AF782859

Engineering Mechanics

Statics and Dynamics

ABOUT THE AUTHOR

A Nelson, did his B Tech (Hons.) from IIT Kharagpur and went on to complete his ME in Civil Structures from University of Tokyo, Japan. He also has an added honour of being a silver medalist during his undergraduation. He has vast years of experience in teaching for Engineering as well as MBA students at P R Engineering College, Thanjavur and Aurora's Engineering College, Hyderabad. During his teaching tenure, he was rated by students as a teacher who gave importance even to the slightest details. He has also served in managerial capacity in several of Indian based international organisations, namely, DPM-South Pacific, Structures America Innovative and Deegit Inc. Currently, he is into design field, having worked in US and Indian construction. At present, he is working as a team leader in Akshaya Infra Consultants Ltd, Hyderabad. He is also a life member of Indian Society of Technical Education and Institution of Engineers.

Engineering Mechanics

Statics and Dynamics

A Nelson

General Manager
Structures America Innovative Ltd., Hyderabad
Formerly with Dept. of Mechanical Engineering,
Aurora's Engineering College, Hyderabad

Mc Graw Hill

McGraw Hill Education (India) Private Limited

Published by McGraw Hill Education (India) Private Limited,
444/1, Sri Ekambara Naicker Industrial Estate, Alapakkam, Porur, Chennai 600116, Tamil Nadu, India.

Engineering Mechanics: Statics and Dynamics

20th reprint 2019
RXBRRRDRTDZAB

ISBN (13 digits): 978-0-07-014614-3
ISBN (10 digits): 0-07-014614-4

21 22 23 24 25 AVO 23 22 21 20 19

Managing Director: *Lalit Singh*
Senior Manager: Content Development & Production Services: *Shalini Jha*
General Manager—Production: *Rajender P Ghansela*

Typeset at Bharati Composers, D-6/159A, Sector VI, Rohini, Delhi 110085, and,
Printed and Bound in India at Avon Printers, Plot No. 16, Main Loni Road, Shahdara, Delhi 110094.

Visit us at: www.mheducation.co.in ; (Toll free in India): **1800 103 5875**
Write to us at: info.india@mheducation.com

CIN: U80302TN2010PTC111532

Preface

I am delighted to present this treatise on "Engineering Mechanics" to admirers of the subject. Engineering mechanics is the fundamental subject for many engineering disciplines like civil, mechanical, electrical, chemical, aeronautical, naval, etc. A thorough understanding of this subject is a prerequisite for pursuing these disciplines as well as for other disciplines in their first year course as followed by most of the Indian universities.

It is a general trend that students have a dislike for this subject because of its vastness and numerical problem oriented nature. This book has been written with an aim to help the student community to overcome this fear to learn the basic concepts in an easy-to-understand manner and to apply them in numerical problems. A large number of solved examples have been provided to clearly explain the concepts under each topic. This book also addresses the need of the teachers who wish to teach the subject in depth. The theory has been provided in a lucid manner with figurative description wherever possible to clearly explain the concepts. Summary, objective type questions, theoretical questions and numerical problems with answers are provided at the end of each chapter.

The content of the book is so arranged that it would suit the standard syllabus of any university. The chapters of the book have been so written as to introduce the student to the underlying concepts, develop relevant theory, illustrate with examples and relate the subject to real world situations.

Vector approach, being widely accepted, has been used in this text. An introduction to vector algebra has been presented in the beginning chapters to know the basics. The problems in static's and dynamics can be solved by expressing the physical problems in terms of idealized models. Analysis of these models under the action of forces involves drawing free-body diagrams as the first step. For drawing of these free body diagrams, various cases that we normally come across have been dealt with in detail devoting an entire chapter.

Throughout the book, SI units have been used for all calculations. However, in dimensioning certain members, CGS system of units are also considered. I have attempted to treat the subject matter in the book in a self sufficient and integrated manner to minimize if not eliminate the need for reference to other texts as far as possible.

This book is accompanied by a comprehensive website http://www.mhhe.com/nelson/em designed to provide valuable resources. Instructors can access a solution manual and PowerPoint presentation with figures. Among the many resources that students can access are web links for further reading, additional objective type questions and sample question papers with solutions.

I thank my God, who has given me the strength and wisdom to write this book. I wish to express my sincere thanks to Dr U M Chaudhari, my Principal and HOD for giving me valuable guidance and suggestions. He took the pain of going through the entire manuscript and did the proof correction. The book wouldn't be in the current form had not his encouragement been there right from the beginning till its completion. It is in vain to attempt to thank one's spouse for anything, for her contribution is much more than can be acknowledged by mere words. However, if I could do it, I wish to thank my wife for her help in typing the material and her encouragement throughout the completion of this book. I also wish to remember my mother, family members and colleagues who were of real encouragement to me throughout my work.

A note of acknowledgement is due to the following reviewers for their valuable inputs.

S K Chakraborty
IIT Kanpur, Uttar Pradesh

Ahmad Ali Khan
Aligarh Muslim University, Uttar Pradesh

K R Viswanathan
Anna University, Tamil Nadu

Saravana Sankar
Arulmigu Kalasalingam College of Engineering, Tamil Nadu

N P N Murthy
Acharya Nagarjuna University, Andhra Pradesh

K V Ramana Reddy
Osmania University, Andhra Pradesh

Dipak J Choudhary
Mumbai University, Maharashtra

V Y Barve
Mumbai University, Maharashtra

Meeta Gandhi
Mumbai University, Maharashtra

Abhay D Kakirde
Truba Institute of Science & Technology, Indore

Ashok Kumar Roy
Birla Institute of Technology, Mesra

B G Shukla
Shree Institute of Science & Technology, Bhopal

Bharat Chhede
Mahakal Institute of Technology and Science, Ujjain

S S S Govil
ABES engineering college, Ghaziabad

Suresh B
Globus Engineering College, Bhopal

P S Bharti
Guru Govind Singh Inderprastha University, Delhi

Ranadip Kumar Das
Indian School of Mines, Dhanbad

Vivek Sharma
College of Engineering Roorkee (COER)

I would be looking forward to suggestions for further improvement of the book. You may contact me at the following email id—tmh.corefeedback@gmail.com (*kindly mention the title and author name in the subject line*).

A Nelson

Contents

Guided Tour

Introduction

The introduction, at the beginning of each chapter, gives an overview of what is going to be discussed in that chapter. It also forms a bridge between the preceding chapter and the current chapter.

5 Equilibrium of a System of Forces

5.1 INTRODUCTION

In the previous two chapters, we discussed various systems of forces and the methods to find their resultants. If the forces in the system are *concurrent*, they can be replaced by a resultant force acting at the point of concurrency. The effect of such a force system is to *translate* the body in the direction of the resultant. If the forces are *non-concurrent*, they can be replaced by a resultant force acting at a common point and a moment about the same point. The effect of such a force system would be to *translate* and *rotate* it as well.

In this chapter, we will discuss a *special* case that arises when the resultant **force** and **moment** turn out to be **zero**. If the resultant **force** of a system of forces is **zero**, the body will remain at *rest* or move with *constant velocity*, if it was already moving with constant velocity, i.e., its acceleration will be zero; and if the **moment** is also **zero**, then there will not be any rotational motion. Such a condition is called **static equilibrium**.

To analyze a body in equilibrium, we must first identify the forces and moments acting on it and check if the resultant force and moment are zero or not. In the previous two chapters, the system of forces acting on a body was given **explicitly** and hence, we could determine its resultant. However, in real situations, one must determine this force system in order to check the equilibrium condition. If one can master the way to determine these forces then one can solve any type of problem in mechanics. Therefore, the student is advised to go through the following section very carefully as this forms the basis for solving problems in statics as well as dynamics.

5.2 FREE-BODY DIAGRAM

The system of forces acting on a body tends to translate it or rotate it or do both. In general, the translational and rotational motions can be resolved into **six** components, namely, **three translations** along X, Y and Z directions, and **three rotations** about X, Y and Z directions. To represent these, we require six independent variables—three for translational motion and three for rotational motion. Hence, we say that bodies have **six degrees of freedom.** Whenever a body is restricted to move in any of these directions due to its attachments with the surroundings, the body is said to be **constrained**. To investigate the equilibrium of a constrained body, we must first isolate it from all its attachments with its surroundings.

10.6 MASS MOMENT OF INERTIA OF SOLIDS

The results derived in the previous section for mass moments of inertia of thin plates can be used to determine mass moments of inertia of solids as explained below for various regular shaped bodies.

10.6.1 Solid Cylinder

Consider a cylinder of radius R, length L and mass density ρ. The coordinate axes are chosen about the centroid as shown in Fig.10.13. Suppose we cut a circular disc of infinitesimal thickness dz perpendicular to the Z-axis at a distance z from the origin, its mass is given as

$$dm = \rho \pi R^2 dz \qquad (10.33)$$

Therefore, its mass moment of inertia about the Z-axis is

$$(dI_{zz})_{\text{mass}} = dm R^2/2$$
$$= \rho(\pi R^4/2)dz \qquad (10.34)$$

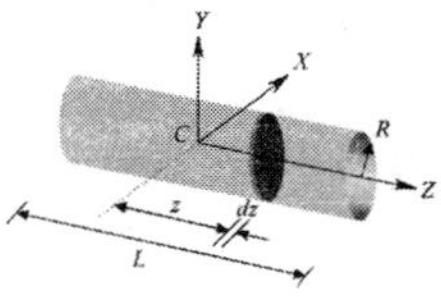

Fig. 10.13 Section of a cylinder

Sections and Sub-sections

Each chapter has been divided into sections and sub-sections to present the basic concepts in a lucid manner.

Illustrations

Illustrations are a way of presenting the theory for easy and better understanding. Pictures, line diagrams, sketches in two dimensional (plane) and three dimensional (space), tables and charts have been used liberally throughout this book.

8.1 INTRODUCTION

The forces that we have so far dealt with were all treated as *concentrated* forces. However, in reality, these forces are *distributed* in nature. Forces can be distributed over the entire volume of a body as in the case of force of gravity (i.e., weight of a body) or distributed over a surface area in contact as in the case of contact forces such as normal reaction or pressure distribution of water against a dam gate, load distribution in beams, stress distribution in deformable bodies, etc.

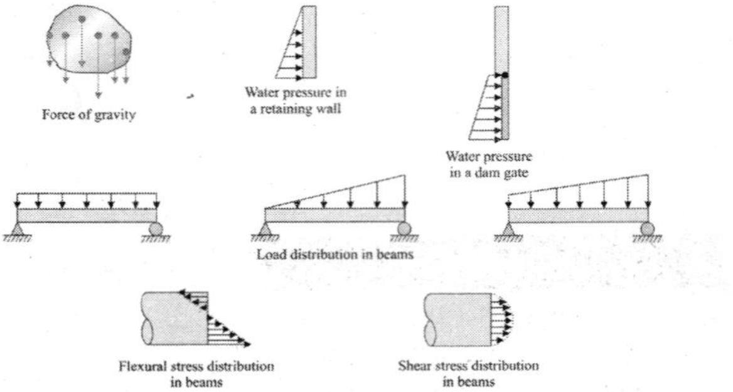

Fig. 8.1 Distributed forces

Though these forces discussed above are all distributed, for *analytical* purposes while applying the conditions of equilibrium, it is customary to replace them by a *single resultant force*, which would produce the same effect as that of the distributed forces. As these forces are *parallel*, we must determine the point of application of the resultant force, which is a point at which the forces are assumed to be

Example 14.34 A block of mass m is at rest at the topmost point of a hemispherical shell. If the block begins to slide over the hemispherical shell, determine the position of a point on the hemisphere at which the block loses contact with the shell.

Solution When the block begins to slide, it undergoes circular motion in vertical plane. The forces acting on the block are its weight mg, normal reaction R exerted by the shell on the block. Resolving the motion along normal direction, we get

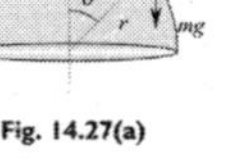

Fig. 14.27

$$mg\cos\theta - R = \frac{mv^2}{r}$$

or

$$R = mg\cos\theta - \frac{mv^2}{r}$$

We see from the figure that $\cos\theta = (r-h)/r$ and as the block slides from rest, its velocity after falling though a height h is $v = \sqrt{2gh}$. Therefore,

$$R = m\left[g\left(\frac{r-h}{r}\right) - \frac{2gh}{r}\right]$$

$$= mg\left[\left(\frac{r-h}{r}\right) - \frac{2h}{r}\right]$$

$$= \frac{mg}{r}[r-h-2h]$$

$$= \frac{mg}{r}[r-3h]$$

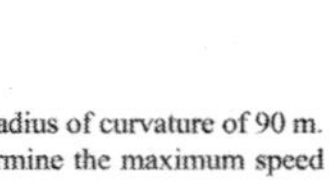

Fig. 14.27(a)

We see that when the block loses contact with the shell, its reaction R is zero. Therefore,

$$0 = \frac{mg}{r}[r-3h]$$

$$\Rightarrow \quad h = \frac{r}{3}$$

Example 14.35 A motorcyclist travels along a level curved track with a radius of curvature of 90 m. If the coefficient of friction between the wheels and the road is 0.25, determine the maximum speed with which he can travel without skidding.

Solved Examples

Solved examples are provided at the end of each major section within a chapter to explain the application of the theory to real world situations. There are as high as 500 solved examples in total.

Solution To draw the free-body diagrams, we detach the cylinder, plank and string separately. Then they are drawn to scale.

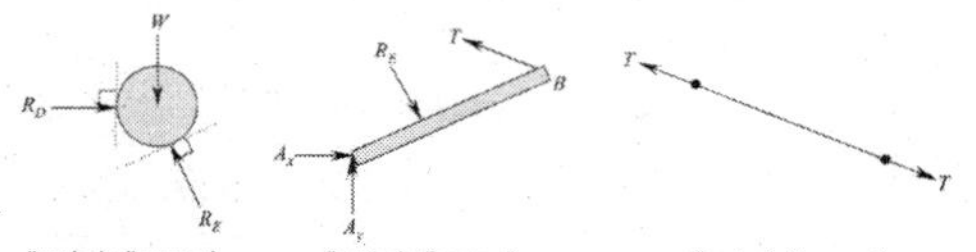

Free-body diagram of cylinder | Free-body diagram of plank | Free-body diagram of string

Fig. 5.23(a)

As the cylinder is in contact with the wall at D and with the plank at E, normal reactions R_D (exerted by the wall on the cylinder) and R_E (exerted by the plank on the cylinder) are shown in its free-body diagram. Also, the weight W of the cylinder is placed at the centre of gravity and directed vertically downwards. Since R_E is the force exerted by the plank on the cylinder, the force exerted by the cylinder on the plank will be equal and in the opposite direction. Hence, R_E for the plank is shown in the opposite direction. In addition, at point A, as it is a hinge, the reactions A_x and A_y are shown. At point B, there will be a tension T as it is connected by a string. Note that here we have neglected the weight of the plank. If its weight is to be included, it must be placed at the centre of gravity of the plank. The free-body diagram of the string will have a tension T at both of its ends. Normally, strings are assumed to be weightless.

Free Body Diagrams

Free body diagrams are essential to understand the forces acting in members of a structure or mechanical system. Nearly every solved example is explained with separate free body diagrams for each member constituting a system.

: ladder at the end B; e contact point C with ım along the X and Y directions at the point of impending motion,

$\sum F_x = 0 \Rightarrow$

$$N_C \sin\theta - F_C \cos\theta - F_B = 0$$

$$N_C \sin\theta - \mu N_C \cos\theta - \mu N_B = 0$$

$$\therefore \quad N_C[\sin\theta - \mu \cos\theta] - \mu N_B = 0 \qquad \text{(a)}$$

$\sum F_y = 0 \Rightarrow$

$$N_C \cos\theta + F_C \sin\theta + N_B - W = 0$$

$$N_C \cos\theta + \mu N_C \sin\theta + N_B - W = 0$$

$$\therefore \quad N_C [\cos\theta + \mu \sin\theta] + N_B = W \qquad \text{(b)}$$

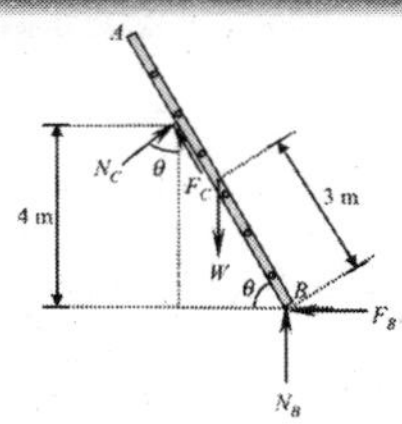

Fig. 6.37(a)

SUMMARY

Any motion that repeats itself after equal intervals of time is termed *periodic* motion. The motion of water waves in seas under the action of wind can be cited as an example. As water waves are free to move, they displace from one place to the other. On the other hand, real civil structures such as transmission cable, diving board in swimming pool, bridge, tall towers, etc., and even mechanical systems such as simple pendulum, string of a musical instrument due to their attachment with the surroundings do not displace from one place to the other upon the action of external forces. Instead, they move back and forth over the *same path*. Such types of periodic motions, which trace the same path in a cyclic manner, are termed *vibratory* or *oscillatory* motions.

For all analytical purposes, we assume all resistance to vibration to be eliminated and such vibrations are termed *undamped free vibrations*. However, this is an ideal condition as we normally observe these vibrations to die out after some time and such vibrations are termed *damped free vibrations*. If the external force continues to act on the structure periodically then the resulting vibrations are termed *forced vibrations*.

Simple Harmonic Motion

Simple harmonic motion is a special case of *rectilinear* motion with *variable* acceleration, in which the acceleration of the particle is proportional to the displacement from the origin and is always directed towards the origin. Most of the structures and mechanical systems are observed to execute this motion under small displacements from the equilibrium position.

The equations of motion as functions of time and displacement for ready reference are shown below:

	As function of displacement	*As function of time*	
		t = 0, at mean position	*t = 0, at extreme position*
Displacement	x	$x = A \sin \omega t$	$x = A \cos \omega t$
Velocity	$v = \pm\omega\sqrt{A^2 - x^2}$	$v = A\omega \cos \omega t$	$v = -A\omega \sin \omega t$
Acceleration	$a = -\omega^2 x$	$a = -A\omega^2 \sin \omega t$	$a = -A\omega^2 \cos \omega t$

Summary

The summary at the end of each chapter gives a handy description of all definitions and formulas discussed in that chapter. It would be of great help for preparation just before exams.

Numerical Problems with Answers

Each chapter at the end has numerical problems with answers, totaling 639 in the book. They give practice in applying the theory to real world situations and provide further confidence to solve any kind of problem.

Numerical Problems

18.1 A flywheel of 15 kg mass and a 25 cm radius of gyration rotates at a constant angular speed of 1200 rpm. When the power supply is switched off, if it coasts to rest in 15 seconds, determine retarding torque due to friction in the bearings assuming it to be uniform.

Ans. 7.86 N.m

18.2 The water jet is shut off when a hydraulic turbine is rotating at a speed of 3000 rpm. If the friction in the bearings is uniform exerting a retarding torque of 20 N.m, determine the number of revolutions made by the rotor before coming to rest and the time taken to come to rest. The mass of the rotor is 20 kg and the radius of gyration is 15 cm.

Ans. 176.73 rev, 7.07 s

Numerical Problems

3.1. Determine the magnitude and direction of resultant of two forces acting on a block as shown in Fig. E.3.1 by trigonometry using (i) parallelogram law, and (ii) triangle law.

Ans. $R = 75.5$ N, $\alpha = 25.2°$

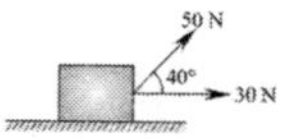

Fig. E.3.1

3.2. Two forces $\vec{F}_1$ and $\vec{F}_2$ act upon a body. If the magnitude of their resultant is equal to that of $\vec{F}_1$ and direction perpendicular to $\vec{F}_1$, then find the magnitude and direction of force $\vec{F}_2$. Take $F_l = 20$ N.

Ans. $F_2 = 28.28$ N, $\theta = 135°$

3.3. Two unequal forces acting at a point at an angle of 150° have a resultant, which is perpendicular to the smaller force. The larger force is 24 N. Find the smaller force and the resultant.

Online Resources

The quickstep solutions to the numerical problems can also be accessed through the publisher's website.

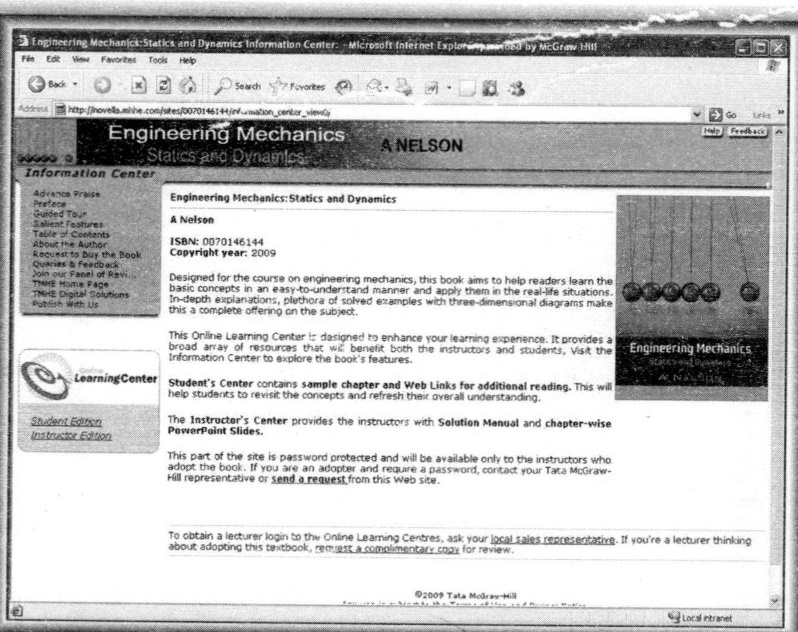

EXERCISES

Objective-type Questions

1. A rigid body can be idealized as a particle
   (a) only when its size is very minute
   (b) only when the body is at rest
   (c) when there is no translational motion involved
   (d) when there is no rotational motion involved
2. In rectilinear motion, all the particles in the body
   (a) have the same displacement (b) have the same velocity
   (c) have the same acceleration (d) all of these
3. Average velocity is defined as
   (a) average of initial and final velocities
   (b) ratio of change in displacement and elapsed time
   (c) ratio of distance travelled and elapsed time
   (d) average of initial and final speeds
4. When a car moves at a constant speed around a curved path, its velocity
   (a) is zero (b) is constant
   (c) changes in magnitude (d) changes in direction
5. A man walks from one town to another and then comes back. State which of the following statements is true concerning his journey:
   (a) his displacement is zero (b) distance traveled is zero
   (c) average speed is zero (d) time taken is zero
6. A man walks from one town to another at a constant speed of 15 kmph and then returns back at a constant speed of 10 kmph. His average speed for the journey is
   (a) 12.5 kmph (b) 12 kmph (c) 2.5 kmph (d) 25 kmph
7. A particle can move with constant velocity when motion is
   (a) rectilinear (b) curvilinear (c) rotational (d) general motion
8. Uniform motion implies that
   (a) acceleration is constant (b) velocity is constant
   (c) position is constant (d) time is constant
9. The area under an a–t curve represents
   (a) average acceleration (b) instantaneous acceleration
   (c) change in position of the particle (d) change in velocity of the particle
10. The area under a v–t curve represents
    (a) average velocity of the particle (b) instantaneous velocity of the particle
    (c) distance travelled by the particle (d) acceleration of the particle

Objective-type Questions

Objective type questions presented at the end of each chapter are useful not only in competitive exams like GATE but also in the online internal exams of regular engineering courses as followed by some of the universities. There are around 220 questions in total.

Short-answer Questions

1. Distinguish between statics and dynamics.
2. Distinguish between kinematics and kinetics.
3. Distinguish between particle and rigid body.
4. Explain the types of motion with suitable examples.
5. Define position vector and displacement vector.
6. Distinguish between displacement vector and distance travelled.
7. Define velocity of a particle.
8. Define average velocity and instantaneous velocity.
9. Under what conditions is average velocity equal to instantaneous velocity?
10. Define average acceleration and instantaneous acceleration.
11. If a particle moves with constant speed but changes in direction, can there be acceleration?
12. Distinguish between rectilinear motion and curvilinear motion.
13. State the differential equations of motion.
14. Distinguish between uniform motion and uniformly accelerated motion.
15. Derive the x–t, v–t and a–t relationships for uniformly accelerated motion.
16. What are motion curves? What are they used for?
17. Define free fall.
18. What are the assumptions made in free fall?

Short-answer Questions

Each chapter has ample short answer questions for practicing the basic concepts learnt in the theory. There are a total of 255 questions in this book.

1 Introduction

1.1 INTRODUCTION TO MECHANICS

Mechanics is the oldest of *physical sciences,* which deals with the state of rest or motion of bodies under the action of forces. As we know that matter can exist in three different states—solids, liquids and gases—and that their behaviours under the action of forces vary, we study them separately under different headings. Hence, mechanics can be classified into **mechanics of solids** and **mechanics of fluids**. Mechanics of fluids or fluid mechanics deals with the study of liquids and gases (together called fluids) at rest or in motion. This will be covered in books on fluid mechanics and will not be covered in this study. In this book, we shall study mechanics of solids only.

Mechanics of solids can further be classified into **mechanics of deformable bodies** and **mechanics of non-deformable** or **rigid bodies**. Mechanics of deformable bodies is the study dealing with internal force distribution and the deformation developed in actual engineering structures and machine components. It is also called *strength of materials* or *mechanics of materials* and will be covered in books on such topics. Hence, it will not be covered in this book.

Here, we shall study mechanics of *non-deformable* or *rigid* bodies. As such, *no* body is perfectly rigid but does *deform* under the action of forces. However, when we are concerned only with the *external effect* of forces on solid bodies, we can generally idealize them to be rigid for analytical purposes without appreciable error unless we deal with extremely flexible or low modulus materials. The study of solid bodies assuming such an idealized condition is termed *mechanics of rigid bodies.*

Mechanics of rigid bodies can further be classified into **statics** and **dynamics**. Statics is the study of distribution and effect of forces on bodies which are at *rest* and remain at rest. It also includes the case of bodies moving with uniform velocity and experiencing no acceleration. Here, we derive the *conditions of equilibrium* for bodies at rest or moving with constant velocity. Dynamics is the study of *motion* of bodies and their correlation with the forces causing them. The current study is divided into two parts—in the first part comprising the first eleven chapters, we deal with statics, and in the next part covering the remaining chapters, we deal with dynamics.

Dynamics can further be classified into **kinematics** and **kinetics**. Kinematics is the study of motion of bodies *without* considering the forces causing the motion. It deals with the relationship between displacement, velocity and acceleration, and their variation with time. Kinetics is the study of motion of bodies *together* with the forces causing the motion. It relates the force and the acceleration by the laws of motion. The following chart depicts various branches of mechanics discussed so far.

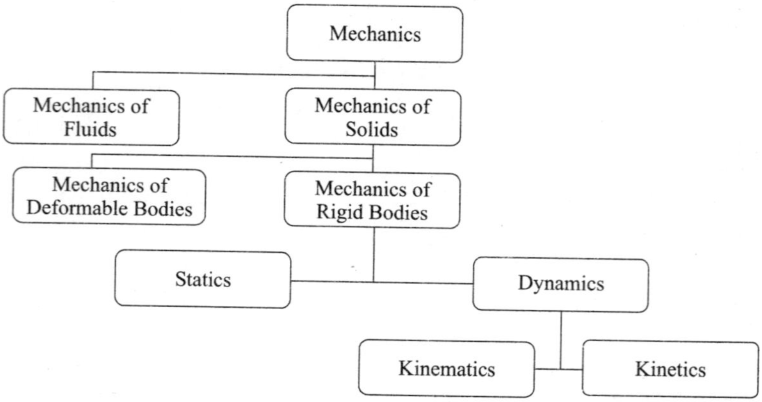

Chart 1.1 Branches of mechanics

1.2 AN OVERVIEW OF THE TEXT

As engineering mechanics forms the foundation for most engineering fields such as civil, mechanical, electrical, chemical, aeronautical, naval, etc., a thorough understanding of this subject is an essential prerequisite for the students of these disciplines. An overview of the text is given below.

In this chapter, we will discuss some of the basic concepts underlying the field of mechanics. As vector approach is widely used in this text, an introduction to vector algebra is dealt with in Chapter 2. Chapters 3 to 11 are about statics. In Chapters 3 and 4, we will discuss the definition of force, various systems of forces and the methods to determine their resultants. As different methods are followed for different types of force systems, we will deal with each type separately for better understanding of the effect of various force systems on bodies. In Chapter 5, we will introduce the concept of free-body diagrams, which will be applied to solving problems in statics as well as in dynamics. Finally, in Chapters 6 and 7, we will also discuss the effect of friction on the motion of bodies and their applications in machines and mechanisms.

Chapters 8 and 9 deal with centre of gravity and centroids of solid bodies, plane areas and lines, and area moments of inertia. These find application in the study of strength of materials. Chapter 10 deals with mass moments of inertia and it finds application in the study of rigid-body dynamics. Chapter 11 deals with an alternate way of solving equilibrium of particles, called *virtual work method.*

Chapters 12 to 19 discuss the dynamics part. Here we will deal with kinematics and kinetics separately. Kinematics deals with geometry of the motion, while kinetics deals with the correlation between the motion and the cause of motion. Finally, in Chapter 19, we will discuss vibrations, which we normally encounter in civil structures and mechanical systems.

1.3 DEVELOPMENT OF THE FIELD OF MECHANICS

The ancient civilizations give evidence that man had constructed various structures like towers, houses, etc., for his living. Hence, the subject of mechanics can be said to be as old as the human civilization itself. However, contributions to the field of mechanics in the form of writings were not made until a few centuries before Christ. These earlier contributions were towards the field of statics. Aristotle (384–322 BC), a Greek philosopher, worked on problems of lever and the concept of centre of gravity. Archimedes (287–212 BC) contributed towards buoyancy and equilibrium of forces acting on a lever, which is fundamental for the realization of all machines.

Aristotle had made important assertions regarding the *motion* of bodies also. He had stated that *heavier* bodies fall at a faster rate than *lighter* bodies, if dropped from a height. This was believed to be true and it was a commonly held view for many centuries until Galileo Galilei (1564–1642), an Italian scientist, refuted this assertion by experiments. He later wrote a treatise entitled *Dialogues Concerning Two New Sciences* in which he detailed his experimental studies of motion. This treatise marked the beginning of the field of dynamics.

Isaac Newton, born in England in the year of Galileo's death, followed the ideas of Galileo and others who preceded him. He related the forces acting on bodies and the resulting motion by stating his famous laws of motion (in 1686) in his *Principia Mathematica Philosophiae Naturalis*. He also invented the principles of calculus to prove his point and formulated the law of universal gravitation. Newton's principles were later expressed in modified forms by d'Alembert, Lagrange, Euler and Hamilton.

Leonardo Da Vinci (1452–1519), who lived much before Newton, i.e., before the concept of force was fully developed, formulated the static and kinetic laws of friction. Varignon, Laplace and others were the subsequent contributors to the development of mechanics.

Newtonian mechanics, also called *classical mechanics*, is found to hold good while describing motion of gross objects such as planets, jet planes, baseballs, etc. However, it is found to fail when applied to the motion of small particles like atoms and in dynamical astronomy where speeds nearing the speed of light are attained. This was stated by Einstein in 1905 in his theory called the *special theory of relativity*. In our study, we limit ourselves only to classical mechanics or mechanics of gross objects whose speeds are much below the speed of light.

1.4 IDEALIZATION OF BODIES

While studying the effect of forces on bodies either at rest or in motion, the bodies are normally *idealized* to simplify the problem without affecting the actual results. We know that bodies are composed of molecules and there is *intermolecular attraction* between these molecules. Moreover, there are *voids* or *spaces* in between these molecules. The degree of intermolecular attraction and the amount of voids present depend upon the *state* of matter contained in the bodies. For instance, in the case of solids, this intermolecular attraction is stronger and voids are lesser as compared to that of fluids. However, for all analytical purposes, bodies, whether they be solids or fluids are always idealized to be in **continuum**, i.e., they have a continuous distribution of matter with no voids or empty spaces.

Solid bodies upon the application of external loads may expand, contract, bend or twist causing internal stresses. These internal stresses arise due to resistance to deformation offered by the molecules. If we are interested only in the external effect of forces on solid bodies, usually these deformations can be neglected compared to the sizes of the bodies. In such cases, the bodies can be idealized to be **rigid**. A rigid body can be defined as one having a specific amount of matter in which all the particles are *fixed* in position relative to one another.

Further, consider the motion of a cricket ball in its trajectory. The ball may spin during its motion along the trajectory. However, if we are interested only in the translational motion of the ball, this spinning effect can be neglected. This assumption is justified as the size of the ball is small compared to its trajectory. Under such a condition, instead of considering the motion of the ball as a whole, we may consider the motion of a *particle* in the body to describe its motion. This is true because when there is no rotational motion involved, the displacement of every particle in the body will be same. This is called idealization of a body to be a **particle**. Normally, the centre of mass of the body is chosen to be that particle. Mathematically, a particle is treated as a point, a body without extent that rotational consideration is not involved and at the same time, it does not significantly affect the solution of the problems. Based on this idealization, in the case of planetary motion, the planets are considered particles even though they are of huge sizes.

1.5 PHYSICAL QUANTITIES—UNITS AND DIMENSIONS

1.5.1 Fundamental and Derived Quantities

To define the state of rest or motion of bodies, we must specify the physical characteristics of the bodies by certain measurable quantities. These quantities can be classified into **fundamental quantities** and **derived quantities**. Fundamental quantities are those quantities based upon which all other quantities are derived. Hence, fundamental quantities are independent. Derived quantities are those that are derived from fundamental quantities.

Now we are faced with a question: What should be the minimum number of fundamental or independent quantities required to express every other physical quantity in mechanics? It has been found that three fundamental quantities are sufficient to express every other quantity in mechanics. They are chosen to be Length (L), Mass (M) and Time (T). In some countries instead of mass, force is also chosen as the fundamental quantity. The first letter of these quantities, namely, [L], [M] and [T] are used to express every other quantity and their expressions are termed **dimensions**. For instance, area is expressed as a product of two lengths. Hence, its dimension is $[L]^2$. Similarly, force is expressed as $[M][L][T]^{-2}$. These will be discussed later in detail. Alternatively, we can also define dimension of a physical quantity as the powers to which the fundamental quantities are raised to define the derived quantity.

1.5.2 Units and Systems of Units

Unlike mathematics where we deal with mere numbers, in engineering, we must assign **units** to each physical quantity apart from their numerical values. For instance, to express a measurement of length, we may say that it is '*n*' times a standard length. Such types of standard measurements are termed *units*. Thus, unit may be defined as those standards in terms of which the physical quantities are measured. These standards also enable comparison of two different measurements of the same category. Different systems of units are followed for this purpose by different countries. The most common ones are mks system of units, cgs system of units and fps system of units.

Mks System of Units In mks system of units, metre (m) is the unit assigned to length, kilogram (kg) to mass and second (s) to time. It is also represented as metre–kilogram–second. In this system of units, unit force (1 newton) is defined as that force which will cause a unit acceleration ($1 m/s^2$) upon a body of unit mass (1 kg).

Cgs System of Units In cgs system of units, centimetre (cm) is the unit assigned to length, gram (g) to mass and second (s) to time. It can also be represented as centimetre–gram–second. It is also called Gaussian system of units. Here, unit force (1 dyne) is defined as that force which will cause a unit acceleration (1 cm/s^2) upon a body of unit mass (1 g).

We must understand that though we may use different systems of units to express the same physical quantity, they are *not* altogether very different. There always exists a relationship between various systems of units followed. Thus, a physical quantity in one system of unit can be converted to another by using *conversion factors*. The relationship between cgs system of units and mks system of units is given below:

$$1 \text{ cm} = 10^{-2} \text{ m (or) } 100 \text{ cm} = 1 \text{ m}$$

$$1 \text{ g} = 10^{-3} \text{ kg (or) } 1000 \text{ g} = 1 \text{ kg}$$

Also, $$1 \text{ dyne} = 1 \text{ g} \times 1 \text{ cm/s}^2 = 10^{-3} \text{ kg} \times 10^{-2} \text{ m/s}^2 = 10^{-5} \text{ N}$$

or $$10^5 \text{ dyne} = 1 \text{ N}$$

Fps System of Units In fps system of units, foot (ft) is the unit assigned to length, pound (lb) to force and second (s) to time. Here, force is chosen as the fundamental quantity and mass is a derived quantity. Unit mass is defined as that mass of a body whose acceleration is 1 ft/s^2 when acted upon by a force of 1 lb. Mathematically, it is expressed as:

$$m = F/a, \text{ where } F \text{ is force and } a \text{ is acceleration.}$$

The unit of mass is *slug*. The relationships between foot and centimetre, slug and kilogram, and that between pound and newton are as follows:

$$1 \text{ ft } (= 12 \text{ in}); \; 1 \text{ in} = 2.54 \text{ cm}$$

$$1 \text{ lb} \cong 4.448222 \text{ N}$$

Also, $$1 \text{ slug} = 0.45359237 \text{ kg}$$

This system of units is followed by British and American engineers and it can be represented as foot–pound–second.

SI System of Units As different systems of units are used by different countries of the world, it leads to confusion. Hence, a universal standard for units was sought for at the Eleventh General Conference of Weights and Measures held in Paris in 1960. The system of units proposed during that time is called Système Internationale d'Unités, abbreviated as SI units. In English, it is referred as International System of units.

1.5.3 Standards in SI Units

The fundamental quantities, length, mass and time, must be standardized to enable comparison of two different measurements of the same category. The following are the standards for these quantities in SI units.

Standard Length Standard length is defined as the distance between two fine lines engraved on gold plugs near the ends of a bar of platinum–iridium alloy called the standard metre kept at the International Bureau of Weights and Measures at Sevres near Paris, France. The bar must be at 0.00°C and supported mechanically in a prescribed way.

Due to problems encountered in accessing this standard length from different parts of the world and also as the accuracy in reproducing it is not high enough to meet the current requirements, an alternate standard was thought of, which can be easily available everywhere and also invariable. As a result, a metre is now defined as 1 650 763.73 wavelengths of a particular orange radiation emitted by atoms of a particular isotope of krypton (Kr^{86}) in an electrical discharge.

Standard Mass Mass is a measure of quantity of matter contained in a body. It is also referred as a measure of resistance to change in velocity, which is termed as *inertia* of a body. Standard kilogram is the mass of a particular platinum cylinder carefully preserved at the International Bureau of Weights and Measures at Sevres near Paris, France.

Standard Time One second was defined originally as the fraction 1/86 400 of a mean solar day. This standard required astronomical observations for a certain period of time. It did not take into account the irregularities in the rotation of the earth and hence the required accuracy could not be achieved. Later, a secondary terrestrial clock was sought and it also lacked precision. The second is now defined more precisely as the duration of 9 192 631 770 periods of radiation corresponding to the transition between two hyperfine levels of the ground state of the cesium-133 atom.

1.5.4 Derived Quantities

Once fundamental quantities and their units and dimensions are chosen, we can define every other physical quantity in terms of these quantities. For instance, area can be expressed as a product of lengths. Hence, its dimension and unit are

$$[\mathrm{L}][\mathrm{L}] = \mathrm{L}^2 \text{ and } \mathrm{m}^2$$

Similarly, volume can be expressed as a product of three lengths. Hence, its dimension and unit are

$$[\mathrm{L}][\mathrm{L}][\mathrm{L}] = \mathrm{L}^3 \text{ and } \mathrm{m}^3$$

Velocity is defined as rate of change of displacement. Mathematically,

$$\text{velocity} = \frac{\text{displacement}}{\text{time}}$$

Hence, its dimension and unit can be expressed in terms of fundamental quantities as

$$\frac{[\mathrm{L}]}{[\mathrm{T}]} \text{ (or) } \mathrm{LT}^{-1} \text{ and m/s}$$

Similarly, acceleration is defined as the rate of change of velocity. Mathematically,

$$\text{acceleration} = \frac{\text{velocity}}{\text{time}}$$

Its dimension and unit can be expressed as

$$\frac{[\mathrm{L}][\mathrm{T}]^{-1}}{[\mathrm{T}]} \text{ (or) } \mathrm{LT}^{-2} \text{ and m/s}^2$$

In this way, we can express the dimension and unit of every quantity in mechanics in terms of fundamental quantities. These are summarized in Table 1.1. As these quantities are derived based upon the fundamental quantities, they are termed derived quantities.

Table 1.1 Fundamental and derived quantities in mechanics

S.No.	*Quantity*	*Dimension*	*Unit*
1	Length	L	m
2	Mass	M	kg
3	Time	T	s
4	Area	L^2	m^2
5	Volume	L^3	m^3
6	Velocity	LT^{-1}	m/s
7	Acceleration	LT^{-2}	m/s^2
8	Momentum	$M LT^{-1}$	N.s
9	Force	$M LT^{-2}$	N
10	Work done	$M L^2T^{-2}$	N.m (or) J
11	Energy	$M L^2T^{-2}$	N.m (or) J
12	Moment of force	$M L^2T^{-2}$	N.m
13	Torque	$M L^2T^{-2}$	N.m
14	Power	$M L^2T^{-3}$	N.m/s (or) W
15	Impulse	$M LT^{-1}$	N.s
16	Area moment of inertia	L^4	m^4
17	Mass moment of inertia	ML^2	$kg.m^2$
18	Mass density	ML^{-3}	kg/m^3
19	Weight density	$ML^{-2}T^{-2}$	N/m^3
20	Frequency	T^{-1}	s^{-1} (or) Hz
21	Angular displacement		rad*
22	Angular velocity	T^{-1}	rad/s
23	Angular acceleration	T^{-2}	rad/s^2

(*rad-radian is measurement of plane angle and it is used as a supplementary unit)

Certain units are not part of the International System of units, that is, they are outside the SI units, but are important and widely used. They are

$$\text{minute, } 1 \text{ min} = 60 \text{ s}$$

$$\text{hour, } 1 \text{ h} = 60 \text{ min} = 3600 \text{ s}$$

$$\text{degree, } 1° = \pi/180 \text{ rad}$$

$$\text{metric ton, } 1 \text{ t} = 1000 \text{ kg}$$

It is worthwhile mentioning here that there are other physical quantities too, such as temperature, pressure, current, stress, etc., but as these are not covered in our study, we have not mentioned them. These may be found in books on related topics.

To express multiples and submultiples of quantities, prefixes are used as shown in Table 1.2.

Table 1.2 Prefixes for multiplication factors

Multiplication Factor	*Prefix*	*Symbol*
10^{-12}	pico-	p
10^{-9}	nano-	n
10^{-6}	micro-	μ
10^{-3}	milli-	m
10^{-2}	centi-	c
10^{-1}	deci-	d
10^{1}	deca-	da
10^{2}	hecto-	h
10^{3}	kilo-	k
10^{6}	mega-	M
10^{9}	giga-	G
10^{12}	tera-	T

For instance,

1000 N is normally expressed as 10^3 N (or) 1 kN
1000 000 W is expressed as 10^6 W (or) 1 MW
0.001 m is expressed as 10^{-3} m (or) 1 mm
200 000 N is expressed as 200 kN and not as 0.2 MN

1.5.5 Dimensional Homogeneity

While expressing physical quantities in terms of other quantities using mathematical equations, the following rule must be checked. That is, the dimensions of the terms on both sides of the equation must be equal. This is known as **dimensional homogeneity**. For instance, the velocity of a body in rectilinear motion (which we will learn later) is given by the equation,

$$v = v_0 + at$$

where v_0 is the initial velocity, a is the acceleration and t is the time. The dimensions of the terms on both sides of the equation can be written as

$$\mathrm{LT^{-1} = LT^{-1} + LT^{-2}T}$$

or

$$\mathrm{LT^{-1} = LT^{-1} + LT^{-1}}$$

We see that the dimensions of the terms on both sides of the equation are equal. Hence, this equation is said to be dimensionally *homogeneous*. Consider now the following equation,

$$s = v_0 + (1/2)at^2$$

where s is the displacement, and v_0, a and t are as defined before. The dimensions of the terms can be written as

$$\mathrm{L = LT^{-1} + LT^{-2}T^2}$$

or

$$\mathrm{L = LT^{-1} + L}$$

We see that the dimensions of the terms on both sides of the equation are *not* equal. Hence, this equation is said to be dimensionally *non-homogeneous* and it can be seen that the correct equation must be

$$s = v_0 t + (1/2)at^2$$

for it to be dimensionally homogeneous.

1.6 SCALARS AND VECTORS

The physical quantities alone will not suffice to describe the state of rest or of motion of a body. In addition to these quantities, which describe only the magnitude, the direction should also need to be specified. It is for this reason that we introduce the concept of **scalars** and **vectors**.

Scalars are those quantities that have magnitude only (specified by a number and unit). The examples are mass, length, time, density, energy, etc. Vectors are those quantities that have both magnitude and direction and combine according to the parallelogram law. The examples are displacement, velocity, acceleration, momentum, force, etc. As vectors are used throughout this book, an introduction to vector algebra is dealt with in Chapter 2.

1.7 LAWS OF MECHANICS

The mechanics of bodies is governed by two basic laws: the laws of motion and the force laws. Laws of motion were stated by Isaac Newton.

First Law *Every body continues in its state of rest or of uniform motion in a straight line unless it is compelled to change that state by forces acting on it.*

Stated in an alternate way, to cause a change in velocity or in other words, to accelerate a body, an external force must act on it.

Second Law *If the resultant force acting on a particle is not zero, the particle will have acceleration proportional to the magnitude of the resultant and in the direction of this resultant force.*

The mathematical statement of this law is given as

$$\vec{F} = m\vec{a}$$

where $\vec{F}$ represents the resultant of a system of forces acting on a particle, m mass of the particle and $\vec{a}$ its acceleration.

Third Law *To every action there is always an equal and opposite reaction.*

Forces always occur in pairs of active and reactive forces. A single isolated force is an impossibility. When one body exerts a force on a second body, the second body always exerts a force on the first.

Newton's Law of Universal Gravitation *Any two particles of mass, m_1 and m_2, are mutually attracted along the line connecting them with equal and opposite forces of magnitude F proportional to the product of their masses and inversely proportional to the square of the distance r between them.*

Mathematically, $$F = \frac{Gm_1m_2}{r^2},$$

where G is the universal gravitational constant and r is the distance between the two particles.

$$G = 6.673 \times 10^{-11}\ \text{N.m}^2/\text{kg}^2$$

Consider a body of mass m located on the surface of the earth, whose mass is M and radius R. Then the force of attraction between the two bodies is given as

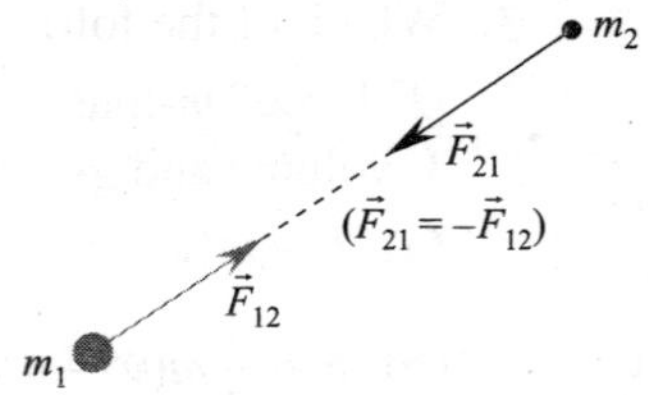

Fig. 1.1 Gravitational force of attraction

$$F = \frac{GMm}{R^2} = m\left[\frac{GM}{R^2}\right]$$

We can see that the terms within the bracket are constants, hence we can write

$$F = mg$$

where g is a constant called the **acceleration due to gravity**. Here, the force F is also called the **force of gravity**. As the earth is not perfectly spherical, the value of R varies for different latitudes and for different altitudes. However, if we consider smaller altitudes, the value of g can be assumed constant. Hence, for all calculation purposes, a value of g equal to 9.81 m/s^2 can be used.

Mass vs. Weight Mass is an intrinsic property of a body. It is a scalar quantity. Weight of a body is defined as the gravitational force exerted by the earth on the body. As it is a force, weight is a vector quantity. Its magnitude is given as a product of mass and acceleration due to gravity.

1.8 PRECISION AND SIGNIFICANT FIGURES

While measuring the physical quantities, absolute precision is not possible. There may be some uncertainty or error. Hence, measurements are normally represented by the measured value followed by a ± symbol and a second number. For instance, the diameter of a rod can be written as 23.56 mm ± 0.02 mm. The second number represents the uncertainty or error in measurement. In this example, it implies that the true value must be somewhere in between 23.54 and 23.58 mm.

When quantities with errors are used to calculate other quantities, these errors will be carried over and the calculated values also will be uncertain. Hence, in engineering calculations instead of always looking for the precise values, we do allow some errors. In general, no numerical result can have more significant figures than the numbers from which it is computed. The number of meaningful digits in a number is called the number of significant figures. An accuracy of three significant figures is considered satisfactory for the majority of engineering calculations. In this book, in all of the numerical calculations, an accuracy of two and at the most three significant digits have been considered.

EXERCISES

Objective-type Questions

1. N.cm expressed in SI units is
   (a) 100 N.m (b) 0.01 N.m (c) 10 N.m (d) 0.001 N.m
2. g/cm^3 expressed in SI units is
   (a) 10 kg/m^3 (b) 100 kg/m^3 (c) 1000 kg/m^3 (d) 0.1 kg/m^3

3. Which of the following equation is dimensionally homogeneous?
[F-force, m-mass, v_0 and v-initial and final velocities, s-displacement, t-time, ρ-density, V-volume and g-acceleration due to gravity]
(a) $F = m(v - v_o)$
(b) $F = \frac{1}{2}m(v^2 - v_o^2)$
(c) $F = ms/t$
(d) $F = \rho Vg$

4. The unit of power is
(a) N.m/s (b) Watt (c) kg.m^2/s^3 (d) all of these

5. ML^2T^{-2} is the dimension for which of the following physical quantities?
(a) Work done (b) Energy (c) Moment of force (d) All of these

6. The dimension of universal gas constant G in the expression $F = \frac{Gm_1m_2}{r^2}$ is
(a) $M^2L^3T^{-2}$ (b) $M^{-2}L^{3T^{-2}}$ (c) $M^{-1}L^3T^{-2}$ (d) $M^{-2}L^2T^{-2}$

7. 1/100th of a second can also be expressed as
(a) 10 ms (b) 10 μs (c) 10 ks (d) 10 ps

8. 50 000 J is expressed in SI units as
(a) 0.05 MJ (b) 50 kJ (c) 0.05 μJ (d) 500 kJ

9. 24 000 cm is expressed in SI units as
(a) 2.4 m (b) 24 m (c) 240 m (d) 2400 m

Answers

1. (b) 2. (c) 3. (d) 4. (d) 5. (d) 6. (c) 7. (a) 8. (b)
9. (c)

Short-answer Questions

1. Distinguish between mechanics of deformable bodies and mechanics of non-deformable bodies.
2. Distinguish between statics and dynamics.
3. Define the following terms: continuum, rigid body and particle.
4. Define fundamental and derived quantities.
5. Define units and dimensions.
6. What are the different systems of units followed? Explain.
7. Discuss on the standards followed in SI units.
8. What is meant by dimensional homogeneity? Give examples.
9. Distinguish between scalars and vectors. Give examples.
10. Distinguish between mass and weight.
11. State Newton's laws of motion and the law of universal gravitation.

2 Vector Algebra

2.1 INTRODUCTION

Certain physical quantities can be specified *completely* by stating their magnitude (a number and unit) only. Such physical quantities are termed **scalars**. The examples are mass, length, time, density, energy, power, etc.

10 kg of mass; 2.3 m length; 35 s duration; 1000 kg/m^3 mass density

50 J of energy; 10 kW of power

As these are merely numbers with units, they follow the normal rules of arithmetic and algebra. For instance, if the times taken for two related activities are respectively 3 s and 5 s, then the total time taken for the two activities is given by normal arithmetic as $3 + 5 = 8$ s.

There are certain other physical quantities, which cannot be specified completely by stating their magnitude alone. Hence, in addition to the magnitude, we need to specify the direction in which they act. Such quantities, which are specified completely by their magnitude and direction and in addition, which obey the parallelogram law of addition (which we will learn later in the chapter) are termed **vectors**. The examples are displacement, velocity, acceleration, momentum, force, etc.

a displacement of *10 m towards south*

a car moving with a velocity of *20 m/s along the eastern direction*

gravitational acceleration of *9.81 m/s^2 acting towards the centre of the earth*

50 N weight acting vertically downwards

As these quantities are specified with direction as well as magnitude, the normal rules of mathematical operations cannot be performed on them. Hence, we must establish new rules for operation of vectors. Thus, for this reason we discuss the basics of vector algebra in this chapter, which will form the foundation for the succeeding chapters, as vectors are widely used in the solution of statics and dynamics problems.

2.2 REPRESENTATION OF VECTORS

Graphically, a vector is represented by a line segment drawn from a point called the **point of application** of the vector such that its length is proportional to the magnitude of the vector. Its inclination with a reference line represents the **direction** of the vector. The arrowhead indicates **sense** of the vector, i.e., whether it is acting away or towards the point of application.

Fig. 2.1

For instance, a body moving with a velocity of 10 m/s along NE direction is shown graphically in Fig. 2.2.

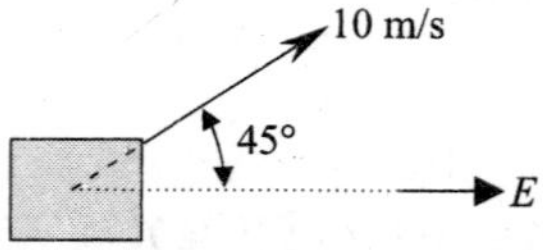

Fig. 2.2 Graphical representation

Mathematically, a vector is represented by an **arrow** placed over the symbol used to denote the vector like $\vec{a}$. The **magnitude** of a vector is a positive quantity corresponding to the length of the vector. It is denoted by $|\vec{a}|$ or simply '*a*.'

2.3 CLASSIFICATION OF VECTORS

Vectors are classified into three types, namely, fixed, sliding and free vectors.

Fixed Vectors Some vectors have a fixed point of application, i.e., they cannot be moved without affecting the conditions of the problem. Such vectors are termed **fixed vectors**. They have specific magnitude, direction and their line of action passes through a *unique* point in space. The example for such type of vectors is *moment of a force* about a point.

Sliding Vectors Some vectors are not fixed and can be moved along their lines of action. Such vectors are termed **sliding vectors**. Their magnitude, direction and sense remain the same. The point of application can be anywhere along their line of action. The example for such type of vectors is a *force* acting on a *rigid body*. In Fig. 2.3, a force $\vec{F}$ acting on a rigid body can be moved anywhere along its line of action, without affecting the conditions of the problem.

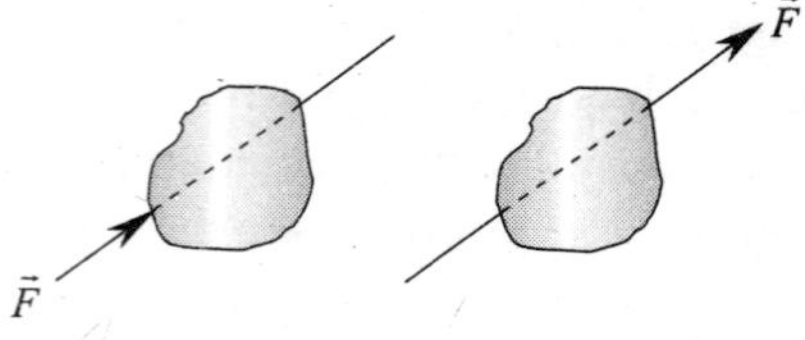

Fig. 2.3 Sliding vector

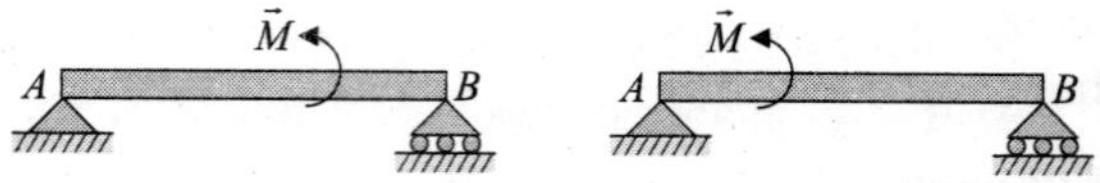

Fig. 2.4 Free vector

Free Vectors Some vectors can be moved over in space. Such vectors are termed **free vectors**. They have a specific magnitude, direction and sense but their lines of action do not pass through a unique point in space. The example for such type of vectors is a *couple*. In Fig. 2.4, the couple whose direction is perpendicular to the plane of the paper can be placed anywhere on the beam without affecting its conditions.

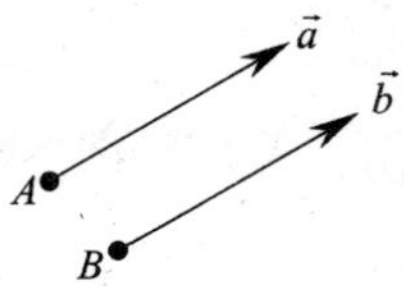

Fig. 2.5 Equal vectors

Equality of Vectors Two vectors $\vec{a}$ and $\vec{b}$, are said to be **equal** if they have the *same* magnitude and direction. However, they can have *different* points of application [refer Fig. 2.5].

Equivalent or Equipollent Vectors Two vectors are said to be **equivalent**, if not only are they *equal*, but also produce the *same effects* on the body. A force P acting on a beam will produce different deflections, if placed at different points on the beam as shown in Fig. 2.6. Hence, the forces though equal cannot be said to be equivalent, as they do not produce the same effect on the body.

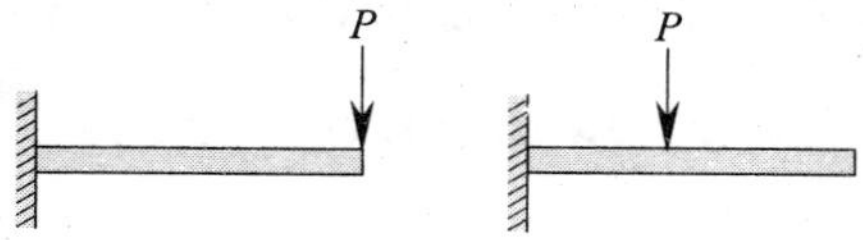

Fig. 2.6

Unit Vector A vector whose magnitude is *unity* is called a **unit vector**. It is normally represented by a circumflex placed over the letter. For example: $\hat{n}$. Any vector can be represented as a product of its *magnitude* and *unit vector* along its direction, i.e.,

$$\vec{a} = |\vec{a}|\hat{n} = a\hat{n} \tag{2.1}$$

Here the length of the vector is specified by its magnitude and the direction by the unit vector along its direction. The unit vector along the direction of any vector can be obtained as

$$\hat{n} = \frac{\vec{a}}{|\vec{a}|} \tag{2.2}$$

The unit vectors that are of importance to us are the unit vectors along the rectangular coordinate axes. For a rectangular coordinate system, the unit vectors along X, Y and Z axes are normally represented respectively as $\hat{i}$, $\hat{j}$, and $\hat{k}$, or simply $\vec{i}$, $\vec{j}$, and $\vec{k}$.

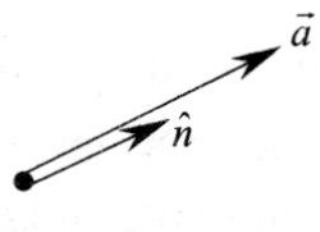

Fig. 2.7 Unit vector along the direction of any vector

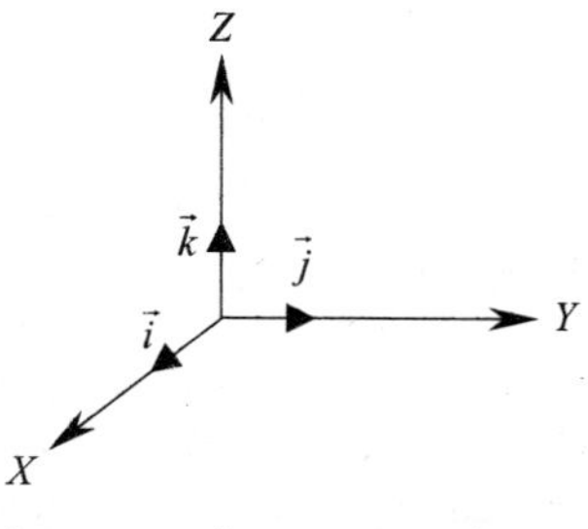

Fig. 2.8 Unit vectors along rectangular coordinate axes

There are other unit vectors too pointing along the path of the particle, $\hat{e}_t$, and normal to the path, $\hat{e}_n$. Though their magnitudes are unity, their directions, unlike the above rectangular unit vectors, keep on changing with respect to time.

Null Vector A vector whose magnitude is *zero* is called a **null vector**. It is similar to zero in scalars, but it is differentiated by a vector sign as $\vec{O}$.

Negative of a Vector A vector having the same magnitude, but direction opposite to that of a given vector is called the **negative** of the vector. Suppose the given vector is $\vec{a}$, then its negative vector is denoted as $-\vec{a}$. Graphically, it is represented as shown in Fig. 2.9.

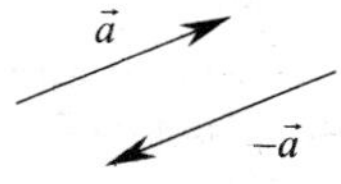

Fig. 2.9 Negative of a vector

2.4 MATHEMATICAL OPERATIONS OF VECTORS

2.4.1 Addition of Vectors

The addition of vectors can be determined graphically using the **parallelogram law**. It states that, when two concurrent vectors $\vec{a}$ and $\vec{b}$ are represented by two adjacent sides of a parallelogram, then the diagonal passing through their point of concurrency represents the sum of vectors $\vec{a} + \vec{b}$ in magnitude and direction.

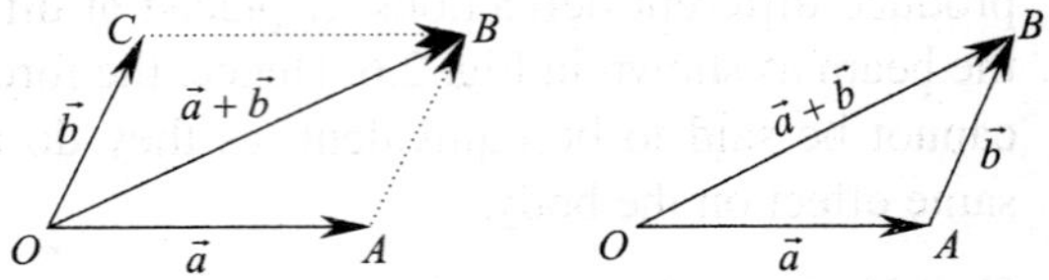

Fig. 2.10 Addition of vectors

Alternatively, treating $\vec{b}$ as a free vector, we can construct the triangle *OAB* and the summation can be determined using **triangle law**. It states that if two vectors $\vec{a}$ and $\vec{b}$ can be represented by the two sides of a triangle (in magnitude and direction) taken in order, then the third side (closing side) represents the sum of the two vectors in the opposite order.

The addition of vectors is independent of the order in which the vectors are selected. Hence, we say that vectors follow **commutative law**. Also, the vectors can be grouped without any restriction. Hence, vectors also follow **associative law** as shown in Fig. 2.11.

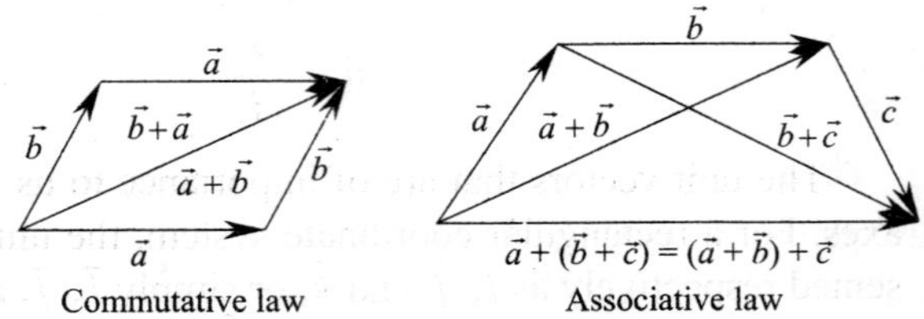

Fig. 2.11

$$\vec{a} + \vec{b} = \vec{b} + \vec{a} \qquad \text{(Commutative law)} \tag{2.3}$$

$$(\vec{a} + \vec{b}) + \vec{c} = \vec{a} + (\vec{b} + \vec{c}) = \vec{a} + \vec{b} + \vec{c} \qquad \text{(Associative law)} \tag{2.4}$$

Thus, we should note that vectors, because of their directions do not add up algebraically, but follow the parallelogram law of addition.

Corollary It is worthwhile mentioning at this point that some physical quantities like *angular rotation*, though having magnitude and direction do not obey the parallelogram law. Hence, it is for this reason we included in the definition of vectors at the beginning that vectors not only must have magnitude and direction but also that they should obey the parallelogram law of addition.

2.4.2 Subtraction of Vectors

The difference of two vectors is obtained by adding the first vector with the negative of the second vector, i.e., $\vec{a} + (-\vec{b})$. The negative of $\vec{b}$ is given by $\overrightarrow{AC}$. Hence, $\overrightarrow{OC}$ gives the difference of $\vec{a}$ and $\vec{b}$.

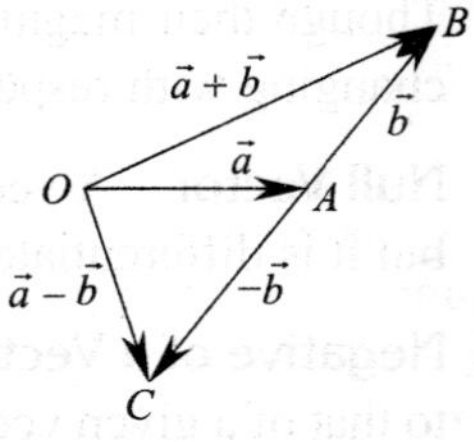

Fig. 2.12 Subtraction of vectors

The graphical methods so far discussed become tedious to work with when the number of vectors are more and when the vectors are three-dimensional or in space. Hence, we introduce the analytical method, called the *resolution of vectors*, to simplify such problems.

2.4.3 Resolution of a Vector

This method is the inverse of the parallelogram law. That is, just as two vectors can be added to give a single vector, any vector can be resolved into two components along a pair of given axes. As there could be infinite number of such pairs of axes, we will have infinite number of components to a given vector.

However, we are interested only in a particular case, where the components are taken about a pair of perpendicular axes. Such axes are called *orthogonal* or *rectangular* axes and the components are called *rectangular components*.

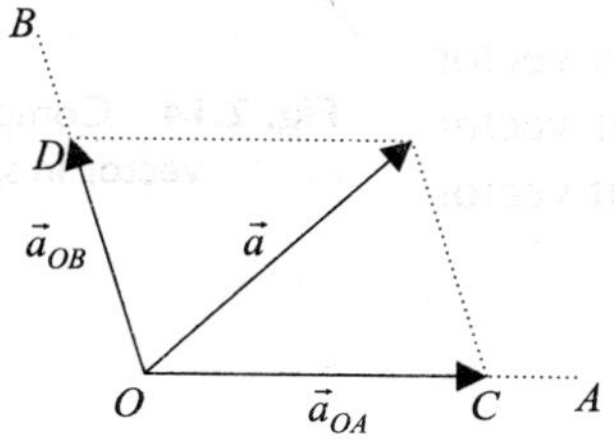

Fig. 2.13(a) Resolution of a vector

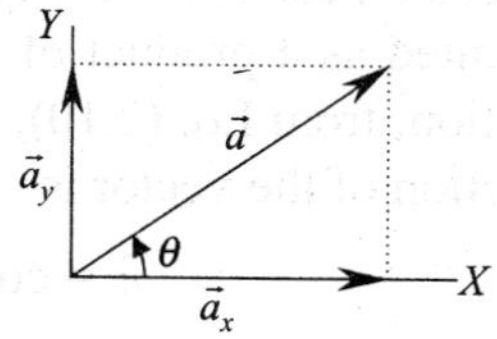

Fig. 2.13(b) Rectangular components of a vector

The rectangular components of a vector are obtained by projecting the vector onto *X* and *Y* axes as shown in Fig. 2.13(b). Hence, the vector can be written as

$$\vec{a} = \vec{a}_x + \vec{a}_y \tag{2.5}$$

where $\vec{a}_x$ and $\vec{a}_y$ are called **vector components** of $\vec{a}$.

As we already saw that any vector could be represented as a product of its magnitude and the unit vector along its direction, we can also express the vector $\vec{a}$ in Eq. 2.5 as

$$\vec{a} = a_x\vec{i} + a_y\vec{j} \tag{2.6}$$

where a_x and a_y are called **scalar components** or simply **components** of the vector, and $\vec{i}$ and $\vec{j}$ are the unit vectors along *X* and *Y* axes respectively. If θ is the inclination of $\vec{a}$ with respect to the *X*-axis, then

$$\vec{a} = a\cos\theta\vec{i} + a\sin\theta\vec{j} \tag{2.7}$$

The magnitude and direction of the vector can be expressed in terms of its components as

$$|\vec{a}| = a = \sqrt{a_x^2 + a_y^2} \tag{2.8}$$

$$\theta = \tan^{-1}\left[\frac{a_y}{a_x}\right] \tag{2.9}$$

The same procedure can be extended to a vector in three-dimensional space with the addition of a component along the *Z*-direction. Hence, a vector in space can be expressed in terms of its components as

$$\begin{aligned}\vec{a} &= \vec{a}_x + \vec{a}_y + \vec{a}_z \\ &= a_x\vec{i} + a_y\vec{j} + a_z\vec{k} \\ &= a\cos\theta_x\vec{i} + a\cos\theta_y\vec{j} + a\cos\theta_z\vec{k} \\ &= a\,[\cos\theta_x\vec{i} + \cos\theta_y\vec{j} + \cos\theta_z\vec{k}]\end{aligned} \tag{2.10}$$

where θ_x, θ_y and θ_z are the angles made by the vector with respect to the X, Y and Z axes respectively.

Its magnitude and direction are given in terms of the components as

$$|\vec{a}| = a = \sqrt{a_x^2 + a_y^2 + a_z^2} \tag{2.11}$$

$$\cos\theta_x = \frac{a_x}{a},\ \cos\theta_y = \frac{a_y}{a} \text{ and } \cos\theta_z = \frac{a_z}{a} \tag{2.12}$$

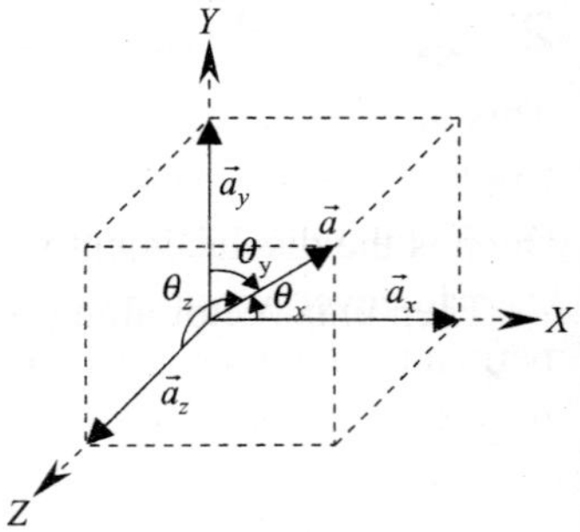

Fig. 2.14 Components of a vector in space

where cos θ_x, cos θ_y, cos θ_z are called **direction cosines**. As any vector can be represented as a product of its magnitude and the unit vector along its direction, from Eq. (2.10), we readily see that the unit vector along the direction of the vector is

$$\hat{n} = \cos\theta_x\vec{i} + \cos\theta_y\vec{j} + \cos\theta_z\vec{k} \tag{2.13}$$

Since the magnitude of the unit vector is unity,

$$\cos^2\theta_x + \cos^2\theta_y + \cos^2\theta_z = 1 \tag{2.14}$$

If the inclinations of the vector with respect to any two orthogonal axes are known, then its inclination with the third axis can be determined from the above expression.

Addition of Vectors Analytically If two vectors can be represented in terms of their components as

$$\vec{a} = a_x\vec{i} + a_y\vec{j} + a_z\vec{k}$$

and

$$\vec{b} = b_x\vec{i} + b_y\vec{j} + b_z\vec{k}$$

then their summation is obtained by adding the corresponding components along the X, Y and Z directions, i.e.,

$$\vec{a} + \vec{b} = (a_x + b_x)\vec{i} + (a_y + b_y)\vec{j} + (a_z + b_z)\vec{k} \tag{2.15}$$

and their difference is given as

$$\vec{a} - \vec{b} = (a_x - b_x)\vec{i} + (a_y - b_y)\vec{j} + (a_z - b_z)\vec{k} \tag{2.16}$$

The physical application of summation and difference of two vectors is to determine the **displacement vector**. Suppose a body is at point A at an instant of time, then its **position vector** with respect to a coordinate system is obtained by drawing a line segment from the origin to the point A. If (x_1, y_1, z_1) be the coordinates of point A, then its position vector is given as

$$\overrightarrow{OA} = x_1\vec{i} + y_1\vec{j} + z_1\vec{k}$$

At a later instant of time, if it is at point B, whose coordinates are (x_2, y_2, z_2), then its position vector is given as

$$\overrightarrow{OB} = x_2\vec{i} + y_2\vec{j} + z_2\vec{k}$$

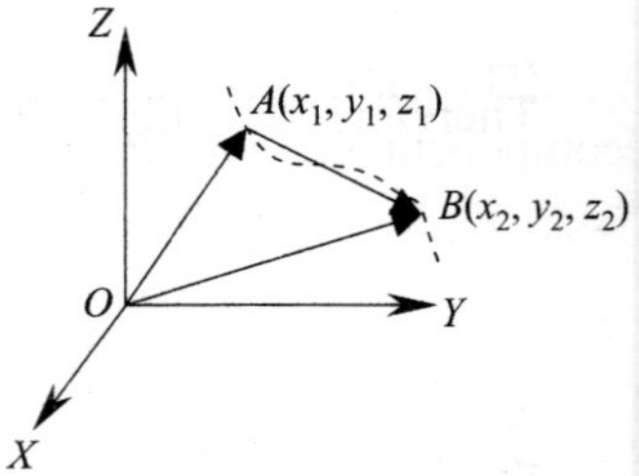

Fig. 2.15 Displacement vector

Hence, in this time interval, the body has displaced from A to B and the displacement is obtained by drawing a line segment from A to B. Note that the actual path traveled may not be a straight line. Using the addition of vectors, we know that,

$$\overrightarrow{OB} = \overrightarrow{OA} + \overrightarrow{AB}$$

$$\therefore \qquad \overrightarrow{AB} = \overrightarrow{OB} - \overrightarrow{OA}$$

which is the displacement vector.

Then unit vector along AB is given as

$$\hat{n}_{AB} = \frac{\overrightarrow{AB}}{|\overrightarrow{AB}|}$$

Example 2.1 Find the magnitude of the vector $5\vec{i} + 6\vec{j} - 2\vec{k}$ and unit vector along its direction.

Solution Let $\vec{a} = 5\vec{i} + 6\vec{j} - 2\vec{k}$. Its magnitude is given as

$$|\vec{a}| = a = \sqrt{5^2 + 6^2 + (-2)^2} = \sqrt{65} \text{ units}$$

Therefore, unit vector along its direction is obtained as

$$\hat{n} = \frac{\vec{a}}{|\vec{a}|} = \frac{5\vec{i} + 6\vec{j} - 2\vec{k}}{\sqrt{65}}$$

Example 2.2 Determine the unit vector parallel to the sum of vectors $(2\vec{i} + 4\vec{j} - 5\vec{k})$ and $(\vec{i} + 2\vec{j} + 3\vec{k})$.

Solution Let $\vec{a} = 2\vec{i} + 4\vec{j} - 5\vec{k}$ and $\vec{b} = \vec{i} + 2\vec{j} + 3\vec{k}$, then

$$\begin{aligned}\vec{a} + \vec{b} &= (2\vec{i} + 4\vec{j} - 5\vec{k}) + (\vec{i} + 2\vec{j} + 3\vec{k}) \\ &= 3\vec{i} + 6\vec{j} - 2\vec{k}\end{aligned}$$

Therefore, unit vector parallel to $\vec{a} + \vec{b}$ is

$$\begin{aligned}\hat{n} &= \frac{3\vec{i} + 6\vec{j} - 2\vec{k}}{\sqrt{3^2 + 6^2 + (-2)^2}} \\ &= \frac{3\vec{i} + 6\vec{j} - 2\vec{k}}{7}\end{aligned}$$

Example 2.3 Find the direction cosines of the vector $2\vec{i} + 3\vec{j} + 4\vec{k}$.

Solution Let $\vec{a} = 2\vec{i} + 3\vec{j} + 4\vec{k}$, then

$$|\vec{a}| = a = \sqrt{2^2 + 3^2 + 4^2} = \sqrt{29} \text{ units}$$

Therefore, direction cosines are

$$\cos\theta_x = \frac{a_x}{a} = \frac{2}{\sqrt{29}},$$

$$\cos\theta_y = \frac{a_y}{a} = \frac{3}{\sqrt{29}}$$

and

$$\cos\theta_z = \frac{a_z}{a} = \frac{4}{\sqrt{29}}$$

Example 2.4 If the position vectors of two points A and B are $2\vec{i}-9\vec{j}-4\vec{k}$ and $6\vec{i}-3\vec{j}-8\vec{k}$ respectively, then find $\overrightarrow{AB}$ and its magnitude.

Solution Given that $\overrightarrow{OA} = 2\vec{i} - 9\vec{j} - 4\vec{k}$ and $\overrightarrow{OB} = 6\vec{i} - 3\vec{j} - 8\vec{k}$,

then
$$\overrightarrow{AB} = \overrightarrow{OB} - \overrightarrow{OA}$$
$$= (6\vec{i} - 3\vec{j} - 8\vec{k}) - (2\vec{i} - 9\vec{j} - 4\vec{k})$$
$$= 4\vec{i} + 6\vec{j} - 4\vec{k}$$

Therefore, its magnitude is
$$|\overrightarrow{AB}| = \sqrt{4^2 + 6^2 + (-4)^2}$$
$$= \sqrt{68} \text{ units}$$

2.4.4 Multiplication of Vectors

Multiplication of a Vector by a Scalar If m is a scalar, then the product of a vector $\vec{a}$ by the scalar m is a vector whose magnitude is $m|\vec{a}|$. Its direction is *same* as $\vec{a}$ if m is *positive* and *opposite* to that of $\vec{a}$ if m is *negative*.

Hence,
$$\text{if } m = 1 \Rightarrow 1(\vec{a}) = \vec{a}$$
$$\text{if } m = -1 \Rightarrow -1(\vec{a}) = -\vec{a}$$

In general, if
$$\vec{a} = a_x\vec{i} + a_y\vec{j} + a_z\vec{k}, \text{ then}$$
$$m\vec{a} = m(a_x\vec{i} + a_y\vec{j} + a_z\vec{k})$$
$$= ma_x\vec{i} + ma_y\vec{j} + ma_z\vec{k} \qquad (2.17)$$

Scalar (or) Dot Product Scalar product of two vectors is defined as *the product of their magnitudes and the cosine of the included angle between them.* If $\vec{a}$ and $\vec{b}$ are two vectors, then the scalar product is expressed as

$$\vec{a}.\vec{b} = ab\cos\theta \qquad (2.18)$$

Since a and b are scalars, and $\cos\theta$ is a pure number, the scalar product of two vectors is a **scalar**. Because of the notation used, scalar product is also called **dot product** and it is read as $\vec{a}$ dot $\vec{b}$. Given two vectors, then the angle between them can be obtained as

$$\cos\theta = \frac{\vec{a}.\vec{b}}{ab} \qquad (2.19)$$

Geometrical Interpretation of Scalar Product From Fig. 2.16, we can see that the dot product of two vectors can also be written as

$$\vec{a}.\vec{b} = ab\cos\theta$$
$$= a\,[\text{projection of } \vec{b} \text{ on } \vec{a}] \qquad (2.20)$$

Also,
$$\vec{a}.\vec{b} = b\,[\text{projection of } \vec{a} \text{ on } \vec{b}] \qquad (2.21)$$

Hence, the dot product of two vectors is the product of magnitude of one of the vectors and the projection of the other vector on the first vector.

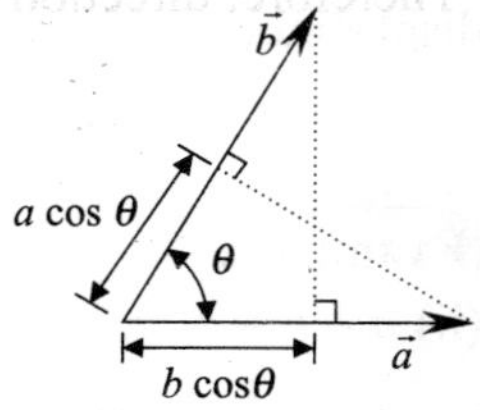

Fig. 2.16 Geometrical interpretation of dot product

Therefore, projection of $\vec{b}$ on $\vec{a} = b\cos\theta = \dfrac{\vec{a}.\vec{b}}{a}$ (2.22)

Similarly, projection of $\vec{a}$ on $\vec{b} = a\cos\theta = \dfrac{\vec{a}.\vec{b}}{b}$ (2.23)

Dot Product of Unit Vectors The dot product of two *collinear* or *parallel* vectors is equal to the product of their magnitudes as the included angle θ is zero or $\cos\theta = 1$. In the same way, the dot product of two *perpendicular* vectors is zero as the included angle θ is 90° or $\cos\theta = 0$. Based on this, we can determine the dot product of unit vectors:

$$\vec{i}.\vec{i} = \vec{j}.\vec{j} = \vec{k}.\vec{k} = 1 \tag{2.24}$$

$$\vec{i}.\vec{j} = \vec{j}.\vec{i} = \vec{j}.\vec{k} = \vec{k}.\vec{j} = \vec{k}.\vec{i} = \vec{i}.\vec{k} = 0 \tag{2.25}$$

In general, if $\vec{a} = a_x\vec{i} + a_y\vec{j} + a_z\vec{k}$ and

$$\vec{b} = b_x\vec{i} + b_y\vec{j} + b_z\vec{k}$$

then, $\vec{a}.\vec{b} = a_xb_x + a_yb_y + a_zb_z$ (2.26)

If $\vec{a} = \vec{b}$, then

$$\begin{aligned}\vec{a}.\vec{a} &= a_xa_x + a_ya_y + a_za_z\\ &= a_x^2 + a_y^2 + a_z^2\\ &= \left[\sqrt{a_x^2 + a_y^2 + a_z^2}\right]^2\\ &= a^2\end{aligned} \tag{2.27}$$

Dot product is a product of two scalars. As scalars follow the commutative law, the dot product of vectors also follows **commutative law**, i.e.,

$$\vec{a}.\vec{b} = ab\cos\theta = ba\cos\theta = \vec{b}.\vec{a} \tag{2.28}$$

Associative property, i.e.,

$$(\vec{a}.\vec{b}).\vec{c} \tag{2.29}$$

cannot apply to vectors as $\vec{a}.\vec{b}$ is a scalar and dot product of a scalar and a vector is *meaningless*.

Scalar product follows **distributive law**, i.e.,

$$\vec{a}.(\vec{b} + \vec{c}) = \vec{a}.\vec{b} + \vec{a}.\vec{c} \tag{2.30}$$

A number of important physical quantities can be described as the scalar product of two vectors. For instance, **work done** is defined as a product of force component in the direction of displacement and the displacement. Hence, work done can be expressed as

$$\text{Work done} = \vec{F}.d\vec{r} \tag{2.31}$$

Example 2.5 If $\vec{a} = 3\vec{i} - \vec{j} - 4\vec{k}$, $\vec{b} = -2\vec{i} + 4\vec{j} - 7\vec{k}$ and $\vec{c} = \vec{i} + 2\vec{j} + \vec{k}$, find (i) $\vec{a} + 2\vec{b} + 3\vec{c}$, (ii) $\vec{a} - \vec{b} - \vec{c}$, iii) $3\vec{a} - 2\vec{b} + 4\vec{c}$. Also, find the magnitude of the vector in each case.

Solution

(i)
$$\vec{a}+2\vec{b}+3\vec{c} = (3\vec{i}-\vec{j}-4\vec{k})+2(-2\vec{i}+4\vec{j}-7\vec{k})+3(\vec{i}+2\vec{j}+\vec{k})$$
$$= 2\vec{i}+13\vec{j}-15\vec{k}$$
$$|\vec{a}+2\vec{b}+3\vec{c}| = \sqrt{2^2+13^2+(-15)^2} = \sqrt{398} \text{ units}$$

(ii)
$$\vec{a}-\vec{b}-\vec{c} = (3\vec{i}-\vec{j}-4\vec{k})-(-2\vec{i}+4\vec{j}-7\vec{k})-(\vec{i}+2\vec{j}+\vec{k})$$
$$= 4\vec{i}-7\vec{j}+2\vec{k}$$
$$|\vec{a}-\vec{b}-\vec{c}| = \sqrt{4^2+(-7)^2+2^2} = \sqrt{69} \text{ units}$$

(iii)
$$3\vec{a}-2\vec{b}+4\vec{c} = 3(3\vec{i}-\vec{j}-4\vec{k})-2(-2\vec{i}+4\vec{j}-7\vec{k})+4(\vec{i}+2\vec{j}+\vec{k})$$
$$= 17\vec{i}-3\vec{j}+6\vec{k}$$
$$|3\vec{a}-2\vec{b}+4\vec{c}| = \sqrt{17^2+(-3)^2+(6)^2} = \sqrt{334} \text{ units}$$

Example 2.6 If $\vec{a}=2\vec{i}+\vec{j}+2\vec{k}$ and $\vec{b}=\vec{i}-3\vec{j}+\vec{k}$, then find $\vec{a}.\vec{b}$ and the angle between them and the projection of $\vec{b}$ on $\vec{a}$.

Solution The dot product of two vectors is given as
$$\vec{a}.\vec{b} = (2\vec{i}+\vec{j}+2\vec{k}).(\vec{i}-3\vec{j}+\vec{k})$$
$$= 2-3+2 = 1$$

Also,
$$|\vec{a}| = a = \sqrt{2^2+1^2+2^2} = 3$$

and
$$|\vec{b}| = b = \sqrt{1^2+(-3)^2+1^2} = \sqrt{11}$$

Therefore, the angle between the two vectors is given as
$$\cos\theta = \frac{\vec{a}.\vec{b}}{ab} = \frac{1}{3\sqrt{11}} = 0.101$$

$\Rightarrow$
$$\theta = 84.23°$$

Projection of $\vec{b}$ on $\vec{a} = \dfrac{\vec{a}.\vec{b}}{a} = \dfrac{1}{3}$

Example 2.7 If $\vec{a}=2\vec{i}+2\vec{j}+2\vec{k}$ and $\vec{b}=\vec{i}+2\vec{j}+\vec{k}$, then find the angle between $(\vec{a}+\vec{b})$ and $(\vec{a}-\vec{b})$.

Solution
$$\vec{a}+\vec{b} = (2\vec{i}+2\vec{j}+2\vec{k})+(\vec{i}+2\vec{j}+\vec{k}) = 3\vec{i}+4\vec{j}+3\vec{k}$$

Hence, its magnitude is
$$|\vec{a}+\vec{b}| = \sqrt{3^2+4^2+3^2} = \sqrt{34} \text{ units}$$

Similarly,
$$\vec{a}-\vec{b} = (2\vec{i}+2\vec{j}+2\vec{k})-(\vec{i}+2\vec{j}+\vec{k}) = \vec{i}+\vec{k}$$

and its magnitude is
$$|\vec{a}-\vec{b}| = \sqrt{1^2+1^2} = \sqrt{2} \text{ units}$$

Therefore, the angle between the two vectors is given as

$$\cos\theta = \frac{(\vec{a}+\vec{b}).(\vec{a}-\vec{b})}{|\vec{a}+\vec{b}||\vec{a}-\vec{b}|} = \frac{(3\vec{i}+4\vec{j}+3\vec{k}).(\vec{i}+\vec{k})}{\sqrt{34}\sqrt{2}} = \frac{3+3}{\sqrt{68}}$$

$\Rightarrow$ $$\theta = 43.31°$$

Example 2.8 Find the angle, which $\vec{a} = \vec{i} + 2\vec{j} + 3\vec{k}$ makes with the Z-axis.

Solution The magnitude of $\vec{a}$ is given as

$$|\vec{a}| = \sqrt{1^2 + 2^2 + 3^2} = \sqrt{14} \text{ units}$$

Since the unit vector along the Z-axis is $\vec{k}$, we take the dot product of $\vec{a}$ with $\vec{k}$

$$\vec{a}.\vec{k} = (\vec{i} + 2\vec{j} + 3\vec{k}).(\vec{k}) = 3$$

Since dot product of two vectors is also given as $\vec{a}.\vec{k} = (a)(1)\cos\theta$

$$(a)(1)\cos\theta = 3$$

$$\sqrt{14}.1\cos\theta = 3$$

$\Rightarrow$ $$\cos\theta = 3/\sqrt{14}$$

$\therefore$ $$\theta = 36.7°$$

Example 2.9 The dot product of a vector with vectors $(3\vec{i} - 5\vec{k})$, $(2\vec{i} + 7\vec{j})$ and $(\vec{i} + \vec{j} + \vec{k})$ are respectively –1, 6 and 5. Find the vector.

Solution Assume the unknown vector to be $x\vec{i} + y\vec{j} + z\vec{k}$. Taking dot product of this vector with each of the given vectors,

$$3x - 5z = -1 \quad \text{(a)}$$

$$2x + 7y = 6 \quad \text{(b)}$$

$$x + y + z = 5 \quad \text{(c)}$$

From equations (a) and (b), we get the values of y and z in terms of x

$$z = \left[\frac{1+3x}{5}\right] \quad \text{(d)}$$

$$y = \left[\frac{6-2x}{7}\right] \quad \text{(e)}$$

Substituting these values in equation(c)

$$x + y + z = 5$$

$$x + \left[\frac{6-2x}{7}\right] + \left[\frac{1+3x}{5}\right] = 5$$

$$\frac{35x + 30 - 10x + 7 + 21x}{35} = 5$$

$$46x + 37 = 175$$

$$46x = 138$$

$$\Rightarrow \quad x = \frac{138}{46} = 3$$

From equation (e),

$$y = \left[\frac{6-2x}{7}\right] = \left[\frac{6-2(3)}{7}\right] = 0$$

From equation (d),

$$z = \left[\frac{1+3x}{5}\right] = \left[\frac{1+3(3)}{5}\right] = \frac{10}{5} = 2$$

Therefore, the vector is

$$x\vec{i} + y\vec{j} + z\vec{k} = 3\vec{i} + 2\vec{k}$$

Vector (or) Cross Product Vector product of two vectors is a **vector**, whose magnitude is the product of magnitudes of the two vectors and sine of their included angle; its direction is determined by the right-hand screw rule. If $\vec{a}$ and $\vec{b}$ are two vectors, then their vector product is expressed as

$$\vec{a} \times \vec{b} = ab \sin\theta \hat{n} \tag{2.32}$$

where θ is the smaller of the angles between the vectors, i.e., $\theta \leq 180°$. Hence, $\sin\theta$ is always positive.

Because of the notation used, vector product is also called **cross product** and it is read as $\vec{a}$ cross $\vec{b}$. Its magnitude is given as

$$|\vec{a} \times \vec{b}| = ab \sin\theta \tag{2.33}$$

and its direction is specified by right-hand screw rule. Suppose a right-hand screw with its axis perpendicular to the plane formed by the two vectors and passing through their point of concurrency is rotated from $\vec{a}$ to $\vec{b}$ through the smaller angle between them, then the direction of advance of the screw gives the direction of the vector product. Alternatively, if we curl the fingers of our right hand around an axis perpendicular to the plane formed by the two vectors and passing through the point of concurrency while moving from $\vec{a}$ to $\vec{b}$, then the extended thumb specifies the sense of the cross product.

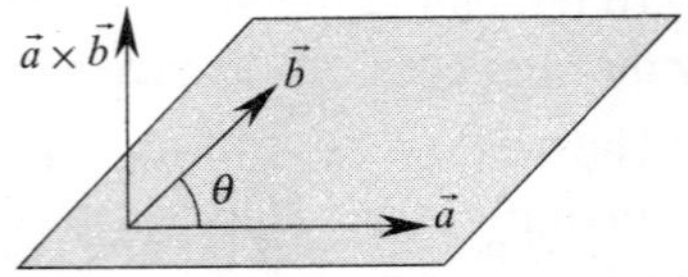

Fig. 2.17 Graphical representation of cross product

Note: Once we define the direction of the *X* and *Y* axes, to define the *Z*-direction in a right-hand triad, we must follow the right-hand screw rule as explained above. When we curl the fingers of our right hand from the *X*-axis to *Y*-axis through the smaller angle, then the direction of the extended thumb indicates the direction of the *Z*-axis. The following figures will clarify this.

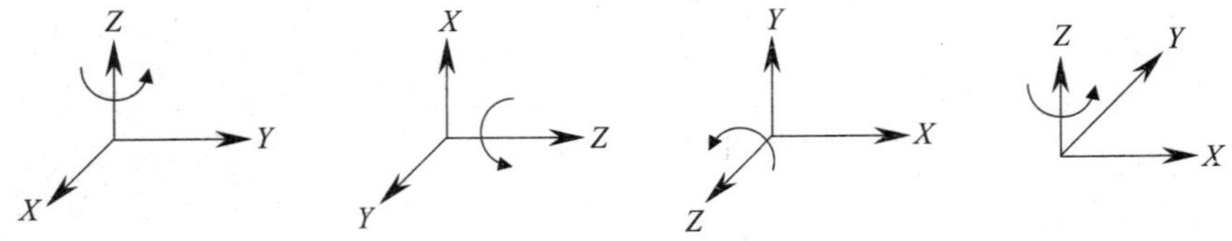

Fig. 2.18 Right-hand triad

Geometrical Interpretation of Cross Product Construct a parallelogram $OABC$ with the two vectors $\vec{a}$ and $\vec{b}$ as its adjacent sides. From Fig. 2.19, we see that $b \sin\theta$ is the perpendicular height DC of the parallelogram.

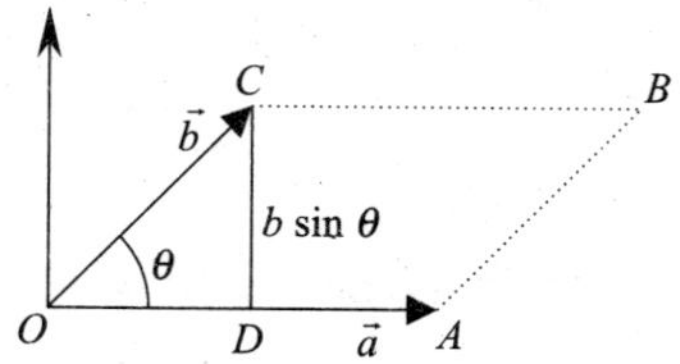

Fig. 2.19 Geometrical interpretation of cross product

Then magnitude of the cross product of two vectors can be written as

$$\begin{aligned}|\vec{a} \times \vec{b}| &= ab \sin\theta \\ &= [a][b \sin\theta] \\ &= \text{base} \times \text{perpendicular height of parallelogram} \\ &= \text{area of parallelogram } OABC \end{aligned} \tag{2.34}$$

Hence, the magnitude of the cross product of two vectors is the *area of the parallelogram formed by the two vectors as adjacent sides.*

Cross Product of Rectangular Unit Vectors When two vectors are *parallel*, then the angle between them is zero or $\sin\theta = 0$. Then

$$\vec{i} \times \vec{i} = \vec{j} \times \vec{j} = \vec{k} \times \vec{k} = \vec{O} \tag{2.35}$$

When two vectors are perpendicular to each other, then the angle between them is 90° or $\sin\theta = 1$. Then

$$\begin{aligned} \vec{i} \times \vec{j} &= \vec{k} & \vec{j} \times \vec{i} &= -\vec{k} \\ \vec{j} \times \vec{k} &= \vec{i} & \vec{k} \times \vec{j} &= -\vec{i} \\ \vec{k} \times \vec{i} &= \vec{j} & \vec{i} \times \vec{k} &= -\vec{j} \end{aligned} \tag{2.36}$$

To remember the above cross product of two rectangular unit vectors, we follow the sign convention as explained in Fig. 2.20.

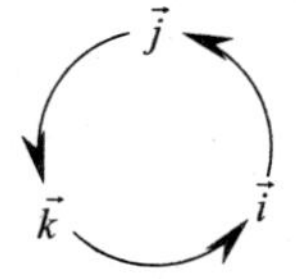

Fig. 2.20 Positive sign convention sequence

The unit vectors $\vec{i}$, $\vec{j}$ and $\vec{k}$ are arranged in an anti-clockwise sequence around the circumference of a circle. While taking cross product of two unit vectors, if we move in the direction of the sequence, then it is taken as **positive**, otherwise, it is **negative**.

In general, if

$$\vec{a} = a_x\vec{i} + a_y\vec{j} + a_z\vec{k} \text{ and}$$

$$\vec{b} = b_x\vec{i} + b_y\vec{j} + b_z\vec{k}$$

then

$$\begin{aligned}\vec{a} \times \vec{b} &= (a_x\vec{i} + a_y\vec{j} + a_z\vec{k}) \times (b_x\vec{i} + b_y\vec{j} + b_z\vec{k}) \\ &= (a_xb_y - a_yb_x)\vec{k} + (a_zb_x - a_xb_z)\vec{j} + (a_yb_z - a_zb_y)\vec{i} \\ &= (a_yb_z - a_zb_y)\vec{i} - (a_xb_z - a_zb_x)\vec{j} + (a_xb_y - a_yb_x)\vec{k} \end{aligned} \tag{2.37}$$

which can also be written in determinant form as

$$\vec{a} \times \vec{b} = \begin{vmatrix} \vec{i} & \vec{j} & \vec{k} \\ a_x & a_y & a_z \\ b_x & b_y & b_z \end{vmatrix} \tag{2.38}$$

Vector product is **not commutative**, i.e.,

$$\vec{a} \times \vec{b} \neq \vec{b} \times \vec{a} \tag{2.39}$$

Instead, $$\vec{a} \times \vec{b} = -(\vec{b} \times \vec{a}) \tag{2.40}$$

Vector product is **not associative**, i.e.,

$$(\vec{a} \times \vec{b}) \times \vec{c} \neq \vec{a} \times (\vec{b} \times \vec{c}) \tag{2.41}$$

Vector product is **distributive** with respect to vector addition, i.e.,

$$\vec{a} \times (\vec{b} + \vec{c}) = (\vec{a} \times \vec{b}) + (\vec{a} \times \vec{c}) \tag{2.42}$$

A number of important physical quantities can be described as vector product of two vectors. For example, **moment of a force** about the origin is defined as the cross product of the position vector of the point of application of the force and the force vector.

Example 2.10 If $\vec{a} = 3\vec{i} - \vec{j} + 2\vec{k}$ and $\vec{b} = 2\vec{i} + \vec{j} - \vec{k}$, compute $\vec{a} \times \vec{b}$.

Solution The cross product of the two vectors is given as

$$\vec{a} \times \vec{b} = \begin{vmatrix} \vec{i} & \vec{j} & \vec{k} \\ 3 & -1 & 2 \\ 2 & 1 & -1 \end{vmatrix}$$

$$= \vec{i}(1 - 2) - \vec{j}(-3 - 4) + \vec{k}(3 + 2)$$

$$= -\vec{i} + 7\vec{j} + 5\vec{k}$$

Example 2.11 Find the area of the parallelogram with adjacent sides $2\vec{i} - 3\vec{j}$ and $3\vec{i} - \vec{k}$.

Solution The area of the parallelogram with the given vectors as adjacent sides is given by the magnitude of their cross product. Let $\vec{a} = 2\vec{i} - 3\vec{j}$, $\vec{b} = 3\vec{i} - \vec{k}$, then

$$\vec{a} \times \vec{b} = \begin{vmatrix} \vec{i} & \vec{j} & \vec{k} \\ 2 & -3 & 0 \\ 3 & 0 & -1 \end{vmatrix}$$

$$= \vec{i}(3 - 0) - \vec{j}(-2 - 0) + \vec{k}(0 + 9)$$

$$= 3\vec{i} + 2\vec{j} + 9\vec{k}$$

Hence, $$|\vec{a} \times \vec{b}| = \sqrt{3^2 + 2^2 + 9^2}$$

$$= 9.7 \text{ units}$$

Example 2.12 A line passes through the points (5 m, 4 m, 0) and (–1 m, –4 m, –2 m). Determine the perpendicular distance d from the line to the origin of the system of coordinates.

Solution Let the two points be A and B respectively, i.e., A (5, 4, 0) and B (–1, –4, –2). Then

$$\overrightarrow{OA} = 5\vec{i} + 4\vec{j}$$

$$\overrightarrow{OB} = -\vec{i} - 4\vec{j} - 2\vec{k}$$

$$\overrightarrow{BA} = \overrightarrow{OA} - \overrightarrow{OB} = 6\vec{i} + 8\vec{j} + 2\vec{k}$$

Taking cross product of the vectors $\overrightarrow{BO}$ and $\overrightarrow{BA}$,

$$\overrightarrow{BO} \times \overrightarrow{BA} = \begin{vmatrix} \vec{i} & \vec{j} & \vec{k} \\ 1 & 4 & 2 \\ 6 & 8 & 2 \end{vmatrix}$$

$$= \vec{i}(8 - 16) - \vec{j}(2 - 12) + \vec{k}(8 - 24)$$

$$= -8\vec{i} + 10\vec{j} - 16\vec{k}$$

$$|\overrightarrow{BO} \times \overrightarrow{BA}| = \sqrt{(-8)^2 + (10)^2 + (-16)^2} = 20.49 \qquad \text{(a)}$$

Also, we know,

$$|\overrightarrow{BO} \times \overrightarrow{BA}| = (BO)(BA)\sin\theta = (BA)d \qquad \text{(b)}$$

where d is the perpendicular distance from the line to the origin. Equating the two equations (a) and (b), we get

$$(BA)d = 20.49$$

$$\Rightarrow \qquad d = \frac{20.49}{BA}$$

$$= \frac{20.49}{\sqrt{6^2 + 8^2 + 2^2}}$$

$$= 2.01 \text{ m}$$

Example 2.13 Find 't' if $4\vec{i} + (2t/3)\vec{j} + t\vec{k}$ is parallel to $\vec{i} + 2\vec{j} + 3\vec{k}$.

Solution When two vectors are parallel, then their cross product is zero.

$$\vec{a} \times \vec{b} = \begin{vmatrix} \vec{i} & \vec{j} & \vec{k} \\ 4 & 2t/3 & t \\ 1 & 2 & 3 \end{vmatrix}$$

$$\vec{O} = \vec{i}(2t - 2t) - \vec{j}(12 - t) + \vec{k}(8 - 2t/3)$$

$$= (t - 12)\vec{j} + (8 - 2t/3)\vec{k}$$

$$\Rightarrow \qquad t = 12$$

Scalar Triple Product For a set of three vectors $\vec{a}$, $\vec{b}$ and $\vec{c}$, the scalar triple product is defined as

$$(\vec{a} \times \vec{b}).\vec{c} \qquad (2.43)$$

We know,

$$\vec{a} \times \vec{b} = \begin{vmatrix} \vec{i} & \vec{j} & \vec{k} \\ a_x & a_y & a_z \\ b_x & b_y & b_z \end{vmatrix}$$

$$= (a_y b_z - a_z b_y)\vec{i} - (a_x b_z - a_z b_x)\vec{j} + (a_x b_y - a_y b_x)\vec{k}$$

Therefore, $(\vec{a} \times \vec{b}).\vec{c} = (a_y b_z - a_z b_y)(c_x) - (a_x b_z - a_z b_x)(c_y) + (a_x b_y - a_y b_x)(c_z)$

On rearranging,

$$(\vec{a} \times \vec{b}).\vec{c} = a_x(b_y c_z - b_z c_y) - a_y(b_x c_z - b_z c_x) + a_z(b_x c_y - b_y c_x)$$

which can also be represented in determinant form as

$$(\vec{a} \times \vec{b}).\vec{c} = \begin{vmatrix} a_x & a_y & a_z \\ b_x & b_y & b_z \\ c_x & c_y & c_z \end{vmatrix} \quad (2.44)$$

Geometrical Interpretation of Scalar Triple Product If we assume $\vec{a}$ and $\vec{b}$ to lie on the X–Y plane, then $\vec{a} \times \vec{b}$ is a vector, whose magnitude, i.e., $|\vec{a} \times \vec{b}|$ is the area of the parallelogram $OABC$ and its direction is perpendicular to X–Y plane or along Z direction. Then

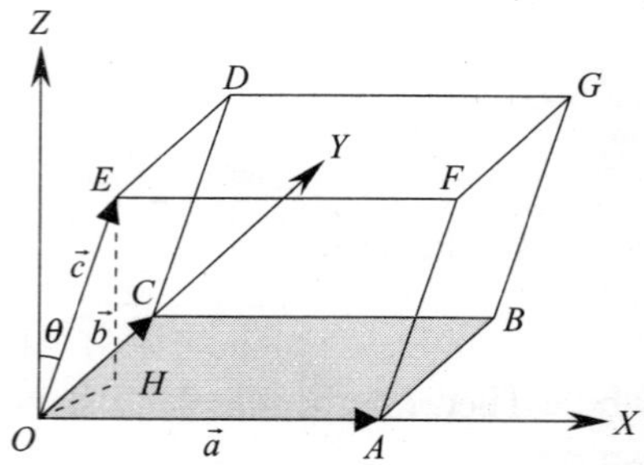

Fig. 2.21

$$(\vec{a} \times \vec{b}).\vec{c} = |\vec{a} \times \vec{b}|\,|\vec{c}| \cos\theta$$

$= [\text{magnitude of } \vec{a} \times \vec{b}][\text{magnitude of } \vec{c}][\cos\theta]$

$= [\text{area of parallelogram } OABC]\ [\text{height } EH]$

$= \text{volume of parallelepiped with sides } \vec{a}, \vec{b} \text{ and } \vec{c}$ (2.45)

Note: The scalar triple product is *zero*, if all the three vectors are *coplanar*. We know that cross product of two vectors will be a vector perpendicular to the plane formed by the two vectors. If its dot product with a third vector is zero, then the angle between the cross product of the first two vectors and the third vector is 90° or in other words, all the three vectors are coplanar. This is useful in checking if three given vectors lie in the same plane or not.

Scalar triple product is commutative

$$(\vec{a} \times \vec{b}).\vec{c} = \vec{a}.(\vec{b} \times \vec{c}) \quad (2.46)$$

Also, $\vec{a}.(\vec{b} \times \vec{c}) = \vec{b}.(\vec{c} \times \vec{a}) = \vec{c}.(\vec{a} \times \vec{b})$, because the volume is constant. (2.47)

Vector Triple Product For a set of three vectors $\vec{a}$, $\vec{b}$ and $\vec{c}$, the vector triple product is defined as

$$\vec{a} \times (\vec{b} \times \vec{c}) \quad (2.48)$$

It is used in dynamics. Instead of finding the determinant in two steps, vector triple product can be found out from the simple expression

$$\vec{a} \times (\vec{b} \times \vec{c}) = \vec{b}(\vec{a}.\vec{c}) - \vec{c}(\vec{a}.\vec{b})$$

The student can easily verify the proof of this expression.

Example 2.14 If $\vec{a} = \vec{i} - 2\vec{j} - 3\vec{k}$, $\vec{b} = \vec{i} + 3\vec{j} + 2\vec{k}$ and $\vec{c} = 2\vec{i} - 4\vec{j} + \vec{k}$, find $\vec{a}.(\vec{b} \times \vec{c})$ and $\vec{a} \times (\vec{b} \times \vec{c})$.

Solution The scalar triple product is given as

$$\vec{a}.(\vec{b} \times \vec{c}) = \begin{vmatrix} 1 & -2 & -3 \\ 1 & 3 & 2 \\ 2 & -4 & 1 \end{vmatrix}$$

$$= 1(3 + 8) - (-2)(1 - 4) - 3(-4 - 6) = 11 - 6 + 30 = 35 \text{ units}$$

The vector triple product is given as

$$\vec{a} \times (\vec{b} \times \vec{c}) = \vec{b}(\vec{a}.\vec{c}) - \vec{c}(\vec{a}.\vec{b})$$

$$\vec{a}.\vec{c} = (\vec{i} - 2\vec{j} - 3\vec{k}).(2\vec{i} - 4\vec{j} + \vec{k})$$

$$= 2 + 8 - 3 = 7 \text{ units}$$

$$\vec{a}.\vec{b} = (\vec{i} - 2\vec{j} - 3\vec{k}).(\vec{i} + 3\vec{j} + 2\vec{k})$$

$$= 1 - 6 - 6 = -11 \text{ units}$$

Therefore,

$$\vec{a} \times (\vec{b} \times \vec{c}) = \vec{b}(\vec{a}.\vec{c}) - \vec{c}(\vec{a}.\vec{b})$$

$$= (\vec{i} + 3\vec{j} + 2\vec{k})(7) - (2\vec{i} - 4\vec{j} + \vec{k})(-11)$$

$$= 29\vec{i} - 23\vec{j} + 25\vec{k}$$

Example 2.15 Given the vectors $\vec{a} = 3\vec{i} - \vec{j} + \vec{k}$, $\vec{b} = 4\vec{i} + b_y\vec{j} - 2\vec{k}$ and $\vec{c} = 2\vec{i} - 2\vec{j} + 2\vec{k}$, determine the value of b_y for which the three vectors are coplanar.

Solution Considering the scalar triple product,

$$(\vec{a} \times \vec{b}).\vec{c} = \begin{vmatrix} 3 & -1 & 1 \\ 4 & b_y & -2 \\ 2 & -2 & 2 \end{vmatrix}$$

$$= 3(2b_y - 4) - (-1)(8 + 4) + 1(-8 - 2b_y)$$

$$= 6b_y - 12 + 12 - 8 - 2b_y = 4b_y - 8$$

The three vectors are coplanar if their scalar triple product is zero. Therefore,

$$0 = 4b_y - 8 \quad \Rightarrow \quad b_y = 2$$

SUMMARY

Scalars and Vectors

Those physical quantities which can be specified completely by stating their magnitude (a number and unit), are termed *scalars*. The examples are mass, length, time, density, energy, power, etc. Those physical quantities which are specified completely by their magnitude and direction and in addition, which obey the parallelogram law of addition are termed *vectors*. The examples are displacement, velocity, acceleration, momentum, force, etc.

Types of Vectors

Fixed vectors are those vectors which have a fixed point of application.

Sliding vectors are those vectors which are not fixed and can be moved along their lines of action.

Free vectors are those vectors which can be moved over in space.

Equal vectors are two vectors having the same magnitude and direction. However, they can have different points of application.

Equivalent vectors are two vectors that are not only equal, but also produce the same effect.

Unit vector is a vector whose magnitude is *unity*.

Null vector is a vector whose magnitude is *zero*.

Negative of a vector is a vector having the same magnitude, but direction opposite to that of the given vector.

Addition of Vectors by Graphical Methods

The addition of vectors can be determined graphically using the *parallelogram law* or *triangle law*. The parallelogram law states that when two concurrent vectors $\vec{a}$ and $\vec{b}$ are represented by two adjacent sides of a parallelogram, then the diagonal passing through their point of concurrency represents the sum of the vectors $\vec{a} + \vec{b}$ in magnitude and direction. The triangle law states that if two vectors $\vec{a}$ and $\vec{b}$ can be represented by the two sides of a triangle (in magnitude and direction) taken in order, then the third side (closing side) represents the sum of the two vectors in the opposite order.

Addition of Vectors by Analytical Method

If two vectors can be represented in terms of their components as

$$\vec{a} = a_x\vec{i} + a_y\vec{j} + a_z\vec{k} \quad \text{and} \quad \vec{b} = b_x\vec{i} + b_y\vec{j} + b_z\vec{k}$$

then their summation is obtained by adding the corresponding components along *X*, *Y* and *Z* directions, i.e.,

$$\vec{a} + \vec{b} = (a_x + b_x)\vec{i} + (a_y + b_y)\vec{j} + (a_z + b_z)\vec{k}$$

and their difference is given as

$$\vec{a} - \vec{b} = (a_x - b_x)\vec{i} + (a_y - b_y)\vec{j} + (a_z - b_z)\vec{k}$$

A vector in space can also be represented as

$$\vec{a} = a[\cos\theta_x\vec{i} + \cos\theta_y\vec{j} + \cos\theta_z\vec{k}]$$

where $\cos\theta_x$, $\cos\theta_y$ and $\cos\theta_z$ are called *direction cosines* of the vector.

Multiplication of a Vector by a Scalar

If m is a scalar, then the product of a vector $\vec{a}$ by the scalar m is a vector whose magnitude is $m|\vec{a}|$. Its direction is *same* as $\vec{a}$ if k is *positive* and *opposite* to that of $\vec{a}$ if k is *negative.*

Scalar (or) Dot Product

Scalar product of two vectors is defined as *the product of their magnitudes and the cosine of the included angle between them.*

In general, if $\vec{a} = a_x\vec{i} + a_y\vec{j} + a_z\vec{k}$ and $\vec{b} = b_x\vec{i} + b_y\vec{j} + b_z\vec{k}$

then $\vec{a}.\vec{b} = a_xb_x + a_yb_y + a_zb_z$

Vector (or) Cross Product

Vector product of two vectors is a **vector**, whose magnitude is the product of magnitudes of the two vectors and sine of their included angle; its direction is determined by right-hand screw rule.

In general, if $\vec{a} = a_x\vec{i} + a_y\vec{j} + a_z\vec{k}$ and $\vec{b} = b_x\vec{i} + b_y\vec{j} + b_z\vec{k}$, then

$$\vec{a} \times \vec{b} = \begin{vmatrix} \vec{i} & \vec{j} & \vec{k} \\ a_x & a_y & a_z \\ b_x & b_y & b_z \end{vmatrix}$$

Scalar Triple Product and Vector Triple Product

Scalar triple product is given as

$$(\vec{a} \times \vec{b}).\vec{c}$$

Vector triple product is given as

$$\vec{a} \times (\vec{b} \times \vec{c}) = \vec{b}(\vec{a}.\vec{c}) - \vec{c}(\vec{a}.\vec{b})$$

EXERCISES

Short-answer Questions

1. Distinguish between scalars and vectors.
2. Can any physical quantity with direction as well as magnitude be treated as a vector? Discuss.
3. Distinguish between equal and equivalent vectors.
4. Can equal vectors be always considered as equivalent vectors?
5. Define unit vector and null vector.
6. How can we represent a vector in terms of a unit vector along its direction?
7. Define the negative of a vector.
8. State the commutative and associative laws of vector addition.
9. Describe how to subtract a vector from another vector.
10. Define the resolution of a vector.
11. Define scalar and vector components of a vector.
12. What are direction cosines?
13. Describe how to determine the addition and subtraction of vectors by analytical method?
14. Define position vector and displacement vector.
15. What are the geometrical interpretations of dot product and cross product?
16. Distinguish between scalar triple product and vector triple product.

Numerical Problems

2.1 Find the magnitude of the vector $3\vec{i} + \vec{j} + 4\vec{k}$ and unit vector along its direction.

Ans. $\sqrt{26}$ units, $\dfrac{3\vec{i} + \vec{j} + 4\vec{k}}{\sqrt{26}}$

2.2 If $\vec{a} = 5\vec{i} + 2\vec{j} + 3\vec{k}$, $\vec{b} = 2\vec{i} + 4\vec{j} + \vec{k}$, $\vec{c} = 3\vec{i} + \vec{j} + 2\vec{k}$, then find (i) $2\vec{a} + \vec{b} + 3\vec{c}$, (ii) $\vec{a} - 2\vec{b} + \vec{c}$ and (iii) $3\vec{a} - \vec{b} - \vec{c}$.

Ans. (i) $21\vec{i} + 11\vec{j} + 13\vec{k}$, (ii) $4\vec{i} - 5\vec{j} + 3\vec{k}$ and (iii) $10\vec{i} + \vec{j} + 6\vec{k}$

2.3 If $\vec{a} = 4\vec{i} + 2\vec{j} + \vec{k}$, $\vec{b} = 2\vec{i} + \vec{j} + 2\vec{k}$, then find the unit vector parallel to (i) $\vec{a} + \vec{b}$ and (ii) $\vec{a} - \vec{b}$.

Ans. $\dfrac{6\vec{i} + 3\vec{j} + 3\vec{k}}{\sqrt{54}}$, $\dfrac{2\vec{i} + \vec{j} - \vec{k}}{\sqrt{6}}$

2.4 What vector should be added to $2\vec{i} + \vec{j} + 3\vec{k}$ such that the resultant is unit vector $\vec{k}$?

Ans. $-2\vec{i} - \vec{j} - 2\vec{k}$

2.5 Find a vector of magnitude 10 units, which is parallel to the vector $3\vec{i} + 4\vec{j}$.

Ans. $6\vec{i} + 8\vec{j}$

2.6 Find the direction cosines of the vector: $3\vec{i} + 2\vec{j} + \vec{k}$.

Ans. $\dfrac{3}{\sqrt{14}}$, $\dfrac{2}{\sqrt{14}}$ and $\dfrac{1}{\sqrt{14}}$

2.7 If $|\vec{a}| = 3$ and $|\vec{b}| = 5$, and the angle between $\vec{a}$ and $\vec{b}$ is 50°, determine the value of $\vec{a}.\vec{b}$.

Ans. 9.64

2.8 If $|\vec{a}| = 2$, $|\vec{b}| = 3$ and $\vec{a}.\vec{b} = 4$, then find the area of the parallelogram formed by the two vectors as adjacent sides.

Ans. 4.47 units

2.9 For what values of a are $\vec{A} = a\vec{i} - 2\vec{j} + \vec{k}$ and $\vec{B} = 2a\vec{i} + a\vec{j} - a\vec{k}$ perpendicular.

Ans. $a = 3/2$

2.10 If $\vec{a} = \vec{i} + 2\vec{j} + 3\vec{k}$ and $\vec{b} = \vec{i} + 3\vec{j} + \vec{k}$, find (i) $\vec{a}.\vec{b}$, (ii) angle between the vectors, (iii) projection of $\vec{a}$ on $\vec{b}$ and (iv) projection of $\vec{b}$ on $\vec{a}$.

Ans. 10, 36.31°, 3.02, 2.67

2.11 If $a = 3$, $b = 4$ and $\vec{a}.\vec{b} = 2\sqrt{11}$, find $|\vec{a} \times \vec{b}|$.

Ans. 10

2.12 If $\vec{a} = 3\vec{i} - \vec{j} - 2\vec{k}$ and $\vec{b} = 2\vec{i} + 3\vec{j} + \vec{k}$, find $(\vec{a} + 2\vec{b}) \times (2\vec{a} - \vec{b})$.

Ans. $-25\vec{i} + 35\vec{j} - 55\vec{k}$

2.13 Two vectors A and B are given as $\vec{A} = 2\vec{i} + 3\vec{j} + \vec{k}$ and $\vec{B} = 3\vec{i} - 3\vec{j} + 4\vec{k}$. Determine their cross product and the unit vector along it.

Ans. $15\vec{i} - 5\vec{j} - 15\vec{k}, \dfrac{3\vec{i} - \vec{j} - 3\vec{k}}{\sqrt{19}}$

2.14 If $\vec{a} = \vec{i} + 2\vec{j} + 3\vec{k}, \vec{b} = 2\vec{i} + 2\vec{j} + 2\vec{k}$ and $\vec{c} = \vec{i} - \vec{j} - \vec{k}$, find $\vec{a} . (\vec{b} \times \vec{c})$ and $\vec{a} \times (\vec{b} \times \vec{c})$.

Ans. $-4, -20\vec{i} + 4\vec{j} + 4\vec{k}$

2.15 Given the vectors $\vec{a} = 2\vec{i} - 2\vec{j} + 3\vec{k}, \vec{b} = \vec{i} + \vec{j} + 3\vec{k}$ and $\vec{c} = 2\vec{i} + \vec{j} + \vec{k}$, determine whether they are coplanar or not.

Ans. Not coplanar

2.16 Given the vectors $\vec{a} = 2\vec{j} + 6\vec{k}, \vec{b} = \vec{j} + 3\vec{k}$, and $\vec{c} = 2\vec{i} + \vec{k}$, determine whether they are coplanar or not.

Ans. coplanar

2.17 Find the area of the parallelogram formed by two vectors: $\vec{a} = 2\vec{i} + 3\vec{j}$ and $\vec{b} = \vec{j} - 2\vec{k}$.

Ans. $2\sqrt{14}$ units

2.18 $\overrightarrow{OA} = \vec{i} + \vec{j} - 2\vec{k}, \overrightarrow{OB} = 2\vec{i} - \vec{j} + \vec{k}$ are two vectors from the origin. Using vector product, determine the length of the perpendicular from A to the line OB.

Ans. $\sqrt{35/6}$ units

2.19 If $|\vec{a} + \vec{b}| = |\vec{a} - \vec{b}|$, then prove that the vectors $\vec{a}$ and $\vec{b}$ are mutually perpendicular.

2.20 Show that the diagonals of a rhombus are at right angles.

2.21 Prove that the altitudes of a triangle are concurrent.

3 System of Forces and Resultant-I (Concurrent Forces)

3.1 INTRODUCTION

In this chapter, we will introduce the concept of force, its characteristics and its effect on bodies. When more than one force acts on a body, they constitute a *system of forces*. Various types of systems of forces are possible by the relative orientation of forces within the system. As each such type of force system produces different effects on bodies, they are studied separately. In this chapter, we will discuss the effects of concurrent forces and in the following chapter, the effects of non-concurrent forces.

In Sections 3.5–3.6, we will discuss the methods to determine the resultant of concurrent force systems in a plane and their effects upon bodies. In Section 3.7, we will discuss concurrent spatial force systems.

3.2 FORCE

Newton's first law of motion states that every body continues in its state of rest or of uniform motion in a straight line *unless* it is compelled to change that state by *forces* impressed on it. Hence, **force** can be defined as any *action* of a body on another, which tends to *change* the state of rest or of motion of the other body. There are various types of forces that can act upon a body in a given environment: such as gravitational force acting on a body when it is placed in a gravitational field or push and pull exerted by our hand or tractive force of a locomotive on the train which it is pulling, etc.

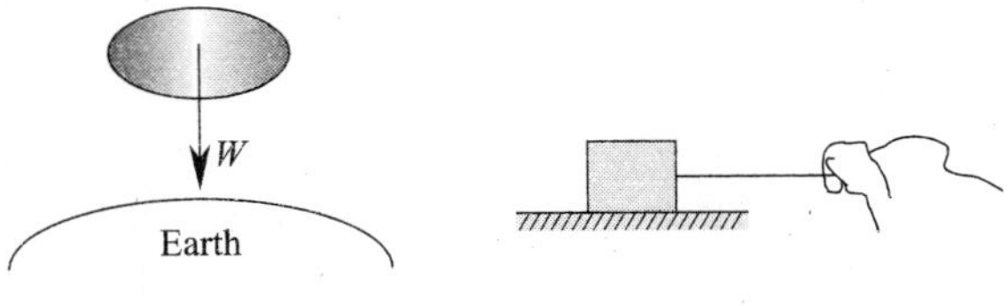

Fig. 3.1 (a) Gravitational force (b) Pull exerted by our hand

To define a force completely, particularly for graphical representation, we must specify its **magnitude**, its **point of application** and its **direction**, which are known as **force characteristics**.

The **magnitude** of a force is obtained by comparing it with a certain *standard force.* Standard force is defined as that force exerted on a standard kilogram body (which is a platinum cylinder preserved at the International Bureau of Weights and Measures at Sevres near Paris) causing uniform acceleration of 1 m/s^2. The magnitude of force can be measured by a spring balance or by a dynamometer. The SI unit of force is given in **newton** (N).

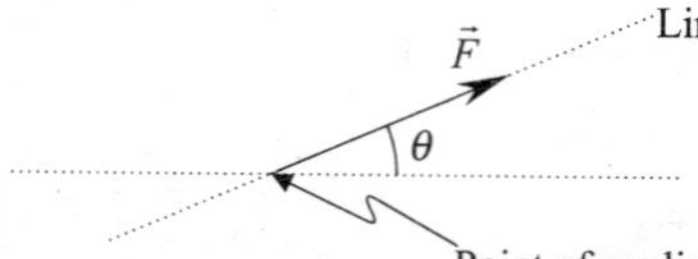

Fig. 3.2 Graphical representation of a force

The **point of application** of a force is that point in the body at which the force can be assumed to be *concentrated.* In reality, forces are never concentrated but are distributed over the entire volume of the body, as in the case of force of gravity or over the entire contact area, as in the case of contact forces. However, for theoretical purposes, these forces may be assumed to be concentrated at a point called the *point of application* of the force without affecting the accuracy of the problem. For example, the weight of a body, even though distributed over its entire volume, is normally applied at its centre of gravity.

The **direction** of a force is that direction in which the acting force *tends to move* the body. It is a straight line passing through the point of application of the force. It is also called the **line of action** of the force. The **sense** of force is indicated by an *arrowhead*. For example, gravity force is always directed towards the centre of the earth or the tension in a string always acts along its length away from the body.

Consider for instance, a block being pulled by a string with a force of 50 N. Then the force acting on the block is represented graphically as shown in Fig. 3.3(a), where its magnitude and sense are also shown. Suppose it is pushed with a force of 50 N. Then it is represented as shown in Fig. 3.3(b). Note that even though this force may be distributed over the entire contact area between the body and the agent causing the motion, it is represented as a concentrated force placed at the centroid of the contact area.

Graphical representation of a force pulling and pushing a block

Fig. 3.3(a) **Fig. 3.3(b)**

From the above discussion, we can readily see that force is a **vector** as it has both **magnitude** and **direction**, and also obeys the **parallelogram law,** which will be explained later in Section 3.6.1.

3.3 EFFECT OF FORCE ON A BODY

An external force acting on a body tends to *deform* the body, causing internal stresses within the body. This internal stress distribution is dependent on the point of application of the force. For example, forces $\vec{F_1}$ and $\vec{F_2}$ applied respectively at ends A and B of a rod cause tension in the rod. The same forces when applied at ends B and A respectively, cause *compression* in the rod.

Hence, in such a case, we must treat the force as a **fixed vector**, i.e., having a *fixed* point of application. Such types of problems are dealt with in the field of strength of materials.

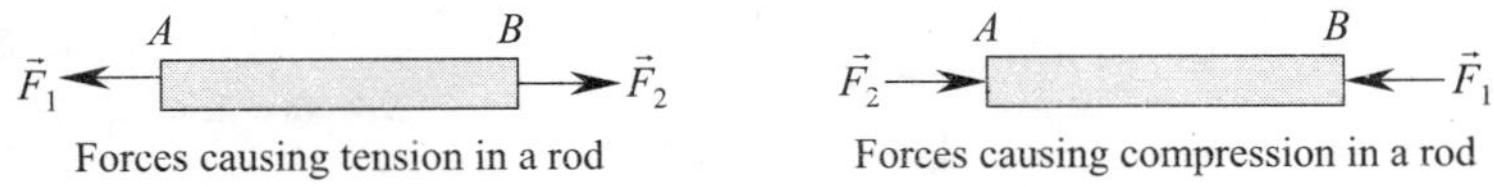

Fig. 3.4

However, our subject of interest is that of the effect of external forces on **rigid bodies**, i.e., bodies that *do not deform* under the action of external forces. In such cases, the external forces only tend to **translate** the body or **rotate** the body as a whole. Consider a block resting on a table. The motion of the block will remain the same, irrespective of whether a force is pulling the block or pushing the block from the other side over the table.

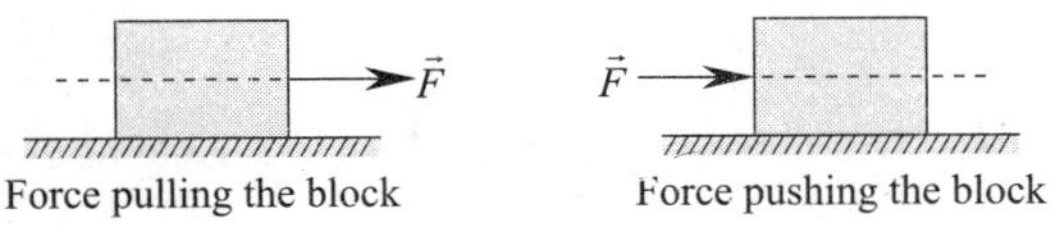

Fig. 3.5

Hence, we can see that forces can be moved to any point along their lines of action on the rigid body. This is known as **principle of transmissibility**. It states that *the conditions of equilibrium or of motion of a rigid body remains unchanged if a force acting at a given point of the rigid body is replaced by a force of the same magnitude and direction, but acting at a different point, provided that the two forces have the same line of action.*

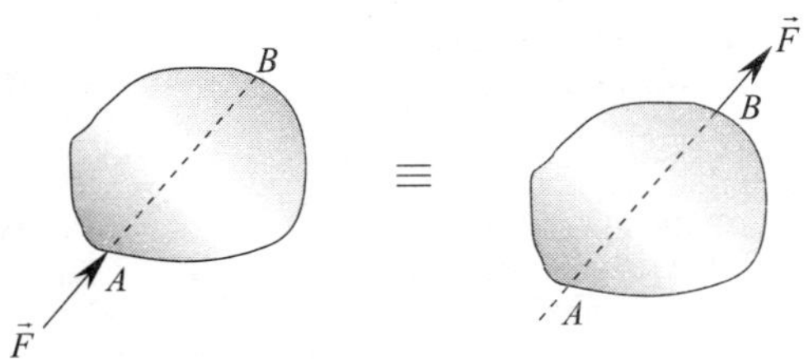

Fig. 3.6 Principle of transmissibility

Thus, the effect of force $\vec{F}$ acting at A will be same as the effect of $\vec{F}$ acting at B along the same line of action. It should be noted that here the force is treated as a *sliding vector*.

3.4 SYSTEM OF FORCES

When more than one force acts on a body at a particular instant, they are said to constitute a **system of forces**. Within the system of forces, all the forces may lie on the same plane or on different planes. If they all lie on the same plane, they are said to be **coplanar** forces. If they lie on different planes, they are said to be **non-coplanar** or **spatial** forces. Further, if the lines of action of all of them (whether coplanar or non-coplanar forces) intersect at a point as in Fig. 3.7(iii), they are termed **concurrent forces**; if not, they are termed **non-concurrent forces** [refer Fig. 3.7(iv)]. Again, if the lines of action of all lie along the same line as in Fig. 3.7(i), they are termed **collinear** forces; and if their lines of action are parallel to each other as in Fig. 3.7(ii), they are termed **parallel** forces. Chart 3.1 below summarises the various types of force systems discussed.

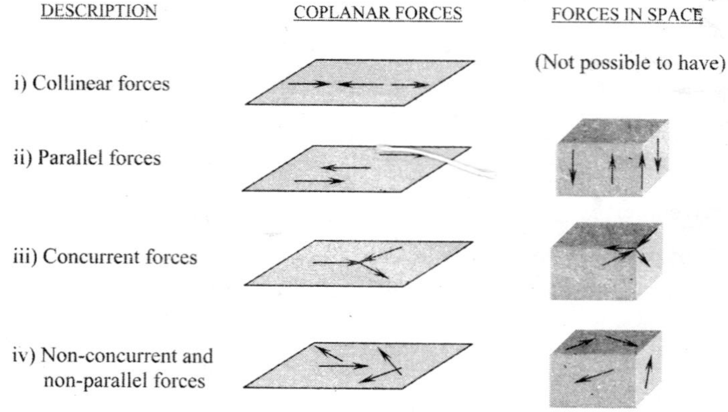

Fig. 3.7 Graphical representation of various systems of forces

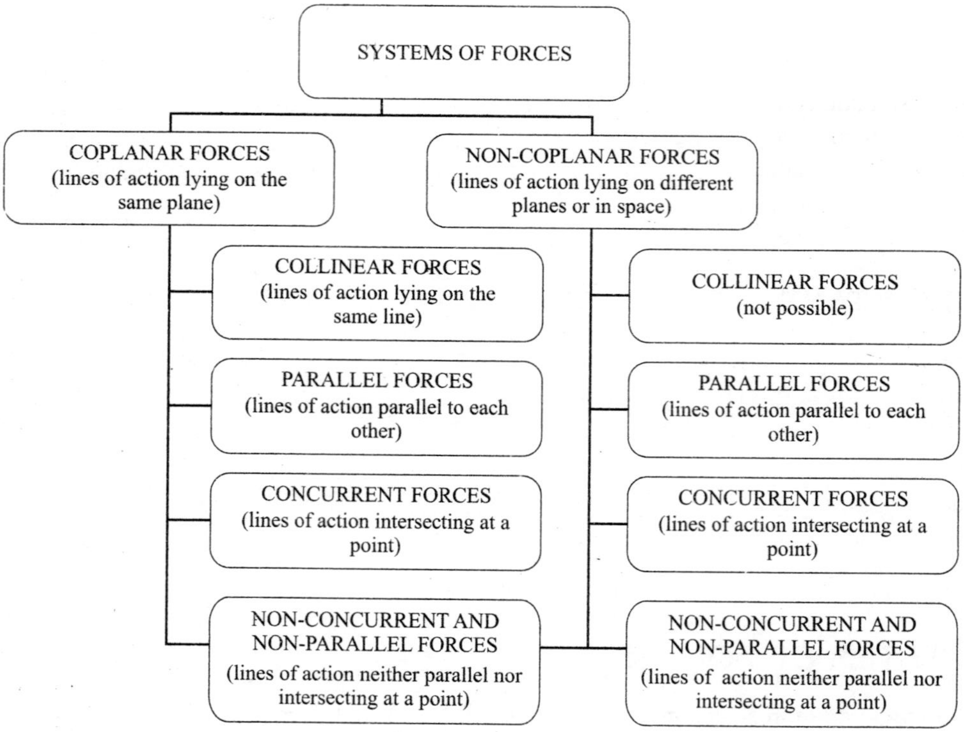

Chart 3.1 Types of systems of forces

Different force systems will have different effects on rigid bodies. For instance, the **concurrent** forces in a plane (Fig. 3.8) and in space (Fig. 3.9) tend to *move* or *translate* the body as a whole. As there is *no rotational motion* involved, the body in such cases can be idealized as a *particle*, i.e., a body without any extent.

However, if the forces are non-concurrent in a plane [Fig. 3.10(a)] and in space [Fig. 3.10(b)], they tend to *rotate* the body in addition to *translating* the body. Hence, in such cases the body can no more be idealized as a particle but treated as a *rigid body* itself.

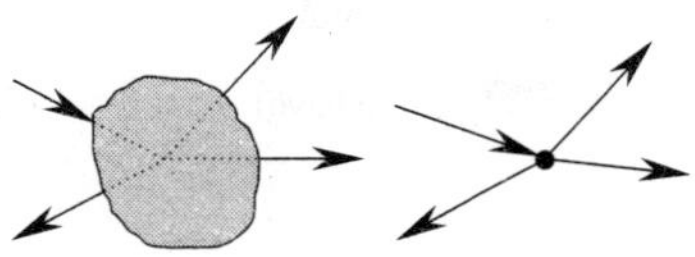

(a) (b)

Fig. 3.8 Concurrent forces in a plane Idealization as a particle

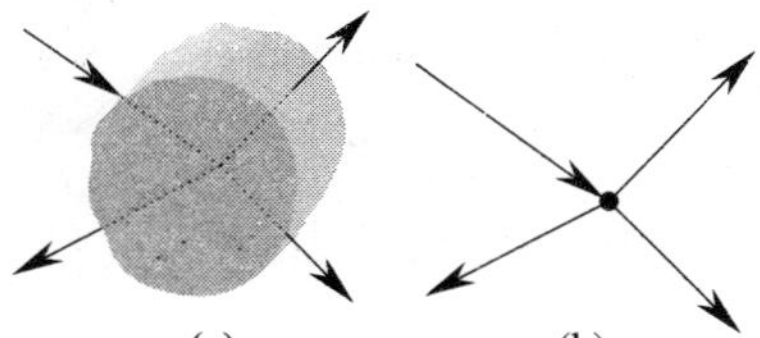

(a) (b)

Fig. 3.9 Concurrent forces in space Idealization as a particle

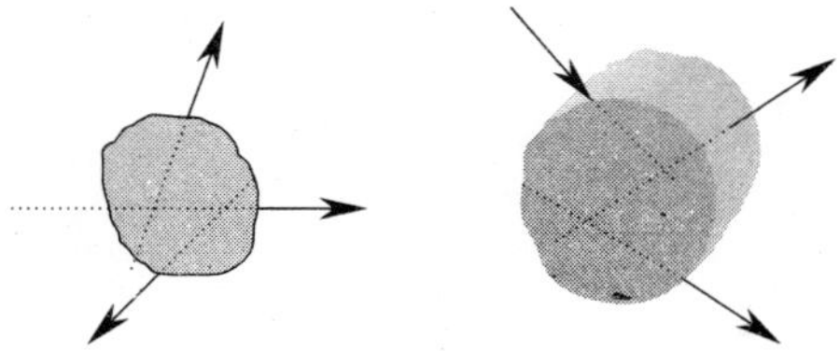

(a) (b)

Fig. 3.10 (a) Non-concurrent forces in a plane (b) Non-concurrent forces in space

For a better understanding of the effect of various force systems, it will be good if we consider each type separately in our study. Hence, we will consider mainly four types of force systems, namely, coplanar concurrent forces, concurrent forces in space, coplanar non-concurrent forces and non-concurrent forces in space. It should be noted that collinear forces and parallel forces are special cases of non-concurrent forces. Hence, these will not be discussed separately.

3.5 RESULTANT OF A FORCE SYSTEM

To study the effect of a system of forces acting on a body, it is customary to replace the system of forces by its **resultant**. It is defined as a **single equivalent force** which produces the same effect on the body as that of all given forces. This resultant is helpful in determining the motion of the body.

For instance, if the resultant of a force system turns out to be *zero*, the body will remain at rest if it was already at rest or move with constant velocity if it was already moving with constant velocity. Hence, we say that the body remains in a state of *equilibrium*. Such a condition is called **statics**. If the resultant turns out to be *non-zero*, the body will have varying state of motion. Such types of problems are discussed under **dynamics**.

As different methods are employed to find the resultant of different types of force systems, we explain them separately in two chapters. In the remaining part of this chapter, we discuss methods to find the resultant of **concurrent** forces in a plane and in space; and in the next chapter, **non-concurrent** forces in a plane and in space.

3.6 RESULTANT OF COPLANAR CONCURRENT FORCES

Various methods are employed to determine the *resultant* of *concurrent forces* in a *plane*. They are described below:

(i) Graphical methods : Parallelogram law, triangle law and polygon law

(ii) Trigonometric methods : Cosine law and sine law
(iii) Analytical method : Vector approach

3.6.1 Parallelogram Law

The parallelogram law states that *when two concurrent forces $\vec{F}_1$ and $\vec{F}_2$ acting on a body are represented by two adjacent sides of a parallelogram, the diagonal passing through their point of concurrency represents the resultant force $\vec{R}$ in magnitude and direction.*

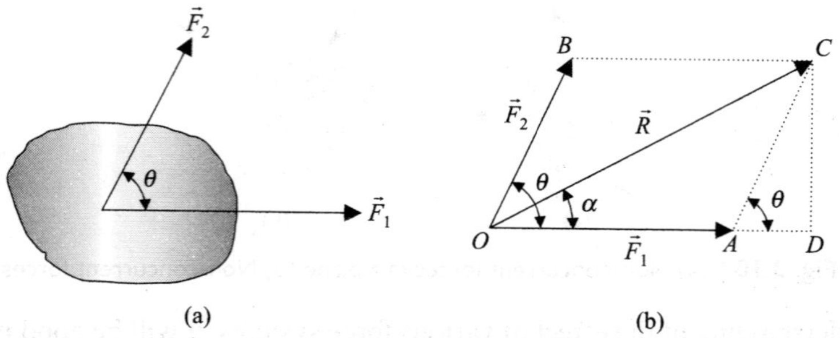

Fig. 3.11 Graphical representation of the parallelogram law

It should be noted that this law is based on experimental evidence (which the student would have come across in school-level physics) and it cannot be proved mathematically.

Graphical Solution To obtain the resultant graphically, from the origin O, draw the two force vectors on a graph to a convenient scale and in the directions specified, i.e., OA and OB respectively. Complete the parallelogram $OACB$ with the two force vectors as adjacent sides. Draw the diagonal passing through the origin. Then the length of the diagonal OC gives the magnitude of the resultant to scale and its inclination α to the reference axis OA gives the direction.

3.6.2 Law of Cosine

The mathematical statement of the parallelogram law is called the **law of cosine**. The magnitude and direction of the resultant can also be determined from Fig. 3.11(b) by trigonometry as follows.

From ΔOCD, we know,

$$OC^2 = (OA + AD)^2 + (CD)^2 \quad (3.1)$$

Hence, the magnitude of the resultant $\vec{R}$ is given by

$$R = OC = \sqrt{(F_1 + F_2 \cos\theta)^2 + (F_2 \sin\theta)^2}$$

$$= \sqrt{F_1^2 + F_2^2 + 2F_1F_2\cos\theta} \quad (3.2)$$

The inclination of $\vec{R}$ with $\vec{F}_1$ is given by

$$\alpha = \tan^{-1}\left[\frac{CD}{(OA + AD)}\right]$$

$$= \tan^{-1}\left[\frac{F_2 \sin\theta}{F_1 + F_2 \cos\theta}\right] \tag{3.3}$$

Corollary

Case I: If $\theta = 90°$, i.e., if the two forces are perpendicular to each other then

$$R = \sqrt{F_1^2 + F_2^2} \text{ and } \alpha = \tan^{-1}\left(\frac{F_2}{F_1}\right) \tag{3.4}$$

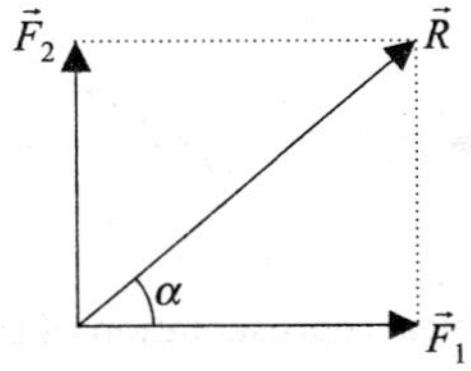

Fig. 3.12(a) Resultant of two concurrent and perpendicular forces

Case II: If $\theta = 0°$, i.e., if the two forces are collinear and act in the same direction then

$$R = F_1 + F_2 \text{ and } \alpha = 0 \tag{3.5}$$

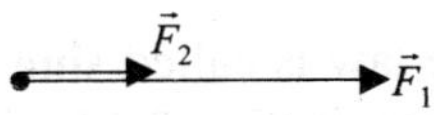

Fig. 3.12(b) Resultant of two collinear forces in the same direction

Case III: If $\theta = 180°$, i.e., if the two forces are collinear but acting in the opposite direction, where $F_1 > F_2$ then

$$R = F_1 - F_2 \text{ and } \alpha = 0 \tag{3.6}$$

Fig. 3.12(c) Resultant of two collinear forces in the opposite direction

Hence, we see that if two forces are **collinear** then their resultant is given by their **algebraic sum**.

When there are more than two concurrent forces in a force system, the resultant can be found out in steps. First, considering any two forces, the resultant can be determined as explained above. The resultant thus obtained is added on to the next force, and so on, until all given forces are added on by the parallelogram law. Thus, the overall resultant can be obtained.

3.6.3 Triangle Law

The resultant can also be determined by the triangle law which states that *if two forces $\vec{F}_1$ and $\vec{F}_2$ acting simultaneously on a body can be represented by the two sides of a triangle (in magnitude and direction) taken in order then the third side (closing side) represents the resultant in the opposite order.*

Graphical Solution From the origin, draw one of the force vectors on a graph to a convenient scale and in the direction specified. With the head of this vector as origin, draw the second vector to scale and in the specified direction. Join the tail of the first vector with the head of the second vector. Then its length gives the magnitude of the resultant to scale and its inclination to the reference axis gives the direction. It should be noted that due to commutative property of vectors, the order in which we construct the forces do not affect the resultant.

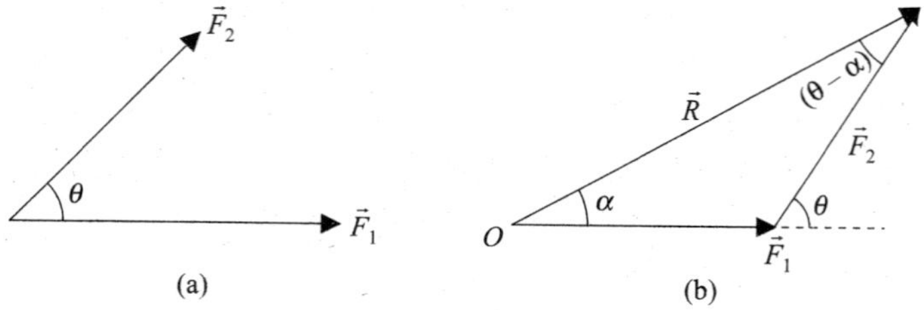

Fig. 3.13 Graphical representation of the triangle law

Though this may appear to be a different law by itself, it is actually a corollary of the parallelogram law explained before. The same triangle in Fig. 3.13(b) can also be obtained by considering the lower part of the parallelogram in Fig. 3.11(b) and treating $\vec{F}_2$ as a free vector.

3.6.4 Sine Law

The mathematical statement of the triangle law is called **sine law**. For a triangle of sides and included angles as shown in Fig. 3.14, sine law can be expressed as

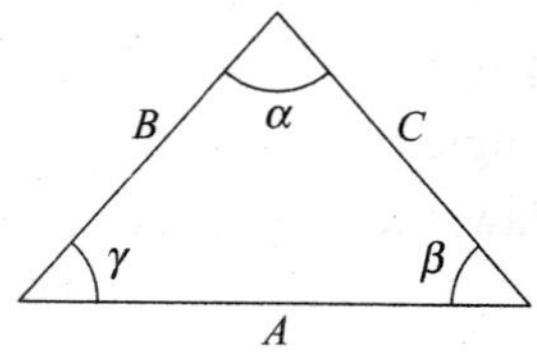

Fig. 3.14

$$\frac{A}{\sin\alpha} = \frac{B}{\sin\beta} = \frac{C}{\sin\gamma} \tag{3.7}$$

where A, B and C are the sides of the triangle and the angles opposite to these sides being respectively α, β and γ.

Applying sine law to Fig. 3.13(b) we get,

$$\frac{F_1}{\sin(\theta - \alpha)} = \frac{F_2}{\sin(\alpha)} = \frac{R}{\sin(180° - \theta)} \tag{3.8}$$

from which the magnitude of the resultant R and its inclination α can be determined. Also, for a triangle, we know

$$A^2 = B^2 + C^2 - 2BC\cos\alpha \tag{3.9}$$

$$B^2 = A^2 + C^2 - 2AC\cos\beta \tag{3.10}$$

$$C^2 = A^2 + B^2 - 2AB\cos\gamma \tag{3.11}$$

Corollary If a number of concurrent forces acting simultaneously on a body are represented in magnitude and direction by the sides of a polygon taken in order then the closing side of the polygon represents the resultant in the opposite order. This is called the **polygon law**, which is an extension of the triangle law to find the resultant of more than two concurrent forces. It should be noted that due to associative property of vectors, the order in which we add the forces do not affect the resultant.

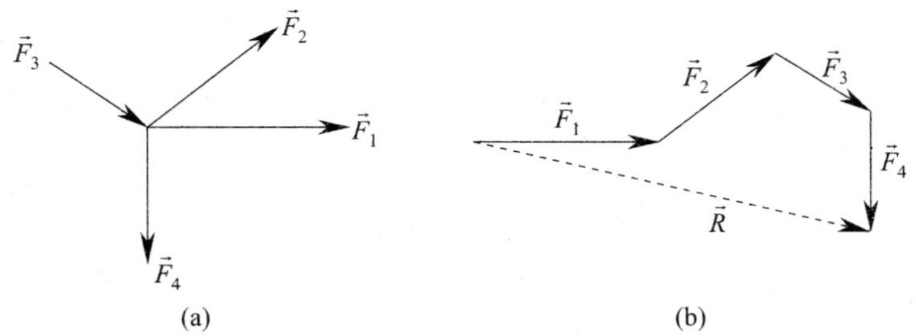

Fig. 3.15 Graphical representation of the polygon law

Example 3.1 Two forces are applied at the point A of a hook support as shown in Fig. 3.16. Determine the magnitude and direction of the resultant force by the trigonometric method using (i) parallelogram law, and (ii) triangle law.

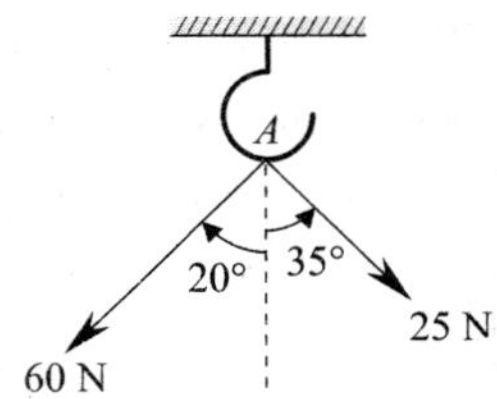

Fig. 3.16

Solution

(i) Parallelogram Law

Let us take the two forces as $|\vec{F}_1| = F_1 = 60$ N, $|\vec{F}_2| = F_2 = 25$ N. The included angle θ between them is $20° + 35° = 55°$. Hence, according to the parallelogram law, the magnitude of the resultant force $\vec{R}$ is given by,

$$R = \sqrt{F_1^2 + F_2^2 + 2F_1F_2 \cos\theta}$$

$$= \sqrt{60^2 + 25^2 + 2(60)(25)\cos 55°} = 77.11 \text{ N}$$

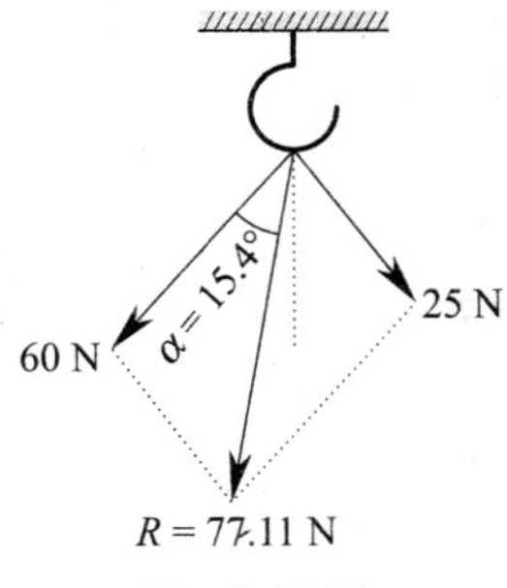

Fig. 3.16(a)

The angle made by the resultant $\vec{R}$ with $\vec{F}_1$ is given by,

$$\alpha = \tan^{-1}\left[\frac{F_2 \sin\theta}{F_1 + F_2 \cos\theta}\right]$$

$$= \tan^{-1}\left[\frac{25 \sin 55°}{60 + 25 \cos 55°}\right] = 15.4°$$

(ii) Triangle Law

Draw the force triangle with $\vec{F}_1$ and $\vec{F}_2$ taken in order. It should be noted that due to associative property of the vectors, the order in which we take the forces would not affect the result. Then the closing side represents the resultant in the opposite order. The included angles between the forces are also shown. While measuring the included angle between the two vectors, we must take both the vectors either pointing towards or pointing away from the point of concurrency. The angle included between the two forces being 55°, we can see that the angle between them in the force triangle is $180° - 55° = 125°$. The angle between the resultant force $\vec{R}$ and the force $\vec{F}_1$ is taken as α. From the properties of a triangle, we know

$$R^2 = 60^2 + 25^2 - 2(60)(25)\cos(180° - 55°)$$
$$= 60^2 + 25^2 + 2(60)(25)\cos 55°$$
$$[\text{since } \cos(180° - \theta) = -\cos\theta]$$

$\therefore$ $$R = 77.11 \text{ N}$$

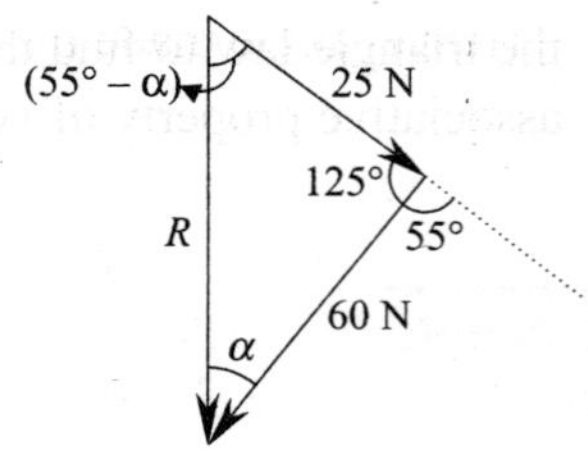

Fig. 3.16(b) Force triangle

By sine law, we have

$$\frac{R}{\sin 125°} = \frac{25}{\sin\alpha} = \frac{60}{\sin(55° - \alpha)}$$

$\Rightarrow$ $$\alpha = 15.4°$$

Example 3.2 The angle between the resultant (35 N) of two forces and one of the forces (15 N) is 38.21°. Find the other force and its inclination with the resultant.

Solution We draw the forces as shown in Fig. 3.17(a), where $F_1 = 15$ N, $R = 35$ N and $\alpha = 38.21°$. Let θ be the included angle between the two forces.

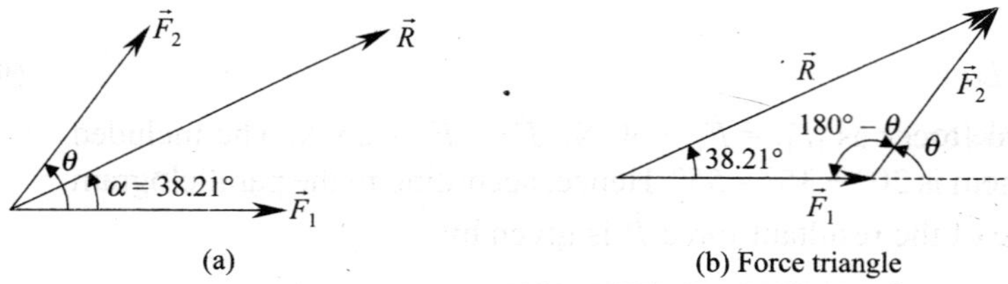

Fig. 3.17

Though this problem can be solved by any of the two methods, it will be found convenient to use the triangle law here as it simplifies the problem. Consider the force triangle for the given system of forces as shown in Fig. 3.17(b). Draw force $\vec{F}_1$ from the origin to scale and in the direction specified. With the head of $\vec{F}_1$ as the origin, draw $\vec{F}_2$ to scale and in the direction specified. Then close the triangle. The unknown force F_2 is then given as

$$F_2 = \sqrt{R^2 + F_1^2 - 2RF_1\cos(38.21°)}$$
$$= \sqrt{35^2 + 15^2 - 2(35)(15)\cos(38.21)°}$$
$$= 25 \text{ N}$$

Also, by sine law, we have

$$\frac{R}{\sin(180° - \theta)} = \frac{F_2}{\sin 38.21°}$$

$$\frac{35}{\sin\theta} = \frac{25}{\sin 38.21°} \quad [\text{since } \sin(180° - \theta) = \sin\theta]$$

$\Rightarrow$ $$\theta = 60°$$

Hence, angle made by the other force with the resultant is

$$= \theta - 38.21°$$

$$= 60° - 38.21° = 21.79°$$

Example 3.3 The resultant of two forces is 400 N. If the forces are inclined at 40° and 60° with the resultant, one on either side, calculate the magnitude of the two forces.

Solution This problem can be solved by constructing the force triangle as shown in Fig. 3.18(b) and applying the sine rule. In Fig. 3.18(a), we see that both the forces $\vec{F}_2$ and the resultant $\vec{R}$ point away from the origin and in the force triangle, both point towards the meeting point. Hence, the included angles between them in the two must be equal, i.e., 60°. The third angle can then be determined as 80°.

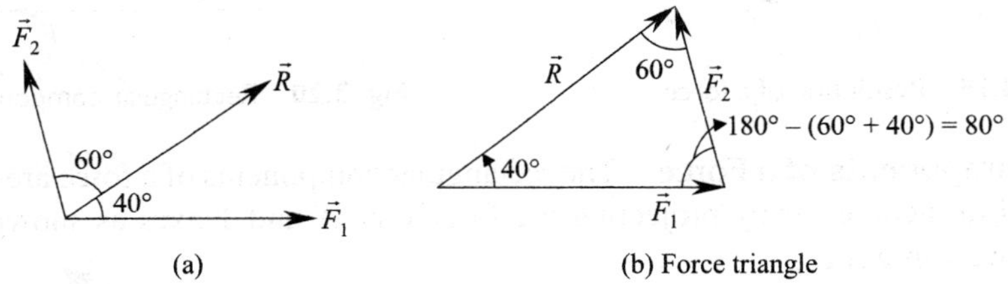

Fig. 3.18

Applying sine rule to the force triangle,

$$\frac{R}{\sin 80°} = \frac{F_1}{\sin 60°} = \frac{F_2}{\sin 40°}$$

$$\Rightarrow \quad F_1 = R\frac{\sin 60°}{\sin 80°}$$

$$= 351.75 \text{ N}$$

Similarly,

$$F_2 = R\frac{\sin 40°}{\sin 80°}$$

$$= 261.08 \text{ N}$$

3.6.5 Analytical Method

When there are more than three concurrent forces acting on a particle then graphical and trigonometric methods become tedious to work with. In such cases, we resort to the analytical method as it can be applied to any number of forces.

Resolution of a Force into Components The parallelogram law states that forces acting on a body can be replaced by a single resultant force. The inverse of this should also be true. That is, a single force acting on a body may be replaced by two or more forces, which produce the same effect on the body. These forces are called **components** of the original force $\vec{F}$.

To graphically determine these components, in Fig. 3.19, from the tip of $\vec{F}$, we draw lines *EC* and *ED* parallel to the arbitrarily given axes *OB* and *OA* respectively. The points *C* and *D* at which they meet the axes define respectively the components $\vec{F}_1$ and $\vec{F}_2$ of the force along the axes. Here we have chosen the directions of axes in an arbitrary way. Hence, there could be infinite number of such axes and hence, infinite number of such components to a single force. However, we are interested only in a particular case, where the components are taken about rectangular or orthogonal axes and the corresponding components are termed **rectangular** or **orthogonal** components.

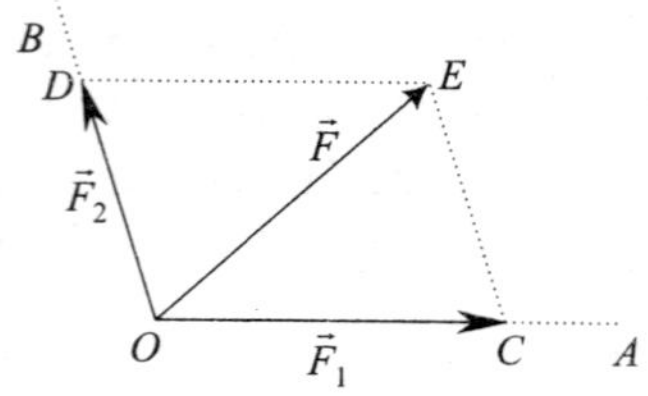

Fig. 3.19 Resolution of a force

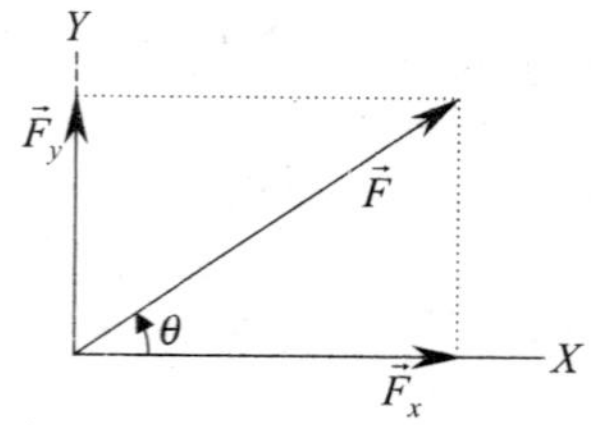

Fig. 3.20 Rectangular components of a force

Rectangular Components of a Force The rectangular components of a force are obtained by drawing lines parallel to the axes or by projecting the force onto X and Y axes as shown in Fig. 3.20. By vector addition, we can write

$$\vec{F} = \vec{F}_x + \vec{F}_y \tag{3.12}$$

where $\vec{F}_x$ and $\vec{F}_y$ are called **vector components** of $\vec{F}$.

Since any vector can be represented as a product of its magnitude and unit vector along its direction, we can also write the force vector as

$$\vec{F} = F_x\vec{i} + F_y\vec{j} \tag{3.13}$$

where F_x and F_y are called the **scalar components** or simply **components** of $\vec{F}$.

If θ is inclination of $\vec{F}$ with respect to the X-axis then the scalar components of $\vec{F}$ are

$$F_x = |\vec{F}|\cos\theta = F\cos\theta \tag{3.14}$$

and

$$F_y = |\vec{F}|\sin\theta = F\sin\theta \tag{3.15}$$

Therefore, force $\vec{F}$ can also be written as

$$\vec{F} = [F\cos\theta]\vec{i} + [F\sin\theta]\vec{j} \tag{3.16}$$

The magnitude and direction of $\vec{F}$ can be expressed in terms of its components as

$$|\vec{F}| = F = \sqrt{F_x^2 + F_y^2} \quad \text{and} \quad \theta = \tan^{-1}\left[\frac{F_y}{F_x}\right] \tag{3.17}$$

Note: This process of substituting a force $\vec{F}$ with its components is called **resolving** the force $\vec{F}$ into components. While expressing the forces graphically, to differentiate between the force and its components, the original force is normally shown in *dotted* lines.

Corollary Depending upon the angle θ made by the force vector with respect to the **positive X-axis**, the sign of the components of force vary as shown in Table 3.1.

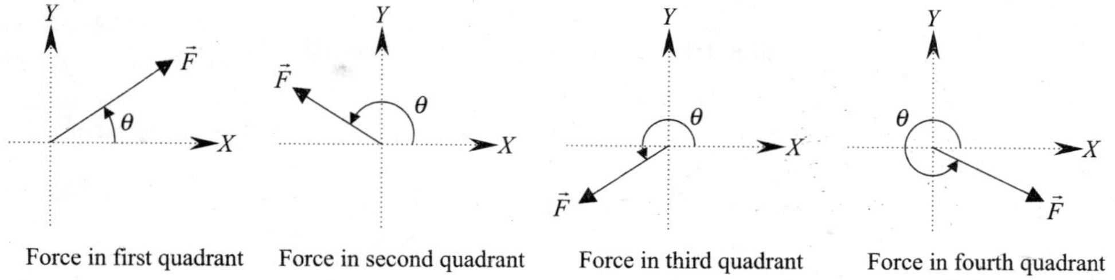

Fig. 3.21

Table 3.1 Sign of Components

Quadrant	θ	*Sign of*	
		x-component	*y-component*
I	0 – 90°	+	+
II	90° – 180°	–	+
III	180° – 270°	–	–
IV	270° – 360°	+	–

However, if we measure θ with respect to the *X-axis* such that it is always an *acute angle* as shown in Fig. 3.22 and consider the above sign convention for the components depending upon the quadrant in which the force lies then the vector approach can be avoided. This will be made clear when we solve some of the problems.

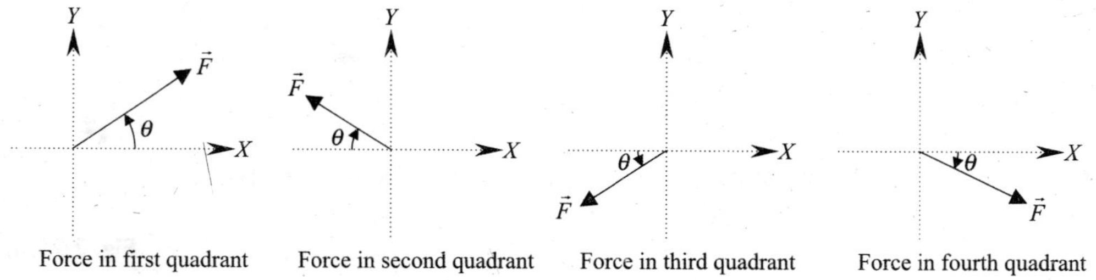

Fig. 3.22

Note: When we resolve a force into orthogonal components, the orientation of the coordinate axes is arbitrary. Hence, in case of inclined plane problems, i.e., bodies lying on an inclined plane as shown in Fig. 3.23, it will be convenient to choose the *X*-axis parallel to the inclined plane rather than the ground plane. [Also, refer Solved Examples 3.6 and 3.7].

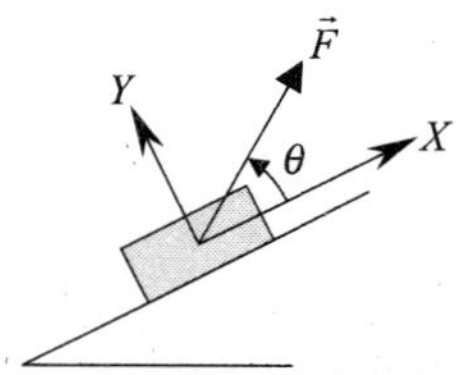

Fig. 3.23 Inclined plane problem

Example 3.4 Find the rectangular components of the force shown in Fig. 3.24.

Solution The given force is resolved into rectangular components along *X* and *Y* axes as shown in Fig. 3.24(a). Then the *x* and *y*-components of the force are

$$F_x = F\cos\theta$$
$$= 100 \times \cos 30°$$
$$= 86.6 \text{ N}$$
$$F_y = F\sin\theta$$
$$= 100 \times \sin 30°$$
$$= 50 \text{ N}$$

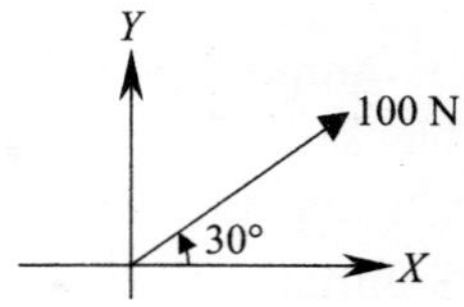

Fig. 3.24

Note that in order to differentiate the given force and its components, the given force is represented with a dashed line or a dotted line. This would avoid confusion as it may imply that there are three forces acting at the origin.

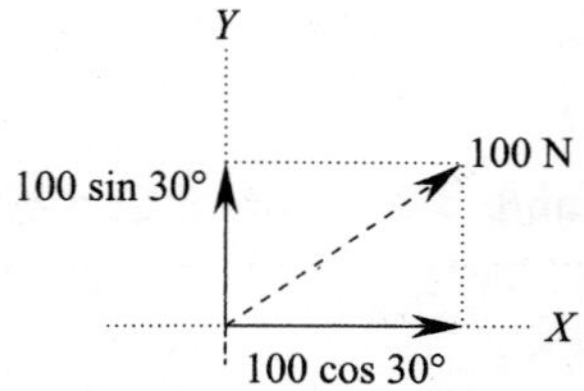

Fig. 3.24(a)

Example 3.5 Find the rectangular components of a force of magnitude 50 N acting as shown in Fig. 3.25.

Solution As the force points towards the second quadrant, we know that the *x*-component is negative and *y*-component is positive. Hence, we can write its components as

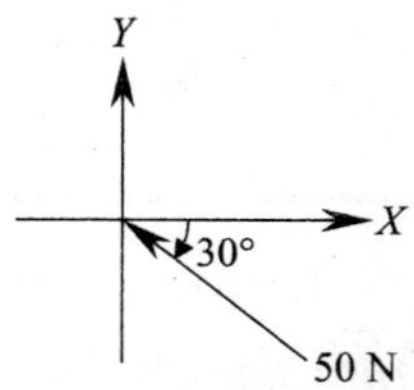

Fig. 3.25

$$F_x = -F\cos\theta$$
$$= -50 \times \cos 30°$$
$$= -43.3 \text{ N}$$

The negative sign indicates that it points along the negative *X*-axis.

$$F_y = F\sin\theta$$
$$= 50 \times \sin 30°$$
$$= 25 \text{ N}$$

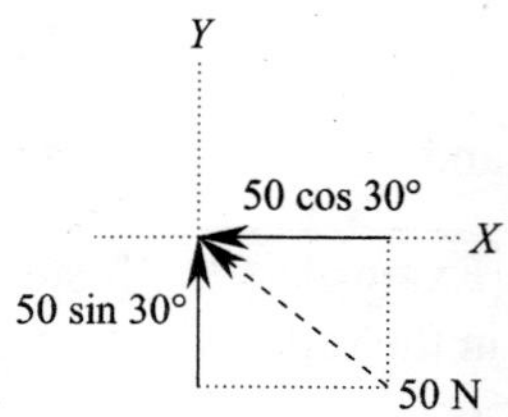

Fig. 3.25(a)

Example 3.6 Determine the components of weight [100 N] of a block resting on an inclined plane along the incline and the normal to the incline.

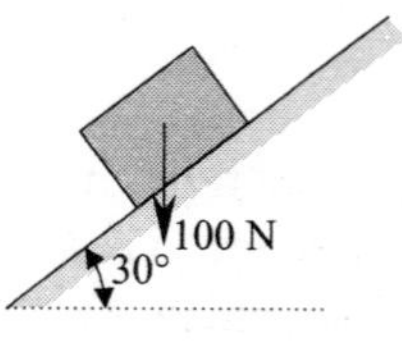

Fig. 3.26

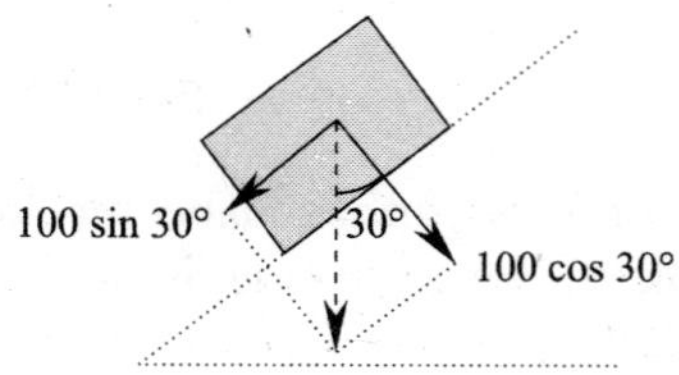

Fig. 3.26(a)

Solution As the block is resting on an inclined plane, it is convenient to resolve the force along and normal to the inclined plane. The inclination of the force with respect to the normal line is 30° as can be seen from the Fig. 3.26(a).

Hence, the components of the force are

$$F_{\text{normal}} = 100\cos 30° = 86.6 \text{ N}$$

and

$$F_{\text{along}} = 100\sin 30° = 50 \text{ N}$$

Example 3.7 Find the components of a force of magnitude 50 N acting on a block as shown in Fig. 3.27, (i) along lines parallel and perpendicular to the inclined plane, and (ii) along the horizontal and vertical axes.

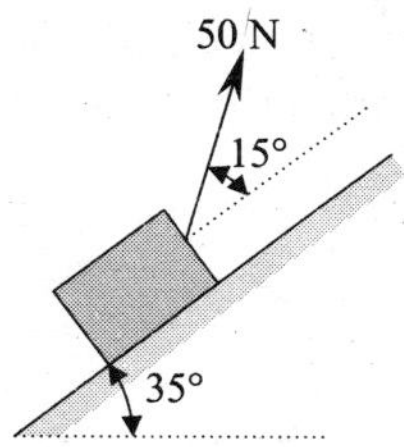

Fig. 3.27

Solution *(i) Components of the force parallel and perpendicular to the inclined plane*

As the inclination of the force with respect to the inclined plane is 15°, its components parallel and perpendicular to the inclined plane are

$$F_{par} = 50 \cos 15° = 48.3 \text{ N}$$

and
$$F_{per} = 50 \sin 15° = 12.94 \text{ N}$$

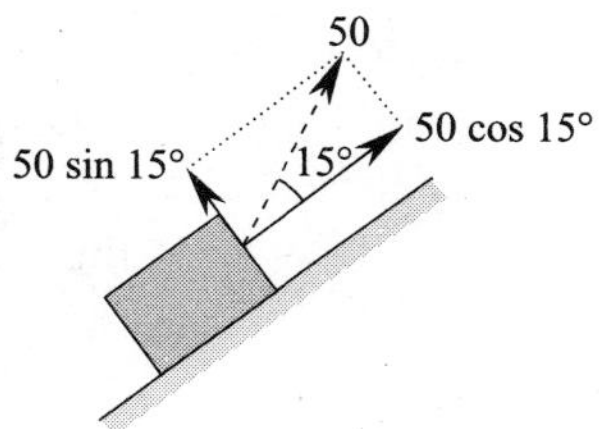

Fig. 3.27(a)

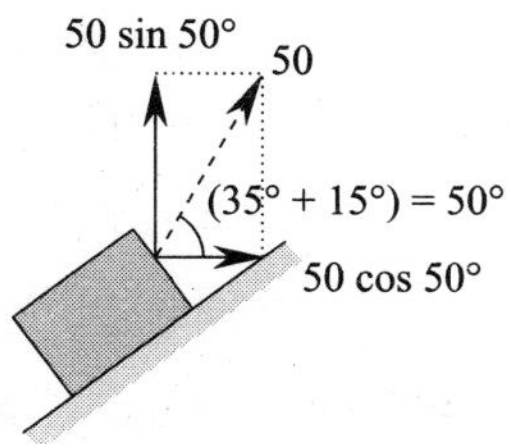

Fig. 3.27(b)

(ii) Components of the force along horizontal and vertical axes

The inclination of the force with respect to the horizontal axis is 50°. Hence, its components along the horizontal and vertical axes are

$$F_x = 50 \cos 50° = 32.14 \text{ N}$$

and
$$F_y = 50 \sin 50° = 38.3 \text{ N}$$

Example 3.8 A steel beam is lifted as shown in Fig. 3.28. Determine the components of the tension T in the rope, parallel and perpendicular to the axis of the beam.

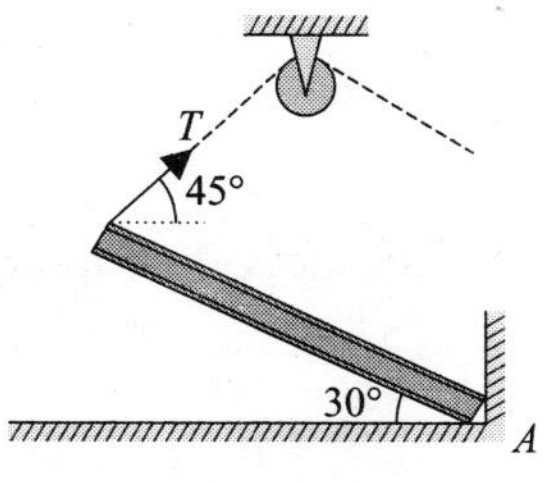

Fig. 3.28

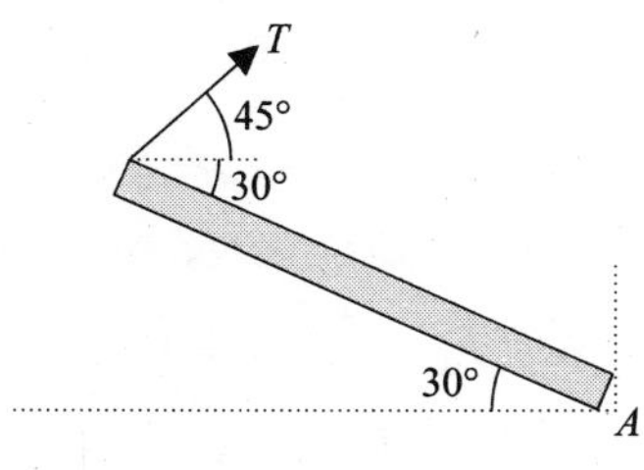

Fig. 3.28(a)

Solution From Fig. 3.28(a), we see that the inclination of the rope with respect to the beam is 45° + 30° = 75°. [The other details in the figure have been omitted for simplicity]. Hence, the components of the tension parallel and perpendicular to the axis of the beam are

$$T_{par} = T \cos 75°$$

and
$$T_{per} = T \sin 75°$$

Example 3.9 A man opens the door by applying a 10 N force at 10° to the direction of the y-axis as shown in Fig. 3.29. Determine the components of the force along the door and normal to the door.

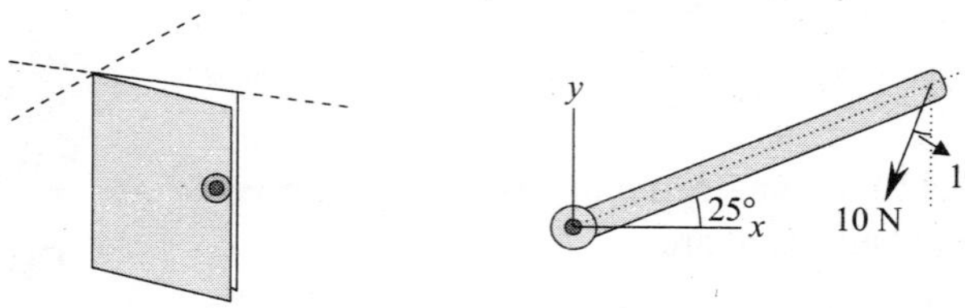

Fig. 3.29

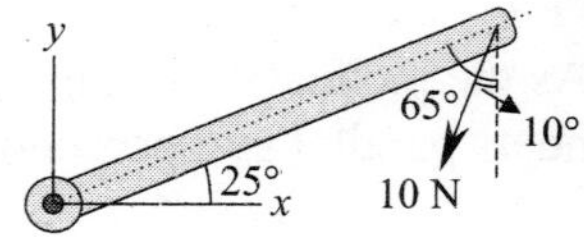

Fig. 3.29(a)

Solution From Fig. 3.29(a), we see that the inclination of the applied force with respect to the axis of the door is 65° – 10° = 55°. Therefore, its components along the door and normal to the door are

$$F_{\text{along}} = 10 \cos 55° = 5.74 \text{ N}$$

and

$$F_{\text{normal}} = 10 \sin 55° = 8.19 \text{ N}$$

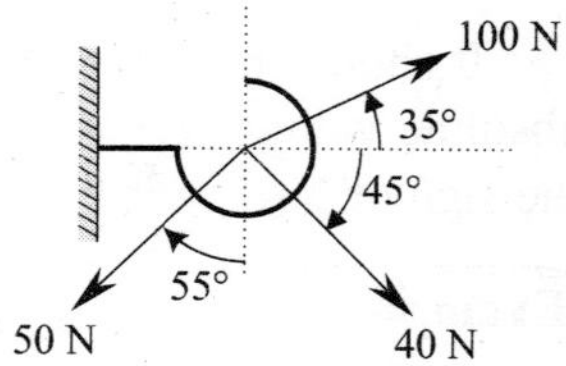

Fig. 3.30

Example 3.10 Find the components of each force shown in Fig. 3.30.

Solution The calculations are summarized in the table below:

Force	*Magnitude*	*Inclination with respect to X-axis*	*X-component (N)*	*Y-component (N)*
$\vec{F}_1$	100 N	35°	100 cos 35° = 81.92	100 sin 35° = 57.36
$\vec{F}_2$	40 N	45°	40 cos 45° = 28.28	–40 sin 45° = –28.28
$\vec{F}_3$	50 N	35°	–50 cos 35° = –40.96	–50 sin 35° = –28.68

Resultant of Several Concurrent Forces Consider 'n' number of concurrent forces, namely, $\vec{F}_1, \vec{F}_2, ..., \vec{F}_n$ acting on a particle. Representing each force in the system by its components, we have,

$$\vec{F}_1 = F_{1x}\vec{i} + F_{1y}\vec{j}$$

$$\vec{F}_2 = F_{2x}\vec{i} + F_{2y}\vec{j}$$

$$\vdots \quad \vdots \quad \vdots$$

and

$$\vec{F}_n = F_{nx}\vec{i} + F_{ny}\vec{j}$$

Therefore, the resultant is obtained by the vector addition of all individual forces:

$$\begin{aligned}\vec{R} &= \vec{F}_1 + \vec{F}_2 + .. + \vec{F}_n \\ &= (F_{1x}\vec{i} + F_{1y}\vec{j}) + (F_{2x}\vec{i} + F_{2y}\vec{j}) + \ldots + (F_{nx}\vec{i} + F_{ny}\vec{j}) \\ &= \Sigma(F_x)_i\vec{i} + \Sigma(F_y)_i\vec{j}\end{aligned} \tag{3.18}$$

If R_x and R_y are x and y components of the resultant then

$$R_x\vec{i} + R_y\vec{j} = \Sigma(F_x)_i\vec{i} + \Sigma(F_y)_i\vec{j} \tag{3.19}$$

which implies that

$$R_x = \Sigma(F_x)_i \text{ and } R_y = \Sigma(F_y)_i \tag{3.20}$$

From the above expressions, we can conclude that the x-component of the resultant of a system of forces is the *algebraic sum* of x-components of individual forces and the y-component of the resultant of a system of forces is the *algebraic sum* of the y-components of individual forces.

The magnitude and direction of the resultant are given by

$$|\vec{R}| = \sqrt{R_x^2 + R_y^2}$$

$$= \sqrt{\Sigma(F_x)_i^2 + \Sigma(F_y)_i^2} \quad (3.21)$$

and $$\tan \alpha = \frac{R_y}{R_x} = \frac{|\Sigma(F_y)_i|}{|\Sigma(F_x)_i|} \quad (3.22)$$

It should be noted that while determining the angle α made by the resultant with the X-axis, the absolute values of $\Sigma(F_y)_i$ and $\Sigma(F_x)_i$ are taken. The exact quadrant in which it lies is determined from the sign of $\Sigma(F_y)_i$ and $\Sigma(F_x)_i$. Here 'α' will always be acute angle (i.e., $\alpha \leq 90°$).

Example 3.11 Determine the resultant of four forces concurrent at the origin as shown in Fig. 3.31.

Solution Let us number the forces $\vec{F_1}$ to $\vec{F_4}$, starting from 400 N and moving in the anticlockwise direction. For the first force $\vec{F_1}$, its magnitude and inclination with respect to the X-axis are

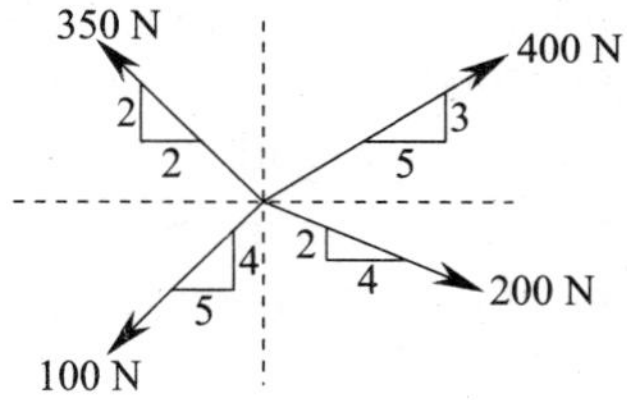

Fig. 3.31

$$|\vec{F_1}| = F_1 = 400 \text{ N}$$

$$\theta_1 = \tan^{-1}(3/5) = 30.96°$$

Therefore, its x- and y-components are

$$F_{1x} = F_1 \cos \theta_1$$

$$= 400 \times \cos(30.96°) = 343.01 \text{ N}$$

$$F_{1y} = F_1 \sin \theta_1$$

$$= 400 \times \sin(30.96°) = 205.78 \text{ N}$$

Similarly, for the other forces, the components can be determined and they are summarised in the following table:

Force	*Magnitude of force (N)*	*Inclination of force with X-axis*	$(F_x)_i$ N	$(F_y)_i$ N
$\vec{F_1}$	400	30.96°	343.01	205.78
$\vec{F_2}$	350	$\tan^{-1}(2/2) = 45°$	$-350 \cos 45° = -247.49$	$350 \sin 45° = 247.49$
$\vec{F_3}$	100	$\tan^{-1}(4/5) = 38.66°$	$-100 \cos 38.66° = -78.09$	$-100 \sin 38.66° = -62.47$
$\vec{F_4}$	200	$\tan^{-1}(2/4) = 26.57°$	$200 \cos 26.57° = 178.88$	$-200 \sin 26.57° = -89.46$
$\Sigma =$			196.31	301.34

Therefore, the magnitude of the resultant is given by

$$R = \sqrt{\Sigma(F_x)_i^2 + \Sigma(F_y)_i^2}$$

$$= \sqrt{(196.31)^2 + (301.34)^2}$$

$$= 359.64 \text{ N}$$

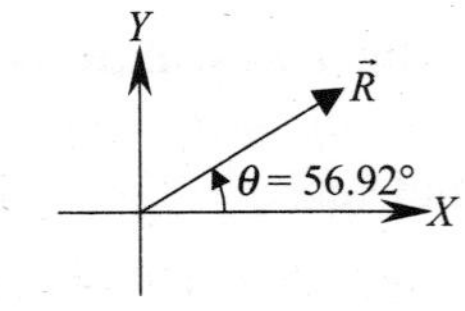

Fig. 3.31(a) Graphical representation of the resultant force

and its inclination with respect to the X-axis is given by

$$\theta = \tan^{-1}\left[\frac{|\Sigma(F_y)_i|}{|\Sigma(F_x)_i|}\right]$$

$$= \tan^{-1}\left[\frac{301.34}{196.31}\right] = 56.92°$$

Since both $\Sigma(F_x)_i$ and $\Sigma(F_y)_i$ are positive, the resultant lies in the *first*-quadrant [see Fig. 3.31(a)].

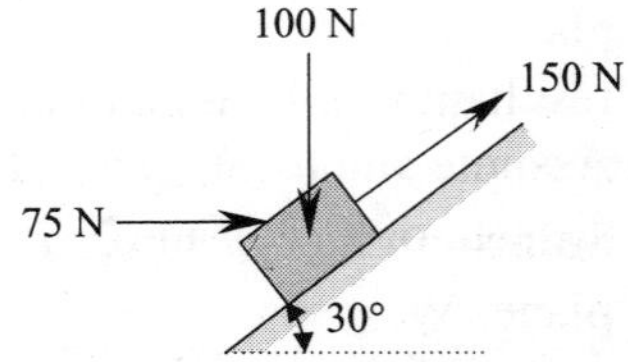

Fig. 3.32

Example 3.12 Find the resultant of a system of forces acting on the block as shown in Fig. 3.32. The 150 N force acts parallel to the incline; the 100 N force acts vertically downwards and the 75 N force acts horizontally. Assume all the forces are concurrent at a point.

Solution As the block is resting on the inclined plane, it will be convenient to resolve the forces acting on it along the incline and normal to the incline. Hence, we must know the inclination of each force with respect to the incline [refer Fig. 3.32(a)]. The calculations are summarised in the table below:

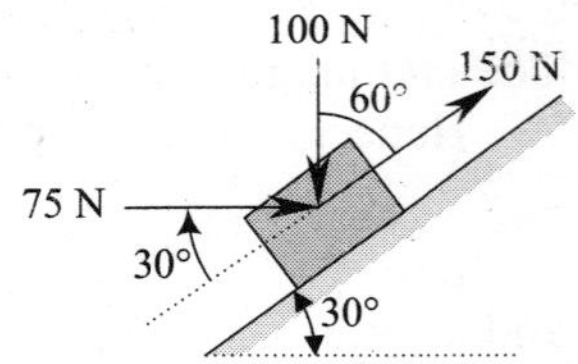

Fig. 3.32(a)

Force	*Magnitude of force (N)*	*Inclination of force with respect to the incline*	$(F_x)_i$ N	$(F_y)_i$ N
$\vec{F}_1$	150	0°	150	0
$\vec{F}_2$	100	60°	–50	–86.6
$\vec{F}_3$	75	30°	64.95	–37.5
$\Sigma =$			164.95	–124.1

[Note that the 100 N force points along *third*-quadrant and hence, both of its components are negative; the 75 N force points along *fourth*-quadrant and hence, its x-component is positive, while its y-component is negative.]

Therefore, the magnitude of the resultant is given by

$$R = \sqrt{\Sigma(F_x)_i^2 + \Sigma(F_y)_i^2}$$

$$= \sqrt{(164.95)^2 + (-124.1)^2}$$

$$= 206.42 \text{ N}$$

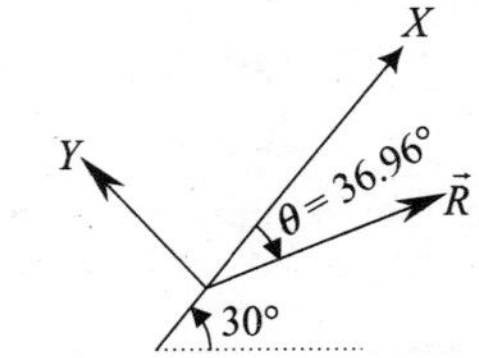

Fig. 3.32(b) Graphical representation of the resultant force

and its inclination with respect to the inclined plane is

$$\theta = \tan^{-1}\left[\frac{|\Sigma(F_y)_i|}{|\Sigma(F_x)_i|}\right]$$

$$= \tan^{-1}\left[\frac{124.1}{164.95}\right] = 36.96°$$

Since $\Sigma(F_x)_i$ is positive and $\Sigma(F_y)_i$ is negative, the resultant lies in the *fourth*-quadrant [see Fig. 3.32(b)].

Example 3.13 The system of forces acting on a block lying on a horizontal plane is shown in Fig. 3.33. Determine the resultant force, if the component of resultant along the y-direction is zero. Also, determine the unknown force 'F.' Assume all the forces are concurrent at a point.

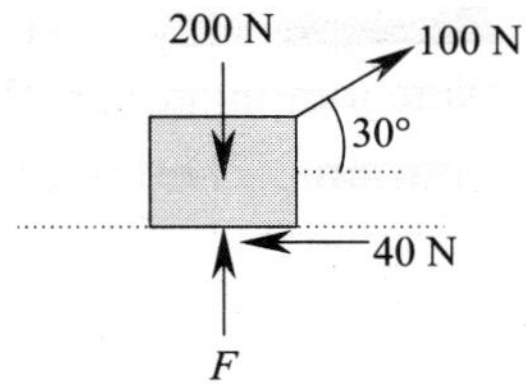

Fig. 3.33

Solution Let us choose the reference axes along and perpendicular to the plane. We see that only the 100 N force is inclined to the axes. Hence, its components along the X and Y axes are respectively:

$$100 \cos 30° = 86.6 \text{ N}$$

and $$100 \sin 30° = 50 \text{ N}$$

Hence, summing up the components of individual forces along X and Y axes, we have

$$\Sigma(F_y)_i = F + 100 \sin 30° - 200 \quad \text{(a)}$$

and $$\Sigma(F_x)_i = 100 \cos 30° - 40 = 46.6 \text{ N} \quad \text{(b)}$$

As the component of the resultant along the y-direction is zero, we get from equation (a),

$$F = 150 \text{ N}$$

Since the component of the resultant along the y-direction is zero, the resultant of the forces is same as x-component, i.e., $R = \Sigma(F_x)_i = 46.6$ N.

Example 3.14 Five forces are acting at corner A of a regular hexagon as shown in Fig. 3.34. Determine the resultant of the system of forces.

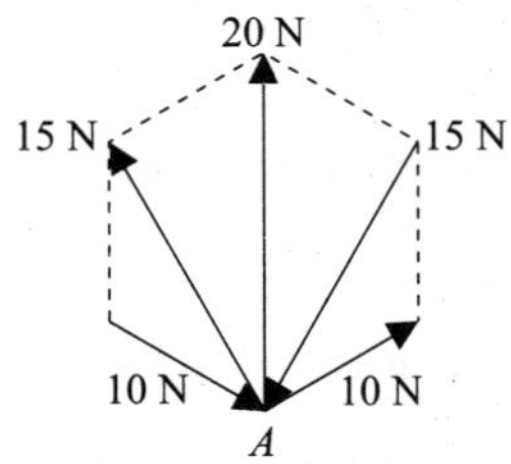

Fig. 3.34

Solution We know that in a regular hexagon, the internal angle at each vertex is 120°. Hence, the angle between the forces can be determined as shown in Fig. 3.34(a).

Force	*Magnitude of force (N)*	*Inclination of force with X-axis*	$(F_x)_i$ N	$(F_y)_i$ N
$\vec{F}_1$	10	30°	10 cos 30°	10 sin 30°
$\vec{F}_2$	15	60°	–15 cos 60°	–15 sin 60°
$\vec{F}_3$	20	90°	0	20
$\vec{F}_4$	15	60°	–15 cos 60°	15 sin 60°
$\vec{F}_5$	10	30°	10 cos 30°	–10 sin 30°
Σ =			2.32	20

Therefore, the magnitude of the resultant is given as

$$R = \sqrt{\sum(F_x)_i^2 + \sum(F_y)_i^2} = 20.13 \text{ N}$$

and its inclination with respect to the X-axis is

$$\alpha = \tan^{-1}\left[\frac{|\sum(F_y)_i|}{|\sum(F_x)_i|}\right] = 83.38°$$

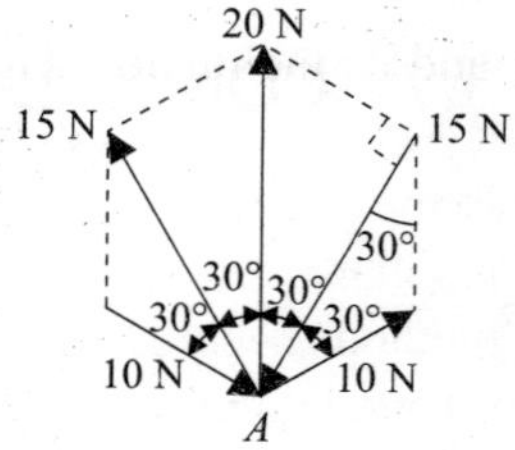

Fig. 3.34(a)

As the x and y components of the resultant are positive, we can understand that the resultant lies in the *first*-quadrant.

Example 3.15 Determine the resultant of two concurrent forces shown in Fig. 3.35 acting at a gusset plate joint in a truss. What force should be applied to make the resultant zero?

Solution The forces acting along the members can be represented in vector form using rectangular components as

$$\vec{F}_1 = 30\vec{i}$$

and $$\vec{F}_2 = -20 \cos 30° \vec{i} - 20 \sin 30° \vec{j}$$

Hence, the resultant of the two forces can be obtained by vector addition:

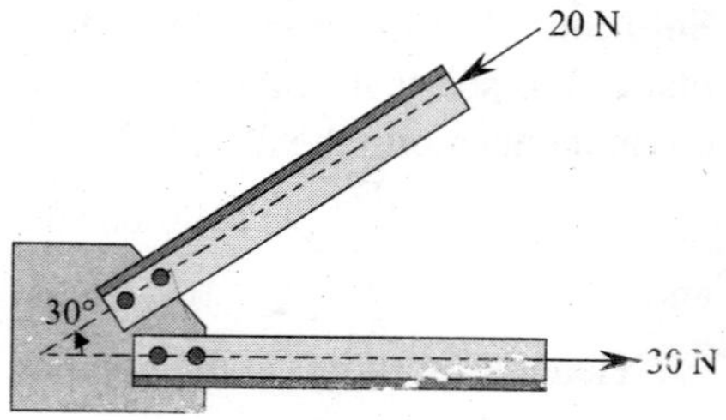

Fig. 3.35

$$\vec{R} = \vec{F}_1 + \vec{F}_2$$

$$= [30 - 20 \cos 30°]\vec{i} - [20 \sin 30°]\vec{j}$$

$$= 12.7\vec{i} - 10\vec{j}$$

Therefore, the force that must be applied at the point of concurrency to make the resultant zero is

$$-\vec{R} = -12.7\vec{i} + 10\vec{j}$$

Example 3.16 Show that the resultant of forces $k_1\,\overrightarrow{OA}$, $k_2\,\overrightarrow{OB}$ is $(k_1 + k_2)\overrightarrow{OC}$, where C is a point on AB such that $k_1\,AC = k_2\,CB$.

Solution Given: $k_1\,AC = k_2\,CB$

Therefore, $$k_1\,\overrightarrow{AC} = k_2\,\overrightarrow{CB}$$

Since $\overrightarrow{AC}$ can be written as $(\overrightarrow{OC} - \overrightarrow{OA})$ and $\overrightarrow{CB}$ can be written as $(\overrightarrow{OB} - \overrightarrow{OC})$,

$$k_1(\overrightarrow{OC} - \overrightarrow{OA}) = k_2(\overrightarrow{OB} - \overrightarrow{OC})$$

On rearranging,

$$(k_1 + k_2)\overrightarrow{OC} = k_1\,\overrightarrow{OA} + k_2\,\overrightarrow{OB}$$

Thus, we can see that the resultant of forces $k_1\,\overrightarrow{OA}$ and $k_2\,\overrightarrow{OB}$ is $(k_1 + k_2)\,\overrightarrow{OC}$.

3.7 CONCURRENT FORCES IN SPACE

In the previous sections, we discussed the coplanar concurrent forces and the methods to determine their resultants. In this section, we will discuss concurrent forces in *space*. Graphical and trigonometric methods become complicated when applied to solve more number of concurrent forces. Still more complicated do they become when applied to forces in space. Hence, we employ *only* analytical method to solve forces in space.

3.7.1 Forces in Space

Consider a force $\vec{F}$ in space acting at the origin. Its components along mutually perpendicular *X*, *Y* and *Z* axes can be determined by the method of resolution. First, resolve the force $\vec{F}$ into components, $\vec{F}_y$ and $\vec{F}_{xz}$ along the vertical plane *OCGF* [Fig. 3.36(a)]. Then resolve $\vec{F}_{xz}$ into components, $\vec{F}_x$ and $\vec{F}_z$ on the *X–Z* plane [Fig. 3.36(b)].

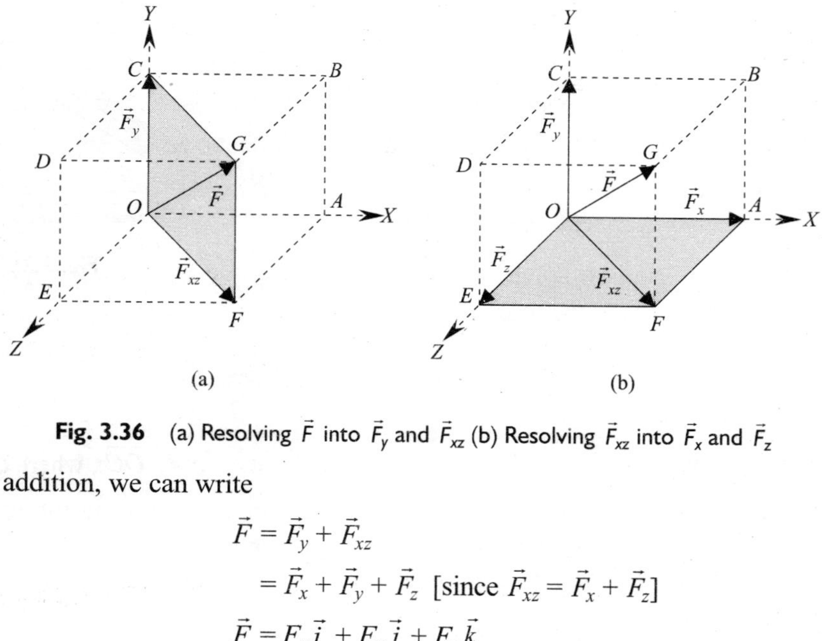

Fig. 3.36 (a) Resolving $\vec{F}$ into $\vec{F}_y$ and $\vec{F}_{xz}$ (b) Resolving $\vec{F}_{xz}$ into $\vec{F}_x$ and $\vec{F}_z$

By vector addition, we can write

$$\vec{F} = \vec{F}_y + \vec{F}_{xz}$$

$$= \vec{F}_x + \vec{F}_y + \vec{F}_z \quad [\text{since } \vec{F}_{xz} = \vec{F}_x + \vec{F}_z] \tag{3.23}$$

or

$$\vec{F} = F_x\vec{i} + F_y\vec{j} + F_z\vec{k} \tag{3.24}$$

where $\vec{i}$, $\vec{j}$ and $\vec{k}$ are unit vectors respectively along *X*, *Y* and *Z* directions.

The magnitude of the force is represented in terms of its components as

$$|\vec{F}| = F = \sqrt{F_x^2 + F_y^2 + F_z^2} \tag{3.25}$$

If θ_x, θ_y and θ_z are angles made by $\vec{F}$ with *X*, *Y* and *Z* axes respectively [Fig. 3.36(c)], then Eq. 3.24 can be written as

$$\vec{F} = F\cos\theta_x\vec{i} + F\cos\theta_y\vec{j} + F\cos\theta_z\vec{k}$$

$$= F[\cos\theta_x\vec{i} + \cos\theta_y\vec{j} + \cos\theta_z\vec{k}] \tag{3.26}$$

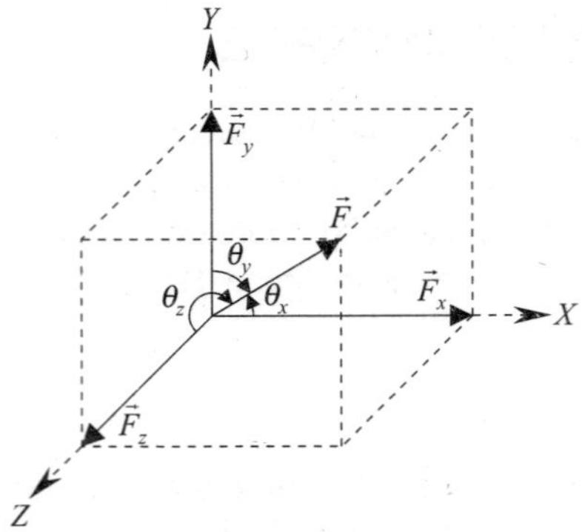

Fig. 3.36(c) Inclination of $\vec{F}$ with respect to X, Y and Z axes

We know that any vector can be expressed as a product of its magnitude and unit vector along its line of action. Hence, force vector can also be written as

$$\vec{F} = F\hat{n} \tag{3.27}$$

Comparing the above two expressions for force [Eqs 3.26 and 3.27], we can readily see that $\hat{n} = \cos\theta_x \vec{i} + \cos\theta_y \vec{j} + \cos\theta_z \vec{k}$ is the unit vector along the line of action of $\vec{F}$. Since the magnitude of $\hat{n}$ is **unity**, we have,

$$\cos^2\theta_x + \cos^2\theta_y + \cos^2\theta_z = 1 \tag{3.28}$$

where $\cos\theta_x$, $\cos\theta_y$ and $\cos\theta_z$ are called **direction cosines** of the force. The direction cosines can be expressed in terms of components of the force as

$$\cos\theta_x = \frac{F_x}{F}, \cos\theta_y = \frac{F_y}{F} \text{ and } \cos\theta_z = \frac{F_z}{F} \tag{3.29}$$

3.7.2 Representation of a Force Passing through any Two Points in Space

In the previous section, we saw how to express a force vector in space passing through the origin. However, in general, forces do not pass through the origin, but pass through any two points in space. Hence, in this section, we will see how to express a force vector passing through any two points in space.

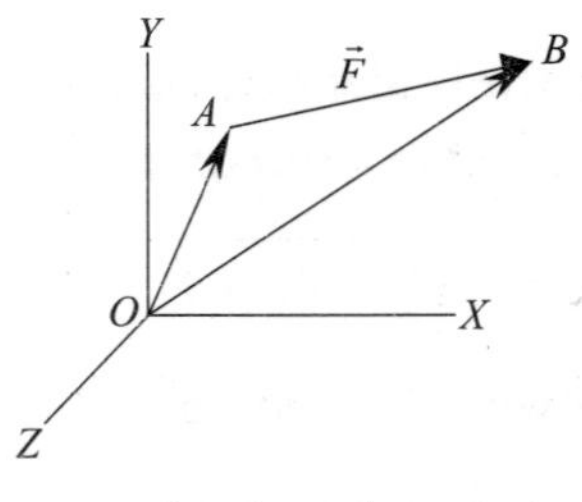

Fig. 3.37 Force passing through any two points in space

Consider a force $\vec{F}$ passing through two points A and B in space, whose respective coordinates are (x_1, y_1, z_1) and (x_2, y_2, z_2). Then the position vectors of the points A and B with respect to the origin are

$$\overrightarrow{OA} = x_1\vec{i} + y_1\vec{j} + z_1\vec{k} \tag{3.30}$$

and

$$\overrightarrow{OB} = x_2\vec{i} + y_2\vec{j} + z_2\vec{k} \tag{3.31}$$

Hence,

$$\overrightarrow{AB} = \overrightarrow{OB} - \overrightarrow{OA}$$

$$= (x_2 - x_1)\vec{i} + (y_2 - y_1)\vec{j} + (z_2 - z_1)\vec{k} \tag{3.32}$$

Then unit vector along AB is given by $\hat{n}_{AB} = \dfrac{\overrightarrow{AB}}{|\overrightarrow{AB}|}$

Hence, the force $\vec{F}$ can be expressed as

$$\vec{F} = F\hat{n}_{AB} \tag{3.33}$$

Example 3.17 Express the force of 100 N passing through the origin and point A as shown in Fig. 3.38 in vector form.

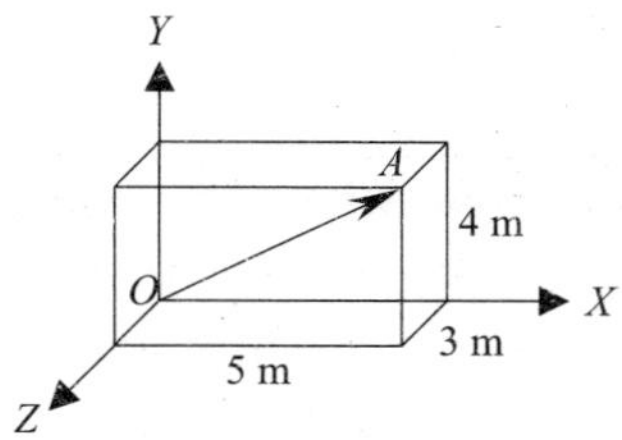

Fig. 3.38

Solution From the figure, we see that the coordinates of A are (5, 4, 3). Therefore,

$$\overrightarrow{OA} = 5\vec{i} + 4\vec{j} + 3\vec{k}$$

Then its magnitude is obtained as

$$|\overrightarrow{OA}| = \sqrt{5^2 + 4^2 + 3^2}$$

$$= \sqrt{50} \text{ m}$$

and unit vector along OA is obtained as

$$\hat{n}_{OA} = \frac{\overrightarrow{OA}}{|\overrightarrow{OA}|}$$

$$= \frac{5\vec{i} + 4\vec{j} + 3\vec{k}}{\sqrt{50}}$$

Therefore, the force can be expressed in vector form as

$$\vec{F} = F\hat{n}_{OA}$$

$$= 100\left[\frac{5\vec{i} + 4\vec{j} + 3\vec{k}}{\sqrt{50}}\right]$$

$$= 70.71 \text{ N}\vec{i} + 56.57 \text{ N}\vec{j} + 42.43 \text{ N}\vec{k}$$

Example 3.18 A force of magnitude 200 N passes through the origin. Its inclinations with respect to the axes are $\theta_x = 30°$, $\theta_y = 75°$. If F_z is positive, then determine the inclination of the force with respect to the Z-axis and express the force in vector form.

Solution We know that the direction cosines are related by the expression

$$\cos^2\theta_x + \cos^2\theta_y + \cos^2\theta_z = 1$$

$$\Rightarrow \quad \cos^2\theta_z = 1 - \cos^2\theta_x - \cos^2\theta_y$$

$$= 1 - \cos^2(30°) - \cos^2(75°) = 0.183$$

$$\therefore \quad \cos\theta_z = \pm 0.428$$

Which of the above two values is to be considered depends upon the sign of F_z. Since F_z is positive, $\cos\theta_z$ will also be positive. [Note that $\cos\theta_z = F_z/F$.]

$$\therefore \quad \cos\theta_z = 0.428$$

$$\Rightarrow \quad \theta_z = 64.66°$$

Therefore, the force can be expressed in vector form as

$$\vec{F} = F[\cos\theta_x\vec{i} + \cos\theta_y\vec{j} + \cos\theta_z\vec{k}]$$

$$= 200[\cos(30°)\vec{i} + \cos(75°)\vec{j} + \cos(64.66°)\vec{k}]$$

$$= 173.21\text{ N}\vec{i} + 51.76\text{ N}\vec{j} + 85.6\text{ N}\vec{k}$$

Example 3.19 The inclinations of a force passing through the origin are $\theta_y = 55.4°$ and $\theta_z = 67.2°$. Determine the angle θ_x, if $F_x = -100$ N. Also, express the force in vector form.

Solution The direction cosines are related by the expression

$$\cos^2\theta_x + \cos^2\theta_y + \cos^2\theta_z = 1$$

$$\Rightarrow \quad \cos^2\theta_x = 1 - \cos^2\theta_y - \cos^2\theta_z$$

$$= 1 - \cos^2(55.4°) - \cos^2(67.2°) = 0.527$$

$$\therefore \quad \cos\theta_x = \pm 0.726$$

Which of the above two values is to be considered depends upon the sign of F_x. Since F_x is negative, $\cos\theta_x$ will also be negative.

$$\therefore \quad \cos\theta_x = -0.726 \quad \Rightarrow \quad \theta_x = 136.55°$$

We know,

$$\cos\theta_x = \frac{F_x}{F}$$

$$\therefore \quad F = \frac{F_x}{\cos\theta_x}$$

$$= \frac{-100}{-0.726}$$

$$= 137.74\text{ N}$$

Therefore, the force vector is given as

$$\vec{F} = F[\cos\theta_x\vec{i} + \cos\theta_y\vec{j} + \cos\theta_z\vec{k}]$$

$$= 137.74[\cos(136.55°)\vec{i} + \cos(55.4°)\vec{j} + \cos(67.2°)\vec{k}]$$

$$= -100\text{ N}\vec{i} + 78.21\text{ N}\vec{j} + 53.38\text{ N}\vec{k}$$

Example 3.20 The line of action of a force of magnitude 50 N passes through points A and B as shown in Fig. 3.39. Express the force in vector form.

Solution From the figure, we see that

$$\overrightarrow{AB} = \overrightarrow{OB} - \overrightarrow{OA}$$
$$= (2\vec{i} - 2\vec{j} + 2\vec{k}) - (4\vec{i} + 5\vec{j} + 6\vec{k})$$
$$= -2\vec{i} - 7\vec{j} - 4\vec{k}$$

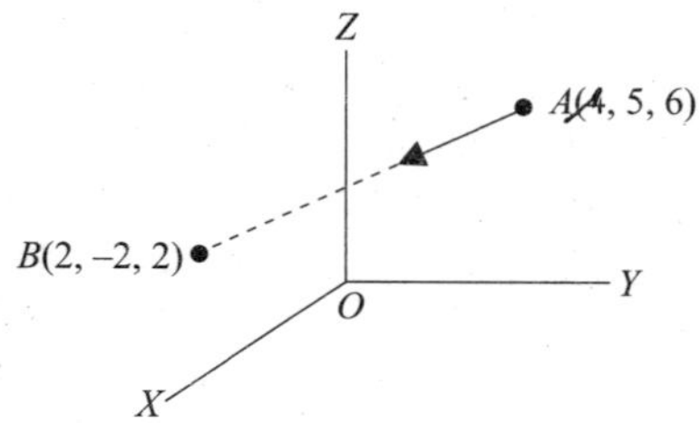

Fig. 3.39

Therefore, unit vector along $\overrightarrow{AB}$ is

$$\hat{n}_{AB} = \frac{-2\vec{i} - 7\vec{j} - 4\vec{k}}{\sqrt{(-2)^2 + (-7)^2 + (-4)^2}}$$
$$= -0.241\vec{i} - 0.843\vec{j} - 0.482\vec{k}$$

Therefore, the force can be expressed in vector form as

$$\vec{F} = F\hat{n}_{AB}$$
$$= 50[-0.241\vec{i} - 0.843\vec{j} - 0.482\vec{k}]$$
$$= -12.05\text{ N}\vec{i} - 42.15\text{ N}\vec{j} - 24.1\text{ N}\vec{k}$$

3.7.3 Resultant of Several Concurrent Forces in Space

Consider 'n' number of concurrent forces, namely, $\vec{F}_1$, $\vec{F}_2$, ..., $\vec{F}_n$ acting in space. Representing each force in the system by its components, we have,

$$\vec{F}_1 = F_{1x}\vec{i} + F_{1y}\vec{j} + F_{1z}\vec{k}$$
$$\vec{F}_2 = F_{2x}\vec{i} + F_{2y}\vec{j} + F_{2z}\vec{k}$$
$$\vdots \quad \vdots \quad \vdots \quad \vdots$$

and
$$\vec{F}_n = F_{nx}\vec{i} + F_{ny}\vec{j} + F_{nz}\vec{k}$$

Then their resultant is obtained by vector addition of all individual forces:

$$\vec{R} = \vec{F}_1 + \vec{F}_2 + \ldots + \vec{F}_n$$
$$= [F_{1x} + F_{2x} + \ldots + F_{nx}]\vec{i} + [F_{1y} + F_{2y} + \ldots + F_{ny}]\vec{j} + [F_{1z} + F_{2z} + \ldots + F_{nz}]\vec{k}$$
$$= \Sigma(F_x)_i\vec{i} + \Sigma(F_y)_i\vec{j} + \Sigma(F_z)_i\vec{k} \quad (3.34)$$

If R_x, R_y and R_z are components of the resultant then

$$R_x\vec{i} + R_y\vec{j} + R_z\vec{k} = \Sigma(F_x)_i\vec{i} + \Sigma(F_y)_i\vec{j} + \Sigma(F_z)_i\vec{k}$$

which implies that

$$R_x = \Sigma(F_x)_i,\ R_y = \Sigma(F_y)_i \text{ and } R_z = \Sigma(F_z)_i \quad (3.35)$$

The magnitude and direction of the resultant are given by

$$|\vec{R}| = \sqrt{R_x^2 + R_y^2 + R_z^2}$$

$$= \sqrt{\sum(F_x)_i^2 + \sum(F_y)_i^2 + \sum(F_z)_i^2} \tag{3.36}$$

$$\cos\theta_x = \frac{R_x}{R} = \frac{\sum(F_x)_i}{R}$$

$$\cos\theta_y = \frac{R_y}{R} = \frac{\sum(F_y)_i}{R}$$

and

$$\cos\theta_z = \frac{R_z}{R} = \frac{\sum(F_z)_i}{R} \tag{3.37}$$

It should be noted that unlike two-dimensional, i.e., coplanar system of forces, here algebraic values of $\Sigma(F_x)_i$, $\Sigma(F_y)_i$ and $\Sigma(F_z)_i$ are considered for finding the direction cosines and not absolute values.

Example 3.21 Determine the resultant of a system of three concurrent forces passing through the origin and points (10, –5, 6), (–5, 5, 7) and (6, –4, –3) respectively. The respective magnitudes of the forces are 1500 N, 2500 N and 2000 N.

Solution Let the points through which the forces pass be denoted as *A*, *B* and *C* respectively. Then their position vectors are given as

$$\overrightarrow{OA} = 10\vec{i} - 5\vec{j} + 6\vec{k}$$

$$\overrightarrow{OB} = -5\vec{i} + 5\vec{j} + 7\vec{k}$$

$$\overrightarrow{OC} = 6\vec{i} - 4\vec{j} - 3\vec{k}$$

Then unit vectors along $\overrightarrow{OA}$, $\overrightarrow{OB}$ and $\overrightarrow{OC}$ can be determined as follows

$$\hat{n}_{OA} = \frac{10\vec{i} - 5\vec{j} + 6\vec{k}}{\sqrt{(10)^2 + (-5)^2 + (6)^2}}$$

$$= \frac{10\vec{i} - 5\vec{j} + 6\vec{k}}{\sqrt{161}}$$

$$\hat{n}_{OB} = \frac{-5\vec{i} + 5\vec{j} + 7\vec{k}}{\sqrt{(-5)^2 + (5)^2 + (7)^2}}$$

$$= \frac{-5\vec{i} + 5\vec{j} + 7\vec{k}}{\sqrt{99}}$$

$$\hat{n}_{OC} = \frac{6\vec{i} - 4\vec{j} - 3\vec{k}}{\sqrt{(6)^2 + (-4)^2 + (-3)^2}}$$

$$= \frac{6\vec{i} - 4\vec{j} - 3\vec{k}}{\sqrt{61}}$$

Hence, the forces can be expressed in vector form as

$$\vec{F}_{OA} = F_{OA}\,\hat{n}_{OA}$$

$$= 1500\left(\frac{10\vec{i} - 5\vec{j} + 6\vec{k}}{\sqrt{161}}\right)$$

$$= 1182.17\vec{i} - 591.08\vec{j} + 709.3\vec{k}$$

$$\vec{F}_{OB} = F_{OB}\,\hat{n}_{OB}$$

$$= 2500\left(\frac{-5\vec{i} + 5\vec{j} + 7\vec{k}}{\sqrt{99}}\right)$$

$$= -1256.3\vec{i} + 1256.3\vec{j} + 1758.82\vec{k}$$

$$\vec{F}_{OC} = F_{OC}\,\hat{n}_{OC}$$

$$= 2000\left(\frac{6\vec{i} - 4\vec{j} - 3\vec{k}}{\sqrt{61}}\right)$$

$$= 1536.44\vec{i} - 1024.3\vec{j} - 768.22\vec{k}$$

Therefore, the resultant force $\vec{R}$ is given as

$$\vec{R} = \vec{F}_{OA} + \vec{F}_{OB} + \vec{F}_{OC}$$

$$= 1462.31\text{ N}\vec{i} - 359.08\text{ N}\vec{j} + 1699.9\text{ N}\vec{k}$$

Its magnitude is given by

$$|\vec{R}| = \sqrt{(1462.31)^2 + (-359.08)^2 + (1699.9)^2}$$

$$= 2270.89\text{ N}$$

Example 3.22 Determine the resultant of the tension forces acting at point A of the transmission tower. The magnitudes of tensions along cables AB, AC and AD are respectively 1000 N, 2000 N and 1800 N.

Solution The coordinates of the points A, B, C and D are

$$A(0, 0, 10) \text{ or } \overrightarrow{OA} = 10\vec{k}$$

$$B(-5, -2, 0) \text{ or } \overrightarrow{OB} = -5\vec{i} - 2\vec{j}$$

$$C(4, -3, 0) \text{ or } \overrightarrow{OC} = 4\vec{i} - 3\vec{j}$$

$$D(-3, 3, 0) \text{ or } \overrightarrow{OD} = -3\vec{i} + 3\vec{j}$$

Calculation of unit vectors along AB, AC and AD

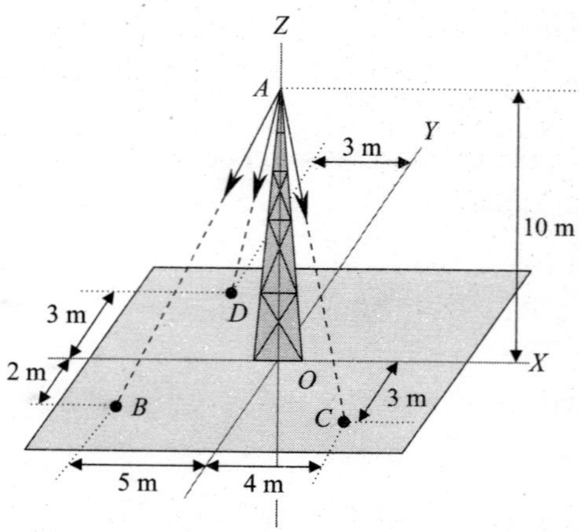

Fig. 3.40

$\hat{n}_{AB}$: $\overrightarrow{AB} = \overrightarrow{OB} - \overrightarrow{OA}$

$$= -5\vec{i} - 2\vec{j} - 10\vec{k}$$

$$|\overrightarrow{AB}| = \sqrt{(-5)^2 + (-2)^2 + (-10)^2} = \sqrt{129}$$

$$\therefore \quad \hat{n}_{AB} = \frac{-5\vec{i} - 2\vec{j} - 10\vec{k}}{\sqrt{129}}$$

$\hat{n}_{AC}$: $\overrightarrow{AC} = \overrightarrow{OC} - \overrightarrow{OA}$

$$= 4\vec{i} - 3\vec{j} - 10\vec{k}$$

$$|\overrightarrow{AC}| = \sqrt{(4)^2 + (-3)^2 + (-10)^2} = \sqrt{125}$$

$$\therefore \quad \hat{n}_{AC} = \frac{4\vec{i} - 3\vec{j} - 10\vec{k}}{\sqrt{125}}$$

$\hat{n}_{AD}$: $\overrightarrow{AD} = \overrightarrow{OD} - \overrightarrow{OA}$

$$= -3\vec{i} + 3\vec{j} - 10\vec{k}$$

$$|\overrightarrow{AD}| = \sqrt{(-3)^2 + (3)^2 + (-10)^2} = \sqrt{118}$$

$$\therefore \quad \hat{n}_{AD} = \frac{-3\vec{i} + 3\vec{j} - 10\vec{k}}{\sqrt{118}}$$

Calculation of force vectors

$$\vec{T}_{AB} = T_{AB}\,\hat{n}_{AB}$$

$$= 1000\left(\frac{-5\vec{i} - 2\vec{j} - 10\vec{k}}{\sqrt{129}}\right)$$

$$= -440.23\vec{i} - 176.09\vec{j} - 880.45\vec{k}$$

$$\vec{T}_{AC} = T_{AC}\,\hat{n}_{AC}$$

$$= 2000\left(\frac{4\vec{i} - 3\vec{j} - 10\vec{k}}{\sqrt{125}}\right)$$

$$= 715.54\vec{i} - 536.66\vec{j} - 1788.85\vec{k}$$

$$\vec{T}_{AD} = T_{AD}\,\hat{n}_{AD}$$

$$= 1800\left(\frac{-3\vec{i} + 3\vec{j} - 10\vec{k}}{\sqrt{118}}\right)$$

$$= -497.11\vec{i} + 497.11\vec{j} - 1657.03\vec{k}$$

Hence, the resultant force is given by

$$\vec{R} = \vec{T}_{AB} + \vec{T}_{AC} + \vec{T}_{AD}$$

$$= -221.8\vec{i} - 215.64\vec{j} - 4326.33\vec{k}$$

Its magnitude is

$$R = \sqrt{(-221.8)^2 + (-215.64)^2 + (-4326.33)^2}$$

$$= 4337.38 \text{ N}$$

and its inclinations with respect to the axes are

$$\theta_x = \cos^{-1}\left[\frac{R_x}{R}\right] = \cos^{-1}\left[\frac{-221.8}{4337.38}\right] = 92.93°$$

$$\theta_y = \cos^{-1}\left[\frac{R_y}{R}\right] = \cos^{-1}\left[\frac{-215.64}{4337.38}\right] = 92.85°$$

$$\theta_z = \cos^{-1}\left[\frac{R_z}{R}\right] = \cos^{-1}\left[\frac{-4326.33}{4337.38}\right] = 175.91°$$

SUMMARY

Force is defined as any action of a body on another, which tends to change the state of rest or of motion of the other body. To completely define a force, we must specify its *magnitude*, its *point of application* and its *direction*, which are known as *force characteristics*.

The *magnitude* of a force is obtained by comparing it with a certain standard force. The unit of force is given in newton (N). The *point of application* of the force is that point in the body at which the force can be assumed to be concentrated. The *direction* or *line of action* of the force is that direction in which the acting force tends to move the body. It is a straight line passing through the point of application of the force. The *sense* of the force is indicated by an *arrowhead*.

The principle of transmissibility states that the conditions of equilibrium or of motion of a rigid body remains unchanged if a force acting at a given point of the rigid body is replaced by a force of the same magnitude and direction, but acting at a different point, provided that the two forces have the same line of action.

System of forces: A number of forces acting on a body at a particular instant constitute a system of forces.

Coplanar forces: When all the forces in the system of forces lie on the *same* plane.

Non-coplanar or spatial forces: When all the forces lie on *different* planes.

Concurrent forces: When the lines of action of all the forces intersect at a *point.*

Non-concurrent forces: When the lines of action of all the forces do not intersect at a point.

Collinear forces: When the lines of action of all the forces lie along the *same line.*

Parallel forces: When the lines of action of all the forces are *parallel* to each other.

The resultant of a force system is a *single equivalent force* which produces the *same* effect on the body as that of all given forces.

The parallelogram law states that when two forces $\vec{F}_1$ and $\vec{F}_2$ acting on a body are represented by two adjacent sides of a parallelogram then the diagonal passing through their point of concurrency represents the resultant force $\vec{R}$ in magnitude and direction. Mathematically, the magnitude and direction of the resultant force are given as

$$R = \sqrt{F_1^2 + F_2^2 + 2F_1F_2 \cos\theta}$$

and $$\alpha = \tan^{-1}\left[\frac{F_2 \sin\theta}{F_1 + F_2 \cos\theta}\right]$$

The triangle law states that if two forces $\vec{F}_1$ and $\vec{F}_2$ acting simultaneously on a body can be represented by the two sides of a triangle (in magnitude and direction) taken in order then the third side (closing side) represents the resultant in the opposite order. Mathematically, it is stated as

$$\frac{F_1}{\sin\alpha} = \frac{F_2}{\sin\beta} = \frac{R}{\sin(180° - \alpha - \beta)}$$

This is also called *law of sines.*

The polygon law states that if a number of concurrent forces acting simultaneously on a body are represented in magnitude and direction by the sides of a polygon taken in order then the closing side of the polygon represents the resultant in the opposite order. This is an extension of the triangle law.

Resolution of a Force into Components (Analytical Method)

A single force acting on a body may be replaced by two forces along perpendicular axes, which produce the same effect on the body. These forces are called *rectangular* or *orthogonal components* of the original force $\vec{F}$. Mathematically,

$$\vec{F} = \vec{F}_x + \vec{F}_y = F_x\vec{i} + F_y\vec{j} = [F\cos\theta]\vec{i} + [F\sin\theta]\vec{j}$$

where $\vec{F}_x$ and $\vec{F}_y$ are called *vector components* of $\vec{F}$, and F_x and F_y are called *scalar components* of $\vec{F}$; θ is the angle made by the resultant with the X-axis.

Resultant of Several Concurrent Forces

The resultant of a system of forces is obtained by vector addition of all the individual forces:

$$\vec{R} = \vec{F}_1 + \vec{F}_2 + \ldots + \vec{F}_n$$

or $$R_x\vec{i} + R_y\vec{j} = \Sigma(F_x)_i\vec{i} + \Sigma(F_y)_i\vec{j}$$

The x-component of the resultant of a system of forces is the algebraic sum of the x-components of individual forces and the y-component of the resultant of a system of forces is the algebraic sum of the y-components of individual forces.

Concurrent Forces in Space

By resolving the force in space along three mutually perpendicular axes, force vector can be written as

$$\vec{F} = \vec{F}_x + \vec{F}_y + \vec{F}_z$$
$$= F_x\vec{i} + F_y\vec{j} + F_z\vec{k}$$
$$= F[\cos\theta_x\vec{i} + \cos\theta_y\vec{j} + \cos\theta_z\vec{k}]$$

where θ_x, θ_y and θ_z are the angles made by $\vec{F}$ with *X*, *Y* and *Z* axes respectively. They are related by the expression,

$$\cos^2\theta_x + \cos^2\theta_y + \cos^2\theta_z = 1$$

Resultant of Several Concurrent Forces in Space

The resultant of several concurrent forces in space can be determined by vector addition as

$$\vec{R} = \vec{F}_1 + \vec{F}_2 + \ldots + \vec{F}_n$$

or

$$R_x\vec{i} + R_y\vec{j} + R_z\vec{k} = \Sigma(F_x)_i\vec{i} + \Sigma(F_y)_i\vec{j} + \Sigma(F_z)_i\vec{k}$$

EXERCISES

Objective-type Questions

1. Principle of transmissibility can be applied only when the body is treated as
(a) a particle (b) a rigid body (c) deformable (d) a continuum

2. In the study of strength of materials, the forces acting on the bodies are treated as
(a) free vectors (b) sliding vectors (c) fixed vectors (d) unit vectors

3. The weight of a body is a
(a) body force (b) surface force (c) line force (d) reactive force

4. The system of forces shown in the figure is

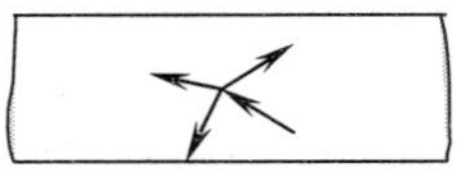

(a) coplanar non-concurrent forces (b) coplanar collinear forces
(c) coplanar concurrent forces (d) coplanar parallel forces

5. Collinear forces are those which
(a) are concurrent at a point (b) are parallel to each other
(c) lie on the same line (d) act on different planes

6. The magnitude of the resultant of two concurrent forces $\vec{F}_1$ and $\vec{F}_2$ with the included angle θ is
(a) $\sqrt{F_1^2 + F_2^2 + 2F_1F_2\cos\theta}$ (b) $\sqrt{F_1^2 + F_2^2 - 2F_1F_2\cos\theta}$
(c) $\sqrt{F_1^2 + F_2^2 + 2F_1F_2\sin\theta}$ (d) $\sqrt{F_1^2 + F_2^2 - 2F_1F_2\sin\theta}$

7. If $|\overrightarrow{AC}| = 50$ N, and $\overrightarrow{AB}$ and $\overrightarrow{BC}$ are equal in magnitude, then the magnitude of either of them is

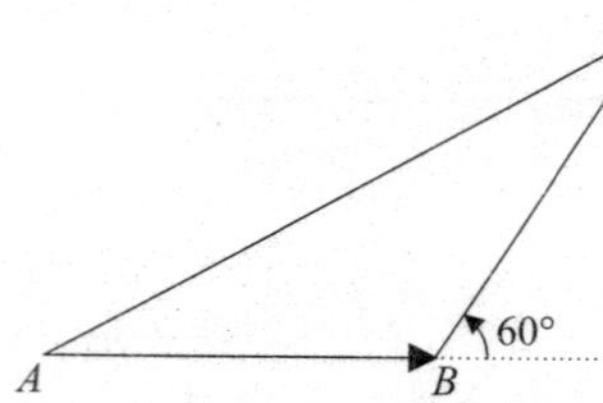

(a) 25.88 N (b) 50 N (c) 28.87 N (d) 25 N

8. The mathematical statement of the parallelogram law is called
   (a) sine law (b) cosine law (c) triangle law (d) polygon law
9. The mathematical statement of the triangle law is called
   (a) sine law (b) cosine law (c) parallelogram law (d) polygon law
10. State which of the following statement is true?
    (a) The parallelogram law can be proved mathematically.
    (b) The parallelogram law can be proved experimentally.
    (c) The parallelogram law is applicable for non-concurrent forces.
    (d) The parallelogram law cannot be applied to spatial concurrent forces.
11. An example for body force is
    (a) contact force (b) force of gravity (c) tensile force (d) support reaction
12. An example for surface force is
    (a) contact force (b) force of gravity (c) tensile force (d) support reaction
13. Force couple is a
    (a) fixed vector (b) sliding vector (c) free vector (d) unit vector
14. When a force vector is resolved into rectangular components, the components are ________ quantities.
    (a) scalar (b) vector (c) variable (d) zero
15. The rectangular components of the force are respectively

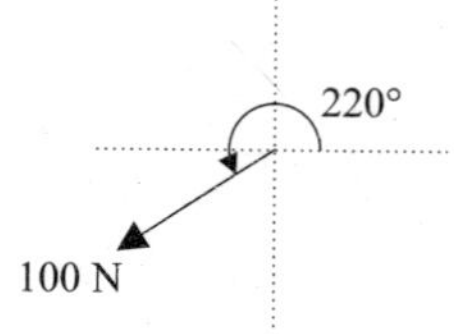

    (a) –100 sin 220°, –100 cos 220° (b) –100 sin 40°, –100 cos 40°
    (c) –100 cos 220°, –100 sin 220° (d) –100 cos 40°, –100 sin 40°
16. If the resultant of two concurrent forces is zero then it implies that the two forces are
    (a) equal in magnitude
    (b) equal in magnitude and direction
    (c) equal in magnitude and opposite in direction
    (d) equal in magnitude and perpendicular to each other
17. When resolving a force into its components
    (a) only one component is possible

(b) only two components are possible
(c) only three components are possible
(d) infinite number of components are possible

18. A 50 N force acts from point A (0, 0, 0) to point B (1, 1, 1); then the force is represented as

(a) $\overrightarrow{50}$ (b) $50(\vec{i} + \vec{j} + \vec{k})$ (c) $\frac{50}{\sqrt{3}}(\vec{i} + \vec{j} + \vec{k})$ (d) 50 N

Answers

1. (b)	**2.** (c)	**3.** (a)	**4.** (c)	**5.** (c)	**6.** (a)	**7.** (c)	**8.** (b)
9. (a)	**10.** (b)	**11.** (b)	**12.** (a)	**13.** (c)	**14.** (b)	**15.** (d)	**16.** (c)
17. (d)	**18.** (c)						

Short-answer Questions

1. Define force and force characteristics.
2. Why is force treated as a vector quantity?
3. Distinguish between particle and rigid body.
4. State the principle of transmissibility of a force.
5. Define system of forces and classify the system of forces with neat sketches.
6. Define resultant of a system of forces. What are the various methods to determine the resultant of concurrent forces?
7. State the parallelogram law of forces.
8. State the triangle law and polygon law to determine the resultant of concurrent forces.
9. State the laws of cosines and sines.
10. Distinguish between vector and scalar components of a force vector.
11. Discuss the analytical method to find the resultant of a force system.
12. Discuss on the sign convention to be followed while finding the components of a force.
13. Explain how to resolve a force in space to determine its components.

Numerical Problems

3.1. Determine the magnitude and direction of resultant of two forces acting on a block as shown in Fig. E.3.1 by trigonometry using (i) parallelogram law, and (ii) triangle law.

Ans. $R = 75.5$ N, $\alpha = 25.2°$

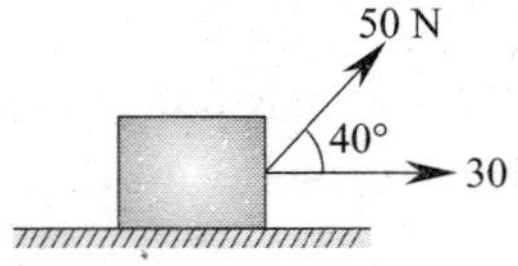

Fig. E.3.1

3.2. Two forces $\vec{F}_1$ and $\vec{F}_2$ act upon a body. If the magnitude of their resultant is equal to that of $\vec{F}_1$ and direction perpendicular to $\vec{F}_1$, then find the magnitude and direction of force $\vec{F}_2$. Take $F_1 = 20$ N.

Ans. $F_2 = 28.28$ N, $\theta = 135°$

3.3. Two unequal forces acting at a point at an angle of 150° have a resultant, which is perpendicular to the smaller force. The larger force is 24 N. Find the smaller force and the resultant.

Ans. 20.78 N, 12 N

3.4. The resultant of two forces acting at a point is 75.71 kN, where one force is double that of the other, and if the direction of one is reversed, the resultant becomes 57.17 kN. Find the magnitudes of the two forces and the angle between them.

Ans. 60 kN, 30 kN and 70°

3.5. Determine x and y components of the force in each case shown in Figs. E.3.5 (i)-(iv).

Ans. (i) $F \cos 35°$, $F \sin 35°$; (ii) $F \sin 30°$, $F \cos 30°$; (iii) $F \sin 40°$, $-F \cos 40°$; (iv) $-F \sin 20°$, $F \cos 20°$

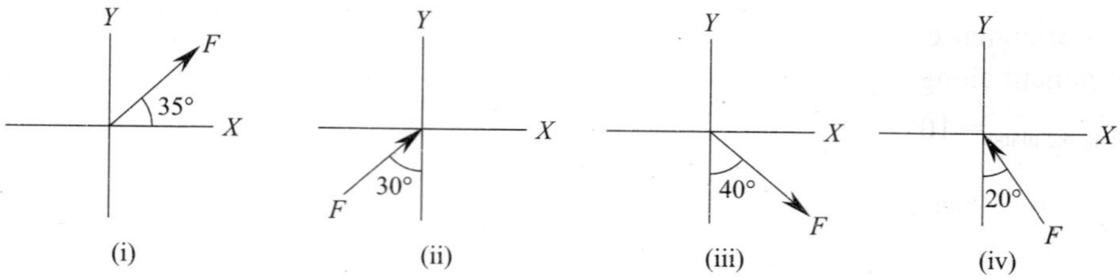

Fig. E.3.5

3.6. The 100 N force shown in Fig. E.3.6 is resolved along OA and OB. If the component along OA is 120 N, find (i) the component along OB, and (ii) its inclination with respect to OA.

Ans. 156.7 N, 90.3°

3.7. Find the components of a 25 N force shown in Fig. E.3.7 acting on a block, (i) along the horizontal and vertical directions, and (ii) along the inclined plane and normal to the plane.

Ans. (i) 17.7 N, 17.7 N; (ii) 24.15 N, 6.5 N

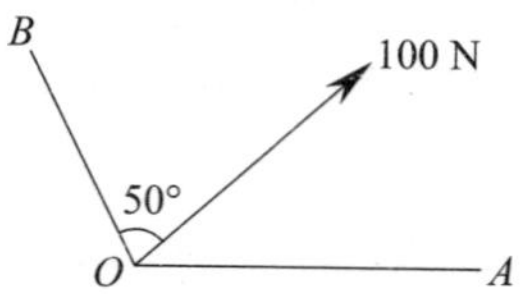

Fig. E.3.6

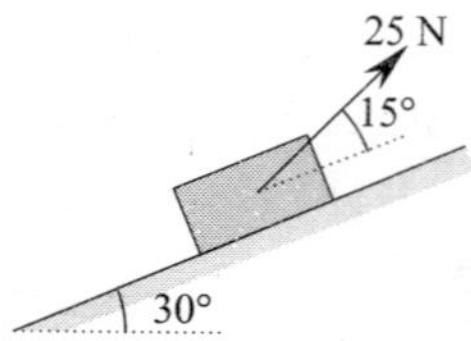

Fig. E.3.7

3.8. A body of weight 100 N rests on a rough inclined plane as shown in Fig. E.3.8. Find the components of its weight along the plane and normal to the plane.

Ans. 50 N, 86.6 N

3.9. A reaction force R acts on the wedge surface AB normal to that surface as shown in Fig. E.3.9. Determine its horizontal and vertical components.

Ans. horizontal component = $-R \sin 10°$, vertical component = $R \cos 10°$

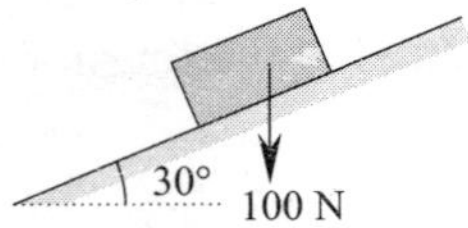

Fig. E.3.8

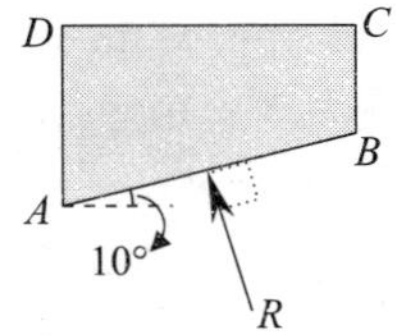

Fig. E.3.9

3.10. Determine the components of a push of 100 N acting on a block along and normal to the plane. Refer Fig. E.3.10.

Ans. 81.9 N, 57.4 N into the plane.

3.11. The horizontal component of a force acting as shown in Fig. E.3.11 is 150 N. Determine its component along the plane and normal to the plane.

Ans. $F_{\text{along the plane}} = 109.8$ N, $F_{\text{normal to the plane}} = 109.8$ N

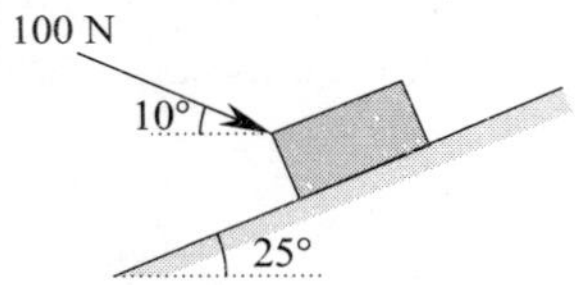

Fig. E.3.10

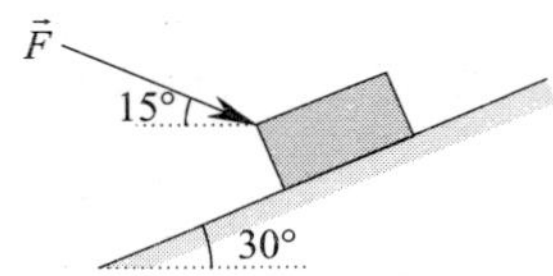

Fig. E.3.11

3.12. Find the resultant of an 800 N force acting towards the eastern direction and a 500 N force acting towards the north-eastern direction. Also, determine its direction.

Ans. $R = 1206.5$ N, 17.04° towards north of eastern direction

3.13. Two concurrent forces 20 N and 30 N are acting as shown in Fig. E.3.13. Determine the resultant force in magnitude and direction by analytical method.

Ans. 45.5 N, 44.7° to X-axis

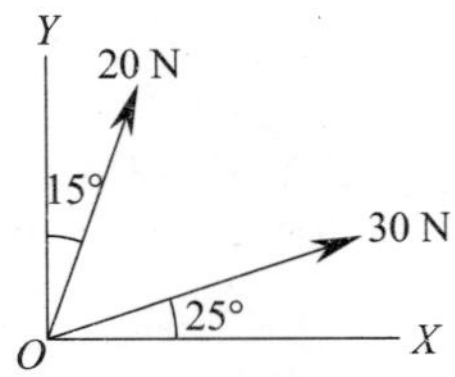

Fig. E.3.13

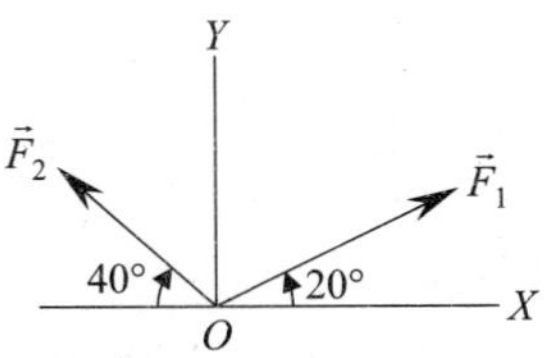

Fig. E.3.14

3.14. Two forces $\vec{F}_1$ and $\vec{F}_2$ are applied at the origin as shown in Fig. E.3.14. If their resultant is 500 N vertical, determine the two forces.

Ans. $F_1 = 442.3$ N, $F_2 = 542.5$ N

3.15. If the resultant of two forces is 1000 N horizontal then determine the individual forces. Refer Fig. E.3.15.

Ans. $F_1 = 662.3$ N, $F_2 = 488$ N

3.16. Three forces are concurrent at the origin as shown in Fig. E.3.16. Determine (a) the resultant of the three forces, and (b) the magnitude and direction of the fourth force, which must be added to make the resultant zero.

Ans. (a) –48.6 N$\vec{i}$ + 42.7 N$\vec{j}$; (b) 64.7 N, 41.3° to X-axis lying in *fourth* quadrant

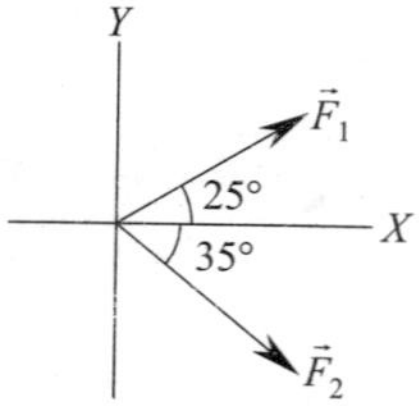

Fig. E.3.15

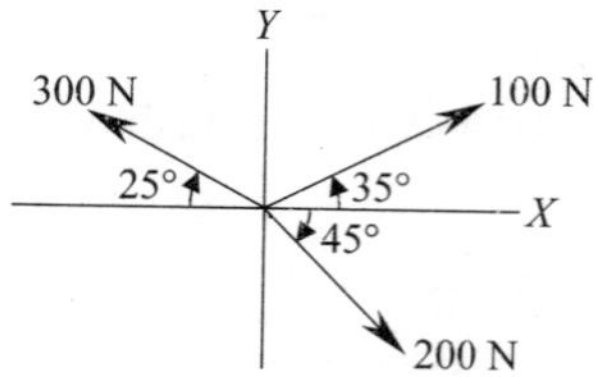

Fig. E.3.16

3.17. A 100 kg box is shifted by two persons, one pulling it exerting a force of 200 N inclined at 20° to the horizontal and another pushing it from behind by exerting a force of 150 N inclined at 10° to the horizontal. Determine the resultant force acting on the box. Refer Fig. E.3.17.

Ans. 338.3 N, 7.2°

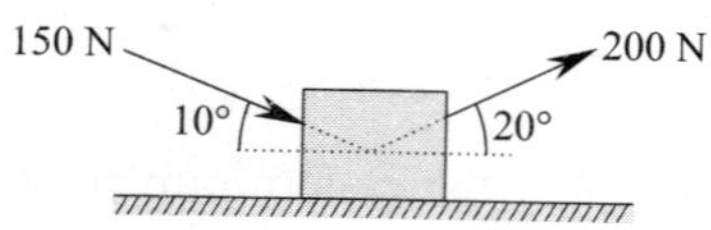

Fig. E.3.17

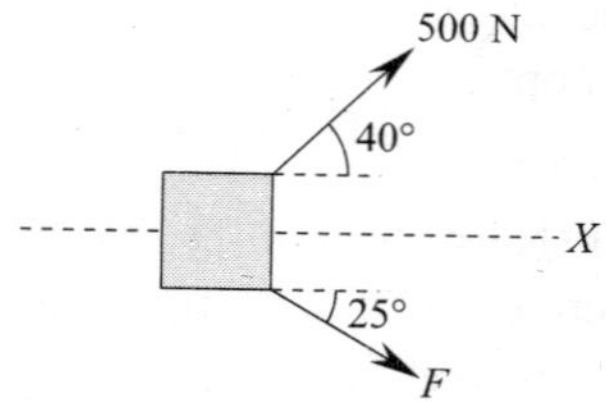

Fig. E.3.18

3.18. Two men (not shown) are pulling a block as shown in Fig. E.3.18, such that the block moves along the X-axis (i.e., the direction of resultant). If the first man exerts a force of 500 N, determine the force exerted by the other man and the resultant.

Ans. 760.5 N, 1072.3 N

3.19. Two men (not shown) are pulling a rope passing over a peg as shown in Fig. E.3.19. If they exert a force of 35 N and 50 N respectively, determine the force exerted by the rope on the peg.

Ans. 74 N, 35.8° to 35 N force

3.20. The tensions on both ends of a string passing over a smooth pulley are equal to 50 N. Determine the resultant force exerted by the string on the pulley bearing. Refer Fig. E.3.20.

Ans. 86.6 N at 30° to the vertical

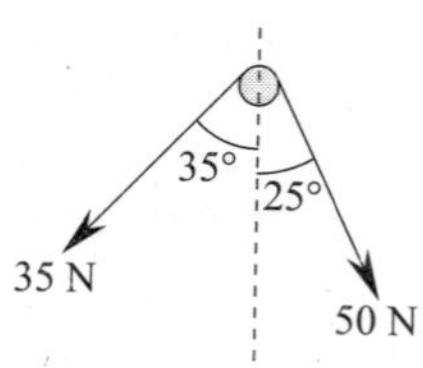

Fig. E.3.19

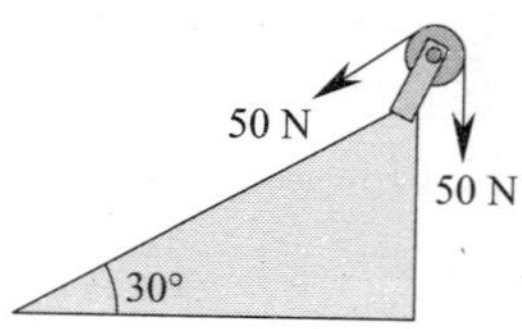

Fig. E.3.20

3.21. Determine the resultant of two concurrent forces of magnitudes 10 N and 20 N respectively as shown in Fig. E.3.21.

Ans. 29.8 N, 40° to *X*-axis

3.22. A log of wood is lifted using two ropes as shown in Fig. E.3.22. If the tensions in the ropes are 1.5 kN and 1.2 kN acting away from point *C* and their respective inclinations with respect to the vertical are 25.1° and 32° then determine the resultant force (in magnitude and direction) at point *C*.

Ans. 2.4 kN, acting vertically downward

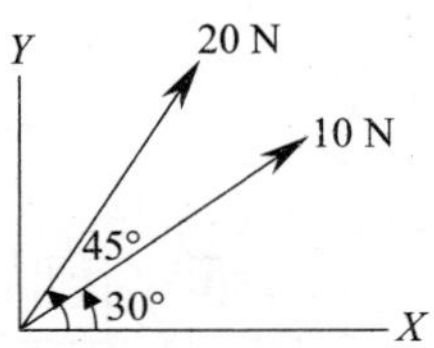

Fig. E.3.21

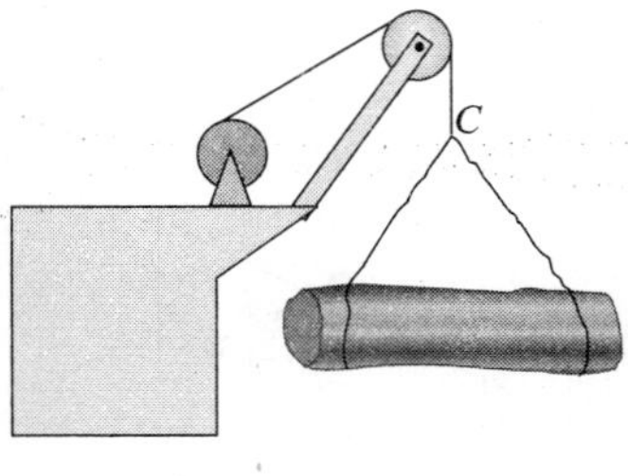

Fig. E.3.22

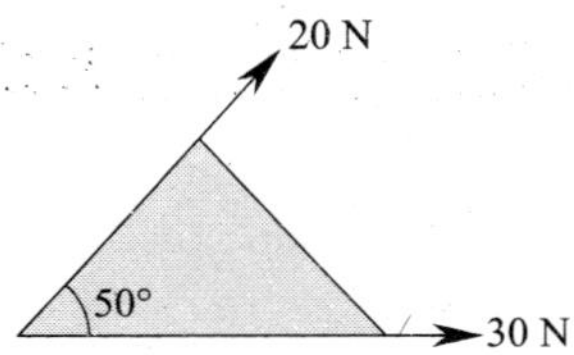

Fig. E.3.23

3.23. Two forces act along the sides of a triangular plate as shown in Fig. E.3.23. Determine the force that must be applied to keep the plate in equilibrium.

Ans. $-42.9\text{ N}\vec{i} - 15.3\text{ N}\vec{j}$

3.24. A force of 300 N passing through the origin forms an angle of 60° and 45° with respect to *X* and *Y* axes respectively. If the angle made with respect to the *Z*-axis is obtuse, find its components and express it as a vector.

Ans. $150\text{ N}\vec{i} + 212.13\text{ N}\vec{j} - 150\text{ N}\vec{k}$

3.25. A force $\vec{F}$ of magnitude 200 N acts at origin and directed as shown in Fig. E.3.25. Express the force in vector form.

Ans. $\vec{F} = 64.8\text{ N}\vec{i} + 138.9\text{ N}\vec{j} + 128.6\text{ N}\vec{k}$

3.26. The inclination of a force passing through the origin is θ_x = 48.6° and θ_z = 65.6°. Determine the angle θ_y, if $F_y = -250$ N. Also, express the force in vector form.

Ans. $\theta_y = 128.8°$; $263.9\text{ N}\vec{i} - 250\text{ N}\vec{j} + 164.8\text{ N}\vec{k}$

3.27. The inclination of a force passing through the origin is $\theta_y = 50°$ and $\theta_z = 70°$. Determine the angle θ_x, if $F_x = -125$ N. Also, express the force in vector form.

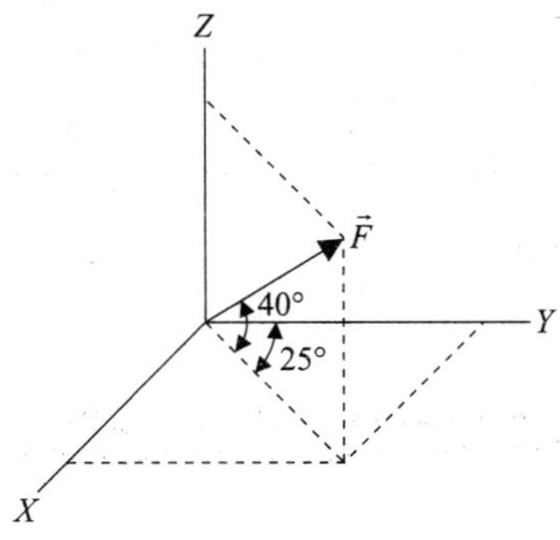

Fig. E.3.25

Ans. $\theta_x = 133.2°$; $-125\text{ N}\vec{i} + 117.4\text{ N}\vec{j} + 62.5\text{ N}\vec{k}$

3.28. A force of magnitude 300 N passes through the origin. The *x* and *y* components of the force are F_x = 100 N, F_y = −200 N. Determine the *z*-component of the force and the inclination of the force with respect to the axes.

Ans. $F_z = 200$ N, $\theta_x = 70.5°$, $\theta_y = 131.8°$, $\theta_z = 48.2°$

3.29. A force of magnitude 200 N has a component of 120 N on the x–y plane and an inclination of 60° with respect to the X-axis. Express the force in vector form.

Ans. $100\text{ N}\vec{i} + 66.33\text{ N}\vec{j} + 160\text{ N}\vec{k}$

3.30. A force of magnitude 200 N acts along a line passing through two points A and B in space as shown in Fig. E.3.30. Express the force in vector form.

Ans. $74.3\text{ N}\vec{i} + 111.4\text{ N}\vec{j} + 148.6\text{ N}\vec{k}$

3.31. A force of magnitude 90 N acts along a line passing through two points A and B in space as shown in Fig. E.3.31. Express the force in vector form.

Ans. $-30\sqrt{3}\text{ N}(\vec{i} + \vec{j} + \vec{k})$

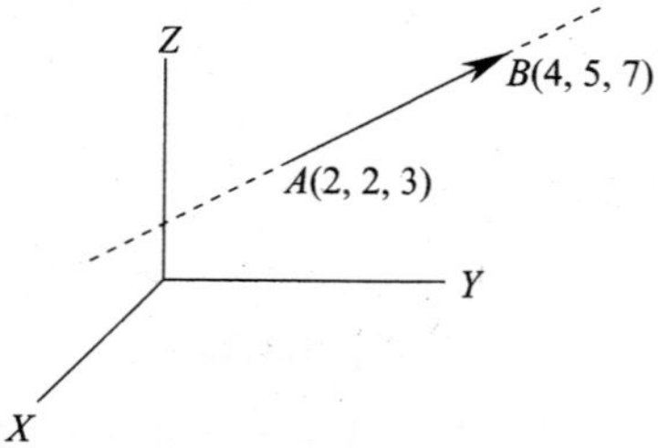

Fig. E.3.30

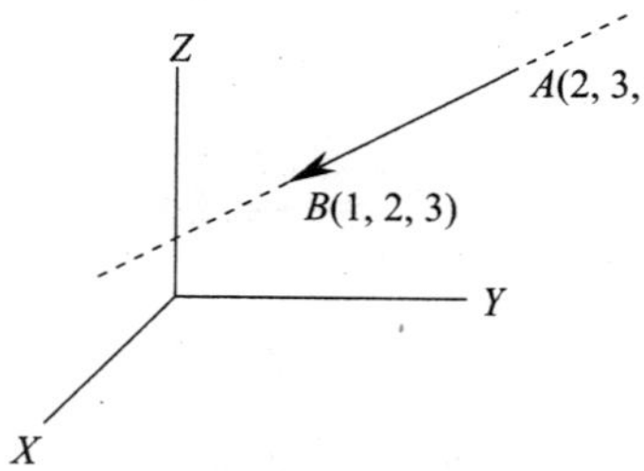

Fig. E.3.31

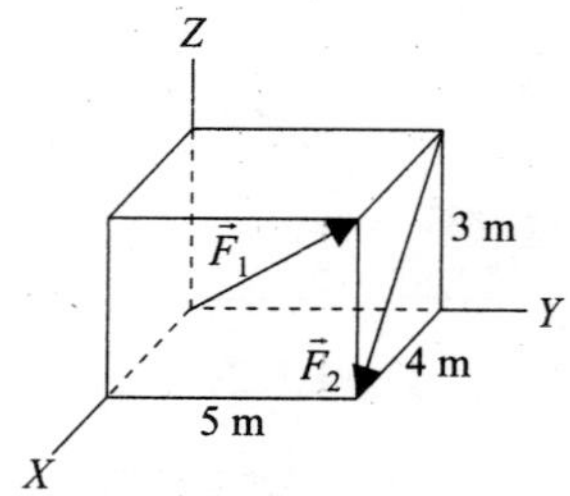

Fig. E.3.32

3.32. In Fig. E.3.32, express the two forces $\vec{F}_1$ and $\vec{F}_2$ of magnitudes 100 N and 50 N respectively in vector form.

Ans. $\vec{F}_1 = 56.6\text{ N}\vec{i} + 70.7\text{ N}\vec{j} + 42.4\text{ N}\vec{k}$; $\vec{F}_2 = 40\text{ N}\vec{i} - 30\text{ N}\vec{k}$

3.33. Express the forces acting along the diagonals as shown in Fig. E.3.33, in vector form.

Ans. 100 N force : $38.4\text{ N}\vec{i} - 51.2\text{ N}\vec{j} - 76.8\text{ N}\vec{k}$; 200 N force: $-76.8\text{ N}\vec{i} + 102.4\text{ N}\vec{j} - 153.6\text{ N}\vec{k}$

3.34. Express the force of magnitude 500 N acting along the diagonal as shown in Fig. E.3.34 in vector form.

Ans. $371.4\text{ N}\vec{i} + 278.5\text{ N}\vec{j} - 185.7\text{ N}\vec{k}$

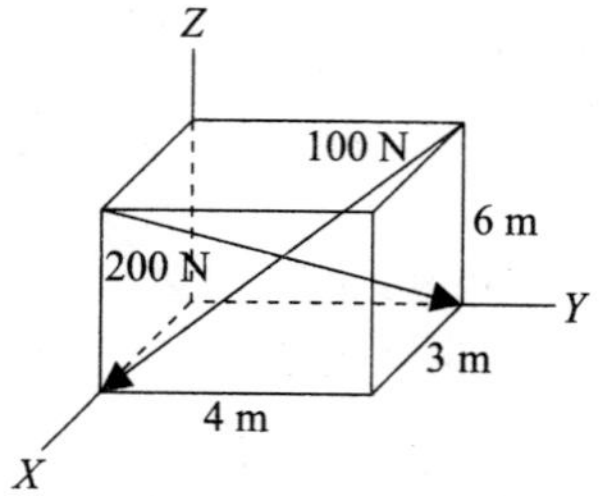

Fig. E.3.33

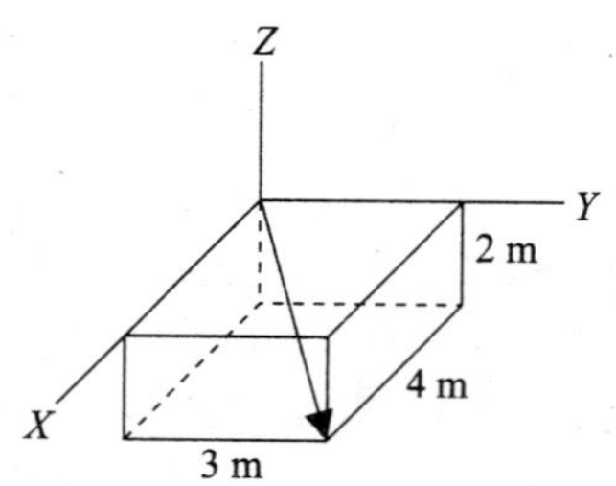

Fig. E.3.34

3.35. Determine the resultant of a system of three concurrent forces passing through the origin and the points (8, 4, 0), (–5, –3, 6) and (4, –5, 3) respectively. The respective magnitudes of the forces are 2000 N, 1600 N and 1500 N.

Ans. $\vec{R} = 1681.2\text{ N}\vec{i} - 739.9\text{ N}\vec{j} + 1783.8\text{ N}\vec{k}$

3.36. Find the resultant of the tension forces concurrent at A. The tensions along cables AB, AC and AD are 1200 N, 1500 N and 1000 N respectively. Refer Fig. E.3.36.

Ans. $-344.0\vec{i} - 463.4\vec{j} - 3196.6\vec{k}$

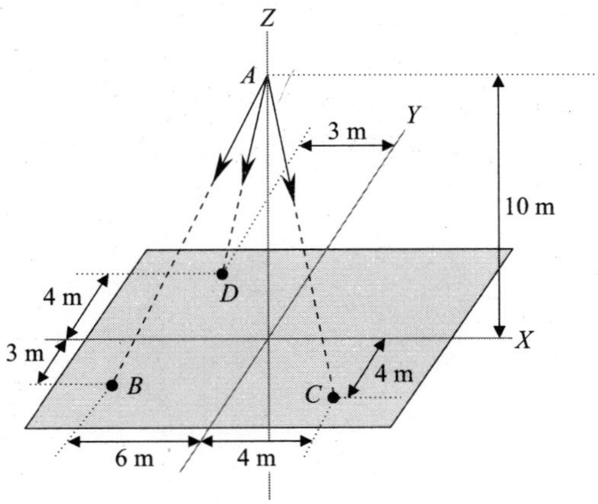

Fig. E.3.36

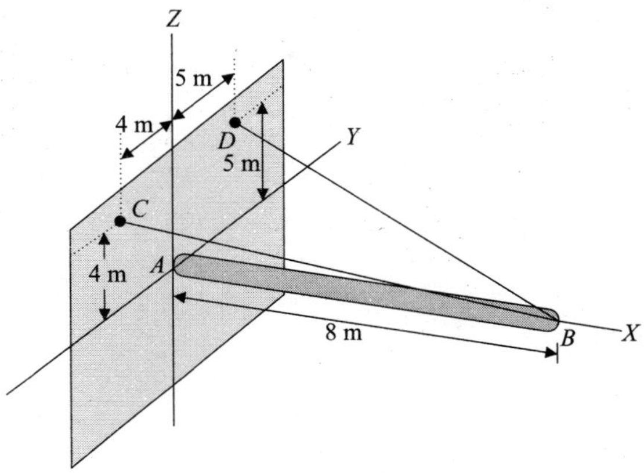

Fig. E.3.37

3.37. Find the resultant of the tension forces along the cables BC and BD acting at the end B of a boom held horizontal by the cables. The tensions in the cables are respectively 500 N and 800 N. Refer Fig. E.3.37.

Ans. $-1007.7\vec{i} + 170.5\vec{j} + 578.8\vec{k}$

3.38. A mast of height h is supported by three cables AB, AC and AD placed equidistant around the circumference of a circle of radius r on the ground level. Determine the resultant of the tensions at A if the tension in each of the cable is T. Refer Fig. E.3.38.

Ans. $3T\dfrac{h}{\sqrt{r^2 + h^2}}$ acting vertically downwards along the axis of the mast

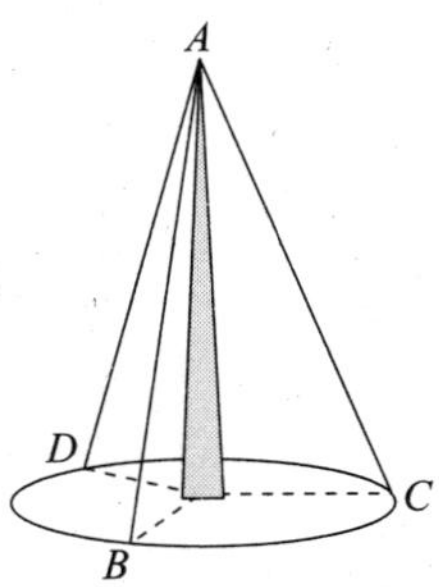

Fig. E.3.38

4 System of Forces and Resultant-II (Non-Concurrent Forces)

4.1 INTRODUCTION

In the previous chapter, we have seen how to determine the resultant of concurrent forces in a plane and in space. Since the forces were *concurrent*, we could treat the body as a *particle* as there is only translational motion. However, when a body is subjected to a *non-concurrent* force system, the body will have **rotational motion** in addition to the translational motion. Hence, in such cases, the body can no more be idealized as a particle but treated as a **rigid body** itself. Before we proceed to determine the resultant of non-concurrent forces in a plane and in space, we must understand two new concepts, namely, the **moment of a force** and the **moment of a couple**. These are discussed in the following few sections. Here, unlike the previous chapter, we will discuss forces in space first and then treat the coplanar system as a special case of the spatial system.

4.2 MOMENT OF A FORCE

Consider a door hinged about one edge for instance. While closing or opening the door, we apply a force at the other edge. Since this force is applied away from the hinge and not at the hinge, its effect is to **rotate** the door about the hinge. A similar kind of effect is observed while using a pipe wrench in fixing pipes. This type of rotational effect of a force is measured by a physical quantity known as the **moment of a force**. The moment of a force about a point is defined as *a measure of the tendency of the force to rotate a body about that point*.

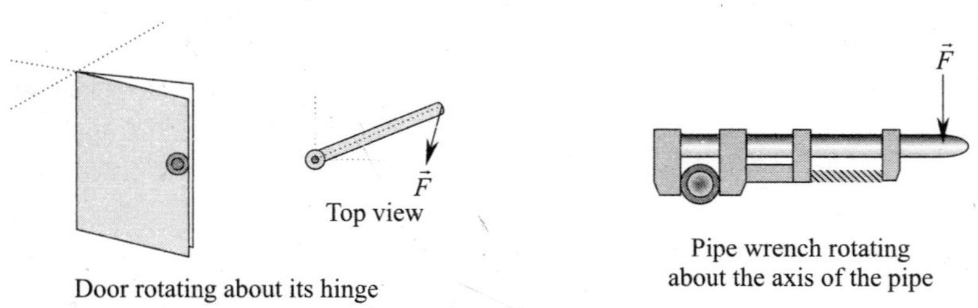

Fig. 4.1

4.2.1 Moment of a Force About the Origin

Consider a force $\vec{F}$ as shown in Fig. 4.2 acting at point A, whose position vector is $\vec{r}$. The effect of such a force will be to rotate the body about the origin, assuming the body is free to rotate about the origin. This rotational effect of $\vec{F}$ about the origin is determined by a physical quantity called *moment of the force*. Mathematically, the moment of the force about point O is defined as the cross product of the *position vector* $\vec{r}$ of the *point of application* of the force and the *force vector* $\vec{F}$:

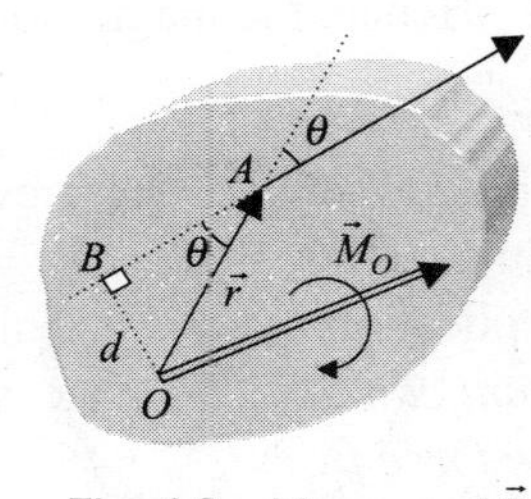

Fig. 4.2 Moment of $\vec{F}$ about the origin O

$$\vec{M}_O = \vec{r} \times \vec{F} \tag{4.1}$$

Since the cross product of two vectors is a *vector*, the moment of a force is also a *vector*. It has both magnitude and direction, and adds according to the parallelogram law of addition. Its magnitude is given as

$$M_O = rF \sin\theta = (r \sin\theta)F \tag{4.2}$$

where θ is the angle made by the positive sense of $\vec{r}$ with the positive sense of $\vec{F}$. From Fig. 4.2, we can see that $r \sin\theta = d$. Hence, the magnitude of the moment can also be expressed as

$$M_O = Fd \tag{4.3}$$

where d is the perpendicular distance from the origin O to the line of action of the force. As the magnitude of the moment of a force is the product of force and perpendicular distance, its SI unit is **N.m.** Its direction is perpendicular to the plane formed by $\vec{r}$ and $\vec{F}$ and whether it acts into that plane or out of that plane is determined by the right-hand screw rule as we curl our fingers from $\vec{r}$ to $\vec{F}$.

The point O about which the force tends to rotate the body is termed **moment centre** and the perpendicular distance d from the moment centre to the line of action of the force is termed **moment arm**. The line perpendicular to the plane containing the force and its point of application, and passing through the moment centre O is called **axis of the moment**.

Note: The moment vector is normally denoted by a **double-line arrow** or by a **curved arrow** about the moment axis as shown in Fig. 4.2. As vector product is not commutative, the order of vectors should be noted while taking the cross product, i.e., $\vec{M}_O = \vec{r} \times \vec{F}$ and not otherwise.

Corollary In the above discussion, we considered the position vector $\vec{r}$ of the point of application of the force, namely, A. Now consider some other point, C for instance (refer Fig. 4.3), on the line of action of the force, having position vector $\vec{r}_1$.

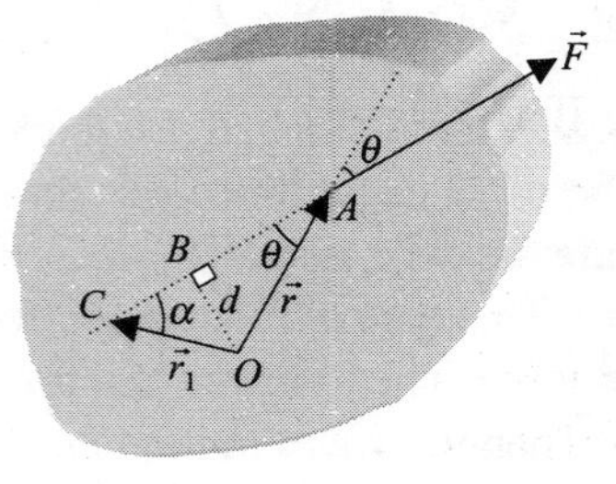

Fig. 4.3 Moment of $\vec{F}$ about O; point of application of the force anywhere along its line of action

Since force is a sliding vector in our case, its moment about O is also given as

$$\vec{M}_O = \vec{r}_1 \times \vec{F} \tag{4.4}$$

From the Fig. 4.3, we see that

$$r_1 \sin\alpha = d$$

Hence, the magnitude of the moment can be expressed as

$$M_O = (r_1 \sin\alpha)F = Fd \tag{4.5}$$

which is same as the previous expression (4.3), considering the position vector of point A. Hence, we can conclude that the magnitude of the moment of a force remains the *same* irrespective of the point of application of the force along its line of action. Hence, the *magnitude of the moment of a force about a point can be defined as the product of the force and the perpendicular distance of the line of action of the force from that point.*

If the perpendicular distance of the line of action of the force is known then we can directly find the magnitude of the moment as explained above. However, for forces in space, determination of perpendicular distance may not be possible. Hence, we take the cross product of the position vector and force vector from which the magnitude of the moment may be determined. If the position vector of the point of application of the force and the force vector are given in rectangular components as

$$\vec{r} = x\vec{i} + y\vec{j} + z\vec{k} \tag{4.6}$$

and

$$\vec{F} = F_x\vec{i} + F_y\vec{j} + F_z\vec{k} \tag{4.7}$$

then the moment vector can be determined as

$$\begin{aligned}\vec{M}_O &= \vec{r} \times \vec{F} = (x\vec{i} + y\vec{j} + z\vec{k}) \times (F_x\vec{i} + F_y\vec{j} + F_z\vec{k})\\ &= xF_y\vec{k} - xF_z\vec{j} - yF_x\vec{k} + yF_z\vec{i} + zF_x\vec{j} - zF_y\vec{i}\end{aligned} \tag{4.8}$$

On rearranging,

$$\vec{M}_O = (yF_z - zF_y)\vec{i} - (xF_z - zF_x)\vec{j} + (xF_y - yF_x)\vec{k} \tag{4.9}$$

which can also be written in determinant form as

$$\vec{M}_O = \begin{vmatrix} \vec{i} & \vec{j} & \vec{k} \\ x & y & z \\ F_x & F_y & F_z \end{vmatrix} \tag{4.10}$$

If M_x, M_y and M_z are components of $\vec{M}_O$ then

$$M_x = yF_z - zF_y \tag{4.11}$$

$$M_y = zF_x - xF_z \tag{4.12}$$

and

$$M_z = xF_y - yF_x \tag{4.13}$$

Then the magnitude of the moment is given as

$$M_O = \sqrt{M_x^2 + M_y^2 + M_z^2} \tag{4.14}$$

4.2.2 Varignon's Theorem (or) Principle of Moments

In this section, we will introduce a theorem known as Varignon's theorem, which is useful in calculating the moment of a force when its moment arm cannot be easily determined. The theorem states that *the moment about a given point O of the resultant of several concurrent forces is equal to the sum of the moments of individual forces about the same point O.*

Proof Consider n number of forces $\vec{F}_1, \vec{F}_2, ..., \vec{F}_n$ concurrent at a point A; then the resultant force is given as the vector sum of individual forces:

$$\vec{R} = \vec{F}_1 + \vec{F}_2 + ... + \vec{F}_n \tag{4.15}$$

Then the moment of the resultant about the origin O is

$$\vec{M}_O = \vec{r} \times \vec{R}$$

$$= \vec{r} \times (\vec{F}_1 + \vec{F}_2 + ... + \vec{F}_n) \tag{4.16}$$

By the distributive property of vector multiplication, we can write it as

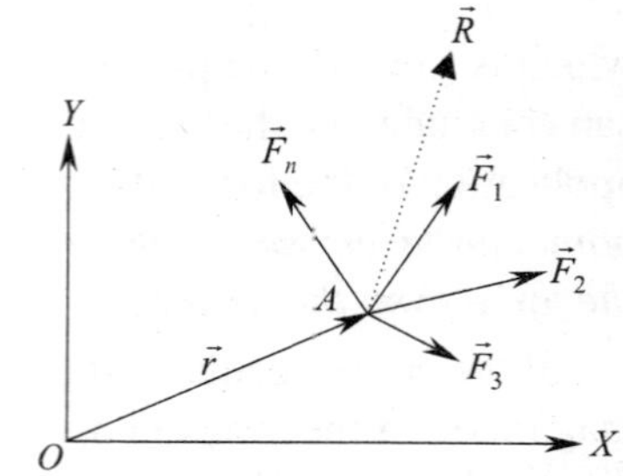

Fig. 4.4 Moment of a system of concurrent forces

$$\vec{M}_O = (\vec{r} \times \vec{F}_1) + (\vec{r} \times \vec{F}_2) + ... + (\vec{r} \times \vec{F}_n)$$

$$= \sum(\vec{r} \times \vec{F}_i) \tag{4.17}$$

= sum of the moment of individual forces about point O

Thus, we see that the moment of the resultant of several concurrent forces about a point is equal to the sum of the moments of individual forces about the same point.

Corollary This theorem is useful when the moment arm of a force cannot be easily determined, which arises usually in *coplanar* problems. In such cases, the force is resolved into orthogonal components and the moment can be determined by applying Varignon's theorem. Refer to the solved examples 4.5 to 4.12.

4.2.3 Moment of a Force About a Point Other Than The Origin

Suppose the moment of a force about a point other than the origin is required then we need to consider the vector from the point about which the moment is required to the given point of application of the force. For instance, if the moment of the force acting at A is required about a point B (refer Fig. 4.5) then the position of point A with respect to point B is

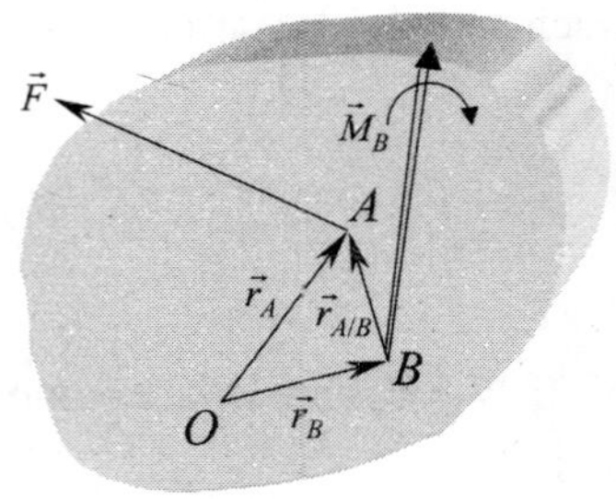

Fig. 4.5 Moment of a force about a point other than the origin

$$\vec{r}_{A/B} = \overrightarrow{BA} = \overrightarrow{OA} - \overrightarrow{OB} = \vec{r}_A - \vec{r}_B \tag{4.18}$$

Therefore, the moment of the force about point B is

$$\vec{M}_B = \vec{r}_{A/B} \times \vec{F}$$

$$= (\vec{r}_A - \vec{r}_B) \times \vec{F} \tag{4.19}$$

This moment is perpendicular to the plane formed by $\overrightarrow{BA}$ and $\vec{F}$.

4.2.4 Moment of a Force About an Axis

Further, if the moment of a force about a point (say B) is known then the component of this moment about an axis passing through that point, say BC, is given as a dot product of the unit vector along BC and $\vec{M}_B$, i.e.,

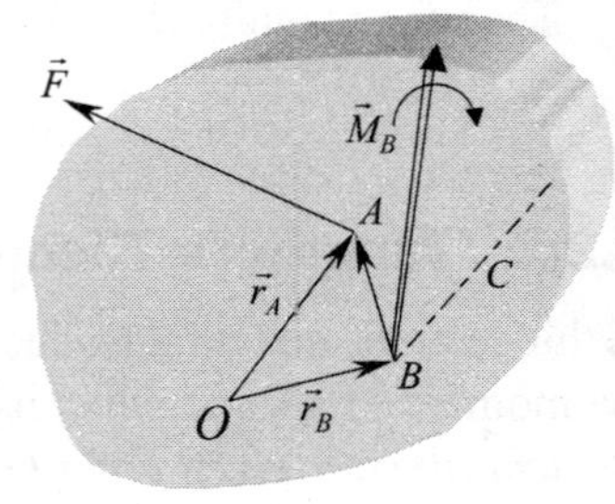

Fig. 4.6 Moment of a force about an axis

$$|\vec{M}_{BC}| = \hat{n}_{BC} \cdot \vec{M}_B \tag{4.20}$$

It should be noted that the moment of a force about an axis is a *scalar* quantity.

Example 4.1 Find the moment of a 100 N force acting between points A and B as shown in Fig. 4.7, (i) about the origin O, (ii) about point C, and (iii) about axis CD.

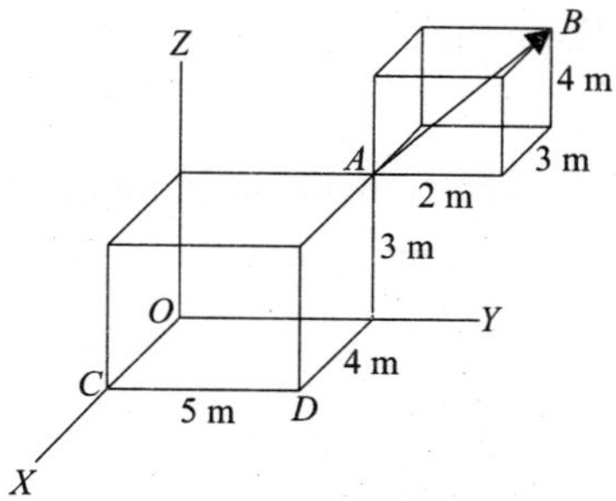

Fig. 4.7

Solution From the given figure, we see that

$$\overrightarrow{AB} = -3\vec{i} + 2\vec{j} + 4\vec{k}$$

Therefore, unit vector along AB is

$$\hat{n}_{AB} = \frac{-3\vec{i} + 2\vec{j} + 4\vec{k}}{\sqrt{(-3)^2 + (2)^2 + (4)^2}} = \frac{-3\vec{i} + 2\vec{j} + 4\vec{k}}{\sqrt{29}}$$

Since the magnitude of the force is 100 N, the force can be represented in vector form as

$$\vec{F} = F\hat{n}_{AB} = 100\left[\frac{-3\vec{i} + 2\vec{j} + 4\vec{k}}{\sqrt{29}}\right] = \frac{-300\vec{i} + 200\vec{j} + 400\vec{k}}{\sqrt{29}}$$

(i) *Determination of moment of the force about the origin O*

The position vector of point of application A of the force is

$$\vec{r}_A = \overrightarrow{OA} = 5\vec{j} + 3\vec{k}$$

Hence, the moment of the force about the origin is given as

$$\vec{M}_O = \vec{r}_A \times \vec{F}$$

$$= (5\vec{j} + 3\vec{k}) \times \left[\frac{-300\vec{i} + 200\vec{j} + 400\vec{k}}{\sqrt{29}}\right]$$

$$= \frac{1400\vec{i} - 900\vec{j} + 1500\vec{k}}{\sqrt{29}}$$

(ii) *Determination of moment of the force about point C*

Here we must consider the vector from point C to the point of application A of the force, i.e.,

$$\vec{r}_{A/C} = \overrightarrow{CA} = \overrightarrow{OA} - \overrightarrow{OC}$$

$$= (5\vec{j} + 3\vec{k}) - (4\vec{i})$$

$$= -4\vec{i} + 5\vec{j} + 3\vec{k}$$

Therefore, the moment of the force about point C is given as

$$\vec{M}_C = \vec{r}_{A/C} \times \vec{F}$$

$$= (-4\vec{i} + 5\vec{j} + 3\vec{k}) \times \left[\frac{-300\vec{i} + 200\vec{j} + 400\vec{k}}{\sqrt{29}}\right]$$

$$= \frac{1400\vec{i} + 700\vec{j} + 700\vec{k}}{\sqrt{29}}$$

(iii) *Determination of moment of the force about axis CD*

From the given figure, we see that,

$$\overrightarrow{CD} = 5\vec{j}$$

The component of moment about axis CD is obtained by taking the dot product between the unit vector along CD and $\vec{M}_C$. Therefore,

$$M_{CD} = \hat{n}_{CD} \cdot \vec{M}_C$$

$$= [\vec{j}] \cdot \frac{1400\vec{i} + 700\vec{j} + 700\vec{k}}{\sqrt{29}} = \frac{700}{\sqrt{29}} \text{ N.m}$$

Example 4.2 A force $\vec{F} = (50\text{ N})\vec{i} + (75\text{ N})\vec{j} + (100\text{ N})\vec{k}$ acts as shown in Fig. 4.8. Determine the moment of the force about X, Y and Z axes.

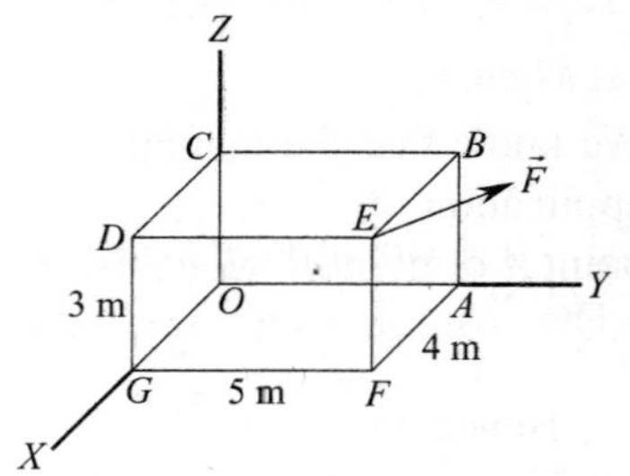

Fig. 4.8

Solution The position vector of point of application of the force is

$$\overrightarrow{OE} = 4\vec{i} + 5\vec{j} + 3\vec{k}$$

Therefore, the moment of the force about the origin is given as

$$\vec{M}_O = \overrightarrow{OE} \times \vec{F}$$

$$= (4\vec{i} + 5\vec{j} + 3\vec{k}) \times (50\vec{i} + 75\vec{j} + 100\vec{k})$$

$$= 25\begin{vmatrix} \vec{i} & \vec{j} & \vec{k} \\ 4 & 5 & 3 \\ 2 & 3 & 4 \end{vmatrix}$$

$$= 25[\vec{i}(20 - 9) - \vec{j}(16 - 6) + \vec{k}(12 - 10)]$$

$$= 275\vec{i} - 250\vec{j} + 50\vec{k}$$

Hence, the moment of the force about X, Y and Z axes are given by taking dot product of the unit vectors along the respective directions and the moment vector, i.e.,

$$M_x = \vec{i} \cdot \vec{M}_O$$

$$= 275 \text{ N.m}$$

$$M_y = \vec{j} \cdot \vec{M}_O$$

$$= -250 \text{ N.m}$$

$$M_z = \vec{k} \cdot \vec{M}_O$$

$$= 50 \text{ N.m}$$

Example 4.3 The line of action of a 50 N force passes through points A (2, 7, 5) and B (6, 2, 3), the coordinates being given in metres. What is the moment of this force (a) about the origin, and (b) about a point C (3, 5, 4)?

Solution We know that $\overrightarrow{AB} = \overrightarrow{OB} - \overrightarrow{OA}$

$$= (6\vec{i} + 2\vec{j} + 3\vec{k}) - (2\vec{i} + 7\vec{j} + 5\vec{k})$$

$$= 4\vec{i} - 5\vec{j} - 2\vec{k}$$

and $$|\overrightarrow{AB}| = \sqrt{4^2 + (-5)^2 + (-2)^2} = \sqrt{45} \text{ m}$$

Therefore, unit vector along *AB* is

$$\hat{n}_{AB} = \frac{4\vec{i} - 5\vec{j} - 2\vec{k}}{\sqrt{45}}$$

Hence, the force vector can be written as

$$\vec{F} = F\hat{n}_{AB}$$

$$= 50\left[\frac{4\vec{i} - 5\vec{j} - 2\vec{k}}{\sqrt{45}}\right] \text{N}$$

(a) *Moment of the force about the origin*

We know that the magnitude of the moment of a force remains the same irrespective of the point of application of the force along its line of action. Hence, we can consider the position vector of either point *A* or point *B*. Considering point *A* then,

$$\vec{r} = \overrightarrow{OA} = 2\vec{i} + 7\vec{j} + 5\vec{k}$$

Hence, the moment of the force about the origin is given as

$$\vec{M}_O = \vec{r} \times \vec{F} = \frac{50}{\sqrt{45}}\begin{vmatrix} \vec{i} & \vec{j} & \vec{k} \\ 2 & 7 & 5 \\ 4 & -5 & -2 \end{vmatrix}$$

$$= \left[\frac{550\vec{i} + 1200\vec{j} - 1900\vec{k}}{\sqrt{45}}\right] \text{N.m}$$

(b) *Moment of the force about point C (3, 5, 4)*

Here we consider the position of any point on the line of action of the force with respect to point *C*. Therefore,

$$\vec{r}_{A/C} = \overrightarrow{CA} = \overrightarrow{OA} - \overrightarrow{OC}$$

$$= (2\vec{i} + 7\vec{j} + 5\vec{k}) - (3\vec{i} + 5\vec{j} + 4\vec{k})$$

$$= -\vec{i} + 2\vec{j} + \vec{k}$$

Hence, the moment of the force about the point *C* is given as

$$\vec{M}_C = \vec{r}_{A/C} \times \vec{F} = \frac{50}{\sqrt{45}}\begin{vmatrix} \vec{i} & \vec{j} & \vec{k} \\ -1 & 2 & 1 \\ 4 & -5 & -2 \end{vmatrix} = \left[\frac{50\vec{i} + 100\vec{j} - 150\vec{k}}{\sqrt{45}}\right] \text{N.m}$$

Example 4.4 A force $\vec{F}$ of magnitude 200 N is directed from A (1, 2, 3) to B (5, 6, 2), the coordinates being given in metres. Determine (a) the moment of $\vec{F}$ about the origin and the magnitude of the moment; (b) the moment of $\vec{F}$ about point C (4, 5, 6) and its moment arm from C; and (c) the moment of $\vec{F}$ about an axis passing through the origin and point C.

Solution We know that

$$\overrightarrow{AB} = \overrightarrow{OB} - \overrightarrow{OA}$$

$$= (5\vec{i} + 6\vec{j} + 2\vec{k}) - (\vec{i} + 2\vec{j} + 3\vec{k})$$

$$= 4\vec{i} + 4\vec{j} - \vec{k}$$

and

$$|\overrightarrow{AB}| = \sqrt{(4)^2 + (4)^2 + (-1)^2} = \sqrt{33} \text{ m}$$

Therefore, unit vector along AB is

$$\hat{n}_{AB} = \frac{4\vec{i} + 4\vec{j} - \vec{k}}{\sqrt{33}}$$

Hence, the force vector can be written as

$$\vec{F} = 200\left[\frac{4\vec{i} + 4\vec{j} - \vec{k}}{\sqrt{33}}\right]$$

(a) *Moment of the force about the origin*

Here

$$\vec{r} = \overrightarrow{OA} = \vec{i} + 2\vec{j} + 3\vec{k}$$

$\therefore$

$$\vec{M}_O = \vec{r} \times \vec{F} = \frac{200}{\sqrt{33}}\begin{vmatrix} \vec{i} & \vec{j} & \vec{k} \\ 1 & 2 & 3 \\ 4 & 4 & -1 \end{vmatrix}$$

$$= \frac{200}{\sqrt{33}}[-14\vec{i} + 13\vec{j} - 4\vec{k}]$$

Hence, the magnitude of the moment is

$$|\vec{M}_O| = \frac{200}{\sqrt{33}}(19.52) = 679.6 \text{ N.m}$$

(b) *Moment of the force about the point C (4, 5, 6)*

Here

$$\vec{r}_{A/C} = \overrightarrow{CA} = \overrightarrow{OA} - \overrightarrow{OC}$$

$$= (\vec{i} + 2\vec{j} + 3\vec{k}) - (4\vec{i} + 5\vec{j} + 6\vec{k})$$

$$= -3\vec{i} - 3\vec{j} - 3\vec{k}$$

$\therefore$

$$\vec{M}_C = \vec{r}_{A/C} \times \vec{F} = \frac{200}{\sqrt{33}}\begin{vmatrix} \vec{i} & \vec{j} & \vec{k} \\ -3 & -3 & -3 \\ 4 & 4 & -1 \end{vmatrix} = \frac{200}{\sqrt{33}}(15\vec{i} - 15\vec{j})$$

Hence, the magnitude of the moment is

$$|\vec{M}_C| = \frac{200}{\sqrt{33}}(21.21) = 738.44 \text{ N.m}$$

We know that $|\vec{M}_C| = |\vec{F}|d$

Therefore, moment arm, $d = \frac{|\vec{M}_C|}{|\vec{F}|} = \frac{738.44}{200} = 3.692$ m

(c) *Moment of the force about an axis passing through the origin and the point C*

To determine this moment, we must determine the unit vector along CO

$$\overrightarrow{CO} = -\overrightarrow{OC} = -4\vec{i} - 5\vec{j} - 6\vec{k}$$

$$|\overrightarrow{CO}| = \sqrt{(-4)^2 + (-5)^2 + (-6)^2} = 8.775 \text{ m}$$

$$\therefore \qquad \hat{n}_{CO} = \frac{-4\vec{i} - 5\vec{j} - 6\vec{k}}{8.775}$$

Hence, the moment of the force about the axis CO is given by the component of the moment about C along the axis CO:

$$|\vec{M}_{CO}| = \hat{n}_{CO} \cdot \vec{M}_C = \frac{-4\vec{i} - 5\vec{j} - 6\vec{k}}{8.775} \cdot \frac{200}{\sqrt{33}}(15\vec{i} - 15\vec{j})$$

$$= \frac{200}{8.775\sqrt{33}}[-60 + 75] = 59.51 \text{ N.m}$$

4.3 COPLANAR NON-CONCURRENT FORCES

For forces in space, the moment calculations involve the cross product of two vectors, as explained in the previous section. However, these calculations can be much simplified for a planar problem. Consider a force $\vec{F}$ acting on a body in X–Y plane as shown in Fig. 4.9. If $\vec{r}$ is the position vector of point of application of the force then the moment of the force about the origin O is:

$$\vec{M}_O = \vec{r} \times \vec{F} \qquad (4.21)$$

Its magnitude is given as

$$M_O = (r \sin\theta)F$$
$$= Fd \qquad (4.22)$$

Since $\vec{r}$ and $\vec{F}$ lie on the plane, the moment vector $\vec{M}$ is always perpendicular to the plane of the paper. Whether it points into the plane or out of the plane is determined by the right-hand screw rule as explained below.

Let us resolve the force into rectangular components F_x and F_y as shown in Fig. 4.10. Then we can write the position vector of point of application of the force and the force vector in rectangular components as: $\vec{r} = x\vec{i} + y\vec{j}$ and $\vec{F} = F_x\vec{i} + F_y\vec{j}$. Therefore, the moment vector can be written as

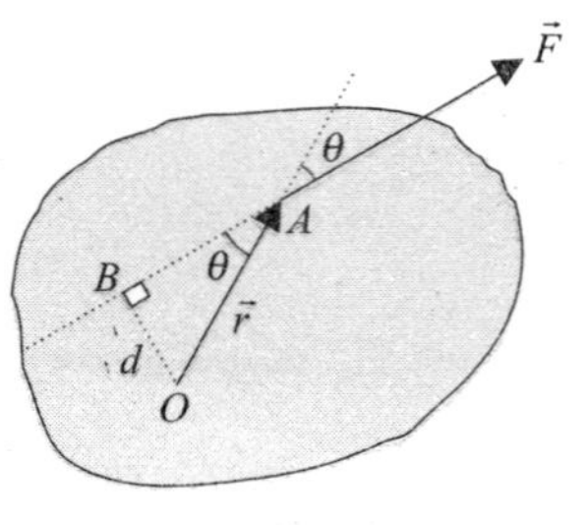

Fig. 4.9 Moment of a coplanar force

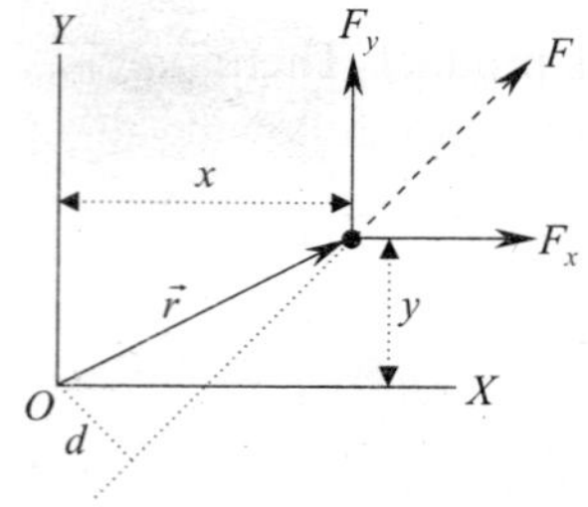

Fig. 4.10 Moment of the components of a force about the origin

$$\vec{M}_O = \vec{r} \times \vec{F} = (x\vec{i} + y\vec{j}) \times (F_x\vec{i} + F_y\vec{j})$$
$$= (xF_y - yF_x)\vec{k} \tag{4.23}$$

It can be seen that for planar problems, the moment always acts perpendicular to the plane or in other words along the Z direction. It can be either into the plane or out of the plane. The vector approach in such cases can be avoided if we follow the below-mentioned sign convention.

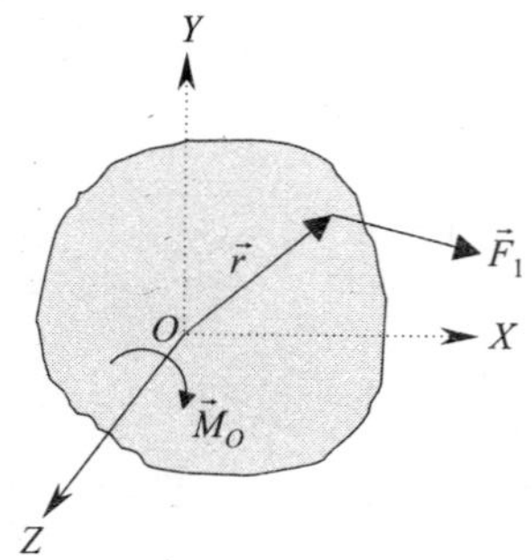

Fig. 4.11(a) Clockwise moment

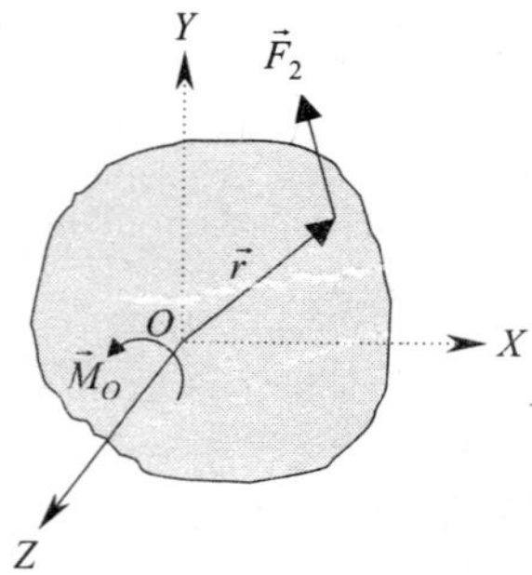

Fig. 4.11(b) Anti-clockwise moment

Consider two forces $\vec{F}_1$ [Fig. 4.11(a)] and $\vec{F}_2$ [Fig. 4.11(b)] acting on a body as shown. It can be seen that the effect of $\vec{F}_1$ is to rotate the body about O in the **clockwise direction** and that of $\vec{F}_2$ is to rotate the body about O in the **anti-clockwise direction**. In other words, the moment of $\vec{F}_1$ is in the clockwise sense and that of $\vec{F}_2$ is in the anti-clockwise sense. As we curl the fingers of our right hand from the positive sense of $\vec{r}$ to the positive sense of $\vec{F}$, we can see that the thumb points into the plane in Fig. 4.11(a) or in other words along the **negative** Z-direction; and out of the plane in Fig. 4.11(b) or in other words along the **positive** Z-direction. Thus, we can avoid the vector approach by considering **clockwise moments** as **negative** and **anti-clockwise** moments as **positive**.

In Fig. 4.10, we see that F_x creates clockwise moment and F_y creates anticlockwise moment. Hence, considering the sign convention as discussed above, the moment of the force about O can be directly written as:

$$M_O = -yF_x + xF_y \tag{4.24}$$

It can be seen that it is exactly same as that obtained using the vector approach. Hence, using the components of the force, we can directly determine the moment of the force without vector calculation of the cross product. Therefore, in the coplanar problems, we will avoid the vector approach by following the sign convention. The following examples will clarify this.

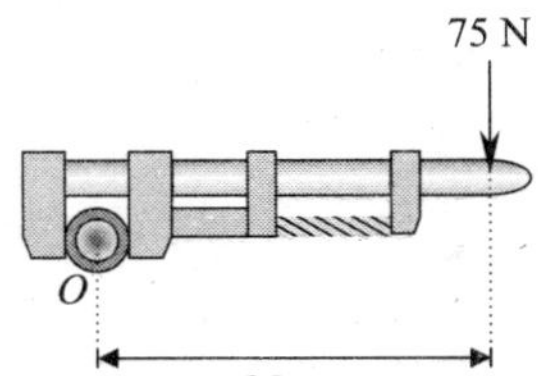

Fig. 4.12

Example 4.5 Determine the moment of a 75 N force applied on a pipe wrench as shown in Fig. 4.12 about centre O of the pipe.

Solution We know that the moment of a force about a point is given by the product of the force and the perpendicular distance of the line of action of the force from that point. Hence,

$$M_O = -75 \times 0.25$$
$$= -18.75 \text{ N.m}$$

The negative sign indicates that the force tends to rotate the wrench in the *clockwise* direction.

Example 4.6 While opening a butterfly valve in a liquid storage tank, a force of 100 N is applied as shown in Fig. 4.13. Determine the moment of the applied force about the hinge O.

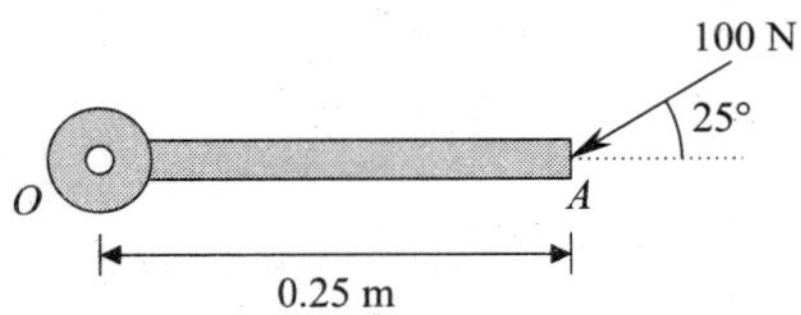

Fig. 4.13

Solution We know that the moment of a force about a point is given by the product of the force and the perpendicular distance of the line of action of the force from that point. However, here as the force is inclined to the axis of the handle, we cannot readily determine this perpendicular distance. Hence, we solve this problem by applying Varignon's theorem, which states that the moment of a force about a point is the same as the moment of its components about the same point. Hence, resolving the force along X and Y directions, we have $F_x = 100 \cos 25°$ and $F_y = 100 \sin 25°$ as shown in Fig. 4.13(a). Note that the original force is shown in the dotted line to avoid the confusion, as it may imply that there are three forces acting.

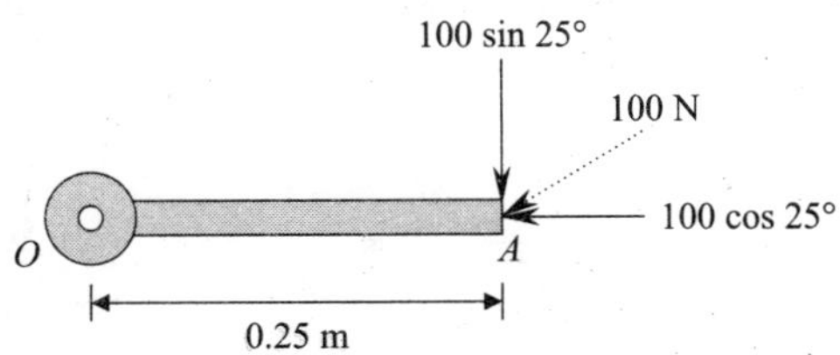

Fig. 4.13(a)

According to the sign convention, treating clockwise moments as *negative* and anti-clockwise moments as *positive*, the sum of moments of the components about O is given as

$$M_O = -100 \sin 25°(0.25) + 100 \cos 25°(0)$$
$$= -10.57 \text{ N.m}$$

Note that as the x-component of the force passes through the origin O, its moment arm is zero.

Alternative method:

The above problem can also be solved by determining the perpendicular distance of the line of action of the force from O. Referring to Fig. 4.13(b), we see that this perpendicular distance is: 0.25 sin 25°.

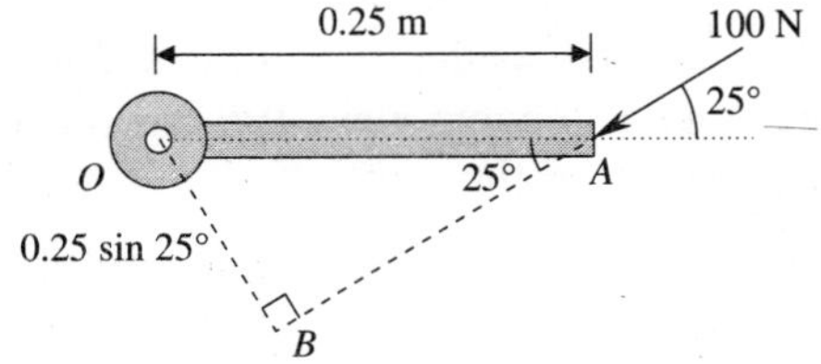

Fig. 4.13(b)

Hence, the moment of the force about O is given as the product of the force and the perpendicular distance:

$$M_O = -(100)(0.25 \sin 25°)$$
$$= -10.57 \text{ N.m}$$

Thus, we see that the same result is obtained as that of the previous method using Varignon's theorem. Though the latter method looks simpler, it must be noted that the determination of perpendicular distance will not always be as easy as in this problem.

Example 4.7 A 100 N force is applied at the end A of a lever OA (i) horizontally, (ii) vertically, and (iii) perpendicular to the lever as shown in the figures below. Determine the moment of the force about O in each case. What do you infer from the three moments?

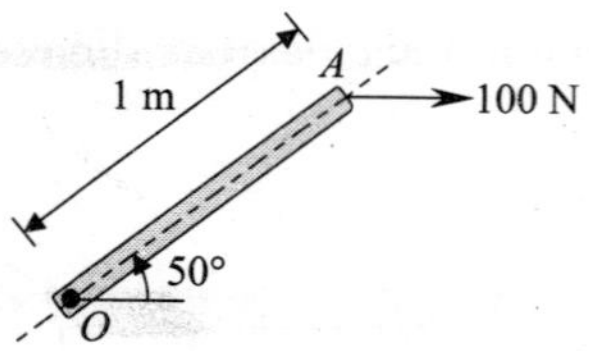

Fig. 4.14(a)

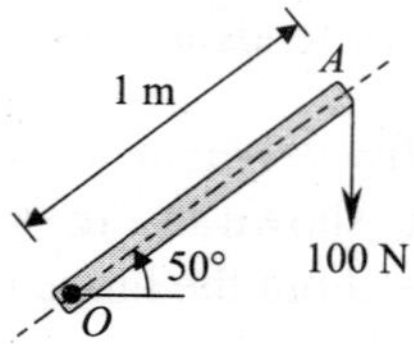

Fig. 4.14(b)

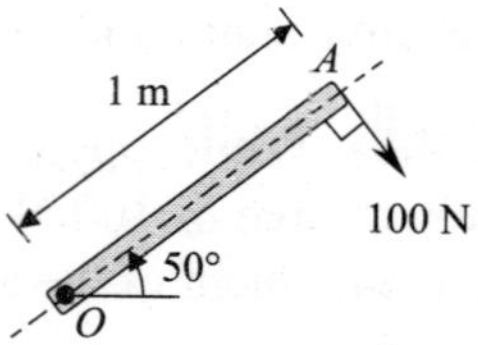

Fig. 4.14(c)

Solution The applied forces are resolved into components along the axis of the lever and perpendicular to the lever as shown in the figures below:

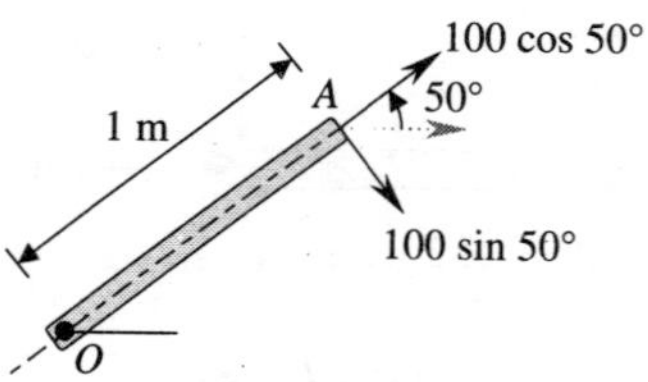

Fig. 4.14(d)

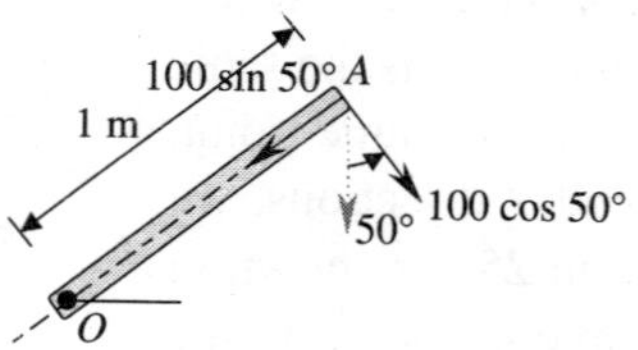

Fig. 4.14(e)

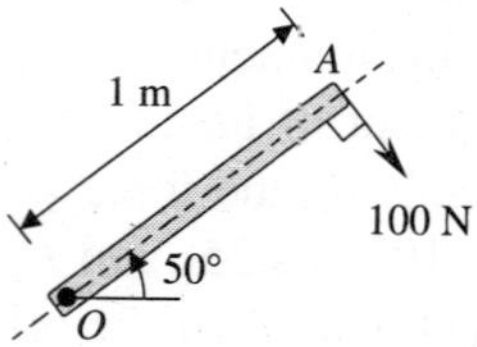

Fig. 4.14(f)

(i) *When the force is applied horizontally*

Taking moment about the hinge O, we have

$$M_O = -100 \sin 50° \times 1$$
$$= -76.6 \text{ N.m}$$

[Note that as the moment is in the clockwise direction, as per the sign convention, it is considered negative. The component 100 cos 50° does not contribute to the moment as its line of action passes through the origin.]

(ii) *When the force is applied vertically*

Similarly, we have

$$M_O = -100 \cos 50° \times 1$$
$$= -64.28 \text{ N.m}$$

[Note that as the moment is in the clockwise direction, as per the sign convention, it is considered negative. The component 100 sin 50° does not contribute to the moment as its line of action passes through the origin.]

(iii) *When the force is applied perpendicular to the lever*

$$M_O = -100(1)$$
$$= -100 \text{ N.m}$$

From the moments obtained in the above three cases, we see that the magnitude of the moment is **maximum** when the force is applied **perpendicular** to the lever.

Example 4.8 A force of 50 N is applied on a bench vice as shown in Fig. 4.15. Determine the moment of this force about O.

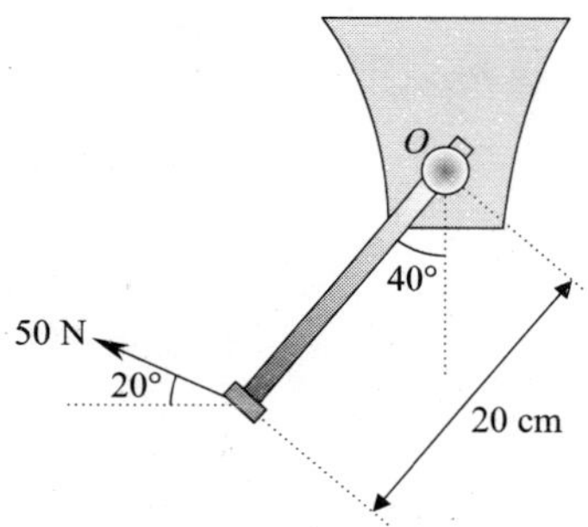

Fig. 4.15

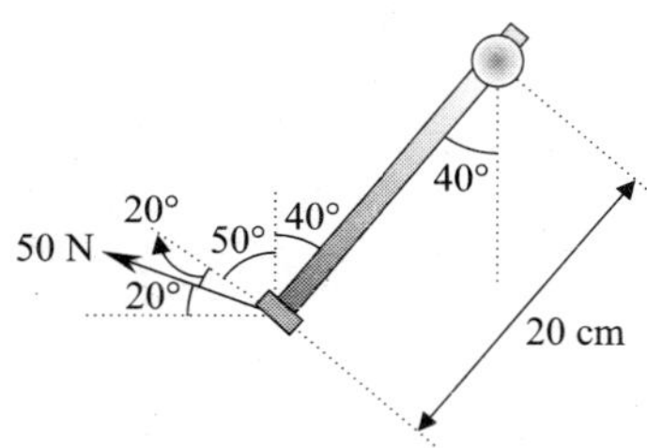

Fig. 4.15(a)

Solution As the applied force is inclined to the axis of the handle, we must determine its inclination and the components along the axis and normal to the axis of the handle. From Fig. 4.15(a), we can see that its inclination normal to the handle is 20°. Hence, the components of the force along and normal to the axis are respectively 50 sin 20° and 50 cos 20°. It should be noted that the first component does not contribute to the moment as its line of action passes through the origin. Hence, the moment of the applied force about O is

$$M_O = 50 \cos 20° \times 0.2$$
$$= 9.4 \text{ N.m (clockwise)}$$

Example 4.9 Find the moment of a 50 N force applied on a bent handle as shown in Fig. 4.16 about the base O.

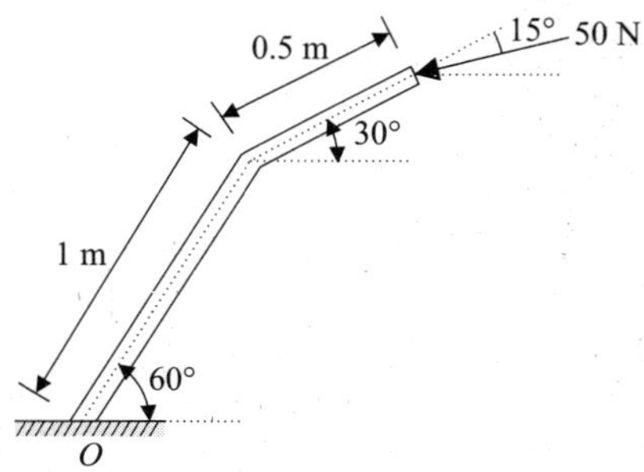

Fig. 4.16

Solution From the above figure, we can see that the inclination of the force with respect to the horizontal or X-axis is 30° – 15° = 15°. Hence, resolving the force into rectangular components along X and Y axes, we have $F_x = 50 \cos 15°$ and $F_y = 50 \sin 15°$. The perpendicular distances of the components from the origin O are shown in Fig. 4.16(a).

Taking the summation of the moments about O considering the sign convention, we have

$$M_O = 50 \cos 15° \,[(0.5) \sin 30° + 1 \sin 60°] - 50 \sin 15° \,[(0.5) \cos 30° + 1 \cos 60°]$$

$$= 41.83 \text{ N.m}$$

The positive sign indicates that the moment is in the anti-clockwise direction.

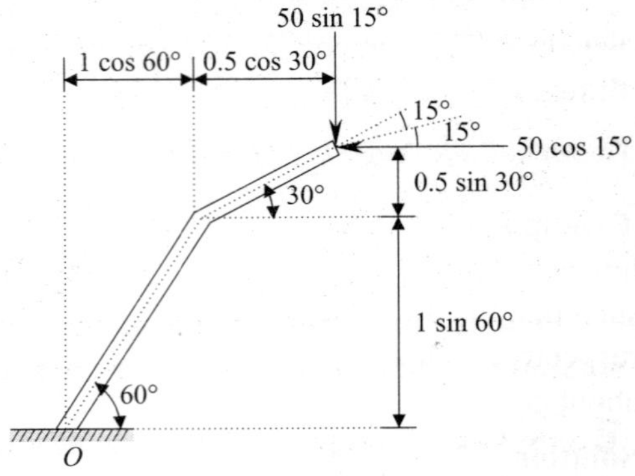

Fig. 4.16(a)

Example 4.10 Compute the moment of a 100 N force applied on a cantilever beam as shown in Fig. 4.17 about the fixed end A.

Solution Resolving the force along X and Y directions, we have $F_x = 100 \cos 30°$ and $F_y = 100 \sin 30°$ as shown in Fig. 4.17(a).

Taking the moment about the fixed end A,

$$M_A = 100 \cos 30° \,[0.25] - 100 \sin 30° \,[3 + 0.5]$$

$$= -153.35 \text{ N.m}$$

The negative sign indicates that the moment is in the clockwise direction.

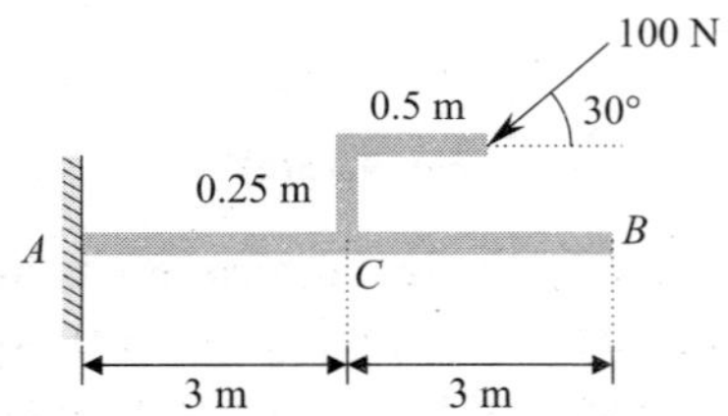

Fig. 4.17

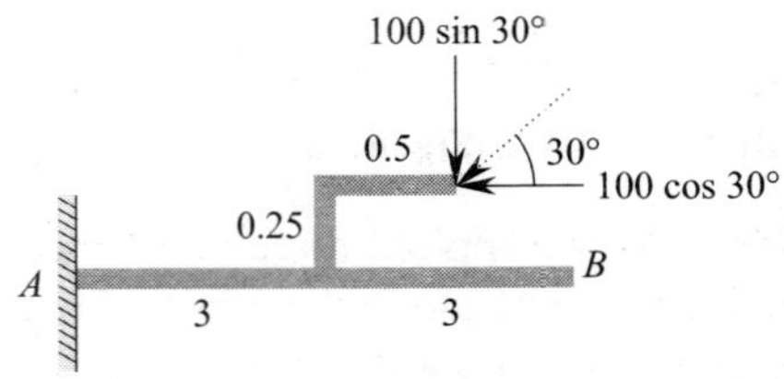

Fig. 4.17(a)

Example 4.11 Compute the moment of a 200 N force applied as shown in Fig. 4.18 about points A and B.

Solution Resolve the force along X and Y directions as shown in Fig. 4.18(a). Therefore,

$$M_B = -(200 \cos 45°) \times 4$$

$$= -565.69 \text{ N.m}$$

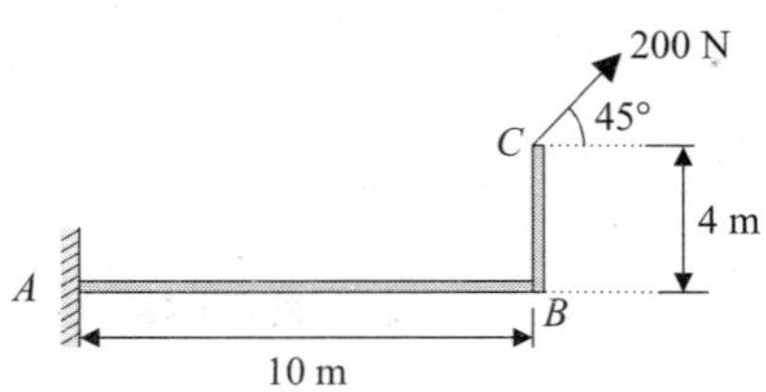

Fig. 4.18

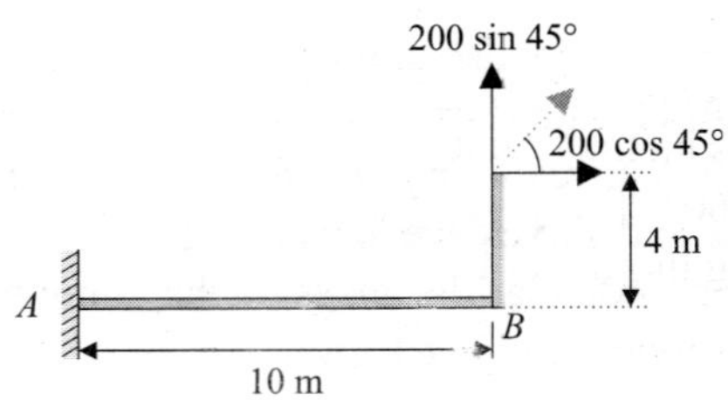

Fig. 4.18(a)

Note that the y-component of the force passes through point B and hence it does not contribute to the moment. Similarly,

$$M_A = [-(200 \cos 45°) \times 4] + [(200 \sin 45°) \times 10]$$
$$= 848.53 \text{ N.m}$$

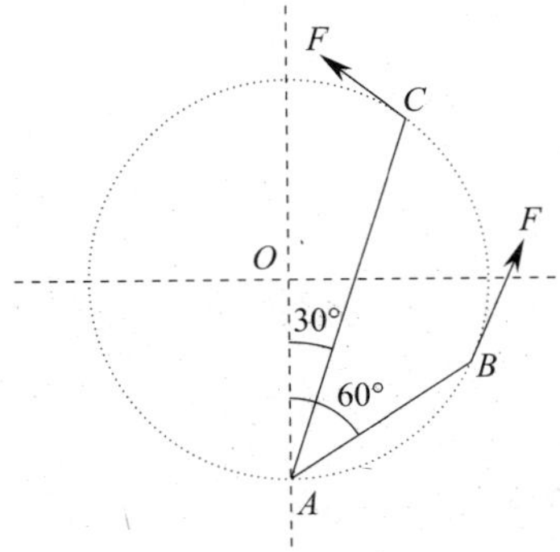

Fig. 4.19

Example 4.12 Forces of magnitudes F are applied on the circumference of a circle of radius r at points B and C as shown in Fig. 4.19, such that the direction of the forces are tangential to the circle at the respective points. Determine the moment of each of these forces about point A.

Solution *Force acting at B:* Since the force acting at B is tangential to the circle, we can see from Fig. 4.19(a) that its inclination to the radial line OB must be 90°. In addition, we can see that the triangle OAB must be an equilateral triangle, i.e., $\overline{OA} = \overline{OB} = \overline{AB} = r$ and the included angle is 60°. Hence, the inclination of the force with respect to the vertical line is 30°. Therefore, the horizontal and vertical components of the force are $F \sin 30°$ and $F \cos 30°$. Taking moment of the components about A, we have

$$M_A = -F \sin 30° \, [\overline{AB} \cos 60°] + F \cos 30° \, [\overline{AB} \sin 60°]$$
$$= -F \sin 30°[r \cos 60°] + F \cos 30°[r \sin 60°]$$
$$= -Fr \sin^2 30° + Fr \cos^2 30°$$
$$= 0.5Fr$$

[Note that sin 30° = cos 60° and cos 30° = sin 60°]

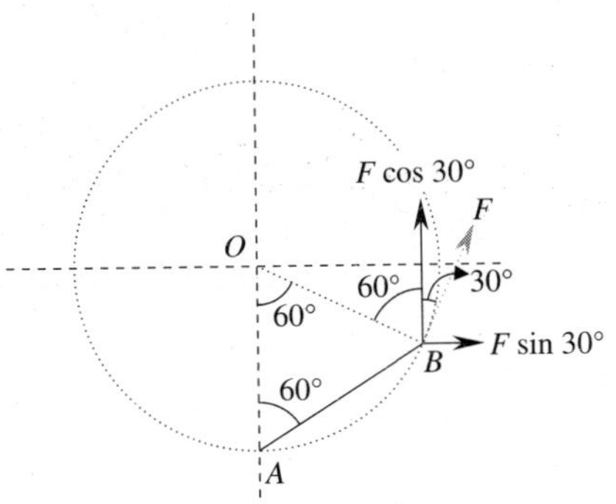

Fig. 4.19(a)

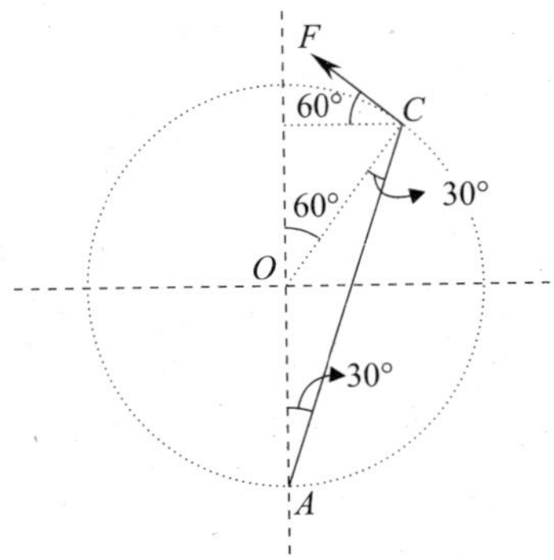

Fig. 4.19(b)

Force acting at C: Since the force acting at C is tangential to the circle, we can see from Fig. 4.19(b) that its inclination to the radial line OC must be 90°. Further, its inclination with respect to the horizontal is 60°. Therefore, the horizontal and vertical components of the force are $F \cos 60°$ and $F \sin 60°$. Taking moment of the components about A, we have

$$M_A = F \cos 60° \, [\overline{AO} + \overline{OC} \cos 60°] + F \sin 60° \, [\overline{OC} \sin 60°]$$
$$= F \cos 60° \, [r + r \cos 60°] + F \sin 60° \, [r \sin 60°]$$

$$= Fr\,[\sin^2 60° + \cos^2 60° + \cos 60°]$$
$$= Fr\,[1 + \cos 60°]$$
$$= 1.5Fr$$

4.4 COUPLE

When two forces $\vec{F}$ and $-\vec{F}$ having the same magnitude, parallel lines of action and opposite sense act on a body, then they are said to form a **couple**. Their *only* effect is to *rotate* the body about the axis perpendicular to the plane of the couple. For instance, while driving a vehicle and taking a turn, a couple is applied on the steering as shown in Fig. 4.20, as well when a screw is tightened using a screwdriver.

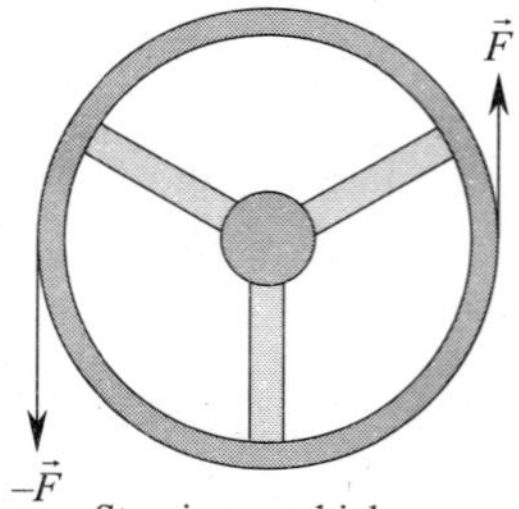

Steering a vehicle

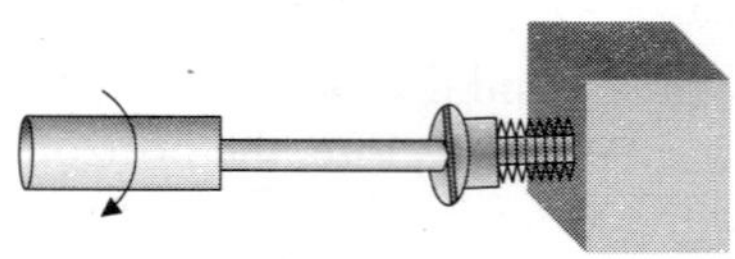

Tightening a screw

Fig. 4.20

4.4.1 Moment of a Couple

Consider two equal and opposite forces acting at points A and B, whose respective position vectors are $\vec{r}_A$ and $\vec{r}_B$ as shown in Fig. 4.21. Then the sum of the moments of these two forces about the origin O is given as

$$\vec{M}_O = [\vec{r}_A \times \vec{F}] + [\vec{r}_B \times (-\vec{F})]$$
$$= (\vec{r}_A - \vec{r}_B) \times \vec{F}$$
$$= \vec{r}_{BA} \times \vec{F} \qquad (4.25)$$

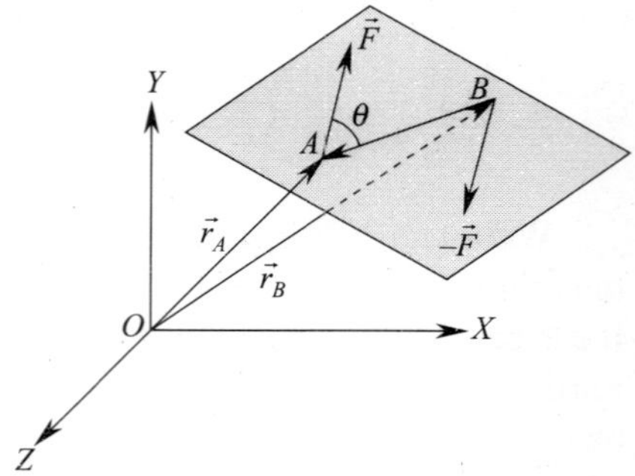

Fig. 4.21 Moment of a couple

The moment vector thus obtained is called **moment of a couple**. Its magnitude is

$$|\vec{M}_O| = Fr_{BA} \sin (180° - \theta)$$

Note that while taking the cross product of two vectors, the angle between their positive senses need to be considered. Thus, the angle between $\vec{r}_{BA}$ and $\vec{F}$ is $(180° - \theta)$.

$\therefore$
$$|\vec{M}_O| = Fr_{BA} \sin \theta$$
$$= Fd \qquad (4.26)$$

Note that $r_{BA} \sin \theta = d$, where d is the perpendicular distance between the two forces.

Hence, the moment of a couple can be defined as *the product of the magnitude of one of the forces and the perpendicular distance between them*. Its direction is perpendicular to the plane containing the two forces.

Note: Since $\vec{r}_{BA}$, the difference between the two position vectors is independent of the choice of the origin O of the coordinate axes, the moment of the couple remains the *same* irrespective of the reference point chosen. Hence, $\vec{M}$ is a **free vector,** i.e., it can be applied at any point. It can also be rotated through any angle.

4.5 FORCE–COUPLE SYSTEM

Consider a force $\vec{F}$ acting on a body at point A, whose position vector is $\vec{r}$. According to the principle of transmissibility, the force can be moved to any point along its line of action, as it produces the same effect on the body. However, if we want to move the force to a point not lying on its line of action then we must introduce a *couple* such that it produces the same effect as the force.

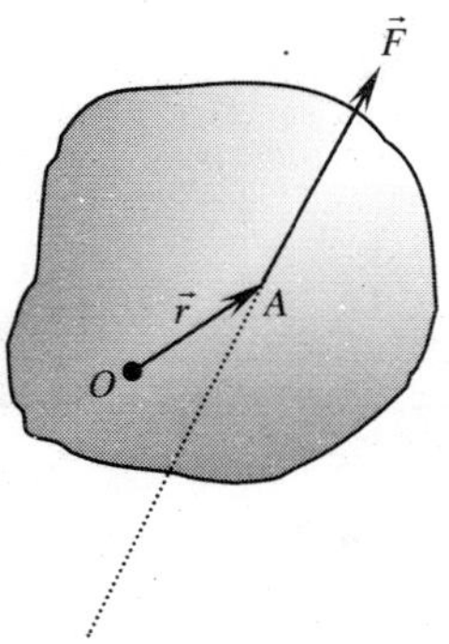

Fig. 4.22(a)

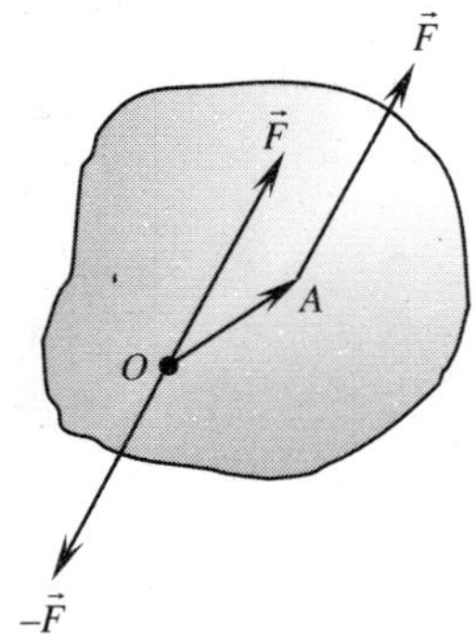

Fig. 4.22(b)

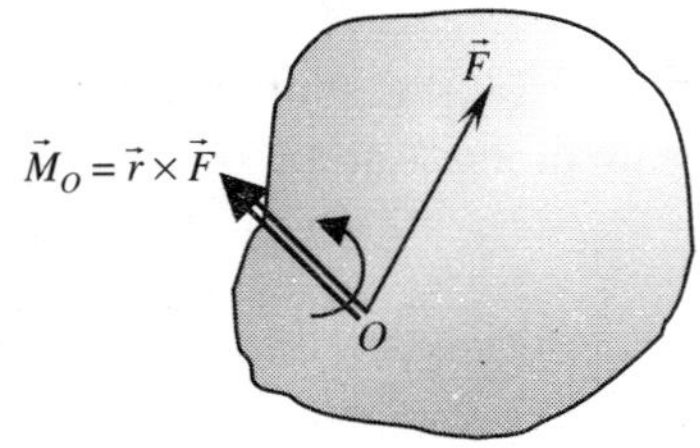

Fig. 4.22(c) Equivalent force–couple

Without affecting the effect of the force on the rigid body, let us introduce two forces $\vec{F}$ and $-\vec{F}$ at the origin as shown in Fig. 4.22(b). The forces $\vec{F}$ acting at A and $-\vec{F}$ acting at O together form a *couple*. It can be readily seen that the moment of this couple is same as the moment of $\vec{F}$ about O in Fig. 4.22(a). Further, it can be moved to any point, as it is a free vector. Hence, any force $\vec{F}$ acting on a rigid body can be moved to an arbitrary point O provided that a couple is added whose moment is equal to the moment of $\vec{F}$ about O. This is known as **force–couple** system.

4.6 RESULTANT OF NON-CONCURRENT FORCES

If there are n numbers of non-concurrent forces acting on a rigid body, then all those forces can be moved to a *common* point by introducing force–couple for each individual force. Once the forces are *concurrent* at a point, their resultant can be obtained by vector addition as explained in the previous chapter for concurrent forces. Similarly, the resultant of all couples can be added vectorially.

Hence, the **resultant** of non-concurrent forces is a single **force–couple system**, whose **force** is equal to *the summation of all individual forces* and the **couple** is equal to *the summation of all individual couples*. Mathematically, it is written as

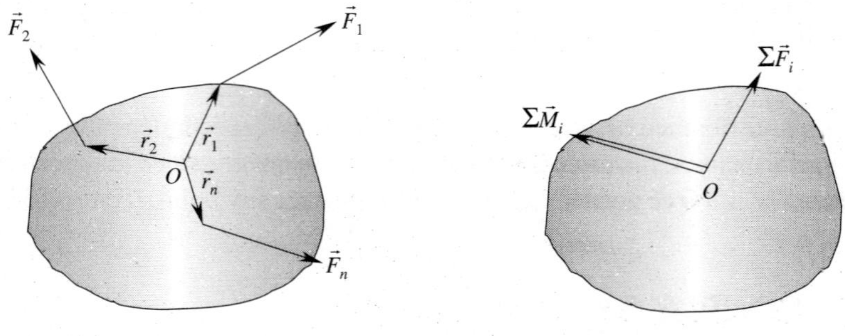

Fig. 4.23 Resultant of non-concurrent forces

$$\vec{R} = \vec{F}_1 + \vec{F}_2 + ... + \vec{F}_n = \sum \vec{F}_i \tag{4.27}$$

and
$$\vec{M}_O = (\vec{r}_1 \times \vec{F}_1) + (\vec{r}_2 \times \vec{F}_2) + ... + (\vec{r}_n \times \vec{F}_n) = \sum \vec{r}_i \times \vec{F}_i \tag{4.28}$$

Example 4.13 In Fig. 4.24 shown, reduce the given system of forces acting on beam *AB* to (i) an equivalent force–couple system at *A*, and (ii) an equivalent force–couple system at *B*.

Solution Taking summation of all the forces acting on the beam along *X* and *Y* directions:

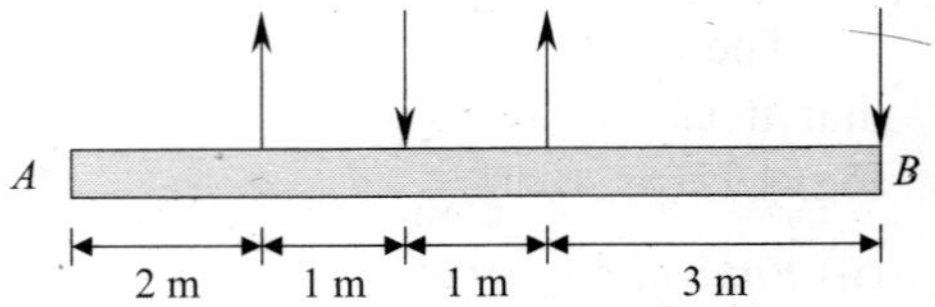

Fig. 4.24

$$\sum F_x = 0$$

and
$$\sum F_y = 100 - 500 + 750 - 600$$
$$= -250 \text{ N}$$

The negative sign indicates that the resultant points in the *negative Y*-direction.

(i) *Equivalent force–couple system at A*

Taking summation of the moments of all the forces about *A*,

$$\sum M_A = (100 \times 2) - (500 \times 3) + (750 \times 4) - (600 \times 7) = -2500 \text{ N.m}$$

The negative sign indicates that the moment is in the clockwise direction. Figure 4.24(a) shows the equivalent force–couple system at *A*.

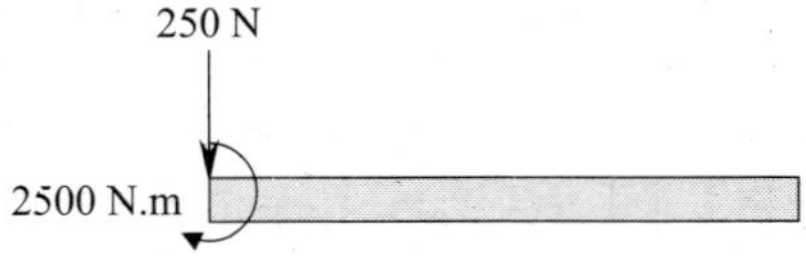

Fig. 4.24(a)

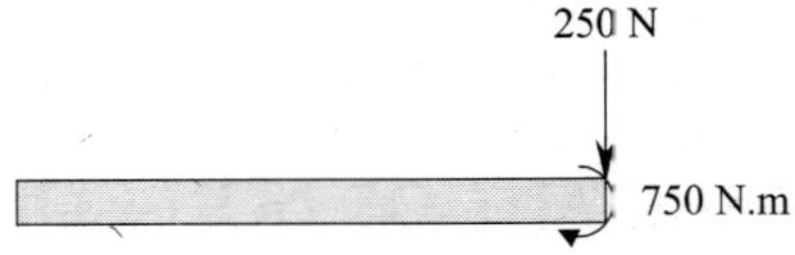

Fig. 4.24(b)

(ii) *Equivalent force–couple system at B*

Taking summation of the moments of all the forces about *B*,

$$\sum M_B = -(100 \times 5) + (500 \times 4) - (750 \times 3) + (600 \times 0) = -750 \text{ N.m}$$

The negative sign indicates clockwise moment. Figure 4.24(b) shows the equivalent force–couple system at *B*.

Example 4.14 In the Fig. 4.25 shown, reduce the given system of forces acting on beam AB to (i) an equivalent force–couple system at A, and (ii) an equivalent force–couple system at B.

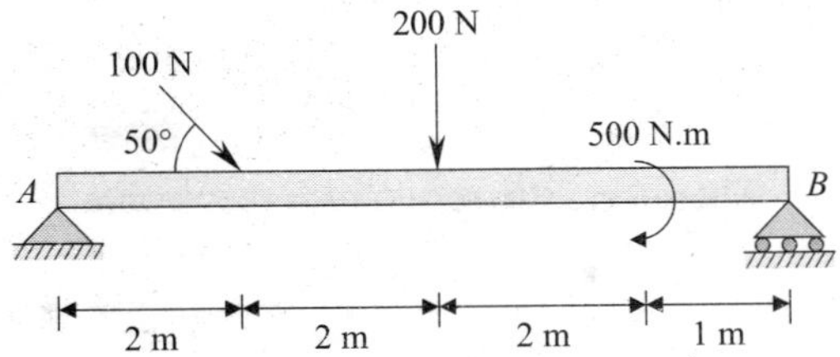

Fig. 4.25

Solution Taking summation of all the forces along X and Y-directions,

$$\sum F_x = 100 \cos 50°$$
$$= 64.28 \text{ N}$$

$$\sum F_y = -100 \sin 50° - 200$$
$$= -276.6 \text{ N}$$

The negative sign indicates that the y-component of the resultant is directed along the *negative* Y-direction.

(i) *Equivalent force–couple system at A*

Taking summation of the moments of all the forces about A

$$\sum M_A = -100 \sin 50°(2) - 200(4) - 500 = -1453.21 \text{ N.m}$$

The negative sign indicates clockwise moment. Note that the couple 500 N.m is a free vector and that it can be placed anywhere on the beam. The equivalent force–couple system is shown in Fig. 4.25(a).

(ii) *Equivalent force–couple system at B*

Taking summation of the moments of all the forces about B

$$\sum M_B = 100 \sin 50°(5) + 200(3) - 500 = 483.02 \text{ N.m}$$

The positive sign indicates that it is in the anticlockwise direction. The equivalent force–couple system is shown in Fig. 4.25(b).

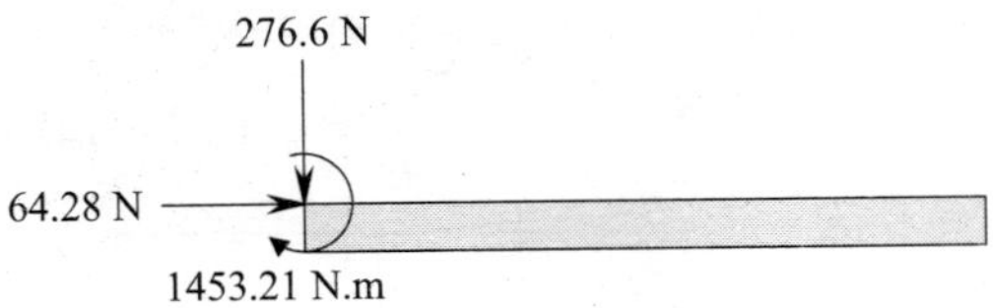

Fig. 4.25(a)

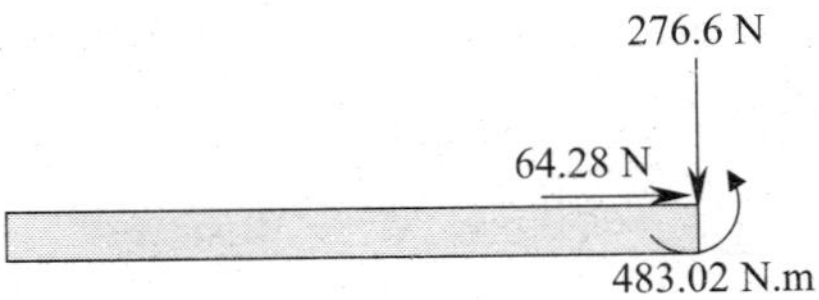

Fig. 4.25(b)

Example 4.15 Four forces and a couple are applied to a rectangular plate as shown in Fig. 4.26. Replace the forces and the couple by an equivalent force–couple system at the bolt 1.

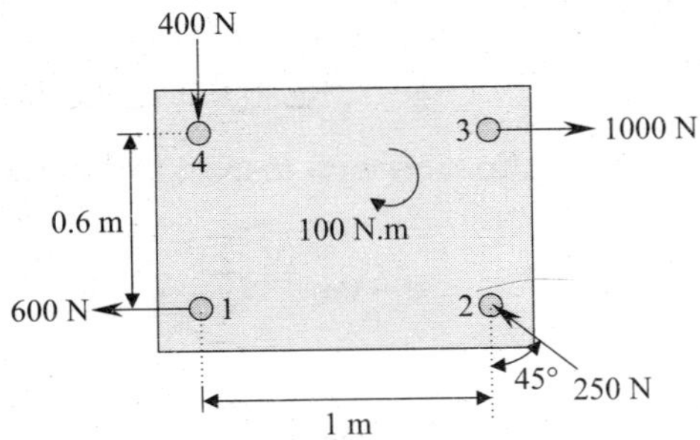

Fig. 4.26

Solution Taking summation of all the forces along X and Y directions,

$$\sum F_x = -600 - 250 \sin 45° + 1000$$
$$= 223.22 \text{ N}$$

$$\sum F_y = -400 + 250 \cos 45°$$
$$= -223.22 \text{ N}$$

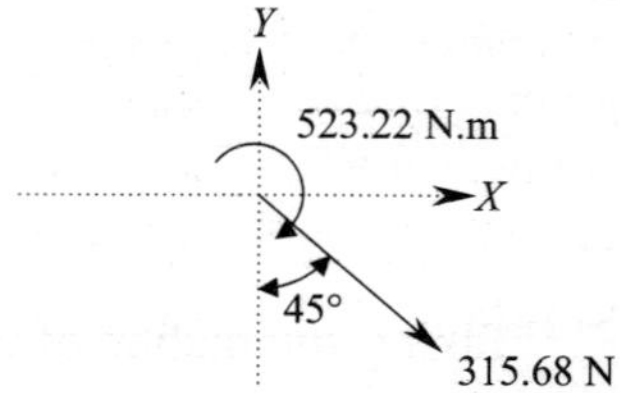

Fig. 4.26(a)

Therefore, the resultant force is obtained as

$$R = \sqrt{\left(\sum F_x\right)^2 + \left(\sum F_y\right)^2} = 315.68 \text{ N}$$

Its inclination with respect to the X-axis is obtained as

$$\theta = \tan^{-1}\left(\frac{\left|\sum F_y\right|}{\left|\sum F_x\right|}\right)$$

$$= \tan^{-1}\left(\frac{223.22}{223.22}\right) = 45°$$

Since $\sum F_x$ is positive and $\sum F_y$ is negative, we know that the resultant lies in the fourth quadrant. Taking summation of the moments of all the forces about the bolt 1,

$$\sum M_1 = -100 + (250 \cos 45° \times 1) - (1000 \times 0.6)$$
$$= -523.22 \text{ N.m}$$

Note that the couple of 100 N.m is a free vector and hence, it can be moved to any point. Also, note that the forces 600 N, 400 N and the x-component of 250 N do not contribute to the moment, as their lines of action pass through the bolt 1. The equivalent force–couple system is shown in Fig. 4.26(a).

Example 4.16 An equilateral triangular plate of side 3 m is acted on by three forces as shown in Fig. 4.27. Replace them by an equivalent force–couple system at A.

Solution The given forces are resolved into rectangular components as shown in Fig. 4.27(a).

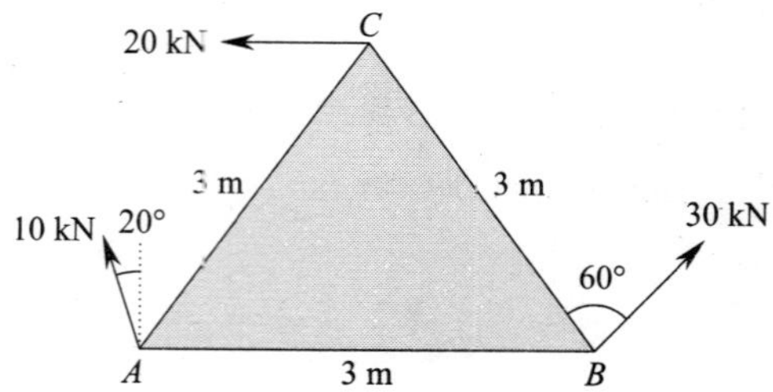

Fig. 4.27

Taking summation of the forces along X and Y axes:

$$\sum F_x = -10 \sin 20° - 20 + 30 \cos 60°$$
$$= -8.42 \text{ kN}$$

$$\sum F_y = 10 \cos 20° + 30 \sin 60°$$
$$= 35.38 \text{ kN}$$

Therefore, the resultant force is obtained as

$$R = \sqrt{\left(\sum F_x\right)^2 + \left(\sum F_y\right)^2} = 36.37 \text{ kN}$$

Its inclination with respect to the X-axis is

$$\theta = \tan^{-1}\left(\frac{\left|\sum F_y\right|}{\left|\sum F_x\right|}\right)$$

$$= \tan^{-1}\left(\frac{35.38}{8.42}\right) = 76.61°$$

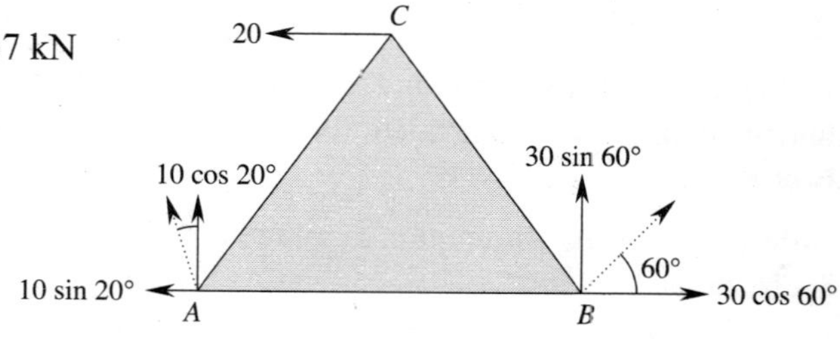

Fig. 4.27(a)

Since $\sum F_x$ is negative and $\sum F_y$ is positive, we know that the resultant lies in the second quadrant. From Fig. 4.27(b), we know that

$$\overline{CD} = \sqrt{3^2 - (1.5)^2} = \sqrt{6.75} \text{ m}$$

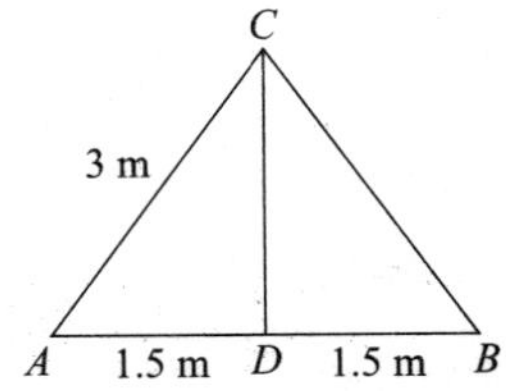

Fig. 4.27(b)

Taking summation of the moments of all the forces about A,

$$M_A = 20 \times \sqrt{6.75} + 30 \sin 60° \times 3$$
$$= 129.9 \text{ N.m}$$

Note that the forces 10 kN acting at A and horizontal component of 30 kN acting at B do not contribute to the moment as their lines of action pass through A.

Example 4.17 Replace the tension T in the cable AB by an equivalent force–couple system at O. Take the magnitude of the tension to be 200 N.

Solution From the figure, we see that the coordinates of A and B are:

$$A\ (0, 0, 6) \text{ and } B\ (-2, 4, 0)$$

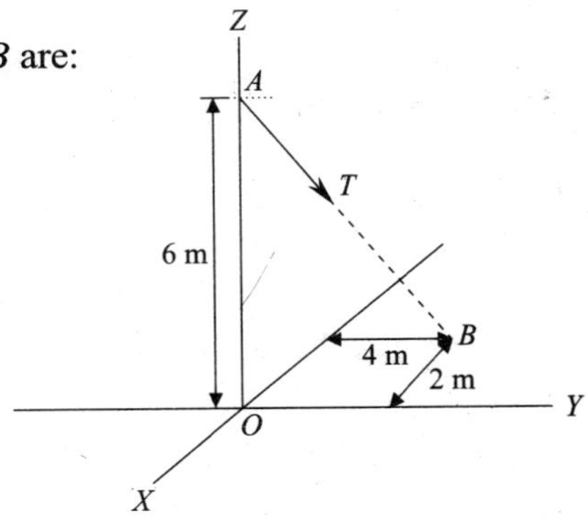

Fig. 4.28

Therefore,

$$\overrightarrow{AB} = \overrightarrow{OB} - \overrightarrow{OA}$$
$$= (-2\vec{i} + 4\vec{j}) - (6\vec{k})$$
$$= -2\vec{i} + 4\vec{j} - 6\vec{k}$$

Then unit vector along AB is obtained as

$$\hat{n}_{AB} = \frac{-2\vec{i} + 4\vec{j} - 6\vec{k}}{\sqrt{(-2)^2 + (4)^2 + (-6)^2}}$$
$$= \frac{-2\vec{i} + 4\vec{j} - 6\vec{k}}{\sqrt{56}}$$

Hence, the tension can be written in vector form as

$$\vec{T}_{AB} = T_{AB}\hat{n}_{AB}$$
$$= 200\left[\frac{-2\vec{i} + 4\vec{j} - 6\vec{k}}{\sqrt{56}}\right]$$
$$= \frac{-400\vec{i} + 800\vec{j} - 1200\vec{k}}{\sqrt{56}}$$

The tension can be replaced by an equivalent force–couple system, whose force is equal to the tension and the moment of the couple is equal to the moment of the tension about O. Moment of the tension about O is given by

$$\vec{M}_O = (\overrightarrow{OA}) \times \vec{T}_{AB}$$
$$= 6\vec{k} \times \left[\frac{-400\vec{i} + 800\vec{j} - 1200\vec{k}}{\sqrt{56}}\right] = \frac{-4800\vec{i} - 2400\vec{j}}{\sqrt{56}}$$

Example 4.18 Reduce the two couples acting as shown in Fig. 4.29 to a single couple. The forces of one of the couples are represented by two diagonals of opposite faces of a cube, and those of the other by the edges, which do not meet those diagonals. Let the forces along the edges be of magnitude F and that along the diagonals be of magnitude $\sqrt{2}F$ and the side of the cube be a.

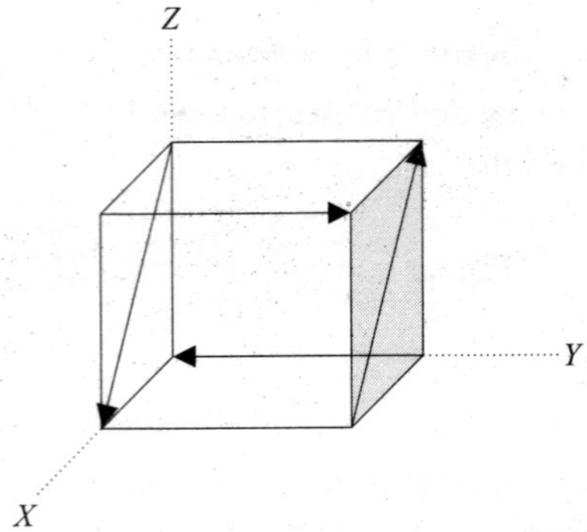

Fig. 4.29

Solution *Moment of the forces along the diagonal*
Consider the two forces of magnitude $\sqrt{2}F$ acting along the diagonals of opposite faces of the cube. Then the force on the right face [see Fig. 4.29(a)] can be represented as

$$\vec{F}_1 = \sqrt{2}F[\hat{n}]$$

$$= \sqrt{2}F\left[\frac{-a\vec{i} + a\vec{k}}{\sqrt{2}a}\right] = F[-\vec{i} + \vec{k}]$$

The position vector of the force with respect to the force on the left face is

$$\vec{r}_1 = (a\vec{i} + a\vec{j}) - (a\vec{k}) = a\vec{i} + a\vec{j} - a\vec{k}$$

Therefore, the moment of the couple is obtained as

$$\vec{M}_1 = (\vec{r}_1 \times \vec{F}_1) = \begin{vmatrix} \vec{i} & \vec{j} & \vec{k} \\ a & a & -a \\ -F & 0 & F \end{vmatrix}$$

$$= \vec{i}(Fa - 0) - \vec{j}(Fa - Fa) + \vec{k}(0 + Fa)$$

$$= Fa[\vec{i} + \vec{k}]$$

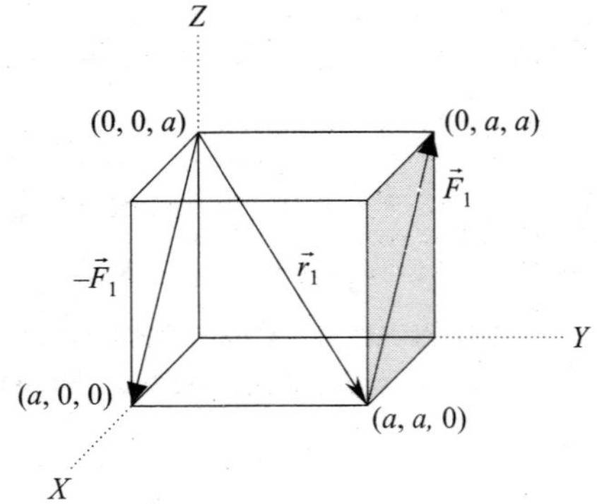

Fig. 4.29(a)

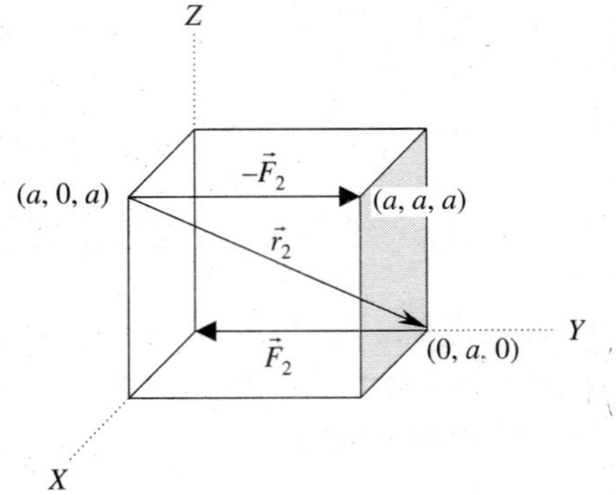

Fig. 4.29(b)

Moment of the forces along the edges

Consider the two forces of magnitude F acting along the edges as shown in Fig. 4.29(b). Then the force on the farther side can be represented as

$$\vec{F}_2 = -F\vec{j} \qquad \text{[Note that it points along the negative } Y\text{-axis.]}$$

Its position vector with respect to the force acting along the nearer edge is

$$\vec{r}_2 = (a\vec{j}) - (a\vec{i} + a\vec{k}) = -a\vec{i} + a\vec{j} - a\vec{k}$$

Therefore, the moment of the couple is obtained as

$$\vec{M}_2 = (\vec{r}_2 \times \vec{F}_2) = \begin{vmatrix} \vec{i} & \vec{j} & \vec{k} \\ -a & a & -a \\ 0 & -F & 0 \end{vmatrix}$$

$$= \vec{i}[0 - Fa] - \vec{j}[0] + \vec{k}[Fa - 0]$$

$$= Fa[-\vec{i} + \vec{k}]$$

Therefore, the two couples can be reduced to a single couple by vector addition:

$$\vec{M}_1 + \vec{M}_2 = 2Fa\vec{k}$$

that is the magnitude of the resultant couple is $2Fa$ directed along the Z direction.

4.7 REDUCTION OF A FORCE–COUPLE INTO A SINGLE FORCE

In the previous section, we saw that the resultant of non-concurrent forces is a *single* equivalent force–couple system placed at an arbitrary point. The resultant is obtained by summation of all individual forces and the moment of the couple is obtained by summation of the moments of all individual forces about that point. In general, the resultant force and couple would not be *perpendicular* to each other. However, a special case that is of interest to us arises, where the *force* and *the moment of the couple* are *perpendicular* to each other. This happens when all the forces lie on the **same plane**, i.e., they are coplanar or when the forces are **parallel** to each other. These two cases are discussed below in detail.

4.7.1 Coplanar Non-Concurrent Forces

When the forces are *coplanar*, we can see that the *resultant* will also lie on the *same plane*. Moreover, the moments of the individual forces about O will be perpendicular to the plane. Hence, the resultant *couple moment* will also be *perpedicular* to that plane. In such a case, the resultant force–couple system can be reduced to a **single force**, by moving the resultant force to a new line of action on the same plane such that its moment about O is equal to the moment of the couple, i.e.,

$$Rd = \sum M_O$$

where d is the perpendicular distance from the origin to the new line of action of the resultant. The value of d can be determined as

$$d = \frac{\sum M_O}{R} \tag{4.29}$$

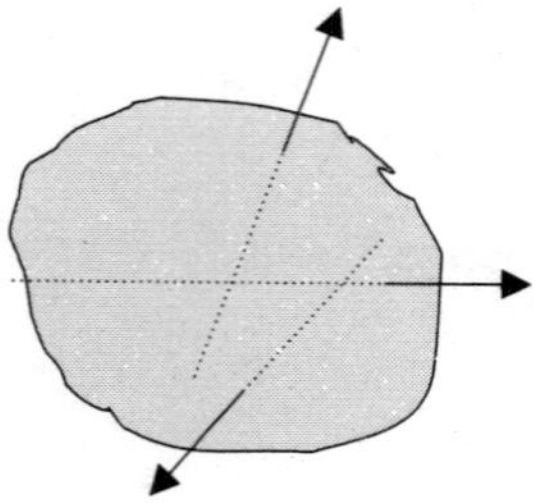

Fig. 4.30(a) Coplanar non-concurrent forces

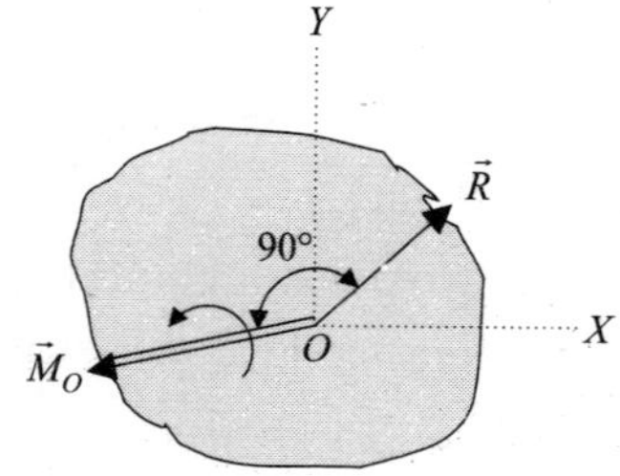

Fig. 4.30(b) Equivalent force–couple system

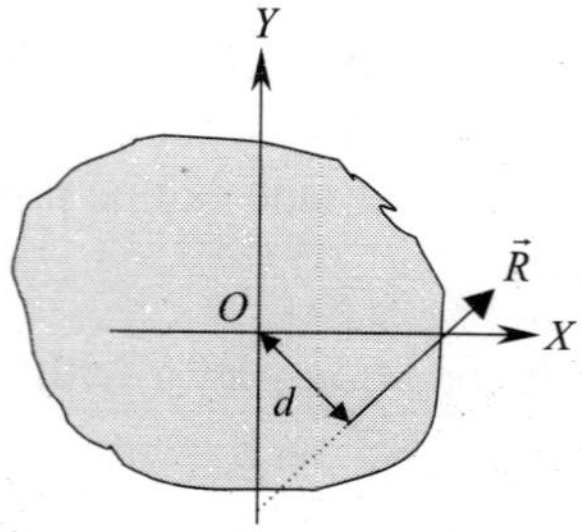

Fig. 4.30(c) Equivalent force

If instead of the perpendicular distance d, we are interested in x and y intercepts of the resultant with the coordinate axes, they can be determined as follows:

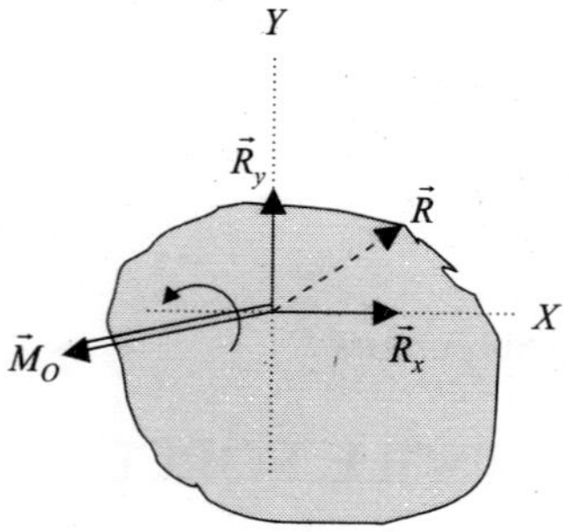

Fig. 4.31(a) Equivalent force–couple system

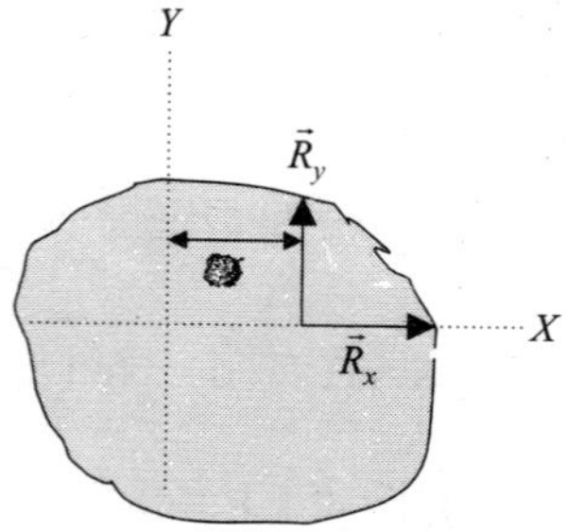

Fig. 4.31(b) x-intercept of equivalent force

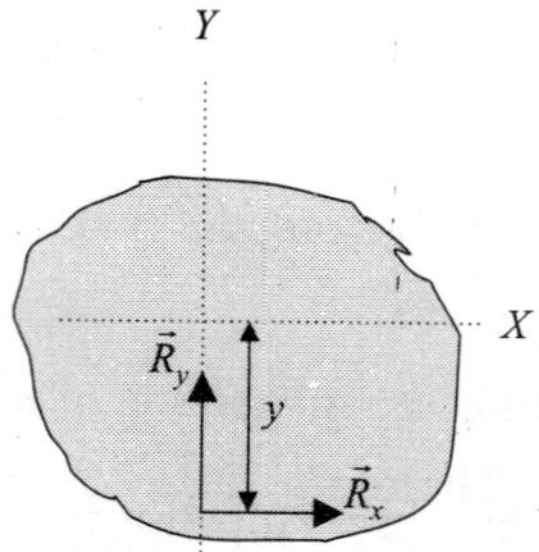

Fig. 4.31(c) y-intercept of equivalent force

If (x, y) be the coordinates of the point of application of the resultant such that its moment about the origin is equal to the moment of the couple then

$$(x\vec{i} + y\vec{j}) \times (\vec{R}) = \sum \vec{M}_O$$

$$(x\vec{i} + y\vec{j}) \times \left(\sum F_x\vec{i} + \sum F_y\vec{j}\right) = \sum \vec{M}_O$$

$$\Rightarrow \qquad \left(x\sum F_y - y\sum F_x\right)\vec{k} = \left(\sum M_O\right)\vec{k} \tag{4.30}$$

If $y = 0$, the resultant force will lie on the X-axis [refer Fig. 4.31(b)] and its location is given by

$$x = \frac{\sum M_O}{\sum F_y} = \frac{\sum M_O}{R_y} \tag{4.31}$$

If $x = 0$, the resultant force will lie on the Y-axis [refer Fig. 4.31(c)] and its location is given by

$$y = -\frac{\sum M_O}{\sum F_x} = -\frac{\sum M_O}{R_x} \tag{4.32}$$

Example 4.19 Two forces are acting as shown in Fig. 4.32. Determine a single equivalent force and its x and y intercepts.

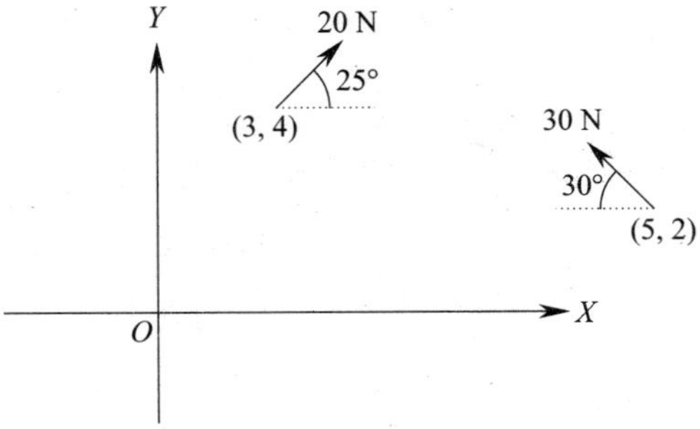

Fig. 4.32

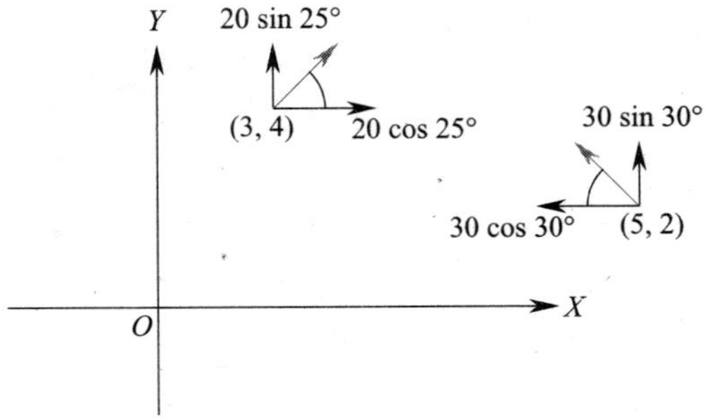

Fig. 4.32(a)

Solution The two forces are resolved into components along X and Y directions as shown in Fig. 4.32(a).

Taking summation of the forces along X and Y directions:

$$\sum F_x = 20 \cos 25° - 30 \cos 30°$$

$$= -7.85 \text{ N}$$

$$\sum F_y = 20 \sin 25° + 30 \sin 30°$$

$$= 23.45 \text{ N}$$

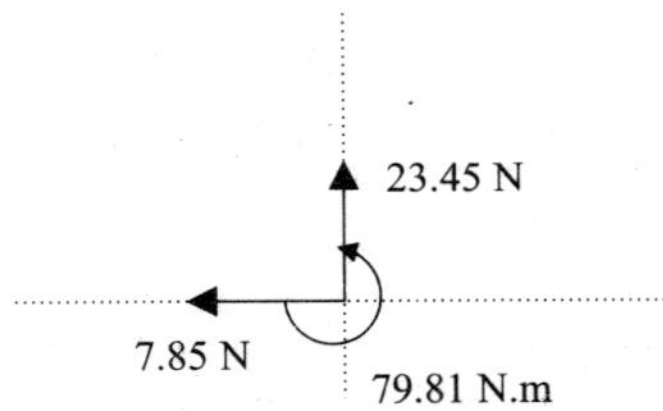

Fig. 4.32(b)

Therefore, the magnitude of the resultant force is given as

$$R = \sqrt{\left(\sum F_x\right)^2 + \left(\sum F_y\right)^2}$$

$$= \sqrt{(-7.85)^2 + (23.45)^2} = 24.73 \text{ N}$$

Its inclination with the horizontal axis is

$$\tan \theta = \frac{\left|\sum F_y\right|}{\left|\sum F_x\right|}$$

$$\Rightarrow \qquad \theta = \tan^{-1}\left[\frac{23.45}{7.85}\right] = 71.49°$$

Taking summation of the moments of all the forces about the origin O,

$$\sum M_O = -(20 \cos 25° \times 4) + (20 \sin 25° \times 3) + (30 \cos 30° \times 2) + (30 \sin 30° \times 5)$$

$$= 79.81 \text{ N.m}$$

The positive sign indicates that the moment is pointing along the positive Z-axis. The equivalent force–couple system is shown in Fig. 4.32(b).

Determination of x and y intercepts

$$\bar{x} = \frac{\sum M_O}{\sum F_y} = \frac{79.81}{23.45} = 3.4 \text{ m}$$

$$\bar{y} = -\frac{\sum M_O}{\sum F_x}$$

$$= -\left[\frac{79.81}{-7.85}\right] = 10.17 \text{ m}$$

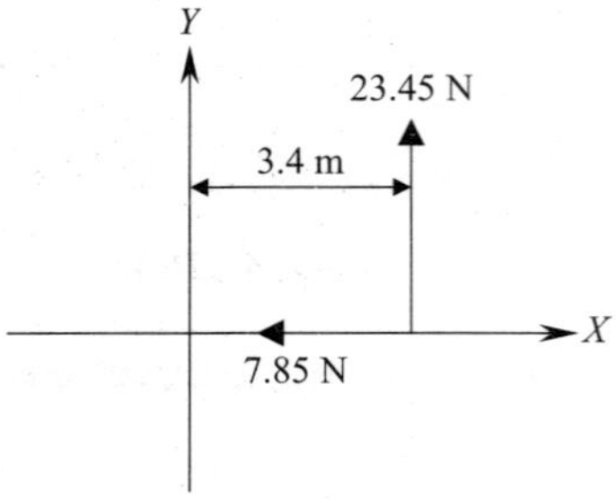

Fig. 4.32(c) *x*-intercept

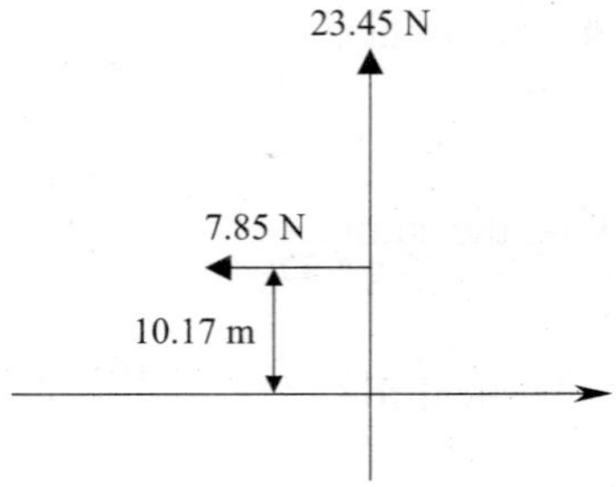

Fig. 4.32(d) *y*-intercept

Alternative method—vector approach

As we already know the components of the forces, we can write them in vector form as

$$\vec{F}_1 = 20 \cos 25° \vec{i} + 20 \sin 25° \vec{j}$$

and

$$\vec{F}_2 = -30 \cos 30° \vec{i} + 30 \sin 30° \vec{j}$$

In addition, their points of application can be written as

$$\vec{r}_1 = 3\vec{i} + 4\vec{j} \quad \text{and} \quad \vec{r}_2 = 5\vec{i} + 2\vec{j}$$

Therefore, the resultant of the two forces and their moments about the origin can be determined as follows:

$$\vec{R} = \vec{F}_1 + \vec{F}_2$$

$$= (20 \cos 25° - 30 \cos 30°)\vec{i} + (20 \sin 25° + 30 \sin 30°)\vec{j}$$

$$= -7.85\vec{i} + 23.45\vec{j}$$

$$\vec{M}_O = (\vec{r}_1 \times \vec{F}_1) + (\vec{r}_2 \times \vec{F}_2)$$

$$= (3\vec{i} + 4\vec{j}) \times (20 \cos 25° \vec{i} + 20 \sin 25° \vec{j})$$

$$+ (5\vec{i} + 2\vec{j}) \times (-30 \cos 30° \vec{i} + 30 \sin 30° \vec{j})$$

$$= 60 \sin 25° \vec{k} - 80 \cos 25° \vec{k} + 150 \sin 30° \vec{k} + 60 \cos 30° \vec{k}$$

$$= 79.81\vec{k}$$

Let $(x\vec{i} + y\vec{j})$ be the coordinates of the location of the resultant. We know that the moment of the resultant about the origin must be same as the moment of the couple. Hence,

$$(x\vec{i} + y\vec{j}) \times (-7.85\vec{i} + 23.45\vec{j}) = 79.81\vec{k}$$

$$(23.45x + 7.85y)\vec{k} = 79.81\vec{k}$$

When $y = 0$, $x = \dfrac{79.81}{23.45} = 3.4$ m

When $x = 0$, $y = \dfrac{79.81}{7.85} = 10.17$ m

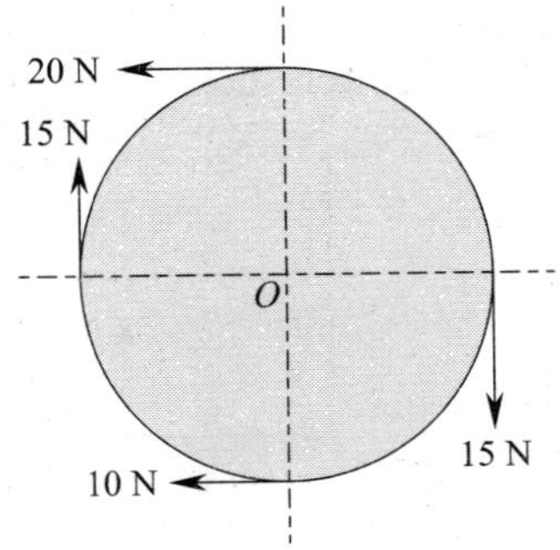

Fig. 4.33

Example 4.20 A circular disc of radius 1 m is acted upon by four forces as shown in Fig. 4.33. Replace the forces by a single equivalent force.

Solution Taking the summation of all the forces along X and Y axes:

$$\Sigma F_x = -20 - 10 = -30 \text{ N}$$

As the two vertical forces are equal in magnitude and opposite to each other, they form a couple. Hence,

$$\Sigma F_y = 0$$

Taking the summation of the moments of all the forces about the origin O,

$$\Sigma M_O = 20 \times 1 - 10 \times 1 - 15 \times 2 = -20 \text{ N.m}$$

Note that the moment of the couple is equal to the product of one of the forces and the perpendicular distance between them. The equivalent *force–couple* system is shown in Fig. 4.33(a).

Determination of x and y intercepts

As ΣF_y is zero, x-intercept does not exist. Therefore,

$$y = -\frac{\Sigma M_O}{\Sigma F_x}$$

$$= -\left[\frac{-20}{-30}\right] = -0.67 \text{ m}$$

The *equivalent force* is shown in Fig. 4.33(b).

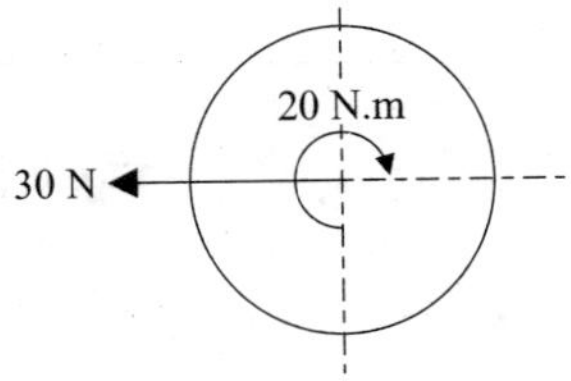

Fig. 4.33(a)

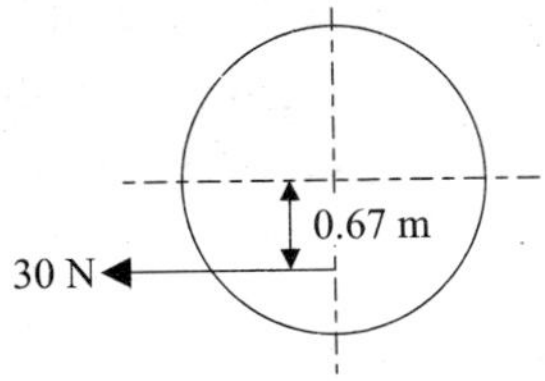

Fig. 4.33(b)

4.7.2 Parallel Forces

When the forces are **parallel** to each other (say directed along vertical axis) then we can see that the *resultant* will also lie along the *vertical axis*. Moreover, the moments of individual forces about O will lie on the horizontal plane. Hence, the resultant *couple moment* will also lie on the same plane. In such a case, the resultant force–couple can be reduced to a **single force**, by moving the force to a new parallel line of action such that its moment about O is equal to the moment of the couple.

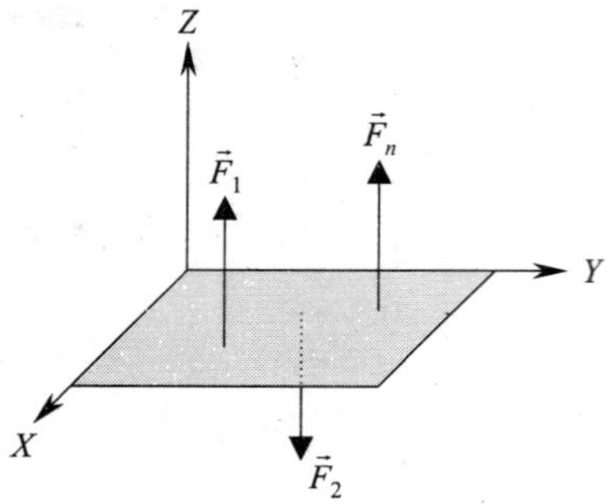

Fig. 4.34(a) Parallel forces

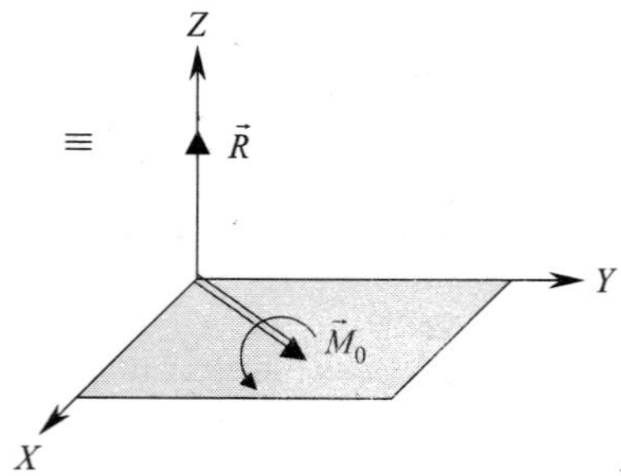

Fig. 4.34(b) Equivalent force–couple system

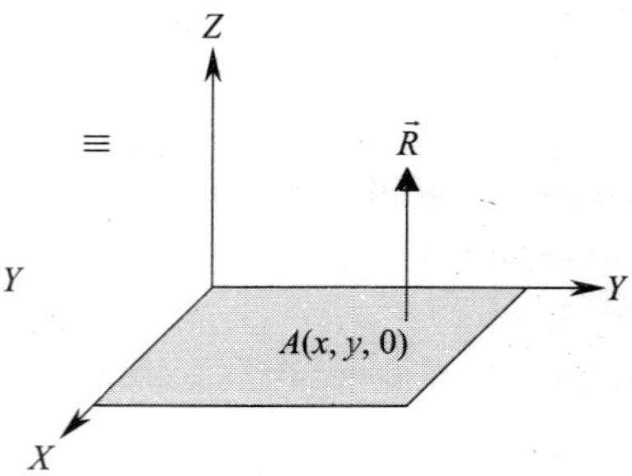

Fig. 4.34(c) Equivalent force

If $(x, y, 0)$ be the location of the resultant then

$$(x\vec{i} + y\vec{j}) \times \vec{R} = \vec{M}_O \tag{4.33}$$

If M_x and M_y be the components of $\vec{M}_O$ then

$$(x\vec{i} + y\vec{j}) \times R\vec{k} = M_x\vec{i} + M_y\vec{j}$$

$$-xR\vec{j} + yR\vec{i} = M_x\vec{i} + M_y\vec{j} \tag{4.34}$$

Note that as the moment of the couple lies on the X–Y plane, its z-component is zero. Equating the coefficients of $\vec{i}$ and $\vec{j}$, we get

$$x = -\frac{M_y}{R} \tag{4.35}$$

and

$$y = \frac{M_x}{R} \tag{4.36}$$

Example 4.21 A uniformly distributed load of intensity w N/m is shown in Fig. 4.35. Replace it by an equivalent force.

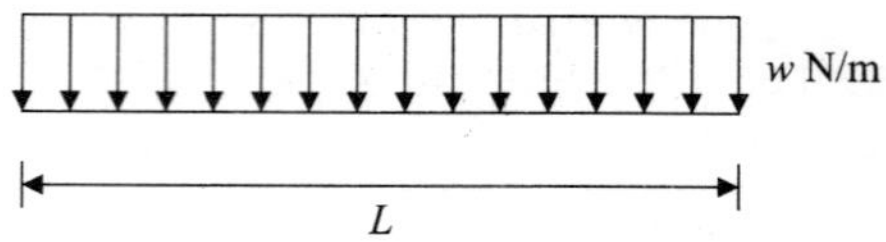

Fig. 4.35

Solution Consider an infinitesimally small load [shaded portion in Fig. 4.35(a)] at a distance x from the left end. Its magnitude is

$$dF = \text{(load/unit length) (length)}$$

$$= (w)dx \tag{a}$$

The moment of this infinitesimally small load about the left end is

$$dM = (dF) \times \text{perpendicular distance}$$

$$= [w dx]x$$

$$= wx dx$$

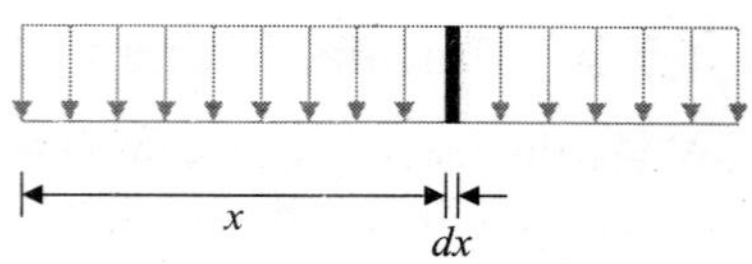

Fig. 4.35(a)

Hence, the total moment of all such loads about the left end is obtained by integrating between the limits,

$$M = \int_0^L dM = \int_0^L wx\,dx$$

$$= \frac{wL^2}{2}$$

Also, from equation (a), we can determine the total load acting as

$$F = \int_0^L dF = \int_0^L w\,dx$$

$$= wL$$

Hence, the location of the equivalent force is given as

$$\bar{x} = \frac{M}{F} = \frac{wL^2/2}{wL} = \frac{L}{2}$$

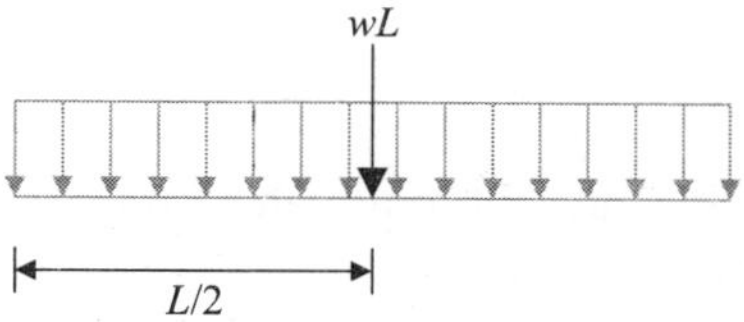

Fig. 4.35(b)

Thus, the equivalent force is wL located at $L/2$ from the left end or at the mid-point of the load distribution as shown in Fig. 4.35(b).

Example 4.22 A uniformly varying load varying from zero to w N/m is shown in Fig. 4.36. Replace it by an equivalent force.

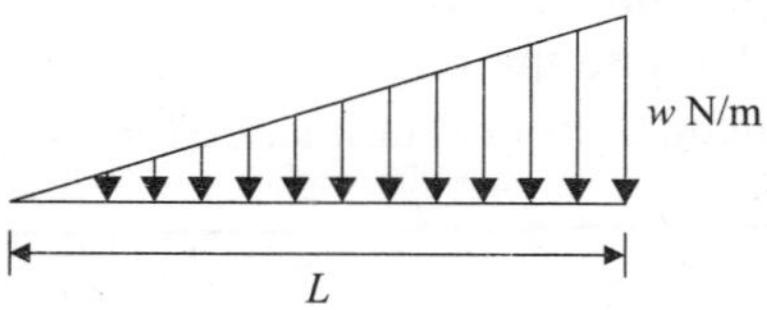

Fig. 4.36

Solution Consider an infinitesimally small load [shaded portion in Fig. 4.36(a)] at a distance x from the left end. The magnitude of the load per unit length at distance x is obtained from similar triangles as:

$$\frac{w}{L}x$$

Therefore, the infinitesimally small load is given as

$$dF = \frac{w}{L}x\,dx \qquad \text{(a)}$$

The moment of this infinitesimally small load about the left end is

$$dM = (dF) \times \text{perpendicular distance}$$

$$= \left[\frac{w}{L}x\,dx\right]x$$

$$= \frac{w}{L}x^2\,dx$$

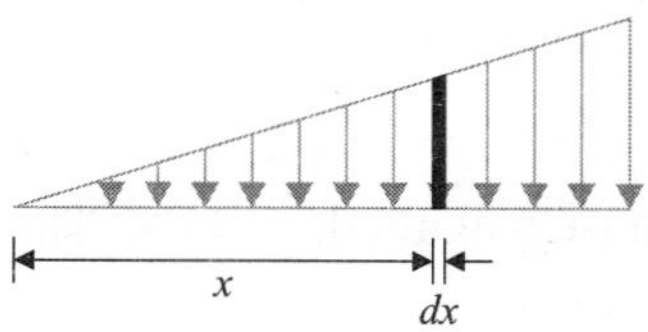

Fig. 4.36(a)

Hence, the total moment of all such loads about the left end is obtained by integrating between the limits.

$$M = \int_0^L dM = \int_0^L \frac{w}{L}x^2\,dx$$

$$= \frac{wL^2}{3}$$

Also, from equation (a), we can determine the total load acting as

$$F = \int_0^L dF = \int_0^L \frac{w}{L} x\,dx$$

$$= \frac{wL}{2}$$

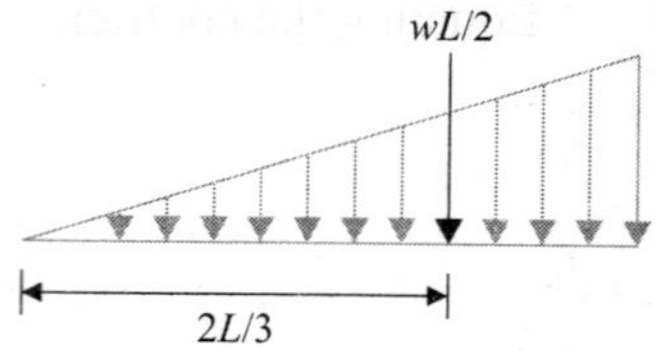

Fig. 4.36(b)

Hence, the location of the equivalent force is given as

$$\bar{x} = \frac{M}{F} = \frac{wL^2/3}{wL/2} = \frac{2}{3}L$$

Thus, the equivalent force is $wL/2$ located at $2L/3$ from the left end of the load as shown in Fig. 4.36(b).

Example 4.23 Four forces are acting perpendicular to the *X–Y* plane as shown in Fig. 4.37. The lines of action of all the forces are parallel to the *Z*-axis. The *x* and *y* coordinates of the points of action of these forces are given in metres. Determine (i) the magnitude of the resultant force, and (ii) the location of the resultant.

Solution Representing each force and the corresponding point of application in vector form, we can determine the resultant of the forces and the moment of the individual forces about the origin. The calculations are summarized in tabular form:

$\vec{r}_i$	$\vec{F}_i$	$\vec{M}_O = \vec{r}_i \times \vec{F}_i$
$4\vec{i} + 3\vec{j}$	$25\vec{k}$	$75\vec{i} - 100\vec{j}$
$6\vec{i} + 4\vec{j}$	$-30\vec{k}$	$-120\vec{i} + 180\vec{j}$
$2\vec{i} + 6\vec{j}$	$10\vec{k}$	$60\vec{i} - 20\vec{j}$
$\vec{i} + 7\vec{j}$	$-40\vec{k}$	$-280\vec{i} + 40\vec{j}$
$\Sigma =$	$-35\vec{k}$	$-265\vec{i} + 100\vec{j}$

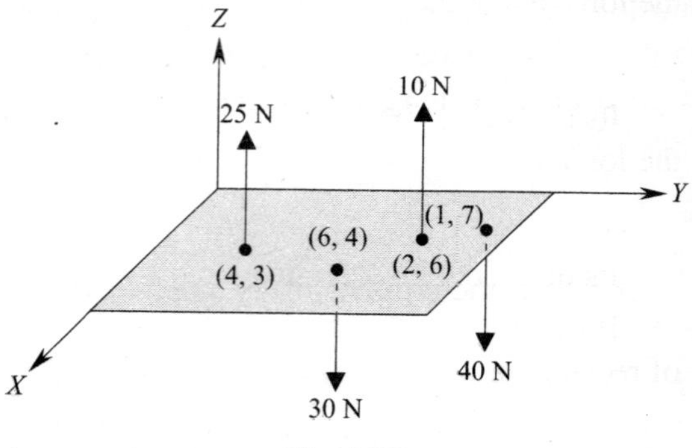

Fig. 4.37

Therefore, the resultant force is

$$\vec{R} = \sum \vec{F}_i = -35\vec{k}$$

whose magnitude is 35 N. The summation of the moments of the individual forces about the origin is

$$\sum \vec{M}_O = -265\vec{i} + 100\vec{j}$$

Since $\vec{R}$ and $\Sigma \vec{M}_O$ are perpendicular, the force–couple system can be reduced further to a single force $\vec{R}$. Let $x\vec{i} + y\vec{j}$ be the new location of the resultant force. Then

$$\vec{r} \times \vec{R} = \sum \vec{M}_O$$

$$(x\vec{i} + y\vec{j}) \times (-35\vec{k}) = -265\vec{i} + 100\vec{j}$$

$$35x\vec{j} - 35y\vec{i} = -265\vec{i} + 100\vec{j}$$

Equating the coefficients of $\vec{i}$ and $\vec{j}$, we get the x and y intercepts of the resultant force:

$$y = \frac{-265}{-35} = 7.57 \text{ m}$$

and
$$x = \frac{100}{35} = 2.86 \text{ m}$$

SUMMARY

Moment of a Force

Forces which are concurrent at a common point tend to translate the body as a whole, while forces that are non-concurrent tend to rotate the body as a whole in addition to translating it. This rotational effect of a force is measured by a physical quantity known as *the moment of the force*. The moment of a force about a point is defined as *a measure of the tendency of the force to rotate a body about that point*. These types of rotational effect of forces acting on bodies is observed while closing or opening a door, turning a pipe using a wrench, and so on.

Moment of a Force about the Origin

If $\vec{r}$ is position vector of the point of application of a force and $\vec{F}$ is the force vector then the moment of the force about the origin is defined as

$$\vec{M}_O = \vec{r} \times \vec{F}$$

Its magnitude is given as the product of the force and the perpendicular distance of line of action of the force from the origin, i.e.,

$$M_O = rF \sin \theta = (r \sin \theta)F = Fd$$

Its direction is perpendicular to the plane formed by $\vec{r}$ and $\vec{F}$.

If the position vector of the point of application of the force and the force vector are given in terms of rectangular components as:

$$\vec{r} = x\vec{i} + y\vec{j} + z\vec{k} \quad \text{and} \quad \vec{F} = F_x\vec{i} + F_y\vec{j} + F_z\vec{k}$$

then the moment vector is given as

$$\vec{M}_O = \vec{r} \times \vec{F}$$

$$= \begin{vmatrix} \vec{i} & \vec{j} & \vec{k} \\ x & y & z \\ F_x & F_y & F_z \end{vmatrix}$$

The moment of a force (acting at a point A) about a point B, other than the origin is given as

$$\vec{M}_B = \vec{r}_{A/B} \times \vec{F} = \overrightarrow{BA} \times \vec{F}$$

$$= (\vec{r}_A - \vec{r}_B) \times \vec{F}$$

Varignon's Theorem (or) Principle of Moments

The theorem states that the moment about a given point O of the resultant of several concurrent forces is equal to the sum of the moments of individual forces about the same point O.

Moment of a Force About an Axis

The component of moment $\vec{M}_B$ of a force about an axis BC is given as the dot product of the unit vector along that axis and the moment of the force about point B.

$$|\vec{M}_{BC}| = \hat{n}_{BC} \cdot \vec{M}_B$$

Coplanar Non-Concurrent Forces

For planar problems, the vector approach can be avoided if we follow the sign convention, treating *clockwise* moments as *negative* and *anti-clockwise* moments as *positive*. Accordingly, moment of a force acting at a point (x, y) about the origin is given as

$$M_O = -yF_x + xF_y$$

Moment of a Couple

When two forces $\vec{F}$ and $-\vec{F}$ having the same magnitude, parallel lines of action and opposite sense act on a body, they are said to form a *couple*. Their only effect is to rotate the body about the axis perpendicular to the plane of the couple. The moment of a couple is defined as the product of the magnitude of one of the forces and the perpendicular distance between them. Its direction is perpendicular to the plane containing the two forces. Mathematically, it is given as

$$|\vec{M}_O| = Fd$$

Force–Couple System

Any force $\vec{F}$ acting on a rigid body can be moved to an arbitrary point O provided that a couple is added whose moment is equal to the moment of $\vec{F}$ about O. This is known as *force–couple system*.

Resultant of Non-Concurrent Forces

The resultant of non-concurrent forces is a single force–couple system, whose force is equal to the summation of all individual forces and the couple is equal to the summation of all individual couples. Mathematically, it is written as

$$\vec{R} = \vec{F}_1 + \vec{F}_2 + \dots + \vec{F}_n = \sum \vec{F}_i$$

$$\vec{M}_O = (\vec{r}_1 \times \vec{F}_1) + (\vec{r}_2 \times \vec{F}_2) + \dots + (\vec{r}_n \times \vec{F}_n) = \sum \vec{r}_i \times \vec{F}_i$$

Reduction of a Force–Couple Into a Single Force

In a force–couple system, the force and couple will be perpendicular to each other when all the individual forces lie on the same plane, i.e., they are *coplanar* or they are *parallel* to each other. In such cases, the resultant force–couple system can be reduced to a single force by moving the force to a new line of action such that its moment about O is equal to the moment of the couple.

Coplanar Non-Concurrent Forces

If (x, y) be the coordinates of the resultant such that its moment about the origin is equal to the moment of the couple, then x and y intercepts of the resultant are given as

$$x = \frac{\sum M_O}{\sum F_y} \quad \text{and} \quad y = -\frac{\sum M_O}{\sum F_x}$$

Parallel Forces

If $(x, y, 0)$ be the coordinates of the location of the resultant of forces parallel to the Z-axis such that the moment of the resultant about the origin is equal to the moment of the couple then

$$(x\vec{i} + y\vec{j}) \times R\vec{k} = M_x\vec{i} + M_y\vec{j}$$

$$x = -\frac{M_y}{R} \quad \text{and} \quad y = \frac{M_x}{R}$$

EXERCISES

Objective-type Questions

1. State in which of the following actions the applied force *does not* produce a moment.
   (a) Pedalling a bicycle (b) Stretching a spring
   (c) Opening a water tap (d) Opening a door
2. The magnitude of the moment of a force about a point is equal to the product of the force and perpendicular distance between ___________ and the point.
   (a) the magnitude of the force (b) the line of action of the force
   (c) the sense of the force (d) the point of application of the force
3. The unit of moment is
   (a) N/m (b) N.s (c) N.m (d) N/m^2
4. State which of the following statements is true?
   Two forces can produce the same moment about a point
   (a) only when the two forces are equal
   (b) only when the moment arms of the forces are equal
   (c) when the two forces are concurrent
   (d) when the product of the force and the moment arm are equal
5. Varignon's theorem is applicable only when the forces are
   (a) coplanar (b) concurrent (c) non-concurrent (d) parallel
6. Which of the following is a scalar quantity?
   (a) Moment of a force about the origin
   (b) Moment of a force about a point other than the origin
   (c) Moment of a force about an axis
   (d) Moment of a couple
7. Anti-clockwise moments are considered positive
   (a) merely for the sake of convenience
   (b) since they point along the $-Z$ direction
   (c) since they point along the $+Z$ direction
   (d) none of the above

8. The magnitude of the moment is ____________ when a force is applied perpendicular to a lever.
   (a) maximum (b) minimum (c) zero (d) negative
9. The magnitude of the moment is maximum when a force is applied ____________ the lever.
   (a) parallel to (b) inclined to (c) perpendicular to (d) at the hinge of
10. State in which of the following actions is a couple *not* required:
    (a) Opening a pipe using a wrench (b) Steering a vehicle while taking a turn
    (c) Opening a water tap (c) Tightening a screw with a screw driver
11. Moment of a couple is a
    (a) free vector (b) fixed vector (c) sliding vector (d) null vector
12. A force–couple system can be reduced to a single force only when the resultant force and couple are ____________ to each other.
    (a) parallel (b) perpendicular (c) inclined at 45° (d) inclined at 135°
13. Which of the following system of forces *cannot* be reduced to a single force?
    (a) Non-concurrent forces in space (b) Non-concurrent forces in a plane
    (c) Parallel forces in space (d) Parallel forces in a plane

Answers

1. (b) 2. (b) 3. (c) 4. (d) 5. (b) 6. (c) 7. (c) 8. (a)
9. (c) 10. (a) 11. (a) 12. (b) 13. (a).

Short-answer Questions

1. Define moment of a force. What are its unit and dimension?
2. Discuss the rotational effect of force on a body with suitable examples.
3. Define moment centre, moment arm and axis of moment.
4. The magnitude of the moment of a force remains the same irrespective of the point of application of the force along its line of action. Discuss.
5. State and prove Varignon's theorem.
6. Distinguish between moment about a point and moment about an axis.
7. What are the sign conventions for clockwise and anti-clockwise moments in planar problems? Explain.
8. Define a force couple and moment of a couple.
9. Distinguish between moment of a force and moment of a couple.
10. Explain how a system of non-concurrent forces can be reduced to an equivalent force–couple system.
11. Under what conditions can a force–couple system be reduced to a single force. Discuss.

Numerical Problems

4.1 A steel column though designed for an axial load, due to misalignment while assembling is subjected to an eccentric load as shown in Fig. E.4.1. Determine the moment of the load about the midpoint *O*.

Ans. 60 N.m anti-clockwise

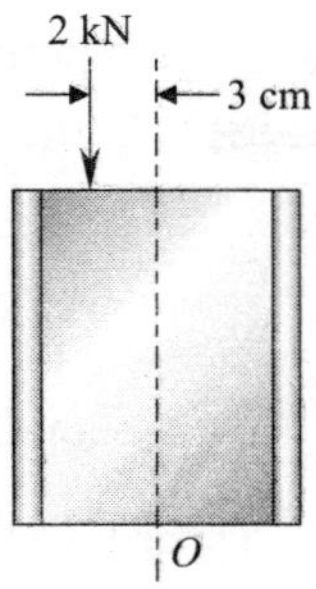

Fig. E.4.1

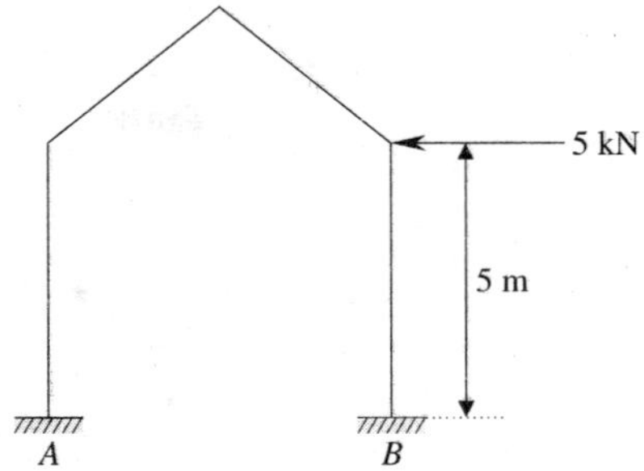

Fig. E.4.2

4.2 A portal frame is subjected to a horizontal wind load as shown in Fig. E.4.2. Determine its moment about the base *A*.

Ans. 25 kN.m anti-clockwise

4.3 In the foot lever used in spot welding, a vertical force of 60 N is applied as shown in Fig. E.4.3. Determine the moment of this force about an axis passing through the point *O* and normal to the plane of the paper.

Ans. 31.8 N.m clockwise

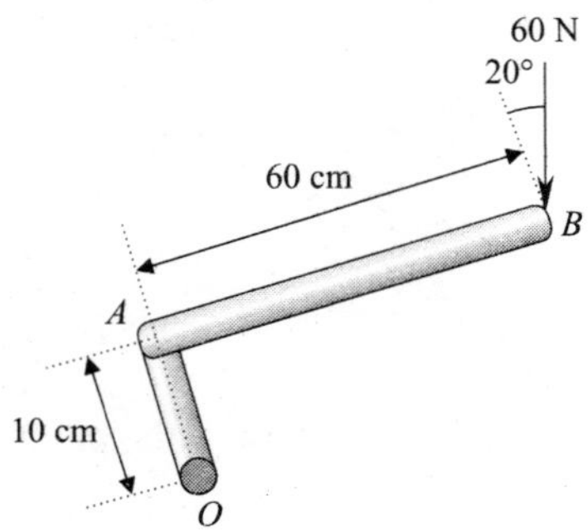

Fig. E.4.3

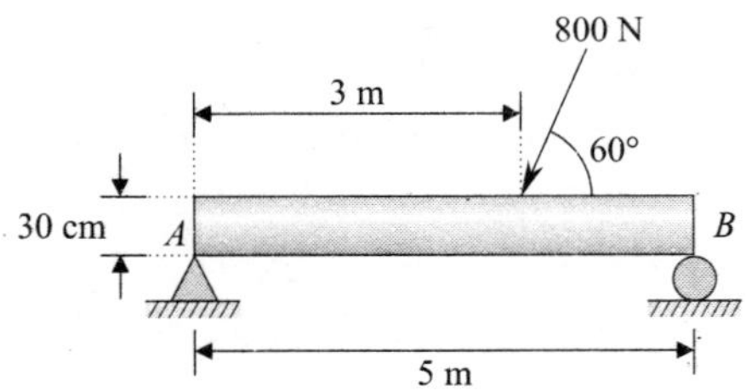

Fig. E.4.4

4.4 In Fig. E.4.4, determine the moment of the 800 N force acting on a simply supported beam *AB* about supports *A* and *B*.

Ans. 1958.5 N.m (clockwise), 1505.6 N.m (anti-clockwise)

4.5 A force of 500 N is applied as shown in Fig. E.4.5. Determine the moment of this force (i) about point *A*, and (ii) about origin *O*.

Ans. (i) 250 N.m clockwise; (ii) 899.5 N.m clockwise

4.6 A 6 m steel column to be erected is pulled up as shown in Fig. E.4.6 using a cable by applying a force *T* of 1500 N. Determine the moment of the tension about point *A* on the ground.

Ans. 8.69 kN.m

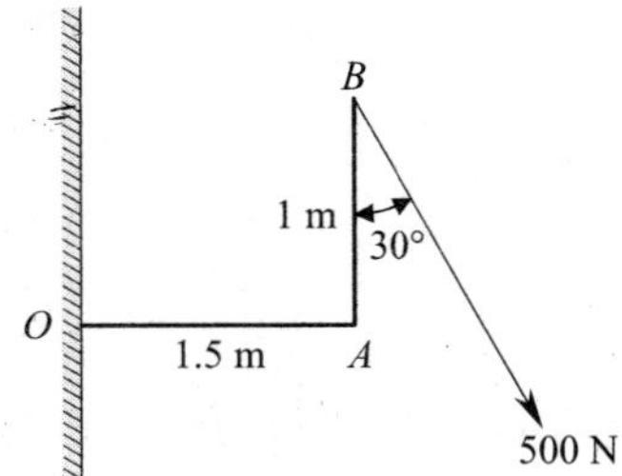

Fig. E.4.5

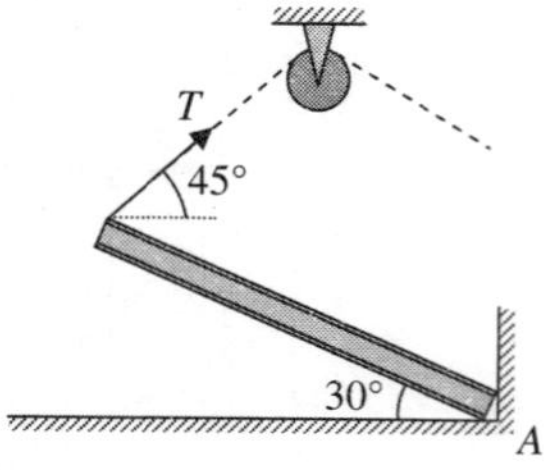

Fig. E.4.6

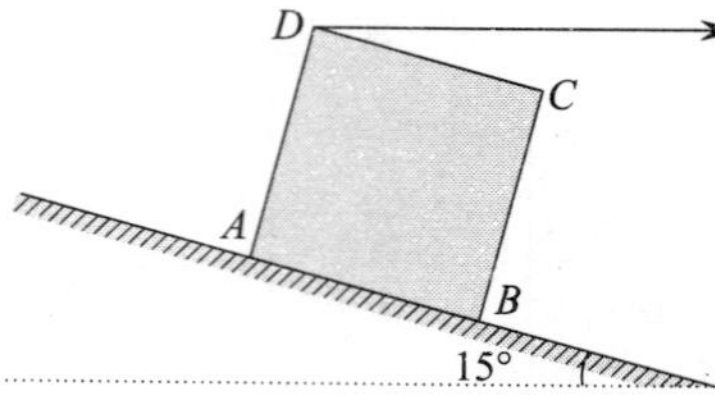

Fig. E.4.7

4.7 A square box *ABCD* of side *a* is resting on an inclined plane as shown in Fig. E.4.7. It is to be pulled down by toppling. A horizontal force *F* is applied at the corner *D*. Determine the moment of this force about the corner *B*.

Ans. 1.225 Fa

4.8 A jet of water strikes centrally a flat plate *AB* of height 1.5 m hinged at its upper edge *A* as shown in Fig. E.4.8. Due to the force exerted by the jet, the plate swings through 30°. Determine the moment of this force about the hinge at *A*. (Hint: even though the jet of water strikes centrally at the initial position, at the inclined position it will not be so).

Ans. 0.75 F

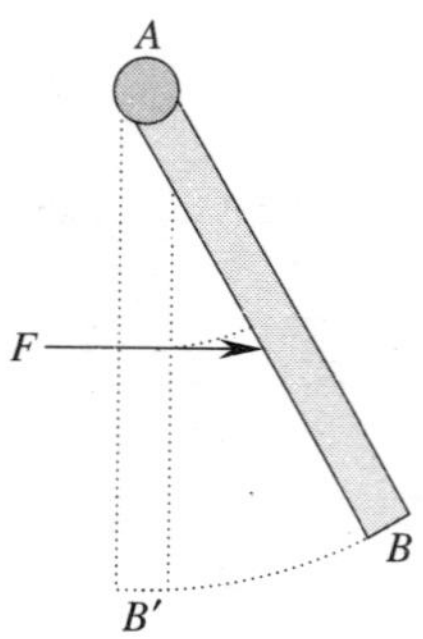

Fig. E.4.8

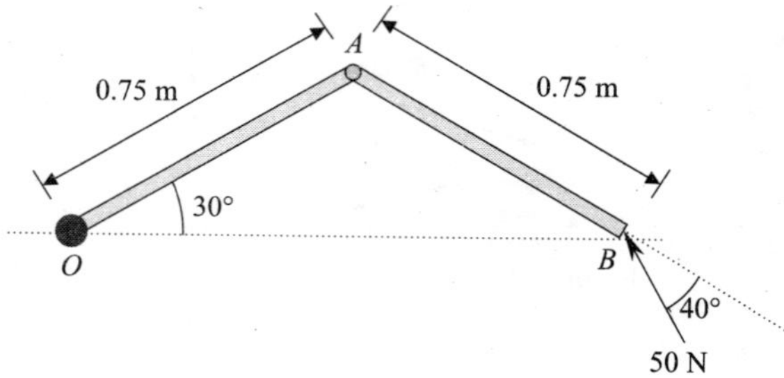

Fig. E.4.9

4.9 The top view of an iron gate, which has two halves hinged together at *A*, is shown in Fig. E.4.9. Determine the moment of a 50 N force exerted at the point *B* about *O*.

Ans. 61.03 N.m

4.10 A force of 1 kN acts at the joint *F* of a fink truss as shown in Fig. E.4.10. Determine the moment of this force about supports (i) *A*, and (ii) *B*. The construction of the truss is such that *AD* = *DE* = *EB* = *DC* = *EC* = *a*, where *a* is in metres; *GD* is perpendicular to *AC* and *EF* is perpendicular to *BC*.

Ans. (i) 1.732*a* kN.m clockwise; (ii) 0.866*a* kN.m anti-clockwise

4.11 An adjustable roof truss in a stadium is suspended by cables as shown in Fig. E.4.11. If the tension in the cable *CE* is 4000 N, when the truss is held in a position such that *AB* is horizontal, determine the moment of this tension about the point *A*.

Ans. 20.74 kN.m

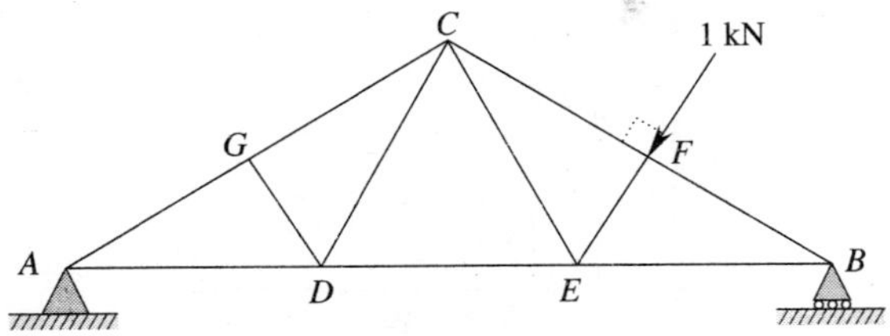

Fig. E.4.10

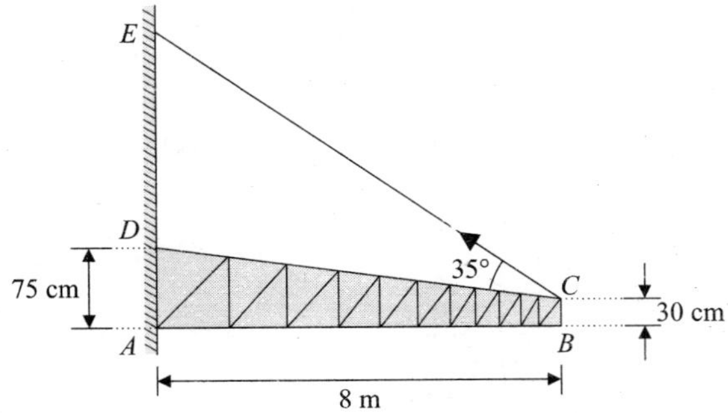

Fig. E.4.11

4.12 A part of a cable-stayed bridge is shown in Fig. E.4.12. Determine the moment of the tensions in cables C_1, C_2 and C_3 about the point D, the base of the pillar. The respective tensions in the cable are 2500 N, 2000 N and 1000 N.

Ans. 18.62 kN.m, 11.74 kN.m, 4 kN.m

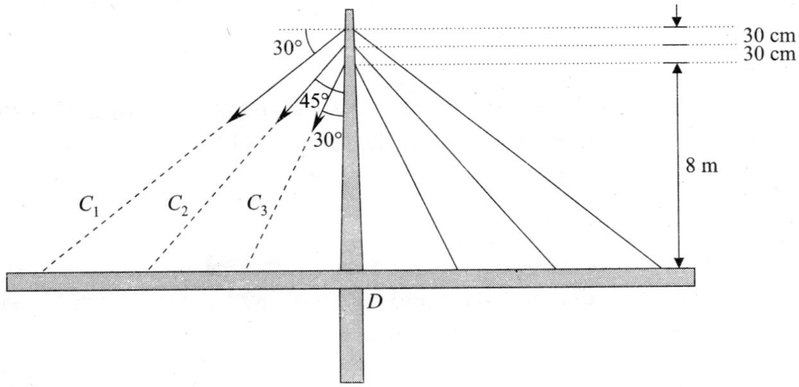

Fig. E.4.12

4.13 A force of magnitude 500 N is applied vertically downwards at B of a structural member lying in the X–Y plane as shown in Fig. E.4.13. Determine its moment about the origin O.

Ans. $500(-3\vec{i} + 4\vec{j})$

4.14 In Fig. E.4.14, find the moment of the force $\vec{F}$ of magnitude 200 N (i) about the origin O, (ii) about the point A, and (iii) about the X-axis.

Ans. $\dfrac{-600\vec{i} + 400\vec{j} + 600\vec{k}}{\sqrt{43}}$, $\dfrac{-600\vec{i} + 4000\vec{j} - 3000\vec{k}}{\sqrt{43}}$, $\dfrac{-600}{\sqrt{43}}$ N.m

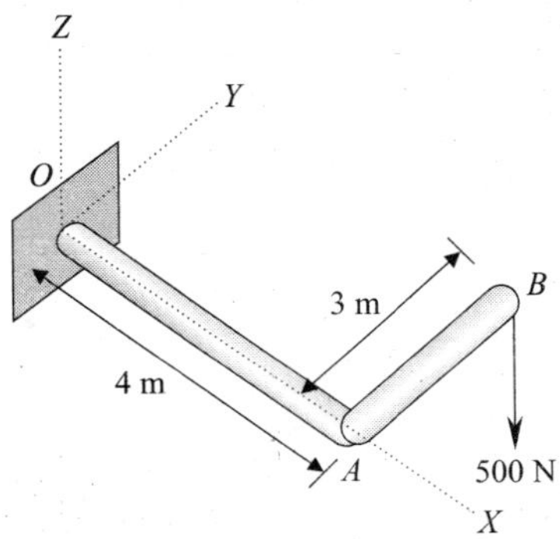

Fig. E.4.13

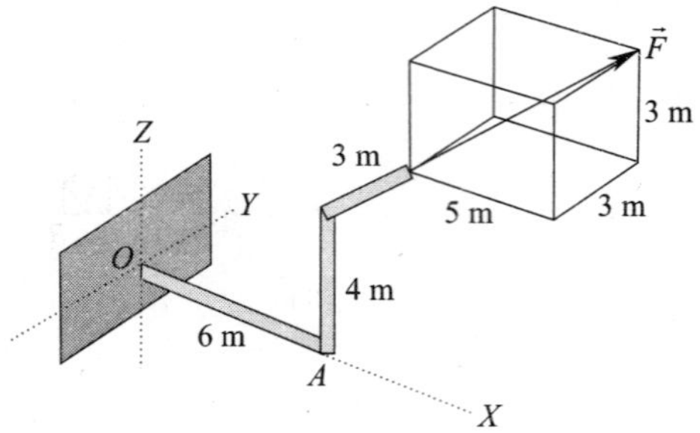

Fig. E.4.14

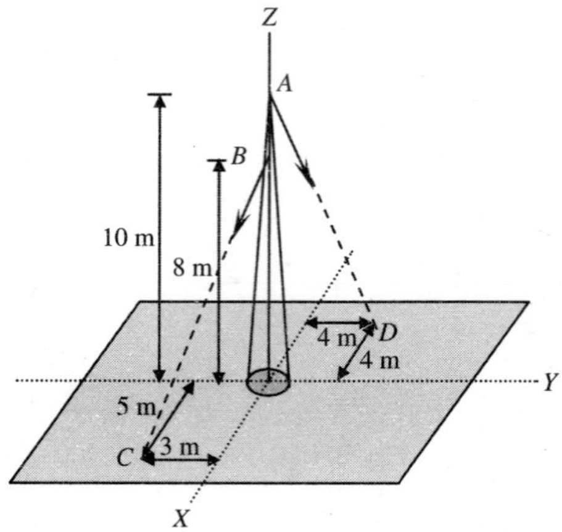

Fig. E.4.15

4.15 In Fig. E.4.15, find the moment of the tensions in cables supporting a tower about the base of the tower. Take tensions in the cables *AD* and *BC* to be respectively 100 N and 150 N.

Ans. $\dfrac{-4000\vec{i} - 4000\vec{j}}{\sqrt{132}}, \dfrac{3600\vec{i} + 6000\vec{j}}{\sqrt{98}}$

4.16 A force of magnitude 200 N acts at a point *A* (2, 3, 4) and passes through a point *B* (5, 6, 7), the coordinates being given in metres. What is the moment of this force about the origin and about a point *C* (4, 5, 5)?

Ans. $\dfrac{-200\vec{i} + 400\vec{j} - 200\vec{k}}{\sqrt{3}}, \dfrac{-200\vec{i} + 200\vec{j}}{\sqrt{3}}$

4.17 A 100 N force acts at a point (5, 2, 4) m and passes through a point in space, whose coordinates are (–2, 4, 3) m. Find the moment of this force about the *Z*-axis.

Ans. $\dfrac{2400}{\sqrt{54}}$ N.m

4.18 A force of 100 N acts at a point *A* as shown in Fig. E.4.18. Determine the magnitude of the force acting at *B*, which would produce an equal and opposite moment of the force acting at *A* about the origin.

Ans. 214.5 N

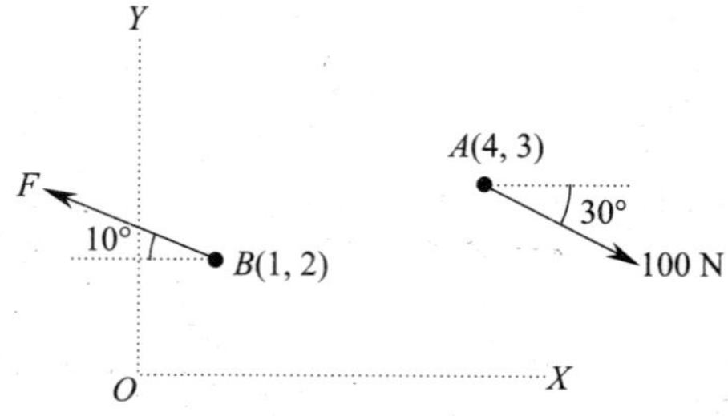

Fig. E.4.18

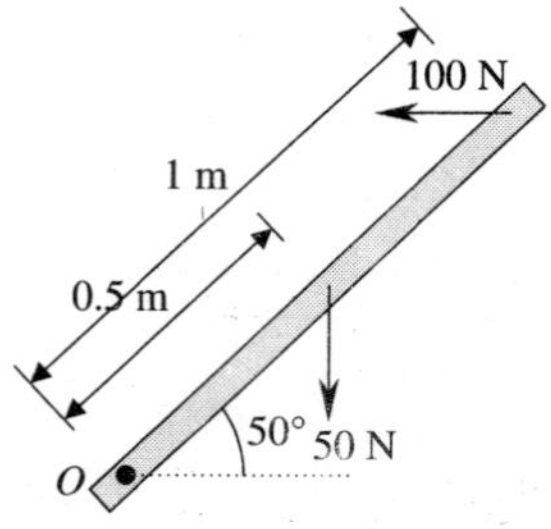

Fig. E.4.19

4.19 A horizontal force of 100 N is applied at the free end of a lever and a vertical force of 50 N is applied at its midpoint as shown in Fig. E.4.19. Find the resultant moment of the forces about the hinge *O*.

Ans. 60.5 N.m (anti-clockwise)

4.20 While taking a left turn, the driver of an automobile exerts 20 N forces of equal magnitude on the steering (of outer radius of 15 cm) as shown in Fig. E.4.20. Determine the moment of the couple about the axis of rotation.

Ans. 6 N.m (anti-clockwise)

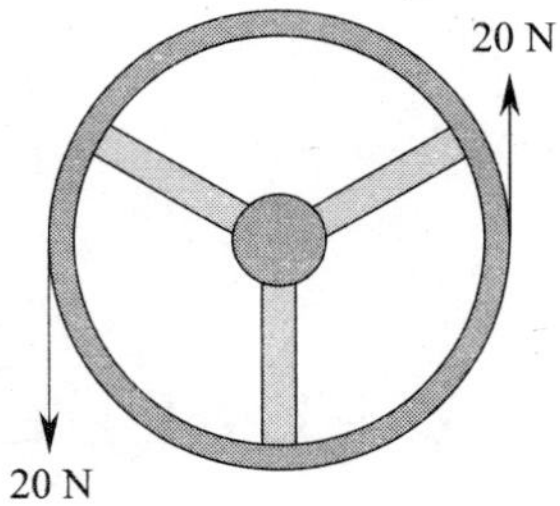

Fig. E.4.20

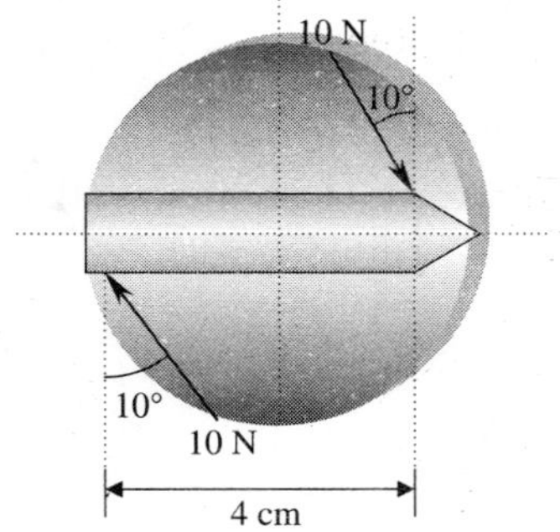

Fig. E.4.21

4.21 While turning the knob of a gas stove, forces are exerted by the fingers as shown in Fig. E.4.21. Determine the moment of the forces about the axis of rotation.

Ans. 0.394 N.m

4.22 A force of magnitude 100 N and a couple of 50 N.m are applied on a lever as shown in Fig. E.4.22. Find the resultant moment about the hinge *O*.

Ans. 18.4 N.m (anti-clockwise)

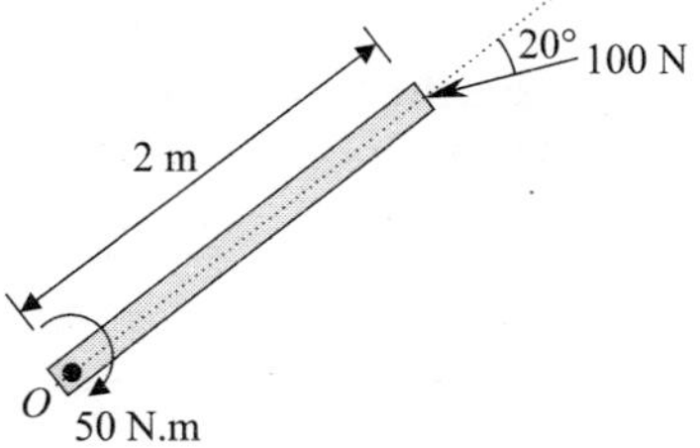

Fig. E.4.22

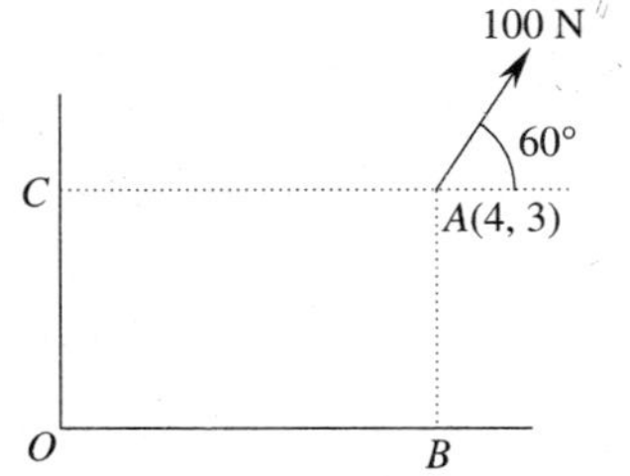

Fig. E.4.23

4.23 Replace the force shown in Fig. E.4.23 by an equivalent force–couple system (i) at the origin *O*, (ii) at the point *B*, and (iii) at the point *C*.

Ans. (i) 100 N force at 60° to *X*-axis, 196.4 N.m anti-clockwise, (ii) 100 N force at 60° to *X*-axis, 150 N.m clockwise, (iii) 100 N force at 60° to *X*-axis, 346.4 N.m anti-clockwise

4.24 Replace the system of forces shown in Fig. E.4.24 by an equivalent force.

Ans. 30 N acting vertically downwards at 3.33 m from left

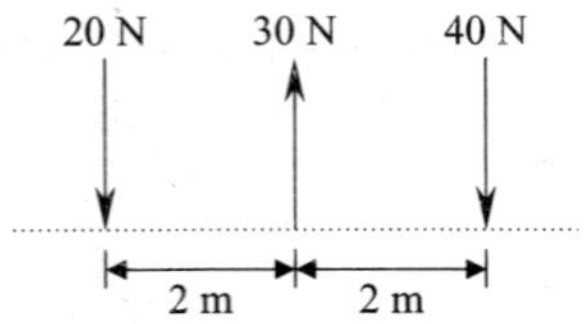

Fig. E.4.24

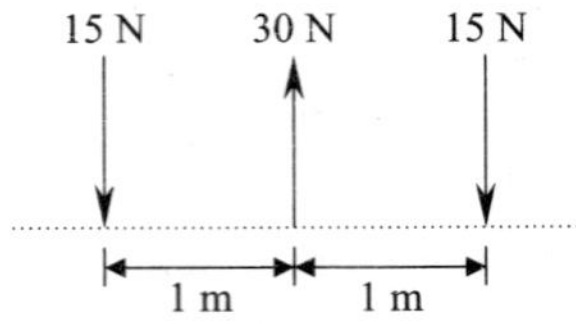

Fig. E.4.25

4.25 Replace the system of forces shown in Fig. E.4.25 by an equivalent force.

Ans. 30 N acting vertically upwards at the right extreme

4.26 A beam *AB* of span *L* is subjected to a concentrated load *P* and a couple *M* as shown in Fig. E.4.26. Replace the system of forces by an equivalent force–couple system at supports *A* and *B*.

Ans. (i) $P, M + \frac{PL}{4}$ (clockwise), (ii) $P, \frac{3PL}{4} - M$ (anti-clockwise)

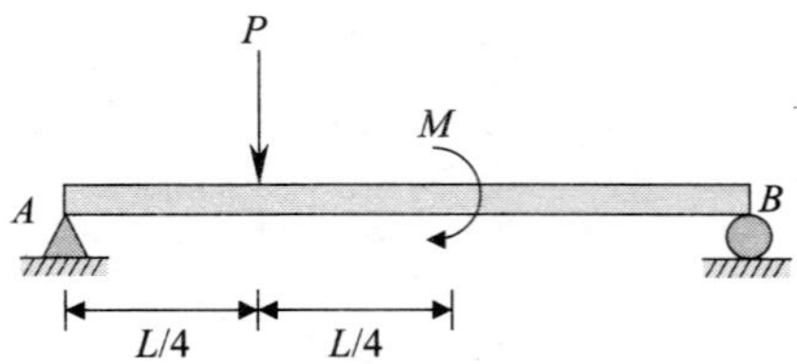

Fig. E.4.26

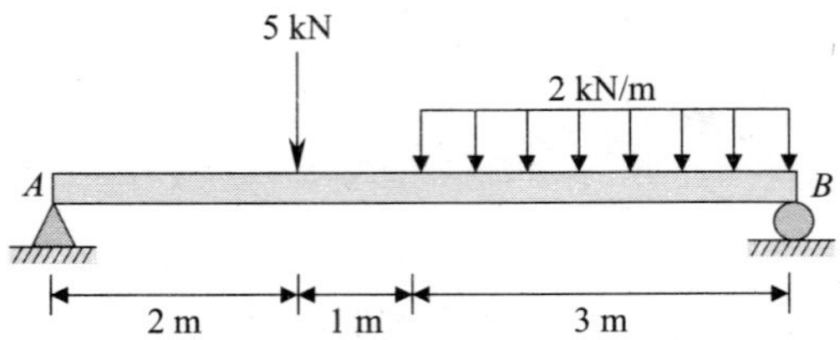

Fig. E.4.27

4.27 A beam *AB* of 6 m span is subjected to a concentrated load and a distributed load as shown in Fig. E.4.27. Replace the system of forces by an equivalent force–couple system at supports *A* and *B*.

Ans. (i) 11 kN, 37 kN.m (clockwise), (ii) 11 kN, 29 kN.m (anti-clockwise)

4.28 Replace the system of forces shown in Fig. E.4.28 by (i) an equivalent force–couple system at the origin *O*. The size of each square in the mesh is 10 cm × 10 cm.

Ans. $F_x = 98.8$ N, $F_y = 202.4$ N, $M = 2.98$ N.m anti-clockwise

4.29 Forces are acting on a mesh as shown in Fig. E.4.29. Replace the system of forces by an equivalent force–couple system at the corner *A*. One square is 1 cm × 1 cm.

Ans. $F_x = 10.61$ N, $F_y = 4.04$ N, $M_A = -30.81$ N.cm

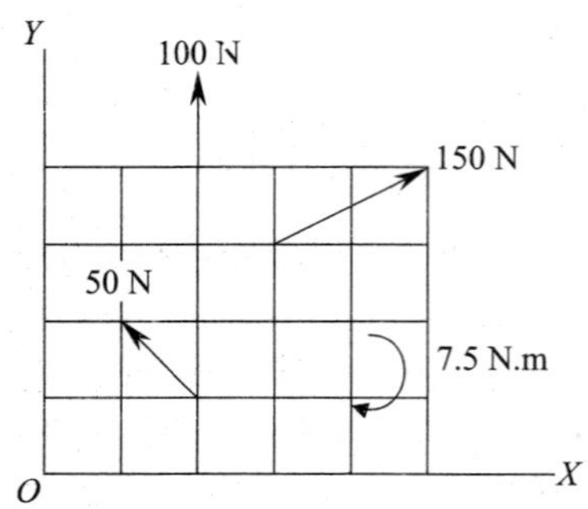

Fig. E.4.28

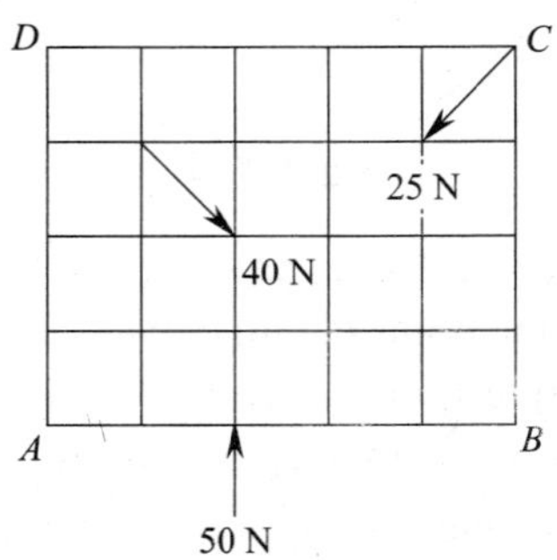

Fig. E.4.29

4.30 A pulley of 30 cm diameter mounted on a shaft is acted upon by its own weight of 300 N and tensions on either side of the belt passing over it. Determine the equivalent force acting on the pulley. Refer Fig. E.4.30.

Ans. 1824.8 N, $\bar{x} = 10$ cm, $\bar{y} = 1.67$ cm

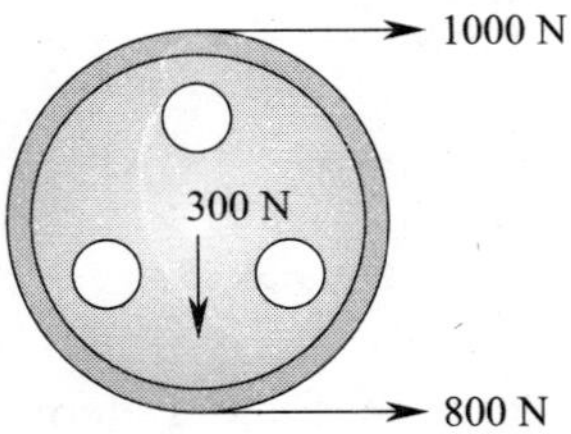

Fig. E.4.30

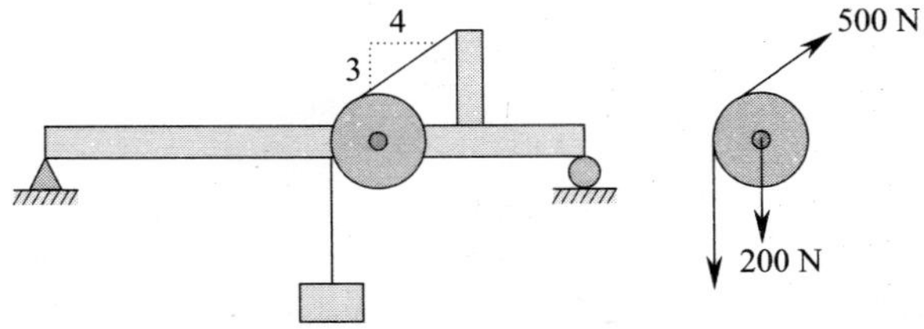

Fig. E.4.31

4.31 A frictionless pulley mounted on a beam as shown in Fig. E.4.31 is subjected to forces: its own weight of 200 N and tensions of 500 N in the rope on either side. Replace the system of forces by an equivalent force.

Ans. $400\sqrt{2}$ N acting at the axis of the pulley at 45° down towards right

4.32 A uniformly varying load varying from w_1 to w_2 N/m is shown in Fig. E.4.32. Replace it by an equivalent force.

Ans. $\dfrac{w_1 + w_2}{2}L, \; \dfrac{L}{3}\left[\dfrac{w_1 + 2w_2}{w_1 + w_2}\right]$ from the left side

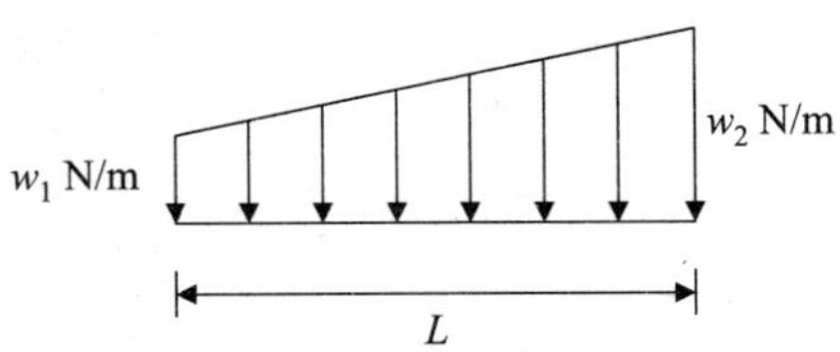

Fig. E.4.32

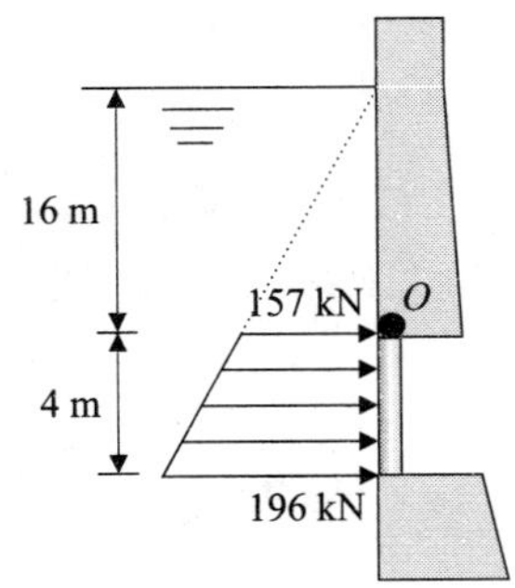

Fig. E.4.33

4.33 Using the answer of the previous problem, determine the moment of the hydrostatic force on the dam gate about the hinge *O*. Refer Fig. E.4.33.

Ans. 1464 kN.m

4.34 Reduce the system of forces shown in Fig. E.4.34 to an equivalent force and determine its magnitude and location with respect to *A*.

Ans. 16.91 kN, 4.15 m from *A*

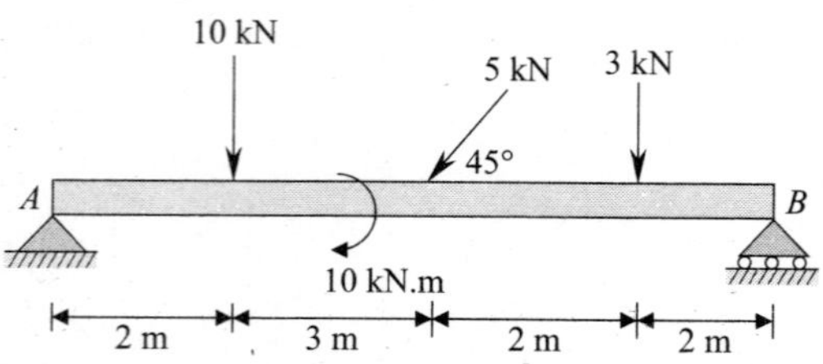

Fig. E.4.34

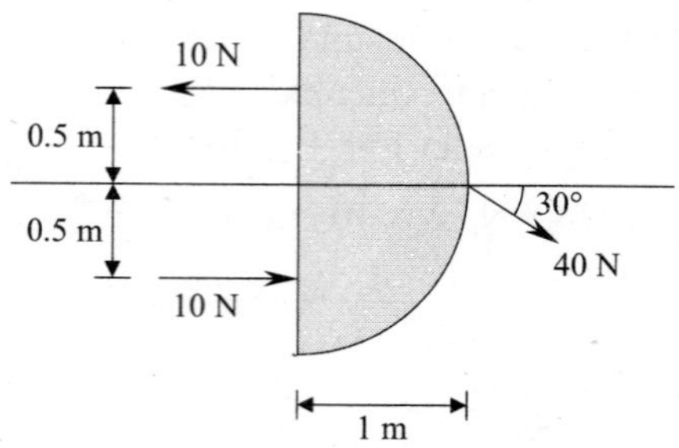

Fig. E.4.35

4.35 Reduce the system of forces shown in Fig. E.4.35 to an equivalent force. Determine its magnitude, and x and y intercepts.

Ans. $\Sigma F_x = 34.64$ N; $\Sigma F_y = -20$ N; $x = 0.5$ m, $y = 0.29$ m

4.36 In Fig. E.4.36, replace the tension T in the cable by an equivalent force–couple system at O. Take the magnitude of the tension to be 100 N.

Ans. $\dfrac{-400\vec{i} + 500\vec{j} - 800\vec{k}}{\sqrt{105}}, \dfrac{-4000\vec{i} - 3200\vec{j}}{\sqrt{105}}$

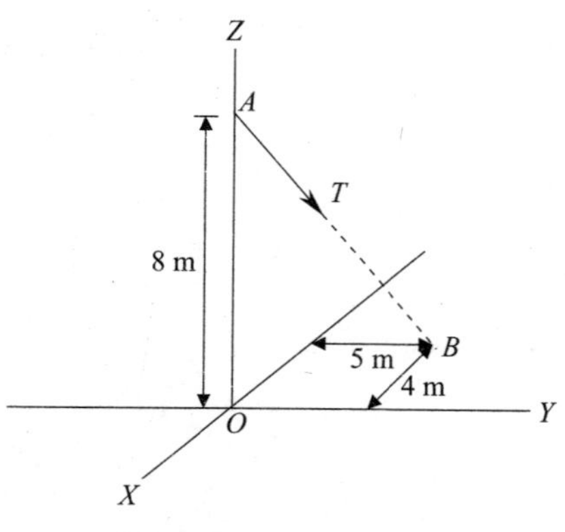

Fig. E.4.36

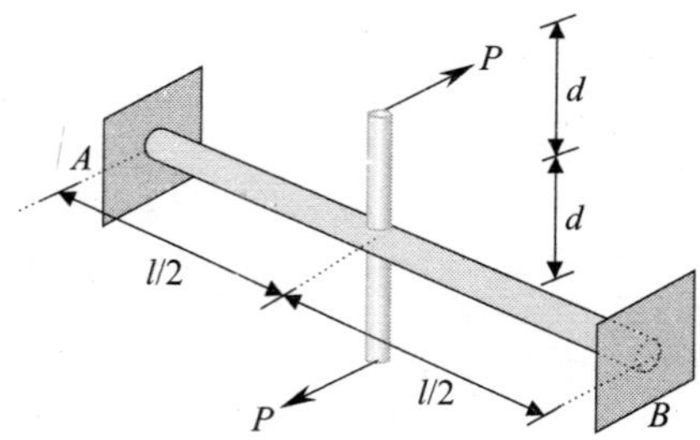

Fig. E.4.37

4.37 A shaft AB supported at both the ends is subjected to forces as shown in Fig. E.4.37. Replace the system of forces by an equivalent force–couple system at A and at B.

Ans. $2Pd$ directed along BA, $2Pd$ directed along BA

4.38 Three forces are acting perpendicular to the X–Y plane as shown in Fig. E.4.38. The lines of action of all the forces are parallel to the Z-axis. The x and y coordinates of the point of action of these forces are given in metres. Determine (i) the magnitude of the resultant force, and (ii) location of the resultant.

Ans. (i) $-15\vec{k}$, (ii) $x = 4$ m, $y = 10$ m

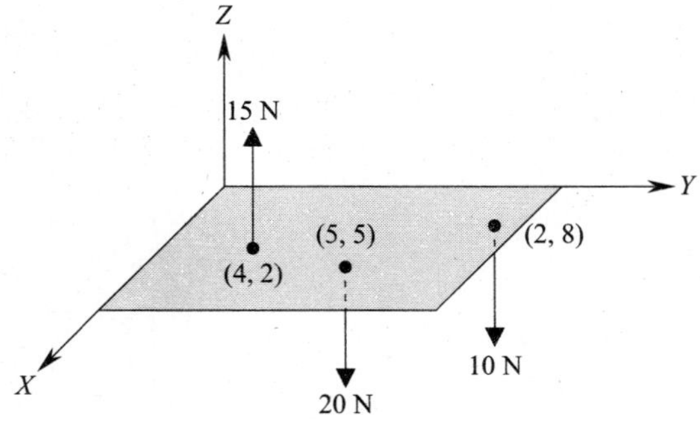

Fig. E.4.38

5 Equilibrium of a System of Forces

5.1 INTRODUCTION

In the previous two chapters, we discussed various systems of forces and the methods to find their resultants. If the forces in the system are *concurrent*, they can be replaced by a resultant force acting at the point of concurrency. The effect of such a force system is to *translate* the body in the direction of the resultant. If the forces are *non-concurrent*, they can be replaced by a resultant force acting at a common point and a moment about the same point. The effect of such a force system would be to *translate* and *rotate* it as well.

In this chapter, we will discuss a *special* case that arises when the resultant **force** and **moment** turn out to be **zero**. If the resultant **force** of a system of forces is **zero,** the body will remain at *rest* or move with *constant velocity*, if it was already moving with constant velocity, i.e., its acceleration will be zero; and if the **moment** is also **zero**, then there will not be any rotational motion. Such a condition is called **static equilibrium**.

To analyze a body in equilibrium, we must first identify the forces and moments acting on it and check if the resultant force and moment are zero or not. In the previous two chapters, the system of forces acting on a body was given **explicitly** and hence, we could determine its resultant. However, in real situations, one must determine this force system in order to check the equilibrium condition. If one can master the way to determine these forces then one can solve any type of problem in mechanics. Therefore, the student is advised to go through the following section very carefully as this forms the basis for solving problems in statics as well as dynamics.

5.2 FREE-BODY DIAGRAM

The system of forces acting on a body tends to translate it or rotate it or do both. In general, the translational and rotational motions can be resolved into **six** components, namely, **three translations** along X, Y and Z directions, and **three rotations** about X, Y and Z directions. To represent these, we require six independent variables—three for translational motion and three for rotational motion. Hence, we say that bodies have **six degrees of freedom**. Whenever a body is restricted to move in any of these directions due to its attachments with the surroundings, the body is said to be **constrained**. To investigate the equilibrium of a constrained body, we must first isolate it from all its attachments with its surroundings.

A body thus isolated from all other members, which are connected to it or in contact with it, is called a **free-body**. A sketch of the isolated body, showing all the **forces** acting on it by vectors is called a **free-body diagram**. Normally, two kinds of forces must be shown to act on a free-body; they are **external forces** or **active forces**, and **reactive forces** replacing the attachments and supports.

5.2.1 External Forces or Active Forces

The *external or active forces* acting on the free body depend upon the *environment* in which it is placed. For example, if a body is placed in a gravitational, magnetic or electrical field then a force of *attraction* or *repulsion* acts as the case may be. The study of magnetic and electrical forces involves electrons, whereas in our study, we deal only with gross bodies. Hence, in our study we will limit ourselves only to force due to gravity.

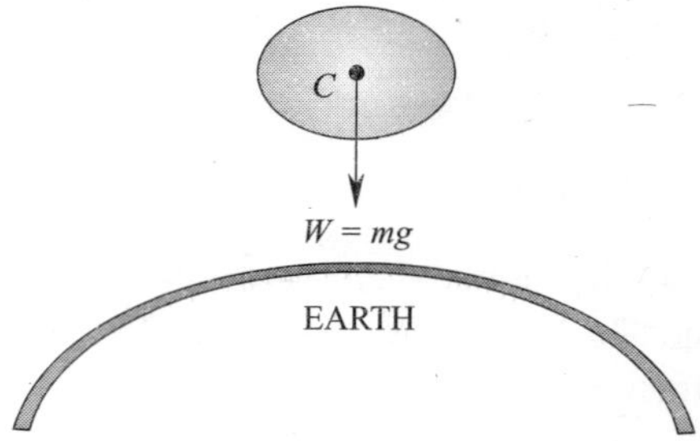

Fig. 5.1 Force of gravity

Consider a body left in a uniform gravitational field; then it is attracted by the earth. Hence, we say a **force of gravity** acts on the body and this force always acts vertically downwards towards the centre of the earth. If m is the mass of the body and g is the acceleration due to gravity then **gravitational force** (also termed as **weight** of the body) is given as $W = mg$. This force is placed at the **centre of gravity** (C) of the body. [The centre of gravity is a point in the body at which the entire weight is assumed to be concentrated.]

Hence, the first force that must be shown acting on the free-body diagram of any solid body must be its own weight acting vertically downwards and placed at its centre of gravity. In some cases, the weight of the body may be **negligible** in comparison with other bodies in the system, and in that case, this force of gravity or its weight can be neglected.

Corollary By Newton's law of gravitation, we know that each body is attracted to another with a force known as *force of gravity*. Hence, in the above case, the earth is also attracted by the body under consideration. But the earth being so massive, this force is not expected to impart a noticeable acceleration to the earth.

5.2.2 Reactive Forces

Reactive forces arise whenever the motion of a body in a particular direction is *restrained* by another body attached to it or in contact with it. By Newton's third law of motion, we know that when one body exerts a force on a second body, the second body always exerts a force on the first. Furthermore, we know that these forces are equal in magnitude but opposite in direction. A single isolated force is, therefore, an *impossibility*. Also, the action and reaction forces lie along the line joining the bodies.

There are many types of reactive forces that act upon a body due to its attachment with its surroundings. These are discussed below in detail.

(A) Normal Reaction Normal reaction arises whenever two bodies with *smooth* surfaces are in *contact* with each other. For example, consider a block resting on a smooth horizontal table as shown in Fig. 5.2(a). The block is *free* to move along the table, but its motion downwards, due to pull of gravity, is *restrained* by the table. Hence, we say that the table exerts a force R vertically upwards (opposite to the direction of motion of the body) to restrain its motion downwards.

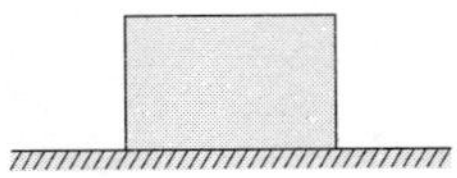

Fig. 5.2(a) A block resting on a smooth table

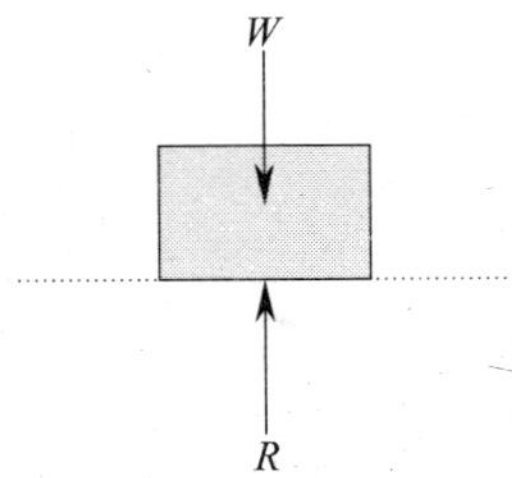

Fig. 5.2(b) Free-body diagram of the block

Therefore, when we draw the free-body diagram of the block, we must show this force R exerted by the table on the block [refer Fig. 5.2(b)]. Since this force acts normal to the plane of the table, it is called **normal reaction**.

We should note that by Newton's third law of motion, the block also tries to exert a force on the table that is equal and opposite to the normal reaction. As we are not interested in the free-body diagram of the table, we do not consider this force.

Next, consider a block lying on a *smooth* inclined plane [Fig. 5.3(a)]. It is free to move along the plane, but it is restrained from moving normal *into* the plane. Hence, we say a normal reaction R is exerted by the plane on the block. Similar reactions will exist even if a ball is resting on a horizontal plane [Fig. 5.3(b)] or in a trough [Fig. 5.3(c)].

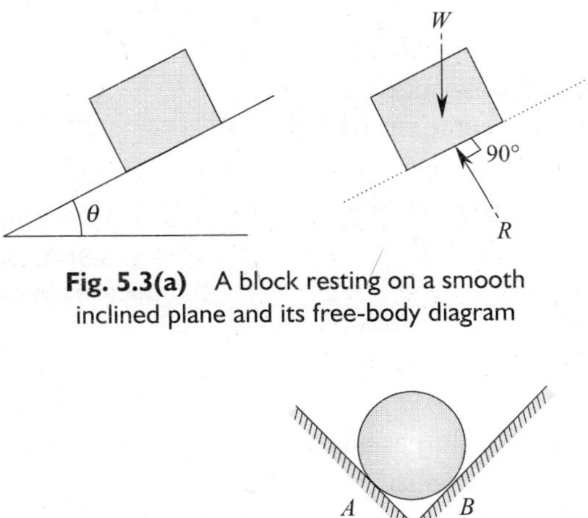

Fig. 5.3(a) A block resting on a smooth inclined plane and its free-body diagram

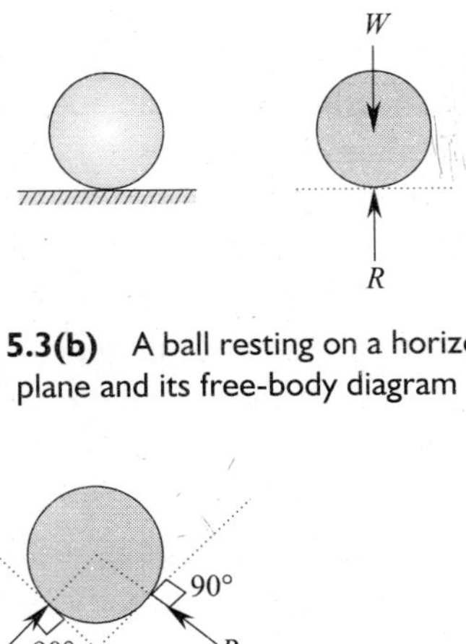

Fig. 5.3(b) A ball resting on a horizontal plane and its free-body diagram

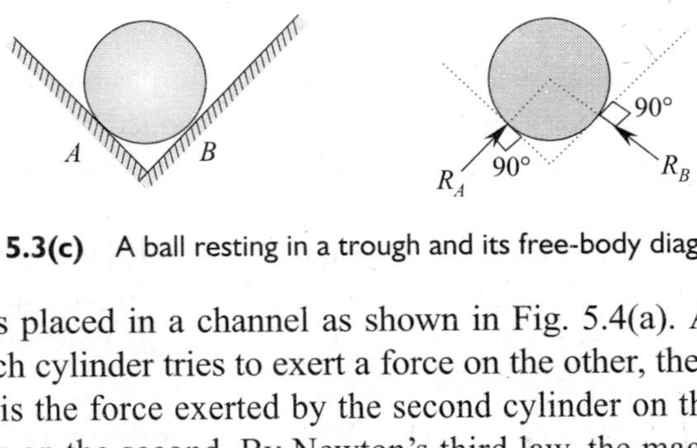

Fig. 5.3(c) A ball resting in a trough and its free-body diagram

Consider two cylinders placed in a channel as shown in Fig. 5.4(a). As the two cylinders are in contact with each other, each cylinder tries to exert a force on the other, the two forces being equal but opposite to each other. R_{21} is the force exerted by the second cylinder on the first, and R_{12} is the force exerted by the first cylinder on the second. By Newton's third law, the magnitudes of these two forces are equal, i.e.,

$$R_{12} = R_{21}$$

and they act normal to the common tangential plane between them.

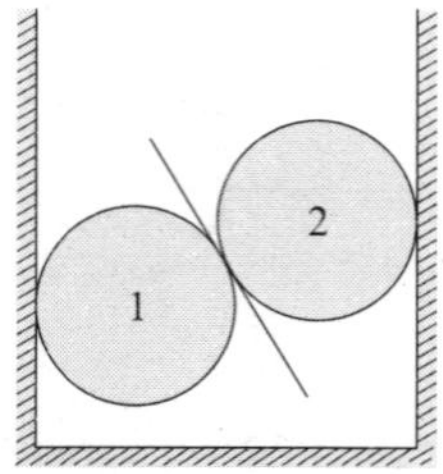

Fig. 5.4(a) Two cylinders placed in a channel

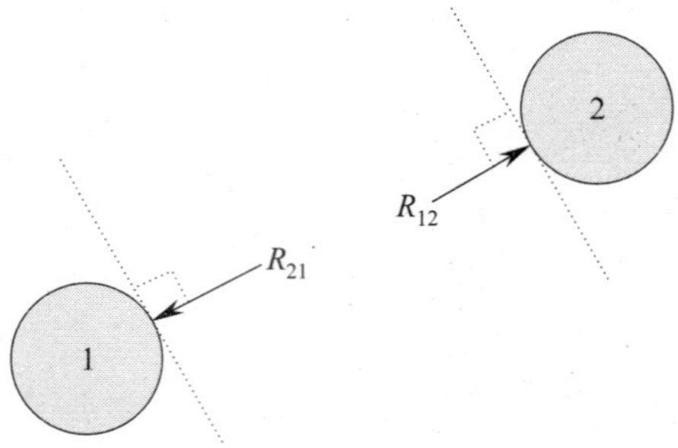

Fig. 5.4(b) Reactive forces exerted by each cylinder on the other

> Hence, we can conclude that whenever two bodies with smooth surfaces are in contact with each other, each body will exert a force on the other and this force always acts normal to the plane of contact between them and at the point of contact.

Corollary Here we have considered bodies with smooth surfaces. However, in reality, no surface is smooth. Hence, when the contact surfaces are *rough*, in addition to the *normal* reaction, a *tangential* frictional force will arise at the contact surfaces. This will be discussed later when we deal with friction in Chapter 6.

(B) Tensile Pull Whenever a body is attached to an inextensible cord, a rope or a cable, a **tensile pull** acts on the body. Consider a ball suspended by a string attached to the ceiling as shown in Fig. 5.5(a). Due to its weight, the ball tries to pull the string downwards, which causes a **tension** in the string. By Newton's third law of motion, the string also tries to exert an equal and opposite reaction on the ball which prevents its motion downwards. Hence, when we draw the free-body diagram of the ball [Fig. 5.5(b)], we must show this reaction force **T** of the string exerted on the ball. Its *magnitude* is equal to the *tension* in the string and it always acts **along** the string and **away** from the body as shown in the figure.

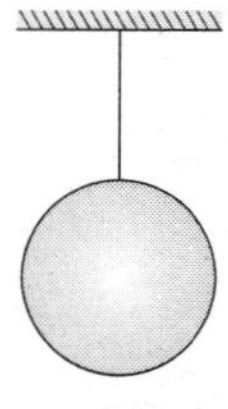

Fig. 5.5(a) A ball suspended by a string

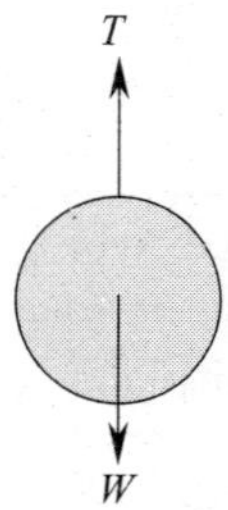

Fig. 5.5(b) Free-body diagram of the ball

Next, consider an inextensible string, a rope or a cable passing over a *smooth* pulley and supporting a body of weight W by an applied force P as shown in Fig. 5.6(a). The free-body diagram of the string is shown in Fig. 5.6(b). If the mass of the string is neglected and the pulley being frictionless, the *tension* at both ends of the string will be **equal**, i.e., $P = T$. However, in practice, we never find a frictionless pulley. For a rough pulley, the tension at both ends of the string passing over the pulley will *not* be equal. This will be discussed later when we deal with friction in Chapter 7.

The tensile pull exerted by the string or the cable in other forms may be as shown in Fig. 5.7.

> Hence, whenever a body is attached to a string, a rope or a cable, then a force of tension is exerted on the body and it acts along the string, rope or cable and away from the body.

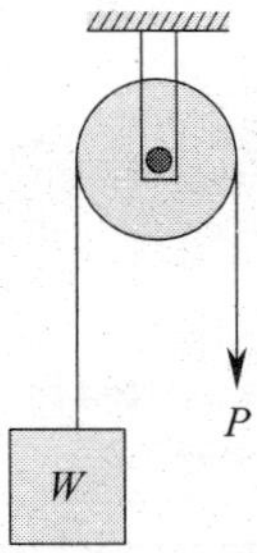

Fig. 5.6(a) Block and pulley arrangement

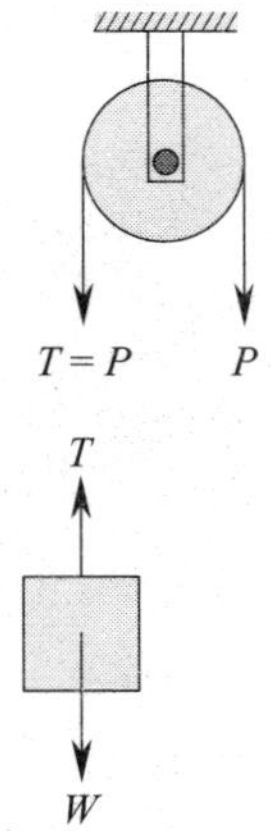

Fig. 5.6(b) Free-body daigarms of string and block

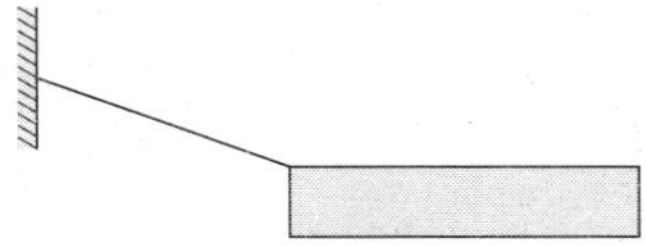

Fig. 5.7(a)

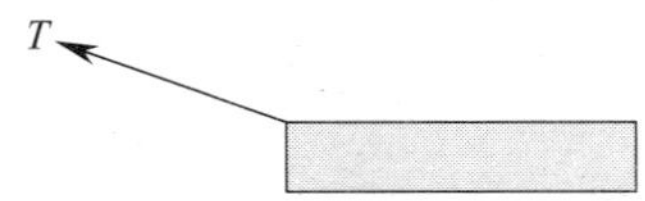

Fig. 5.7(b)

(C) Restoring Force in a Spring Consider a body attached to a spring of unstretched length l and resting on a smooth table. If the body is displaced to the right, it will pull the spring also with it to the right, thus causing *elongation* in the spring. By Newton's third law, the spring restricts this elongation and tries to exert an equal and opposite force on the block acting towards the left. The magnitude of this restoring force is given by Hooke's law as

$$F = k\Delta l$$

where Δl is the change in length of the spring and k is the spring constant.

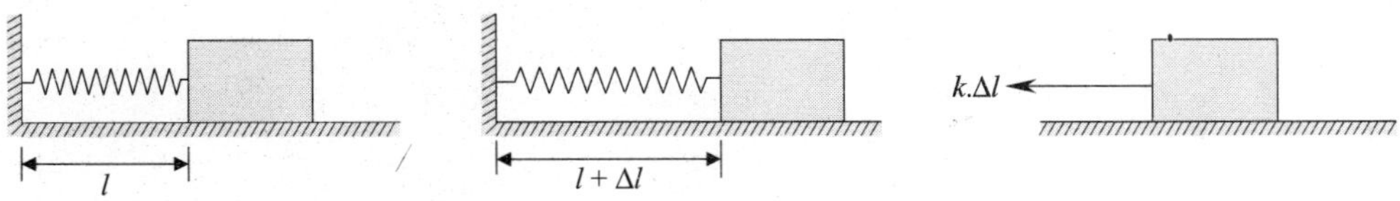

Block attached to a spring and at rest on a smooth horizontal table

Block displaced to the right

Force exerted by the spring on the block

Fig. 5.8

Hence, whenever a body is attached to a spring, a restoring force is exerted by the spring on the body. Its magnitude is $k\Delta l$ and it acts along the spring and in the direction opposite to the direction of displacement of the body.

(D) Tension or Compression Members Sometimes members like angles or rods are hinged at both the ends as shown in Fig. 5.9(a). Such types of connections are very common in bridge trusses and link-mechanisms. The forces in such members act along the axis of the members. Hence, they are either in **tension** or in **compression** depending upon the forces acting on them. Normally, the direction of these forces in the members are unknown, i.e., whether they are in tension or in compression. As a rule, we may assume them to be under **tension** to begin with. If the answer turns out to be of **negative** value then it implies that our assumption is wrong and that it should be in the opposite direction. In other words, the member is under **compression**.

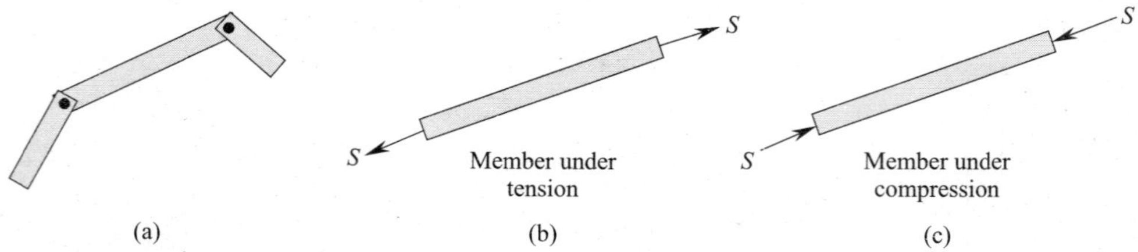

Fig. 5.9

Hence, whenever forces act on a member hinged at both the ends as in a link mechanism, the member is in tension or in compression and this force acts along the axis of the member.

(E) Support Reactions In the case of rigid bodies, particularly beams, to prevent not only translational motion but also rotational motion, these are normally held by various supports. These induce reactions on the bodies and these are discussed below:

(i) Hinge or pin support In this type of support, the body is restrained from moving along the *XY* plane. The restraining force can be resolved into two reactions R_x and R_y along *X* and *Y* directions respectively acting on the body at the point of support. At the same time, it should be noted that the body is free to rotate about the *Z*-axis and hence, no restraining moment acts on the body.

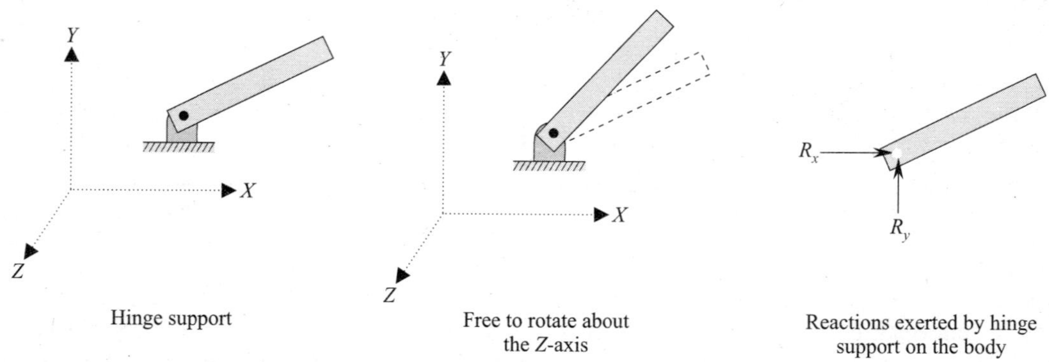

Fig. 5.10

This type of support may also be represented as shown below:

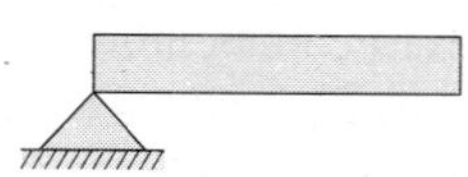

Fig. 5.11 Alternate representation of hinge support

Thus, a hinge support exerts two reactions in the plane of the body along X and Y axes at the point of support.

(ii) Roller or frictionless support In this type of support, the body is not free to move downwards along the negative Y direction, but is free to move along the X direction. Hence, a reaction force R_y acts on the body at the support preventing its motion downwards along the Y direction.

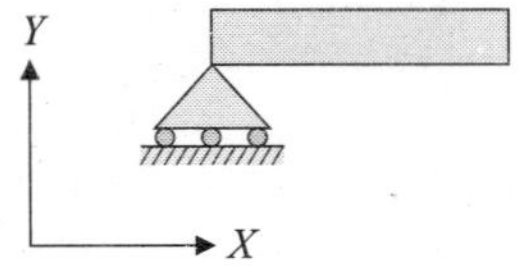

Fig. 5.12(a) Roller support

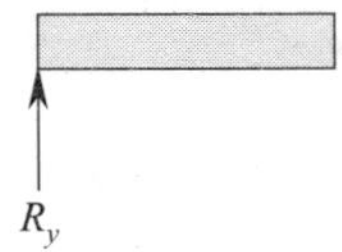

Fig. 5.12(b) Reaction exerted by roller support

This type of support may also be represented as shown below:

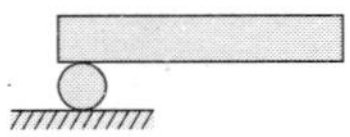

Fig. 5.13 Alternate representation of roller support

Thus, a roller support exerts one normal reaction at the point of support.

(iii) Fixed or built-in support In this type of support, the body is restrained from moving along the XY plane and also from rotating about the Z-axis. Hence, two reactions R_x and R_y and a restraining moment M_z act on the body at the support.

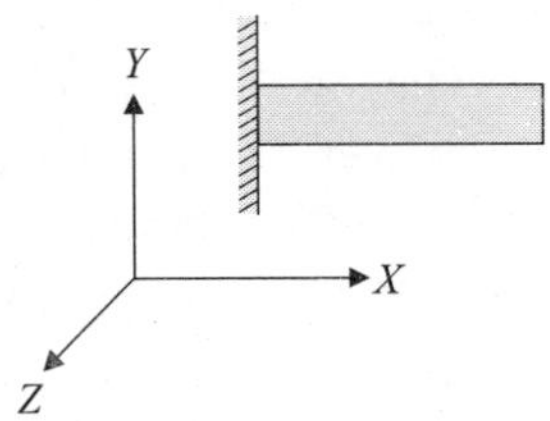

Fig. 5.14(a) Fixed end

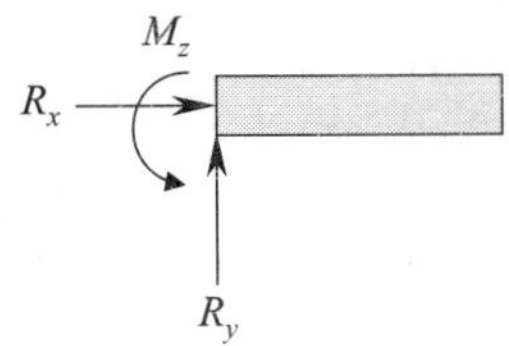

Fig. 5.14(b) Reactions at fixed end

Hence, a fixed or built-in support exerts two normal reactions along the X and Y directions and one restraining moment along the Z-direction.

(iv) Ball-and-socket joint This type of support is same as the hinge support in three dimensions. Here, the body is restrained from moving in all three directions, but is free to rotate in any direction. Hence, reactions R_x, R_y and R_z act on the free-body diagram of the body.

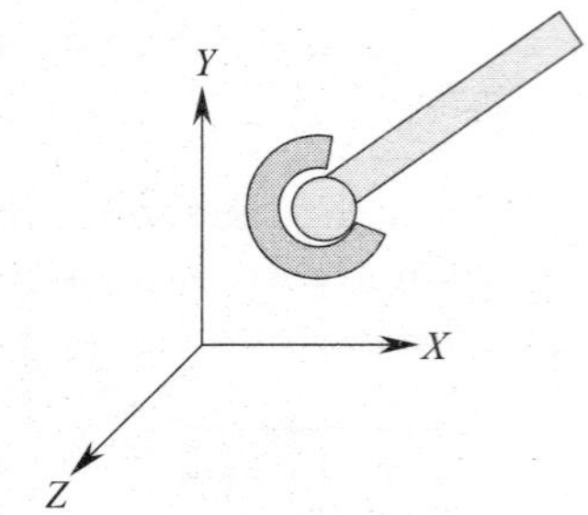

Fig. 5.15(a) Ball-and-socket joint

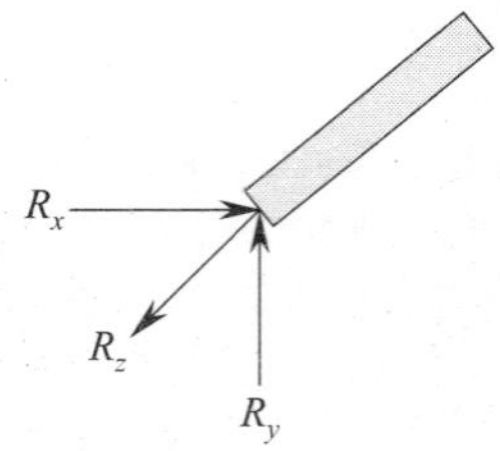

Fig. 5.15(b) Reactions at ball-and-socket joint

> Thus, a ball-and-socket joint exerts three reactions at the point of support along the *X*, *Y* and *Z* directions.

Corollary While finding the *unknown reactions* and *moments* at the supports, they are normally assumed to act in the *positive* directions. If after calculations, they turn out to be of *negative values* then we conclude that the assumed directions are wrong and that they should be in the opposite direction.

There are other reactions too, but these will suffice to understand the basics of static equilibrium. Even though we have discussed various forces that may act on a body in a structure or mechanical system, we must remember that *not* all of them are going to act on the body at a time.

5.2.3 Steps to be Followed While Drawing Free-Body Diagrams

The following steps can be followed while drawing free-body diagrams in a structure or mechanical system:

Step 1 Any structure or mechanical system is made up of a number of bodies, which are connected together or are in contact with other members. Free-body diagrams can be drawn for all individual elements in a structure or mechanical system. However, for analyzing a structure or mechanical system, we consider only the free-body diagrams of those bodies that are of interest to us. Hence, the first step is to **identify** the body or bodies for which free-body diagrams must be drawn in order to analyze the structure or mechanical system.

Step 2 **Isolate** the body from all its attachments in the structure or mechanical system.

Step 3 Draw a **sketch** of the body to scale and show all the forces (active and reactive forces) acting on the body by **vectors**.

Step 4 The first force that must be shown in the free-body diagram is the **weight** of the body. It is drawn vertically downwards at the centre of gravity of the body. In some cases, the weight of the body may be negligible; hence, it need not be shown.

Step 5 If the body is in contact with a table, a wall or any other body then the normal reaction must be shown at the point of contact and perpendicular to the common tangential plane.

Step 6 If the body is attached by a string, a rope or a cable then a force of tension must be shown at that point acting along the string or rope and away from the body.

Step 7 If the body is attached to a spring then a restoring force must be shown acting on the body in the direction opposite to that in which the body tries to move.

Step 8 If the body is attached to any support, as in beams then support reactions must be shown accordingly. If the support reactions are unknown then normally they are assumed to be positive.

Step 9 When all the forces, active as well as reactive, are shown in the free-body diagram, we then proceed to analyze its equilibrium.

Note: The forces shown in the free-body diagram are the forces exerted **on** the body by the members attached to it or in contact with it, and *not* the forces exerted **by** the body on the other members.

The following illustrative examples explain the step-by-step procedure for drawing free-body diagrams for various cases, which the student would come across throughout this book.

Example 5.1 Draw free-body diagrams for the following cases:

(i) A block or a ball resting on a smooth horizontal plane

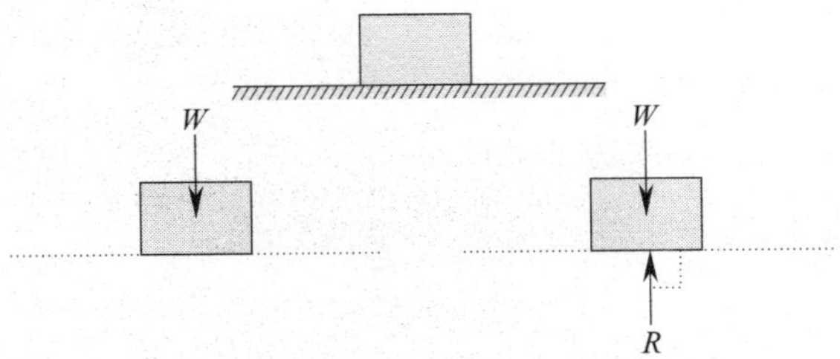

Fig. 5.16(a) A block resting on a smooth horizontal plane

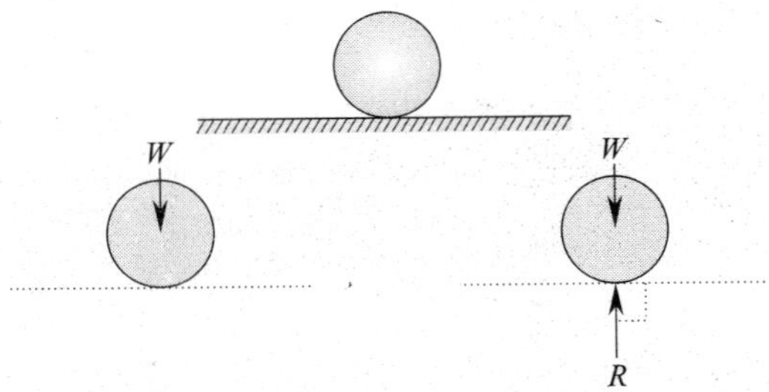

Fig. 5.16(b) A ball resting on a smooth horizontal plane

Step 1 Detach the block or the ball from its attachment with the surrounding, i.e., the horizontal plane.

Step 2 Draw a sketch of the block or ball.

Step 3 Draw its weight W acting at the centre of the block or ball directed vertically downwards.

Step 4 Draw the normal reaction R at the point of contact between the block or ball and the horizontal plane.

(ii) A block on a smooth inclined plane is restrained from moving downwards by a string attached to it

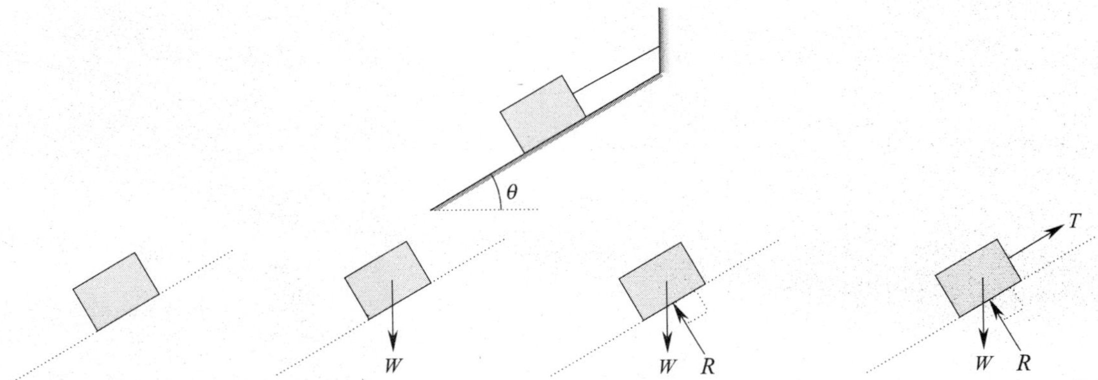

Fig. 5.17

Step 1 Detach the block from its attachment with the surrounding, i.e., the inclined plane and the string.

Step 2 Draw a sketch of the block.

Step 3 Draw its weight W acting at the centre of the block directed vertically downwards.

Step 4 Draw the normal reaction R at the point of contact between the block and the inclined plane.

Step 5 Draw the tensile pull exerted by the string acting along the string and away from the block.

(iii) A sphere on a smooth inclined plane is restrained from moving downwards by a string attached to the sphere whose other end is attached to the inclined plane

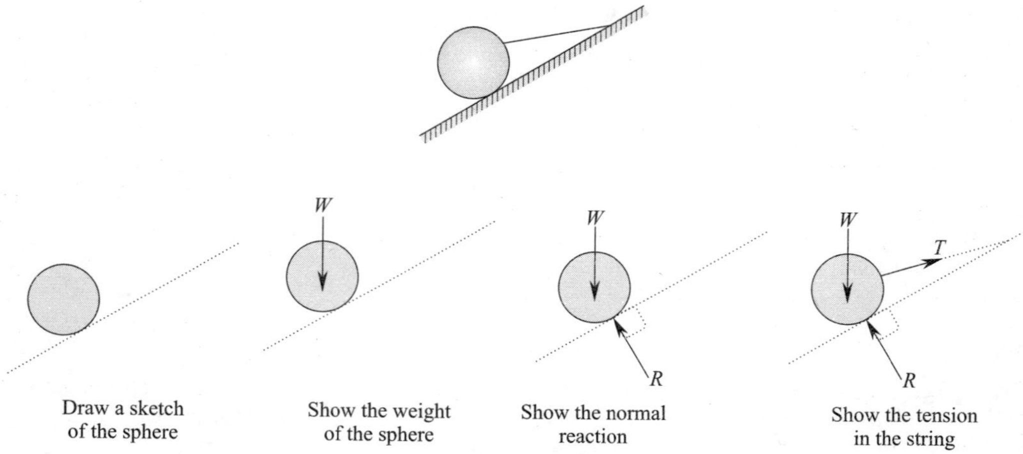

Fig. 5.18

(iv) A sphere on a smooth inclined plane is restrained from moving downwards by a vertical plane

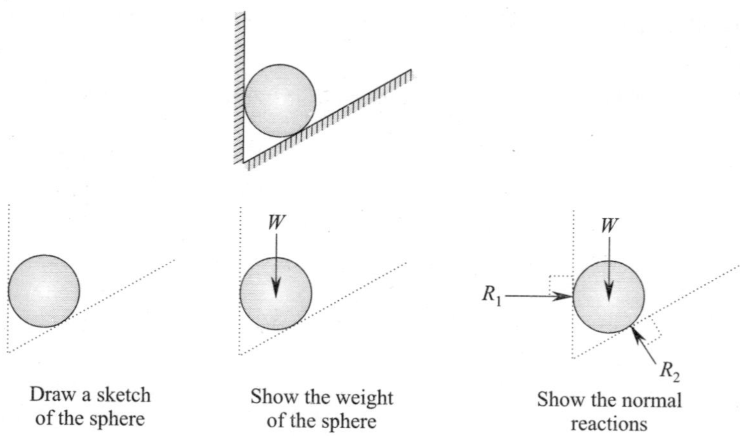

Fig. 5.19

(v) A sphere resting in a trough

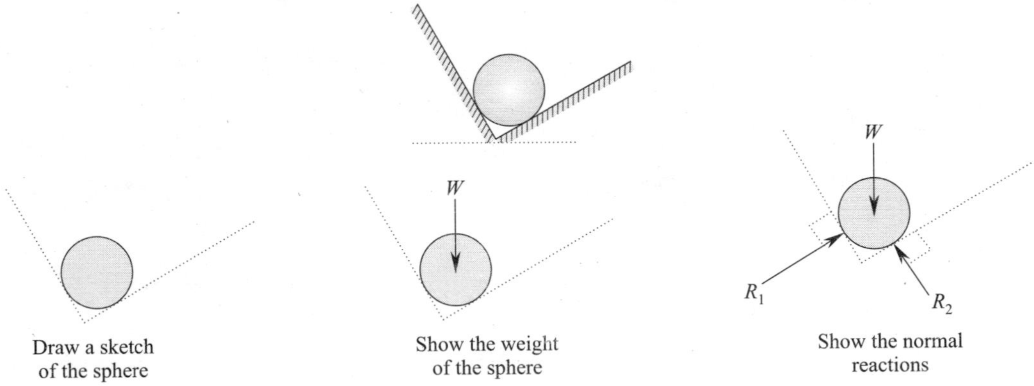

Fig. 5.20

Example 5.2 Sketch the free-body diagram of a ball suspended by a string and a horizontal force F applied to it as shown in Fig. 5.21.

Solution Detach the ball from its attachment with the string. Draw a sketch of the ball to scale and show all the forces acting on it by vectors. The free-body diagram is shown in Fig. 5.21(a). It must include weight of the ball W placed at its centre of gravity and directed vertically downwards; tension T as it is suspended by the string (acting along the string and away from the body) and the applied force F.

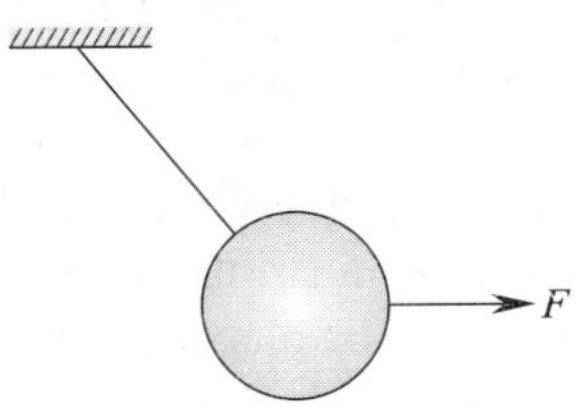

Fig. 5.21

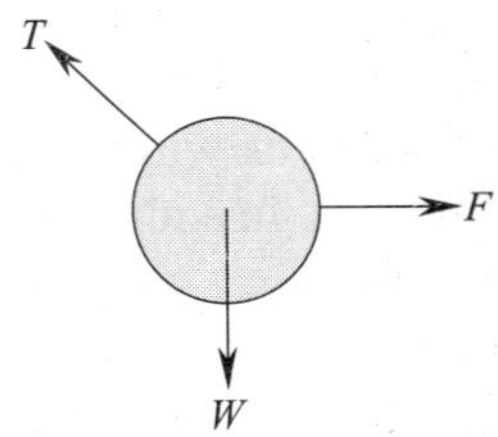

Fig. 5.21(a)

Example 5.3 Draw the free-body diagram of a cylinder resting in a channel as shown in Fig. 5.22. Assume all contact surfaces to be smooth.

Solution Detach the cylinder from all its attachments with the surroundings. Draw the cylinder to scale and show all the forces acting on it by vectors. The free-body diagram is shown in Fig. 5.22(a). The free-body diagram must include the weight W of the cylinder placed at its centre of gravity and directed downwards. As the cylinder is in contact with the planes at A and B, the normal reactions R_A and R_B exerted by the planes on the cylinder must also be shown. Note that these reactions must be perpendicular to the tangential planes of contact.

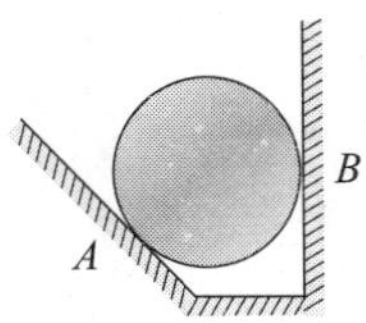

Fig. 5.22

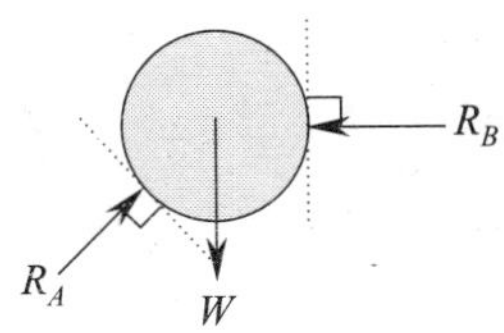

Fig. 5.22(a)

Example 5.4 A cylinder of weight W is supported on a plank AB hinged at A. The end B of the plank is tied by a string BC to the wall. Draw free-body diagrams of the cylinder, plank and string separately. Assume all contact surfaces to be smooth and the plank to be weightless.

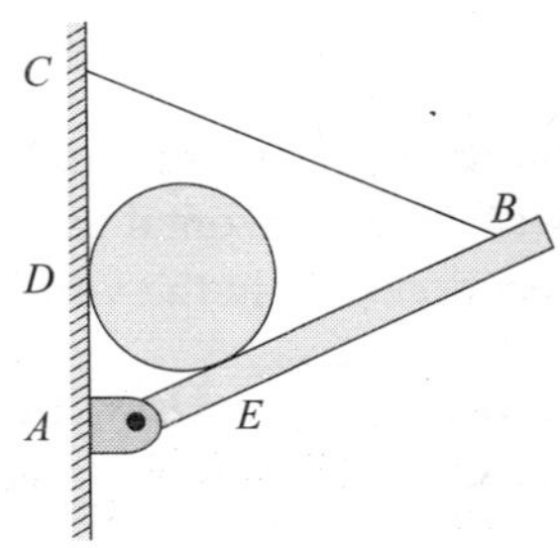

Fig. 5.23

Solution To draw the free-body diagrams, we detach the cylinder, plank and string separately. Then they are drawn to scale.

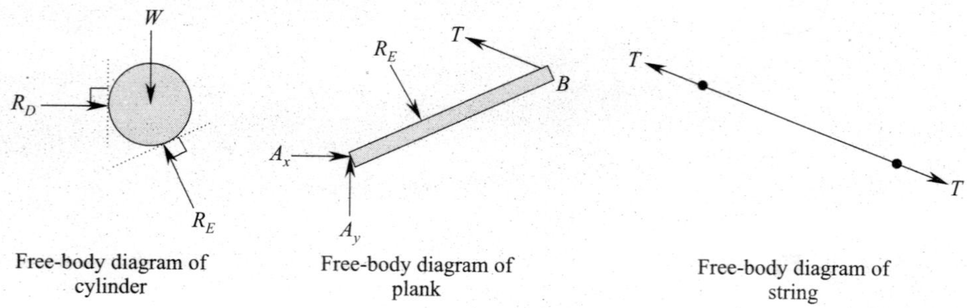

Fig. 5.23(a)

As the cylinder is in contact with the wall at D and with the plank at E, normal reactions R_D (exerted by the wall on the cylinder) and R_E (exerted by the plank on the cylinder) are shown in its free-body diagram. Also, the weight W of the cylinder is placed at the centre of gravity and directed vertically downwards. Since R_E is the force exerted by the plank on the cylinder, the force exerted by the cylinder on the plank will be equal and in the opposite direction. Hence, R_E for the plank is shown in the opposite direction. In addition, at point A, as it is a hinge, the reactions A_x and A_y are shown. At point B, there will be a tension T as it is connected by a string. Note that here we have neglected the weight of the plank. If its weight is to be included, it must be placed at the centre of gravity of the plank. The free-body diagram of the string will have a tension T at both of its ends. Normally, strings are assumed to be weightless.

Example 5.5 Draw free-body diagram of the truss arrangement shown.

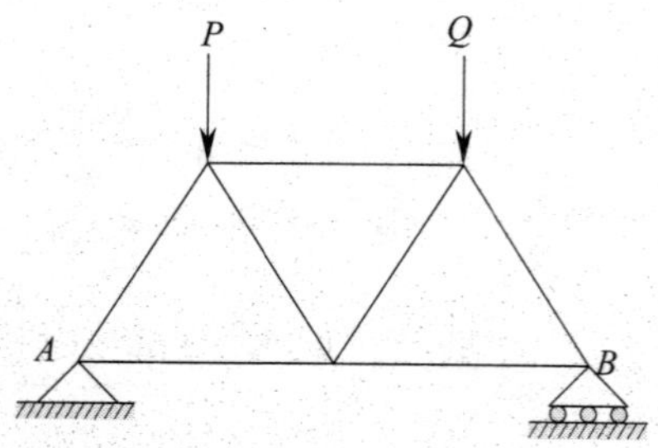

Fig. 5.24

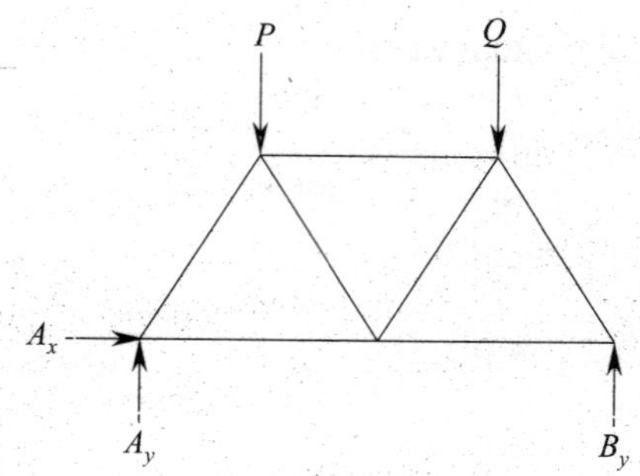

Fig. 5.24(a)

Solution The truss is supported by a hinge at A and a roller at B. At the hinge support A, the reaction will have horizontal and vertical components, namely, A_x and A_y; at the roller support B, there will be reaction only in the vertical direction, i.e., B_y. The external loads P and Q are also shown.

Example 5.6 Draw free-body diagrams of pulleys, block and string in the arrangement shown. The weights of the pulleys can be neglected.

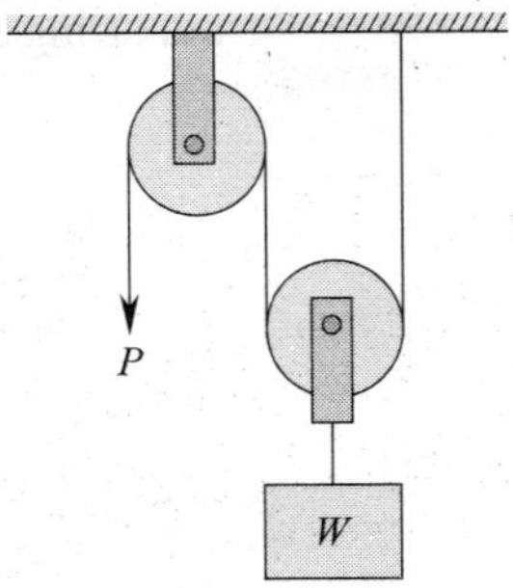

Fig. 5.25

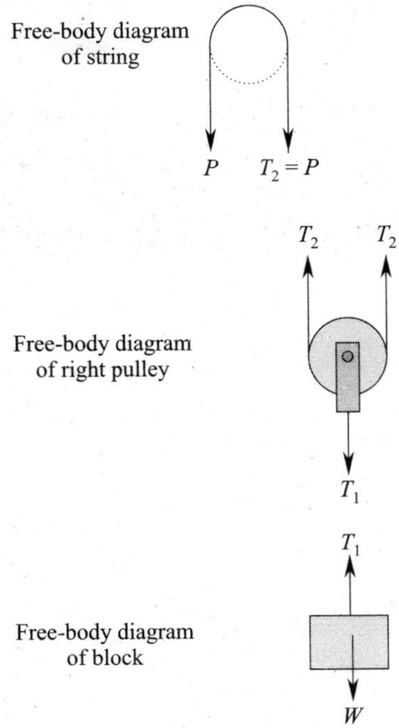

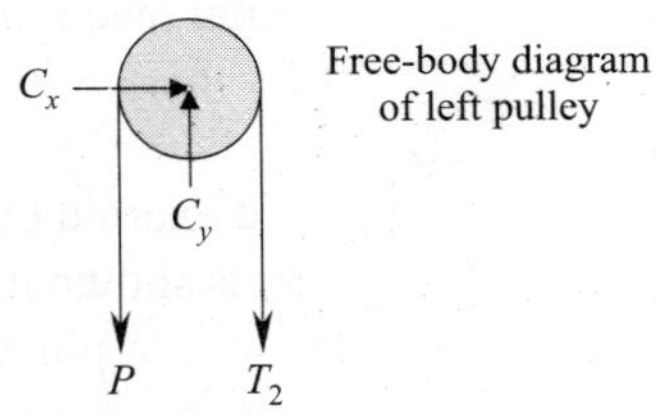

Fig. 5.25(b)

Fig. 5.25(a)

Solution To draw the free-body diagrams, we detach the pulleys, string and block separately, and draw them to scale. Since the pulleys are frictionless and also as the mass of the string can be neglected, the tension at both the ends of the string will be equal.

Let T_2 be the tension in the string passing over the pulleys. Then for the string passing over the left pulley, $T_2 = P$; for the string passing over the right pulley, the two tensions will be equal to T_2. In the free-body diagram of the block, two forces are shown acting on it: (i) its weight W, and (ii) tension T_1. The free-body diagram of the left pulley includes: (i) the tensions in the string, and (ii) reactions C_x and C_y as it is hinged at its centre [refer Fig. 5.25(b)].

Example 5.7 A block of mass m_1 is resting on a smooth horizontal plane. It is attached to another block of mass m_2 by a string passing over a smooth and massless pulley as shown in Fig. 5.26. The block is also restrained from moving by a string attached to the wall. Draw free-body diagrams of the two blocks and the pulley.

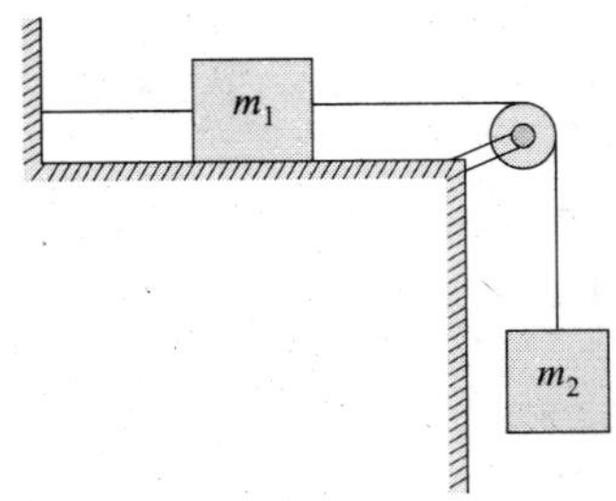

Fig. 5.26

Solution The free-body diagrams of the two blocks and the pulley are shown in Figs 5.26(a). The forces acting on the block resting on the plane are its weight m_1g, normal reaction N and tensions T_1 and T_2 due to its attachment with the strings on the left and right sides respectively. As the pulley is frictionless, the tensions on both the ends of the string are shown equal. The pulley is assumed to be massless. Hence, its weight has been neglected. In the free-body diagram of the hanging block, the forces shown are its weight m_2g and tension T_2 in the string.

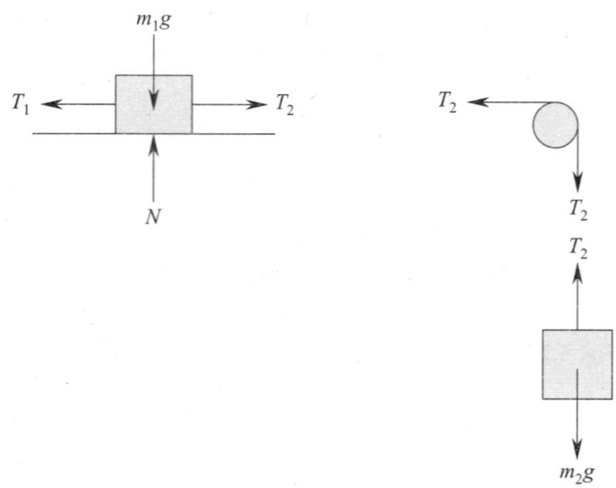

Fig. 5.26(a)

Example 5.8 A ladder AB of length l and weight W rests against a smooth vertical wall and floor. To prevent it from sliding, end B is connected to the wall by a string BC. If a girl of weight P is standing on it at a distance of $l/3$ from the ground, draw the free-body diagram of the ladder.

Solution The free-body diagram of the ladder is shown in Fig. 5.27(a). The forces acting on it are its weight W, normal reactions R_W and R_F at the contact points between the ladder and the wall and between the ladder and the floor respectively. As the girl is standing on it, the weight P of the girl must be shown acting vertically downwards at the point where she is standing. Also, as the ladder is prevented from sliding by its attachment with the string, tension T is shown in the free-body diagram.

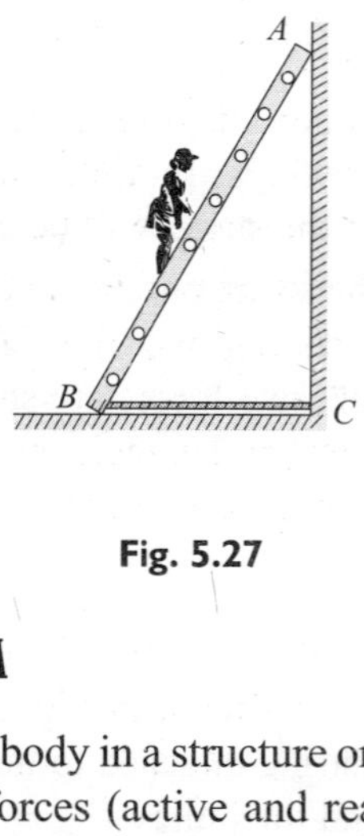

Fig. 5.27

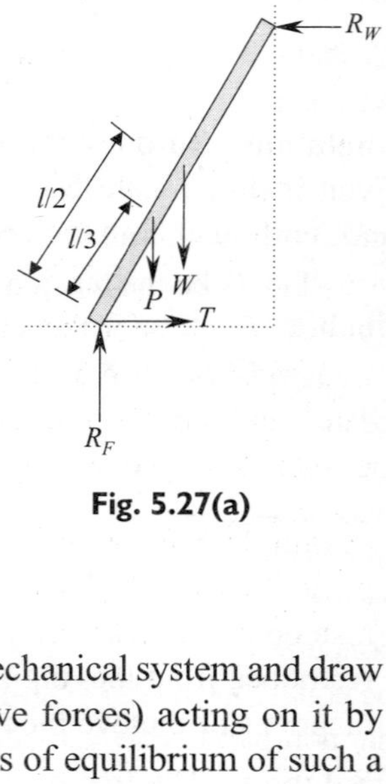

Fig. 5.27(a)

5.3 CONDITIONS OF EQUILIBRIUM

In the previous section, we discussed how to isolate a body in a structure or mechanical system and draw a sketch of its free-body diagram showing all the forces (active and reactive forces) acting on it by vectors. In this section, we will go a step further and determine the conditions of equilibrium of such a body.

The conditions of equilibrium of a body are different for different force systems. Hence, as a first step, we must check if the lines of action of all the forces acting on the free body are **concurrent** or not. If they are *concurrent*, we can idealize the body as a particle as already discussed in Chapter 3. Hence, we can replace the system of forces by a resultant force acting at the point of concurrency. The effect of a concurrent force system will be to translate the particle in the direction of the resultant. Thus, to keep the particle in equilibrium, the **resultant force** acting on it must be **zero**; graphically speaking, the *force polygon* must *close*. Mathematically, the condition for equilibrium of a particle can be expressed as

$$\vec{R} = \sum \vec{F} = \vec{O} \tag{5.1}$$

i.e., vectorial summation of all the forces acting on the free body is a *null* vector. This is termed *condition of equilibrium of a particle.*

Suppose, the lines of action of all the forces acting on the free body are *not concurrent*. Then we must treat it as rigid body itself. We already saw in Chapter 4 that forces on a rigid body can be replaced by a resultant force acting at a common point and a moment about the same point. The effect of such a force system on the body would be to translate as well as to rotate it. Thus, to keep a rigid body in equilibrium, the **resultant force** and **moment** must be **zero**. Mathematically, the condition for equilibrium of a rigid body can be expressed as

$$\vec{R} = \sum \vec{F} = \vec{O} \tag{5.2}$$

and

$$\vec{M} = \sum(\vec{r} \times \vec{F}) = \vec{O} \tag{5.3}$$

i.e., vectorial summation of individual forces about a common point is a null vector and vectorial summation of moments of the forces about the same point is also a null vector.

Corollary Two forces acting on a body will keep it in equilibrium, if the two forces are *equal*, *opposite* and *collinear*. Even if they are equal, but not opposite and collinear then equilibrium is not possible. This is termed **two-force equilibrium**. Figure 5.28 illustrates this point.

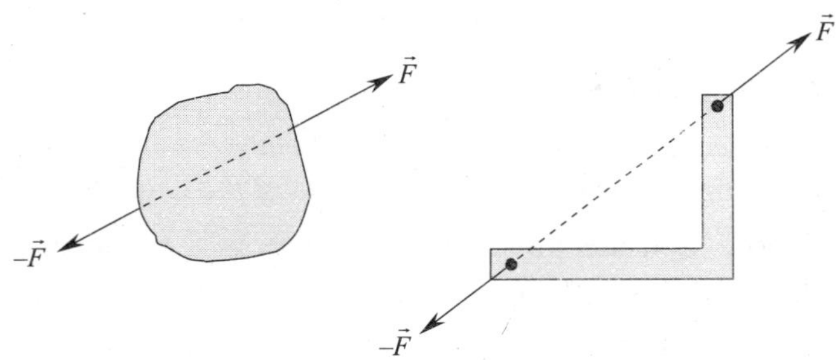

Fig. 5.28 Two-force equilibrium

Three forces acting on a body will keep it in equilibrium if the three forces are *coplanar* and *concurrent*. Consider two non-parallel forces on a plane. Then their lines of action will intersect at some point since they lie on the same plane. Hence, their resultant can be determined which will pass through their point of concurrency. For the equilibrium of three forces, the third force must be equal, opposite and collinear to the resultant of the first two coplanar non-parallel forces found out as above. This is possible only if the third force passes through the point of concurrency of the other two forces and also that it lies on the same plane as the other two forces. This is termed **three-force equilibrium**.

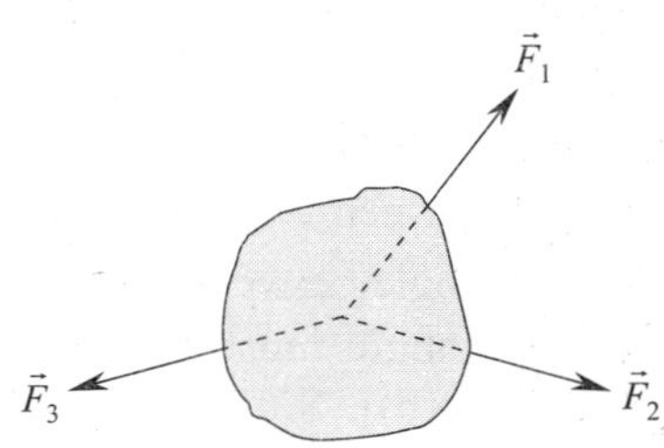

Fig. 5.29 Three-force equilibrium

In the subsequent sections, we will consider each type of force system separately and discuss the equilibrium conditions in detail along with solved examples.

5.3.1 Equilibrium of Coplanar Concurrent Forces

For a *coplanar concurrent force* system, expressing the resultant in terms of rectangular components of individual forces, the equilibrium condition can be stated as

$$\vec{R} = (\Sigma F_x)\vec{i} + (\Sigma F_y)\vec{j} = \vec{O} \tag{5.4}$$

This is satisfied only when both the coefficients of the unit vectors are zero. Hence, the necessary and sufficient conditions for equilibrium are

$$\Sigma F_x = 0 \tag{5.5}$$

and

$$\Sigma F_y = 0 \tag{5.6}$$

Stated in words, a particle is said to be in equilibrium, if summation of the X components of all the forces acting on it is zero and also summation of the Y components is zero. We know from mathematics that from a system of two independent simultaneous equations, a *maximum* of only *two* unknowns can be solved. Hence, we can solve for only two unknowns from the equilibrium Eqs 5.5 and 5.6. These unknowns may be magnitudes of two reactions whose directions are known or may be one reaction and one inclination angle.

This method of finding the components to check the condition of equilibrium is applicable to any number of concurrent forces in the system as well as forces in space. However, the calculations can be much simplified if there are only three concurrent forces. This is done by applying Lami's theorem, which makes use of graphical and trigonometric methods.

Lami's Theorem

The theorem states that *if three coplanar concurrent forces acting on a body keep it in equilibrium then each force is proportional to the sine of the angle between the other.*

Proof Consider three concurrent forces $\vec{F}_1$, $\vec{F}_2$ and $\vec{F}_3$ as shown in Fig. 5.30(a). Then we can construct the force triangle by taking the forces in order [refer Fig. 5.30(b)]. Draw $\vec{F}_1$ to scale parallel to its given direction. Then from the head of force $\vec{F}_1$, draw force $\vec{F}_2$ to scale and parallel to its direction. Finally, from the head of force $\vec{F}_2$, draw force $\vec{F}_3$ to scale and parallel to its direction to close the triangle. It should be noted that the force triangle must close, as the resultant is zero.

The inclination of each force must also be shown accordingly. As the angle between $\vec{F}_1$ and $\vec{F}_2$ is γ, the included angle between $\vec{F}_1$ and $\vec{F}_2$ in the force triangle must be $180° - \gamma$ [see Fig. 5.30(b)]. Similarly, as the angle between $\vec{F}_1$ and $\vec{F}_3$ is β, the included angle between $\vec{F}_1$ and $\vec{F}_3$ in the force triangle must be $180° - \beta$. Therefore, the third angle in the force triangle is given as

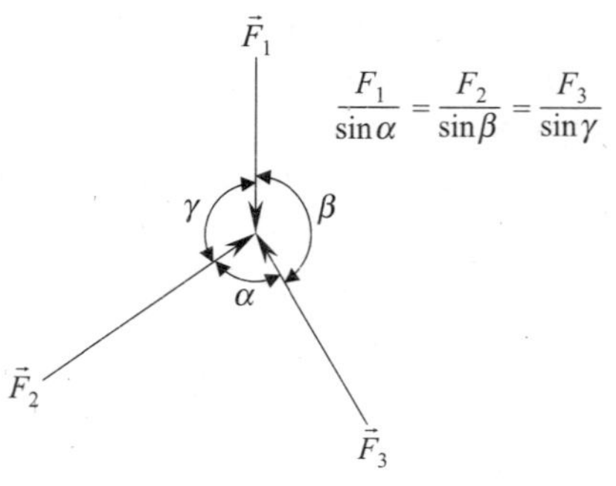

Fig. 5.30(a)

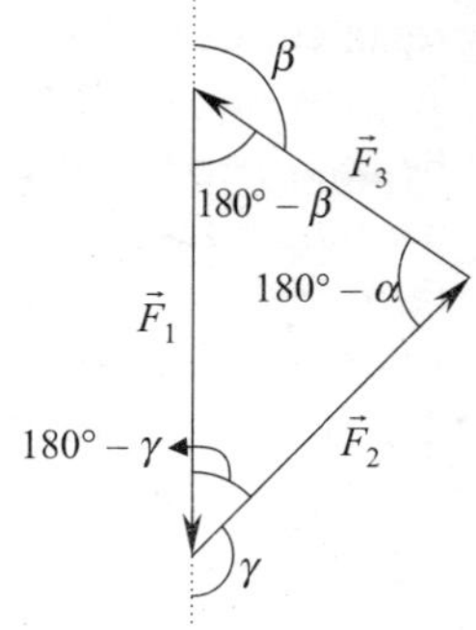

Fig. 5.30(b)

$$180° - (180° - \beta) - (180° - \gamma)$$
$$= \beta + \gamma - 180°$$
$$= (360° - \alpha) - 180° \quad [\text{since } \beta + \gamma = 360° - \alpha]$$
$$= 180° - \alpha \tag{5.7}$$

We saw in Chapter 3 that the sine law states

$$\frac{A}{\sin(180° - \alpha)} = \frac{B}{\sin(180° - \beta)} = \frac{C}{\sin(180° - \gamma)} \tag{5.8}$$

Hence, applying sine law to the force triangle, we get

$$\frac{F_1}{\sin\alpha} = \frac{F_2}{\sin\beta} = \frac{F_3}{\sin\gamma} \quad [\text{since } \sin(180° - \theta) = \sin\theta] \tag{5.9}$$

Example 5.9 A strut AB supporting a block of weight 500 N is held by a cable BC as shown in Fig. 5.31. Find the tension T in the cable BC and the force S in the strut AB. Neglect weight of the strut.

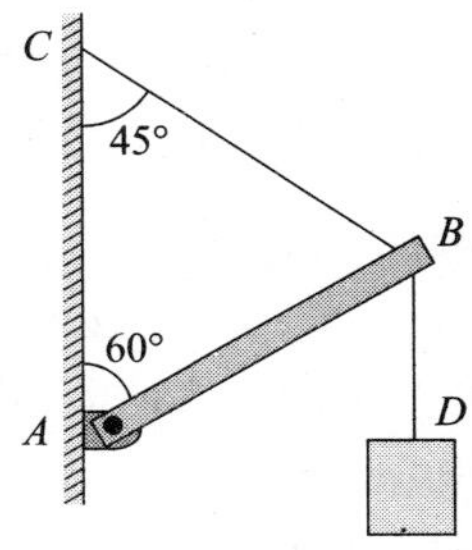

Fig. 5.31

Solution For a better understanding of the problem, let us consider the free-body diagrams of the strut, block and point B separately as shown in Fig. 5.31(a). To begin with, let us assume the strut to be under tension. Hence, the direction of the force S acting on the strut is shown as acting away from the member.

Considering the free body diagram of the block and applying the condition of equilibrium along the Y direction, i.e.,

$$\Sigma F_y = 0 \Rightarrow$$
$$T_{BD} - 500 = 0$$
$$\therefore \quad T_{BD} = 500 \text{ N} \tag{a}$$

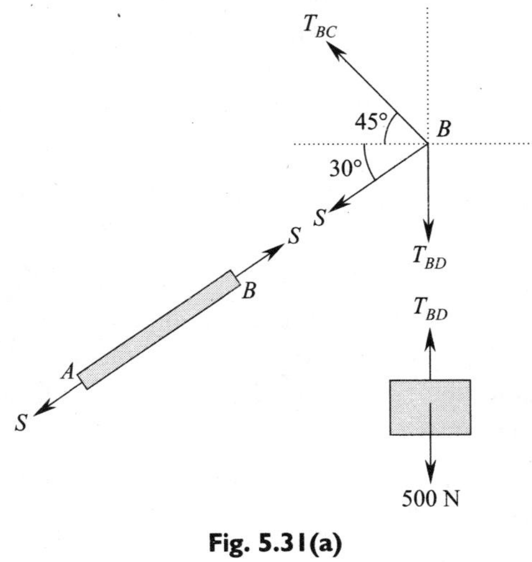

Fig. 5.31(a)

The forces acting at point B are then (i) tension T_{BC} along the cable BC, (ii) force S along the strut BA (since we do not know the direction of this force, we have assumed the strut to be under tension) and (iii) tension T_{BD} along BD. Since these three forces are concurrent at point B, we can consider equilibrium of point B.

Applying the conditions of equilibrium of a particle (point B) along the X and Y directions,

$$\Sigma F_x = 0 \Rightarrow$$
$$-S\cos 30° - T_{BC}\cos 45° = 0$$
$$\therefore \quad S\cos 30° = -T_{BC}\cos 45° \tag{b}$$

$$\Sigma F_y = 0 \Rightarrow$$
$$T_{BC}\sin 45° - S\sin 30° - T_{BD} = 0$$
$$\therefore \quad S\sin 30° = T_{BC}\sin 45° - 500 \quad [\text{since } T_{BD} = 500 \text{ N}] \tag{c}$$

Note that there are two unknowns S and T_{BC} and these can be solved from the two equations of equilibrium (b) and (c) available.

Dividing equation (c) by equation (b), we get

$$\tan 30° = \frac{T_{BC} \sin 45° - 500}{-T_{BC} \cos 45°}$$

$$-T_{BC} \cos 45° \tan 30° = T_{BC} \sin 45° - 500$$

$$\Rightarrow \quad T_{BC} = 448.29 \text{ N}$$

Substituting the value of T_{BC} in equation (b),

$$S \cos 30° = -T_{BC} \cos 45°$$

$$S \cos 30° = -(448.29) \cos 45°$$

$$\Rightarrow \quad S = -366.03 \text{ N}$$

The negative sign in the value of S indicates that our initial assumption that the member is under tension is wrong. Hence, it is actually under compression.

Example 5.10 In Fig. 5.32, find the tension T in the cable BC and the force S in the strut AB. The weight of the block suspended is 1000 N. Neglect the weight of the strut.

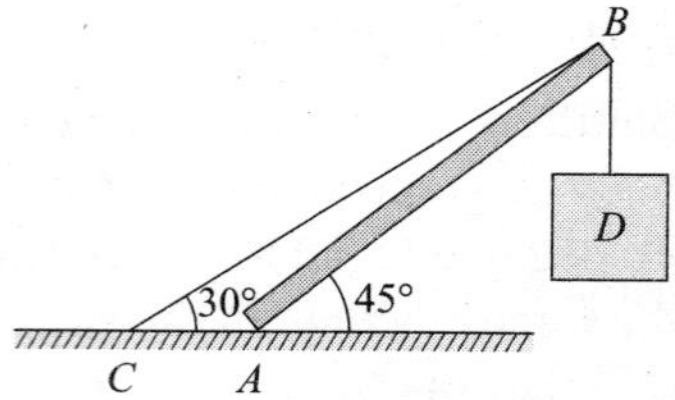

Fig. 5.32

Solution For a better understanding of the problem, let us consider free-body diagrams of the strut, block and the point B separately as shown in Fig. 5.32(a). To begin with, let us assume the strut to be under tension. Hence, the direction of the force S acting on the strut is shown as acting away from the member.

Considering the free-body diagram of the block and applying the condition of equilibrium along the Y direction, i.e.,

$$\sum F_y = 0 \Rightarrow$$

$$T_{BD} - 1000 = 0$$

$$\therefore \quad T_{BD} = 1000 \text{ N}$$

The forces acting at the point B are then (i) tension T_{BC} along the cable BC, (ii) force S along the strut BA, and (iii) tension T_{BD} along the cable BD. Since these three forces are concurrent at the point B, we can consider equilibrium of the point B.

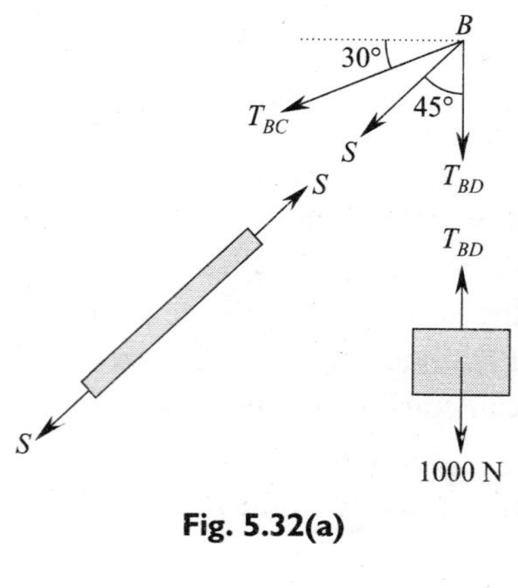

Fig. 5.32(a)

Applying the conditions of equilibrium along the X and Y directions,

$$\sum F_x = 0 \Rightarrow$$

$$-S \sin 45° - T_{BC} \cos 30° = 0$$

$$\therefore \quad S \sin 45° = -T_{BC} \cos 30° \qquad \text{(a)}$$

$$\Sigma F_y = 0 \Rightarrow$$

$$-T_{BC} \sin 30° - S \cos 45° - T_{BD} = 0$$

$$S \cos 45° = -T_{BC} \sin 30° - T_{BD}$$

$$= -T_{BC} \sin 30° - 1000 \qquad \text{(b)}$$

Dividing equation (a) by equation (b),

$$\tan 45° = \frac{-T_{BC} \cos 30°}{-T_{BC} \sin 30° - 1000}$$

Since tan 45° = 1,

$$T_{BC} \sin 30° + 1000 = T_{BC} \cos 30°$$

$$\Rightarrow \qquad T_{BC} = 2732.05 \text{ N}$$

Substituting the value of T_{BC} in equation (a),

$$S \sin 45° = - T_{BC} \cos 30°$$

$$S \sin 45° = -(2732.05) \cos 30°$$

$$\Rightarrow \qquad S = -3346.06 \text{ N}$$

The negative sign in the value of S indicates that our initial assumption that the member is under tension is wrong. Hence, it is actually under compression.

Example 5.11 Two persons lift a mass of 100 kg by cables passing over two pulleys as shown in Fig. 5.33. Determine the forces P and Q that must be applied by the two persons if the body is in equilibrium at the position shown.

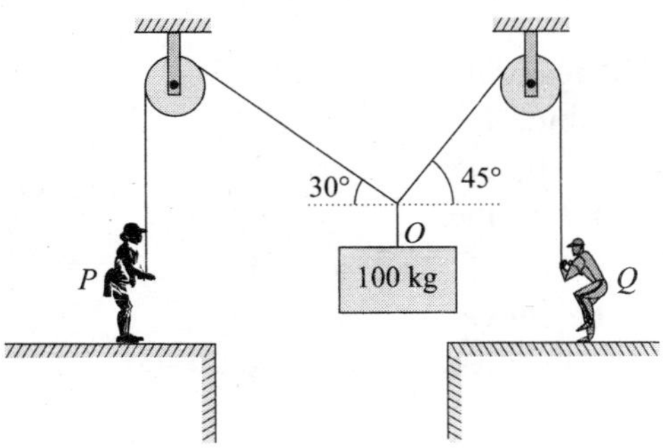

Fig. 5.33

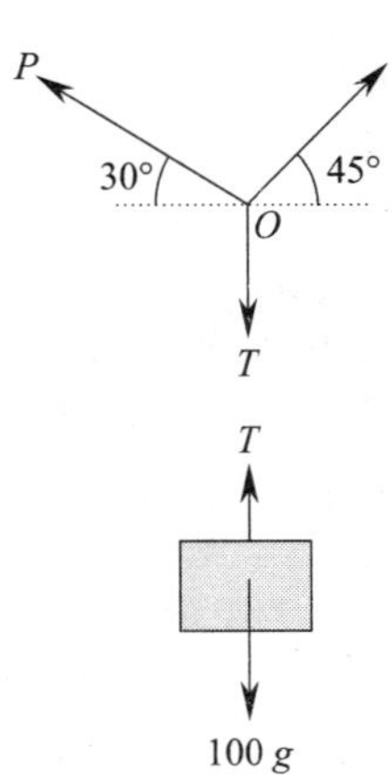

Fig. 5.33(a)

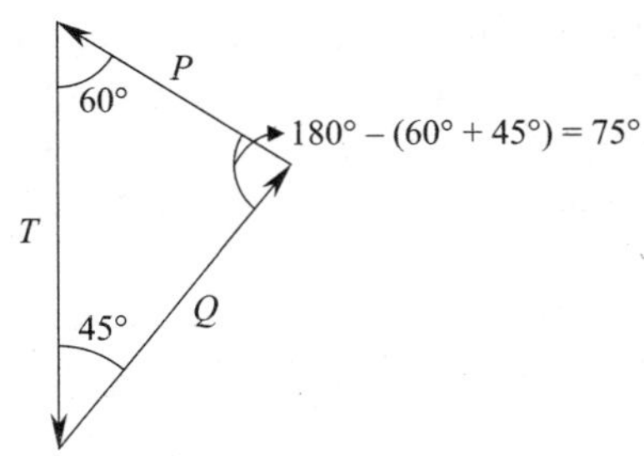

Fig. 5.33(b)

Solution The free-body diagrams of the particle (point O) and block are shown in Fig. 5.33(a).

Considering the free-body diagram of the block and applying the conditions of equilibrium along vertical direction, i.e.,

$\sum F_y = 0 \Rightarrow$

$$T - (100g) = 0$$

$$\therefore \quad T = 100g$$

$$= 100 \times 9.81 = 981 \text{ N}$$

The forces acting at point O are tensions T, P and Q. Since it is in equilibrium under the action of three concurrent forces, we can apply Lami's theorem to the triangle of forces shown in Fig. 5.33(b). The three forces are drawn to scale parallel to the given directions. The inclination of force Q with respect to horizontal or vertical is 45°; and that of force P is 30° to the horizontal or 60° to the vertical. The third angle can then be determined as 75°. Applying Lami's theorem then to the force triangle:

$$\frac{T}{\sin 75°} = \frac{P}{\sin 45°} = \frac{Q}{\sin 60°}$$

$$\Rightarrow \quad P = T\frac{\sin 45°}{\sin 75°}$$

$$= (981)\frac{\sin 45°}{\sin 75°} = 718.14 \text{ N}$$

Also,

$$Q = T\frac{\sin 60°}{\sin 75°}$$

$$= (981)\frac{\sin 60°}{\sin 75°} = 879.54 \text{ N}$$

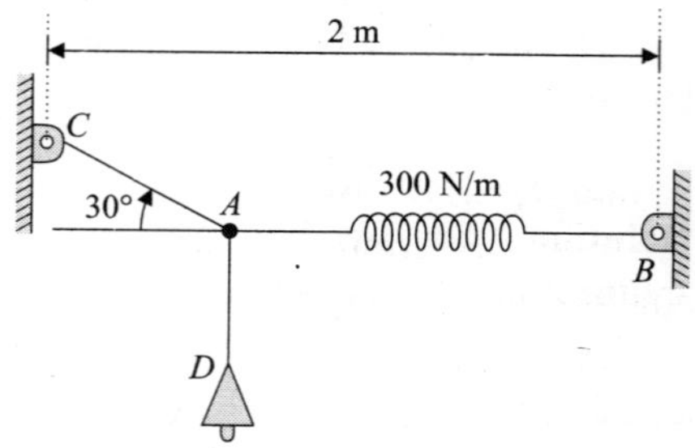

Fig. 5.34

Example 5.12 Determine the length of cord AC in Fig. 5.34, which will keep the 8 kg lamp, suspended in the position shown. The undeformed length of the spring AB is 0.4 m, and the spring has a stiffness of $k_{AB} = 300$ N/m.

Solution Let Δl be the change in length of the spring; then the restoring force exerted by the spring is $k\Delta l$. The free-body diagram of the point A is shown in Fig. 5.34(a). The forces acting at the point A are then: (i) weight W of the lamp, (ii) restoring force of the spring, and (iii) tension T in the cord AC.

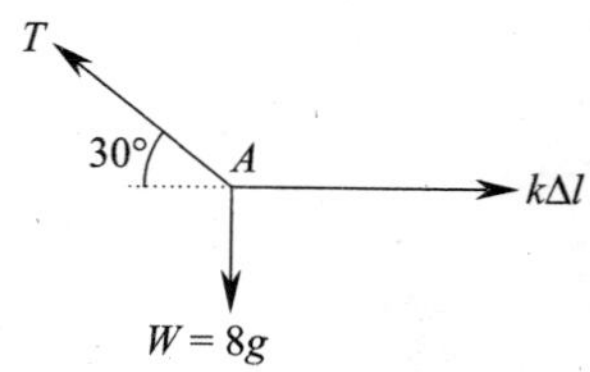

Fig. 5.34(a)

Here even though there are three concurrent forces, it will be convenient to use the method of resolution rather than Lami's theorem to solve the unknowns. Applying the conditions of equilibrium,

$\sum F_x = 0 \Rightarrow$

$$k(\Delta l) - T\cos 30° = 0 \qquad \text{(a)}$$

$\sum F_y = 0 \Rightarrow$

$$T\sin 30° - 8g = 0 \quad \Rightarrow \quad T = 156.96 \text{ N} \qquad \text{(b)}$$

Substituting the value of T in equation (a), we get,

$$\Delta l = T \cos 30°/k = 0.453 \text{ m}$$

Hence, the deformed length of the spring = $BA = l + \Delta l = 0.4 + 0.453 = 0.853$ m

Hence, the remaining horizontal length between points B and C is

$$= 2 - 0.853 = 1.147 \text{ m}$$

Therefore, length, $AC = (1.147)/(\cos 30°) = 1.324$ m

Example 5.13 A smooth cylinder of weight W rests in a trough as shown in Fig. 5.35. Determine the contact forces at points A and B. Assume the inclined surfaces also to be smooth.

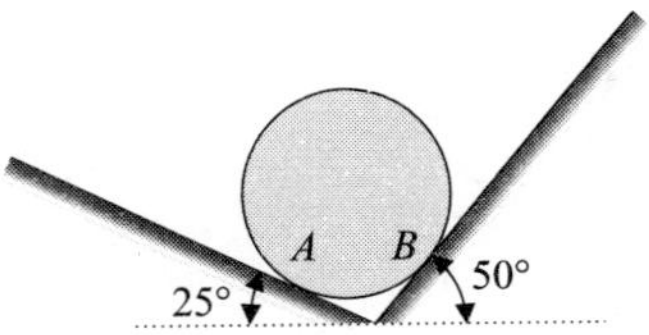

Fig. 5.35

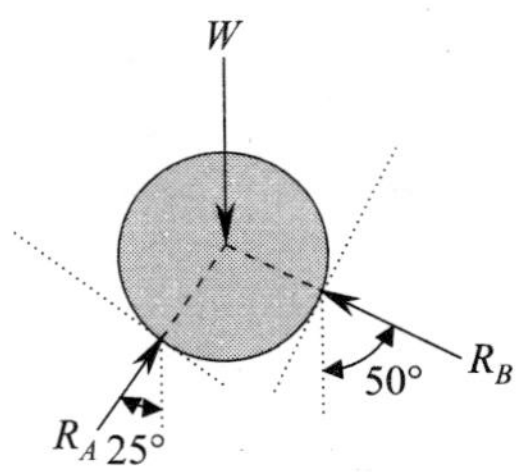

Fig. 5.35(a)

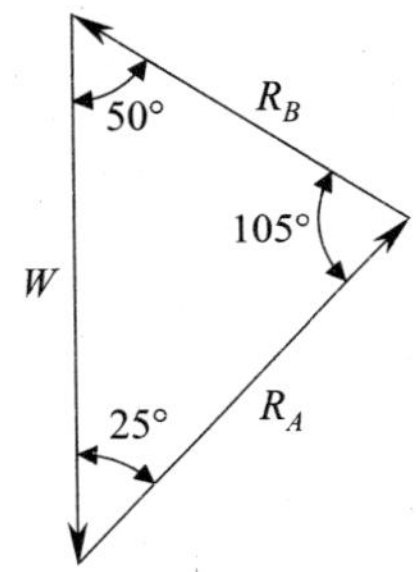

Fig. 5.35(b)

Solution The free-body diagram of the cylinder is shown in Fig. 5.35(a). The forces acting on the cylinder are: (i) its weight W, (ii) reactions R_A and R_B exerted by the inclined planes.

As the reactive forces act normal to the tangential planes, the three forces will be concurrent at the centre of the cylinder. Hence, we can solve the unknowns by applying Lami's theorem to the force triangle [Fig. 5.35(b)].

$$\frac{W}{\sin 105°} = \frac{R_A}{\sin 50°} = \frac{R_B}{\sin 25°}$$

$$\Rightarrow \qquad R_A = W\frac{\sin 50°}{\sin 105°} = 0.793 \text{ W}$$

Also,

$$R_B = W\frac{\sin 25°}{\sin 105°} = 0.438 \text{ W}$$

Example 5.14 Three smooth cylinders, each of radius r and weight W are placed in a rectangular channel of width $5r$ as shown in Fig. 5.36. Determine the reactions at all contact surfaces.

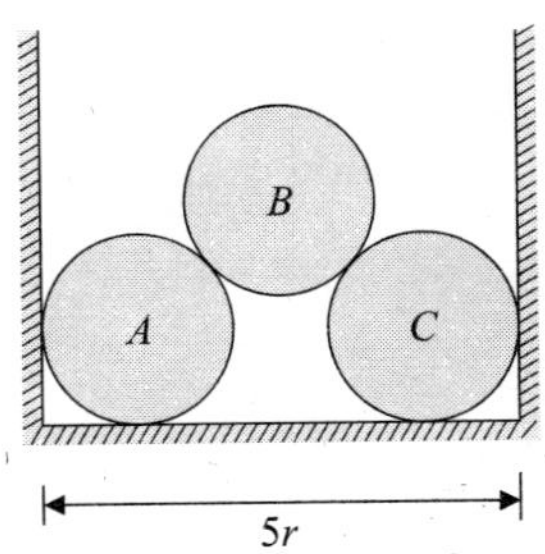

Fig. 5.36

Solution As the width of the channel is given as $5r$, we can readily determine the distances between the centres A, B and C of the cylinders as shown in Fig. 5.36(a). Construct triangle OAB joining the centres of the cylinders A and B [refer Fig. 5.36(b)]. From the triangle OAB we know,

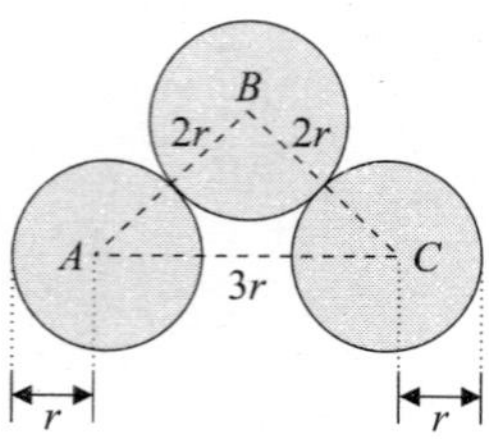

Fig. 5.36(a)

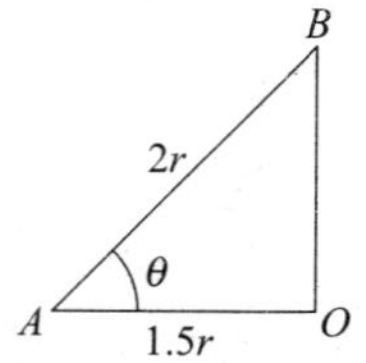

Fig. 5.36(b)

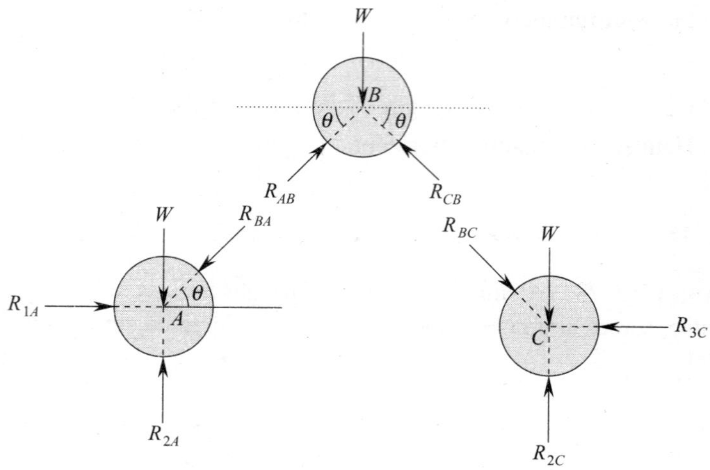

Fig. 5.36(c)

$$\cos\theta = \frac{1.5r}{2r} = \frac{3}{4}$$

$\Rightarrow$ $$\theta = \cos^{-1}(3/4) = 41.41°$$

The free-body diagrams of the cylinders are shown in Fig. 5.36(c): R_{1A} is the force exerted by the left wall on the cylinder A, R_{2A} is the force exerted by the base on the cylinder A, and so on.

Cylinder A

Applying the conditions of equilibrium for the cylinder A,

$\Sigma F_x = 0 \Rightarrow$

$$R_{1A} - R_{BA}\cos\theta = 0$$

$\therefore$ $$R_{1A} = R_{BA}\cos\theta \quad \text{(a)}$$

$\Sigma F_y = 0 \Rightarrow$

$$R_{2A} - W - R_{BA}\sin\theta = 0$$

$\therefore$ $$R_{2A} = W + R_{BA}\sin\theta \quad \text{(b)}$$

Cylinder B

As three concurrent forces are acting on the cylinder B, we can apply Lami's theorem to the force triangle shown in Fig. 5.36(d). Note that due to symmetry, the inclination of the reactions R_{AB} and R_{CB} must be same equal to θ.

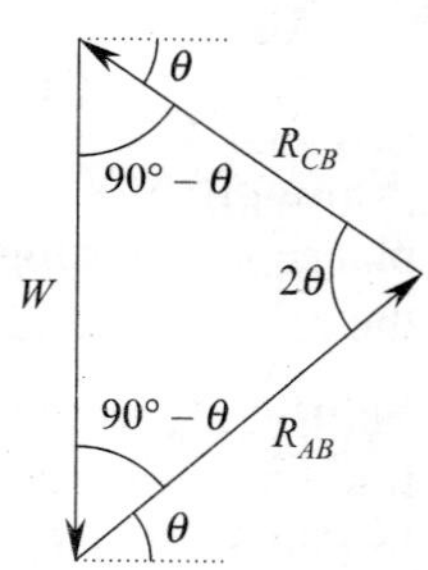

Fig. 5.36(d)

$$\frac{W}{\sin 2\theta} = \frac{R_{AB}}{\sin(90° - \theta)} = \frac{R_{CB}}{\sin(90° - \theta)}$$

$$\frac{W}{\sin 2\theta} = \frac{R_{AB}}{\cos\theta} = \frac{R_{CB}}{\cos\theta}$$

Due to symmetry, we can see that $R_{AB} = R_{CB}$. Hence, we have

$$R_{AB} = R_{CB} = W\frac{\cos 41.41°}{\sin(2 \times 41.41°)} = 0.756\ \text{W} \tag{c}$$

By Newton's third law, we know that $R_{AB} = R_{BA}$. Substituting the value of R_{AB} from equation (c) in equations (a) and (b), we get,

$$R_{1A} = R_{BA}\cos\theta$$
$$= (0.756\ W)\cos 41.41°$$
$$= 0.567\ W$$

and

$$R_{2A} = W + R_{BA}\sin\theta$$
$$= W + (0.756\ W)\sin 41.41°$$
$$= 1.5\ \text{W}$$

Due to symmetry, we know that

$$R_{1A} = R_{3C},\ R_{AB} = R_{CB} \text{ and } R_{2A} = R_{2C}$$

Example 5.15 Two smooth cylinders are placed in a channel as shown in Fig. 5.37. Their respective diameters are indicated in the figure. The weight of the smaller cylinder is W and that of the larger cylinder is $3W$. Determine contact forces at points A, B, C and D. Take $W = 10$ kN.

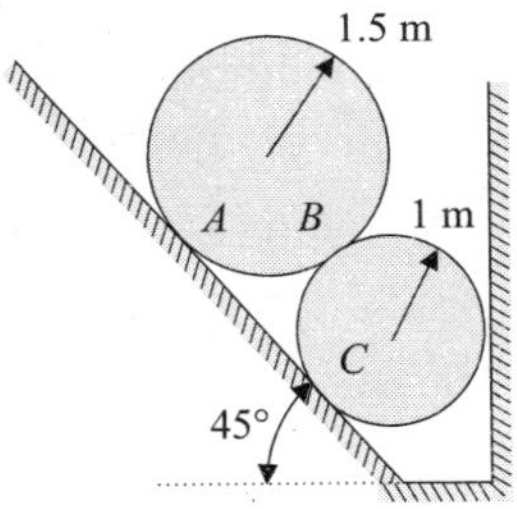

Fig. 5.37

Solution The geometry of the cylinders is shown in Fig. 5.37(a). From A and C draw perpendiculars AE and CF respectively to the inclined plane. Draw FG parallel to AC. Its inclination with the horizontal is same as the inclination of AC, i.e., 45°. Next, we must determine the inclination of EF (i.e., line joining the centres of the two cylinders) with the horizontal. We know,

$$EF = \text{sum of radii of the two cylinders}$$
$$= 1 + 1.5 = 2.5\ \text{m}$$
$$GE = AE - AG = AE - CF \quad [\text{since } AG = CF]$$
$$= 1.5 - 1 = 0.5\ \text{m}$$
$$\angle EFG = \sin^{-1}(GE/EF)$$
$$= \sin^{-1}(0.5/2.5) = 11.54°$$

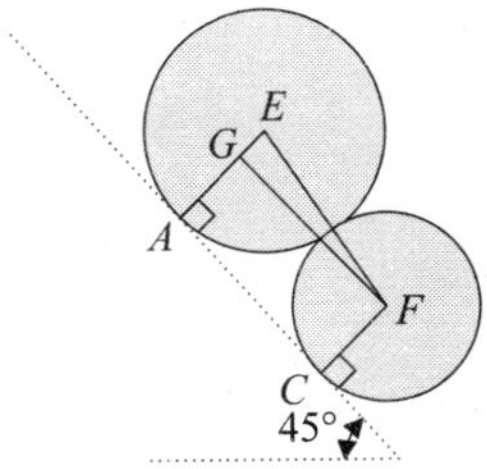

Fig. 5.37(a)

Therefore, inclination of EF with the horizontal is 45° + 11.54° = 56.54°

The free-body diagrams of the cylinders can then be drawn as shown below.

As the number of unknowns is smaller in the case of larger cylinder than that of the smaller cylinder, we proceed with the equilibrium of larger cylinder first.

Larger cylinder

As the forces are concurrent at the centre of the cylinder, we can apply Lami's theorem to the force triangle [see Fig. 5.37(c)].

$$\frac{3W}{\sin 101.54°} = \frac{R_A}{\sin 33.46°} = \frac{R_B}{\sin 45°}$$

$$\Rightarrow \qquad R_A = 3W\left[\frac{\sin 33.46°}{\sin 101.54°}\right]$$

$$= 3(10)\frac{\sin 33.46°}{\sin 101.54°}$$

$$= 16.88 \text{ kN}$$

$$\Rightarrow \qquad R_B = 3W\frac{\sin 45°}{\sin 101.54°}$$

$$= 3(10)\frac{\sin 45°}{\sin 101.54°}$$

$$= 21.65 \text{ kN}$$

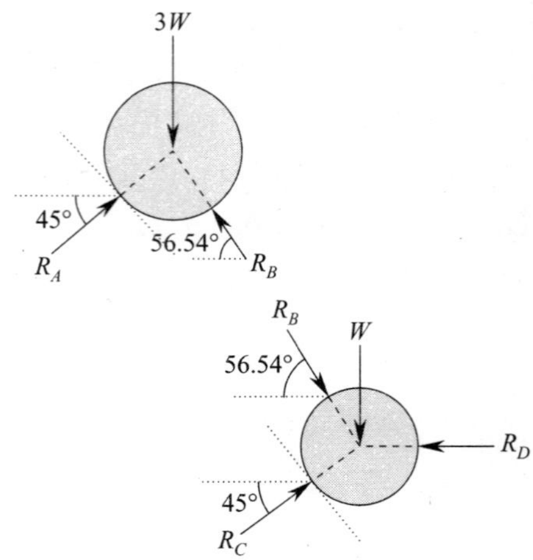

Fig. 5.37(b)

Smaller cylinder

Since there are four forces concurrent at the centre of the cylinder, we use equilibrium conditions to solve the unknowns. Applying the conditions of equilibrium along X and Y directions,

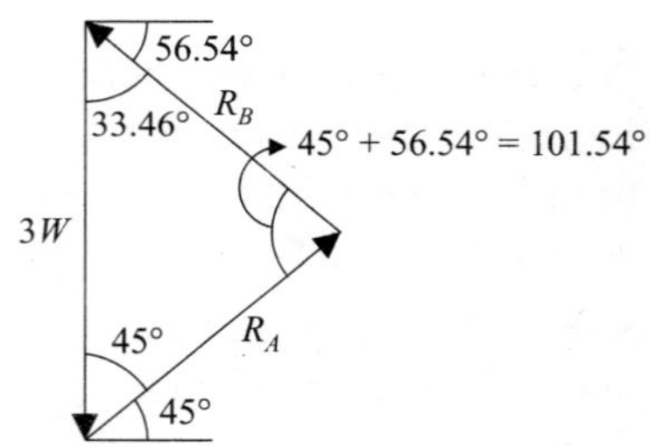

Fig. 5.37(c)

$\sum F_x = 0 \Rightarrow$

$$R_B \cos 56.54° + R_C \cos 45° - R_D = 0$$

$$\therefore \qquad R_D - R_C \cos 45° = R_B \cos 56.54°$$

$$= 11.94 \text{ kN} \qquad \text{(a)}$$

$\sum F_y = 0 \Rightarrow$

$$R_C \sin 45° - W - R_B \sin 56.54° = 0$$

$$\therefore \qquad R_C = [W + R_B \sin 56.54°]/\sin 45°$$

$$= [10 + (21.65) \sin 56.54°]/\sin 45°$$

$$= 39.69 \text{ kN}$$

Substituting this value in equation (a), we get

$$R_D - R_C \cos 45° = 11.94 \text{ kN}$$

$$\Rightarrow \qquad R_D = (39.69) \cos 45° + 11.94 = 40 \text{ kN}$$

Example 5.16 Two identical smooth cylinders each of weight W and radius r are placed in a quarter circular cross-sectional channel of radius R as shown in Fig. 5.38, such that they just fit in the channel. Determine the reactions at the contact surfaces A, B, C and D.

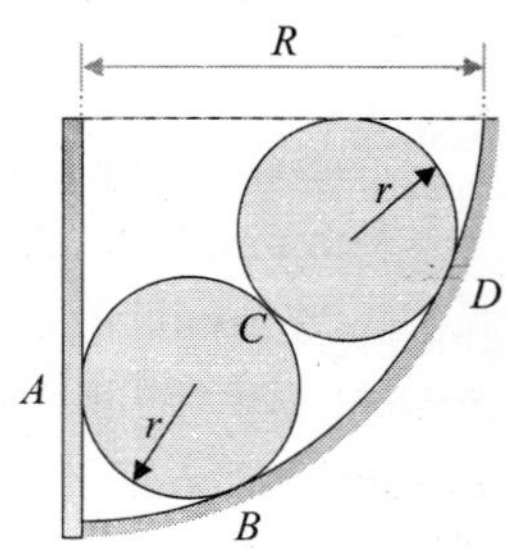

Fig. 5.38

Solution Due to symmetry, we see that the angle made by the radial line OB with the vertical wall is 90°/4 =22.5° as shown in Fig. 5.38(a).

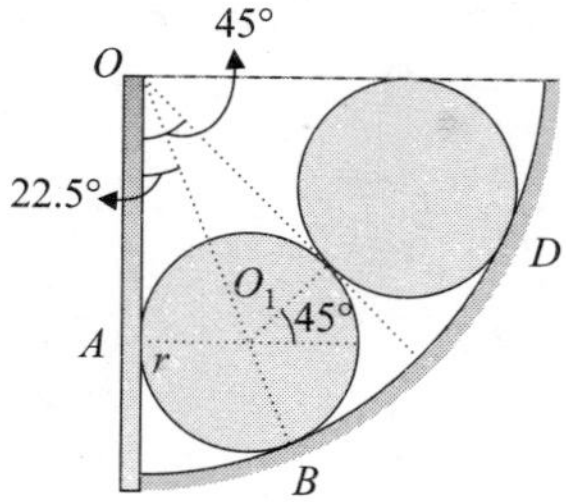

Fig. 5.38(a)

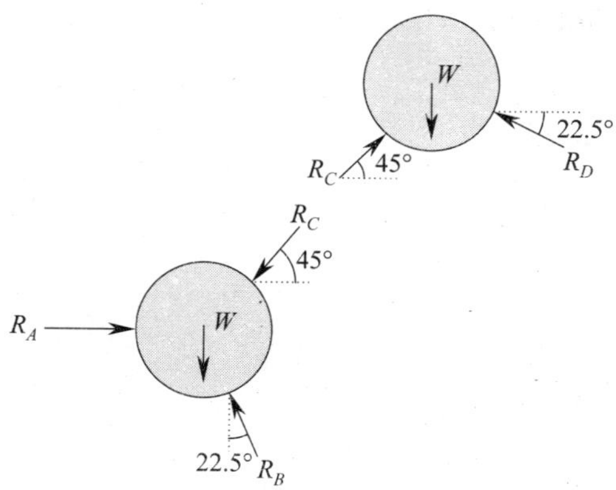

Fig. 5.38(b)

$$\sin 22.5° = \frac{O_1A}{OO_1} = \frac{O_1A}{OB - O_1B} = \frac{r}{R - r}$$

$$\Rightarrow \qquad R - r = \frac{r}{\sin 22.5°}$$

$$\therefore \qquad R = 3.613r$$

As the number of unknowns is smaller in the case of upper cylinder than that of the lower cylinder, we proceed with the equilibrium of upper cylinder first.

Upper cylinder

Applying the conditions of equilibrium along the X and Y directions,

$\sum F_y = 0 \Rightarrow$

$$R_C \sin 45° + R_D \sin 22.5° - W = 0 \tag{a}$$

$\sum F_x = 0 \Rightarrow$

$$R_C \cos 45° - R_D \cos 22.5° = 0 \tag{b}$$

Solving for R_C and R_D from the above two equations, we get

$$R_D = 0.765\ W \quad \text{and} \quad R_C = W$$

Lower cylinder

Applying the conditions of equilibrium along the X and Y directions,

$\sum F_y = 0 \Rightarrow$

$$R_B \cos 22.5° - W - R_C \sin 45° = 0 \tag{c}$$

$\sum F_x = 0 \Rightarrow$

$$R_A - R_C \cos 45° - R_B \sin 22.5° = 0 \tag{d}$$

Substituting the value of R_C in the above two equations, we get

$$R_B = 1.848\ W \quad \text{and} \quad R_A = 1.414\ W$$

Example 5.17 Two inclined planes are placed on a smooth horizontal plane, and a sphere rests upon them. The inclined planes are prevented from separating by a string connecting them at the lowest points. Show that the ratio of the weight of the sphere to the tension of the string is the sum of the cotangents of the inclinations of the planes.

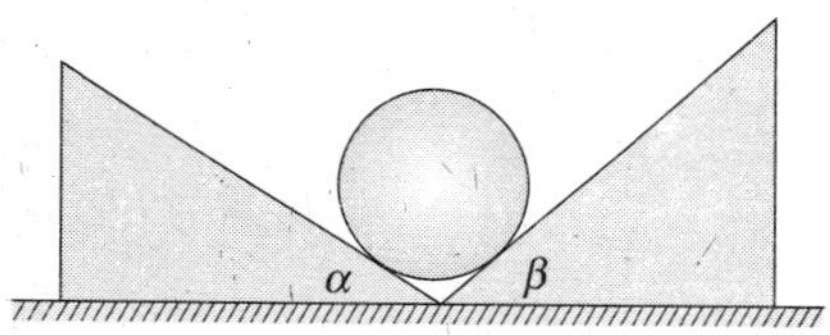

Fig. 5.39

Solution The free-body diagrams of the sphere and the inclined planes are shown in Fig. 5.39(a).

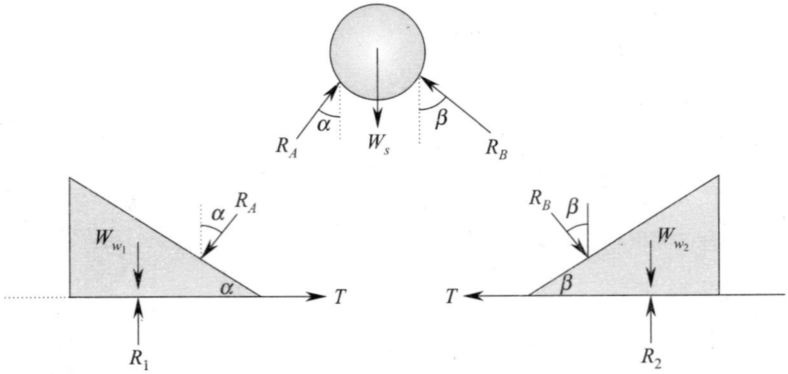

Fig. 5.39(a)

Sphere

Applying the conditions of equilibrium along the X and Y directions,

$\sum F_x = 0 \Rightarrow$

$$R_A \sin\alpha - R_B \sin\beta = 0$$

$\therefore$
$$R_A \sin\alpha = R_B \sin\beta$$

$\sum F_y = 0 \Rightarrow$

$$R_A \cos\alpha + R_B \cos\beta - W_s = 0$$

$\therefore$
$$R_A \cos\alpha + R_B \cos\beta = W_s \qquad \text{(a)}$$

Inclined planes

Applying the condition of equilibrium along the X direction for the left inclined plane,

$\sum F_x = 0 \Rightarrow$

$$T - R_A \sin\alpha = 0$$

$\therefore$
$$R_A = T/\sin\alpha \qquad \text{(b)}$$

Similarly, applying the condition of equilibrium along the X direction for the right inclined plane,

$$R_B \sin\beta - T = 0$$

$\therefore$
$$R_B = T/\sin\beta \qquad \text{(c)}$$

Substituting the values of R_A and R_B respectively from equations (b) and (c) in equation (a), we have

$$\frac{T}{\sin\alpha}\cos\alpha + \frac{T}{\sin\beta}\cos\beta = W_s$$

$$T[\cot\alpha + \cot\beta] = W_s$$

$\Rightarrow$
$$\frac{W_s}{T} = \cot\alpha + \cot\beta$$

5.3.2 Equilibrium of Concurrent Forces in Space

For a *concurrent force* system in *space*, expressing the resultant in terms of orthogonal components of individual forces, the equilibrium condition can be stated as

$$\vec{R} = (\sum F_x)\vec{i} + (\sum F_y)\vec{j} + (\sum F_z)\vec{k} = \vec{O} \tag{5.10}$$

Hence, the necessary and sufficient conditions for equilibrium are

$$\sum F_x = 0 \tag{5.11}$$

$$\sum F_y = 0 \tag{5.12}$$

and

$$\sum F_z = 0 \tag{5.13}$$

Since there are three independent equations, a *maximum* of only *three* unknowns can be solved.

Example 5.18 A vertical load of 50 kg is supported by three rods positioned as shown in Fig. 5.40. Determine the force in each rod.

Solution From the figure, we see that the coordinates of points *A*, *B*, *C* and *D* are

$$A(-4, 1, 0),\ B(3, -3, 0),\ C(3, 2, 0) \text{ and } D(0, 0, 6)$$

Determination of unit vectors along DA, DB and DC

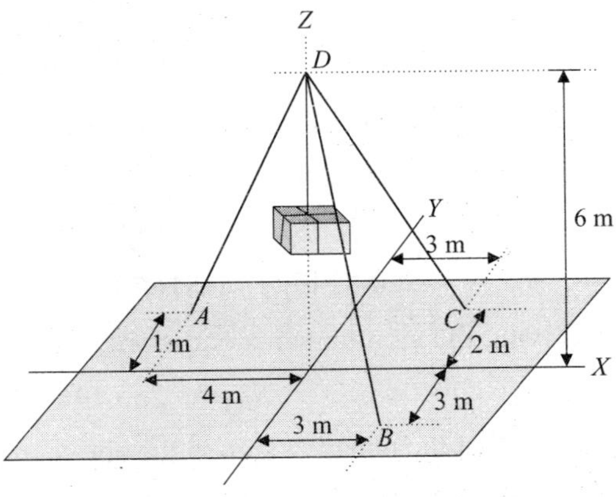

Fig. 5.40

$$\overrightarrow{DA} = \overrightarrow{OA} - \overrightarrow{OD}$$

$$= -4\vec{i} + \vec{j} - 6\vec{k}$$

$$\therefore \quad \hat{n}_{DA} = \frac{-4\vec{i} + \vec{j} - 6\vec{k}}{\sqrt{(-4)^2 + (1)^2 + (-6)^2}} = \frac{-4\vec{i} + \vec{j} - 6\vec{k}}{\sqrt{53}}$$

$$= -0.549\vec{i} + 0.137\vec{j} - 0.824\vec{k}$$

$$\overrightarrow{DB} = \overrightarrow{OB} - \overrightarrow{OD}$$

$$= 3\vec{i} - 3\vec{j} - 6\vec{k}$$

$$\therefore \quad \hat{n}_{DB} = \frac{3\vec{i} - 3\vec{j} - 6\vec{k}}{\sqrt{(3)^2 + (-3)^2 + (-6)^2}} = \frac{3\vec{i} - 3\vec{j} - 6\vec{k}}{\sqrt{54}}$$

$$= 0.408\vec{i} - 0.408\vec{j} - 0.816\vec{k}$$

$$\overrightarrow{DC} = \overrightarrow{OC} - \overrightarrow{OD}$$

$$= 3\vec{i} + 2\vec{j} - 6\vec{k}$$

$$\therefore \quad \hat{n}_{DC} = \frac{3\vec{i} + 2\vec{j} - 6\vec{k}}{\sqrt{(3)^2 + (2)^2 + (-6)^2}} = \frac{3\vec{i} + 2\vec{j} - 6\vec{k}}{7}$$

$$= 0.429\vec{i} + 0.286\vec{j} - 0.857\vec{k}$$

Let S_{DA}, S_{DB} and S_{DC} be the magnitudes of the forces in the rods DA, DB and DC respectively. Then the force vectors can be represented as

$$\vec{S}_{DA} = S_{DA}\hat{n}_{DA}$$
$$= S_{DA}[-0.549\vec{i} + 0.137\vec{j} - 0.824\vec{k}]$$
$$\vec{S}_{DB} = S_{DB}\hat{n}_{DB}$$
$$= S_{DB}[0.408\vec{i} - 0.408\vec{j} - 0.816\vec{k}]$$
$$\vec{S}_{DC} = S_{DC}\hat{n}_{DC}$$
$$= S_{DC}[0.429\vec{i} + 0.286\vec{j} - 0.857\vec{k}]$$

The load at D can be represented in vector form as

$$\vec{W} = 50 \times 9.81[-\vec{k}] \quad \text{(Note that the weight is directed along the negative } Z\text{-axis.)}$$
$$= -490.5\vec{k}$$

Since the forces are concurrent at the point D, the resultant of the system of forces is given as

$$\vec{R} = \vec{S}_{DA} + \vec{S}_{DB} + \vec{S}_{DC} + \vec{W}$$
$$= [-0.549S_{DA} + 0.408S_{DB} + 0.429S_{DC}]\vec{i}$$
$$+ [0.137S_{DA} - 0.408S_{DB} + 0.286S_{DC}]\vec{j}$$
$$+ [-0.824S_{DA} - 0.816S_{DB} - 0.857S_{DC} - 490.5]\vec{k}$$

Applying the condition of equilibrium, $\vec{R} = \vec{O}$, we get three independent simultaneous equations:

$$-0.549S_{DA} + 0.408S_{DB} + 0.429S_{DC} = 0$$
$$0.137S_{DA} - 0.408S_{DB} + 0.286S_{DC} = 0$$
$$-0.824S_{DA} - 0.816S_{DB} - 0.857S_{DC} = 490.5$$

Solving the simultaneous equations, we get

$$S_{DA} = -255.14 \text{ N}, S_{DB} = -188.9 \text{ N and } S_{DC} = -147.02 \text{ N}$$

(Note that the negative sign in the above values indicates that the members are under compression.)

Example 5.19 A tripod supports a load of 30 kg as shown in Fig. 5.41. Determine the forces in the legs of the tripod, if the length of each leg is 5 m and the ends touching the ground are at an equal distance of 3 m from one another.

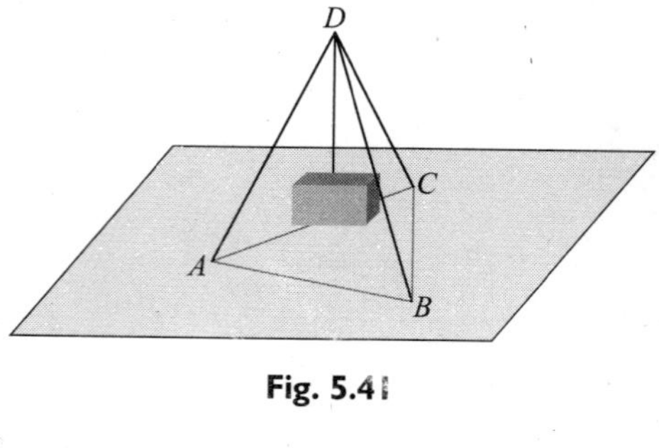

Fig. 5.41

Solution Since the structure is symmetric, we can apply simple trigonometry instead of vector notation. Let θ be the angle made by each leg with the vertical [refer Fig. 5.41(a)].

Since the forces in the legs and the weight W are concurrent at point D, we can apply the condition of equilibrium along vertical direction at point D,

$$S_{DA}\cos\theta + S_{DB}\cos\theta + S_{DC}\cos\theta + W = 0$$

Due to symmetry, the forces in the legs are equal. Therefore,

$$3S\cos\theta = -W$$

$$\Rightarrow \quad S = -\frac{W}{3\cos\theta} \qquad \text{(a)}$$

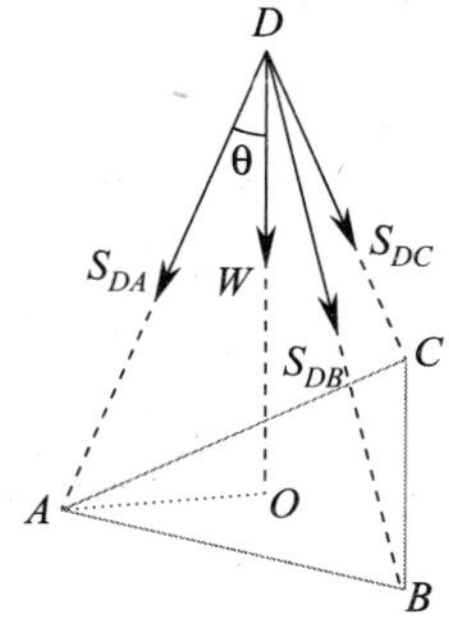

Fig. 5.41(a)

Determination of angle θ

Consider the triangle *ABC*. Draw perpendiculars *DC*, *EA* and *FB* to sides *AB*, *BC* and *CA* respectively. They will intersect at point *O*.

$$AE = \sqrt{(AB)^2 - (BE)^2}$$

$$= \sqrt{(3)^2 - (1.5)^2} = 2.6 \text{ m}$$

From the property of the triangle, we know that

$$OA : OE = 2 : 1$$

Therefore, $$OA = \left(\frac{2}{3}\right) AE = 1.73 \text{ m}$$

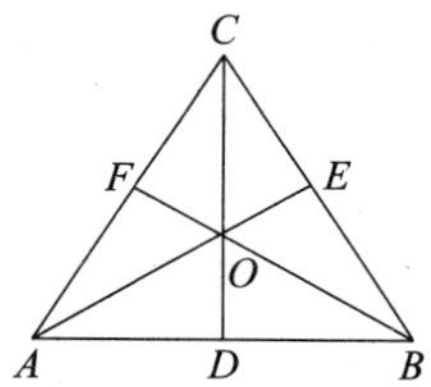

Fig. 5.41(b)

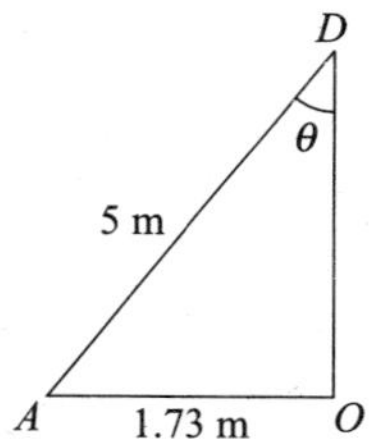

Fig. 5.41(c)

From Fig. 5.41(c), $$OD = \sqrt{(5)^2 - (1.73)^2} = 4.69 \text{ m}$$

Therefore, $$\cos\theta = 4.69/5 \qquad \text{(b)}$$

Substituting this value in equation (a), we have

$$S = -\frac{W}{3\cos\theta} = -\frac{30 \times 9.81}{3 \times (4.69/5)}$$

$$= -104.58 \text{ N}$$

The negative sign indicates that the members are under compression.

Example 5.20 A box of mass 200 kg is supported by three cables as shown in Fig. 5.42. Determine the tension in each cable.

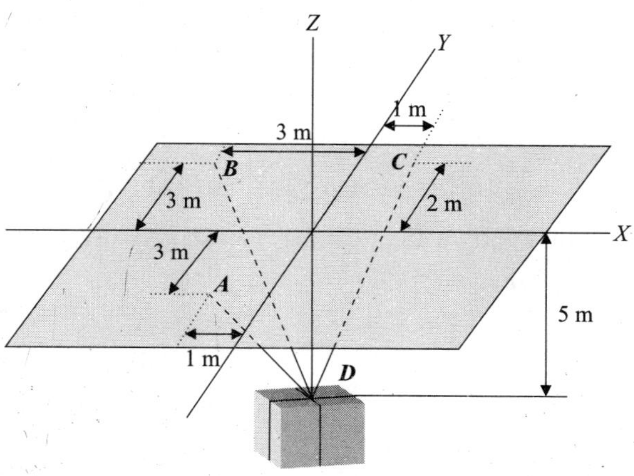

Fig. 5.42

Solution From the figure, we see that the coordinates of points A, B, C and D are

$$A(-1, -3, 0),\ B(-3, 3, 0),\ C(1, 2, 0) \text{ and } D(0, 0, -5)$$

Determination of unit vectors along DA, DB and DC

$$\overrightarrow{DA} = \overrightarrow{OA} - \overrightarrow{OD}$$

$$= -\vec{i} - 3\vec{j} + 5\vec{k}$$

$$\therefore \quad \hat{n}_{DA} = \frac{-\vec{i} - 3\vec{j} + 5\vec{k}}{\sqrt{(-1)^2 + (-3)^2 + (5)^2}} = \frac{-\vec{i} - 3\vec{j} + 5\vec{k}}{\sqrt{35}}$$

$$= -0.169\vec{i} - 0.507\vec{j} + 0.845\vec{k}$$

$$\overrightarrow{DB} = \overrightarrow{OB} - \overrightarrow{OD}$$

$$= -3\vec{i} + 3\vec{j} + 5\vec{k}$$

$$\therefore \quad \hat{n}_{DB} = \frac{-3\vec{i} + 3\vec{j} + 5\vec{k}}{\sqrt{(-3)^2 + (3)^2 + (5)^2}} = \frac{-3\vec{i} + 3\vec{j} + 5\vec{k}}{\sqrt{43}}$$

$$= -0.457\vec{i} + 0.457\vec{j} + 0.762\vec{k}$$

$$\overrightarrow{DC} = \overrightarrow{OC} - \overrightarrow{OD}$$

$$= \vec{i} + 2\vec{j} + 5\vec{k}$$

$$\therefore \quad \hat{n}_{DC} = \frac{\vec{i} + 2\vec{j} + 5\vec{k}}{\sqrt{(1)^2 + (2)^2 + (5)^2}} = \frac{\vec{i} + 2\vec{j} + 5\vec{k}}{\sqrt{30}}$$

$$= 0.183\vec{i} + 0.365\vec{j} + 0.913\vec{k}$$

Let T_{DA}, T_{DB} and T_{DC} be the tensions in the cables DA, DB and DC respectively. Then the tension vectors can be represented as

$$\vec{T}_{DA} = T_{DA}\hat{n}_{DA}$$

$$= T_{DA}[-0.169\vec{i} - 0.507\vec{j} + 0.845\vec{k}] \qquad \text{(a)}$$

$$\vec{T}_{DB} = T_{DB}\hat{n}_{DB}$$

$$= T_{DB}[-0.457\vec{i} + 0.457\vec{j} + 0762\vec{k}] \qquad \text{(b)}$$

$$\vec{T}_{DC} = T_{DC}\hat{n}_{DC}$$

$$= T_{DC}[0.183\vec{i} + 0.365\vec{j} + 0.913\vec{k}] \qquad \text{(c)}$$

As the mass of the box is 200 kg, its weight can be represented in vector form as

$$\vec{W} = 200 \times 9.81[-\vec{k}]$$

$$= -1962\vec{k} \qquad \text{(d)}$$

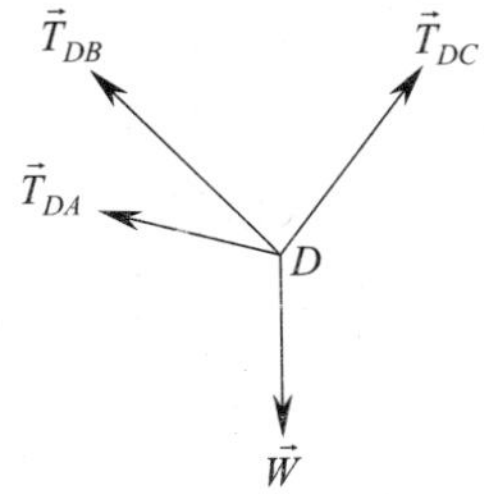

Fig. 5.42(a)

Note that the weight is directed along the negative Z-axis. Since the tension and weight vectors are concurrent at point D, the resultant of the system of forces is given as the vector addition of individual forces.

$$\vec{R} = \vec{T}_{DA} + \vec{T}_{DB} + \vec{T}_{DC} + \vec{W}$$
$$= [-0.169T_{DA} - 0.457T_{DB} + 0.183T_{DC}]\vec{i}$$
$$+ [-0.507T_{DA} + 0.457T_{DB} + 0.365T_{DC}]\vec{j}$$
$$+ [0.845T_{DA} + 0.762T_{DB} + 0.913T_{DC} - 1962]\vec{k}$$

Applying the condition of equilibrium, $\vec{R} = \vec{O}$, we get three independent simultaneous equations:

$$-0.169T_{DA} - 0.457T_{DB} + 0.183T_{DC} = 0 \quad \text{(e)}$$
$$-0.507T_{DA} + 0.457T_{DB} + 0.365T_{DC} = 0 \quad \text{(f)}$$
$$0.845T_{DA} + 0.762T_{DB} + 0.913T_{DC} = 1962 \quad \text{(g)}$$

Solving the above simultaneous equations, we get

$$T_{DA} = 950.02 \text{ N}, T_{DB} = 117.26 \text{ N and } T_{DC} = 1171.92 \text{ N}$$

5.3.3 Equilibrium of Coplanar Non-Concurrent Forces

If the forces acting on the free body are *non-concurrent* then the equivalent resultant will be a single force acting at a common point and a moment about the same point. The effect of such a force system will be to *translate* the body as well as to *rotate* it. Hence, for equilibrium to exist, both the force and the moment must be *null* vectors, i.e.,

$$\vec{R} = \sum\vec{F} = \vec{O} \quad (5.14)$$

and

$$\vec{M} = \sum(\vec{r} \times \vec{F}) = \vec{O} \quad (5.15)$$

When the forces acting on a body lie in the same plane (say X–Y plane) but are non-concurrent, the body will have rotational motion perpendicular to the plane in addition to the translational motion along the plane. Hence, the necessary and sufficient conditions for static equilibrium are

$$\sum F_x = 0 \quad (5.16)$$
$$\sum F_y = 0 \quad (5.17)$$

and

$$\sum M_z = 0 \quad (5.18)$$

where M_z is the moment about the perpendicular Z-axis.

Beams and Types of Beams One of the structural member that we come across in this section is a **beam**. It is a horizontal structural member that is designed to resist forces transverse to its axis. It is held in position by various supports. Depending upon the nature of the supports, beams can be classified as follows:

Simply supported beam Simply supported beam is a beam that is supported by a hinge at one end and a roller at the other end.

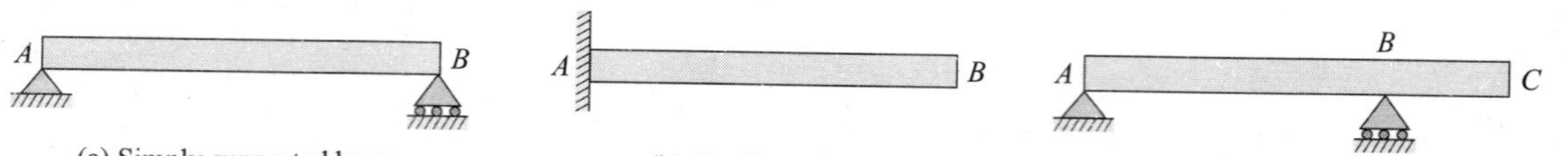

(a) Simply supported beam (b) Cantilever beam (c) Overhanging beam

Fig. 5.43 Types of beams

Cantilever beam Cantilever beam is a beam that is supported by a built-in or fixed support at one end and the other end free.

Overhanging beam The beam is simply supported at A and B, but it also projects beyond the support to the point C, which is a free end.

The corresponding support reactions in each case were already discussed in Section 5.2.2(E).

Types of Loading There are various types of loading which we come across in beams. These loads may be *dead loads* due to their own weight and *live loads* due to people, movable items, etc.

Concentrated (or) point loads These are forces acting at a particular point on a beam, say a man standing on a beam. His weight may be assumed to be concentrated and acting at a point.

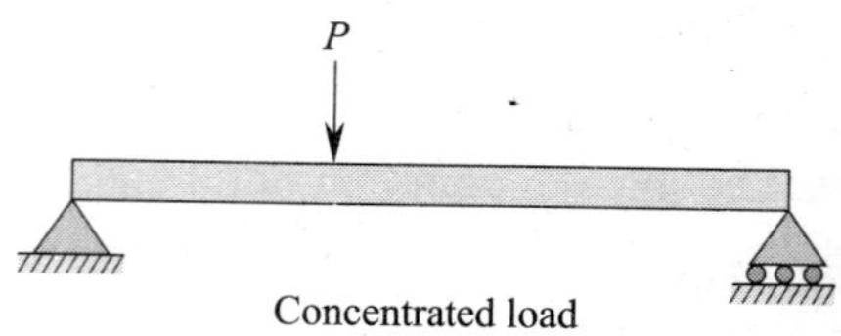

Concentrated load

Fig. 5.44

Distributed loads Distributed loads are those loads that are distributed over a distance. The force distribution is expressed in terms of *force per unit length*. If the load is uniformly distributed over the length, it is termed **uniformly distributed load**; if it is linearly varying over the length it acts, it is termed **linearly varying load**. However, for calculation purposes, we can replace them by an *equivalent concentrated load* such that its magnitude is same as the distributed load and its location is determined from moment considerations. For instance, a uniformly distributed load shown in Fig. 5.45 can be replaced by a point load of magnitude wl and placed from the fixed end at a distance of $a + l/2$. Similarly, a uniformly varying load shown in Fig. 5.46 can be replaced by a point load of magnitude $wl/2$ and placed from the right end of the load at a distance of $l/3$. [Refer to Solved Examples 4.21 and 4.22 in Chapter 4].

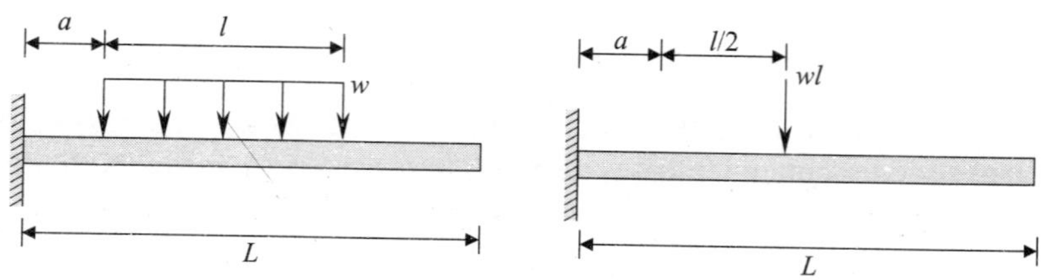

Uniformly distributed load Equivalent point load

Fig. 5.45

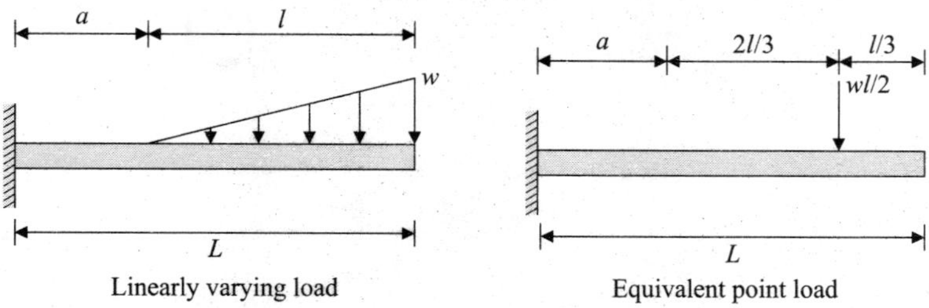

Fig. 5.46

Example 5.21 Determine the support reactions for the cantilever beam AB shown:

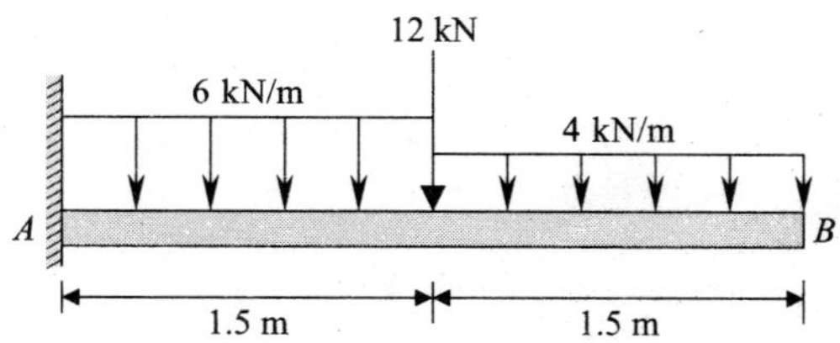

Fig. 5.47

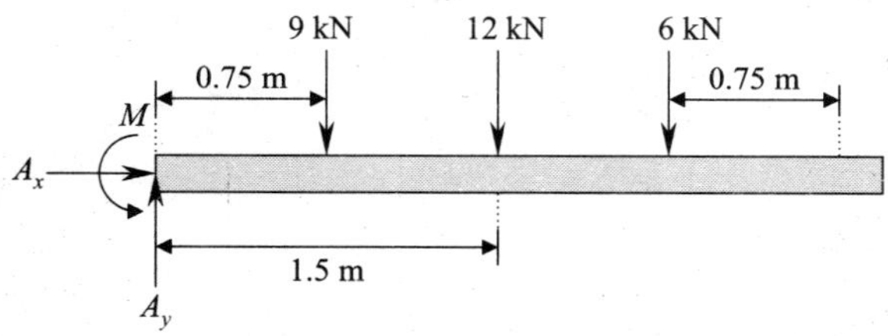

Fig. 5.47(a)

Solution The distributed loads can be replaced by equivalent point loads. The equivalent point load for distributed load with load intensity 6 kN/m is

$$6 \times 1.5 = 9 \text{ kN}$$

and it acts at a distance of $1.5/2 = 0.75$ m from the end A. Similarly, the equivalent point load for distributed load with load intensity 4 kN/m is

$$4 \times 1.5 = 6 \text{ kN}$$

and it acts at a distance of $1.5/2 = 0.75$ m from the end B.

The free-body diagram of the beam is shown in Fig. 5.47 (a). Since the beam is fixed at the end A, the reactions at A are A_x, A_y and moment M along the Z direction.

Applying the conditions of equilibrium for the beam,

$\Sigma F_x = 0 \Rightarrow$

$$A_x = 0$$

$\Sigma F_y = 0 \Rightarrow$

$$A_y - 9 - 12 - 6 = 0$$

$$\therefore \quad A_y = 27 \text{ kN}$$

As the forces are non-concurrent, in addition, we must also take summation of the moments about a point and equate it to zero for equilibrium. Though moment can be taken about any point on the beam, it can be seen that more number of unknowns can be eliminated if we take the moment about the point A. Hence, taking summation of the moments about A and equating it to zero,

$$\Sigma M_A = 0 \Rightarrow$$

$$M - (9 \times 0.75) - (12 \times 1.5) - (6 \times 2.25) = 0$$

$$\therefore \quad M = 38.25 \text{ kN.m}$$

Example 5.22 Find the support reactions for the cantilever beam shown in Fig. 5.48. The load increases linearly from zero to w N/m in the first half of the length and in the second half it decreases from w to zero.

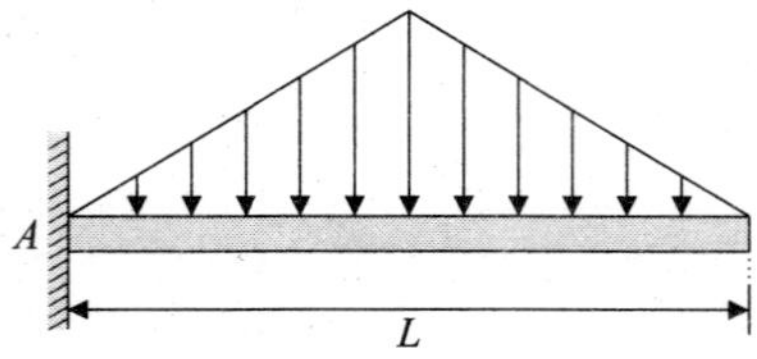

Fig. 5.48

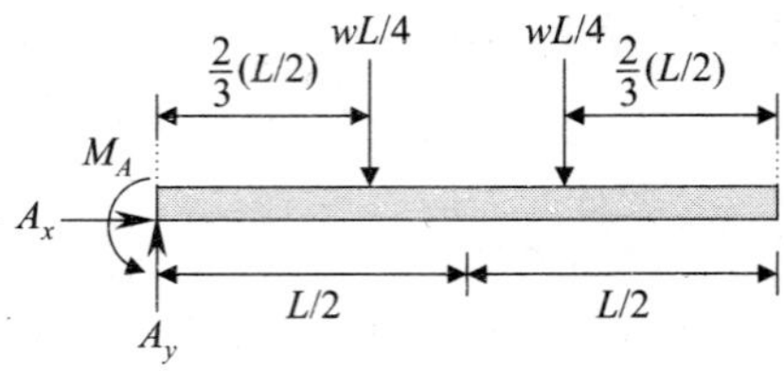

Fig. 5.48(a)

Solution Even though the load is *distributed* over the span of the beam, for finding the support reactions, we replace them by equivalent concentrated loads. From Example 4.22, we know that the equivalent load for a triangular distribution is $wl/2$ placed at $2l/3$ from the vertex. Hence, the equivalent load is shown in Fig. 5.48(a).

Applying the conditions of equilibrium for the beam,

$$\Sigma F_x = 0 \Rightarrow$$

$$A_x = 0$$

$$\Sigma F_y = 0 \Rightarrow$$

$$A_y - \frac{wL}{4} - \frac{wL}{4} = 0$$

$$\therefore \quad A_y = \frac{wL}{2}$$

As the forces are non-concurrent, in addition we must also take summation of the moments about a point and equate it to zero for equilibrium. Hence, taking summation of the moments about A and equating it to zero,

$$\Sigma M_A = 0 \Rightarrow$$

$$M_A - \frac{wL}{4}\frac{L}{3} - \frac{wL}{4}\left(L - \frac{L}{3}\right) = 0$$

$$\therefore \quad M_A = \frac{wL^2}{4}$$

Example 5.23 Find the support reactions for the cantilever beam loaded as shown in Fig. 5.49.

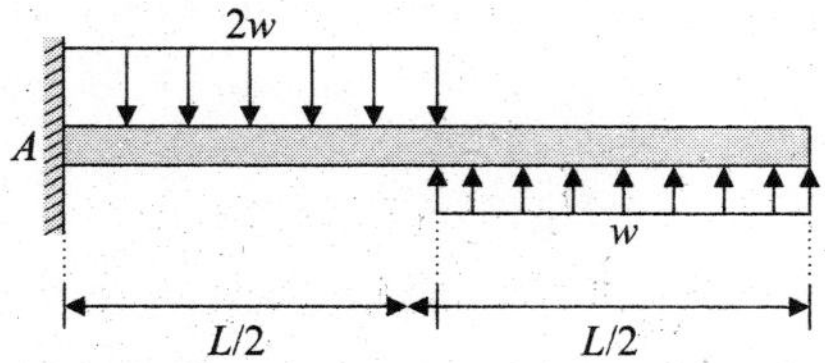

Fig. 5.49

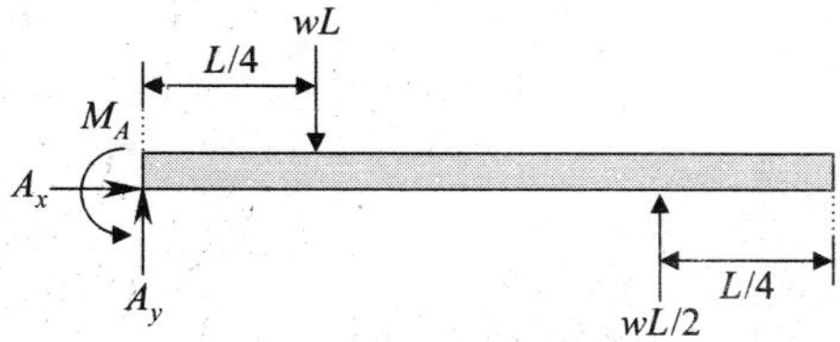

Fig. 5.49(a)

Solution The equivalent point loads for the load distributions are shown in Fig. 5.49(a).

Applying the conditions of equilibrium for the beam,

$\sum F_x = 0 \Rightarrow$

$$A_x = 0$$

$\sum F_y = 0 \Rightarrow$

$$A_y - wL + \frac{wL}{2} = 0$$

$$\therefore \quad A_y = \frac{wL}{2}$$

Taking summation of the moments about A and equating it to zero,

$\sum M_A = 0 \Rightarrow$

$$M_A - wL\frac{L}{4} + \frac{wL}{2}\left(L - \frac{L}{4}\right) = 0$$

$$M_A - \frac{wL^2}{4} + \frac{3wL^2}{8} = 0$$

$$\Rightarrow \quad M_A = -\frac{wL^2}{8}$$

Example 5.24 Determine the support reactions for the beam AB loaded as shown in Fig. 5.50. Neglect the weight of the beam.

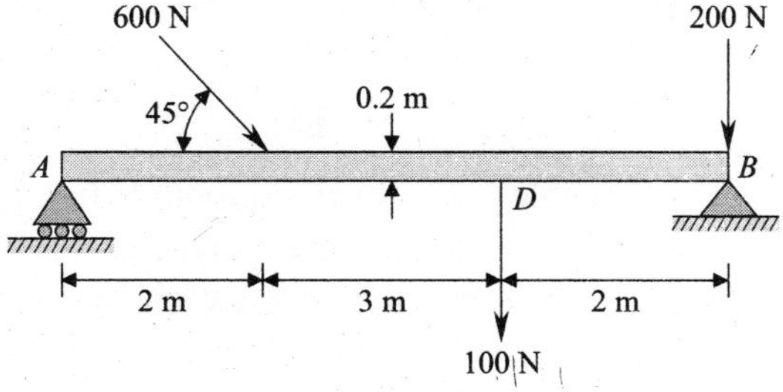

Fig. 5.50

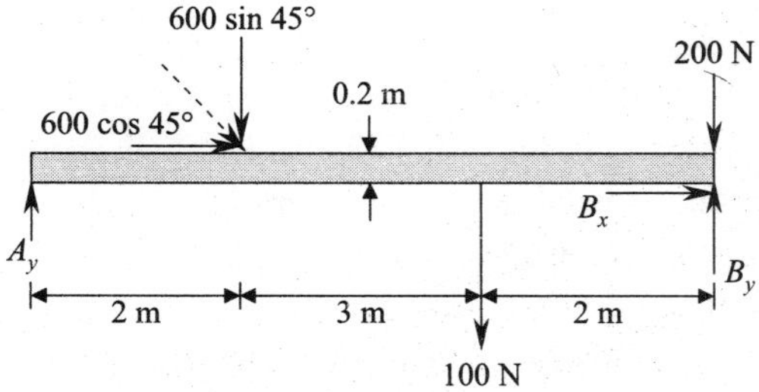

Fig. 5.50(a)

Solution The free-body diagram of the beam is shown in Fig. 5.50(a). Since the end A is a roller support, one vertical reaction A_y is shown; the end B is hinged and hence two reactions B_x and B_y along X and Y axes respectively are shown. The 600 N force is resolved into components as shown.

Applying the conditions of equilibrium,

$\sum F_x = 0 \Rightarrow$

$$B_x + 600\cos 45° = 0$$

$$\therefore \quad B_x = -424.26 \text{ N}$$

$\sum F_y = 0 \Rightarrow$

$$A_y + B_y - 100 - 200 - 600\sin 45° = 0$$

$$\therefore \quad A_y + B_y = 100 + 200 + 600 \sin 45°$$

$$= 300 + 600 \sin 45°$$

$$= 724.26 \text{ N} \qquad \text{(a)}$$

Taking summation of the moments about the point B (as it eliminates more number of unknowns) and equating it to zero,

$\sum M_B = 0 \Rightarrow$

$$-(A_y \times 7) - (600 \cos 45° \times 0.2) + (600 \sin 45° \times 5) + (100 \times 2) = 0$$

$$\Rightarrow \quad (A_y \times 7) = (600 \sin 45° \times 5) + (100 \times 2) - (600 \cos 45° \times 0.2)$$

$$\therefore \quad A_y = 319.5 \text{ N} \qquad \text{(b)}$$

Substituting this value in equation (a),

$$A_y + B_y = 724.26 \text{ N}$$

$$\Rightarrow \quad B_y = 724.26 - 319.5 = 404.76 \text{ N}$$

Example 5.25 A beam AC hinged at A is held in a horizontal position by a cable attached at end C and passing over a smooth pulley as shown in Fig. 5.51. The free end of the cable is connected to a weight 2000 N that rests on the beam. Determine the reaction at A and tension in the cable. Neglect the weight of the beam.

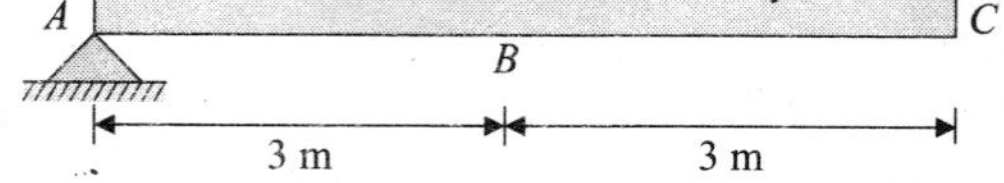

Fig. 5.51

Solution Since a weight is resting on the beam, we can get a clear picture if we consider free-body diagrams for beam and weight separately as shown in Fig. 5.51(a). The forces acting on the weight resting on the beam are its own weight, tension T due to its attachment with the cable and normal reaction R exerted by the beam. Note that as the cable passes over a smooth pulley, the tension at both the ends will be equal to T. The forces acting on the beam are the support reactions at the point A, normal reaction R exerted by the weight on the beam and tension T due to its attachment with the cable. Note that T has been resolved into components along X and Y directions.

Weight W

Applying the equilibrium condition for the weight,

$$\sum F_y = 0 \Rightarrow$$

$$T + R - 2000 = 0$$

$$\therefore \quad T + R = 2000 \text{ N} \qquad \text{(a)}$$

Beam

Similarly, applying the conditions of equilibrium for the beam,

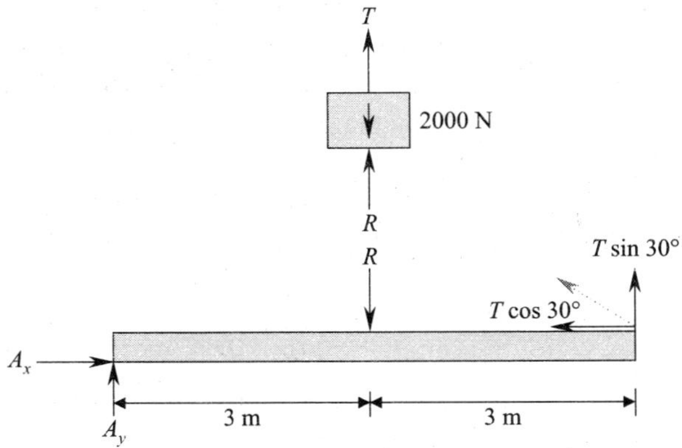

Fig. 5.51(a)

$$\sum F_x = 0 \Rightarrow$$

$$A_x - T\cos 30° = 0$$

$$\therefore \quad A_x = T\cos 30° \qquad \text{(b)}$$

$$\sum F_y = 0 \Rightarrow$$

$$A_y + T\sin 30° - R = 0$$

$$\Rightarrow \quad A_y = R - T\sin 30°$$

$$= R - 0.5T \qquad \text{(c)}$$

Taking moment about the hinge A and equating it to zero,

$$\sum M_A = 0 \Rightarrow$$

$$T\sin 30°(6) - R(3) = 0$$

$$\Rightarrow \quad T = R \qquad \text{(d)}$$

Substituting this in the equation (a), we get,

$$T = R = 1000 \text{ N}$$

From equation (b), we get

$$A_x = T\cos 30°$$

$$= (1000)\cos 30°$$

$$= 866.03 \text{ N}$$

From equation (c), we get

$$A_y = R - 0.5T$$

$$= (1000) - 0.5(1000)$$

$$= 500 \text{ N}$$

Example 5.26 A smooth pulley supporting a load of 3000 N is mounted at B on a horizontal beam AC as shown in Fig. 5.52. If the beam weighs 1000 N, find the support reactions at A and C. Neglect the weight and size of the pulley.

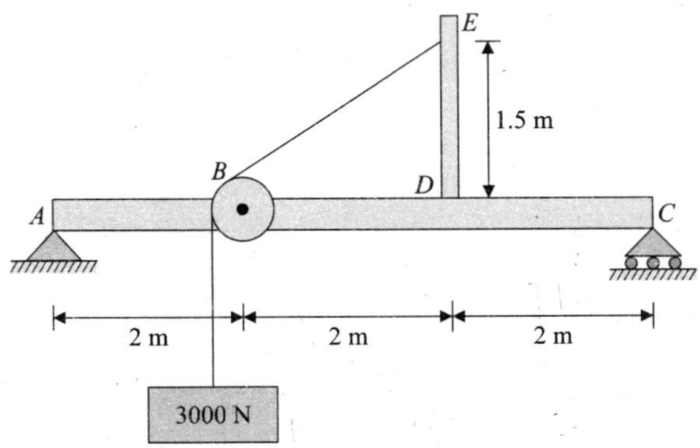

Fig. 5.52

Solution For a clear understanding of the problem, let us isolate the bodies and analyze the free-body diagrams separately. The forces shown in the free-body diagram of the pulley are reactions B_x and B_y as B is a hinge point and equal tension T on both ends of the string as the pulley is frictionless. The forces shown in the free-body diagram of the beam are its weight placed at its centre, reactions A_x and A_y at the support point A, reaction C_y at the support point C, reactions B_x and B_y shown at the point B in the direction opposite to that shown in the pulley and tension T in the string.

The inclination of the string is given as:

$$\tan\theta = 1.5/2 = 0.75$$

Therefore, $$\sin\theta = 0.6$$

and $$\cos\theta = 0.8$$

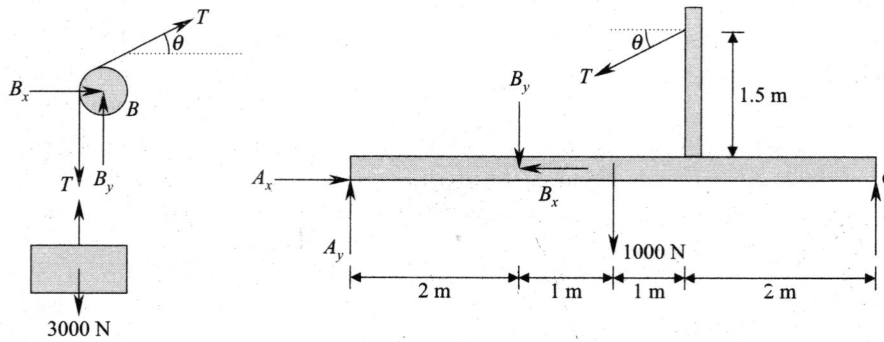

Fig. 5.52(a)

Block

From the free-body diagram of the load, we see that,

$\Sigma F_y = 0 \Rightarrow$

$$T - 3000 = 0$$

$\therefore$ $$T = 3000 \text{ N} \tag{a}$$

Pulley

Applying the conditions of equilibrium to the free-body diagram of the pulley,

$\Sigma F_x = 0 \Rightarrow$

$$B_x + T\cos\theta = 0$$

$\therefore$ $$B_x = -3000 \times 0.8 = -2400 \text{ N}$$

(The negative sign indicates that the force acts in the direction opposite to that of what we have assumed.)

$\Sigma F_y = 0 \Rightarrow$

$$B_y + T\sin\theta - T = 0$$

$$\therefore \quad B_y = 3000 - (3000 \times 0.6)$$
$$= 1200 \text{ N}$$

Beam

Applying the conditions of equilibrium to the free-body diagram of the beam,

$$\Sigma F_x = 0 \Rightarrow$$
$$A_x - B_x - T\cos\theta = 0$$
$$\therefore \quad A_x = B_x + T\cos\theta$$
$$= -2400 + (3000 \times 0.8) = 0$$

$$\Sigma F_y = 0 \Rightarrow$$
$$A_y + C_y - B_y - T\sin\theta - 1000 = 0$$
$$\Rightarrow \quad A_y + C_y = B_y + T\sin\theta + 1000$$
$$= 1200 + 3000(0.6) + 1000$$
$$= 4000 \text{ N} \quad \text{(b)}$$

As the forces are non-concurrent, in addition we also take summation of the moments about A and equate it to zero,

$$\Sigma M_A = 0 \Rightarrow$$
$$[C_y \times 6] + [T\cos\theta \times 1.5] - [T\sin\theta \times 4] - [B_y \times 2] - [1000 \times 3] = 0$$
$$C_y \times 6 = -[3000 \times 0.8 \times 1.5] + [3000 \times 0.6 \times 4] + [1200 \times 2] + [1000 \times 3]$$
$$\therefore \quad C_y = 1500 \text{ N} \quad \text{(c)}$$

Substituting the value of C_y in equation (b),

$$A_y + C_y = 4000$$
$$\Rightarrow \quad A_y = 2500 \text{ N}$$

Example 5.27 A beam AB hinged at A is supported in a horizontal position by a rope passing over a pulley arrangement hinged at C as shown in Fig. 5.53. The free end of the rope supports a load of 1000 N. The weight of the beam is 2000 N and that of the pulley hinged at C is 600 N. Determine the tension in the rope assuming the pulleys to be frictionless and the reaction at A.

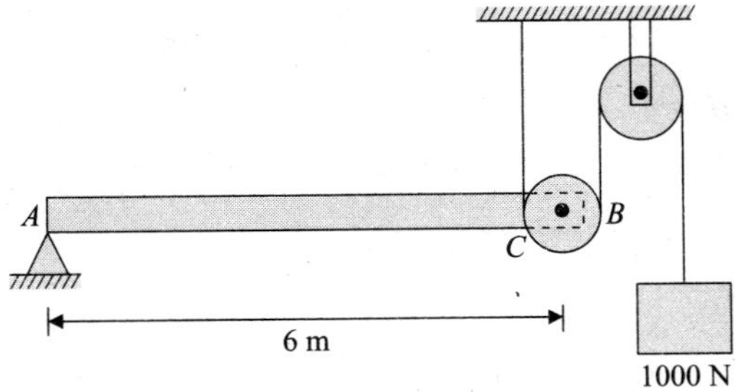

Fig. 5.53

Solution The free-body diagrams are shown below:

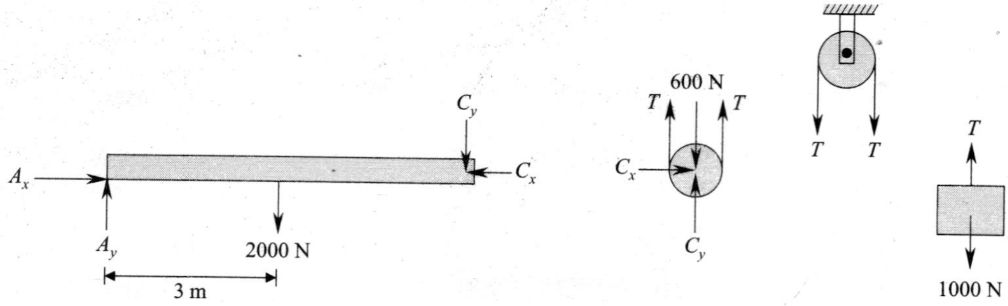

Fig. 5.53(a)

Block

From the free-body diagram of the block suspended at the end of the rope,

$\Sigma F_y = 0 \Rightarrow$

$$T - 1000 = 0$$

$$\therefore \quad T = 1000 \text{ N} \tag{a}$$

Since the pulleys are frictionless, the tension remains the same in the rope.

Left Pulley

Applying the conditions of equilibrium to the free-body diagram of the left pulley:

$\Sigma F_x = 0 \Rightarrow$

$$C_x = 0$$

$\Sigma F_y = 0 \Rightarrow$

$$C_y + 2\text{T} - 600 = 0$$

$$\Rightarrow \quad C_y = 600 - 2T$$

$$= 600 - 2(1000) = -1400 \text{ N}$$

The negative sign indicates that the direction of C_y is opposite to the direction assumed.

Beam

Applying the conditions of equilibrium to the free-body diagram of the beam:

$\Sigma F_x = 0 \Rightarrow$

$$A_x - C_x = 0$$

$$A_x = C_x = 0 \quad [\text{since } C_x = 0]$$

$\Sigma F_y = 0 \Rightarrow$

$$A_y - C_y - 2000 = 0$$

$$\Rightarrow \quad A_y = C_y + 2000$$

$$= -1400 + 2000 = 600 \text{ N}$$

5.3.4 Equilibrium of Non-Concurrent Forces in Space

If all the forces acting on the body are non-concurrent in space, the necessary and sufficient conditions for equilibrium will be same as that for coplanar non-concurrent forces with the addition of the Z-component for $\vec{R}$, and X and Y components for $\vec{M}$.

$$\Sigma F_x = 0 \qquad \Sigma M_x = 0$$

$$\Sigma F_y = 0 \qquad \Sigma M_y = 0$$

$$\Sigma F_z = 0 \qquad \Sigma M_z = 0 \tag{5.19}$$

Example 5.28 A boom AC supporting a load of 10 kN at end C is held in a horizontal position by a ball-and-socket joint at A and by two cables at B as shown in Fig. 5.54. Determine the tension in each cable and the reaction at A. Neglect the weight of the boom.

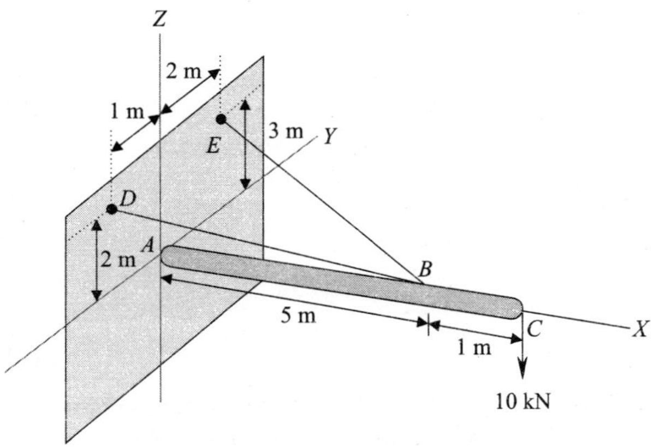

Fig. 5.54

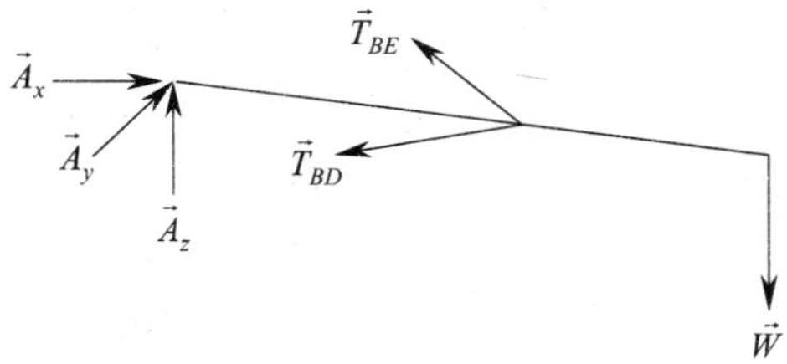

Fig. 5.54(a)

Solution From the figure, we see that the coordinates of points A, B, C, D and E are respectively:

$A(0, 0, 0)$, $B(5, 0, 0)$, $C(6, 0, 0)$, $D(0, -1, 2)$ and $E(0, 2, 3)$.

Determination of unit vectors along BD and BE

$$\overrightarrow{BD} = \overrightarrow{AD} - \overrightarrow{AB}$$

$$= -5\vec{i} - \vec{j} + 2\vec{k}$$

$$\therefore \qquad \hat{n}_{BD} = \frac{-5\vec{i} - \vec{j} + 2\vec{k}}{\sqrt{(-5)^2 + (-1)^2 + (2)^2}}$$

$$= \frac{-5\vec{i} - \vec{j} + 2\vec{k}}{\sqrt{30}}$$

$$= -0.913\vec{i} - 0.183\vec{j} + 0.365\vec{k}$$

$$\overrightarrow{BE} = \overrightarrow{AE} - \overrightarrow{AB}$$
$$= -5\vec{i} + 2\vec{j} + 3\vec{k}$$

$\therefore$
$$\hat{n}_{BE} = \frac{-5\vec{i} + 2\vec{j} + 3\vec{k}}{\sqrt{(-5)^2 + (2)^2 + (3)^2}}$$
$$= \frac{-5\vec{i} + 2\vec{j} + 3\vec{k}}{\sqrt{38}}$$
$$= -0.811\vec{i} + 0.324\vec{j} + 0.487\vec{k}$$

The reaction at A is given in vector form as

$$\vec{R}_A = A_x\vec{i} + A_y\vec{j} + A_z\vec{k} \tag{a}$$

Let T_{BD} and T_{BE} be the magnitudes of tensions in the cables BD and BE respectively. Then the tension vectors can be represented as

$$\vec{T}_{BD} = T_{BD}\hat{n}_{BD}$$
$$= T_{BD}[-0.913\vec{i} - 0.183\vec{j} + 0.365\vec{k}] \tag{b}$$
$$\vec{T}_{BE} = T_{BE}\hat{n}_{BE}$$
$$= T_{BE}[-0.811\vec{i} + 0.324\vec{j} + 0.487\vec{k}] \tag{c}$$

The load at C can be represented in vector form as

$$\vec{W} = 10[-\vec{k}]$$
$$= -10\vec{k} \tag{d}$$

Note that the weight is directed along the negative Z-axis. Since the tension and weight vectors are non-concurrent, we take the moment about A and apply the moment condition for equilibrium,

$$\sum \vec{M}_A = \vec{O} \Rightarrow$$
$$\{5\vec{i} \times T_{BD}[-0.913\vec{i} - 0.183\vec{j} + 0.365\vec{k}]\} + \{5\vec{i} \times T_{BE}[-0.811\vec{i} + 0.324\vec{j} + 0.487\vec{k}]\}$$
$$+ \{6\vec{i} \times [-10\vec{k}]\} = \vec{O}$$

$$\Rightarrow \quad [-1.825T_{BD} - 2.435T_{BE} + 60]\vec{j} + [-0.915T_{BD} + 1.62\, T_{BE}]\vec{k} = \vec{O}$$

Equating the coefficients of $\vec{j}$ and $\vec{k}$ on both sides,

$$-1.825T_{BD} - 2.435T_{BE} + 60 = 0 \tag{e}$$

and
$$-0.915T_{BD} + 1.62T_{BE} = 0 \tag{f}$$

From the above two equations, solving for T_{BD} and T_{BE}, we get

$$T_{BD} = 18.75 \text{ kN} \quad \text{and} \quad T_{BE} = 10.59 \text{ kN} \tag{g}$$

We take the summation of the forces and apply the condition for equilibrium,

$$\vec{R} = \vec{R}_A + \vec{T}_{BD} + \vec{T}_{BE} + \vec{W} = \vec{O}$$
$$= \{A_x\vec{i} + A_y\vec{j} + A_z\vec{k}\} + \{T_{BD}[-0.913\vec{i} - 0.183\vec{j} + 0.365\vec{k}]\}$$
$$+ \{T_{BE}[-0.811\vec{i} + 0.324\vec{j} + 0.487\vec{k}]\} - 10\vec{k} = \vec{O}$$

Equating the coefficients of $\vec{i}$, $\vec{j}$ and $\vec{k}$ to zero,

$$A_x - 0.913T_{BD} - 0.811T_{BE} = 0 \quad \text{(h)}$$

$$A_y - 0.183T_{BD} + 0.324T_{BE} = 0 \quad \text{(i)}$$

$$A_z + 0.365T_{BD} + 0.487T_{BE} - 10 = 0 \quad \text{(j)}$$

Substituting the values of T_{BD} and T_{BE} in the above equations, we get the support reactions at A as

$$A_x = 25.71 \text{ kN}, A_y = 0.09 \text{ N} \quad \text{and} \quad A_z = -2 \text{ kN}$$

Example 5.29 A boom AD supporting a load of 15 kN at the end D is held in a horizontal position by a ball-and-socket joint at A and by two cables BE and CF as shown. Determine the tension in each cable and the reaction at A. Neglect the weight of the boom.

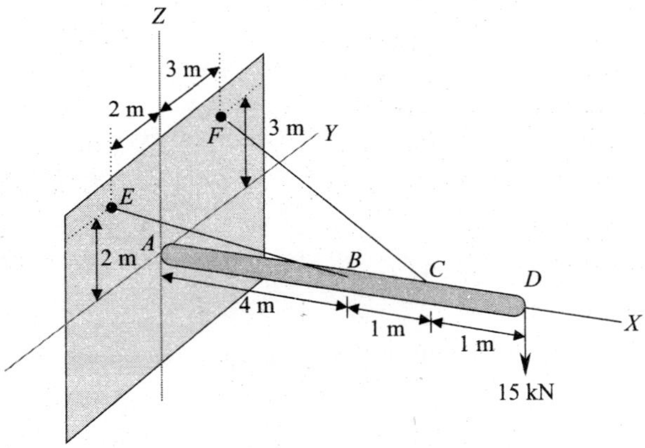

Fig. 5.55

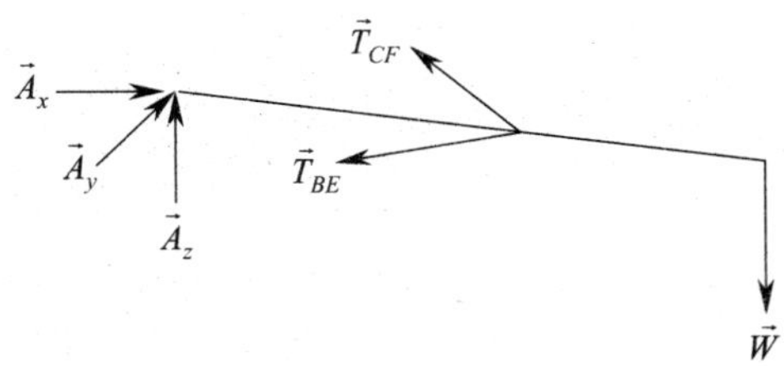

Fig. 5.55(a)

Solution From the figure, we see that the coordinates of points A, B, C, D, E and F are:

$$A(0, 0, 0), B(4, 0, 0), C(5, 0, 0), D(6, 0, 0), E(0, -2, 2) \text{ and } F(0, 3, 3)$$

Determination of unit vectors along BE and CF

$$\overrightarrow{BE} = \overrightarrow{AE} - \overrightarrow{AB}$$

$$= -4\vec{i} - 2\vec{j} + 2\vec{k}$$

$$\therefore \quad \hat{n}_{BE} = \frac{-4\vec{i} - 2\vec{j} + 2\vec{k}}{\sqrt{(-4)^2 + (-2)^2 + (2)^2}}$$

$$= \frac{-4\vec{i} - 2\vec{j} + 2\vec{k}}{\sqrt{24}}$$

$$= -0.816\vec{i} - 0.408\vec{j} + 0.408\vec{k}$$

$$\overrightarrow{CF} = \overrightarrow{AF} - \overrightarrow{AC}$$

$$= -5\vec{i} + 3\vec{j} + 3\vec{k}$$

$$\therefore \qquad \hat{n}_{CF} = \frac{-5\vec{i} + 3\vec{j} + 3\vec{k}}{\sqrt{(-5)^2 + (3)^2 + (3)^2}}$$

$$= \frac{-5\vec{i} + 3\vec{j} + 3\vec{k}}{\sqrt{43}}$$

$$= -0.762\vec{i} + 0.457\vec{j} + 0.457\vec{k}$$

Let T_{BE} and T_{CF} be the magnitudes of tensions in the cables BE and CF respectively. Then the tension vectors can be represented as

$$\vec{T}_{BE} = T_{BE}\hat{n}_{BE}$$

$$= T_{BE}[-0.816\vec{i} - 0.408\vec{j} + 0.408\vec{k}] \qquad \text{(a)}$$

$$\vec{T}_{CF} = T_{CF}\hat{n}_{CF}$$

$$= T_{CF}[-0.762\vec{i} + 0.457\vec{j} + 0.457\vec{k}] \qquad \text{(b)}$$

The reaction at A can be represented in vector form as

$$\vec{R}_A = A_x\vec{i} + A_y\vec{j} + A_z\vec{k} \qquad \text{(c)}$$

The load at D can be represented in vector form as

$$\vec{W} = 15[-\vec{k}] = -15\vec{k} \qquad \text{(d)}$$

Note that the weight is directed along the negative Z-axis.

Since the tensions, load and reaction vectors are non-concurrent, we take the moment about A and apply the condition for moment equilibrium.

$$\sum \vec{M}_A = \vec{O} \Rightarrow$$

$$\{4\vec{i} \times T_{BE}[-0.816\vec{i} - 0.408\vec{j} + 0.408\vec{k}]\} + \{5\vec{i} \times T_{CF}[-0.762\vec{i} + 0.457\vec{j} + 0.457\vec{k}]\}$$
$$+ \{6\vec{i} \times [-15\vec{k}]\} = \vec{O}$$

$$\Rightarrow \qquad [-1.632T_{BE} - 2.285T_{CF} + 90]\vec{j} + [-1.632T_{BE} + 2.285T_{CF}]\vec{k} = \vec{O}$$

Equating the coefficients of $\vec{j}$ and $\vec{k}$ on both sides,

$$-1.632T_{BE} - 2.285T_{CF} + 90 = 0 \qquad \text{(e)}$$

$$-1.632T_{BE} + 2.285T_{CF} = 0 \qquad \text{(f)}$$

Solving for T_{BE} and T_{CF}, we get

$$T_{BE} = 27.57 \text{ kN} \quad \text{and} \quad T_{CF} = 19.69 \text{ kN} \qquad \text{(g)}$$

We take the summation of the forces and apply the condition for equilibrium,

$$\vec{R} = \vec{R}_A + \vec{T}_{BE} + \vec{T}_{CF} + \vec{W} = \vec{O}$$

$$= [A_x\vec{i} + A_y\vec{j} + A_z\vec{k}] + T_{BE}[-0.816\vec{i} - 0.408\vec{j} + 0.408\vec{k}]$$
$$+ T_{CF}[-0.762\vec{i} + 0.457\vec{j} + 0.457\vec{k}] - 15\vec{k} = \vec{O}$$

Equating the coefficients of $\vec{i}$, $\vec{j}$ and $\vec{k}$ to zero,

$$A_x - 0.816T_{BE} - 0.762\,T_{CF} = 0 \qquad \text{(h)}$$

$$A_y - 0.408T_{BE} + 0.457T_{CF} = 0 \quad \text{(i)}$$

$$A_z + 0.408T_{BE} + 0.457T_{CF} - 15 = 0 \quad \text{(j)}$$

Substituting the values of T_{BE} and T_{CF} in the above equations, we get the support reactions at A as

$$A_x = 37.5 \text{ kN}, A_y = 2.25 \text{ kN} \quad \text{and} \quad A_z = -5.25 \text{ kN}$$

5.4 STATICALLY INDETERMINATE STRUCTURES

All the structures that we have so far come across had as many number of *unknowns* in the free-body diagrams to as many *equations of equilibrium* available. Hence, the unknowns could be determined by solving the equilibrium equations. Such types of structures are termed **statically determinate** structures. However, in some structures it may so happen that the free bodies of the members have more number of unknown forces than the equations of equilibrium available. Such types of structures are termed **statically indeterminate** structures. Structures of such type can be analyzed by supplementing the equilibrium equations with additional equations pertaining to displacements of the structure, which is beyond the scope of our study.

Consider for instance, coplanar concurrent forces shown in Fig. 5.56(a). Here the tensions in the two strings AC and BC are unknown and they can be determined from the equilibrium equations, namely, two for concurrent forces. However, if one more string is attached as shown in Fig. 5.56(b), then the structure becomes statically indeterminate as the number of unknowns (three in this case) is greater than the number of equilibrium equations (two for concurrent forces) available.

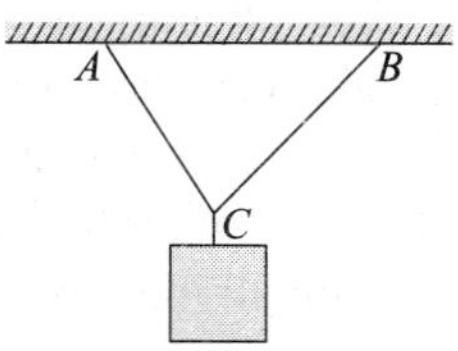

Fig. 5.56(a) Statically determinate structure

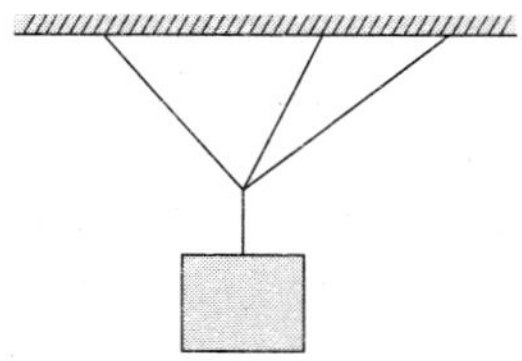

Fig. 5.56(b) Statically indeterminate structure

Next, consider concurrent forces in space as shown in Fig. 5.57(a). Here the tensions in the three cables OA, OB and OC can be determined using the equations of equilibrium, namely, three for concurrent forces in space. However, if one more cable is added as shown in Fig. 5.57(b), then the structure becomes statically indeterminate as the number of unknowns (four in this case) is greater than the number of equilibrium equations available.

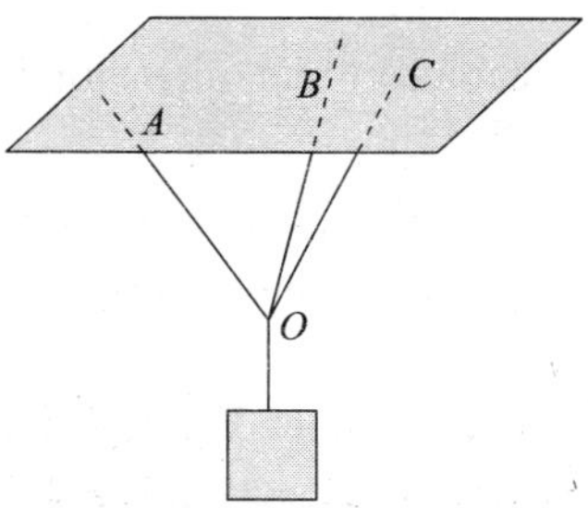

Fig. 5.57(a) Statically determinate structure

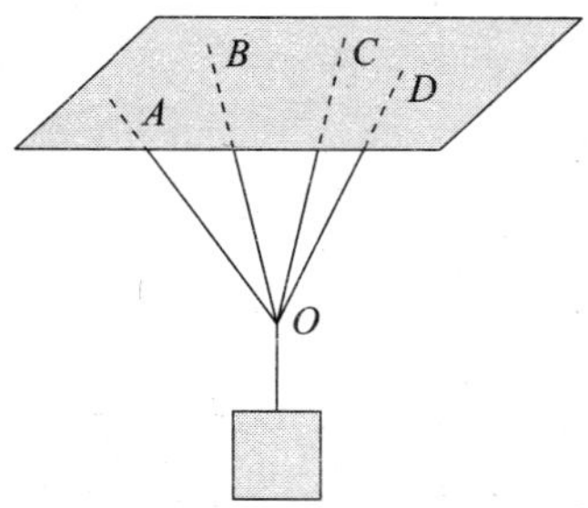

Fig. 5.57(b) Statically indeterminate structure

Next, consider non-concurrent coplanar forces shown in Fig. 5.58(a). Here there are two support reactions and one moment at fixed support and these can be determined from the three equations of equilibrium. However, if one more support shown in Fig. 5.58(b) is added like roller support, then the number of unknowns exceeds the number of independent equations available. Hence, it is a statically indeterminate beam.

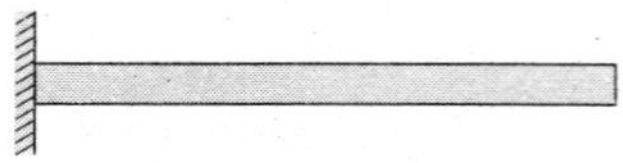

Fig. 5.58(a) Statically determinate beam

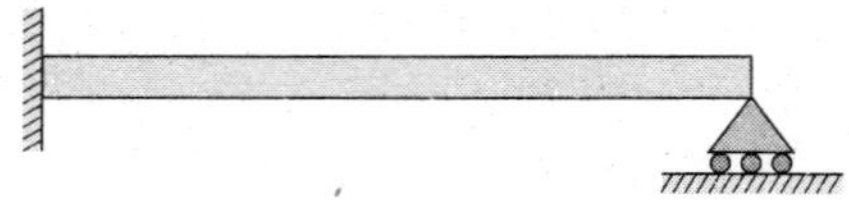

Fig. 5.58(b) Statically indeterminate beam

SUMMARY

Any force system acting on a body can be replaced by an equivalent resultant force acting at a common point and a moment about the same point. When the resultant *force* of a system of forces is *zero* then the body will remain at rest or move with constant velocity, if it was already moving with constant velocity, i.e., its acceleration will be zero; and if the *moment* is also *zero*, there will not be any rotational motion. Such a condition is called *static equilibrium*.

Free-Body Diagram

The individual members in a structure or mechanical system, due to their attachments or contact with other members are restricted to move in translational directions and rotational directions. Such bodies are said to be *constrained*. To investigate the equilibrium of a constrained body, we must first isolate it from all its attachments with its surroundings. A body thus, isolated from all other members, which are connected to it is called a *free-body*. A sketch of the isolated body showing all the *forces* acting on it by vectors is called a *free-body diagram*. Normally, two kinds of forces must be shown to act on a free body; they are *external forces* or *active forces* and *reactive forces* replacing the attachments and supports.

External Forces or Active Forces Gross bodies when placed in a uniform gravitational field are subjected to a force of gravity $W = mg$. This force is placed at the *centre of gravity* of the body.

Normal Reaction Whenever two bodies with *smooth* surfaces are in contact with each other, then each body will exert a force on the other and this force always acts *normal* to the plane of contact between them and at the point of contact.

Tensile Pull Whenever a body is attached to a string, a rope or a cable, a force of tension is exerted on the body and it acts along the string, rope or cable and away from the body.

Restoring Force in a Spring Whenever a body is attached to a spring, a restoring force is exerted by the spring on the body. Its magnitude is $k\Delta l$ and it acts along the spring and in the direction opposite to the direction of displacement of the body.

Tension or Compression Members Whenever forces act on a member hinged at both the ends as in link mechanism, the member is in tension or in compression and this force acts along the axis of the member.

Support Reactions

(i) Hinge or pin support: A hinge support exerts two reactions along X and Y axes at the point of support.

(ii) Roller or frictionless support: A roller support exerts one normal reaction at the point of support.

(iii) Fixed or built-in support: A fixed or built-in support exerts two normal reactions along X and Y directions and one moment along Z-direction.

(iv) Ball-and-socket joint: A ball-and-socket joint exerts three reactions at the point of support along X, Y and Z directions.

Conditions of Equilibrium

(A) Coplanar concurrent forces

$$\sum F_x = 0 \quad \text{and} \quad \sum F_y = 0$$

(B) Concurrent forces in space

$$\sum F_x = 0, \quad \sum F_y = 0 \quad \text{and} \quad \sum F_z = 0$$

(C) Coplanar non-concurrent forces

$$\sum F_x = 0, \quad \sum F_y = 0 \quad \text{and} \quad \sum M_z = 0$$

where M_z is the moment about the perpendicular Z-axis.

(D) Non-concurrent forces in space

$$\sum F_x = 0 \qquad \sum M_x = 0$$
$$\sum F_y = 0 \qquad \sum M_y = 0$$
$$\sum F_z = 0 \qquad \sum M_z = 0$$

Lami's Theorem: The theorem states that if three coplanar concurrent forces acting on a body keep it in equilibrium, then each force is proportional to the sine of the angle between the other two.

$$\frac{F_1}{\sin\alpha} = \frac{F_2}{\sin\beta} = \frac{F_3}{\sin\gamma}$$

Two force equilibrium: Two forces acting on a body will keep it in equilibrium if the two forces are *equal, opposite* and *collinear.*

Three-force equilibrium: Three forces acting on a body will keep it in equilibrium if the three forces are *coplanar* and *concurrent.*

Beams and Types of Beams

Simply supported beam: It is a beam supported with a hinge support at one end and a roller support at the other end.

Cantilever beam: It is a beam that is supported by a built-in or fixed support at one end and the other end free.

Overhanging beam: It is a simply supported beam with overhanging portions on either or both sides.

Types of Loading

Concentrated (or) point loads: These are forces acting at a particular point on a beam.

Distributed loads: Such types of loads are distributed over a distance. If the load is uniformly distributed over a length then it is termed *uniformly distributed load*; if it is linearly varying over the length it acts then it is termed *linearly varying load*. The force distribution is expressed in terms of *force per unit length*.

EXERCISES

Objective-type Questions

1. A reactive force arises
   (a) whenever motion is restrained in a direction
   (b) whenever motion is free in a direction
   (c) whenever bodies are free from its surroundings
   (d) due to intermolecular attraction
2. If a body is at rest, it implies that
   (a) the forces acting on it are always zero
   (b) the resultant of the forces acting on it are zero
   (c) the moment of the forces acting on it are zero
   (d) both the resultant force and moment are zero
3. If a body is in static equilibrium then it implies that the body
   (a) is at rest
   (b) is at rest or moving with constant velocity
   (c) is moving with constant acceleration
   (d) is oscillating about a fixed point
4. A free-body diagram is a diagram
   (a) drawn by free hand
   (b) of a body suspended freely in air
   (c) of a body in vacuum free from any influence from the surroundings
   (d) drawn by detaching the body from its attachments with the surroundings and replacing the attachments with force vectors
5. The forces shown in the free-body diagram are
   (a) the forces exerted by the body under consideration upon the surrounding bodies
   (b) the forces exerted by the surrounding bodies upon the body under consideration
   (c) the internal forces resisting deformation of the body
   (d) the forces exerted in free space
6. A rigid body has ____________ degree(s) of freedom
   (a) one (b) two (c) four (d) six

7. Which of the following statement is true?
   When forces acting on a particle keep it in equilibrium then
   (a) the force polygon closes
   (b) the resultant force is zero
   (c) the acceleration of the particle is zero
   (d) all of these
8. A hinge support constrains
   (a) translational motion along the x direction
   (b) translational motion along the xy plane
   (c) rotational motion perpendicular to the axis
   (d) translational motion along the xy plane and rotational motion perpendicular to the axis
9. A body is resting on an inclined plane. Which of the following figure represents the correct weight vector?

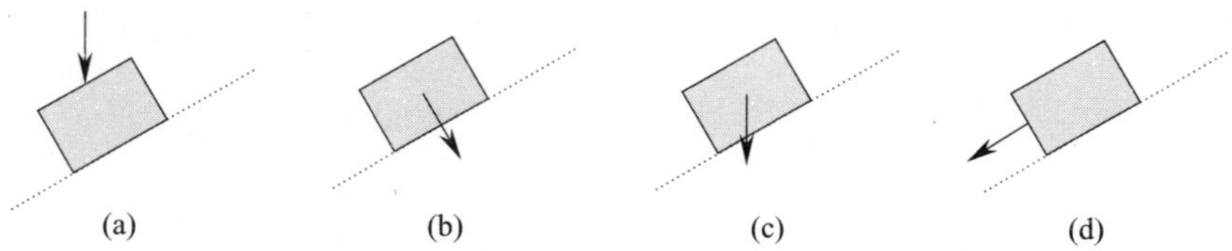

10. The conditions of equilibrium for non-concurrent coplanar force systems are
    (a) $\sum F_x = 0,\ \sum F_y = 0$
    (b) $\sum F_x = 0,\ \sum F_y = 0,\ \sum F_z = 0$
    (c) $\sum F_x = 0,\ \sum F_y = 0,\ \sum M_z = 0$
    (d) $\sum F_x = 0,\ \sum F_y = 0,\ \sum F_z = 0,\ \sum M_x = 0,\ \sum M_y = 0,\ \sum M_z = 0$
11. Three forces acting on a body can keep it in equilibrium, only when they are
    (a) collinear
    (b) coplanar and concurrent
    (c) coplanar and parallel
    (d) coplanar and non-concurrent

Answers

1. (a) **2.** (d) **3.** (b) **4.** (d) **5.** (b) **6.** (d) **7.** (d) **8.** (b)
9. (c) **10.** (c) **11.** (b)

Short-answer Questions

1. Define static equilibrium of a body.
2. Define a free body and free-body diagram.
3. Differentiate between active and reactive forces.
4. Explain the reactive force between two smooth surfaces in contact with each other.
5. Discuss on various support reactions.
6. What are the steps to be followed while drawing free-body diagrams?
7. Define two-force equilibrium and three-force equilibrium.
8. State and prove Lami's theorem.

9. State the conditions of equilibrium for: (i) coplanar concurrent forces; (ii) concurrent forces in space; (iii) coplanar non-concurrent forces, and (iv) non-concurrent forces in space.

10. Distinguish between statically determinate structures and statically indeterminate structures.

11. Define beams and types of beams.

12. What are the various types of loading in beams?

Numerical Problems

5.1 A rope AC is fixed horizontally between two walls 20 m apart. If a mass of 50 kg is suspended as shown in Fig. E.5.1 at its midpoint B, then it sags by 3 m. Determine the tensions in the sections AB and BC of the rope, neglecting its weight.

Ans. $T_{BA} = T_{BC} = 853.5$ N

5.2 A rope is tied horizontally between two buildings, 30 m apart. A weight of 200 N is attached to it at its midpoint. Determine the angle the rope would make with the horizontal if the tension developed in the rope is three times the weight of the body. Assume the string to be inextensible and neglect its weight. Also, determine by how much the weight would sag.

Ans. 9.6°, 2.54 m

5.3 A body of 50 kg mass is suspended as shown in Fig. E.5.3. Determine the tensions in the cable portions AC and BC, neglecting the weight of the cable.

Ans. $T_{AC} = T_{BC} = 717.1$ N

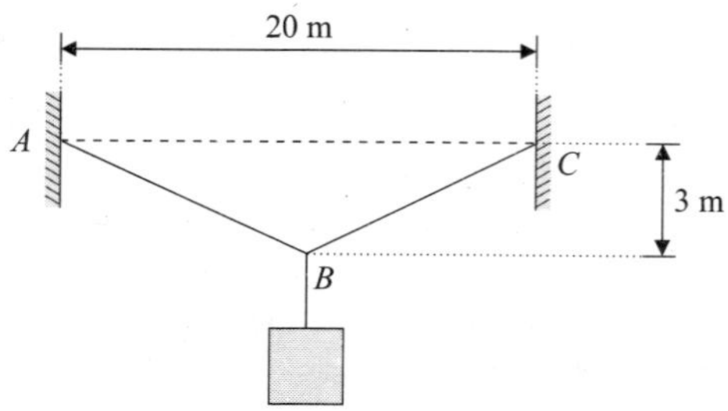

Fig. E.5.1

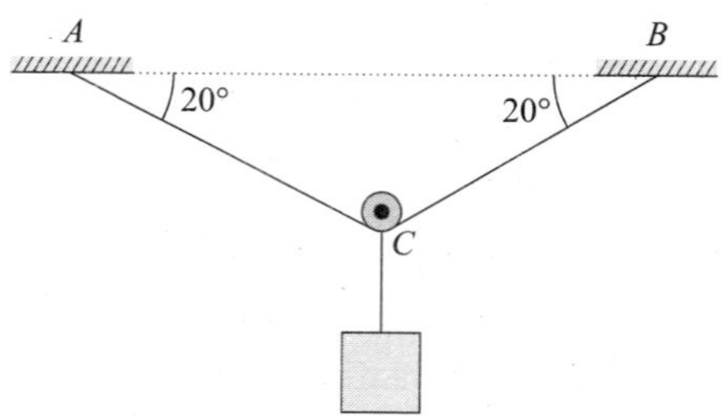

Fig. E.5.3

5.4 A body of 40 kg mass is suspended as shown in Fig. E.5.4. Determine the tensions in the cable portions AC and BC.

Ans. $T_{AC} = 287.3$ N, $T_{BC} = 351.8$ N

5.5 In the arrangement shown in Fig. E.5.5, determine the angle θ, which the string would make with the horizontal, when weights of the blocks A and B are such that (a) $W_A = W_B = W$ and (b) $W_A = W$; $W_B = 2W$

Ans. (a) 30° (b) 14.5°

5.6 In Fig. E.5.6, determine the force P required to hold a block of 50 kg mass in equilibrium position.

Ans. 283.2 N

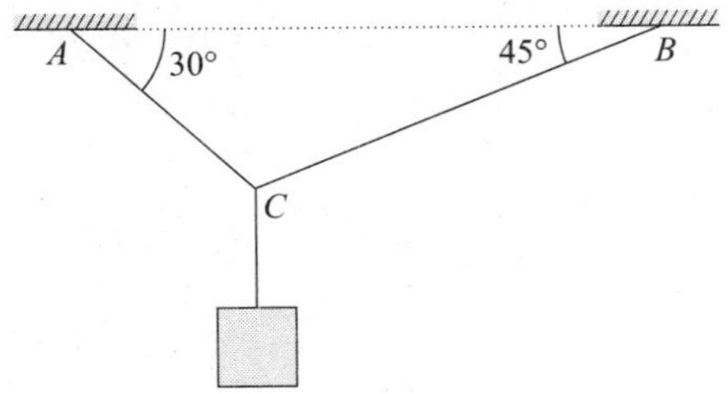

Fig. E.5.4

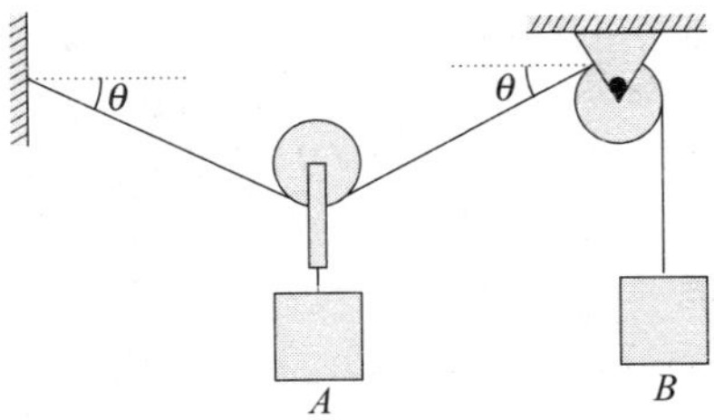

Fig. E.5.5

5.7 A 10 m long inextensible rope is tied to a flag post at its top and bottom points, respectively *A* and *B*. It is pulled aside at its midpoint by a force of 30 N such that it is taut. If the pull at the midpoint is 1 m, determine the tensions in parts *AC* and *BC* of the rope. Refer Fig. E.5.7.

Ans. $T_{CA} = T_{CB} = 75$ N

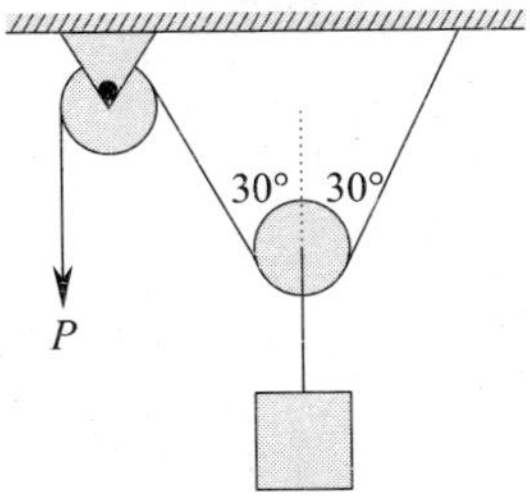

Fig. E.5.6

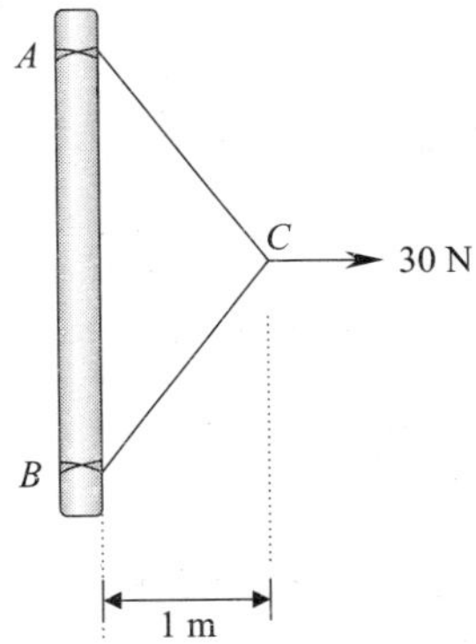

Fig. E.5.7

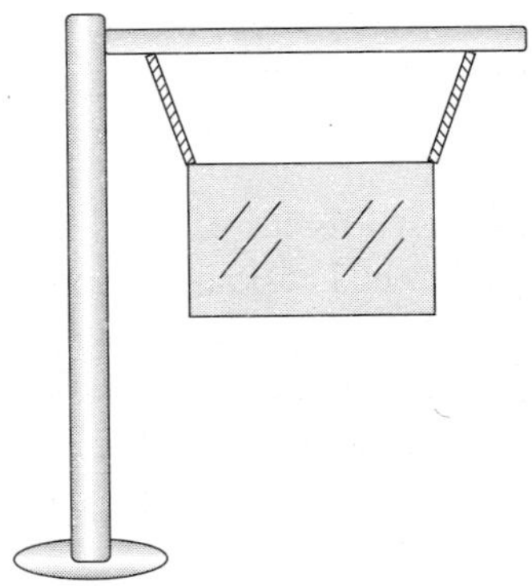

Fig. E.5.8

5.8 A signboard is hung as shown in Fig. E.5.8, by two ropes attached to its upper corners. If the two ropes make the same angle with the vertical, determine the angle when the tension in each rope is equal to the weight of the signboard.

Ans. 60°

5.9 A cable *ABCD* supports two loads 1 kN and 2 kN at points *B* and *C* respectively. Determine the tension in each portion of the cable if the portion *BC* remains horizontal. Also, determine the distance *x* for which equilibrium can be maintained. Refer Fig. E.5.9.

Ans. T_{AB} = 1.25 kN, T_{BC} = 0.75 kN, T_{CD} = 2.14 kN, $x = 0.15$ m

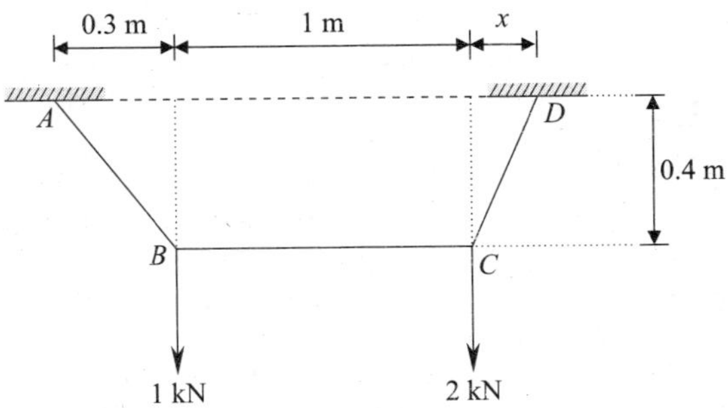

Fig. E.5.9

5.10 A body of 50 kg mass is suspended as shown in Fig. E.5.10. Determine the tensions in the string portions *AC* and *BC*.

Ans. $T_{CA} = 490.5$ N, $T_{CB} = 693.7$ N

5.11 A block of weight 1000 N is suspended as shown in Fig. E.5.11. Determine the tension in each cable.

Ans. $T_{BA} = 2732.05$ N, $T_{BC} = 3346.06$ N, $T_{BD} = 1000$ N

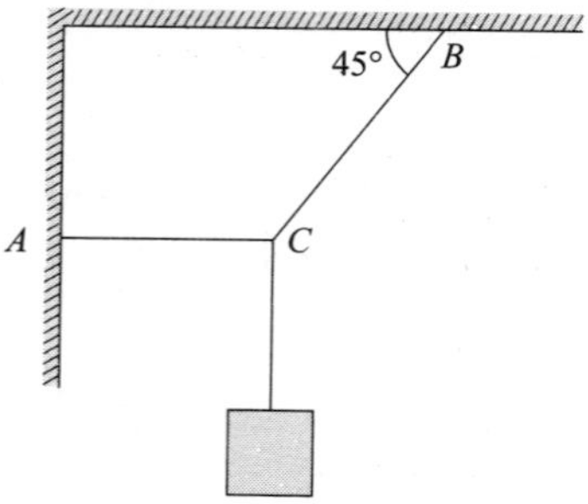

Fig. E.5.10

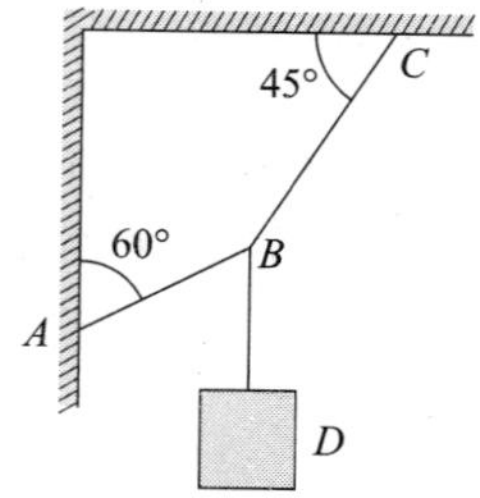

Fig. E.5.11

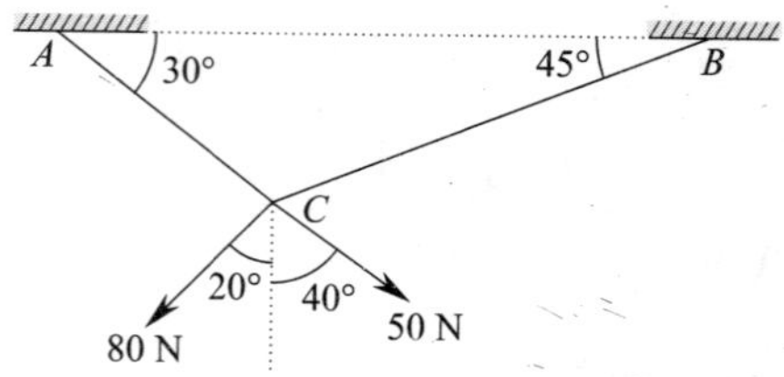

Fig. E.5.12

5.12 Two forces of 80 N and 50 N act at a point *C* of a string *ACB* attached at points *A* and *B*. Determine the tension in each portion of the string. Refer Fig. E.5.12.

Ans. $T_{CA} = 86.6$ N, $T_{CB} = 99.3$ N

5.13 A sphere of 10 kg mass is suspended as shown in Fig. E.5.13. Determine the tension in the strings *AC* and *BC*.

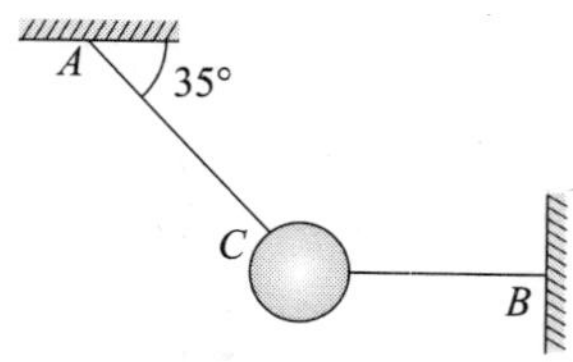

Fig. E.5.13

Ans. $T_{CA} = 171.03$ N, $T_{CB} = 140.1$ N

5.14 A long string carrying a weight of 300 g at each end passes over two smooth pegs 80 cm apart so that the string is horizontal. What mass must be attached to the string midway between the pegs to depress the point by 30 cm?

Ans. 360 g

5.15 A smooth wire is bent into the form of a circle, and is supported by a small ring sliding on it and attached by a string fast to a vertical wall, against which it rests. Determine the inclination of the string to the wall when the tension is double the weight of the circle.

Ans. 60°

5.16 A sphere of weight *W* is suspended by a string attached to a vertical wall as shown in Fig. E.5.16. Determine the inclination of the string if the tension in it is one-and-a-half times that of the weight of the sphere. Also, determine the reaction between the sphere and the wall.

Ans. $\theta = 48.2°$, $R = 1.12$ W

5.17 In Fig. E.5.17, find the tension in the cable *BC* and the force in the strut *AB*. The weight of the block suspended at end *B* is 500 N. Neglect the weight of the strut.

Ans. $T_{BC} = 866.03$ N, $S_{BA} = 1000$ N (compression)

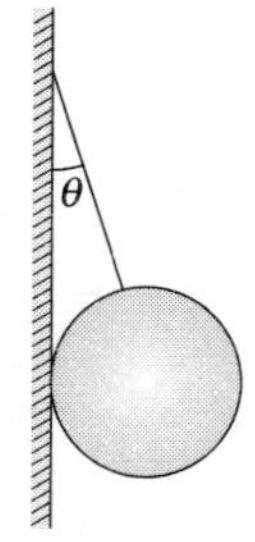

Fig. E.5.16

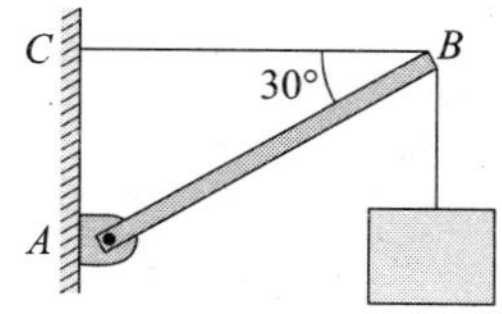

Fig. E.5.17

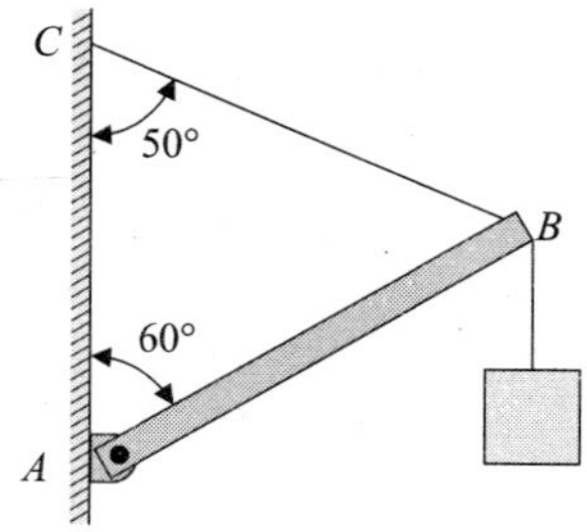

Fig. E.5.18

5.18 A bar *AB* of 6 m length and 100 N weight is hinged at *A*. It supports a load of 25 kg at point *B*. The end *B* of the bar is connected to the wall by a string *BC*. Determine the tension in the string and the reaction at *A*. Refer Fig. E.5.18.

Ans. $T = 272.1$ N, $A_x = 208.4$ N, $A_y = 170.4$ N

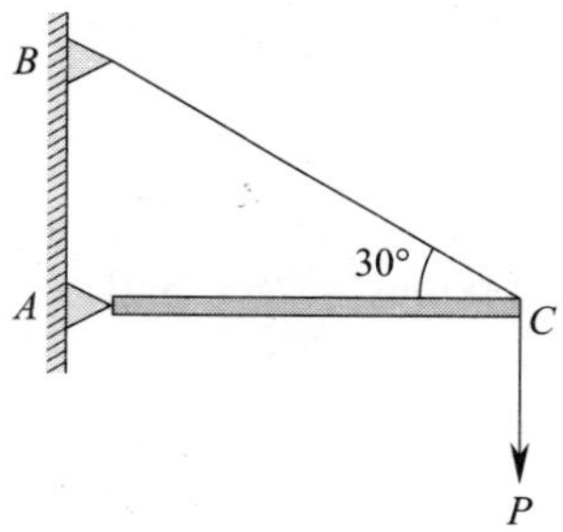

Fig. E.5.19

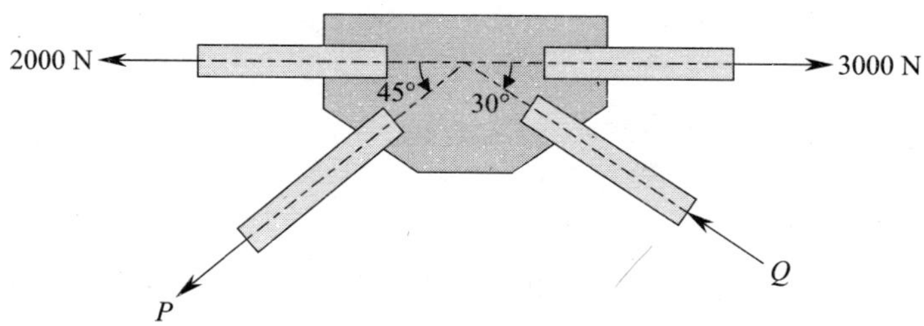

Fig. E.5.20

5.19 A strut *AC* hinged at *A* is held in the horizontal position by a string passing over its free end as shown in Fig. E.5.19. One end of the string is attached to the wall and the other end supports a weight *P*. Determine the tension in the string and the force in the strut. Neglect the weight of the strut.

Ans. $T = 2P$; $S = \sqrt{3}P$ (compression)

5.20 A gusset plate arrangement at a joint in a truss is shown in Fig. E.5.20. Determine the values of *P* and *Q* in order to maintain equilibrium at the joint.

Ans. $P = 517.64$ N, $Q = 732.05$ N

5.21 A smooth sphere of diameter $2r$ and weight W is resting in a groove as shown in Fig. E.5.21. If the width of the groove is r, determine the contact forces at *A* and *B*.

Ans. $W/\sqrt{3}$

5.22 A cylinder of weight W rests in a trough as shown in Fig. E.5.22. Determine the reactions at contact points *A* and *B*.

Ans. $R_A = W/2$, $R_B = \sqrt{3}\,W/2$

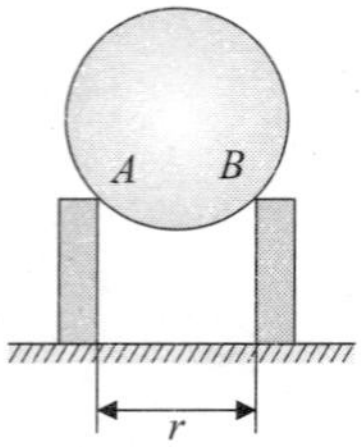

Fig. E.5.21

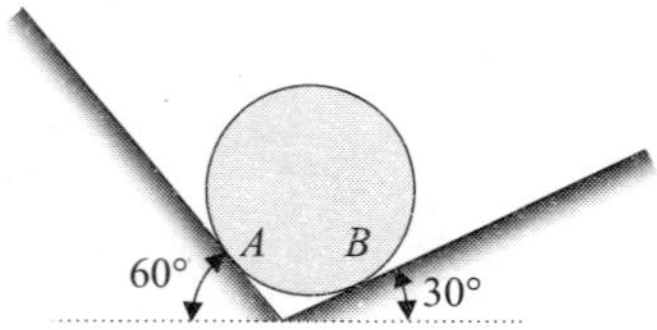

Fig. E.5.22

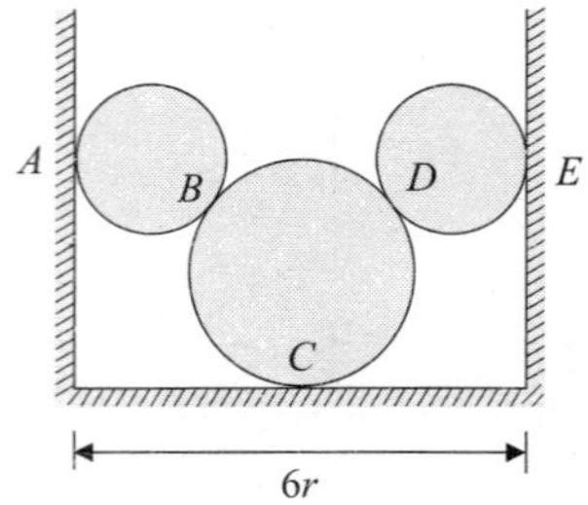

Fig. E.5.23

5.23 Three smooth cylinders are placed in a rectangular channel as shown in Fig. E.5.23. Determine the reactions at all contact surfaces. The weights of smaller cylinders are W and that of the larger one is 2 W. The corresponding radii are respectively r and $2r$.

Ans. $R_A = R_E = 0.895\ W$, $R_B = R_D = 1.342\ W$, $R_C = 4\ W$

5.24 Two smooth spheres each of weight W and diameter $2r$ rest in a hemispherical shell of diameter $6r$ as shown in Fig. E.5.24. Determine the contact forces.

Ans. $R_A = R_C = 2\ W/\sqrt{3}$, $R_B = W/\sqrt{3}$

5.25 A smooth sphere of weight 50 N and a smooth block of weight 150 N are placed in a smooth trough as shown in Fig. E.5.25. Determine the reaction forces at points A, B, C and D.

Ans. $R_A = 115.5$ N, $R_B = 101.04$ N, $R_C = 75$ N, $R_D = 129.9$ N

5.26 Two smooth cylinders, of equal radii, just fit in between two parallel vertical walls, and rest on a horizontal plane, without pressing against the walls. If a third were placed on top of them, find the resulting pressure against either wall.

Ans. $\dfrac{W}{2\sqrt{3}}$

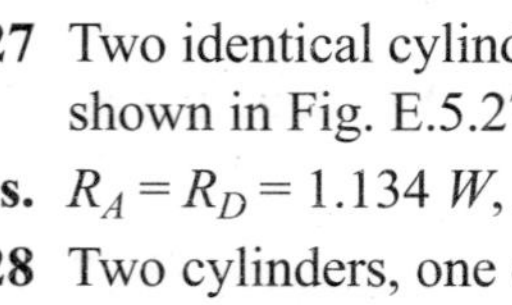

Fig. E.5.24

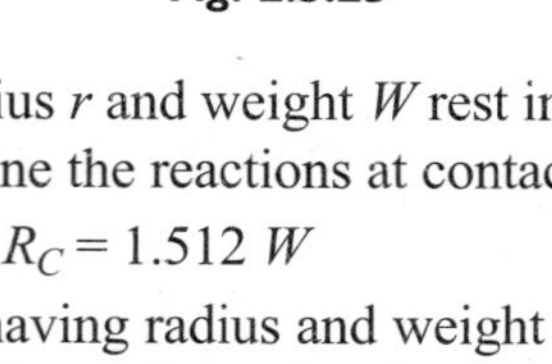

Fig. E.5.25

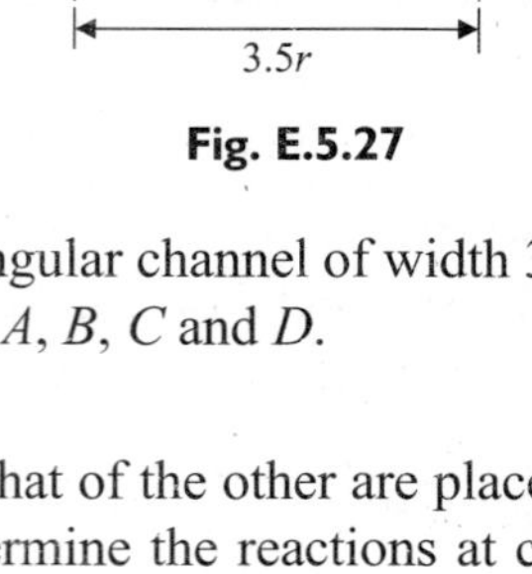

Fig. E.5.27

5.27 Two identical cylinders of radius r and weight W rest in a rectangular channel of width $3.5r$ as shown in Fig. E.5.27. Determine the reactions at contact points A, B, C and D.

Ans. $R_A = R_D = 1.134\ W$, $R_B = 2\ W$, $R_C = 1.512\ W$

5.28 Two cylinders, one of which having radius and weight double that of the other are placed in a rectangular channel of width $5r$ as shown in Fig. E.5.28. Determine the reactions at contact points A, B, C and D. Take weight and radius of the smaller cylinder as W and r respectively.

Ans. $R_A = R_D = 1.79\ W$, $R_B = 3\ W$, $R_C = 2.68\ W$

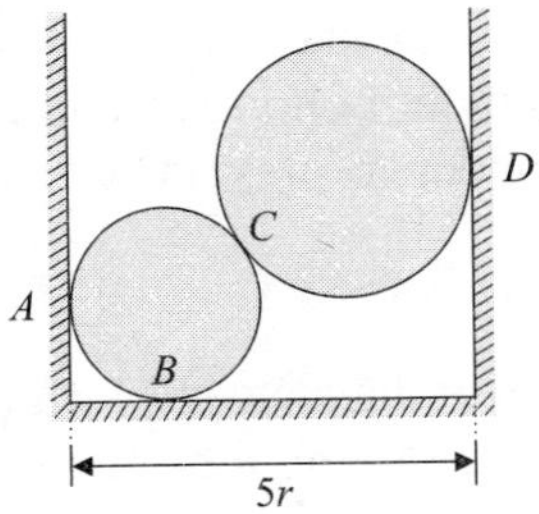

Fig. E.5.28

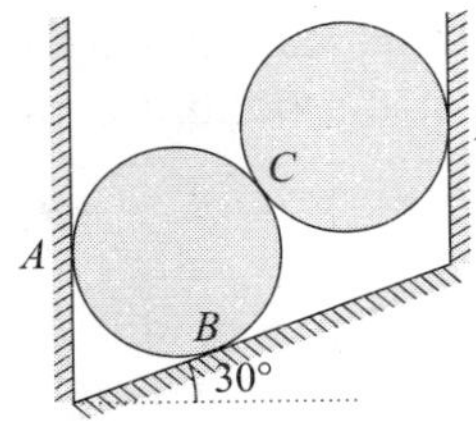

Fig. E.5.29

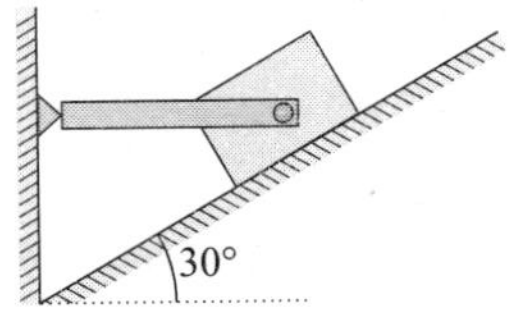

Fig. E.5.30(a)

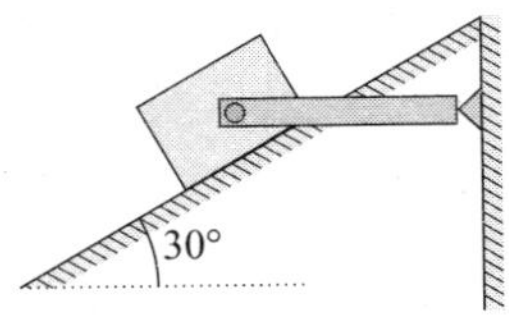

Fig. E.5.30(b)

5.29 Two identical cylinders of radius r and weight W rest in a channel with inclined base as shown in Fig. E.5.29. Determine the reactions at contact points A, B, C and D. The base width is $3.5r$ in the horizontal direction and its inclination is 30°.

Ans. $R_A = 2.29\ W$, $R_B = 2.31\ W$, $R_C = 1.512\ W$, $R_D = 1.134\ W$

5.30 A block placed on a smooth inclined plane is restrained from moving downwards by a horizontal rod hinged as shown in Fig. E.5.30 (a) and (b). Determine the reaction and the force acting on the rod in the two cases and comment on the values.

Ans. $R = \dfrac{W}{\cos 30°}$, $S = W \tan 30°$ (In the first case, the member is under compression and in the second case, it is under tension.)

5.31 A homogeneous 6 m steel column to be erected at a construction site is lifted up as shown in Fig. E.5.31. Determine the force P required to hold the column under equilibrium in the position shown. Also, determine the reaction at A. The mass of the column is 28.5 kg/m.

Ans. 790 N, $A_x = 632$ N, $A_y = 1203.5$ N

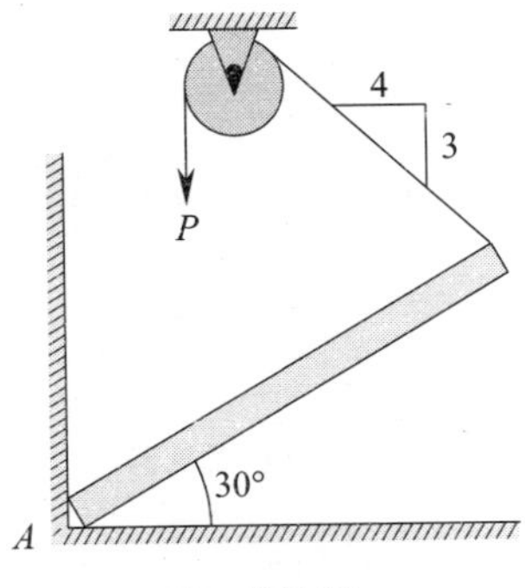

Fig. E.5.31

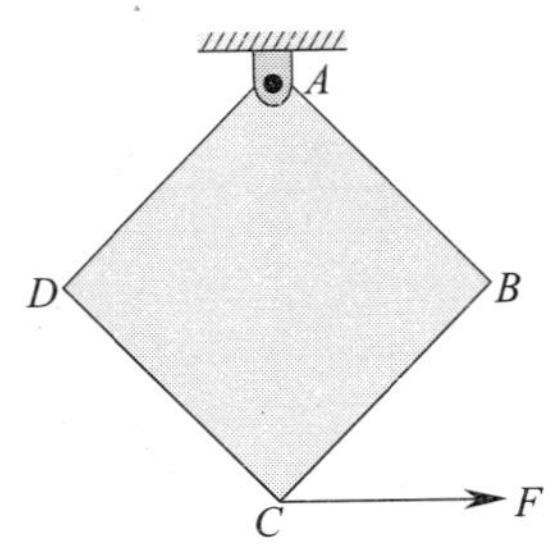

Fig. E.5.32

5.32 A homogenous square plate of 2 kg mass is suspended as shown in Fig. E.5.32. Determine the horizontal force F required to be applied at corner C to keep the plate in equilibrium with side AB held horizontal.

Ans. 9.81 N

5.33 A 6 m long ladder weighing 100 N rests against a smooth vertical wall at an angle of 30° to the wall. If a man of 750 N weight climbs up the ladder and stays at 4 m from the bottom, determine the horizontal force required to be applied at the bottom of the ladder to prevent it from slipping.

Ans. 317.5 N

5.34 A quarter circular plate is supported as shown in Fig. E.5.34. The mass of the plate is 2 kg and a force of 50 N is applied vertically downwards. Determine the force acting on the spring in pulling it. Assume the weight of the plate to act at $\frac{4}{3\pi}$ times its radius from the hinge.

Ans. 58.33 N

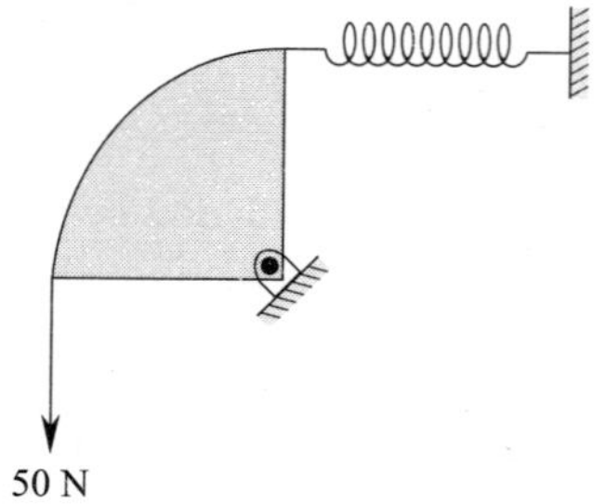

Fig. E.5.34

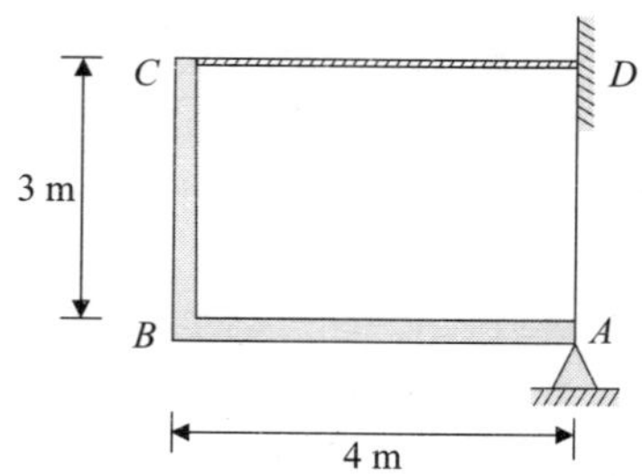

Fig. E.5.35

5.35 A member *ABC* hinged at *A* is held such that *AB* is in the horizontal position by a cable attached at *C*, whose other end is fixed to the wall. Determine the tension in the cable under the weight of the member. Assume the weight of the member to be 2 kN/m. Refer Fig. E.5.35.

Ans. 13.33 kN

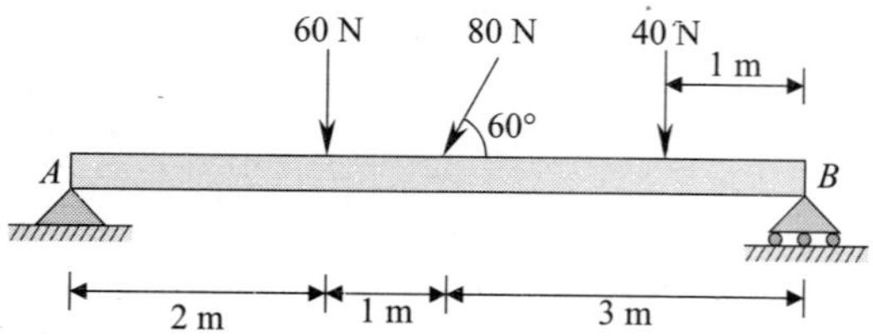

Fig. E.5.36

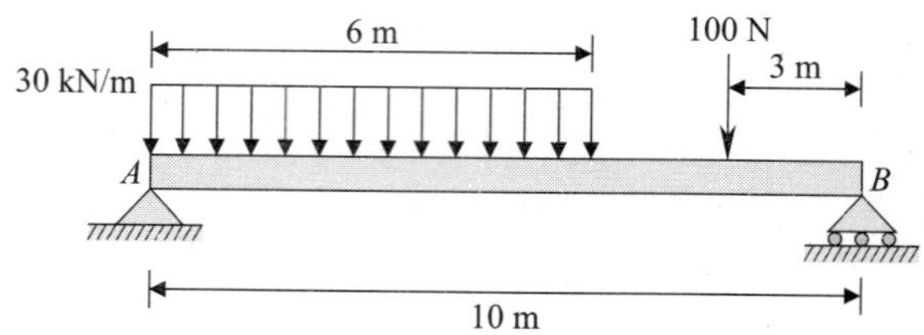

Fig. E.5.37

5.36 Find the support reactions for the simply supported beam *AB* loaded as shown in Fig. E.5.36.

Ans. $A_x = 40$ N, $A_y = 81.3$ N and $B_y = 88$ N

5.37 Find the support reactions for the simply supported beam *AB* loaded as shown in Fig. E.5.37.

Ans. $A_y = 156$ N, $B_y = 124$ N

5.38 Find the support reactions for the beam loaded as shown in Fig. E.5.38.

Ans. $A_x = 18.75$ kN, $A_y = 16.25$ kN, $R_B = 26.52$ kN

5.39 Find the support reactions for the simply supported beam *AB* loaded as shown in Fig. E.5.39.

Ans. $A_x = 0$, $A_y = 7.2$ kN, $B_y = 14.8$ kN

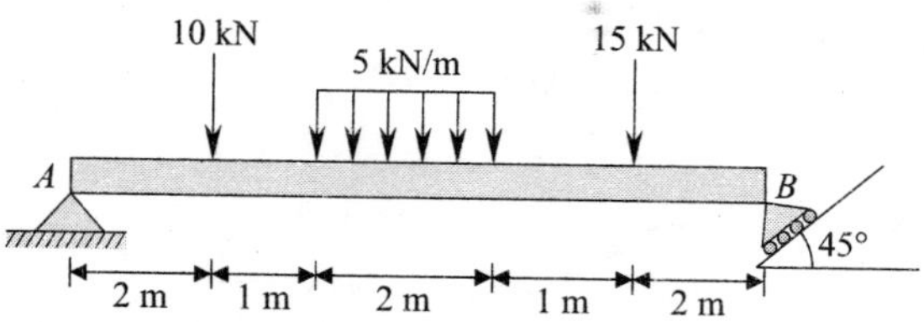

Fig. E.5.38

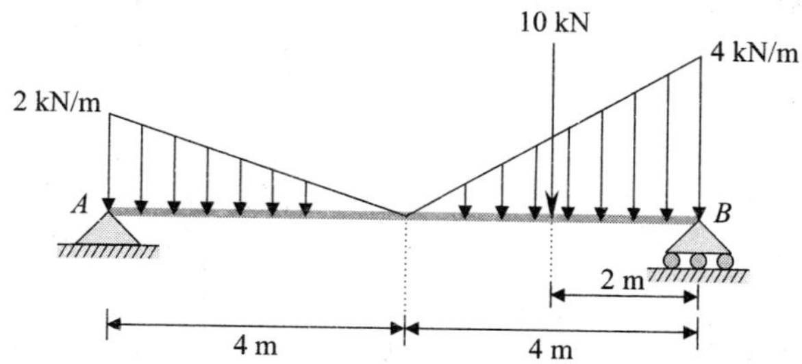

Fig. E.5.39

5.40 A pulley supporting a load of 4000 N is mounted at C on a horizontal beam as shown in Fig. E.5.40. If the beam weighs 800 N, find the support reactions at A and B. Neglect the weight of the pulley.

Ans. $A_x = 0$, $A_y = -3120$ N, $B_y = 7920$ N

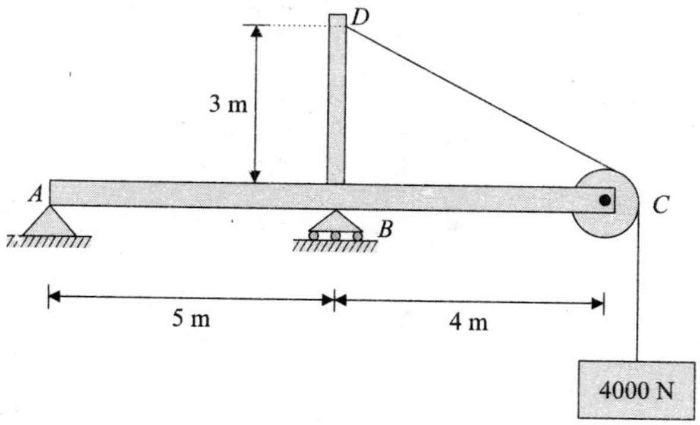

Fig. E.5.40

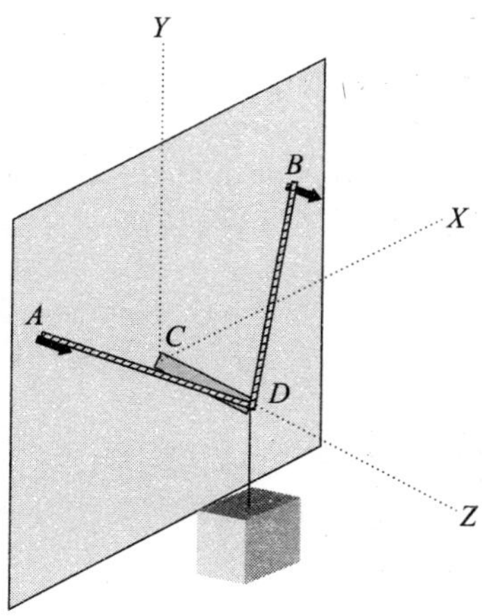

Fig. E.5.41

5.41 A block of 50 kg mass is suspended by two ropes attached to the wall at points A and B as shown in Fig. E.5.41. The block is held away from the wall by a horizontal strut CD. Determine the tensions in the two portions of the rope and the force in the strut. The coordinates of the points A, B, C and D are respectively (–4, 3, 0), (3, 3, 0), (0, 0, 0) and (0, 0, 1).

Ans. $T_{DA} = 357.5$ N, $T_{DB} = 407.4$ N, $R = 163.4$ N (compression)

5.42 A boom AC supporting a load of 20 kN at the end C is held in a horizontal position by a ball-and-socket joint at A and by two cables BD and BE as shown in Fig. E.5.42. Determine the tension in each cable and the reaction at A.

Ans. $T_{BD} = 21.4$ kN, $T_{BE} = 16.95$ kN, $A_x = 26.5$ kN, $A_y = 6.3$ N, $A_z = -2.51$ kN

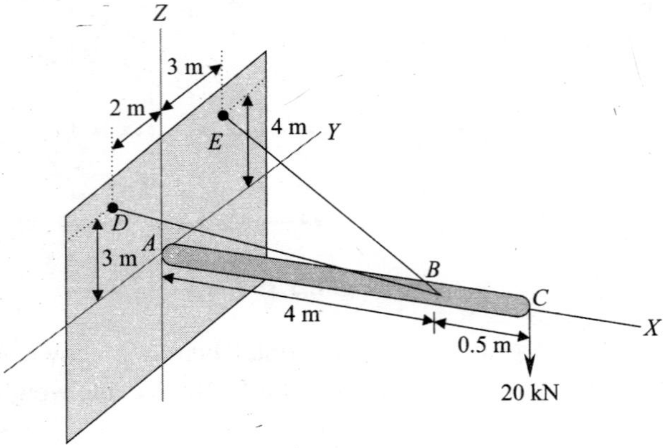

Fig. E.5.42

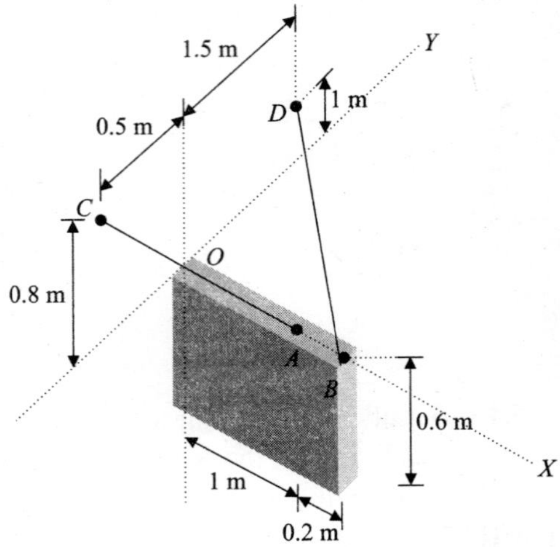

Fig. E.5.43

5.43 A rectangular signboard of 100 kg mass is supported by a ball-and-socket joint at its corner *O* and by cables *AC* and *BD* connected to a wall. Determine the tension in the cables and the reaction at the ball-and-socket joint. Refer Fig. E.5.43.

Ans. $T_{AC} = 714.1$ N, $T_{BD} = 312.4$ N, $O_x = 692.2$ N, $O_y = 43.4$ N, $O_z = 421.1$ N

6 Friction

6.1 INTRODUCTION

In the previous chapter, where we discussed the *equilibrium* of bodies, we assumed the surfaces of two bodies in contact with each other to be **perfectly smooth** in nature. In which case, the bodies are *free* to slide *tangentially* past one another, but are *restricted* from moving *normal* into the plane of contact. Hence, only a normal reaction N acts on each body as exerted by the other. However, this is just an *ideal* condition, as we do not find any surface to be perfectly smooth in nature. Even a highly polished steel surface is found to have some *irregularities* on its surface when viewed under magnification. Because of these irregularities, when one body *slides* or *tends to slide* over another, resistance is always offered to its motion in the **tangential** direction. This tangential resistive force is termed as **frictional force** or **force of friction**.

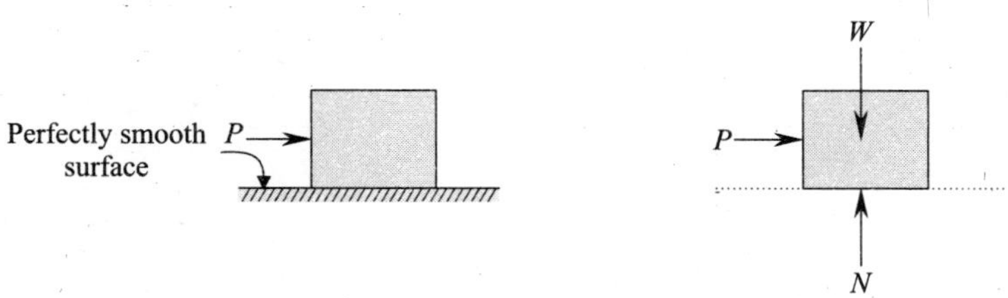

Fig. 6.1 No resistance offered when the surfaces are smooth

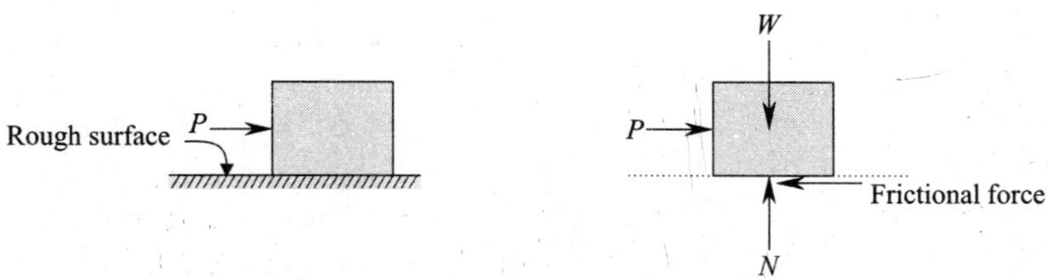

Fig. 6.2 Resistance is offered when the surfaces are rough

Hence, we see that whenever the surface of one body slides or tends to slide over another, each body exerts a **tangential** frictional force on the other apart from the **normal** reaction. This frictional force tries to prevent the motion of one surface with respect to the other. However, these frictional forces are found to be *limited* in *magnitude* and will not be able to prevent the relative motion when sufficiently large external forces are applied.

Because of the resistive nature of frictional force, it causes undesirable effects like wear and tear of mechanical parts in machines when they slide over one another and the overall efficiency of machines is reduced too. On the other hand, without friction we cannot do things in our daily lives as we do, as we cannot walk or drive a car or hold a pen and write, and so on. Hence, we see that friction causes favourable as well as undesirable effects.

Even though friction cannot be avoided in machines, it should be noted that it could be reduced to a large extent by application of **lubricants**, which form a thin film in between the contact surfaces. Due to the viscous nature of these lubricant fluids, resistance is developed between sliding layers of fluid moving at different velocities and the wear and tear of mechanical parts can be reduced to a large extent. This type of fluid friction is studied in the field of fluid mechanics and will not be covered in this study. Our study is limited only to that of dry friction, i.e., dry unlubricated surfaces.

In this chapter, we will introduce the basic laws governing dry friction and solve simple plane problems, and in the following chapter, we will discuss the specific application of this friction in mechanisms and machines.

6.2 LIMITING FRICTION AND IMPENDING MOTION

The existence of friction and its effect on motion of bodies can best be understood from the following experiment. Consider a block of weight W resting on a horizontal surface. If no external forces act on it, then its free-body diagram is as shown in Fig. 6.3(a), where the weight of the block W and the normal reaction N exerted by the plane on the block are shown.

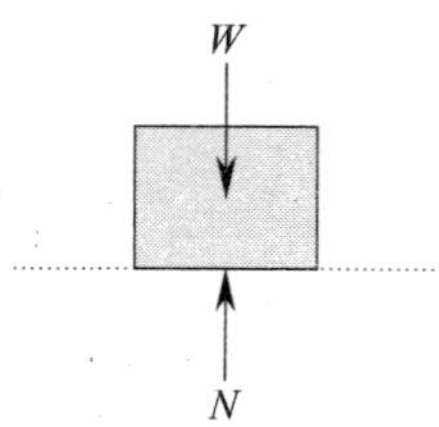

Free-body diagram of block at rest

Fig. 6.3(a)

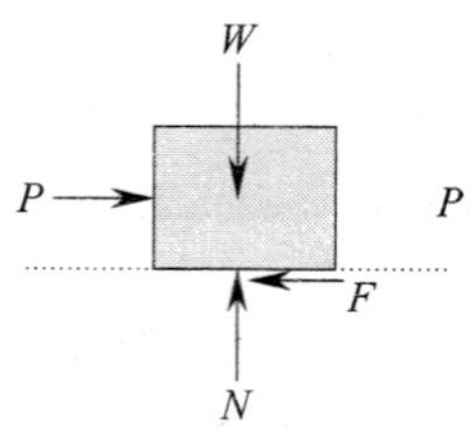

$P > 0$, but very small

Equilibrium is maintained; no motion

Fig. 6.3(b)

Suppose a horizontal force P be applied on the block as shown in Fig. 6.3(b). Then we see that the block will not move even though we apply a small force. Hence, to maintain equilibrium, the applied force P must be balanced by an equal and opposite force. This force arises from the surface irregularities and it tries to oppose the relative motion. This tangential resistive force is termed as **static frictional force**. Further, it is not a single concentrated force, but the resultant of a number of distributed forces acting over the entire surface of contact. However, for analytical purposes, they can be assumed to be concentrated without much appreciable error.

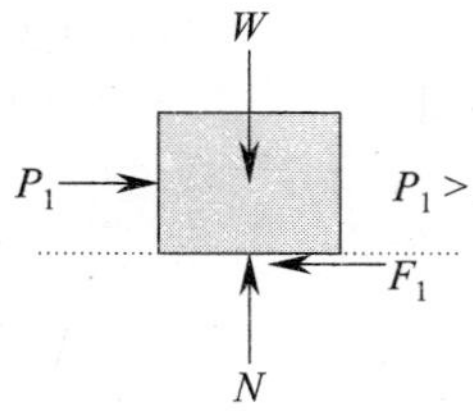

Fig. 6.3(c) Equilibrium is maintained; no motion

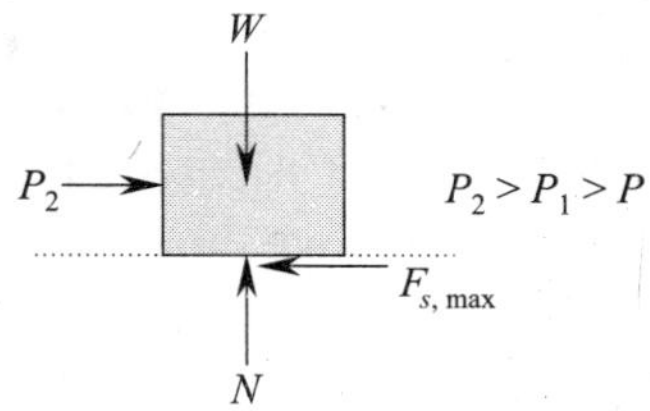

Fig. 6.3(d) Point of impending motion

If P is gradually increased to P_1, then F also increases correspondingly to F_1 to maintain equilibrium as in Fig. 6.3(c). However, there is a **limit** up to which the frictional force can increase. The maximum or limiting value of friction is termed as **limiting static friction** and it is denoted as $F_{s,\,\max}$ [refer Fig. 6.3(d)]. At this point, the block is on the verge of sliding and this instant is known as point of **impending motion**.

If P is increased further, the frictional force cannot balance it and motion occurs. Once motion has started, the frictional forces acting between the surfaces usually decrease so that a smaller force is necessary to maintain uniform motion. The frictional force acting between surfaces in relative motion is termed as **force of kinetic friction**.

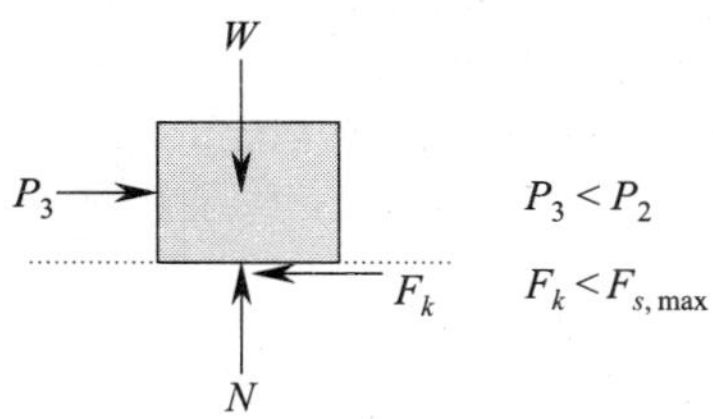

Fig. 6.3(e) Body under motion

The results of this experiment can be depicted in a graph as shown in Fig. 6.4, plotting magnitude of frictional force (F) against the externally applied force P. Up to the point of impending motion, the frictional force (F) increases linearly with the applied force P. The slope of the line being 45°, as F is equal to P until the limiting value is reached. Beyond the point of impending motion, the frictional force drops to force of kinetic friction (F_k) and it almost remains constant. The *sudden drop* in frictional force from $F_{s,\max}$ to F_k can be observed when we try to push or pull a chair on the floor; there will be a *jerk* just before the motion begins.

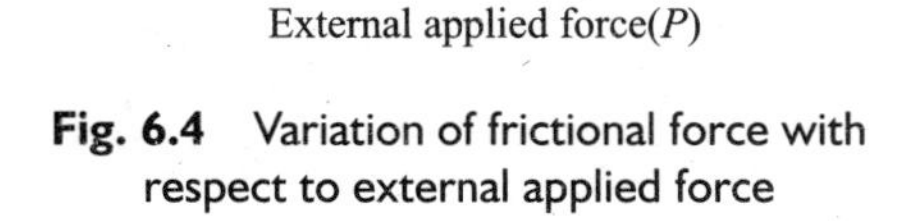

Fig. 6.4 Variation of frictional force with respect to external applied force

Depending upon the value of the frictional force developed between the contact surfaces, we come across three different cases:

Case I When $F < F_{s,\,\max}$, no motion occurs. (6.1)

Case II When $F = F_{s,\,\max}$, motion impends. (6.2)

Case III When $F > F_{s,\,\max}$, body is under motion. (6.3)

6.3 COULOMB'S LAWS OF DRY FRICTION

In 1781, a French scientist named Coulomb conducted many such experiments as discussed above to investigate the frictional forces between dry surfaces. His results are collectively termed as **Coulomb's laws of dry friction** and they can be applied to solve problems on dry friction. These are summarized below:

1. The frictional force always acts such as to *oppose* the tendency of one surface to slide relative to the other. It acts tangential to the surfaces in contact.
2. The magnitude of frictional force is exactly equal to the tangential component of the force, which tends to move the body until the limiting value is reached.
3. The maximum force of friction is independent of the area of contact between the two sliding surfaces and depends on the nature of surfaces in contact.
4. The magnitude of limiting static friction is proportional to the normal reaction between the two sliding surfaces. Mathematically, it is expressed as

$$F_{s,\max} \propto N \tag{6.4}$$

Introducing a constant of proportionality,

$$F_{s,\max} = \mu_s N \tag{6.5}$$

where the constant μ_s is called the **coefficient of static friction**.

5. For low relative velocities between sliding bodies, frictional force is independent of the relative speed with which the surfaces move over each other.
6. The magnitude of kinetic friction is proportional to the normal reaction between the two surfaces. Mathematically, it is expressed as

$$F_k \propto N \tag{6.6}$$

Introducing a constant of proportionality,

$$F_k = \mu_k N \tag{6.7}$$

where the constant μ_k is called the **coefficient of kinetic friction**.

Note: It should be noted that frictional force equal to $\mu_s N$, is applicable *only* at the point of impending motion. If the body does not move under the action of external forces then the frictional force developed is less than the maximum value of static friction. Once motion has started, the frictional force is equal to kinetic friction, which is less than the maximum static friction.

6.4 COEFFICIENTS OF FRICTION

From Eqs (6.5) and (6.7), we can write

$$\mu_s = \frac{F_{s\max}}{N} \text{ and } \mu_k = \frac{F_k}{N} \tag{6.8}$$

Thus, the ratio of magnitude of the maximum force of static friction to the magnitude of normal force is termed the **coefficient of static friction** [μ_s] for the surfaces involved. Similarly, the ratio of magnitude of the force of kinetic friction to the magnitude of normal force is termed the **coefficient of kinetic friction** [μ_k] for the surfaces involved.

Both μ_s and μ_k are **dimensionless** constants as they are ratios of forces. Their values depend on the nature of *both* the surfaces in contact. The following table gives the coefficients of static friction between various pairs of surfaces in contact. In general, for a given pair of surfaces, the coefficient of static friction is greater than that of the kinetic friction, i.e., $\mu_s > \mu_k$. Normally, μ_k is 25% less than μ_s.

Table 6.1 Coefficients of static friction for various pairs of contact surfaces

Contact surfaces	μ_s
Wood on wood	0.2–0.5
Wood on leather	0.2–0.5
Metal on metal	0.15–0.25
Metal on wood	0.2–0.6
Metal on stone	0.3–0.7
Metal on leather	0.3–0.5
Stone on stone	0.4–0.7
Earth on earth	0.2–1.0
Rubber on concrete	0.6–0.9
Rope on wood	0.5–0.7

6.5 ANGLE OF FRICTION

We have seen that the frictional force F acts *tangential* to the plane of contact and the normal reaction N acts *normal* to the plane of contact. Sometimes it will be convenient to express these two orthogonal forces by their resultant R as it simplifies the solution to the problem, particularly if we use **graphical** method. Let φ be the angle of inclination of the resultant R with respect to the normal reaction N. If the external horizontal force P applied is zero, then F is also zero. Hence, R will be same as N and in which case $\varphi = 0$. If the horizontal force increases to P_1 [refer Fig. 6.5(a)], then F also increases to F_1 to balance the external force and φ also increases correspondingly. At the point of impending motion, F reaches the maximum value, $F_{s,\,max}$ and φ also reaches its maximum value, $\varphi_{s,\,max}$ or simply φ_s. This maximum value of φ is termed as **angle of static friction** [refer Fig. 6.5(b)].

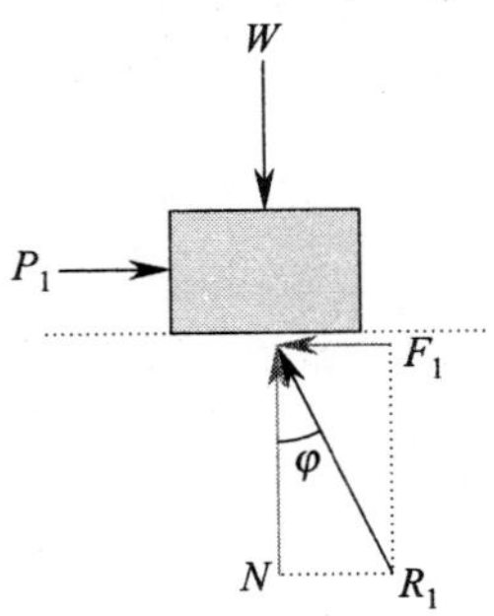

Fig. 6.5(a)

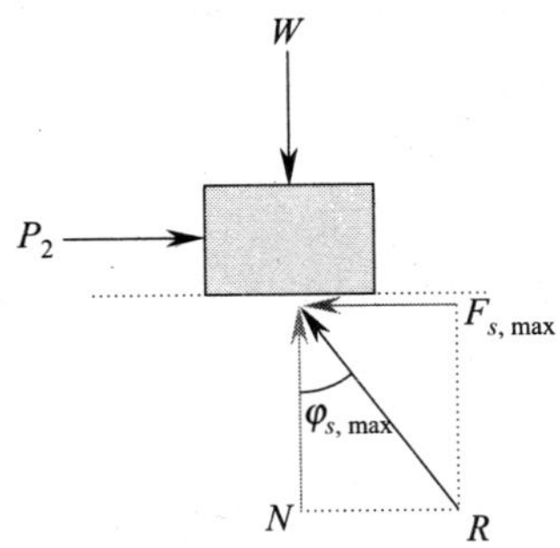

Fig. 6.5(b)

The angle of static friction can thus be defined as *the inclination of the resultant of force of friction and normal reaction to the normal reaction at the point of impending motion.* From the geometry of the triangle in Fig. 6.5(b), we see that,

$$\tan \varphi_s = \frac{F_{s,max}}{N} \tag{6.9}$$

Since $\quad F_{s,\max} = \mu_s N,$ (6.10)

$$\tan \varphi_s = \mu_s \tag{6.11}$$

Thus, the tangent of the angle of static friction is equal to the coefficient of static friction.

Similarly, if the block is in motion, we can determine the angle of kinetic friction as

$$\tan \varphi_k = \mu_k \tag{6.12}$$

6.6 DETERMINATION OF COEFFICIENT OF FRICTION

The coefficient of friction between any two contact surfaces can be determined experimentally as follows: Consider a block resting on a horizontal plane as shown in Fig. 6.6(a). If by some mechanism, we can gradually increase the angle of inclination θ of the plane with the horizontal then at a particular inclination, the block will be at the point of sliding down the incline under its own weight. This angle of inclination corresponding to impending motion is termed the **angle of repose**.

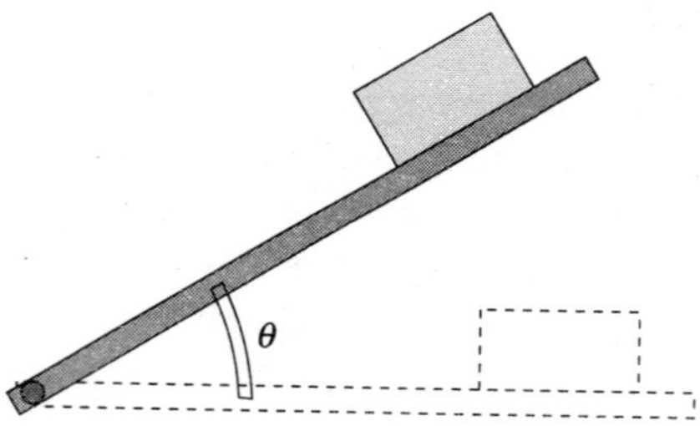

Fig. 6.6(a) Experimental apparatus to determine coefficient of friction

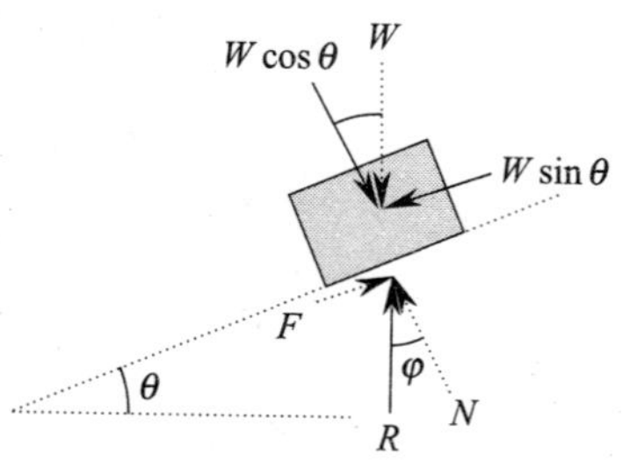

Fig. 6.6(b)

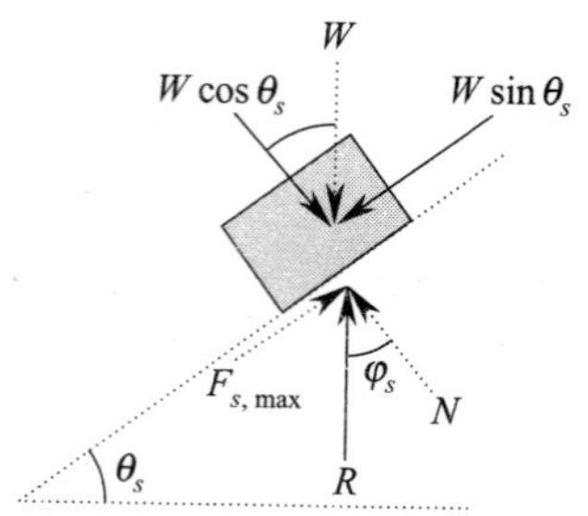

Fig. 6.6(c)

To maintain equilibrium until limiting condition is reached, we can see that the resultant R must always act vertically upwards to balance the weight of the block. Hence, from the geometry of the Fig. 6.6(b), we can readily see that the inclination of the plane, θ at any instant must be equal to φ, as N always acts normal to the inclined plane. At the point of impending motion, as φ reaches its maximum value, φ_s [refer Fig. 6.6(c)], the angle of repose, that is, the inclination at which motion impends, must be equal to φ_s. Hence, by determining the angle of inclination θ at which the body is at the point of impending motion, the angle of friction φ_s can be determined, i.e., $\theta = \varphi_s$. Once φ_s is known, μ_s can be determined from Eq. 6.11.

6.7 CONE OF FRICTION

From the above discussion, we can see that for coplanar problems, if the resultant force (R) of the normal reaction and frictional force lies within the triangle AOB having an inner angle $2\varphi_s$ then the body will be at rest. For non-coplanar problems, it must lie within the cone generated by revolving AO about the normal axis by one complete revolution. The cone thus generated is known as the **cone of friction**.

Triangle of friction   Cone of friction

Fig. 6.7

Example 6.1 A man lifts a carton of mass m by applying a force P laterally on either side of the opposite faces as shown in Fig. 6.8. If the coefficient of friction between his hands and the carton is μ, determine what minimum force he should apply to lift the carton.

Fig. 6.8

Solution Since his hands exert a force P normal to the face of the box on either side, the frictional force between his hands and the faces is equal to μP at the limiting condition. As the box tries to fall downwards due to pull of gravity, the frictional forces act such as to oppose this motion and hence they act upwards. At the point of impending motion when equilibrium is maintained, we can write

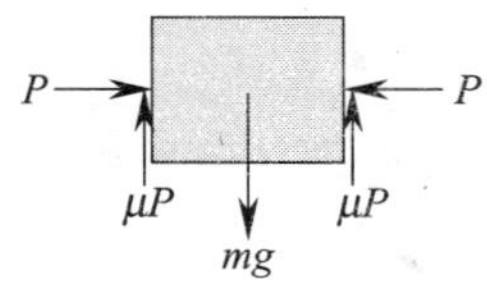

Fig. 6.8(a)

$$2\,\mu P - mg = 0$$

$$\Rightarrow \qquad P = \frac{mg}{2\mu}$$

Example 6.2 A sweeper while mopping the floor exerts a force of 50 N along the axis of the handle as shown in Fig. 6.9. Determine the coefficient of friction between the contact surfaces, if the mop is on the verge of sliding. The weight of the mop can be neglected.

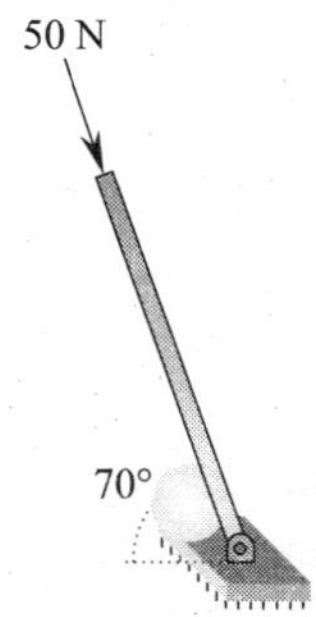

Fig. 6.9

Solution As the weight of the mop is negligible, the forces acting on it are the force exerted by the sweeper, normal reaction N and the frictional force F. Applying the conditions of equilibrium at the point of impending motion,

$$\sum F_y = 0 \Rightarrow$$

$$N - 50 \sin 70° = 0$$

$$\therefore \qquad N = 50 \sin 70°$$

$\sum F_x = 0 \Rightarrow$

$$50 \cos 70° - F = 0$$

$$50 \cos 70° - \mu\, 50 \sin 70° = 0 \quad [\text{since } F = \mu N]$$

$$\Rightarrow \qquad \mu = \frac{1}{\tan 70°} = 0.364$$

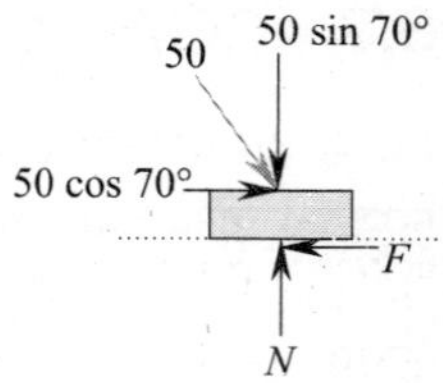

Fig. 6.9(a)

Example 6.3 A block of 50 kg mass rests on a rough horizontal plane as shown in Fig. 6.10. If the coefficients of static and kinetic friction between the block and the plane are respectively 0.2 and 0.15, describe the resulting motion, (i) when a horizontal force of 75 N is applied, and (ii) when a horizontal force of 120 N is applied. Also, determine the force required to cause motion to impend.

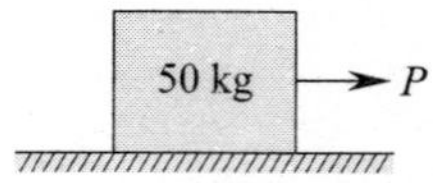

Fig. 6.10

Solution The free-body diagram of the block is shown in Fig. 6.10(a). The forces acting on the block are its weight W, normal reaction N, force of friction F acting to oppose the motion and the applied force P.

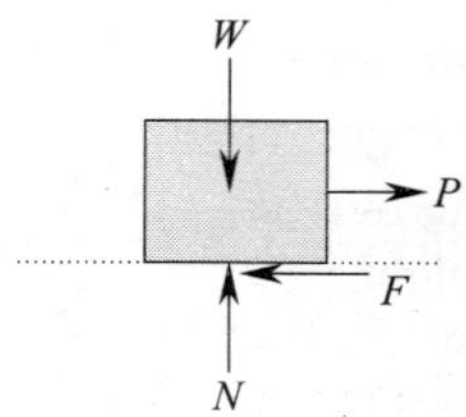

Fig. 6.10(a)

Applying the conditions of equilibrium along the X and Y directions,

$\sum F_y = 0 \Rightarrow$

$$N - W = 0$$

$$\therefore \qquad N = W$$

$$= 50 \times g = 490.5 \text{ N} \qquad \text{(a)}$$

$\sum F_x = 0 \Rightarrow$

$$P - F = 0 \qquad \text{(b)}$$

The maximum force of friction possible is equal to the limiting force of static friction, i.e.,

$$F_{max} = F_{s,\,max} = \mu_s N$$

$$= 0.2 \times 490.5 = 98.1 \text{ N} \qquad \text{(c)}$$

Force required to cause motion to impend is equal to the maximum force of static friction, i.e., $P = 98.1$ N.

(i) When a horizontal force of 75 N is applied

When $P = 75$ N, from equation (b), we can see that the frictional force to maintain equilibrium is

$$F = P = 75 \text{ N}$$

Since this frictional force is less than $F_{s,\,max}$, the block is in *equilibrium* or in other words, it is at rest.

(ii) When a horizontal force of 120 N is applied

When $P = 120$ N, from equation (b), we can see that the frictional force to maintain equilibrium is equal to 120 N. Since this frictional force is greater than $F_{s,\,max}$, which cannot be, the block is actually under motion. Therefore, the actual force of friction is equal to the force of kinetic friction whose magnitude is given as

$$F = F_k = \mu_k \text{ N}$$
$$= 0.15 \times 490.5 = 73.58 \text{ N}$$

Example 6.4 A block of 200 kg mass rests on a rough horizontal plane. Find the force required in each of the following cases: (i) to just pull the block by a horizontal force P, (ii) to just pull the block by an inclined force P inclined at 30° to the horizontal, and (iii) to just push the block by a horizontal force P. The coefficient of static friction between the contact surfaces is 0.3.

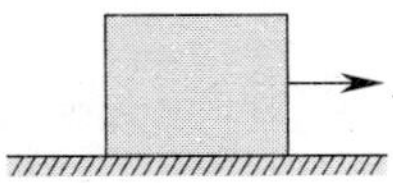

Fig. 6.11(a)

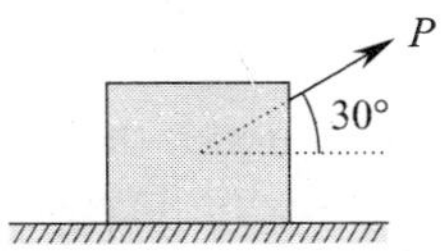

Fig. 6.11(b)

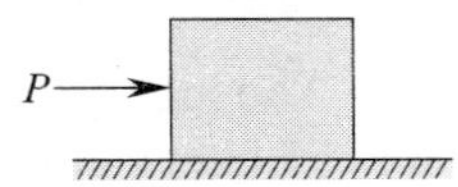

Fig. 6.11(c)

Solution

(i) Force required to just pull the block by a horizontal force [P]

The free-body diagram of the block is shown in Fig. 6.11(d). The forces acting on the block are its weight, normal reaction N, applied force P and frictional force F. At the point of impending motion towards right, the frictional force acts in the opposite direction and its magnitude is the limiting value equal to μ_sN. Applying the conditions of equilibrium along the X and Y directions:

$\sum F_y = 0 \Rightarrow$

$$N - 200 \text{ g} = 0$$

$\therefore$ $$N = 200\, g = 200 \times 9.81 = 1962 \text{ N}$$

$\sum F_x = 0 \Rightarrow$

$$P - F = 0$$

$\therefore$ $$P = F = \mu_s N = 0.3 \times 1962 = 588.6 \text{ N}$$

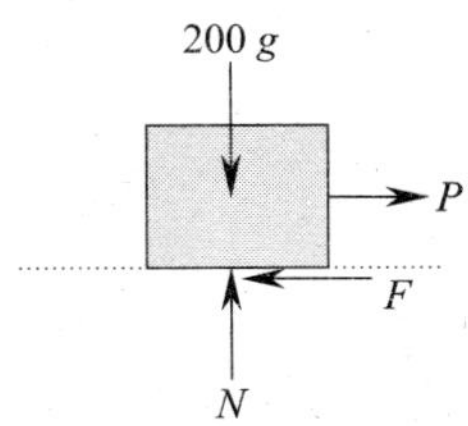

Fig. 6.11(d)

(ii) Force required to just pull the block by an inclined force [P]

The free-body diagram of the block is shown in Fig. 6.11(e). The applied force P has been resolved into rectangular components. Applying the conditions of equilibrium along the X and Y directions:

$\sum F_y = 0 \Rightarrow$

$$N + P \sin 30° - 200 \text{ g} = 0$$

$\therefore$ $$N = 200 \text{ g} - P \sin 30° = 1962 - 0.5\, P$$

$\sum F_x = 0 \Rightarrow$

$$P \cos 30° - F = 0$$

Since $F = \mu_s N$ at the point of impending motion,

$$P \cos 30° - \mu_s N = 0$$

$$P \cos 30° - (0.3)[1962 - 0.5\, P] = 0$$

or, $$1.016\, P = 588.6$$

$\therefore$ $$P = 579.33 \text{ N}$$

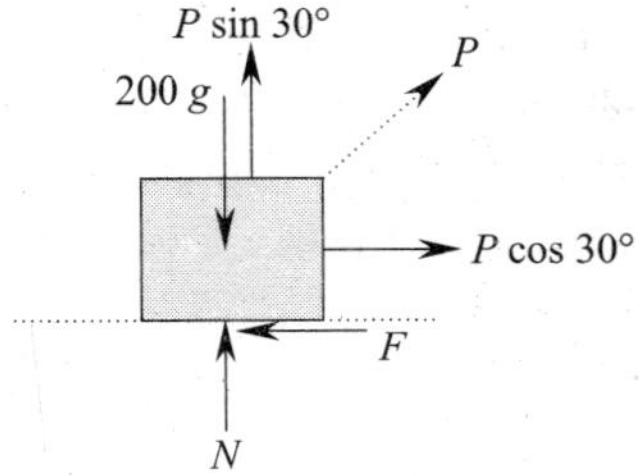

Fig. 6.11(e)

(iii) Force required to just push the block by a horizontal force [P]

Since the force P pushes the block to the right, the frictional force F acts towards the left. Hence, the value of P will be the same as that obtained for the force required to just pull the block by a horizontal force.

Example 6.5 A block of 50 kg mass is pushed up an inclined plane by a force of 100 N acting parallel to the incline as shown in Fig. 6.12. Determine whether the block shown is in equilibrium or not. Also, find the magnitude and direction of the frictional force. The coefficients of friction between the block and the plane are $\mu_s = 0.25$ and $\mu_k = 0.2$.

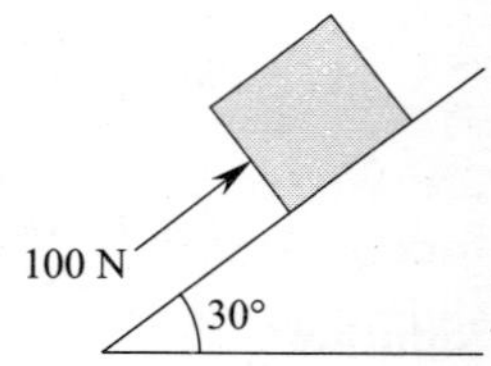

Fig. 6.12

Solution The impending motion of the block under the action of an external force may be either upwards or downwards. Accordingly, the frictional force would act such as to oppose the motion. To begin with, let us assume that the block is at the point of moving upwards. Then the frictional force will act along the downward direction as shown in Fig. 6.12(a). Applying the equilibrium condition along normal to the plane,

$$\Sigma F_y = 0 \Rightarrow$$

$$N - 50\,g \cos 30° = 0$$

$$\therefore \quad N = 50 \times 9.81 \times \cos 30° = 424.79 \text{ N}$$

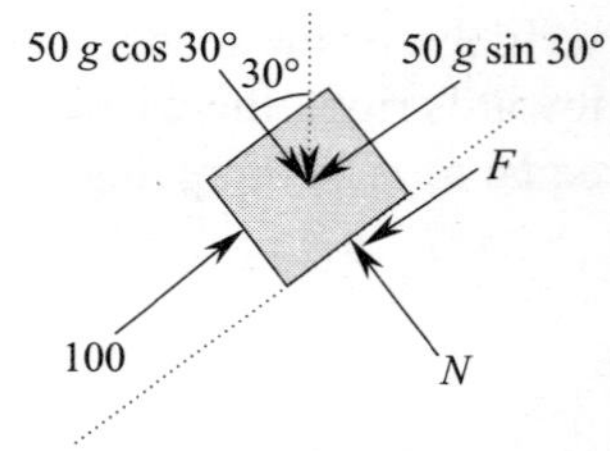

Fig. 6.12(a)

The maximum or the limiting static force of friction is then obtained as:

$$F_{s,\,\max} = \mu_s \text{N}$$

$$= 0.25 \times 424.79 = 106.2 \text{ N}$$

The frictional force required to maintain equilibrium along the inclined plane is obtained as follows:

$$\Sigma F_x = 0 \Rightarrow$$

$$100 - 50\,g \sin 30° - F = 0$$

$$\therefore \quad F = -145.25 \text{ N}$$

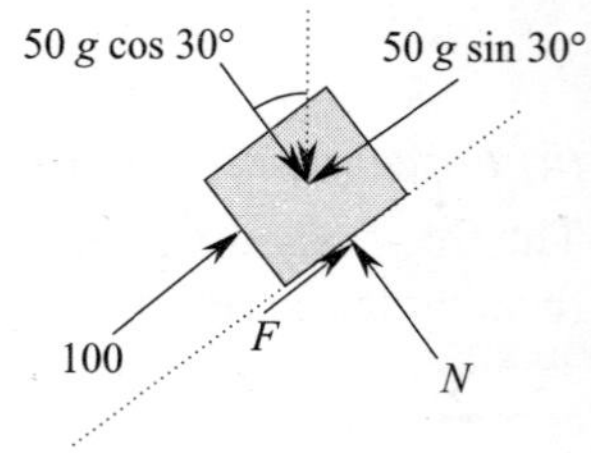

Fig. 6.12(b)

Since the above value is negative, the assumed direction is wrong and the block actually slides downwards. Hence, the frictional force required to maintain equilibrium is a force of magnitude 145.25 N acting upwards as shown in Fig. 6.12(b). We know that the maximum force of friction possible is $F_{s,\,\max}$. Since the absolute value of F is greater than $F_{s,\,\max}$, which cannot be, the block is in motion. Hence, we must use μ_k to determine the actual value of frictional force, i.e.,

$$F = F_k = \mu_k \text{N}$$

$$= (0.2)\ 424.79 = 84.96 \text{ N (acting upwards)}$$

Example 6.6 A horizontal force P is applied on a 50 kg block placed on an inclined plane as shown in Fig. 6.13. Determine the value of P required (i) to just push the block up the incline, (ii) to just prevent motion down the incline. The coefficient of static friction between the contact surfaces is 0.2.

Solution Since the applied force is not parallel to the inclined plane, it will be convenient to express the normal reaction and the frictional force by their resultant as it simplifies the problem. Hence, we must determine the angle of static friction between the contact surfaces to know the inclination of the resultant with respect to the normal reaction. We know,

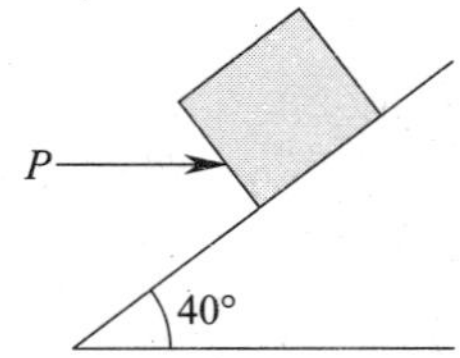

Fig. 6.13

$$\tan \varphi_s = \mu_s$$

$$\Rightarrow \quad \varphi_s = \tan^{-1}(\mu_s) = \tan^{-1}(0.2) = 11.31°$$

We see that the inclination of the plane with respect to the horizontal is greater than the angle of friction. Hence, the block would slide down under its own weight. Therefore, to prevent the block from sliding down or to just move it up the plane, we must apply a force. The two cases are different as the direction of frictional force depends upon the direction of impending motion.

(i) When the block is at the point of moving up the plane

Here the force of friction will act in the downward direction to prevent motion up the plane. Therefore, the free-body diagram is as shown in Fig. 6.13(a):

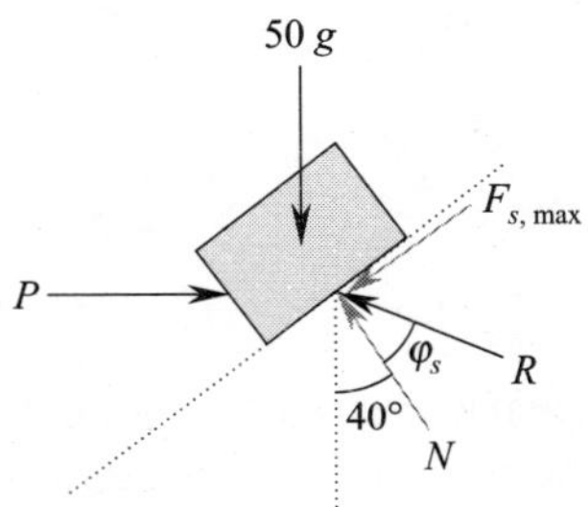

Fig. 6.13(a)

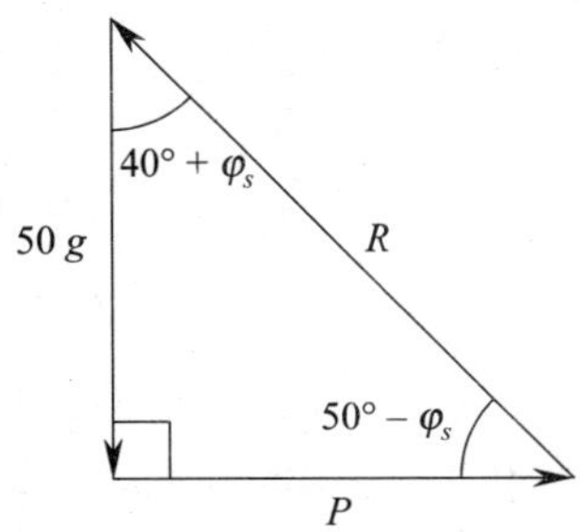

Fig. 6.13(b)

At the point of impending motion, the resultant force R will be inclined to the normal reaction N at an angle of φ_s. Since there are three forces acting on the block and they keep it in equilibrium, we can construct the force triangle as shown in Fig. 6.13(b). Applying sine law to the force triangle, we have

$$\frac{P}{\sin(40° + \varphi_s)} = \frac{50g}{\sin[90° - (40° + \varphi_s)]}$$

$$\frac{P}{\sin(40° + 11.31°)} = \frac{50 \times 9.81}{\sin[50° - 11.31°]}$$

$$\frac{P}{\sin(51.31°)} = \frac{50 \times 9.81}{\sin(38.69°)}$$

$$\therefore \quad P = 612.46 \text{ N}$$

(ii) When the block is at the point of moving down the plane

Here the force of friction will act in the upward direction to prevent motion down the plane. The free-body diagram is as shown in Fig. 6.13(c). The inclination of R with respect to the normal reaction is the same as φ_s. However, its inclination with respect to the vertical is different and it is equal to $(40° - \varphi_s)$.

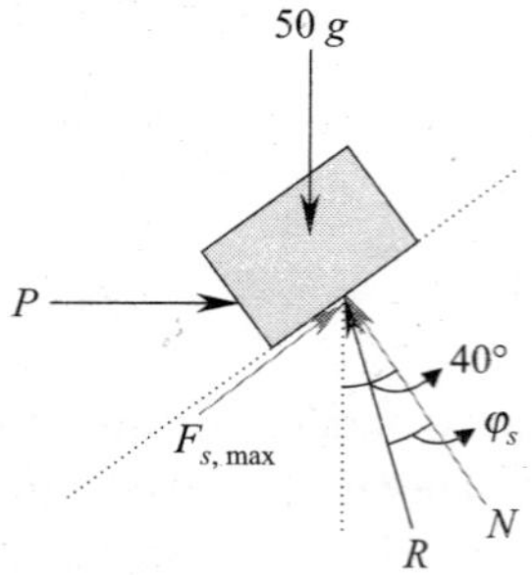

Fig. 6.13(c)

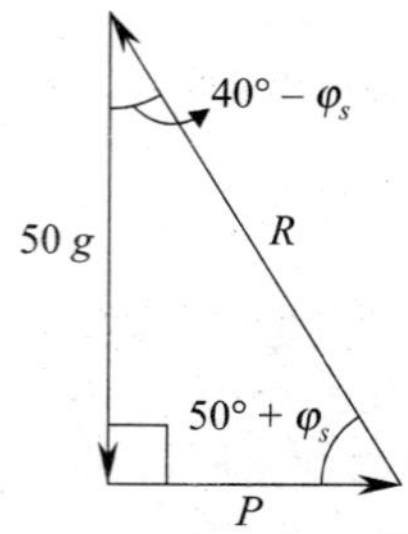

Fig. 6.13(d)

The force triangle can be constructed as shown in Fig. 6.13(d). Applying sine law to the force triangle,

$$\frac{P}{\sin(40° - 11.31°)} = \frac{50 \times 9.81}{\sin[50° + 11.31°)]}$$

$$\frac{P}{\sin 28.69°} = \frac{50 \times 9.81}{\sin(61.31°)}$$

$\therefore$ $$P = 268.43 \text{ N}$$

Example 6.7 A body of mass m is to be lowered down by a rope passing over a smooth pulley and the free end is held by a man as shown in Fig. 6.14. Determine the force he has to exert to prevent the downward motion of the block.

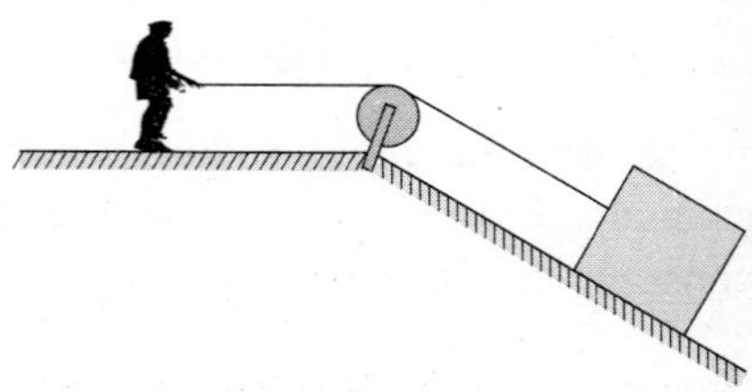

Fig. 6.14

Solution The free-body diagram of the block is shown in Fig. 6.14(a). The forces acting on it are (i) its weight mg, resolved into components $mg \sin\theta$ along the incline and $mg \cos\theta$ normal to the incline, (ii) normal reaction N, (iii) force of friction F acting upward as the block is about to slide downward, and (iv) applied force P.

Applying the equilibrium conditions normal to the plane and along the plane,

$\sum F_y = 0 \Rightarrow$

$$N - mg\cos\theta = 0$$

$\therefore$ $$N = mg\cos\theta$$

$\sum F_x = 0 \Rightarrow$

$$mg\sin\theta - F - P = 0$$

$\therefore$ $$P = mg\sin\theta - F$$

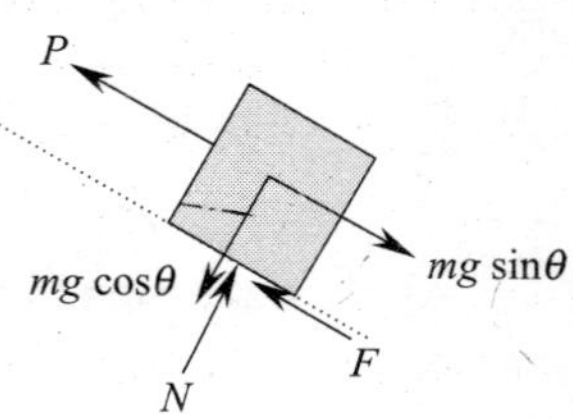

Fig. 6.14(a)

Since $F = \mu_s N$ at the point of impending motion,

$$P = mg \sin\theta - \mu mg \cos\theta$$
$$= mg\,[\sin\theta - \mu \cos\theta]$$

Example 6.8 A block of mass m is resting in an inclined right-angled trough as shown in Figs 6.15(a) and (b). If the coefficient of friction between the contact surfaces is 0.3, at what angle of inclination θ will the motion be impending?

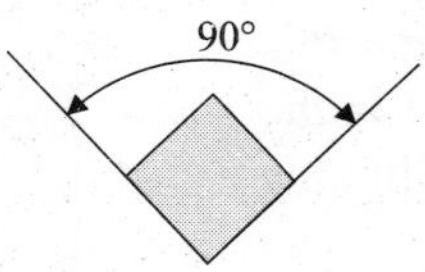

Block in trough

Fig. 6.15(a)

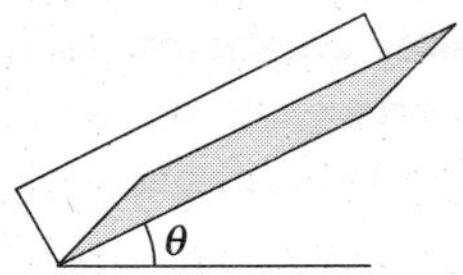

Inclined right-angled trough

Fig. 6.15 (b)

Solution As it is difficult to visualize this three-dimensional problem, we will draw two different diagrams: 6.15 (c) and (d). In Fig. 6.15(c), we see that the normal component of the weight of the block is $mg \cos\theta$. This must be balanced by the normal reactions N exerted by the plane on the two faces of the block [refer Fig. 6.15(d)]. Note that as it is a right-angled trough and symmetrical to the vertical axis, the normal reactions exerted by the planes are equal.

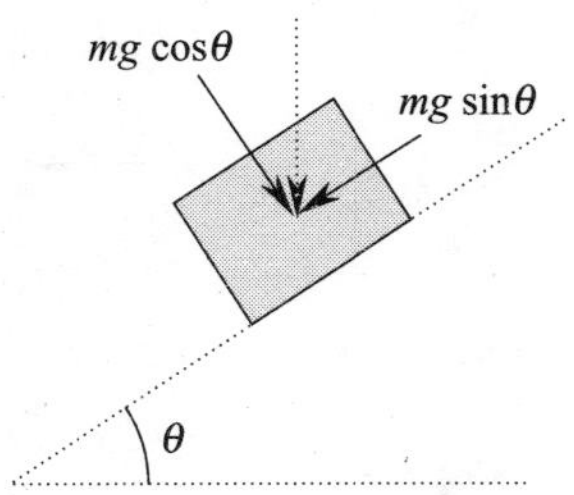

Fig. 6.15(c)

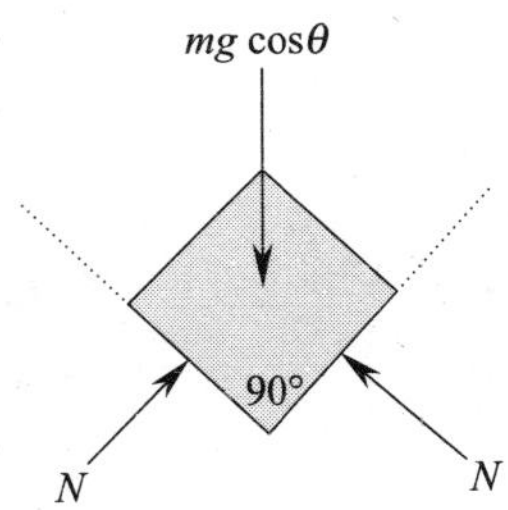

Fig. 6.15(d)

Hence, applying the conditions of equilibrium along the normal direction to the incline,

$\sum F_y = 0 \Rightarrow$

$$2N \cos 45^\circ - mg \cos\theta = 0$$

$\Rightarrow$
$$N = \frac{1}{\sqrt{2}}\, mg \cos\theta$$

$\therefore$
$$F = \mu N = \frac{\mu}{\sqrt{2}}\, mg \cos\theta$$

Applying the conditions of equilibrium parallel to the incline,

$\sum F_x = 0 \Rightarrow$

$$mg \sin\theta - F - F = 0$$

(Note that as the block is in contact with two planes, two forces of friction are considered.)

$$mg\sin\theta - 2\frac{\mu}{\sqrt{2}}mg\cos\theta = 0$$

$\Rightarrow$ $$\tan\theta = \sqrt{2}\,\mu$$

$\therefore$ $$\theta = \tan^{-1}(\sqrt{2}\times 0.3) = 23°$$

Example 6.9 A rectangular block of 100 kg mass rests on a rough horizontal floor having coefficient of friction 0.25. What minimum force P is required to just move the block? If the base width is 40 cm, what is the maximum height at which the force P can be applied without causing the block to tip?

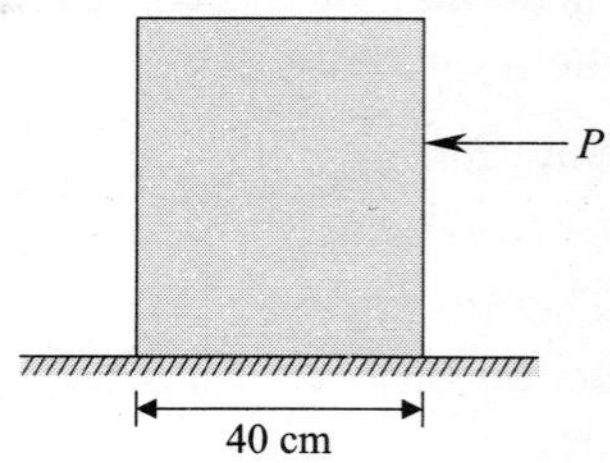

Fig. 6.16

Solution

(i) When the block is at the point of impending motion

The free-body diagram of the block at impending motion is shown in Fig. 6.16(a).

Applying the conditions of equilibrium,

$\sum F_y = 0 \Rightarrow$

$$N - 100\,g = 0$$

$\therefore$ $$N = 100\,g \qquad \text{(a)}$$

$\sum F_x = 0 \Rightarrow$

$$F - P = 0$$

$\therefore$ $$P = F = \mu N$$

$$= 0.25 \times 100 \times 9.81$$

$$= 245.25 \text{ N} \qquad \text{(b)}$$

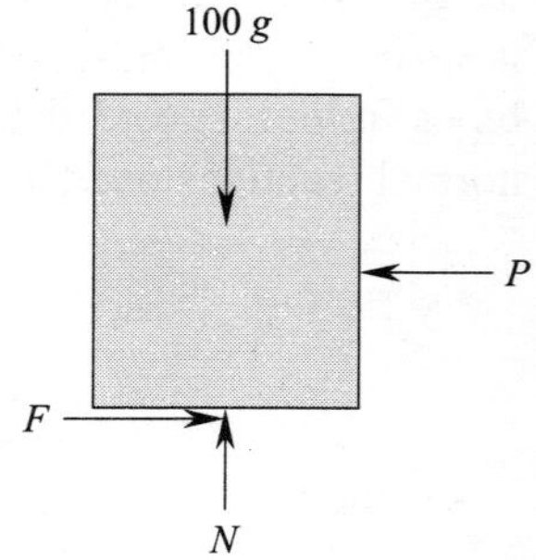

Fig. 6.16(a)

(ii) When the block is at the point of tipping

At the point of tipping, the block will have contact only at the left corner A. Hence, the frictional force and normal reaction are shown only at the left corner [refer Fig. 6.16(b)]. Taking the summation of moments about A and applying the condition of equilibrium,

$\sum M_A = 0 \Rightarrow$

$$P \times h - 100 \times g \times 0.2 = 0$$

$\Rightarrow$ $$h = \frac{100 \times g \times 0.2}{P}$$

$$= \frac{100 \times 9.81 \times 0.2}{245.25} = 0.8 \text{ m} \qquad \text{(c)}$$

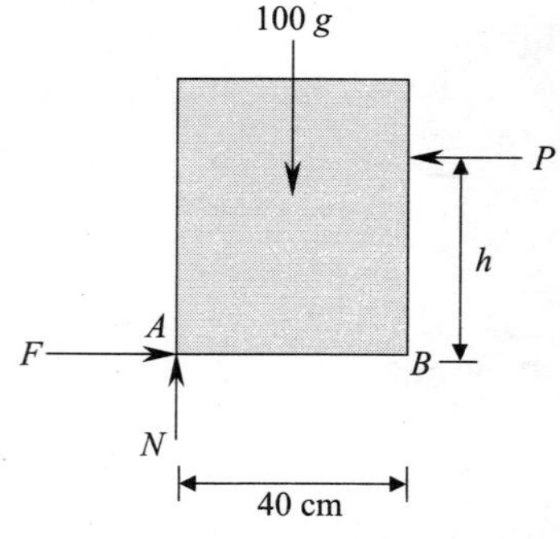

Fig. 6.16(b)

Example 6.10 A table of 1 m length, 0.75 m height and 25 kg weight is to be pulled on a rough surface ($\mu_s = 0.3$) as shown in Fig. 6.17. Determine (i) the horizontal force required to just pull it to the right, (ii) the normal reactions at the front and rear legs. Suppose the table is pushed with a force of 200 N, determine the maximum height at which the force can be applied without causing the table to tip. Assume the centre of gravity of the table to be midway between the front and rear legs.

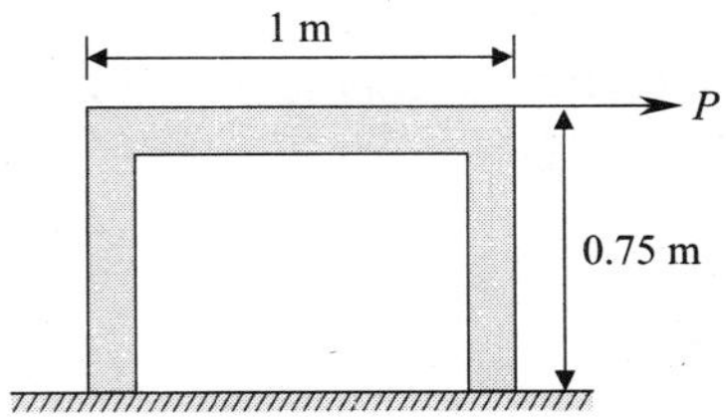

Fig. 6.17

Solution

(i) Horizontal force required to pull the table to the right

The free-body diagram of the table is shown in Fig. 6.17(a). The forces acting on it are its weight, normal reactions, frictional forces at the front and rear legs and the applied force.

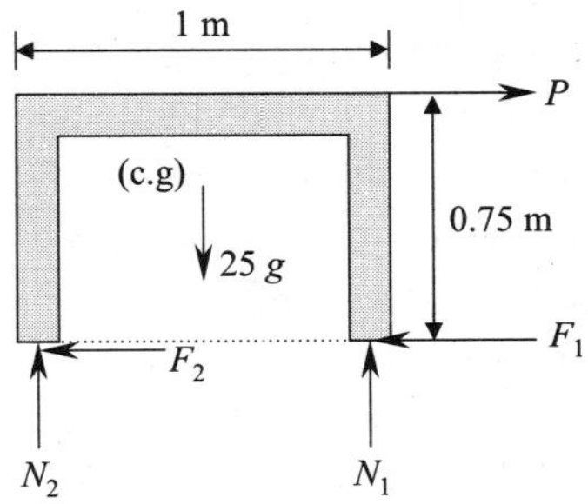

Fig. 6.17(a)

At the point of impending motion, applying the conditions of equilibrium along the X and Y directions,

$$\sum F_y = 0 \Rightarrow$$

$$N_1 + N_2 - 25 \times 9.81 = 0$$

$$\Rightarrow \quad N_1 + N_2 = 245.25 \text{ N} \quad \text{(a)}$$

$$\sum F_x = 0 \Rightarrow$$

$$P - F_1 - F_2 = 0$$

At the point of impending motion, we know that $F_1 = \mu N_1$ and $F_2 = \mu N_2$. Therefore,

$$P = F_1 + F_2$$

$$= \mu N_1 + \mu N_2 = \mu(N_1 + N_2) = 73.58 \text{ N}$$

(ii) Normal reactions at the front and rear legs

Taking moment about the front leg and applying the equilibrium condition,

$$\sum M_{FL} = 0 \Rightarrow$$

$$-P \times 0.75 - N_2 \times 1 + (25 \times 9.81 \times 0.5) = 0$$

$$\Rightarrow \quad N_2 = 67.44 \text{ N}$$

From equation (a), we get

$$N_1 = 245.25 - 67.44 = 177.81 \text{ N}$$

(iii) Maximum height at which the force can be applied without causing the table to tip

When a force is pushing the table, at the point of tipping, the table will have contact only at the left leg. Hence, the frictional force and normal reaction are shown only at the left leg.

Taking the summation of moments about the left leg and applying the condition of equilibrium,

$\sum M_{LL} = 0 \Rightarrow$

$$200 \times h - 25 \times g \times 0.5 = 0$$

$$\Rightarrow \quad h = \frac{25 \times 9.81 \times 0.5}{200} = 0.613 \text{ m}$$

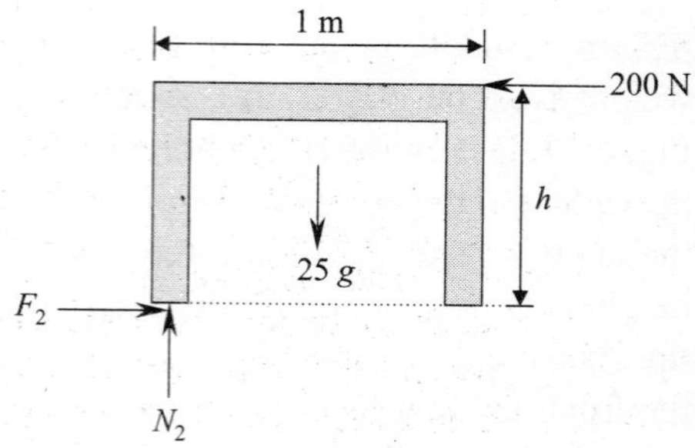

Fig. 6.17(b)

Example 6.11 Determine the vertical force P to be applied at the pin joint B of a link mechanism shown in Fig. 6.18 to cause motion to impend in 20 kg block at C. The coefficient of friction between the block and the horizontal plane is 0.2. Solve for angle $\theta = 30°$.

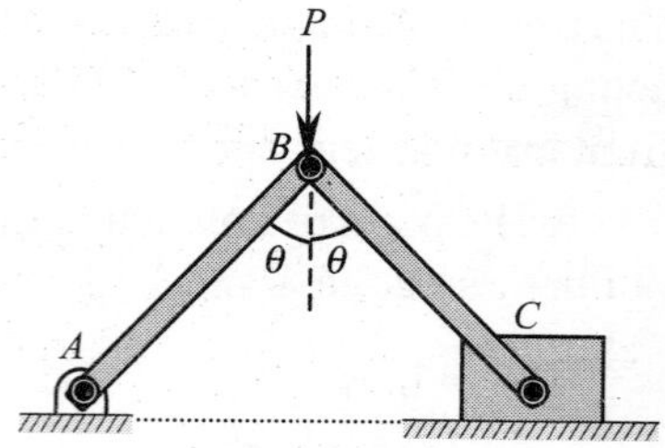

Fig. 6.18

Solution Due to symmetry, we know that the forces in the members AB and BC are equal and let that force be S. Hence, the forces acting on the block are the force S exerted by the member BC, weight mg, normal reaction N and frictional force F. Applying the conditions of equilibrium at the point of impending motion,

$\sum F_y = 0 \Rightarrow$

$$N - mg - S \cos 30° = 0$$

$$\therefore \quad N = mg + S \cos 30°$$

$\sum F_x = 0 \Rightarrow$

$$S \sin 30° - F = 0$$

$$S \sin 30° - \mu N = 0$$

$$S \sin 30° - \mu[mg + S \cos 30°] = 0$$

$$\Rightarrow \quad S = \frac{\mu mg}{\sin 30° - \mu \cos 30°}$$

$$= \frac{(0.2)(20)(9.81)}{\sin 30° - (0.2)\cos 30°} = 120.08 \text{ N}$$

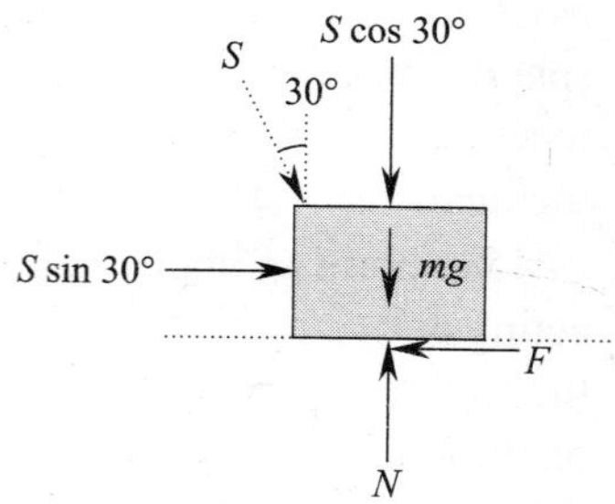

Fig. 6.18(a)

Therefore, applying the condition of equilibrium at point B, we have

$$P = 2S \cos 30°$$

$$= \sqrt{3}\, S = 208 \text{ N}$$

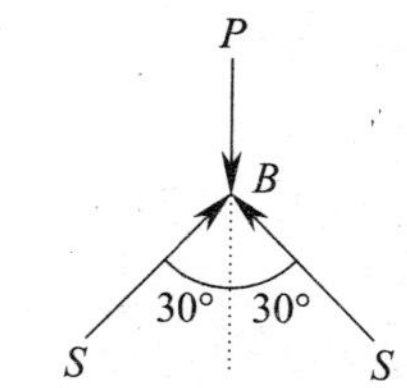

Fig. 6.18(b)

Example 6.12 In the above problem, if the force P is applied horizontally as shown in Fig. 6.19, determine its value to cause motion to impend in block at C.

Solution Here unlike the previous problem, the forces in the members AB and BC are not equal. Let the force in the member AB be S_1 and that in BC be S_2. As before, the force in the member BC can be determined by considering the equilibrium of the block and we can get

$$S_2 = 120.08 \text{ N}$$

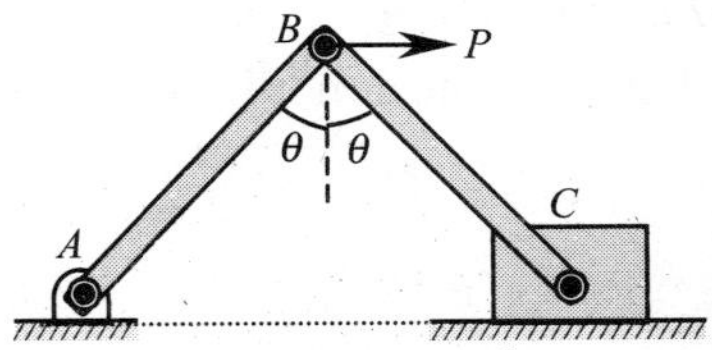

Fig. 6.19

Applying the conditions of equilibrium at point B,

$\Sigma F_y = 0 \Rightarrow$

$$S_1 \cos 30° + S_2 \cos 30° = 0$$

$\therefore$ $$S_1 + S_2 = 0$$

$\Sigma F_x = 0 \Rightarrow$

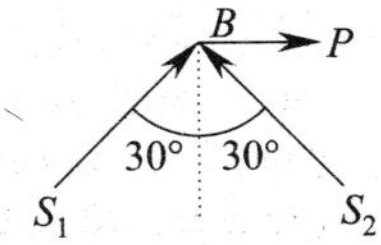

Fig. 6.19(a)

$$P + S_1 \sin 30° - S_2 \sin 30° = 0$$

$$P - S_2 \sin 30° - S_2 \sin 30° = 0 \text{ [since } S_1 = -S_2]$$

$\therefore$ $$P = 2S_2 \sin 30° = S_2 = 120.08 \text{ N}$$

Example 6.13 A smooth sphere rests against a smooth vertical plane, and upon a wedge that rests on a horizontal plane supposed rough. If the coefficient of friction between the wedge and the horizontal plane is μ, determine the angle of the wedge when there is equilibrium.

Solution Let the weights of the sphere and wedge be respectively W_s and W_w. Let the angle of the wedge be θ. The free-body diagrams of the sphere and wedge are shown in Fig. 6.20(a). The forces acting on the free-body diagram of the sphere are its weight W_s, reaction R_1 at the contact point with the wall and reaction R_2 exerted by the wedge on the sphere. It should be noted that as the sphere is smooth, there is no frictional force at these contact points. Hence, the reaction R_2 acts normal to the common plane of contact or in other words, it is inclined at wedge angle θ to the vertical.

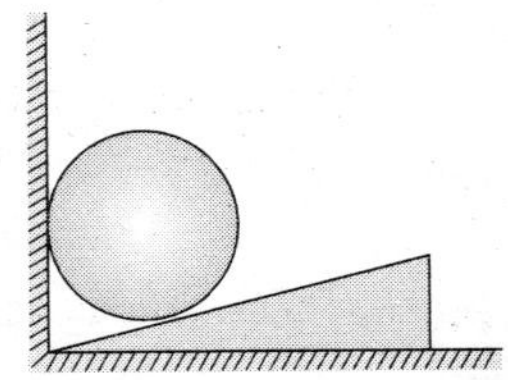

Fig. 6.20

Similarly, in the free-body diagram of the wedge, the forces acting on it are its weight W_w, reaction R_2 exerted by the sphere on the wedge, being equal and opposite to that exerted by the wedge on the sphere, reaction R_3 exerted by the horizontal rough plane on the wedge. As this contact point is rough, there will be frictional force in addition to the normal reaction and hence, their resultant R_3 is inclined at an angle φ_s to the normal direction. It should be noted that $\tan \varphi_s = \mu$.

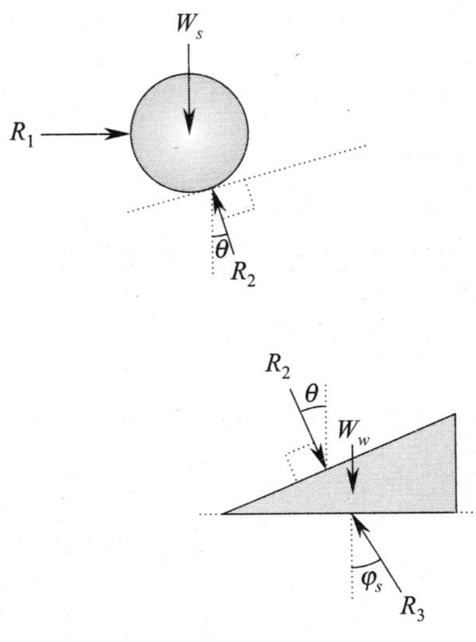

Fig. 6.20(a)

Equilibrium of sphere

Applying the conditions of equilibrium of the sphere along the X and Y directions,

$\Sigma F_x = 0 \Rightarrow$

$$R_1 - R_2 \sin \theta = 0$$

$\therefore$ $$R_1 = R_2 \sin \theta \quad \text{(a)}$$

$\Sigma F_y = 0 \Rightarrow$

$$R_2 \cos \theta - W_s = 0$$

$\therefore$ $$W_s = R_2 \cos \theta \quad \text{(b)}$$

Equilibrium of wedge

Applying the conditions of equilibrium of wedge along the X and Y directions,

$\sum F_y = 0 \Rightarrow$

$$R_3 \cos \varphi_s - R_2 \cos \theta - W_w = 0$$

$$\therefore \quad R_3 \cos \varphi_s = R_2 \cos \theta + W_w \quad \text{(c)}$$

$\sum F_x = 0 \Rightarrow$

$$R_2 \sin \theta - R_3 \sin \varphi_s = 0$$

$$\therefore \quad R_3 \sin \varphi_s = R_2 \sin \theta \quad \text{(d)}$$

Dividing equation (d) by equation (c), we have

$$\tan \varphi_s = \frac{R_2 \sin\theta}{R_2 \cos\theta + W_w}$$

Substituting the value of $R_2 \cos \theta$ from equation (b) and noting that $\tan \varphi_s = \mu$, we have

$$R_2 \sin \theta = \mu[W_s + W_w] \quad \text{(e)}$$

From equations (b) and (e), we have

$$\frac{W_s}{\cos \theta} \sin \theta = \mu[W_s + W_w]$$

$$\Rightarrow \quad \tan \theta = \mu\left[\frac{W_s + W_w}{W_s}\right]$$

or

$$\theta = \tan^{-1} [\mu(W_s + W_w)/W_s]$$

Example 6.14 A cylinder of weight W and radius r rests in a right-angled trough as shown in Fig. 6.21. Determine the moment M that must be applied in the clockwise direction to just turn the cylinder.

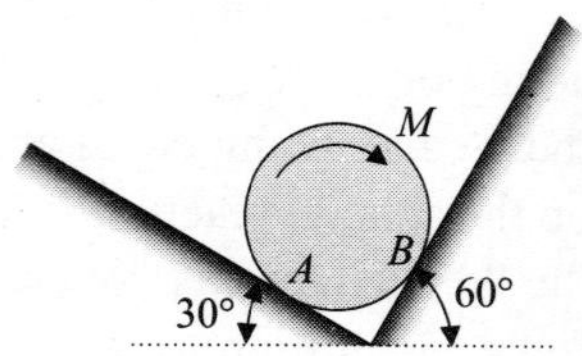

Fig. 6.21

Solution

Normal reactions at the contact points A and B

The free-body diagram of the cylinder is shown in Fig. 6.21(a). The forces acting on the cylinder are: (i) its weight W, (ii) normal reactions N_A and N_B exerted by the inclined planes.

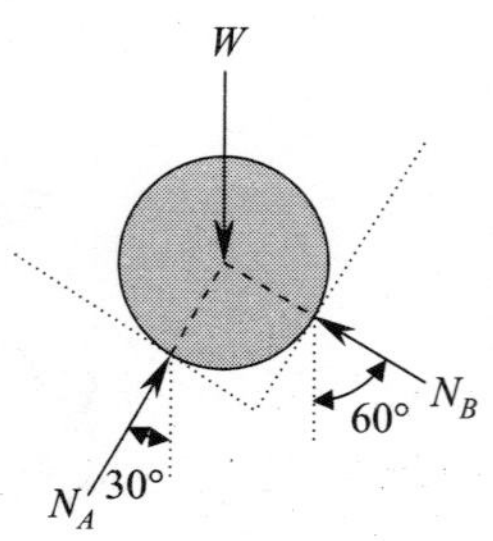

Fig. 6.21(a)

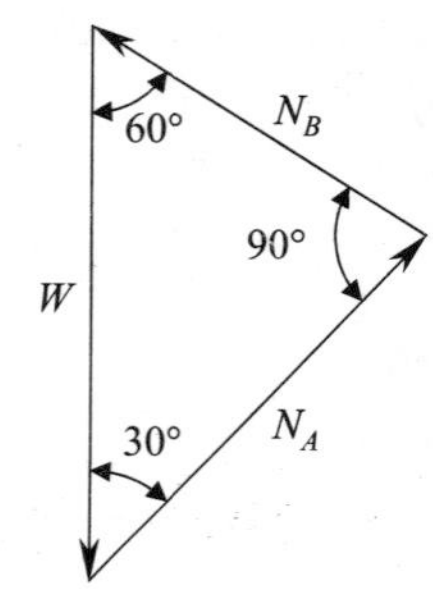

Fig. 6.21(b)

As the reactive forces act normal to the tangential planes, the three forces will be concurrent at the centre of the cylinder. Hence, we can solve the unknowns by applying Lami's theorem to the force triangle [Fig. 6.21(b)].

$$\frac{W}{\sin 90^\circ} = \frac{N_A}{\sin 60^\circ} = \frac{N_B}{\sin 30^\circ}$$

$$\Rightarrow \qquad N_A = W\frac{\sin 60^\circ}{\sin 90^\circ} = 0.866\ W$$

Also,

$$N_B = W\frac{\sin 30^\circ}{\sin 90^\circ} = 0.5\ W$$

Frictional forces at contact points A and B

Since the external moment is applied in the clockwise direction, the frictional forces developed at the contact points oppose this rotation and hence, they act in the opposite direction as shown in Fig. 6.21(c). At the point of impending motion, the frictional forces at the contact points A and B are:

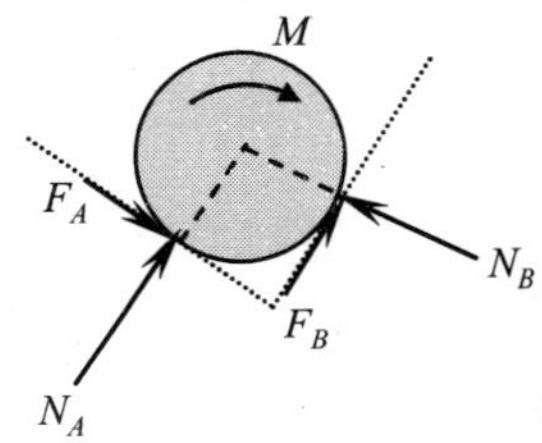

Fig. 6.21(c)

$$F_A = \mu N_A = 0.866\ \mu W$$
$$F_B = \mu N_B = 0.5\ \mu W$$

Considering the equilibrium of the cylinder at the point of impending motion,

$$\sum M_O = 0 \Rightarrow$$

$$F_A r + F_B r - M = 0$$

$$\Rightarrow \qquad M = (F_A + F_B)r$$
$$= 1.366\ \mu Wr$$

6.8 IMPENDING MOTION OF CONNECTED BODIES

In all of the previous examples, we considered individual bodies at the point of impending motion. We also encounter situations in which more than one body is connected together by a string or a strut and the system is at the point of impending motion. In such cases, we consider each body in the system to be at the point of impending motion and solve for the unknowns.

Example 6.15 In the system of blocks shown in Fig. 6.22, if the weight of the block B is 150 N, determine the range of the weight of the block A for which motion is impending. Take coefficient of friction for all contact surfaces to be 0.25. Assume the pulley at the top to be frictionless.

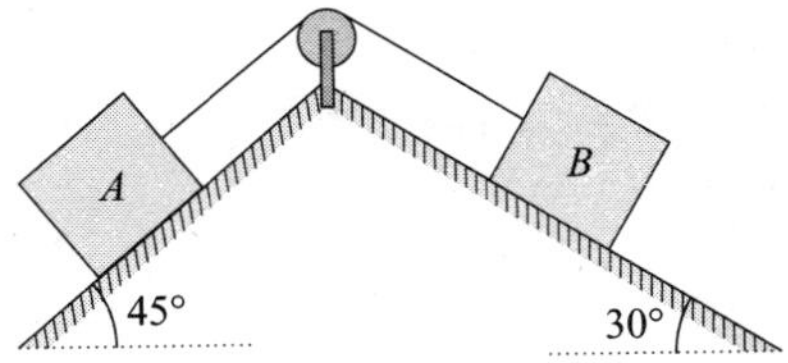

Fig. 6.22

Solution Before we draw the free-body diagrams of the blocks, we must know the direction of impending motion as the direction of frictional force acts opposite to the direction of impending motion. As the weight of the block A is unknown, the motion could be in either direction, i.e., the block A may move downwards or the block B may move downwards. Hence, we must consider the two cases separately to determine the range of weight of the block A, for which motion is impending.

Case I Consider the motion of the block B to be downwards; then the frictional forces act in the direction opposite to the direction of motion. Also, we should note that at the point of impending motion, the frictional forces acting on both the blocks would reach their maximum values. The free-body diagrams of the blocks are shown below:

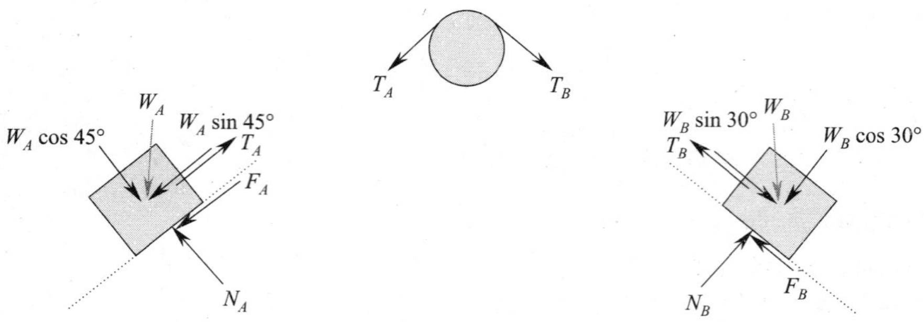

Fig. 6.22(a)

As the weight of the block A is unknown, we proceed from the equilibrium of the block B. Applying the conditions of equilibrium for the block B along the inclined plane and normal to the plane,

$\sum F_y = 0 \Rightarrow$

$$N_B - W_B \cos 30° = 0$$

$$\therefore \quad N_B = W_B \cos 30° = 150 \times \cos 30° = 129.9 \text{ N}$$

$\sum F_x = 0 \Rightarrow$

$$W_B \sin 30° - T_B - F_B = 0$$

$$\therefore \quad T_B = W_B \sin 30° - F_B$$

$$= W_B \sin 30° - \mu N_B \text{ [at the point of impending motion, } F_B = \mu N_B]$$

$$= (150 \times \sin 30°) - (0.25 \times 129.9) = 42.53 \text{ N}$$

Since the pulley is frictionless, the tensions on both the ends of the string are equal, i.e.,

$$T_A = T_B = 42.53 \text{ N}$$

Applying the conditions of equilibrium for the block A along the inclined plane and normal to the plane,

$\sum F_y = 0 \Rightarrow$

$$N_A - W_A \cos 45° = 0$$

$$\therefore \quad N_A = W_A \cos 45°$$

$\sum F_x = 0 \Rightarrow$

$$T_A - F_A - W_A \sin 45° = 0$$

$$\therefore \quad W_A \sin 45° = T_A - F_A$$

$$= T_A - \mu N_A \text{ [at the point of impending motion, } F_A = \mu N_A]$$

$$= 42.53 - (0.25)W_A \cos 45°$$

$$\Rightarrow \quad W_A = 48.12 \text{ N}$$

Case II Consider motion of the block *A* to be downwards; then the frictional forces act in the direction opposite to the direction of motion. Also, we should note that at the point of impending motion, the frictional forces acting on both the blocks would reach their maximum values. The free-body diagrams of the blocks are shown below:

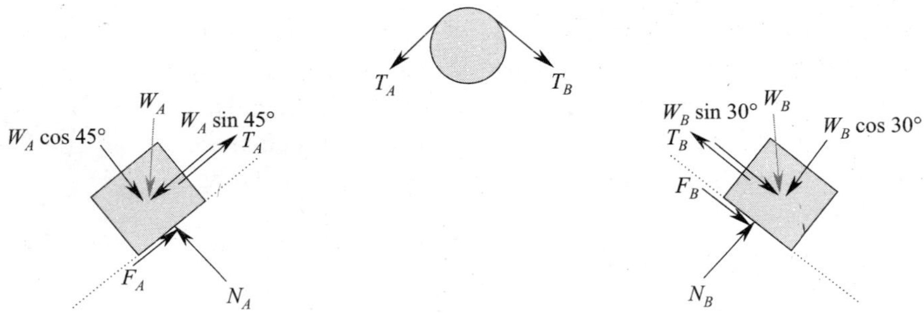

Fig. 6.22(b)

Applying the conditions of equilibrium for the block *B* along the inclined plane and normal to the plane,

$\sum F_y = 0 \Rightarrow$

$$N_B - W_B \cos 30° = 0$$

$$\therefore \quad N_B = W_B \cos 30° = 129.9 \text{ N}$$

$\sum F_x = 0 \Rightarrow$

$$W_B \sin 30° - T_B + F_B = 0$$

$$\therefore \quad T_B = W_B \sin 30° + F_B$$

$$= W_B \sin 30° + \mu N_B$$

$$= (150 \times \sin 30°) + (0.25 \times 129.9) = 107.48 \text{ N}$$

Since the pulley is frictionless, the tensions on both the ends of the string are equal, i.e.,

$$T_A = T_B = 107.48 \text{ N}$$

Applying the conditions of equilibrium for the block *A* along the inclined plane and normal to the plane,

$\sum F_y = 0 \Rightarrow$

$$N_A - W_A \cos 45° = 0$$

$$\therefore \quad N_A = W_A \cos 45°$$

$\sum F_x = 0 \Rightarrow$

$$W_A \sin 45° - T_A - F_A = 0$$

$$\therefore \quad W_A \sin 45° = T_A + F_A$$
$$= T_A + \mu N_A$$
$$= 107.48 + (0.25) W_A \cos 45°$$
$$\Rightarrow \quad W_A = 202.67 \text{ N}$$

From the two cases, we conclude that the range of weight of the block A for which motion is impending is 48.12 N to 202.67 N

Example 6.16 Two blocks A and B of weights W_A and W_B respectively rest on a rough inclined plane and are connected by a short piece of string as shown in Fig. 6.23. If the coefficients of friction between the blocks and the plane are respectively $\mu_A = 0.2$ and $\mu_B = 0.3$, find (a) the angle of inclination of the plane for which sliding will impend, and (b) tension in the string. Take $W_A = W_B = 20$ N.

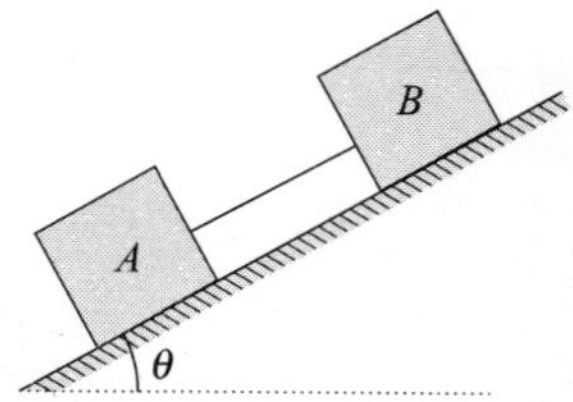

Fig. 6.23

Solution The free-body diagrams of the blocks A and B are shown in Figs 6.23(a) and (b) respectively. At the point of impending motion, the frictional forces acting on both the blocks will reach their maximum values.

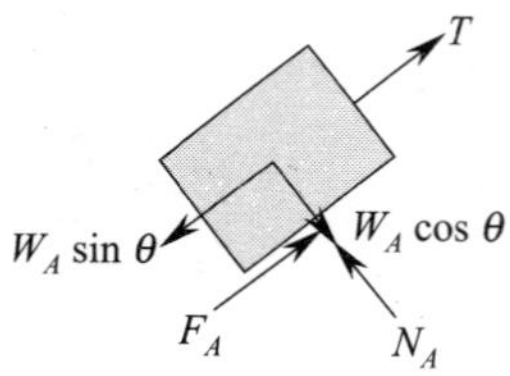

Fig. 6.23(a)

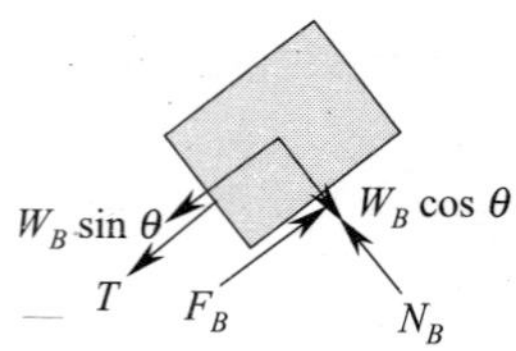

Fig. 6.23(b)

Block A

Applying the conditions of equilibrium at the point of impending motion along the incline and normal to the incline,

$$\Sigma F_y = 0 \Rightarrow$$

$$N_A - W_A \cos\theta = 0$$
$$\therefore \quad N_A = W_A \cos\theta$$

At the point of impending motion, we know

$$F_A = \mu_A N_A = \mu_A W_A \cos\theta$$

$$\Sigma F_x = 0 \Rightarrow$$

$$T + F_A - W_A \sin\theta = 0$$
$$\therefore \quad T = W_A \sin\theta - \mu_A W_A \cos\theta$$
$$= W_A (\sin\theta - \mu_A \cos\theta) \qquad \text{(a)}$$

Block B

Applying the conditions of equilibrium at the point of impending motion along the incline and normal to the incline,

$\sum F_y = 0 \Rightarrow$

$$N_B - W_B \cos\theta = 0$$

$$\therefore \quad N_B = W_B \cos\theta$$

Again, at the point of impending motion, we know

$$F_B = \mu_B N_B = \mu_B W_B \cos\theta$$

$\sum F_x = 0 \Rightarrow$

$$-T - W_B \sin\theta + F_B = 0$$

$$\Rightarrow \quad T = F_B - W_B \sin\theta$$

$$= \mu_B W_B \cos\theta - W_B \sin\theta$$

$$= W_B(\mu_B \cos\theta - \sin\theta) \qquad \text{(b)}$$

(a) Angle of inclination of the plane for which sliding will impend

Equating the value of tension T in the two equations (a) and (b),

$$W_A(\sin\theta - \mu_A \cos\theta) = W_B(\mu_B \cos\theta - \sin\theta)$$

Since $W_A = W_B$, we get

$$2\sin\theta = \cos\theta(\mu_A + \mu_B)$$

$$\tan\theta = \frac{(\mu_A + \mu_B)}{2} = \frac{0.2 + 0.3}{2} = 0.25$$

$$\theta = 14.04° \qquad \text{(c)}$$

b) Tension in the string

Substituting the value of θ in either of the equations for T, we get

$$T = W_A(\sin\theta - \mu_A \cos\theta)$$

$$= 20[\sin(14.04°) - 0.2\cos(14.04°)] = 0.971 \text{ N} \qquad \text{(d)}$$

Example 6.17 Block B in Fig. 6.24 weighs 100 N. Determine the maximum weight of the block A for which the system will be in equilibrium. The coefficient of static friction between the block B and the table is 0.20.

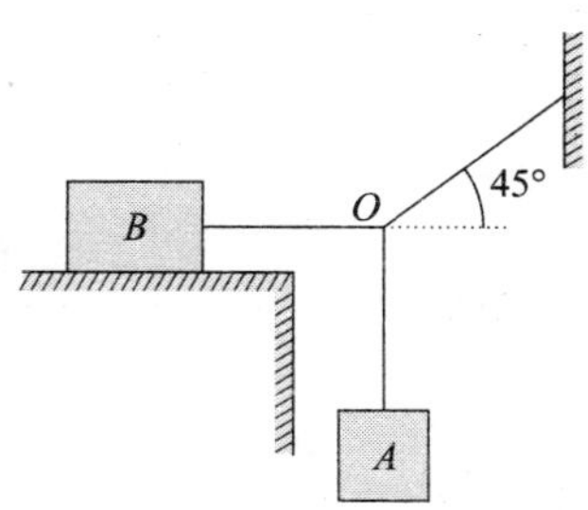

Fig. 6.24

Solution The free-body diagrams of the members in the system are shown below. The forces acting on the block A are its weight W and tension T_A in the string. The forces acting on the block B are its weight, normal reaction N_B, frictional force F_B acting opposite to the direction of impending motion and tension T_B in the string. As the weight of the block A is unknown, we proceed from the equilibrium of the block B.

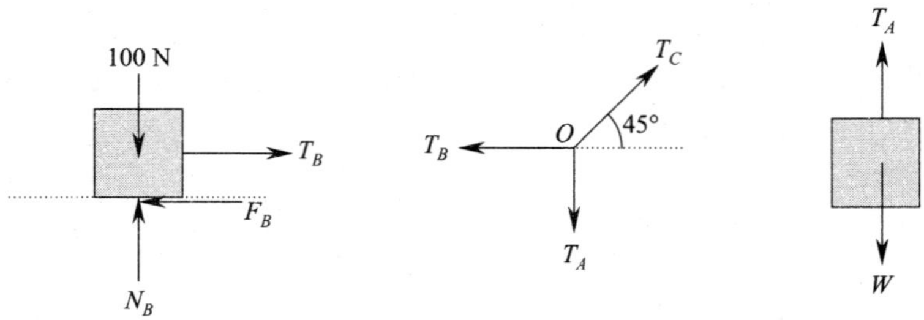

Fig. 6.24(a)

Equilibrium of the block B

At the point of impending motion, applying the conditions of equilibrium along the *X* and *Y* directions,

$\Sigma F_y = 0 \Rightarrow$

$$N_B - 100 = 0$$

$\therefore$ $$N_B = 100 \text{ N}$$

$\Sigma F_x = 0 \Rightarrow$

$$T_B - F_B = 0$$

At the point of impending motion, we know that the frictional force reaches the maximum value and hence,

$$T_B = F_B = \mu_B N_B$$

$$= 0.2 \times 100 = 20 \text{ N} \quad \text{(a)}$$

Equilibrium of point O

Since the three tensions T_A, T_B and T_C are concurrent at point *O*, we can apply the conditions of equilibrium along the *X* and *Y* directions,

$\Sigma F_x = 0 \Rightarrow$

$$T_C \cos 45° - T_B = 0$$

$\therefore$ $$T_C \cos 45° = T_B \quad \text{(b)}$$

$\Sigma F_y = 0 \Rightarrow$

$$T_C \sin 45° - T_A = 0$$

$\therefore$ $$T_C \sin 45° = T_A \quad \text{(c)}$$

Dividing equation (c) by equation (b), we get

$$\tan 45° = \frac{T_A}{T_B}$$

$\Rightarrow$ $$T_A = T_B = 20 \text{ N [since } \tan 45° = 1]$$

Equilibrium of block A

Applying the condition of equilibrium for the block A along the Y-direction,

$$\Sigma F_y = 0 \Rightarrow$$

$$T_A - W = 0$$

$$\therefore \quad W = T_A = 20 \text{ N}$$

Example 6.18 A heavy rope hangs over the edge of a rough horizontal table; find how much of it can hang over without slipping.

Solution Let L be the total length of the rope and its weight per unit length be w.

If x be the length of the rope hanging over the edge of the table, then its weight is

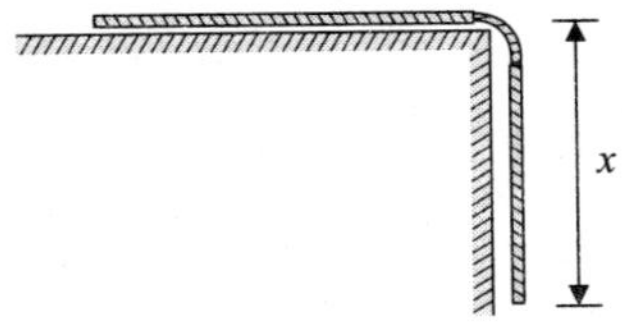

Fig. 6.25

$$w\,x$$

Similarly, the weight of the portion lying over the table is

$$w\,(L - x)$$

Therefore, normal reaction N acting on the portion of the rope lying over the table is

$$N = w\,(L - x)$$

At the point of impending motion, the force of friction reaches the maximum value. Hence, we can write

$$F = \mu N = \mu w (L - x)$$

When the rope is under equilibrium, the weight of the rope hanging over the table must be balanced by the frictional force. Hence,

$$w\,x = \mu w (L - x)$$

$$\Rightarrow \quad x = \frac{\mu L}{(1 + \mu)}$$

6.9 RELATIVE MOTION

In all of the previous examples, we saw the effect of a single body or connected bodies about to slide over a plane surface, which is at **rest**. However, at times, we may come across two bodies in contact with each other, which **move** relative to one another. Hence, while drawing the free body diagram, we must show the direction of frictional force in the right sense. Also, the frictional force exerted by each body on the other will be equal and opposite.

For illustration, let us consider two bodies, A and B, which slide over one another as shown in Fig. 6.26(a). In order to determine the correct sense of frictional force, we must analyze the relative motion of each body with respect to the other. From the figure, we can see that the body A moves to the right relative to B. Since the frictional force always acts opposite to the direction of motion, the frictional force acting on the body A is shown as directed towards the left [refer Fig. 6.26(b)]. Next, we see that the body B moves to the left relative to A. Hence, the frictional force is shown as directed towards the right on its top and bottom surfaces. The following examples will clarify this concept.

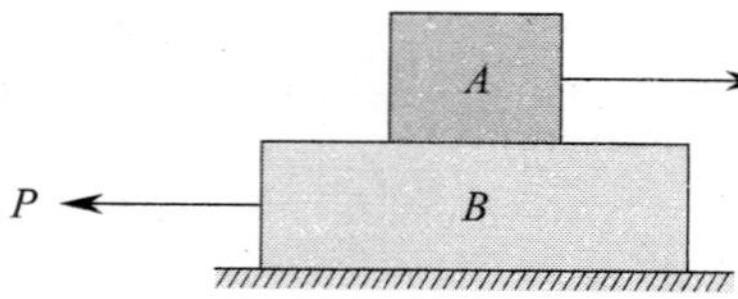

Fig. 6.26(a)

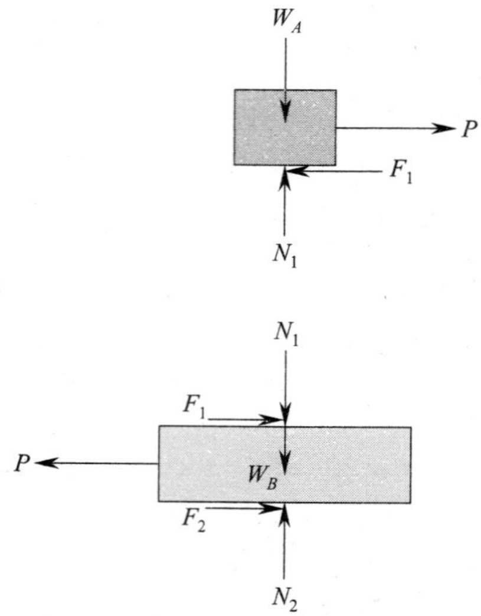

Fig. 6.26(b)

Example 6.19 A block of 100 kg mass is to be pulled to the right by a horizontal force *P*. Over the block, another block of 50 kg mass is placed which is attached to the wall by a string. If the coefficient of friction between all contact surfaces is 0.25, determine the value of *P*, for which motion is impending. Also, determine the tension in the string connecting the upper block with the wall.

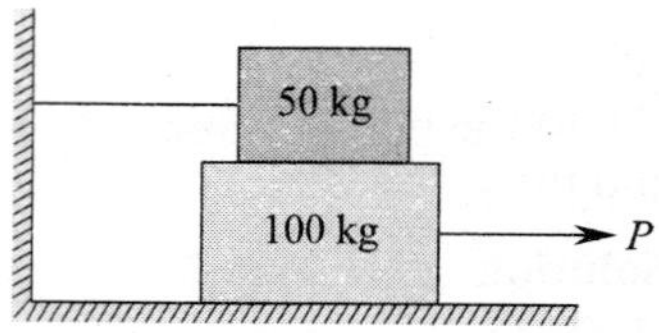

Fig. 6.27

Solution Since the lower block is to be pulled to the right, the force of friction on both of its faces act in the opposite direction, i.e., towards the left. By Newton's third law, the frictional force exerted by the lower block on the upper block then must be in the opposite direction, i.e., towards the right. The free-body diagrams of the two blocks are shown in Fig. 6.27(a).

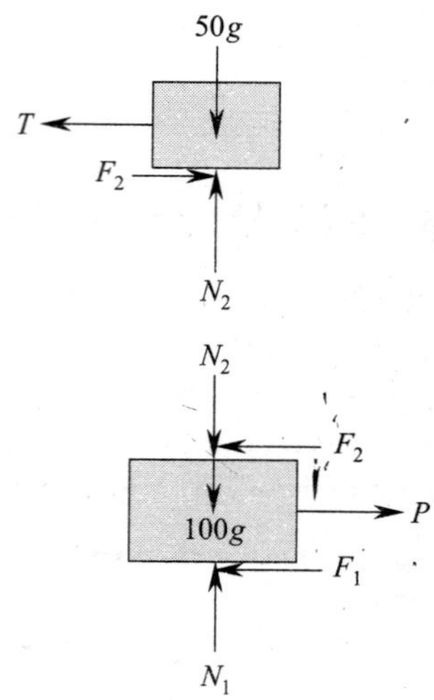

Fig. 6.27(a)

As we must determine the value of *P* applied to the lower block, we proceed from the equilibrium of the upper block.

Upper block

At the point of impending motion, applying the conditions of equilibrium along the *X* and *Y* directions,

$\sum F_y = 0 \Rightarrow$

$$N_2 - 50\,g = 0$$

$\therefore$
$$N_2 = 50\,g = 490.5 \text{ N}$$

$\sum F_x = 0 \Rightarrow$

$$F_2 - T = 0$$

$\therefore$ $$T = F_2 = \mu N_2$$
$$= (0.25)\ (490.5) = 122.63 \text{ N}$$

Lower block

At the point of impending motion, applying the conditions of equilibrium along the X and Y directions,

$\sum F_y = 0 \Rightarrow$

$$N_1 - N_2 - 100\, g = 0$$

$\therefore$ $$N_1 = N_2 + 100\, g$$
$$= 490.5 + (100 \times 9.81) = 1471.5 \text{ N}$$

$\sum F_x = 0 \Rightarrow$

$$P - F_1 - F_2 = 0$$

$\therefore$ $$P = F_1 + F_2$$

At the point of impending motion, we know that $F_1 = \mu N_1$ and $F_2 = \mu N_2$. Therefore,

$$P = \mu N_1 + \mu N_2$$
$$= 0.25\ (1471.5 + 490.5) = 490.5 \text{ N}$$

Example 6.20 In Fig. 6.28, determine the horizontal force P applied to the lower block to just pull it to the right. The coefficient of friction between the blocks is 0.2 and that between the lower block and the plane is 0.25. Assume the pulley to be frictionless.

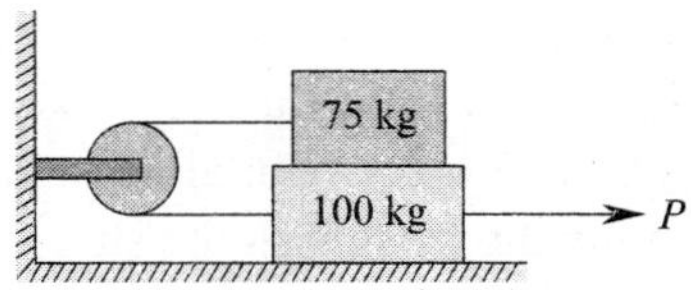

Fig. 6.28

Solution The free-body diagrams of the blocks are shown below. Here, unlike the previous problem, the tension acting on the upper block acts on the lower block also as the pulley is frictionless. As the lower block is pulled to the right, the frictional forces acting on both of its faces opposing the motion are directed towards the left. By Newton's third law, the frictional force exerted by the lower block on the upper block then must be in the opposite direction, i.e., towards the right.

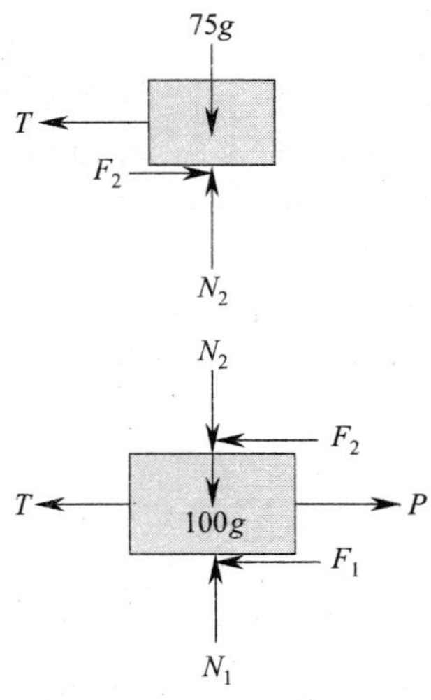

Fig. 6.28(a)

Upper block

At the point of impending motion, applying the conditions of equilibrium along the X and Y directions,

$\sum F_y = 0 \Rightarrow$

$$N_2 - 75\, g = 0$$

$\therefore$ $$N_2 = 75\, g = 735.75 \text{ N}$$

$\sum F_x = 0 \Rightarrow$

$$F_2 - T = 0$$

$\therefore$ $$T = F_2 = \mu_2 N_2$$
$$= (0.2)\ (735.75) = 147.15 \text{ N}$$

[Note that at the point of impending motion, $F_2 = \mu_2 N_2$].

Lower block

At the point of impending motion, applying the conditions of equilibrium along the X and Y directions,

$\sum F_y = 0 \Rightarrow$

$$N_1 - N_2 - 100\,g = 0$$

$$\therefore \quad N_1 = N_2 + (100\,g)$$

$$= (735.75) + (100 \times 9.81) = 1716.75 \text{ N}$$

$\sum F_x = 0 \Rightarrow$

$$P - F_1 - F_2 - T = 0$$

$$\therefore \quad P = F_1 + F_2 + T$$

$$= \mu_1 N_1 + \mu_2 N_2 + T$$

$$= (0.25 \times 1716.75) + (0.2 \times 735.75) + (147.15)$$

$$= 723.5 \text{ N}$$

Example 6.21 In Fig. 6.29 shown, a block A of 15 kg mass is connected to another block B of 10 kg mass by a string passing over a frictionless pulley. Determine the minimum mass of the block C, (connected to the wall by a string CD) which must be placed over the block A to keep it from sliding. Take coefficient of friction between all contact surfaces to be 0.25. Also, determine the tension in the string CD.

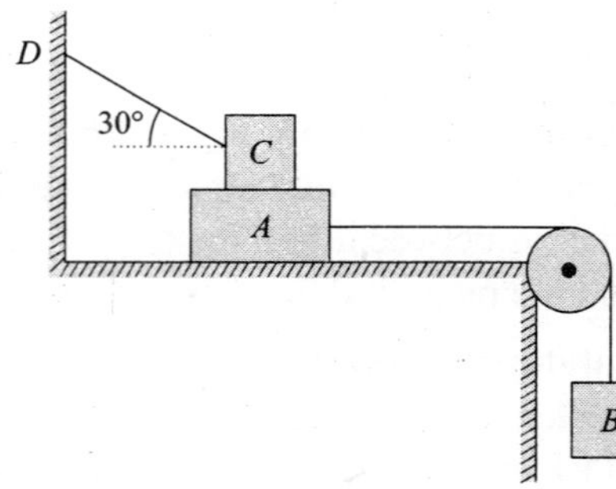

Fig. 6.29

Solution The free-body diagrams of the individual members are shown below. As the block A moves towards the right, the frictional forces act on it towards left on both of its faces. Hence, the frictional force acting on the block C must be in the opposite direction, i.e., towards the right.

As we have to determine the minimum mass of the block C to maintain equilibrium, we proceed from the equilibrium of the block B.

Block B

Applying the conditions of equilibrium along the Y direction,

$\sum F_y = 0 \Rightarrow$

$$T_B - W_B = 0$$

$$\therefore \quad T_B = W_B = 10 \times 9.81 = 98.1 \text{ N}$$

Since the pulley is frictionless, the tensions at both ends of the string passing over it are equal. Therefore,

$$T_A = T_B = 98.1 \text{ N}$$

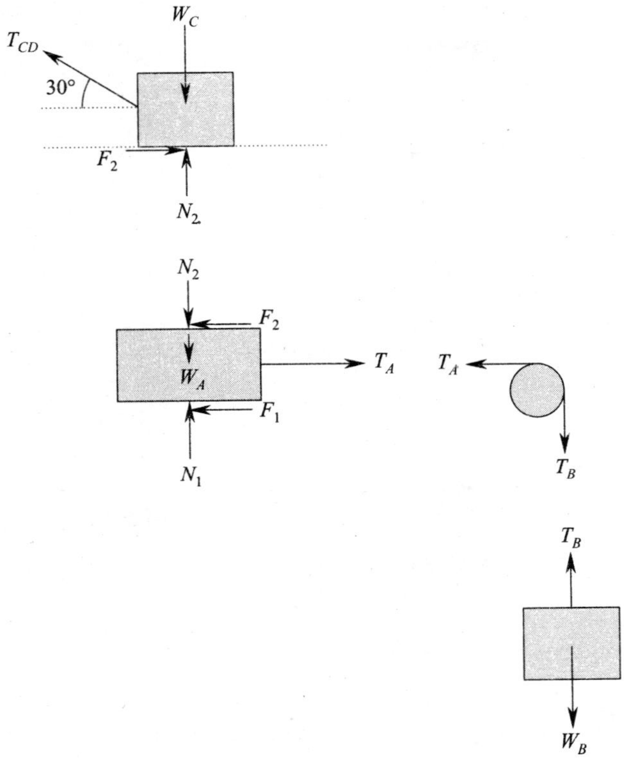

Fig. 6.29(a)

Block A

Applying the conditions of equilibrium along the *X* and *Y* directions,

$\sum F_x = 0 \Rightarrow$

$$T_A - F_1 - F_2 = 0$$

∴ $$T_A = F_1 + F_2$$

Since $F_1 = \mu_1 N_1$ and $F_2 = \mu_2 N_2$ at the point of impending motion, we can write,

$$T_A = \mu_1 N_1 + \mu_2 N_2$$

or, $$N_1 + N_2 = T_A/\mu \quad [\text{since } \mu_1 = \mu_2 = \mu]$$

$$= 98.1/0.25 = 392.4 \text{ N} \tag{a}$$

$\sum F_y = 0 \Rightarrow$

$$N_1 - N_2 - W_A = 0$$

∴ $$N_1 - N_2 = W_A = 15 \times 9.81 = 147.15 \text{ N} \tag{b}$$

Solving for N_1 and N_2 from equations (a) and (b), we get

$$N_1 = 269.78 \text{ N} \quad \text{and} \quad N_2 = 122.62 \text{ N}$$

Block C

Applying the conditions of equilibrium along the X and Y directions,

$\sum F_x = 0 \Rightarrow$

$$F_2 - T_{CD} \cos 30° = 0$$

$$\therefore \quad T_{CD} = F_2/\cos 30°$$

$$= \mu_2 N_2/\cos 30°$$

$$= (0.25 \times 122.62)/\cos 30° = 35.4 \text{ N}$$

$\sum F_y = 0 \Rightarrow$

$$N_2 + T_{CD} \sin 30° - W_C = 0$$

$$\therefore \quad W_C = N_2 + T_{CD} \sin 30°$$

$$= 122.62 + (35.4 \times \sin 30°) = 140.32 \text{ N}$$

Therefore, the mass of the block C is obtained as

$$m_C = W_C/g$$

$$= (140.32)/(9.81) = 14.3 \text{ kg}$$

Example 6.22 Suppose in the previous problem, the block C is not connected to the wall by a string, but merely rests on the block A, determine the mass of the block C for the motion to be impending.

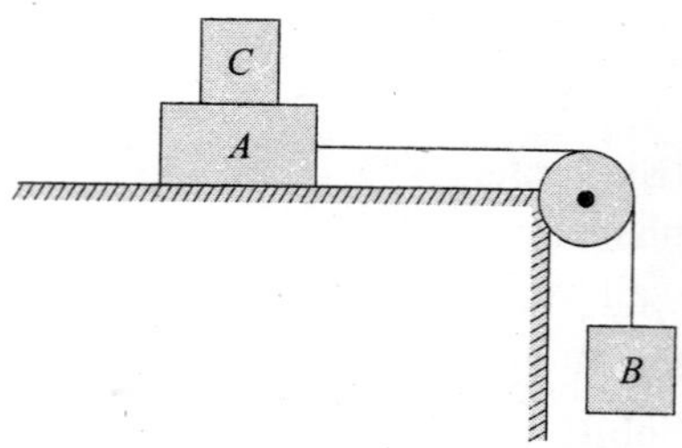

Fig. 6.30

Solution The free-body diagrams of the individual blocks are shown below:

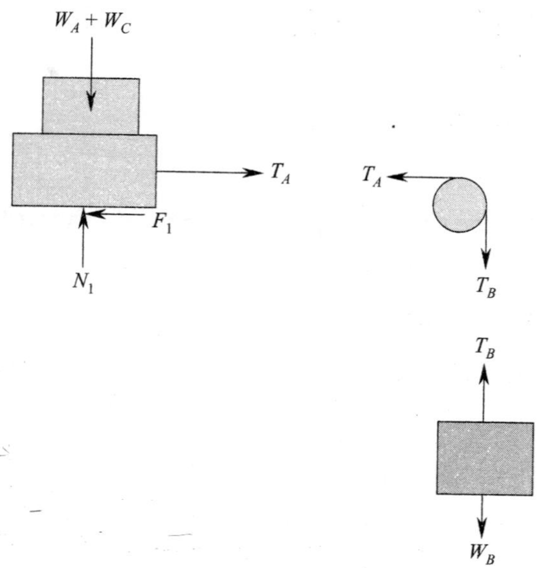

Fig. 6.30(a)

Just as in the previous problem, we can apply the conditions of equilibrium for the block B and obtain $T_A = 98.1$ N as before.

Unlike the previous problem, since the block C is resting on the block A, there is no relative motion between them and hence, they move together as a single body. Applying the equilibrium condition to the two blocks at the point of impending motion,

$\sum F_x = 0 \Rightarrow$

$$T_A - F_1 = 0$$

$\therefore$ $$F_1 = T_A = 98.1 \text{ N}$$

Since $$F_1 = \mu N_1,$$

$$N_1 = F_1/\mu$$

$$= 98.1/0.25 = 392.4 \text{ N}$$

$\sum F_y = 0 \Rightarrow$

$$N_1 - (W_A + W_C) = 0$$

$\therefore$ $$W_C = N_1 - W_A$$

$$= 392.4 - 147.15 = 245.25 \text{ N}$$

Therefore, the mass of the block C is obtained as

$$m_C = W_C/g$$

$$= (245.25)/(9.81) = 25 \text{ kg}$$

Example 6.23 Block A of weight W resting on an inclined plane is prevented from moving down the plane by a plank B of same weight W placed over it as shown in Fig. 6.31. The plank is attached to the wall by a string CD parallel to the incline. If the coefficient of friction is same for all contact surfaces, determine its value at which the motion is impending. Also, determine the tension in the string CD.

Solution The free-body diagrams for both the block and the plank are shown below:

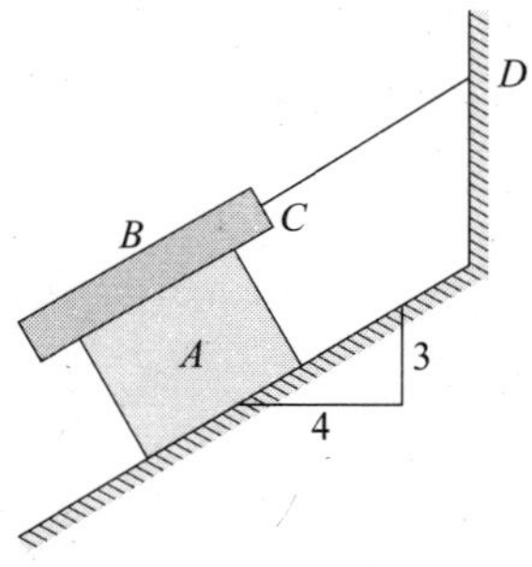

Fig. 6.31

Block A

Applying the conditions of equilibrium at the point of impending motion,

$\sum F_y = 0 \Rightarrow$

$$N_1 - N_2 - W\cos\theta = 0 \qquad \text{(a)}$$

$\sum F_x = 0 \Rightarrow$

$$F_1 + F_2 - W\sin\theta = 0 \qquad \text{(b)}$$

Since $F_1 = \mu N_1$ and $F_2 = \mu N_2$ at the point of impending motion, equation (b) can be written as

$$\mu N_1 + \mu N_2 - W\sin\theta = 0 \qquad \text{(c)}$$

Plank B

Applying the conditions of equilibrium at the point of impending motion,

$\sum F_x = 0 \Rightarrow$

$$T - F_2 - W\sin\theta = 0 \qquad \text{(d)}$$

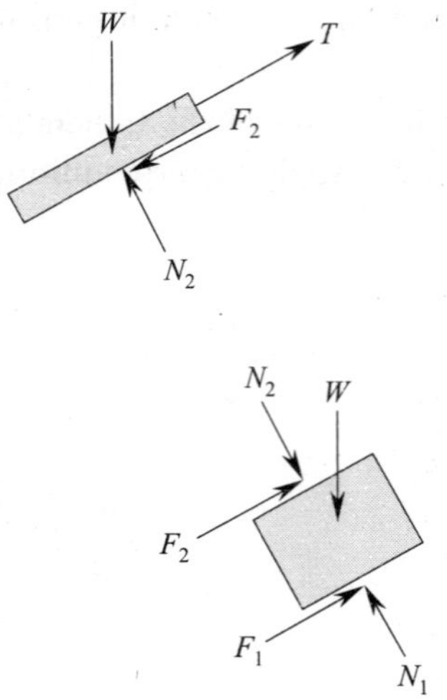

Fig. 6.31(a)

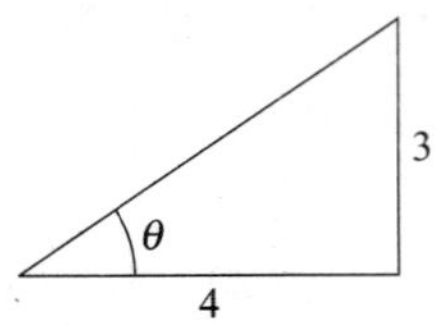

$\tan\theta = 3/4$
$\therefore \sin\theta = 3/5$
$\cos\theta = 4/5$

Fig. 6.31(b)

$\Sigma F_y = 0 \Rightarrow$

$$N_2 - W\cos\theta = 0$$

$\therefore$
$$N_2 = W\cos\theta \qquad \text{(e)}$$

Substituting this value in equation (a), we get,

$$N_1 = 2\,W\cos\theta \qquad \text{(f)}$$

Substituting the values of N_1 and N_2 from equations (f) and (e) in equation (c), we get,

$$\mu N_1 + \mu N_2 - W\sin\theta = 0$$

$$\mu(2W\cos\theta) + \mu(W\cos\theta) - W\sin\theta = 0$$

$\Rightarrow$
$$\mu = \frac{\sin\theta}{3\cos\theta}$$

$$= \frac{\tan\theta}{3} = \frac{3/4}{3} = 0.25$$

From equation (d), we get,

$$T = F_2 + W\sin\theta$$

$$= \mu N_2 + W\sin\theta$$

$$= \mu W\cos\theta + W\sin\theta \text{ [from equation (e)]}$$

$$= (0.25)W(4/5) + W(3/5) = 0.8\text{ W}$$

6.10 LADDER FRICTION

In this section, we will introduce another type of sliding friction, which is seen in **ladders**. Here, the forces acting on the member are *non-concurrent*, unlike the previous sections. Hence, we have to apply the moment equilibrium in addition to the force equilibrium to solve the unknowns. Consider a ladder *AB* resting on a *rough* horizontal surface and leaning against a *rough* wall. Then by its own weight or when a man starts climbing upon it, it tends to slide down. At the point of impending motion, we know that the point *A* tends to slide downwards and the point *B* tends to slide away from the wall. Hence, apart from normal reactions at these points, frictional forces act in the directions as indicated trying to oppose its motion.

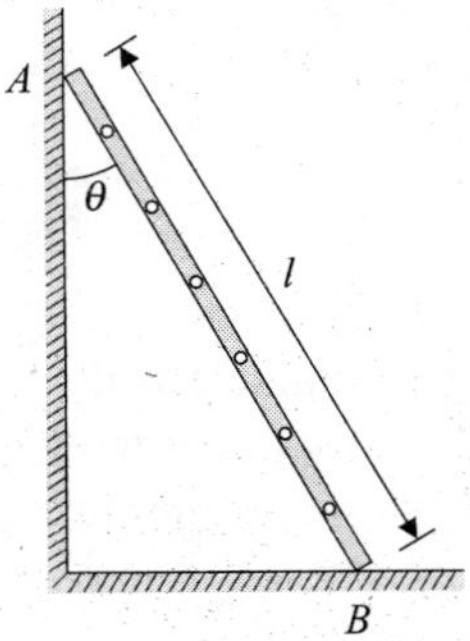

Fig. 6.32(a)

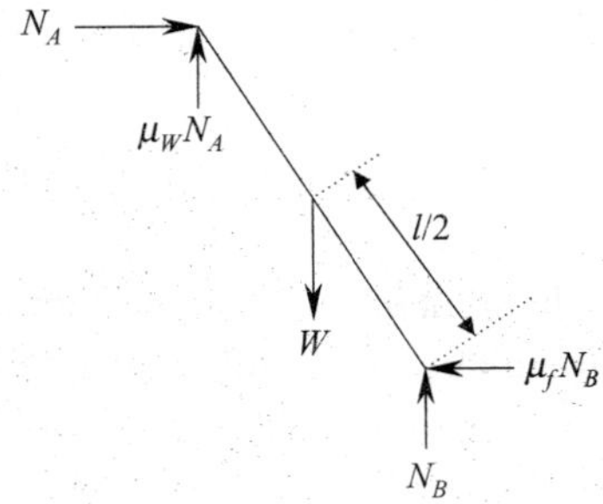

Fig. 6.32(b)

In the Fig. 6.32(b) above, W is the weight of the ladder; μ_w and μ_f are the coefficients of friction between the ladder and the wall, and between the ladder and the floor respectively. The force of friction at the wall acts upwards preventing the downward motion of the end A and its magnitude is equal to $\mu_w N_A$ at the point of impending motion. Similarly, the force of friction at the floor acts towards the wall preventing the motion of the end B away from the wall and its magnitude is equal to $\mu_f N_B$ at the point of impending motion. If a man climbs up the ladder, his weight must also be shown in the free-body diagram. At the point of impending motion, we can apply the conditions of equilibrium:

$$\sum F_x = 0 \tag{6.13}$$

$$\sum F_y = 0 \tag{6.14}$$

and

$$\sum M = 0 \tag{6.15}$$

to solve the unknowns. While taking moment summation, it can be taken about any point on the ladder. However, we should take such points that would eliminate more number of unknowns. The following examples will clarify this.

Example 6.24 A uniform ladder weighing 800 N and 10 m long is resting on a rough horizontal floor and inclined at an angle of 30° with the vertical wall. The ladder would just slip if a man of 1000 N weight reaches a point that is 8 m from the lower end of the ladder. If the coefficient of friction between the wall and the ladder is 0.4, determine the coefficient of friction between the ladder and the floor.

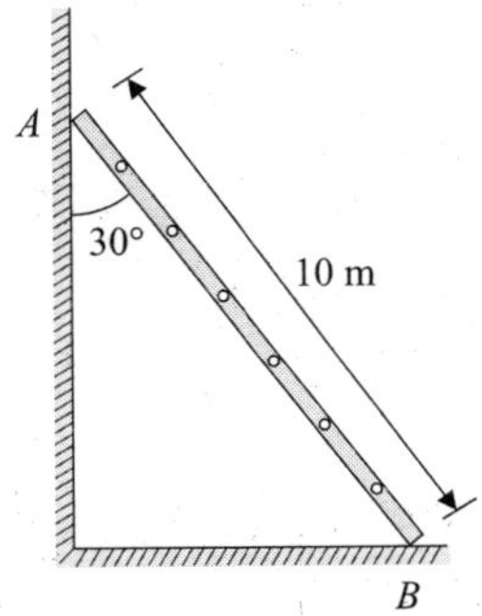

Fig. 6.33

Solution When the man reaches a point that is 8 m from the lower end of the ladder, it is about to slip. Accordingly, we can draw the free-body diagram of the ladder considering the direction of friction at the point of impending motion as shown in Fig. 6.33(a). The forces acting on the ladder are its weight acting at mid-length, weight of man at 8 m from bottom end, normal reactions N_A and N_B respectively at ends A and B, frictional forces acting upwards at the end A and towards the wall at the end B.

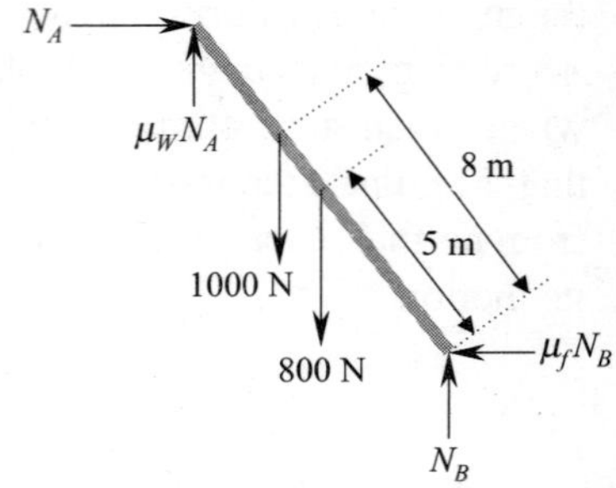

Fig. 6.33(a)

Applying the conditions of equilibrium along the X and Y directions,

$\sum F_x = 0 \Rightarrow$

$$N_A - \mu_f N_B = 0 \quad \text{(a)}$$

$\sum F_y = 0 \Rightarrow$

$$\mu_w N_A + N_B - 1000 - 800 = 0$$

$$0.4\, N_A + N_B = 1800 \quad \text{(b)}$$

Though moment can be taken about any point on the ladder, we should consider a point where more number of unknowns are eliminated. We see that at the point A, the only unknown is N_A, whereas at the point B, there are two unknowns, N_B and μ_B. Hence, taking the moment about B and applying the condition of equilibrium,

$\sum M_B = 0 \Rightarrow$

$$-N_A\,(10 \cos 30°) - \mu_w\, N_A\,(10 \sin 30°) + 800(5) \sin 30° + 1000(8) \sin 30° = 0$$

$$\Rightarrow \quad N_A = 562.84 \text{ N} \quad \text{(c)}$$

Substituting the value of N_A in equation (b), we get

$$N_B = 1574.86 \text{ N} \quad \text{(d)}$$

Hence, from equation (a), we get,

$$\mu_f = N_A/N_B$$
$$= 562.84/1574.86 = 0.36$$

Example 6.25 A 10 m long ladder rests on a horizontal floor and leans against a vertical wall. If the coefficients of friction between the ladder and the floor, and between the ladder and the wall are respectively $\mu_f = 0.3$ and $\mu_w = 0.15$, determine the angle of inclination of the ladder with the floor at the point of impending motion.

Solution The free-body diagram of the ladder is shown below:

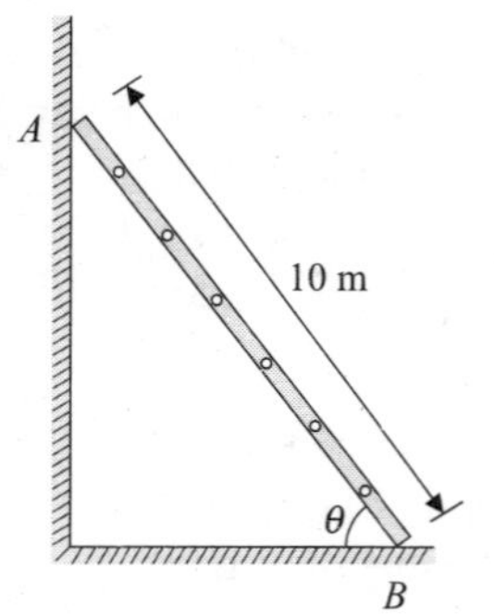

Fig. 6.34(a)

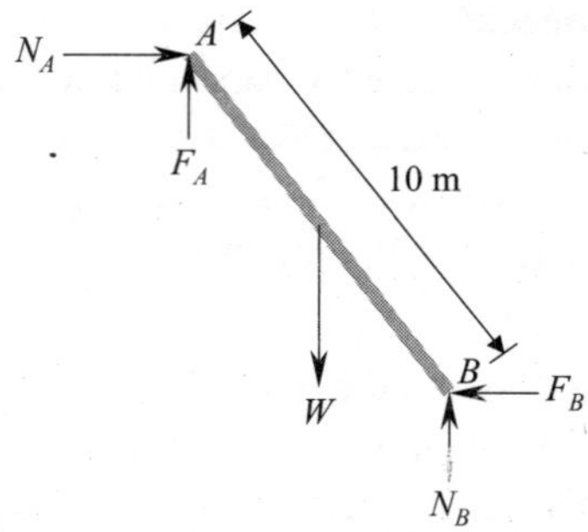

Fig. 6.34(b)

At the point of impending motion, applying the equations of equilibrium along the X and Y directions,

$\sum F_x = 0 \Rightarrow$

$$N_A - F_B = 0$$

Since $F_B = \mu_f N_B$,

$$N_A - \mu_f N_B = 0$$

$$N_A - 0.3\, N_B = 0 \qquad \text{(a)}$$

$\sum F_y = 0 \Rightarrow$

$$F_A + N_B - W = 0$$

Since $F_A = \mu_w N_A$,

$$\mu_w N_A + N_B - W = 0$$

$$0.15\, N_A + N_B - W = 0 \qquad \text{(b)}$$

Substituting the value of N_B from equation (a) in the above equation, we get

$$W = 0.15\, N_A + (N_A / 0.3)$$

$$= 3.48\, N_A \qquad \text{(c)}$$

We can take the moment about either A or B as both have only one unknown. Taking the moment about B and applying the condition of equilibrium,

$\sum M_B = 0 \Rightarrow$

$$-[F_A \times 10 \cos\theta] - [N_A \times 10 \sin\theta] + [W \times 5 \cos\theta] = 0$$

$$-[\mu_w N_A \times 10 \cos\theta] - [N_A \times 10 \sin\theta] + [3.48\, N_A \times 5 \cos\theta] = 0$$

$$-1.5 \cos\theta - 10 \sin\theta + 17.4 \cos\theta = 0$$

$$15.9 \cos\theta = 10 \sin\theta$$

$$\tan\theta = 1.59$$

$$\Rightarrow \qquad \theta = 57.83°$$

Example 6.26 An 8 m long uniform ladder weighing 500 N is resting on a rough horizontal floor and inclined at an angle of 30° with a vertical wall. A man weighing 750 N climbs the ladder. At what position will he induce slipping? The coefficient of friction between the ladder and the wall is 0.3 and that between the ladder and the floor is 0.2.

Solution Let the man be at a point x metres from the floor along the ladder at the point of impending motion. Accordingly, we draw the free-body diagram as shown below.

Applying the conditions of equilibrium along the X and Y directions,

$\sum F_x = 0 \Rightarrow$

$$N_A - \mu_f N_B = 0$$

$$\therefore \qquad N_A - 0.2\, N_B = 0 \qquad \text{(a)}$$

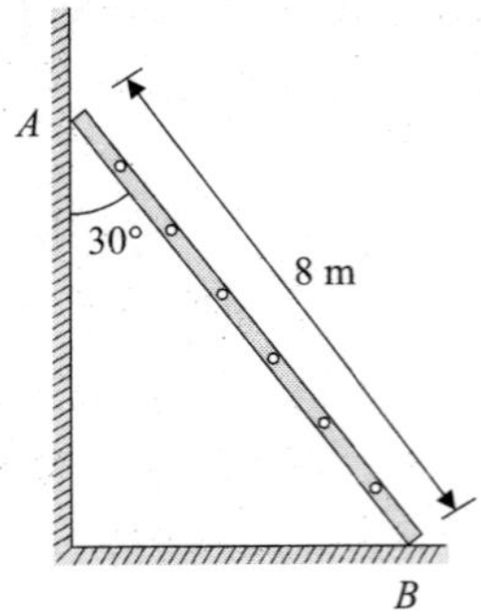

Fig. 6.35(a)

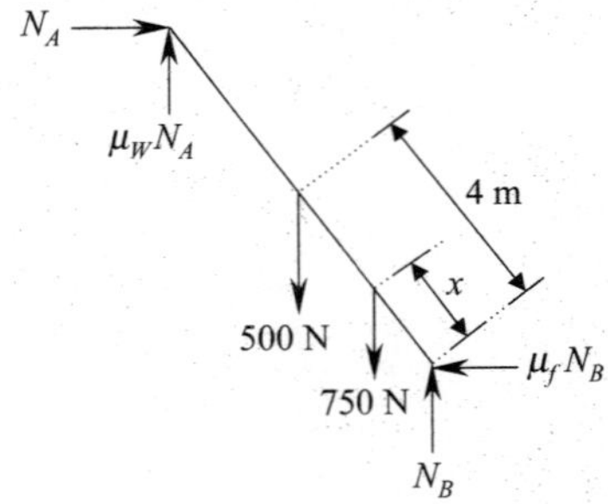

Fig. 6.35(b)

$\sum F_y = 0 \Rightarrow$

$$\mu_w N_A + N_B - 750 - 500 = 0$$

$$0.3 N_A + N_B = 1250 \qquad \text{(b)}$$

Solving for N_A and N_B from the above two equations (a) and (b), we get

$$N_A = 235.85 \text{ N} \qquad \text{(c)}$$

and $$N_B = 1179.25 \text{ N} \qquad \text{(d)}$$

Taking the moment about B and applying the condition of equilibrium,

$\sum M_B = 0 \Rightarrow$

$$-N_A (8 \cos 30°) - \mu_w N_A (8 \sin 30°)$$
$$+ 500 (4 \sin 30°) + 750 (x \sin 30°) = 0$$
$$-(235.85)(8 \cos 30°) - (0.3)(235.85)(8 \sin 30°)$$
$$+ 500(4 \sin 30°) + 750(x \sin 30°) = 0$$

$\Rightarrow$ $$x = 2.45 \text{ m}$$

Example 6.27 A 10 m long uniform ladder weighing 500 N is resting on a rough horizontal floor and inclined at an angle of 30° with the vertical wall. A man weighing 800 N climbs the ladder up to 4 m point, from the ground and along the ladder. The ladder is held in position by another man standing on the ground by applying a horizontal force P at a vertical height, 1.2 m from ground level. Determine the force he must apply if the coefficients of friction between the ladder and wall and that between the ladder and floor are 0.2.

Solution The free-body diagram of the ladder is shown below.

Applying the conditions of equilibrium along the X and Y directions,

$\sum F_x = 0 \Rightarrow$

$$\mu_f N_B + P - N_A = 0$$

$\Rightarrow$ $$N_A = \mu_f N_B + P$$

$$= 0.2 N_B + P \qquad \text{(a)}$$

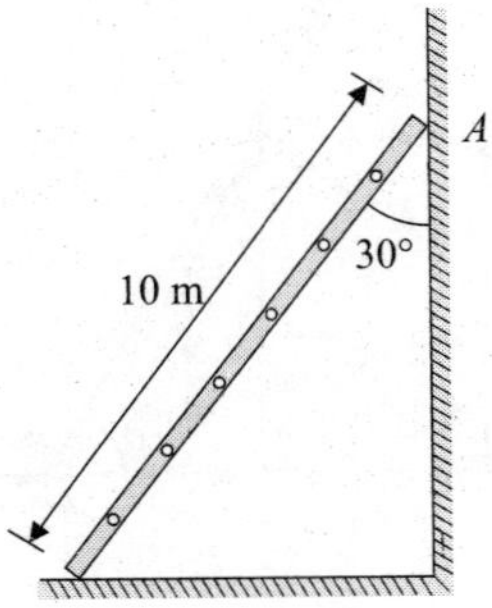

Fig. 6.36(a)

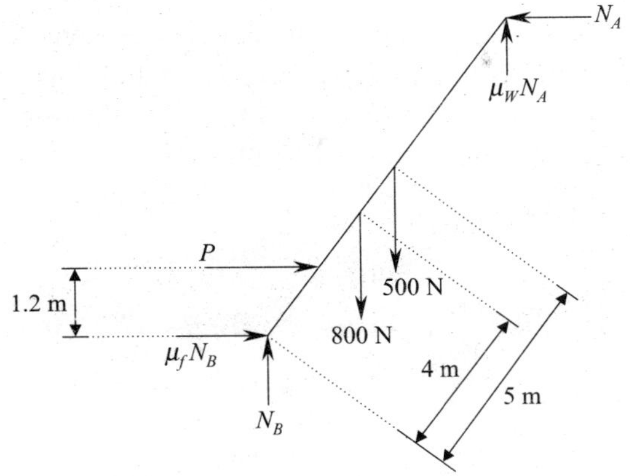

Fig. 6.36(b)

$\sum F_y = 0 \Rightarrow$

$$N_B + \mu_w N_A - 800 - 500 = 0$$

$$0.2\, N_A = 1300 - N_B \qquad \text{(b)}$$

Dividing equation (b) by equation (a),

$$\frac{0.2 N_A}{N_A} = \frac{1300 - N_B}{0.2 N_B + P}$$

On rearranging,

$$1.04\, N_B + 0.2\, P = 1300 \qquad \text{(c)}$$

Taking the summation of moments about A and applying the condition of equilibrium,

$\sum M_A = 0 \Rightarrow$

$$[-N_B \times 10 \sin 30°] + [\mu_f N_B \times 10 \cos 30°]$$
$$+ [P \times (10 \cos 30° - 1.2)] + [500 \times 5 \sin 30°] + [800 \times 6 \sin 30°] = 0$$

On simplification,

$$3.27\, N_B - 7.46\, P = 3650 \qquad \text{(d)}$$

Solving the simultaneous equations (c) and (d), we get,

$$P = 54.1 \text{ N}$$

Example 6.28 A 6 m long ladder AB leans against a wall as shown in Fig. 6.37. If the coefficient of friction between the ladder and the wall and that between the ladder and the floor are same and equal to 0.3, determine the minimum inclination of the ladder with respect to the floor for which equilibrium is maintained.

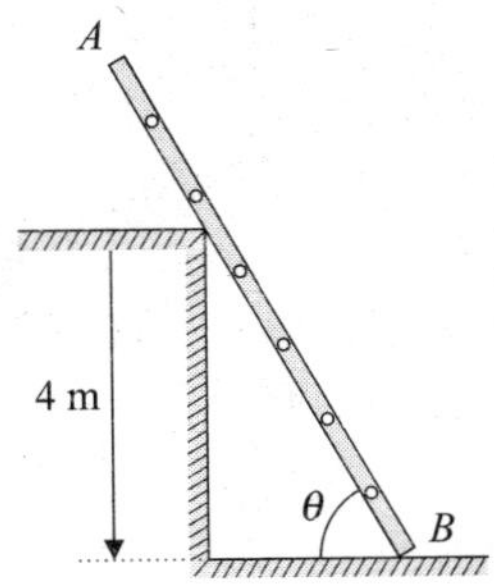

Fig. 6.37

Solution The free-body diagram of the ladder is shown in Fig. 6.37(a). The forces acting on it are its weight placed at its centre; normal reaction N_B and frictional force F_B opposing the motion of the ladder at the end B; normal reaction N_C and frictional force F_C at the contact point C with the wall. Applying the conditions of equilibrium along the X and Y directions at the point of impending motion,

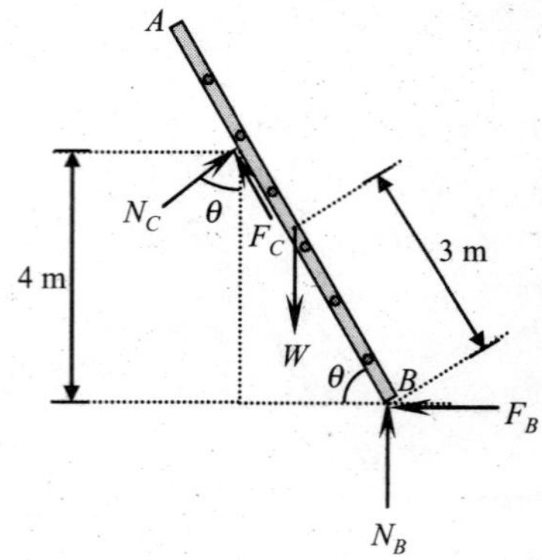

Fig. 6.37(a)

$\sum F_x = 0 \Rightarrow$

$$N_C \sin\theta - F_C \cos\theta - F_B = 0$$

$$N_C \sin\theta - \mu N_C \cos\theta - \mu N_B = 0$$

$$\therefore \quad N_C[\sin\theta - \mu \cos\theta] - \mu N_B = 0 \qquad \text{(a)}$$

$\sum F_y = 0 \Rightarrow$

$$N_C \cos\theta + F_C \sin\theta + N_B - W = 0$$

$$N_C \cos\theta + \mu N_C \sin\theta + N_B - W = 0$$

$$\therefore \quad N_C[\cos\theta + \mu \sin\theta] + N_B = W \qquad \text{(b)}$$

From the Fig. 6.37(a), we can see that if we take moment about the point B, then more number of unknowns can be eliminated and hence taking summation of the moments of all the forces about the point B and equating it to zero,

$\sum M_B = 0 \Rightarrow$

$$-N_C \frac{4}{\sin\theta} + W\,3\cos\theta = 0$$

$$\therefore \quad N_C = \frac{3}{4} W \sin\theta \cos\theta \qquad \text{(c)}$$

Multiplying equation (b) by μ and adding it with equation (a), we get

$$N_C[\sin\theta - \mu\cos\theta] + \mu N_C[\cos\theta + \mu\sin\theta] = \mu W$$

$$N_C[\sin\theta - \mu\cos\theta + \mu\cos\theta + \mu^2\sin\theta] = \mu W$$

$$N_C \sin\theta[1 + \mu^2] = \mu W$$

Substituting the value of N_C from equation (c),

$$\frac{3}{4} W \sin^2\theta \cos\theta[1 + \mu^2] = \mu W$$

$$\Rightarrow \quad \sin^2\theta\cos\theta = \frac{4}{3}\frac{\mu}{1+\mu^2}$$

$$= \frac{4}{3}\frac{[0.3]}{[1+(0.3)^2]}$$

$$= 0.367$$

By trial and error method, we get $\theta = 61.8°$.

SUMMARY

Frictional Force

Whenever the surface of a body slides or tends to slide over another, each body exerts a *tangential* frictional force on the other body apart from the *normal* reaction. This tangential frictional force tries to prevent motion of one surface with respect to the other. However, these frictional forces are found to be *limited* in magnitude and will not be able to prevent relative motion when sufficiently large external forces are applied.

The frictional forces arise from surface irregularities. They are the resultant of a number of distributed forces acting over the entire surface of contact. However, for analytical purposes, they can be assumed to be concentrated without much appreciable error.

Friction causes *favourable* as well as *undesirable* effects. Because of the resistive nature of frictional force, it causes undesirable effects like wear and tear of mechanical parts in machines. On the other hand, without friction we cannot do things in our daily lives as we do, as we cannot walk or drive a car or hold a pen and write, and so on.

Impending Motion and Forces of Friction

When the tangential component of the resultant external force acting on a body increases, the frictional resistance also increases correspondingly to maintain *equilibrium* of the body. The frictional force acting between surfaces at rest with respect to each other is called *force of static friction.* The maximum or limiting value of friction is termed *limiting static friction* and it is denoted as $F_{s,\,\max}$. When the frictional force reaches the maximum value, the body is on the verge of *sliding* and this instant is known as point of *impending motion.*

If the resultant tangential force increases beyond the limiting friction, the frictional force cannot balance it and motion occurs. Once motion has started, the frictional forces acting between the surfaces usually decrease so that a smaller force is necessary to maintain uniform motion. The frictional force acting between surfaces in relative motion is termed *force of kinetic friction.*

Depending upon the value of the frictional force, we come across three different cases:

Case I When $F < F_{s,\,\max}$, no motion occurs

Case II When $F = F_{s,\,\max}$, motion impends

Case III When $F > F_{s,\,\max}$, the body is under motion.

Coulomb's laws of dry friction

The frictional force always acts such as to *oppose* the relative motion of bodies. Its magnitude is exactly equal to the force, which tends to move the body until the limiting value is reached. The maximum force of friction is independent of the area of contact between two sliding surfaces and depends on the nature of surfaces in contact. The magnitude of limiting or maximum static friction is proportional to the normal reaction between the two surfaces, i.e.,

$$F_{s,\,\max} \propto N \text{ (or) } F_{s,\,\max} = \mu_s N$$

where the constant μ_s is called *coefficient of static friction.* When the body is under motion, the force of kinetic friction is constant and it is proportional to the normal reaction between the two surfaces, i.e.,

$$F_k \, \alpha N \text{ (or) } F_k = \mu_k N$$

where the constant μ_k is called *coefficient of kinetic friction.*

Coefficient of static friction [μ_s] is defined as the ratio of the magnitude of maximum force of static friction to the magnitude of normal reaction.

Coefficient of kinetic friction [μ_k] is defined as the ratio of the magnitude of force of kinetic friction to the magnitude of normal reaction.

$$\mu_s = \frac{F_s}{N} \text{ and } \mu_k = \frac{F_k}{N}$$

Both μ_s and μ_k are dimensionless constants as they are ratios of forces.

Angle of static friction [φ_s] is defined as the inclination of the resultant force R of the frictional force F and normal reaction N with the normal reaction at the point of impending motion. Mathematically,

$$\tan \varphi_s = \frac{F_{s,\max}}{N} = \mu_s$$

Similarly, if the block is motion,

$$\tan \varphi_k = \mu_k$$

Angle of repose is defined as the angle of inclination of the plane on which a body is resting corresponding to impending motion. At the point of impending motion, this angle is equal to the angle of static friction, φ_s.

Cone of Friction

For coplanar problems, if the resultant R lies within a triangle subtending an internal angle of $2\varphi_s$, then the body will be at rest. For non-coplanar problems, it must lie within the cone generated by revolving the maximum resultant vector about the normal axis by one complete revolution. The cone thus generated is known as *cone of friction.*

EXERCISES

Objective-type Questions

1. When a body is resting on a rough horizontal plane,
   (a) frictional force exists and it is maximum
   (b) frictional force exists and it is minimum
   (c) frictional force comes into play only when external force is applied tangential to the plane
   (d) frictional force is equal to the weight of the body

2. When a body resting on a rough plane is acted upon by gradually increasing tangential force,
   (a) frictional force is zero
   (b) frictional force remains constant
   (c) frictional force increases indefinitely
   (d) there is a limit up to which frictional force can increase

3. Frictional force acts __________ the surfaces in contact.
   (a) tangential to (b) normal to (c) inclined to (d) away from
4. Coulomb's laws of friction can be applied to
   (a) fluid friction
   (b) fluid-structure interaction
   (c) dry friction between solid bodies
   (d) lubricated surfaces
5. Impending motion of a body refers to a
   (a) body at rest
   (b) body about to move
   (c) body moving with uniform speed
   (d) body moving with uniform acceleration
6. At the point of impending motion, the static frictional force is
   (a) zero (b) maximum (c) minimum (d) infinite
7. At the point of impending motion,
   (a) the body is on the verge of moving
   (b) the frictional force reaches the maximum value
   (c) the frictional force is equal to the tangential applied force
   (d) all of these
8. Force required to start motion is
   (a) less than that required for keeping it in motion
   (b) more than that required for keeping it in motion
   (c) same as the force required for keeping it in motion
   (d) zero, while force required for keeping it in motion is non-zero
9. When a body is acted on by external forces tangential to the plane of contact and if no motion occurs then it implies that
   (a) the force of friction is zero
   (b) the force of friction is equal to $\mu_s N$
   c) the force of friction is equal to $\mu_k N$
   (d) the force of friction is equal to the externally applied tangential force
10. Force of friction depends upon
    (a) area of contact surfaces
    (b) normal reaction between the contact surfaces
    (c) dimension of the body
    (d) shape of the body
11. The tangent of the angle of friction is __________.
    (a) angle of repose
    (b) coefficient of friction
    (c) cone of friction
    (d) limiting friction
12. The angle made by the resultant of normal reaction and frictional force with the normal reaction at the point of impending motion is called
    (a) angle of inclination
    (b) angle of repose
    (c) angle of friction
    (d) normal angle

13. Angle of friction is given as

(a) $\sin^{-1}(\mu)$ (b) $\cos^{-1}(\mu)$ (c) $\tan^{-1}(\mu)$ (d) $\cot^{-1}(\mu)$

14. Which of the following statements is correct?

(a) Static friction force is greater than $F_{s,\,\max}$. (b) Kinetic friction force is greater than $F_{s,\,\max}$.

(c) Kinetic friction force is less than $F_{s,\,\max}$. (d) Kinetic friction force is equal to $F_{s,\,\max}$.

15. The unit of coefficient of friction is

(a) Newton (b) radian (c) metre (d) dimensionless

16. Frictional force

(a) is always zero

(b) increases linearly with the applied force up to a limit

(c) increases parabolically with the applied force up to a limit

(d) decreases with the applied force

17. Frictional forces arise

(a) due to surface irregularities

(b) due to large area of contact

(c) due to elastic property of the materials

(d) due to temperature change

18. A ball and a block of same mass are placed on a rough surface and are pulled by a force. When they are about to slide, the force of friction in the case of the ball is

(a) greater than the force of friction in the case of the block

(b) less than the force of friction in the case of the block

(c) equal to the force of friction in the case of the block

(d) not existent

19. When a block of weight W resting on a rough inclined plane of inclination θ does not slide, then the frictional force acting on it is

(a) $w \sin\theta$ (b) $w \cos\theta$ (c) $\mu w \sin\theta$ (d) $\mu w \cos\theta$

Answers

1. (c)	**2.** (d)	**3.** (a)	**4.** (c)	**5.** (b)	**6.** (b)	**7.** (d)	**8.** (b)
9. (d)	**10.** (b)	**11.** (b)	**12.** (c)	**13.** (c)	**14.** (c)	**15.** (d)	**16.** (b)
17. (a)	**18.** (c)	**19.** (a)					

Short-answer Questions

1. Define friction.
2. Explain the types of friction with examples.
3. What are surface irregularities? How do they affect the relative motion of bodies over one another?
4. Is it possible to eliminate friction completely from mechanical parts in machines? Discuss.
5. Why are lubricants used in machines? What is the theory behind their application?

6. Define limiting friction and impending motion.
7. What is meant by coefficient of static friction?
8. Can we always use the formula $F = \mu_s N$ to determine the frictional force? If not, state why.
9. State Coulomb's laws of dry friction.
10. What are the dimensions of coefficients of friction?
11. Figure 6.38 shows different shapes of bodies made of same material and having the same mass. If they are pulled by a constant force P on a rough surface with their bottom surfaces in contact with the plane, comment on how the maximum frictional force varies in each case.

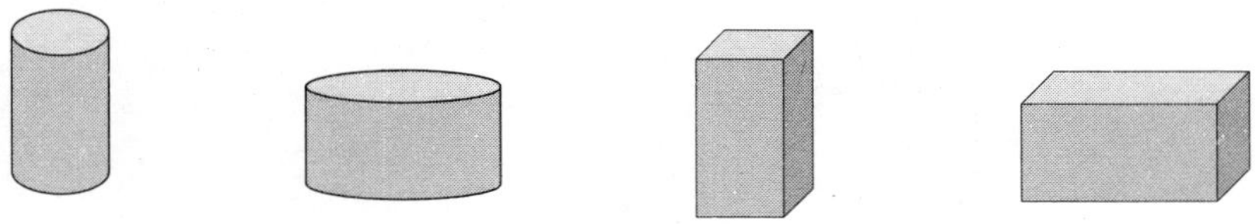

Fig. 6.38

12. Differentiate between static friction and kinetic friction.
13. Define angle of friction, angle of repose and cone of friction.
14. Explain how to determine the correct sense of frictional forces when two bodies move relative to one another.
15. Define ladder friction and discuss the sense of frictional forces acting at the contact points.

Numerical Problems

6.1 A block of 200 N weight must be held against a wall as shown in Fig. E.6.1 by applying a force P normal to the contact surface. If the coefficient of friction between the surfaces is 0.3, determine the minimum force required.

Ans. 666.67 N

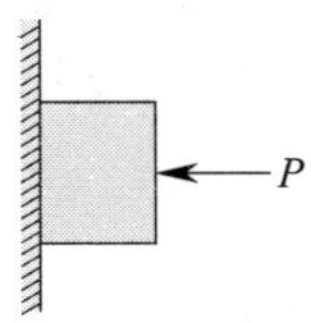

Fig. E.6.1

6.2 A block of 50 kg mass rests on a 35° incline. Find the horizontal force required to just push the block up the plane. The coefficient of friction between contact surfaces is 0.2.

Ans. 513.5 N

6.3 If the ratio of the greatest to the least force which acting parallel to a rough inclined plane, can support a weight on it, is equal to that of the weight to the pressure on the plane then prove that the coefficient of friction is $\tan \alpha . \tan^2 \dfrac{\alpha}{2}$, where α is the inclination of the plane to the horizontal.

6.4 A block weighing 200 N must be pushed up a plane inclined at 25° to the horizontal as shown in Fig. E.6.4. Find the value of P required if it is inclined at 40° to the horizontal. The coefficient of friction between the contact surfaces is 0.2.

Ans. 500.4 N

6.5 Determine the state of the block shown in Fig. E.6.5, i.e., whether it is in equilibrium or in motion; find the magnitude and direction of the frictional force. The mass of the block is 15 kg and the coefficients of friction are $\mu_s = 0.2$ and $\mu_k = 0.15$.

Ans. Under motion up the plane, 14 N acting down the plane

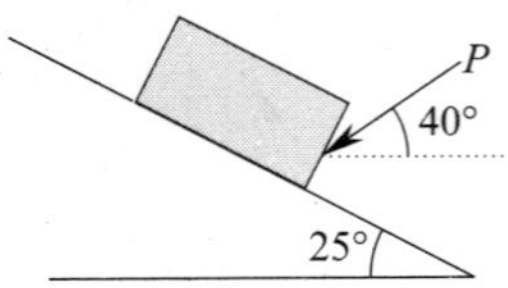

Fig. E.6.4

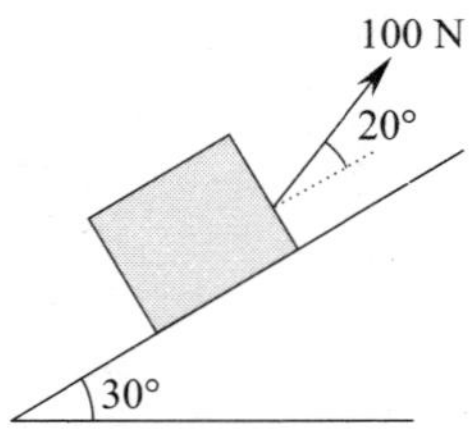

Fig. E.6.5

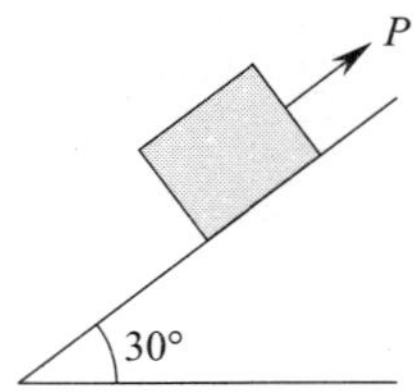

Fig. E.6.6

6.6 A force P of 360 N just pulls a body of 600 N weight up on a rough plane inclined at 30° to the horizontal as shown in Fig. E.6.6. The force is applied parallel to the plane. If the weight of the body is increased to 840 N, determine the additional pull to be applied parallel to the plane to move the body up the plane.

Ans. 143.7 N

6.7 A block of 200 kg mass resting on a horizontal floor must be moved. Determine the force P required (i) to pull the body, and (ii) to push the body as shown in Fig. E.6.7. The coefficient of friction between contact surfaces is 0.25

Ans. (i) 495 N, (ii) 662 N

6.8 A block of 75 kg mass is pulled up an inclined plane as shown in Fig. E.6.8. If the coefficient of friction between the contact surfaces is 0.2, determine the force P required to just make the block move.

Ans. 440.7 N

6.9 In the previous problem, suppose the block is to be pushed by a force P as shown in Fig. E.6.9, determine the value of P to start the motion of the block.

Ans. 697 N

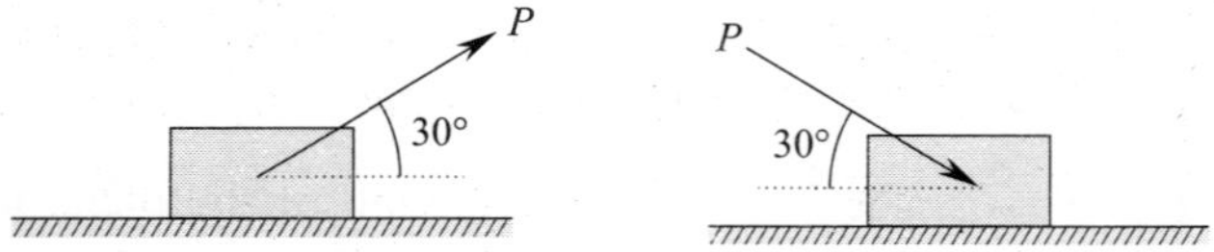

Fig. E.6.7

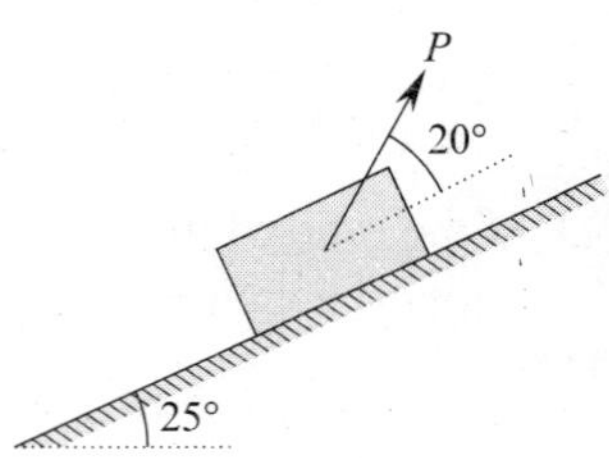

Fig. E.6.8

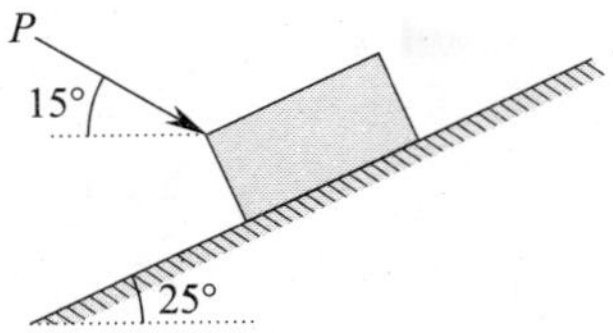

Fig. E.6.9

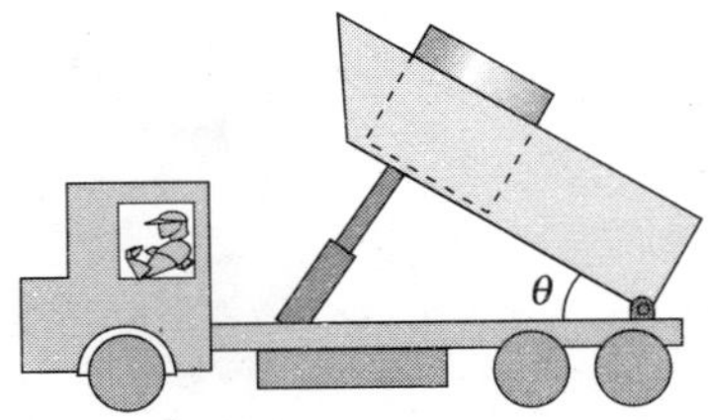

Fig. E.6.10

6.10 The driver of a truck decides to unload a heavy box of 800 kg mass by tilting the rear bed as shown in Fig. E.6.10. Determine what should be the minimum angle θ of the inclination at which the box begins to slide. The coefficient of friction between the box and the base is 0.35. Also, determine the force acting on it causing it to slide.

Ans. 19.29°, 2.6 kN

6.11 A heavy block of 200 kg mass is unloaded from a truck using a plank at the rear side. The plank is kept inclined at 15° to the ground as shown in Fig. E.6.11. Determine the horizontal force the man has to exert on the block to make it to slide. Take $\mu = 0.35$.

Ans. 147.2 N

6.12 A heavy block of 200 kg mass is unloaded from a truck using a plank at the rear side. The plank is kept inclined at 30° to the ground as shown in Fig. E.6.12. Determine the horizontal force the man has to exert on the block to prevent it from sliding down. Take $\mu = 0.35$.

Ans. 371.1 N

Fig. E.6.11

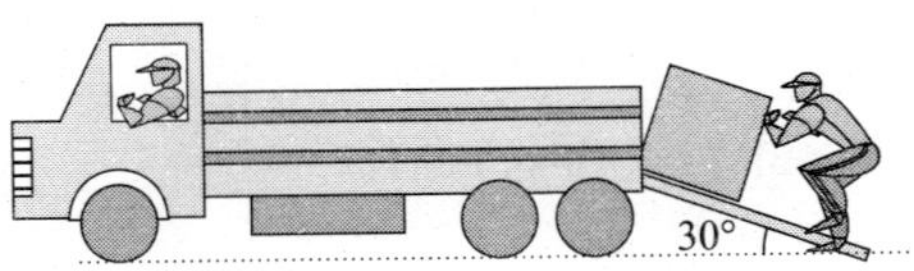

Fig. E.6.12

6.13 A body of dimensions as shown in Fig. E.6.13 must be pushed up the inclined plane. What must be the maximum height at which a force P parallel to the incline must be applied to cause it to slide without tipping over? Take the coefficient of friction as 0.2.

Ans. 60 cm

6.14 A cube of 100 kg mass and of 50 cm side rests on a horizontal surface as shown in Fig. E.6.14. Determine the horizontal force P, (i) required to start the motion of the cube (ii) required to push the cube at which it will begin to topple. Take $\mu = 0.3$.

Ans. (i) 294.3 N, (ii) 490.5 N

6.15. A block of mass m rests in a V-shaped channel as shown in figures E.6.15 (a) and (b). Determine what horizontal force P must be applied for the motion to impend.

Ans. $P = \sqrt{2}\,\mu mg$

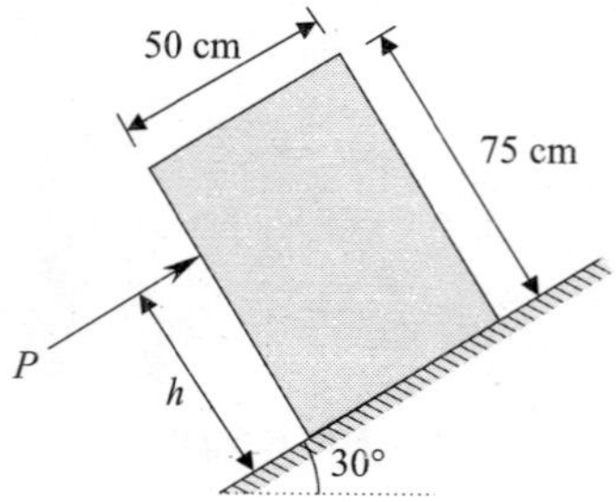

Fig. E.6.13

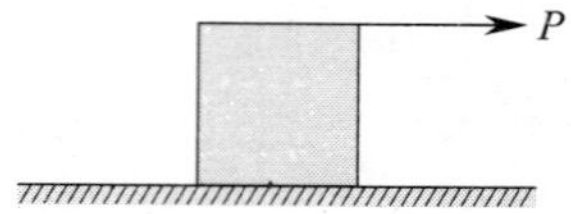

Fig. E.6.14

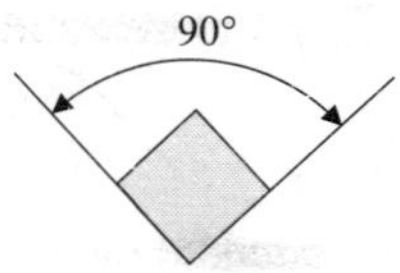

Fig. E.6.15(a)

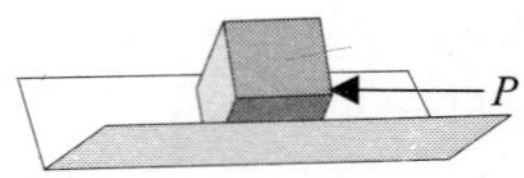

Fig. E.6.15(b)

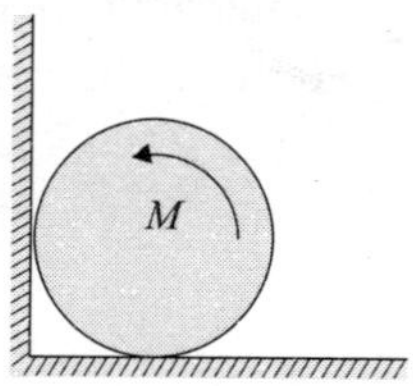

Fig. E.6.16

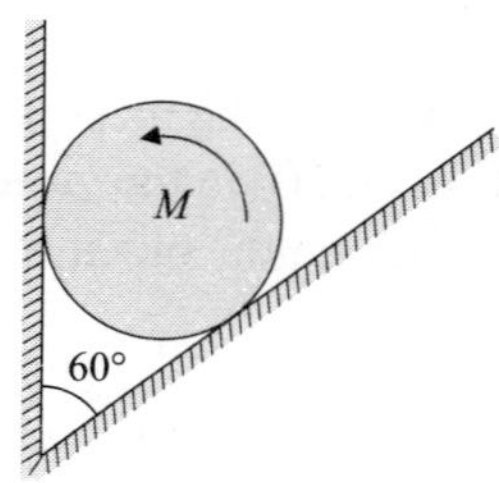

Fig. E.6.17

6.16 Determine the moment required to be applied to the cylinder of weight W and radius r resting as shown in Fig. E.6.16 to cause motion to impend. The coefficient of friction between all contact surfaces is μ.

Ans. $\dfrac{\mu Wr(1+\mu)}{(1+\mu^2)}$

6.17. Determine the moment required to be applied to the cylinder of weight W and radius r resting in a trough as shown in Fig. E.6.17 to cause motion to impend. The coefficient of friction between all contact surfaces is μ.

Ans. $\sqrt{3}\,\mu Wr$

6.18 A cylinder of 20 kg mass and 30 cm diameter rests on a rough horizontal surface. A plate AB is tightly held over it by two springs, each of spring constant 1 kN/m as shown in Fig. E.6.18. If the plate AB is of negligible weight, the coefficient of friction between plate and cylinder is 0.15, and that between horizontal surface and cylinder is 0.2, determine the moment M required to cause the motion to impend in cylinder. The extension in each spring is 5 cm from the unstretched position.

Ans. 11.1 N.m

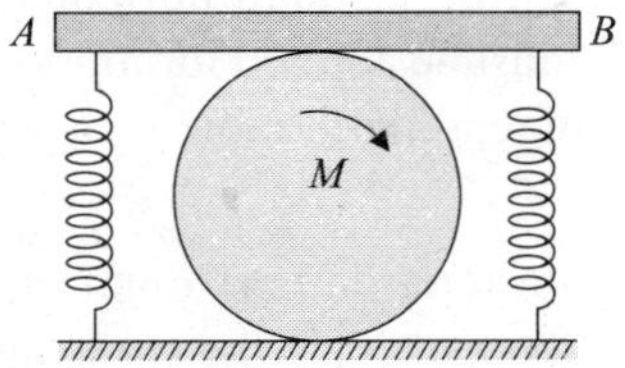

Fig. E.6.18

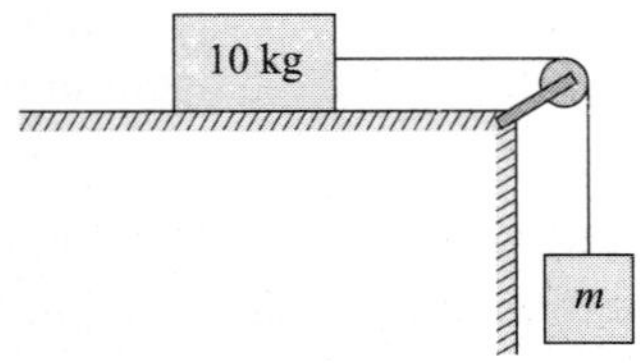

Fig. E.6.20

6.19 Two equal bodies A and B of weight W each are placed on a rough inclined plane. The bodies are connected by a light string. If $\mu_A = 1/2$ and $\mu_B = 1/3$, show that the bodies will be both at the point of impending motion when the plane is inclined at $\tan^{-1}$ (5/12).

6.20 In Fig. E.6.20, determine the value of mass m required to start the 10 kg block to slide. The coefficient of static friction between the contact surfaces is 0.2.

Ans. 2 kg

6.21 In Fig. E.6.21, determine the range of values of W for which the 50 kg block will neither move up nor slide down the plane. The coefficient of static friction between the contact surfaces is 0.2. Assume the pulley to be frictionless

Ans. 28.3 to 42.43 kg

6.22 If the block A in Fig. E.6.22 is on the verge of sliding downwards, determine the coefficient of static friction assuming it the same at all contact surfaces. Mass of the block A is 100 kg and that of the block B is 50 kg. Block B is held by a string attached to the wall such that it is parallel to the incline. Also, determine the tension in the string.

Ans. 0.29, 368 N

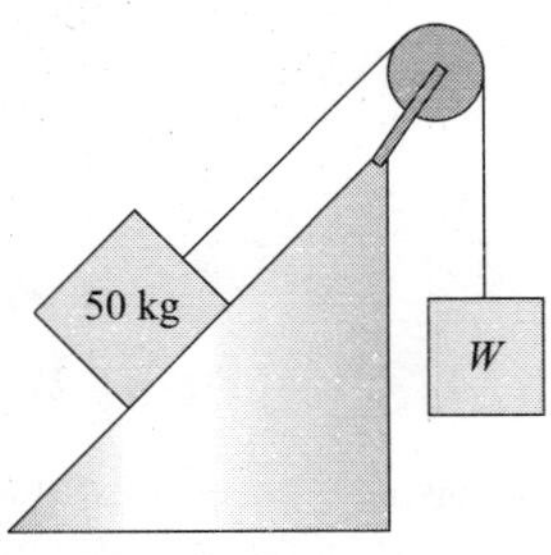

Fig. E.6.21

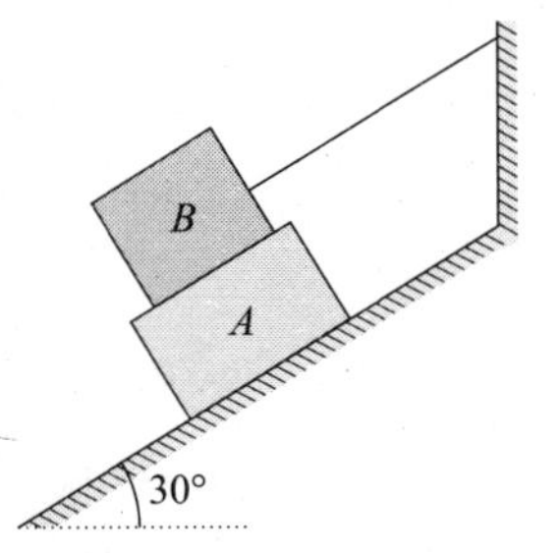

Fig. E.6.22, Fig. E.6.23

6.23 In the previous problem, suppose the coefficient of friction between the two blocks is 0.3 and that between the lower block and the inclined plane is 0.33. Determine the horizontal force P applied parallel to the plane required to start the 100 kg block moving downwards. Refer Fig. E.6.23

Ans. 57.5 N

6.24 Two blocks of equal masses M connected by a string rest on a rough horizontal surface as shown in Fig. E.6.24. Determine what horizontal force P must be applied for motion to impend. Also, determine the tension in the string at the point of impending motion.

Ans. $2\,\mu Mg$, μMg

6.25 In the above problem, if the force P is applied at an angle θ to the horizontal, determine the force required for the motion to impend. Also, determine the tension in the string at the point of impending motion. Refer Fig. E.6.25.

Ans. $\dfrac{2\mu Mg}{\cos\theta + \mu \sin\theta}$, μMg

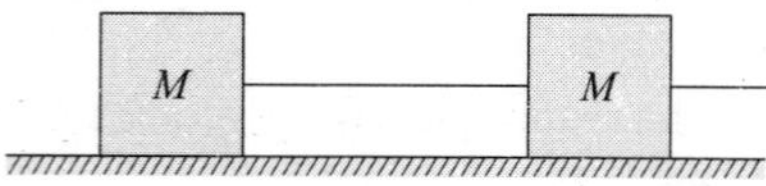

Fig. E.6.24

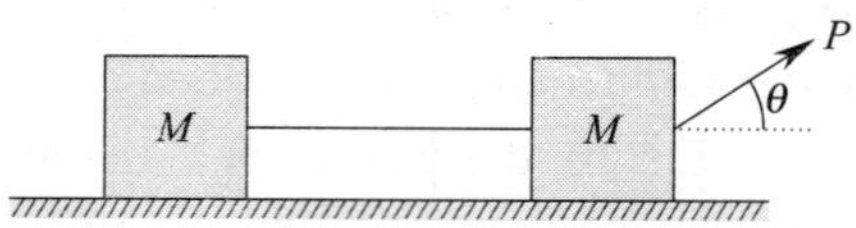

Fig. E.6.25

6.26 Block B of 75 kg mass is placed over a block A of 100 kg mass resting on a rough horizontal plane as shown in Fig. E.6.26. The two blocks are connected by a string. A third block C of 50 kg mass is placed over the block B and it is hinged by a horizontal rod. Determine the horizontal force P required to pull the block A to the left, taking coefficient of static friction for all contact surfaces to be 0.25.

Ans. 1287.6 N

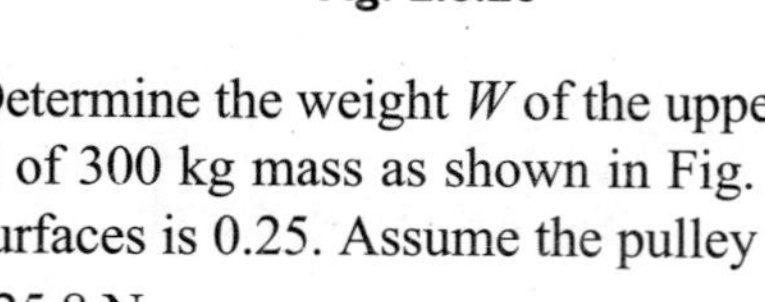

Fig. E.6.26

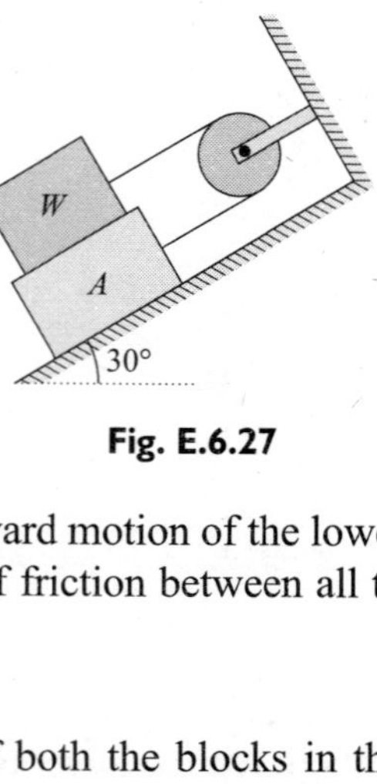

Fig. E.6.27

6.27. Determine the weight W of the upper block to prevent downward motion of the lower block of A of 300 kg mass as shown in Fig. E.6.27. The coefficient of friction between all the contact surfaces is 0.25. Assume the pulley to be frictionless.

Ans. 725.8 N

6.28. Determine the vertical load W that would initiate motion of both the blocks in the position shown in Fig. E.6.28. The coefficient of friction between the 150 kg block and plane is 0.2. Also, determine the coefficient of friction between the 200 kg block and the plane.

Ans. 402 N, 0.184

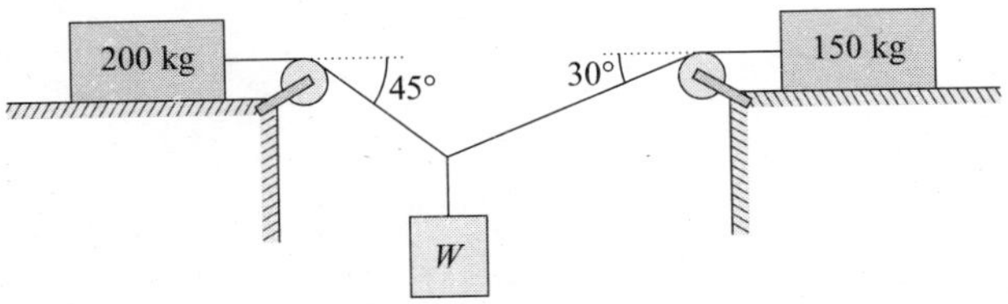

Fig. E.6.28

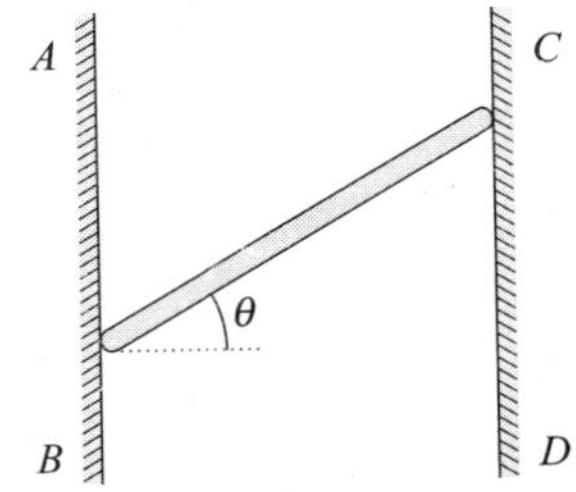

Fig. E.6.29

6.29. A homogeneous bar of length l is placed in between two parallel rough walls AB and CD as shown in Fig. E.6.29. Determine the angle of inclination of the rod with respect to the horizontal for which the rod is in equilibrium. The coefficients of friction between the rod and wall AB and that between the rod and the wall CD are respectively μ_1 and μ_2.

Ans. $\tan^{-1}\left[\frac{\mu_1 - \mu_2}{2}\right]$

6.30 A uniform bar of length l rests on a rough horizontal floor and a rough inclined wall as shown in Fig. E.6.30. Determine the inclination of the wall with respect to the horizontal for which equilibrium can be maintained. Take coefficient of friction for all contact surfaces to be 0.3.

Ans. 53.5°

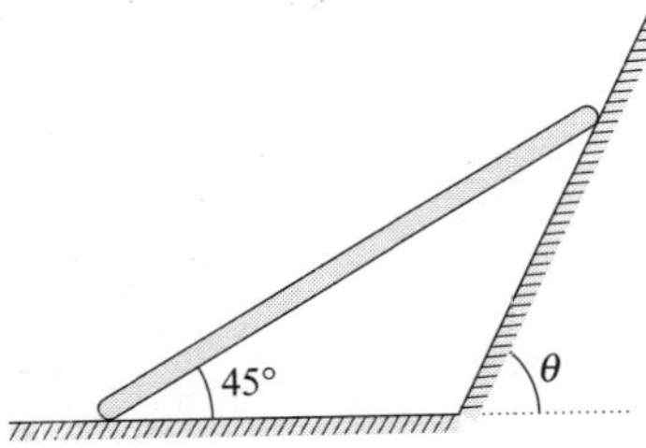

Fig. E.6.30

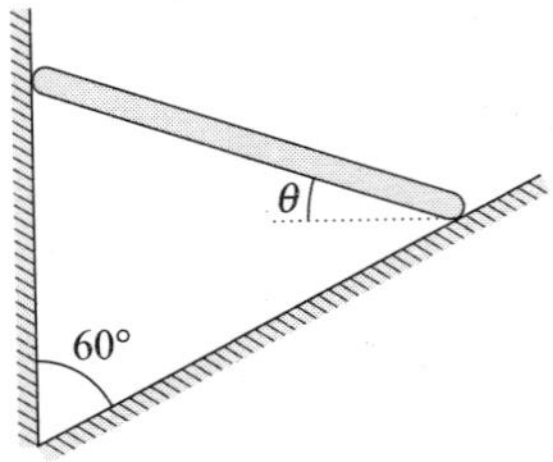

Fig. E.6.31

6.31 A uniform bar of length l rests between a rough vertical wall and a rough incline as shown in Fig. E.6.31. Determine the minimum inclination of the bar with respect to the horizontal for which the equilibrium can be maintained. Take coefficient of friction for all contact surfaces to be 0.2.

Ans. 25.1°

6.32 A rod of length $2r$ is placed in a hemispherical bowl of diameter $2r$ as shown in Fig. E.6.32. If the maximum inclination of the rod possible is 40° with respect to the horizontal, determine the coefficient of friction assuming it the same for all contact points.

Ans. 0.234

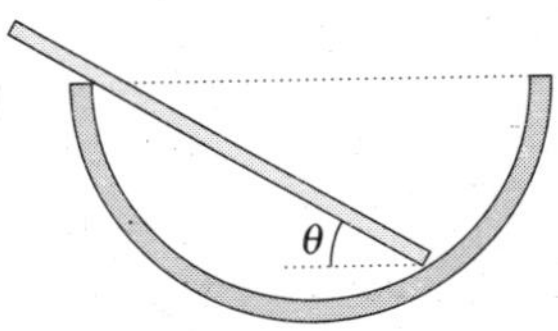

Fig. E.6.32

7 Application of Friction

7.1 INTRODUCTION

A number of machines and mechanisms make use of the friction developed between contact surfaces when one part slides or tends to slide over another to provide fastening, grips, transmission of power, absorption of power between machine components and for lifting heavy loads. These include wedges, screw jacks, belt drives, band brakes, etc. The theory and numerical problems related to them are discussed in detail in this chapter. Wedges are discussed in Section 7.2, screw friction in Section 7.3 and belt friction in Section 7.4.

7.2 WEDGES

In steel structures, prior to bolting or welding the members, they need to be adjusted *slightly* to bring them to exact alignment; or heavy loads may have to be moved *slightly* for proper positioning; or some tightening may have to be provided for keys in shafts. To serve these purposes, **triangular** or **trapezoidal** shaped blocks as shown in Fig. 7.1 are used, which are termed **wedges**. These are simple machines made with a very small sloping angle called **wedge angle**.

Fig. 7.1

Figure 7.2 (a) shows a two-wedge arrangement for lifting a steel column by applying a horizontal force P on both the wedges and Fig. 7.2(b) shows a wedge arrangement for pushing a heavy block to the right by applying a vertical force P. When forces are applied to these wedges, due to friction developed between the contact surfaces, they help in lifting or moving heavy loads. Normally, the weights of the wedges are *very small* when compared to the weight of the bodies they lift. Hence, their weights can be *neglected* in the calculations. Wedges are so shaped that when the applied forces are removed, they remain under **self-locking**. Friction helps in holding the bodies in the shifted position and prevents them from slipping back. Using wedges, large weights can be moved using relatively smaller forces.

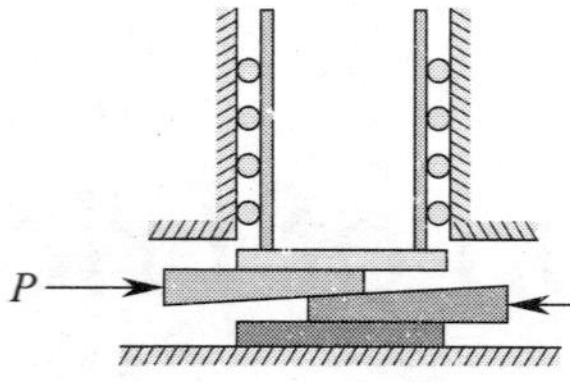

Lifting a column

Fig. 7.2(a)

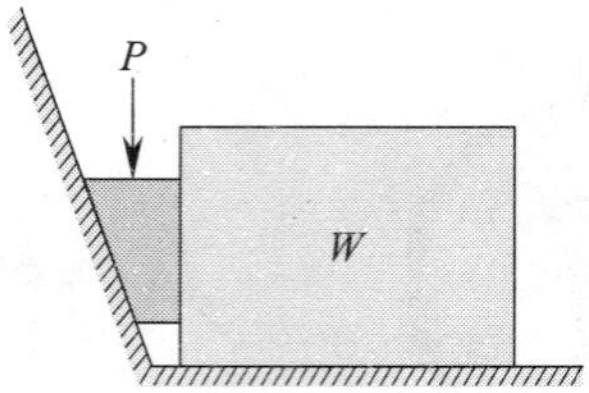

Pushing a heavy block

Fig. 7.2(b)

The force P required to be applied on the wedges can be determined by considering free-body diagrams of the block and the wedge. For illustration, let us consider the example shown in Fig. 7.2(b). The free-body diagrams of the block and the wedge are shown in the figures below. One must carefully analyze the direction of friction between contact planes while drawing the free-body diagrams. As the wedge tries to move downwards, the frictional forces F_1 and F_2 at the two contact surfaces will be directed such as to oppose this downward movement. The respective normal reactions N_1 and N_2 are also shown acting at the contact surfaces.

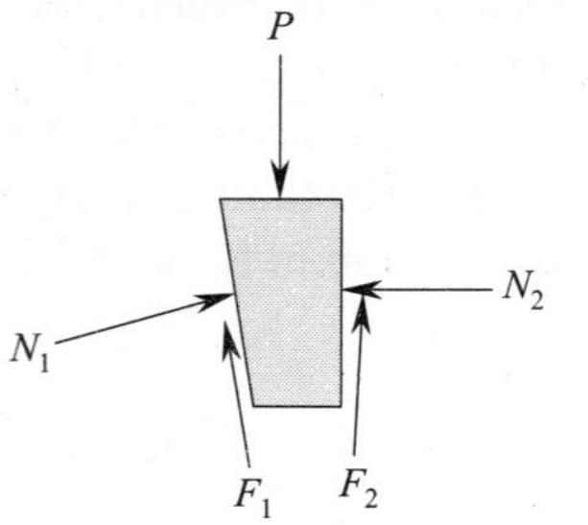

Fig. 7.3(a) Free-body diagram of wedge

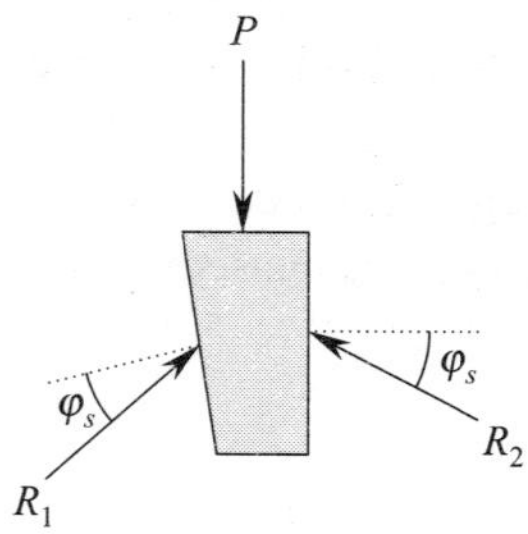

Fig. 7.3(b) Free-body diagram of wedge

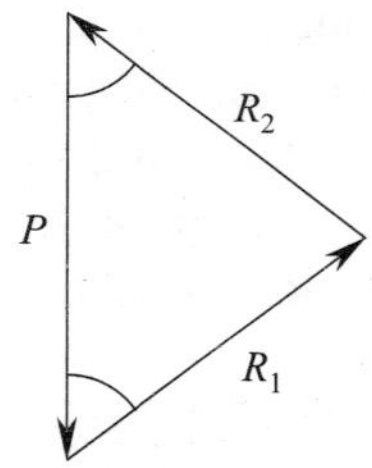

Fig. 7.3(c) Force triangle

However, in wedge problems it will be found convenient if we replace the normal reaction N and the frictional force F by their resultant R as shown in Fig. 7.3(b). As there are three forces P, R_1 and R_2 acting on the wedge and they keep it in equilibrium, we can draw the force triangle as shown in Fig. 7.3(c) and apply Lami's theorem to determine the unknown forces.

Similarly, the free-body diagram of the block is shown in Fig. 7.4. The reaction exerted by the wedge on the block must be equal and opposite to that exerted by the block on the wedge. Hence, R_2 is shown in the opposite direction. The force triangle can be constructed for the block also and the unknown forces can be determined by applying Lami's theorem.

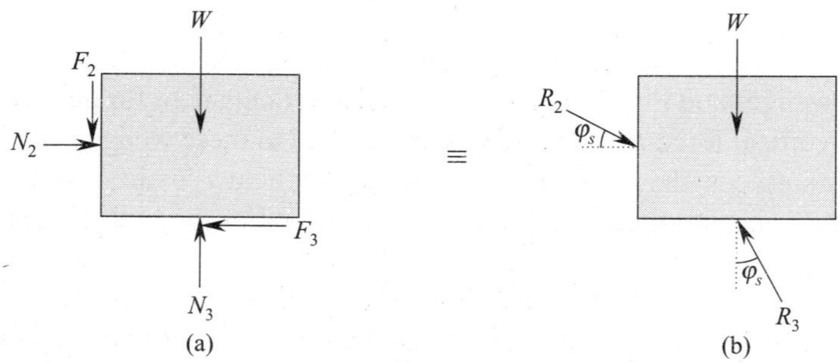

Fig. 7.4 Free-body diagram of block

Example 7.1 A 2000 kg block must be raised using two similar wedges as shown in Fig. 7.5. What horizontal force P must be applied on both the wedges to raise the block? Take coefficient of friction at all contact surfaces as 0.2.

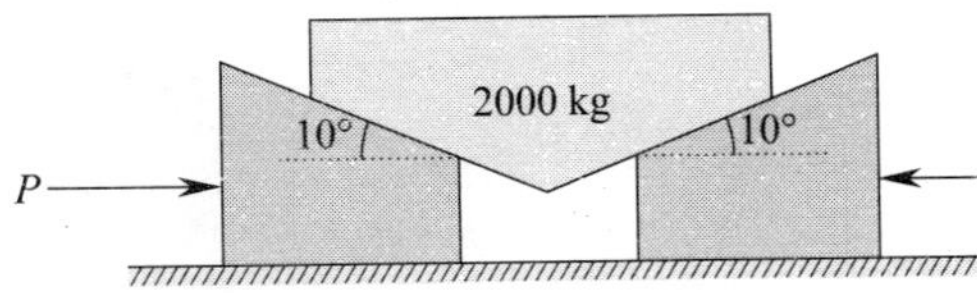

Fig. 7.5

Solution Given that the coefficient of friction between all contact surfaces as 0.2, the angle of friction is obtained as

$$\varphi_s = \tan^{-1}(\mu_s)$$
$$= \tan^{-1}(0.2) = 11.31°$$

The free-body diagrams of the wedges and the block are shown below. The weights of the wedges being very small as compared to the weight of the block, they have been neglected. Since the wedges move inwards, the frictional forces acting on them will be to oppose this motion. Hence, the direction of resultant R for the wedges should be as shown inclined at an angle of φ_s to the normal to the plane [Fig. 7.5(a)]. Note that the reaction R_1 between the left wedge and the block is inclined at an angle φ_s to the normal line and further that the normal line is inclined at 10° to the vertical. It should be noted that due to symmetry, the reactions on both sides of the block are equal, i.e., $R_1 = R_3$.

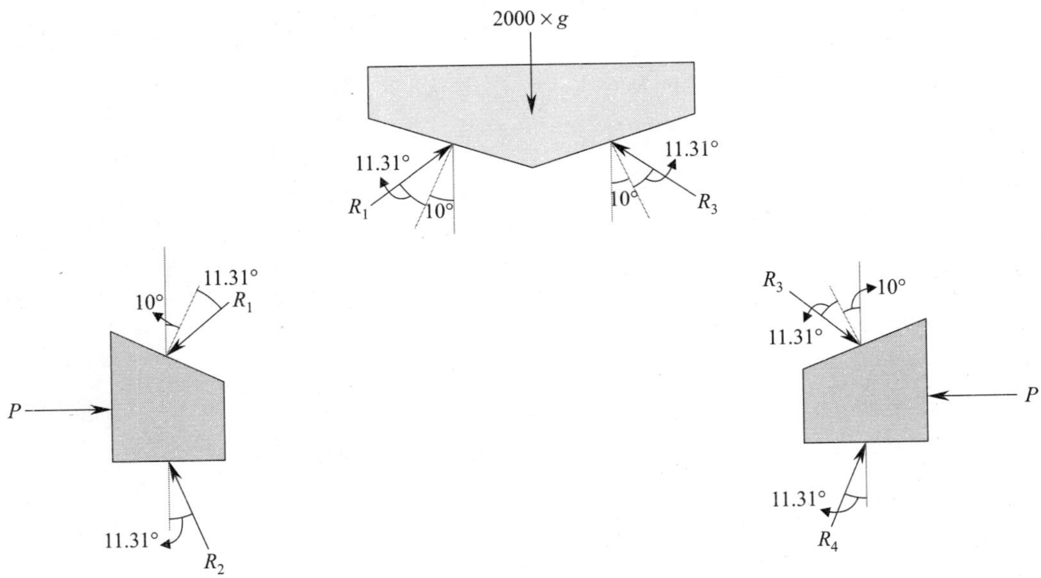

Fig. 7.5(a) Free-body diagrams of block and wedges

We see that there are three forces acting on each free-body and they keep each of them in equilibrium. From Chapter 5, we know that three coplanar forces can be in equilibrium only when they are concurrent. Hence, the three forces can be represented by the sides of a closed triangle taken in order and we can solve for the unknowns by applying Lami's theorem to the force triangles, which are drawn for the block and the wedges.

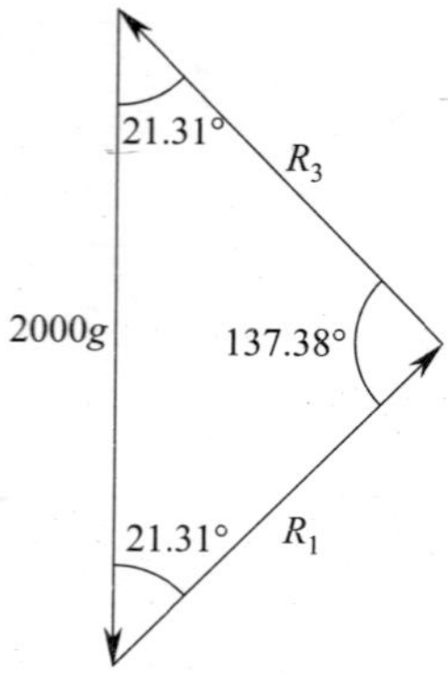

Fig. 7.5(b) Force triangle for block

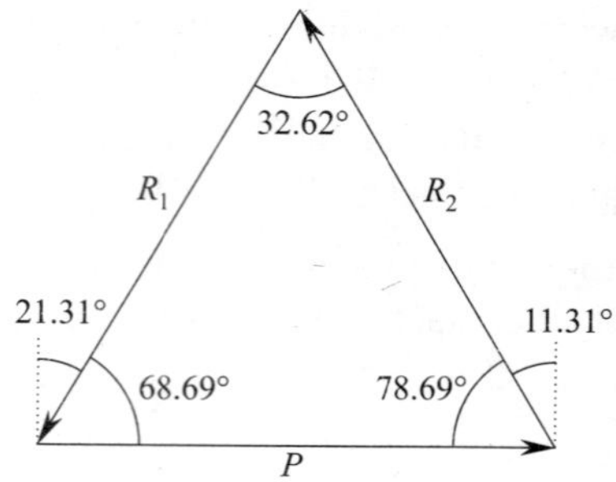

Fig. 7.5(c) Force triangle for wedges

Applying Lami's theorem to the force triangle for the block,

$$\frac{2000 \times g}{\sin 137.38°} = \frac{R_1}{\sin 21.31°} = \frac{R_3}{\sin 21.31°}$$

$$\Rightarrow \qquad R_1 = R_3 = 10.53 \text{ kN}$$

Similarly, applying Lami's theorem to the force triangle for the wedges,

$$\frac{P}{\sin 32.62°} = \frac{R_1}{\sin 78.69°}$$

$$\Rightarrow \qquad P = 5.79 \text{ kN}$$

Example 7.2 A heavy block of mass 500 kg is to be adjusted horizontally using an 8° wedge by applying a vertical force *P*. If the coefficient of static friction for both the contact surfaces of the wedge is 0.25 and that between the block and the horizontal surface is 0.5, determine the least force *P* required to move the block.

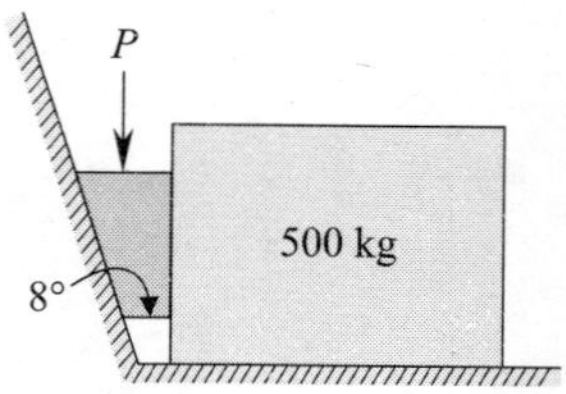

Fig. 7.6

Solution

Determination of angles of friction

For wedge surfaces, it is given that coefficient of static friction is 0.25. Hence,

$$\tan \varphi_1 = \mu_w = 0.25$$

$$\therefore \qquad \varphi_1 = \tan^{-1}(0.25) = 14.04°$$

For block and horizontal surface, it is given that coefficient of static friction is 0.5. Hence,

$$\tan \varphi_2 = \mu_b = 0.5$$

$$\therefore \qquad \varphi_2 = \tan^{-1}(0.5) = 26.57°$$

The free-body diagrams of the block and the wedge are shown in Fig. 7.6(a). Since the wedge is pushed downwards, the frictional forces acting on the two faces of the wedge are directed in the opposite direction, i.e., upwards. Hence, the resultant forces R_1 and R_2 are shown in the correct sense. Since the block is pushed by the wedge towards right, the frictional force between the floor and the block must be directed in the opposite direction, i.e., towards left. Hence, the direction of resultant R_3 is shown accordingly in the figure.

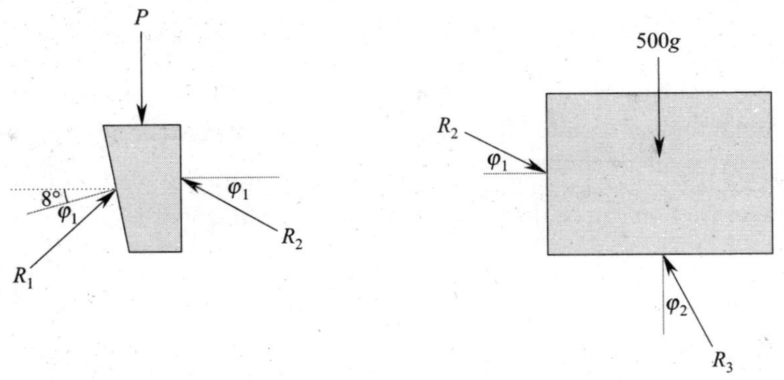

Fig. 7.6(a)

Since there are three forces acting on each free body and they keep each of them in equilibrium, we can conclude that the three forces must be concurrent. Hence, the three forces can be represented by the sides of a closed triangle taken in order and we can solve for the unknowns by applying Lami's theorem to the force triangles shown below.

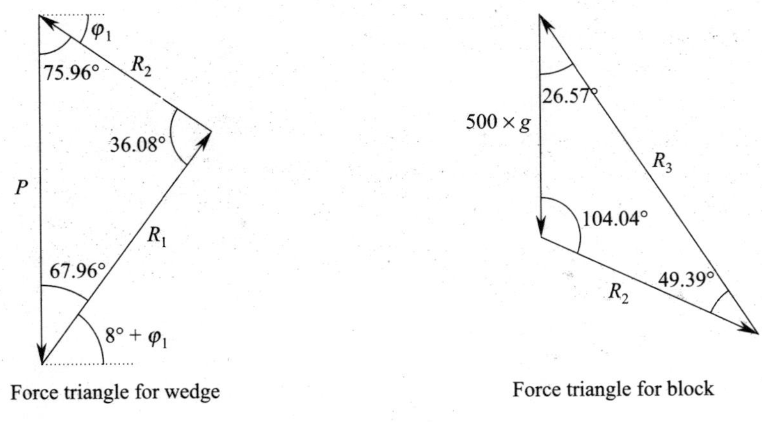

Fig. 7.6(b)

Since we must determine the value of the force P applied to the wedge, we begin with the free-body diagram of the block. Applying Lami's theorem to the force triangle for the block,

$$\frac{500 \times 9.81}{\sin (49.39)°} = \frac{R_2}{\sin (26.57)°}$$

$$\Rightarrow \quad R_2 = 2890 \text{ N}$$

Similarly, applying Lami's theorem to the force triangle for the wedge,

$$\frac{P}{\sin(36.08)^\circ} = \frac{R_2}{\sin(67.96)^\circ}$$

$$\Rightarrow \qquad P = 1836.14 \text{ N}$$

Example 7.3 Two 8° wedges are used to lift and position a 500 kg block. Determine the horizontal force P required to be applied at the top wedge taking coefficient of friction between all contact surfaces as 0.15.

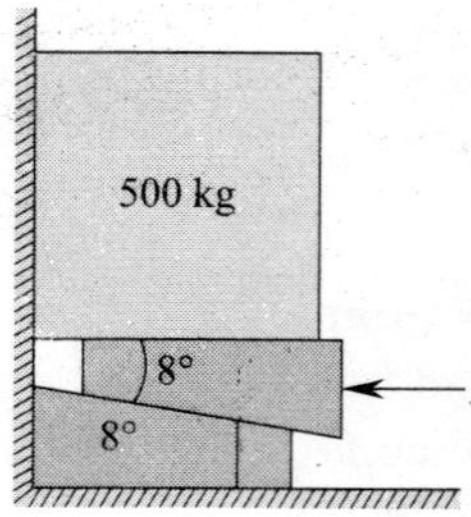

Fig. 7.7

Solution The angle of friction can be determined from the coefficient of friction as follows

Given that $\tan \varphi_s = \mu_s = 0.15$

$$\therefore \qquad \varphi_s = \tan^{-1}(0.15) = 8.53^\circ$$

The free-body diagrams of the block and the top wedge are shown below. As the lower wedge cannot move at all, we do not consider its free-body diagram. Since the top wedge is pushed towards the left side, the frictional forces acting on the two faces of the wedge are directed in the opposite direction, i.e., towards right. Hence, the resultant forces R_2 and R_3 are shown in the correct sense. Since the block is pushed by the wedge upwards, the frictional force between the wall and the block must be directed downwards. Hence, the direction of resultant R_1 is shown accordingly in the figure.

Since three forces keep the body in equilibrium, Lami's theorem is applicable.

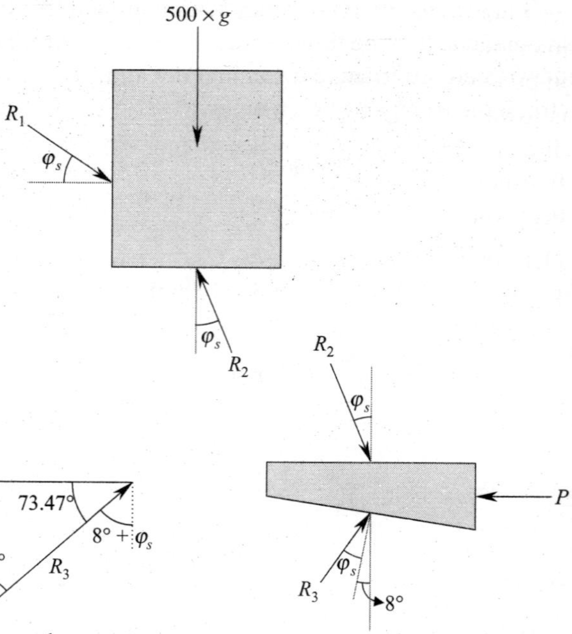

Fig. 7.7(a)

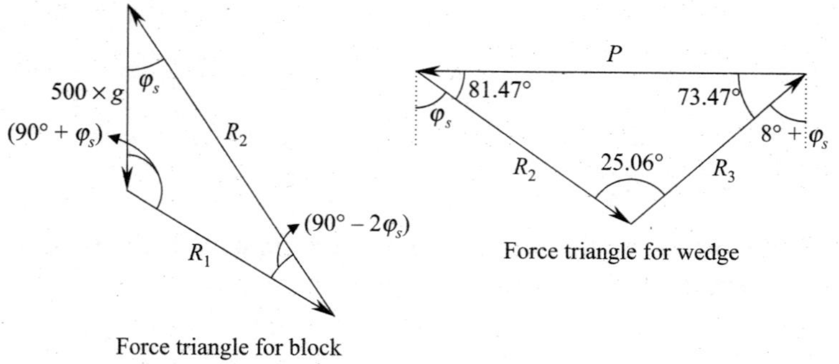

Fig. 7.7(b)

Applying Lami's theorem to the force triangle for the block

$$\frac{500 \times g}{\sin[90 - 2(8.53)]^\circ} = \frac{R_2}{\sin[90 + 8.53]^\circ}$$

$\Rightarrow$ $$R_2 = 5.07 \text{ kN}$$

Similarly, applying Lami's theorem to the force triangle for the wedge

$$\frac{P}{\sin 25.06^\circ} = \frac{R_2}{\sin 73.47^\circ}$$

$\Rightarrow$ $$P = 2.24 \text{ kN}$$

Example 7.4 In Fig.7.8, the block A supports a weight of 4000 N and it is to be prevented from sliding down by applying a horizontal force P on the block B. If the coefficient of friction at all surfaces of contact is 0.2, determine the smallest force P required to maintain equilibrium. Neglect the weights of the blocks.

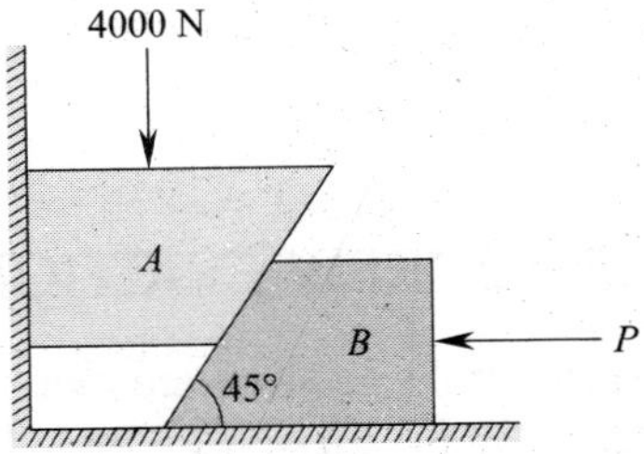

Fig. 7.8

Solution Unlike the previous problem, where the external force P was applied to raise a block, here the force is applied merely to maintain equilibrium. As a result, we can see that the direction of impending motion of the block A is downwards and that of the block B is towards the right. Accordingly, we must show the direction of frictional forces acting on the blocks. The free-body diagrams of the blocks are shown in Fig. 7.8(a).

Determination of angle of friction

For the contact surfaces, it is given that coefficient of static friction is 0.2. Hence,

$$\tan \varphi_s = \mu_s = 0.2$$

$$\therefore \quad \varphi_s = \tan^{-1}(0.2) = 11.31^\circ$$

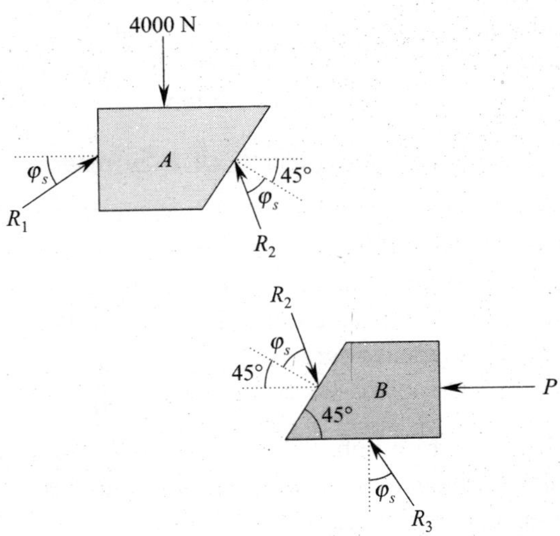

Fig. 7.8(a)

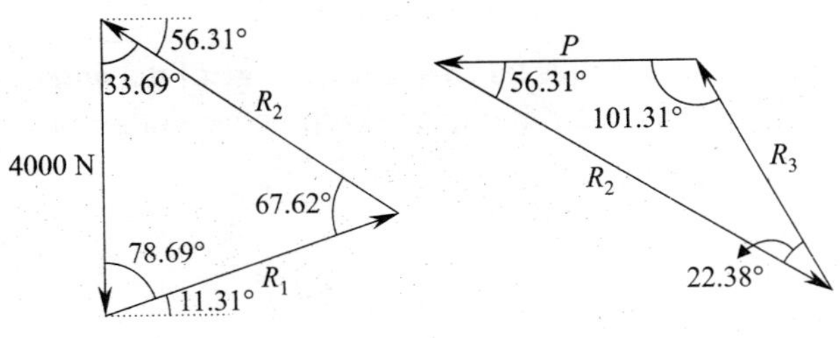

Fig. 7.8(b)

Since we must determine the value of the force P applied to the block B, we proceed from the free-body diagram of the block A. Applying Lami's theorem to the force triangle for the block A,

$$\frac{4000}{\sin(67.62)^\circ} = \frac{R_1}{\sin(33.69)^\circ} = \frac{R_2}{\sin(78.69)^\circ}$$

$$\Rightarrow \quad R_2 = 4000\frac{\sin 78.69°}{\sin 67.62°}$$

$$= 4241.82 \text{ N}$$

Similarly, applying the conditions of equilibrium for the block B

$$\frac{P}{\sin (22.38)°} = \frac{R_2}{\sin (101.31)°} = \frac{R_3}{\sin (56.31)°}$$

$$\Rightarrow \quad P = R_2\frac{\sin 22.38°}{\sin 101.31°}$$

$$= 1647 \text{ N (or) } 1.65 \text{ kN}$$

7.3 SCREW FRICTION

Screws, bolts and nuts are used in various machines and structures for *fastening* purposes and for *lifting heavy loads*. These have threads, which are cut around their bodies in *helical* fashion. There are two types of screw threads that are normally cut; they are ***V-threads*** and ***square threads***. V-threads are used for fastening purposes and square threads for lifting heavy loads. In our study, we will limit ourselves only to square threads. When these screws rotate, resistance is offered by the machine parts to which they are fastened. This resistance is utilized in lifting loads as in ***screw jacks*** or clamping two bodies as in ***vices***. Before we discuss them, we must understand certain terminologies related to screws, which are discussed in the following section.

7.3.1 Lead and Pitch of a Screw

When a screw rotates about its axis, the distance through which the screw advances in one turn along the axis is called the ***lead*** [L] of the screw. The distance between two consecutive threads is termed the ***pitch*** [p]. The screw threads cut on the body of the screw may be *single threaded* or *multiple threaded*. Accordingly, the lead of the screw varies. For a single-threaded screw, the lead is equal to the pitch and for multi-threaded screws,

$$\text{lead} = n \times \text{pitch}$$

where n is the number of threads. The ***mean radius*** (r) of the thread is taken as the average of the ***outer radius*** and ***root*** (or ***inner***) ***radius*** of the thread. For square threads, the relationship between the root radius and outer radius is given as

$$\text{root (or inner) radius} = \text{outer radius} - p/2$$

7.3.2 Lead Angle and Friction Angle

If the square thread is imagined to be unwound then we can visualize it to form an inclined plane as shown in Fig. 7.9(b). The slope θ, called the ***lead angle*** or ***helix angle*** of the inclined plane can be determined from the portion of the inclined plane for one revolution. For one revolution, we know that the screw advances in the vertical direction a distance equal to the *lead* [L] of the screw and the horizon-

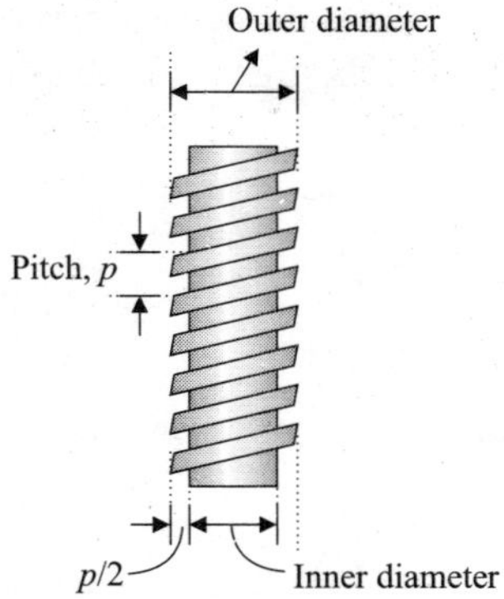

Fig. 7.9(a) Section of square thread

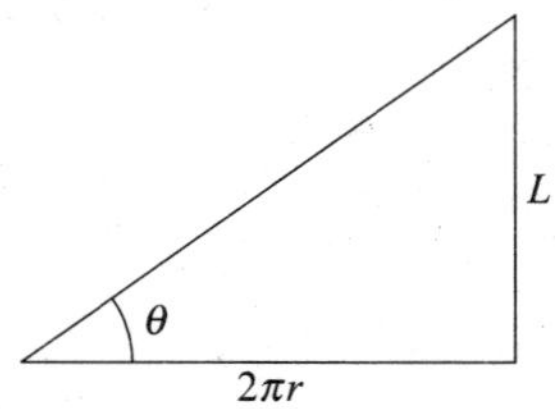

Fig. 7.9(b) Inclined plane

tal distance traveled is equal to the *circumference* [$2\pi r$] of the circle with mean radius (r) of the screw. Therefore,

$$\tan\theta = \frac{L}{2\pi r} \tag{7.1}$$

Since screw threads, when imagined to be unwound, form an inclined plane, screw friction problems can be solved by just treating them as inclined plane problems, which were discussed in the previous chapter. The coefficient of friction between the screw and the surrounding metal part is given as

$$\tan\varphi_s = \mu_s \tag{7.2}$$

where φ_s is the angle of static friction.

7.3.3 Screw Jack

A screw jack is a mechanical device used for lifting and holding heavy loads with little effort. It consists of a spindle around which square threads are cut in helical fashion. The spindle rotates in a nut, which has internal square threads. The nut is fitted in a metallic frame, which is a tapering cylinder with an enlarged base to provide a large bearing area. Provision is made on the head of the screw spindle to insert a handle for rotating it during operation. The load W to be lifted or to be lowered is placed over the head of the spindle.

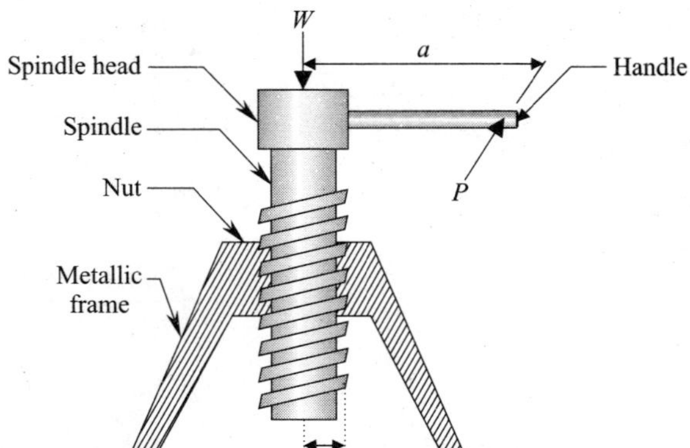

Fig. 7.10 Screw jack

The friction developed between the threads of the spindle and the nut help in lifting or lowering the loads. As this friction is *independent* of the area of contact, the load may be assumed to be concentrated on a small element of the thread at the mean radius (r) of the thread. When a force P is applied at the handle, the equivalent force Q acting on the concentrated weight at the mean radius (r) of the thread can be obtained by taking moment about the axis of the screw. Hence,

$$Pa = Qr \tag{7.3}$$

Effort Required to Raise a Load Let us consider a load being raised using a screw jack. As the screw-jack problem can be treated as an inclined-plane problem, the free-body diagram is as shown in Fig. 7.11(a).

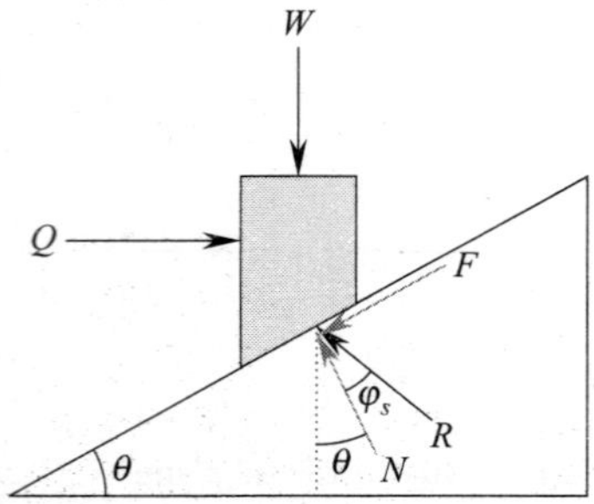

Fig. 7.11(a) Inclined-plane problem

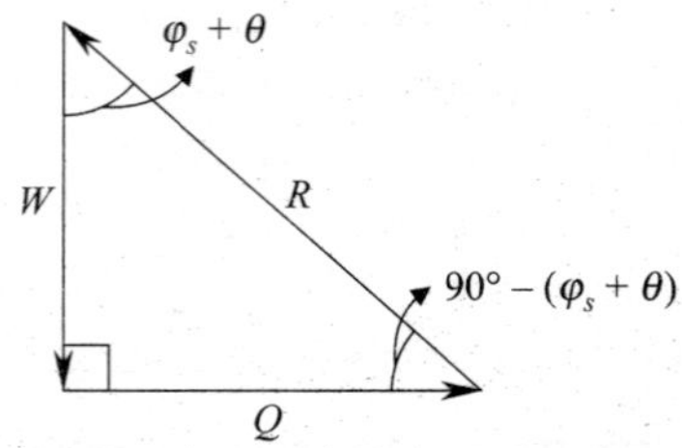

Fig. 7.11(b) Force triangle

Here, the force of friction acts downwards to prevent upward motion of the load. Hence, the direction of the resultant force R of the frictional force and normal reaction is as shown in the figure. At the point of *impending motion*, the forces Q, W and R are in equilibrium. We know that three-force equilibrium is possible only if the forces are concurrent. Hence, the three forces must be concurrent at a point and can be represented by the sides of a closed triangle taken in order. We must note that inclination of the resultant R with respect to the normal reaction is φ_s and that the inclination of the normal reaction with respect to the vertical line is the inclination θ of the incline. Hence, the inclination of the resultant with respect to the vertical line is $\varphi_s + \theta$ [refer Fig. 7.11(a) and (b)].

Applying Lami's theorem to the triangle of forces [refer Fig. 7.11(b)],

$$\frac{Q}{\sin(\varphi_s + \theta)} = \frac{W}{\sin[90° - (\varphi_s + \theta)]}$$

$$\therefore \quad Q = W\frac{\sin(\varphi_s + \theta)}{\sin[90° - (\varphi_s + \theta)]}$$

$$= W\frac{\sin(\varphi_s + \theta)}{\cos(\varphi_s + \theta)}$$

$$= W\tan(\varphi_s + \theta) \qquad (7.4)$$

From Eqs 7.3 and 7.4, we can see that to raise a load, the effort required to be applied at the handle is

$$P = \frac{Wr}{a}\tan(\varphi_s + \theta) \qquad (7.5)$$

or the torque required to raise the load is

$$\tau = Pa = Wr\tan(\varphi_s + \theta) \qquad (7.6)$$

Once motion has started, *little* effort is required to keep the screw jack turning and it is given by

$$P = \frac{Wr}{a} \tan (\varphi_k + \theta) \tag{7.7}$$

where φ_k is the coefficient of kinetic friction.

If the lead angle θ is greater than the friction angle φ_s, the screw will ***unwind*** itself under the load even in the absence of any external moment. However, if the lead angle θ is less than the friction angle φ_s then in the absence of any external applied moment, the screw will ***not unwind*** itself under the load. Hence, the screw is said to be ***self-locking***. In order to lower the load in such cases, some external moment must be applied. Its value can be determined as explained below.

Effort Required to Lower the Load Let us consider a load being lowered using a screw jack. Here the force of friction acts upwards to prevent the downward motion of the load. Hence, the direction of the resultant force R of the frictional force and normal reaction is as shown in Fig. 7.12(a). At the point of impending motion, the forces Q, W and R are in equilibrium. Hence, the three forces must be concurrent at a point and can be represented by the sides of a closed triangle taken in order. In this case, it can be seen that the inclination of the resultant R with respect to the vertical line is: $\varphi_s - \theta$.

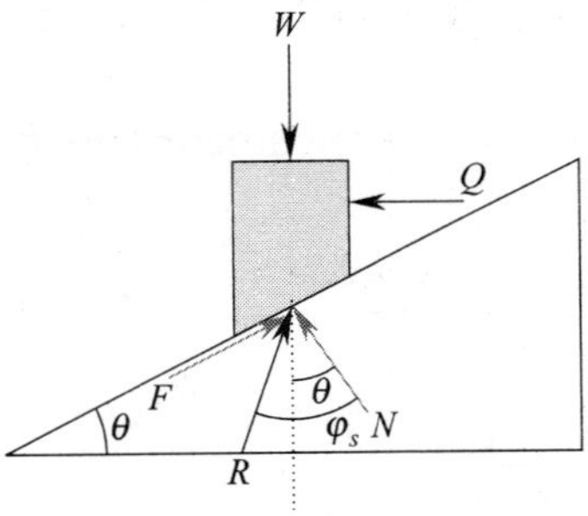

Fig. 7.12(a) Inclined-plane problem

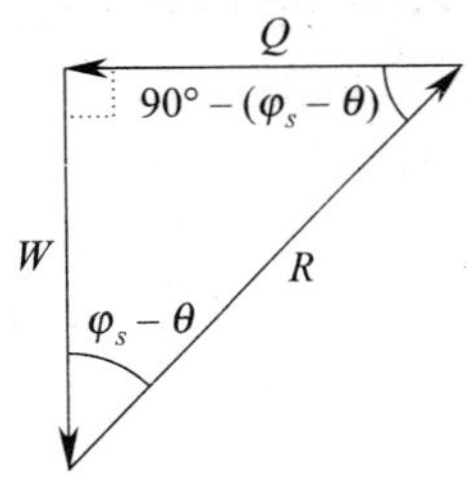

Fig. 7.12(b) Force triangle

Applying Lami's theorem to the triangle of forces [refer Fig. 7.12(b)],

$$\frac{Q}{\sin (\varphi_s - \theta)} = \frac{W}{\sin [90^\circ - (\varphi_s - \theta)]}$$

$$\therefore \quad Q = W \frac{\sin (\varphi_s - \theta)}{\sin [90^\circ - (\varphi_s - \theta)]}$$

$$= W \frac{\sin (\varphi_s - \theta)}{\cos (\varphi_s - \theta)}$$

$$= W \tan (\varphi_s - \theta) \tag{7.8}$$

From Eqs 7.3 and 7.8, we can see that to lower a load, the effort required to be applied at the handle is

$$P = \frac{Wr}{a} \tan (\varphi_s - \theta) \tag{7.9}$$

Similarly, the torque required to lower the load is:

$$\tau = Pa = Wr \tan(\varphi_s - \theta) \tag{7.10}$$

7.3.4 Efficiency of a Screw Jack

When we assume no friction between the surfaces of contact then the static angle of friction $\varphi_s = 0$. Therefore, the ideal effort required to lift a load of weight W is given by,

$$P = \frac{Wr}{a} \tan\theta \tag{7.11}$$

The ***efficiency*** of a screw jack is defined as the ratio of the ***effort*** required under ***frictionless condition*** to that of the ***actual effort*** required to raise a load, i.e.,

$$\eta = \frac{\frac{Wr}{a}\tan\theta}{\frac{Wr}{a}\tan(\varphi_s + \theta)}$$

Hence,

$$\eta = \frac{\tan\theta}{\tan(\varphi_s + \theta)} \tag{7.12}$$

7.3.5 Condition for Maximum Efficiency

Since the static angle of friction φ_s is *constant* for a particular screw, we can differentiate η with respect to θ in the Eq. 7.12. Therefore,

$$\frac{d\eta}{d\theta} = \frac{\tan(\varphi_s + \theta)\sec^2\theta - \tan\theta\sec^2(\varphi_s + \theta)}{\tan^2(\varphi_s + \theta)}$$

$$= \frac{\sec^2\theta}{\tan(\varphi_s + \theta)} - \frac{\tan\theta\sec^2(\varphi_s + \theta)}{\tan^2(\varphi_s + \theta)} \tag{7.13}$$

The condition for maximum efficiency is given by equating the above equation to zero, i.e.,

$$\frac{d\eta}{d\theta} = 0 \Rightarrow$$

$$\frac{\sec^2\theta}{\tan(\varphi_s + \theta)} = \frac{\tan\theta\sec^2(\varphi_s + \theta)}{\tan^2(\varphi_s + \theta)}$$

Since, $\sec^2\theta = \dfrac{1}{\cos^2\theta}$ and $\tan\theta = \dfrac{\sin\theta}{\cos\theta}$, we get

$$\frac{\cos(\varphi_s + \theta)}{\cos^2\theta\sin(\varphi_s + \theta)} = \frac{\sin\theta}{\cos\theta\sin^2(\varphi_s + \theta)}$$

$$\sin(\varphi_s + \theta)\cos(\varphi_s + \theta) = \sin\theta\cos\theta \tag{7.14}$$

Since $\sin 2\theta = 2\sin\theta\cos\theta$, we can write the above equation as

$$\sin 2(\varphi_s + \theta) = \sin 2\theta \tag{7.15}$$

Therefore, the angles on two sides must be equal, i.e.,

$$2(\varphi_s + \theta) = 2\theta$$

$$\Rightarrow \qquad \varphi_s = 0, \text{ which is not possible.}$$

Since $\sin 2\theta$ is also equal to $\sin (180° - 2\theta)$, the Eq. 7.15 can be written as

$$2(\varphi_s + \theta) = 180° - 2\theta$$

$$\Rightarrow \qquad \theta = 45° - \varphi_s/2 \tag{7.16}$$

which is the condition for maximum efficiency.

7.3.6 Maximum Efficiency

Substituting the condition for maximum efficiency in the expression for η in the Eq. 7.12, we get

$$\eta_{max} = \frac{\tan\theta}{\tan(\varphi_s + \theta)}$$

$$= \frac{\tan(45° - \varphi_s/2)}{\tan(45° + \varphi_s/2)}$$

$$= \frac{\sin(45° - \varphi_s/2)}{\cos(45° - \varphi_s/2)} \, \frac{\cos(45° + \varphi_s/2)}{\sin(45° + \varphi_s/2)} \tag{7.17}$$

Since $\sin(\alpha \pm \beta) = \sin\alpha\cos\beta \pm \cos\alpha\sin\beta$

and $\cos(\alpha \pm \beta) = \cos\alpha\cos\beta \mp \sin\alpha\sin\beta$

$$\eta_{max} = \frac{[\cos(\varphi_s/2) - \sin(\varphi_s/2)]}{[\cos(\varphi_s/2) + \sin(\varphi_s/2)]} \, \frac{[\cos(\varphi_s/2) - \sin(\varphi_s/2)]}{[\cos(\varphi_s/2) + \sin(\varphi_s/2)]}$$

[It should be noted that since sin 45° = cos 45°, they have been cancelled in the numerator and denominator].

$$= \frac{[\cos(\varphi_s/2) - \sin(\varphi_s/2)]^2}{[\cos(\varphi_s/2) + \sin(\varphi_s/2)]^2}$$

$$= \frac{1 - 2\sin(\varphi_s/2)\cos(\varphi_s/2)}{1 + 2\sin(\varphi_s/2)\cos(\varphi_s/2)} \qquad [\text{since } \sin^2\theta + \cos^2\theta = 1]$$

Therefore,

$$\eta_{max} = \frac{1 - \sin\varphi_s}{1 + \sin\varphi_s} \tag{7.18}$$

Example 7.5 A single-threaded screw jack has a pitch of 12 mm and a mean diameter of 75 mm. The coefficient of static friction between the screw and nut is 0.2 and that of kinetic friction is 0.1. Determine the force P to be applied at the end of a 500 mm long lever (i) to just lift a weight of 25 kN, and (ii) to keep the screw jack turning.

Solution *Given data*

pitch(p) = 12 mm

mean radius(r) = 75/2 = 37.5 mm

coefficient of static friction (μ_s) = 0.2

coefficient of kinetic friction (μ_k) = 0.1

lever length (a) = 500 mm

weight to be lifted (W) = 25 kN

Since the screw is *single threaded*, lead (L) = pitch (p) = 12 mm

Determination of helix angle

We know, $$\tan\theta = \frac{L}{2\pi r} = \frac{12}{2\pi(37.5)} = 0.051$$

$$\Rightarrow \quad \theta = \tan^{-1}(0.051) = 2.92° \qquad \text{(a)}$$

(a) Force required to just lift a weight of 25 kN

We know, $$\tan\varphi_s = \mu_s$$

$$\Rightarrow \quad \varphi_s = \tan^{-1}(0.2) = 11.31°$$

$$\therefore \quad \varphi_s + \theta = 11.31° + 2.92° = 14.23°$$

$$\tan(\varphi_s + \theta) = 0.254$$

Therefore, force required to just raise the load is given as

$$P = \frac{Wr}{a}\tan(\varphi_s + \theta)$$

$$= \frac{25000 \times 0.0375}{0.5}(0.254) = 476.25 \text{ N}$$

(b) Force required to keep the screw jack turning

Once motion has started, we must replace coefficient of static friction with coefficient of kinetic friction. Hence, we determine the angle of kinetic friction as follows:

$$\tan\varphi_k = \mu_k$$

$$\Rightarrow \quad \varphi_k = \tan^{-1}(0.1) = 5.71°$$

$$\therefore \quad \varphi_k + \theta = 5.71° + 2.92° = 8.63°$$

$$\tan(\varphi_k + \theta) = 0.152$$

Therefore, force required to keep the screw jack turning is given as

$$P = \frac{Wr}{a}\tan(\varphi_k + \theta)$$

$$= \frac{25000 \times 0.0375}{0.5}(0.152) = 285 \text{ N}$$

[Note that the force required to keep the screw jack turning, once the motion has started, is lesser than that required to start lifting the load].

Example 7.6 A single square-threaded screw jack has a pitch of 16 mm and a mean radius of 50 mm. Determine the force that must be applied at the end of a 60 cm lever to raise a weight of 100 kN and the efficiency of the jack. State whether it is self-locking or not? If yes, determine the force that must be applied to lower the same weight. Take coefficient of static friction as 0.2.

Solution *Given data*

Pitch (p) = 16 mm

mean radius (r) = 50 mm

coefficient of friction (μ_s) = 0.2

weight to be lifted (W) = 100 kN

lever arm (a) = 60 cm

Determination of lead angle and angle of friction

Since the screw is *single threaded*, lead (L) = pitch (p) = 16 mm. Therefore,

$$\tan\theta = \frac{L}{2\pi r} = \frac{16}{2\pi(50)} = 0.051$$

$$\Rightarrow \qquad \theta = 2.92°$$

We know, $$\tan\varphi_s = \mu_s$$

$$\therefore \qquad \varphi_s = \tan^{-1}(0.2) = 11.31°$$

Force required to raise a weight of 100 kN

$$P = \frac{Wr}{a}\tan(\varphi_s + \theta)$$

$$= \frac{100\,000 \times 0.05}{0.6}\tan(11.31 + 2.92)° = 2113.3 \text{ N (or) } 2.11 \text{ kN}$$

Efficiency of jack

The efficiency of a screw jack is given as

$$\eta = \frac{\tan\theta}{\tan(\varphi_s + \theta)}$$

$$= \frac{\tan(2.92)°}{\tan(11.31 + 2.92)°} = 0.2011 \text{ (or) } 20.11\%$$

Condition for self-locking

Since $\varphi_s > \theta$, it is under ***self-locking***. Hence, force is required to lower the load.

Force required to lower the same weight of 100 kN

$$P = \frac{Wr}{a}\tan(\varphi_s - \theta)$$

$$= \frac{100\,000 \times 0.05}{0.6}\tan(11.31 - 2.92)° = 1229.07 \text{ N (or) } 1.23 \text{ kN}$$

Example 7.7 A square threaded spindle of a screw jack has an outer diameter of 50 mm and a pitch of 10 mm. If coefficient of friction between the screw and nut is 0.25, determine (i) the force required to be applied at the screw to raise a load of 5000 N, (ii) efficiency of screw jack, (iii) the force required to be applied at pitch radius to lower the same load of 5000 N (iv) the pitch required for maximum efficiency of the screw and the corresponding maximum efficiency. Neglect friction between the nut and the collar.

Solution *Given data*

pitch (p) = 10 mm

outer radius (r) = 25 mm

coefficient of static friction (μ_s) = 0.25

load to be lifted (W) = 5000 N

Determination of mean radius

Root radius of thread = outer radius– $p/2$ = 25 – 10/2 = 20 mm

Therefore, mean radius of thread [r] = (1/2) (outer radius + root radius)

= (1/2) (25 + 20) = 22.5mm

Determination of helix angle and angle of friction

As nothing is said about the number of threads in the screw, we assume that it is single threaded. Therefore, lead of the screw is equal to the pitch.

We know, $$\tan\theta = \frac{L}{2\pi r} = \frac{10}{2\pi(22.5)} = 0.071$$

$$\Rightarrow \quad \theta = \tan^{-1}(0.071) = 4.06°$$

Also, $$\tan\varphi_s = \mu_s$$

$$\Rightarrow \quad \varphi_s = \tan^{-1}(0.25) = 14.04°$$

$$\therefore \quad \varphi_s + \theta = 14.04° + 4.06° = 18.1°$$

(i) Force required to be applied at the screw to raise a load of 5000 N

$$Q = W\tan(\varphi_s + \theta)$$

$$= 5000\tan(18.1)° = 1634.25 \text{ N}$$

(ii) Efficiency of screw jack while raising the load

$$\eta = \frac{\tan\theta}{\tan(\varphi_s + \theta)} = \frac{\tan(4.06°)}{\tan(18.1°)} = 0.2172 \text{ (or) } 21.72\%$$

(iii) Force required to be applied at pitch radius to lower the same load of 5000 N

$$Q = W\tan(\varphi_s - \theta)$$

$$= 5000\tan(14.04 - 4.06)°$$

$$= 5000\tan(9.98)° = 879.84 \text{ N}$$

(iv) Maximum efficiency of screw and pitch for maximum efficiency

The condition for maximum efficiency is

$$\theta = 45° - \varphi_s/2$$
$$= 45° - 14.04°/2 = 37.98°$$

$$\tan\theta = \frac{L}{2\pi r}$$

$$\Rightarrow \quad L = \tan\theta\,(2\pi r) = 110.4 \text{ mm}$$

Hence, the pitch for maximum efficiency is 110.4 mm and the corresponding maximum efficiency is given as

$$\eta_{\max} = \frac{1 - \sin\varphi_s}{1 + \sin\varphi_s}$$

$$= \frac{1 - \sin 14.04°}{1 + \sin 14.04°} = 0.6095 \text{ (or) } 60.95\%$$

Example 7.8 The mean radius of the screw spindle is 40 mm and the pitch of the thread is 16 mm. If coefficient of friction between the screw and nut is 0.15, find the torque required to raise a load of 5000 N and the efficiency of the screw jack. Also, find the torque required to lower the load.

Solution *Given data*

pitch (p) = 16 mm

mean radius (r) = 40 mm

coefficient of static friction (μ_s) = 0.15

load (W) = 5000 N

Determination of helix angle and angle of friction

$$\tan\theta = \frac{L}{2\pi r} = \frac{16}{2\pi(40)} = 0.0637$$

$$\Rightarrow \quad \theta = 3.64°$$

$$\tan\varphi_s = \mu_s$$

$$\Rightarrow \quad \varphi_s = \tan^{-1}(0.15) = 8.53°$$

(i) Torque required to raise the load

The torque required to raise the load is given as

$$\tau = Pa = Wr\tan(\varphi_s + \theta)$$
$$= 5000 \times 0.04 \times \tan(8.53 + 3.64)° = 43.13 \text{ N.m}$$

(ii) Efficiency of screw jack

The efficiency of screw jack is given as

$$\eta = \frac{\tan\theta}{\tan(\varphi_s + \theta)}$$

$$= \frac{\tan (3.64)^\circ}{\tan (8.53 + 3.64)^\circ} = 0.295 \text{(or) } 29.5\%$$

(iii) Torque required to lower the load

Since $\varphi_s > \theta$, it is under ***self-locking***. Hence, torque is required to lower the load and it is given by

$$\tau = Pa = Wr \tan (\varphi_s - \theta)$$

$$= 5000 \times 0.04 \times \tan (8.53 - 3.64)^\circ = 17.11 \text{ N.m}$$

Example 7.9 A C-clamp is used to compress two wooden boards. The thread of the clamp is a double square thread with a 40 mm mean diameter and a pitch of 6 mm. The coefficient of static friction is 0.25. If a maximum torque of 50 N.m is applied in tightening the clamp, determine the torque required to loosen the clamp.

Solution *Given data*

pitch (p) = 6 mm

mean radius (r) = 20 mm

coefficient of static friction (μ_s) = 0.25

maximum torque applied (τ) = 50 N.m

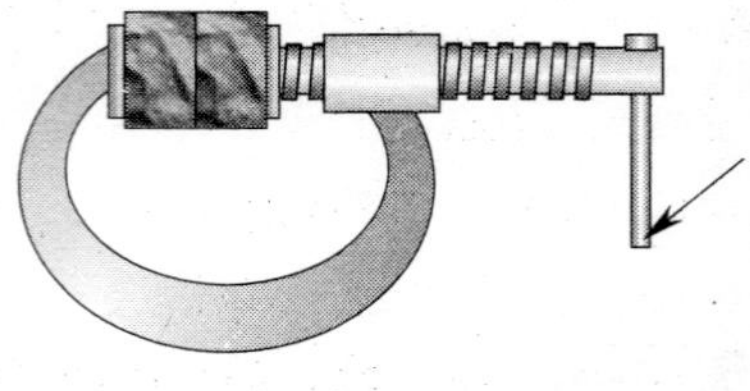

Fig. 7.13

Determination of helix angle and angle of friction

Since the screw is double-threaded, lead[L] = 2 × pitch = 12 mm. Hence,

$$\tan \theta = \frac{L}{2\pi r} = \frac{12}{2\pi(20)} = 0.095$$

$$\Rightarrow \quad \theta = 5.43^\circ$$

$$\tan \varphi_s = \mu_s$$

$$\Rightarrow \quad \varphi_s = \tan^{-1}(0.25) = 14.04^\circ$$

The torque exerted on the boards while tightening is given by

$$\tau = Wr \tan (\varphi_s + \theta)$$

$$50 = W (0.02) \tan (14.04 + 5.43)^\circ$$

$$\Rightarrow \quad W = 7.07 \text{ kN}$$

Therefore, the torque required to loosen the clamp

$$\tau = Wr \tan(\varphi_s - \theta)$$

$$= (7.07 \times 10^3)(0.02) \tan (14.04 - 5.43)^\circ = 21.41 \text{ N.m}$$

Example 7.10 The distance between adjacent threads of a triple-threaded screw jack is 16 mm. The mean radius is 50 mm. The coefficient of static friction is 0.10. What load can be raised by exerting a moment of 500 N.m?

Solution *Given data*

pitch (p) = 16 mm

mean radius (r) = 50 mm

coefficient of friction (μ_s) = 0.10

moment (M) = 500 N.m

Determination of lead angle and angle of friction

Since the screw is triple threaded, lead (L) = $3 \times p$ = 48 mm

$$\tan\theta = \frac{L}{2\pi r} = \frac{48}{2\pi(50)} = 0.153$$

$$\Rightarrow \quad \theta = 8.7°$$

$$\tan\varphi_s = \mu_s$$

$$\Rightarrow \quad \varphi_s = \tan^{-1}(0.10) = 5.71°$$

$$\varphi_s + \theta = 5.71° + 8.7° = 14.41°$$

$$\tan(\varphi_s + \theta) = 0.257$$

Determination of load that can be raised

$$M = Wr\tan(\varphi_s + \theta)$$

$$\Rightarrow \quad W = \frac{M}{r\tan(\varphi_s + \theta)}$$

$$= \frac{500 \text{ N.m}}{50 \times 0.257 \text{ mm}} = 38.91 \text{ kN}$$

7.3.7 Differential Screw Jack

A differential screw jack has two spindles, one within the other. The outer spindle with a larger diameter has a hollow cylinder in which square threads are cut, both outside as well as inside. The external threads glide over the nut while the internal threads glide over the external threads of the smaller diameter spindle. The load to be lifted is placed on the head of the inner spindle. The outer spindle can be rotated by turning a handle inserted over its head. The inner spindle does not rotate, but displaces in the vertical direction only.

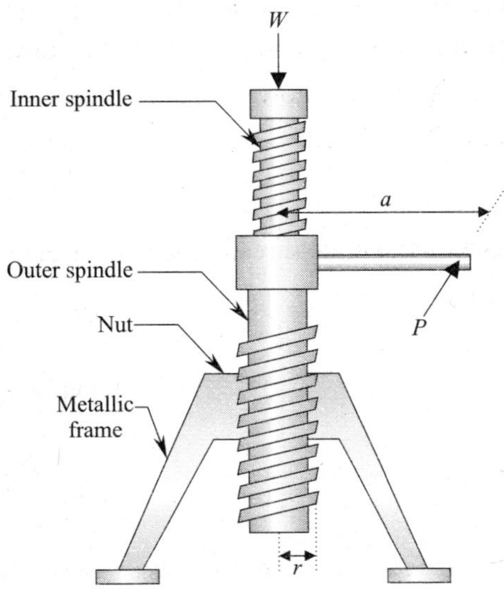

Fig. 7.14 Differential screw jack

Let d_1 and d_2 be the mean diameters of the larger (outer) and smaller (inner) screws respectively. Let p_1 and p_2 be their respective pitches. Hence, when the handle is rotated once, the larger screw will move ***up*** by p_1 while the smaller screw on whose head the load W is placed will move ***down*** by p_2. Hence, the net upward displacement of the load is

$$p_1 - p_2 \tag{7.19}$$

A differential screw jack allows finer movements compared to normal screw jacks as the distance moved per rotation reduces from p_1 to $(p_1 - p_2)$. When the outer spindle rotates, friction is developed between its inner thread and the outer thread of the smaller spindle and also between its outer thread and the nut. Hence, the external torque applied must be the sum of the torques required to rotate the outer screw with respect to the nut and the inner screw with respect to the external screw of the smaller spindle.

The torque required to rotate the outer screw with respect to the nut is equivalent to the effort in raising a load, i.e., $\tau_1 = r_1 W \tan(\varphi_1 + \theta_1)$; and that required to rotate the inner screw with respect to the outer screw of the smaller spindle is equivalent to the effort in lowering a load, i.e., $\tau_2 = r_2 W \tan(\varphi_2 - \theta_2)$. Hence, the torque required to raise the load is given as

$$\tau = \tau_1 + \tau_2$$

$$\therefore \qquad \tau = r_1 W \tan(\varphi_1 + \theta_1) + r_2 W \tan(\varphi_2 - \theta_2) \tag{7.20}$$

where φ_1 = friction angle for surfaces between outer screw of the larger spindle and the nut

φ_2 = friction angle for contact surfaces between inner screw of the larger spindle and outer screw of the smaller spindle

θ_1 and θ_2 are lead angles of the outer and inner screws respectively.

If the lever arm is a then the effort to be applied at the end of the handle is

$$P = \tau / a \tag{7.21}$$

Example 7.11 Determine the torque to be applied in a differential screw jack to lift a load of 10 kN. The pitches of the outer and inner threads are 10 mm and 6 mm respectively. The mean radius of the outer thread is 30 mm and that of the inner thread is 20 mm, coefficient of friction for both threads is 0.1.

Solution We know, $\tan \varphi_s = \mu_s$

$$\Rightarrow \qquad \varphi_s = \tan^{-1}(0.1) = 5.71°$$

Outer screw

Assuming single-threaded screw for outer spindle,

$$L_1 = p_1 = 10 \text{ mm}$$

$$\therefore \qquad \theta_1 = \tan^{-1}\left[\frac{L_1}{2\pi r_1}\right]$$

$$= \tan^{-1}\left[\frac{10}{2\pi(30)}\right] = 3.04°$$

Hence, $\tan(\varphi_s + \theta_1) = \tan(5.71 + 3.04)° = 0.154$

Inner screw

Assuming single-threaded screw for inner spindle,

$$L_2 = p_2 = 6 \text{ mm}$$

$$\therefore \quad \theta_2 = \tan^{-1}\left[\frac{L_2}{2\pi r_2}\right] = \tan^{-1}\left[\frac{6}{2\pi(20)}\right] = 2.73°$$

Hence, $\tan(\varphi_s - \theta_2) = \tan(5.71 - 2.73)° = 0.052$

Hence, the torque to be applied to raise a load of 10 kN is given as

$$\tau = r_1 W \tan(\varphi_1 + \theta_1) + r_2 W \tan(\varphi_2 - \theta_2)$$
$$= W[r_1 \tan(\varphi_1 + \theta_1) + r_2 \tan(\varphi_2 - \theta_2)]$$
$$= 10 \times 10^3 [(0.03 \times 0.154) + (0.02 \times 0.052)] = 56.6 \text{ N.m}$$

7.4 BELT FRICTION

In Chapter 5, we saw that whenever a flexible member like a string, a rope, a cable or a belt passes over a **smooth** pulley or a cylindrical drum, as there is no frictional force developed between the two contact surfaces, the tension at both the ends of the member will be equal [refer Fig. 7.15(a)]. However, we know that no surface is perfectly smooth.

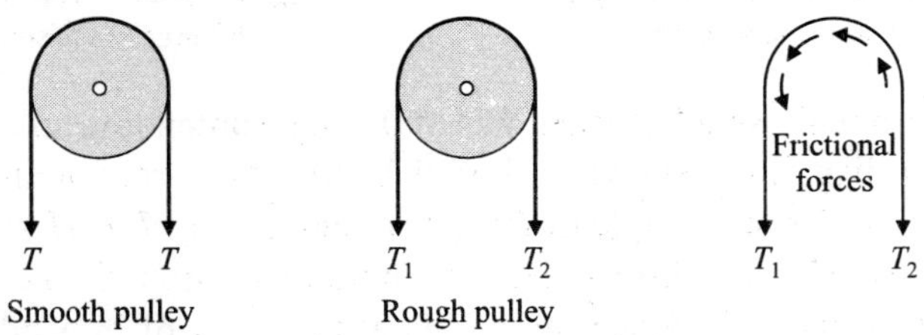

Fig. 7.15(a) **Fig. 7.15(b)** **Fig. 7.15(c)**

Hence, whenever a flexible member like a string, a rope, a cable or a belt *moves* or *tends to move* over a pulley or a cylindrical drum, frictional forces are always developed between the contact surfaces [Fig. 7.15(b) and (c)], which tend to oppose the relative motion. This frictional force is termed as ***rope*** or ***belt friction***.

Unlike sliding friction, the belt friction is found *not* to be constant but to vary *exponentially* throughout the length of contact surface. As a result, the tension at both the ends of the rope or belt will not be equal. For impending motion of rope or belt in the *clockwise* direction, the tension on the right end will be greater than that on the left end, i.e., $T_2 > T_1$. The side of the belt with *greater* tension is termed as ***tight side*** and that with *lesser* tension is termed as ***slack side***. Throughout this book, we consider T_2 to be the tension on the tight side and T_1 to be the tension on the slack side. Hence, $T_2 > T_1$. The relationship between these two tensions can be derived as explained below in the Section 7.4.1.

This belt friction is utilized in ***transmitting power*** as in the case of ***belt drives*** and ***power absorption*** as in the case of *band brakes*. In belt drives, both the belt and the pulley rotate by friction developed between contact surfaces. As a result, power is transmitted from one pulley to another through the belt.

This is discussed in Sections 7.4.2–7.4.10. In the case of band brakes, the band remains fixed while the drum rotates. Due to the friction developed between the band and the drum, the drum can be stopped by tightening the band. This is discussed in Section 7.5.

7.4.1 Relationship Between Tensions on Tight Side and Slack Side

Consider a flat belt passing over a stationary cylindrical drum [Fig. 7.16(a)]. Let the arc length AB be the portion of the belt in contact with the drum subtending an angle β at the centre O of the drum. The angle β is also called ***contact*** or ***lap angle***.

Let the belt be at the point of impending motion in the clockwise direction. Hence, we know that tension on the tight side is greater than the tension on the slack side, i.e., $T_2 > T_1$.

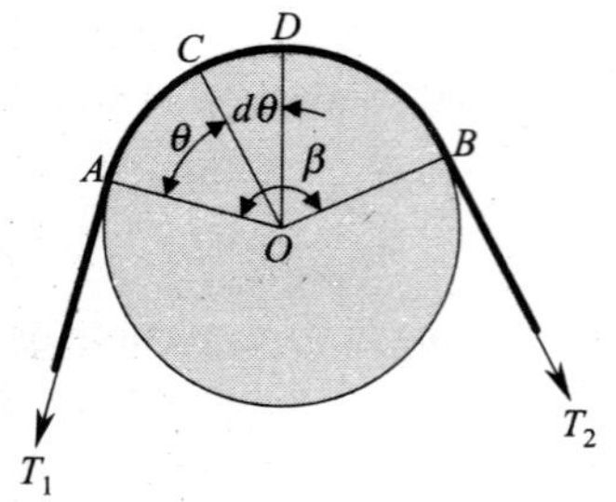

Fig. 7.16(a) Flat belt passing over a stationary drum

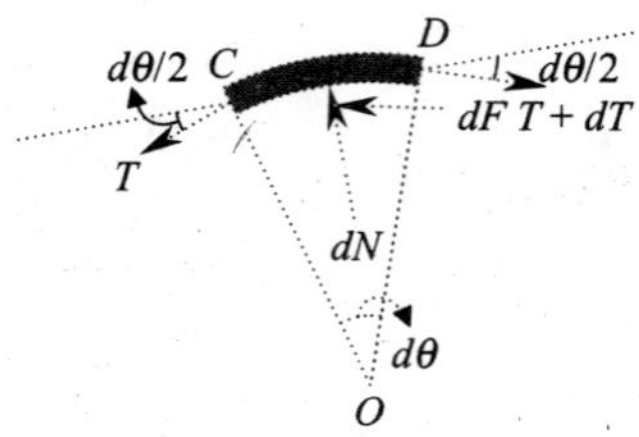

Fig. 7.16(b) Free-body diagram of infinitesimally small element

Consider an infinitesimally small element CD of the belt subtending an angle $d\theta$ at the centre O of the drum. Its free-body diagram is shown in Fig. 7.16(b). The forces acting on the element will be tension T at point C (point C is at an angle of θ from A) and tension $T + dT$ at the point D (point D is at an angle of $\theta + d\theta$ from A). Also, the drum will exert a normal reaction dN and frictional force dF against the direction of motion of the belt. Since the belt is at the point of impending motion, applying the conditions of equilibrium along the normal and tangential directions to the element, we get

$\Sigma F_t = 0 \Rightarrow$

$$(T + dT)\cos\frac{d\theta}{2} - T\cos\frac{d\theta}{2} - dF = 0 \tag{7.22}$$

For an infinitesimally small element, we can assume the contact surfaces to be plane and we can apply the coulomb's laws of dry friction. Hence, $dF = \mu_s dN$. Therefore, the above equation can be written as

$$(T + dT)\cos\frac{d\theta}{2} - T\cos\frac{d\theta}{2} - \mu_s dN = 0$$

$$dT\cos\frac{d\theta}{2} - \mu_s dN = 0 \tag{7.23}$$

$\Sigma F_n = 0 \Rightarrow$

$$dN - (T + dT)\sin\frac{d\theta}{2} - T\sin\frac{d\theta}{2} = 0$$

$$dN - (2T + dT)\sin\frac{d\theta}{2} = 0 \tag{7.24}$$

Substituting the value of dN from the Eq. 7.24 in the Eq. 7.23, we get

$$dT\cos\frac{d\theta}{2} - \mu_s(2T + dT)\sin\frac{d\theta}{2} = 0 \tag{7.25}$$

We know that as $d\theta \to 0$, $\cos\frac{d\theta}{2} \to 1$ and $\sin\frac{d\theta}{2} \to \frac{d\theta}{2}$. Hence,

$$dT - \mu_s(2T + dT)\frac{d\theta}{2} = 0 \tag{7.26}$$

Neglecting the product of differentials $dTd\theta$ as compared to the first-order differentials,

$$dT - \mu_s T d\theta = 0$$

or,

$$\frac{dT}{T} = \mu_s d\theta \tag{7.27}$$

Upon integration between limits,

$$\int_{T_1}^{T_2}\frac{dT}{T} = \mu_s\int_0^{\beta} d\theta$$

$$\ln T_2 - \ln T_1 = \mu_s\beta$$

or,

$$\ln\left(\frac{T_2}{T_1}\right) = \mu_s\beta \tag{7.28}$$

The above equation can also be written in the form

$$\frac{T_2}{T_1} = e^{\mu_s\beta} \tag{7.29}$$

Corollary Though we have derived the above formulas considering a ***flat belt*** passing over a ***fixed*** drum, they apply equally well to problems involving belts passing over ***moving*** drum or ***ropes*** wrapped around a ***post*** or a ***capstan***. As the frictional resistance depends on the angle of contact β, the frictional resistance can be increased by winding the rope with more than one turn as in the case of capstans. If the rope makes n number of revolutions, the contact angle β is $n(2\pi)$ radian.

Note: The following points must be kept in mind while applying the formulas (7.28 and 7.29):

(i) The tension on the tight side T_2 is always greater than the tension T_1 on the slack side, i.e., $T_2 > T_1$.

(ii) The contact angle β is always expressed in ***radians***.

(iii) Since $\frac{T_2}{T_1}$ is independent of the radius of the pulley, the formula (7.29) is applicable to non-circular sections also with the total contact angle being β.

Example 7.12 Find the angle of contact in the following cases:

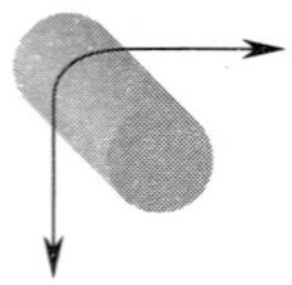

Fig. 7.17(a)

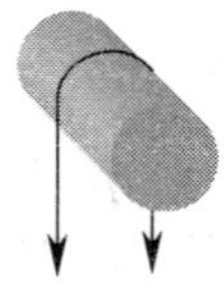

Fig. 7.17(b)

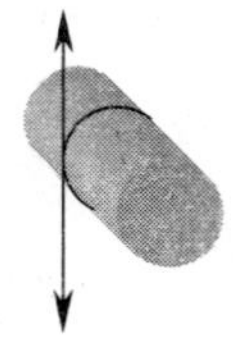

Fig. 7.17(c)

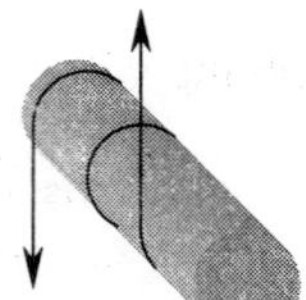

Fig. 7.17(d)

Solution We know that for one complete revolution, the angle traced is 2π radian. Hence, in the four cases mentioned, we can determine the angle of contact as follows:

(a) Since the rope is in contact with the cylinder for ¼th of a revolution, the angle of contact, $\beta = \frac{\pi}{2}$ radian.

(b) Here, the rope is in contact with the cylinder for ½ of a revolution. Hence, $\beta = \pi$ radian.

(c) Here, the rope is in contact with the cylinder for a full revolution. Hence, $\beta = 2\pi$ radian.

(d) Here, the rope makes two complete revolutions. Hence, $\beta = 4\pi$ radian.

Example 7.13 A rope is wrapped three and a half times around a cylinder. Determine the force T exerted on the free end of the rope that is required to just support a 5 kN weight. The coefficient of friction between the rope and the cylinder is 0.2.

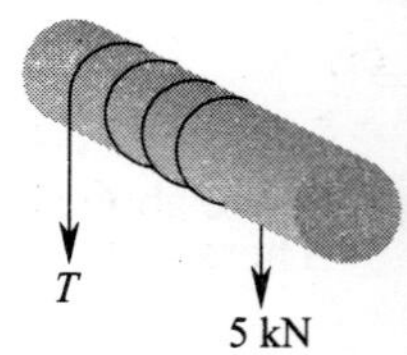

Fig. 7.18

Solution *Given data*

Coefficient of friction, $\mu_s = 0.2$

Number of turns, $n = 3.5$

Contact angle, $\beta = 3.5(2\pi) = 7\pi$ rad

Since T is the force required *to just support* the weight, we know that this force is less than the weight supported, i.e., $T < 5$ kN. Therefore, we take by the convention mentioned:

$$T_2 = 5 \text{ kN} \quad \text{and} \quad T_1 = T$$

At the point of impending motion, the ratio of tensions on the tight and slack sides is given as

$$\frac{T_2}{T_1} = e^{\mu_s \beta}$$

$$\frac{5}{T} = e^{(0.2)(7\pi)}$$

$$\Rightarrow \quad T = 0.0615 \text{ kN (or) } 61.5 \text{ N}$$

Example 7.14 A 150 kg block is supported by a rope, which is wrapped 1½ times around a horizontal rod. If the coefficient of static friction between the rope and the rod is 0.16, determine the range of values of P for which equilibrium is maintained.

Solution *Given data*

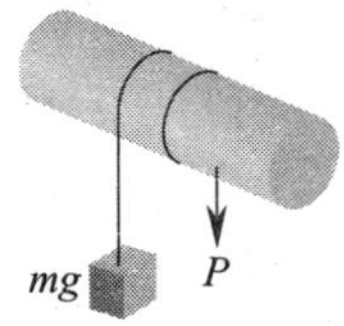

Fig. 7.19

Mass of block, $m = 150$ kg

Contact angle, $\beta = 1.5 \times 2\pi = 3\pi$ rad

Coefficient of static friction, $\mu_s = 0.16$

We encounter two different cases, namely, one in which the block is about to move downwards and the other in which the block is about to move upwards. In the former case, since mg is greater, we take T_2 = mg and $T_1 = P$; and in the latter case, P is greater. Hence, we take $T_2 = P$ and $T_1 = mg$.

We know that at the point of impending motion,

$$\frac{T_2}{T_1} = e^{\mu_s \beta}$$

$$= e^{(0.16)(3\pi)} = 4.52$$

Case I $\quad T_2 = mg$

$\Rightarrow \quad T_1 = P = \dfrac{T_2}{e^{\mu_s \beta}}$

$$= \frac{150 \times 9.81}{4.52} = 325.55 \text{ N}$$

Case II $\quad T_1 = \text{mg}$

$\Rightarrow \quad T_2 = P = T_1\, e^{\mu_s \beta}$

$$= 150 \times 9.81 \times 4.52 = 6651.18 \text{ N}$$

Therefore, range of P for which equilibrium is maintained is

$$325.55 \le P \le 6651.18 \text{ N}$$

Example 7.15 A block of weight 3000 N is prevented from sliding down an inclined plane by another block of weight W placed over it. The top block is tied to the bottom block by a rope passing over a fixed drum. If coefficient of friction between the rope and the fixed drum and that between all contact surfaces is 0.15, determine the minimum value of W to prevent sliding.

Solution The free-body diagrams of the blocks and the pulley with rope are shown below. As the bottom block tries to move downwards, the frictional forces acting on both of its faces are such as to oppose this motion. Hence, they are shown acting upwards. For the top block, it will act downwards as per Newton's third law of motion. From the Fig. 7.20, we see that the inclination of the plane is $\tan\theta$ = 3/4. Therefore, $\sin\theta = 3/5$ and $\cos\theta = 4/5$.

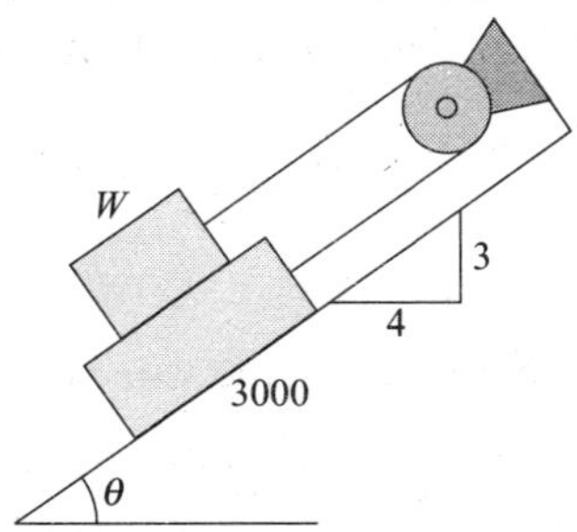

Fig. 7.20

For the top block

At the point of impending motion, applying the conditions of equilibrium along X and Y directions,

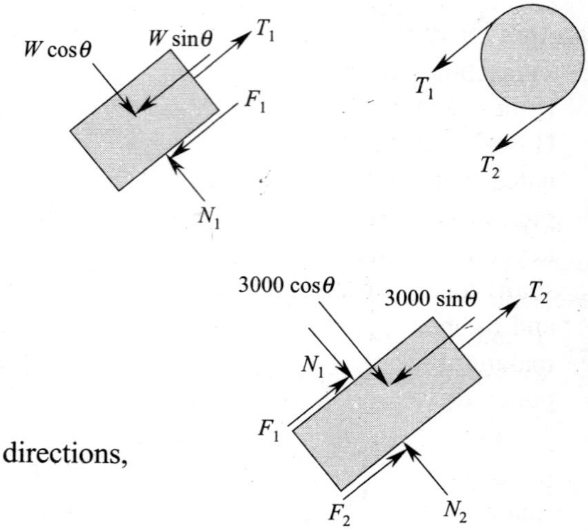

Fig. 7.20(a)

$\Sigma F_y = 0 \Rightarrow$

$$N_1 - W\cos\theta = 0$$

$\therefore$ $$N_1 = W\cos\theta \quad \text{(a)}$$

Hence, $F_1 = \mu N_1 = \mu W\cos\theta$ (b)

$\Sigma F_x = 0 \Rightarrow$

$$T_1 - W\sin\theta - F_1 = 0$$

or, $$T_1 = W(\sin\theta + \mu\cos\theta)$$

$$= W[3/5 + (0.15)(4/5)] = 0.72\ W \quad \text{(c)}$$

For the bottom block

Applying the conditions of equilibrium along X and Y directions,

$\Sigma F_y = 0 \Rightarrow$

$$N_2 - N_1 - 3000\cos\theta = 0$$

$\therefore$ $N_2 = (W + 3000)\cos\theta$ [from the equation (a)]

Hence, $F_2 = \mu N_2 = \mu(W + 3000)\cos\theta$ (d)

$\Sigma F_x = 0 \Rightarrow$

$$T_2 + F_1 + F_2 - 3000\sin\theta = 0$$

$\Rightarrow$ $$T_2 = 3000\sin\theta - F_1 - F_2$$

Substituting the values of F_1 and F_2 from equations (b) and (d), we have

$$T_2 = 3000\sin\theta - \mu W\cos\theta - \mu(W + 3000)\cos\theta$$

$$= 3000(\sin\theta - \mu\cos\theta) - 2\mu W\cos\theta$$

$$= 3000[3/5 - (0.15)(4/5)] - 2(0.15) \times W \times (4/5)$$

$$= 1440 - 0.24W \quad \text{(e)}$$

For pulley and rope

Since the 3000 N block is about to move downwards, the impending motion of the rope is from top to bottom. Therefore, the bottom end is tighter and the top end is slacker. Hence, $T_2 > T_1$. Therefore,

$$\frac{T_2}{T_1} = e^{\mu\beta}$$

$$= e^{(0.15)(\pi)} = 1.602 \quad \text{(the contact angle, } \beta = \pi \text{ radian)} \quad \text{(f)}$$

Substituting the values of T_1 and T_2 from equations (c) and (e) in the above equation,

$$\frac{1440 - 0.24W}{0.72W} = 1.602$$

$\Rightarrow$ $$1.393\,W = 1440$$

$$W = 1033.74 \text{ N}$$

7.4.2 Belt Drives

In machines, usually we need to transmit power and rotary motion from ***engine shaft*** to ***machine shaft***. If these shafts are close to each other then gears can be used for transmission of power and motion. However, if they are not close to each other then flexible belts made of leather, fabric, rubber-impregnated fabric or synthetics are used for transmitting power and motion. These belts normally pass over pulleys keyed to the shafts. The pulley keyed to the engine shaft is called the ***driver pulley*** and that keyed to the machine shaft is called the ***driven pulley***. As the driver pulley (connected to the engine shaft) rotates, it carries with it the belt embracing over it due to friction developed between the pulley and the belt. The belt in turn carries with it the driven pulley (connected to the machine shaft) thus making it rotate. This way, power is transmitted from the engine shaft to the machine shaft. Belt and pulley drives are also used to either increase or decrease the rotational speed of the driven shafts.

Figures 7.21 and 7.22 show simple belt drives. Depending upon the direction in which the driven pulley is required to rotate, we can have two types of simple drives. If the driven pulley is required to rotate in the same direction as that of the driver pulley then it is termed an ***open belt drive*** (refer Fig. 7.21) and if it is required to rotate in the opposite direction then it is termed a ***cross-belt drive*** (refer Fig. 7.22).

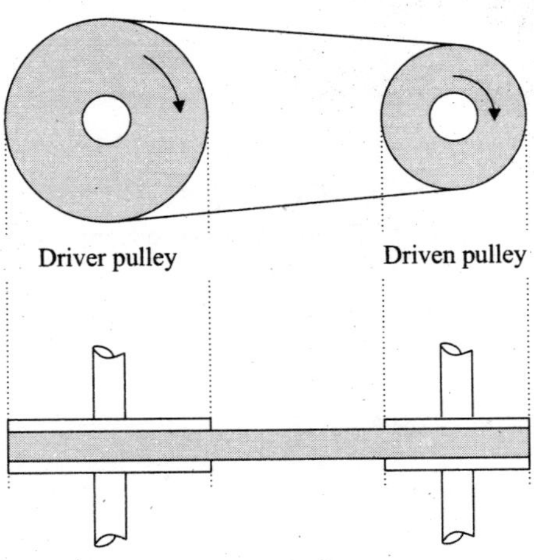

Fig. 7.21 Simple belt drive: open belt

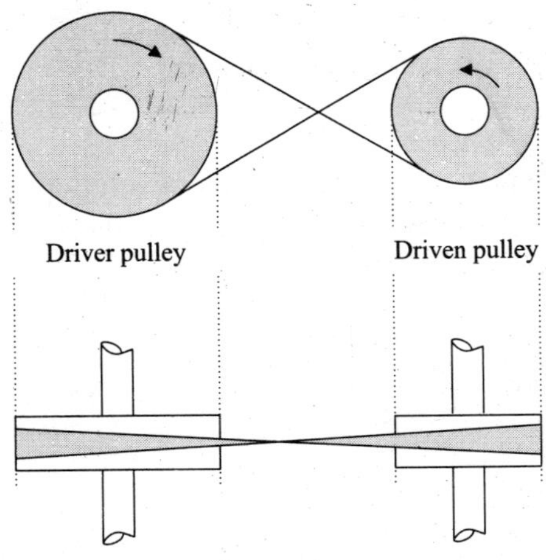

Fig. 7.22 Simple belt drive: cross belt

If the engine shaft and machine shaft are quite far away then intermediate shafts are provided, known as ***secondary shafts***. The adjacent shafts are connected by simple drives as shown in Fig. 7.23. Such a type of drive is known as **compound drive**.

7.4.3 Speed Ratio of Driver Pulley and Driven Pulley

Simple Belt Drive Consider a simple belt drive. Let d_1 and d_2 be diameters of the driver and driven pulleys respectively. Let their respective speeds be N_1 and N_2 rpm (revolutions per minute). Since the angle traced in one revolution is 2π radian, the respective angular velocities ω_1 and ω_2 of the driver and driven pulleys in rad/s are given as

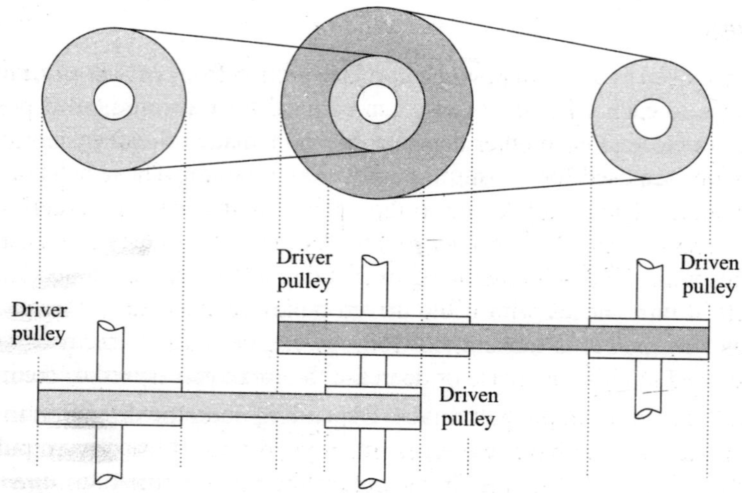

Fig. 7.23 Compound belt drive

$$\omega_1 = \frac{2\pi N_1}{60} \tag{7.30}$$

and

$$\omega_2 = \frac{2\pi N_2}{60} \tag{7.31}$$

Note that as N is expressed in revolution per minute, it is divided by 60 to convert it to per **second**. The tangential velocity of the pulley at its outer radius (rim) is given by the relation $r\omega$, where r is radius of the pulley. Therefore,

$$v_1 = \frac{d_1}{2}\omega_1 = \frac{d_1}{2}\frac{2\pi N_1}{60} = \frac{\pi d_1 N_1}{60} \tag{7.32}$$

and

$$v_2 = \frac{d_2}{2}\omega_2 = \frac{d_2}{2}\frac{2\pi N_2}{60} = \frac{\pi d_2 N_2}{60} \tag{7.33}$$

No slip If a firm grip exists between pulley and belt then there will not be any *slippage* between them during motion. Hence, the tangential velocity of the belt leaving the driver pulley will be the same as the velocity of the driver pulley, i.e., v_1. Similarly, the velocity (v_2) of the driven pulley will be the same as the velocity of the belt passing over it. Thus,

$$v_1 = v_2 \tag{7.34}$$

$$\Rightarrow \quad \frac{\pi d_1 N_1}{60} = \frac{\pi d_2 N_2}{60}$$

$$\therefore \quad \frac{N_2}{N_1} = \frac{d_1}{d_2}$$

or,

$$N_2 = \frac{d_1}{d_2}N_1 \tag{7.35}$$

From the above expression, we see that *the speed is inversely proportional to the diameter.* Hence, the diameter of the driven pulley is normally kept smaller than the diameter of the driver pulley to increase the speed of the machine shaft.

Thickness of belt In the above derivation, we considered the velocity of the outer radius (rim) of the pulley neglecting thickness of the belt. However, if the thickness of the belt is also considered, then we must determine the speed of the belt at its mean radius, i.e.,

$$v_1 = \left(\frac{d_1}{2} + \frac{t}{2}\right)\omega_1 \tag{7.36}$$

and

$$v_2 = \left(\frac{d_2}{2} + \frac{t}{2}\right)\omega_2 \tag{7.37}$$

Therefore,

$$\frac{N_2}{N_1} = \frac{d_1 + t}{d_2 + t} \tag{7.38}$$

With slip If a firm grip does not exist between pulley and belt then there is a possibility of slippage occurring between them. As a result, when the driver pulley rotates, it does not carry along with it the belt. Hence, the tangential velocity of the belt will be less than that of the driver pulley. This difference is termed as *slip* and it is expressed as a percentage. If p_1 is the percentage slip of the belt embracing the driver pulley then the speed of the belt leaving the driver pulley is given as:

$$v_1 = \frac{\pi d_1 N_1}{60}\left(1 - \frac{p_1}{100}\right) \tag{7.39}$$

Similarly, the belt will move forward without carrying the driven pulley along with it. As a result, the speed of the driven pulley will be less than that of the belt. If p_2 is the percentage slip of the belt at the driven pulley then the speed of the driven pulley will be further reduced:

$$v_2 = v_1\left(1 - \frac{p_2}{100}\right)$$

$$= \frac{\pi d_1 N_1}{60}\left(1 - \frac{p_1}{100}\right)\left(1 - \frac{p_2}{100}\right)$$

$$= \frac{\pi d_1 N_1}{60}\left[1 - \frac{(p_1 + p_2)}{100}\right] \quad \text{(approximately)} \tag{7.40}$$

Since $v_2 = \dfrac{\pi d_2 N_2}{60}$, we can write

$$\frac{\pi d_2 N_2}{60} = \frac{\pi d_1 N_1}{60}\left[1 - \frac{(p_1 + p_2)}{100}\right]$$

Therefore,

$$\frac{N_2}{N_1} = \frac{d_1}{d_2}\left[1 - \frac{(p_1 + p_2)}{100}\right] \tag{7.41}$$

where $(p_1 + p_2)$ is termed as ***percentage slip*** in the ***drive***.

Compound Belt Drive Consider a compound belt drive (refer Fig. 7.23) in which the first belt passes over the driver pulley mounted over the engine shaft and driven pulley mounted over the secondary shaft. Let the respective diameters of the pulleys be d_1 and d_2. Then their speed ratios are given by

$$N_1 d_1 = N_2 d_2$$

or,
$$\frac{N_2}{N_1} = \frac{d_1}{d_2} \tag{7.42}$$

The second belt passes over the driver pulley mounted over the secondary shaft and driven pulley mounted over the machine shaft. Let the respective diameters of the pulleys be d_3 and d_4. Then their speed ratios are given by

$$N_3 d_3 = N_4 d_4$$

or,
$$\frac{N_4}{N_3} = \frac{d_3}{d_4} \tag{7.43}$$

Since the speed of the pulleys mounted on the secondary shaft are same, i.e., $N_2 = N_3$,

$$\frac{N_4}{N_1} = \frac{N_2}{N_1}\frac{N_4}{N_3} = \frac{d_1}{d_2}\frac{d_3}{d_4} \tag{7.44}$$

From the above expression, we see that diameters of the driver pulleys appear in the numerator and that of the driven pulleys appear in the denominator. Hence, in general, for a compound drive with any number of secondary shafts, we can write that

$$\frac{\text{speed of machine shaft}}{\text{speed of engine shaft}} = \frac{\text{product of diameters of all driver pulleys}}{\text{product of diameters of all driven pulleys}}$$

If slip between belt and pulley is considered then

$$N_2 = \frac{d_1}{d_2} N_1 \left(1 - \frac{p_1}{100}\right) \tag{7.45}$$

and
$$N_4 = \frac{d_3}{d_4} N_3 \left(1 - \frac{p_2}{100}\right) \tag{7.46}$$

where p_1 and p_2 are percentage slips in the two simple drives.

Since $N_2 = N_3$,

$$N_4 = \frac{d_1}{d_2}\frac{d_3}{d_4} N_1 \left(1 - \frac{p_1}{100}\right)\left(1 - \frac{p_2}{100}\right)$$

$$\approx \frac{d_1}{d_2}\frac{d_3}{d_4} N_1 \left(1 - \frac{p}{100}\right) \tag{7.47}$$

where $p = p_1 + p_2$ is the percentage slip of the compound drive.

Example 7.16 An engine drives a shaft by means of a belt. The diameter of driver pulley is 1.5 m and that of driven pulley is 1 m. If the speed of the engine shaft is 200 r.p.m, determine the speed of the machine shaft (i) when there is no slip between the belt and the pulleys, and (ii) when there is a slip of 3%.

Solution *Given data*

$$d_1 = 1.5 \text{ m} \quad N_1 = 200 \text{ r.p.m}$$
$$d_2 = 1 \text{ m}$$

(i) *When there is no slip*

When there is no slip, the speed ratio of the pulleys is given as

$$\frac{N_2}{N_1} = \frac{d_1}{d_2}$$

$$\Rightarrow \quad N_2 = \frac{d_1}{d_2} N_1 = \frac{1.5}{1} \times 200 = 300 \text{ r.p.m}$$

(ii) *When there is a slip of 3%*

When there is a slip between the belt and the pulleys, the speed ratio of the pulleys is given as

$$\frac{N_2}{N_1} = \frac{d_1}{d_2}\left(1 - \frac{p}{100}\right)$$

$$\Rightarrow \quad N_2 = 291 \text{ r.p.m}$$

Example 7.17 An engine drives a shaft by means of a belt of thickness 6 mm. The driving pulley of the engine is 600 mm and that of the shaft is 400 mm in diameter. If the engine runs at 200 r.p.m, determine the speed of the shaft (i) neglecting belt thickness, and (ii) considering the belt thickness.

Solution *Given data*

$$d_1 = 600 \text{ mm} \quad N_1 = 200 \text{ r.p.m}$$
$$d_2 = 400 \text{ mm} \quad t = 6 \text{ mm}$$

(i) *Neglecting belt thickness*

The speed ratio of the engine and shaft is given as

$$\frac{N_2}{N_1} = \frac{d_1}{d_2}$$

$$\Rightarrow \quad N_2 = \frac{d_1}{d_2} N_1 = \frac{600}{400} \times 200 = 300 \text{ r.p.m}$$

(ii) *Considering the belt thickness*

The speed ratio of the engine and shaft is given as

$$\frac{N_2}{N_1} = \frac{d_1 + t}{d_2 + t}$$

$$\Rightarrow \quad N_2 = \left[\frac{600 + 6}{400 + 6}\right] 200 = 298.52 \text{ r.p.m}$$

Example 7.18 An engine shaft running at 200 r.p.m drives a secondary shaft by means of a belt. The diameters of the engine pulley and driven pulley are respectively 600 mm and 400 mm. Another belt connects a pulley of 500 mm diameter mounted on the same secondary shaft to the machine shaft whose diameter is 300 mm. Determine the speed of the machine shaft (i) when there is no slip, and (ii) when there is a slip of 3%.

Solution *Given data*

Diameters of driver pulleys, $d_1 = 600$ mm and $d_3 = 500$ mm

Diameters of driven pulleys, $d_2 = 400$ mm and $d_4 = 300$ mm

Speed of engine shaft, $N_1 = 200$ r.p.m

(i) *When there is no slip*

Speed of machine shaft is given as

$$N_4 = \frac{d_1}{d_2}\frac{d_3}{d_4}N_1$$

$$= \frac{(600)(500)}{(400)(300)}200 = 500 \text{ r.p.m}$$

(ii) *When there is a slip of 3%*

Speed of machine shaft is given as

$$N_4 = \frac{d_1}{d_2}\frac{d_3}{d_4}N_1\left(1 - \frac{p}{100}\right)$$

$$= 500\left(1 - \frac{3}{100}\right) = 485 \text{ r.p.m}$$

7.4.4 Length of Belt Required

Once the diameters of the pulleys of the driver and driven shafts, and the distance between the shafts are decided, we must determine the length of belt required to embrace the pulleys. Since this length will be different for open and cross belt drives, they are explained below separately.

Open Belt Drive Consider an open belt drive as shown in Fig. 7.24. Let r_1 and r_2 be the radii of the driver and driven pulleys respectively and l be the distance between their centres.

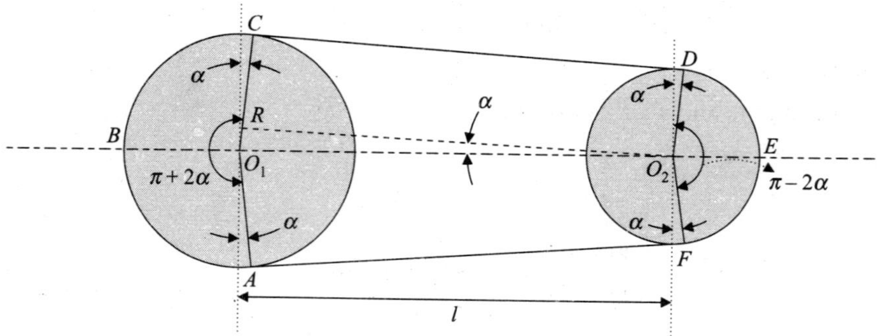

Fig. 7.24 Length of belt in an open belt drive

Since the pulleys are circular, the belt will be tangential to the pulleys at points A, C, D and F, where the belt leaves contact with the pulleys. Draw O_2R parallel to CD. Let $\angle RO_2O_1 = \alpha$. Then from the geometry, we see that the contact angle for driver pulley is

$$\angle AO_1C = (\pi + 2\alpha) \text{ radian} \tag{7.48}$$

and that for driven pulley is

$$\angle DO_2F = (\pi - 2\alpha) \text{ radian} \tag{7.49}$$

Since $O_1O_2 = l$, $O_2R = CD = l \cos \alpha$ (7.50)

The total length L of the belt required is then given as

$$\begin{aligned} L &= \text{arc length } ABC + (\overline{CD}) + \text{arc length } DEF + (\overline{FA}) \\ &= \text{arc length } ABC + 2(\overline{CD}) + \text{arc length } DEF \quad [\text{since } \overline{CD} = \overline{FA}] \\ &= (\pi + 2\alpha)r_1 + 2l\cos\alpha + (\pi - 2\alpha)r_2 \\ &= \pi(r_1 + r_2) + 2\alpha(r_1 - r_2) + 2l\cos\alpha \end{aligned} \tag{7.51}$$

For small values of α, we know,

$$\begin{aligned} \alpha = \sin\alpha &= \frac{O_1R}{O_1O_2} \\ &= \frac{O_1C - RC}{O_1O_2} = \frac{O_1C - O_2D}{O_1O_2} = \frac{r_1 - r_2}{l} \end{aligned} \tag{7.52}$$

Also,

$$\begin{aligned} \cos\alpha &= 1 - \frac{\alpha^2}{2!} + \frac{\alpha^4}{4!} - \ldots \\ &\approx 1 - \frac{\alpha^2}{2!}, \text{ since } \alpha \text{ is a small quantity.} \end{aligned}$$

Substituting the value of α from the Eq. 7.52, we get

$$\cos\alpha = 1 - \frac{\alpha^2}{2} = 1 - \frac{(r_1 - r_2)^2}{2l^2} \tag{7.53}$$

Therefore, the total length of the belt can be written as

$$\begin{aligned} L &= \pi(r_1 + r_2) + 2\left(\frac{r_1 - r_2}{l}\right)(r_1 - r_2) + 2l\left[1 - \frac{(r_1 - r_2)^2}{2l^2}\right] \\ &= \pi(r_1 + r_2) + 2\frac{(r_1 - r_2)^2}{l} + 2l - \frac{(r_1 - r_2)^2}{l} \end{aligned}$$

$$L = 2l + \pi(r_1 + r_2) + \frac{(r_1 - r_2)^2}{l} \tag{7.54}$$

Cross-Belt Drive Consider a cross-belt drive as shown in Fig. 7.25. Let r_1 and r_2 be the radii of the driver and driven pulleys respectively and l be the distance between their centres.

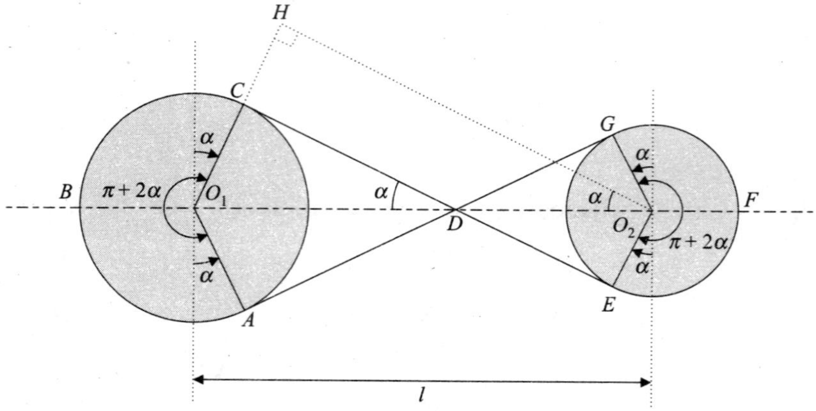

Fig. 7.25 Length of belt in a cross-belt drive

Since the pulleys are circular, the belt will be tangential to the pulleys at points A, C, E and G, where the belt leaves contact with the pulleys. Draw O_2H parallel to EC. If $\angle CDO_1 = \alpha$ then $\angle HO_2O_1 = \alpha$. From the geometry, we can see that the contact angle for driver pulley is

$$\angle AO_1C = (\pi + 2\alpha) \text{ radian} \tag{7.55}$$

and that for driven pulley is

$$\angle EO_2G = (\pi + 2\alpha) \text{ radian} \tag{7.56}$$

Since $O_1O_2 = l$, $CE = O_2H = l\cos\alpha$ (7.57)

The total length L of the belt required is then given as

$$L = \text{arc length } ABC + (\overline{CE}) + \text{arc length } EFG + (\overline{GA})$$

$$= \text{arc length } ABC + 2(\overline{CE}) + \text{arc length } EFG \quad [\text{since } \overline{CE} = \overline{GA}]$$

$$= (\pi + 2\alpha)r_1 + 2l\cos\alpha + (\pi + 2\alpha)r_2$$

$$= \pi(r_1 + r_2) + 2\alpha(r_1 + r_2) + 2l\cos\alpha \tag{7.58}$$

For small values of α, we know that,

$$\alpha = \sin\alpha = \frac{O_1H}{O_1O_2}$$

$$= \frac{O_1C + CH}{O_1O_2} = \frac{O_1C + O_2E}{O_1O_2} = \frac{r_1 + r_2}{l} \tag{7.59}$$

Also,

$$\cos\alpha = 1 - \frac{\alpha^2}{2!} + \frac{\alpha^4}{4!} - \dots$$

$$\approx 1 - \frac{\alpha^2}{2}, \text{ since } \alpha \text{ is a small quantity.}$$

Substituting the value of α from the Eq. 7.59,

$$\cos\alpha = 1 - \frac{\alpha^2}{2} = 1 - \frac{(r_1 + r_2)^2}{2l^2} \tag{7.60}$$

Therefore, the total length can be written as

$$L = \pi(r_1 + r_2) + 2\left(\frac{r_1 + r_2}{l}\right)(r_1 + r_2) + 2l\left[1 - \frac{(r_1 + r_2)^2}{2l^2}\right]$$

$$= \pi(r_1 + r_2) + 2\frac{(r_1 + r_2)^2}{l} + 2l - \frac{(r_1 + r_2)^2}{l}$$

$$L = 2l + \pi(r_1 + r_2) + \frac{(r_1 + r_2)^2}{l} \tag{7.61}$$

Corollary The length of the cross belt depends only on the ***sum*** of radii of the pulleys [i.e., $(r_1 + r_2)$], whereas in the case of open belts, the length depends on the ***sum*** and ***difference*** of the radii [i.e., $(r_1 + r_2)$ and $(r_1 - r_2)$]. Hence, for stepped pulleys (explained in the next section) it will be easy to determine the length of the belt required if a cross belt arrangement is used than an open belt arrangement.

Example 7.19 Two pulleys mounted respectively on engine shaft and machine shaft have diameters of 60 cm and 40 cm. If the shafts are 2.2 m apart, determine the angle of contact and the length of belt required to connect the two pulleys for (a) an open belt arrangement, and (b) a cross-belt arrangement.

Solution *Given data*

Distance between centres of the pulleys, $l = 2.2$ m

Radius of larger pulley, $r_1 = 0.3$ m

Radius of smaller pulley, $r_2 = 0.2$ m

(a) *Open belt arrangement*

We know that for an open belt drive,

$$\sin\alpha = \frac{r_1 - r_2}{l} = \frac{0.3 - 0.2}{2.2}$$

$$\Rightarrow \qquad \alpha = 2.61°$$

Therefore, the contact angle is

for driver pulley, $\pi + 2\alpha = 180° + 2(2.61°) = 185.22°$

and for driven pulley, $\pi - 2\alpha = 180° - 2(2.61°) = 174.78°$

The length of belt required for an open belt drive is given as

$$L = 2l + \pi(r_1 + r_2) + \frac{(r_1 - r_2)^2}{l}$$

$$= 2(2.2) + \pi(0.3 + 0.2) + \frac{(0.3 - 0.2)^2}{2.2} = 5.975 \text{ m}$$

(b) *Cross-belt arrangement*

We know that for a cross belt drive,

$$\sin\alpha = \frac{r_1 + r_2}{l} = \frac{0.3 + 0.2}{2.2}$$

$$\Rightarrow \qquad \alpha = 13.14°$$

Therefore, the contact angle for both the pulleys is $\pi + 2\alpha = 180° + 2(13.14°) = 206.28°$.

The length of belt required for a cross belt drive is given as

$$L = 2l + \pi(r_1 + r_2) + \frac{(r_1 + r_2)^2}{l}$$

$$= 2(2.2) + \pi(0.3 + 0.2) + \frac{(0.3 + 0.2)^2}{2.2} = 6.084 \text{ m}$$

7.4.5 Stepped Pulleys and Speed Cones

Sometimes the machine shafts may be required to run at different speeds. For which, coupling it with different engine shafts is not possible. To overcome this difficulty, a number of pulleys of different diameters are mounted in stepped fashion over the same shaft and such an arrangement is known as ***stepped pulley*** or ***cone pulley***. This way the machine shaft can be run at different speeds by coupling it with the same engine shaft. Again, this requires different lengths of belts to run at different speeds. This can also be overcome by providing stepped pulleys for the engine shaft also, but in the opposite direction such that the largest pulley of the machine shaft can be coupled with the smallest pulley of the engine shaft and vice versa.

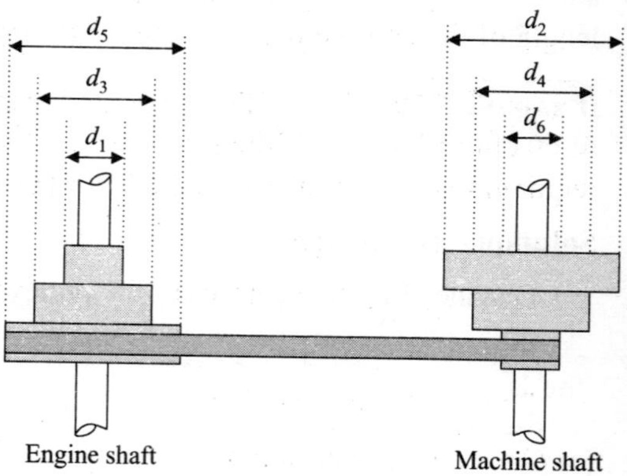

Fig. 7.26 Stepped pulleys

Let the stepped diameters of the pulleys mounted on the engine shaft be d_1, d_3 and d_5; and the corresponding diameters of the stepped pulleys mounted on the machine shaft be d_2, d_4 and d_6. Suppose the speed of the engine shaft is N r.p.m. Then the variable speeds of the machine shaft can be determined as follows:

(i) When pulleys with diameters d_1 and d_2 are coupled then

$$(N_2)_1 = \frac{d_1}{d_2}N \qquad (7.62)$$

(ii) When pulleys with diameters d_3 and d_4 are coupled then

$$(N_2)_2 = \frac{d_3}{d_4}N \qquad (7.63)$$

(iii) When pulleys with diameters d_5 and d_6 are coupled then

$$(N_2)_3 = \frac{d_5}{d_6}N \qquad (7.64)$$

Corollary Looking at the expression for length of belt required in a cross-belt drive [refer Eq. 7.61], we can see that the same belt can be used for all the speeds, provided the sums of radii or diameters for each coupling remain constant, i.e.,

$$(d_1 + d_2) = (d_3 + d_4) = (d_5 + d_6) \tag{7.65}$$

However, this cannot be achieved in the case of open belt drives, as the length does not depend only on the sum of the radii but also on their difference [refer Eq. 7.54].

Example 7.20 A shaft, which rotates at a constant speed of 150 r.p.m, is connected by belting to a parallel shaft, 2 m apart, which has to run at 70, 90 and 110 r.p.m. The smallest pulley on the driver shaft is 140 mm in radius. Determine the remaining radii of the two stepped pulleys for (i) a crossed belt arrangement, and (ii) an open belt arrangement. Neglect belt thickness and slip.

Solution *Given data*

Speed of driver shaft, $N_1 = 150$ r.p.m

Radius of smallest pulley on driver shaft, $r_1 = 140$ mm

Distance between the shafts, $l = 2$ m

Speeds at which driven shaft must be run,

$$(N_2)_1 = 70 \text{ r.p.m}$$

$$(N_2)_2 = 90 \text{ r.p.m}$$

$$(N_2)_3 = 110 \text{ r.p.m}$$

Note that the driven shaft has to be run at speeds less than the speed of the driver shaft.

(i) *Cross-belt drive*

As the length of the belt remains constant, the smallest pulley on the driver shaft must be connected to the largest pulley of the driven shaft. In which case, the speed of the driven shaft is the least, i.e., 70 r.p.m.

$$\frac{(N_2)_1}{N_1} = \frac{d_1}{d_2} = \frac{r_1}{r_2}$$

$$\Rightarrow \quad \frac{r_1}{r_2} = \frac{70}{150}$$

$$\therefore \quad r_2 = 300 \text{ mm}$$

Hence, the sum of radii of the two pulleys is

$$r_1 + r_2 = 440 \text{ mm}$$

For a cross-belt drive, the sums of radii of the pulleys coupled remain constant. Hence,

$$r_1 + r_2 = r_3 + r_4 = r_5 + r_6 = 440 \text{ mm}$$

For the second pair of pulleys,

$$\frac{(N_2)_2}{N_1} = \frac{d_3}{d_4} = \frac{r_3}{r_4}$$

$$\Rightarrow \qquad \frac{r_3}{r_4} = \frac{90}{150} = 0.6$$

$$\therefore \qquad r_3 = 0.6r_4 \qquad \text{(a)}$$

Also, $\qquad r_3 + r_4 = 440 \qquad$ (b)

Solving for r_3 and r_4 from equations (a) and (b), we get

$$r_3 = 165 \text{ mm}$$

and $$r_4 = 275 \text{ mm}$$

For the third pair of pulleys,

$$\frac{(N_2)_3}{N_1} = \frac{d_5}{d_6} = \frac{r_5}{r_6}$$

$$\Rightarrow \qquad \frac{r_5}{r_6} = \frac{110}{150} = \frac{11}{15}$$

$$\therefore \qquad r_5 = \frac{11}{15}r_6 \qquad \text{(c)}$$

Also, $\qquad r_5 + r_6 = 440 \qquad$ (d)

Solving for r_5 and r_6 from equations (c) and (d), we get

$$\frac{11}{15}r_6 + r_6 = 440$$

$$\therefore \qquad r_6 = 253.85 \text{ mm}$$

and $$r_5 = 186.15 \text{ mm}$$

(ii) *Open belt drive*

For a cross-belt drive, the length of the belt is the same provided the sums of radii of coupled pulleys remain constant, whereas for an open belt drive, it is not so. Hence, we must determine the length of the belt and accordingly, the radii of stepped pulleys must be determined. The smallest radius of the driver pulley is 140 mm and the corresponding radius of the driven pulley is same as that determined for a cross-belt drive, i.e.,

$$r_2 = 300 \text{ mm}$$

The length L required for an open belt drive is obtained as

$$L = 2l + \pi(r_1 + r_2) + \frac{(r_1 - r_2)^2}{l}$$

$$= 2(2) + \pi(0.14 + 0.3) + \frac{(0.14 - 0.3)^2}{2} = 5.395 \text{ m}$$

For the second pair of pulleys, we know from the equation (a), $r_3 = 0.6r_4$. Since the length of the belt should remain the same for this pair of radii also,

$$L = 2l + \pi(r_3 + r_4) + \frac{(r_3 - r_4)^2}{l}$$

$$5.395 = 2(2) + \pi(0.6r_4 + r_4) + \frac{(0.6r_4 - r_4)^2}{2}$$

$$0.08r_4^2 + 5.027r_4 - 1.395 = 0$$

Solving the quadratic equation, we get

$$r_4 = 0.2763 \text{ m (or) } 276.3 \text{ mm}$$

and $$r_3 = 165.8 \text{ mm}$$

For the third pair of pulleys, we know from the equation (c), $r_5 = \frac{11}{15}r_6$. Since the length of the belt should remain the same for this pair of radii also,

$$L = 2l + \pi(r_5 + r_6) + \frac{(r_5 - r_6)^2}{l}$$

$$5.395 = 2(2) + \pi\left(\frac{11}{15}r_6 + r_6\right) + \frac{[(11/15)r_6 - r_6]^2}{2}$$

$$0.0356r_6^2 + 5.445r_6 - 1.395 = 0$$

Solving the quadratic equation, we get

$$r_6 = 0.2558 \text{ m (or) } 255.8 \text{ mm}$$

and $$r_5 = 187.6 \text{ mm}$$

7.4.6 Power Transmitted by a Belt in a Belt Drive

The relationship between the tensions on the tight and slack sides as derived in the Eq. 7.29 is also applicable to belt drives, even though we considered the drum to be stationary while deriving the equation. Hence:

$$\frac{T_2}{T_1} = e^{\mu_s \beta} \qquad (7.66)$$

where μ_s is the coefficient of static friction.

Points to consider while applying the above formula

1. When the driver pulley rotates in the clockwise direction, due to friction, it pulls the lower portion of the belt and delivers it to the upper side. Hence, the tension on the lower side will be more than that on the upper side. Accordingly, the tension on lower side (tight side) is taken as T_2 and that on the upper side (slack side) is taken as T_1.
2. Since the lap angle for the smaller-diameter pulley is less than that for the larger-diameter pulley, slippage is likely to occur at this pulley. Hence, β is always chosen as the lap angle for the *smaller*-diameter pulley. [Note that the lap angle for a cross belt drive is the same for both the pulleys].

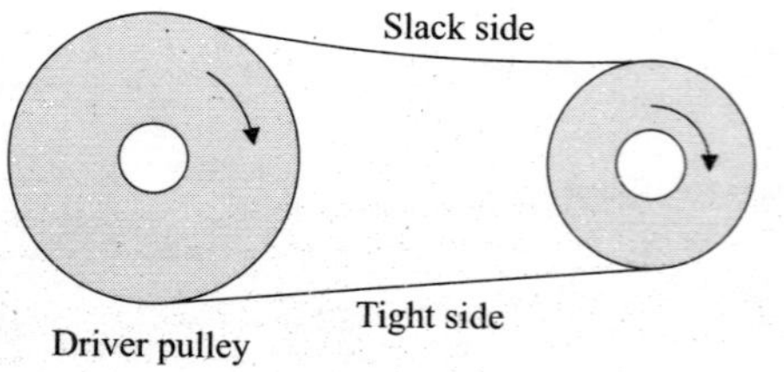

Fig. 7.27(a) Open belt

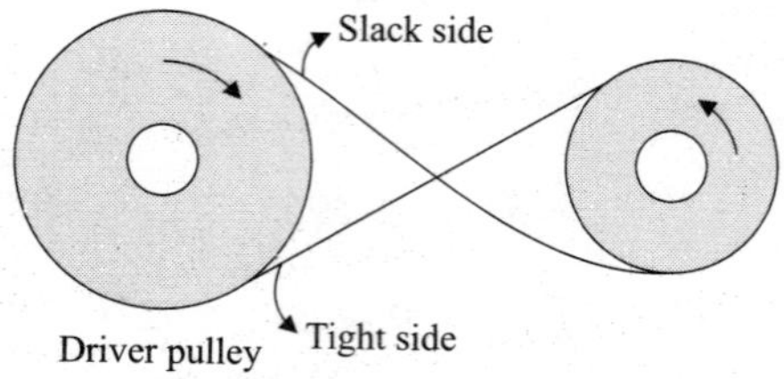

Fig. 7.27(b) Cross belt

Torque exerted on the driver pulley is

$$\tau_1 = T_2 r_1 - T_1 r_1 = (T_2 - T_1) r_1 \tag{7.67}$$

Similarly, torque exerted on the driven pulley is

$$\tau_2 = T_2 r_2 - T_1 r_2 = (T_2 - T_1) r_2 \tag{7.68}$$

where r_1 and r_2 are radii of the driver and driven pulleys respectively.

Power transmitted by the belt is given as the product of net force acting on the pulley and the speed of the belt, i.e.,

$$P = (T_2 - T_1)v \quad \text{Watt} \tag{7.69}$$

Example 7.21 The contact angle of a belt passing over a pulley of 600 mm diameter rotating at 250 r.p.m is 175°. If the coefficient of friction between the belt and the pulley is 0.25 and the maximum tension in the belt is 1500 N, determine (i) the torque exerted, and (ii) the power transmitted by the belt.

Solution *Given data*

Maximum tension, $T_2 = 1500$ N

Coefficient of friction, $\mu_s = 0.25$

Diameter of pulley, $d = 0.6$ m

Speed of pulley, $N = 250$ rpm

Contact angle, $\beta = \dfrac{175}{180}\pi = 3.05$ rad

Tangential velocity v at the rim of the pulley is given as

$$v = \pi d \frac{N}{60} = \pi(0.6)\frac{250}{60} = 7.85 \text{ m/s}$$

We know, $$\frac{T_2}{T_1} = e^{\mu_s \beta}$$

$$\therefore \quad \frac{1500}{T_1} = e^{0.25(3.05)}$$

$$\Rightarrow \quad T_1 = 699.75 \text{ N}$$

Torque exerted on the pulley is given as

$$\tau = (T_2 - T_1)r$$
$$= (1500 - 699.75) \times 0.3 = 240.1 \text{ N.m}$$

Power transmitted is given as

$$\text{Power} = (T_2 - T_1)v$$
$$= (1500 - 699.75) \times 7.85 = 6281.96 \text{ W (or) } 6.28 \text{ kW}$$

7.4.7 Centrifugal Tension

As the belt embracing the pulleys undergoes circular motion about the axis of the pulley, a centrifugal force is induced upon it. Due to this force, it tries to move away from the centre of the pulleys. This results in an increase in tension of the belt on both tight and slack sides. This increase in tension caused by the centrifugal force is called the ***centrifugal tension***.

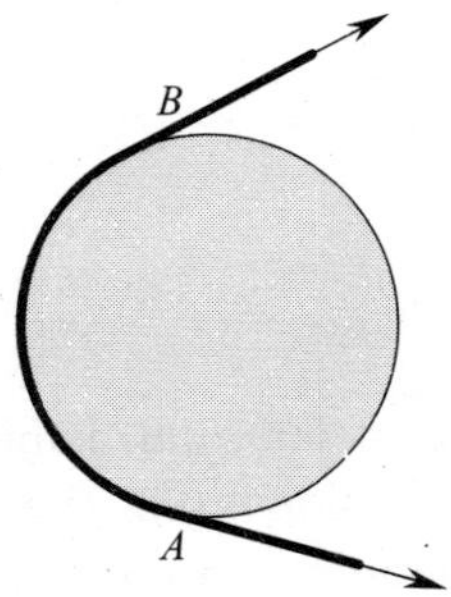

Fig. 7.28(a)

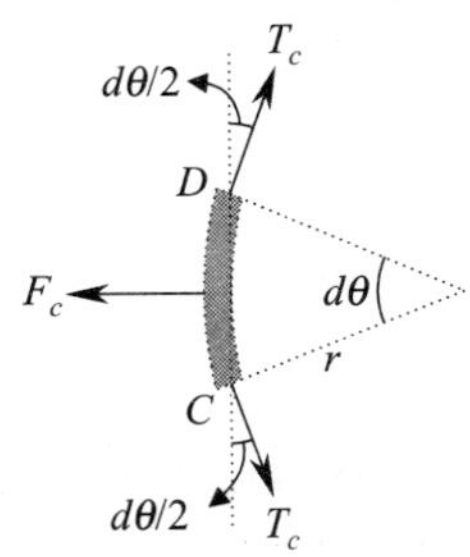

Fig. 7.28(b) Element of belt

Consider an infinitesimally small element *CD* of the belt in contact with the pulley. If *m* be mass of the belt per unit length, then mass of the element *CD* is given as

$$dm = (\text{mass/unit length})\ (\text{arc length } CD) = m\,r\,d\theta \tag{7.70}$$

If *v* is the linear velocity of the belt then the centrifugal force on the element *CD* is given as

$$F_c = dm\frac{v^2}{r} = (m\,r\,d\theta)\frac{v^2}{r} = m\,d\theta\,v^2 \tag{7.71}$$

It acts radially outwards. For equilibrium condition, this force must be balanced by the tension in the belt at both the ends *C* and *D*. As we are considering the tension only due to the centrifugal force and not due to friction, the two tensions will be equal. Let it be T_c. Taking the summation of forces along the radial direction, we get

$$mv^2 d\theta = 2T_c \sin\left(\frac{d\theta}{2}\right) \tag{7.72}$$

Since the element is infinitesimally small, $d\theta \to 0$. Therefore, $\sin\left(\frac{d\theta}{2}\right) = \frac{d\theta}{2}$

Hence, the above equation reduces to

$$mv^2 d\theta = 2T_c \frac{d\theta}{2}$$

or,

$$T_c = mv^2 \tag{7.73}$$

Hence, considering centrifugal tension,

$$\text{tension on tight side} = T_2 + T_c$$
$$= T_2 + mv^2 \tag{7.74}$$
$$\text{tension on slack side} = T_1 + T_c$$
$$= T_1 + mv^2 \tag{7.75}$$

7.4.8 Maximum Tension in the Belt

Consider a belt of width b and thickness t. If σ_{max} is the maximum allowable stress developed then the maximum allowable tension in the belt is given as

$$T_{max} = \sigma_{max}\, bt \tag{7.76}$$

Since the maximum tension occurs at the tight side, we have

$$T_{max} = \sigma_{max}\, bt = T_2 + mv^2 \tag{7.77}$$

Example 7.22 A belt transmits 10 kW of power from a pulley of 600 mm diameter running at 240 r.p.m. The angle of contact of the belt is 165° and coefficient of friction is 0.2. The thickness of the belt is 6 mm and the density of the belt material is 2000 kg/m^3. Determine minimum width of the belt required if stress in the belt is not to exceed 3 N/mm^2.

Solution *Given data*

$P = 10$ kW $\quad \mu_s = 0.2$

$d = 600$ mm $\quad t = 6$ mm

$N = 240$ rpm $\quad \rho = 2000$ kg/m^3

$\beta = 165° = \frac{165}{180} \times \pi = 2.88$ rad $\quad \sigma_{max} = 3$ N/mm^2

Determination of maximum allowable tension

$$v = \frac{\pi d N}{60}$$
$$= \frac{\pi(0.6)(240)}{60} = 7.54 \text{ m/s}$$

We know,

$$P = (T_2 - T_1)v$$

Therefore,

$$T_2 - T_1 = \frac{P}{v} = \frac{10 \times 10^3}{7.54} = 1326.26 \text{ N} \tag{a}$$

Also, $$\frac{T_2}{T_1} = e^{\mu_s \beta}$$

$$= e^{(0.2)(2.88)} = 1.78 \quad \text{(b)}$$

From the two equations (a) and (b), solving for T_1 and T_2, we get

$$T_1 = 1700.33 \text{ N}$$

$$T_2 = 3026.59 \text{ N}$$

Neglecting centrifugal tension

Since T_2 is the maximum tension if centrifugal force is neglected,

$$T_{max} = T_2 = \sigma_{max} \times b \times t$$

$$\Rightarrow \quad b = \frac{3026.59}{3 \times 6} = 168.14 \text{ mm}$$

Considering centrifugal tension

$$T_{max} = \sigma_{max} \times b \times t = T_2 + T_c$$

or, $$\sigma_{max} \times b \times t = T_2 + mv^2$$

$$3 \times 10^6 \times b \times 0.006 = 3026.59 + (2000) \times b \times 0.006 \times (7.54)^2$$

$$\Rightarrow \quad b = 0.175\,m \text{ (or) } 175 \text{ mm}$$

Example 7.23 The engine shaft and machine shaft separated by 2 metres are connected by a belt of 120 mm width and 6 mm thickness. The belt transmits 10 kW of power when the shafts rotate at 150 rpm and 300 rpm respectively. If the diameter of the driven pulley is 400 mm and coefficient of friction is 0.2, determine the stress in the belt in (i) an open belt arrangement, and (ii) a cross-belt arrangement.

Solution *Given data*

$N_1 = 150$ rpm $\qquad \mu_s = 0.2$

$N_2 = 300$ rpm $\qquad l = 2$ m

$d_2 = 400$ mm $\qquad b = 120$ mm

$P = 10$ kW $\qquad t = 6$ mm

Determination of speed of belt

The speed ratio of the pulleys is given as

$$\frac{N_2}{N_1} = \frac{d_1}{d_2}$$

$$\Rightarrow \quad d_1 = 800 \text{ mm}$$

Therefore, tangential speed of the belt is given as

$$v = \frac{\pi d_1 N_1}{60} = 6.28 \text{ m/s}$$

(a) *Open belt arrangement*

Determination of lap angle

$$\sin\alpha = \left(\frac{r_1 - r_2}{l}\right) = \frac{0.4 - 0.2}{2} = 0.1$$

$$\Rightarrow \quad \alpha = 5.74°$$

Since lap angle is considered for the smaller-diameter pulley, lap angle is given as

$$\beta = 180° - 2\alpha = 168.52° = 2.94 \text{ rad}$$

Determination of maximum tension

We know,

$$\frac{T_2}{T_1} = e^{\mu_s \beta}$$

$$= e^{(0.2)(2.94)} = 1.8 \quad \text{(a)}$$

Also,

$$P = (T_2 - T_1)v$$

$$\Rightarrow \quad T_2 - T_1 = \frac{10 \times 10^3}{6.28} = 1592.36 \text{ N} \quad \text{(b)}$$

Solving for tensions from the above two equations (a) and (b),

$$1.8T_1 - T_1 = 1592.36$$

$$\therefore \quad T_1 = 1990.45 \text{ N}$$

and

$$T_2 = 3582.81 \text{ N}$$

Maximum allowable stress

Neglecting the effect due to centrifugal force, the maximum allowable stress is given as

$$\sigma_{max} = \frac{T_2}{b \times t} = \frac{3582.81}{120 \times 6} = 4.98 \text{ N/mm}^2$$

(b) *Cross-belt arrangement*

Determination of lap angle

$$\sin\alpha = \left(\frac{r_1 + r_2}{l}\right) = \frac{0.4 + 0.2}{2} = 0.3$$

$$\Rightarrow \quad \alpha = 17.46°$$

Since the lap angle is same for both the pulleys,

$$\beta = 180° + 2\alpha = 214.92° = 3.75 \text{ rad}$$

Determination of maximum tension

$$\frac{T_2}{T_1} = e^{\mu_s \beta}$$

$$= e^{(0.2)(3.75)} = 2.12 \quad \text{(c)}$$

Also, $$T_2 - T_1 = \frac{P}{v} = \frac{10 \times 10^3}{6.28} = 1592.36 \text{ N} \quad \text{(d)}$$

From equations (c) and (d),

$$2.12T_1 - T_1 = 1592.36 \text{ N}$$

$\Rightarrow$ $$T_1 = 1421.75 \text{ N}$$

and $$T_2 = 3014.11 \text{ N}$$

Maximum allowable stress

Neglecting the effect due to centrifugal force, the maximum allowable stress is given as

$$\sigma_{max} = \frac{T_2}{b \times t} = \frac{3014.11}{120 \times 6} = 4.19 \text{ N/mm}^2$$

7.4.9 Condition for Maximum Power Transmitted

Power transmitted in a belt drive is given as

$$P = (T_2 - T_1)v \quad (7.78)$$

Since $\frac{T_2}{T_1} = e^{\mu_s \beta}$, we can also write the above expression as

$$P = \left(T_2 - \frac{T_2}{e^{\mu_s \beta}} \right) v$$

$$= T_2 \left(1 - \frac{1}{e^{\mu_s \beta}} \right) v \quad (7.79)$$

Also, considering centrifugal tension, $T_{max} = T_2 + mv^2$. Therefore,

$$P = (T_{max} - mv^2) \left(1 - \frac{1}{e^{\mu_s \beta}} \right) v$$

$$= (T_{max}\, v - mv^3) \left(1 - \frac{1}{e^{\mu_s \beta}} \right) \quad (7.80)$$

We know that the power transmitted is maximum, when

$$\frac{dP}{dv} = 0$$

$\Rightarrow$ $$T_{max} - 3mv^2 = 0 \quad (7.81)$$

Since $T_c = mv^2$,

$$T_{max} - 3T_c = 0$$

Therefore, $$T_{max} = 3T_c \quad (7.82)$$

Also, from the Eq. 7.81,

$$v = \sqrt{\frac{T_{max}}{3\,m}} \tag{7.83}$$

The above two expressions are the conditions for maximum power transmission.

Example 7.24 The maximum permissible stress of a belt of 150 mm × 10 mm and weighing 1200 kg/m^3 is 2.5 N/mm^2. Determine the maximum power it can transmit if the ratio of the tensions developed on either side is two.

Solution *Given data*

$\sigma_{max} = 2.5$ N/mm^2 $\quad$ $T_2/T_1 = 2$

$b = 150$ mm $\quad$ $\rho = 1200$ kg/m^3

$t = 10$ mm

The maximum allowable tension is

$$T_{max} = \sigma_{max} \times b \times t$$

$$= 2.5 \times 150 \times 10 = 3750 \text{ N}$$

The condition for maximum power transmission is

$$T_c = T_{max}/3 = 1250 \text{ N}$$

Therefore, tension on tight side is given as

$$T_2 = T_{max} - T_c$$

$$= 3750 - 1250 = 2500 \text{ N}$$

Therefore, tension on slack side is

$$T_1 = \frac{T_2}{2} = 1250 \text{ N}$$

Speed of belt for maximum power is

$$v = \sqrt{\frac{T_{max}}{3\,m}}$$

$$= \sqrt{\frac{3750}{3(1200 \times 0.15 \times 0.01)}} = 26.35 \text{ m/s}$$

Therefore, maximum power transmitted is given as

$$P_{max} = [(T_2 + T_c) - (T_1 + T_c)]v$$

$$= (T_2 - T_1)v$$

$$= (2500 - 1250)\ 26.35 = 32.94 \text{ kW}$$

7.4.10 Initial Tension in the Belt

To ensure a firm grip between the belt and the pulleys, an ***initial tension*** is induced in the belt by tightening it. Otherwise, the tension on the slack side would cease to exist, the drive would not operate satisfactorily and the belt would be prone to slip off the pulleys. This tension is present in the belt even when the belt is stationary. Let T_i be the initial tension in the belt. When the belt is under motion, its tension on the tight side is T_2 and that on the slack side is T_1. Therefore, increase in tension on the tight side is

$$= T_2 - T_i$$

If α is the coefficient of increase in the belt length per unit force then increase in length of belt on the tight side is

$$= \alpha(T_2 - T_i)$$

Similarly, decrease in tension on the slack side is

$$= T_i - T_1$$

Therefore, decrease in length of belt on the slack side is

$$= \alpha(T_i - T_1)$$

If the belt length is assumed to be constant, increase in length on the tight side must be equal to decrease in the length on the slack side. Therefore,

$$\alpha(T_2 - T_i) = \alpha(T_i - T_1)$$

$$T_2 - T_i = T_i - T_1$$

$$\Rightarrow \qquad T_i = \frac{T_1 + T_2}{2} \tag{7.67}$$

If centrifugal tension is taken into consideration then

$$T_i = \frac{(T_1 + T_c) + (T_2 + T_c)}{2} = \frac{T_1 + T_2}{2} + T_c \tag{7.68}$$

Example 7.25 An open flat belt drive connects two parallel shafts, 1.2 m apart. The driving and driven shafts rotate at 140 and 350 rpm respectively and the driven pulley is 400 mm in diameter. The belt is 5 mm thick and 80 mm wide. The coefficient of friction between the belt and the pulleys is 0.3 and the maximum permissible stress in the belting is 1.4 N/mm^2. Determine (a) the maximum power that may be transmitted by the belting, and (b) the required initial tension in the belt. Neglect centrifugal tension.

Solution *Given data*

$l = 1.2$ m $\qquad t = 5$ mm

$N_1 = 140$ rpm $\qquad b = 80$ mm

$N_2 = 350$ rpm $\qquad \mu_s = 0.3$

$d_2 = 400$ mm $\qquad \sigma_{max} = 1.4$ N/mm^2

Diameter of driver pulley

$$\frac{N_2}{N_1} = \frac{d_1}{d_2}$$

$$\Rightarrow \quad d_1 = \frac{350}{140}(400) = 1000 \text{ mm}$$

(i) *Maximum power transmitted*

Maximum allowable tension,

$$T_{\max} = \sigma_{\max} \times b \times t$$

$$= 1.4 \times 5 \times 80 = 560 \text{ N}$$

We know for open belt drive,

$$\sin \alpha = \left(\frac{r_1 - r_2}{l}\right) = \frac{0.5 - 0.2}{1.2} = 0.25$$

$$\Rightarrow \quad \alpha = 14.48°$$

Therefore, contact angle is given as

$$\beta = 180° - 2\alpha = 151.04° = 2.64 \text{ rad}$$

We know that at the point of impending motion,

$$\frac{T_2}{T_1} = e^{\mu_s \beta} = e^{(0.3)(2.64)} = 2.21$$

Since $T_2 > T_1$, we take $T_2 = T_{\max} = 560$ N [since centrifugal tension can be neglected]

$$\Rightarrow \quad T_1 = 253.39 \text{ N}$$

Velocity of belt,

$$v = \frac{\pi d_1 N_1}{60}$$

$$= \frac{\pi(1)(140)}{60} = 7.33 \text{ m/s}$$

Therefore, maximum power transmitted is given as

$$P = (T_2 - T_1)v$$

$$= (560 - 253.39) \times 7.33 = 2247.4 \text{ W}$$

(ii) *Initial tension*

Neglecting centrifugal tension,

$$T_i = \frac{T_1 + T_2}{2} = \frac{560 + 253.39}{2} = 406.7 \text{ N}$$

Example 7.26 A driving pulley of 600 mm diameter, rotating at 240 rpm is connected to a driven pulley of 400 mm diameter by a cross-belt arrangement. The power transmitted between the shafts separated by 3 m is 8 kW. If the thickness of the belt is 6 mm and the permissible load on it is 20 N/mm width, determine (i) the width of the belt, and (ii) the necessary initial tension in the belt. Take coefficient of friction between the belt and the pulleys as 0.25.

Solution *Given data*

$l = 3$ m	$P = 8$ kW
$d_1 = 600$ mm	Permissible load = 20 N/mm width
$d_2 = 400$ mm	$t = 6$ mm
$N_1 = 240$ rpm	$\mu_s = 0.25$

Velocity of the belt

The velocity of the belt is given as

$$v = \frac{\pi d_1 N_1}{60}$$

$$= \frac{\pi(0.6)(240)}{60} = 7.54 \text{ m/s}$$

Determination of allowable tension

Power, $$P = (T_2 - T_1)v$$

$$8000 = (T_2 - T_1)(7.54)$$

$\Rightarrow$ $$(T_2 - T_1) = 1061.01 \text{ N} \qquad \text{(a)}$$

Since a cross belt is used,

$$\sin\alpha = \left(\frac{r_1 + r_2}{l}\right) = \frac{0.3 + 0.2}{3} \quad \Rightarrow \quad \alpha = 9.6°$$

Therefore, lap angle is given as

$$\beta = 180° + 2\alpha$$

$$= 199.2° = 3.48 \text{ rad}$$

$\therefore$ $$\frac{T_2}{T_1} = e^{\mu_s \beta} = e^{(0.25)(3.48)} = 2.39 \qquad \text{(b)}$$

From the above two equations (a) and (b), we get

$$2.39T_1 - T_1 = 1061.01 \text{ N}$$

$\Rightarrow$ $$T_1 = 763.3 \text{ N}$$

and $$T_2 = 1824.3 \text{ N}$$

(i) *Width of belt*

Since permissible load is 20 N/mm width,

$$T_2 = 20 \times b$$

$$\Rightarrow \qquad b = 91.2 \text{ mm}$$

(ii) *Necessary initial tension*

Neglecting centrifugal tension,

$$T_i = \frac{T_1 + T_2}{2} = \frac{763.3 + 1824.3}{2} = 1293.8 \text{ N}$$

7.5 BAND BRAKES

A band brake is a simple mechanism for power absorption in a rotating drum. A flexible band passes over a drum and its free ends are connected to a lever, which is hinged at a point as shown in Fig. 7.29. Unlike the belt drive, here the drum rotates and the band remains stationary. By applying a force P at the end of the lever, the band gets tightened and due to friction developed between the band and the drum, the drum is brought to a stop.

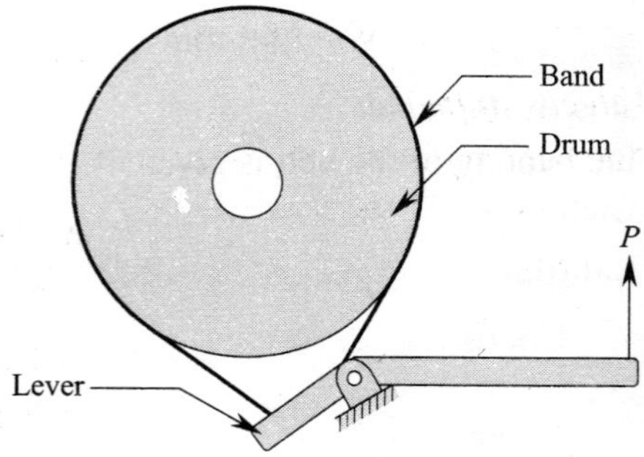

Fig. 7.29 Band brake

The free-body diagrams of the drum, band and lever are shown below separately. Suppose the drum is rotating in the clockwise direction. Then the frictional force exerted by the band in contact with the drum trying to stop it will be such as to oppose it. Hence, it will act in the anticlockwise direction [refer Fig. 7.30(a)]. It should be noted that this frictional force is distributed over the entire contact angle and only the average value is shown in the figure.

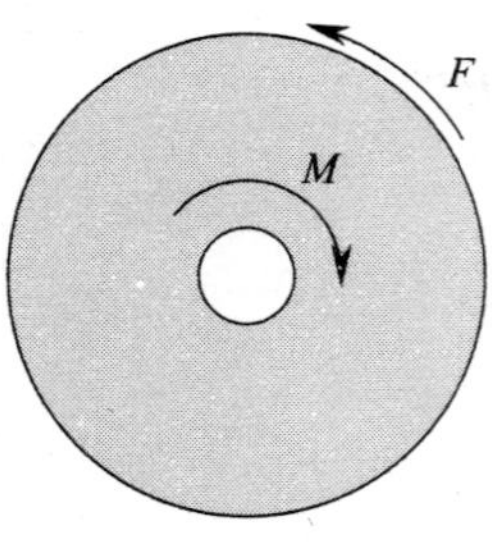

Fig. 7.30(a) Free-body diagram of drum

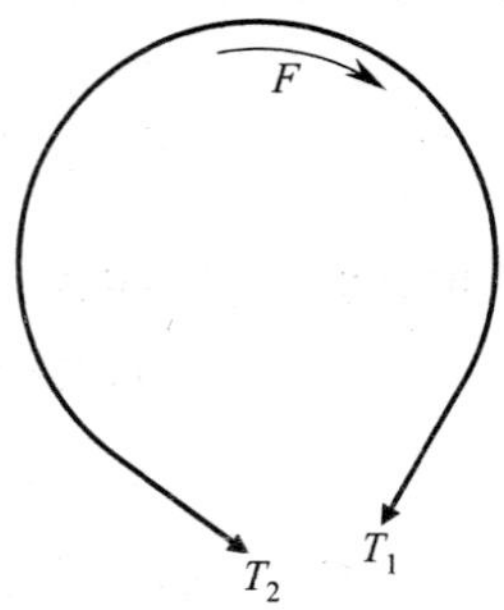

Fig. 7.30(b) Free-body diagram of band

By Newton's third law of motion, the frictional force exerted on the band by the drum will act in the direction opposite to that it acts over the drum as shown in Fig. 7.30(b). Since the frictional force acts in the clockwise direction on the band, we can readily see that the left end of the band is at a higher tension than that at the right end. Hence, we name the tensions as T_2 on the left end and T_1 on the right end as per the convention described before, i.e., $T_2 > T_1$. The ratio of these tensions is given by the same expression as derived for belt drive, i.e.,

$$\frac{T_2}{T_1} = e^{\mu_k \beta} \tag{7.69}$$

Note that unlike the belt drive, slippage between the band and drum will always be present. Hence, coefficient of kinetic friction μ_k is used.

Once T_1 and T_2 are known, by considering the free-body diagram of the lever and applying moment about any point on it, the force P required to be applied at the end of the lever can be determined.

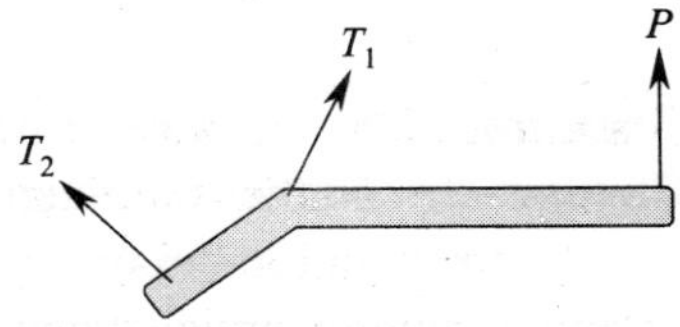

Fig. 7.30(c) Free body diagram of lever

Example 7.27 A force P acting at the end of the lever shown (hinged at point B) produces a torque of 200 N.m on the rotating brake drum. If the band embraces the drum for an angle of 240° and the coefficient of friction is 0.2, determine the required value of P.

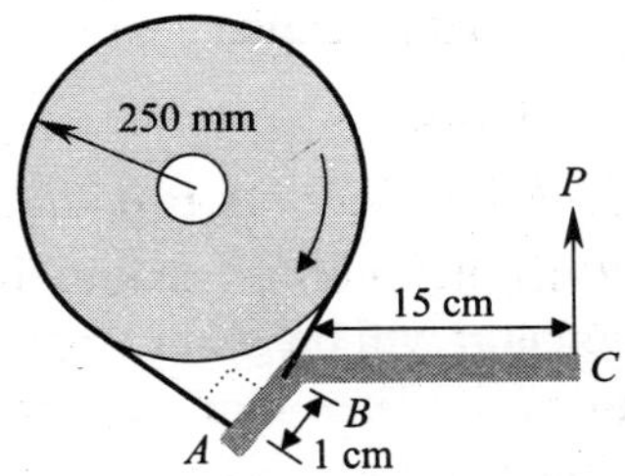

Fig. 7.31

Solution *Given data*

Coefficient of friction, $\mu_k = 0.2$

Contact angle, $\beta = \frac{240}{180}\pi = 4.189$ rad

Torque, $\tau = 200$ N.m

Since the drum is rotating in the clockwise direction, the frictional resistance acting on the drum will be in the anticlockwise direction. Hence, the frictional force exerted by the drum on the band will be in the clockwise direction. Thus, tension T_2 will act on the left end of the band and tension T_1 at the right end (refer the discussion in Section 7.5).

We know, $\tau = (T_2 - T_1)r$

$$200 = (T_2 - T_1)(0.25)$$

$$\Rightarrow \quad T_2 - T_1 = 800 \text{ N} \tag{a}$$

Also, $\frac{T_2}{T_1} = e^{\mu_k \beta}$

$$= e^{(0.2)(4.189)} = 2.31$$

or

$$T_1 = 0.433 T_2 \tag{b}$$

Substituting this value in the equation (a), we get

$$T_2 = \frac{800}{(1 - 0.433)} = 1410.93 \text{ N}$$

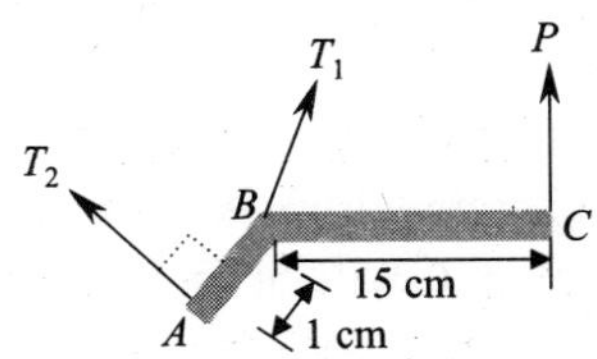

Fig. 7.31(a)

Taking the summation of moments about the point B and equating it to zero for equilibrium,

$\sum M_B = 0 \Rightarrow$

$$-T_2 \times 1 + P \times 15 = 0$$

$$\therefore \quad P = \frac{1410.93 \times 1}{15} = 94.06 \text{ N}$$

Example 7.28 A brake band arrangement is shown in Fig. 7.32. The force applied at the end of the lever is 50 N and the coefficient of friction between the band and drum is 0.3. The diameter of the drum is 300 mm. Find the braking moment on the drum (a) if the drum is rotating clockwise, and (b) if the drum is rotating anticlockwise.

Solution *Given data*

Coefficient of friction, $\mu_k = 0.3$

Contact angle, $\beta = \pi$ rad

Force, $P = 50$ N

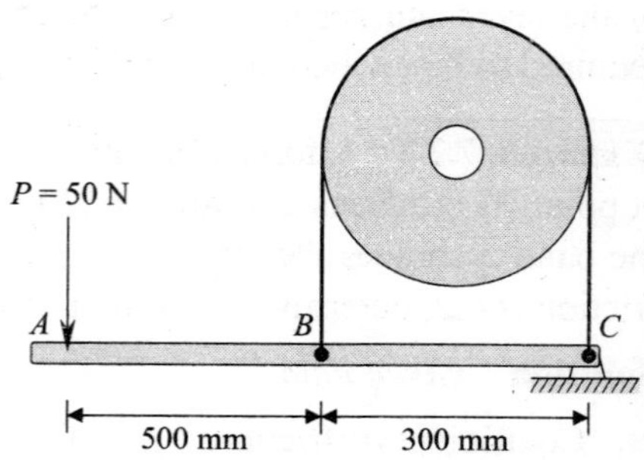

Fig. 7.32

(a) *When the drum is rotating in the clockwise direction*

When the drum is rotating in the clockwise direction, the tension at B will be greater than that at C (refer discussion in Section 7.5). Therefore, the free-body diagram of the lever is as shown in Fig. 7.32(a).

Taking summation of moments about C and applying the condition of equilibrium,

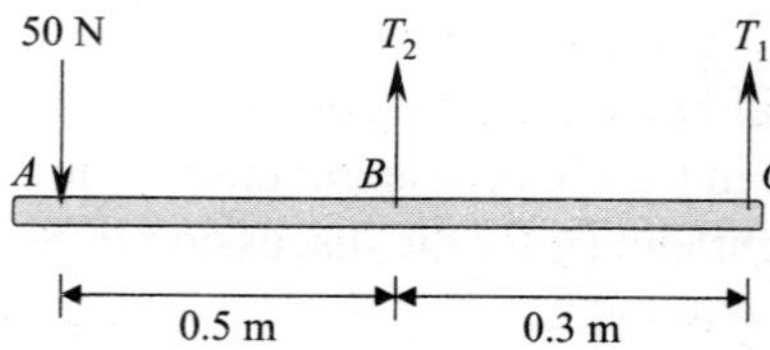

Fig. 7.32(a)

$\sum M_C = 0 \Rightarrow$

$$50 \times (0.8) - T_2(0.3) = 0$$

$$\therefore \quad T_2 = 133.33 \text{ N}$$

Also, we know,

$$\frac{T_2}{T_1} = e^{\mu_k \beta}$$

$$= e^{(0.3)\pi} = 2.57$$

$$\therefore \quad T_1 = \frac{T_2}{2.57} = 51.88 \text{ N}$$

Therefore, braking moment on the drum is obtained as

$$M = (T_2 - T_1)r$$

$$= (133.33 - 51.88)(0.15) = 12.22 \text{ N.m}$$

(b) *When the drum is rotating in the anticlockwise direction*

When the drum is rotating in the anticlockwise direction, the tension at C will be greater than that at B (refer discussion in Section 7.5) and accordingly the free-body diagram of the lever is as shown in Fig. 7.32(b).

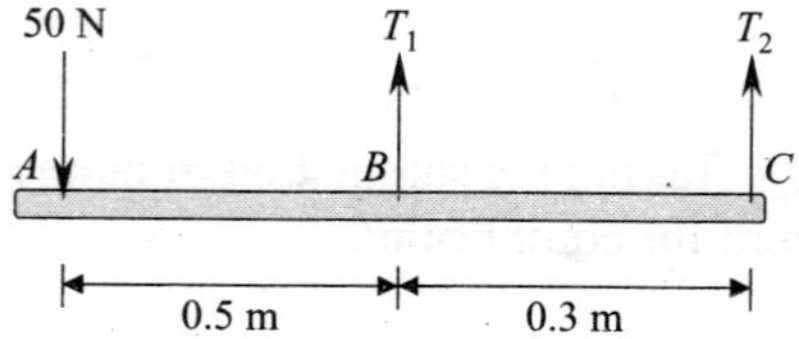

Fig. 7.32(b)

Taking summation of moments about C and equating it to zero for equilibrium,

$\sum M_C = 0 \Rightarrow$

$$50 \times (0.8) - T_1(0.3) = 0$$

$$\Rightarrow \quad T_1 = 133.33 \text{ N}$$

Also, we know,

$$\frac{T_2}{T_1} = e^{\mu_k \beta}$$

$$= e^{(0.3)\pi} = 2.57$$

$$\therefore \quad T_2 = 2.57\,T_1 = 342.66 \text{ N}$$

Therefore, braking moment on the drum is obtained as

$$M = (T_2 - T_1)r$$

$$= (342.66 - 133.33)(0.15) = 31.4 \text{ N.m}$$

SUMMARY

Wedges

Wedges are simple machines used to lift or move heavy loads for slight adjustments in their positions. These are normally *triangular* or *trapezoidal*-shaped blocks made with a very small sloping angle called *wedge angle*. Wedges are so shaped that when the applied forces are removed, they remain under *self-locking*.

Screw Friction

Square threaded screws are used for lifting heavy loads. When these screws rotate, resistance is offered by the machine parts to which they are fastened. This resistance is utilized in lifting loads as in screw jacks or clamping two bodies as in vices.

Lead [L] of a screw It is defined as the distance through which the screw advances in one turn along the axis, when rotated about its axis.

Pitch [p] of a screw It is the distance between two consecutive threads.

For single threaded screws, lead = pitch; and for multithreaded screws, lead = $n \times$ pitch.

Mean radius (r) of the thread It is taken as the average of the outer radius and root (or inner) radius of the thread. For a square thread, the relationship between root radius and outer radius is given as

root (or inner) radius = outer radius – $p/2$

Lead angle or helix angle $\theta = \tan^{-1}\left[\dfrac{L}{2\pi r}\right]$, where L is the lead of the screw and r is mean radius of the screw.

Screw Jack

A screw jack is a mechanical device used for lifting and holding heavy loads with relatively little effort. The effort required to be applied at the handle (lever arm = a) to raise a load W is

$$P = \frac{Wr}{a} \tan(\varphi_s + \theta),$$ where φ_s is coefficient of static friction.

Once motion has started, little effort is required to keep the screw-jack turning and it is given by

$$P = \frac{Wr}{a} \tan(\varphi_k + \theta),$$ where φ_k is the coefficient of kinetic friction.

If the lead angle is greater than static angle of friction, i.e., $\theta > \varphi_s$ then the screw will *unwind* itself under the load, when the external moment is removed. If the lead angle is less than static angle of friction, i.e., $\theta < \varphi_s$ then the screw will *self-lock* and effort is required to lower it. It is given as

$$P = \frac{Wr}{a} \tan(\varphi_s - \theta)$$

Efficiency of screw jack It is defined as the ratio of the effort required under frictionless condition to that of the actual effort required to raise a load.

$$\eta = \frac{\tan\theta}{\tan(\varphi_s + \theta)}$$

The condition for maximum efficiency is

$$\theta = 45° - \varphi_s/2$$

and the maximum efficiency is given as

$$\eta_{\max} = \frac{1 - \sin\varphi_s}{1 + \sin\varphi_s}$$

Belt Friction

Whenever a flexible member like a string, a rope, a cable or a belt moves or tends to move over a pulley or a cylindrical drum, frictional forces are always developed between the contact surfaces, which tend to oppose the relative motion. As a result, the tension at both the ends of the rope or belt will not be equal. The relationship between the two tensions is given as

$$\frac{T_2}{T_1} = e^{\mu_s \beta}$$

where T_2 is the tension on the tight side and T_1 is the tension on the slack side; β is the contact or lap angle made by the belt with the drum and it is always expressed in *radians*.

Belt Drives

The speed ratio of engine and machine shafts:

For a simple belt drive

$$N_2 = \frac{d_1}{d_2} N_1 \quad \text{(without slip)}$$

$$\frac{N_2}{N_1} = \frac{d_1}{d_2}\left[1 - \frac{(p_1 + p_2)}{100}\right] \quad \text{(with slip)}$$

$$\frac{N_2}{N_1} = \frac{d_1 + t}{d_2 + t} \quad \text{(with thickness of belt considered and without pull)}$$

For a compound belt drive

$$\frac{\text{speed of machine shaft}}{\text{speed of engine shaft}} = \frac{\text{product of diameters of all driver pulleys}}{\text{product of diameters of all driven pulleys}}$$

Length of Belt Required

For an open belt drive: $L = 2l + \pi(r_1 + r_2) + \dfrac{(r_1 - r_2)^2}{l}$

For a cross belt drive: $L = 2l + \pi(r_1 + r_2) + \dfrac{(r_1 + r_2)^2}{l}$

Power transmitted by the belt is given as

$P = (T_2 - T_1)v$ watts, where v is velocity of the belt.

Centrifugal Tension

The increase in tension caused by the centrifugal force is called the *centrifugal tension*. It is given as

$$T_c = mv^2$$

Therefore, tension on the tight side $= T_2 + T_c = T_2 + mv^2$

tension on the slack side $= T_1 + T_c = T_1 + mv^2$

Maximum tension in the belt is given as

$$T_{max} = \sigma_{max}\, bt$$

Since the maximum tension occurs at the tight side,

$$T_{max} = T_2 + mv^2$$

Conditions for Maximum Power Transmitted

The conditions for maximum power transmission are

$$T_{max} = 3T_c \quad \text{and} \quad v = \sqrt{\frac{T_{max}}{3m}}$$

Initial Tension in the Belt

Initial tension in the belt is given as

$$T_i = \frac{T_1 + T_2}{2} \quad \text{(without centrifugal tension)}$$

$$T_i = \frac{T_1 + T_2}{2} + T_c \quad \text{(with centrifugal tension)}$$

Band Brakes

A band brake is a simple mechanism for power absorption in a rotating drum. Here the drum rotates and the band remains stationary. As there will be slippage between band and drum, the ratio of the tensions is given as

$$\frac{T_2}{T_1} = e^{\mu_k \beta}$$

Note that if the drum rotates in the clockwise direction then the tension on the left end will be greater than that on the right end; if it rotates in the anticlockwise direction the tension on the right end will be greater than that on the left end.

EXERCISES

Objective-type Questions

1. In a square threaded screw of pitch p, the relationship between root radius and outer radius is
   (a) root radius = outer radius – $p/3$
   (b) root radius = outer radius – $p/2$
   (c) root radius = outer radius – p
   (d) root radius = outer radius – $2p$
2. In a multithreaded screw of n threads, lead is equal to
   (a) $(n - 1) \times$ pitch
   (b) $(n + 1) \times$ pitch
   (c) $n \times$ pitch
   (d) pitch
3. Pitch of a screw is defined as
   (a) distance between two consecutive threads
   (b) distance through which the screw advances in one turn along the axis
   (c) the number of threads in a screw
   (d) ratio of outer and inner radii of the screw
4. Lead of a screw is defined as
   (a) distance between two consecutive threads
   (b) distance through which the screw advances in one turn along the axis
   (c) the number of threads in a screw
   (d) ratio of outer and inner radii of the screw
5. Lead angle in a screw is also called
   (a) angle of static friction
   (b) angle of repose
   (c) helix angle
   (d) angle of kinetic friction
6. For a screw of lead L, mean radius r and angle of static friction ϕ_s, lead or helix angle θ is
   (a) $\tan^{-1} \dfrac{L}{2\pi\varphi_s}$
   (b) $\tan^{-1} \dfrac{p}{2\pi\varphi_s}$
   (c) $\tan^{-1} \dfrac{L}{2\pi r}$
   (d) $\tan^{-1} \dfrac{p}{2\pi r}$
7. The ratio of efforts required to raise a load and to lower the same load is
   (a) $\dfrac{\tan(\varphi_s)}{\tan(\theta)}$
   (b) $\dfrac{\tan(\theta)}{\tan(\varphi_s)}$
   (c) $\dfrac{\tan(\varphi_s + \theta)}{\tan(\varphi_s - \theta)}$
   (d) $\dfrac{\tan(\varphi_s - \theta)}{\tan(\varphi_s + \theta)}$
8. State which of the following statement is correct:
   In a screw jack, when lead angle θ is greater than the angle of static friction φ_s,
   (a) the screw jack will self-lock
   (b) effort is required to lower a load

(c) the screw will unwind itself
(d) more effort is required to lower than to raise a load

9. A screw jack with lead angle [θ] and friction angle [φ_s] is said to be in self-locking, if
(a) $\theta > \varphi_s$ (b) $\theta < \varphi_s$ (c) $\theta = \varphi_s$ (d) $\varphi_s = 0$

10. The efficiency of screw jack is defined as the ratio of
(a) effort required under frictionless condition and actual effort
(b) actual effort and effort required under frictionless condition
(c) efforts required to raise and to lower a load
(d) efforts required to lower and to raise a load

11. The condition for maximum efficiency in a screw jack is
(a) $\theta = 45° - \varphi_s/2$ (b) $\theta = 45° + \varphi_s/2$ (c) $\varphi_s = 45° - \theta/2$ (d) $\varphi_s = 45° + \theta/2$

12. The maximum efficiency η_{max} in a screw jack is given as
(a) $\frac{1+\cos\varphi_s}{1-\cos\varphi_s}$ (b) $\frac{1-\cos\varphi_s}{1+\cos\varphi_s}$ (c) $\frac{1+\sin\varphi_s}{1-\sin\varphi_s}$ (d) $\frac{1-\sin\varphi_s}{1+\sin\varphi_s}$

13. Effort required to raise a load as compared to that required at impending motion is
(a) lesser (b) greater (c) equal (d) nil effort

14. In an open belt drive with unequal diameters of pulleys, the contact angle [β]
(a) is that of smaller ϕ pulley (b) is that of larger ϕ pulley
(c) can be any of the two diameters (d) is the average of both the pulleys

15. The abbreviation rpm stands for
(a) radians per minute (b) revolutions per minute
(c) radians per metre (d) revolutions per metre

16. For given diameters of driver and driven pulleys and spacing between them, the length required for cross-belt arrangement is __________ that required for open-belt arrangement.
(a) lesser than (b) greater than (c) equal to (d) half

17. The tangential velocity of a belt embracing a pulley of diameter d and rotating at N rpm is
(a) πdN (b) $\frac{\pi dN}{60}$ (c) $\pi \frac{d}{2}\frac{N}{60}$ (d) $\pi \frac{d}{2} N$

Answers

1. (b) 2. (c) 3. (a) 4. (b) 5. (c) 6. (c) 7. (c) 8. (c)
9. (b) 10. (a) 11. (a) 12. (d) 13. (a) 14. (a) 15. (b) 16. (b)
17. (b)

Short-answer Questions

1. What is a wedge and what is its application?
2. What are the various shapes of wedges used and what is wedge angle?
3. Discuss the method of solving wedge problems.
4. What is a square thread? Where is it used?

5. Define lead and pitch of a screw. How are they different for multithreaded screws?
6. What is the relationship between mean radius, outer radius and inner radius of a square threaded screw?
7. Distinguish between lead angle and angle of friction in a screw.
8. Derive the expressions for efforts required to raise and to lower loads.
9. Define efficiency of a screw jack.
10. What is the condition in terms of efficiency for a machine to be self-locking?
11. Derive the condition for maximum efficiency in a screw jack.
12. Describe the advantage of a differential screw jack over a normal screw jack.
13. Deduce the effort required to raise a load in a differential screw jack.
14. Classify and explain the types of belt drives with neat sketches.
15. Define slip of the belt.
16. Distinguish between open and cross-belt drives.
17. Derive the expression for the length of belt required in (a) an open belt drive, and (b) a cross belt drive.
18. Derive the expression for the ratio of tensions in a belt drive.
19. Define initial tension in a belt.
20. Distinguish between initial tension and centrifugal tension in a belt.
21. Obtain the conditions for the maximum power transmitted by a belt from one pulley to another.

Numerical Problems

7.1 A 15° wedge is used to raise a 1000 kg block as shown in Fig. E.7.1. Determine the horizontal force P that must be applied on the wedge to raise it. Take the angle of friction at all contact surfaces to be 0.2.

Ans. 7.6 kN

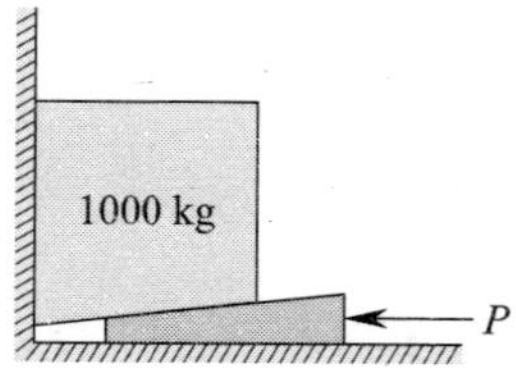

Fig. E.7.1

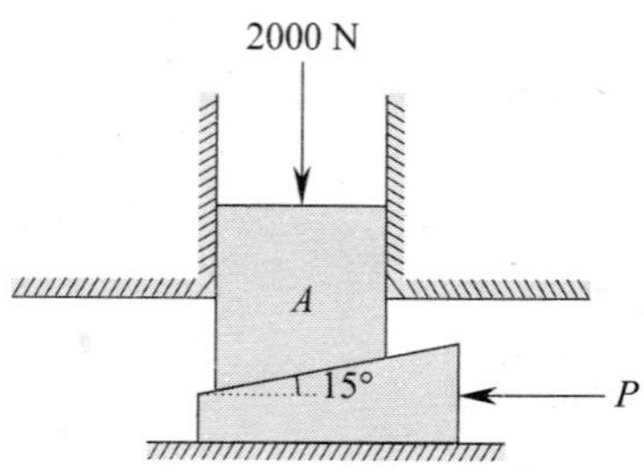

Fig. E.7.2

7.2 In Fig. E.7.2, the block A supports a weight of 2000 N and it is to be raised by forcing the wedge under it. Determine the horizontal force P required to just lift it. Assume the block and wedge to be of negligible weight and the angle of contact friction for all surfaces to be 10°.

Ans. 1.4 kN

7.3 A 12° wedge is pushed inside a gap in between two blocks by a vertical force P as shown in Fig. E.7.3. Determine the value of P to just move the 1000 kg block to the left. Also, determine the value of W to maintain equilibrium. The coefficient of friction at all contact surfaces is 0.25.

Ans. 2064.3 N, 10.5 kN

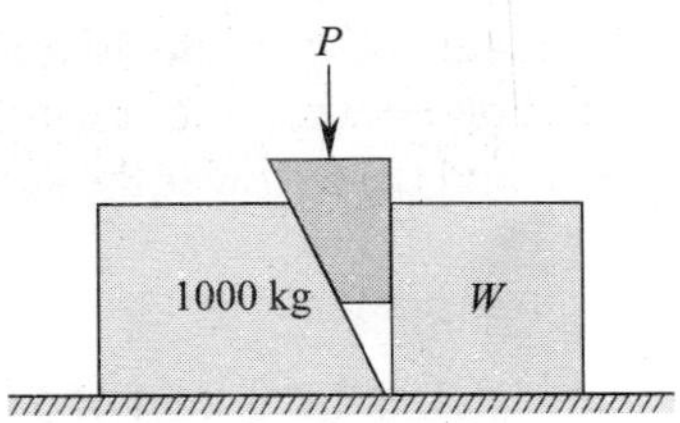

Fig. E.7.3

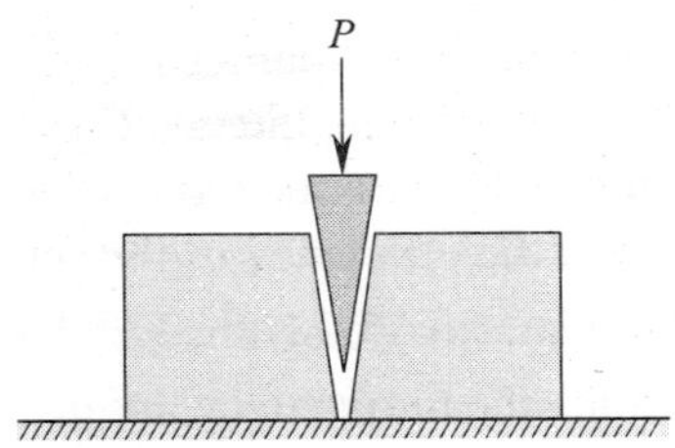

Fig. E.7.4

7.4 A 15° triangular wedge is pushed between two blocks of equal mass of 500 kg as shown in Fig. E.7.4. Determine the minimum force P acting vertically to make the blocks move away. The coefficient of friction at all contact surfaces is 0.2.

Ans. 717.2 N

7.5 A single-threaded screw jack has square threads of a mean diameter of 75 mm and a pitch of 10 mm. Determine the force that must be applied to the end of the 500 mm lever to raise a weight of 30 kN, and the efficiency of the jack. State whether it is under self-locking or not? If yes, determine the force that must be applied to lower the same load. Take $\mu = 0.15$.

Ans. 435.7 N, 21.9%, self-locking, 240.5 N

7.6 The mean radius of a single threaded screw jack is 50 mm and the pitch of the thread is 16 mm. If the coefficient of friction between the screw and nut is 0.2, determine the torque required to raise a load of 1 kN and the efficiency of the screw jack. Also, determine the torque required to lower the load.

Ans. 12.7 N.m, 20.1%, 7.4 N.m

7.7 The distance between adjacent threads of a double-threaded screw jack is 10 mm; mean radius is 60 mm; coefficient of friction is 0.10. What load can be raised by exerting a moment of 100 N.m?

Ans. 10.8 kN

7.8 A square-threaded spindle of a screw jack has a mean diameter of 50 mm and pitch of 10 mm. If the coefficient of friction between the screw and the nut is 0.4, determine (i) force required to be applied at the screw to raise a load of 5 kN, (ii) efficiency of the screw jack, (iii) what should be the pitch for the maximum efficiency of the screw and the corresponding maximum efficiency. Neglect friction between the nut and collar.

Ans. (i) 2374 N, (ii) 13.25%, (iii) 106.4 mm, 45.8%

7.9 A C-clamp is used to compress two wooden boards. The thread of the clamp is a single square thread of 42 mm outer diameter and a 6 mm pitch. The coefficient of static friction is 0.2. If a maximum torque of 25 N.m is applied in tightening the clamp, determine the torque required to loosen the clamp.

Ans. 16.6 N.m

7.10 A steam valve of 15 cm diameter has a pressure of 3 MPa acting on it. If it is closed by means of square-threaded screw of 60 mm external diameter and a 6 mm pitch, determine the torque required to be exerted on the handle of the valve. The coefficient of friction is 0.2.

Ans. 355.2 N.m

7.11 The outer and inner diameters of the spindle in a screw jack are 60 mm and 40 mm respectively. If the screw is single threaded, and the coefficient of friction between the screw and nut is 0.2, determine (i) the torque required to raise a load of 20 kN, and (ii) to lower the same load. Also, determine the efficiency of the screw jack.

Ans. (i) 168 N.m, (ii) 35.4 N.m, 37.92%

7.12 Determine the torque to be applied in a differential screw jack to lift a load of 5 kN, given that the pitch of the outer and inner threads are 12 mm and 8 mm respectively. The mean radius of the outer thread is 40 mm and that of the inner thread is 25 mm; coefficient of friction for both threads is 0.1.

Ans. 35.8 N.m

7.13 Find how many coils must be taken by a string round a rough cylinder in order that a weight may support another *n* times as great.

Ans. $\dfrac{\ln n}{2\pi\mu}$

7.14 Block *B* of 100 kg mass is supported by a rope, which is wrapped around three rough horizontal rods, such that the rope is vertical in all portions as shown in Fig. E.7.14. Knowing that the coefficient of static friction between the rope and the rods is 0.2, determine the range of values of the mass of the block *A* for which equilibrium is maintained.

Ans. 15.2 to 658.6 kg

7.15 Block *B* of 100 kg mass is supported by a rope, which is wrapped around three rough horizontal rods as shown in Fig. E.7.15. Knowing that the coefficient of static friction between the rope and the rods is 0.2, determine the range of values of the mass of the block *A* for which equilibrium is maintained.

Ans. 28.5 to 351.4 kg

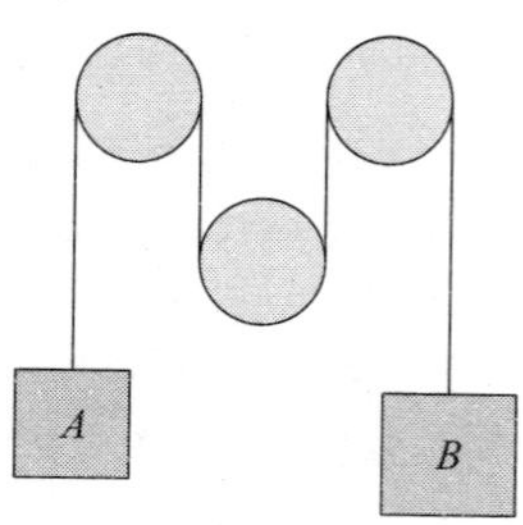

Fig. E.7.14

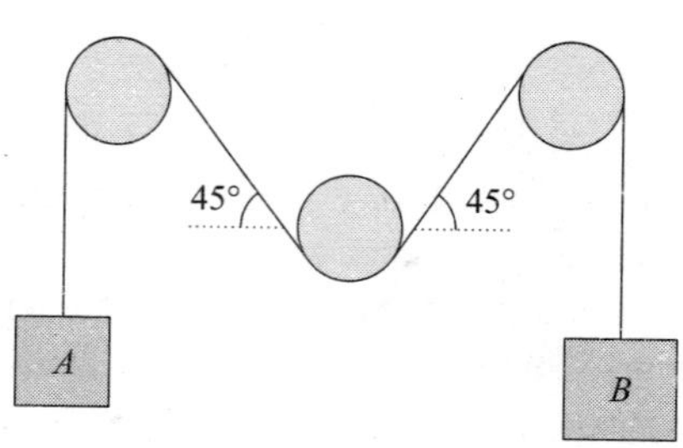

Fig. E.7.15

7.16 A block of 50 kg mass is supported by a rope, which is wrapped around two horizontal cylinders of 8 cm diameters as shown in Fig. E.7.16. Knowing that the coefficient of static friction between the rope and the cylinders is 0.2, determine the range of values of the force P applied at the free end of the rope for which equilibrium is maintained.

Ans. 211 to 1141 N

7.17 A block of 100 kg mass is supported by a rope, which is wrapped around two horizontal cylinders as shown in Fig. E.7.17; the diameter of the upper cylinder being 8 cm and that of the lower cylinder being 10 cm. Knowing that the coefficient of static friction between the rope and the cylinders is 0.2, determine the range of values of the force P applied at the free end of the rope for which equilibrium is maintained.

Ans. 399 to 2413 N

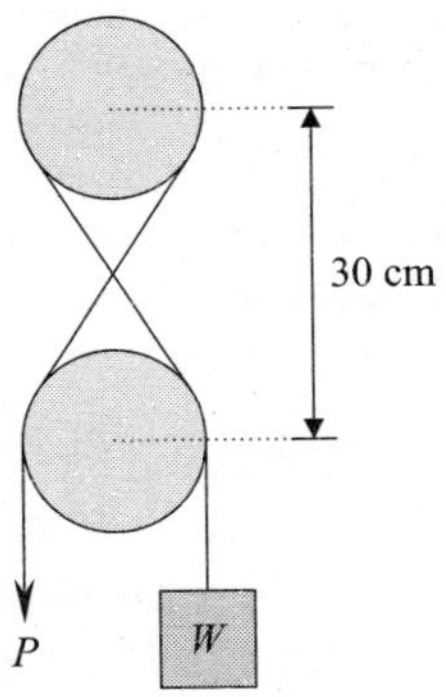

Fig. E.7.16

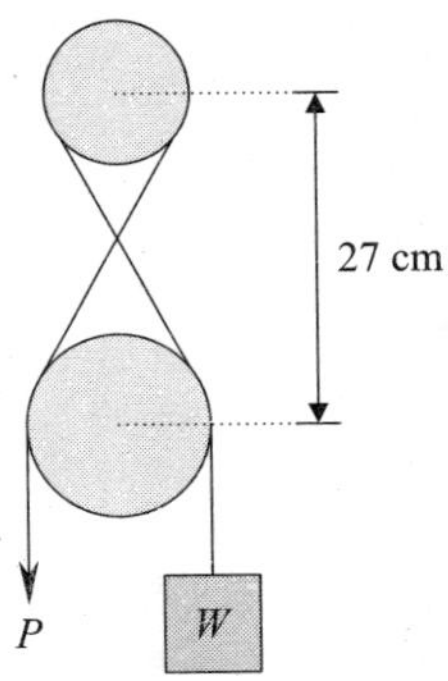

Fig. E.7.17

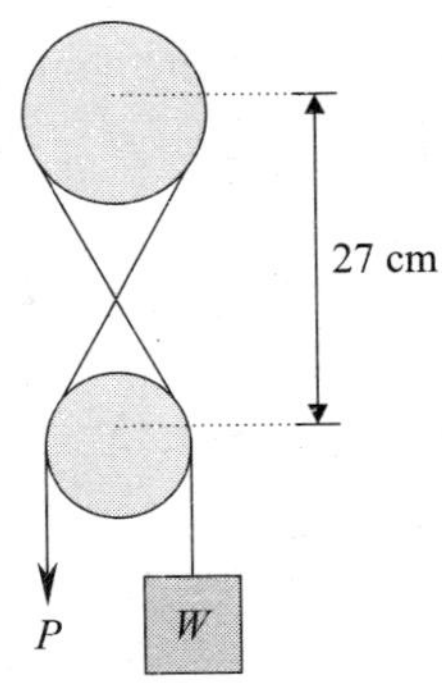

Fig. E.7.18

7.18 Solve the above problem with the upper cylinder diameter being 10 cm and that of the lower cylinder being 8 cm. Refer Fig. E.7.18.

Ans. 399 to 2413 N

7.19 Determine the weight W of the hanging block required to make the 200 kg block resting on a rough horizontal plane to just move. The coefficient of friction between all contact surfaces is 0.2. Refer Fig. E.7.19.

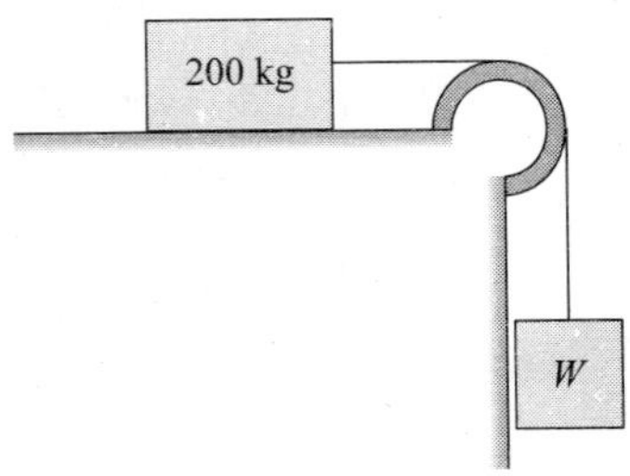

Fig. E.7.19

Ans. 537.2 N

7.20 A homogeneous rope of length l is to be suspended by passing over a cylindrical drum as shown in Fig. E.7.20. What maximum length can be suspended on a side without the rope falling down, if the coefficient of friction between the rope and the drum is 0.2?

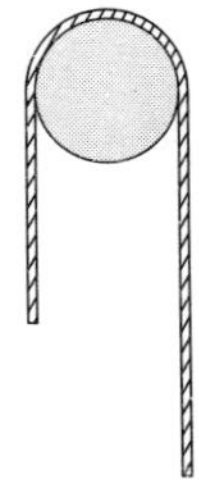

Fig. E.7.20

Ans. 0.65l

7.21 Two pulleys mounted on an engine shaft and machine shaft have respective diameters of 600 mm and 400 mm. If the speed of the driven shaft is 300 rpm, determine the speed of the driver pulley.

Ans. 200 rpm

7.22 An engine drives a shaft by means of a belt. The driving pulley of the engine is 2 m and that of the shaft is 1 m in diameter. If the engine runs at 200 rpm, what will be the speed of the shaft when (i) there is no slip, (ii) there is a slip of 4%.

Ans. 400 rpm, 384 rpm

7.23 An engine shaft running at 500 rpm drives a secondary shaft by means of a belt. The diameter of the engine pulley and driven pulley are 800 mm and 500 mm respectively. Another belt connects a pulley of 600 mm diameter mounted on the same secondary shaft to the machine shaft whose diameter is 400 mm. Determine the speed of the machine shaft when (i) there is no slip, (ii) there is a slip of 3%.

Ans. 1200 rpm, 1164 rpm

7.24 An engine drives a shaft by means of a belt of 6 mm thickness. The driving pulley of the engine is 5 m and that of the shaft is 3 m in diameter. If the engine runs at 200 rpm, what will be the speed of the shaft (i) neglecting belt thickness, (ii) considering the belt thickness?

Ans. 333.33 rpm, 333.07 rpm

7.25 The diameters of driver and driven pulleys in a belt drive are respectively 500 mm and 300 mm. If they are spaced at 1.8 m, determine the angle of contact of the belt over the two pulleys in (i) an open belt arrangement, and (ii) a cross-belt arrangement.

Ans. (i) driver pulley = 186.4°, driven pulley = 173.6°, (ii) driver and driven pulleys = 205.6°

7.26 Two pulleys mounted respectively on an engine shaft and a machine shaft have diameters of 40 cm and 30 cm. The shafts are 60 cm apart. If the coefficient of friction is 0.15, determine the length of belt required to connect the two pulleys for (a) open belt, and (b) cross-belt arrangements.

Ans. 2.3 m, 2.5 m

7.27 A shaft, which rotates at a constant speed of 110 rpm, is connected by belting to a parallel shaft, 2.5 m apart, which has to run at different speeds of 50, 70 and 90 rpm. The smallest pulley on the driver shaft is 160 mm in diameter. Determine the remaining radii of the two stepped pulleys for (i) a crossed-belt arrangement, and (ii) an open belt arrangement. Neglect belt thickness and slip.

Ans. cross-belt: 80 mm – 176 mm; 99.55 mm – 156.45 mm; 115.2 mm – 140.8 mm;

open belt: 80 mm – 176 mm; 101.4 mm – 159.3 mm; 117.4 mm – 143.5 mm

7.28 A shaft, which rotates at a constant speed of 160 rpm, is connected by belting to a parallel shaft, 1800 mm apart, which has to run at 60, 80 and 100 rpm. The smallest pulley on the driver shaft is 150 mm in diameter. Determine the remaining radii of the two stepped pulleys for (i) a crossed belt, and (ii) an open belt. Neglect belt thickness and slip.

Ans. cross belt: 75 mm–200 mm; 91.67 mm – 183.33 mm; 105.75 mm – 169.25 mm

Open belt: 75 mm–200 mm; 92.13 mm – 184.26 mm; 106.6 mm – 170.6 mm

7.29 Find (i) the torque exerted, and (ii) the power transmitted by a belt running over a pulley of 600 mm diameter at 300 r.p.m. The coefficient of friction between the belt and the pulley is 0.2. Angle of lap or contact is 165° and maximum tension in the belt is 2000 N.

Ans. 263 N.m, 8.3 kW

7.30 A leather belt of size 150 mm × 10 mm and weighing 1200 kg/m^3 is used to transmit 10 kW of power from a driver pulley of 600-mm diameter to a driven pulley of 400 mm diameter running at 400 r.p.m. The pulleys are spaced at 2 m, coefficient of friction being 0.25. Determine the permissible stress in the belt (i) neglecting centrifugal tension, and (ii) considering centrifugal tension.

Ans. (i) 1.49 N/mm^2, (ii) 1.58 N/mm^2

7.31 Solve the above problem, considering a cross-belt arrangement.

Ans. (i) 1.33 N/mm^2, (ii) 1.41 N/mm^2

7.32 A torque of 400 N.m acts on the brake drum. One end of the brake band is attached at *B* of the brake lever where the lever is connected to the fulcrum. The other end of the band is attached to the lever at *A*. If the brake band is in contact with the brake drum through 250° and the coefficient of kinetic friction is 0.3, determine the force *P* applied at the end of the brake lever when the drum rotates in the clockwise direction. Refer Fig. E.7.32.

Ans. 342.6 N

7.33 Solve the above problem , if the drum rotates in the anticlockwise direction.

Ans. 92.6 N

7.34 A simple band brake is applied to a rotating drum of 600 mm diameter. A torque of 300 N.m acts on the drum. If the coefficient of friction between the band and the drum is 0.2, determine the value of *P* applied at the end of the lever, when the drum is rotating in the clockwise direction. Refer Fig. E.7.34.

Ans. 363.9 N

7.35 Solve the above problem, if the drum is rotating in the anticlockwise direction.

Ans. 488.9 N

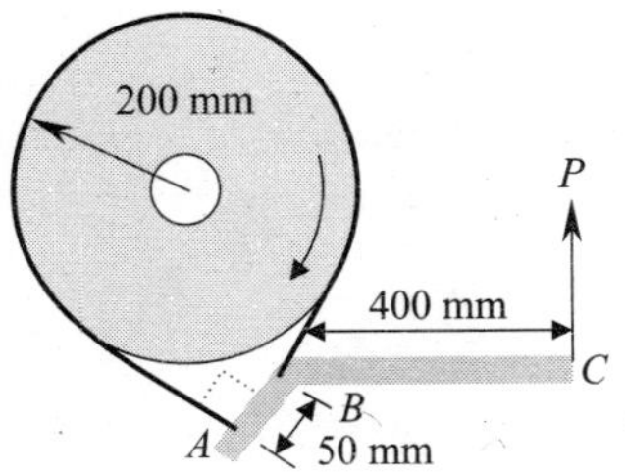

Fig. E7.32, E7.33

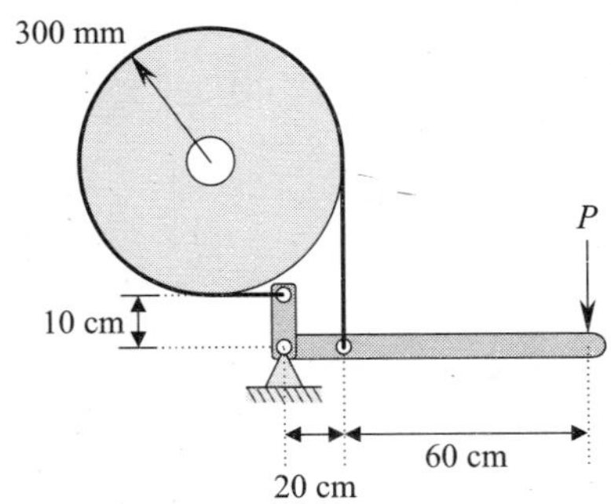

Fig. E.7.34, E.7.35

8 Centre of Gravity and Centroid

8.1 INTRODUCTION

The forces that we have so far dealt with were all treated as *concentrated* forces. However, in reality, these forces are *distributed* in nature. Forces can be distributed over the entire volume of a body as in the case of force of gravity (i.e., weight of a body) or distributed over a surface area in contact as in the case of contact forces such as normal reaction or pressure distribution of water against a dam gate, load distribution in beams, stress distribution in deformable bodies, etc.

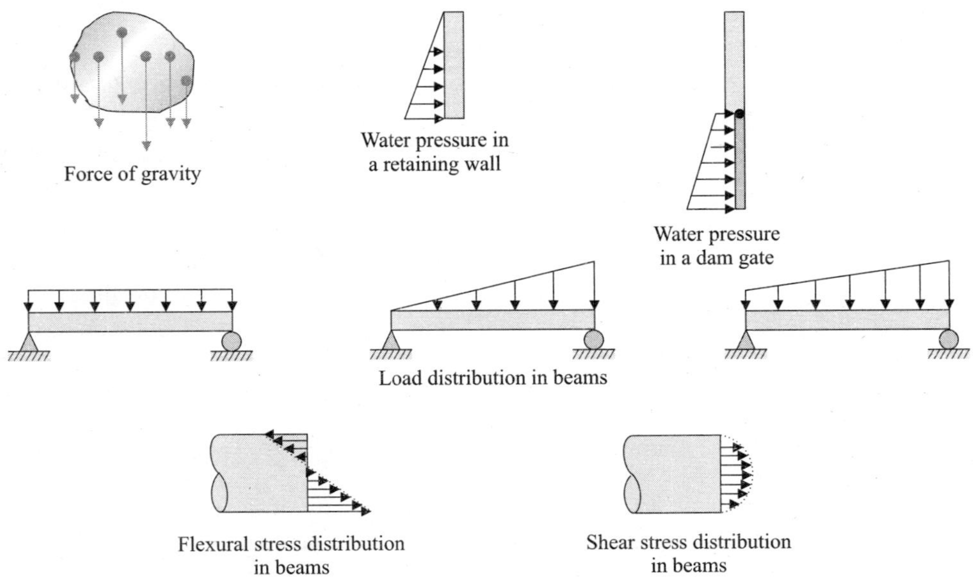

Fig. 8.1 Distributed forces

Though these forces discussed above are all distributed, for *analytical* purposes while applying the conditions of equilibrium, it is customary to replace them by a *single resultant force*, which would produce the same effect as that of the distributed forces. As these forces are *parallel*, we must determine the point of application of the resultant force, which is a point at which the forces are assumed to be concentrated. In the case of gravitational forces, the point of application is termed as **centre of gravity**, a point at which the entire weight is assumed to be concentrated. However, when we deal with merely the geometry of a body as in the case of volume, area or line, this point is termed as **centroid**.

The student will be introduced to determination of expressions for coordinates of centre of gravity in the following section. From this expression, we will derive the expressions for centre of mass and centroid in the subsequent sections. In Sections 8.4–8.7, we will discuss the centroids of lines, areas and volumes. In Section 8.8, we will discuss the centre of mass for various solid bodies and in Section 8.6, we will discuss two theorems, which are helpful in determining surface areas and volumes of revolution of solid bodies.

8.2 CENTRE OF GRAVITY

One of the forces encountered in mechanics is the *force of gravity*. Actually, for an extended body, this is not just one force but the *resultant* of a great many forces. Each particle in the body is acted on by a gravitational force. This force acts vertically downwards towards the centre of the earth.

Suppose we divide the body into n number of elements; then the gravitational force on the ith element is given as $W_i = m_i g$ (see Fig. 8.2), where m_i is mass of the ith element and g is acceleration due to gravity. The value of g is not a constant, but varies from geographical region to region and also with altitude. Since the dimensions of the bodies that we deal in mechanics are small as compared to the distances over which g changes appreciably, we can assume that g is *uniform or constant* over the body.

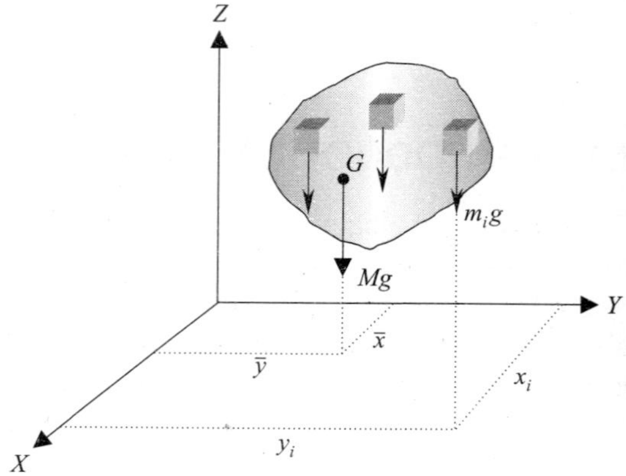

Fig. 8.2 Centre of gravity of a solid body

Further, the sizes of the bodies that we encounter in mechanics are very small compared to the size of the earth. Hence, the gravitational forces acting on these elements, though all of them are directed radially towards the centre of the earth, can be assumed to be **parallel**. Thus, they constitute a system of **non-coplanar parallel forces**. We already saw in Chapter 4 that a system of non-coplanar parallel forces could be replaced by a resultant force. The individual forces being parallel, i.e., having the same direction, the magnitude of this resultant is given by the summation of all individual forces. Hence, we replace them by a resultant whose magnitude is given by $W = \sum_{i=1}^{n} W_i = Mg$ [where M is the mass of the entire body] acting vertically downwards directed towards the centre of the earth. The location (G) of this resultant force is found out by the *principle of moments*, that is the summation of moments of all individual forces about an axis should be same as that of the moment produced by the resultant force.

If (x_i, y_i) is the position of the ith element then the **first moment** of W_i (weight of the ith element) about Y-axis is given as

$$W_i x_i \tag{8.1}$$

(The first moment of weight is defined as the product of the elemental weight multiplied by the perpendicular distance of the element from the reference axis). This is termed as the first moment to differentiate it from the second moment, which we will come across in the next chapter.

Hence, the moment of all n elements about the Y-axis is obtained by summing up:

$$M_y = \sum_{i=1}^{n} W_i x_i \tag{8.2}$$

We know that the sum of the moments of the individual forces about an axis is equal to the moment of the resultant about the same axis. Hence, if $(\bar{x}, \bar{y})$ are coordinates of location of the resultant force $W = Mg$ (refer Fig. 8.2), then

$$\left[\sum_{i=1}^{n} W_i\right]\bar{x} = \sum_{i=1}^{n} W_i x_i \tag{8.3}$$

$$\therefore \qquad \bar{x} = \frac{\sum_{i=1}^{n} W_i x_i}{\sum_{i=1}^{n} W_i} = \frac{M_y}{\sum_{i=1}^{n} W_i} \tag{8.4}$$

If we consider infinitesimally small elements then we can replace the summation sign by integral sign, i.e.,

$$\bar{x} = \frac{\int x\,dW}{\int dW} \tag{8.5}$$

Similarly, it can be shown that y-coordinate of location of the resultant force W is

$$\bar{y} = \frac{\int y\,dW}{\int dW} \tag{8.6}$$

The point of application of the resultant of gravitational forces acting on all the particles of a body is called the **centre of gravity**. In other words, it is **a point** in the body where the entire weight is assumed to be **concentrated**. It remains the same for all orientations of the body. The term centre of gravity is applied to three-dimensional bodies with weight.

8.3 RELATIONSHIP BETWEEN CENTRE OF GRAVITY, CENTRE OF MASS AND CENTROID

As weight can be expressed as a product of mass and acceleration due to gravity, we can write the Eq. 8.5 as

$$\bar{x} = \frac{\int x\,dW}{\int dW} = \frac{\int x\,d(mg)}{\int d(mg)} \tag{8.7}$$

Suppose the acceleration due to gravity is the same for all the particles in the body (which is true always, as the sizes of the bodies that we encounter in mechanics are smaller as compared to the size of the earth) then acceleration due to gravity can be taken outside the integral sign. Canceling g in the numerator and denominator, we can write

$$\bar{x} = \frac{g\int x\,dm}{g\int dm} = \frac{\int x\,dm}{\int dm} \tag{8.8}$$

The above expression locates the coordinate of the **centre of mass**. It is defined as that point in the body at which the entire *mass* is assumed to be *concentrated*. The determination of centre of mass is discussed in Section 8.8. It can be seen that the centre of mass and centre of gravity *coincide* when the acceleration due to gravity is *same* throughout the body.

Further, the mass of a body can be written as a product of its volume and its density. Hence, we can write the Eq. 8.8 as

$$\bar{x} = \frac{\int x\,dm}{\int dm} = \frac{\int x\,d(V\rho)}{\int d(V\rho)} \tag{8.9}$$

If the density ρ of the body is uniform throughout, i.e., if the body is **homogeneous** then it can be taken outside the integral sign. Hence, we can write

$$\bar{x} = \frac{\int x\,dV}{\int dV} \tag{8.10}$$

Thus, we see that when the body is *homogeneous*, the location of the centre of gravity of the body no more depends upon the material of the body, but only on its volume. Hence, to differentiate this point from the centre of gravity, we term it **centroid**, which deals only with the *geometry* of the body. Thus, *centroid* of a *volume* is defined as that point in the body at which the entire *volume* is assumed to be *concentrated*. The determination of the centroid of volume is discussed in Section 8.7.

Further, the volume can be expressed as a product of cross-sectional area and thickness and hence we can write

$$\bar{x} = \frac{\int x\,d(At)}{\int d(At)} \tag{8.11}$$

Again, if the thickness or depth of the body is uniform throughout then we can take it outside the integral sign. Therefore,

$$\bar{x} = \frac{\int x\,dA}{\int dA} \tag{8.12}$$

The above expression describes the location of **centroid** of an **area**. It is defined as that point in the body at which the entire *area* is assumed to be *concentrated*. The location of the centroid of an area is of great importance in the field of strength of materials. In order to have uniform stress distribution in a structural member, the external loads must be applied such that the line of action of their resultant coincides with the centroid of cross section of the member. The determination of location of centroid of an area is discussed in Section 8.5.

Further, if we consider a **homogeneous** wire of uniform cross-sectional area, we can treat the wire as a line. Thus, we can write its centroid as

$$\bar{x} = \frac{\int x\,dL}{\int dL} \tag{8.13}$$

As the above expression is dependent only on the length of the wire, it is termed **centroid** of a **line**. This is discussed in the following section.

8.4 CENTROID OF A LINE

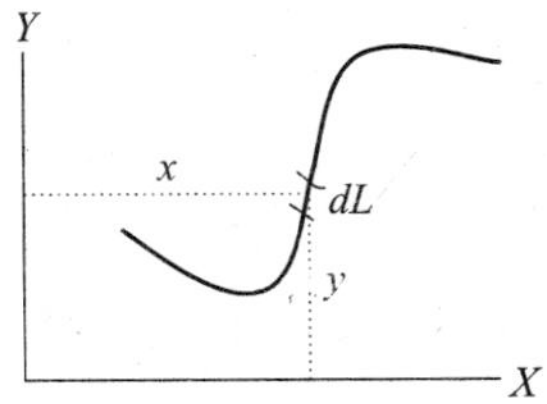

Fig. 8.3 Centroid of a line

Consider a *homogenous* wire of uniform cross-sectional area A, total length L and density ρ. If we divide it into infinitesimally small elements then the weight of an element of length dL is given as

$$dW = \rho A(dL)g \quad (8.14)$$

Hence, the weight of the entire wire is obtained by integrating the above expression over the length,

$$W = \rho AgL \quad (8.15)$$

The first moment of weight of the infinitesimally small element about the X-axis is given as the weight multiplied by the perpendicular distance, i.e., $\rho Ag(dL)y$. Using the principle of moments, the y-coordinate of location of centre of gravity of the entire wire is determined as

$$\bar{y}W = \int \rho Ag(dL)y$$

or

$$\bar{y}\rho AgL = \int \rho Ag(dL)y \quad (8.16)$$

Since the density ρ and cross-sectional area A are constant throughout the length of the wire, they can be taken outside the integral sign. Therefore,

$$\bar{y} = \frac{\int y\,dL}{L} \quad (8.17)$$

Similarly, the x-coordinate of location of centre of gravity of the wire can be determined as

$$\bar{x} = \frac{\int x\,dL}{L} \quad (8.18)$$

We see that the above expressions do not depend upon the material of the wire or its cross-sectional area. Hence, we can conclude that *the centre of gravity of a homogenous wire of uniform cross-sectional area is the same as the centroid of its centre line*. Thus, we can treat the wire as a line while solving for the location of its centroid. The determination of centroids for the various simple curves is discussed below.

8.4.1 Centroid of a Straight Line

Consider a straight line of length L along the X-axis. If we take an infinitesimally small length dx at a distance x from the origin then its first moment about the Y-axis is

$$dM_y = x\,dx$$

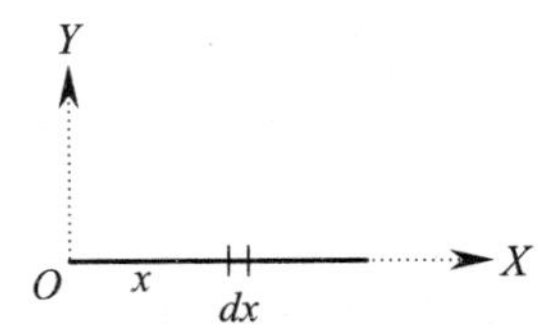

Fig. 8.4 Centroid of a straight line

Therefore, the first moment of the entire length about the Y-axis is

$$M_y = \int_0^L x\,dx = \frac{L^2}{2} \quad (8.19)$$

Therefore, the x-coordinate of the centroid is given as

$$\bar{x} = \frac{M_y}{L}$$

$$= \frac{L^2/2}{L} = L/2 \tag{8.20}$$

From the figure, we can readily see that as the line is along the X-axis, $\bar{y} = 0$. Therefore, we can conclude that the centroid of a straight line lies at the **mid-point** of the line.

8.4.2 Centroid of an Arc of a Circle

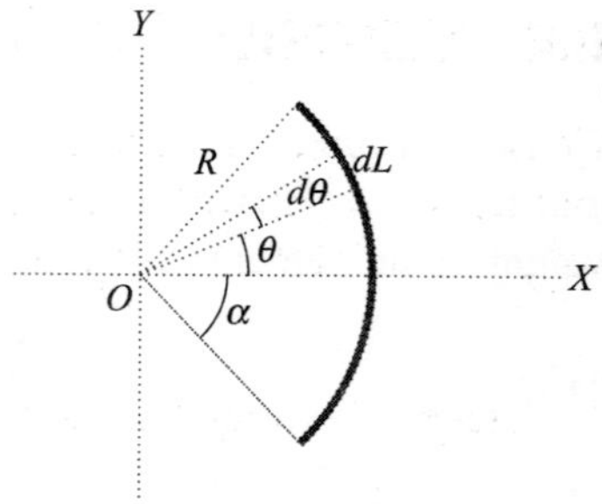

Fig. 8.5 Centroid of an arc of a circle

Consider an arc of a circle symmetric about the X-axis as shown in Fig. 8.5. Let R be the radius of the arc and 2α be the subtended angle. Consider an infinitesimally small length dL such that the radius to the length makes an angle θ with the X-axis. Then its length dL is given as

$$dL = R\,d\theta$$

Therefore, the total length of the arc is

$$L = \int_{-\alpha}^{\alpha} R\,d\theta = 2\alpha R$$

The first moment of the infinitesimally small length about the Y-axis is

$$dM_y = x\,dL = (R\cos\theta)(R\,d\theta) = R^2\cos\theta\,d\theta$$

Hence, the first moment of the entire arc about the Y-axis is given as

$$M_y = \int_{-\alpha}^{\alpha} R^2\cos\theta\,d\theta$$

$$= R^2[\sin\theta]_{-\alpha}^{\alpha} = 2R^2\sin\alpha$$

Therefore, the x-coordinate of centroid of the arc is given as

$$\bar{x} = \frac{M_y}{L} = \frac{2R^2\sin\alpha}{2\alpha R} = \frac{R\sin\alpha}{\alpha} \tag{8.21}$$

From the Fig. 8.5, we can see that due to symmetry of the arc about X-axis, $\boxed{\bar{y} = 0}$.

For a **semicircular arc**, θ varies from $-\pi/2$ to $\pi/2$; hence, the location of its centroid is obtained by substituting $\alpha = \pi/2$ in the above equation:

$$\bar{x} = 2R/\pi \quad \text{and} \quad \bar{y} = 0 \tag{8.22}$$

Note: The centroid of a circular arc due to symmetry about the X and Y axes must lie at the centre of the circle.

The following table summarizes the centroids of various simple curves that we normally encounter:

Table 8.1 Centroids of simple curves

S.No.	*Shape*	*Figure*	*Length*	$\bar{x}$	$\bar{y}$
1.	Straight line	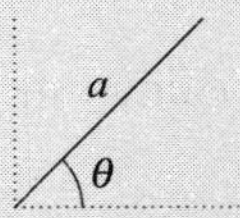	a	$\frac{a}{2}\cos\theta$	$\frac{a}{2}\sin\theta$
2.	Straight line	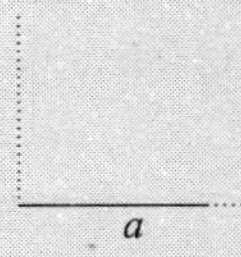	a	$\frac{a}{2}$	0
3.	Straight line	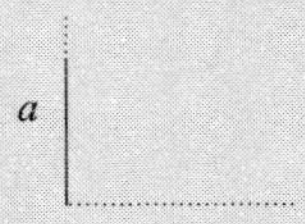	a	0	$\frac{a}{2}$
4.	Semicircular arc	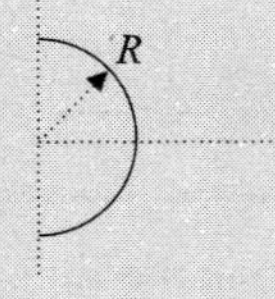	πR	$\frac{2R}{\pi}$	0
5.	Semicircular arc	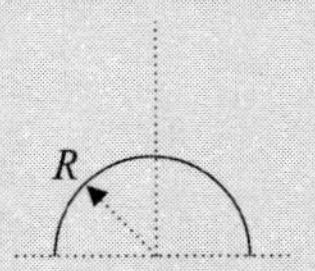	πR	0	$\frac{2R}{\pi}$
6.	Quarter circular arc	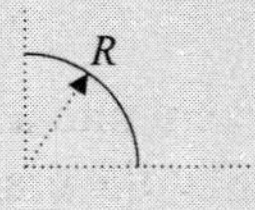	$\frac{\pi R}{2}$	$\frac{2R}{\pi}$	$\frac{2R}{\pi}$
7.	Arc of a circle	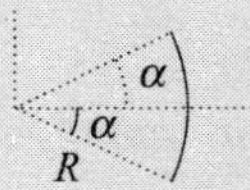	$2\alpha R$	$\frac{R\sin\alpha}{\alpha}$	0

8.4.3 Centroid of a Composite Line

In general, a given curve may not be of regular shape as shown in table above; then in that case, it is divided into finite segments of regular shapes for which positions of centroids are readily known. Let L_i be the length of a segment for which the centroid is known and $(\bar{x}_i, \bar{y}_i)$ be the location of its centroid. Then the centroid of the composite line is given by

$$\bar{x} = \frac{\sum L_i \bar{x}_i}{L} \tag{8.23}$$

and

$$\bar{y} = \frac{\sum L_i \bar{y}_i}{L} \tag{8.24}$$

where $L = \Sigma L_i$. Note that as the number of elements is finite, we use the summation sign. If the length is *added* on, we use a *positive* sign and if it is *removed*, we use a *negative* sign.

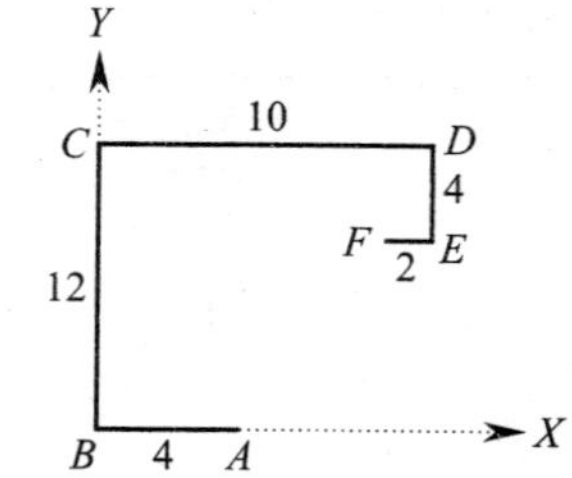

Fig. 8.6

Example 8.1 Find the centroid of a wire bent as shown in Fig. 8.6. All dimensions are in cm.

Solution Divide the wire into straight-line segments *AB, BC, CD, DE* and *EF* for which positions of centroids are readily known. The centroid calculations are summarized below in tabular form:

S.No	Elements	L_i	$\bar{x}_i$	$\bar{y}_i$	$L_i\bar{x}_i$	$L_i\bar{y}_i$
1.	*AB*	4	2	0	8	0
2.	*BC*	12	0	6	0	72
3.	*CD*	10	5	12	50	120
4.	*DE*	4	10	12 – (4/2) = 10	40	40
5.	*EF*	2	10 – (2/2) = 9	12 – 4 = 8	18	16
	$\Sigma =$	32			116	248

Therefore, the location of the centroid of the composite line is given as:

$$\bar{x} = \frac{\sum L_i\bar{x}_i}{\sum L_i} = \frac{116}{32} = 3.625 \text{ cm} \qquad \bar{y} = \frac{\sum L_i\bar{y}_i}{\sum L_i} = \frac{248}{32} = 7.75 \text{ cm}$$

Example 8.2 A slender homogeneous wire *ABCD* bent into a three quarter circle is suspended from a hinge at *A* as shown in Fig. 8.7. Determine the angle that the diameter *AC* would make with the vertical in the suspended position.

Solution We know that the centre of gravity of a *thin homogeneous* wire is the same as the centroid of its centre line. The composite wire can be considered to be made up of three arcs *AB, BC* and *CD*, for which positions of centroids are readily known. An alternate way is to consider the whole circle *ABCDA*, from which we remove the arc *AD*. It should be noted that the length of arc *AD* then must have a *negative* sign. We follow the latter one as it simplifies the calculations. Since the reference axes are not mentioned in the given problem, we choose them conveniently as shown in Fig. 8.7(a).

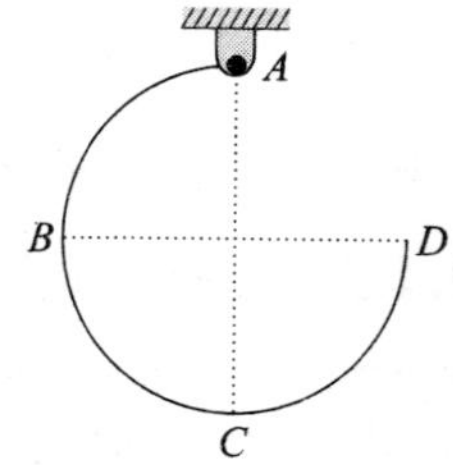

Fig. 8.7

S.No.	Elements	L_i	$\bar{x}_i$	$\bar{y}_i$	$L_i\bar{x}_i$	$L_i\bar{y}_i$
1.	Circle *ABCDA*	$2\pi r = 6.28r$	r	0	$6.28r^2$	0
2.	Quarter circular arc *AD*	$-\pi r/2 = -1.57r$	$r - 2r/\pi$	$2r/\pi$	$-0.57r^2$	$-r^2$
	$\Sigma =$	$4.71r$			$5.71r^2$	$-r^2$

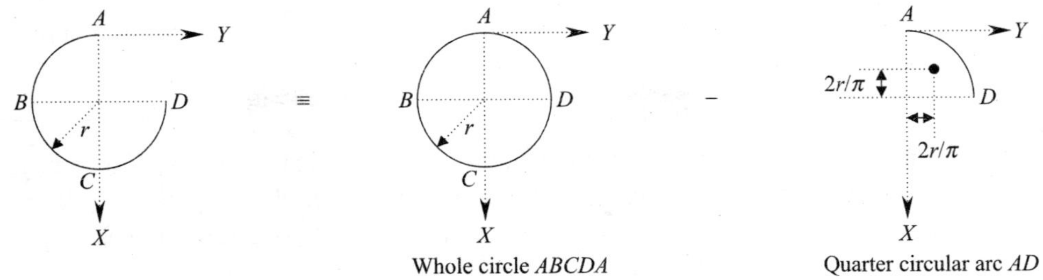

Fig. 8.7(a)

Therefore, the location of the centroid of the wire is given as

$$\bar{x} = \frac{\sum L_i \bar{x}_i}{\sum L_i} \qquad \bar{y} = \frac{\sum L_i \bar{y}_i}{\sum L_i}$$

$$= \frac{5.71r^2}{4.71r} = 1.212r \qquad = \frac{-r^2}{4.71r} = -0.212r$$

From the above result, we see that in Fig. 8.7(b), $\overline{AE} = 1.212r$ and $\overline{EG} = 0.212r$. For stable equilibrium, the centre of gravity (G) must lie below the point of support A, or in other words, points A and G must lie on the same vertical line. Hence, from Fig. 8.7(c), we see that

$$\tan\theta = \left|\frac{\overline{GE}}{\overline{AE}}\right| = \frac{0.212r}{1.212r}$$

$$\Rightarrow \qquad \theta = 9.92°$$

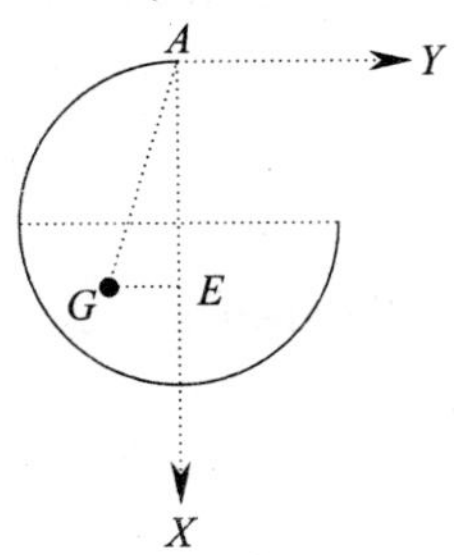

Fig. 8.7(b)

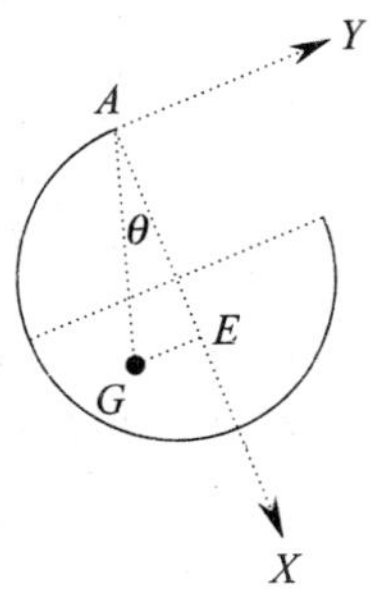

Fig. 8.7(c)

8.5 CENTROID OF AN AREA

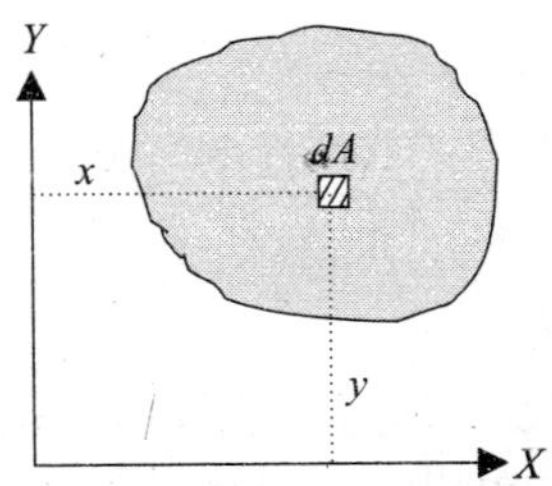

Fig. 8.8 First moment of an area

In this section, we will discuss the expressions to locate the centroid of a plane lamina or area. Consider a homogeneous plate of uniform thickness t, area A and density ρ. If we take an infinitesimally small area dA (refer Fig. 8.8) then its weight is

$$dW = (\rho t\, dA)g \tag{8.25}$$

Hence, the weight of the entire plate is obtained by integrating the above expression over the entire area,

$$W = \rho t g A \tag{8.26}$$

The first moment of weight of this infinitesimally small element about the X-axis is given as the product of its weight and perpendicular distance from the X-axis, i.e., $dM_x = (\rho t g\, dA)y$. By the principle of moments, the y-coordinate of the location of the centre of gravity is determined as

$$\bar{y}\, W = \int \rho t g\, (dA) y$$

or

$$\bar{y} \rho t g A = \int \rho t g\, (dA) y \tag{8.27}$$

Since the density and thickness are constant, they can be taken outside the integral sign.

Therefore,

$$\bar{y} = \frac{\int y\, dA}{A} \tag{8.28}$$

Similarly, it can be shown that

$$\bar{x} = \frac{\int x\, dA}{A} \tag{8.29}$$

We see that the above expressions do not depend upon the material of the plate or its thickness. Hence, we can conclude that *the centre of gravity of a homogenous plate of uniform thickness is the same as the centroid of its surface area.* Thus, we can treat the plate as a plane area while solving for the location of its centroid. The determination of centroid for the various sections is discussed below.

8.5.1 Centroid of a Rectangle

Consider a rectangle of base length b and height h. If we take a thin strip parallel to the X-axis at a distance y from the X-axis and of infinitesimally small thickness dy then its area is given as $dA = b\, dy$.

Hence, the area of the rectangle is

$$A = \int_0^h dA = \int_0^h b\, dy = bh$$

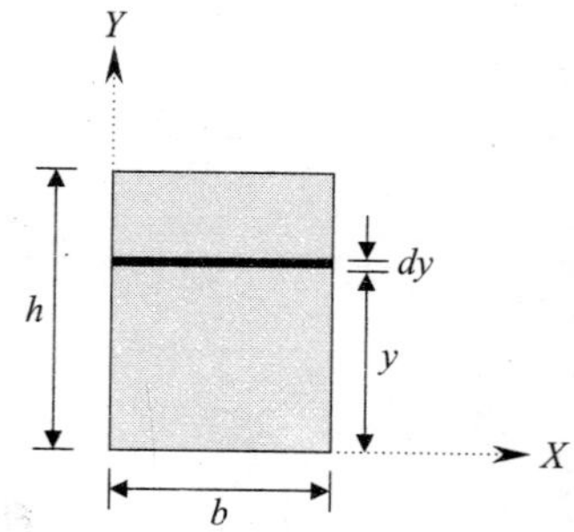

Fig. 8.9 Centroid of a rectangle

As each point on this strip is at the same distance y from the X-axis, we can take moment of area of the strip about the X-axis as

$$dM_x = y dA = yb\, dy$$

Therefore, the first moment of the entire area about the X-axis is

$$M_x = \int_0^h y(b\, dy) = \frac{bh^2}{2} \tag{8.30}$$

Hence, the y-coordinate of the centroid of the rectangle is given as

$$\bar{y} = \frac{M_x}{A} = \frac{h}{2} \tag{8.31}$$

In a similar manner, we can consider a vertical strip at a distance x from the Y-axis and of infinitesimally small thickness dx, and obtain the x-coordinate of the centroid as

$$\bar{x} = \frac{b}{2} \tag{8.32}$$

Thus, we can see that the centroid of a rectangle lies at the midpoint or in other words, at the intersection of its two diagonals.

8.5.2 Centroid of a Right-angled Triangle

Consider a right-angled triangle of base b and height h. If we take a thin strip parallel to the base at a distance y from the X-axis and of infinitesimally small thickness dy then its area is $dA = b'\,dy$, where b' is the width of the strip. From similar triangles, we know that

$$\frac{b'}{h-y} = \frac{b}{h} \Rightarrow b' = \frac{b}{h}(h-y)$$

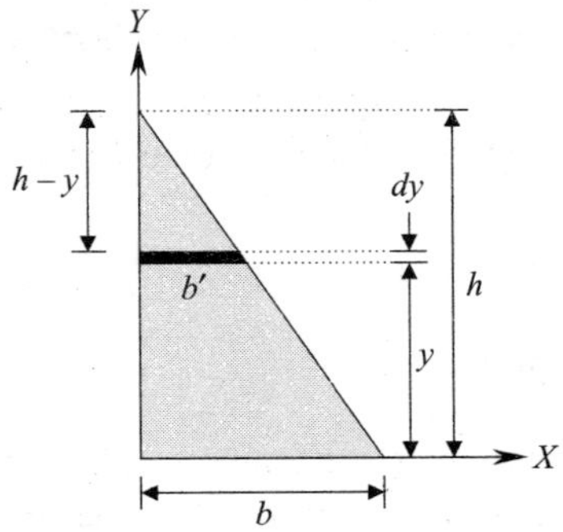

Fig. 8.10 Centroid of a right-angled triangle

Therefore,

$$dA = \frac{b}{h}(h-y)dy$$

Then area of the triangle is obtained as

$$A = \frac{b}{h}\int_0^h (h-y)dy$$

$$= \frac{b}{h}\left[hy - \frac{y^2}{2}\right]_0^h = \frac{bh}{2} \tag{8.33}$$

The first moment of the strip with respect to the X-axis is

$$dM_x = ydA = y\left[\frac{b}{h}(h-y)\right]dy$$

Therefore, the first moment of the entire area about the X-axis is given as

$$M_x = \int_0^h y\,dA = \int_0^h y\frac{b}{h}(h-y)dy$$

$$= \frac{b}{h}\int_0^h (hy - y^2)dy$$

$$= \frac{b}{h}\left[h\frac{y^2}{2} - \frac{y^3}{3}\right]_0^h$$

$$= \frac{bh^2}{6} \tag{8.34}$$

Therefore, the y-coordinate of the centroid is given as

$$\bar{y} = \frac{M_x}{A} = \frac{bh^2/6}{bh/2} = \frac{h}{3} \tag{8.35}$$

In a similar manner, we can consider a vertical strip of area dA parallel to the Y-axis and obtain the x-coordinate of the centroid as

$$\bar{x} = M_y/A = \frac{b}{3} \tag{8.36}$$

Corollary Even though the centroid is a fixed point on the area, its coordinates with respect to the reference axes will be different for **different orientations** of the area.

For instance, if the right-angled triangle is oriented as in the adjacent Fig. 8.11 then the coordinates of centroid are

$$\bar{x} = \frac{2}{3}b \tag{8.37}$$

and

$$\bar{y} = \frac{h}{3} \tag{8.38}$$

which is different from the coordinates for the orientation of triangle considered in the above derivation.

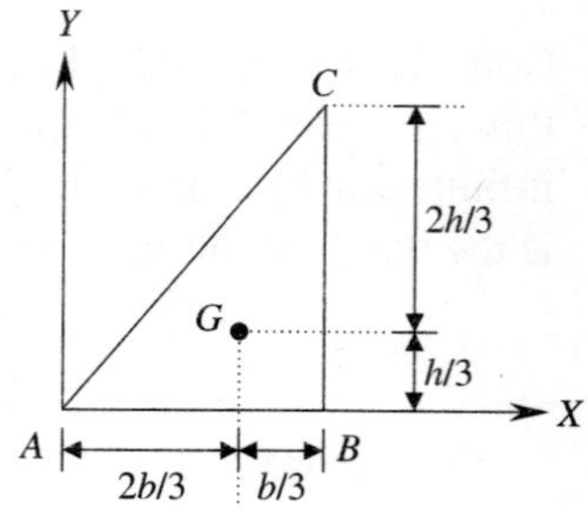

Fig. 8.11

We can also observe from Fig. 8.11 that if the x coordinate of the centroid is measured from the vertex A then it is $2b/3$, whereas if measured from B, it is $b/3$. Similarly, if the y-coordinate of the centroid is measured from the vertex B then it is $h/3$, whereas if measured from C, it is $2h/3$. Thus, we see that the coordinates of the centroid will be different for different orientations of the triangle. Since it is not possible to remember these coordinates for different orientations, the student may follow this general rule: If the coordinates are measured from the vertex with a slant side then it should be 2/3rd of base or height and if it is measured from a vertex with a vertical side then it is 1/3rd of the base or height.

8.5.3 Centroid of a Triangle in General

If (x_1, y_1), (x_2, y_2) and (x_3, y_3) are the coordinates of vertices of a triangle in general, then its centroid is given as

$$\bar{x} = \frac{x_1 + x_2 + x_3}{3} \quad \text{and} \tag{8.39}$$

$$\bar{y} = \frac{y_1 + y_2 + y_3}{3} \tag{8.40}$$

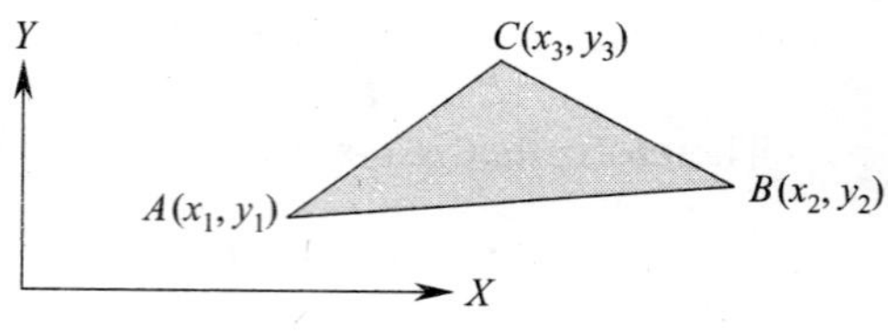

Fig. 8.12 Centroid of a triangle

Proof Suppose we divide the triangle into rectangular strips of infinitesimally small thickness parallel to side AB as shown in Fig. 8.13(a).

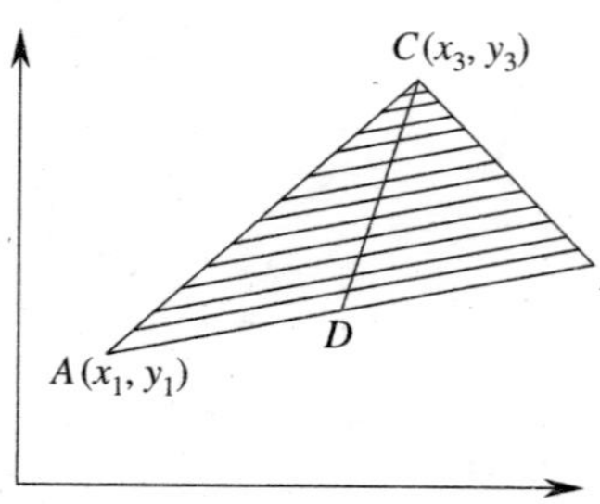

Fig. 8.13(a)

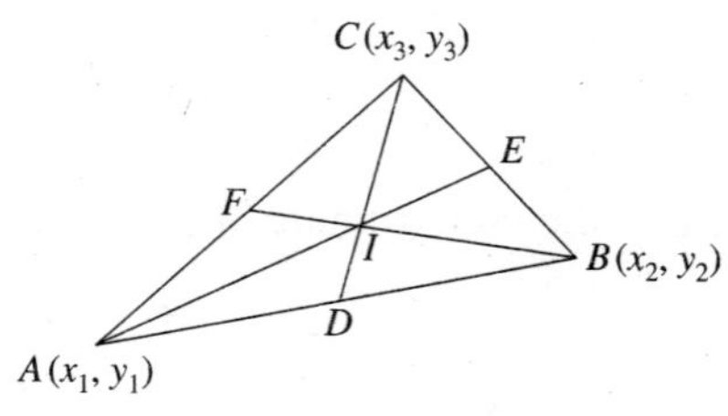

Fig. 8.13(b)

Since the centroid of a rectangular area lies at the midpoint, the centroid of the triangle must lie on the line joining the midpoints of these rectangular strips, i.e., *CD*. Since *D* is the midpoint on side *AB*, its coordinates are given by

$$\left(\frac{x_1 + x_2}{2}\right) \text{ and } \left(\frac{y_1 + y_2}{2}\right) \tag{8.41}$$

Similarly, we can consider strips parallel to sides *BC* and *AC*. Then the centroid will lie on the lines *AE* and *BF* respectively, where *E* and *F* are midpoints of sides *BC* and *AC* respectively. Hence, we can conclude that the centroid of the triangle must lie at the intersection point *I* of these three lines *CD*, *AE* and *BF*. We know from the properties of the triangle, that this point divides the line *CD*, in the ratio of 2:1, i.e., $CI = 2\,DI$.

Horizontal length of line CD = (x coordinate of C) – (x coordinate of D)

$$= x_3 - \left(\frac{x_1 + x_2}{2}\right)$$

Therefore, the x-coordinate of I can be determined as

$$\bar{x} = (x\text{-coordinate of } D) + (1/3)\,(\text{horizontal length of } CD)$$

$$= \left(\frac{x_1 + x_2}{2}\right) + \frac{1}{3}\left[x_3 - \left(\frac{x_1 + x_2}{2}\right)\right]$$

$$= \frac{x_1 + x_2 + x_3}{3} \tag{8.42}$$

Similarly, its y-coordinate can be determined as

$$\bar{y} = \frac{y_1 + y_2 + y_3}{3} \tag{8.43}$$

8.5.4 Centroid of Area of a Circular Sector

Consider an area of a circular sector of radius R with subtended angle 2α and symmetric about the X-axis. If we take an element of area *OCD* at an angle θ from the X-axis then its area can be determined as follows:

OCD can be considered as a triangle and its area is then given as

$$dA = 1/2\,R\,Rd\theta = \frac{R^2}{2}d\theta$$

The centroid of this triangle lies at a distance of 2/3 R from O. Hence, the x and y coordinates of the centroid are

$$x = \frac{2}{3}R\cos\theta \quad \text{and} \quad y = \frac{2}{3}R\sin\theta \tag{8.44}$$

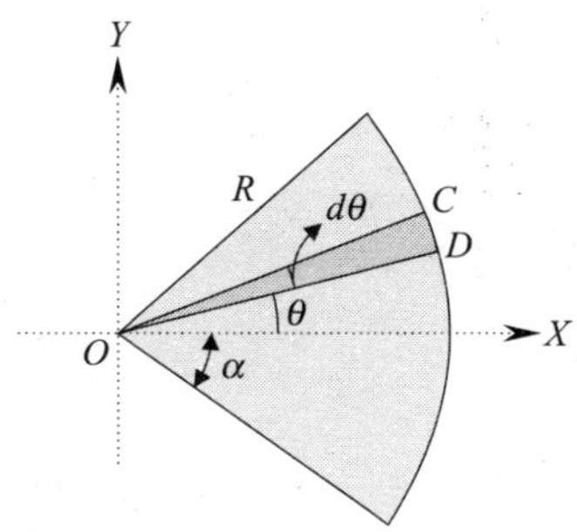

Fig. 8.14 A circular sector

Area of the entire circular sector is obtained by integrating the expression for dA between limits, i.e.,

$$A = \int_{-\alpha}^{\alpha} \frac{R^2}{2}\, d\theta = R^2 \alpha \tag{8.45}$$

Taking the first moment of the triangle OCD about the Y-axis,

$$dM_y = x\, dA = \frac{2}{3} R \cos\theta \frac{R^2}{2} d\theta$$

Therefore, the first moment of the entire area about the Y-axis is

$$M_y = \int x\, dA$$

$$= \int_{-\alpha}^{\alpha} \frac{2}{3} R \cos\theta \frac{R^2}{2} d\theta$$

$$= \frac{R^3}{3}\left[\sin\theta\right]_{-\alpha}^{\alpha} = \frac{2R^3 \sin\alpha}{3}$$

Therefore, the x-coordinate of the centroid is

$$\bar{x} = M_y/A = \frac{2}{3}\frac{R \sin\alpha}{\alpha} \tag{8.46}$$

and the y-coordinate of the area is $\bar{y} = 0$ because of the symmetry of the sector with respect to the X-axis.

Corollary For a semicircular area, we know that θ varies from $-\pi/2$ to $\pi/2$. Hence, its centroid is obtained by substituting $\alpha = \pi/2$ in the above expression for $\bar{x}$. Therefore, we get

$$\bar{x} = \frac{4R}{3\pi} \quad \text{and} \quad \bar{y} = 0$$

Similarly, if the area is oriented as shown in Fig. 8.15(b) then the centroidal coordinates are

$$\bar{x} = 0 \quad \text{and} \quad \bar{y} = \frac{4R}{3\pi}$$

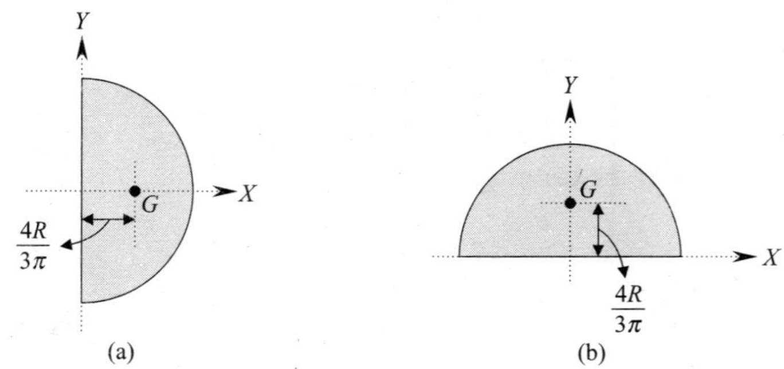

Fig. 8.15 Centroid of semicircular area

8.5.5 Centroid of a Parabola

Consider a shaded area bounded by a parabola of equation $y = kx^2$, X-axis and line $x = b$ as shown in Fig. 8.16. Then we see that at $x = 0$, $y = 0$ and at $x = b$, $y = h$. Therefore,

$$k = \frac{h}{b^2}$$

Hence, we can write the equation of the curve as

$$y = \frac{h}{b^2}x^2$$

Consider a vertical strip parallel to the Y-axis at a distance x from the origin and of infinitesimally small thickness dx as shown in the figure. Then its elemental area is given as $dA = y\,dx = (h/b^2)x^2\,dx$. Therefore, the area under the entire curve is

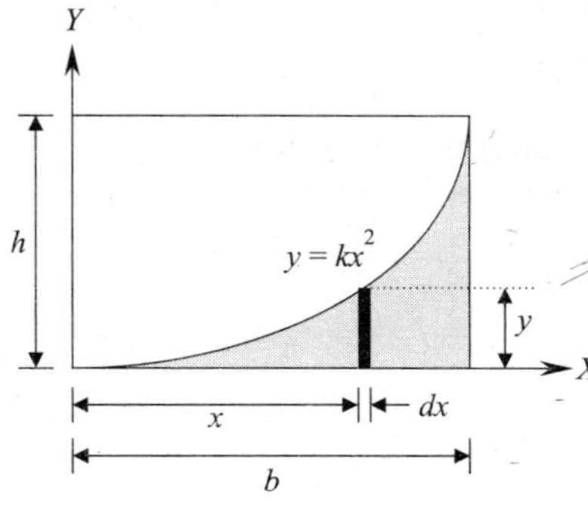

Fig. 8.16

$$A = \int_0^b \left(\frac{h}{b^2}\right) x^2\,dx$$

$$= \frac{h}{b^2}\frac{b^3}{3} = \frac{bh}{3} \tag{8.47}$$

We see that the area of the curve is 1/3rd of the area of the enclosed rectangle. The first moment of the area about the Y-axis is given as

$$M_y = \int x\,dA$$

$$= \int_0^b x\frac{h}{b^2}x^2\,dx$$

$$= \frac{h}{b^2}\frac{b^4}{4} = \frac{b^2h}{4}$$

Therefore, the x-coordinate of the centroid is given as

$$\bar{x} = \frac{M_y}{A} = \frac{b^2h/4}{bh/3} = \frac{3}{4}b \tag{8.48}$$

In a similar manner, we can consider a thin strip parallel to the X-axis and of infinitesimally small thickness dy as shown in Fig. 8.16(a) and obtain the y-coordinate of the centroid as follows:

The elemental area is given as $dA = (b - x)dy$. Therefore, the first moment of the area about the X-axis is given as

$$M_x = \int y\,dA = \int_0^h y(b-x)\,dy$$

$$= \int_0^h y\left(b - \frac{b}{h^{1/2}}y^{1/2}\right)dy$$

$$= \left[b\frac{y^2}{2} - \frac{b}{h^{1/2}}\frac{y^{5/2}}{5/2}\right]_0^h = \frac{bh^2}{10}$$

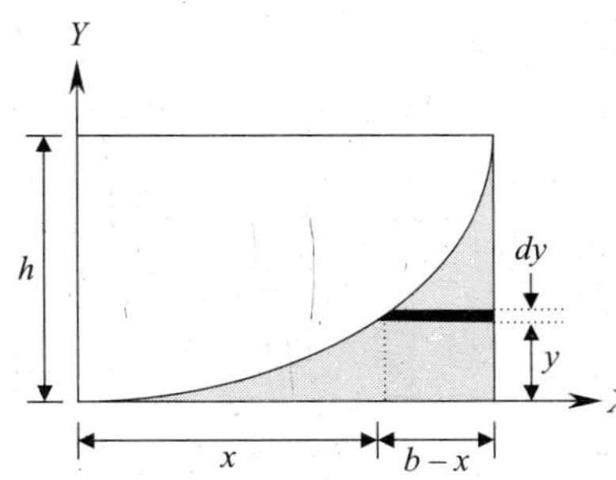

Fig. 8.16(a)

Therefore, the y-coordinate of the centroid is given as

$$\bar{y} = \frac{M_x}{A} = \frac{bh^2/10}{bh/3} = \frac{3}{10}h \tag{8.49}$$

The centroidal coordinates of regular geometrical shapes can be determined in a like manner and the results are summarized below in tabular form.

Table 8.2 Centroids of regular geometrical shapes

S.No	*Shape*	*Figure*	*Area*	$\bar{x}$	$\bar{y}$
1.	Rectangle	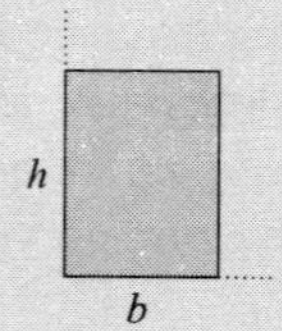	bh	$\frac{b}{2}$	$\frac{h}{2}$
2.	Right-angled triangle	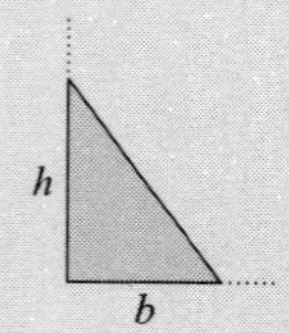	$\frac{1}{2}bh$	$\frac{b}{3}$	$\frac{h}{3}$
3.	Right-angled triangle	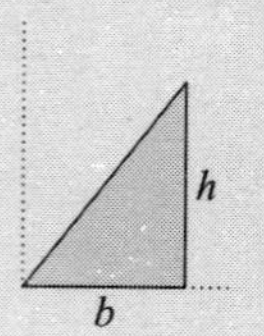	$\frac{1}{2}bh$	$\frac{2}{3}b$	$\frac{h}{3}$
4.	Semicircle	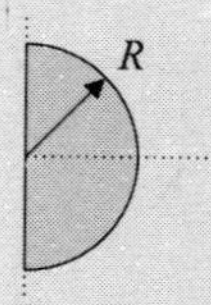	$\frac{\pi R^2}{2}$	$\frac{4R}{3\pi}$	0
5.	Semicircle	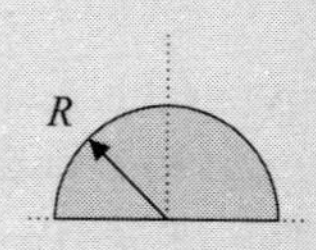	$\frac{\pi R^2}{2}$	0	$\frac{4R}{3\pi}$
6.	Quadrant	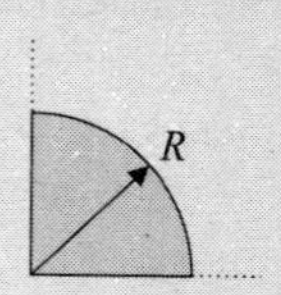	$\frac{\pi R^2}{4}$	$\frac{4R}{3\pi}$	$\frac{4R}{3\pi}$
7.	Circular sector	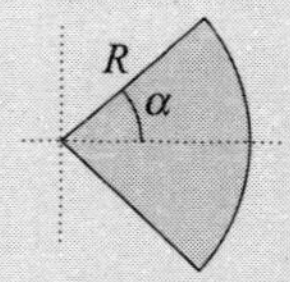	αR^2	$\frac{2R\sin\alpha}{3\alpha}$	0

Contd.

8.	Semi-elliptical area	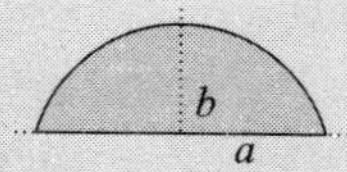	$\frac{\pi ab}{2}$	0	$\frac{4b}{3\pi}$
9.	Parabola	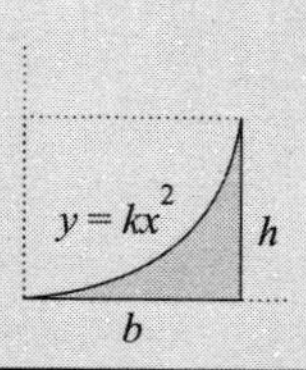	$\frac{1}{3}bh$	$\frac{3b}{4}$	$\frac{3h}{10}$

8.5.6 Axis of Symmetry

If an area has an axis of symmetry (say Y–Y axis) then its centroid will lie on that axis. This can be proved by looking at the Fig. 8.17. For every element of area dA on the positive side of the axis, there is a mirror element dA with moment arm $-x$. Hence, taking the first moment of the area about the Y-axis, while summing up, these two values cancel out. Since there will be such pairs of areas throughout the entire area, the first moment of the entire area about the Y-axis will be zero, i.e., $M_y = 0$. Hence, $\bar{x} = 0$; or in other words, the centroid lies on the Y–Y axis. The examples for such areas are an isosceles triangle, a semi-circle, etc.

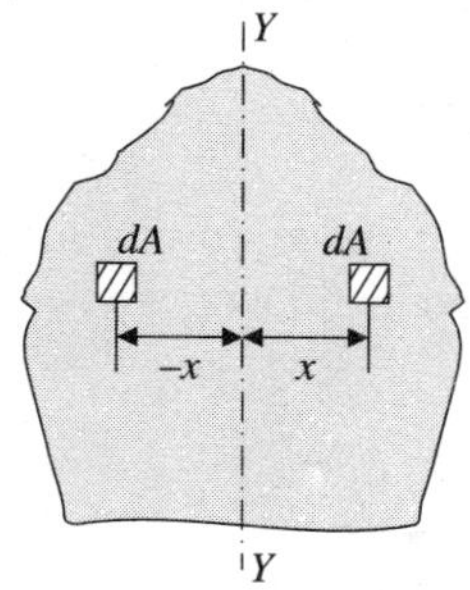

Fig. 8.17 Area with axis of symmetry

Similarly, if the area has two axes of symmetry like rectangle, circle, equal flange I-section, etc., then the centroid will lie at the intersection of the two axes.

Example 8.3 By direct integration, find the location of the centroid of the area of a spandrel bounded by the X-axis, the line $x = b$ and $y = kx^n$ where $n \geq 0$.

Solution From Fig. 8.18, we see that at $x = 0$, $y = 0$ and at $x = b$, $y = h$

$$\Rightarrow \qquad k = \frac{h}{b^n}$$

Therefore, the equation of the curve can be written as $y = \frac{h}{b^n}x^n$

Consider a thin strip parallel to the Y-axis at a distance x from the origin and of infinitesimal thickness dx. Its area dA is then given as $dA = y\,dx$

Hence, the area of the spandrel is obtained by integrating the infinitesimal area between the limits:

$$A = \int_0^b y\,dx$$

$$= \int_0^b \frac{h}{b^n}x^n\,dx$$

$$= \frac{h}{b^n}\left[\frac{x^{n+1}}{n+1}\right]_0^b = \frac{bh}{n+1}$$

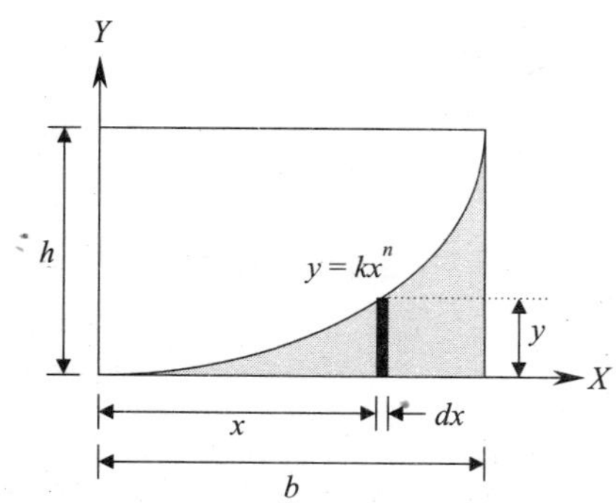

Fig. 8.18

The first moment of the strip about the Y-axis is

$$dM_y = x\,dA$$

Therefore, the first moment of the entire area about the Y-axis is obtained as

$$M_y = \int_0^b x\,dA = \int_0^b \frac{h}{b^n} x^{n+1}\,dx$$

$$= \frac{h}{b^n}\left[\frac{x^{n+2}}{n+2}\right]_0^b = \frac{hb^2}{n+2}$$

Therefore, the x-coordinate of the centroid is obtained as

$$\bar{x} = \frac{M_y}{A} = \frac{hb^2}{n+2}\,\frac{n+1}{bh} = b\frac{n+1}{n+2}$$

In a similar manner, we can consider a thin horizontal strip and obtain the y-coordinate of the centroid as

$$\bar{y} = h\frac{n+1}{4n+2}$$

Example 8.4 Determine the centroid of quadrant of an ellipse, whose equation is $\frac{x^2}{a^2} + \frac{y^2}{b^2} = 1$.

Solution From the equation of the ellipse, we get the expression for y as

$$y = \frac{b}{a}\sqrt{a^2 - x^2}$$

Consider a thin strip of infinitesimal thickness dx parallel to the Y-axis at a distance x from the origin. Its area dA is then given as

$$dA = y\,dx = \frac{b}{a}\sqrt{a^2 - x^2}\,dx$$

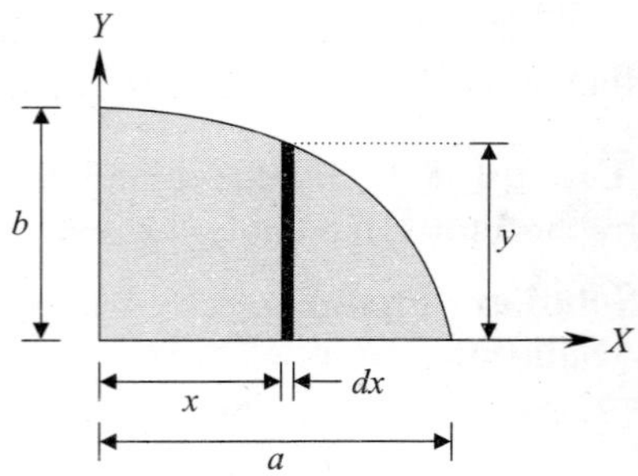

Fig. 8.19

Hence, the area of the quadrant of the ellipse is obtained by integrating the infinitesimal area between the limits:

$$A = \int_0^a dA = \int_0^a y\,dx$$

$$= \int_0^a \frac{b}{a}\sqrt{a^2 - x^2}\,dx$$

Upon integration by parts, we get

$$= \frac{b}{a}\left[\frac{x\sqrt{a^2 - x^2}}{2} + \frac{a^2}{2}\sin^{-1}\frac{x}{a}\right]_0^a$$

$$= \frac{\pi ab}{4}$$

Taking the first moment of the infinitesimal area about the Y-axis, we have

$$dM_y = x\,dA = x\frac{b}{a}\sqrt{a^2 - x^2}\,dx$$

Hence, the first moment of the entire area about the Y-axis is obtained as

$$M_y = \int_0^a x\,dA = \frac{b}{a}\int_0^a x\sqrt{a^2 - x^2}\,dx$$

$$= -\frac{b}{a}\left[\frac{(a^2 - x^2)^{3/2}}{3}\right]_0^a$$

$$= \frac{a^2 b}{3}$$

Therefore, the x-coordinate of the centroid of the quadrant of the ellipse is given as

$$\bar{x} = \frac{M_y}{A}$$

$$= \frac{a^2 b/3}{\pi ab/4} = \frac{4a}{3\pi}$$

In a similar manner, we can consider a thin horizontal strip and obtain the y-coordinate of the centroid as

$$\bar{y} = \frac{4b}{3\pi}$$

Example 8.5 Find the centroid of the area bounded by $y = \sin x$ between $x = 0$ and $x = \pi$, and the X-axis.

Solution Consider a thin strip of infinitesimal thickness dx parallel to the Y-axis at a distance x from the origin. Its area dA is then given as

$$dA = y\,dx$$

$$= \sin x\,dx$$

Hence, the area of the shaded region is obtained by integrating the above infinitesimal area between the limits:

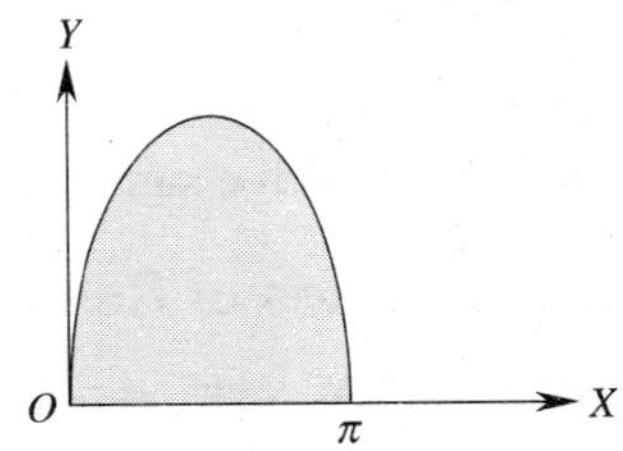

Fig. 8.20

$$A = \int_0^\pi dA = \int_0^\pi \sin x\,dx$$

$$= [-\cos x]_0^\pi$$

$$= -[(-1) - (1)] = 2$$

Since the strip is thin, we can assume it to be a rectangle. Hence, its centroid lies at its midpoint, i.e., at $y/2$ from the X-axis. Therefore, taking the first moment of the infinitesimal area about the X-axis, we have

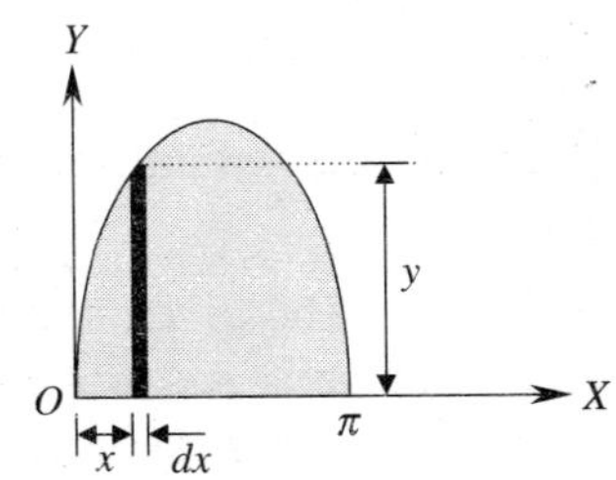

Fig. 8.20(a)

$$dM_x = \frac{y}{2}dA = \frac{\sin x}{2}\sin x\,dx = \frac{\sin^2 x}{2}\,dx$$

Hence, the first moment of the entire area about the X-axis is obtained as

$$M_x = \int_0^{\pi} \frac{\sin^2 x}{2} dx$$

This can be written as

$$M_x = \frac{1}{2}\int_0^{\pi} \frac{(1-\cos 2x)}{2} dx = \frac{1}{4}\left[x - \frac{\sin 2x}{2}\right]_0^{\pi} = \frac{\pi}{4}$$

Therefore, the y-coordinate of the centroid is given as

$$\bar{y} = \frac{M_x}{A} = \frac{\pi/4}{2} = \frac{\pi}{8}$$

Due to symmetry, we know that the x-coordinate of the centroid is

$$\bar{x} = \frac{\pi}{2}$$

8.5.7 Centroid of Composite Figure

In engineering work, we frequently need to locate the centroid of a composite area. Such an area may be composed of regular geometric shapes such as rectangle, triangle, circle, semicircle, quarter circle, etc. In such cases, we divide the given area into regular geometric shapes for which the positions of centroids are readily known. Let A_i be the area of an element and $(\bar{x}_i, \bar{y}_i)$ be the respective centroidal coordinates. Then for the composite area,

$$A\bar{x} = A_1\bar{x}_1 + A_2\bar{x}_2 + \ldots + A_n\bar{x}_n$$

Therefore, $$\bar{x} = \frac{\sum A_i\bar{x}_i}{A} \tag{8.50}$$

Similarly, $$\bar{y} = \frac{\sum A_i\bar{y}_i}{A}, \tag{8.51}$$

where the total area, $A = \Sigma A_i$, in which the areas are added up algebraically. When an area is **added** on, we use a **positive** sign and if it is **removed**, we use a **negative** sign.

8.5.8 Axes of Reference

Although the centroid is a fixed point on the area, its position coordinates depend on the reference axes chosen. Sometimes the reference axes are specified in the problem itself. If they are *not* specified, we have the liberty to choose them such that it simplifies the calculations. Generally, the area is placed on the first quadrant with the lowest line of the area coinciding with the X-axis and the left line of the area coinciding with the Y-axis.

Example 8.6 Find the centroid of the plain lamina shown in Fig. 8.21.

Solution The given section can be considered to be made up of two rectangular sections as shown in the figures below. The reference axes can also be conveniently chosen as shown below.

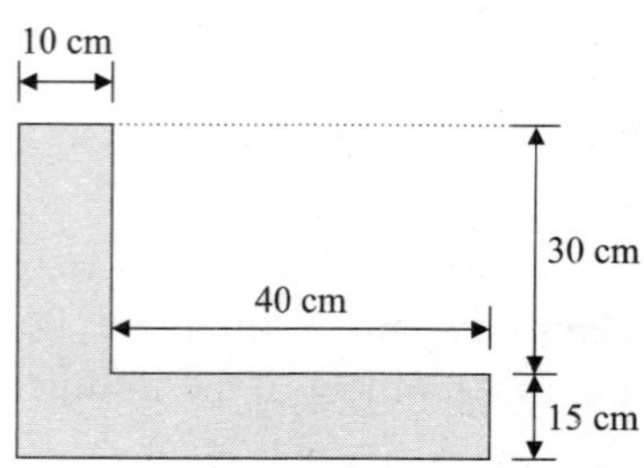

Fig. 8.21

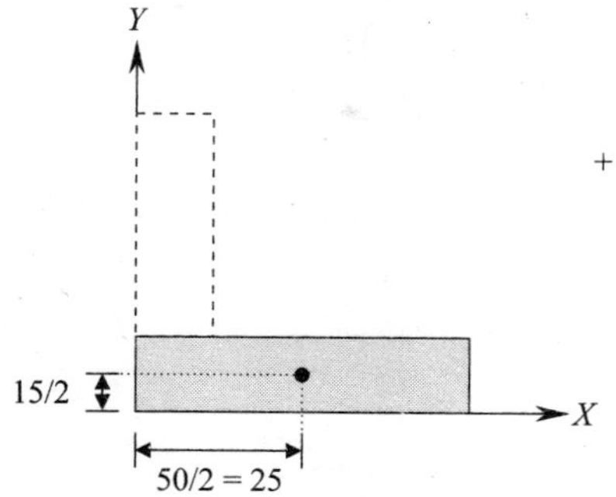

Fig. 8.21(a)

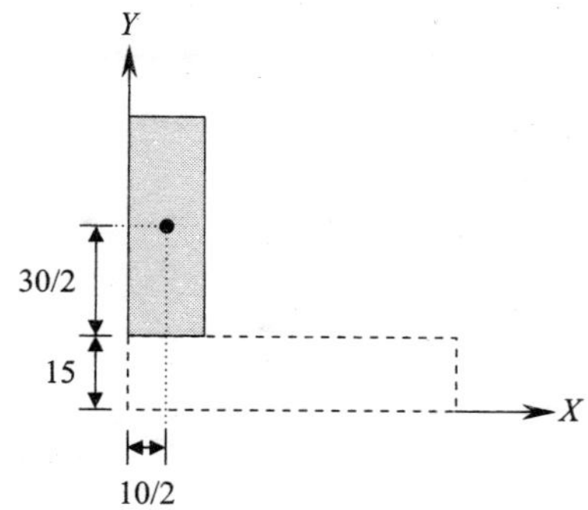

Fig. 8.21(b)

S.No	Element	A_i (cm^2)	$\bar{x}_i$ (cm)	$\bar{y}_i$ (cm)	$A_i\bar{x}_i$ (cm^3)	$A_i\bar{y}_i$ (cm^3)
1.	Rectangle-(1)	$50 \times 15 = 750$	$50/2 = 25$	$15/2 = 7.5$	18 750	5 625
2.	Rectangle-(2)	$10 \times 30 = 300$	$10/2 = 5$	$15 + (30/2) = 30$	1500	9000
$\Sigma =$		1050			20 250	14 625

Therefore, the coordinates of the centroid are

$$\bar{x} = \frac{\sum A_i\bar{x}_i}{\sum A_i} = \frac{20\,250}{1050} = 19.29 \text{ cm} \qquad \bar{y} = \frac{\sum A_i\bar{y}_i}{\sum A_i} = \frac{14\,625}{1050} = 13.93 \text{ cm}$$

Example 8.7 Find the centroid of the plain lamina shown in Fig. 8.22.

Solution Divide the area into three rectangular elements as shown below. To simplify the calculations, the given dimensions have been converted to cm and then solved.

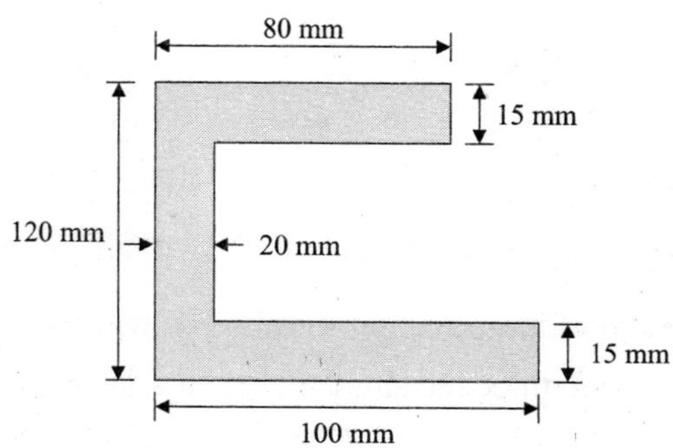

Fig. 8.22

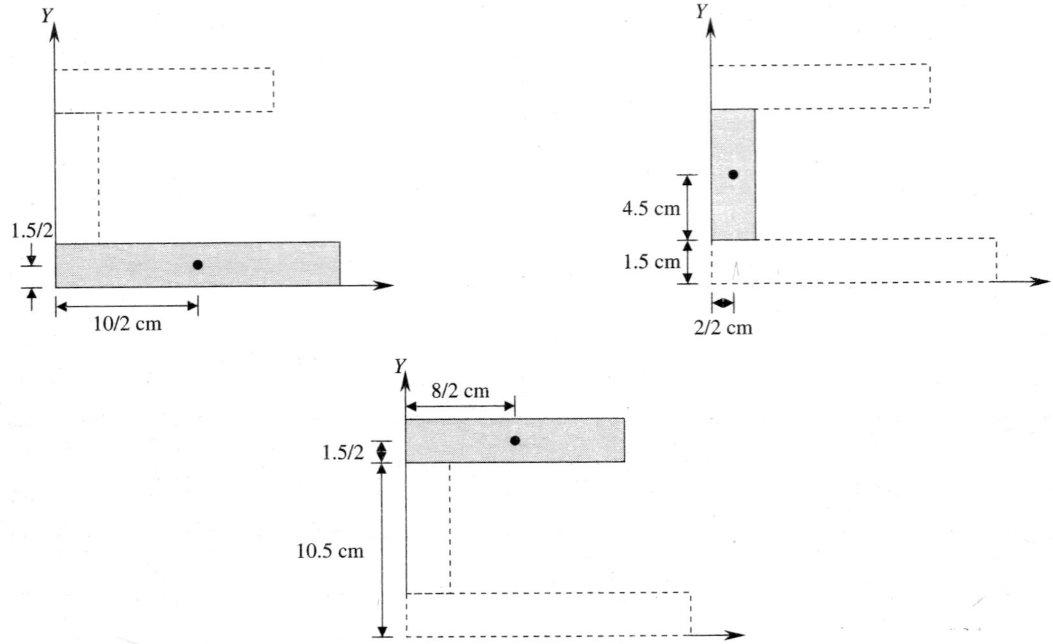

Fig. 8.22(a)

S.No	Element	A_i (cm^2)	$\bar{x}_i$ (cm)	$\bar{y}_i$ (cm)	$A_i\bar{x}_i$ (cm^3)	$A_i\bar{y}_i$ (cm^3)
1.	Rectangle-(1)	$10 \times 1.5 = 15$	$10/2 = 5$	$1.5/2 = 0.75$	75	11.25
2.	Rectangle-(2)	$[12 - 2(1.5)] \times 2 = 18$	$2/2 = 1$	$12/2 = 6$	18	108
3.	Rectangle-(3)	$8 \times 1.5 = 12$	4	$12 - 1.5/2 = 11.25$	48	135
	$\Sigma =$	45			141	254.25

Therefore, the coordinates of the centroid are

$$\bar{x} = \frac{\sum A_i\bar{x}_i}{\sum A_i} = \frac{141}{45} = 3.13 \text{ cm (or) } 31.3 \text{ mm}$$

$$\bar{y} = \frac{\sum A_i\bar{y}_i}{\sum A_i} = \frac{254.25}{45} = 5.65 \text{ cm (or) } 56.5 \text{ mm}$$

Example 8.8 Find the centroid of the *Z*-section shown. The width at all sections is 20 mm.

Solution Divide the *Z*-section into three rectangular elements and choose the reference axes as shown below. To simplify the calculations, the given dimensions have been converted to cm and then solved.

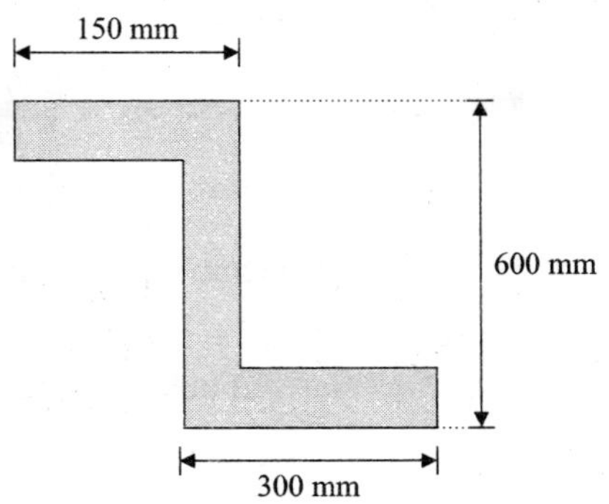

Fig. 8.23

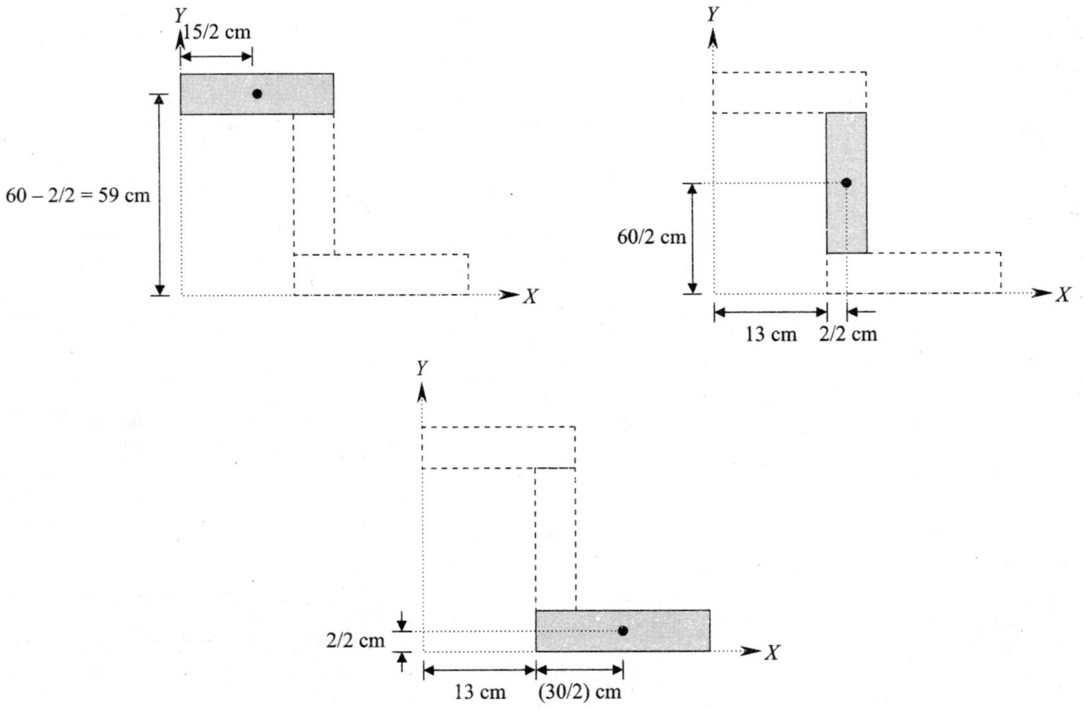

Fig. 8.23(a)

S.No	Elements	A_i (cm^2)	$\bar{x}_i$ (cm)	$\bar{y}_i$ (cm)	$A_i\bar{x}_i$ (cm^3)	$A_i\bar{y}_i$ (cm^3)
1.	Rectangle-(1)	$15 \times 2 = 30$	$15/2 = 7.5$	$60 - 2/2 = 59$	225	1770
2.	Rectangle-(2)	$(60 - 4) \times 2 = 112$	$13 + 2/2 = 14$	$60/2 = 30$	1568	3360
3.	Rectangle-(3)	$30 \times 2 = 60$	$13 + 30/2 = 28$	$2/2 = 1$	1680	60
	$\Sigma =$	202			3473	5190

Therefore, coordinates of location of the centroid are

$$\bar{x} = \frac{\sum A_i \bar{x}_i}{\sum A_i} = \frac{3473}{202} = 17.19 \text{ cm (or)} 171.9 \text{ mm} \qquad \bar{y} = \frac{\sum A_i \bar{y}_i}{\sum A_i} = \frac{5190}{202} = 25.69 \text{ cm (or) } 256.9 \text{ mm}$$

Example 8.9 Find the centroid of the shaded area (trapezium) shown.

Solution The trapezium section can be considered to be made up of a rectangle and two triangles at its ends as shown below. The calculations are summarized below in the table.

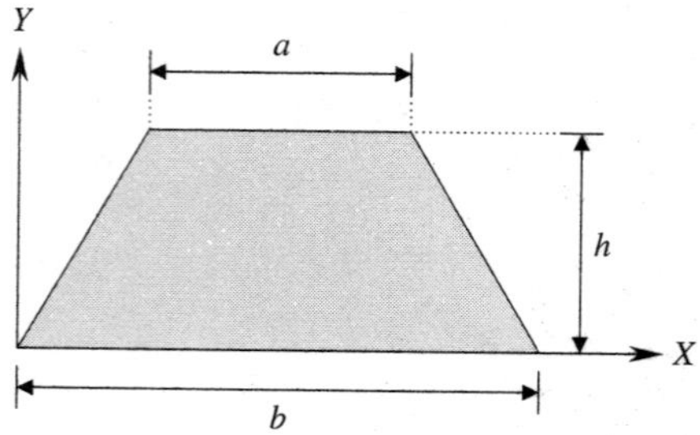

Fig. 8.24

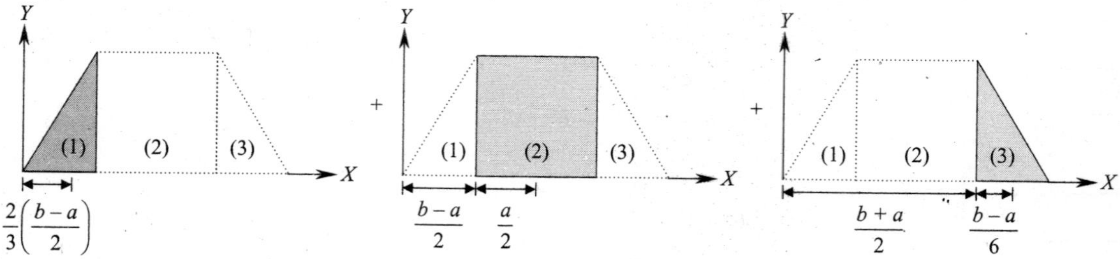

Fig. 8.24(a)

S.No	Element	A_i	$\bar{x}_i$	$\bar{y}_i$	$A_i\bar{x}_i$	$A_i\bar{y}_i$
1.	Triangle-(1)	$\frac{1}{2}\left(\frac{b-a}{2}\right)h$	$\frac{2}{3}\left(\frac{b-a}{2}\right)$	$\frac{h}{3}$	$\frac{(b-a)^2}{12}h$	$\frac{(b-a)h^2}{12}$
2.	Rectangle-(2)	ah	$\frac{b}{2}$	$\frac{h}{2}$	$\frac{abh}{2}$	$\frac{ah^2}{2}$
3.	Triangle-(3)	$\frac{1}{2}\left(\frac{b-a}{2}\right)h$	$\frac{2b+a}{3}$	$\frac{h}{3}$	$\frac{(2b+a)(b-a)h}{12}$	$\frac{(b-a)h^2}{12}$
	$\Sigma =$	$\frac{(b+a)}{2}h$			$\frac{bh}{4}(b+a)$	$\frac{h^2}{6}(b+2a)$

Therefore, the coordinates of location of centroid of the trapezium are

$$\bar{x} = \frac{\sum A_i \bar{x}_i}{\sum A_i} = \frac{b}{2} \qquad \bar{y} = \frac{\sum A_i \bar{y}_i}{\sum A_i} = \frac{h}{3}\left[\frac{b+2a}{b+a}\right]$$

Example 8.10 Determine the centroid of the shaded area shown.

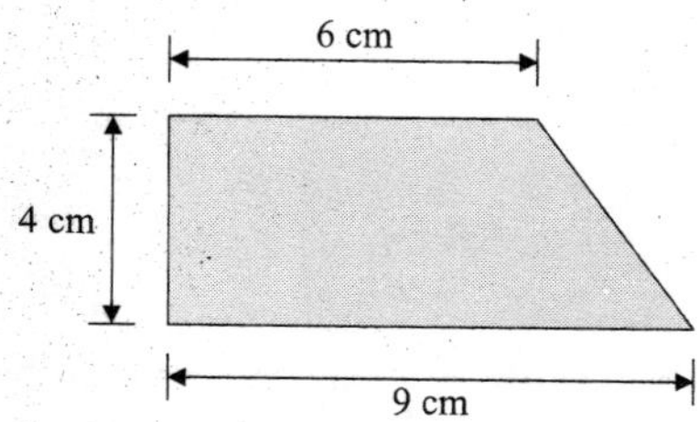

Fig. 8.25

Solution The given section can be considered to be made up of a rectangle and a triangle. The calculations are summarized below in the table.

S.No	Element	A_i (cm^2)	$\bar{x}_i$ (cm)	$\bar{y}_i$ (cm)	$A_i\bar{x}_i$ (cm^3)	$A_i\bar{y}_i$ (cm^3)
1.	Rectangle-(1)	$4 \times 6 = 24$	3	2	72	48
2.	Triangle-(2)	$(1/2)(9-6)(4) = 6$	$6 + (9-6)/3 = 7$	4/3	42	8
	$\Sigma =$	30			114	56

Therefore, location of the centroid is obtained as

$$\bar{x} = \frac{\sum A_i \bar{x}_i}{\sum A_i} = \frac{114}{30} = 3.8 \text{ cm} \qquad \bar{y} = \frac{\sum A_i \bar{y}_i}{\sum A_i} = \frac{56}{30} = 1.87 \text{ cm}$$

Example 8.11 Determine the centroid of the shaded area shown in Fig. 8.26.

Solution The given section can be considered to be made up of a rectangle and a triangle, from which a quarter circular area has been removed. The calculations are summarized below in the table.

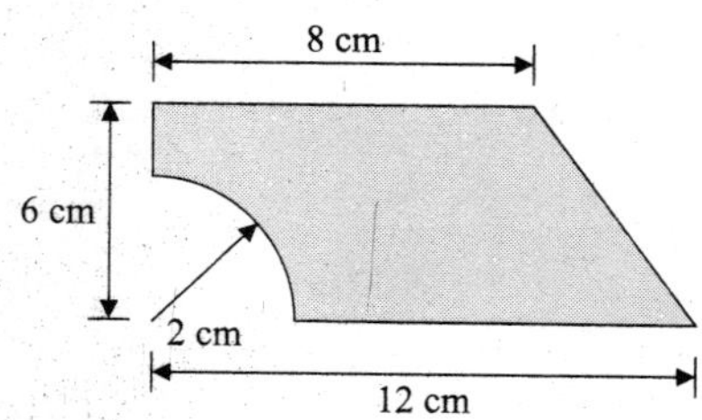

Fig. 8.26

S.No	Elements	A_i (cm^2)	$\bar{x}_i$ (cm)	$\bar{y}_i$ (cm)	$A_i\bar{x}_i$ (cm^3)	$A_i\bar{y}_i$ (cm^3)
1.	Rectangle-(1)	$8 \times 6 = 48$	4	3	192	144
2.	Triangle-(2)	$(1/2)(4)(6) = 12$	$8 + 4/3 = 9.33$	$6/3 = 2$	111.96	24
3.	Quarter circle-(3)	$-\pi(2)^2/4 = -3.142$	$4(2)/3\pi = 0.849$	$4(2)/3\pi = 0.849$	−2.668	−2.668
	$\Sigma =$	56.858			301.292	165.332

Note that as the quartercircular area has been removed, the negative sign has been used in the area of the figure.

$$\bar{x} = \frac{\sum A_i \bar{x}_i}{\sum A_i} = \frac{301.292}{56.858} = 5.3 \text{ cm} \qquad \bar{y} = \frac{\sum A_i \bar{y}_i}{\sum A_i} = \frac{165.332}{56.858} = 2.91 \text{ cm}$$

Example 8.12 Determine the centroid of the shaded area formed by removing a semicircle of diameter r from a quarter circle of radius r.

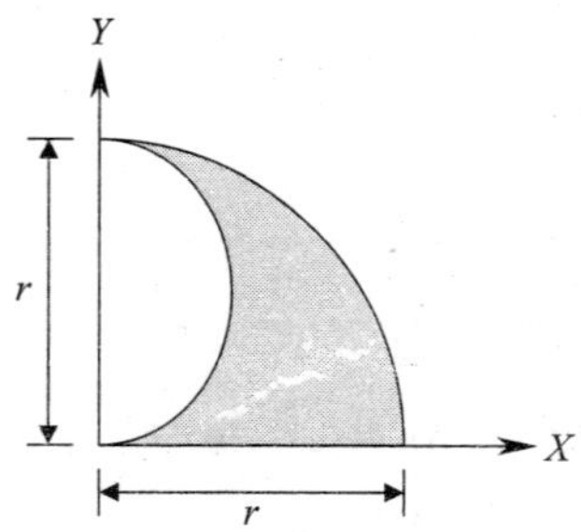

Fig. 8.27

Solution The given area can be considered a quartercircular area of radius r, from which a semicircular area of radius $r/2$ is removed as shown below. The calculations are summarized below in tabular form.

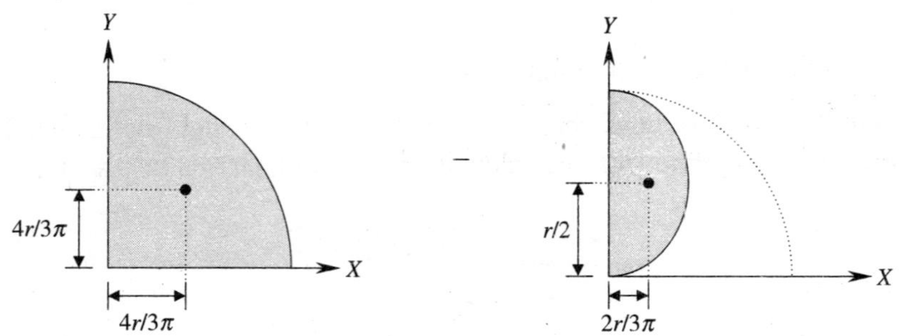

Fig. 8.27(a)

S.No	Elements	A_i	$\bar{x}_i$	$\bar{y}_i$	$A_i\bar{x}_i$	$A_i\bar{y}_i$
1.	Quarter circle	$\pi r^2/4$	$4r/3\pi$	$4r/3\pi$	$r^3/3$	$r^3/3$
2.	Semicircle	$-\pi r^2/8$	$4(r/2)/3\pi = 2r/3\pi$	$r/2$	$-r^3/12$	$-\pi r^3/16$
	$\Sigma =$	$\pi r^2/8$			$r^3/4$	$6.58\, r^3/48$

Therefore, the coordinates of the location of the centroid are

$$\bar{x} = \frac{\sum A_i\bar{x}_i}{\sum A_i} = \frac{r^3/4}{\pi r^2/8} = \frac{2r}{\pi} = 0.637r \qquad \bar{y} = \frac{\sum A_i\bar{y}_i}{\sum A_i} = \frac{6.58r^3/48}{\pi r^2/8} = \frac{6.58r}{6\pi} = 0.349r$$

Example 8.13 A right-angled triangular plate ABC is suspended at corner A as shown in Fig. 8.28. Determine the inclination of the hypotenuse AC with the vertical at the suspended position. The dimensions of the triangle are $\overline{AB} = 8$ cm, $\overline{BC} = 6$ cm and $\overline{AC} = 10$ cm.

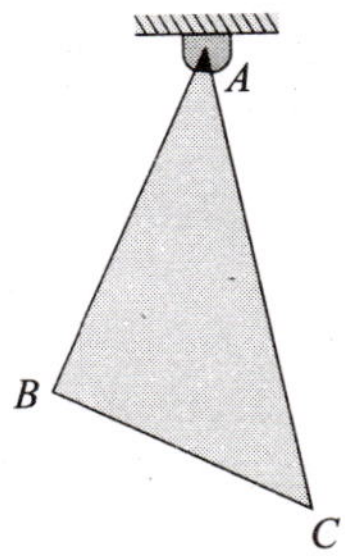

Fig. 8.28

Solution Consider the triangle ABC such that the hypotenuse AC is vertical. Choosing the reference axes as shown in Fig. 8.28(a), with corner A as reference point, the coordinates of the two vertices A and C are readily known:

$$A\,(0, 0),\ C\,(0, 10)$$

To determine the coordinates of B, we proceed as follows:

$$\angle CAB = \tan^{-1}(6/8) = 36.87°$$

Hence, x and y coordinates of corner B are respectively

$$8\sin(36.87)° = 4.8 \text{ cm}$$

and

$$8\cos(36.87)° = 6.4 \text{ cm}$$

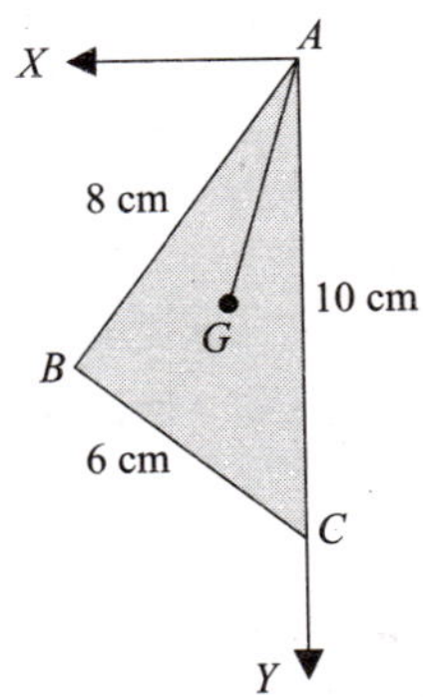

Fig. 8.28(a)

Therefore, the coordinates of the centroid (G) of the triangle are

$$\bar{x} = (x_A + x_B + x_C)/3 = (0 + 4.8 + 0)/3 = 1.6 \text{ cm}$$

and

$$\bar{y} = (y_A + y_B + y_C)/3 = (0 + 6.4 + 10)/3 = 5.47 \text{ cm}$$

The inclination of line AG with respect to side AC is then given as

$$\theta = \tan^{-1}(\bar{x}/\bar{y}) = \tan^{-1}(1.6/5.47) = 16.3°$$

When the triangular plate is suspended, for equilibrium position, the centre of gravity must lie below the point of support, i.e., line AG must be vertical. Hence, the hypotenuse AC will make an angle of 16.3° with respect to the vertical.

8.6 THEOREMS OF PAPPUS AND GULDINUS

Pappus and Guldinus, two mathematicians, developed two theorems, which are useful in determining the surface area and volume of revolution generated by revolving respectively a plane curve and an area about a non-intersecting axis. Both these theorems are based on the determination of the centroid of the generating curve and area discussed in the previous sections. The two theorems can be stated as follows:

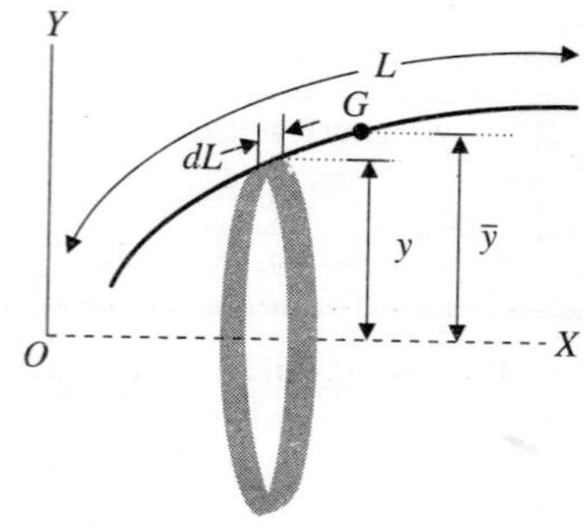

Fig. 8.29 Theorem I

Theorem I *The area of a surface generated by revolving a plane curve about a non-intersecting axis in the plane of the curve is equal to the product of length of the curve and distance travelled by the centroid G of the curve during revolution.*

Proof Consider a curve of length L and let it be revolved about the OX axis through 2π radians. Then an infinitesimally small element of length dL will generate a hoop of area $2\pi y dL$.

Therefore, the total surface area generated by the curve is given as

$$A = \int 2\pi y dL$$

$$= 2\pi \int y dL = 2\pi \bar{y} L \quad \left[\text{Note that } \bar{y} = \frac{\int y dL}{L}\right] \tag{8.52}$$

Depending upon the generating curve, the surface areas generated are different as shown below. A straight line parallel to the axis of revolution generates surface area of a cylinder; an inclined line with one end touching the axis of revolution generates surface area of a cone; a semicircular arc with the ends touching the axis of revolution generates surface area of a sphere and a circular arc away from the axis of revolution generates surface area of a torus.

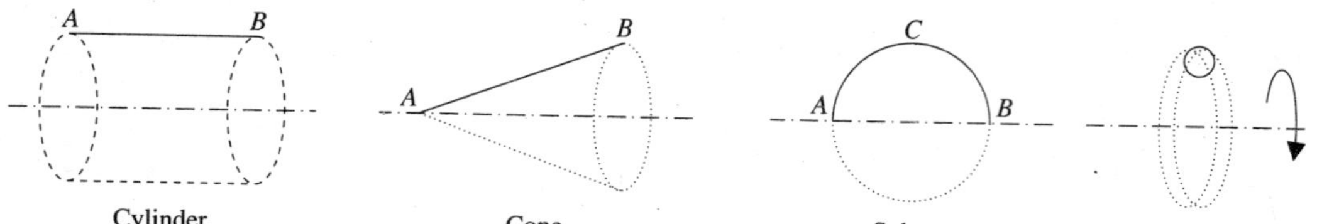

Fig. 8.30 Generation of surface area

Theorem II *The volume of a solid generated by revolving a plane area about a non-intersecting axis in its plane is equal to the product of the area and length of the path travelled by the centroid G of the area during the revolution about the axis.*

Consider a laminar area A and let it be rotated about the horizontal axis through an angle of 2π radian. Then an infinitesimally small area dA will generate a ring of volume $2\pi y dA$. Therefore, the entire volume generated by the area is given as

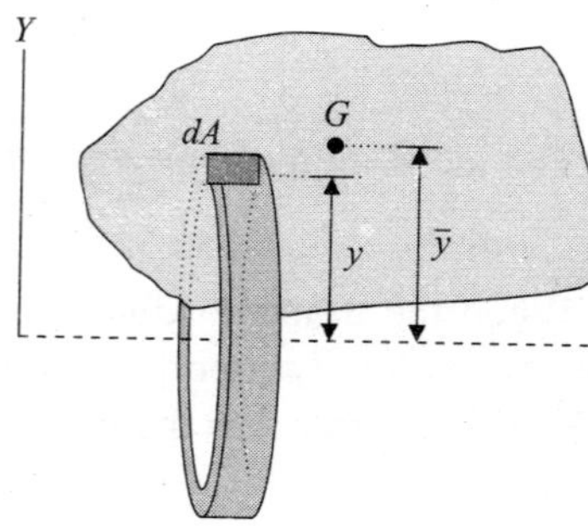

Fig. 8.31 Theorem – II

$$V = \int 2\pi y dA$$

$$= 2\pi \int y dA = 2\pi \bar{y} A \quad \left[\text{Note that } \bar{y} = \frac{\int y dA}{A}\right] \tag{8.53}$$

Depending upon the generating area, the volumes generated are different as shown below. A rectangular area when rotated about one of its sides generates volume of a cylinder; a right-angled triangle when rotated about a side other than the hypotenuse generates volume of a cone; a semicircular area when rotated about its diameter generates volume of a sphere and a circle away from the axis rotated about the axis of rotation generates volume of a torus.

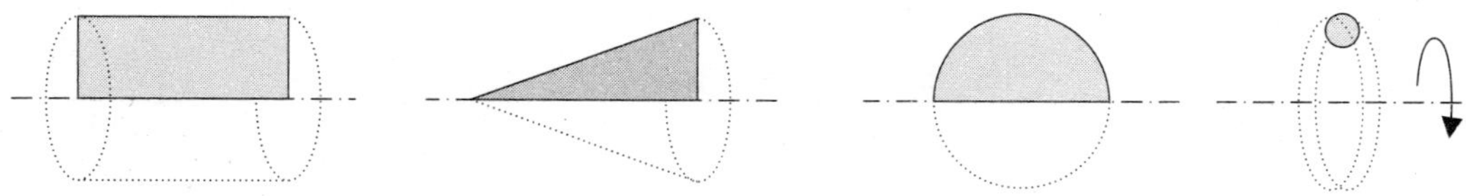

Fig. 8.32 Generation of volume of solid body

Example 8.14 Determine the surface area and volume of a cylinder using the Pappus and Guldinus theorems.

Solution Consider a straight line AB of length H, parallel to the Y-axis at a distance R from the Y-axis as shown in Fig. 8.33. Revolving this line about the Y-axis through 360° will generate surface area of a cylinder of radius R and height H. By Pappus and Guldinus Theorem I, its surface area can be determined as

$$A = (\text{length of curve})(\bar{x})(\theta)$$

$$= H(R)\,2\pi = 2\pi RH$$

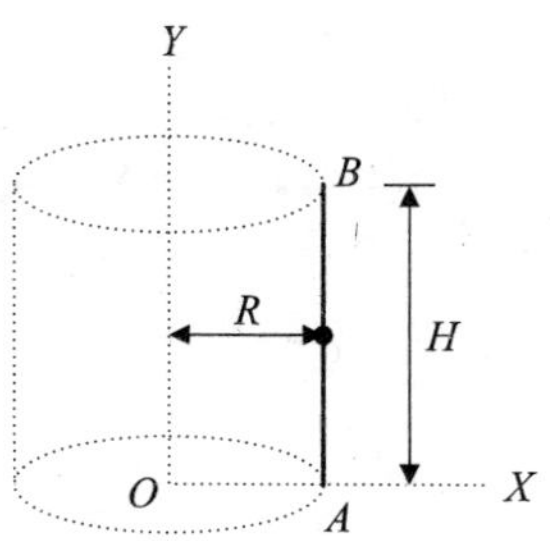

Fig. 8.33

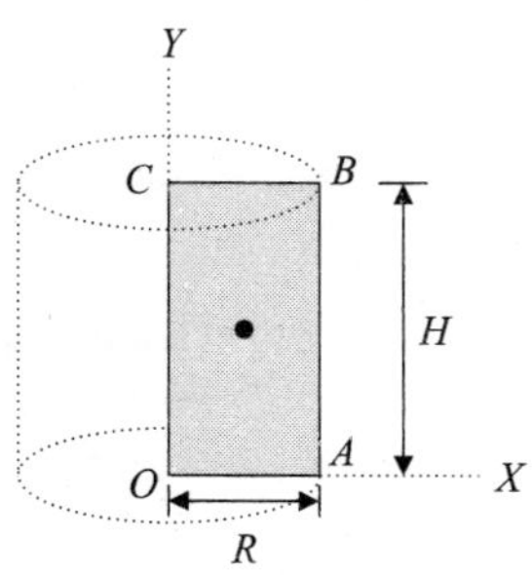

Fig. 8.33(a)

Similarly, considering a rectangle $OABC$ as shown in Fig. 8.33(a) and rotating it about the Y-axis will generate a solid cylinder. Its volume can be determined by using Theorem II as

$$V = (\text{area of the plane})(\bar{x})(\theta)$$

$$= HR\left[\frac{R}{2}\right]2\pi = \pi R^2 H$$

Example 8.15 Determine the surface area and volume of a cone using the Pappus and Guldinus theorems.

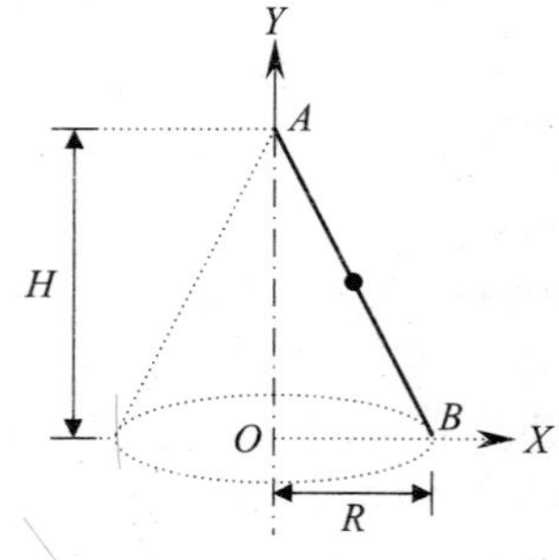

Fig. 8.34

Solution Consider a straight line *AB* of length *L*, inclined as shown in Fig. 8.34. By revolving this through 360° about the *Y*-axis will generate surface area of a cone of base radius *R* and height *H*. By Pappus and Guldinus Theorem I, its surface area can be determined as

$$A = (\text{length of curve})(\bar{x})(\theta)$$

$$= L\left[\frac{R}{2}\right]2\pi$$

$$= \pi RL = \pi R\sqrt{H^2 + R^2}$$

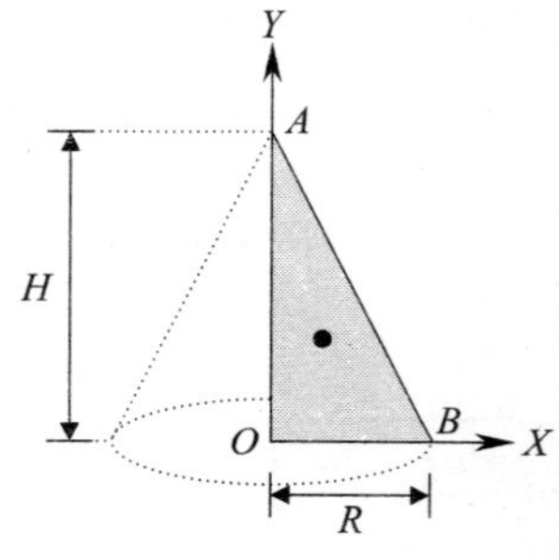

Fig. 8.34(a)

Similarly, considering a triangle *OAB* as shown in Fig. 8.34(a) and rotating it about the *Y*-axis will generate a solid cone. Its volume can be determined by using Theorem II as

$$V = (\text{area of plane})(\bar{x})(\theta)$$

$$= \frac{1}{2}RH\left[\frac{R}{3}\right]2\pi = \frac{1}{3}\pi R^2 H$$

Example 8.16 Determine the surface area and volume of a sphere using the Pappus and Guldinus theorems.

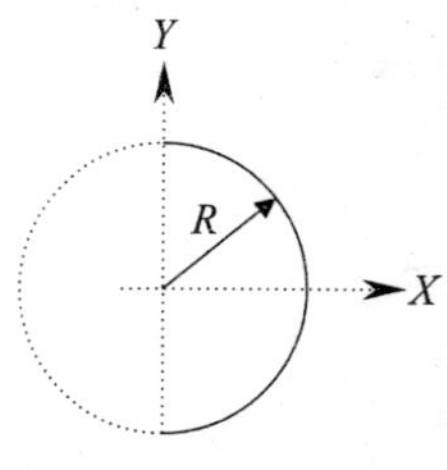

Fig. 8.35

Solution Consider a semicircular arc of radius *R* as shown in Fig. 8.35. By revolving this through 360° about the *Y*-axis will generate a surface area of a sphere of radius *R*. By the Pappus and Guldinus theorem, its surface area can be determined as

$$A = (\text{length of curve})(\bar{x})(\theta)$$

$$= \pi R\left(\frac{2R}{\pi}\right)2\pi = 4\pi R^2$$

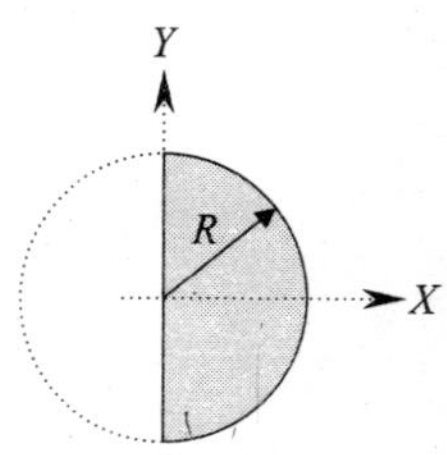

Fig. 8.35(a)

Similarly, considering a semicircular area as shown in Fig. 8.35(a) and rotating it about the *Y*-axis will generate a solid sphere. Its volume can be determined as

$$V = (\text{area of plane})(\bar{x})(\theta)$$

$$= \frac{\pi R^2}{2}\left(\frac{4R}{3\pi}\right)2\pi = \frac{4}{3}\pi R^3$$

Example 8.17 Determine the surface area and volume of a torus of radius R obtained by rotating a circle of radius r about an axis away from the circle.

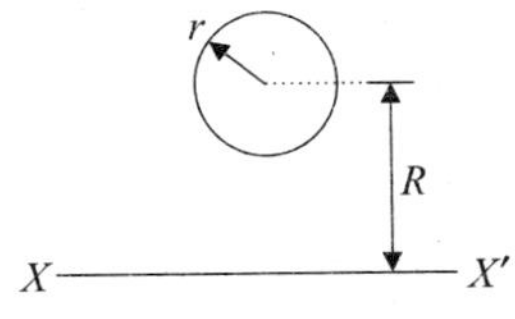

Fig. 8.36

Solution Consider a circular arc of radius r with its centre at a distance of R from X–X axis as shown in Fig. 8.36. By revolving this through 360° about X–X axis will generate surface area of a torus. By Pappus and Guldinus Theorem I, its surface area can be determined as

$$A = (\text{length of curve})(\bar{y})(\theta)$$
$$= 2\pi r\,(R)\,2\pi = 4\pi^2 Rr$$

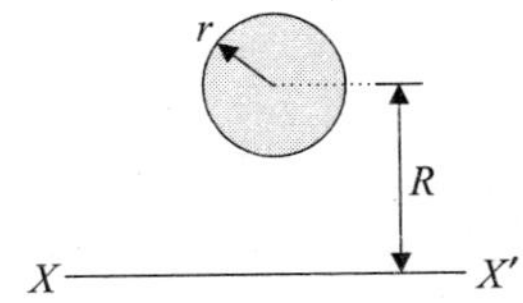

Fig. 8.36(a)

Similarly, considering a circular area as shown in Fig. 8.36(a) and rotating it about the X–X axis will generate a solid torus. Its volume can be determined as

$$V = (\text{area of plane})(\bar{y})(\theta)$$
$$= \pi r^2\,(R)\,2\pi = 2\pi^2 r^2 R$$

Example 8.18 Determine the volume of a frustum of a cone of base radius R, top radius r and height H.

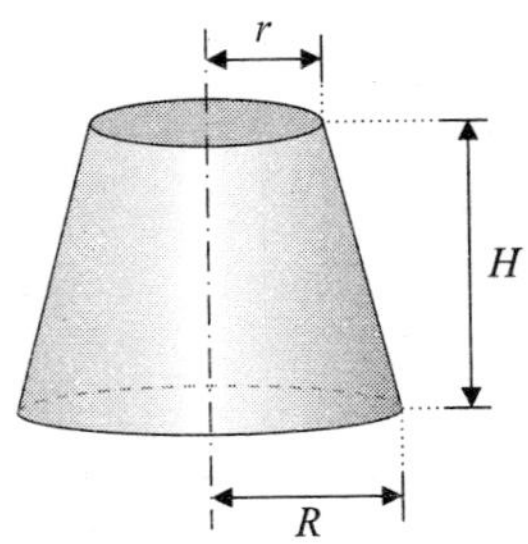

Fig. 8.37

Solution The frustum of a cone can be generated by revolving a trapezoidal area as shown below about the axis of rotation. The position of the centroid with respect to the axis of rotation can be determined as follows:

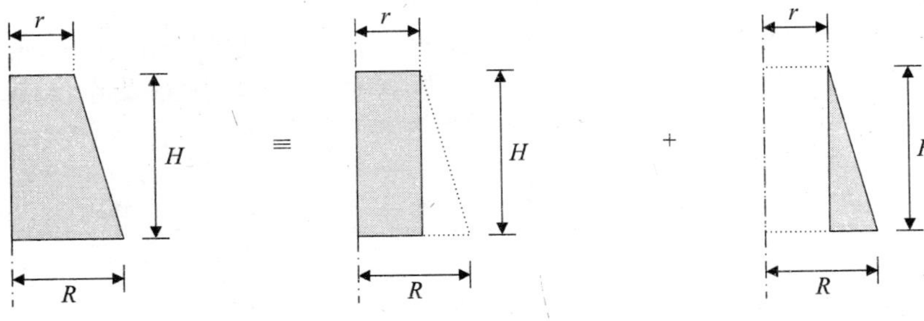

Fig. 8.37(a)

S.No	Element	A_i	$\bar{x}_i$	$A_i\bar{x}_i$
1.	Rectangle	Hr	$r/2$	$Hr^2/2$
2.	Triangle	$\frac{1}{2}(R-r)H$	$r+\frac{(R-r)}{3}$	$\frac{1}{6}(R-r)(R+2r)H$
$\Sigma =$		$\frac{(R+r)H}{2}$		$\frac{H(R^2+r^2+rR)}{6}$

Therefore, the x-coordinate of centroid is obtained as

$$\bar{x} = \frac{\sum A_i\bar{x}_i}{\sum A_i}$$

$$= \frac{H(R^2 + r^2 + rR)}{6} \cdot \frac{2}{(R+r)H}$$

$$= \frac{1}{3} \frac{(R^2 + r^2 + rR)}{(R+r)}$$

Therefore, by applying the Pappus and Guldinus theorem II, the volume of the frustum of cone is obtained as

$$V = 2\pi (A)\bar{x} = \frac{\pi H(R^2 + r^2 + rR)}{3}$$

Example 8.19 Using the Pappus and Guldinus theorems, find the volume of the machine component in which a cylindrical portion is removed as shown.

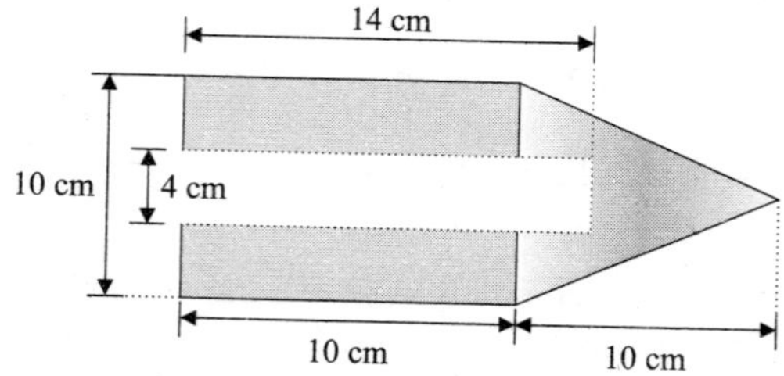

Fig. 8.38

Solution If we consider the plane area shown in Fig. 8.38(a), then rotating this through one complete revolution about the horizontal axis will generate the given volume. The plane area can be considered to be made up of a rectangle and a triangle, from which a rectangle has been removed.

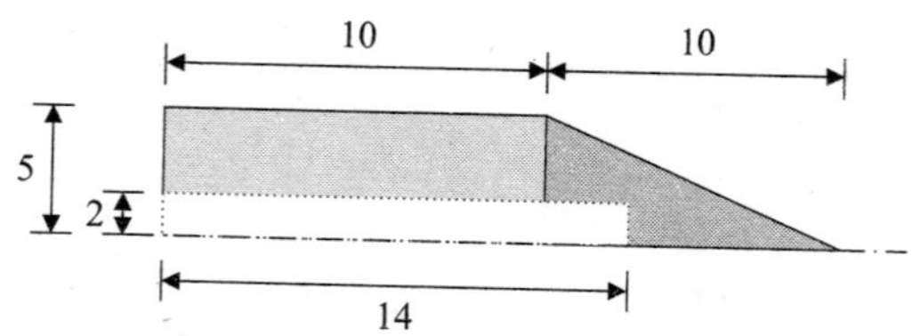

Fig. 8.38(a)

Determination of $\bar{y}$

S.No	Element	$A_i (cm^2)$	$\bar{y}_i (cm)$	$A_i \bar{y}_i (cm^3)$
1.	Rectangle	50	2.5	125
2.	Triangle	25	⅓(5) = 1.67	41.67
3.	Rectangle	–28	1	–28
	Σ =	47		138.67

$$\therefore \quad \bar{y} = \frac{\sum A_i \bar{y}_i}{\sum A_i}$$

$$= \frac{138.67}{47} = 2.95 \text{ cm}$$

Therefore, by applying the Pappus and Guldinus theorem II, the volume of the solid is obtained as

$$V = 2\pi \bar{y} A$$

$$= 2\pi(2.95)(47) = 871.16\text{cm}^3$$

Example 8.20 Find the surface area and volume of the body formed by joining two hemispheres at the end of a cylinder as shown by using Pappus and Guldinus theorems.

Solution

Determination of surface area

By revolving the curve shown in Fig. 8.39(a) about the horizontal axis will generate the surface area of the solid.

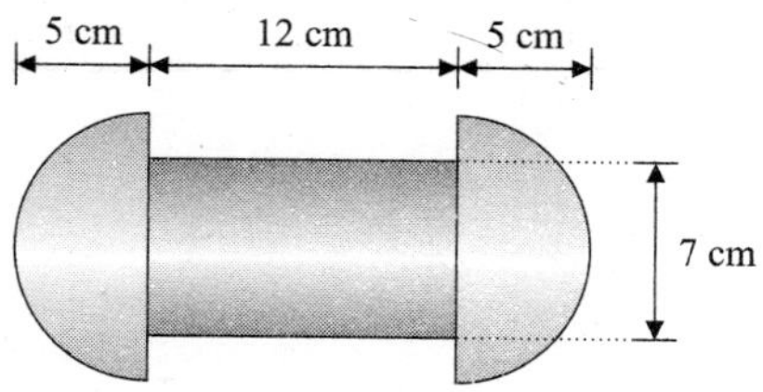

Fig. 8.39

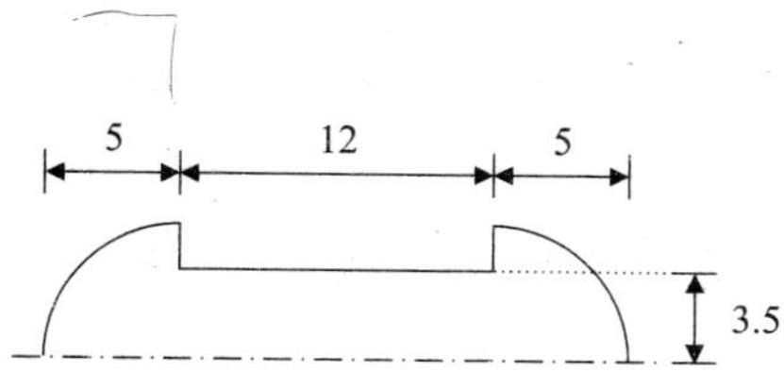

Fig. 8.39(a)

The *y*-coordinate of the centroid of the curve is obtained as below:

S.No	*Element*	$L_i(cm)$	$\bar{y}_i(cm)$	$L_i\bar{y}_i(cm^2)$
1.	Quartercircular arc	$\frac{\pi(5)}{2}$	$\frac{2(5)}{\pi}$	25
2.	Vertical line	5 − 3.5 = 1.5	3.5 + 1.5/2 = 4.25	6.375
3.	Horizontal Line	12	3.5	42
4.	Vertical line	1.5	4.25	6.375
5.	Quartercircular arc	$\frac{\pi(5)}{2}$	$\frac{2(5)}{\pi}$	25
	Σ =	30.71		104.75

$\therefore$

$$\bar{y} = \frac{\sum L_i \bar{y}_i}{\sum L_i}$$

$$= \frac{104.75}{30.71} = 3.41 \text{ cm}$$

Therefore, surface area $= 2\pi \bar{y} L$

$$= 2\pi\,(3.41)30.71 = 658 \text{ cm}^2$$

Determination of volume

By revolving the area shown in Fig. 8.39(b) about the horizontal axis will generate the volume of the solid. The y-coordinate of the centroid of the area is obtained as below:

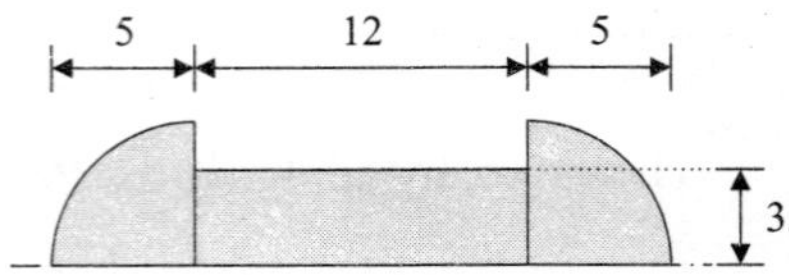

Fig. 8.39(b)

S.No	Element	$A_i(cm^2)$	$\bar{y}_i(cm)$	$A_i\bar{y}_i(cm^3)$
1.	Quartercircular area	$\frac{\pi}{4}(5)^2$	$\frac{4(5)}{3\pi}$	41.67
2.	Rectangle	42	1.75	73.5
3.	Quartercircular area	$\frac{\pi}{4}(5)^2$	$\frac{4(5)}{3\pi}$	41.67
	$\Sigma =$	81.27		156.84

$$\therefore \quad \bar{y} = \frac{\sum A_i \bar{y}_i}{\sum A_i}$$

$$= \frac{156.84}{81.27} = 1.93 \text{ cm}$$

Therefore, volume $= 2\pi\,\bar{y}\,A$

$$= 2\pi(1.93)81.27 = 985.5 \text{ cm}^3$$

Example 8.21 Determine the volume of liquid that can be stored in the vessel shown in Fig. 8.40 using the Pappus and Guldinus theorems. It has a hollow cylindrical top and a hollow hemispherical bottom.

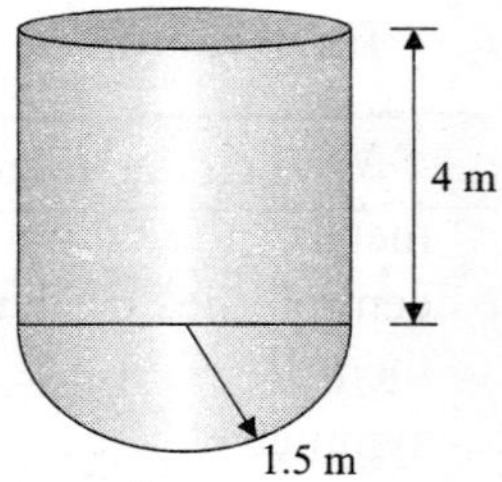

Fig. 8.40

Solution The volume of liquid that can be stored is nothing but the volume of the vessel. The volume of the solid of revolution is obtained by revolving the plane area shown in Fig. 8.40(a) about the vertical axis. The centroid of the plane area is obtained as follows [refer Figs 8.40(b) and (c)]:

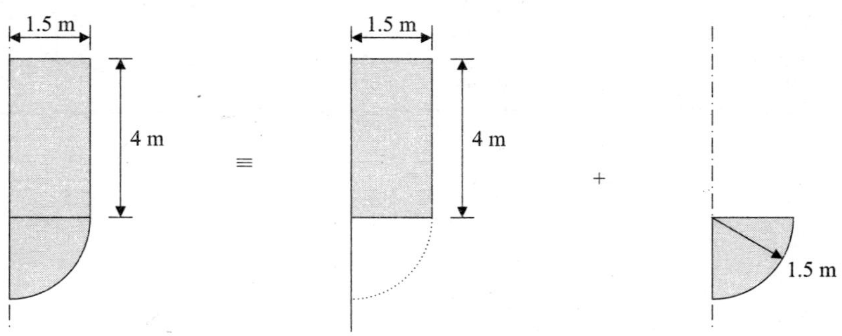

Fig. 8.40(a) Fig. 8.40(b) Fig. 8.40(c)

S.No	Element	$A_i(m^2)$	$\bar{x}_i(m)$	$A_i\bar{x}_i(m^3)$
1.	Rectangle	6	0.75	4.5
2.	Quartercircular area	$\frac{\pi(1.5)^2}{4} = 1.77$	$\frac{4(1.5)}{3\pi}$	$\frac{1}{3}(1.5)^3 = 1.125$
$\Sigma =$		7.77		5.625

$$\therefore \quad \bar{x} = \frac{\sum A_i \bar{x}_i}{\sum A_i}$$

$$= \frac{5.625}{7.77} = 0.724 \text{ m}$$

Hence, the volume of the solid of revolution is obtained as

$$V = 2\pi A\bar{x}$$

$$= 2\pi(7.77)(0.724) = 35.35 \text{ m}^3$$

8.7 CENTROIDS OF VOLUMES

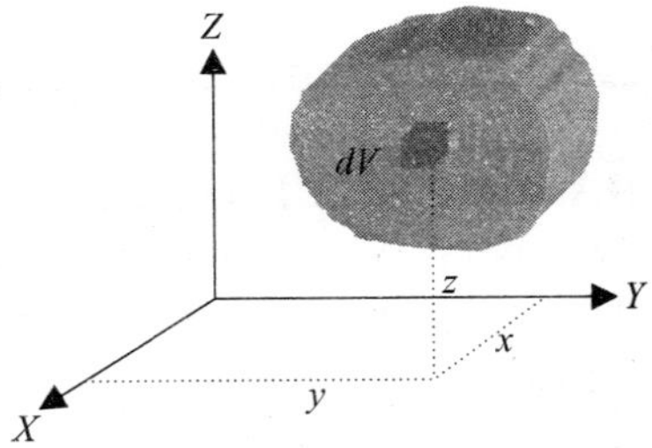

Fig. 8.41

Consider a homogeneous body of volume V. The coordinates of the centroid of the volume were derived in Section 8.3 and they are restated below:

$$\bar{x} = \frac{\int x\,dV}{V}, \; \bar{y} = \frac{\int y\,dV}{V} \quad \text{and} \quad \bar{z} = \frac{\int z\,dV}{V} \qquad (8.54)$$

When the volume possesses a plane of symmetry, the first moment of the volume with respect to that plane is zero. Hence, the centroid lies on that plane. If it possesses two planes of symmetry then the centroid lies on the line of intersection of these two planes. If it possesses three planes of symmetry then it lies at the point of intersection of the three planes.

The centroid of regular geometrical bodies can be determined by the integration methods as explained for lines and plane areas and the results are summarized below in the table. It should be noted that the centroid of a volume of revolution does not coincide with the centroid of its cross section. Hence, the centroid of a hemisphere is different from that of a semicircular area and the centroid of a cone is different from that of a triangle.

Table 8.3 Centroids of Volumes

S.No	Shape	Figure	Volume	Centroid
1.	Cylinder (radius–R, length–H)	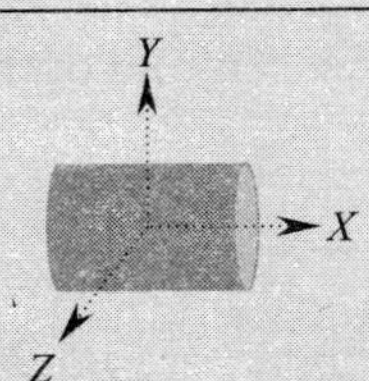	$\pi R^2 H$	At the centre

Contd.

2.	Cone (radius–R, length–H)	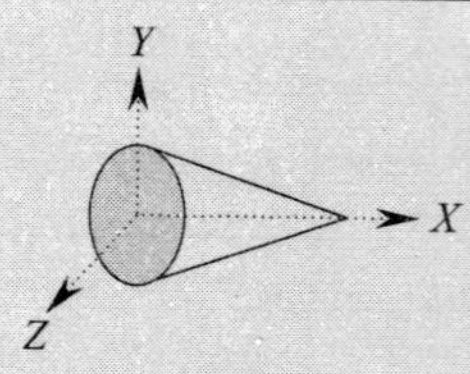	$\frac{1}{3}\pi R^2 H$	$\bar{x} = H/4$
3.	Hemisphere (radius–R)	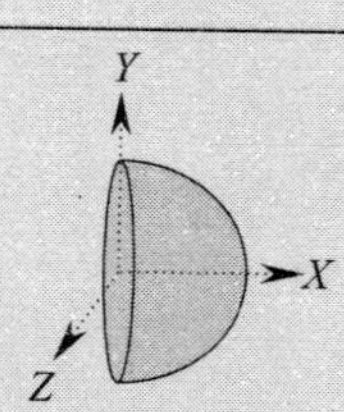	$\frac{2}{3}\pi R^3$	$\bar{x} = 3R/8$
4.	Paraboloid of revolution (length–H)	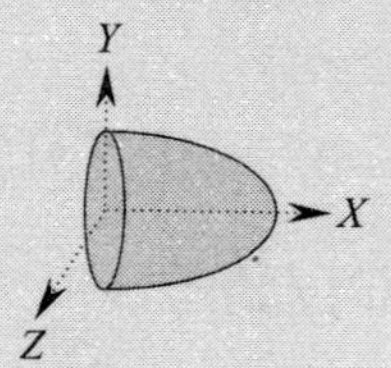	$\frac{1}{2}\pi a^2 H$	$\bar{x} = H/3$

Example 8.22 Find the centroid of volume of a solid formed by a right circular cone of 100 mm base radius and a height of 150 mm placed over a cylinder having the same radius and a 75 mm height.

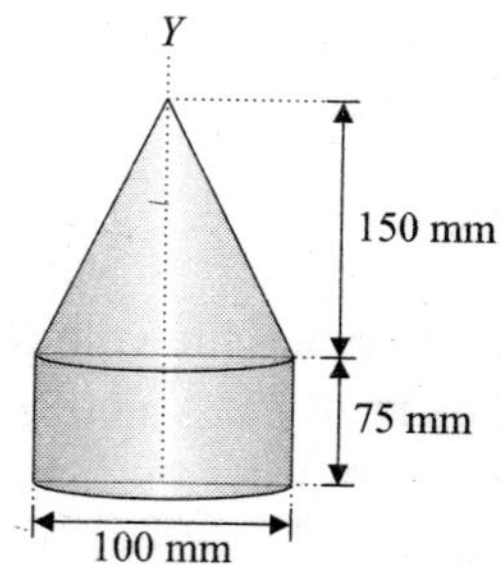

Fig. 8.42

Solution Since Y–Z is a plane of symmetry, the x-coordinate of the centroid is zero, i.e., $\bar{x} = 0$. Also, X–Y is a plane of symmetry and hence the z-coordinate of the centroid is zero, i.e., $\bar{z} = 0$. Thus, we need to find only the y-coordinate of the centroid. The base radius of the cylinder and cone are equal to 0.05 m. The calculations are given in tabular form as follows:

S.No	*Element*	$V_i(m^3)$	$\bar{y}_i(m)$	$V_i\bar{y}_i(m^4)$
1.	Cone	$\frac{1}{3}\pi R^2 H = 3.93 \times 10^{-4}$	$0.075 + \frac{0.15}{4} = 0.1125$	0.442×10^{-4}
2.	Cylinder	$\pi R^2 H = 5.89 \times 10^{-4}$	$\frac{0.075}{2} = 0.0375$	0.221×10^{-4}
	$\Sigma =$	9.82×10^{-4}		0.663×10^{-4}

Therefore, the y-coordinate of the centroid of the composite volume is obtained as

$$\bar{y} = \frac{\Sigma V_i \bar{y}_i}{\Sigma V_i}$$

$$= \frac{0.663 \times 10^{-4}}{9.82 \times 10^{-4}} = 0.0675 \text{ m (or) } 67.5 \text{ mm}$$

Example 8.23 Find the centroid of volume of a solid formed by joining a cone, a cylinder and a hemisphere as shown in Fig. 8.43.

Solution Since the volume has two planes of symmetry, namely, *X–Y* and *Y–Z*, the *z* and *x*-coordinates of the centroid are zero, i.e., $\bar{z} = 0$, $\bar{x} = 0$. Thus, we need to find only the *y*-coordinate of the centroid. The radius *R* of the members is 3 cm as can be seen in the figure. The calculations are given in tabular form as follows:

S.No	Element	$V_i (cm^3)$	$\bar{y}_i (cm)$	$V_i \bar{y}_i (cm^4)$
1.	Cone	$(1/3)\pi R^2 H = 75.4$	$(3/4)8 = 6$	452.4
2.	Cylinder	$\pi R^2 H = 141.37$	$8 + (5/2) = 10.5$	1484.39
3.	Hemisphere	$(2/3)\pi R^3 = 56.55$	$13 + 3(3)/8 = 14.13$	799.05
	$\Sigma =$	273.32		2735.84

Therefore, the *y*-coordinate of the centroid of the composite volume is obtained as

$$\bar{y} = \frac{\Sigma V_i \bar{y}_i}{\Sigma V_i}$$

$$= \frac{2735.84}{273.32} = 10.01 \text{ cm}$$

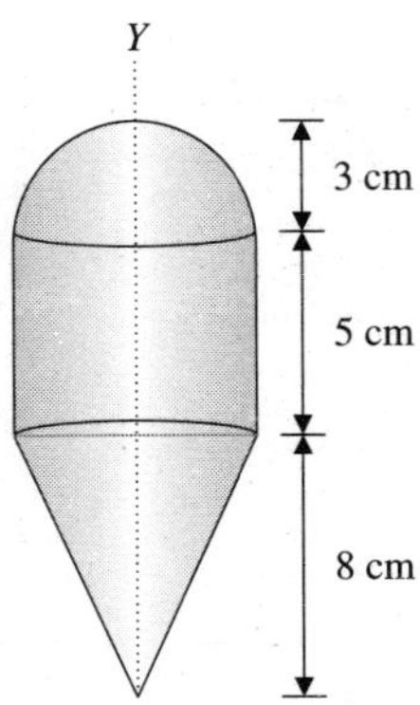

Fig. 8.43

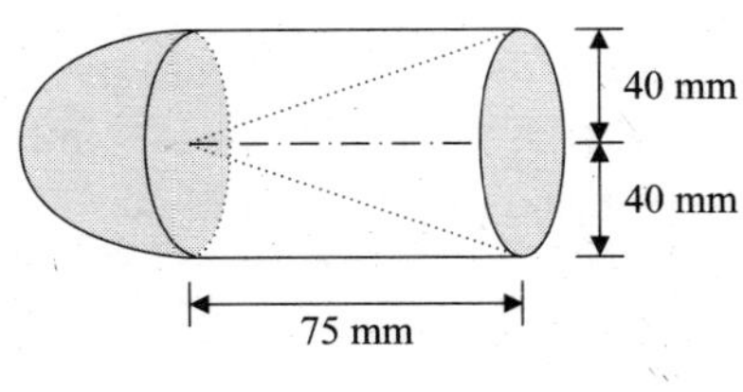

Fig. 8.44

Example 8.24 Find the centroid of volume of a solid obtained by joining a hemisphere and a cylinder, and carving out a cone as shown in Fig. 8.44.

Solution The reference axes are chosen as shown in Fig. 8.44(a). The given solid can be considered to be made up of a hemisphere, a cylinder and a cone removed from it. The radius *R* of each member is 4 cm. Since the volume has two planes of symmetry, namely, *X–Y* and *X–Z*, we need to find only $\bar{x}$.

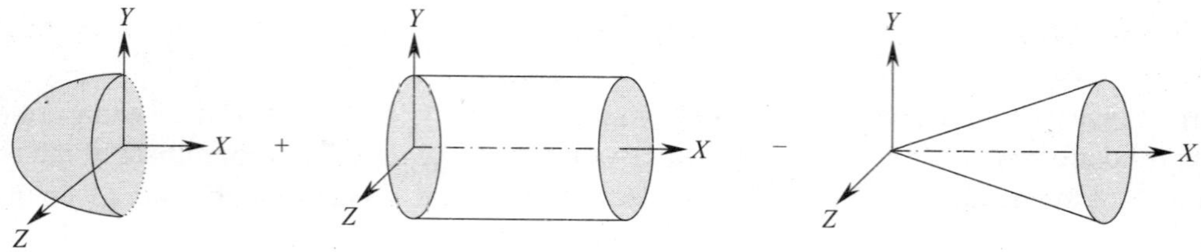

Fig. 8.44(a)

S.No	Element	$V_i(cm^3)$	$\bar{x}_i(cm)$	$V_i\bar{x}_i(cm^4)$
1.	Hemisphere	$(2/3)\pi R^3 = 134.04$	$-3R/8 = -1.5$	-201.06
2.	Cylinder	$\pi R^2 H = 377$	$7.5/2 = 3.75$	1413.75
3.	Cone	$-(1/3)\pi R^2 H = -125.67$	$(3/4)7.5 = 5.625$	-706.89
	$\Sigma =$	385.37		505.8

Note that as the centroid of the hemisphere lies to the left of the origin, it is taken as negative. Hence, the x-coordinate of the centroid of the composite volume is obtained as

$$\bar{x} = \frac{\sum V_i \bar{x}_i}{\sum V_i}$$

$$= \frac{505.8}{385.37} = 1.31 \text{ cm}$$

Example 8.25 Find the centroid of volume of a solid included between two spheres of radii a and b, which touch internally.

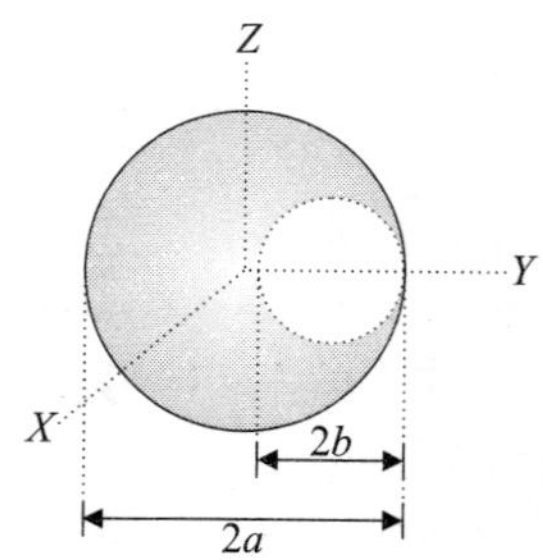

Fig. 8.45

Solution Consider the reference axes about the centre of the outer sphere. Since Y–Z is a plane of symmetry, the x-coordinate of the centroid is zero, i.e., $\bar{x} = 0$. Also, X–Y is a plane of symmetry and hence the z-coordinate of the centroid is zero, i.e., $\bar{z} = 0$. Thus, we need to find only the y-coordinate of the centroid. The calculations are given in tabular form as follows:

S.No	Element	V_i	$\bar{y}_i$	$V_i\bar{y}_i$
1.	Sphere of radius a	$\frac{4}{3}\pi a^3$	0	0
2.	Sphere of radius b	$-\frac{4}{3}\pi b^3$	$(a-b)$	$-\frac{4}{3}\pi b^3(a-b)$
	$\Sigma =$	$\frac{4}{3}\pi(a^3 - b^3)$		$-\frac{4}{3}\pi b^3(a-b)$

Therefore, the y-coordinate of the centroid of the composite volume is obtained as

$$\bar{y} = \frac{\sum V_i \bar{y}_i}{\sum V_i}$$

$$= \frac{-\frac{4}{3}\pi b^3(a-b)}{\frac{4}{3}\pi(a^3-b^3)}$$

Since $$a^3 - b^3 = (a-b)(a^2+ab+b^2)$$

$$\bar{y} = \frac{-b^3}{a^2+ab+b^2}$$

8.8 CENTRE OF MASS

So far, we have seen the procedure to determine the centroids of various geometrical shapes. In this section, we will determine the centre of mass of solid bodies. If the density of the material is *uniform* throughout the body then the centre of mass coincides with the centroid of its volume. However, if the density is *not uniform* then the centre of mass does not coincide with the centroid of its volume. The expressions for coordinates of the centre of mass as derived in Section 8.3 are restated below:

$$\bar{x} = \frac{\int x\,dm}{\int dm}; \quad \bar{y} = \frac{\int y\,dm}{\int dm} \quad \text{and} \quad \bar{z} = \frac{\int z\,dm}{\int dm} \tag{8.55}$$

The centre of mass of a composite body is given as

$$\bar{x} = \frac{\sum m_i \bar{x}_i}{\sum m_i}; \quad \bar{y} = \frac{\sum m_i \bar{y}_i}{\sum m_i} \quad \text{and} \quad \bar{z} = \frac{\sum m_i \bar{z}_i}{\sum m_i} \tag{8.56}$$

Example 8.26 Determine the centre of mass of a composite body formed by placing a brass cone with a base diameter of 8 cm and 12 cm height over a steel cylinder of same diameter and a height of 10 cm. Density of steel is 7850 kg/m^3 and that of brass is 8650 kg/m^3.

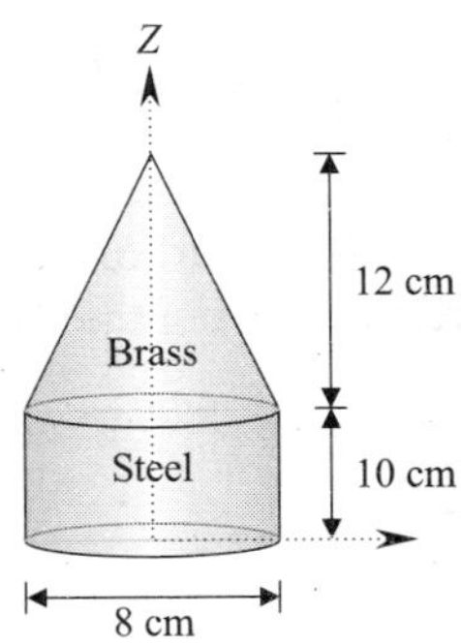

Fig. 8.46

Solution If the solid is homogeneous then we can directly determine the centroid of the volume, as the centre of mass and centroid coincide in such a case. However, in this problem, the two components are of different materials. Hence, to determine the centre of mass of the composite body, we must proceed as follows:

Density of steel, $\rho_1 = 7850$ kg/m^3

Density of brass, $\rho_2 = 8650$ kg/m^3

Mass of steel cylinder,

$$m_1 = \rho_1 V_1 = (7850)(\pi)(0.04)^2(0.1) = 3.946 \text{ kg}$$

Mass of brass cone,

$$m_2 = \rho_2 V_2 = (8650)(\pi/3)(0.04)^2(0.12) = 1.739 \text{ kg}$$

As the body is symmetrical about *X*–*Z* and *Y*–*Z* planes, we need to determine only the *z*-coordinate of the centre of mass. Hence,

$$\bar{z} = \frac{z_1 m_1 + z_2 m_2}{m_1 + m_2}$$

$$= \frac{(5)(3.946) + [10 + (12/4)](1.739)}{3.946 + 1.739} = 7.45 \text{ cm from the base}$$

[Note that within the cylinder or the cone, as the material is uniform the centre of mass coincides with the centroid of the volume].

Example 8.27 In a steel cylinder with a 20 cm base diameter and a 30 cm height, a vertical hole of 4 cm diameter is drilled up to half the depth and the portion is filled with lead, whose density is 11 370 kg/m^3. Determine the centre of mass of the composite body. Take the density of steel as 7850 kg/m^3.

Solution

Given data

Density of steel, $\rho_1 = 7850$ kg/m^3

Density of lead, $\rho_2 = 11370$ kg/m^3

Mass of entire steel cylinder,

$$m_1 = \rho_1 V_1 = (7850)(\pi)(0.1)^2\,(0.3) = 73.98 \text{ kg}$$

If V_2 is the volume of the cylinder removed then mass of the steel cylinder removed is

$$m_2' = \rho_1 V_2$$

Mass of lead filled in that place is then

$$m_3' = \rho_2 V_2$$

Since part of steel is removed and filled with lead, the net mass removed from the steel cylinder is

$$m_2 = m_2' - m_3'$$

$$= (\rho_1 - \rho_2) V_2$$

$$= (7850 - 11370)\,(\pi)\,(0.02)^2\,(0.15)$$

$$= -0.664 \text{ kg}$$

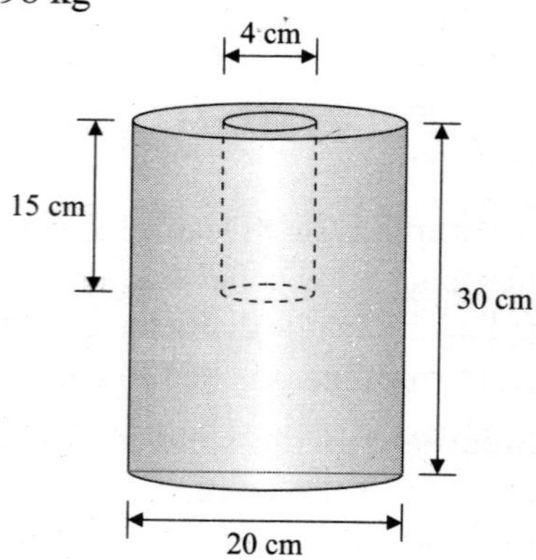

Fig. 8.47

Let us consider the axes with respect to the top surface such that the positive Z-axis is pointing down. Then the z-coordinate of the centroid is given as

$$\bar{z} = \frac{z_1 m_1 - z_2 m_2' + z_3 m_3'}{m_1 - m_2' + m_3'}$$

$$= \frac{z_1 m_1 - z_2 (m_2' - m_3')}{m_1 - (m_2' - m_3')} \quad [\text{since } z_2 = z_3]$$

$$= \frac{z_1 m_1 - z_2 (m_2)}{m_1 - (m_2)}$$

$$= \frac{(15)(73.98) - (7.5)(-0.664)}{73.98 - (-0.664)}$$

$$= 14.93 \text{ cm from the top surface}$$

Example 8.28 If the density of a hemisphere varies as the distance from the bounding plane, show that the distance of the centre of gravity from that plane is 8/15th of its radius.

Solution Consider a hemisphere of radius r oriented as shown such that the YZ plane is the bounding plane. Consider a circular disc ACB at a distance x from the bounding plane and of infinitesimal thickness dx.

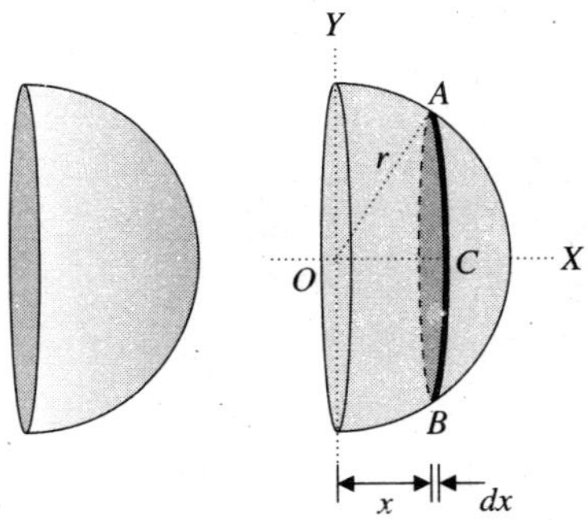

Fig. 8.48

Its radius is given by

$$AC = \sqrt{(OA)^2 - (OC)^2}$$

$$= \sqrt{r^2 - x^2}$$

Therefore, volume of the disc is

$$dV = \pi\left(\sqrt{r^2 - x^2}\right)^2 dx = \pi(r^2 - x^2)\, dx$$

Hence, its weight is

$$dW = (dm)\, g$$

$$= d(\rho V)\, g$$

Since over the disc, the density can be assumed to be constant,

$$dW = (\rho)\,(dV)\, g$$

Since the density varies as the distance from the bounding plane,

$$\rho = kx \text{ (where } k \text{ is a constant)}$$

Hence, we can write

$$dW = [k\,x]\ \pi(r^2 - x^2)\, g\, dx$$

Hence, the total weight of the hemisphere is obtained by integrating the above expression between limits:

$$W = \int_0^r dW = \int_0^r kx\,\pi\,(r^2 - x^2)\, g\, dx$$

$$= k\,\pi\, g \int_0^r (r^2 x - x^3)\, dx$$

$$= k\,\pi\, g\left[\frac{r^2 x^2}{2} - \frac{x^4}{4}\right]_0^r = k\,\pi\, g\left[\frac{r^4}{4}\right]$$

The first moment of the circular disc ACB about the bounding plane is given as

$$x\,dW = k\,\pi\,g(r^2 x^2 - x^4)\,dx$$

Hence, the first moment of the entire hemisphere about the bounding plane is obtained by integrating the above expression between limits,

$$\int_0^r x\,dW = \int_0^r k\,\pi\,g\,(r^2 x^2 - x^4)\,dx$$

$$= k\,\pi\,g\left[\frac{r^2 x^3}{3} - \frac{x^5}{5}\right]_0^r = k\,\pi\,g\left[\frac{2r^5}{15}\right]$$

Therefore, the centre of gravity of the hemisphere is obtained as

$$\bar{x} = \frac{\int_0^r x\,dW}{\int_0^r dW} = \frac{k\pi g\left[\frac{2r^5}{15}\right]}{k\pi g\left[\frac{r^4}{4}\right]} = \frac{8}{15}r$$

SUMMARY

Centre of Gravity

The force of gravity for an extended body is not just one force but the *resultant* of a great many forces. This resultant acts vertically downwards towards the centre of the earth. Its magnitude is given by $W = \int dW = Mg$, where M is the mass of the body and g is the acceleration due to gravity. The location (G) of the resultant is found out by the principle of moments as

$$\bar{x} = \frac{\int x\,dW}{\int dW} \quad \text{and} \quad \bar{y} = \frac{\int y\,dW}{\int dW}$$

The point of application of the resultant of gravitational forces acting on all the particles of a body is called the *centre of gravity*. In other words, it is *a point* on the body where the entire weight is assumed to be *concentrated*. It remains the same for all orientations of the body.

Centroid

The term centre of gravity is applied to three-dimensional bodies with weight. The term centre of gravity, when applied to two-dimensional bodies like homogeneous plate of *infinitesimal thickness* is termed as *centroid*. It is defined as the *point* on the area where the entire area is assumed to be *concentrated*. The term centroid is also used for length of a curve and volume of a body without considering its weight.

Centroid of a Line

The centroid of a line of entire length L is given by

$$\bar{x} = \frac{\int x\,dL}{L} \quad \text{and} \quad \bar{y} = \frac{\int y\,dL}{L}$$

The centre of gravity of a thin homogenous wire is the same as the centroid of its centre line.

Centroid an Area

The centroid of a plane lamina of area A is given as

$$\bar{x} = \frac{\int x\,dA}{A} \quad \text{and} \quad \bar{y} = \frac{\int y\,dA}{A}$$

The centre of gravity of a homogenous plate of infinitesimally small thickness is the same as the centroid of the surface area.

Centroid of a Volume

The centroid of a body of volume V is given as

$$\bar{x} = \frac{\int x\,dV}{V}, \quad \bar{y} = \frac{\int y\,dV}{V} \quad \text{and} \quad \bar{z} = \frac{\int z\,dV}{V}$$

Centre of Mass

The centre of mass of a body of mass M is given as

$$\bar{x} = \frac{\int x\,dm}{M}, \quad \bar{y} = \frac{\int y\,dm}{M} \quad \text{and} \quad \bar{z} = \frac{\int z\,dm}{M}$$

If the body is non-homogeneous then the centre of mass does not coincide with the centroid of its volume.

Axes of Reference

While choosing the coordinate axes, generally, the area is placed on the first quadrant with the lowest line of the area coinciding with the X-axis and the left line of the area coinciding with the Y-axis. If an area has an axis of symmetry then its centroid will lie on that axis. If it has two axes of symmetry like rectangle, equal flange I-section, etc., then the centroid will lie at the intersection of the two axes.

In a similar way, the axes are chosen for solid bodies. When a volume possesses a plane of symmetry then the first moment of volume with respect to that plane is zero. Hence, the centroid lies on that plane. If it possesses two planes of symmetry then the centroid lies on the line of intersection of these two planes. If it possesses three planes of symmetry then it lies at the point of intersection of the three planes.

Theorems of Pappus and Guldinus

Theorem I The area of surface generated by revolving a plane curve about a non-intersecting axis in the plane of the curve is equal to the product of the length of the curve and the distance travelled by the centroid of the curve during revolution.

Theorem II The volume of a solid generated by revolving a plane area about a non-intersecting axis in its plane is equal to the product of the area and the length of the path travelled by the centroid of the area during the revolution about the axis.

EXERCISES

Objective-type Questions

1. The centre of gravity of a body and its centre of mass coincide when ____________ is uniform throughout the body.
   (a) acceleration due to gravity (b) density
   (c) thickness (d) mass
2. The centre of mass of a body and its centroid of volume coincide when ____________ is uniform throughout the body.
   (a) acceleration due to gravity (b) density
   (c) thickness (d) area
3. Centre of gravity of a body
   (a) is a point in the body at which g is constant
   (b) is a point in the body different for different orientations of the body
   (c) is a point in the body at which the entire weight is assumed to be concentrated
   (d) always coincides with the centroid of its volume
4. When a body is suspended about a horizontal axis, its centre of gravity lies
   (a) above the point of suspension (b) anywhere on the body
   (c) vertically below the point of suspension d) at the point of suspension
5. If an area has an axis of symmetry then
   (a) its first moment about that axis is zero (b) its centroid lies on that axis
   (c) both (a) and (b) (d) none of these
6. The centroid of arc of a circle of radius R and symmetric about the X-axis with subtended angle 2α is
   (a) $\dfrac{2R}{\alpha}$ (b) $\dfrac{R\tan\alpha}{\alpha}$ (c) $\dfrac{R\cos\alpha}{\alpha}$ (d) $\dfrac{R\sin\alpha}{\alpha}$
7. The x or y coordinate of the centroid of a quarter-circular arc of radius r is
   (a) $\dfrac{4r}{3\pi}$ (b) $\dfrac{2r}{3\pi}$ (c) $\dfrac{2r}{\pi}$ (d) $\dfrac{r}{\pi}$
8. The x or y coordinate of the centroid of a quadrant of a circular area of radius r is
   (a) $\dfrac{4r}{3\pi}$ (b) $\dfrac{2r}{3\pi}$ (c) $\dfrac{2r}{\pi}$ (d) $\dfrac{r}{\pi}$
9. The centroid of an equilateral triangle of side a with a side parallel to the X-axis is
   (a) $\dfrac{a}{2}, \dfrac{a}{\sqrt{6}}$ (b) $\dfrac{a}{2}, \dfrac{a}{\sqrt{12}}$ (c) $\dfrac{a}{2}, \dfrac{a}{\sqrt{24}}$ (d) $\dfrac{a}{3}, \dfrac{a}{3}$
10. The first moment of an area about the X-axis is
    (a) $\int x\,dA$ (b) $\int y\,dA$ (c) $\int x^2\,dA$ (d) $\int y^2\,dA$

11. The coordinates of the centroid of the right-angled triangle shown is

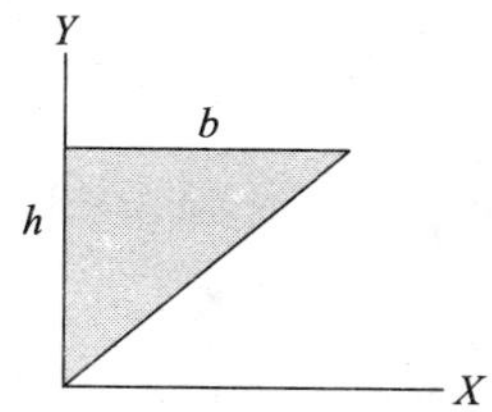

(a) $\left[\frac{b}{3}, \frac{h}{3}\right]$ (b) $\left[\frac{2b}{3}, \frac{h}{3}\right]$ (c) $\left[\frac{b}{3}, \frac{2h}{3}\right]$ (d) $\left[\frac{2b}{3}, \frac{2h}{3}\right]$

12. The coordinates of the centroid of the right-angled triangle shown is

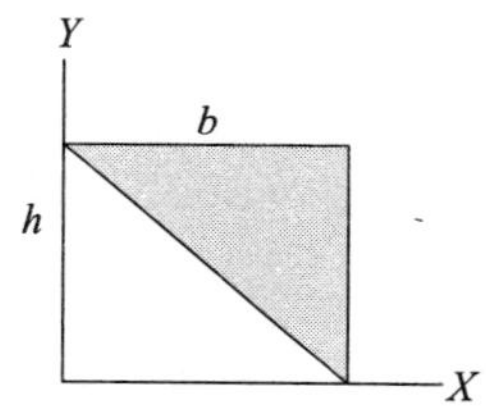

(a) $\left[\frac{b}{3}, \frac{h}{3}\right]$ (b) $\left[\frac{2b}{3}, \frac{h}{3}\right]$ (c) $\left[\frac{b}{3}, \frac{2h}{3}\right]$ (d) $\left[\frac{2b}{3}, \frac{2h}{3}\right]$

13. Assuming a square of side a to be made up of two right-angled triangles then the distance of the centroid of each triangle with respect to the diagonal is

(a) $\frac{a}{\sqrt{2}}$ (b) $\frac{a}{\sqrt{3}}$ (c) $\frac{a}{\sqrt{9}}$ (d) $\frac{a}{\sqrt{18}}$

Answers

1. (a)	**2.** (b)	**3.** (c)	**4.** (c)	**5.** (c)	**6.** (d)	**7.** (c)	**8.** (a)
9. (b)	**10.** (b)	**11.** (c)	**12.** (d)	**13.** (d)			

Short-answer Questions

1. Differentiate between centroid and centre of gravity.
2. Under what condition does the centre of mass coincide with the centre of gravity?
3. Under what condition is the centre of gravity of a wire same as the centroid of its centre line?
4. Under what condition is the centre of gravity of a plate same as the centroid of its surface area?
5. While finding the centroid of a line, an area or a volume, when do we use summation sign and when do we use integral sign?
6. Explain why the first moment of an area with an axis of symmetry is zero.
7. Explain how to choose the axes of reference, while determining the coordinates of the centroid.
8. What are the coordinates of centroid of a triangle in general?
9. If an area has two axes of symmetry then where does the centroid lie?

10. Can the centroid of a volume coincide with the centroid of its cross section? Explain.

11. Define surface of revolution and volume of revolution.

12. State the theorems of Pappus and Guldinus to find out the surface area and the volume of a body.

13. How can you use the Pappus and Guldinus theorems to find the centroid of a line or an area?

Numerical Problems

8.1 Find the centroid of the bent wire shown in Fig. E.8.1. All dimensions are in cm.

Ans. $\bar{x} = 6.2$ cm and $\bar{y} = 6.5$ cm

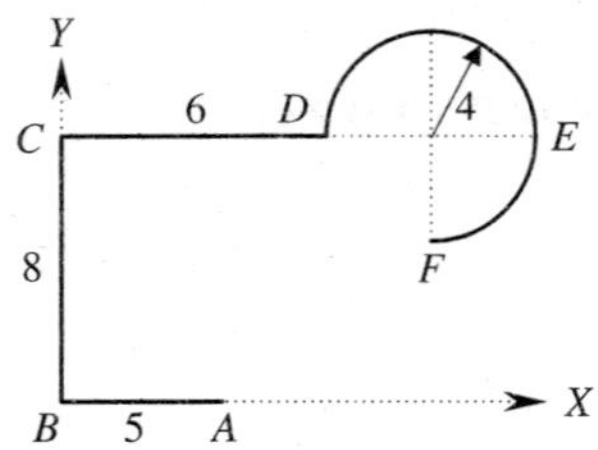

Fig. E.8.1

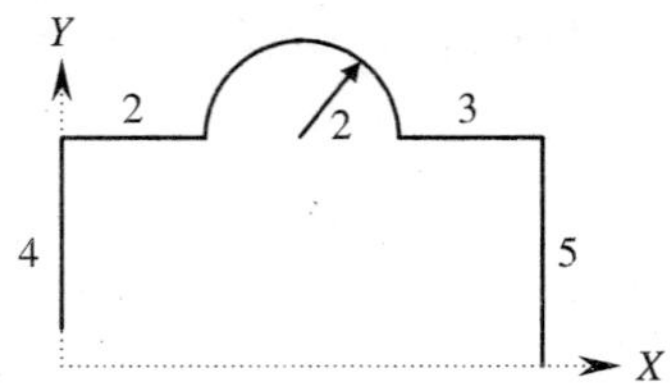

Fig. E.8.2

8.2 Find the centroid of the bent wire shown in Fig. E.8.2. All dimensions are in cm.

Ans. $\bar{x} = 4.7$ cm, $\bar{y} = 4.4$ cm

8.3 Determine the centroid of a thin wire bent into the shapes as shown in Figs E.8.3 (a)–(c) about the given axes.

Ans. (a) $\bar{x} = \bar{y} = \dfrac{3r}{4+\pi}$; (b) $\bar{x} = 0,\ \bar{y} = \dfrac{2r}{2+\pi}$; (c) $\bar{x} = \dfrac{-r}{3\pi+4},\ \bar{y} = \dfrac{r}{3\pi+4}$

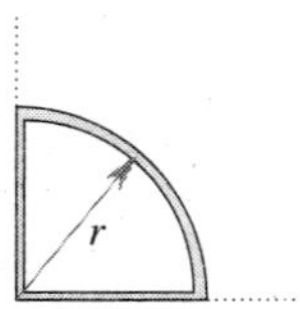

Fig. E.8.3(a)

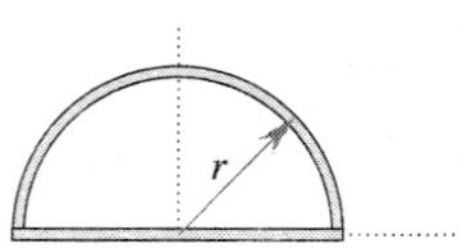

Fig. E.8.3(b)

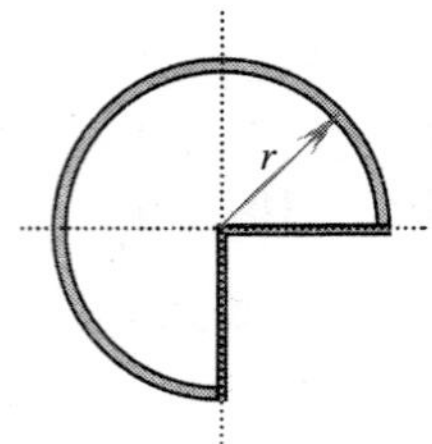

Fig. E.8.3(c)

8.4 A slender homogeneous wire *ABCD* of the shape as shown in Fig. E.8.4 is suspended from a hinge at point *A*. Determine the inclination that the side *AB* would make with the vertical.

Ans. 4.32°

8.5 A slender homogeneous wire of semicircular shape is suspended as shown in Fig. E.8.5. Determine the inclination that the diameter *AB* would make with the vertical.

Ans. 32.5°

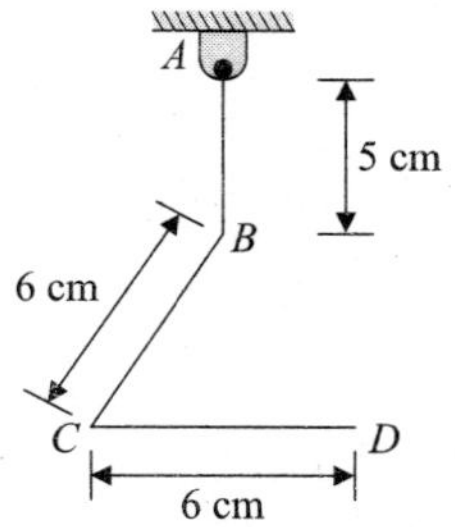

Fig. E.8.4

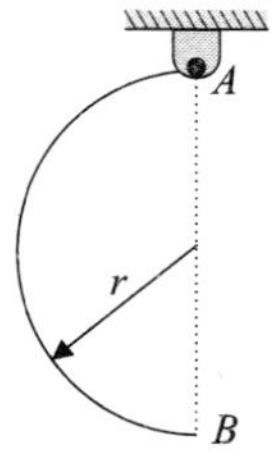

Fig. E.8.5

8.6 A wire bent into a closed triangle *ABC* is suspended about the point *A* as shown in Fig. E.8.6. Determine the inclination that the side *AC* would make with the vertical.

Ans. 15.1°

8.7 A slender homogeneous metallic rod *AB* of mass per unit length *m* is bent into the shape of a semicircular arc of radius *r*. It is hinged at *A* and the end *B* is butting against a wall as shown in Fig. E.8.7. Determine the reaction at *B*.

Ans. *mgr*

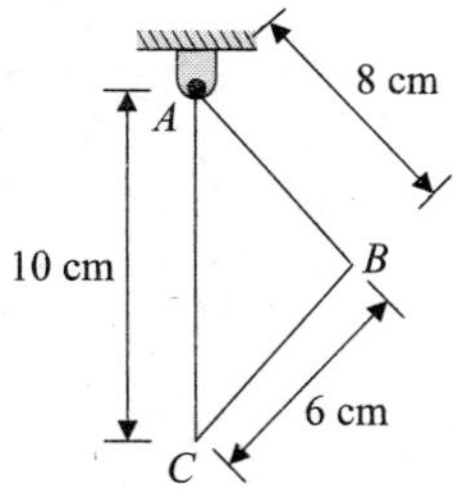

Fig. E.8.6

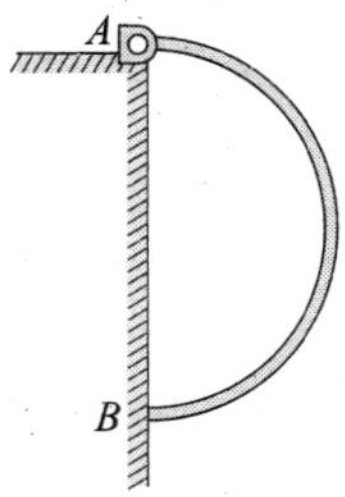

Fig. E.8.7

8.8 Find the centroid of the area included between $ay^2 = x^3$ and $x = b$.

Ans. $\frac{5b}{7}, 0$

8.9 Determine the centroid of the following structural steel sections shown in Figs E.8.9 (a–c).

Ans. (a) $\bar{y}$ = 18 cm from base, (b) $\bar{y}$ = 28.2 cm from base, (c) $\bar{x}$ = 8 cm from extreme left, $\bar{y}$ = 11.1 cm from base

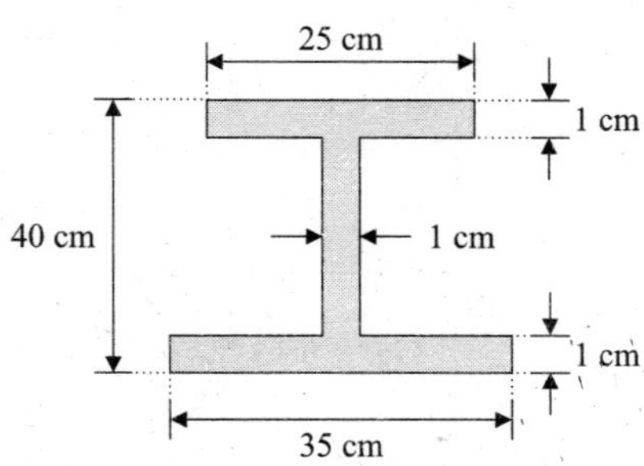

Fig. E.8.9(a)

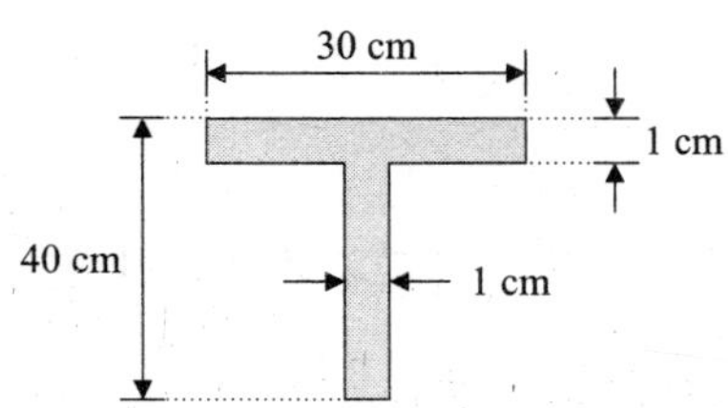

Fig. E.8.9(b)

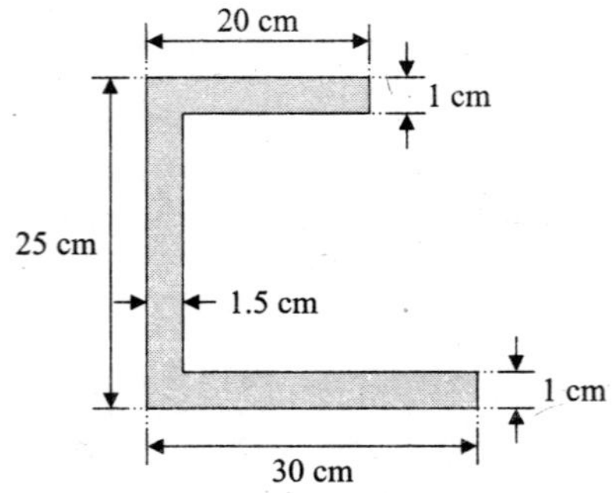

Fig. E.8.9(c)

8.10 Locate the centroid of the shaded area shown in Fig. E.8.10 with respect to the given axes.

Ans. $\bar{x} = 0$ cm and $\bar{y} = 2.6$ cm

8.11 Find the centre of gravity of a thin circular disc of radius r from which is cut out a circle whose diameter is the radius of the disc. Refer Fig. E.8.11. Solve for $r = 5$ cm.

Ans. $\bar{x} = 4.17$ cm and $\bar{y} = 0$ cm

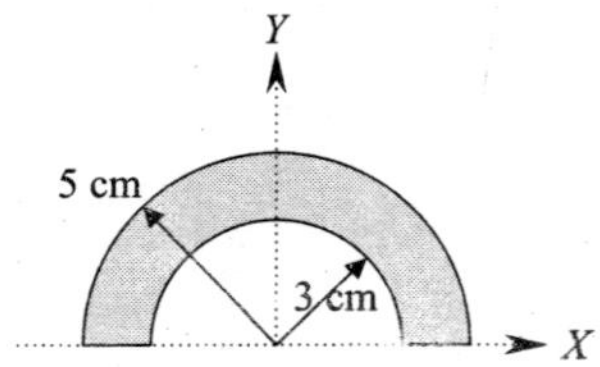

Fig. E.8.10

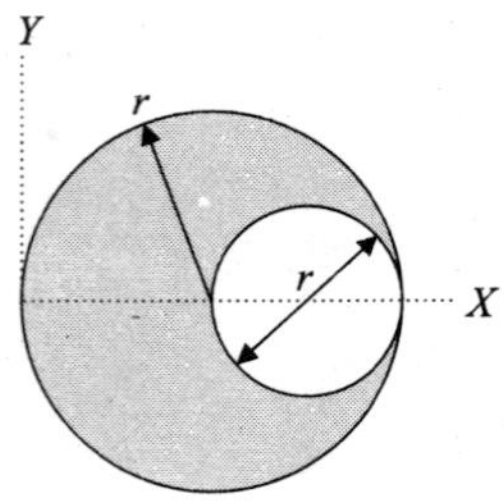

Fig. E.8.11

8.12 Locate the centroid of the cut section shown in Fig. E.8.12. A semicircle of 2-cm radius is cut at the midpoint of the hypotenuse.

Ans. $\bar{x} = 3.6$ cm and $\bar{y} = 1.4$ cm with respect to left corner

8.13 Find the centroid of the shaded area shown in Fig. E.8.13.

Ans. $\bar{x} = 2.33$ cm and $\bar{y} = 0.66$ cm

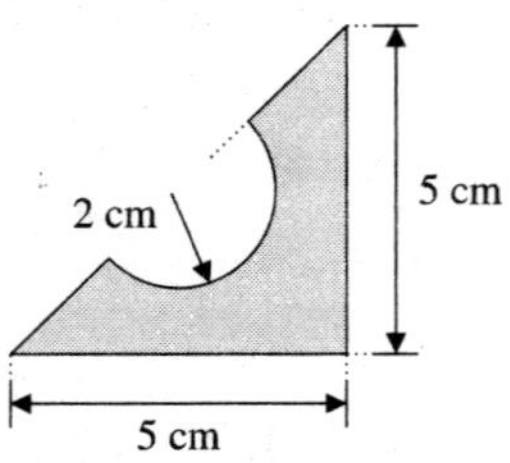

Fig. E.8.12

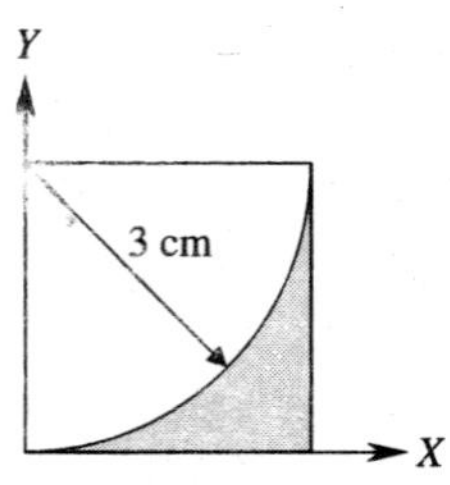

Fig. E.8.13

8.14 Find the centroid of the shaded lamina shown in Fig. E.8.14.

Ans. $\bar{x} = -1.33$ cm and $\bar{y} = 0$

8.15 Find the centroid of the composite section shown in Fig. E.8.15. Consider the curved portions to be semicircular in shape.

Ans. 1.9 cm from base

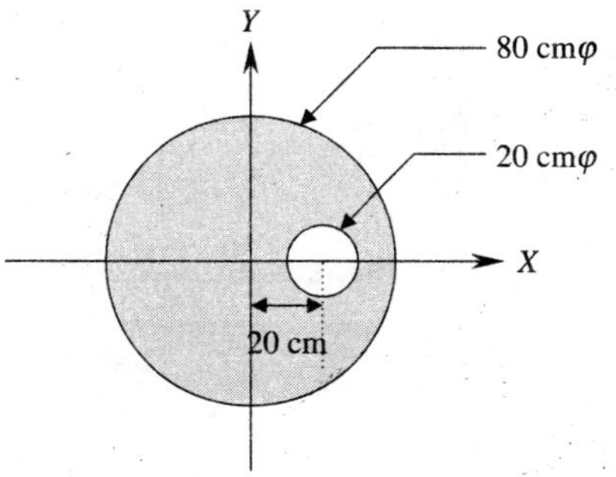

Fig. E.8.14

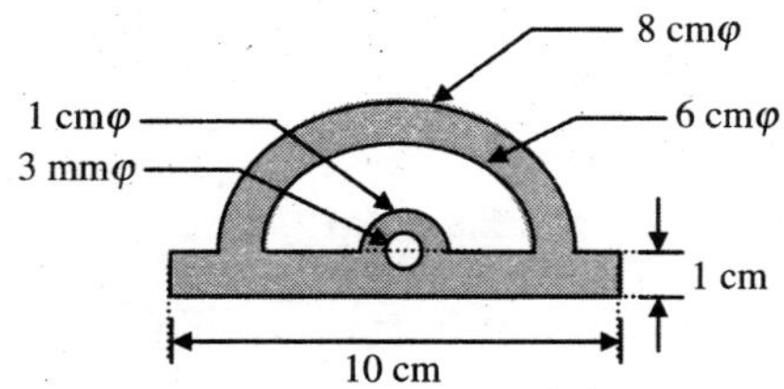

Fig. E.8.15

8.16 Two thin uniform circular discs of 4 cm and 3 cm radii are fixed in the same plane, their edges touching each other externally. Find the centroid of the combination.

Ans. It lies on the line joining their centres at a distance of 1.48 cm from their point of contact towards the larger circular disc.

8.17 Out of a circular lamina of 10 cm radius, a circular portion of 3 cm radius is cut out, the distance between the centres of the given lamina and the portion cut out being 6 cm. Find centroid of the remainder.

Ans. At a distance of 0.593 cm from the centre and away from the cut portion

8.18 Where must a hole of 2 cm diameter be punched out of a uniform circular disc of 10 cm diameter in order to shift the centroid of the remainder by 0.1 cm from the centre of the original disc?

Ans. 2.4 cm

8.19 Find the centroid of a circular disc of radius r, from which is cut out a square whose diagonal length is the radius of the disc. Refer Fig. E.8.19.

Ans. At a distance of $0.095r$ from the centre and away from the cut square

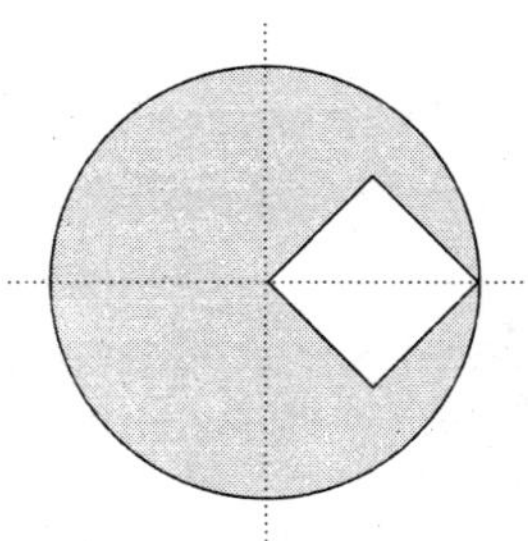

Fig. E.8.19

8.20 Determine the centroid of the composite sections shown in Figs E.8.20 (a) and (b).

Ans. (a) $\bar{y} = 4.5$ cm from base; (b) $\bar{y} = 2.9$ cm from base

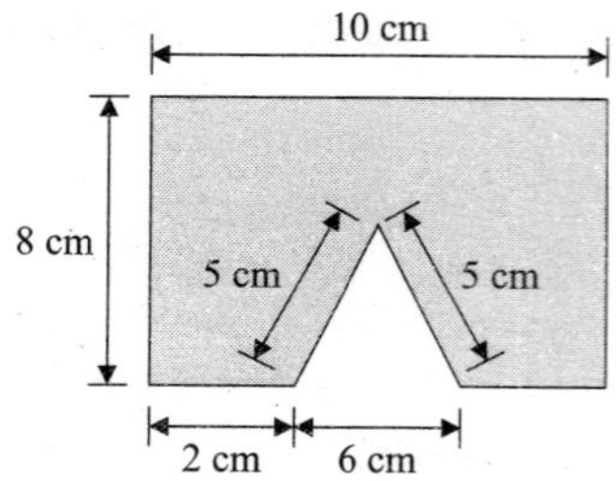

Fig. E.8.20(a)

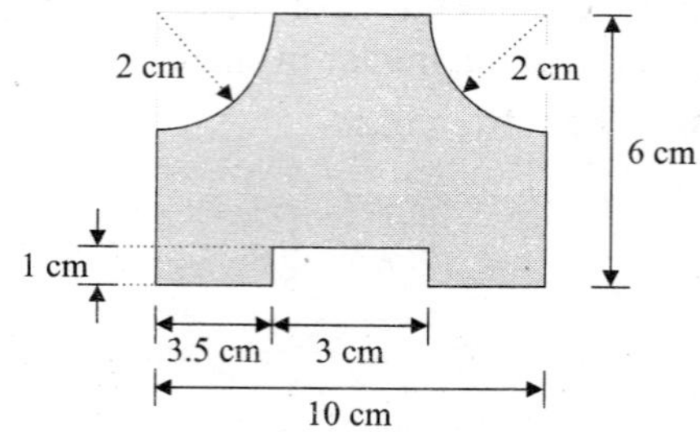

Fig. E.8.20(b)

8.21 Determine the centroid of the composite sections shown in Figs E.8.21 (a) and (b).

Ans. (a) $\bar{x} = 4.5$ cm, $\bar{y} = 0$ with respect to axis of symmetry (b) $\bar{x} = 0$, $\bar{y} = 2.3$ cm with respect to axis of symmetry

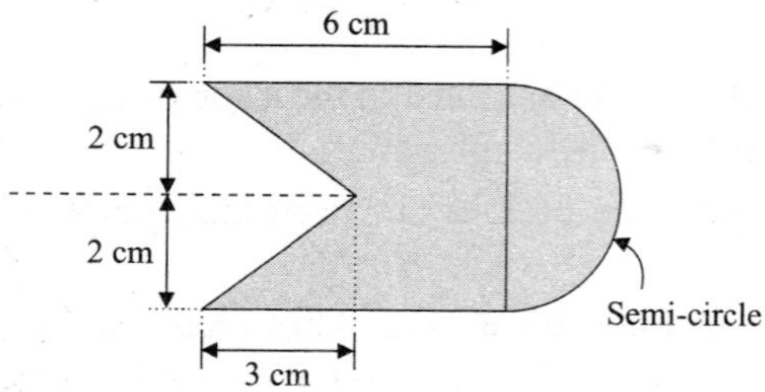

Fig. E.8.21(a)

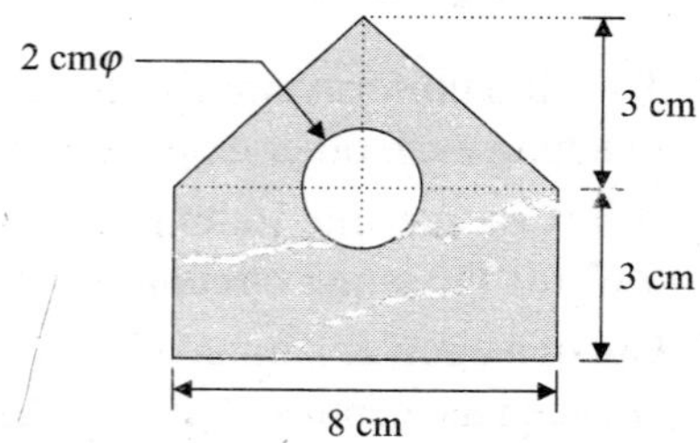

Fig. E.8.21(b)

8.22 Determine the centroid of the composite sections shown in Figs E.8.22 (a) and (b): (a) a rectangular lamina is cut from a semicircular area; (b) a triangular area is cut from a semicircular area.

Ans. (a) $\bar{x} = 0$, $\bar{y} = 2.4$ cm; (b) $\bar{x} = 0$, $\bar{y} = 2.2$ cm

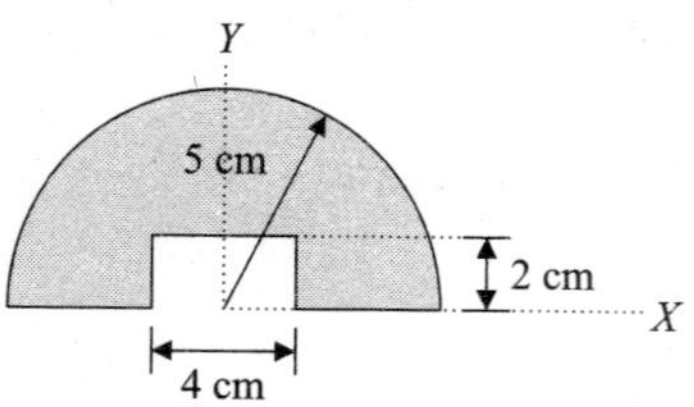

Fig. 8.22(a)

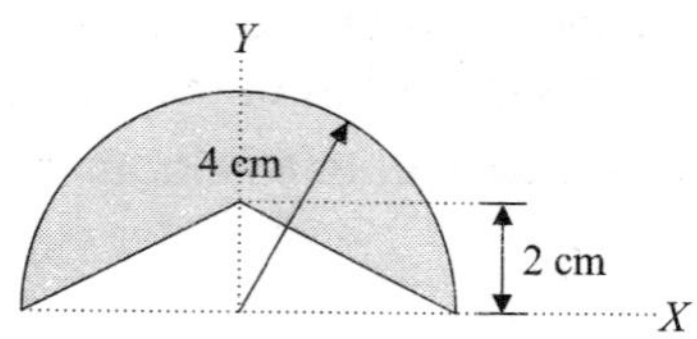

Fig. 8.22(b)

8.23 Determine the centroid of the composite sections shown in Figs E.8.23 (a) and (b). From a quarter-circular area of radius 5 cm, (a) a square of 2 cm side is cut; (b) a quarter-circular area of 3 cm radius is cut.

Ans. (a) $\bar{x} = \bar{y} = 2.41$ cm; (b) $\bar{x} = \bar{y} = 2.6$ cm about centre of the quarter circle

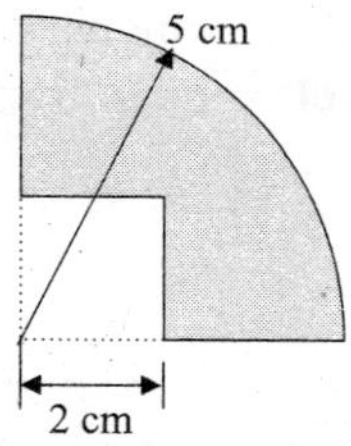

Fig. E.8.23(a)

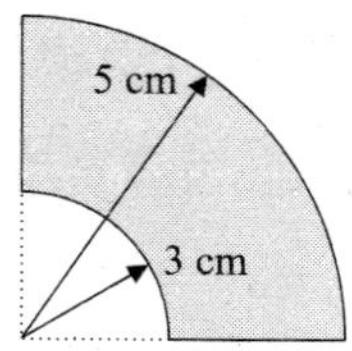

Fig. E.8.23(b)

8.24 Determine the centroid of the composite section shown in Fig. E.8.24.

Ans. $\bar{x} = 8.3$ cm, $\bar{y} = 5.2$ cm

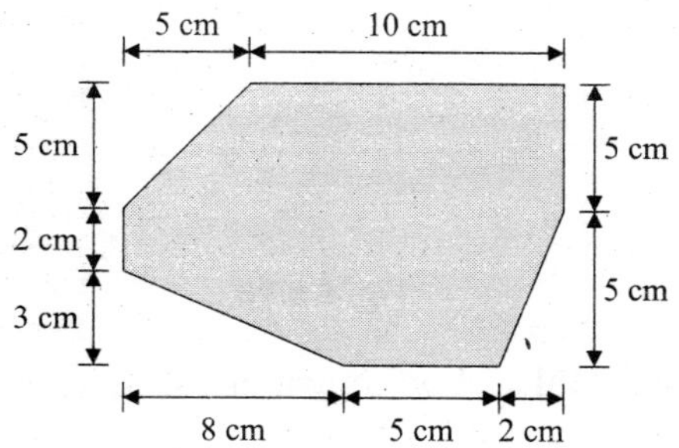

Fig. E.8.24

8.25 A rectangular area of dimensions 5 cm × 2 cm is cut out symmetrically from a quadrant of a circular lamina of 10 cm radius as shown in Fig. E.8.25. Determine the coordinates of centroid of the remaining area.

Ans. $\bar{x} = \bar{y} = 4.4$ cm about O

8.26 Determine the centroid of the composite section formed by removing an equilateral triangle from a semicircle of 5 cm radius. Refer Fig. E.8.26.

Ans. $\bar{x} = 0$, $\bar{y} = 2.17$ cm from base

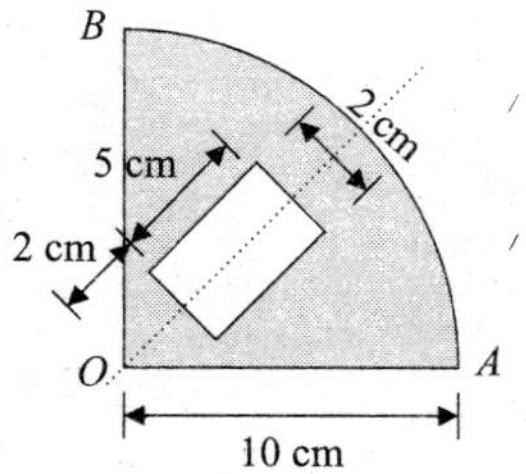

Fig. E.8.25

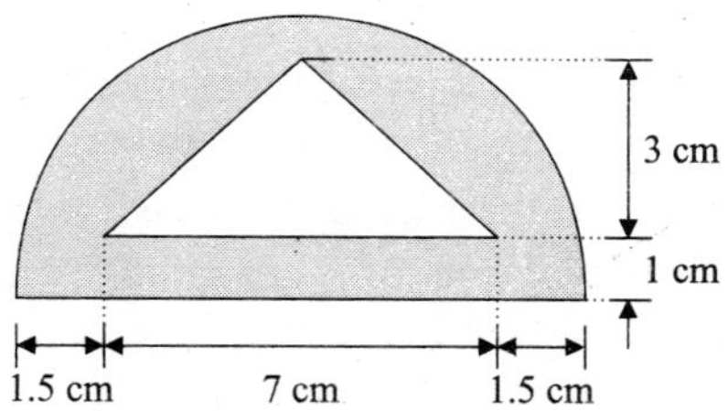

Fig. E.8.26

8.27 From an isosceles right triangle ABC, a semicircle of 2 cm radius is cut at the midpoint of the hypotenuse and a quarter circle of 2.5 cm radius is cut at the corner B as shown in Fig. E.8.27. Determine the centroid of the shaded region.

Ans. $\bar{x} = \bar{y} = 2.8$ cm about corner B

8.28 Determine the centroid of the shaded area shown in Fig. E.8.28, in which a semicircular area is cut from a square lamina such that the diameter of the semicircle is 1 cm from the diagonal AC of the square.

Ans. $\bar{x} = \bar{y} = 3.8$ cm about corner D

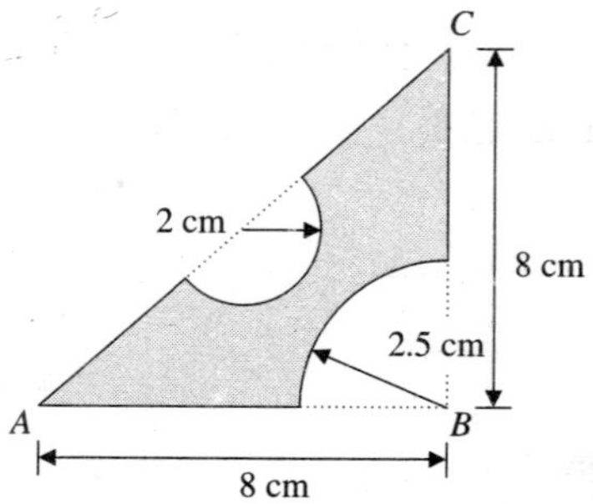

Fig. E.8.27

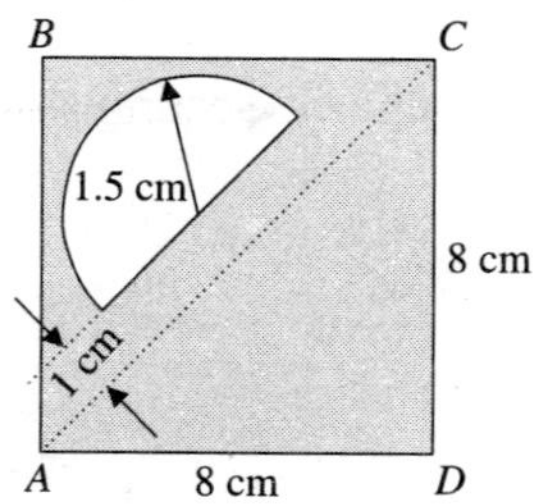

Fig. E.8.28

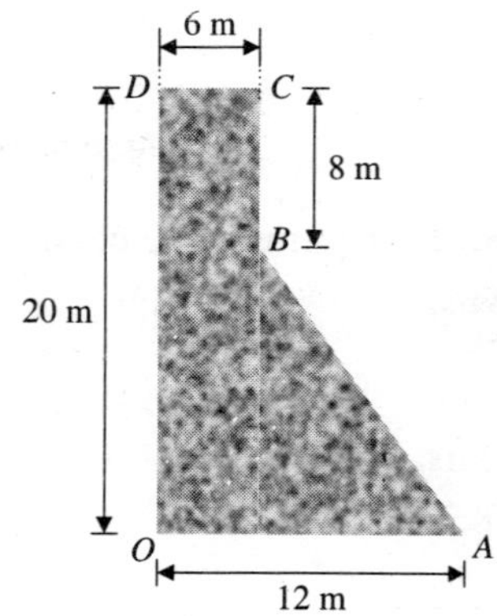

Fig. E.8.29

8.29 Determine the centroid of a dam section shown in Fig. E.8.29 about O.

Ans. $\bar{x} = 4.15$ m, $\bar{y} = 8.62$ m

8.30 Knowing that the surface area of a sphere of radius r is equal to $4\pi r^2$, determine the centroid of a line in the form of semicircular arc.

Ans. $\dfrac{2r}{\pi}$

8.31 Determine the surface area and volume of a solid generated when an equilateral triangle of side a is revolved through 360° about one of its sides.

Ans. S.A $= \sqrt{3}\,\pi a^2$, $V = \pi a^3/4$

8.32 Determine the volume generated by the revolution of a square of side a about one of its diagonals.

Ans. $\dfrac{\pi a^3}{\sqrt{18}}$

8.33 Determine the surface area and volume of a solid generated when a semicircular area of 3 cm radius is revolved through 360° about axis AA as shown in Fig. E.8.33.

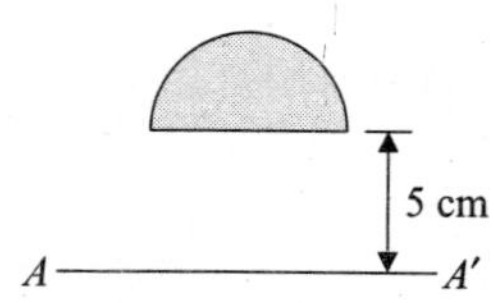

Fig. E.8.33

Ans. S.A $= 597.7$ cm^2, $V = 557.2$ cm^3

8.34 Determine the volume of the solid generated by revolving the area shown in Fig. E.8.34 about the axis O–O'.

Ans. 794.4 cm^3

8.35 Determine the volume of a frustum of cone of 5 cm base radius, 3 cm top radius and 8 cm height, in which a vertical hole of 1 cm radius has been drilled. Refer Fig. E.8.35.

Ans. 386 cm^3

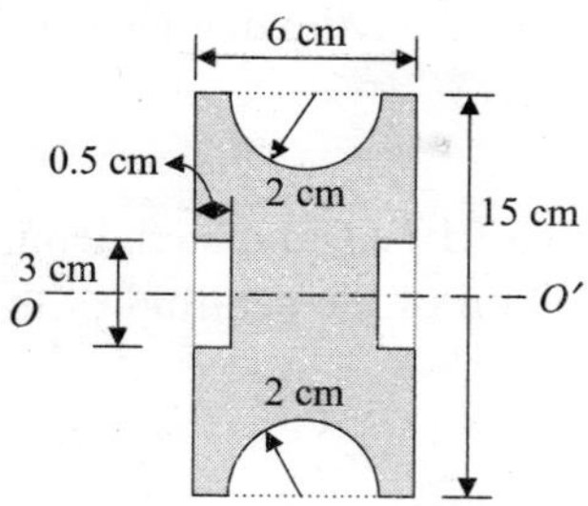

Fig. E.8.34

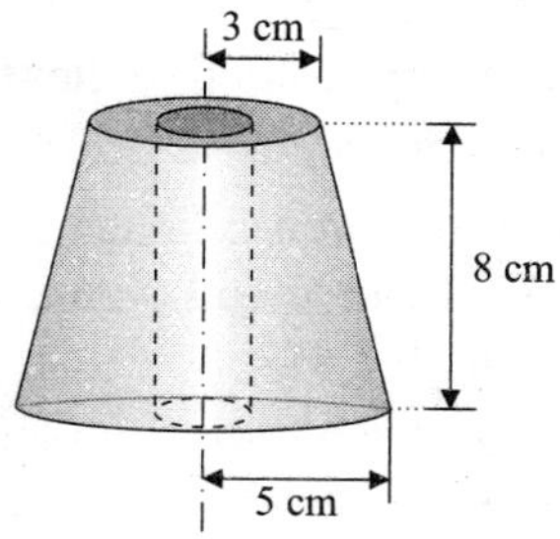

Fig. E.8.35

8.36 Determine the surface area and volume of solid body generated by revolving the plane area shown in Fig. E.8.36 about the axis of revolution O–O'.

Ans. S.A = 668.2 cm^2, V = 1084.4 cm^3

8.37 Using the Pappus and Guldinus theorems, determine the number of steel rivets that can be made from 5 kg mass of steel available, assuming there is no wastage. Take density of steel to be 7850 kg/m^3. Refer Fig. E.8.37.

Ans. approximately 1976 pieces

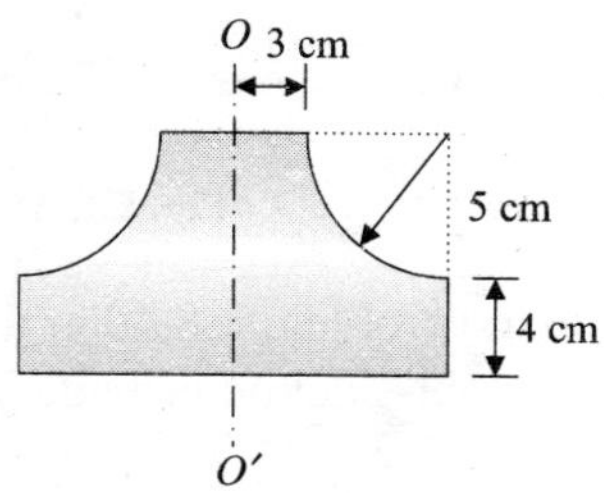

Fig. E.8.36

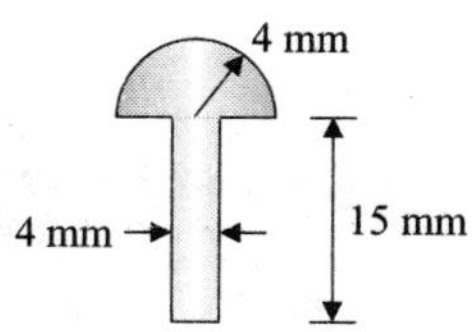

Fig. E.8.37

8.38 Using the Pappus and Guldinus theorems, determine the surface area and volume of a solid cylinder from which a frustum of cone is removed as shown in Fig. E.8.38 (a). The cross-sectional dimensions of the solid are shown in Fig. E.8.38 (b).

Ans. 1448.1 cm^2, 2362.5 cm^3

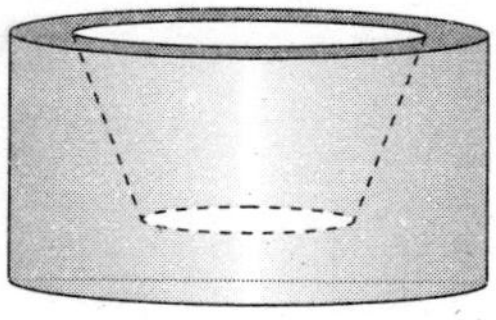

Fig. E.8.38(a)

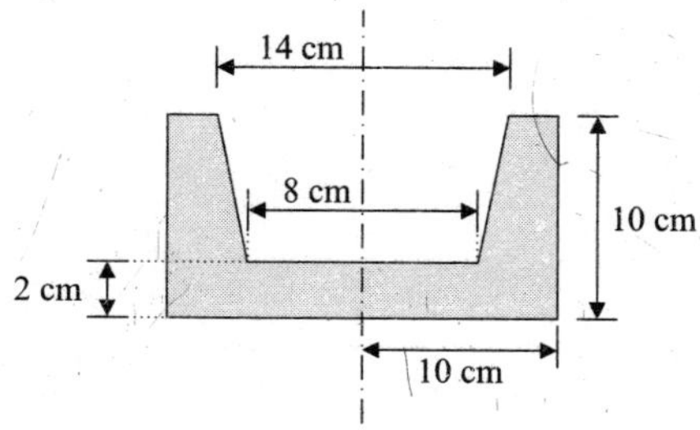

Fig. E.8.38(b)

8.39 Locate the centroid of frustum of a cone of 8 cm height and having diameters of 5 cm and 8 cm at the top and bottom of the frustum of cone respectively.

Ans. $\bar{y} = 3.4$ cm from base

8.40 Two uniform iron cylinders of 2 cm and 1 cm radii and of 10 cm and 8 cm lengths respectively are joined end to end on the same axis. Find the position of the centroid of the combination. Refer Fig. E.8.40.

Ans. 6.5 cm along the axis from left end

8.41 A solid hemisphere of 5 cm radius is placed over a solid cylinder of same base radius and 10 cm height. From the base of the cylinder, a hemisphere of 5 cm radius is cut out. Determine the centroid of volume of the composite body. Refer Fig. E.8.41.

Ans. 8.33 cm from base

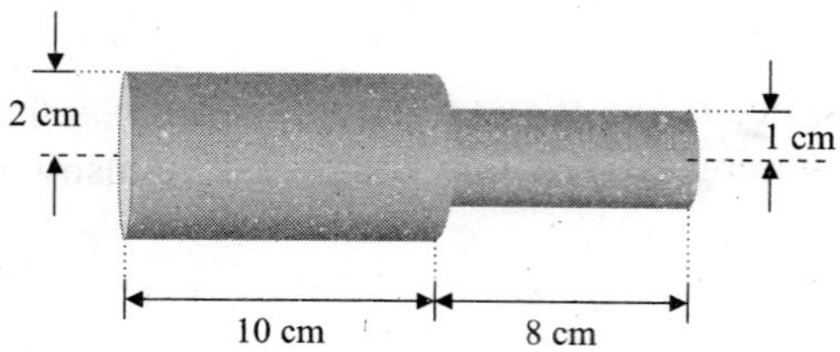

Fig. E.8.40, E.8.42

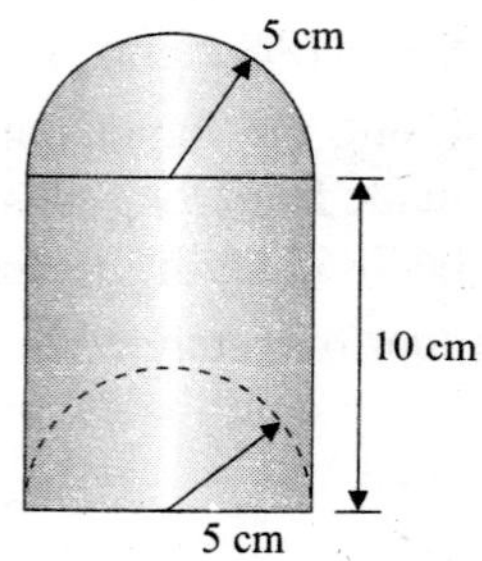

Fig. E.8.41

8.42 Determine the centre of mass for the composite rod in the problem E.8.40, if the left-side rod is made up of steel, whose density is 7850 kg/m^3 and the right-side rod is made up of lead, whose density is 11 370 kg/m^3.

Ans. 7 cm along the axis from left end

9 Moment of Inertia

9.1 INTRODUCTION

In the previous chapter, we discussed how to determine the location of the resultant of distributed forces by using *first moment* of an area, volume, etc. Thus, we saw the methods to determine the centroid, centre of mass and centre of gravity of various bodies. In this chapter, we will discuss the *second moment* of an area across a cross section, which finds application in the design of structural members such as beams, columns, etc.

Beam members when subjected to lateral loading undergo deformation and the flexural or bending stress distribution across a cross section is as shown below:

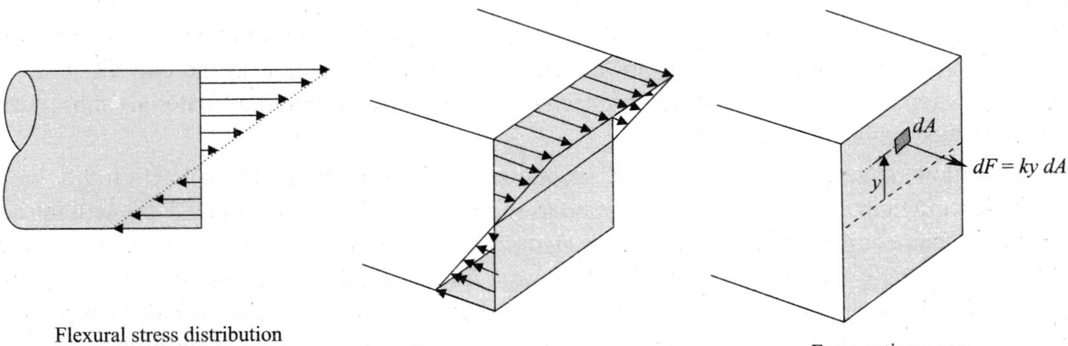

Fig. 9.1

From strength of materials, we know that the stress distribution is not constant across the cross section of the beam, but varies *linearly* reaching maximum values at the top and bottom points. However, there lies an axis on this cross section, where there is no stress at all and this axis is termed as *neutral axis*. The location of this axis is determined from the centroid of the cross section as this axis passes through the centroid in the case of beams subjected to pure bending with no axial forces. Above this axis, the member is subjected to tensile stresses (as in the case of cantilever beams) and below this axis is subjected to compressive stresses. As we know, force acting on the cross section is stress multiplied by cross sectional area, this stress distribution can be replaced by a system of parallel forces. As the stress varies linearly from the neutral axis, the force acting on an infinitesimally small element is proportional to the area dA and distance y from the axis, i.e.,

$$dF = ky\, dA \tag{9.1}$$

The resultant of all these forces can be determined by summing up all the infinitesimally small forces, i.e.,

$$F = k\int y\, dA \tag{9.2}$$

We know that the integral on the right-hand side is the first moment of the area about the neutral axis and it is zero as the neutral axis passes through the centroid. As the forces above the neutral axis and below the neutral axis are of opposite direction, they tend to rotate the beam member about the neutral axis causing bending of the beam about the neutral axis. This bending moment is obtained by taking moment of the elemental force about the neutral axis, i.e.,

$$(ky\, dA)y \tag{9.3}$$

Hence, the moment of all the forces acting over the cross section about the neutral axis is

$$k\int y^2 dA \tag{9.4}$$

This integral is known as **moment of first moment** or **second moment** of the area. As this integral is dependent only on the shape of the cross section, its value is different for different cross sections of the beam member. Its physical significance is that it gives a measure of *resistance* to *bending of beams* under the action of loads. The greater the second moment of an area, the greater the resistance to bending. Hence, it is essential to determine the second moment of cross section of the beams in their designing.

In this chapter, we will discuss how to determine the second moment of an area about an axis for various sections. Though students may not understand its real application in the current study, they would do so when they come to the study of mechanics of solids or strength of materials subsequent to this study in their engineering course.

In the following section, we will define second moment of an area; and in Section 9.6, we will derive the second moments of area for various cross sections. In Sections 9.4 and 9.5, we will introduce two theorems, named respectively *parallel axis theorem* and *perpendicular axis theorem*. The former is useful in determining the second moment of an area about parallel axes, one of the axes being a centroidal axis, while the latter is useful in the study of shafts. In Section 9.7, we will discuss product of inertia, which though does not have any physical significance, is useful in determining the second moment of an area about the inclined axes. The second moment of an area about the inclined axes is discussed in Section 9.8.

9.2 SECOND MOMENT OF AN AREA (OR) MOMENT OF INERTIA

Consider a plane lamina of area A as shown in Fig. 9.2. If we consider a small element of area dA at a distance x and y from the origin, then its first moments with respect to X and Y axes are respectively,

$$dM_x = y dA \quad \text{and} \quad dM_y = x dA \tag{9.5}$$

If we take **moment** of **first moment** of the elemental area dA, i.e.,

$$y(y\,dA) \quad \text{and} \quad x(x\,dA) \tag{9.6}$$

then it is called **second moment** of the elemental area dA about the respective axes. Therefore, the second moment of the entire area A with respect to the X-axis is obtained by integrating the above expression, i.e.,

$$I_{xx} = \int y^2 dA \tag{9.7}$$

Similarly, the second moment of the area A with respect to the Y-axis is

$$I_{yy} = \int x^2\, dA \tag{9.8}$$

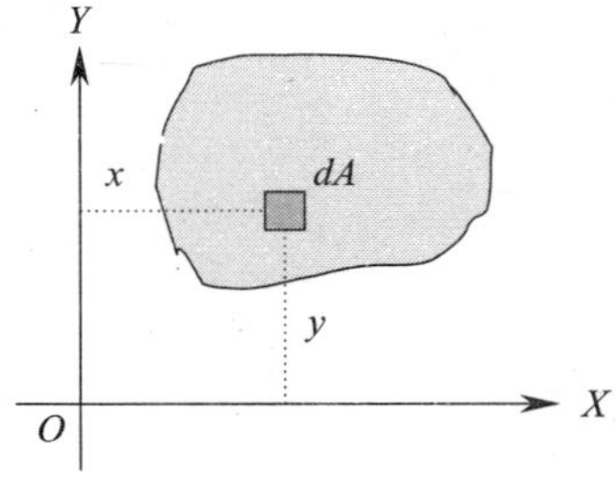

Fig. 9.2 Second moment of an area

The second moment of an area is generally termed as **moment of inertia** and it is denoted by the letter I with a subscript indicating the axes about which the second moment is taken. It finds application in the design of structural members as it gives a measure of resistance to bending in the case of sections or plane areas. Depending upon the distribution of this area [see Fig. 9.3], its resistance to bending varies.

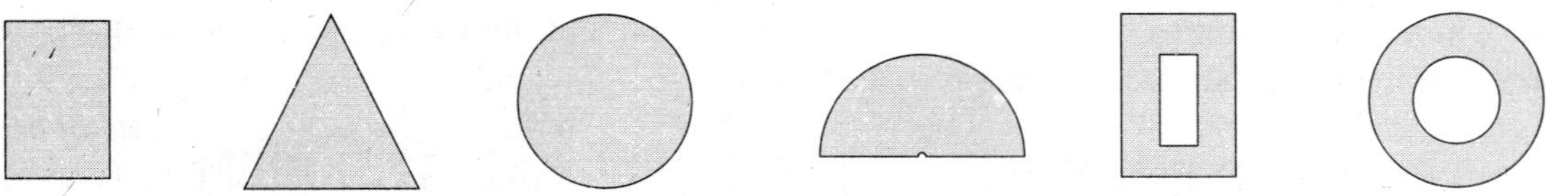

Fig. 9.3 Various cross sections of beams

From Eqs 9.7 and 9.8, we see that moment of inertia is obtained by multiplying the cross-sectional area with square of the distance from the axis. Hence, its dimension and unit are respectively $[\mathbf{L}]^4$ and $\mathbf{m}^4$. It should be noted that the first moment of an area about an axis could be **positive** or **negative** depending upon the sign of the perpendicular distance; and **zero** in the case of areas having an axis of symmetry. Whereas the second moments of area, I_{xx} and I_{yy}, are always **positive** as they involve the **square** of the perpendicular distance.

9.3 RADIUS OF GYRATION

If we can concentrate the entire area A of the lamina into a thin strip parallel to the X-axis [as shown in Fig. 9.4(a)] at a perpendicular distance k_x from the X-axis such that I_{xx} is same for the area, i.e.,

$$I_{xx} = A\,k_x^2 \tag{9.9}$$

then the perpendicular distance k_x is termed as **radius of gyration** of the area with respect to the X-axis. Its value is obtained from the above expression as

$$k_x = \sqrt{\frac{I_{xx}}{A}} \tag{9.10}$$

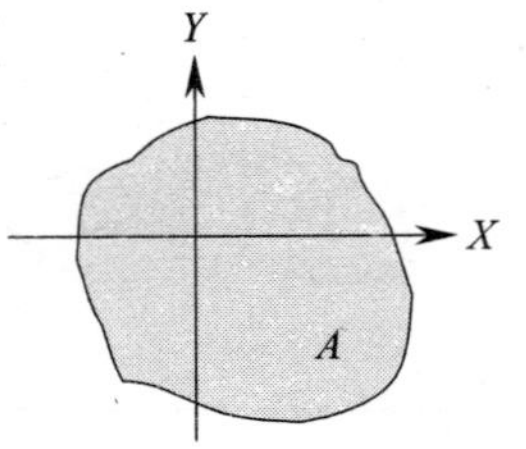

Fig. 9.4.

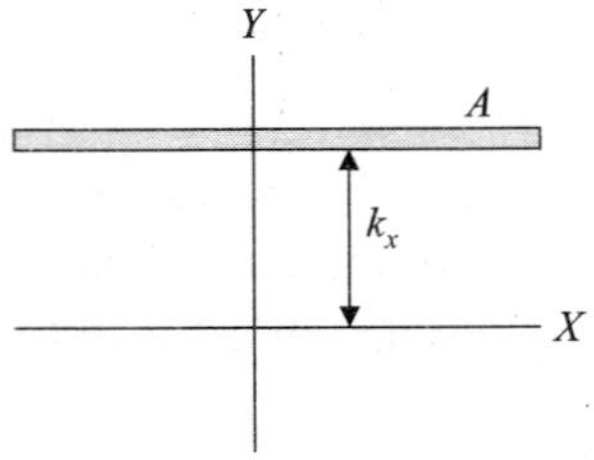

Fig. 9.4(a) Radius of gyration, k_x

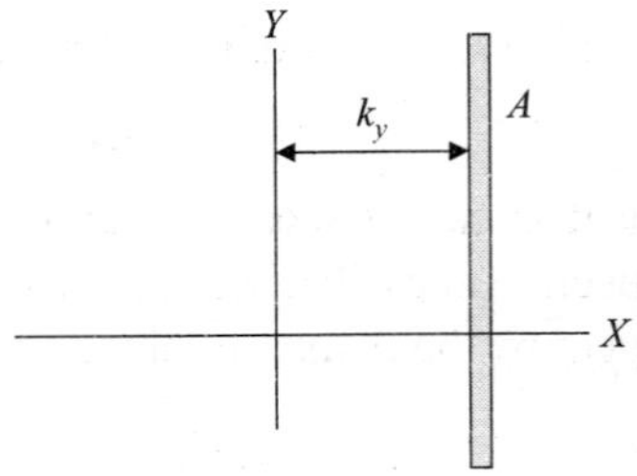

Fig. 9.4(b) Radius of gyration, k_y

Similarly, we can concentrate the entire area A into a thin strip parallel to the Y-axis at a perpendicular distance k_y from the Y-axis such that I_{yy} is same for the area. Then

$$I_{yy} = A\,k_y^2$$

$$\Rightarrow k_y = \sqrt{\frac{I_{yy}}{A}} \tag{9.11}$$

The distance k_y is termed as **radius of gyration** of the area with respect to the Y-axis. Radius of gyration finds application in design of column members. It is defined as the distance from the axis to a point where the entire area of the lamina could be concentrated into a thin strip and have the same moment of inertia with respect to the given axis.

9.4 TRANSFER FORMULA (OR) PARALLEL AXIS THEOREM

Frequently, we come across situations, where we have to determine the moment of inertia of a section about different axes. If we know the moment of inertia about a centroidal axis then we can determine the moment of inertia about a non-centroidal axis, which is *parallel* to the centroidal axis by using the *parallel axis theorem*, also called *transfer formula*. This theorem relates the moment of inertia of an area with respect to any axis in the plane of the area to the moment of inertia with respect to a parallel centroidal axis.

Consider a plane lamina of area A. Let us consider two sets of reference axes, one centroidal axes (X_c–Y_c axes) passing through the centroid C of the area and the other non-centroidal axes (X–Y axes), but parallel to the centroidal axes. If we take a small element of area dA then its coordinates with respect to the centroidal axes are (x, y) and with respect to the non-centroidal axes are $(x + \bar{x}, y + \bar{y})$, where $\bar{x}$ and $\bar{y}$ are coordinates of the centroid (C).

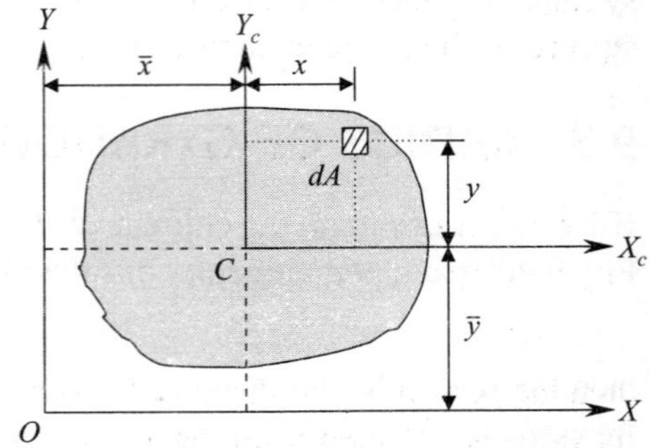

Fig. 9.5 Parallel axis theorem

Then moment of inertia of the entire area with respect to the centroidal X_c-axis is

$$\bar{I}_{xx} = \int y^2 dA$$

[Note that in order to differentiate the moments of inertia about different axes, the moment of inertia about the centroidal axis is shown with a *bar* sign]. Similarly, the moment of inertia with respect to the non-centroidal X-axis is

$$I_{xx} = \int (y + \bar{y})^2 dA$$
$$= \int y^2 dA + \int \bar{y}^2 dA + \int 2y\bar{y} dA$$
$$= \int y^2 dA + \bar{y}^2 \int dA + 2\bar{y} \int y dA$$

We can readily see that the first integral on the right-hand side is the moment of inertia, $\bar{I}_{xx}$ about the centroidal X_c-axis; the second integral is the area A multiplied by $\bar{y}^2$; and the third integral vanishes because it is the first moment of the area and it is zero for the centroidal axes. Therefore, we can write,

$$I_{xx} = \bar{I}_{xx} + A(\bar{y})^2 \tag{9.12}$$

Similarly, we can derive for the Y-axis as

$$I_{yy} = \bar{I}_{yy} + A(\bar{x})^2 \tag{9.13}$$

Hence, the parallel axis theorem can be stated as follows:

Moment of inertia of an area about an axis in the plane of the area is equal to the moment of inertia about an axis passing through the centroid and parallel to the given axis plus the product of the area and the square of the distance between the two parallel axes.

Corollary While using the above two equations, we must understand two things: Firstly, as the second term in the above Eq. 9.12 is always positive (square of $\bar{y}$), I_{xx} is always greater than $\bar{I}_{xx}$. Thus, we can conclude and say that the moment of inertia **increases** as the non-centroidal axis is moved farther from the centroid. Hence, the moment of inertia about the centroidal axis is the **least** moment of inertia of the area.

Secondly, the parallel axis theorem is applicable only if one of the two axes is a centroidal axis. For example, in Fig. 9.6, if $I_{A_1A_1}$ is known then,

$$I_{B_1B_1} \neq I_{A_1A_1} + Ah^2 \tag{9.14}$$

where h is the perpendicular distance between the two axes. In such a case, $I_{B_1B_1}$ is found out in two steps: first, $I_{A_1A_1}$ is transferred to the centroidal X_c axis. While doing so, the moment of inertia decreases as it is moved towards the centroid. In the next step, from the centroidal X_c axis, we transfer it to the axis B_1B_1. While doing so, the moment of inertia increases as it is moved away from the centroid.

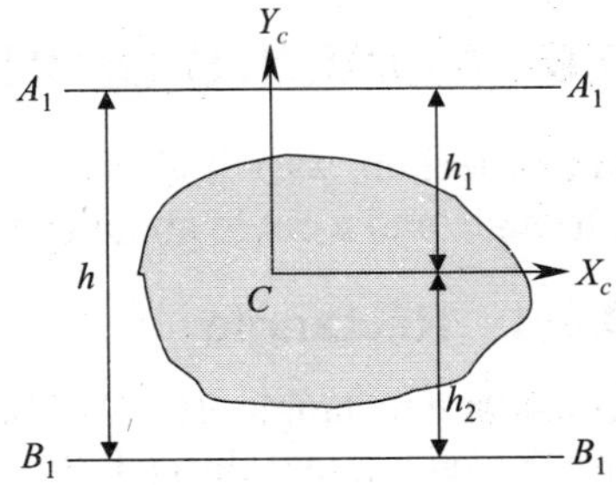

Fig. 9.6

9.5 PERPENDICULAR AXIS THEOREM (OR) POLAR MOMENT OF INERTIA

Just like moments of inertia, I_{xx} and I_{yy}, give a measure of resistance to bending about an axis, we come across another quantity, called **polar moment of inertia** in circular shafts, which gives a measure of resistance to twist about a perpendicular axis to the plane of the area.

Consider a plane lamina of area A. Let the reference axes be chosen as shown in Fig. 9.7 such that Z-axis points outward from the plane of the figure. If we consider an infinitesimally small area dA at a radial distance r from the Z-axis then its moment of inertia about the Z-axis is obtained as a product of area and square of the perpendicular distance, i.e.,

$$d\bar{I}_{zz} = r^2 dA$$

Thus, moment of inertia of the entire area about the perpendicular Z-axis is

$$\bar{I}_{zz} = \int r^2 dA$$

The moment of inertia of a plane lamina about a perpendicular axis is also called the **polar moment of inertia** as it is determined with respect to a pole axis. Since radial distance, r can also be expressed as

$$r^2 = x^2 + y^2$$

the polar moment of inertia can also be determined as follows:

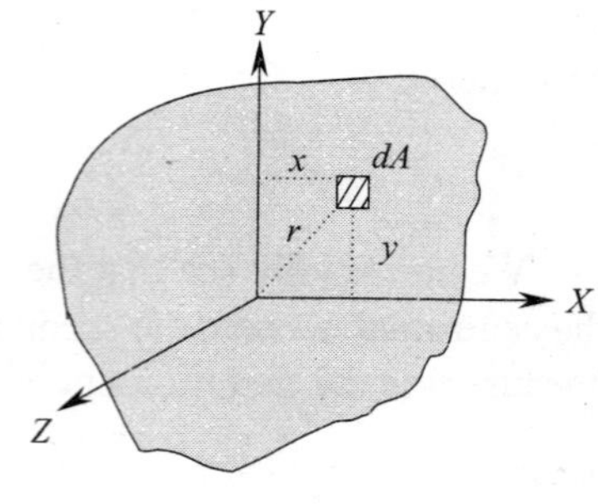

Fig. 9.7 Polar moment of inertia

$$\bar{I}_{zz} = \bar{I}_p = \int (x^2 + y^2)\, dA$$

$$= \int x^2 dA + \int y^2 dA$$

$$= \bar{I}_{xx} + \bar{I}_{yy} \qquad (9.15)$$

Thus, we see that for a plane lamina, polar moment of inertia is equal to the sum of the moments of inertia about two mutually perpendicular axes in the plane of the lamina. The polar moment of inertia is useful in calculating the stresses when a structural member is subjected to torque as in shafts. Also, $\bar{I}_{xx}$ and $\bar{I}_{yy}$ for circular and semicircular areas can be determined using their polar moments of inertia.

Corollary If the polar moment of inertia about the centroid is known then the polar moment of inertia about an axis perpendicular to the plane of the figure but not passing through its centroid can be determined by applying the transfer formula.

$$I_p = \bar{I}_p + Ad^2 \qquad (9.16)$$

where d is the perpendicular distance between the two axes.

9.6 MOMENTS OF INERTIA OF BASIC SHAPES

In this section, we will derive expressions for moments of inertia of regular shapes such as a rectangle, a triangle and a circle about centroidal axes and about non-centroidal axes.

9.6.1 Rectangle

Consider a rectangle $ABCD$ of base b and height h. The reference axes are chosen passing through the centroid of the area, i.e., X_c and Y_c. If we take a thin strip parallel to the X_c-axis at a distance y from the axis and of infinitesimally small thickness dy then its area is given as

$$dA = b\,dy$$

Hence, the moment of inertia of the strip about the X_c-axis is

$$d\bar{I}_{xx} = y^2\, dA = y^2 b\, dy$$

Therefore, moment of inertia of the entire area about the X_c-axis is obtained by integrating the above expression between limits:

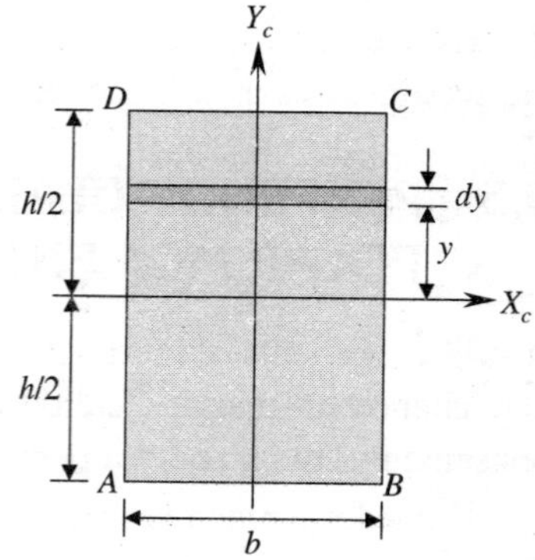

Fig. 9.8 Moment of inertia of a rectangle

$$\bar{I}_{xx} = \int_{-h/2}^{h/2} d\bar{I}_{xx} = \int_{-h/2}^{h/2} y^2\, b\, dy = b\left[\frac{y^3}{3}\right]_{-h/2}^{h/2}$$

Therefore, $$\bar{I}_{xx} = \frac{bh^3}{12} \quad (9.17)$$

In a similar manner, we can use an element of area dA in the form of a vertical strip and obtain the moment of inertia with respect to the Y_c-axis as

$$\bar{I}_{yy} = \frac{hb^3}{12} \quad (9.18)$$

Moment of inertia about non-centroidal axes Once we know the moment of inertia about the centroidal axis, using parallel axis theorem, we can determine the moment of inertia of the rectangle about a non-centroidal axis, say base AB, as follows:

$$I_{AB} = \bar{I}_{xx} + A\left(\frac{h}{2}\right)^2$$

$$= \frac{bh^3}{12} + bh\frac{h^2}{4}$$

Therefore, $$I_{\text{base}} = \frac{bh^3}{3} \quad (9.19)$$

9.6.2 Right-Angled Triangle

Consider a right-angled triangle of base b and height h. If we take a thin strip parallel to the base at a distance y from the base and of infinitesimally small thickness dy then the elemental area is given as

$$dA = b' dy$$

where b' is the width of the strip. From similar triangles, we know that,

$$b'/(h - y) = b/h$$

Therefore,

$$dA = \frac{b}{h}(h - y)dy$$

Here it is easier to find moment of inertia about the base rather than about the centroidal axis. Then by applying transfer formula, we can determine moment of inertia about the centroidal axis. The moment of inertia of the strip about the X-axis is given as

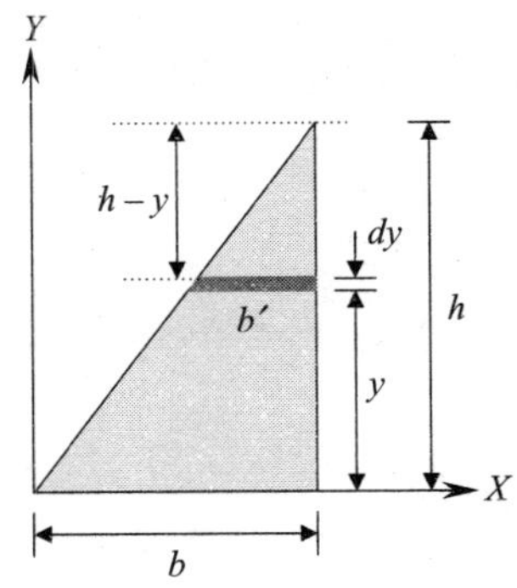

Fig. 9.9 Moment of inertia of a right-angled triangle

$$dI_{xx} = y^2 dA = y^2\,\frac{b}{h}(h - y)dy$$

Therefore, moment of inertia of the entire area about the X-axis is obtained by integrating the above expression between limits:

$$I_{xx} = \int_0^h y^2 dA$$

$$= \int_0^h \frac{b}{h}(h - y)\, y^2 dy$$

$$= \frac{b}{h}\left[\frac{hy^3}{3} - \frac{y^4}{4}\right]_0^h = \frac{bh^3}{12} \tag{9.20}$$

Moment of inertia about centroidal axes By applying the parallel axis theorem, the moment of inertia of the triangle about the centroidal horizontal axis is obtained as

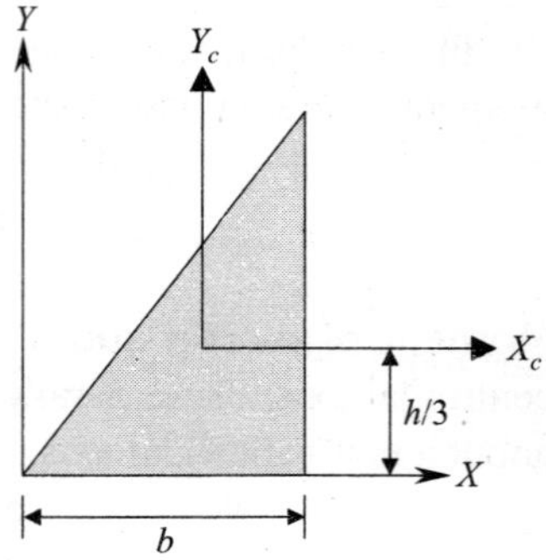

Fig. 9.10

$$\bar{I}_{xx} = I_{xx} - A\left(\frac{h}{3}\right)^2$$

$$= \frac{bh^3}{12} - \frac{1}{2}bh\frac{h^2}{9}$$

$$\therefore \quad \bar{I}_{xx} = \frac{bh^3}{36} \tag{9.21}$$

In a similar manner, we can use an element of area dA in the form of a vertical strip and obtain the moment of inertia with respect to the vertical centroidal axis as

$$\bar{I}_{yy} = \frac{hb^3}{36} \tag{9.22}$$

9.6.3 Circle

Consider a circular area of radius R. If we take a circular ring (darker area) of radius r and thickness dr then its area is: $dA = (2\pi r)(dr)$. Hence, polar moment of inertia of the circular ring, i.e., inertia about perpendicular Z_c-axis is

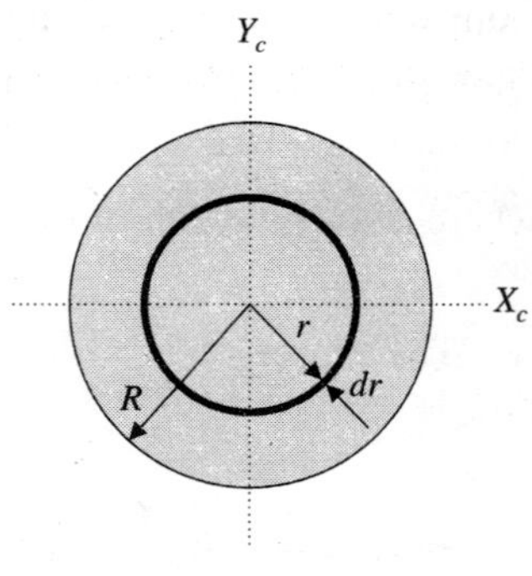

Fig. 9.11

$$d\bar{I}_{zz} = r^2 dA$$

$$= r^2 2\pi r\, dr$$

Therefore, the polar moment of inertia of the entire circular area is

$$\bar{I}_{zz} = \bar{I}_p = \int_0^R r^2 2\pi r\, dr$$

$$= \frac{\pi R^4}{2}$$

Due to symmetry, we can know that for a circular area, $\bar{I}_{xx} = \bar{I}_{yy}$. Hence,

$$\bar{I}_{xx} = \bar{I}_{yy} = \frac{\bar{I}_p}{2} = \frac{\pi R^4}{4} \tag{9.23}$$

Example 9.1 Determine the moments of inertia of a semicircular area of radius R (i) about the diametric axis, and (ii) about its centroidal axes.

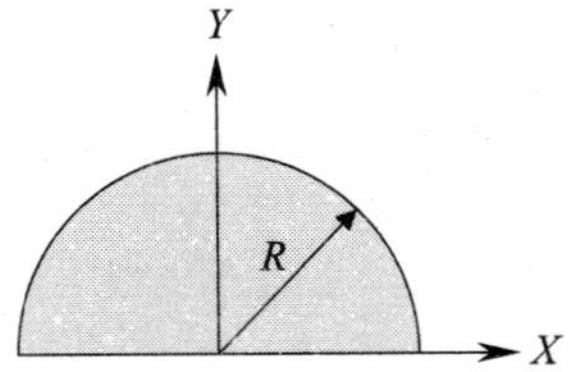

Fig. 9.12

Solution We can consider the semicircular area to be part of the circular area as shown in Fig. 9.12(a). We know that the moment of inertia of a circular area about diameter AB is

$$\pi R^4/4$$

Therefore, the moment of inertia of the *semicircular area* about AB is *half* the moment of inertia of the circular area, i.e.,

$$I_{AB} = \pi R^4/8$$

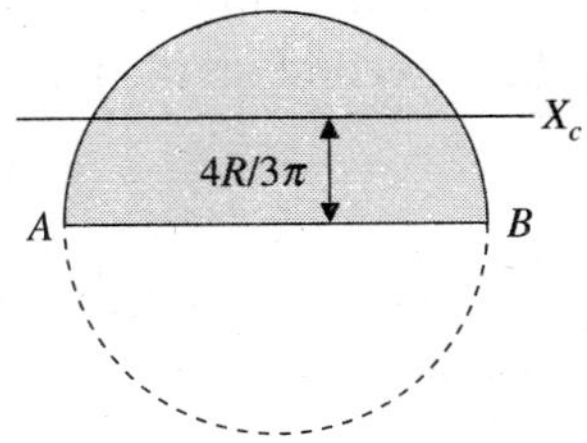

Fig. 9.12(a)

We know that the centroidal axis of the semicircular area is at a perpendicular distance of $4R/3\pi$ from axis AB. Hence, by using parallel axis theorem, we can determine the moment of inertia of the area about the centroidal axis X_c as

$$\bar{I}_{xx} = I_{AB} - A(d)^2 \text{ [where } d \text{ is the perpendicular distance]}$$

$$= \frac{\pi R^4}{8} - \frac{\pi R^2}{2}\left(\frac{4R}{3\pi}\right)^2 = 0.11R^4$$

Example 9.2 Determine the moments of inertia of a quarter-circular area of radius R (i) about the diametric axis, and (ii) about its centroidal axes.

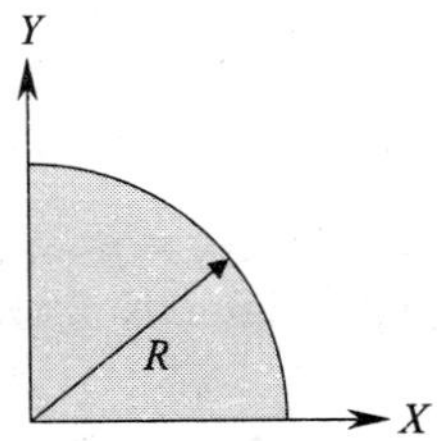

Fig. 9.13

Solution We can consider the quarter-circular area to be part of the circular area as shown in Fig. 9.13(a). We know that the moment of inertia of a circular area about diameter AB is

$$\pi R^4/4$$

Therefore, the moment of inertia of the *quarter-circular area* about AB is *one-fourth* the moment of inertia of the circular area, i.e.,

$$I_{AB} = \pi R^4/16$$

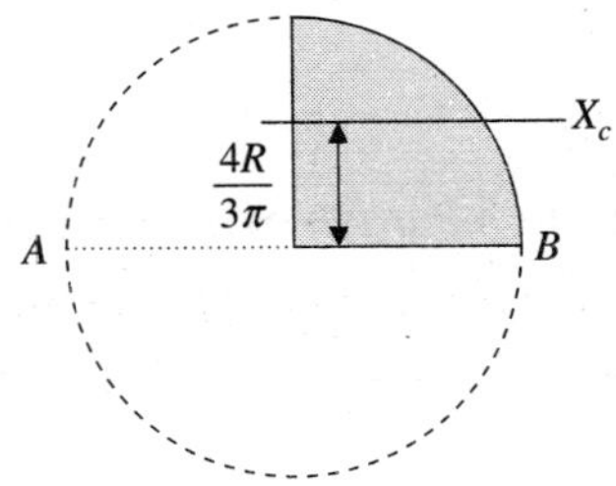

Fig. 9.13(a)

The centroidal axis of the quarter-circular area is at a perpendicular distance of $4R/3\pi$ from the axis AB. Hence, by using parallel axis theorem, we can determine the moment of inertia of the area about the centroidal axis X_c as

$$\bar{I}_{xx} = I_{AB} - A(d)^2$$

$$= \frac{\pi R^4}{16} - \frac{\pi R^2}{4}\left(\frac{4R}{3\pi}\right)^2 = 0.055R^4$$

Example 9.3 Determine the moments of inertia of an isosceles triangle (i) about its base, and (ii) about its centroidal axes.

Solution Consider an isosceles triangle of base b and height h. If we take a thin strip parallel to the X-axis at a distance y from the base and of infinitesimal thickness dy then the area of the strip is given as

$$dA = \overline{DE}\,dy = 2xdy$$

From similar triangles ACB and DCE, we get

$$\frac{2x}{h-y} = \frac{b}{h}$$

Therefore, $$dA = 2x\,dy = \frac{b}{h}(h-y)dy$$

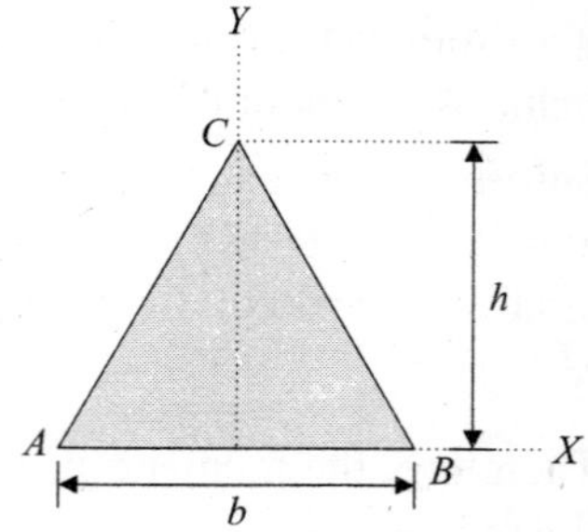

Fig. 9.14

Determination of moment of inertia about the X-axis

Taking second moment of area of the strip about the X-axis

$$dI_{xx} = y^2 dA = \frac{b}{h} y^2 (h-y)dy$$

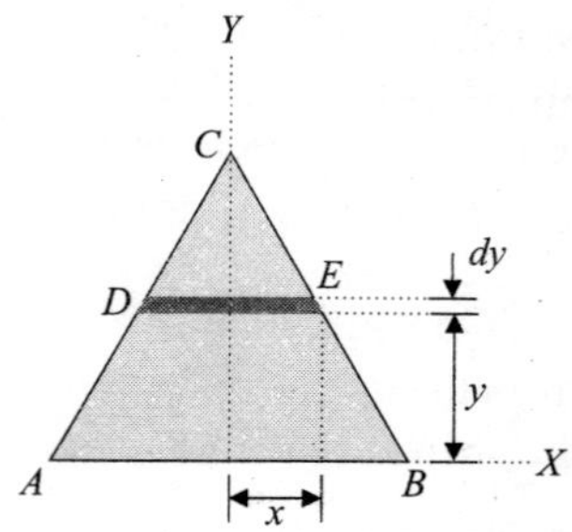

Fig. 9.14(a)

Therefore, moment of inertia of the isosceles triangle about the X-axis is obtained as

$$I_{xx} = \int_0^h dI_{xx} = \frac{b}{h}\int_0^h y^2(h-y)dy$$

$$= \frac{b}{h}\left[h\frac{y^3}{3} - \frac{y^4}{4}\right]_0^h = \frac{b}{h}\left[h\frac{h^3}{3} - \frac{h^4}{4}\right] = \frac{bh^3}{12}$$

Hence, moment of inertia of the isosceles triangle about its horizontal centroidal axis is obtained as

$$\bar{I}_{xx} = I_{xx} - A(h/3)^2$$

$$= \frac{bh^3}{12} - \frac{1}{2}bh\left[\frac{h}{3}\right]^2$$

$$= \frac{bh^3}{36}$$

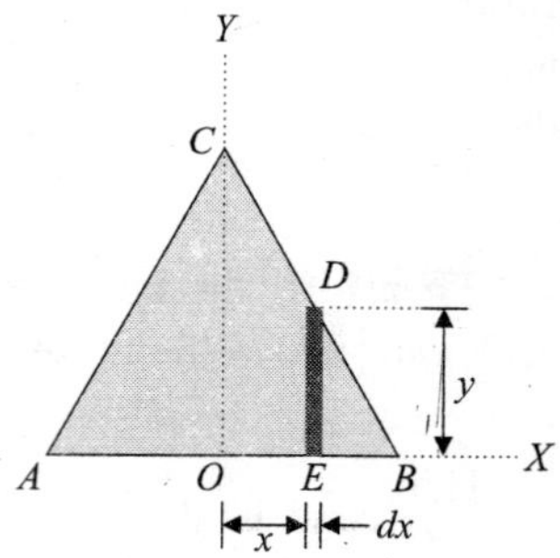

Fig. 9.14(b)

Determination of moment of inertia about the Y-axis

Consider a thin strip parallel to the Y-axis at a distance x from the origin and of infinitesimal thickness dx. Then area of the strip is given as

$$dA = ydx$$

From the similar triangles BED and BOC, we get

$$\frac{b/2}{h} = \frac{b/2 - x}{y} \Rightarrow y = \frac{2h}{b}(b/2 - x)$$

Therefore,

$$dA = ydx = \frac{2h}{b}(b/2 - x)dx$$

Taking second moment of the area of the strip about the Y-axis,

$$d\bar{I}_{yy} = x^2\,dA = \frac{2h}{b}x^2(b/2 - x)\,dx$$

Note that the Y-axis passes through the centroid and hence, a bar sign is used. Therefore, the moment of inertia of the triangle about the Y-axis is given as

$$\bar{I}_{yy} = \int_{-b/2}^{b/2} d\bar{I}_{yy} = \frac{2h}{b}\int_{-b/2}^{b/2} x^2 (b/2 - x)\, dx$$

$$= \frac{4h}{b}\left[\frac{b}{2}\frac{x^3}{3} - \frac{x^4}{4}\right]_0^{b/2} = \frac{4h}{b}\left[\frac{b}{6}\frac{b^3}{8} - \frac{b^4}{64}\right] = \frac{hb^3}{48}$$

Example 9.4 Determine the moment of inertia of an elliptical area about the centroidal axes.

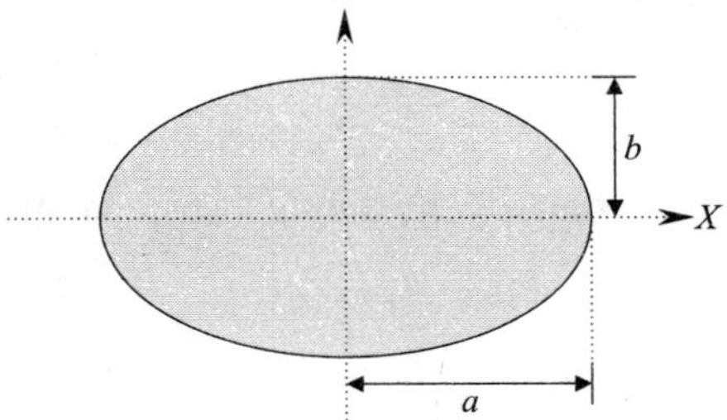

Fig. 9.15

Solution For an ellipse of semi-major axis a and semi-minor axis b, the equation is given as

$$\frac{x^2}{a^2} + \frac{y^2}{b^2} = 1$$

Consider a thin strip parallel to the X-axis at a distance y from the origin and of infinitesimal thickness dy. Then area of the strip is given as

$$dA = 2x\, dy$$

From equation of the ellipse, we get

$$x = \frac{a}{b}\sqrt{b^2 - y^2}$$

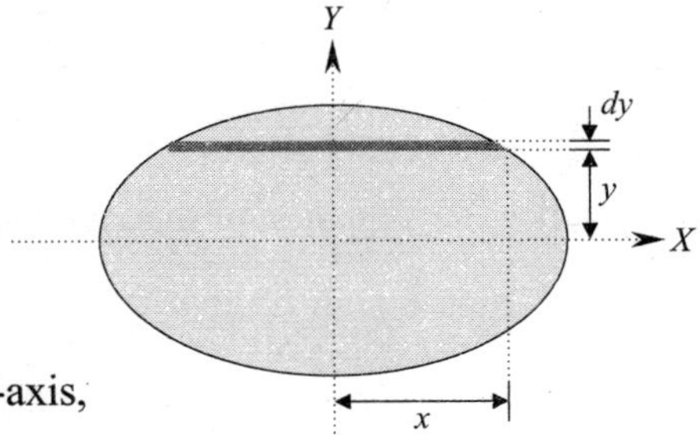

Fig. 9.15(a)

Therefore, $dA = 2\dfrac{a}{b}\sqrt{b^2 - y^2}\, dy$

Taking second moment of the area of the strip about the X-axis,

$$d\bar{I}_{xx} = y^2\, dA = 2\frac{a}{b} y^2 \sqrt{b^2 - y^2}\, dy$$

Hence, moment of inertia of the entire area about the X-axis is obtained as

$$\bar{I}_{xx} = \int_{-b}^{b} d\bar{I}_{xx} = 2\frac{a}{b}\int_{-b}^{b} y^2 \sqrt{b^2 - y^2}\, dy$$

The above integral can be evaluated by a change of variables, explained as follows:

Taking $y = b \sin\theta$, we have $dy = b \cos\theta d\theta$ and the corresponding limits vary from $-\dfrac{\pi}{2}$ to $\dfrac{\pi}{2}$.

Therefore, the integral can be written as

$$\bar{I}_{xx} = 2\frac{a}{b}\int_{-\pi/2}^{\pi/2} b^2 \sin^2\theta \sqrt{b^2 - b^2 \sin^2\theta}\, b \cos\theta d\theta$$

$$= 2ab^3 \int_{-\pi/2}^{\pi/2} \sin^2\theta \cos^2\theta d\theta$$

$$= 2ab^3 \int_{-\pi/2}^{\pi/2} \left[\frac{2\sin\theta\cos\theta}{2}\right]^2 d\theta$$

$$= \frac{1}{2}ab^3 \int_{-\pi/2}^{\pi/2} \sin^2 2\theta\, d\theta \quad [\text{since } 2\sin\theta\cos\theta = \sin 2\theta]$$

$$= \frac{1}{2}ab^3 \int_{-\pi/2}^{\pi/2} \left[\frac{1-\cos 4\theta}{2}\right] d\theta$$

$$= \frac{1}{4}ab^3\left[\theta - \frac{\sin 4\theta}{4}\right]_{-\pi/2}^{\pi/2} = \frac{\pi}{4}ab^3$$

Similarly, by taking a strip parallel to the *Y*-axis, it can be shown that the moment of inertia about the *Y*-axis is $\bar{I}_{yy} = \frac{\pi}{4}ba^3$

Example 9.5 Determine the moment of inertia with respect to the *X* and *Y* axes of the shaded parabolic area by integration method.

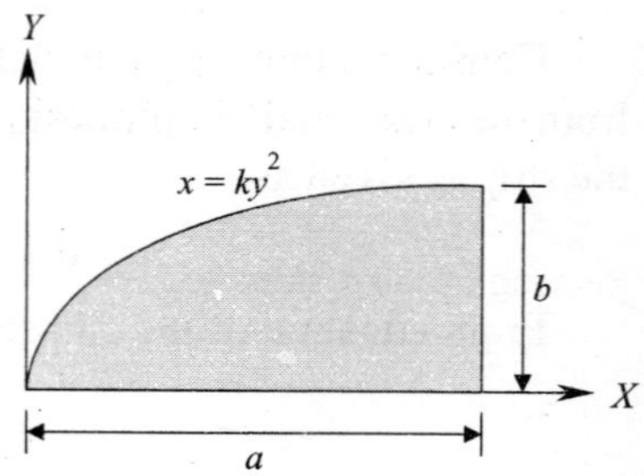

Fig. 9.16

Solution From the Fig. 9.16, we see that at $x = 0$, $y = 0$ and at $x = a$, $y = b$. Therefore,

$$k = \frac{a}{b^2}$$

Hence, the equation of the curve can be written as

$$x = \frac{a}{b^2}y^2 \quad \text{(or)} \quad y = \frac{b}{a^{1/2}}x^{1/2}$$

Consider a vertical strip of infinitesimally small thickness dx at a distance x from the origin. Then the area of the strip is given as $dA = y dx$. Hence, the moment of inertia of the strip about the *Y*-axis is

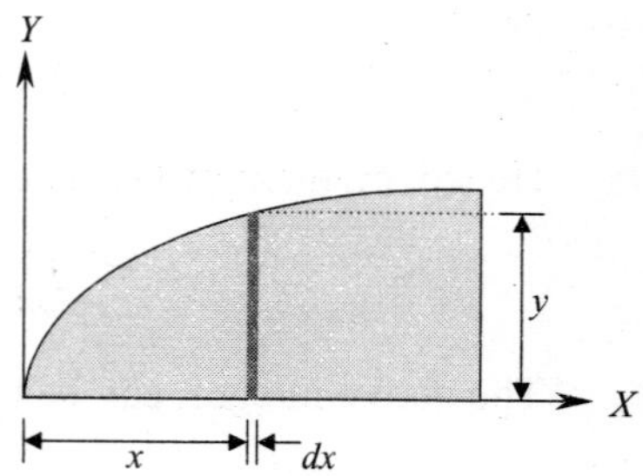

Fig. 9.16(a)

$$dI_{yy} = x^2\, dA$$

Therefore, the moment of inertia of the entire area about the *Y*-axis is obtained by integrating the above expression between limits,

$$I_{yy} = \int_0^a x^2\, dA$$

$$= \int_0^a x^2\, y\, dx$$

$$= \int_0^a x^2 \frac{b}{a^{1/2}} x^{1/2}\, dx = \frac{b}{a^{1/2}} \int_0^a x^{5/2}\, dx$$

$$= \frac{b}{a^{1/2}}\left[\frac{x^{7/2}}{7/2}\right]_0^a = \frac{2}{7}ba^3$$

Similarly, considering a strip parallel to the X-axis, we can determine the moment of inertia of the area about the X-axis as $\quad I_{xx} = \dfrac{2}{15} ab^3$

The following table summarizes the moments of inertia of various regular shaped areas:

Table 9.1 Moments of inertia of regular shapes

Area	*Shape*	$\bar{I}_{xx}$	$\bar{I}_{yy}$	$\bar{I}_{xy}$
Rectangle (about centroidal axes)	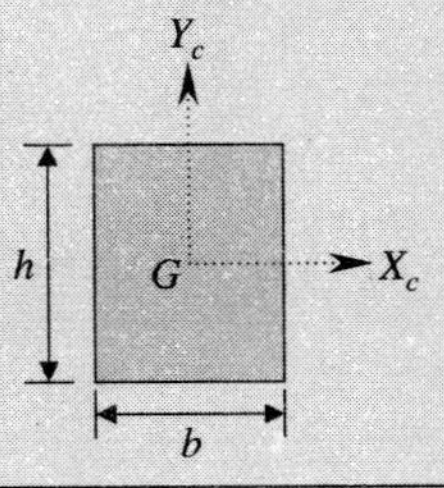	$\dfrac{bh^3}{12}$	$\dfrac{hb^3}{12}$	0
Rectangle (about axes along the sides)	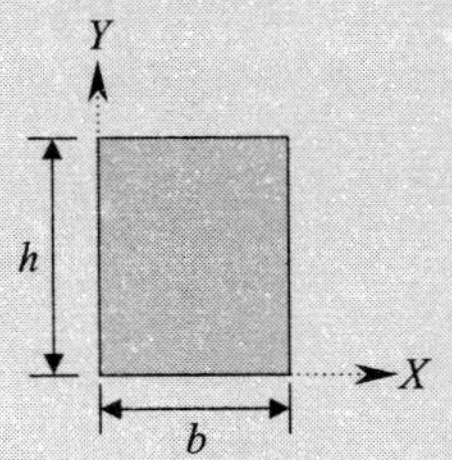	$\dfrac{bh^3}{3}$	$\dfrac{hb^3}{3}$	$\dfrac{b^2h^2}{4}$
Right triangle (about centroidal axes)	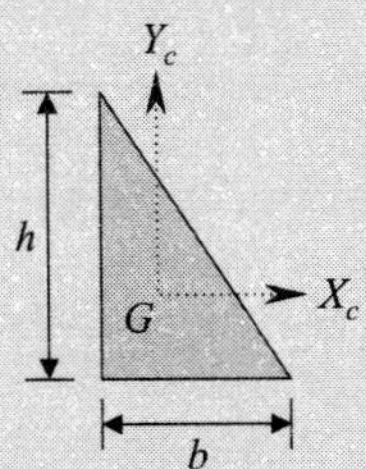	$\dfrac{bh^3}{36}$	$\dfrac{hb^3}{36}$	$-\dfrac{b^2h^2}{72}$
Right triangle (about X–Y axes)	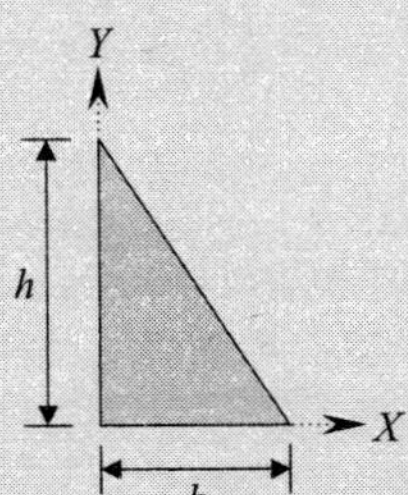	$\dfrac{bh^3}{12}$	$\dfrac{hb^3}{12}$	$\dfrac{b^2h^2}{24}$

Contd.

Isosceles triangle (about *X*–*Y* axes)	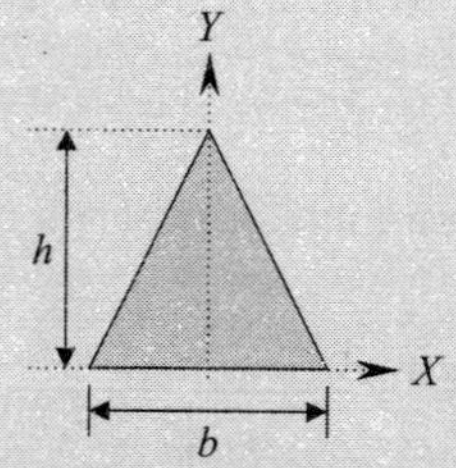	$\frac{bh^3}{12}$	$\frac{hb^3}{48}$	0
Isosceles triangle (about centroidal axes)	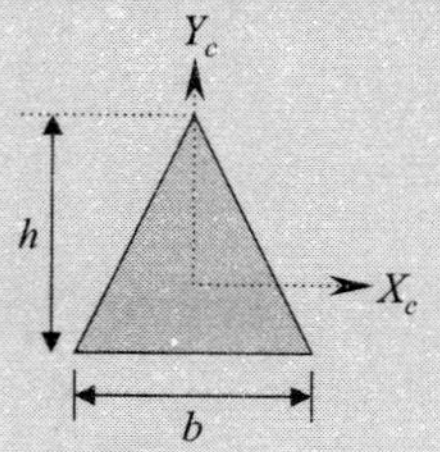	$\frac{bh^3}{36}$	$\frac{hb^3}{48}$	0
Circle (about centroidal axes)	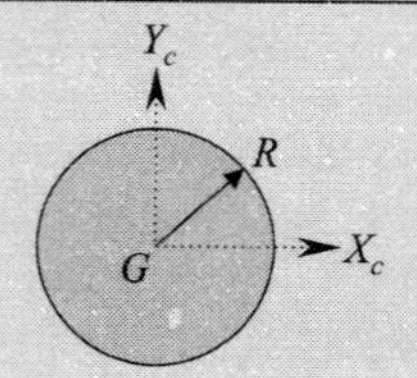	$\frac{\pi R^4}{4}$	$\frac{\pi R^4}{4}$	0
Semicircle (about centroidal axes)	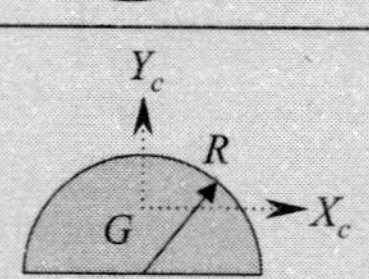	$0.11R^4$	$\frac{\pi R^4}{8}$	0
Semicircle (about diametric axes)	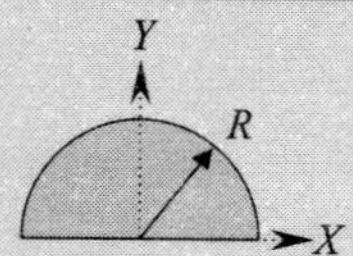	$\frac{\pi R^4}{8}$	$\frac{\pi R^4}{8}$	0
Quarter-circle (about centroidal axes)	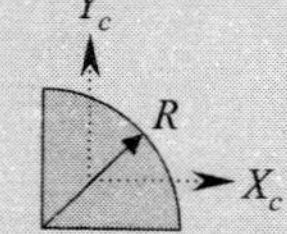	$0.055R^4$	$0.055R^4$	$-0.016R^4$
Quarter-circle (about *X*–*Y* axes)	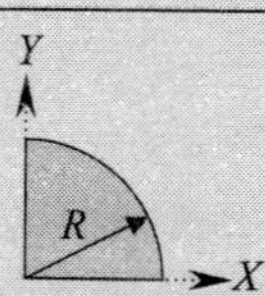	$\frac{\pi R^4}{16}$	$\frac{\pi R^4}{16}$	$\frac{R^4}{8}$
Ellipse (about centroidal axes)	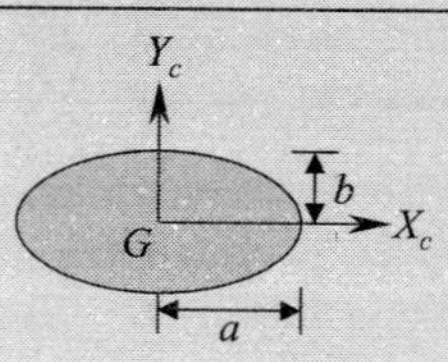	$\frac{\pi ab^3}{4}$	$\frac{\pi ba^3}{4}$	0

9.6.4 Composite Section

In engineering work, we frequently need to find the moment of inertia of an area composed of several parts, each having a familiar geometric shape such as a rectangle, a triangle, etc. In such cases, the following steps can be followed:

Step 1 Divide the given area into regular elements for which moments of inertia are known.

Step 2 Find the coordinates ($\bar{x}$, $\bar{y}$) of the centroid of the composite section.

Step 3 Calculate the moments of inertia of the individual elements about their respective centroidal axes.

Step 4 Apply parallel axis theorem to determine the moments of inertia of the individual elements about the centroidal axes of the composite section as follows

$$(I_{xx})_i = (\bar{I}_{xx})_i + A_i(\bar{y}_i - \bar{y})^2 \tag{9.24}$$

$$(I_{yy})_i = (\bar{I}_{yy})_i + A_i(\bar{x}_i - \bar{x})^2 \tag{9.25}$$

Step 5 The moment of inertia of the composite section is obtained by the algebraic summation of the moments of inertia of individual elements calculated in *Step 4*. When an area is added on, we use a positive sign and if it is removed, we use a negative sign.

$$\bar{I}_{xx} = \Sigma(I_{xx})_i = \Sigma(\bar{I}_{xx})_i + \Sigma A_i\,(\bar{y}_i - \bar{y})^2 \tag{9.26}$$

$$\bar{I}_{yy} = \Sigma(I_{yy})_i = \Sigma(\bar{I}_{yy})_i + \Sigma A_i(\bar{x}_i - \bar{x})^2 \tag{9.27}$$

Step 6 If the moments of inertia of the composite section about non-centroidal reference axes are required, they can be obtained by applying the transfer formula to the above result or else by using the following expressions:

$$I_{xx} = \Sigma(\bar{I}_{xx})_i + \Sigma A_i\,(\bar{y}_i)^2 \tag{9.28}$$

$$I_{yy} = \Sigma(\bar{I}_{yy})_i + \Sigma A_i\,(\bar{x}_i)^2 \tag{9.29}$$

[Note that if formulas 9.28 and 9.29 are adopted then location of centroid of the composite area can be avoided, as it does not come in the calculation].

Example 9.6 In Fig. 9.17, if the moment of inertia of the shaded area (25 cm^2) about 1–1 axis is 7400 cm^4 then determine the moment of inertia of the area about 2–2 axis.

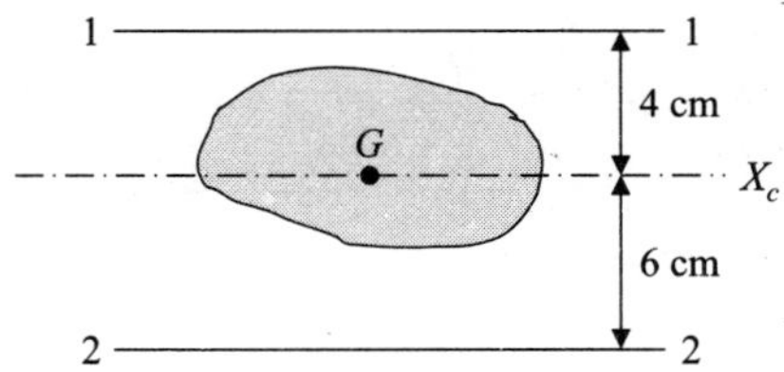

Fig. 9.17

Solution As the parallel axis theorem is applicable only when one of the two axes is the centroidal axis, we cannot directly determine moment of inertia about 2–2 axis from 1–1 axis. Instead, we proceed in two steps as follows:

$$A = 25 \text{ cm}^2$$

$$I_{11} = 7400 \text{ cm}^4$$

First, we determine moment of inertia about the centroidal axis. While doing so, the moment of inertia decreases as we move *towards* the centroidal axis. Hence,

$$\bar{I}_{x_c} = I_{11} - A(d_1)^2$$
$$= 7400 - 25(4)^2 = 7000 \text{ cm}^4$$

Next, we determine moment of inertia about the 2–2 axis. As we do it, the moment of inertia increases as we move *away* from the centroidal axis. Hence,

$$I_{22} = \bar{I}_{x_c} + A(d_2)^2$$
$$= 7000 + 25(6)^2 = 7900 \text{ cm}^4$$

Example 9.7 Find the moments of inertia of the angle section shown about centroidal axes. Also, find the radii of gyration about the same axes.

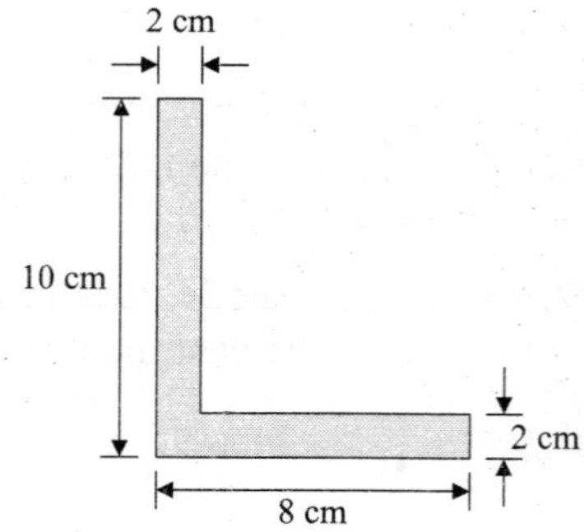

Fig. 9.18

Solution

Determination of centroid

The given angle section can be considered to be made up of two rectangles, both of dimensions 8 cm × 2 cm. Its centroid is determined as follows:

S.No	Element	A_i (cm²)	$\bar{x}_i$ (cm)	$\bar{y}_i$ (cm)	$A_i\bar{x}_i$ (cm³)	$A_i\bar{y}_i$ (cm³)
1.	Rectangle-(1)	8 × 2 = 16	8/2 = 4	2/2 = 1	64	16
2.	Rectangle- (2)	2 × 8 = 16	2/2 = 1	2 + (8/2) = 6	16	96
	Σ =	32			80	112

$\therefore$ $$\bar{x} = \frac{\sum A_i\bar{x}_i}{\sum A_i} = 2.5 \text{ cm} \qquad \bar{y} = \frac{\sum A_i\bar{y}_i}{\sum A_i} = 3.5 \text{ cm}$$

Moments of inertia calculations

S.No	$(\bar{I}_{xx})_i$ cm⁴	$(\bar{I}_{yy})_i$ cm⁴	$A_i(\bar{y}_i - \bar{y})^2$ cm⁴	$A_i(\bar{x}_i - \bar{x})^2$ cm⁴
1.	$(1/12) \times 8 \times 2^3 = 5.33$	$(1/12) \times 2 \times 8^3 = 85.33$	$16(1 - 3.5)^2 = 100$	$16(4 - 2.5)^2 = 36$
2.	$(1/12) \times 2 \times 8^3 = 85.33$	$(1/12) \times 8 \times 2^3 = 5.33$	$16(6 - 3.5)^2 = 100$	$16(1 - 2.5)^2 = 36$
Σ =	90.66	90.66	200	72

Therefore, moments of inertia of the composite section are:

$$\bar{I}_{xx} = \sum(\bar{I}_{xx})_i + \sum A_i(\bar{y}_i - \bar{y})^2$$
$$= 90.66 + 200 = 290.66 \text{ cm}^4$$

$$\bar{I}_{yy} = \sum(\bar{I}_{yy})_i + \sum A_i(\bar{x}_i - \bar{x})^2$$
$$= 90.66 + 72 = 162.66 \text{ cm}^4$$

Radii of gyration

The radii of gyration about the *X* and *Y* axes are determined as follows:

$$k_x = \sqrt{\frac{\bar{I}_{xx}}{A}} \qquad k_y = \sqrt{\frac{\bar{I}_{yy}}{A}}$$

$$= \sqrt{\frac{290.66}{32}} = 3.01 \text{ cm} \qquad = \sqrt{\frac{162.66}{32}} = 2.25 \text{ cm}$$

Example 9.8 In the above problem, determine the moment of inertia of the angle section about the base.

Solution As we know the moment of inertia of the composite section about its centroid, using parallel axis theorem, we can determine the moment of inertia of the composite section about the base as follows:

$$I_{\text{base}} = \bar{I}_{xx} + A(d)^2$$

$$= 290.66 + (32)(3.5)^2 \qquad [\text{Note that } d = \bar{y} = 3.5 \text{ cm}]$$

$$= 682.66 \text{ cm}^4$$

However, if we do not know the moment of inertia of the composite section about its centroidal axis, we can determine the moment of inertia about the base without finding the centroid of the composite section using the Eq 9.28 as follows:

S.No	$(\bar{I}_{xx})_i$ *cm*^{4}	$A_i(\bar{y}_i)^2$ *cm*^{4}
1.	$(1/12) \times 8 \times 2^3 = 5.33$	$16(1)^2 = 16$
2.	$(1/12) \times 2 \times 8^3 = 85.33$	$16(6)^2 = 576$
$\Sigma =$	90.66	592

Therefore,

$$I_{\text{base}} = \sum(\bar{I}_{xx})_i + \sum A_i(\bar{y}_i)^2$$

$$= 90.66 + 592 = 682.66 \text{ cm}^4$$

which is same as the result obtained above.

Example 9.9 Find the moments of inertia of the *I*-section shown about the centroidal axes. Also, find the radii of gyration about the same axes.

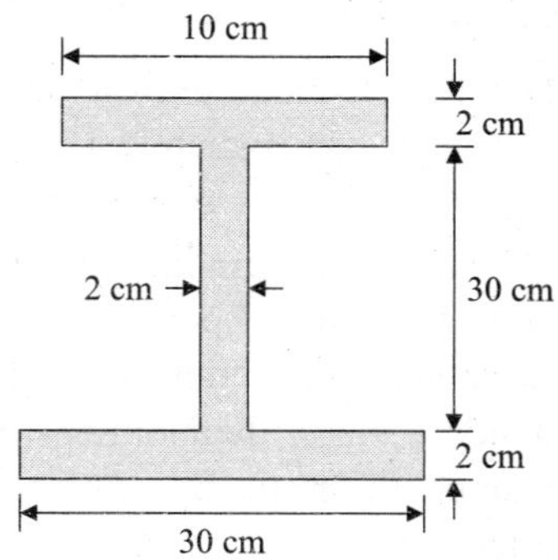

Fig. 9.19

Solution

Determination of centroid

The given *I*-section can be considered to be made up of three rectangles. Due to symmetry, we know that $\bar{x} = 15$ cm from the lower left corner. The *y*-coordinate of the centroid is determined as follows:

S.No	Element	A_i (cm^2)	$\bar{y}_i$ (cm)	$A_i\bar{y}_i$ (cm^3)
1.	Rectangle (1)	$30 \times 2 = 60$	1	60
2.	Rectangle (2)	$30 \times 2 = 60$	17	1020
3.	Rectangle (3)	$10 \times 2 = 20$	33	660
	$\Sigma =$	140		1740

$$\therefore \qquad \bar{y} = \frac{\sum A_i \bar{y}_i}{\sum A_i} = 12.43 \text{ cm}$$

Moments of inertia calculations

S.No	$(\bar{I}_{xx})_i$ cm^4	$(\bar{I}_{yy})_i$ cm^4	$A_i(\bar{y}_i - \bar{y})^2$ cm^4	$A_i(\bar{x}_i - \bar{x})^2$ cm^4
1.	$(1/12) \times 30 \times 2^3 = 20$	$(1/12) \times 2 \times 30^3 = 4500$	$60(1 - 12.43)^2 = 7838.69$	0
2.	$(1/12) \times 2 \times 30^3 = 4500$	$(1/12) \times 30 \times 2^3 = 20$	$60(17 - 12.43)^2 = 1253.09$	0
3.	$(1/12) \times 10 \times 2^3 = 6.67$	$(1/12) \times 2 \times 10^3 = 166.67$	$20(33 - 12.43)^2 = 8462.5$	0
$\Sigma =$	4526.67	4686.67	17 554.28	0

Therefore, moments of inertia of the composite section are

$$\bar{I}_{xx} = \Sigma(\bar{I}_{xx})_i + \sum A_i(\bar{y}_i - \bar{y})^2$$

$$= 4526.67 + 17\,554.28 = 22\,080.95 \text{ cm}^4$$

$$\bar{I}_{yy} = \Sigma(\bar{I}_{yy})_i + \sum A_i\,(\bar{x}_i - \bar{x})^2 = 4686.67 \text{ cm}^4$$

Radii of gyration

The radii of gyration about the *X* and *Y* axes are determined as follows:

$$k_x = \sqrt{\frac{\bar{I}_{xx}}{A}} \qquad\qquad k_y = \sqrt{\frac{\bar{I}_{yy}}{A}}$$

$$= \sqrt{\frac{22\,080.95}{140}} = 12.56 \text{ cm} \qquad\qquad = \sqrt{\frac{4686.67}{140}} = 5.79 \text{ cm}$$

9.6.5 Structural Steel Sections

The steel sections used in structural members such as beams, columns and trusses, etc., are of cross-sectional shapes such as I, C and angle sections. Determination of centroid of such sections were discussed in the previous chapter and the determination of moment of inertia in the previous section.

While determining the centroid or moment of inertia of such sections, we have so far assumed them to be made up of perfect rectangles. However, in practice, such sections are manufactured with little curvature at the ends and at the fillets as shown in Fig. 9.20. As these sections are manufactured in standard sizes, their dimensions and properties of their areas such as cross-sectional area, centroid, moment of inertia, etc., of these sections are readily available in steel section tables and they need not be calculated.

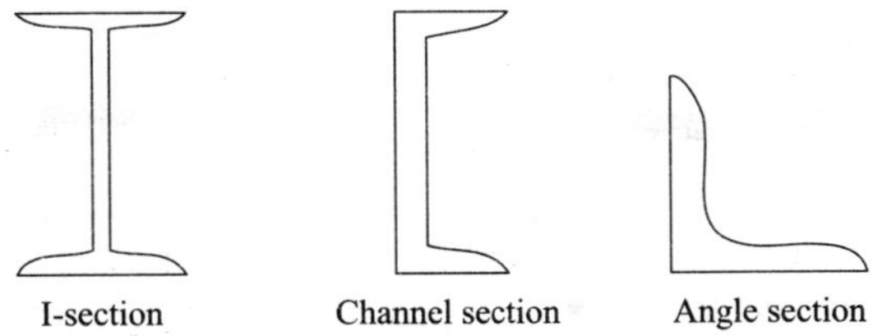

Fig. 9.20

The sections are designated using code, such as ISMB 600, which refers to an *I*-beam section with an overall depth of 600 mm. However, when the structural members have to resist greater moments, even the largest section available may not be sufficient. In such cases, the existing sections are strengthened with extra steel plates and such sections are known as *compound* sections, or plates may be built up to form stronger sections and such sections are known as *built-up* sections.. Determination of moments of inertia of compound sections is discussed below in the following two examples.

Example 9.10 Two channel sections of designation ISMC 400 are laced at a clear distance *d* apart to form a column section such that $I_{xx} = I_{yy}$. Determine the value of spacing *d*. The specifications of the channel section from steel section table are: depth = 400 mm, flange width = 100 mm, area of cross-section = 62.93 cm^2, *x*-coordinate of centroid = 2.42 cm, I_{xx} = 15 082.8 cm^4 and I_{yy} = 504.8 cm^4.

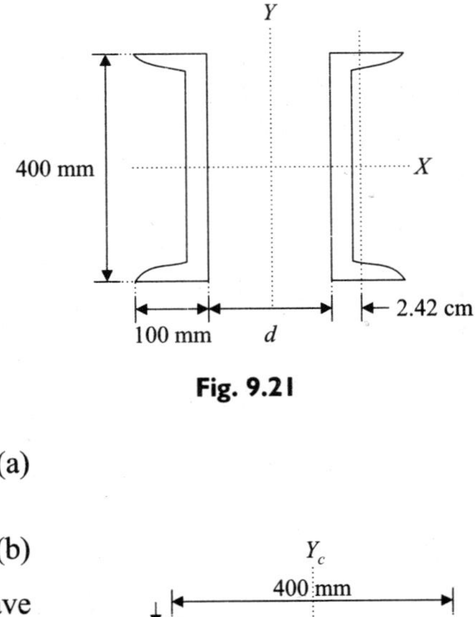

Fig. 9.21

Solution Due to symmetry, we can readily see that the centroid of the composite section lies at the centre of the section. Hence, the moments of inertia of the composite section about the centroidal axes are determined as follows:

$$\bar{I}_{xx} = [15\,082.8] \times 2 = 30165.6 \text{ cm}^4 \qquad \text{(a)}$$

$$\bar{I}_{yy} = 2 \times \{504.8 + (62.93) \times [(d/2) + 2.42]^2\}$$

$$= 1009.6 + 125.86\,[(d/2) + 2.42]^2 \qquad \text{(b)}$$

As the channel sections are spaced such that $\bar{I}_{xx} = \bar{I}_{yy}$, we have

$$30165.6 = 1009.6 + 125.86\,[(d/2) + 2.42]^2$$

$$\Rightarrow \qquad d = 25.6 \text{ cm}$$

Example 9.11 Two ISMC 400 channel sections are welded to a plate, 400 mm × 6 mm, to form a column section as shown in Fig. 9.22. Determine the moment of inertia of the section about horizontal centroidal axis. The specifications of the channel section are same as the previous example.

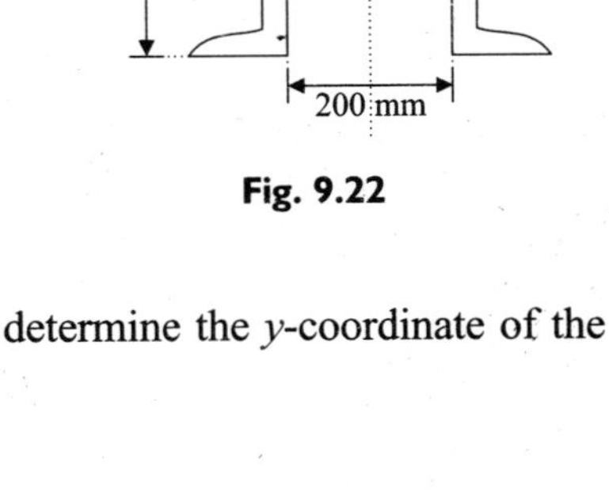

Fig. 9.22

Solution

Determination of location of the centroid of the composite section

As the section is made up of two channel sections and a plate, we determine the *y*-coordinate of the centroid as follows:

$$\bar{y} = \frac{2[62.93 \times 20] + [40 \times 0.6 \times 40.3]}{(2 \times 62.93) \times (40 \times 0.6)}$$

$$= 23.25 \text{ cm}$$

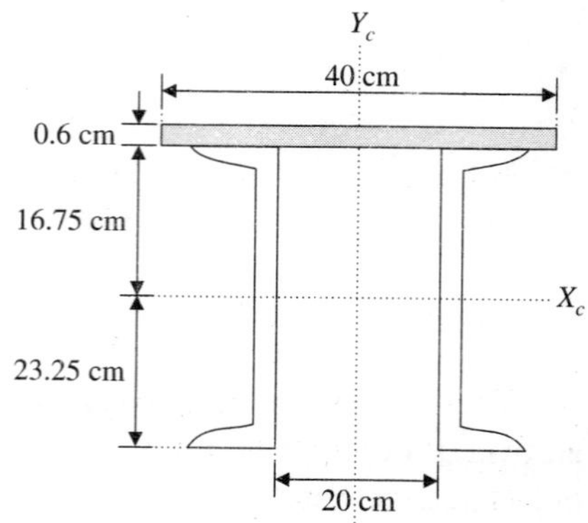

Fig. 9.22(a)

Therefore, moment of inertia of the composite section about centroidal X-axis is obtained as

$$\bar{I}_{xx} = 2[(15\,082.8) + (62.93)\,(40/2 - 23.25)^2] + [(1/12) \times 40 \times (0.6)^3] + (40 \times 0.6)\,(16.75 + 0.3)^2 = 38\,472.58 \text{ cm}^4$$

9.6.6 Moment of Inertia of Composite Sections of Any Shape

In this section, we will discuss the moments of inertia of composite sections of any shape. They can be considered to be made up of regular sections such as a rectangle, circle, triangle, etc., for which moments of inertia are readily known.

Example 9.12 Determine the moment of inertia of a trapezoidal section shown about the base.

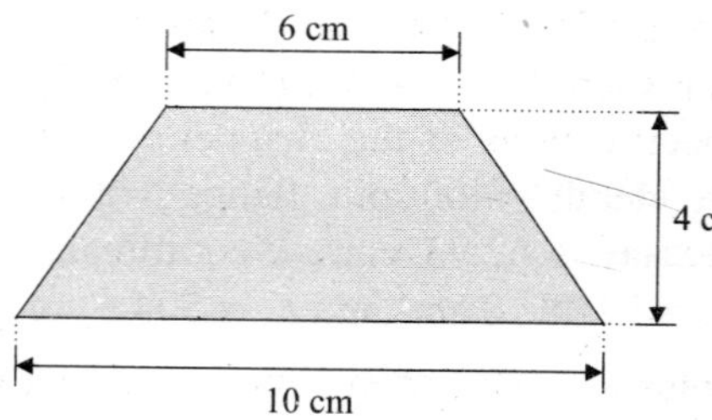

Fig. 9.23

Solution The trapezoidal section can be considered to be made up of a rectangle at the centre and two triangles at the sides as shown in Fig. 9.23(a). Since we know the moments of inertia of a rectangle and a triangle about their bases, we can directly determine the moment of inertia of the composite section about the base as follows:

S.No	*Element*	$(I_{xx})_i$ *cm*^{4}
1.	Triangle	$(1/12) \times 2 \times 4^3 = 10.67$
2.	Rectangle	$(1/3) \times 6 \times 4^3 = 128$
3.	Triangle	$(1/12) \times 2 \times 4^3 = 10.67$
$\Sigma =$		149.34

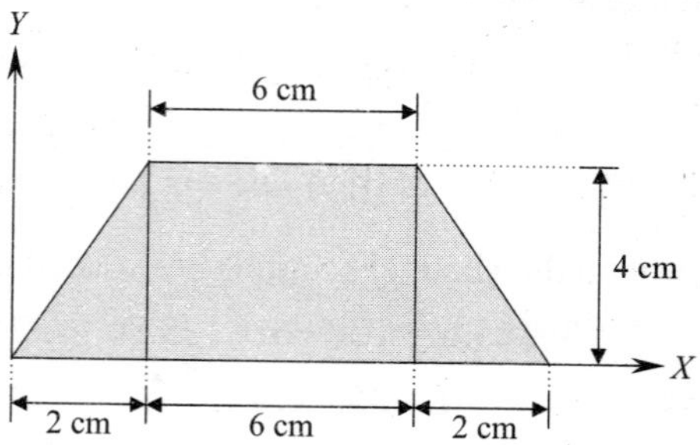

Fig. 9.23(a)

Therefore, the moment of inertia of the composite section about the base is

$$I_{xx} = 149.34 \text{ cm}^4$$

Example 9.13 In the above problem, determine the moment of inertia about the horizontal centroidal axis.

Solution As it is required to determine the moment of inertia about the horizontal centroidal axis, we need to determine the y-coordinate of the centroid of the composite section. We know from the previous chapter that the y-coordinate of the centroid of a trapezium (refer Example 8.9) is

$$\bar{y} = \frac{h}{3}\left[\frac{a+2b}{a+b}\right] = \frac{4}{3}\left[\frac{10+2(6)}{10+6}\right] = 1.83 \text{ cm}$$

and the area of the trapezium is

$$A = \frac{h}{2}[a + b] = \frac{4}{2}[10 + 6] = 32 \text{ cm}^2$$

By parallel axis theorem, we know that

$$I_{xx} = \bar{I}_{xx} + A(\bar{y})^2$$

Therefore,

$$\bar{I}_{xx} = I_{xx} - A(\bar{y})^2$$
$$= 149.34 - (32)(1.83)^2 = 42.18 \text{ cm}^4$$

Example 9.14 Determine $\bar{I}_{yy}$ of the isosceles triangle in Example 9.3 by considering the triangle to be made up of two right-angled triangles.

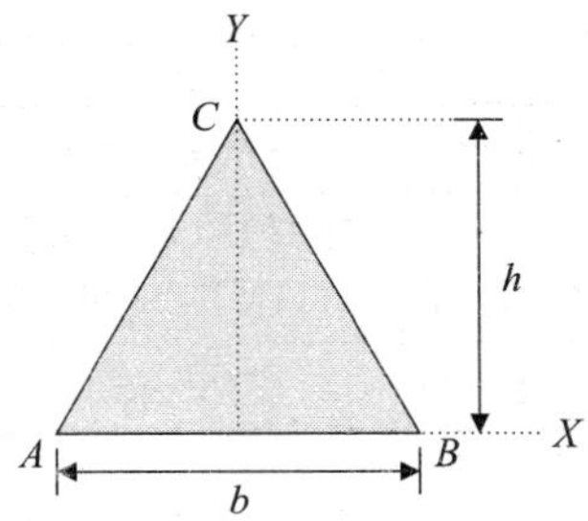

Fig. 9.24

Solution The isosceles triangle can be considered to be made up of two right-angled triangles, for which moments of inertia are readily known. We know that the moment of inertia of a right-angled triangle of base $b/2$ and height h about the Y-axis of the composite section is

$$(I_{yy})_1 = (I_{yy})_2 = \frac{1}{12}h\left[\frac{b}{2}\right]^3 = \frac{hb^3}{96}$$

Therefore, moment of inertia of the entire isosceles triangle about the Y-axis is

$$\bar{I}_{yy} = (I_{yy})_1 + (I_{yy})_2 = \frac{hb^3}{48}$$

Example 9.15 Find the moments of inertia of the shaded area shown about the centroidal axes.

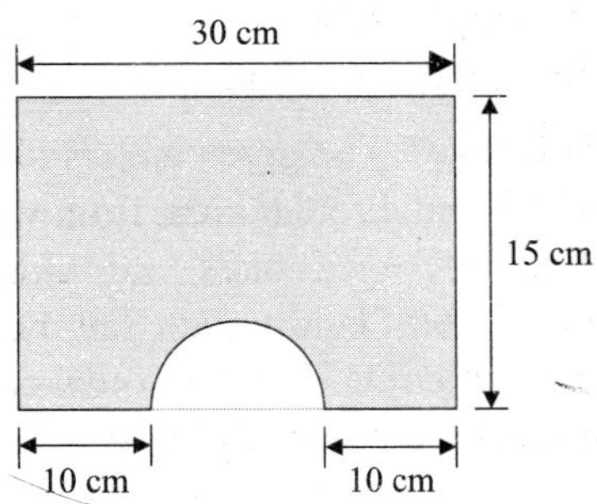

Fig. 9.25

Solution

Centroid calculations

The given area can be considered to be made up of a rectangle, from which a semicircular area has been removed. Due to symmetry, we know that the x-coordinate of the centroid is $\bar{x}$ = 15 cm. The y-coordinate of the centroid is determined as follows:

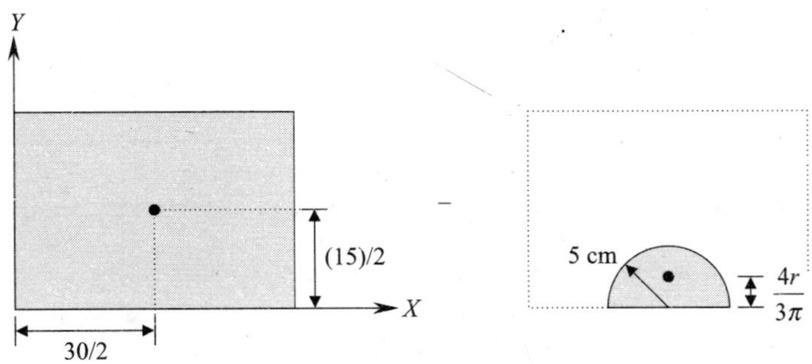

Fig. 9.25(a)

S.No	Element	$A_i (cm^2)$	$\bar{y}_i$ (cm)	$A_i\bar{y}_i (cm^3)$
1.	Rectangle	450	7.5	3375
2.	Semicircle	$-\frac{\pi}{2}(5)^2 = -39.27$	$4(5)/3\pi = 2.12$	−83.25
Σ =		410.73		3291.75

$$\therefore \quad \bar{y} = \frac{\sum A_i\bar{y}_i}{\sum A_i} = 8.01 \text{ cm}$$

Moments of inertia calculations

The moments of inertia of the composite section about its centroid can be determined as follows:

S.No	$(\bar{I}_{xx})_i$ cm^4	$(\bar{I}_{yy})_i$ cm^4	$A_i(\bar{y}_i - \bar{y})^2$ cm^4	$A_i(\bar{x}_i - \bar{x})^2$ cm^4
1.	$(30)(15)^3/12 = 8437.5$	$(15)(30)^3/12 = 33\,750$	$450(7.5 - 8.01)^2 = 117.05$	0
2.	$-0.11(5)^4 = -68.75$	$-\pi(5)^4/8 = -245.44$	$-39.27(2.12 - 8.01)^2 = -1362.36$	0
Σ =	8368.75	33 504.56	−1245.31	0

$$\bar{I}_{xx} = \sum(\bar{I}_{xx})_i + \sum A_i(\bar{y}_i - \bar{y})^2$$

$$= 8368.75 - 1245.31 = 7123.44 \text{ cm}^4$$

$$\bar{I}_{yy} = \sum(\bar{I}_{yy})_i + \sum A_i(\bar{x}_i - \bar{x})^2 = 33\,504.56 \text{ cm}^4$$

Example 9.16 Find the moments of inertia of the shaded area about the centroidal axes.

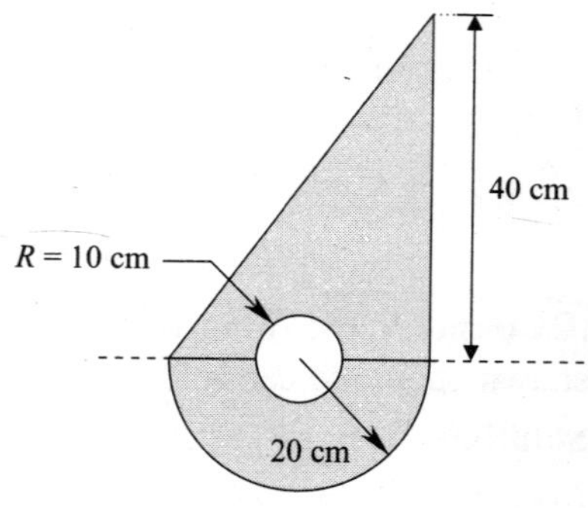

Fig. 9.26

Solution The given area is obtained by joining a triangular area and a semicircular area, from which a circular portion is removed. The reference axes are chosen conveniently as shown in Fig. 9.26(a) below such that the X-axis coincides with diameter of the semicircle. The centroidal coordinates of the composite section are determined as follows:

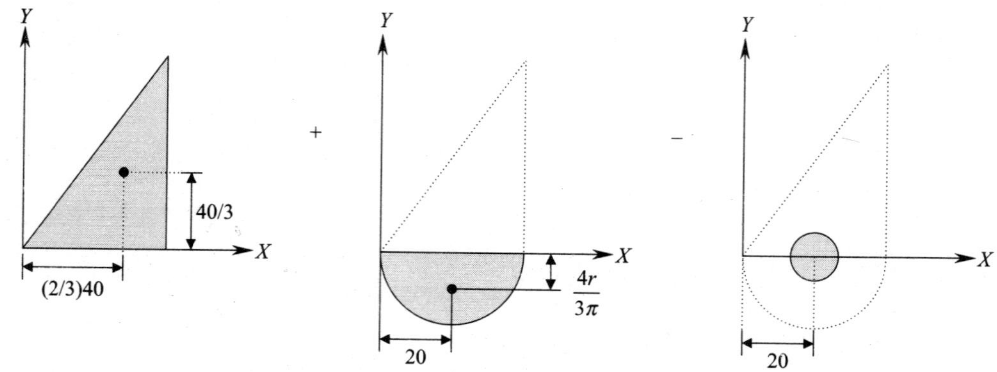

Fig. 9.26(a)

S.No	Element	A_i (cm^2)	$\bar{x}_i$ (cm)	$\bar{y}_i$ (cm)	$A_i\bar{x}_i$ (cm^3)	$A_i\bar{y}_i$ (cm^3)
1.	Triangle	$(1/2) \times 40 \times 40 = 800$	$(2/3)40 = 26.67$	$(1/3)40 = 13.33$	21 336	10 664
2.	Semicircle	$\pi(20)^2/2 = 628.32$	20	$-4(20)/3\pi = -8.49$	12 566.4	–5334.44
3.	Circle	$-\pi(10)^2 = -314.16$	20	0	–6283.2	0
	$\Sigma =$	1114.16			27 619.2	5329.56

$$\therefore \qquad \bar{x} = \frac{\sum A_i\bar{x}_i}{\sum A_i} = 24.79 \text{ cm} \qquad \bar{y} = \frac{\sum A_i\bar{y}_i}{\sum A_i} = 4.78 \text{ cm}$$

The centroidal axes are shown in Fig. 9.26(b) as passing through the centroid of the composite section. The moments of inertia of the composite section about its centroidal axes are calculated as follows:

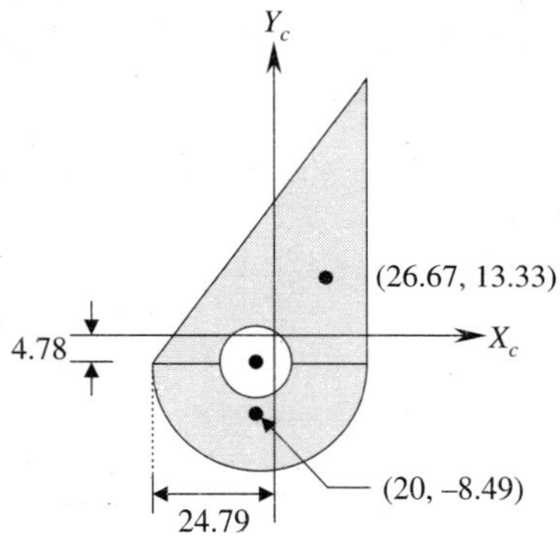

Fig. 9.26(b)

S.No	$(\bar{I}_{xx})_i$ cm^4	$(\bar{I}_{yy})_i$ cm^4	$A_i(\bar{y}_i - \bar{y})^2$ cm^4	$A_i(\bar{x}_i - \bar{x})^2$ cm^4
1.	$40 \times (40)^3/36$ $= 71111.11$	$40 \times (40)^3/36$ $= 71\,111.11$	$800(13.33 - 4.78)^2$ $= 58\,482$	$800(26.67 - 24.79)^2$ $= 2827.52$
2.	$0.11(20)^4$ $= 17\,600$	$\pi(20)^4/8$ $= 62\,831.85$	$628.32(-8.49 - 4.78)^2$ $= 110\,642.69$	$628.32(20 - 24.79)^2$ $= 14\,416.24$
3.	$-\pi(10)^4/4$ $= -7853.98$	$-\pi(10)^4/4$ $= -7853.98$	$-314.16(0 - 4.78)^2$ $= -7178.05$	$-314.16(20 - 24.79)^2$ $= -7208.12$
$\Sigma =$	80 857.13	126 088.98	161 946.64	10 035.64

Therefore, moments of inertia of the composite section about centroidal axes are obtained as

$$\bar{I}_{xx} = \Sigma(\bar{I}_{xx})_i + \sum A_i(\bar{y}_i - \bar{y})^2$$

$$= 80\,857.13 + 161\,946.64 = 242\,803.77 \text{ cm}^4$$

$$\bar{I}_{yy} = \Sigma(\bar{I}_{yy})_i + \sum A_i(\bar{x}_i - \bar{x})^2$$

$$= 126\,088.98 + 10\,035.64 = 136\,124.62 \text{ cm}^4$$

Example 9.17 Find the moments of inertia of the cut section shown about the centroidal axes. Two semicircular portions are cut from a rectangular plate.

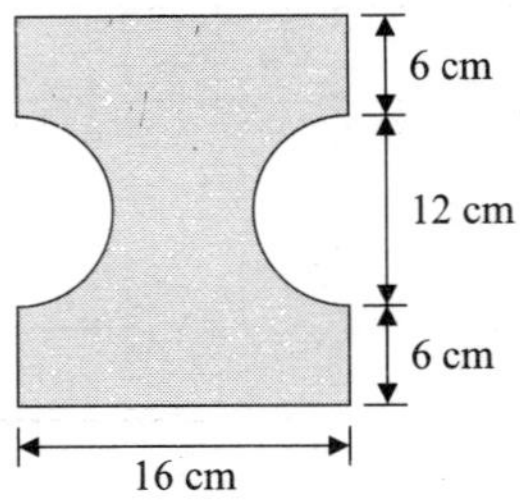

Fig. 9.27

Solution Since the section has two axes of symmetry, we can readily know that its centroid lies at the centre, i.e., $\bar{x} = 8$ cm and $\bar{y} = 12$ cm with respect to the lower left corner. The area of each semicircular portion cut is $\pi(6)^2/2 = 56.55$ cm^2. The centroid of each cut section lies at $4r/3\pi = 4(6)/3\pi = 2.55$ cm from the end of the plate. The moments of inertia calculations are given below:

S.No	$(\bar{I}_{xx})_i$ cm^4	$(\bar{I}_{yy})_i$ cm^4	$A_i(\bar{y}_i - \bar{y})^2$ cm^4	$A_i(\bar{x}_i - \bar{x})^2$ cm^4
1.	$16 \times (24)^3/12 = 18432$	$24 \times (16)^3/12 = 8192$	0	0
2.	$-\frac{\pi}{8}(6)^4 = -508.94$	$-0.11(6)^4 = -142.56$	0	$-56.55(2.55 - 8)^2 = -1679.68$
3.	$-\frac{\pi}{8}(6)^4 = -508.94$	$-0.11(6)^4 = -142.56$	0	$-56.55(16 - 2.55 - 8)^2 = -1679.68$
$\Sigma =$	17 414.12	7906.88	0	−3359.36

Therefore,

$$\bar{I}_{xx} = \sum(\bar{I}_{xx})_i + \sum A_i(\bar{y}_i - \bar{y})^2 = 17\,414.12 \text{ cm}^4$$

$$\bar{I}_{yy} = \sum(\bar{I}_{yy})_i + \sum A_i(\bar{x}_i - \bar{x})^2$$

$$= 7906.88 - 3359.36 = 4547.52 \text{ cm}^4$$

Example 9.18 Compute the second moment of the shaded area shown about the horizontal centroidal axis.

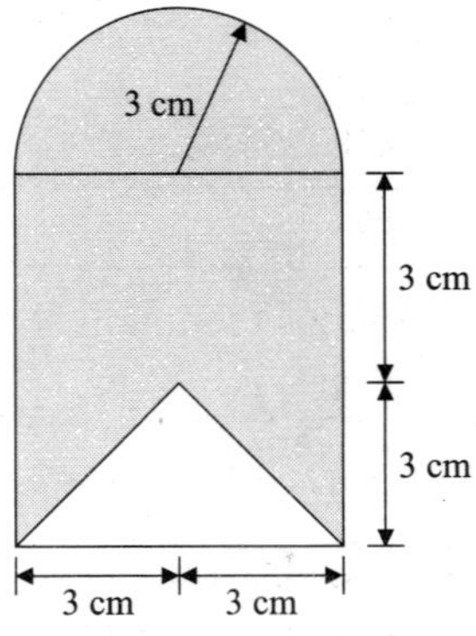

Fig. 9.28

Solution The given area can be considered to be made up of a rectangle and a semicircular area, from which a triangular area has been removed. We see that the composite section has an axis of symmetry and hence we need to determine only the y-coordinate of the centroid.

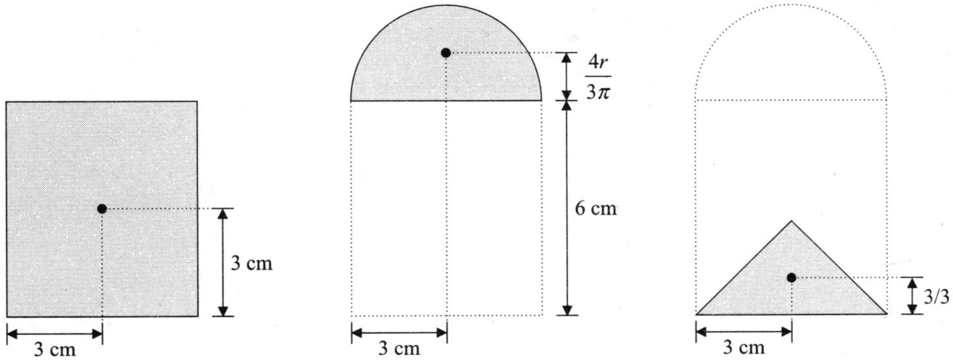

Fig. 9.28(a)

S.No	Element	A_i (cm^2)	$\bar{y}_i$ (cm)	$A_i\bar{y}_i$ (cm^3)
1.	Rectangle	36	3	108
2.	Semicircle	$\pi(3)^2/2 = 14.14$	$6 + [4(3)/3\pi] = 7.27$	102.8
3.	Triangle	–9	1	–9
	$\Sigma =$	41.14		201.8

Therefore, $$\bar{y} = \frac{\sum A_i\bar{y}_i}{\sum A_i} = \frac{201.8}{41.14} = 4.91 \text{ cm}$$

Moment of inertia calculations

The moment of inertia of the composite section about its horizontal centroidal axis is determined as follows:

S.No	$(\bar{I}_{xx})_i$ cm^4	$A_i(\bar{y}_i - \bar{y})^2$ cm^4
1.	$6 \times (6)^3/12 = 108$	$36(3 - 4.91)^2 = 131.33$
2.	$0.11(3)^4 = 8.91$	$14.14(7.27 - 4.91)^2 = 78.75$
3.	$-6 \times (3)^3/36 = -4.5$	$-9(1 - 4.91)^2 = -137.59$
$\Sigma =$	112.41	72.49

$$\therefore \quad \bar{I}_{xx} = \sum(\bar{I}_{xx})_i + \sum A_i(\bar{y}_i - \bar{y})^2$$

$$= 112.41 + 72.49 = 184.9 \text{ cm}^4$$

Example 9.19 Determine the moment of inertia of the quarter-circular spandrel shown in Fig. 9.29 about axes AA and about BB.

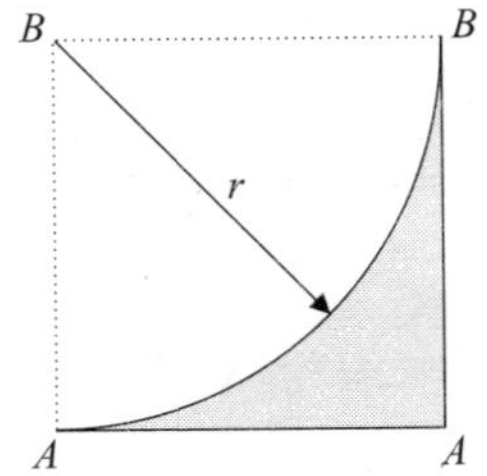

Fig. 9.29

Solution The shaded area can be obtained by removing a quarter-circular area of radius r from a square of side r as shown below:

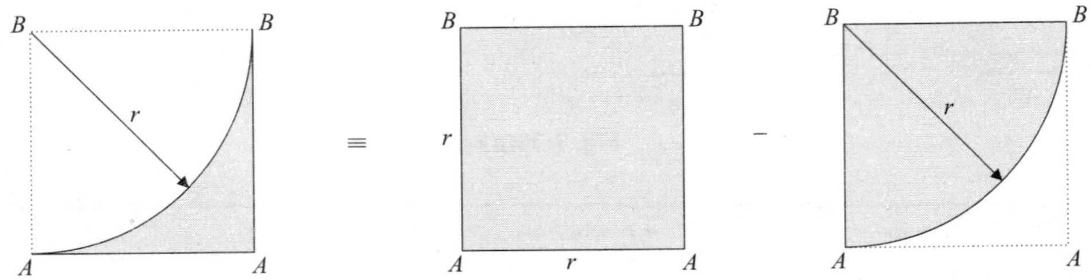

Fig. 9.29(a)

We know that moment of inertia of a square about its base is

$$I_1 = \frac{1}{3}(r)(r)^3 = 0.33r^4$$

The moment of inertia of a quarter-circular area about its centroidal axis is

$$\bar{I}_2 = 0.055(r)^4$$

Therefore, its moment of inertia about the AA axis is

$$I_2 = 0.055(r)^4 + \frac{\pi}{4}r^2\left(r - \frac{4r}{3\pi}\right)^2 = 0.315r^4$$

Hence, the moment of inertia of the spandrel about AA is obtained as

$$\begin{aligned} I_{AA} &= I_1 - I_2 \\ &= 0.33r^4 - 0.315r^4 = 0.015r^4 \end{aligned}$$

Similarly, the moment of inertia of the quarter-circular area about the BB axis is

$$I_2' = 0.055(r)^4 + \frac{\pi}{4}r^2\left(\frac{4r}{3\pi}\right)^2 = 0.196r^4$$

Hence, the moment of inertia of the spandrel about BB is obtained as

$$\begin{aligned} I_{BB} &= I_1 - I_2' \\ &= 0.33r^4 - 0.196r^4 = 0.134r^4 \end{aligned}$$

9.7 PRODUCT OF INERTIA

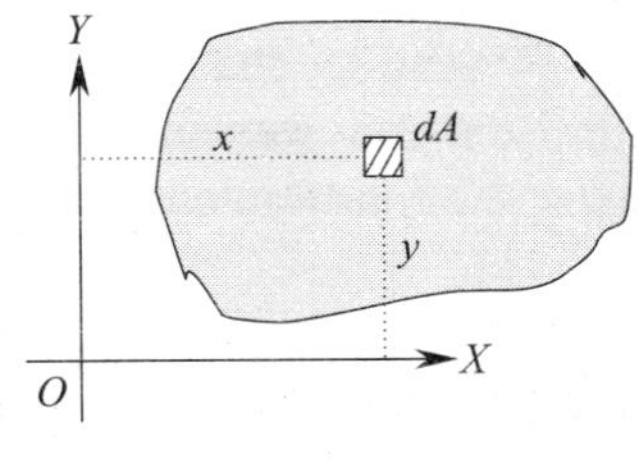

Fig. 9.30 Product of inertia

While finding the moment of inertia of an area, we multiply each element of the area by the **square** of its **perpendicular distance** from the axis. However, if we multiply each element of the area by the **product of its coordinates**, i.e., $xy\,dA$ then it is called **product of inertia**. Hence, product of inertia of the entire area is given as

$$I_{xy} = \int xy\,dA \qquad (9.30)$$

Product of inertia does not have any physical significance except that it is useful in determining the maximum and minimum moments of inertia (discussed in Section 9.8). The dimension and unit of product of inertia are the same as that of the moment of inertia, i.e., $\mathbf{[L]^4}$ and $\mathbf{m^4}$.

Unlike moment of inertia, product of inertia can be **positive** or **negative**, depending upon the position of the area with respect to the reference axes. If the area lies entirely in the first quadrant then it is positive as both x and y coordinates are positive. If it lies entirely in the third quadrant then also it is positive as both x and y coordinates are negative. If it lies entirely in the second or fourth quadrant then it is negative as one of the x and y coordinates is negative. When the area is located in more than one quadrant, the sign of the product of inertia depends upon the distribution of the area within the quadrants.

A special case arises when one of the axes is an *axis of symmetry*. For illustration, consider the area shown in Fig. 9.31, which is symmetric about the Y-axis. Then for each element of area dA on the positive side of the X-axis, there is a symmetric or mirror element dA with the same y-coordinate but with a negative x-coordinate. Hence, while taking the summation, the products of inertia $xydA$ for the two elemental areas cancel out. As there are such pairs of elements throughout the entire area, the product of inertia of the entire area vanishes, i.e.,

$$I_{xy} = 0 \qquad (9.31)$$

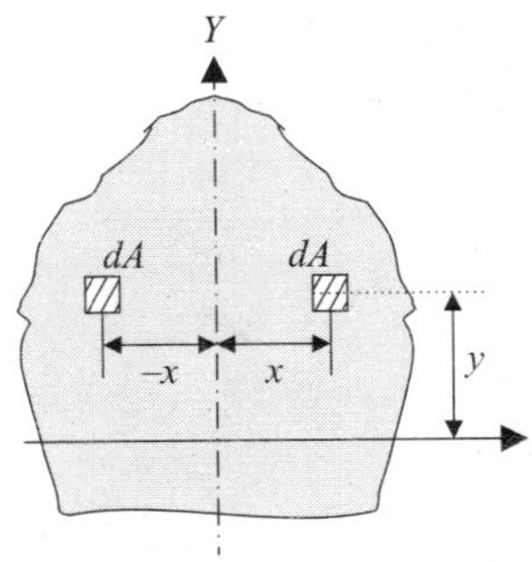

Fig. 9.31 Product of inertia of symmetrical area

Thus, for areas such as a rectangle, circle, semicircle, I-section, etc., which have at least one axis of symmetry, the product of inertia, I_{xy} is zero.

9.7.1 Transfer Theorem for Product of Inertia

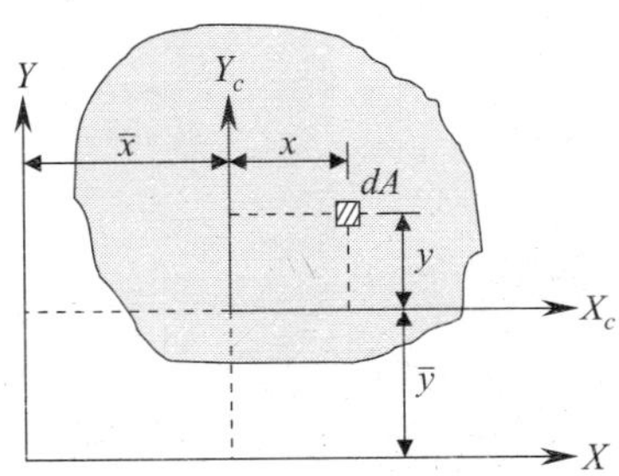

Fig. 9.32 Transfer formula

Just as we derived the transfer formula for moment of inertia, a similar formula can be derived for the product of inertia also. Consider the same figure as before.

The product of inertia about the centroidal axes is

$$\bar{I}_{xy} = \int xy\,dA \qquad (9.32)$$

and the product of inertia about X–Y axes is

$$I_{xy} = \int (x+\bar{x})(y+\bar{y})dA$$

$$= \int xy\,dA + \bar{y}\int x dA + \bar{x}\int y dA + \bar{x}\,\bar{y}\int dA \qquad (9.33)$$

We can see that the first integral on the right-hand side is the product of inertia, $\bar{I}_{xy}$; the second and third integrals vanish because they are the first moments of the area and they are zero for centroidal axes; the fourth integral is the area A multiplied by $\bar{x}\,\bar{y}$. Therefore,

$$I_{xy} = \bar{I}_{xy} + A\bar{x}\,\bar{y} \tag{9.34}$$

9.7.2 Product of Inertia of a Right-Angled Triangle

Consider a right-angled triangle of base b and height h. If we take a thin strip parallel to the base at a distance y from the base and of infinitesimally small thickness dy, then the area of the strip is

$$dA = b'dy$$

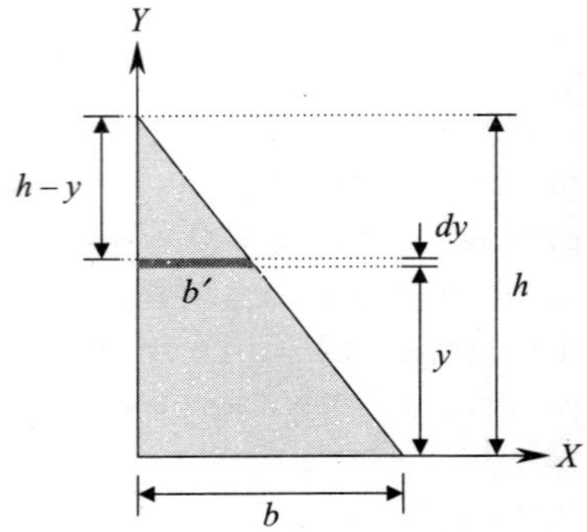

Fig. 9.33

From similar triangles, we know that

$$\frac{b'}{h-y} = \frac{b}{h}$$

Therefore,

$$dA = \frac{b}{h}(h-y)\,dy$$

Then product of inertia of the strip about X–Y axes is

$$dI_{xy} = \left[\frac{b'}{2}\right]y\,dA$$

$$= \left[\frac{b'}{2}\right]y\,\frac{b}{h}(h-y)\,dy$$

$$= \frac{1}{2}\frac{b^2}{h^2}(h-y)^2\,ydy$$

[Note that the strip can be considered to be a rectangle and its centroid lies at its midpoint].

Therefore, the product of inertia of the triangle about the X–Y axes is obtained as

$$I_{xy} = \frac{1}{2}\frac{b^2}{h^2}\int_0^h (h-y)^2\,ydy$$

$$= \frac{1}{2}\frac{b^2}{h^2}\left[\frac{h^2y^2}{2} + \frac{y^4}{4} - \frac{2hy^3}{3}\right]_0^h$$

$$= \frac{b^2h^2}{24} \tag{9.35}$$

By applying the parallel axis theorem for product of inertia, we can obtain the product of inertia of the triangle about the centroidal axes:

$$\bar{I}_{xy} = I_{xy} - A\bar{x}\,\bar{y}$$

$$= \frac{b^2h^2}{24} - \frac{bh}{2}\left(\frac{b}{3}\right)\left(\frac{h}{3}\right) = \frac{-b^2h^2}{72} \tag{9.36}$$

The products of inertia of various regular geometrical shapes are summarized in Table 9.1.

Example 9.20 Find the product of inertia of the channel section shown with respect to centroidal axes.

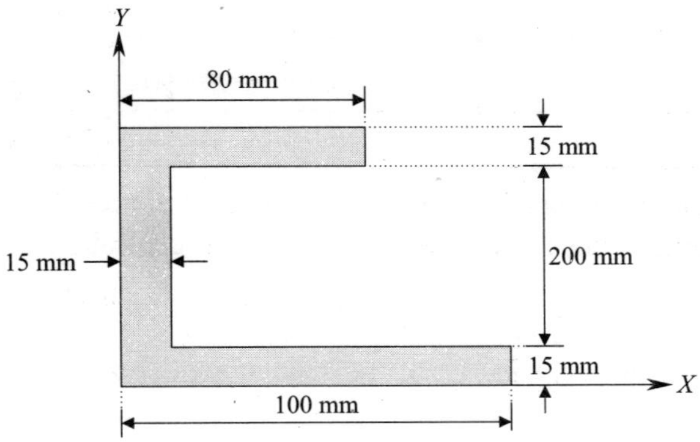

Fig. 9.34

Solution

Centroid Calculations

The given channel section can be considered to be made up of three rectangles. The coordinates of the centroid of the composite section can be determined as follows:

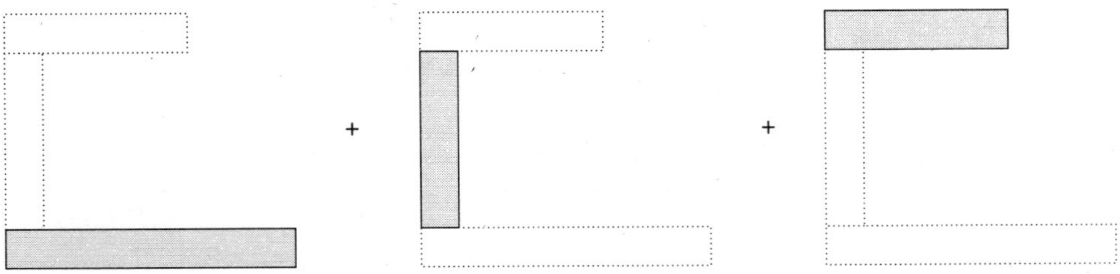

Fig. 9.34(a)

S.No	Element	$A_i(mm^2)$	$\bar{x}_i$ (mm)	$\bar{y}_i$ (mm)	$A_i\bar{x}_i$ (mm^3)	$A_i\bar{y}_i$ (mm^3)
1.	Rectangle (1)	1500	50	15/2 = 7.5	75 000	11 250
2.	Rectangle (2)	3000	7.5	15 + (200/2) = 115	22 500	345 000
3.	Rectangle (3)	1200	40	215 + (15/2) = 222.5	48 000	267 000
Σ =		5700			145 500	623 250

Therefore, coordinates of centroid of the composite section are

$$\bar{x} = \frac{\sum A_i\bar{x}_i}{\sum A_i} = 25.53 \text{ mm} \qquad \bar{y} = \frac{\sum A_i\bar{y}_i}{\sum A_i} = 109.34 \text{ mm}$$

Product of inertia calculations

S.No	$(\bar{I}_{xy})_i$ mm^4	$A_i(\bar{x}_i - \bar{x})(\bar{y}_i - \bar{y})$ mm^4
1.	0	1500(50 – 25.53) (7.5 – 109.34) = –3 738 037.2
2.	0	3000 (7.5 – 25.53) (115 – 109.34) = –306 149.4
3.	0	1200(40 – 25.53) (222.5 –109.34) = 1 964 910.2
Σ =	0	–2 079 276.4

Note that since a rectangle has axes of symmetry, its product of inertia about centroidal axes is zero. Therefore, product of inertia of the composite section about its centroidal axes is determined as

$$\bar{I}_{xy} = \sum(\bar{I}_{xy})_i + \sum A_i(\bar{x}_i - \bar{x})(\bar{y}_i - \bar{y}) = -2\,079\,276.4 \text{ mm}^4$$

Example 9.21 Find the product of inertia of the shaded area shown with respect to the centroidal axes.

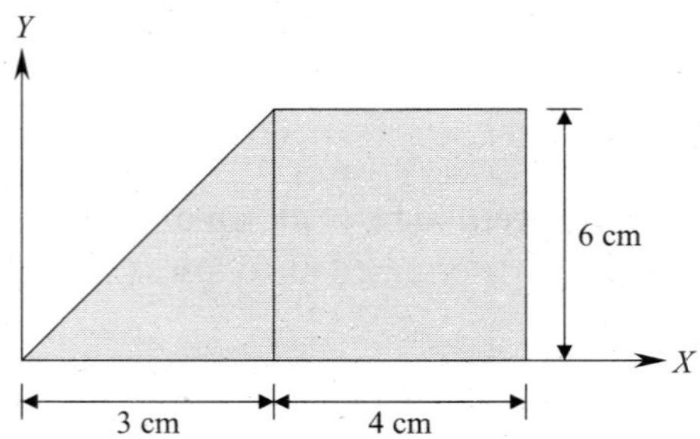

Fig. 9.35

Solution

Centroid Calculations

The given section can be considered to be made up of a triangle and a rectangle. The coordinates of the centroid of the composite section can be determined as follows:

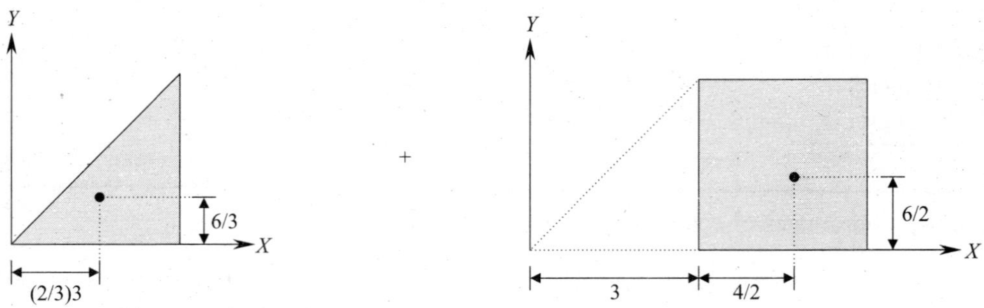

Fig. 9.35(a)

S.No	Element	A_i (cm^2)	$\bar{x}_i$ (cm)	$\bar{y}_i$ (cm)	$A_i\bar{x}_i$ (cm^3)	$A_i\bar{y}_i$ (cm^3)
1.	Triangle	$(1/2)3 \times 6 = 9$	$(2/3) \times 3 = 2$	$6/3 = 2$	18	18
2.	Rectangle	$4 \times 6 = 24$	$3 + (4/2) = 5$	$6/2 = 3$	120	72
$\Sigma =$		33			138	90

Therefore, $\bar{x} = \dfrac{\sum A_i\bar{x}_i}{\sum A_i} = 4.18$ cm $\qquad \bar{y} = \dfrac{\sum A_i\bar{y}_i}{\sum A_i} = 2.73$ cm

Product of inertia calculations

S.No	$(\bar{I}_{xy})_i$ cm^4	$A_i(\bar{x}_i - \bar{x})(\bar{y}_i - \bar{y})$ cm^4
1.	$-(3)^2(6)^2/72 = -4.5$	$9(2 - 4.18)(2 - 2.73) = 14.32$
2.	0	$24(5 - 4.18)(3 - 2.73) = 5.31$
$\Sigma =$	-4.5	19.63

Therefore, $\bar{I}_{xy} = \sum(\bar{I}_{xy})_i + \sum A_i(\bar{x}_i - \bar{x})(\bar{y}_i - \bar{y}) = 15.13$ cm^4

Example 9.22 Determine the product of inertia of the quarter-circular area shown in Fig. 9.36 with respect to the given X and Y axes.

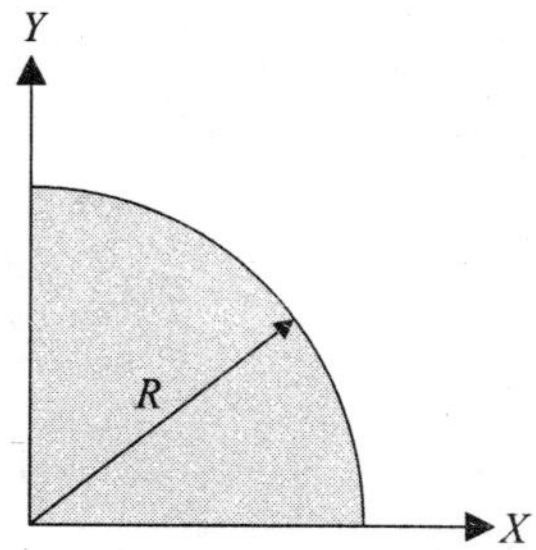

Fig. 9.36

Solution Consider a thin strip parallel to the X-axis at a distance y from the X-axis and of infinitesimally small thickness dy. Then the area of the strip is

$$dA = x dy$$

The product of inertia of this strip is

$$dI_{xy} = (x/2)\, y\, dA = (x/2)\, yx\, dy$$

[Note that the strip can be considered to be a rectangle and its centroid lies at its midpoint].

Therefore, the product of inertia of the entire area about X–Y axes is

$$I_{xy} = \int_0^R (x/2)\, yx\, dy = \int_0^R (1/2)\, x^2 y\, dy$$

We know that the equation of a circle is $x^2 + y^2 = R^2 \Rightarrow x^2 = R^2 - y^2$

Therefore, $I_{xy} = \displaystyle\int_0^R (1/2)(R^2 - y^2)\, y\, dy$

$$= \int_0^R (1/2)(R^2 y - y^3)\, dy$$

$$= \frac{1}{2}\left(\frac{R^4}{2} - \frac{R^4}{4}\right) = \frac{R^4}{8}$$

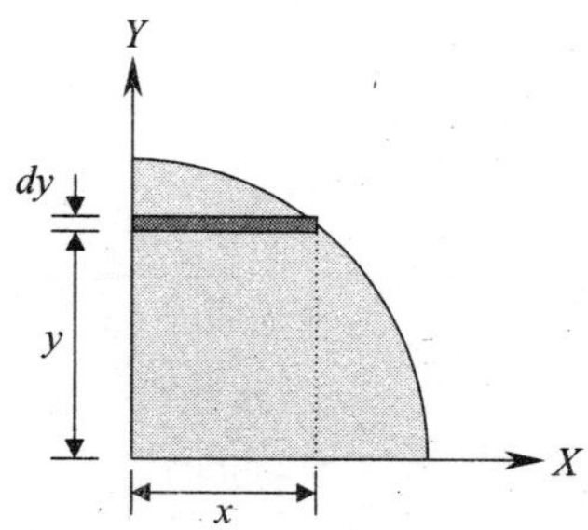

Fig. 9.36(a)

Example 9.23 Determine the product of inertia of the sectioned area shown about the centroidal axes.

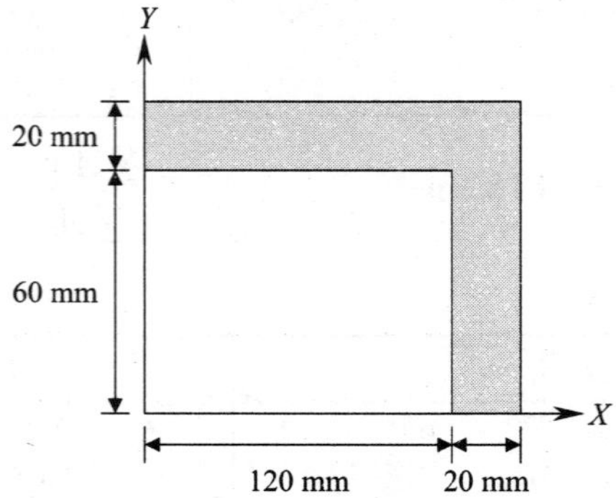

Fig. 9.37

Solution

Centroid Calculations

The given area is divided into two rectangles, 1 and 2, as shown below. To simplify the calculations, the given dimensions have been converted to cm.

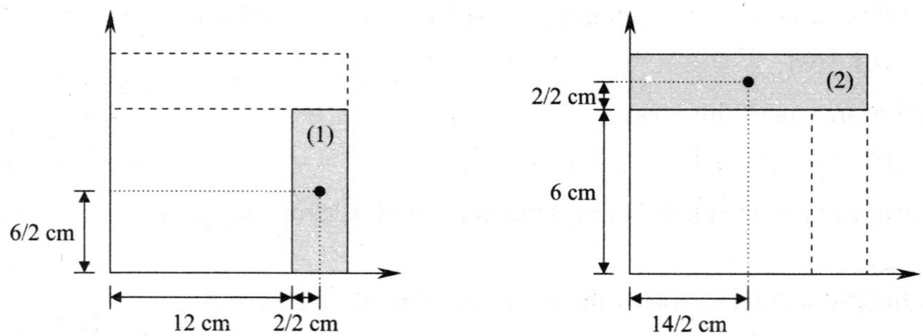

Fig. 9.37(a)

S.No	Element	A_i (cm^2)	$\bar{x}_i$ (cm)	$\bar{y}_i$ (cm)	$A_i\bar{x}_i$ (cm^3)	$A_i\bar{y}_i$ (cm^3)
1.	Rectangle (1)	12	13	3	156	36
2.	Rectangle (2)	28	7	7	196	196
$\Sigma =$		40			352	232

$$\therefore \quad \bar{x} = \frac{\sum A_i\bar{x}_i}{\sum A_i} = 8.8 \text{ cm} \qquad \bar{y} = \frac{\sum A_i\bar{y}_i}{\sum A_i} = 5.8 \text{ cm}$$

Product of inertia calculations

S.No	$(\bar{I}_{xy})_i$ cm^4	$A_i(\bar{x}_i - \bar{x})(\bar{y}_i - \bar{y})$ cm^4
1.	0	12(13 – 8.8) (3 – 5.8) = –141.12
2.	0	28(7 – 8.8)(7 – 5.8) = –60.48
Σ =	0	–201.6

Note that since a rectangle has axes of symmetry, its product of inertia about centroidal axes is zero. Therefore,

$$\bar{I}_{xy} = \sum(\bar{I}_{xy})_i + \sum A_i(\bar{x}_i - \bar{x})(\bar{y}_i - \bar{y})$$

$$= -201.6 \text{ cm}^4$$

$$= -2.02 \times 10^6 \text{ mm}^4$$

9.8 MOMENTS OF INERTIA ABOUT INCLINED AXES

In the previous sections, we saw how to determine the moment of inertia about an axis parallel or perpendicular to the centroidal axes. In this section, we will determine moments of inertia about inclined axes as they find application in the design of beams.

Consider a plane area as shown in Fig. 9.38. Then its moments of inertia about X and Y axes are respectively

$$I_{xx} = \int y^2 dA \qquad I_{yy} = \int x^2\, dA \tag{9.37}$$

and product of inertia about X–Y axes is

$$I_{xy} = \int xy\, dA \tag{9.38}$$

where x and y are coordinates of the element of area dA about the X–Y axes.

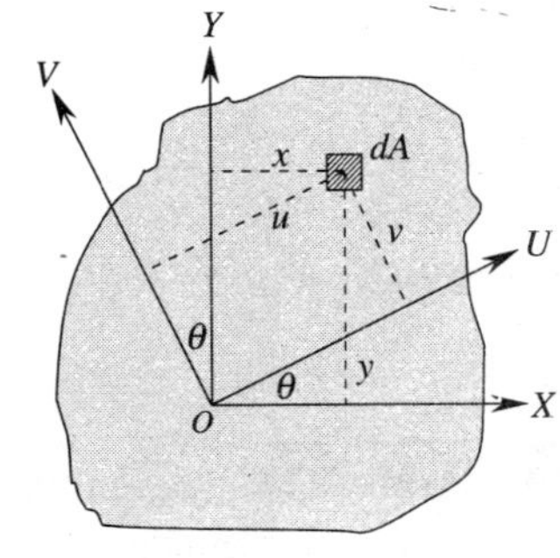

Fig. 9.38

If we consider the inclined axes U–V obtained by rotating about the origin O in the anticlockwise direction through an angle θ then the coordinates of the element dA with respect to the inclined axes can be obtained as

$$u = x\cos\theta + y\sin\theta \text{ and } v = y\cos\theta - x\sin\theta \tag{9.39}$$

or put in matrix form,

$$\begin{bmatrix} u \\ v \end{bmatrix} = \begin{bmatrix} \cos\theta & \sin\theta \\ -\sin\theta & \cos\theta \end{bmatrix} \begin{bmatrix} x \\ y \end{bmatrix} \tag{9.40}$$

Then moment of inertia of the elemental area dA about the U-axis is

$$dI_{uu} = v^2\, dA$$

$$= (y\cos\theta - x\sin\theta)^2\, dA$$

Therefore, moment of inertia of the entire area about the U-axis is

$$I_{uu} = \int (y\cos\theta - x\sin\theta)^2 dA$$

$$= \cos^2\theta \int y^2\, dA + \sin^2\theta \int x^2\, dA - 2\sin\theta\cos\theta \int xy\, dA \tag{9.41}$$

On simplification,

$$I_{uu} = I_{xx}\cos^2\theta + I_{yy}\sin^2\theta - 2I_{xy}\sin\theta\cos\theta \tag{9.42}$$

If we introduce the following trigonometric identities,

$$\cos^2\theta = \frac{1}{2}(1+\cos 2\theta), \quad \sin^2\theta = \frac{1}{2}(1-\cos 2\theta)$$

and $$\sin 2\theta = 2\sin\theta\cos\theta$$

then the Eq. 9.42 reduces to

$$I_{uu} = \left(\frac{I_{xx}+I_{yy}}{2}\right) + \left(\frac{I_{xx}-I_{yy}}{2}\right)\cos 2\theta - I_{xy}\sin 2\theta \tag{9.43}$$

Similarly, the moment of inertia about the V-axis can be obtained as

$$I_{vv} = \int u^2\, dA = I_{xx}\sin^2\theta + I_{yy}\cos^2\theta + 2I_{xy}\sin\theta\cos\theta$$

$$= \left(\frac{I_{xx}+I_{yy}}{2}\right) - \left(\frac{I_{xx}-I_{yy}}{2}\right)\cos 2\theta + I_{xy}\sin 2\theta \tag{9.44}$$

Note that the above expression can also be obtained from the Eq. 9.43 by substituting the angle θ as $90° + \theta$.

The product of inertia about the U–V axes can also be obtained as

$$I_{uv} = \int uv\, dA$$

$$= \sin\theta\cos\theta \int y^2\, dA - \sin\theta\cos\theta\int x^2\, dA + (\cos^2\theta - \sin^2\theta)\int xy\, dA$$

$$= \left(\frac{I_{xx}-I_{yy}}{2}\right)\sin 2\theta + I_{xy}\cos 2\theta \tag{9.45}$$

Example 9.24 Given that moments of inertia of a section are $I_{xx} = I_{yy} = 3 \times 10^4$ mm^4 and $I_{xy} = 1.5 \times 10^4$ mm^4, determine moments of inertia of the section about inclined axes inclined at 30° to the X-axis in the anticlockwise direction.

Solution From the given data, we can determine

$$\frac{I_{xx}+I_{yy}}{2} = 3 \times 10^4 \text{ mm}^4$$

and $$\frac{I_{xx}-I_{yy}}{2} = 0$$

Therefore,

$$I_{uu} = \left(\frac{I_{xx}+I_{yy}}{2}\right) + \left(\frac{I_{xx}-I_{yy}}{2}\right)\cos 2\theta - I_{xy}\sin 2\theta$$

$$= 3 \times 10^4 + 0 - (1.5 \times 10^4 \sin 60°) = 1.7 \times 10^4 \text{ mm}^4$$

$$I_{vv} = \left(\frac{I_{xx}+I_{yy}}{2}\right) - \left(\frac{I_{xx}-I_{yy}}{2}\right)\cos 2\theta + I_{xy}\sin 2\theta$$

$$= 3 \times 10^4 - 0 + (1.5 \times 10^4 \sin 60°) = 4.3 \times 10^4 \text{ mm}^4$$

$$I_{uv} = \left(\frac{I_{xx} - I_{yy}}{2}\right) \sin 2\theta + I_{xy} \cos 2\theta$$

$$= 0 + (1.5 \times 10^4 \cos 60°) = 0.75 \times 10^4 \text{ mm}^4$$

9.8.1 Principal Axes and Principal Moments of Inertia

From Eqs 9.43 and 9.44, we can see that as the inclination of the axes vary, the moments of inertia about the inclined axes also vary and at a particular angle θ, they reach maximum and minimum values. It can be shown that when the moment of inertia about one of the axes reaches a *maximum* value, the moment of inertia about the perpendicular axis in the plane of the figure reaches a *minimum* value. The condition for maxima or minima can be determined by differentiating the Eq. 9.43 or 9.44 with respect to θ and equating it to zero:

$$\frac{dI_{uu}}{d\theta} = 0 \tag{9.46}$$

$\Rightarrow$
$$\tan 2\theta = \frac{-2I_{xy}}{I_{xx} - I_{yy}} \tag{9.47}$$

The above expression gives the inclination of the axes about which the moment of inertia reaches maximum or minimum value. The axes about which moment of inertia is maximum or minimum are termed **principal axes** and the corresponding moments of inertia about these principal axes are termed **principal moments of inertia**. By substituting the condition for maxima or minima in Eqs 9.43 and 9.44, we can get the principal moments of inertia, which can be expressed in compact form as

$$I_{max} = \left(\frac{I_{xx} + I_{yy}}{2}\right) + R \quad \text{(or)} \quad I_{max} = I_{ave} + R \tag{9.48}$$

and
$$I_{min} = \left(\frac{I_{xx} + I_{yy}}{2}\right) - R \quad \text{(or)} \quad I_{min} = I_{ave} - R \tag{9.49}$$

where,
$$R = \sqrt{\left(\frac{I_{xx} - I_{yy}}{2}\right)^2 + (I_{xy})^2} \tag{9.50}$$

and
$$I_{ave} = \frac{I_{xx} + I_{yy}}{2} \tag{9.51}$$

Example 9.25 Find the maximum and minimum moments of inertia for the angle section shown in Fig. 9.39, given that

$$\bar{x} = 25 \text{ mm and } \bar{y} = 35 \text{ mm}$$

$$\bar{I}_{xx} = 290.67 \times 10^4 \text{ mm}^4$$

and
$$\bar{I}_{yy} = 162.67 \times 10^4 \text{ mm}^4$$

All dimensions are in mm.

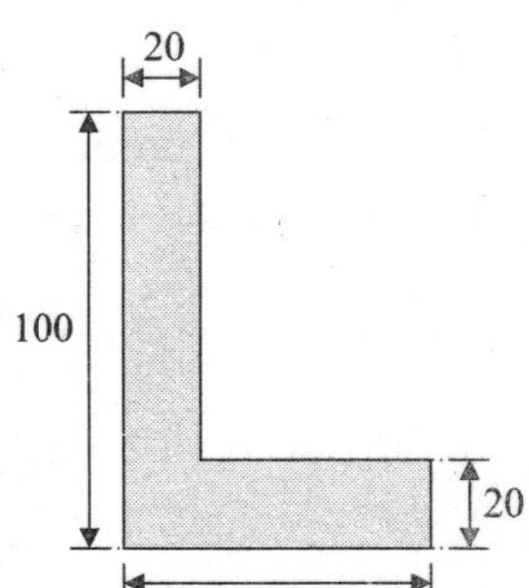

Fig. 9.39

Solution

Determination of product of inertia

To calculate the maximum and minimum moments of inertia, we need to find the product of inertia. The composite section can be considered to be made up of two rectangles. The centroidal coordinates of the two rectangles can be determined as (40,10) and (10, (20 + 80/2 = 60)); the coordinates being with respect to lower left corner. The calculations are summarized below:

S.No	$(\bar{I}_{xy})_i$ mm^4	$A_i(\bar{x}_i - \bar{x})(\bar{y}_i - \bar{y})$ mm^4
1.	0	1600(40 –25)(10 – 35) = – 600 000
2.	0	1600(10 – 25) (60 –35) = – 600 000
Σ =	0	–1200 000

$$\therefore \quad \bar{I}_{xy} = \sum(\bar{I}_{xy})_i + \sum A_i(\bar{x}_i - \bar{x})(\bar{y}_i - \bar{y})$$

$$= -1200\,000 = -120 \times 10^4 \text{ mm}^4$$

Determination of principal moments of inertia

We know,

$$R = \sqrt{\left(\frac{\bar{I}_{xx} - \bar{I}_{yy}}{2}\right)^2 + (\bar{I}_{xy})^2}$$

$$= \sqrt{\left(\frac{290.67 - 162.67}{2}\right)^2 \times 10^8 + (-120 \times 10^4)^2} = 136 \times 10^4 \text{ mm}^4$$

$$\bar{I}_{\text{ave}} = \frac{\bar{I}_{xx} + \bar{I}_{yy}}{2} = 226.67 \times 10^4 \text{ mm}^4$$

Therefore,

$$\bar{I}_{\max} = \bar{I}_{\text{ave}} + R$$

$$= (226.67 + 136) \times 10^4 = 362.67 \times 10^4 \text{ mm}^4$$

and

$$\bar{I}_{\min} = \bar{I}_{\text{ave}} - R$$

$$= (226.67 - 136) \times 10^4 = 90.67 \times 10^4 \text{ mm}^4$$

Example 9.26 Determine moments of inertia and product of inertia about centroidal *U–V* axes for the triangle shown in Fig. 9.40. Also, determine the principal moments of inertia and the inclination of the principal axes.

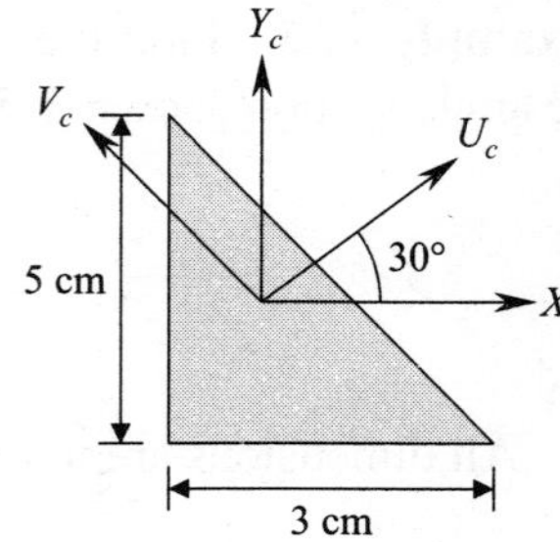

Fig. 9.40

Solution The moments of inertia of the triangle about the horizontal and vertical centroidal axes are

$$\bar{I}_{xx} = \frac{bh^3}{36} = \frac{3(5)^3}{36} = 10.42 \text{ cm}^4 \qquad \bar{I}_{yy} = \frac{hb^3}{36} = \frac{5(3)^3}{36} = 3.75 \text{ cm}^4$$

and the product of inertia is

$$\bar{I}_{xy} = -\frac{b^2h^2}{72} = -3.125\ \text{cm}^4$$

From the above values, we can determine

$$\frac{\bar{I}_{xx} + \bar{I}_{yy}}{2} = \frac{10.42 + 3.75}{2} = 7.085\ \text{cm}^4$$

and

$$\frac{\bar{I}_{xx} - \bar{I}_{yy}}{2} = \frac{10.42 - 3.75}{2} = 3.335\ \text{cm}^4$$

Moments of inertia about inclined axes inclined at $\theta = 30°$

Moments of inertia about axes inclined at $\theta = 30°$ to the X-axis in the anticlockwise direction are

$$\bar{I}_{uu} = \left(\frac{\bar{I}_{xx} + \bar{I}_{yy}}{2}\right) + \left(\frac{\bar{I}_{xx} - \bar{I}_{yy}}{2}\right)\cos 2\theta - \bar{I}_{xy}\sin 2\theta$$

$$= 7.085 + 3.335\cos 60° - (-3.125)\sin 60° = 11.46\ \text{cm}^4$$

$$\bar{I}_{vv} = \left(\frac{\bar{I}_{xx} + \bar{I}_{yy}}{2}\right) - \left(\frac{\bar{I}_{xx} - \bar{I}_{yy}}{2}\right)\cos 2\theta + \bar{I}_{xy}\sin 2\theta$$

$$= 7.085 - 3.335\cos 60° + (-3.125)\sin 60° = 2.711\ \text{cm}^4$$

$$\bar{I}_{uv} = \left(\frac{\bar{I}_{xx} - \bar{I}_{yy}}{2}\right)\sin 2\theta + \bar{I}_{xy}\cos 2\theta$$

$$= 3.335\sin 60° + (-3.125)\cos 60° = 1.33\ \text{cm}^4$$

Principal moments of inertia

$$R = \sqrt{\left(\frac{\bar{I}_{xx} - \bar{I}_{yy}}{2}\right)^2 + (\bar{I}_{xy})^2}$$

$$= \sqrt{(3.335)^2 + (-3.125)^2} = 4.57\ \text{cm}^4$$

Therefore,

$$\bar{I}_{\max} = \left(\frac{\bar{I}_{xx} + \bar{I}_{yy}}{2}\right) + R$$

$$= 7.085 + 4.57 = 11.66\ \text{cm}^4$$

$$\bar{I}_{\min} = \left(\frac{\bar{I}_{xx} + \bar{I}_{yy}}{2}\right) - R$$

$$= 7.085 - 4.57 = 2.52\ \text{cm}^4$$

Inclination of principal axes

$$\tan 2\theta = \frac{-2\bar{I}_{xy}}{\bar{I}_{xx} - \bar{I}_{yy}} = \frac{-\bar{I}_{xy}}{(\bar{I}_{xx} - \bar{I}_{yy})/2} = \frac{-(-3.125)}{3.335}$$

$\Rightarrow$ $$2\theta_1 = 43.14° \text{ and } 2\theta_2 = (180° + 43.14°) = 223.14°$$

Note that $\tan 2\theta = \tan(180° + 2\theta)$. Hence, we obtain two values for θ. Therefore,

$$\theta_1 = 21.57° \text{ and } \theta_2 = 111.57°$$

The angle θ_1 is the inclination of the maximum moment of inertia with respect to the X-axis and θ_2 is the inclination of the minimum moment of inertia with respect to the X-axis. It can be seen that they differ by 90°.

Example 9.27 Show that the moment of inertia of the area of a square about any axis X' through its centre is the same as that about a centroidal axis parallel to a side.

Solution We know that for a square, the moments of inertia about the centroidal axes parallel to its sides are:

$$\bar{I}_{xx} = \bar{I}_{yy} = \frac{1}{12} a\, a^3 = \frac{a^4}{12}$$

and product of inertia is zero due to symmetry, i.e.,

$$\bar{I}_{xy} = 0$$

If θ is the inclination of the X' axis with respect to the X-axis, then

$$\bar{I}_{x'x'} = \left(\frac{\bar{I}_{xx} + \bar{I}_{yy}}{2}\right) + \left(\frac{\bar{I}_{xx} - \bar{I}_{yy}}{2}\right)\cos 2\theta - \bar{I}_{xy}\sin 2\theta$$

$$= \frac{a^4}{12} + 0 - 0$$

$$= \frac{a^4}{12}$$

From the above expression, we see that the moment of inertia about any inclined axis is independent of the inclination θ. Hence, the moment of inertia remains the same for any orientation of the axes.

SUMMARY

Second Moment of Area (or) Moment of Inertia

The second moment of a lamina of area A with respect to the X-axis is defined as

$$I_{xx} = \int y^2\, dA$$

Similarly, the second moment of the area with respect to the Y-axis is

$$I_{yy} = \int x^2\, dA$$

The second moment of an area is also termed the *moment of inertia*. It finds application in the design of structural members and it gives a measure of resistance to bending in the case of sections or plane areas. Depending upon the distribution of this area, its resistance to bending varies. The dimension and unit of moment of inertia are $[L]^4$ and m^4.

Radius of Gyration

It is defined as the distance from the axis to a point where the entire area of the lamina could be concentrated into a thin strip and have the same moment of inertia with respect to the given axis. It finds application in the design of column members.

Transfer formula (or) Parallel Axis Theorem

The theorem states that the moment of inertia of an area about an axis in the plane of the area is equal to the moment of inertia about an axis passing through the centroid and parallel to the given axis plus the product of the area and the square of the distance between the two parallel axes. Mathematically, it is given as

$$I_{xx} = \bar{I}_{xx} + A(\bar{y})^2 \qquad \text{and} \qquad I_{yy} = \bar{I}_{yy} + A(\bar{x})^2$$

It should be noted that the moment of inertia about the centroidal axis is the *least* moment of inertia of the area and that it increases as the axis is moved away from the centroid. Also, the parallel axis theorem is applicable only if one of the two axes is a centroidal axis.

Perpendicular axis Theorem (or) Polar Moment of Inertia

For a plane lamina, moment of inertia about an axis perpendicular to the lamina is equal to the sum of the moments of inertia about two mutually perpendicular axes in the plane of the lamina.

$$\bar{I}_{zz} \text{ (or) } \bar{I}_p = \bar{I}_{xx} + \bar{I}_{yy}$$

Moment of Inertia of Composite Section

While finding the moment of inertia of a composite section, we divide the given section into elements like rectangle, triangle, etc. for which moments of inertia are known. Using parallel axis theorem, the moments of inertia of individual elements are determined about the centroidal axes of the composite section.

$$(I_{xx})_i = (\bar{I}_{xx})_i + A_i(\bar{y}_i - \bar{y})^2$$

$$(I_{yy})_i = (\bar{I}_{yy})_i + A_i(\bar{x}_i - \bar{x})^2$$

From which we determine the moments of inertia of the composite section about the centroidal axes:

$$\bar{I}_{xx} = \sum(\bar{I}_{xx})_i + \sum A_i(\bar{y}_i - \bar{y})^2$$

$$\bar{I}_{yy} = \sum(\bar{I}_{yy})_i + \sum A_i(\bar{x}_i - \bar{x})^2$$

Product of Inertia

The product of inertia for an area A is given as

$$I_{xy} = \int xy \, dA$$

It does not have any physical significance but it is useful in determining the maximum and the minimum moments of inertia. The dimension and unit of product of inertia are same as that of moment of inertia, i.e., $[L]^4$ and m^4.

It should be noted that the first moment of an area about an axis could be *positive* or *negative* depending upon the sign of the perpendicular distance; and *zero* in the case of areas having an axis of symmetry. Whereas the second moments of area, I_{xx} and I_{yy}, are always *positive* as they involve the *square* of the perpendicular distance. Unlike moment of inertia, the product of inertia can be positive or negative depending upon the position of the area with respect to the reference axes. If an area has an axis of symmetry such as a rectangle, circle, semicircle, etc. the product of inertia is zero.

Transfer Theorem for product of Inertia

It is mathematically given as

$$I_{xy} = \bar{I}_{xy} + A\bar{x}\,\bar{y}$$

Moments of Inertia about Inclined Axes

The moments of inertia about inclined axes are given as

$$I_{uu} = \left(\frac{I_{xx} + I_{yy}}{2}\right) + \left(\frac{I_{xx} - I_{yy}}{2}\right)\cos 2\theta - I_{xy}\sin 2\theta$$

$$I_{vv} = \left(\frac{I_{xx} + I_{yy}}{2}\right) - \left(\frac{I_{xx} - I_{yy}}{2}\right)\cos 2\theta + I_{xy}\sin 2\theta$$

and

$$I_{uv} = \left(\frac{I_{xx} - I_{yy}}{2}\right)\sin 2\theta + I_{xy}\cos 2\theta$$

Principal Axes and Principal Moments of Inertia

The condition for maximum or minimum moment of inertia is

$$\frac{dI_{uu}}{d\theta} = 0 \Rightarrow \tan 2\theta = \frac{-2I_{xy}}{I_{xx} - I_{yy}}$$

The axes about which moment of inertia is maximum or minimum are termed *principal axes* and the corresponding moments of inertia about these principal axes are termed *principal moments of inertia*. They can be mathematically expressed in compact form as

$$I_{max} = \left(\frac{I_{xx} + I_{yy}}{2}\right) + R \quad \text{and} \quad I_{min} = \left(\frac{I_{xx} + I_{yy}}{2}\right) - R$$

where,

$$R = \sqrt{\left(\frac{I_{xx} - I_{yy}}{2}\right)^2 + (I_{xy})^2}$$

EXERCISES

Objective-type Questions

1. The moment of inertia of a triangle of base b and height h about its base is

(a) $\frac{bh^3}{36}$ (b) $\frac{bh^3}{24}$ (c) $\frac{bh^3}{12}$ (d) $\frac{bh^3}{3}$

2. The moment of inertia of a rectangle of base b and height h about its base is

(a) $\dfrac{bh^3}{36}$ (b) $\dfrac{bh^3}{24}$ (c) $\dfrac{bh^3}{12}$ (d) $\dfrac{bh^3}{3}$

3. The polar moment of inertia of a circular area of diameter D is

(a) $\dfrac{\pi D^4}{64}$ (b) $\dfrac{\pi D^4}{32}$ (c) $\dfrac{\pi D^4}{16}$ (d) $\dfrac{\pi D^4}{8}$

4. The moment of inertia of a semicircular area of radius R about its diametric axis is

(a) $\dfrac{\pi R^4}{16}$ (b) $\dfrac{\pi R^4}{8}$ (c) $\dfrac{\pi R^4}{4}$ (d) $\dfrac{\pi R^4}{2}$

5. The moment of inertia of a semicircular area of radius R about a centroidal axis parallel to its diameter is:

(a) $0.11\ R^4$ (b) $0.055\ R^4$ (c) $\dfrac{\pi R^4}{4}$ (d) $\dfrac{\pi R^4}{8}$

6. The moment of inertia of a quarter-circular area of radius R about a centroidal axis parallel to its diameter is

(a) $0.11\ R^4$ (b) $0.055\ R^4$ (c) $\dfrac{\pi R^4}{4}$ (d) $\dfrac{\pi R^4}{8}$

7. The product of inertia of a right-angled triangle of base b and height h about its centroidal axes is

(a) $\dfrac{b^2h^2}{36}$ (b) $-\dfrac{b^2h^2}{36}$ (c) $\dfrac{b^2h^2}{72}$ (d) $-\dfrac{b^2h^2}{72}$

8. If the product of inertia of a right-angled triangle of base b and height h about its centroidal axes is $-b^2h^2/72$ then its product of inertia about horizontal and vertical axes passing through its sides are

(a) $\dfrac{b^2h^2}{24}$ (b) $\dfrac{b^2h^2}{36}$ (c) $\dfrac{b^2h^2}{48}$ (d) $\dfrac{b^2h^2}{72}$

9. State which of the following statement is true concerning a plane lamina:
   (a) The moment of inertia can be positive or negative.
   (b) The polar moment of inertia can be positive or negative.
   (c) The product of inertia can be positive or negative.
   (d) The maximum principal moment of inertia can be positive or negative.
10. If an area has an axis of symmetry then
   (a) its centroid lies on that axis (b) its first moment about that axis is zero
   (c) its product of inertia is zero (d) all of these
11. State which of the following statements is true:
   (a) The moment of inertia about the centroidal axis is the maximum.
   (b) The moment of inertia decreases as the axis is moved away from the centroid.

(c) Parallel axis theorem can be applied only when one of the axes is a centroidal axis.
(d) Parallel axis theorem can be applied between any two parallel axes.

12. Which of the following physical quantities can be positive or negative?
(a) I_{xx} (b) I_{yy} (c) I_{xy} (d) I_p

Answers

1. (c)	2. (d)	3. (b).	4. (b)	5. (a)	6. (b)	7. (d)	8. (a)
9. (c)	10. (d)	11. (c)	12. (c)				

Short-answer Questions

1. Differentiate between first moment and second moment of an area.
2. Explain why the first moment of an area can be positive or negative while the second moment of area is always positive.
3. Moment of inertia gives a measure of resistance to bending in the case of sections or plane areas. Discuss.
4. Explain the term radius of gyration. What is its application?
5. State and prove the parallel axis theorem or transfer formula.
6. State and prove the perpendicular axis theorem or polar moment of inertia.
7. Derive an expression for the polar moment of inertia of a circular area having radius r.
8. Determine the second moment of inertia of area of a triangle about its base.
9. Derive an expression to determine the moment of inertia of a semicircular area about its diametric base.
10. Differentiate between polar moment of inertia and product of inertia.
11. Explain product of inertia and transfer formula for product of inertia.
12. Explain why moment of inertia is always positive while product of inertia can be positive or negative.
13. Product of inertia for sections with an axis of symmetry is zero. Explain
14. Define principal axes and principal moments of inertia.

Numerical Problems

9.1 Determine the moments of inertia about rectangular axes of a circular sector of radius r and included angle 2α and symmetric about the X-axis.

Ans. $\frac{r^4}{4}(\alpha - \sin\alpha\cos\alpha)$, $\frac{r^4}{4}(\alpha + \sin\alpha\cos\alpha)$

9.2 Determine the moment of inertia of a square of side a about a diagonal assuming the square to be made up of two triangles. Refer Fig. E.9.2.

Ans. $a^4/12$

9.3 A rectangular beam has to be cut out from a log of wood of radius r. Determine the breadth b and height h of the beam that can be cut out such that the moment of inertia of the beam is maximum. Also, determine the maximum moment of inertia. Refer Fig. E.9.3.

Ans. $b = r$, $h = \sqrt{3}\,r$, $I_{max} = \frac{\sqrt{3}}{4}r^4$

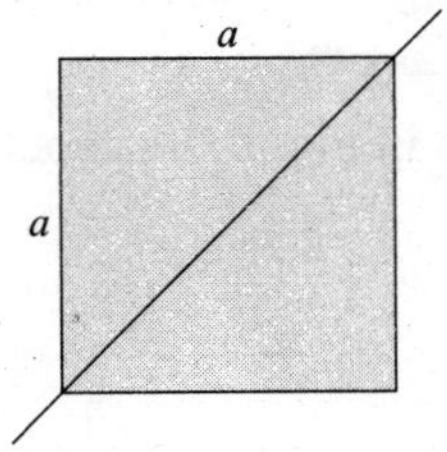

Fig. E.9.2

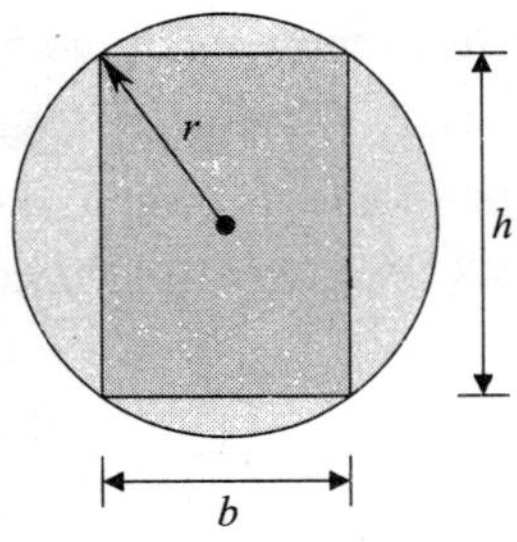

Fig. E.9.3

9.4 Determine the distance d at which the two 2 cm × 8 cm rectangles should be spaced such that $\bar{I}_{xx} = \bar{I}_{yy}$. Refer Fig. E.9.4.

Ans. 2.47 cm

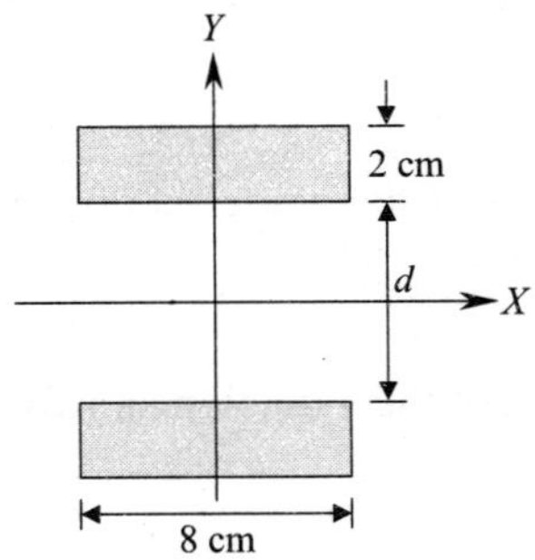

Fig. E.9.4

9.5 Determine moment of inertia of the semicircular area of 5 cm radius shown in Fig. E.9.5 about axis AA.

Ans. 1540.6 cm^4

9.6 Find moment of inertia of the quarter-circular area (i) about the X-axis, and (ii) about the tangent X'-axis. Refer Fig. E.9.6.

Ans. (i) 0.196 R^4, (ii) 0.315 R^4

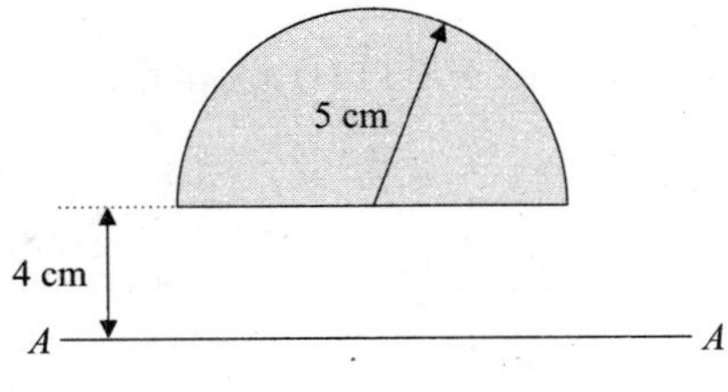

Fig. E.9.5

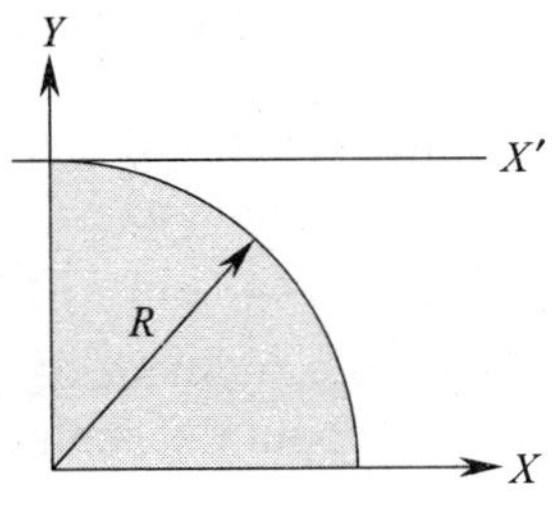

Fig. E.9.6

9.7 Determine moments of inertia of the *I*-section shown in Fig. E.9.7 about centroidal axes.

Ans. 6424.2 cm^4, 121 cm^4

9.8. Find the moments of inertia of the channel section shown in Fig. E.9.8 about centroidal axes.

Ans. 24 275.3 cm^4 and 6831.7 cm^4

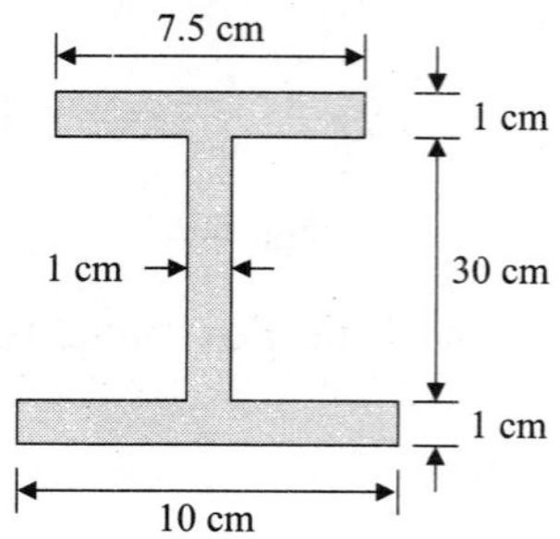

Fig. E.9.7

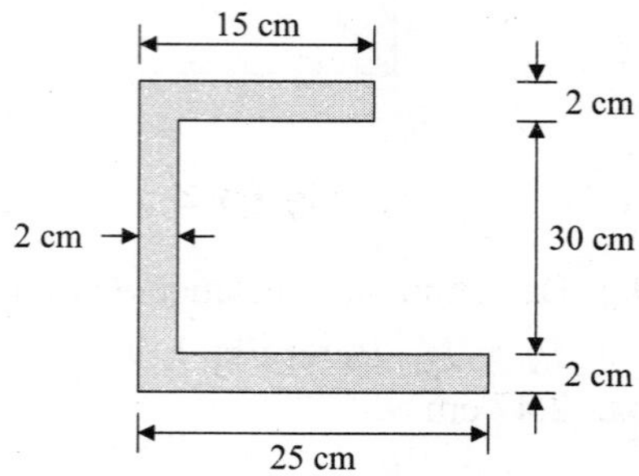

Fig. E.9.8

9.9 Determine moments of inertia of the angle section shown in Fig. E.9.9 about the centroidal axes.

Ans. 1008.7 cm^4, 356.2 cm^4

9.10 Determine moments of inertia of the *T*-section shown in Fig. E.9.10 about centroidal axes.

Ans. 533.33 cm^4, 173.33 cm^4

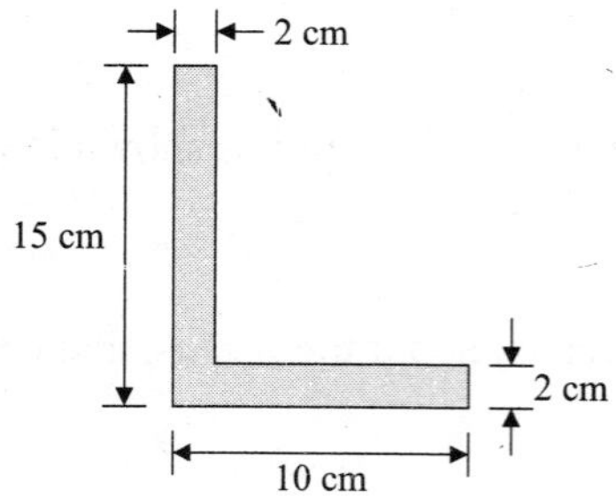

Fig. E.9.9

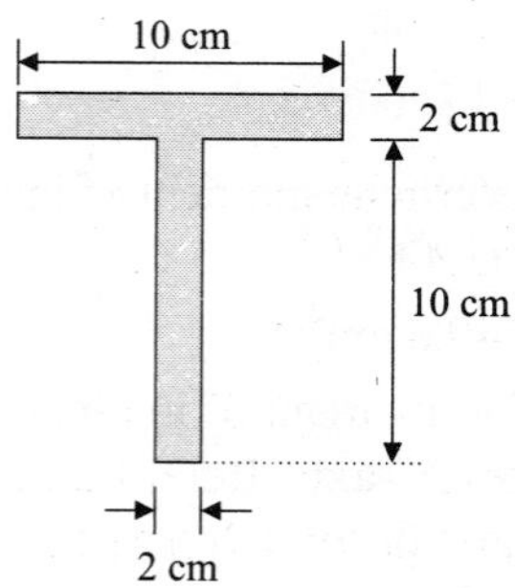

Fig. E.9.10

9.11 Determine moments of inertia of the *Z*-section shown in Fig. E.9.11 (i) about the *X*–*Y* axes, and (ii) about centroidal axes.

Ans. (i) 5438 cm^4, 7448 cm^4 (ii) 2085.3 cm^4, 1357.8 cm^4

9.12 Determine the moment of inertia of the box section formed by four members as shown in Fig. E.9.12 about the centroidal axes.

Ans. 2998.7 cm^4

9.13 An ISMB 600 beam is reinforced by a plate of dimensions 175 mm × 6 mm welded at its top as shown in Fig. E.9.13. Determine the moment of inertia of the composite section about the

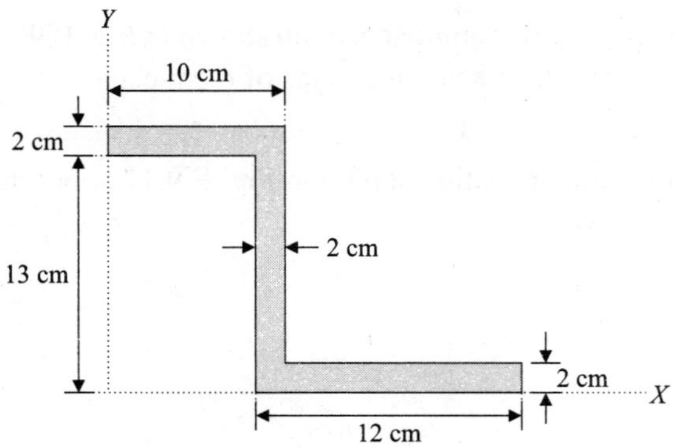

Fig. E.9.11

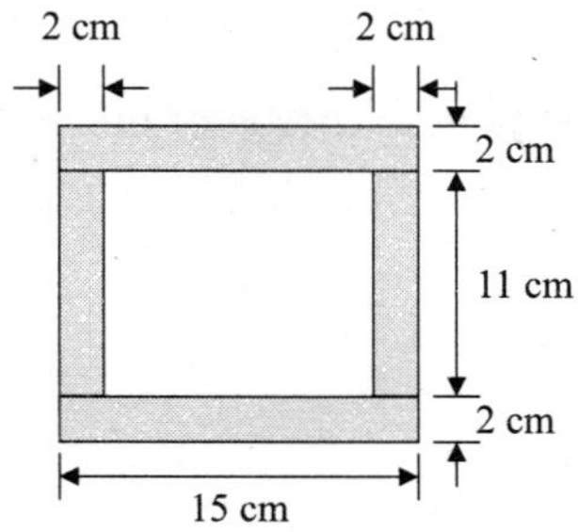

Fig. E.9.12

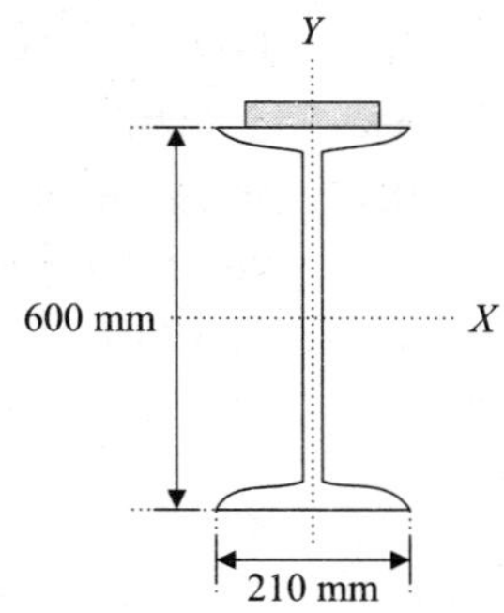

Fig. E.9.13

centriodal X-axis. The specifications of ISMB 600 beam section are beam depth = 600 mm, flange width = 210 mm, area of cross section = 156.21 cm^2, $\bar{I}_{xx}$ = 91 813 cm^4.

Ans. 100 846.1 cm^4

9.14 An ISMB 600 beam is reinforced by an ISMC 400 channel section welded at its top as shown in Fig. E.9.14. Determine the moment of inertia about the centroidal X-axis. The specifications of ISMB 600 beam section are beam depth = 600 mm, flange width = 210 mm, area of cross section = 156.21 cm^2, $\bar{I}_{xx}$ = 91 813 cm^4, $\bar{I}_{yy}$ = 2651 cm^4 and that of ISMC 400 are web depth = 400 mm, flange width = 100 mm, area of cross section = 62.93 cm^2, web thickness = 15.3 mm, centroid = 2.42 cm, $\bar{I}_{xx}$ = 15 082.8 cm^4, $\bar{I}_{yy}$ = 504.8 cm^4.

Ans. 130 330.6 cm^4

9.15 A plate girder is formed by welding a web plate and two identical flange plates together with four ISA 65 mm × 65 mm × 6 mm angles as shown in Fig. E.9.15. The specifications of the angle section are area = 7.44 cm^2, centroid = 1.81 cm, $\bar{I}_{xx}$ = $\bar{I}_{yy}$ = 29.1 cm^4. Determine the moment of inertia about horizontal centroidal axis.

Ans. 62 109.6 cm^4

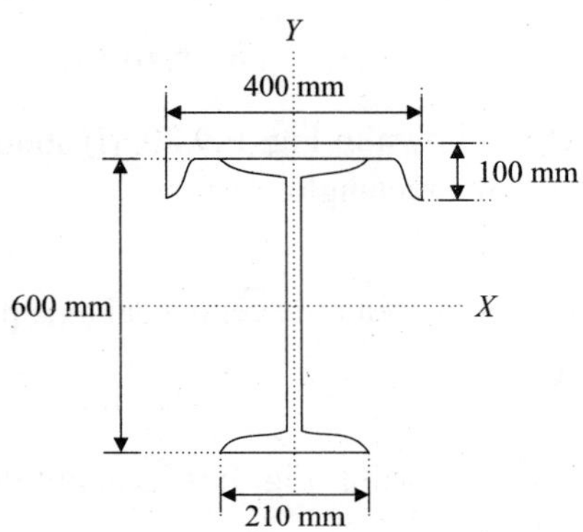

Fig. E.9.14

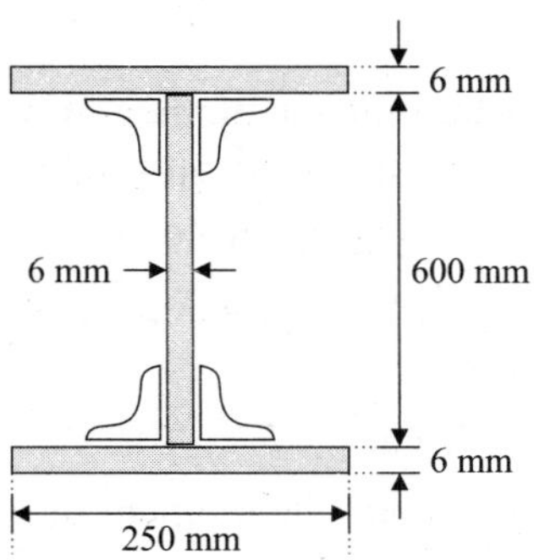

Fig. E.9.15

9.16 Determine the moment of inertia of the hollow isosceles triangular section shown in Fig. E.9.16 about the base *AB*. The inner section has a base length of 8 cm and sides of 6 cm each.

Ans. 71.8 cm^4

9.17 Determine the moment of inertia of the hollow circular section shown in Fig. E.9.17 about the centroidal axes.

Ans. 6836.11 cm^4

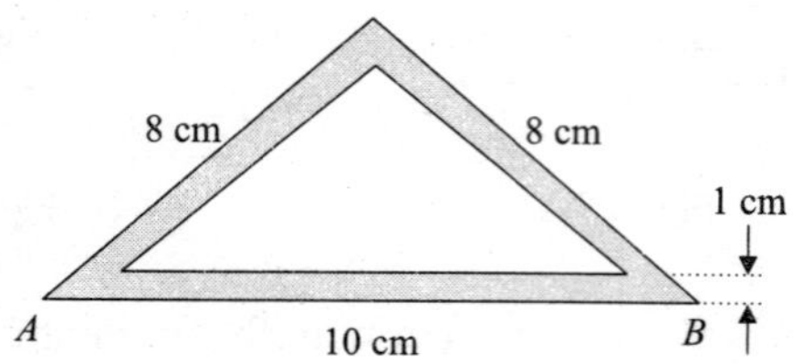

Fig. E.9.16

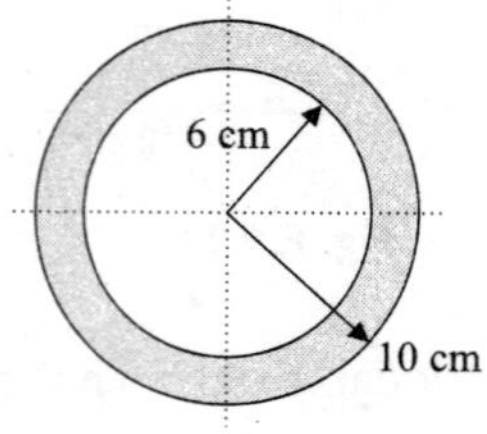

Fig. E.9.17

9.18 Determine the moment of inertia of the diamond section shown in Fig. E.9.18 about the horizontal centroidal axis.

Ans. 248.7 cm^4

9.19 Determine the moment of inertia of the shaded region shown in Fig. E.9.19 about the base.

Ans. 101.33 cm^4

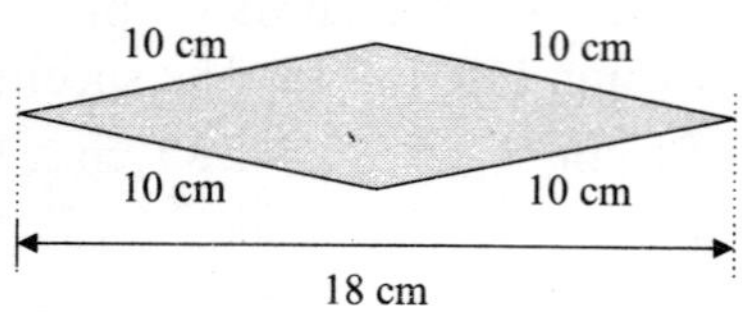

Fig. E.9.18

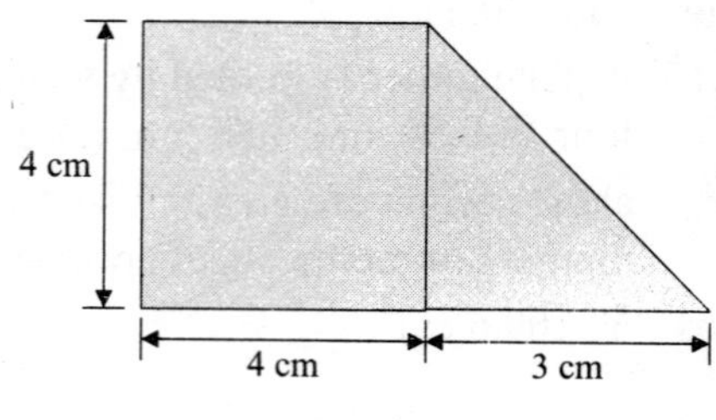

Fig. E.9.19

9.20 Determine the moment of inertia of the cut section shown in Fig. E.9.20, (i) about the bottom side of the rectangle, and (ii) about the top side of the rectangle.

Ans. (i) 1968.2 cm^4 (ii) 914.5 cm^4

9.21 In the above problem, determine the moment of inertia about the centroidal axis parallel to the base.

Ans. 234.2 cm^4

9.22 Determine the moment of inertia of the cut section shown in Fig. E.9.22 about the centroidal axes.

Ans. 165.3 cm^4, 310.9 cm^4

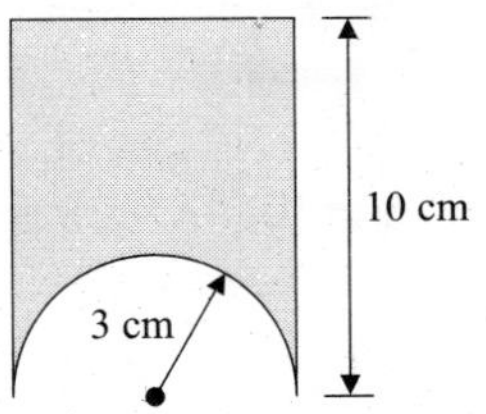

Fig. E.9.20, E.9.21

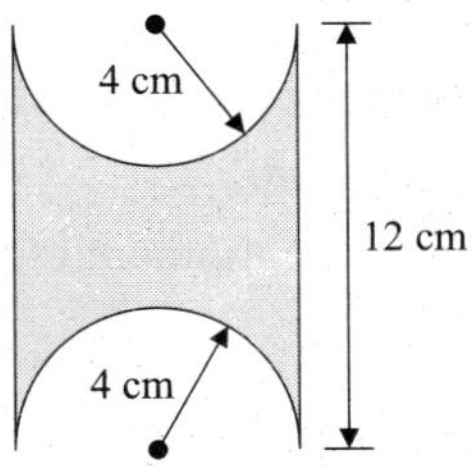

Fig. E.9.22

9.23 Determine the moment of inertia of the semicircular area of 5 cm radius from which a square of 2 cm side has been cut as shown in Fig. E.9.23, (i) about the base *AB*, and (ii) about the horizontal centroidal axis.

Ans. (i) 240.1 cm^4, (ii) 61.5 cm^4

9.24. Determine the moment of inertia of the semicircular area from which a triangle has been cut as shown in Fig. E.9.24, (i) about the base *AB*, and (ii) about the horizontal centroidal axis.

Ans. (i) 222.9 cm^4, (ii) 29.9 cm^4

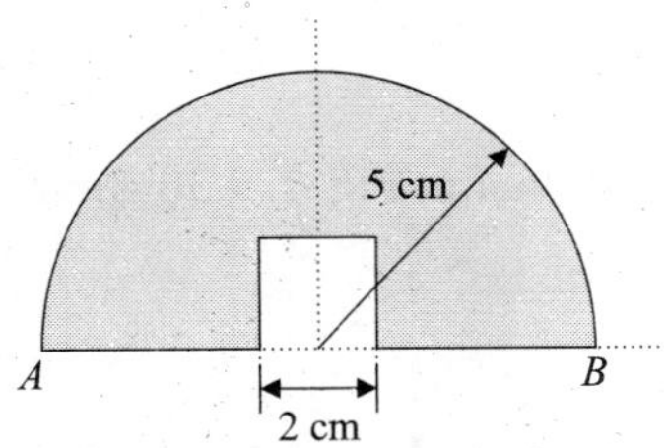

Fig. E.9.23

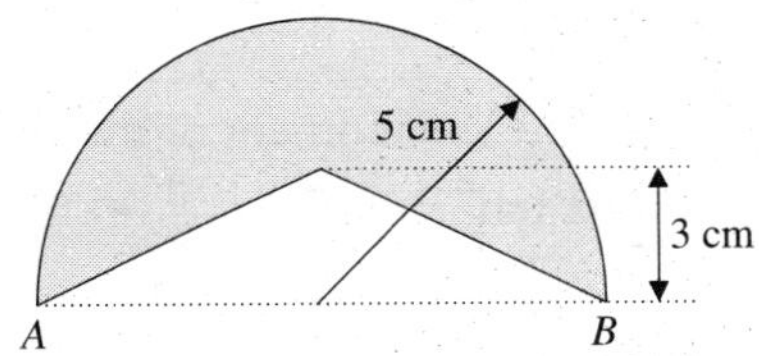

Fig. E.9.24

9.25 Determine the moment of inertia of a square of 10 cm side from which a circle with a diameter of 5 cm has been removed from its centre about (i) its centroidal axis parallel to a side, and (ii) about one of its sides. Refer Fig. E.9.25.

Ans. 802.7 cm^4, 2811.8 cm^4

9.26 Determine the moment of inertia of the triangular area from which a triangle has been cut as shown in Fig. E.9.26, (i) about the base *AB*, and (ii) about the horizontal centroidal axis.

Ans. (i) 180 cm^4, (ii) 26.3 cm^4

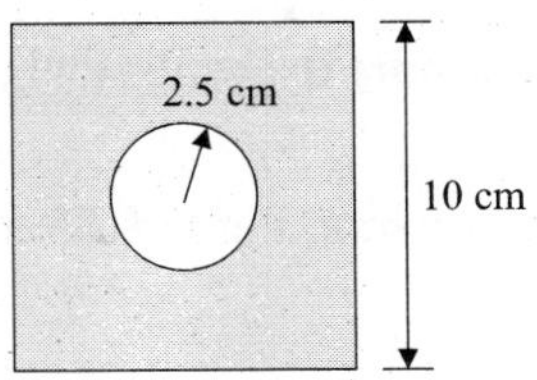

Fig. E.9.25

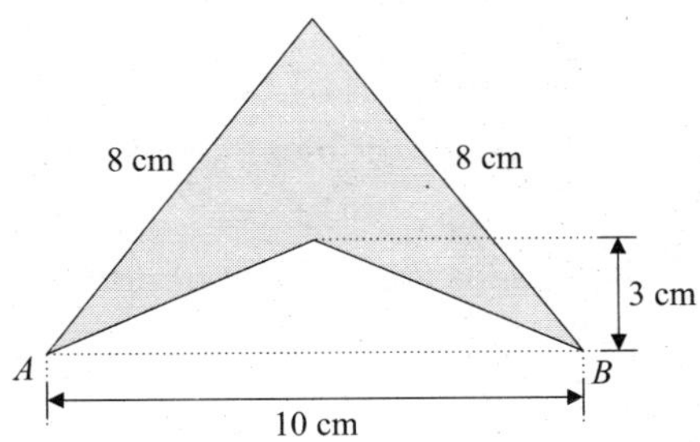

Fig. E.9.26

9.27 Determine the moment of inertia of the triangle shown in Fig. E.9.27, (i) about the base *AB*, and (ii) about the centroidal axis parallel to *AB*.

Ans. 180 cm^4, 60 cm^4

9.28 Determine the moment of inertia of the shaded area shown in Fig. E.9.28 about the axis *AB*.

Ans. 709.33 cm^4

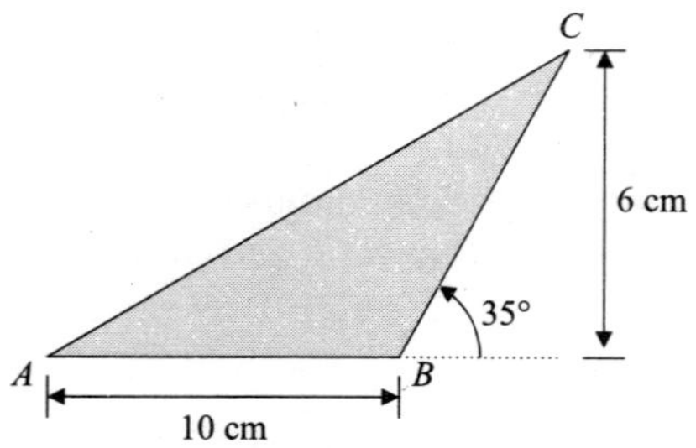

Fig. E.9.27

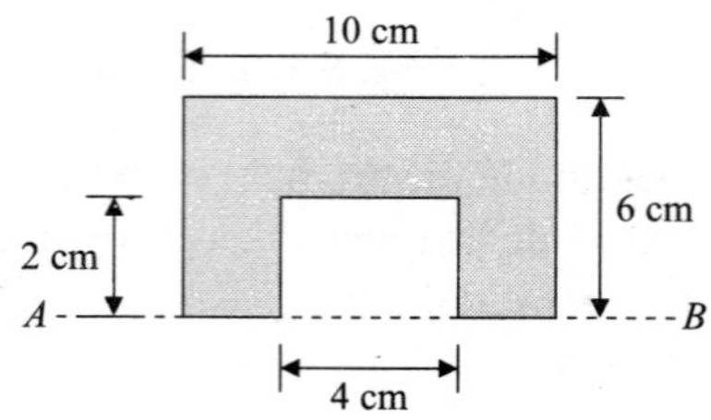

Fig. E.9.28

9.29 Find the moment of inertia of the section shown in Fig. E.9.29 about the *X*-axis. All dimensions are in cm.

Ans. 620 162.5 cm^4

9.30 Determine the moment of inertia of the shaded area shown in Fig. E.9.30 about its base *A–A*.

Ans. 352 691 cm^4

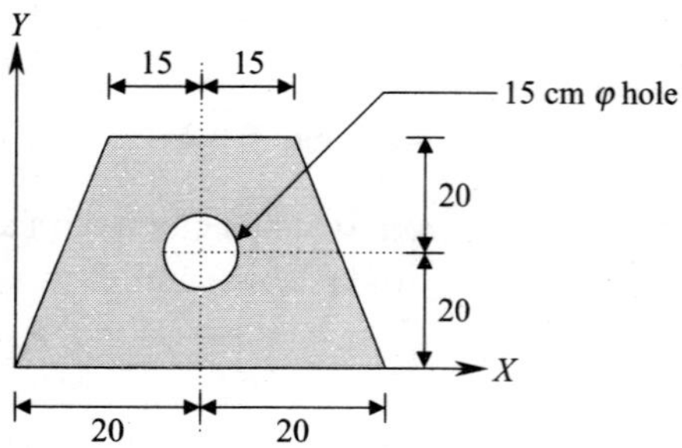

Fig. E.9.29

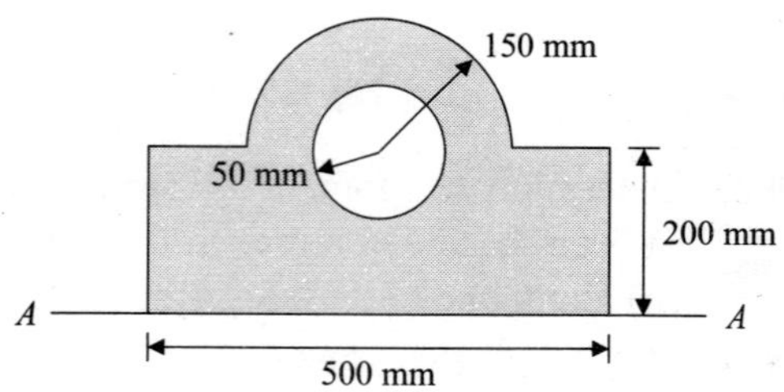

Fig. E.9.30

9.31 Determine moments of inertia of the shaded region shown in Fig. E.9.31 about the *X* and *Y* axes.

Ans. 3015.9 cm^4, 2211.7 cm^4

9.32 In the above problem, determine the moment of inertia about the centroidal axes.

Ans. 3015.9 cm^4, 1943.6 cm^4

9.33 Determine the moment of inertia of the shaded region shown in Fig. E.9.33 about the diameter *AB*.

Ans. 213.6 cm^4

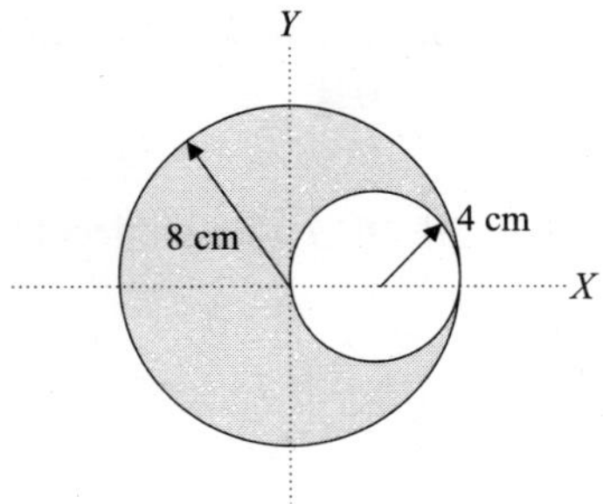

Fig. E.9.31, E.9.32

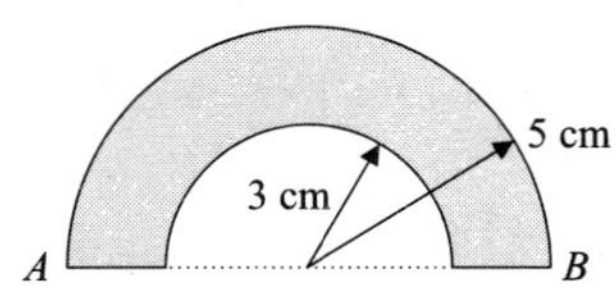

Fig. E.9.33

9.34 Find the moment of inertia and radius of gyration of the shaded area shown in Fig. E.9.34 about the horizontal centroidal axis.

Ans. $\bar{I}_{xx} = 330\,833.33\ \text{cm}^4$; $k_x = 14.4$ cm

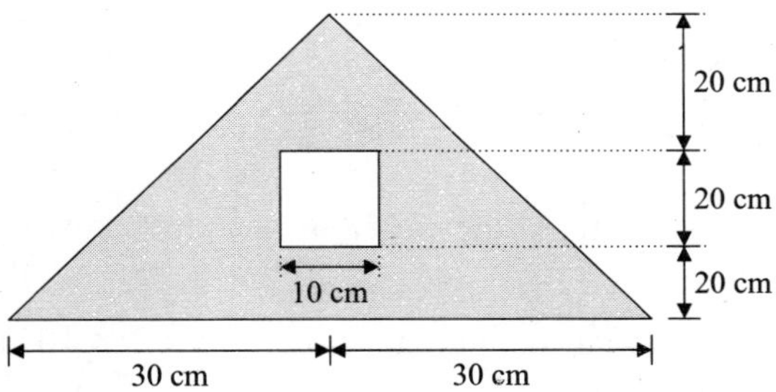

Fig. E.9.34

9.35 Determine the polar moment of inertia of an ellipse of semi-major and semi-minor axes of lengths respectively *a* and *b* units.

Ans. $\dfrac{\pi ab}{4}(a^2 + b^2)$

9.36 Find the polar moment of inertia of the equal flange *T*-section shown in Fig. E.9.36 about the centroidal axes. The thickness at all sections is 2 cm.

Ans. 486.2 cm^4

9.37 Determine the moment of inertia of a regular hexagonal section of side *a* shown in Fig. E.9.37 about the horizontal centroidal axis.

Ans. $\dfrac{5\sqrt{3}}{16}a^4$

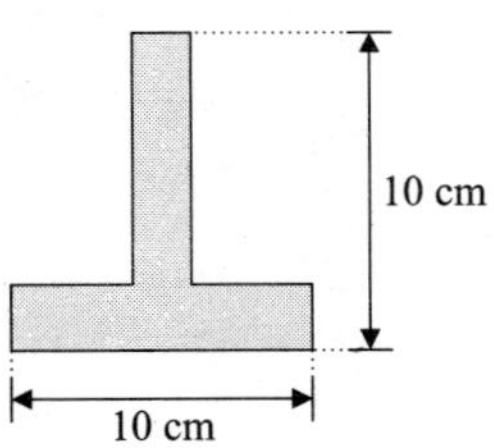

Fig. E.9.36

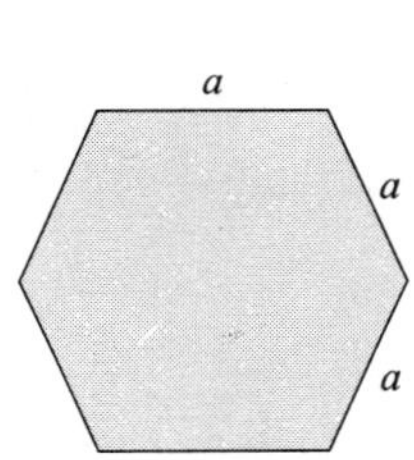

Fig. E.9.37

9.38 Determine the product of inertia of the shaded area shown in Fig. E.9.38 with respect to the given X and Y axes.

Ans. $r^4/12$

9.39 Determine the product of inertia of the area shown in Fig. E.9.39 with respect to the centroidal axes.

Ans. -1897.5 cm^4

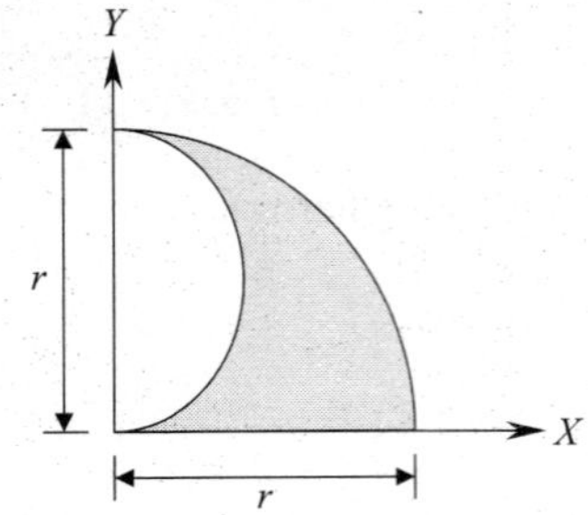

Fig. E.9.38

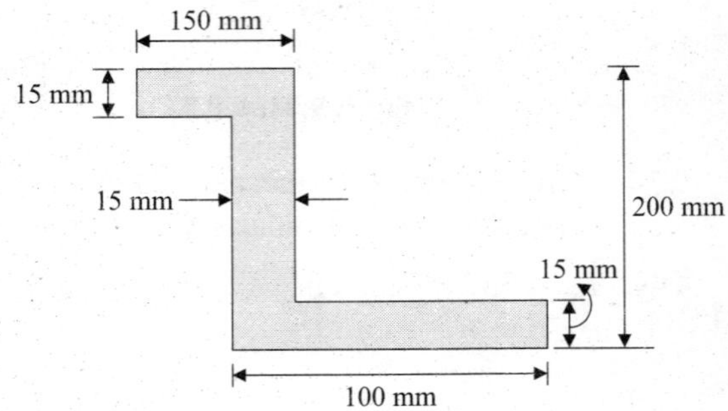

Fig. E.9.39

9.40 Determine the product of inertia of the parabolic spandrel shown in Fig. E.9.40.

Ans. $b^2h^2/12$

9.41 Find moments of inertia about inclined axes, if $I_{xx} = 4.2 \times 10^4\text{ mm}^4$, $I_{yy} = 2.3 \times 10^4\text{ mm}^4$, $I_{xy} = 1.5 \times 10^4\text{ mm}^4$ and $\theta = 40°$.

Ans. $1.94 \times 10^4\text{ mm}^4$ and $4.56 \times 10^4\text{ mm}^4$

9.42 Find the principal moments of inertia, if $I_{xx} = 3.3 \times 10^4\text{ mm}^4$, $I_{yy} = 2.0 \times 10^4\text{ mm}^4$ and $I_{xy} = 1.5 \times 10^4\text{ mm}^4$. Also, find the inclination of the principal axes.

Ans. $I_{max} = 4.28 \times 10^4\text{ mm}^4$ and $I_{min} = 1.02 \times 10^4\text{ mm}^4$; $\theta_1 = 33.3°$ and $\theta_2 = 123.3°$

9.43 Show that the moment of inertia about any rectangular coordinate system, the moment of inertia remains the same irrespective of the angle α. Refer Fig. E.9.43.

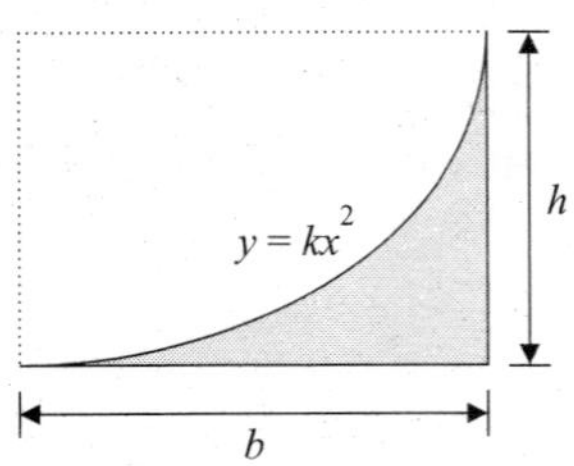

Fig. E.9.40

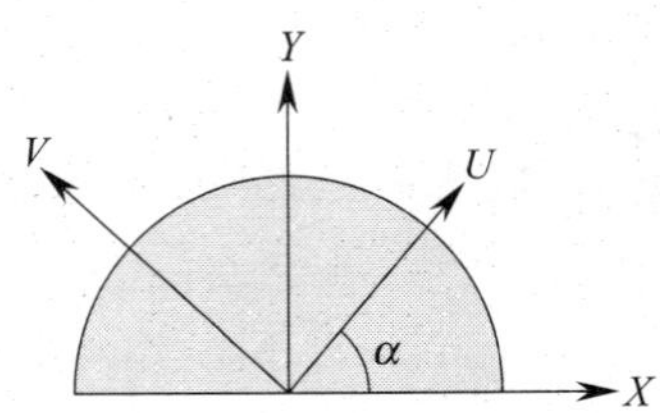

Fig. E.9.43

10 Mass Moment of Inertia

10.1 INTRODUCTION

In the previous chapter, we discussed the second moment of an area or moment of inertia, which finds application in the study of design of structural members. In this chapter, we will extend the same principles developed in the previous chapter for moment of inertia of an *area* to determine moment of inertia of a *solid body*, which we term as **mass moment of inertia**. This finds application in the study of dynamics of rigid bodies as it gives a measure of resistance to angular motion about an axis.

In the following section, we will define mass moment of inertia of a solid body and parallel axis theorem for mass moment of inertia in Section 10.4. In Section 10.5, we will discuss mass moments of inertia of thin plates and their relationship with the area moments of inertia. The results discussed in this section will be utilized in Section 10.6 to determine the mass moments of inertia of solid bodies.

10.2 MASS MOMENT OF INERTIA

Consider a body of mass M. If we take an infinitesimally small element of mass dm then its mass moment of inertia about any axis is defined as the product of the mass ***dm*** and the **square of the perpendicular distance** from the axis. Hence, its mass moment of inertia about the Z-axis is

$$dI_{zz} = r^2 dm \quad (10.1)$$

Therefore, the mass moment of inertia of the entire body about the Z-axis is obtained by integrating the above expression:

$$I_{zz} = \int r^2 \, dm \quad (10.2)$$

From the Fig. 10.1, we see that $r^2 = x^2 + y^2$; hence,

$$I_{zz} = \int (x^2 + y^2) \, dm \quad (10.3)$$

Similarly, it can be shown that,

$$I_{xx} = \int (y^2 + z^2) dm \quad (10.4)$$

and

$$I_{yy} = \int (x^2 + z^2) \, dm \quad (10.5)$$

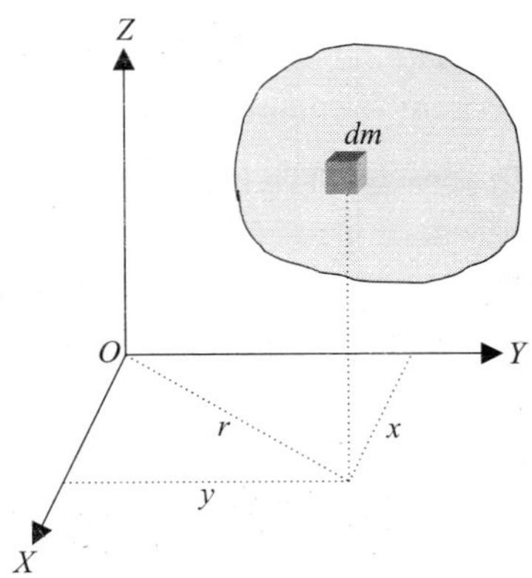

Fig. 10.1 Mass moment of inertia

The mass moment of inertia of a solid body gives a measure of resistance to rotation about an axis. For instance, consider the bodies shown in Fig. 10.2; the resistance to rotation about an axis *A–A* varies depending upon the shape of the body or the distribution of mass with respect to the axis. This property is very important in the study of rigid body dynamics.

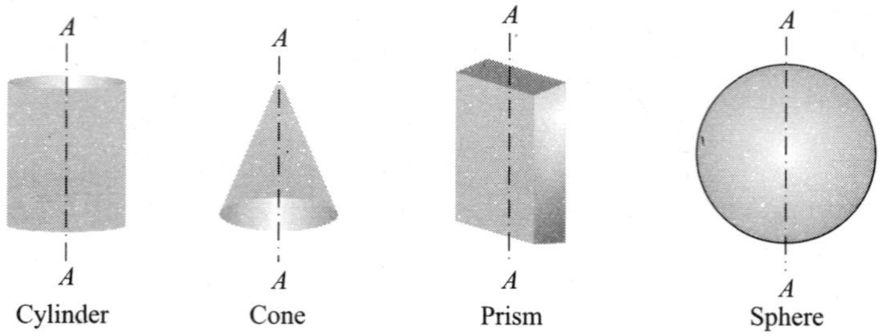

Fig. 10.2

The same symbol *I* is used to denote the mass moment of inertia also; but where confusion arises between area moment of inertia and mass moment of inertia, they are differentiated by the subscripts I_{area} and I_{mass}. As mass moment of inertia is obtained by multiplying the mass of the body with the square of the perpendicular distance from the axis, its dimension and unit are respectively [M] [L]^2 and kg.m^2.

10.3 RADIUS OF GYRATION

The radius of gyration is defined as the distance from the axis of inertia to the point at which the entire mass *M* of the body may be assumed to be concentrated and still have the same moment of inertia. Mathematically, it is given as

$$k = \sqrt{\frac{I}{M}} \tag{10.6}$$

10.4 TRANSFER FORMULA (OR) PARALLEL AXIS THEOREM

Just as we derived the parallel axis theorem for area moment of inertia to determine the moment of inertia about a non-centroidal axis parallel to the centroidal axis, we will derive the parallel axis theorem for mass moment of inertia in this section.

Theorem The mass moment of inertia of a body about an axis at a distance *d* and parallel to the centroidal axis is equal to the sum of the moment of inertia about the centroidal axis and product of mass and square of the perpendicular distance between the parallel axes.

$$I_{zz} = \bar{I}_{zz} + Md^2 \tag{10.7}$$

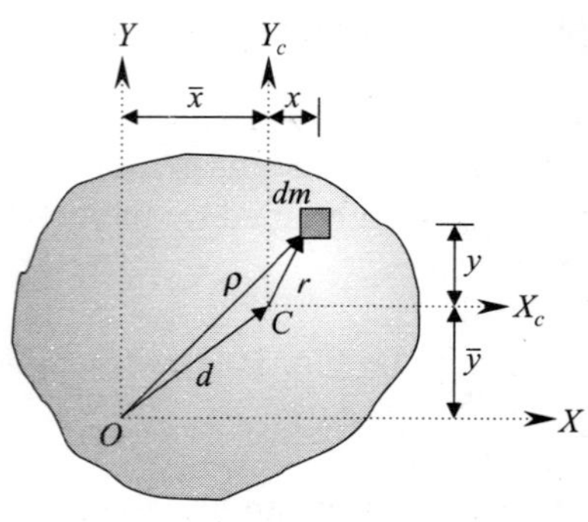

Fig. 10.3 Parallel axis theorem

Proof Let us consider two sets of reference axes, one centroidal (X_c, Y_c) and the other non-centroidal (*X*, *Y*) axes. Figure 10.3 shows the cross section of the mass such that the *Z*-axis is perpendicular to the plane of the paper. If we take a small element of mass *dm* then its coordinates with respect to the centroidal axes are (*x*, *y*) and with respect to the non-centroidal axes are $(x + \bar{x}, y + \bar{y})$, where $\bar{x}$ and $\bar{y}$ are coordinates of the centroid (*C*).

From the Fig.10.3, we know that,

$$\rho^2 = (x + \bar{x})^2 + (y + \bar{y})^2$$
$$= x^2 + \bar{x}^2 + 2x\bar{x} + y^2 + \bar{y}^2 + 2y\bar{y} \quad (10.8)$$

On rearranging,

$$\rho^2 = (x^2 + y^2) + (\bar{x}^2 + \bar{y}^2) + 2x\bar{x} + 2y\bar{y} \quad (10.9)$$

Since $(x^2 + y^2) = r^2$ and $(\bar{x}^2 + \bar{y}^2) = d^2$, the above expression reduces to

$$\rho^2 = r^2 + d^2 + 2x\bar{x} + 2y\bar{y} \quad (10.10)$$

Hence, mass moment of inertia of the entire body about non-centroidal Z-axis is

$$I_{zz} = \int \rho^2\, dm = \int r^2\, dm + \int d^2\, dm + 2\bar{x}\int x dm + 2\bar{y}\int y\, dm \quad (10.11)$$

The first integral on the right-hand side is the moment of inertia about the centroidal Z-axis, $\bar{I}_{zz}$; the second integral is mass M times square of the distance between centroidal and required axes; and the third and fourth integrals vanish because they are the first moments of mass and they are zero for the centroidal axes. Therefore,

$$I_{zz} = \bar{I}_{zz} + Md^2 \quad (10.12)$$

10.5 MASS MOMENT OF INERTIA OF THIN PLATES

In this section, we will derive mass moments of inertia of thin plates and establish a relationship between mass and area moments of inertia. The results derived in this section will be used in the next section to determine the mass moments of inertia of solid bodies.

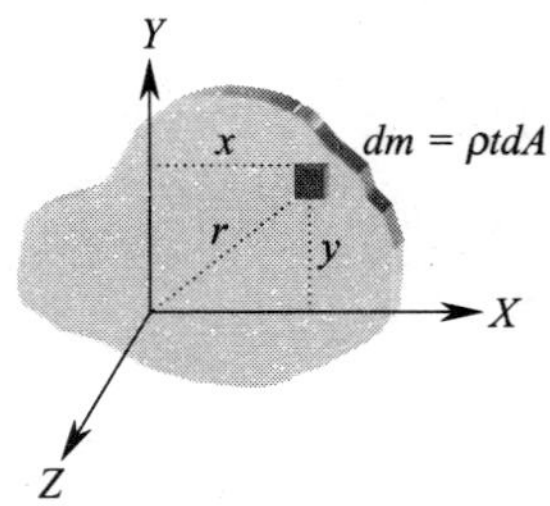

Fig. 10.4 Mass moment of inertia of thin plate

Consider a thin homogeneous plate with constant thickness t and mass density ρ. Assume the plate to be perpendicular to the Z-axis. If we take an infinitesimally small element of mass, $dm = \rho t dA$ then its moment of inertia with respect to the X-axis is $y^2\, dm$. On integration, we get the mass moment of inertia of the entire plate about the X-axis, i.e.,

$$I_{xx} = \int y^2\, dm = \rho t \int y^2 dA$$

As we know that the integral on the right-hand side is the area moment of inertia, the above expression can be written as

$$(I_{xx})_{\text{mass}} = \rho t (I_{xx})_{\text{area}} \quad (10.13)$$

Similarly, $$I_{yy} = \int x^2\, dm = \rho t \int x^2\, dA$$

Hence, $$(I_{yy})_{\text{mass}} = \rho t (I_{yy})_{\text{area}} \quad (10.14)$$

and $$I_{zz} = \int r^2\, dm = \rho t \int (x^2 + y^2)\, dA$$

Hence, $$(I_{zz})_{\text{mass}} = \rho t (I_{zz})_{\text{area}} \quad (10.15)$$

It can be seen from the above expressions that mass moments of inertia of thin plates can be determined if their area moments of inertia are known. These are explained in the following sections for various regular shapes.

10.5.1 Moment of Inertia of a Thin Rectangular Plate

Consider a thin rectangular plate of base b, height h, thickness t and mass density ρ. Then its mass is given as

$$M = \rho\, b\, h\, t \tag{10.16}$$

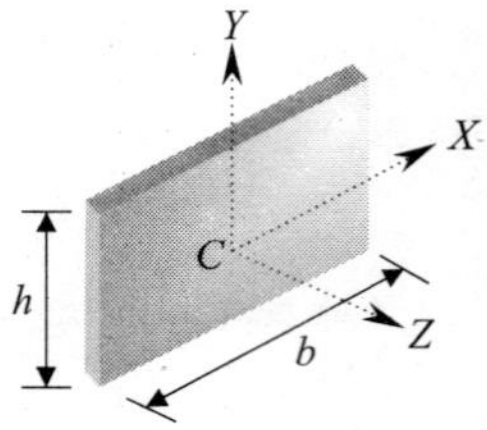

Fig. 10.5

For centroidal axes, we know that the area moments of inertia of a rectangle are

$$\bar{I}_{xx} = bh^3/12,\ \bar{I}_{yy} = hb^3/12,\ \bar{I}_{zz} = bh(b^2 + h^2)/12 \tag{10.17}$$

Therefore, mass moments of inertia are given as

$$(\bar{I}_{xx})_{\text{mass}} = \rho t(\bar{I}_{xx})_{\text{area}} = \rho t\, bh^3/12$$

Since $M = \rho\, b\, h\, t$, $$(\bar{I}_{xx})_{\text{mass}} = \frac{Mh^2}{12} \tag{10.18}$$

Similarly, $$(\bar{I}_{yy})_{\text{mass}} = \frac{Mb^2}{12} \tag{10.19}$$

and $$(\bar{I}_{zz})_{\text{mass}} = \frac{M}{12}[b^2 + h^2] \tag{10.20}$$

10.5.2 Moment of Inertia of a Thin Circular Disc

Consider a thin circular disc of radius R, thickness t and mass density ρ. Then its mass is given as

$$M = \rho \pi R^2 t \tag{10.21}$$

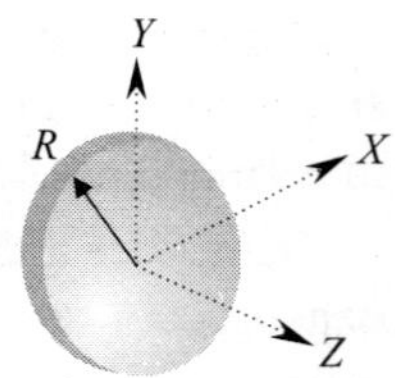

Fig. 10.6

For centroidal axes, we know that the area moments of inertia are

$$\bar{I}_{xx} = \bar{I}_{yy} = \bar{I}_{zz}/2 = \pi R^4/4 \tag{10.22}$$

Therefore, mass moments of inertia are given as

$$(\bar{I}_{zz})_{\text{mass}} = \rho t(\bar{I}_{zz})_{\text{area}} = \rho t\, \pi R^4/2$$

Since $M = \rho \pi R^2 t$, $$(\bar{I}_{zz})_{\text{mass}} = \frac{MR^2}{2} \tag{10.23}$$

Similarly, $$(\bar{I}_{xx})_{\text{mass}} = (\bar{I}_{yy})_{\text{mass}} = MR^2/4 \tag{10.24}$$

10.5.3 Moment of Inertia of a Thin Triangular Plate

Consider a thin isosceles triangular plate of base b, height h and mass density ρ. If t be its thickness then its mass is given as

$$M = \rho \frac{1}{2} bht \tag{10.25}$$

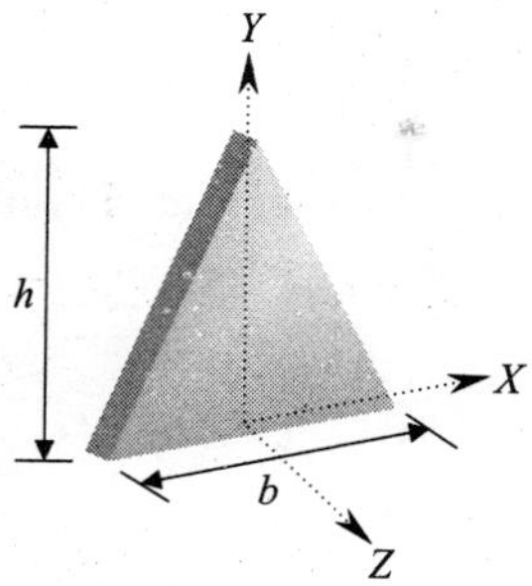

Fig. 10.7(a)

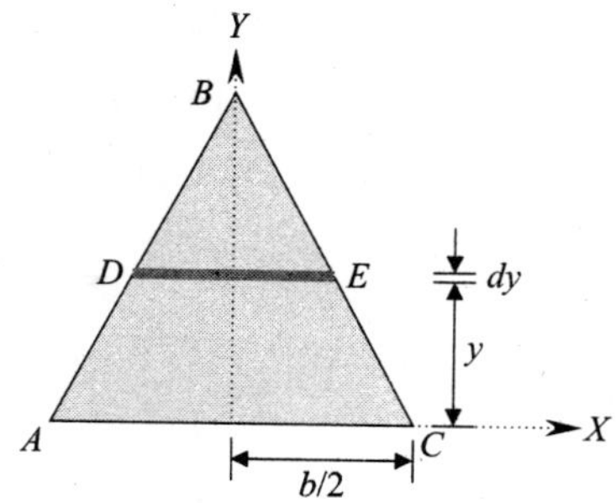

Fig. 10.7(b)

Consider a thin horizontal strip DE at a distance y from the base and of infinitesimal thickness dy. From the similar triangles ABC and DBE, we have

$$\frac{DE}{b} = \frac{h-y}{h} \tag{10.26}$$

Therefore, mass of the strip is

$$dm = \rho(DE)dy\,t = \rho t\frac{b}{h}(h-y)\,dy \tag{10.27}$$

Hence, its second moment about the base is

$$dI_{xx} = dm\,y^2 = \rho t\frac{b}{h}(h-y)y^2 dy \tag{10.28}$$

Therefore, mass moment of inertia of the triangular plate about the base is obtained as

$$I_{xx} = \int_0^h \rho t\frac{b}{h}(hy^2 - y^3)\,dy$$

$$= \rho t\frac{b}{h}\left[\frac{hy^3}{3} - \frac{y^4}{4}\right]_0^h$$

$$= \rho t\frac{b}{h}\left[\frac{h^4}{12}\right] = \frac{\rho tbh^3}{12} \tag{10.29}$$

From the Eq. 10.25, the above equation can be written as

$$I_{xx} = \frac{\rho bht}{2}\frac{h^2}{6} = \frac{Mh^2}{6} \tag{10.30}$$

Hence, mass moment of inertia of the triangular plate about the centroidal axis parallel to the base is given as

$$\bar{I}_{xx} = I_{xx} - M\left[\frac{h}{3}\right]^2$$

$$= \frac{Mh^2}{6} - \frac{Mh^2}{9}$$

$$\bar{I}_{xx} = \frac{Mh^2}{18} \tag{10.31}$$

Similarly, it can be shown that mass moment of inertia of the triangular plate about the centroidal *Y*-axis is

$$\bar{I}_{yy} = \frac{Mb^2}{24} \tag{10.32}$$

Alternative Method

We know that the area moments of inertia of isosceles triangle about the centroidal axes are

$$\bar{I}_{xx} = \frac{bh^3}{36} \text{ and } \bar{I}_{yy} = \frac{hb^3}{48} \quad \text{[refer Example 9.3 in Chapter 9]}$$

Therefore,

$$(\bar{I}_{xx})_{\text{mass}} = \rho t(\bar{I}_{xx})_{\text{area}}$$

$$= \rho t\frac{bh^3}{36}$$

$$= \frac{Mh^2}{18} \text{ [since } M = \rho\frac{1}{2}bht]$$

Similarly, $$(\bar{I}_{yy})_{\text{mass}} = \rho t(\bar{I}_{yy})_{\text{area}} = \frac{Mb^2}{24}$$

The following table summarizes the mass moments of inertia of thin plates:

Table 10.1 Mass moments of inertia of thin plates

Plate	*Shape*	$\bar{I}_{xx}$	$\bar{I}_{yy}$
Rectangular	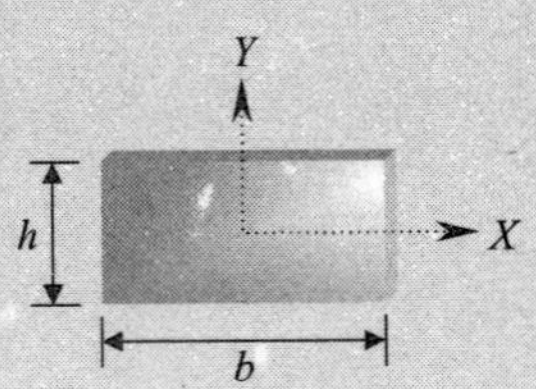	$\frac{Mh^2}{12}$	$\frac{Mb^2}{12}$
Circular	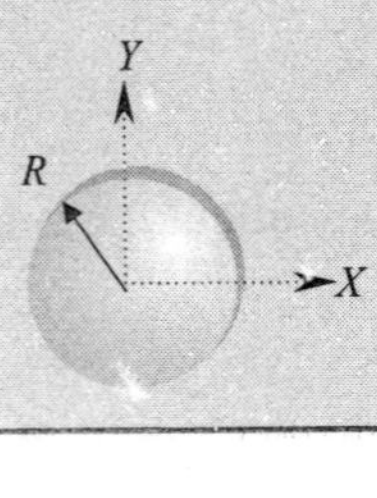	$\frac{MR^2}{4}$	$\frac{MR^2}{4}$

Contd.

Semicircular	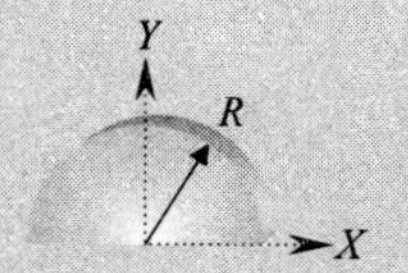	$\frac{MR^2}{4}$ (about base)	$\frac{MR^2}{4}$
Triangular	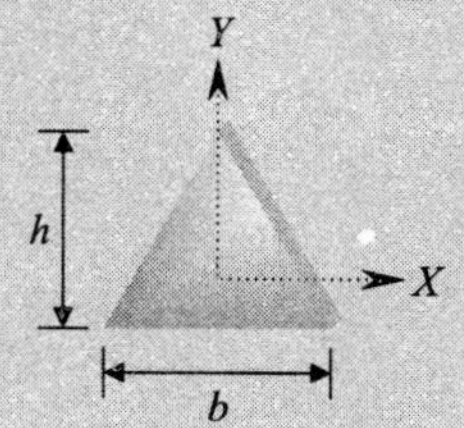	$\frac{Mh^2}{18}$	$\frac{Mb^2}{24}$

Example 10.1 Determine mass moments of inertia of a circular ring or hoop of mass M and radius R about centroidal axes.

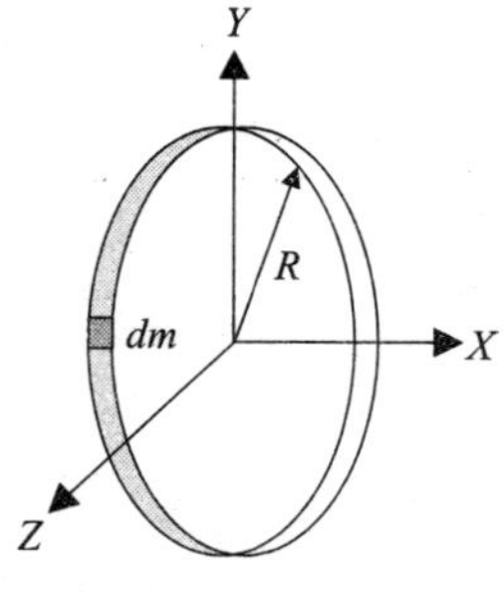

Fig. 10.8

Solution Let the ring be in X–Y plane as shown. If we take an infinitesimally small element of mass dm then its mass moment of inertia about the Z-axis is

$$d\bar{I}_{zz} = dm R^2$$

Hence, the mass moment of inertia of the entire ring about the Z-axis is

$$\bar{I}_{zz} = \int R^2 dm$$

Since every such infinitesimal element lies at the same radial distance R, we can write it as

$$\bar{I}_{zz} = R^2 \int dm$$
$$= MR^2$$

Hence, mass moment of inertia about the X and Y-axes are

$$\bar{I}_{xx} = \bar{I}_{yy} = \frac{\bar{I}_{zz}}{2} = \frac{MR^2}{2}$$

Example 10.2 Determine mass moment of inertia of an annular ring of mass M and, of external and internal radii R_1 and R_2 respectively.

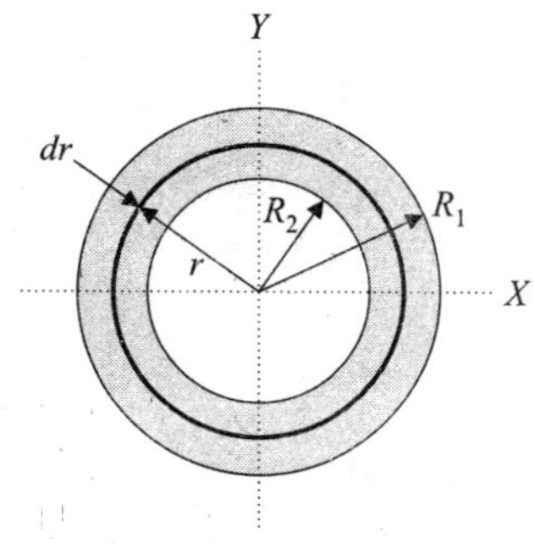

Fig. 10.9

Solution Consider an annular ring of external and internal radii R_1 and R_2 respectively. Let its mass per unit area be m. Consider a ring of radius r and infinitesimal thickness dr. Its area is given as

$$dA = 2\pi r dr$$

Therefore, its mass is

$$dm = m dA = m 2\pi r dr$$

Therefore, its mass moment of inertia about the perpendicular Z-axis is given as

$$d\bar{I}_{zz} = dmr^2 = 2\pi mr^3 dr$$

Hence, the mass moment of inertia of the annular ring is given as

$$\bar{I}_{zz} = \int_{R_2}^{R_1} d\bar{I}_{zz} = \int_{R_2}^{R_1} 2\pi mr^3 dr$$

$$= 2\pi m\left[\frac{r^4}{4}\right]_{R_2}^{R_1}$$

$$= \frac{\pi m}{2}\left[R_1^4 - R_2^4\right]$$

$$= \frac{\pi m}{2}\left[R_1^2 - R_2^2\right]\left[R_1^2 + R_2^2\right]$$

Since the entire area of the annular ring is $\pi\left[R_1^2 - R_2^2\right]$ and if M is the mass of the entire ring, then we can write

$$\bar{I}_{zz} = \frac{M}{2}\left[R_1^2 + R_2^2\right]$$

Example 10.3 Determine the mass moment of inertia of a thin semicircular plate of mass M and radius R about a centroidal axis parallel to the diameter.

Solution We know that the area moment of inertia of a semicircular area about its base is

$$(I_{\text{base}})_{\text{area}} = \frac{\pi R^4}{8}$$

Therefore, mass moment of inertia of a thin semicircular plate of mass M, radius R, thickness t and mass density ρ about its base is given as

$$(I_{\text{base}})_{\text{mass}} = \rho t\frac{\pi R^4}{8}$$

$$= \frac{MR^2}{4} \quad [\text{since } M = \rho t\, \pi R^2/2]$$

Hence, mass moment of inertia of the plate about a centroidal axis parallel to the diameter is obtained by the parallel axis theorem as

$$\bar{I} = I_{\text{base}} - M\left[\frac{4R}{3\pi}\right]^2$$

[It should be noted that as we move towards the centroid, the mass moment of inertia decreases.]

$$\bar{I} = \frac{MR^2}{4} - M\left[\frac{4R}{3\pi}\right]^2 = 0.07\, MR^2$$

Example 10.4 Find the moment of inertia of a thin rectangular plate bent as shown about A–A axis. Density of plate is 7850 kg/m^3 and thickness is 5 mm.

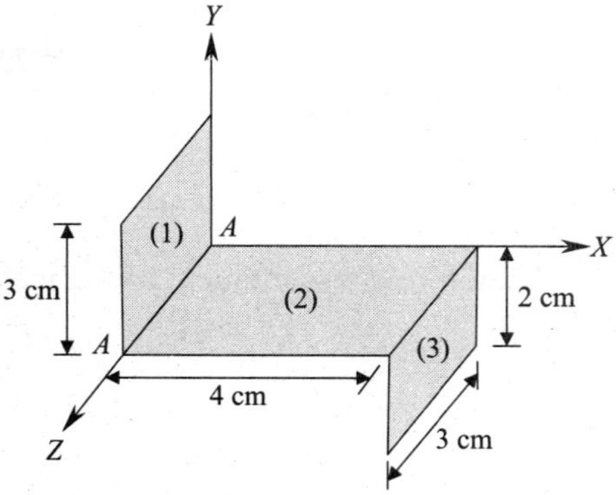

Fig. 10.10

Solution

Given data

density, $\rho = 7850$ kg/m^3

thickness, $t = 5$ mm $= 0.005$ m

Mass Calculations

We know that the mass of thin plate is $\rho A t$. Therefore,

$$M_1 = 7850 \times (0.03 \times 0.03) \times 0.005 = 0.0353 \text{ kg}$$

$$M_2 = 7850 \times (0.04 \times 0.03) \times 0.005 = 0.0471 \text{ kg}$$

$$M_3 = 7850 \times (0.03 \times 0.02) \times 0.005 = 0.0236 \text{ kg}$$

Calculation of mass moments of inertia

For Plate 1

The mass moment of inertia about its centroidal axis parallel to the Z-axis is

$$(\bar{I}_{zz})_1 = \frac{M_1 h_1^2}{12}$$

$$= \frac{0.0353 \times (0.03)^2}{12} = 2.65 \times 10^{-6} \text{ kg.m}^2$$

Therefore, mass moment of inertia about the AA axis is obtained as

$$(I_{AA})_1 = (\bar{I}_{zz})_1 + M_1(d_1)^2$$

$$= 2.65 \times 10^{-6} + [0.0353 \times (0.015)^2] = 1.06 \times 10^{-5} \text{ kg.m}^2$$

Similarly, for the other plates, the mass moments of inertia are determined as follows:

For Plate 2

$$(\bar{I}_{zz})_2 = \frac{M_2 h_2^2}{12}$$

$$= \frac{0.0471 \times (0.04)^2}{12} = 6.28 \times 10^{-6} \text{ kg.m}^2$$

$$(I_{AA})_2 = (\bar{I}_{zz})_2 + M_2(d_2)^2$$

$$= 6.28 \times 10^{-6} + [0.0471 \times (0.02)^2] = 2.51 \times 10^{-5} \text{ kg.m}^2$$

For Plate 3

$$(\bar{I}_{zz})_3 = \frac{M_3 h_3^2}{12}$$

$$= \frac{0.0236 \times (0.02)^2}{12} = 7.87 \times 10^{-7} \text{ kg.m}^2$$

$$(I_{AA})_3 = (\bar{I}_{zz})_3 + M_3(d_3)^2$$

[where $d_3 = \sqrt{(0.04)^2 + (0.01)^2} = \sqrt{1.7 \times 10^{-3}}$ m]

$$\therefore \quad (I_{AA})_3 = 7.87 \times 10^{-7} + 0.0236 \times \left(\sqrt{1.7 \times 10^{-3}}\right)^2 = 4.09 \times 10^{-5} \text{ kg.m}^2$$

Therefore, mass moment of the inertia of the composite plate is given as

$$(I_{AA}) = \sum (I_{AA})_i$$
$$= (1.06 \times 10^{-5}) + (2.51 \times 10^{-5}) + (4.09 \times 10^{-5})$$
$$= 7.66 \times 10^{-5} \text{ kg.m}^2$$

Example 10.5 Determine mass moment of inertia of a thin rectangular plate (thickness = 5 mm) in which a semicircular portion is cut as shown in Fig. 10.11 about the base. The density of the material of the plate is 7850 kg/m^3.

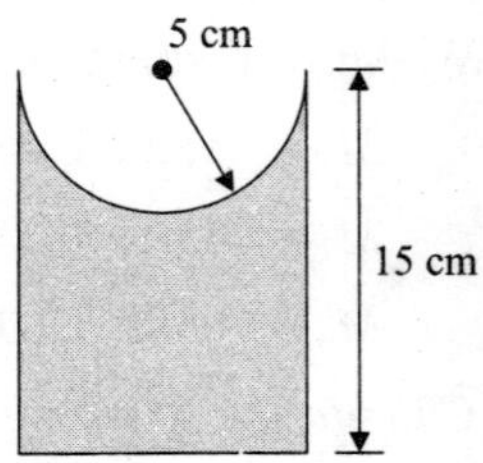

Fig. 10.11

Solution The composite section can be considered to be made up of two sections as shown below.

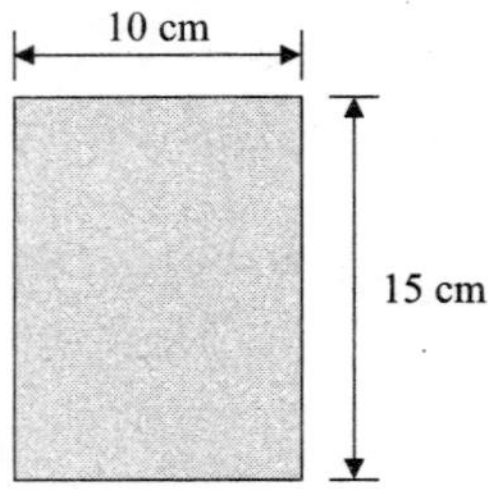

Fig. 10.11(a)

–

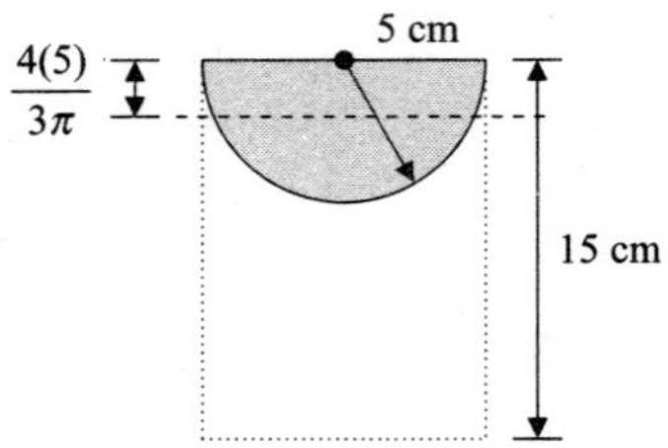

Fig. 10.11(b)

Mass Calculations

Rectangular plate, $M_1 = \rho_1 A_1 t_1$

$= 7850 \times (0.1 \times 0.15) \times 0.005 = 0.589$ kg

Semicircular plate, $M_2 = \rho_2 A_2 t_2$

$= 7850 \times (\pi \times 0.05^2/2) \times 0.005 = 0.154$ kg

Calculation of mass moments of inertia

For the rectangular plate

We know that mass moment of inertia of a thin rectangular plate about its centroidal axis is

$$(\bar{I})_1 = \frac{M_1 h^2}{12}$$

Therefore, mass moment of inertia of the plate about its base is

$$(I)_1 = (\bar{I})_1 + M_1\left[\frac{h}{2}\right]^2$$

$$= \frac{M_1 h^2}{3}$$

$$= \frac{(0.589)\,(0.15)^2}{3} = 4.418 \times 10^{-3}\ \text{kg.m}^2$$

For the semicircular plate

From the result of Example 10.3, we know that the mass moment of inertia of a thin semicircular plate about its centroidal axis is

$$(\bar{I})_2 = 0.07\ M_2 R^2$$

Therefore, the mass moment of inertia of the plate about the base of the composite section is

$$(I)_2 = (\bar{I})_2 + M_2 d^2$$

$$= 0.07(0.154)(0.05)^2 + (0.154)[0.15 - (4 \times 0.05)/(3 \times \pi)]^2$$

$$= 2.58 \times 10^{-3}\ \text{kg.m}^2$$

Therefore, the mass moment of inertia of the composite section is given as

$$I = I_1 - I_2$$

$$= 4.418 \times 10^{-3} - 2.58 \times 10^{-3}$$

$$= 1.838 \times 10^{-3}\ \text{kg.m}^2$$

Example 10.6 A thin plate of mass M is cut in the shape of a parallelogram of thickness t as shown in Fig. 10.12. Determine mass moment of inertia of the plate about the X-axis.

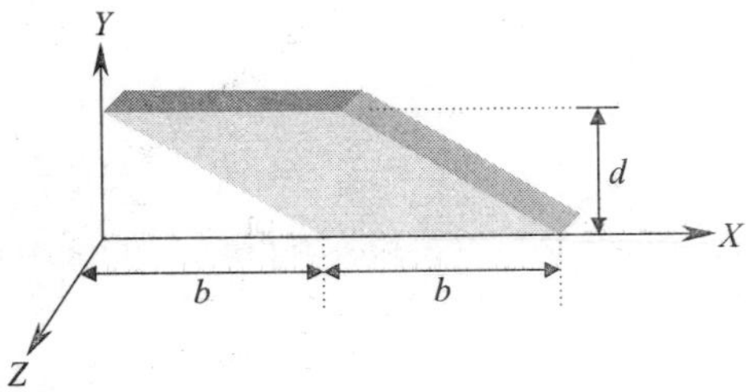

Fig. 10.12

Solution Consider a thin strip in the plane of the figure parallel to the base at a distance y from the base and infinitesimal thickness dy. Since the base length is the same for any strip parallel to the base, we can see that the area moment of inertia of the parallelogram about the centroidal axis parallel to the base is the same as the area moment of inertia of a rectangle of base b and height d, i.e.,

$$(\bar{I}_{xx})_{\text{area}} = \frac{bd^3}{12}$$

Therefore,
$$(\bar{I}_{xx})_{\text{mass}} = \rho t \frac{bd^3}{12}$$

$$= \frac{Md^2}{12}, \text{ where } M = \rho t b d$$

By the parallel axis theorem, the mass moment of inertia of the parallelogram about the X-axis is given as

$$(I_{xx})_{mass} = (\bar{I}_{xx})_{mass} + M(d/2)^2$$
$$= \frac{Md^2}{12} + \frac{Md^2}{4}$$
$$= \frac{Md^2}{3}$$

10.6 MASS MOMENT OF INERTIA OF SOLIDS

The results derived in the previous section for mass moments of inertia of thin plates can be used to determine mass moments of inertia of solids as explained below for various regular shaped bodies.

10.6.1 Solid Cylinder

Consider a cylinder of radius R, length L and mass density ρ. The coordinate axes are chosen about the centroid as shown in Fig.10.13. Suppose we cut a circular disc of infinitesimal thickness dz perpendicular to the Z-axis at a distance z from the origin, its mass is given as

$$dm = \rho\pi R^2 dz \quad (10.33)$$

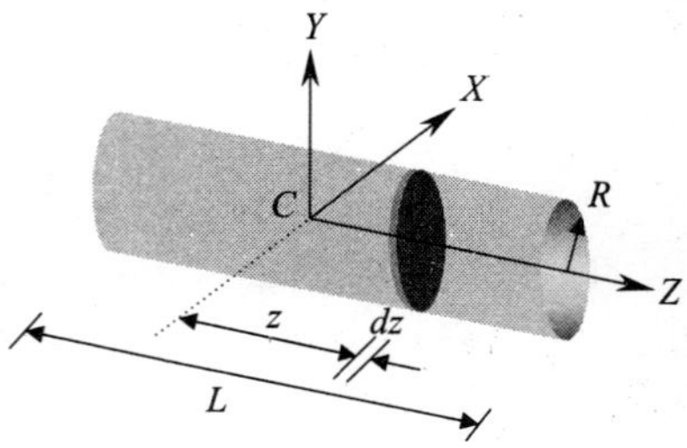

Fig. 10.13 Section of a cylinder

Therefore, its mass moment of inertia about the Z-axis is

$$(dI_{zz})_{mass} = dmR^2/2$$
$$= \rho(\pi R^4/2)dz \quad (10.34)$$

Therefore, the mass moment of inertia of the cylinder about the Z-axis is obtained by integrating the above expression between limits:

$$\bar{I}_{zz} = \int_{-L/2}^{L/2} \frac{\rho\pi R^4}{2} dz$$
$$= \frac{\rho\pi R^4}{2}[z]_{-L/2}^{L/2}$$
$$= \frac{\rho\pi R^4 L}{2} \quad (10.35)$$

We know that volume of the cylinder is $V = \pi R^2 L$ and its mass is $M = \rho\pi R^2 L$.

Hence, $$\bar{I}_{zz} = \frac{MR^2}{2} \quad (10.36)$$

Calculation of $(\bar{I}_{xx})_{mass}$

From the mass moments of inertia of a thin circular disc that we derived in the previous section, we can write the mass moment of inertia of the circular disc about an axis lying on its plane as

$$(dI_{x'x'})_{mass} = dm\, R^2/4$$
$$= \rho(\pi R^4/4)dz \quad (10.37)$$

By transfer of theorem, the mass moment of inertia of the disc about the centroidal X-axis is given as

$$(d\bar{I}_{xx})_{\text{mass}} = \rho(\pi R^4/4)dz + \rho(\pi R^2)dz\, z^2 \tag{10.38}$$

On integration between the limits, we obtain mass moment of inertia of the cylinder about the X-axis as

$$(\bar{I}_{xx}) = \int_{-L/2}^{L/2} \frac{\rho\pi R^4}{4} dz + \int_{-L/2}^{L/2} \rho\pi R^2 z^2 dz$$

$$= \frac{\rho\pi R^4}{4}[z]_{-L/2}^{L/2} + \rho\pi R^2\left(\frac{z^3}{3}\right)_{-L/2}^{L/2}$$

$$= \frac{\rho\pi R^4 L}{4} + \frac{\rho\pi R^2 L^3}{12}$$

$$= \frac{\rho\pi R^2 L}{12}[3R^2 + L^2]$$

Since $\qquad M = \rho\pi R^2 L, \quad \bar{I}_{xx} = \dfrac{M}{12}[3R^2 + L^2]$

As the cylinder is symmetric about X–Z and Y–Z planes,

$$\bar{I}_{xx} = \bar{I}_{yy} = \frac{M}{12}[3R^2 + L^2] \tag{10.39}$$

Corollary

(i) For a **slender rod**, radius $R << L$,

Therefore, $$\bar{I}_{xx} = \bar{I}_{yy} = \frac{ML^2}{12} \tag{10.40}$$

(ii) For a **thin disc**, length $L << R$

Therefore, $$\bar{I}_{xx} = \bar{I}_{yy} = \frac{MR^2}{4} \quad \text{and} \quad \bar{I}_{zz} = \frac{MR^2}{2} \tag{10.41}$$

10.6.2 Prism

Consider a prism of length L, breadth B and height H and let its mass density be ρ. The volume of the prism is then $V = LBH$ and its mass is

$$M = \rho LBH \tag{10.42}$$

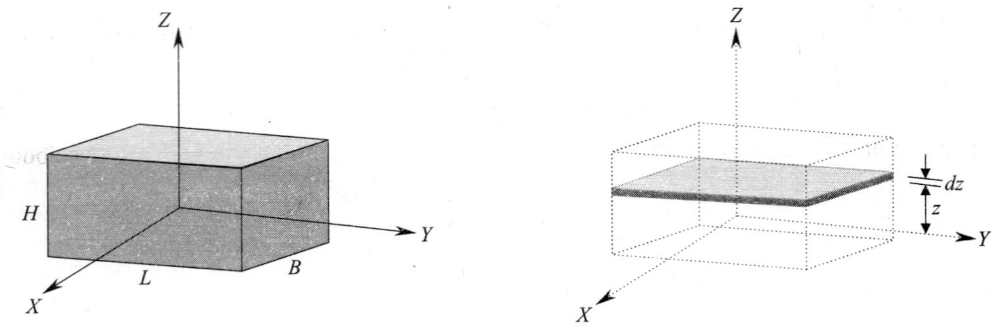

Fig. 10.14 Section of a prism

Suppose we cut a thin plate of thickness dz at a height z from the X–Y plane. Then its mass is given as

$$dm = \rho LB dz \tag{10.43}$$

Therefore, its mass moment of inertia about the Z-axis is given as

$$(d\bar{I}_{zz})_{\text{mass}} = \frac{\rho LB}{12}[L^2 + B^2]dz \tag{10.44}$$

Hence, mass moment of inertia of the prism is obtained by integrating the above expression between the limits:

$$\bar{I}_{zz} = \frac{\rho LB}{12}(L^2 + B^2)\int_{-H/2}^{H/2} dz$$

$$= \frac{\rho LBH}{12}(L^2 + B^2)$$

Since mass, $M = \rho LBH$,

$$\bar{I}_{zz} = \frac{M}{12}(L^2 + B^2) \tag{10.45}$$

Similarly, it can be shown that,

$$\bar{I}_{xx} = \frac{M}{12}[L^2 + H^2] \tag{10.46}$$

and

$$\bar{I}_{yy} = \frac{M}{12}[B^2 + H^2] \tag{10.47}$$

Note: As the orientation of the prism with respect to the axes can be different from the given orientation in Fig. 10.14, an easy way to remember the expressions for mass moments of inertia is the following rule:

Mass moment of inertia of a prism about any axis is equal to $\frac{M}{12}$ times sum of squares of the sides perpendicular to the axis. Thus, from Fig. 10.14, we see that the sides perpendicular to the X-axis are H and L. Hence, $\bar{I}_{xx} = \frac{M}{12}[L^2 + H^2]$ and likewise.

Corollary For a thin plate, height $H \approx 0$

$$\therefore \qquad \bar{I}_{xx} = \frac{ML^2}{12} \quad \bar{I}_{yy} = \frac{MB^2}{12} \tag{10.48}$$

and

$$\bar{I}_{zz} = \frac{M}{12}[L^2 + B^2] \tag{10.49}$$

10.6.3 Sphere

Consider a sphere of radius R and mass density ρ. Suppose we cut a thin circular disc of radius r and infinitesimal thickness dz at a distance z from the X–Y plane (cross section of the sphere shown in the adjacent figure) then its mass is given as

$$dm = \rho \pi r^2 dz \tag{10.50}$$

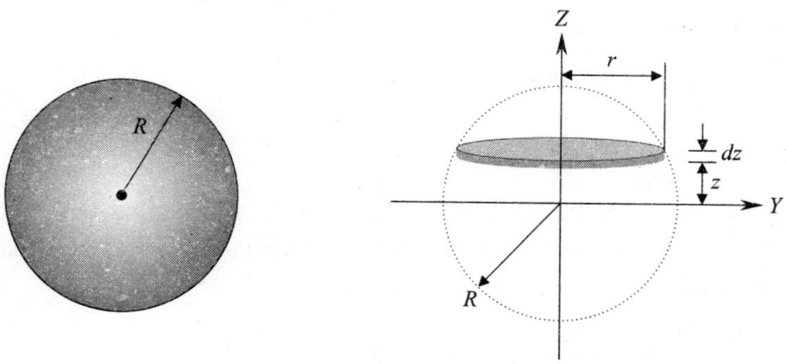

Fig. 10.15 Section of a sphere

Therefore, mass moment of inertia of the disc about the vertical Z-axis is given as

$$(d\bar{I}_{zz})_{\text{mass}} = dm\, r^2/2 = \rho(\pi r^4/2)dz \tag{10.51}$$

Integrating the above expression between limits, we get the mass moment of inertia of the sphere about the Z-axis as

$$\bar{I}_{zz} = \int_{-R}^{R} \rho(\pi r^4/2)dz \tag{10.52}$$

Since R is the radius of the sphere, $r^2 = R^2 - z^2$. Therefore,

$$\bar{I}_{zz} = \int_{-R}^{R} \rho\pi(R^2 - z^2)^2/2\, dz$$

$$= \frac{\rho\pi}{2} \int_{-R}^{R} (R^4 + z^4 - 2R^2z^2)\, dz$$

$$= \frac{\rho\pi}{2}\left[R^4 z + \frac{z^5}{5} - 2R^2\frac{z^3}{3}\right]_{-R}^{R}$$

$$= \frac{8}{15}\rho\pi R^5 \tag{10.53}$$

We know that volume of the sphere is $V = 4/3(\pi R^3)$. Hence, its mass is

$$M = \rho V = 4/3(\rho\pi R^3) \tag{10.54}$$

Substituting this value in the above equation for $\bar{I}_{zz}$, we get

$$\bar{I}_{zz} = \frac{2}{5}MR^2$$

Due to symmetry, the moment of inertia remains the same for any axis passing through the centroid. Hence, in general, we express

$$\bar{I} = \frac{2}{5}MR^2 \tag{10.55}$$

10.6.4 Cone

Consider a cone of base radius R, height H and mass density ρ, oriented with respect to the axes as shown in Fig. 10.16. Suppose we cut a circular disc of radius r and infinitesimal thickness dz at a distance z from the origin. Then its mass is given as

$$dm = \rho\pi r^2 dz \tag{10.56}$$

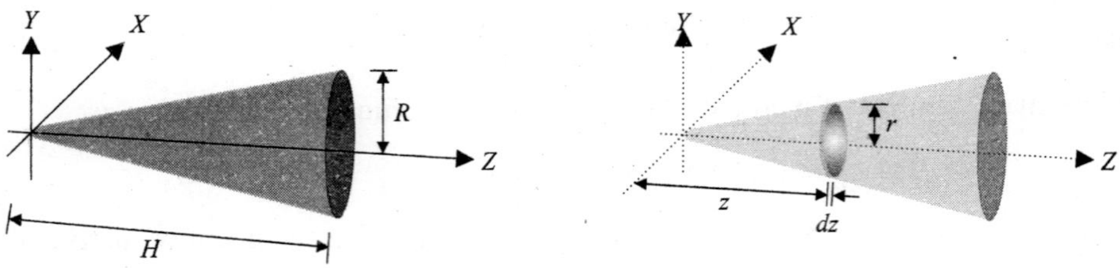

Fig. 10.16 Section of a cone

Therefore, its mass moment of inertia is given as

$$(d\bar{I}_{zz})_{\text{mass}} = dm\, r^2/2 = \rho(\pi r^4/2)dz \tag{10.57}$$

On integration between limits, mass moment of inertia of the cone is obtained as

$$\bar{I}_{zz} = \int_0^H \rho(\pi r^4/2)dz \tag{10.58}$$

By similar triangles, we know, $r/R = z/H$. Therefore,

$$\bar{I}_{zz} = \int_0^H \frac{\rho\pi}{2}\frac{R^4}{H^4}z^4 dz$$

$$= \frac{\rho\pi}{2}\frac{R^4}{H^4}\int_0^H z^4 dz$$

$$= \frac{\rho\pi}{2}\frac{R^4}{H^4}\frac{H^5}{5} = \frac{1}{10}\rho\pi R^4 H \tag{10.59}$$

We know that volume of the cone is $V = 1/3\ \pi R^2 H$. Therefore, its mass is

$$M = \rho V = 1/3 \rho\pi R^2 H \tag{10.60}$$

Substituting this value in the above expression for $\bar{I}_{zz}$, we get

$$\bar{I}_{zz} = \frac{3}{10}MR^2 \tag{10.61}$$

Calculation of $(I_{xx})_{\text{mass}}$

The mass moment of inertia of the disc about an axis lying on its plane is given as

$$(dI_{x'x'})_{\text{mass}} = dm\ r^2/4$$

$$= \rho\pi r^2\ dz\ r^2/4 \tag{10.62}$$

By transfer theorem, the mass moment of inertia of the disc about the X-axis is given as

$$(dI_{xx})_{\text{mass}} = \rho(\pi r^4/4)dz + \rho\pi r^2\ dz\ z^2 \tag{10.63}$$

On integration between the limits, we get the mass moment of inertia of the cone about the X or Y-axis at the vertex as

$$I_{xx} = I_{yy} = \int_0^H \rho(\pi r^4/4)dz + \int_0^H \rho\pi r^2 z^2\ dz$$

$$= \frac{\rho\pi}{4}\frac{R^4}{H^4}\int_0^H z^4 dz + \frac{\rho\pi R^2}{H^2}\int_0^H z^4 dz \quad [\text{since } r/R = z/H]$$

$$= \frac{1}{20}\rho\pi R^4 H + \frac{1}{5}\rho\pi R^2 H^3$$

Since $M = 1/3\rho\pi R^2 H$,

$$I_{xx} = I_{yy} = \frac{3}{5}M\left[\frac{R^2}{4} + H^2\right] \tag{10.64}$$

Corollary If the moment of inertia of the cone about the axes at the base is required, it can be obtained by applying the parallel axis theorem to the above value of I_{xx} or I_{yy}.

The following table summarizes the mass moments of inertia of various solids:

Table 10.2 Mass moments of inertia of solids

Shape	*Figure*	I_{xx}	I_{yy}	I_{zz}
Cylinder	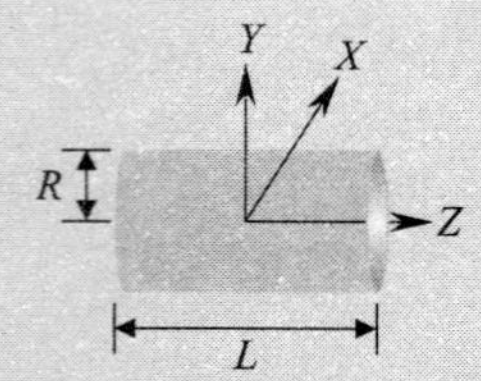	$\frac{M}{12}[3R^2 + L^2]$	$\frac{M}{12}[3R^2 + L^2]$	$\frac{MR^2}{2}$
Slender rod	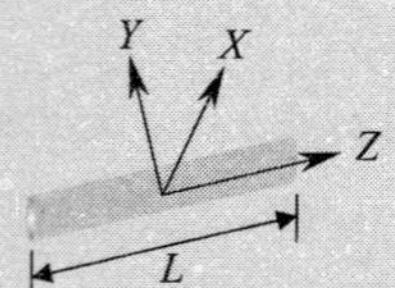	$\frac{ML^2}{12}$	$\frac{ML^2}{12}$	
Prism	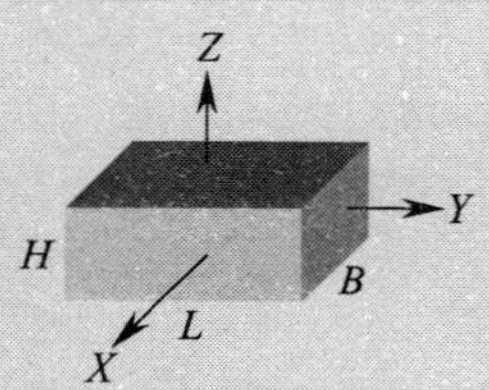	$\frac{M}{12}[L^2 + H^2]$	$\frac{M}{12}[B^2 + H^2]$	$\frac{M}{12}(L^2 + B^2)$
Sphere	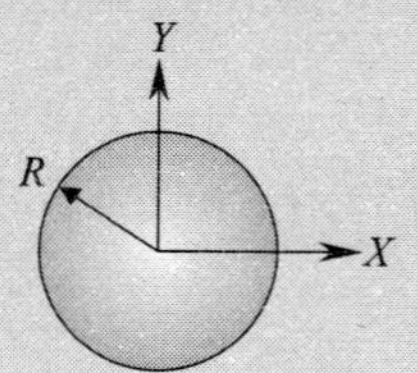	$\frac{2}{5}MR^2$ (about any diametric axis)		
Cone	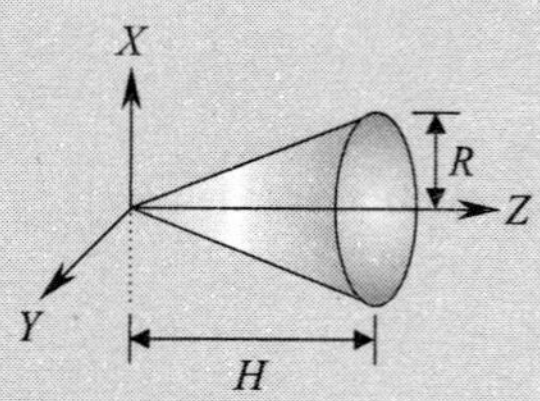	$\frac{3}{5}M\left[\frac{R^2}{4} + H^2\right]$	$\frac{3}{5}M\left[\frac{R^2}{4} + H^2\right]$	$\frac{3}{10}MR^2$

Example 10.7 Derive the expression for mass moment of inertia of a thin-walled spherical shell of mean radius R and thickness t with respect to any diametric axis.

Solution Since R is the mean radius and t is the thickness of the shell, the outer radius R_o and inner radius R_i are given as

$$R_o = R + (t/2), \text{ and}$$

$$R_i = R - (t/2)$$

Hence, the shell can be assumed to be formed by removing a solid sphere of radius R_i from a solid sphere of radius R_o. Thus, mass moment of inertia of the spherical shell is the difference between mass moments of inertia of the outer sphere with radius R_o and inner sphere with radius R_i, i.e.,

$$I_{\text{shell}} = (I)_{\text{outer sphere}} - (I)_{\text{inner sphere}}$$

$$= \frac{2}{5}M_o R_o^2 - \frac{2}{5}M_i R_i^2$$

The masses of the outer and inner spheres are given as

$$M_o = \rho\frac{4}{3}\pi R_o^3$$

and

$$M_i = \rho\frac{4}{3}\pi R_i^3$$

Therefore,

$$I_{\text{shell}} = \frac{2}{5}\left(\rho\frac{4}{3}\pi R_o^3\right)R_o^2 - \frac{2}{5}\left(\rho\frac{4}{3}\pi R_i^3\right)R_i^2$$

$$= \frac{2}{5}\rho\frac{4}{3}\pi(R_o^5 - R_i^5)$$

Since $R_o = R + (t/2)$,

$$R_o^5 = \left(R + \frac{t}{2}\right)^5 = \left(R + \frac{t}{2}\right)\left[\left(R + \frac{t}{2}\right)^2\right]^2$$

$$= \left(R + \frac{t}{2}\right)\left(R^2 + \frac{t^2}{4} + Rt\right)^2$$

$$= \left(R + \frac{t}{2}\right)\left(R^4 + \frac{t^4}{16} + R^2t^2 + \frac{1}{2}R^2t^2 + 2R^3t + \frac{Rt^3}{2}\right)$$

Since the shell is very thin, we can neglect those terms containing t^2, t^3 and t^4. Hence, on simplification,

$$R_o^5 = \left(R + \frac{t}{2}\right)(R^4 + 2R^3t) = R^5 + 2R^4t + \frac{R^4t}{2} + R^3t^2$$

Again neglecting the term with t^2

$$R_o^5 = R^5 + \frac{5}{2}R^4t$$

Similarly, it can be shown that

$$R_i^5 = R^5 - \frac{5}{2}R^4t$$

Therefore,

$$I_{\text{shell}} = \frac{2}{5}\rho\frac{4}{3}\pi\left(R^5 + \frac{5}{2}R^4t - R^5 + \frac{5}{2}R^4t\right)$$

$$= \frac{2}{5}\rho\frac{4}{3}\pi(5R^4t) = \frac{2}{3}(\rho 4\pi R^2t)R^2$$

We know that mass of the shell is the product of surface area with mean radius R and thickness and density, i.e., $M = \rho\, 4\pi R^2 t$. Hence,

$$I_{\text{shell}} = \frac{2}{3} M R^2$$

Example 10.8 Determine mass moment of inertia of a hollow cylinder of length L, and outer and inner radii R_1 and R_2 respectively about the centroidal axes.

Solution

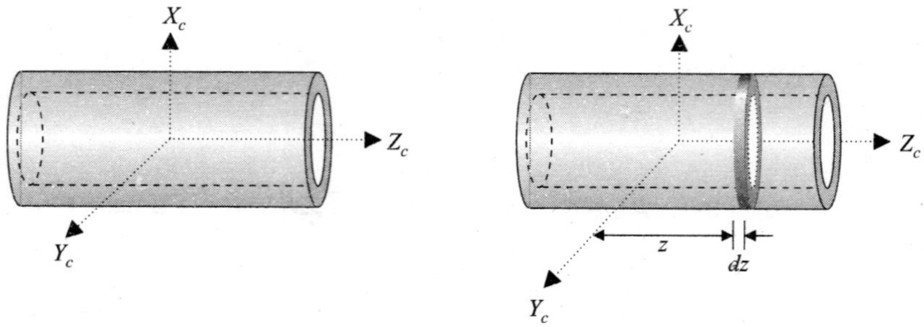

Fig. 10.17

Let m be mass of the cylinder per unit volume. Consider a disc of infinitesimal thickness dz at a distance z from the origin. As already solved in Example 10.2, we know the mass moment of inertia of the annular disc about X, Y and Z axes:

About the Z-axis

Since the Z-axis of the disc is the same as the centroidal Z_c-axis of the cylinder, we have

$$d\bar{I}_{zz} = dI_{zz} = dm\frac{[R_1^2 + R_2^2]}{2}$$

$$= m\pi[R_1^2 - R_2^2]\,dz\,\frac{[R_1^2 + R_2^2]}{2} \qquad \text{(a)}$$

About the X and Y axes

The mass moments of inertia of the disc about the axes in the plane of the disc are

$$dI_{xx} = dI_{yy} = dm\frac{[R_1^2 + R_2^2]}{4}$$

$$= m\pi[R_1^2 - R_2^2]\,dz\,\frac{[R_1^2 + R_2^2]}{4}$$

By the parallel axis theorem, we have mass moment of inertia of the disc about the centroidal axes of the cylinder as

$$d\bar{I}_{xx} = d\bar{I}_{yy} = m\pi[R_1^2 - R_2^2]\,dz\,\frac{[R_1^2 + R_2^2]}{4} + m\pi[R_1^2 - R_2^2]\,dz\,z^2 \qquad \text{(b)}$$

Hence, mass moments of inertia of the hollow cylinder about the centroidal axes are obtained by integrating the expressions (a) and (b):

$$\bar{I}_{zz} = \int_{-L/2}^{L/2} d\bar{I}_{zz} = \int_{-L/2}^{L/2} m\pi[R_1^2 - R_2^2]\,dz\frac{[R_1^2 + R_2^2]}{2}$$

$$= m\pi[R_1^2 - R_2^2]\frac{[R_1^2 + R_2^2]}{2}[z]_{-L/2}^{L/2}$$

$$= m\pi[R_1^2 - R_2^2]\frac{[R_1^2 + R_2^2]}{2}[L]$$

Since $m\pi[R_1^2 - R_2^2]\,[L]$ is the total mass $[M]$ of the hollow cylinder, we can write

$$\bar{I}_{zz} = M\frac{[R_1^2 + R_2^2]}{2}$$

Similarly,
$$\bar{I}_{xx} = \bar{I}_{yy} = \int_{-L/2}^{L/2} m\pi[R_1^2 - R_2^2]\,dz\frac{[R_1^2 + R_2^2]}{4} + \int_{-L/2}^{L/2} m\pi[R_1^2 - R_2^2]\,dz\,z^2$$

$$= M\frac{[R_1^2 + R_2^2]}{4} + m\pi[R_1^2 - R_2^2]\left[\frac{z^3}{3}\right]_{-L/2}^{L/2}$$

$$= M\frac{[R_1^2 + R_2^2]}{4} + M\left[\frac{2L^2}{24}\right]$$

$$= \frac{M}{12}[3(R_1^2 + R_2^2) + L^2]$$

Example 10.9 A cube of 250 mm side has mass density of 4000 kg/m^3. Determine the mass moment of inertia of the cube about one of its edges.

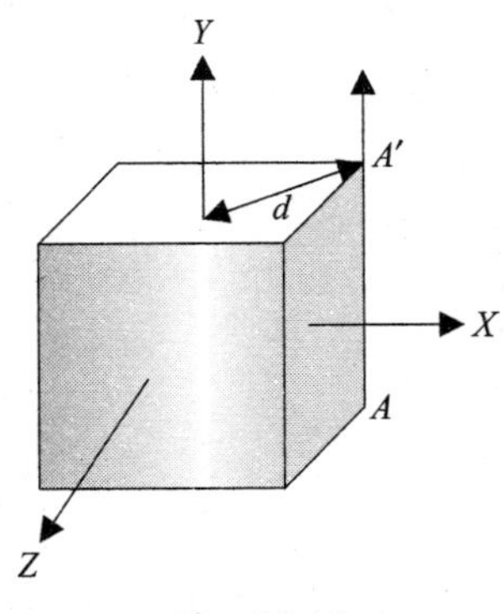

Fig. 10.18

Solution *Given data*

Side, $a = 0.25$ m

Density, $\rho = 4000$ kg/m^3

Therefore, its mass, $M = \rho a^3 = 4000 \times (0.25)^3 = 62.5$ kg

Considering the cube as a prism, the mass moment of inertia about the centroidal Y-axis is obtained as

$$\bar{I}_{yy} = M(a^2 + a^2)/12$$
$$= 62.5[(0.25)^2 + (0.25)^2]/12 = 0.651 \text{ kg.m}^2$$

By transfer formula, moment of inertia about one of the edges AA' is obtained as

$$I_{AA'} = \bar{I}_{yy} + M(d)^2$$

We know, $d^2 = (a/2)^2 + (a/2)^2 = 0.03125 \text{ m}^2$

Therefore, $I_{AA'} = 0.651 + 62.5(0.03125) = 2.6 \text{ kg.m}^2$

Example 10.10 A brass cone having a base diameter of 40 cm and a height of 25 cm is placed on top of a vertical steel cylinder of same diameter and a height of 20 cm. Determine the mass moment of inertia of the composite body about the vertical geometric axis. Take density of brass and steel to be 8400 kg/m^3 and 7850 kg/m^3 respectively.

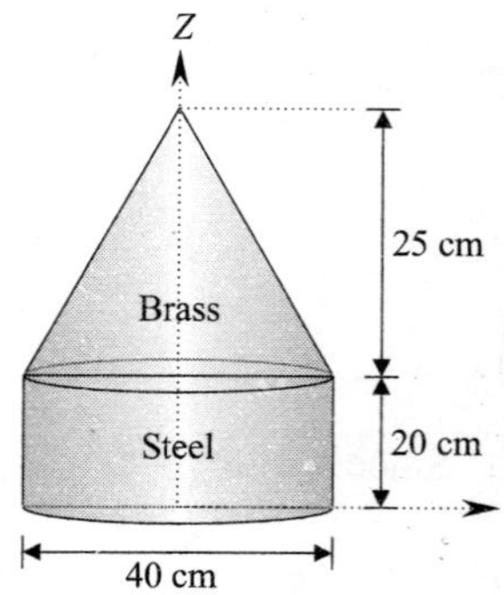

Fig. 10.19

Solution *Given data*

For brass cone

$$R_1 = 20 \text{ cm} = 0.2 \text{ m}$$
$$H_1 = 25 \text{ cm} = 0.25 \text{ m}$$
$$\rho_1 = 8400 \text{ kg/m}^3$$

For steel cylinder

$$R_2 = 20 \text{ cm} = 0.2 \text{ m}$$
$$H_2 = 20 \text{ cm} = 0.2 \text{ m}$$
$$\rho_2 = 7850 \text{ kg/m}^3$$

Mass Calculations

Cone Volume, $V_1 = \dfrac{1}{3}\pi R_1^2 H_1 = \dfrac{1}{3}\pi(0.2)^2\,(0.25) = 10.47 \times 10^{-3} \text{ m}^3$

Mass, $M_1 = \rho_1 V_1 = 8400 \times 10.47 \times 10^{-3} = 87.95$ kg

Cylinder Volume, $V_2 = \pi R_2^2 H_2 = \pi(0.2)^2(0.2) = 25.13 \times 10^{-3} \text{ m}^3$

Mass, $M_2 = \rho_2 V_2 = 7850 \times 25.13 \times 10^{-3} = 197.27$ kg

Mass Moment of Inertia calculations

Cone

$$(\bar{I}_{zz})_1 = \frac{3}{10} M_1 R_1^2$$
$$= \frac{3}{10}(87.95)(0.2)^2 = 1.055 \text{ kg.m}^2$$

Cylinder

$$(\bar{I}_{zz})_2 = \frac{1}{2} M_2 R_2^2$$
$$= \frac{1}{2}(197.27)(0.2)^2 = 3.95 \text{ kg.m}^2$$

Therefore, mass moment of inertia of the composite body about the vertical geometric axis is obtained as

$$\bar{I}_{zz} = (\bar{I}_{zz})_1 + (\bar{I}_{zz})_2$$
$$= 1.055 + 3.95 = 5.005 \text{ kg.m}^2$$

Example 10.11 A uniform steel rod is bent into the shape of an isosceles triangle. Compute the radius of gyration about an axis through O perpendicular to the plane of the figure. The total mass of the steel rod is 10 kg.

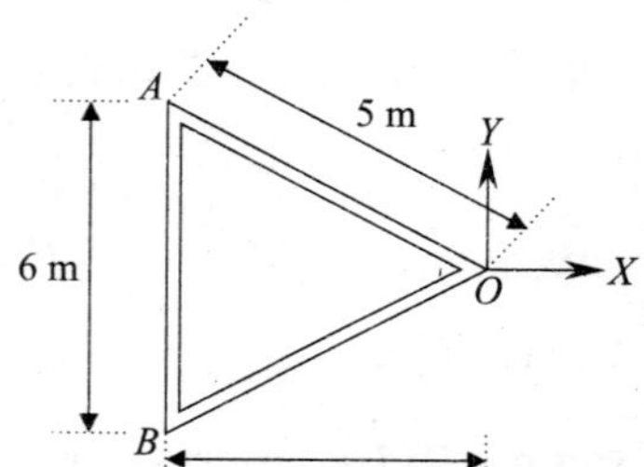

Fig. 10.20

Solution *Mass Calculations*

From the figure, we can see that side $\overline{OB} = \sqrt{4^2 + (6/2)^2} = 5$ m. Therefore, total length of the rod is $\overline{OA} + \overline{AB} + \overline{BO} = 16$ m. Since the total mass of the steel rod is 10 kg, the mass of each side is as follows:

$$M_{OA} = 10\frac{5}{16} = 3.125 \text{ kg}$$

$$M_{AB} = 10\frac{6}{16} = 3.75 \text{ kg}$$

$$M_{OB} = 10\frac{5}{16} = 3.125 \text{ kg}$$

Calculation of mass moments of inertia

For the portion OA

The mass moment of inertia about the centroidal Z-axis is given as

$$(\bar{I}_{zz})_{OA} = \frac{M_{OA}L_{OA}^2}{12}$$

$$= \frac{3.125 \times (5)^2}{12} = 6.51 \text{ kg.m}^2$$

Therefore, the mass moment of inertia of OA about Z-axis of the composite rod is obtained as

$$(I_{zz})_{OA} = (\bar{I}_{zz})_{OA} + M_{OA}(d_{OA})^2$$

$$= 6.51 + [3.125 \times (2.5)^2] = 26.04 \text{ kg.m}^2$$

Similarly, for the other portions, the mass moments of inertia are calculated as follows:

For the portion AB

$$(\bar{I}_{zz})_{AB} = \frac{M_{AB}L_{AB}^2}{12}$$

$$= \frac{3.75 \times (6)^2}{12} = 11.25 \text{ kg.m}^2$$

$$\therefore \quad (I_{zz})_{AB} = (\bar{I}_{zz})_{AB} + M_{AB}(d_{AB})^2$$

$$= 11.25 + [3.75 \times (4)^2] = 71.25 \text{ kg.m}^2$$

For the portion OB

$$(\bar{I}_{zz})_{OB} = \frac{M_{OB}L_{OB}^2}{12}$$

$$= \frac{3.125 \times (5)^2}{12} = 6.51 \text{ kg.m}^2$$

$$\therefore \quad (I_{zz})_{OB} = (\bar{I}_{zz})_{OB} + M_{OB}(d_{OB})^2$$

$$= 6.51 + [3.125 \times (2.5)^2] = 26.04 \text{ kg.m}^2$$

Therefore, mass moment of inertia of the composite rod about the Z-axis is obtained as

$$(I_{zz}) = \sum(I_{zz})_i$$

$$= 26.04 + 71.25 + 26.04 = 123.33 \text{ kg.m}^2$$

Hence, radius of gyration is obtained as

$$k = \sqrt{\frac{I_{zz}}{M}}$$

$$= \sqrt{\frac{123.33}{10}} = 3.51 \text{ m}$$

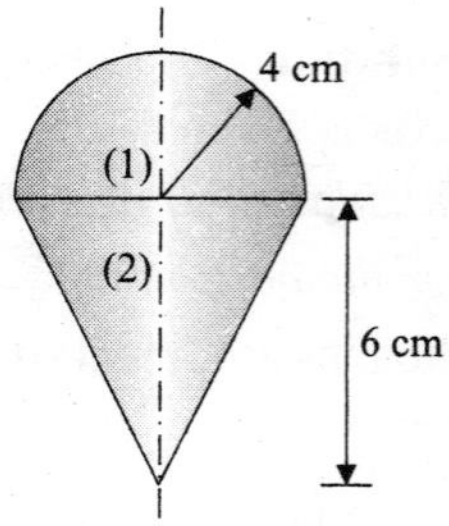

Fig.10.21

Example 10.12 A toy top made up of wood has a hemispherical portion of 8 cm diameter and a cone of 6 cm height as shown. Determine the mass moment of inertia of the top about the axis of revolution, if density of the material is 75 kg/m^3.

Solution *Given data*

Hemisphere

$$R_1 = 4 \text{ cm} = 0.04 \text{ m}$$

Cone

$$R_2 = 4 \text{ cm} = 0.04 \text{ m}$$

$$H_2 = 6 \text{ cm} = 0.06 \text{ m}$$

$$\rho_1 = \rho_2 = 75 \text{ kg/m}^3$$

Mass Calculations

Hemisphere Volume, $V_1 = \frac{2}{3}\pi R_1^3 = \frac{2}{3}\pi(0.04)^3 = 1.34 \times 10^{-4} \text{ m}^3$

Mass, $M_1 = \rho_1 V_1 = 75 \times 1.34 \times 10^{-4} = 10.05 \times 10^{-3}$ kg

Cone Volume, $V_2 = \frac{1}{3}\pi R_2^2 H_2 = \frac{1}{3}\pi(0.04)^2(0.06) = 1.005 \times 10^{-4} \text{ m}^3$

Mass, $M_2 = \rho_2 V_2 = 75 \times 1.005 \times 10^{-4} = 7.54 \times 10^{-3}$ kg

Mass Moment of Inertia calculations

Hemisphere

$$(\bar{I}_{zz})_1 = \frac{2}{5} M_1 R_1^2$$

$$= \frac{2}{5}(10.05 \times 10^{-3})(0.04)^2 = 6.432 \times 10^{-6} \text{ kg.m}^2$$

Cone

$$(\bar{I}_{zz})_2 = \frac{3}{10} M_2 R_2^2$$

$$= \frac{3}{10}(7.54 \times 10^{-3})(0.04)^2 = 3.619 \times 10^{-6} \text{ kg.m}^2$$

Therefore, mass moment of inertia of the composite body about the axis of revolution is obtained as

$$\bar{I}_{zz} = (\bar{I}_{zz})_1 + (\bar{I}_{zz})_2$$

$$= (6.432 \times 10^{-6}) + (3.619 \times 10^{-6}) = 10.05 \times 10^{-6} \text{ kg.m}^2$$

Example 10.13 Determine the mass moment of inertia of the composite solid shown about the axis of rotation. The solid is made up of two identical spheres each of 2 kg mass and 3 cm radius attached at the end of a slender rod of 400 g mass and 15 cm length.

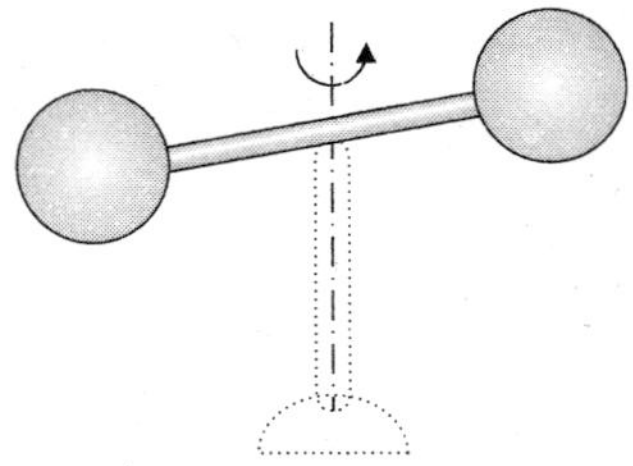

Fig. 10.22

Solution Mass moment of inertia of the slender rod about its centroidal axis is given as

$$\bar{I}_1 = \frac{M_1 l_1^2}{12}$$
$$= \frac{0.4 \times (0.15)^2}{12} = 7.5 \times 10^{-4} \text{ kg.m}^2$$

Mass moment of inertia of a sphere about its centroidal axis is given as

$$\bar{I}_2 = \frac{2}{5} M_2 R_2^2$$
$$= \frac{2}{5}(2) \times (0.03)^2 = 7.2 \times 10^{-4} \text{ kg.m}^2$$

Therefore, the mass moment of inertia of the sphere about the axis of rotation is obtained by transfer formula as

$$I_2 = \bar{I}_2 + M_2 d^2$$
$$= 7.2 \times 10^{-4} + (2)[0.03 + (0.15/2)]^2 = 0.0228 \text{ kg.m}^2$$

Hence, the mass moment of inertia of the composite solid about the axis of rotation is obtained as

$$\bar{I} = \bar{I}_1 + 2I_2$$
$$= 7.5 \times 10^{-4} + [2 \times (0.0228)] = 4.64 \times 10^{-2} \text{ kg.m}^2$$

Example 10.14 A hollow cylinder made of steel has an outer diameter of 8 cm and inner diameter of 6 cm. Its height is 20 cm. Determine its mass moment of inertia about the longitudinal axis. Take density of steel to be 7850 kg/m^3.

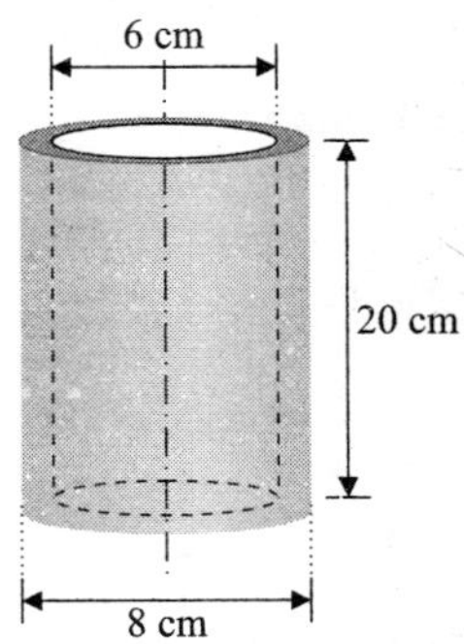

Fig. 10.23

Solution *Given data*

Outer radius, $R_1 = 4 \text{ cm} = 0.04 \text{ m}$

Inner radius, $R_2 = 3 \text{ cm} = 0.03 \text{ m}$

Height, $H = 20 \text{ cm} = 0.2 \text{ m}$

Density, $\rho = 7850 \text{ kg/m}^3$

Mass Calculations

Outer solid cylinder, $M_1 = \rho V_1 = \rho \pi R_1^2 H = 7850 \times \pi(0.04)^2 (0.2)$
$= 7.892 \text{ kg}$

Inner solid cylinder, $M_2 = \rho V_2 = \rho \pi R_2^2 H = 7850 \times \pi(0.03)^2 (0.2)$
$= 4.439 \text{ kg}$

Moment of inertia calculations about the longitudinal Z-axis

$$(\bar{I}_{zz})_1 = \frac{M_1 R_1^2}{2}$$

$$= \frac{7.892 \times (0.04)^2}{2} = 6.314 \times 10^{-3} \text{ kg.m}^2$$

$$(\bar{I}_{zz})_2 = \frac{M_2 R_2^2}{2}$$

$$= \frac{4.439 \times (0.03)^2}{2} = 1.998 \times 10^{-3} \text{ kg.m}^2$$

Therefore, mass moment of inertia of the composite solid about the vertical Z-axis is obtained as

$$\bar{I}_{zz} = (\bar{I}_{zz})_1 - (\bar{I}_{zz})_2$$

$$= (6.314 - 1.998) \times 10^{-3} = 4.316 \times 10^{-3} \text{ kg.m}^2$$

Example 10.15 From a cylinder with a base diameter of 10 cm and height of 20 cm, a cone of same base diameter and 10 cm height is cut out as shown in Fig. 10.24. Determine the mass moment of inertia of the remaining solid about the vertical axis. The density of the material is 7850 kg/m³.

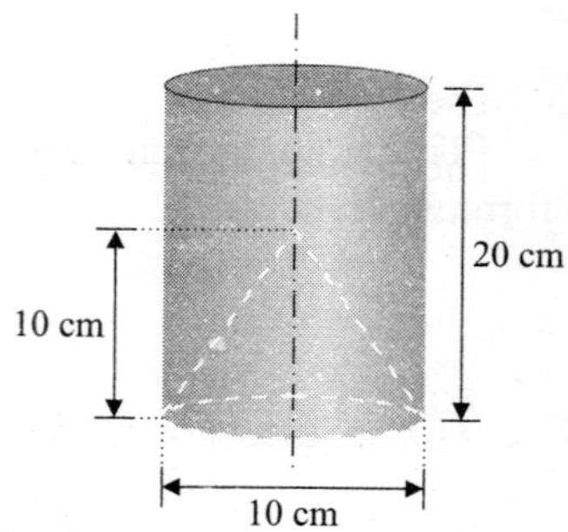

Fig. 10.24

Solution *Given data*

Base radius, $R_1 = R_2 = 5 \text{ cm} = 0.05 \text{ m}$

Height of cylinder, $H_1 = 20 \text{ cm} = 0.2 \text{ m}$

Height of cone, $H_2 = 10 \text{ cm} = 0.1 \text{ m}$

Density, $\rho = 7850 \text{ kg/m}^3$

Mass Calculations

Cylinder $M_1 = \rho V_1 = \rho \pi R_1^2 H_1 = 7850 \times \pi (0.05)^2 (0.2) = 12.33 \text{ kg}$

Cone $M_2 = \rho V_2 = \rho (1/3) \pi R_2^2 H_2 = 7850 \times (1/3) \pi (0.05)^2 (0.1) = 2.06 \text{ kg}$

Moment of inertia calculations about the vertical axis

$$(\bar{I}_{zz})_1 = \frac{M_1 R_1^2}{2}$$

$$= \frac{12.33 \times (0.05)^2}{2} = 15.41 \times 10^{-3} \text{ kg.m}^2$$

$$(\bar{I}_{zz})_2 = \frac{3}{10} M_2 R_2^2$$

$$= \frac{3}{10}(2.06) \times (0.05)^2 = 1.545 \times 10^{-3} \text{ kg.m}^2$$

Therefore, the mass moment of inertia of the composite solid about the vertical axis is obtained as

$$\bar{I}_{zz} = (\bar{I}_{zz})_1 - (\bar{I}_{zz})_2$$

$$= (15.41 - 1.545) \times 10^{-3} = 1.39 \times 10^{-2} \text{ kg.m}^2$$

Example 10.16 In the above problem, determine the mass moment of inertia about an axis in the plane of the base.

Solution The moment of inertia of a cylinder about a centroidal axis parallel to the base is

$$\bar{I}_1 = \frac{M_1}{12}[3R_1^2 + H_1^2]$$

Therefore, the moment of inertia about a parallel axis in the plane of the base is

$$I_1 = \bar{I}_1 + M_1[d]^2$$

$$= \frac{M_1}{12}[3R_1^2 + H_1^2] + M_1[H_1/2]^2$$

$$= M_1\left[\frac{R_1^2}{4} + \frac{H_1^2}{3}\right]$$

$$= 12.33\left[\frac{(0.05)^2}{4} + \frac{(0.2)^2}{3}\right] = 0.172 \text{ kg.m}^2$$

The moment of inertia of the cone about the base can be determined in two steps:

The moment of inertia of a cone about an axis passing through the vertex and parallel to the base is:

$$I_2' = \frac{3}{5}M_2\left[\frac{R_2^2}{4} + H_2^2\right]$$

Hence, the moment of inertia of the cone about a parallel centroidal axis is

$$\bar{I}_2 = \frac{3}{5}M_2\left[\frac{R_2^2}{4} + H_2^2\right] - M_2\left[\frac{3H_2}{4}\right]^2$$

[Note that as we move towards the centroid, the moment of inertia decreases.]

Therefore, the moment of inertia of the cone about a parallel axis in the plane of the base is given as

$$I_2 = \frac{3}{5}M_2\left[\frac{R_2^2}{4} + H_2^2\right] - M_2\left[\frac{3H_2}{4}\right]^2 + M_2\left[\frac{H_2}{4}\right]^2$$

$$= \frac{M_2}{10}[1.5\,R_2^2 + H_2^2]$$

$$= \frac{2.06}{10}[1.5(0.05)^2 + (0.1)^2] = 2.833 \times 10^{-3} \text{ kg.m}^2$$

Therefore, the mass moment of inertia of the composite section about an axis in the plane of the base is obtained as

$$I = I_1 - I_2$$

$$= 0.172 - 2.833 \times 10^{-3} = 0.169 \text{ kg.m}^2$$

Example 10.17 From a prism of dimensions 40 cm × 30 cm × 10 cm, a block of dimensions 10 cm × 15 cm × 10 cm is removed as shown. Determine the mass moment of inertia of the remaining block about axes CC_1 and AA_1. Take density of material to be 1250 kg/m^3.

Solution Let the entire block be named 1 and the removed block, 2.

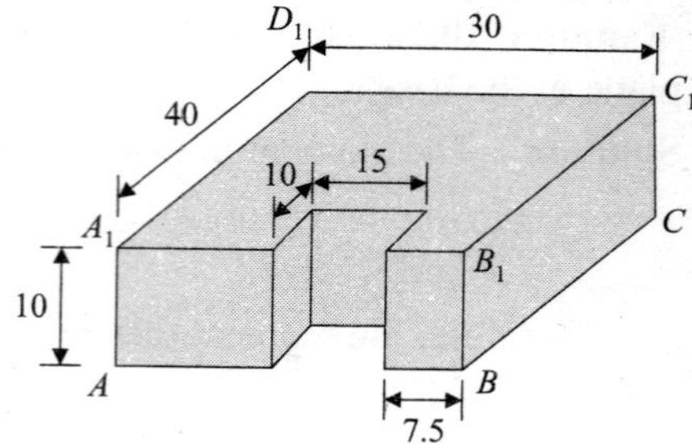

Fig. 10.25

Mass calculations

Mass of the entire block,

$$M_1 = \rho V_1 = \rho L_1 B_1 H_1 = 1250 \times 0.4 \times 0.3 \times 0.1 = 15 \text{ kg}$$

and mass of the removed block,

$$M_2 = \rho V_2 = \rho L_2 B_2 H_2 = 1250 \times 0.1 \times 0.15 \times 0.1 = 1.875 \text{ kg}$$

Mass moment of inertia calculations

We know that mass moment of inertia of a prism about the *centroidal* vertical axis is

$$\bar{I}_{yy} = \frac{M}{12}[L^2 + B^2] \quad \text{where } L \text{ and } B \text{ are length and breadth of the prism}$$

Hence, $$(\bar{I}_{yy})_1 = \frac{15}{12}[(0.4)^2 + (0.3)^2] = 0.3125 \text{ kg.m}^2$$

Similarly, for the element 2

$$(\bar{I}_{yy})_2 = \frac{1.875}{12}[(0.1)^2 + (0.15)^2] = 5.078 \times 10^{-3} \text{ kg.m}^2$$

(i) Mass moment of inertia about CC_1 axis

Therefore, mass moment of inertia of the element 1 about the given CC_1 axis is

$$(I_{yy})_1 = (\bar{I}_{yy})_1 + M_1(d_1)^2 \quad \text{where } d_1 \text{ is the perpendicular distance between the two axes}$$

Since $$d_1 = \sqrt{(0.2)^2 + (0.15)^2} = 0.25 \text{ m}$$

$$(I_{yy})_1 = 0.3125 + 15(0.25)^2 = 1.25 \text{ kg.m}^2$$

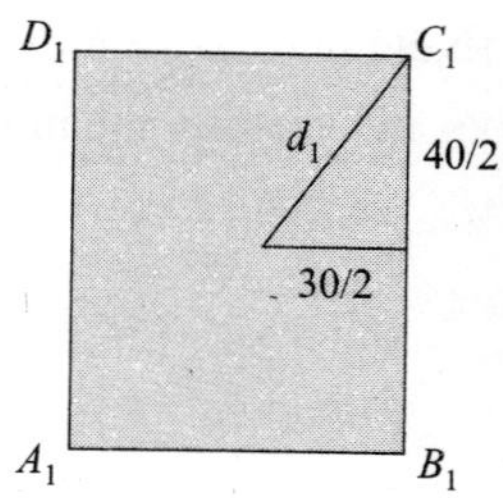

Fig. 10.25(a)

Similarly, moment of inertia of the element 2 about the given CC_1 axis is

$$(I_{yy})_2 = (\bar{I}_{yy})_2 + M_2\,(d_2)^2$$

where d_2 is the perpendicular distance between the two axes.

Since $$d_2 = \sqrt{(0.35)^2 + (0.15)^2} = \sqrt{0.145} \text{ m}$$

$$(I_{yy})_2 = 5.078 \times 10^{-3} + 1.875\left(\sqrt{0.145}\right)^2$$
$$= 0.277 \text{ kg.m}^2$$

Therefore, the moment of inertia of the composite solid about CC_1 axis is obtained as

$$I_{yy} = (I_{yy})_1 - (I_{yy})_2$$
$$= 1.25 - 0.277 = 0.973 \text{ kg.m}^2$$

(ii) Mass moment of inertia about AA_1 axis

For the element 1, the mass moment of inertia is same as before as the perpendicular distance remains the same. Hence,

$$(I_{yy})_1 = 1.25 \text{ kg.m}^2$$

For the element 2

$$(\bar{I}_{yy})_2 = 5.078 \times 10^{-3} \text{ kg.m}^2$$

Since $$d_2 = \sqrt{(0.05)^2 + (0.15)^2} = \sqrt{0.025} \text{ m}$$

$$(I_{yy})_2 = 5.078 \times 10^{-3} + 1.875\left(\sqrt{0.025}\right)^2 = 5.2 \times 10^{-2} \text{ kg.m}^2$$

Therefore, the moment of inertia of the composite solid about the AA_1 axis is obtained as

$$I_{yy} = (I_{yy})_1 - (I_{\bar{y}y})_2$$
$$= 1.25 - 5.2 \times 10^{-2} = 1.198 \text{ kg.m}^2$$

Example 10.18 A hemispherical shell with an outer radius of 5 cm and an inner radius of 4 cm is shown in Fig. 10.26. Determine the mass moment of inertia about the vertical Z-axis. Take density of material to be 7850 kg/m^3.

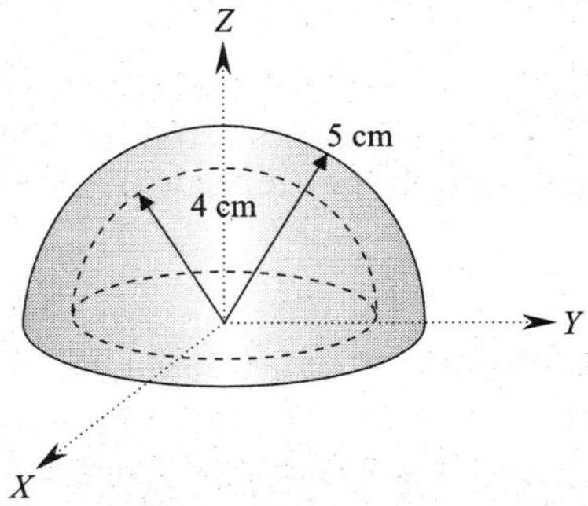

Fig. 10.26

Solution We know that the mass moment of inertia of a sphere about any axis passing through its centre is

$$\bar{I} = \frac{2}{5} MR^2$$

Hence, for a hemisphere, the mass moment of inertia will be half of that for the sphere, i.e.,

$$\bar{I} = \frac{2}{5}\frac{MR^2}{2} = \frac{2}{5}(\text{mass of hemisphere})R^2$$

Since $M/2$ is the mass of the hemisphere, we see that the mass moment of inertia of the hemisphere has the same form as that for the sphere.

Mass Calculations

Outer solid, $$M_1 = \rho \frac{2}{3} \pi R_1^3$$

$$= (7850)\frac{2}{3}\pi(0.05)^3 = 2.055 \text{ kg}$$

Inner solid, $$M_2 = \rho \frac{2}{3} \pi R_2^3$$

$$= (7850)\frac{2}{3}\pi(0.04)^3 = 1.052 \text{ kg}$$

Calculation of mass moments of inertia

$$\bar{I}_1 = \frac{2}{5} M_1 R_1^2$$

$$= \frac{2}{5}(2.055)\,(0.05)^2 = 2.055 \times 10^{-3} \text{ kg.m}^2$$

$$\bar{I}_2 = \frac{2}{5} M_2 R_2^2$$

$$= \frac{2}{5}(1.052)(0.04)^2 = 6.733 \times 10^{-4} \text{ kg.m}^2$$

Therefore, the mass moment of inertia of the hemispherical shell about the vertical Z-axis is obtained as

$$\bar{I} = \bar{I}_1 - \bar{I}_2$$

$$= 2.055 \times 10^{-3} - 6.733 \times 10^{-4} = 1.382 \times 10^{-3} \text{ kg.m}^2$$

Example 10.19 Determine the mass moment of inertia about the horizontal centroidal axis of a solid of revolution obtained by rotating an ellipse of semi-major axis a and semi-minor axis b about the X–X axis.

Solution Consider an ellipse in the X–Y plane with semi-major axis a and semi-minor axis b. Then the equation of the ellipse is

$$\frac{x^2}{a^2} + \frac{y^2}{b^2} = 1$$

The solid of revolution is obtained by revolving the ellipse about the X–X axis. Then we can see that X–Z, Y–Z and X–Y are planes of symmetry for the solid. If we consider a thin disc perpendicular to the X-axis then it can be seen that it will be circular in shape. Consider one such disc at a distance x from the origin and of infinitesimally small thickness dx as shown in Fig. 10.27(b).

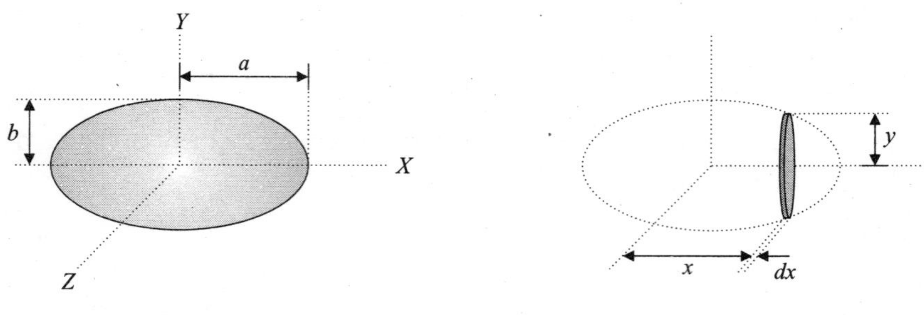

Fig. 10.27(a) **Fig. 10.27(b)**

Since the radius of the disc is y, its area is

$$dA = \pi y^2$$

and its volume is

$$dV = \pi y^2 dx$$

If the density of the material is ρ then its mass is

$$dm = \rho \pi y^2 dx \qquad \text{(a)}$$

Therefore, mass moment of inertia of the disc about the X-axis is

$$d\bar{I}_{xx} = \frac{dm\, y^2}{2}$$

Therefore, the mass moment of inertia of the entire solid of revolution about the X-axis is obtained as

$$\bar{I}_{xx} = \int_{-a}^{a} \frac{dm\, y^2}{2}$$

$$= \frac{1}{2}\pi\rho \int_{-a}^{a} y^4 dx$$

From the equation of the ellipse, we know that,

$$y^2 = \frac{b^2}{a^2}(a^2 - x^2)$$

Hence,

$$\bar{I}_{xx} = \frac{\pi\rho}{2}\frac{b^4}{a^4}\int_{-a}^{a}(a^4 + x^4 - 2a^2x^2)\, dx$$

$$= \frac{\pi\rho}{2}\frac{b^4}{a^4}\left[a^4x + \frac{x^5}{5} - 2a^2\frac{x^3}{3}\right]_{-a}^{a}$$

$$= \frac{\pi\rho}{2}\frac{b^4}{a^4}\left[2a^5 + \frac{2a^5}{5} - \frac{2}{3}a^2(2a^3)\right]$$

$$= \frac{\pi\rho}{2}\frac{b^4}{a^4}a^5\left[2 + \frac{2}{5} - \frac{4}{3}\right]$$

$$= \frac{8\pi\rho}{15}b^4a \tag{b}$$

By integrating the expression for dm in the equation (a), we can get the mass of the solid of revolution as

$$M = \int_{-a}^{a}\pi\rho y^2 dx$$

$$= \pi\rho\frac{b^2}{a^2}\left[a^2x - \frac{x^3}{3}\right]_{-a}^{a}$$

$$= \frac{4}{3}\pi\rho b^2a$$

Therefore, the equation (b) can be written as

$$\bar{I}_{xx} = \frac{4\pi\rho}{3}b^2a\left[\frac{2b^2}{5}\right]$$

$$= \frac{2}{5}Mb^2$$

Example 10.20 In the above problem, determine the mass moment of inertia of the solid about the *Y* and *Z* axes.

Solution Due to symmetry, the moments of inertia about the *Y* and *Z* axes will be equal. The mass moment of inertia of the circular disc about the *Y* or *Z* axis is

$$d\bar{I}_{yy} = d\bar{I}_{zz} = \frac{dm\,y^2}{4} + dm\,x^2$$

Therefore, the mass moment of inertia of the entire solid of revolution about the *Y* or *Z* axis is given as

$$\bar{I}_{yy} = \bar{I}_{zz} = \int \frac{dm\,y^2}{4} + \int dm\,x^2$$

$$= \frac{1}{4}\pi\rho\int y^4 dx + \pi\rho\int y^2 x^2 dx$$

$$= \frac{\pi\rho b^4}{4a^4}\int_{-a}^{a} (a^4 + x^4 - 2a^2x^2)\,dx + \frac{\pi\rho b^2}{a^2}\int_{-a}^{a} (a^2x^2 - x^4)\,dx$$

$$= \frac{\pi\rho b^4}{4a^4}\left[a^4 x + \frac{x^5}{5} - 2a^2\frac{x^3}{3}\right]_{-a}^{a} + \frac{\pi\rho b^2}{a^2}\left[a^2\frac{x^3}{3} - \frac{x^5}{5}\right]_{-a}^{a}$$

$$= \frac{4\pi\rho}{15}b^4 a + \frac{4\pi\rho}{15}b^2 a^3$$

$$= \frac{4\pi\rho}{15}b^2 a(b^2 + a^2)$$

$$= \frac{1}{5}M(a^2 + b^2)$$

SUMMARY

Mass Moment of Inertia

The mass moment of inertia about any axis is defined as the product of *mass* **dm** and the *square of the perpendicular distance* from the axis, i.e.,

$$I = \int r^2\,dm$$

Hence, the mass moment of inertia about the *X*, *Y* and *Z* axes are

$$I_{xx} = \int (y^2 + z^2)\,dm, \quad I_{yy} = \int (x^2 + z^2)\,dm \quad \text{and} \quad I_{zz} = \int (x^2 + y^2)\,dm$$

The mass moment of inertia gives a measure of resistance to rotation about an axis. Its dimension and unit are respectively [M] $[L]^2$ and kg.m^2.

Radius of Gyration

The radius of gyration is defined as the distance from the axis of inertia to the point at which the entire mass *M* of the body may be assumed to be concentrated and still have the same moment of inertia. Mathematically, it is given as

$$k = \sqrt{\frac{I}{M}}$$

Transfer Formula or Parallel Axis Theorem

It states that *the mass moment of inertia of a body about an axis at a distance d and parallel to the centroidal axis is equal to the sum of moment of inertia about the centroidal axis and product of mass and square of the distance between the parallel axes.*

$$I_{zz} = \bar{I}_{zz} + Md^2$$

Mass Moment of Inertia of a Thin Plate

Mass moment of inertia of a thin homogeneous plate of density ρ and thickness t can be determined from the corresponding area moment of inertia by the relation:

$$(I_{xx})_{\text{mass}} = \rho t (I_{xx})_{\text{area}}$$

$$(I_{yy})_{\text{mass}} = \rho t (I_{yy})_{\text{area}}$$

$$(I_{zz})_{\text{mass}} = \rho t (I_{zz})_{\text{area}}$$

For a thin rectangular plate of base b, height h and mass M,

$$(\bar{I}_{xx})_{\text{mass}} = \frac{Mh^2}{12}, \quad (\bar{I}_{yy})_{\text{mass}} = \frac{Mb^2}{12} \quad \text{and} \quad (\bar{I}_{zz})_{\text{mass}} = \frac{M}{12}[b^2 + h^2]$$

For a thin circular plate of radius R and mass M,

$$\bar{I}_{xx} = \bar{I}_{yy} = \frac{MR^2}{4} \quad \text{and} \quad \bar{I}_{zz} = \frac{MR^2}{2}$$

For a thin isosceles triangular plate of base b, height h and mass M,

$$(\bar{I}_{xx})_{\text{mass}} = \frac{Mh^2}{18}, \quad (\bar{I}_{yy})_{\text{mass}} = \frac{Mb^2}{24} \quad \text{and} \quad (\bar{I}_{zz})_{\text{mass}} = \frac{M}{72}[3b^2 + 4h^2]$$

Mass Moment of Inertia of Solids

The mass moment of inertia of solids can be determined as follows:

The given solid is divided into thin plates for which mass moments of inertia are known and they are integrated between the limits.

For a solid cylinder of length L, radius R and mass M about the centroidal axes,

$$\bar{I}_{xx} = \bar{I}_{yy} = \frac{M}{12}[3R^2 + L^2] \quad \text{and} \quad \bar{I}_{zz} = \frac{MR^2}{2}$$

For a slender rod, i.e., R = 0,

$$\bar{I}_{xx} = \bar{I}_{yy} = \frac{ML^2}{12}$$

For a thin disc, i.e., L= 0,

$$\bar{I}_{xx} = \bar{I}_{yy} = \frac{MR^2}{4} \quad \text{and} \quad \bar{I}_{zz} = \frac{MR^2}{2}$$

For a prism of length L, breadth B, height H and mass M about the centroidal axes:

about centroidal axis parallel to breadth, $\bar{I}_{xx} = \dfrac{M}{12}[L^2 + H^2]$

about centroidal axis parallel to length, $\bar{I}_{yy} = \dfrac{M}{12}[B^2 + H^2]$

about centroidal axis parallel to height, $\bar{I}_{zz} = \dfrac{M}{12}[L^2 + B^2]$

For a sphere of radius R and mass M about any diametric axis,

$$\bar{I} = \frac{2}{5}MR^2$$

For a cone of radius R, height H and mass M about the axes at the vertex,

$$I_{zz} = \frac{3}{10}MR^2$$

$$I_{xx} = I_{yy} = \frac{3}{5}M\left[\frac{R^2}{4} + H^2\right]$$

EXERCISES

Objective-type Questions

1. The mass moment of inertia gives a measure of
   (a) resistance to rotation about an axis
   (b) resistance to bending about an axis
   (c) resistance to twisting about an axis
   (d) resistance to elongation
2. The unit of mass moment of inertia is
   (a) m^4 (b) $kg.m^4$ (c) $kg.m^2$ (d) kg.m
3. Mass moment of inertia of a thin plate is ____________ times the area moment of inertia.
   (a) mass (b) density × thickness (c) density (d) weight × thickness
4. Mass moment of inertia of a thin rectangular plate of mass *M*, base *b* and height *h* about its base is
   (a) $\dfrac{Mh^2}{12}$ (b) $\dfrac{M(h^2 + b^2)}{12}$ (c) $\dfrac{Mh^2}{3}$ (d) $\dfrac{Mb^2}{3}$
5. Mass moment of inertia of a thin hoop of mass *M* and radius *R* about an axis perpendicular to its plane is
   (a) MR^2 (b) $\dfrac{MR^2}{2}$ (c) $\dfrac{MR^2}{3}$ (d) $\dfrac{MR^2}{4}$

6. Mass moment of inertia of an annular ring of mass M and external and internal radii R_1 and R_2 respectively is

(a) $\frac{M(R_1^2 - R_2^2)}{2}$ (b) $\frac{M(R_1^2 + R_2^2)}{2}$ (c) $\frac{MR_1^2R_2^2}{2}$ (d) $\frac{M(R_1^2/R_2^2)}{2}$

Answers

1. (a) **2.** (c) **3.** (b) **4.** (c) **5.** (a) **6.** (b)

Short-answer Questions

1. Define mass moment of inertia and explain transfer formula for mass moment of inertia.
2. The mass moment of inertia gives a measure of resistance to rotation about an axis. Discuss.
3. How does the unit of the mass moment of inertia differ from that of the area moment of inertia?
4. Define radius of gyration for mass moment of inertia.
5. State the relationship between the area moment of inertia and mass moment of inertia for a thin uniform plate.
6. Derive the expression for the moment of inertia of a cylinder of length l, radius r and density ρ about the horizontal centroidal axis and about the centroidal transverse axis.
7. Determine mass moment of inertia of a slender rod of length L about its centroidal axis normal to the axis of the rod.
8. Derive the expression for the moment of inertia of a homogeneous sphere of radius r, mass density ρ with reference to its diameter.
9. Derive the expression for the moment of inertia of a homogeneous right circular cone of mass m, base radius r and altitude h with respect to its geometric axis.
10. Show that the moment of inertia of a thin circular ring of mass M and mean radius R with respect to its geometric axis is MR^2.

Numerical Problems

10.1 Determine the mass moment of inertia of a thin semicircular disc of mass M and radius R about diametric axes and axis perpendicular to the plane.

Ans. $\bar{I}_{xx} = \bar{I}_{yy} = \frac{MR^2}{4}, \bar{I}_{zz} = \frac{MR^2}{2}$

10.2 Determine the mass moment of inertia of a thin quarter circular plate of mass M and radius R (i) about its diameter, and (ii) about the centroidal axis parallel to its diameter.

Ans. $\frac{MR^2}{4}$, $0.07\,MR^2$

10.3 Find the mass moment of inertia of a thin isosceles triangular plate of mass M, base b and height h about its base.

Ans. $\frac{1}{6}Mh^2$

10.4 Determine the mass moment of inertia of an equilateral triangular plate of mass M and side a about one of its sides.

Ans. $\dfrac{Ma^2}{8}$

10.5 Determine the mass moment of inertia of a square plate of mass M and side a about its diagonal.

Ans. $\dfrac{Ma^2}{12}$

10.6 Determine the mass moment of inertia of a thin aluminium plate of 3 mm thickness shown in Fig. E.10.6 about its base. The plate is made up of a semicircular plate fitted over a square plate. Take density of aluminium to be 2770 kg/m^3.

Ans. 2.9×10^{-4} kg.m^2

10.7 Determine the mass moments of inertia of a thin homogenous plate of 3 mm thickness bent into a shape as shown in Fig. E.10.7 about the centroidal axes. The ends are of semicircular shapes of 6 cm diameter. Take density of material as 7850 kg/m^3.

Ans. $I_{xx} = 6.3 \times 10^{-5}$ kg.m^2, $I_{yy} = 1.74 \times 10^{-4}$ kg.m^2, $I_{zz} = 2.09 \times 10^{-4}$kg.m^2

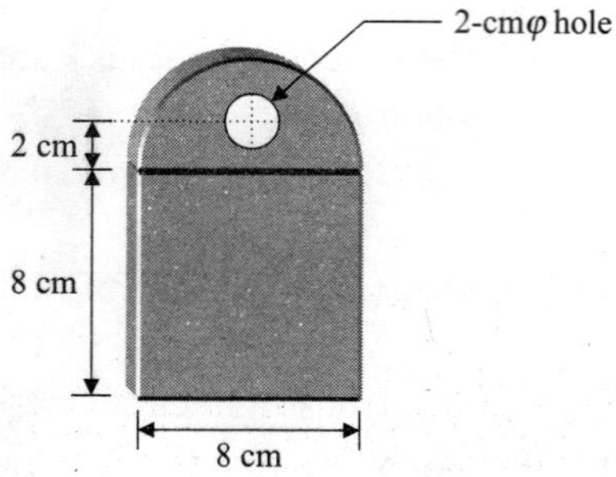

Fig. E.10.6

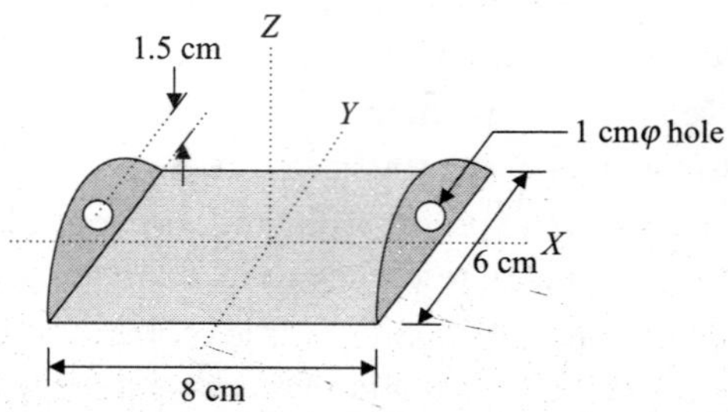

Fig. E.10.7

10.8 Determine the mass moments of inertia of a thin steel plate of 4 mm thickness shown in Fig. E.10.8 about the centroidal axes. Take density of steel to be 7850 kg/m^3.

Ans. $\bar{I}_{xx} = 2.5 \times 10^{-4}$ kg.m^2, $\bar{I}_{yy} = 2.94 \times 10^{-4}$kg.m^2, $\bar{I}_{zz} = 5.44 \times 10^{-4}$ kg.m^2

10.9 Determine the mass moment of inertia of a thin steel plate of 4 mm thickness shown in Fig. E.10.9 about its base. Take density of steel to be 7850 kg/m^3.

Ans. 6.1×10^{-4} kg.m^2

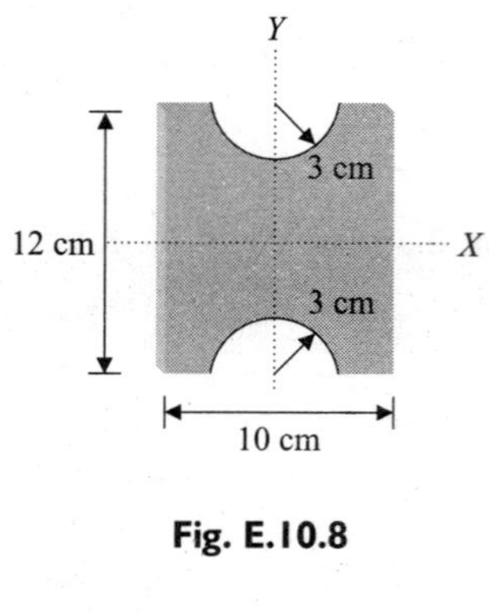

Fig. E.10.8

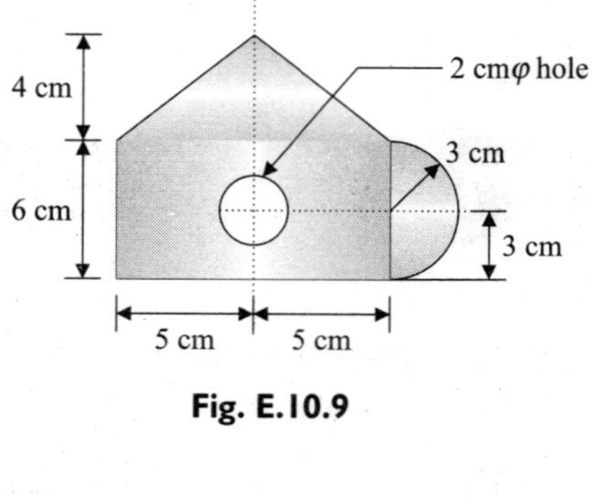

Fig. E.10.9

10.10 Find the mass moment of inertia of a solid sphere of mass M, radius R about a tangent axis.

Ans. $\frac{7}{5}MR^2$

10.11 A thin annular ring of radius R and mass M is mounted on a knife-edge as shown in Fig. E.10.11. Determine its mass moment of inertia about the knife-edge.

Ans. $2MR^2$

10.12 Two uniform slender rods of lengths l_1 and l_2 and mass per unit length m are welded together to form a T-shape as shown in Fig. E.10.12. Determine the mass moment of inertia of the composite rod about the axis of suspension.

Ans. $\frac{M}{12}[4l_1^3 + l_2^3 + 12\, l_2 l_1^2]$

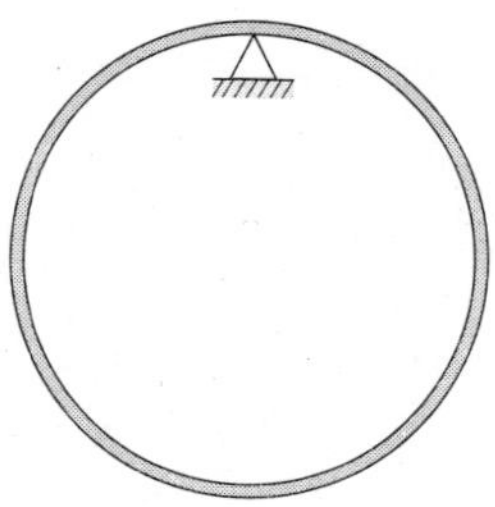

Fig. E.10.11

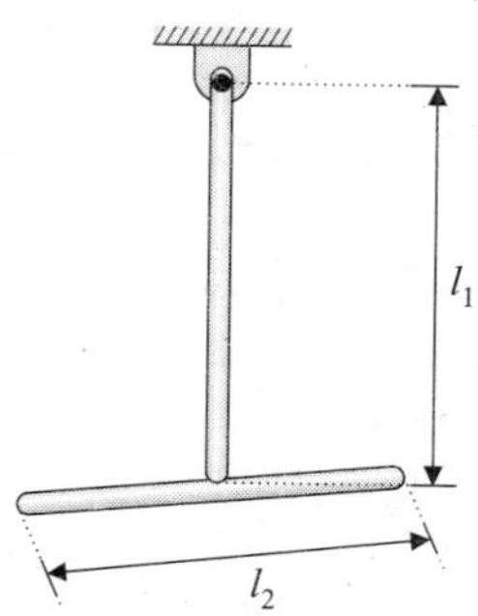

Fig. E.10.12

10.13 Determine the mass moment of inertia of a cylinder of radius R, height H and mass M about the axis of rotation when it is suspended (i) vertically, and (ii) horizontally. Refer Fig. E.10.13.

Ans. $M\left[\frac{R^2}{4} + \frac{H^2}{3}\right], \frac{3}{2}MR^2$

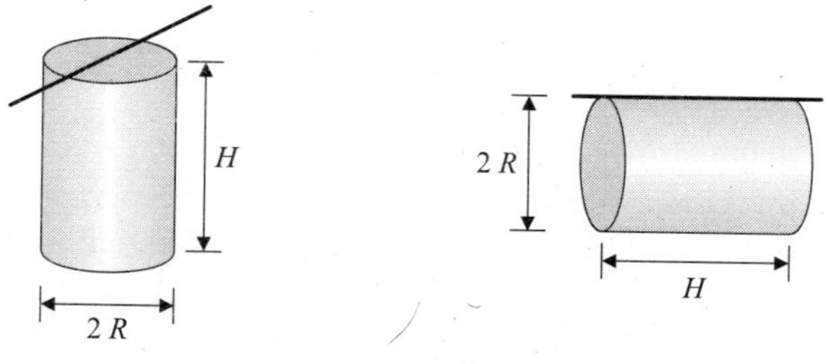

Fig. E.10.13

10.14 A cylinder of 400 mm diameter and 1000 mm height rests vertically. Over this, a cone of 400 mm base diameter and 500 mm height is placed such that the axis of the cone coincides with the axis of the cylinder. Find out the mass moment of inertia of this composite solid about a line which passes through the vertex of the cone and which is parallel to the base of the cylinder if the mass density is 4000 kg/m^3.

Ans. 562.6 kg.m^2

10.15 A square prism of 200 mm × 200 mm cross section and 400 mm height stands vertically over a cylinder of 300 mm diameter and 500 mm height. Calculate the mass moment of inertia of the composite solid about the vertical axis of symmetry if mass density of the material is 2000 kg/m^3.

Ans. 1.008 kg.m^2

10.16 A right circular cone made of steel (density 7850 kg/m^3) has an altitude of 30 cm and a base diameter of 20 cm. A 5 cm deep hole and of 8-cm diameter is drilled from the centre of the base of the cone and filled with a material of density 11,375 kg/m^3. Determine the mass moment of inertia of the resulting solid with respect to the geometric axis.

Ans. 7.5×10^{-2} kg.m^2

10.17 A uniform slender rod weighing 0.15 kg is bent into the shape as shown in Fig. E.10.17. The leg *AB* is in the *XZ* plane and the leg *CD* is in the *XY* plane. Compute the moment of inertia about *X*, *Y* and *Z* axes. All dimensions are in centimetres.

Ans. $I_{xx} = 9 \times 10^{-5}$ kg.m^2, $I_{yy} = 4.05 \times 10^{-4}$ kg.m^2, $I_{zz} = 4.05 \times 10^{-4}$ kg.m^2

10.18 A composite body is made of a cone of 4 cm base diameter and 10 cm height, and a sphere of 3-cm diameter placed over the base of the cone such that the axes of the two solids coincide as shown in Fig. E.10.18. Determine the mass moment of inertia of the composite solid about the *AA* axis. Take density of the solid as 7850 kg/m^3.

Ans. 3.5×10^{-3} kg.m^2

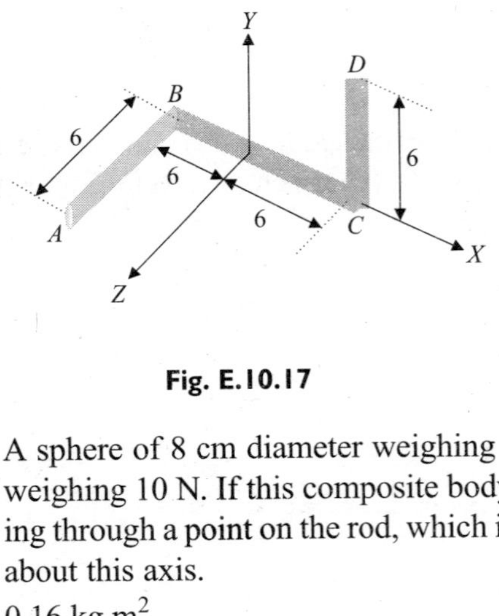

Fig. E.10.17

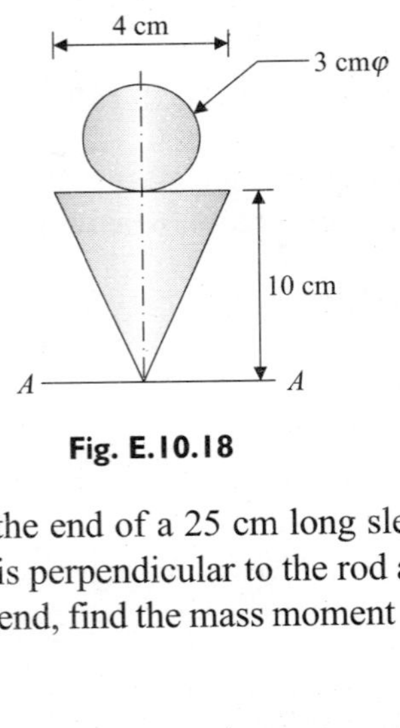

Fig. E.10.18

10.19 A sphere of 8 cm diameter weighing 25 N is attached to the end of a 25 cm long slender rod weighing 10 N. If this composite body rotates about an axis perpendicular to the rod and passing through a point on the rod, which is 5 cm from the free end, find the mass moment of inertia about this axis.

Ans. 0.16 kg.m^2

10.20 A homogeneous block with a mass density of 7850 kg/m^3 with the dimensions as shown in Fig. E.10.20 has a hole of 30 mm diameter drilled at its centre. Determine its mass moment of inertia about the vertical axis.

Ans. 1.54×10^{-3} kg.m^2

10.21 A homogeneous cylinder with mass density of 7850 kg/m^3 has two cylindrical holes of 2 cm diameter cut in it as shown in Fig. E.10.21. Determine its mass moment of inertia about the vertical axis.

Ans. 2.95×10^{-3} kg.m^2

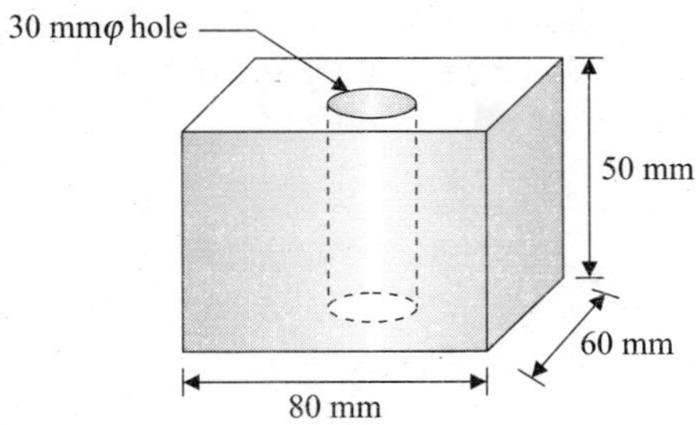

Fig. E.10.20

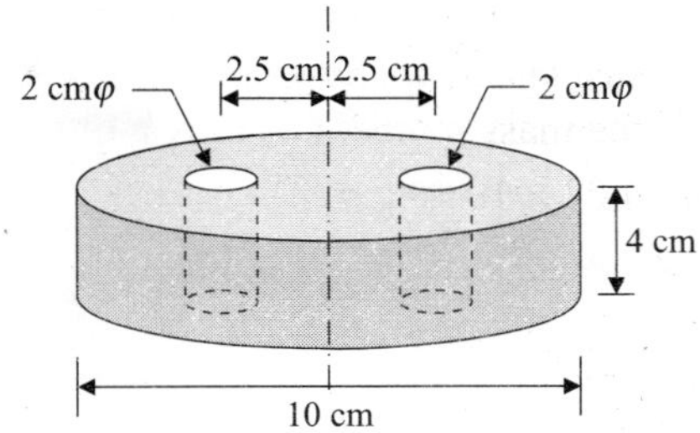

Fig. E.10.21

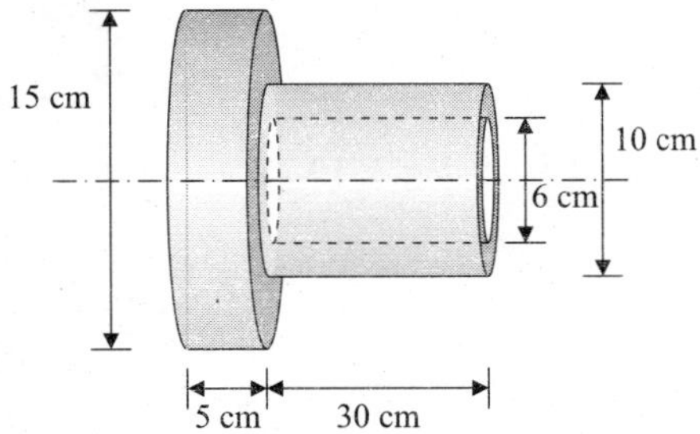

Fig. E.10.22

10.22 Determine the mass moment of inertia about the axis of rotation of the machine component shown in Fig. E.10.22, in which a hollow shaft is welded to a circular plate. Take density of material to be 7850 kg/m^3.

Ans. 3.96×10^{-2} kg.m^2

10.23 Determine the mass moment of inertia of the turntable shown in Fig. E.10.23 about the axis of rotation. Take density of material to be 3000 kg/m^3.

Ans. 2.56×10^{-2} kg.m^2

10.24 Determine the mass moment of inertia of the machine component shown in Fig. E.10.24 about the axis of rotation. The dimensions of the plate are 15 cm × 6 cm × 5 mm and the cylinder is of diameter 6 cm and height 5 cm. Take density of plate material to be 7850 kg/m^3 and of cylinder to be 6400 kg/m^3.

Ans. 1.1×10^{-2} kg.m^2

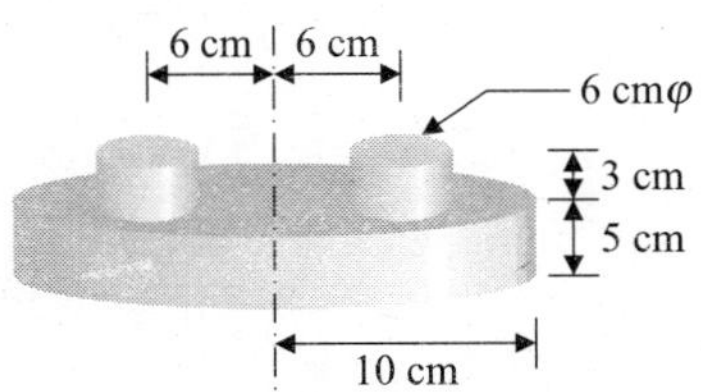

Fig. E.10.23

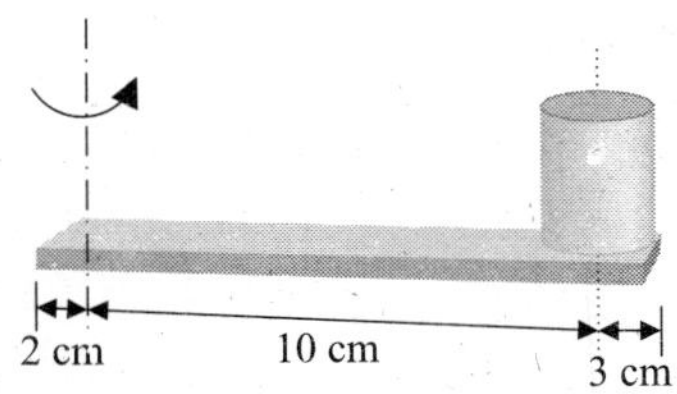

Fig. E.10.24

10.25 A compound pendulum consists of a circular metallic disc welded to a thin metallic rod and suspended as shown in Fig. E.10.25. If their respective masses are 0.4 kg and 0.2 kg, determine the mass moment of inertia of the pendulum about the axis of rotation.

Ans. 4.07×10^{-2} kg.m^2

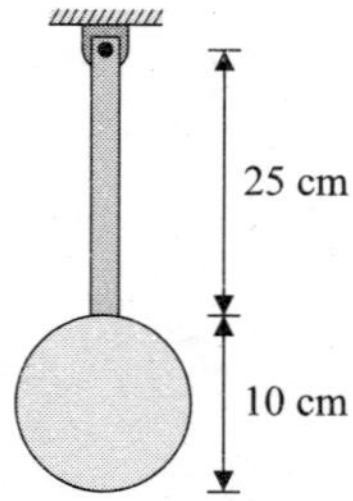

Fig. E.10.25

11 Virtual Work

11.1 INTRODUCTION

In Chapter 5, we analysed the equilibrium of particles and rigid bodies by applying the *equations of equilibrium* to solve the unknown forces. In this chapter, we will introduce an alternative method, known as **virtual work** principle, to analyse the same type of equilibrium of particles and rigid bodies. This method is particularly useful to solve the unknown forces when there are more than one member linked together to form a mechanical system or mechanism. Prior to applying the principle of virtual work, we must understand work done on a particle and a rigid body, which we will discuss in the following few sections.

11.2 WORK DONE ON A PARTICLE

When a force acts on a particle, which is not constrained to move, it causes a *displacement* of the particle. The force is then said to have done **work** on the particle. This definition of work is quite different from our daily usage of the word 'work,' which we refer to any activity involving muscular or mental effort. Consider a force $\vec{F}$ acting on a particle at A causing a displacement $\vec{s}$ (from the point A to B) in the direction of the force. We then define work done on the particle as a **product** of magnitudes of the **force** and the **displacement**. Mathematically, we can write this as

$$W = Fs \tag{11.1}$$

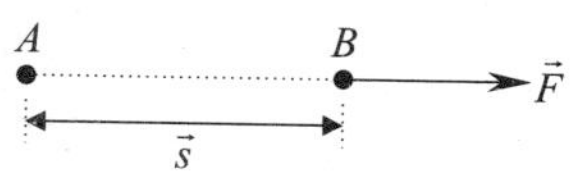

Fig. 11.1 Work done by a force displacing a particle in the direction of the force

A ball freely falling under gravity or a block being pulled on a smooth plane by a horizontal force (refer Fig. 11.2) are examples in which the motion of the body is in the direction of the applied force. (As the line of action of the force passes through the centre of gravity of each body, we can idealize each body as a particle). However, we should note that due to *constraints* involved, the displacement of a particle under the action of forces would not always occur in the direction of the force. Consider for instance, a block sliding down a smooth inclined plane due to pull of gravity or a block being pulled on a smooth plane by an inclined force (refer Fig. 11.3). In these two cases, we observe that the direction of motion is *different* to the direction of the force, i.e., inclined at an angle θ to the direction of the force.

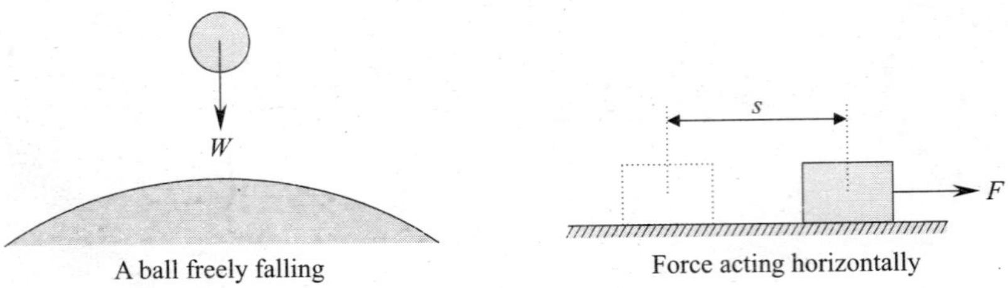

Fig. 11.2

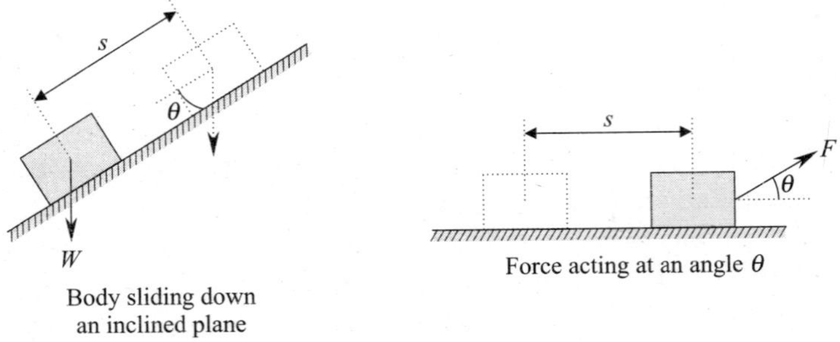

Fig. 11.3

Hence, we can define work done in general, as a product of the *component* of the force in the direction of motion and the displacement. Mathematically, we can write this as

$$W = (F \cos \theta)s \tag{11.2}$$

As the above expression can also be written as

$$W = (F)(s \cos \theta) \tag{11.3}$$

work done can also be defined as a product of force and the *component* of displacement in the direction of the force.

When a system of forces act on a particle, the work done by the system of forces is given by the *algebraic sum* of works done by individual forces. As work done adds up algebraically, it is a *scalar* quantity, that is having magnitude, but not direction. However, we see that force and displacement are vectors. The product of two vectors resulting in a scalar quantity can best be represented by the dot product of two vectors. Thus, work done can be written in vector form as

$$W = \vec{F}.\vec{s} = Fs \cos \theta \tag{11.4}$$

If force and displacement are in the same direction, i.e., $\theta = 0°$ then $\cos \theta = 1$ and hence work done is $W = Fs$, which is same as Equation 11.1. If force and displacement are in the opposite direction, i.e., $\theta = 180°$ then $\cos \theta = -1$ and hence work done is $W = -Fs$. We saw in Chapter 6 that the force of friction always acts in the direction *opposite* to that of the motion; hence, work done by the force of friction is always **negative**. In general, if the component of force is in the *direction* of displacement, then θ is an *acute* angle and $\cos\theta$ is *positive*. Hence, work done in such a case is *positive*. If the component of force is in the *direction opposite* to that of displacement then θ is an *obtuse* angle and $\cos\theta$ is *negative*. Hence, work done in such a case is *negative*.

The work done is **zero** in the following cases:

(i) When the displacement [s] is zero Even though forces may act on a particle, if there is **no** displacement of the particle then *no* work is done on the particle. Consider a block resting on a table. In its free-body diagram, we see that even though its weight W and normal reaction R are acting, they do *no* work on the block, as there is *no* displacement of the block. Similarly, in the case of a ball suspended by a string and a beam supported at one end by a hinge support, no work is done by the forces acting, as there is no displacement involved.

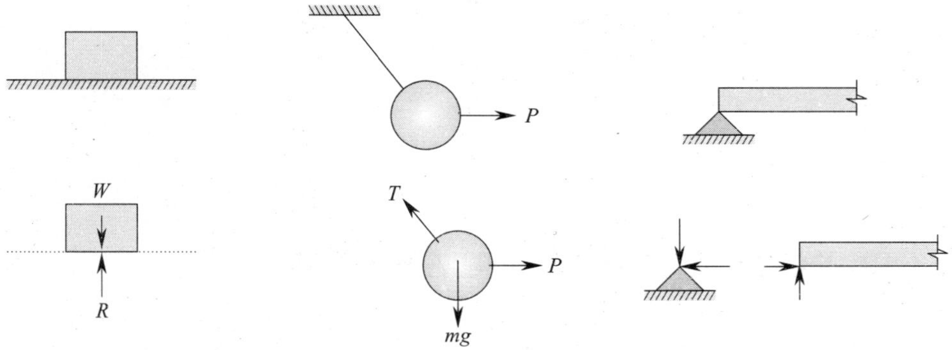

Fig. 11.4 No work is done by forces as there is no displacement

(ii) When the motion is at right angles to the direction of the force When the motion is at right angles to the direction of the force, we see that $\theta = 90°$ and hence, $\cos\theta = 0$. Thus, work done is **zero**. Consider a block moving along a horizontal plane as shown in Fig. 11.5. Since the displacement is at right angles to the direction of the forces, its weight and normal reaction at the contact surfaces do *no* work on the block.

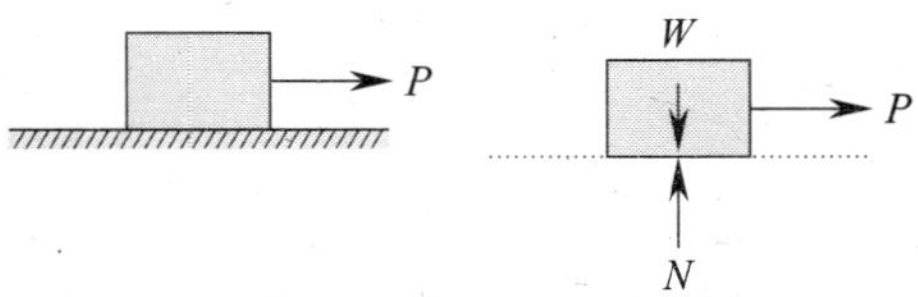

Fig. 11.5 No work is done by forces acting at right angles to the direction of motion

(iii) Total work done is zero When a particle is in *static equilibrium* then the resultant force acting on it is **zero**. Hence, the total work done by the system of forces is also **zero**. Consider a system of concurrent forces $\vec{F}_1, \vec{F}_2, \ldots \vec{F}_n$ acting on a particle causing a displacement $\Delta\vec{s}$. Then we can write the total work done on the particle as

$$
\begin{aligned}
W &= \vec{F}_1 .\, \Delta\vec{s} + \vec{F}_2 . \Delta\vec{s} + \ldots + \vec{F}_n . \Delta\vec{s} \\
&= [\vec{F}_1 + \vec{F}_2 + \ldots + \vec{F}_n] . \Delta\vec{s} \\
&= \sum \vec{F} . \Delta\vec{s} \\
&= 0 \qquad \text{[as the resultant of the forces is zero]} \qquad (11.5)
\end{aligned}
$$

(iv) Work done by internal forces Consider a system of bodies, such as interconnected bodies or mechanisms as shown in Fig. 11.6. The internal force such as tension in the string in the system of blocks shown, acts on the blocks in the opposite directions and hence the net work done for any displacement is zero. Similarly, in the reactions at the pin joint at B in the mechanism shown, no work is done by the reactions as they act in the opposite directions on each link and thus cancel out. Thus, this method is useful in solving the equilibrium of such systems or mechanisms without the need to draw separate free-body diagrams for individual members.

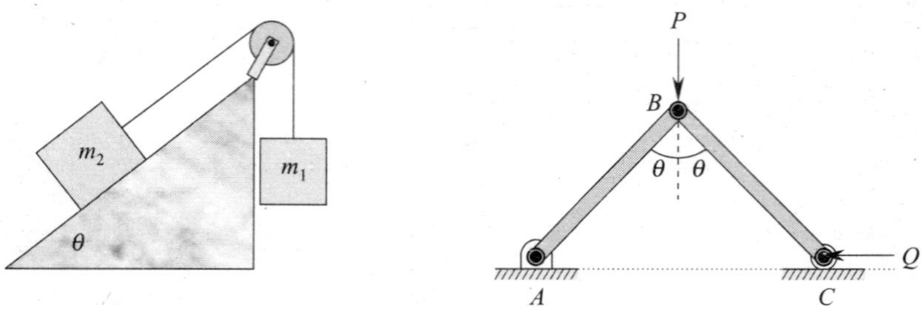

Fig. 11.6 Net work done by internal forces is zero

11.3 WORK DONE IN STRETCHING A SPRING

In the previous section, we assumed the force to be constant, i.e., constant in magnitude and direction, and thus we defined work done by the force. In this section, we will discuss work done by a *varying force*, varying in magnitude but constant in direction, acting on a particle. A **spring** whose one end is fixed and the other end stretched is an example for a variable force constant **in direction** but varying in magnitude.

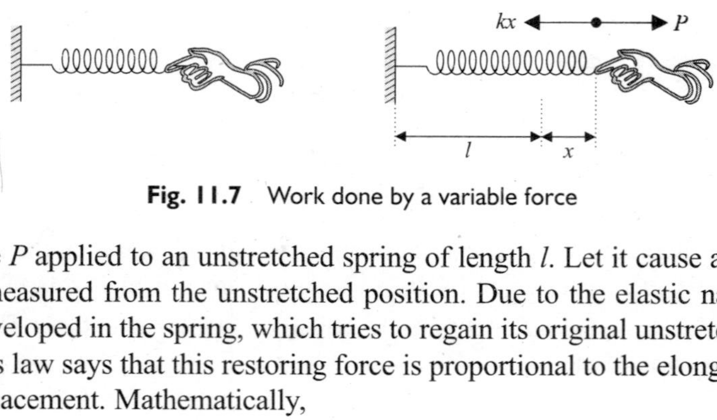

Fig. 11.7 Work done by a variable force

Consider a force P applied to an unstretched spring of length l. Let it cause an elongation x of the spring, where x is measured from the unstretched position. Due to the elastic nature of the spring, a restoring force is developed in the spring, which tries to regain its original unstretched position. Within elastic limit, Hooke's law says that this restoring force is proportional to the elongation and opposite to the direction of displacement. Mathematically,

$$F \propto -x \tag{11.6}$$

Introducing a constant of proportionality, we have

$$F = -kx \tag{11.7}$$

where the constant of proportionality k is called the **stiffness** of the spring or the **spring constant**.

If the spring is stretched slowly, such that it is not accelerated, then the restoring force must be equal and opposite to the applied force for equilibrium to be maintained, i.e.,

$$P = kx \tag{11.8}$$

The relationship between the applied force P and elongation x can be represented graphically as shown in Fig. 11.8.

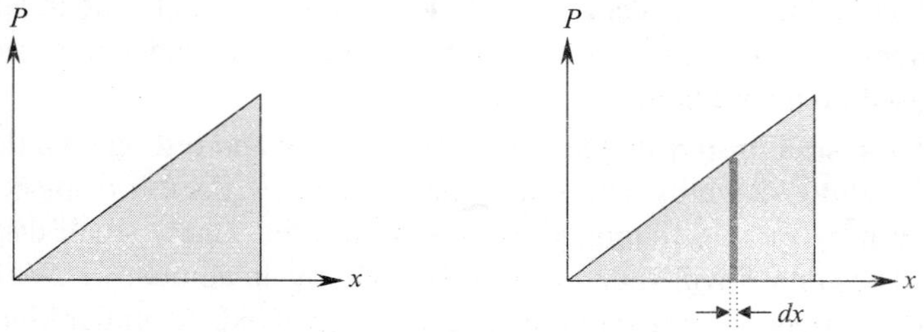

Fig. 11.8 Graphical representation of applied force and elongation

We see that the force and the elongation vary linearly. Suppose we consider an infinitesimally small elongation dx. Then over this infinitesimally small elongation, we can assume the force acting on the spring to be *constant*. Hence, work done over this infinitesimally small displacement is given as

$$dW = Pdx \tag{11.9}$$

Therefore, work done in stretching the spring to an elongation of x_o from its unstretched position is obtained by integrating the above expression between limits.

$$W = \int_0^{x_o} dW = \int_0^{x_o} Pdx$$

$$= \int_0^{x_o} kx\, dx$$

$$= \frac{1}{2}kx_o^2 \tag{11.10}$$

We see that the work done in stretching a spring is given by the area under the curve shown in Fig. 11.8.

11.4 WORK DONE ON A RIGID BODY

We know that a rigid body is subjected to moments in addition to the forces. Just as the forces cause *linear* displacements, moments cause *angular* displacements. In a similar manner as we defined work done on a particle, we can define work done on a rigid body. If a moment M acting on a rigid body causes an angular displacement θ then work done by the moment is defined as product of moment and angular displacement, i.e.,

$$W = M\theta \tag{11.11}$$

In the following discussion, we will refer the term displacement to a translation or a rotation and the term force to a direct force or a moment.

11.5 VIRTUAL DISPLACEMENT

Consider a particle or a rigid body at rest and in equilibrium under the action of a system of forces. Then work done by each individual force is **zero** as there is *no* displacement. As a result, the *total* work done by the system of forces is also zero. The displacement of a particle or a rigid body in equilibrium is *not* at all possible. However, we can assume an *imaginary* displacement to occur, particularly if the system is **partially** constrained. As this displacement is not actually occurring, but an imaginary one, we call it a **virtual displacement**. It is defined as any arbitrary infinitesimal change in position, consistent with the constraints imposed on the motion.

Consider a ladder in equilibrium, and leaning against a wall and resting on a floor as shown. Let us assume that the end A slides down by an infinitesimal amount δy. As this displacement is not actually occurring, we denote it by δy to differentiate it from an infinitesimally small displacement dy which may actually occur. Then end B will slide towards the right by an amount δx. We know that these two virtual displacements are dependent due to the constraints involved. A similar kind of virtual displacement can be assumed in the mechanism also as shown.

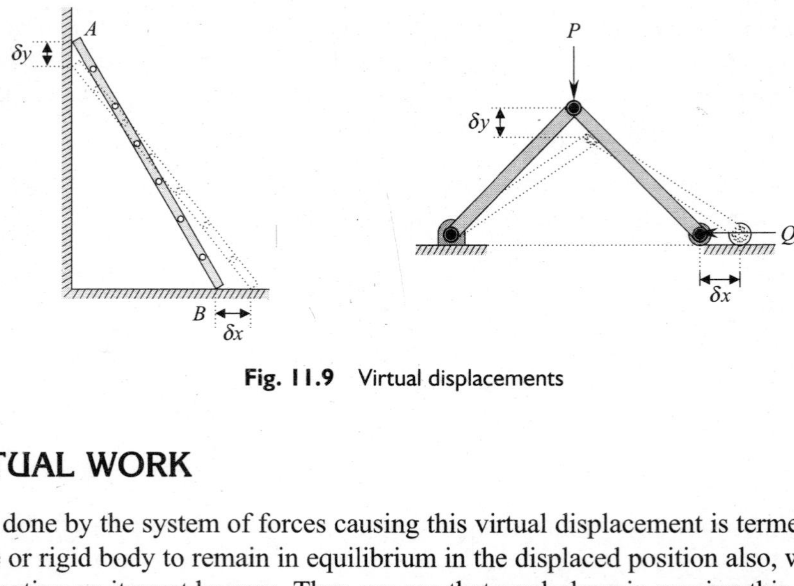

Fig. 11.9 Virtual displacements

11.6 VIRTUAL WORK

The total work done by the system of forces causing this virtual displacement is termed **virtual work**. For the particle or rigid body to remain in equilibrium in the displaced position also, we know that the resultant force acting on it must be zero. Thus, we see that work done in causing this virtual displacement is also zero. This is known as **principle of virtual work**. This was first stated forth by Johann (Jean) Bernoulli (1667–1748).

For a system of concurrent forces $\vec{F}_1, \vec{F}_2, \ldots \vec{F}_n$, the virtual work done is given as

$$\delta U = \vec{F}_1.\delta\vec{r} + \vec{F}_2.\delta\vec{r} + \ldots + \vec{F}_n.\delta\vec{r}$$
$$= [\vec{F}_1 + \vec{F}_2 + \ldots + \vec{F}_n].\delta\vec{r}$$
$$= \sum\vec{F}.\delta\vec{r} \tag{11.12}$$

As a system of concurrent forces can be replaced by a single resultant force, the virtual work done is equal to the work done by the resultant. For the body to remain in equilibrium in the displaced position, we know that the resultant must be zero. Hence, virtual work done in causing this virtual displacement is also zero, i.e.,

$$\delta U = (\sum \vec{F}).\delta \vec{r} = 0 \tag{11.13}$$

The necessary and sufficient condition for the equilibrium of a particle is zero virtual work done by all external forces acting on the particle during any virtual displacement consistent with the constraints imposed on the particle.

Similarly, for a rigid body, we can write the principle of virtual work as

$$\delta U = \sum F\,\delta r + \sum M\,\delta\theta = 0 \tag{11.14}$$

Example 11.1 A uniform ladder AB of length l and weight W leans against a smooth vertical wall and smooth horizontal floor as shown in Fig. 11.10. By the method of virtual work, determine the horizontal force P required to keep the ladder in equilibrium position. Also, determine the angle of inclination of the ladder with respect to the horizontal when P equal to (i) W, (ii) $W/2$, (iii) $W/4$ is applied to maintain equilibrium.

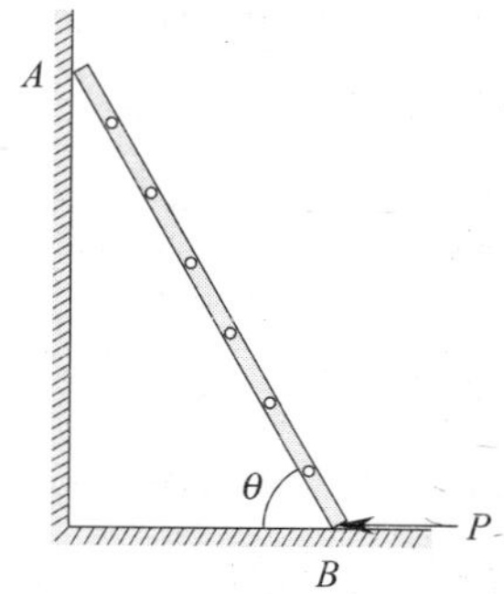

Fig. 11.10

Solution Under its own weight, the ladder tries to slide down, but the horizontal force P holds it in equilibrium. The free-body diagram of the ladder is shown in Fig. 11.10(a).

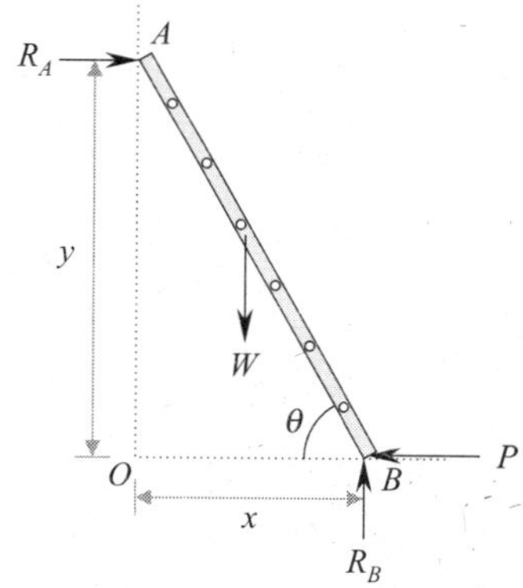

Fig. 11.10(a)

At the equilibrium position, let x and y be the respective positions of the ends B and A of the ladder from the origin O, the meeting point of the wall and floor. Because of the constraints, namely, the wall and the floor, which restrict the motion of the ladder, we can see that the respective positions x and y are related to each other by the relationship

$$x^2 + y^2 = l^2 \tag{a}$$

where l is the length of the ladder. For solving the unknown force P, by using virtual work principle, let us assume that the end A be given a virtual displacement δy in the vertical direction. Then virtual displacement δx of the end B is in the horizontal direction away from the wall. The ladder in the virtually displaced position is shown in Fig. 11.10(b). By differentiating the above expression (a), we get

$$2x\,\delta x + 2y\,\delta y = 0$$

$$\Rightarrow \qquad \delta x = -\frac{y}{x}\,\delta y$$

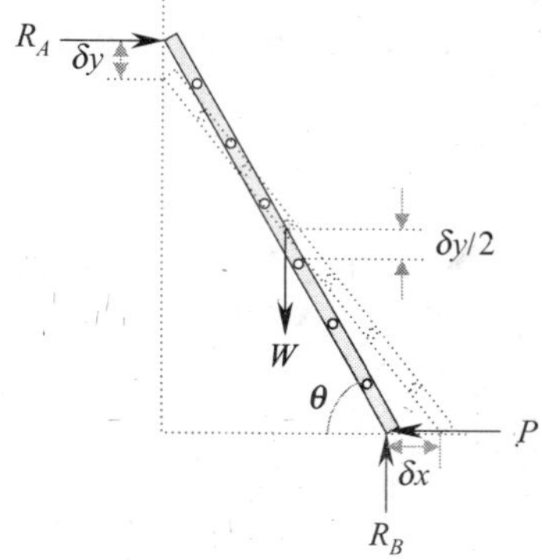

Fig. 11.10(b)

From the figure, we should note that when there is a decrease in the value of y, there is an increase in the value of x. Hence, considering only positive values of virtual displacements, the above expression reduces to

$$\delta x = \frac{y}{x}\,\delta y$$

As the centre of gravity of the *uniform* ladder lies at the centre of its length, we see that its virtual displacement is $\delta y/2$. Applying the principle of virtual work, we have

$$\delta U = 0 \Rightarrow$$

$$-P\,\delta x + W\frac{\delta y}{2} = 0$$

It should be noted that reactions R_A and R_B do no work, as the virtual displacements at the ends A and B are perpendicular to the direction of the forces. Also, the displacement δx is in the direction opposite to that of P and hence the virtual work done by P is negative. Therefore,

$$-P\left[\frac{y}{x}\right]\delta y + W\frac{\delta y}{2} = 0$$

$$\Rightarrow \qquad P = \frac{W}{2}\frac{x}{y}$$

But we know from the triangle OAB that

$$\frac{y}{x} = \tan\theta$$

Hence,
$$P = \frac{W}{2\tan\theta}$$

The angle of inclination of the ladder with respect to the horizontal is given as

$$\theta = \tan^{-1}\left[\frac{W}{2P}\right]$$

(i) When $P = W$, $\theta = 26.57°$
(ii) When $P = W/2$, $\theta = 45°$
(iii) When $P = W/4$, $\theta = 63.43°$

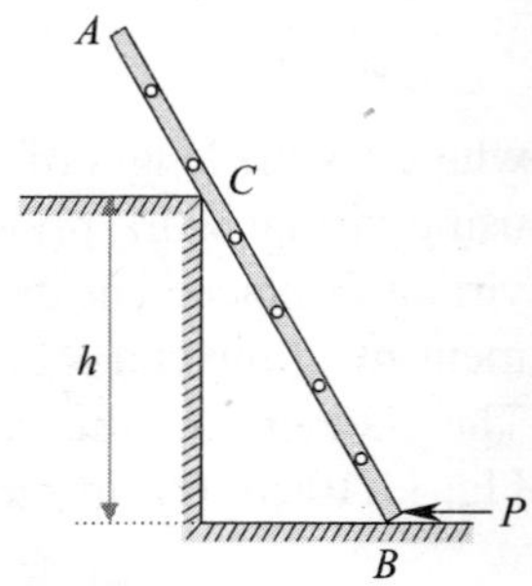

Fig. 11.11

Example 11.2 A uniform ladder AB of length l and weight W leans against a smooth vertical wall and a smooth horizontal floor as shown in Fig. 11.11. By the method of virtual work, determine the horizontal force P required to keep the ladder in equilibrium position.

Solution Under its own weight, the ladder tries to slide down, but the horizontal force P holds it in equilibrium. The free-body diagram of the ladder is shown in Fig. 11.11(a).

Here we cannot establish a relationship between x and y as derived in the previous problem. Instead, we proceed as follows. Let θ be inclination of the ladder with respect to the horizontal. From the geometry of the triangle, we see that the location x of the end B and the location y of the centre of gravity of the ladder with respect to the origin are respectively

$$x = \frac{h}{\tan\theta} \qquad \text{and} \qquad y = \frac{l}{2}\sin\theta$$

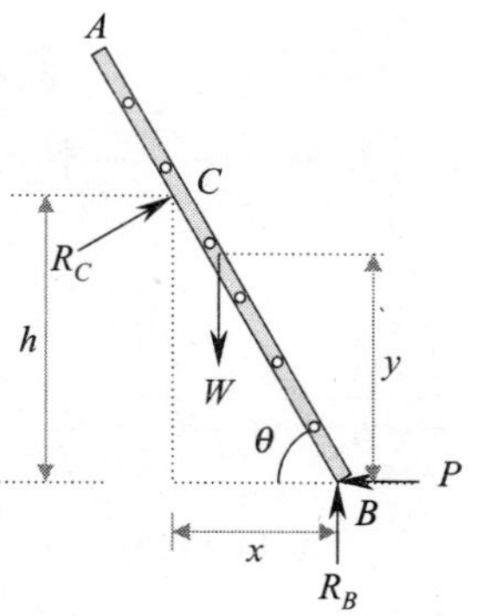

Fig. 11.11(a)

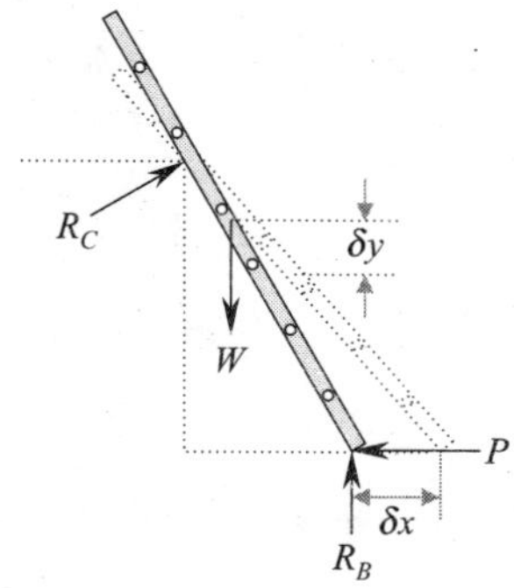

Fig. 11.11(b)

Therefore, the virtual displacements are obtained by differentiating the above expressions,

$$\delta x = -h \operatorname{cosec}^2 \theta \, \delta\theta \qquad \text{and} \qquad \delta y = \frac{l}{2} \cos \theta \, \delta\theta$$

From the figure, we see that as θ decreases, y also decreases but x increases. Hence, considering *only* positive virtual displacements, the above expressions reduce to

$$\delta x = h \operatorname{cosec}^2 \theta \, \delta\theta \qquad \text{and} \qquad \delta y = \frac{l}{2} \cos \theta \, \delta\theta$$

Applying the principle of virtual work,

$\delta U = 0 \Rightarrow$

$$-P\delta x + W\delta y = 0$$

It should be noted that reactions R_B and R_C do no work, as the virtual displacements of the contact points B and C are perpendicular to the direction of the forces. Therefore,

$$-P[h \operatorname{cosec}^2 \theta \, \delta\theta] + W\left[\frac{l}{2} \cos\theta \, \delta\theta\right] = 0$$

$$\Rightarrow \qquad P = \frac{Wl}{2h} \frac{\cos\theta}{\operatorname{cosec}^2\theta}$$

$$= \frac{Wl}{2h} \sin^2 \theta \cos \theta$$

Example 11.3 A parallel rule *ABCD* used in engineering drawing is shown in Fig. 11.12 lying on a table. With link *AB* held fixed, a force *P* is applied at *D* along the *X*-axis and a moment at joint *A*. For the given values of *P* and *M*, determine the angle θ at which equilibrium is maintained.

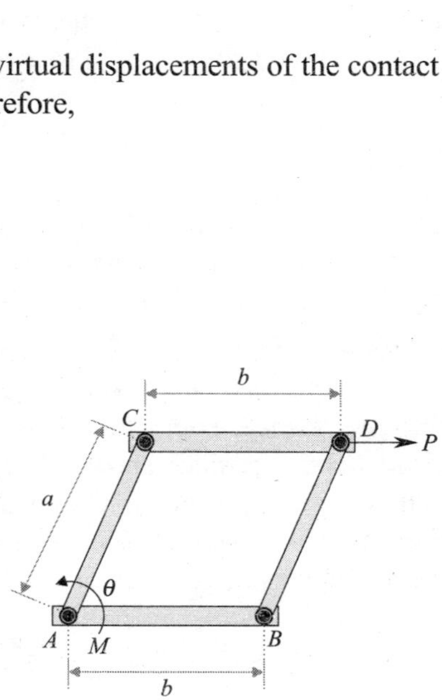

Fig. 11.12

Solution Keeping joint *A* as the origin, we get the position of joint *D* [refer Fig. 11.12(a)] as

$$x = b + a \cos \theta$$

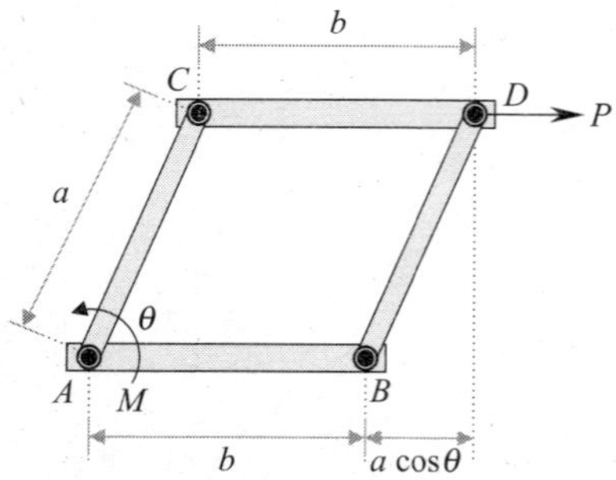

Fig. 11.12(a)

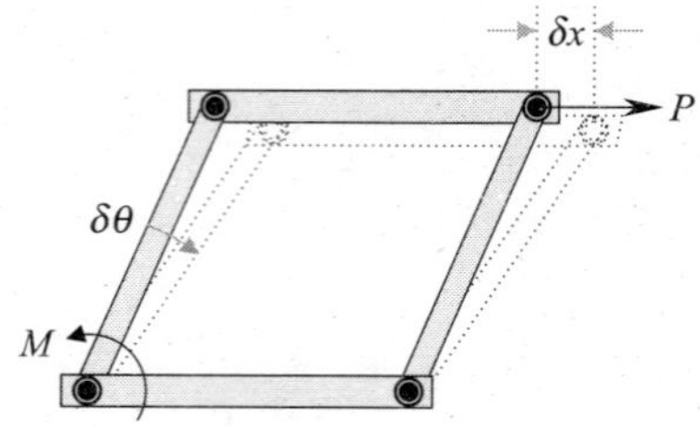

Fig. 11.12(b)

If we give a virtual displacement $\delta\theta$ at joint A then the virtual displacement of joint D is obtained by differentiating the expression for x:

$$\delta x = -a \sin\theta \, \delta\theta$$

From Fig. 11.12(b), we see that as θ decreases, x increases. Hence, considering *only* positive values of virtual displacements, the above expression reduces to

$$\delta x = a \sin\theta \, \delta\theta$$

Therefore, applying the principle of virtual work,

$\delta U = 0 \Rightarrow$

$$-M\delta\theta + P\,\delta x = 0$$

Note that as the virtual displacement δx is in the direction of the applied force P, the virtual work done by P is positive, while as the angular virtual displacement $\delta\theta$ is in the direction opposite to the direction of moment, the virtual work done by M is negative. Therefore,

$$-M\delta\theta + Pa \sin\theta\,\delta\theta = 0$$

$$\Rightarrow \qquad \theta = \sin^{-1}\left[\frac{M}{Pa}\right]$$

Example 11.4 The mechanism shown in Fig. 11.13 is made up of three members, each of equal length l and weight W. Determine the moment M required to be applied to maintain equilibrium in the position shown.

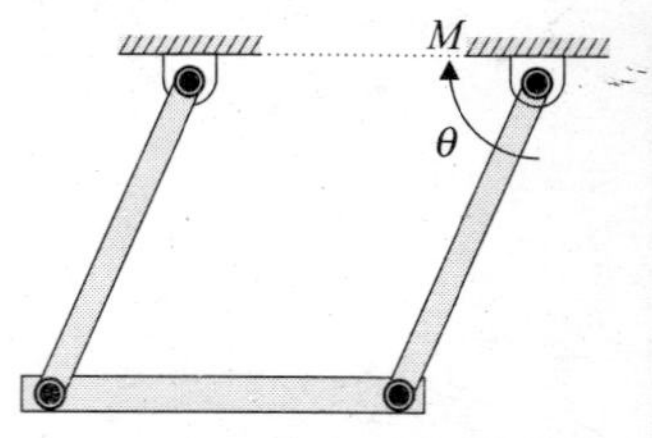

Fig. 11.13

Solution The mechanism under its own weight remains in equilibrium with the end members vertical and the lower member in the horizontal positions. Hence, external moment is required to be applied to displace it to the inclined position shown. The free-body diagram of the mechanism in the displaced position is shown in Fig. 11.13(a). Let θ be the inclination of the vertical members with respect to the horizontal. Then the y-coordinates, y_1, y_2 and y_3 of the centres of gravities of left, right and lower members respectively are

$$y_1 = y_2 = \frac{l}{2}\sin\theta \qquad \text{and} \qquad y_3 = l\sin\theta$$

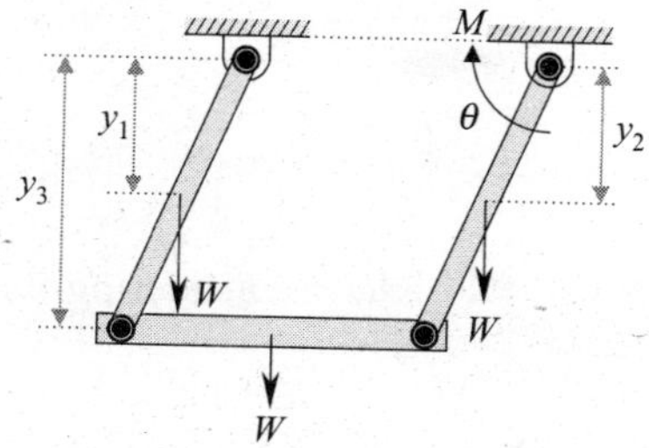

Fig. 11.13(a)

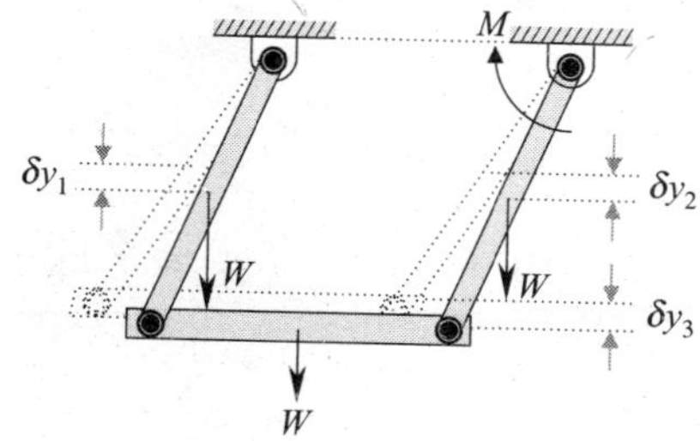

Fig. 11.13(b)

The virtual displacements of the centres of gravities of the members are then given as

$$\delta y_1 = \delta y_2 = \frac{l}{2} \cos\theta\,\delta\theta \qquad \text{and} \qquad \delta y_3 = l\cos\theta\,\delta\theta$$

Therefore, applying the principle of virtual work,

$$-W\delta y_1 - W\delta y_2 - W\delta y_3 + M\delta\theta = 0$$

Note that the virtual works done by the weights of the members are negative as the virtual displacements δy_1, δy_2 and δy_3 are in the direction opposite to that of the weights.

$$-W\frac{l}{2}\cos\theta\,\delta\theta - W\frac{l}{2}\cos\theta\,\delta\theta - Wl\cos\theta\,\delta\theta + M\delta\theta = 0$$

$$\Rightarrow \qquad M = l\cos\theta\left[\frac{W}{2} + \frac{W}{2} + W\right]$$

$$= 2Wl\cos\theta$$

Example 11.5 In the slider crank mechanism shown in Fig. 11.14, determine the horizontal force P that must be applied to hold the system in equilibrium for a given value of moment M applied at the crank.

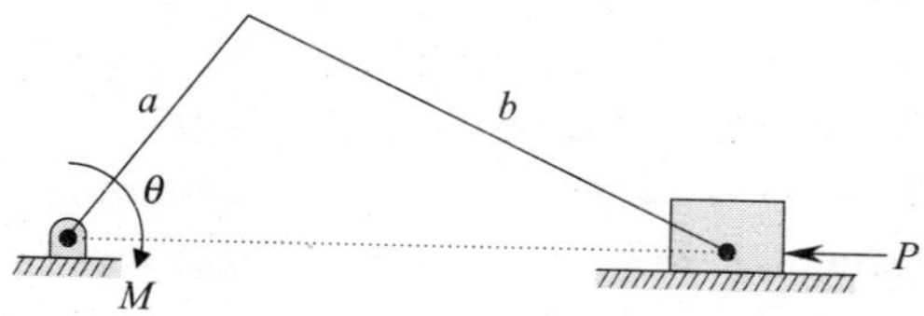

Fig. 11.14

Solution From Fig. 11.14 (a), we see that

$$x = a\cos\theta + \sqrt{b^2 - a^2\sin^2\theta}$$

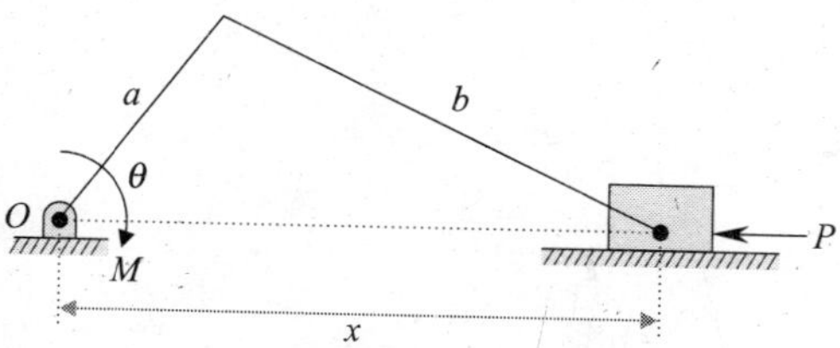

Fig. 11.14(a)

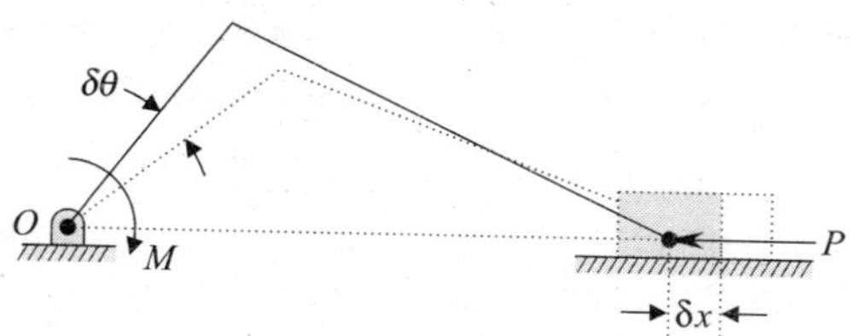

Fig. 11.14(b)

Upon differentiation, we get

$$\delta x = \left[-a \sin\theta + \frac{1}{2}(b^2 - a^2 \sin^2\theta)^{\frac{-1}{2}} (-2a^2 \sin\theta \cos\theta) \right] \delta\theta$$

From Fig. 11.14(b), we see that as θ decreases, x increases and hence considering *only* positive values of virtual displacements, the above expression reduces to

$$\delta x = \left[a \sin\theta + (b^2 - a^2 \sin^2\theta)^{\frac{-1}{2}} (a^2 \sin\theta \cos\theta) \right] \delta\theta$$

Applying the principle of virtual work, we get

$$M\delta\theta - P\delta x = 0$$

Note that the virtual displacement δx is in the direction opposite to that of the force P and hence the virtual work done by P is negative.

$$M\delta\theta - P\left[a \sin\theta + (b^2 - a^2 \sin^2\theta)^{\frac{-1}{2}} (a^2 \sin\theta \cos\theta) \right] \delta\theta = 0$$

$$\Rightarrow \qquad M = P\left[a \sin\theta + a^2 \sin\theta \cos\theta\, (b^2 - a^2 \sin^2\theta)^{\frac{-1}{2}} \right]$$

$$\therefore \qquad P = \frac{M}{a \sin\theta[1 + a \cos\theta/(\sqrt{b^2 - a^2 \sin^2\theta})]}$$

Example 11.6 Two uniform bars of equal lengths l are hinged and supported as shown in Fig. 11.15. For a given vertical force P, determine the value of the horizontal force Q that would hold the system in equilibrium. Neglect weight of the bars.

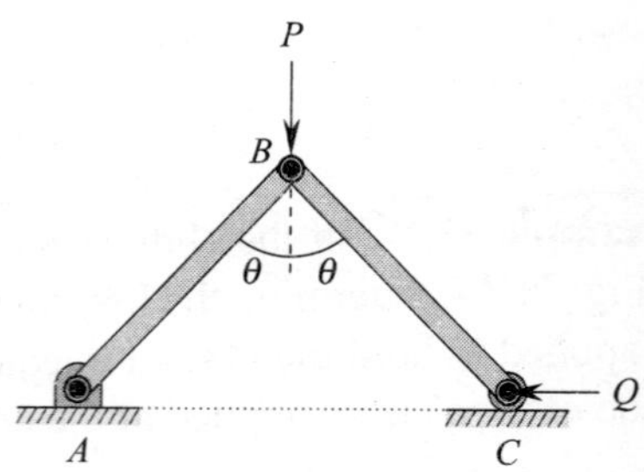

Fig. 11.15

Solution Keeping the hinged support A as the origin, the positions of C and B are given respectively as

$$x = 2l \sin\theta \qquad \text{and} \qquad y = l \cos\theta$$

Hence, the virtual displacements along x and y directions are obtained by differentiating the above expressions:

$$\delta x = 2l \cos\theta\,\delta\theta \qquad \text{and} \qquad \delta y = -l \sin\theta\,\delta\theta$$

Note that as θ increases, x also increases but y decreases. Therefore, considering *only* positive values of virtual displacements, the above expressions reduce to

$$\delta x = 2l \cos\theta\,\delta\theta \qquad \text{and} \qquad \delta y = l \sin\theta\,\delta\theta$$

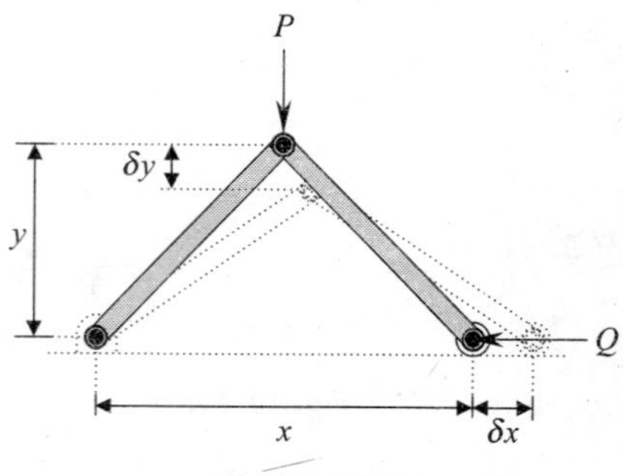

Fig. 11.15(a)

Applying the principle of virtual work, we have

$$P\delta y - Q\delta x = 0$$

$$P[l \sin\theta]\delta\theta - Q[2l \cos\theta]\delta\theta = 0$$

$$P \sin\theta - 2Q \cos\theta = 0$$

$$\Rightarrow \qquad Q = \frac{P \tan\theta}{2}$$

Example 11.7 A two bar pendulum has two identical links, each of length l and weight W. A horizontal force P holds the pendulum in equilibrium in the position shown. Determine the angles of inclination θ_1 and θ_2 of the two links with respect to the vertical

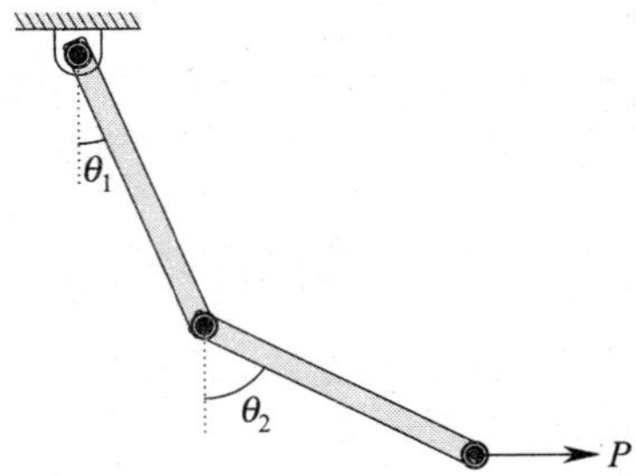

Fig. 11.16

Solution With the point of support as the origin, the locations of the points of application of the forces W and P are

$$y_1 = \frac{l}{2}\cos\theta_1$$

$$y_2 = l\cos\theta_1 + \frac{l}{2}\cos\theta_2$$

$$x = l\sin\theta_1 + l\sin\theta_2$$

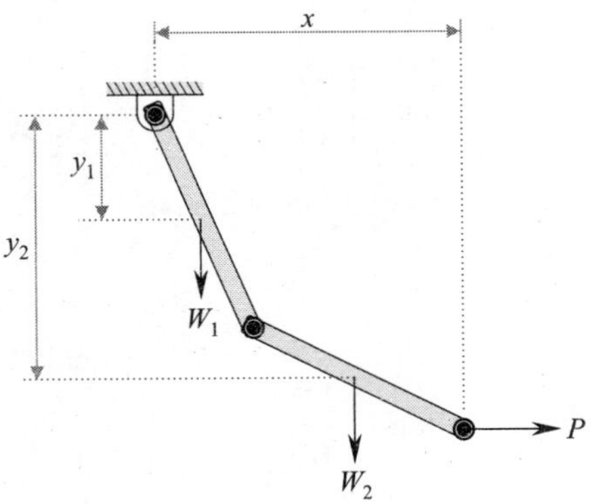

Fig. 11.16(a)

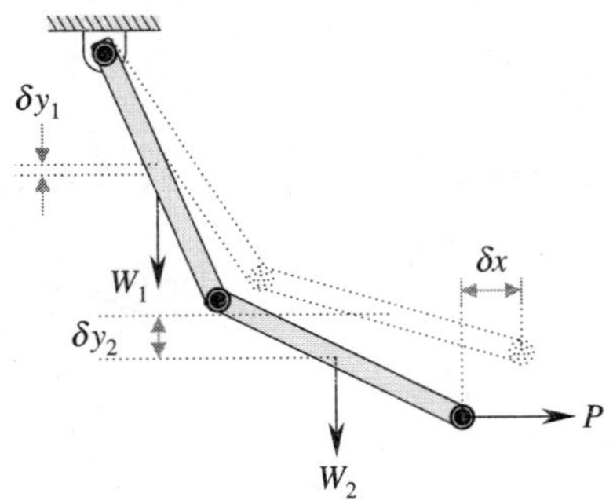

Fig. 11.16(b)

The virtual displacements are obtained by differentiating the above expressions:

$$\delta y_1 = -\frac{l}{2}\sin\theta_1\,\delta\theta_1$$

$$\delta y_2 = -l\sin\theta_1\,\delta\theta_1 - \frac{l}{2}\sin\theta_2\,\delta\theta_2$$

$$\delta x = l\cos\theta_1\,\delta\theta_1 + l\cos\theta_2\,\delta\theta_2$$

Note that as θ_1 and θ_2 increase, y_1 and y_2 decrease, while x increases. Hence, considering *only* positive virtual displacements, the above expressions reduce to

$$\delta y_1 = \frac{l}{2}\sin\theta_1\,\delta\theta_1$$

$$\delta y_2 = l \sin\theta_1\, \delta\theta_1 + \frac{l}{2} \sin\theta_2\, \delta\theta_2$$

$$\delta x = l \cos\theta_1\, \delta\theta_1 + l \cos\theta_2\, \delta\theta_2$$

Applying the principle of virtual work,

$$-W_1\, \delta y_1 - W_2\, \delta y_2 + P\, \delta x = 0$$

Note that as the virtual displacements δy_1 and δy_2 are in the direction opposite to that of the weights of the links, the virtual works done by the weights are negative.

$$-W_1\left[\frac{l}{2} \sin\theta_1\, \delta\theta_1\right] - W_2\left[l \sin\theta_1\, \delta\theta_1 + \frac{l}{2} \sin\theta_2\, \delta\theta_2\right] + P[l \cos\theta_1\, \delta\theta_1 + l \cos\theta_2\, \delta\theta_2] = 0$$

As the virtual displacements are arbitrary, we can equate the coefficients of $\delta\theta_1$ on either side of the equation. Hence,

$$-W_1\left[\frac{l}{2} \sin\theta_1\, \delta\theta_1\right] - W_2[l \sin\theta_1\, \delta\theta_1] + P[l \cos\theta_1\, \delta\theta_1] = 0$$

$$\Rightarrow \quad -\left[\frac{W_1}{2} + W_2\right] [\sin\theta_1] + P[\cos\theta_1] = 0$$

$$\therefore \quad \theta_1 = \tan^{-1}\left[\frac{2P}{W_1 + 2W_2}\right] = \tan^{-1}\left[\frac{2P}{3W}\right]$$

Similarly, equating the coefficients of $\delta\theta_2$ on either side of the equation, we have

$$-W_2\left[\frac{l}{2} \sin\theta_2\, \delta\theta_2\right] + P[l \cos\theta_2\, \delta\theta_2] = 0$$

$$\Rightarrow \quad \theta_2 = \tan^{-1}\left[\frac{2P}{W_2}\right] = \tan^{-1}\left[\frac{2P}{W}\right]$$

Example 11.8 In the mechanism shown, determine the horizontal force P required to be applied to hold the system in equilibrium. The length of each link is l and that of the weight is W.

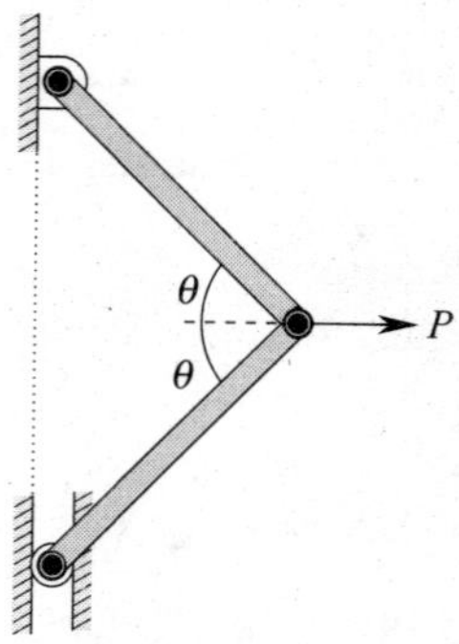

Fig. 11.17

Solution Choosing the hinge point as the origin, the positions of the points of application of the forces W and P are y_1, y_2 and x as shown in Fig. 11.17(a). If the inclination of the links with respect to the horizontal is θ then their respective positions are given as

$$y_1 = \frac{l}{2} \sin\theta$$

$$y_2 = l \sin\theta + \frac{l}{2} \sin\theta = \frac{3l}{2} \sin\theta$$

and

$$x = l \cos\theta$$

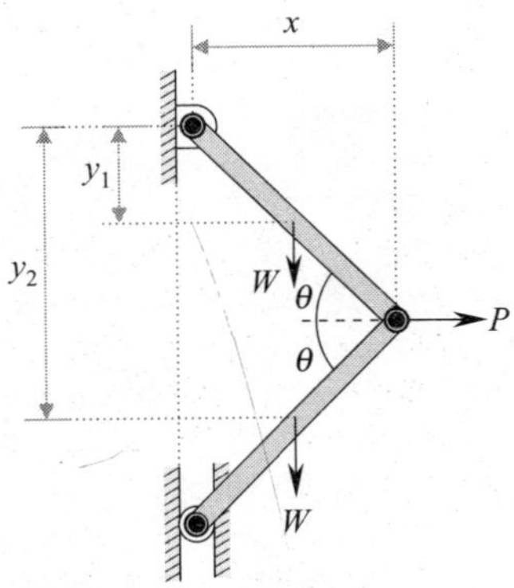

Fig. 11.17(a)

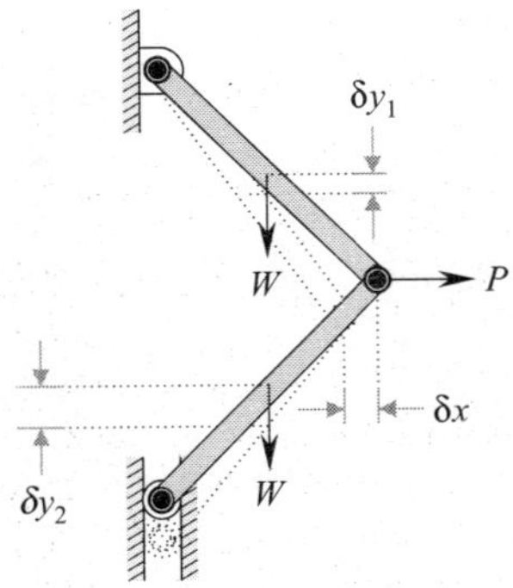

Fig. 11.17(b)

Then virtual displacements are obtained by differentiating the above expressions:

$$\delta y_1 = \frac{l}{2} \cos\theta\, \delta\theta$$

$$\delta y_2 = \frac{3l}{2} \cos\theta\, \delta\theta$$

and $$\delta x = -l \sin\theta\, \delta\theta$$

Note that as θ increases, x decreases and y increases. Hence, the expressions for δy_1, δy_2 do not change, but δx changes to $l \sin\theta\delta\theta$ when we consider *only* positive virtual displacements.

Applying the principle of virtual work,

$$W\delta y_1 + W\delta y_2 - P\delta x = 0$$

Note that as the virtual displacement δx is in the direction opposite to that of the direction of the force P, the virtual work done by P is negative.

$$W\frac{l}{2}\cos\theta\delta\theta + W\frac{3l}{2}\cos\theta\delta\theta - Pl\sin\theta\delta\theta = 0$$

$\Rightarrow$ $$P = \frac{4W\cos\theta}{2\sin\theta} = \frac{2W}{\tan\theta}$$

Example 11.9 A load W supported on platform AB is to be held in equilibrium by applying a moment M as shown in Fig. 11.18. Determine the value of M required to maintain equilibrium. The length of each cross member supporting the load is l. The weights of the links can be neglected.

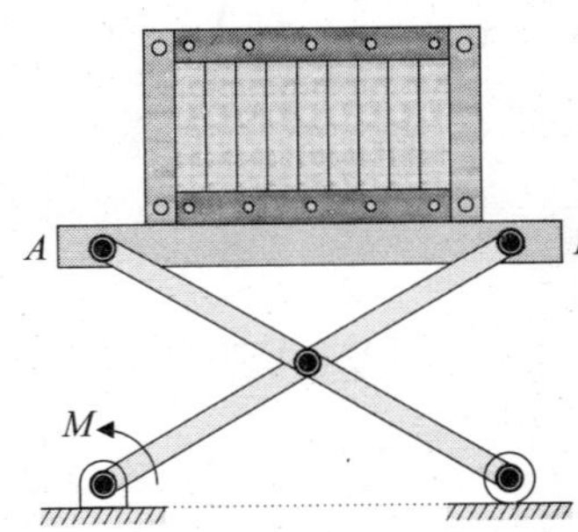

Fig. 11.18

Solution Choosing the hinge point as the origin, the point of application of the load W is y and the inclination of the links with respect to the vertical is θ. Then

$$y = 2\frac{l}{2}\cos\theta = l\cos\theta$$

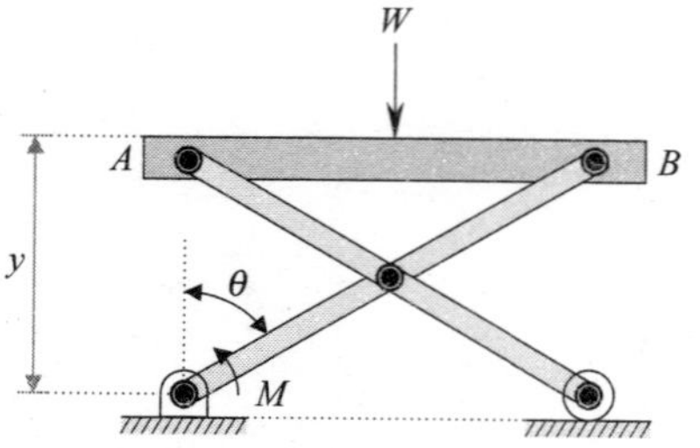

Fig. 11.18(a)

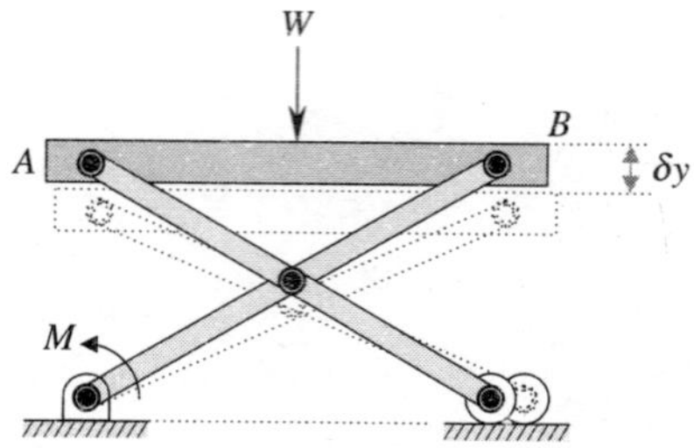

Fig. 11.18(b)

The virtual displacement δy is then obtained by differentiating the above expression:

$$\delta y = -l \sin\theta\, \delta\theta$$

From the figure, we notice that as θ increases, y decreases. Hence, considering *only* positive values of virtual displacements, the above expression becomes

$$\delta y = l \sin\theta\, \delta\theta$$

Applying the principle of virtual work, we have

$$-M\delta\theta + W\delta y = 0$$

Note that as the virtual angular displacement $\delta\theta$ is in the direction opposite to that of the moment, the virtual work done by M is negative.

$$-M\delta\theta + W(l \sin\theta\delta\theta) = 0$$

$$\therefore \quad M = Wl \sin\theta$$

Example 11.10 Using the principle of virtual work, determine the angle θ for which equilibrium is maintained in the mechanism shown for given values of forces P_1 and P_2 applied. Length of the longer links is l and that of the shorter links is $l/2$.

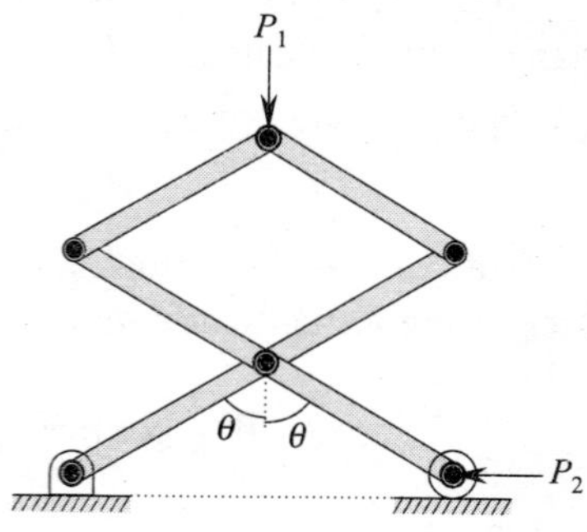

Fig. 11.19

Solution Choosing the hinge point as the origin, the point of application of the forces P_1 and P_2 are respectively y and x. Expressing these positions x and y in terms of θ, we have

$$y = \frac{l}{2}\cos\theta + \frac{l}{2}\cos\theta + \frac{l}{2}\cos\theta$$

$$= 3\frac{l}{2}\cos\theta \qquad \text{(a)}$$

and

$$x = 2\frac{l}{2}\sin\theta = l\sin\theta \qquad \text{(b)}$$

The virtual displacements are obtained by differentiating the above two expressions, (a) and (b):

$$\delta y = -\frac{3l}{2}\sin\theta\,\delta\theta$$

and

$$\delta x = l\cos\theta\,\delta\theta$$

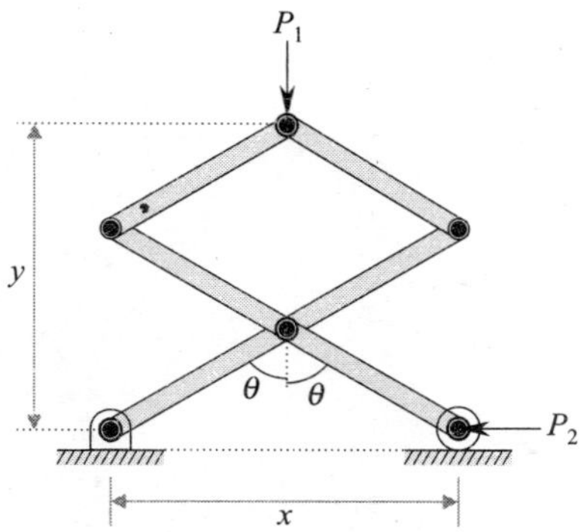

Fig. 11.19(a)

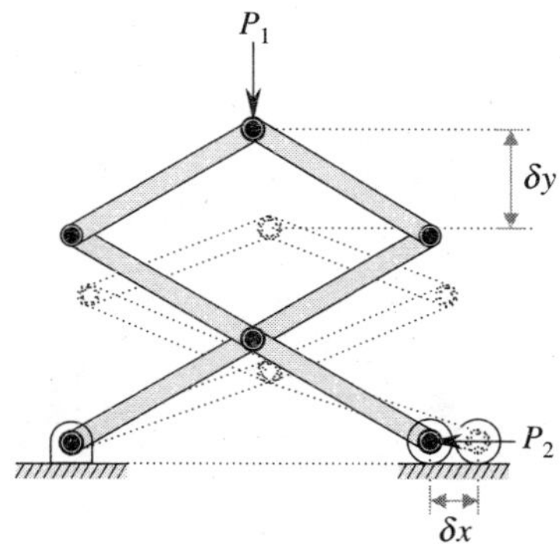

Fig. 11.19(b)

From the figure, we see that as θ increases, x increases while y decreases. Hence, considering *only* positive values of virtual displacements, the above expressions reduce to

$$\delta y = -\frac{3l}{2}\sin\theta\,\delta\theta \qquad \text{and} \qquad \delta x = l\cos\theta\,\delta\theta$$

Applying the principle of virtual work, we have

$$P_1\delta y - P_2\delta x = 0$$

Note that the virtual work done by P_2 is negative as the virtual displacement δx is in the direction opposite to that of the force.

$$P_1\left(\frac{3l}{2}\sin\theta\,\delta\theta\right) - P_2(l\cos\theta\,\delta\theta) = 0$$

$$\frac{3P_1}{2}\sin\theta = P_2\cos\theta$$

$$\Rightarrow \qquad \theta = \tan^{-1}\left[\frac{2P_2}{3P_1}\right]$$

Example 11.11 Two links of equal lengths l are hinged and arranged vertically as shown in Fig. 11.20. They are connected at their lower ends by a spring of unstretched length s. When a vertical force P is applied, determine the spring constant k to maintain equilibrium at the position shown.

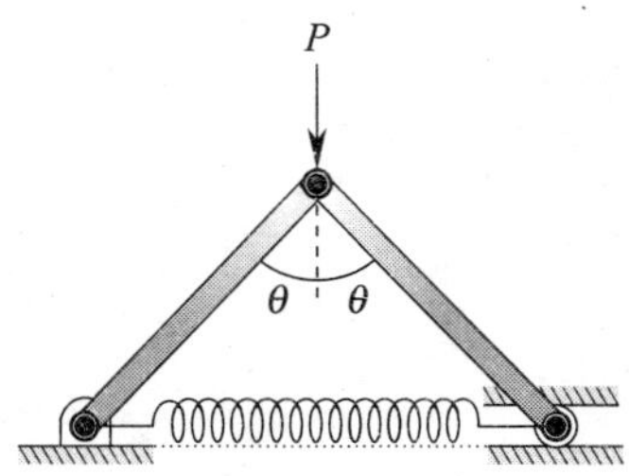

Fig. 11.20

Solution Under the applied force P, the links separate apart at the bottom, thus extending the spring. Let x be the extension in the spring from its unstretched position. Then the restoring force exerted by the spring on the roller support is kx acting such as to oppose this elongation. The free-body diagram of the mechanism is shown in Fig. 11.20(a).

From the figure, we can write x and y values in terms of the angle θ as

$$y = l\cos\theta \qquad \text{and} \qquad x = 2l\sin\theta - s$$

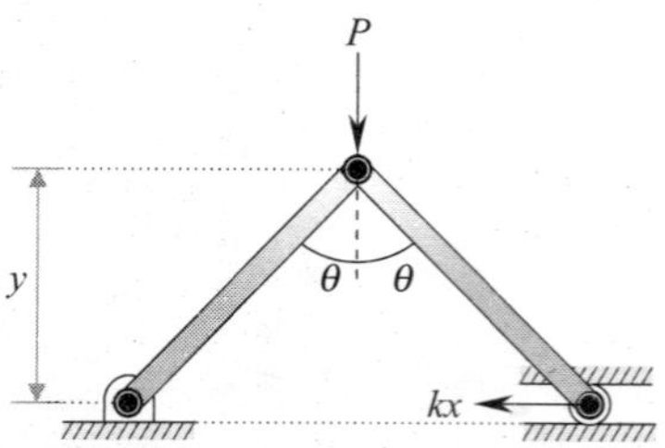

Fig. 11.20(a)

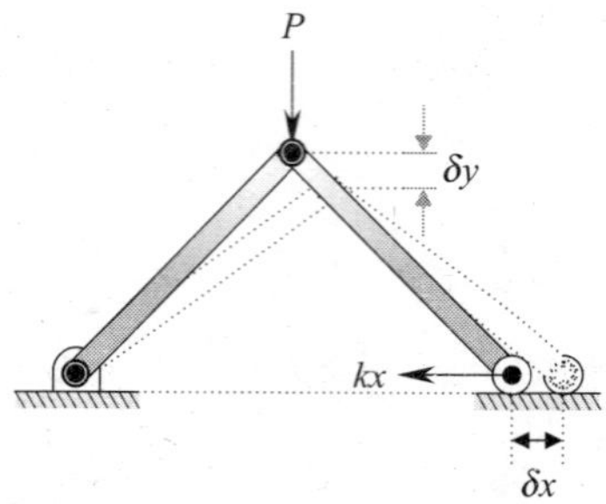

Fig. 11.20(b)

Upon differentiation, we get

$$\delta y = -l \sin\theta\,\delta\theta \qquad \text{and} \qquad \delta x = 2l \cos\theta\,\delta\theta$$

As θ increases, we can see that x increases but y decreases. Hence, for *only* positive values of virtual displacements, the above expressions reduce to

$$\delta y = l \sin\theta\,\delta\theta \qquad \text{and} \qquad \delta x = 2l \cos\theta\,\delta\theta$$

Applying the principle of virtual work, we get

$$P\,\delta y - kx\,\delta x = 0$$

$$P(l \sin\theta\,\delta\theta) - k(2l \sin\theta - s)(2l \cos\theta\,\delta\theta) = 0$$

$$\Rightarrow \qquad k = \frac{P \tan\theta}{2(2l \sin\theta - s)}$$

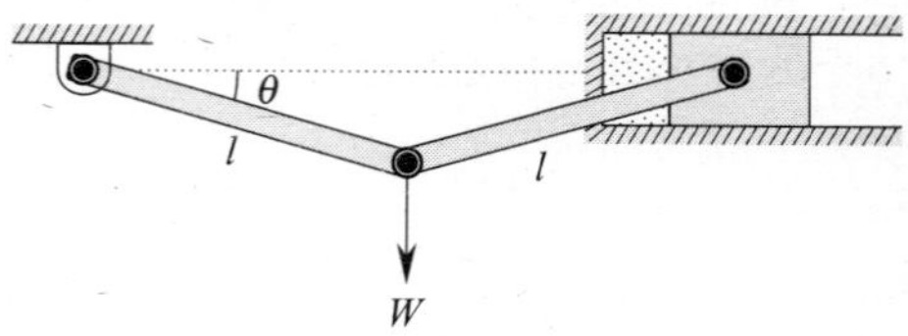

Fig. 11.21

Example 11.12 What pressure p should be applied to the plunger of cross-sectional area A to hold the system in equilibrium in the position shown? Neglect weight of the links.

Solution If p is the pressure inside the cylinder, the thrust exerted on the plunger is pA, where A is the cross-sectional area of the plunger. The free-body diagram of the system is shown in Fig. 11.21(a).

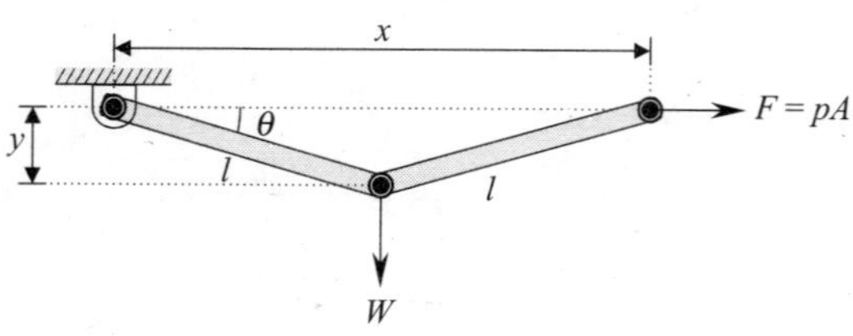

Fig. 11.21(a)

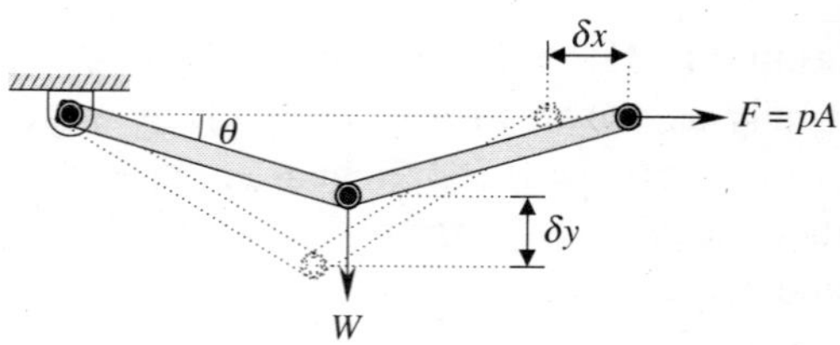

Fig. 11.21(b)

Taking the hinge support as the origin, we can write

$$x = 2(l \cos\theta) \qquad \text{and} \qquad y = l \sin\theta$$

Upon differentiation, we have

$$\delta x = -2l \sin\theta\,\delta\theta \qquad \text{and} \qquad \delta y = l \cos\theta\,\delta\theta$$

As θ increases, y also increases but x decreases. Hence, considering *only* positive values of virtual displacements, the above expressions reduce to

$$\delta x = 2l \sin\theta\delta\theta \qquad \text{and} \qquad \delta y = l \cos\theta\delta\theta$$

Applying the principle of virtual work, we have

$$W\delta y - F\delta x = 0$$

$$Wl \cos\theta\delta\theta - pA(2l \sin\theta\delta\theta) = 0$$

$$\Rightarrow \qquad p = \frac{W}{2A\tan\theta}$$

Example 11.13 By the method of virtual work, determine the effort P required to be applied to hold the block of weight W in equilibrium.

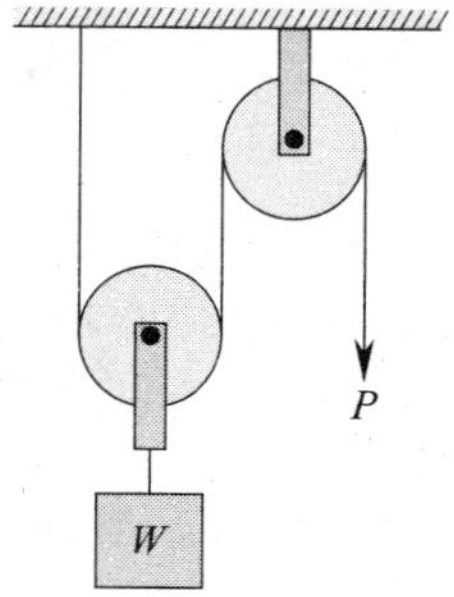

Fig. 11.22

Solution Let us assume that the free end of the string is given a virtual displacement δy in the vertically downward direction. From the figure, we can see that when the free end moves *down* by δy then W moves *up* by $\delta y/2$. Hence, applying the principle of virtual work, we have

$$P\delta y - W\frac{\delta y}{2} = 0$$

$$\Rightarrow \qquad P = \frac{W}{2}$$

Example 11.14 By the method of virtual work, determine the coefficient of friction between the block of mass m_2 and inclined plane to maintain equilibrium for impending motion of the block up the plane. The string connecting the blocks is inextensible.

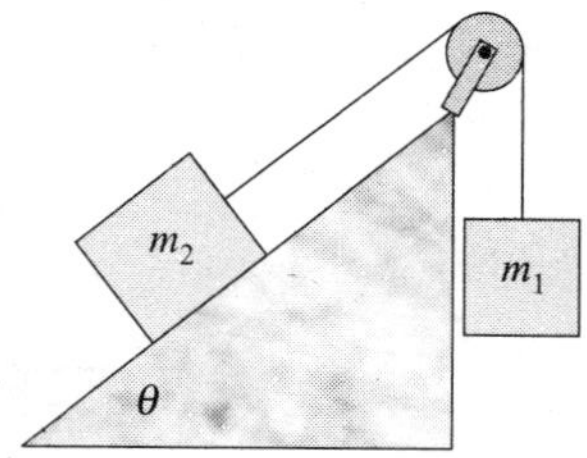

Fig. 11.23

Solution For interconnected bodies, the works done by internal forces cancel out and hence only the external forces are effective in doing work on the system. In the system shown, the internal force is tension T in the string and hence only the other forces are shown acting on the blocks in the system as in Fig. 11.23(a).

If the virtual displacement of the block of mass m_1 is δy in the vertically downward direction then the virtual displacement of the block of mass m_2 up the plane is also the same as the string connecting the two blocks is inextensible. Noting that the forces N_2 and $m_2g\cos\theta$ do not contribute to the work done on the system as they act perpendicular to the displacements and applying the principle of virtual work, we have

$$m_1g\,\delta y - (F_2 + m_2g\sin\theta)\delta y = 0$$

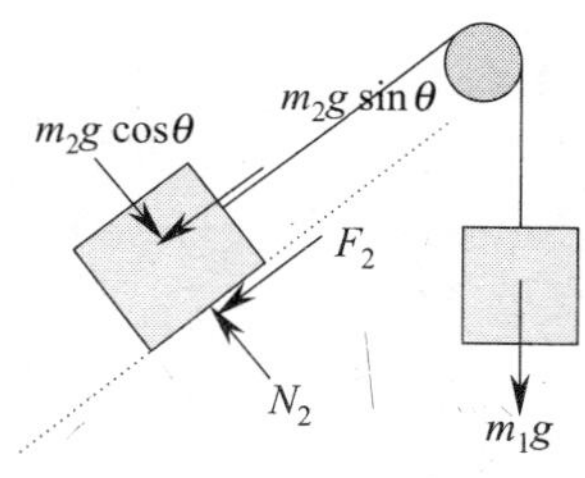

Fig. 11.23(a)

At the point of impending motion, we know that $F_2 = \mu N_2 = \mu m_2 g\cos\theta$ and hence the above equation can be written as

$$m_1g\,\delta y - (\mu m_2g\cos\theta + m_2g\sin\theta)\,\delta y = 0$$

$$\Rightarrow \qquad \mu = \frac{m_1 - m_2\sin\theta}{m_2\cos\theta}$$

Example 11.15 Determine the coefficient of friction between the 100 kg block and floor to maintain equilibrium of the system shown in Fig. 11.24.

Solution As the pulley rotates, the angle through which any radial line moves remains the same. However, the linear distance is dependent upon the radial length. Let the virtual angular displacement of the pulley be $\delta\theta$. Then the virtual linear displacement of any point on the rim of the smaller diameter pulley is $r_1\,\delta\theta$ and hence the vertical virtual displacement of the 75-kg block is

$$\delta y = r_1\,\delta\theta$$

Similarly, the virtual linear displacement any point on the rim of the larger diameter pulley is $r_2\,\delta\theta$ and hence the horizontal virtual displacement of the 100 kg block is

$$\delta x = r_2\,\delta\theta$$

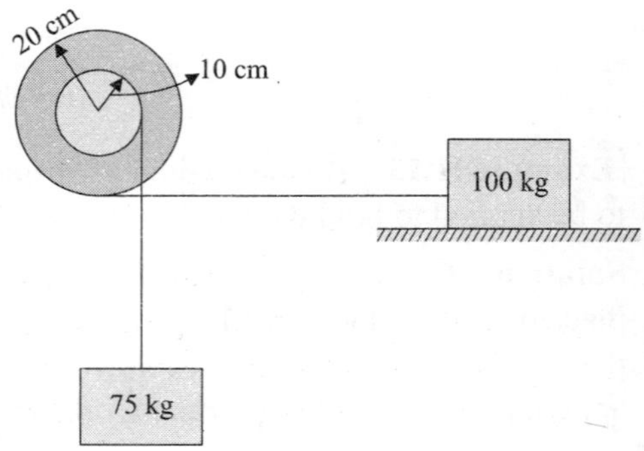

Fig. 11.24

Applying the principle of virtual work,

$$(m_1 g)\delta y - (\mu m_2 g)\delta x = 0$$

Note that the force of friction acts in the direction opposite to the motion of the 100 kg block and hence work done by the force is negative. The other forces acting on the block are perpendicular to the direction of motion of the block and hence do not contribute to the work done.

$\therefore \quad (m_1 g)r_1\,\delta\theta - (\mu m_2 g)r_2\,\delta\theta = 0$

$$\Rightarrow \quad \mu = \frac{m_1 r_1}{m_2 r_2}$$

$$= \frac{(75)(10)}{(100)(20)} = 0.375$$

11.7 COMPLETELY CONSTRAINED BODY

If a body is completely constrained, then even virtual displacement of the body as discussed before is not possible. In such cases, we release one of the constraints by replacing it with the reaction force and at the same time keeping the other constraints undisturbed. Then we assume an imaginary displacement and solve the unknown reaction force as done earlier for partially constrained bodies. Similarly, we can release each of the constraints, while keeping the remaining constraints undisturbed. This way we can solve the unknown forces and reactions.

Example 11.16 Determine the reactions of a simply supported beam shown by applying the virtual work principle.

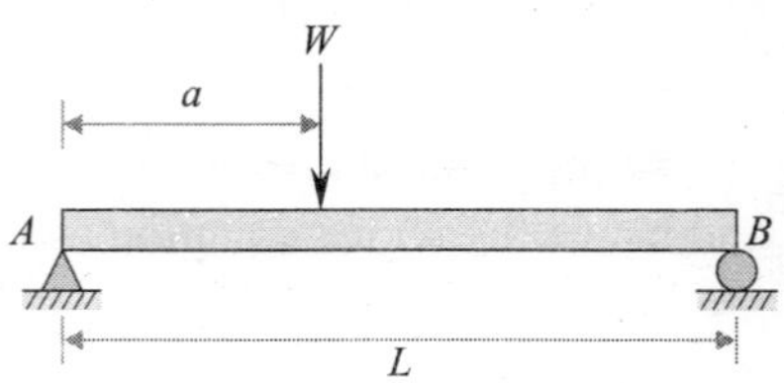

Fig. 11.25

Solution The free-body diagram of the beam is shown in Fig. 11.25(a).

Keeping end A of the beam fixed, let end B be given a virtual displacement δy in the vertical direction. Then from similar triangles ADE and ABC, we see that

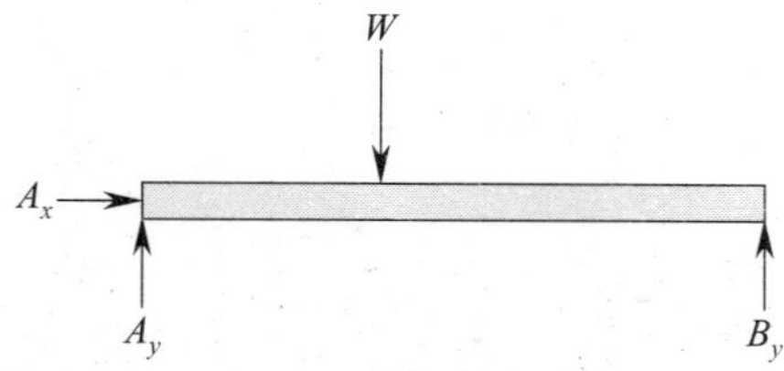

Fig. 11.25(a)

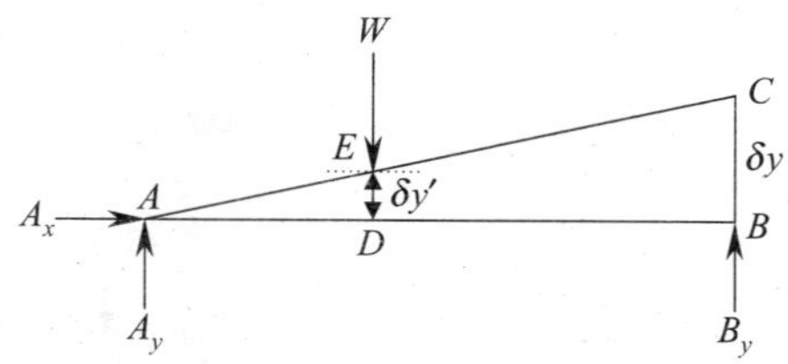

Fig. 11.25(b)

$$\frac{\delta y'}{\delta y} = \frac{a}{L} \quad \Rightarrow \delta y' = \frac{a}{L}\delta y$$

where $\delta y'$ is virtual displacement of the point D at which the external load W is acting. It should be noted that the virtual displacement at D is opposite to the direction of W acting and hence work done by W is negative.

Applying the principle of virtual work,

$$-W\delta y' + B_y\,\delta y = 0$$

$$-W\frac{a}{L}\,\delta y + B_y\,\delta y = 0$$

$$\Rightarrow \qquad B_y = W\frac{a}{L}$$

Similarly, keeping the end B as fixed and giving the end A a virtual displacement, it can be shown that

$$A_y = \frac{W}{L}(L-a)$$

Example 11.17 Determine the reactions of a simply supported beam shown by applying the virtual work principle.

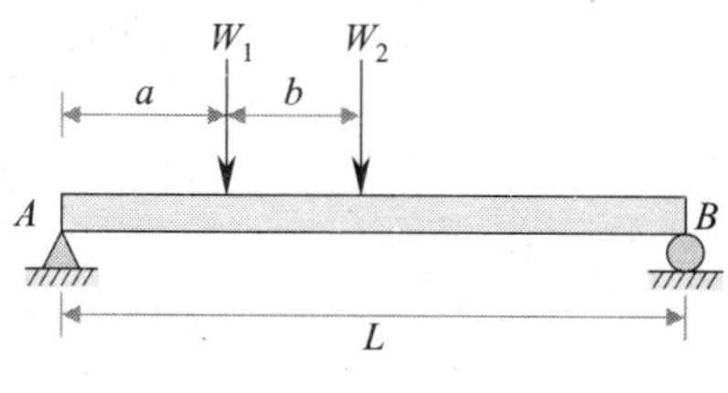

Fig. 11.26

Solution Keeping the end A of the beam fixed, let the end B be given a virtual displacement δy in the vertical direction. Then the corresponding virtual displacements of W_1 and W_2 are respectively δy_1 and δy_2.

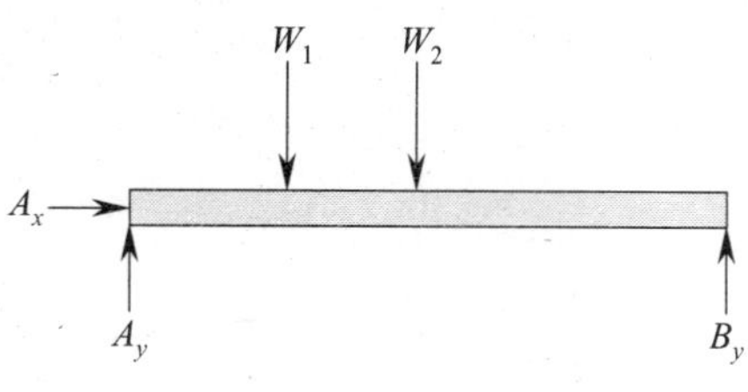

Fig. 11.26(a)

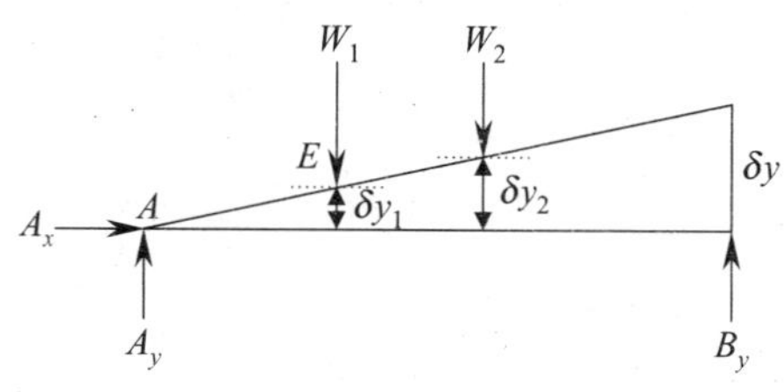

Fig. 11.26(b)

From the geometry of the triangle, we see that

$$\frac{\delta y_1}{\delta y} = \frac{a}{L} \quad \Rightarrow \quad \delta y_1 = \frac{a}{L}\delta y$$

and

$$\frac{\delta y_2}{\delta y} = \frac{a+b}{L} \quad \Rightarrow \quad \delta y_2 = \frac{a+b}{L}\delta y$$

Applying the principle of virtual work,

$$-W_1\,\delta y_1 - W_2\,\delta y_2 + B_y\,\delta y = 0$$

$$-W_1\left(\frac{a}{L}\right)\delta y - W_2\left(\frac{a+b}{L}\right)\delta y + B_y\,\delta y = 0$$

$$\Rightarrow \quad -W_1\left(\frac{a}{L}\right) - W_2\left(\frac{a+b}{L}\right) + B_y = 0$$

$$\therefore \quad B_y = \frac{aW_1 + (a+b)W_2}{L}$$

Similarly, keeping the end B as fixed and giving the end A a virtual displacement, it can be shown that

$$A_y = \frac{W_1(L-a) + W_2(L-a-b)}{L}$$

Example 11.18 Determine the reactions of overhanging beam shown by applying virtual work principle.

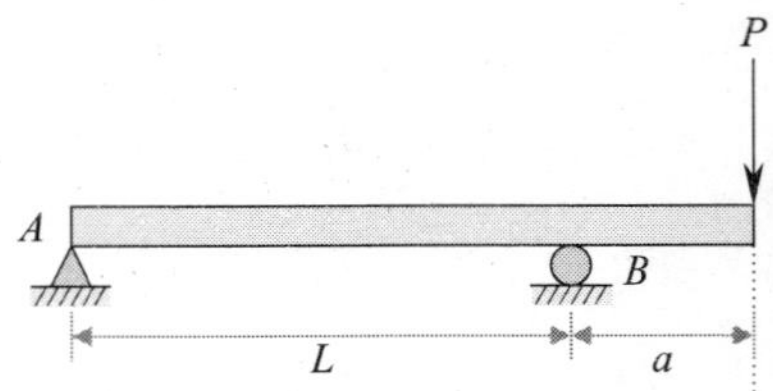

Fig. 11.27

Solution Keeping the end A of the beam fixed, let the roller support B be given a virtual displacement δy in the vertical direction. Then the corresponding virtual displacement of P be $\delta y'$.

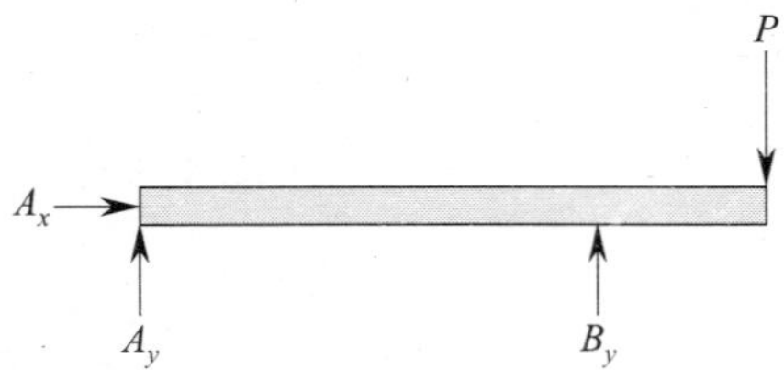

Fig. 11.27(a)

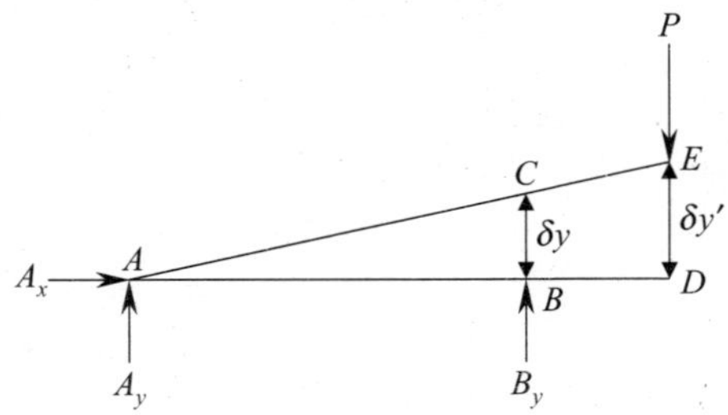

Fig. 11.27(b)

From the geometry of the triangle, we see that

$$\frac{\delta y'}{\delta y} = \frac{a+L}{L} \quad \Rightarrow \quad \delta y' = \frac{a+L}{L}\delta y$$

Applying the principle of virtual work,

$$B_y\,\delta y - P\,\delta y' = 0$$

It should be noted that the virtual displacement at D is opposite to the direction of P acting and hence the work done by P is negative.

$$B_y \delta y - P\left[\frac{a+L}{L}\right]\delta y = 0$$

$$\Rightarrow \qquad B_y = P\left[\frac{a+L}{L}\right]$$

Similarly, keeping the end B as fixed and giving the end A a virtual displacement, it can be shown that

$$A_y = -P\left[\frac{a}{L}\right]$$

Example 11.19 Determine the reactions of a simply supported beam shown by applying the virtual work principle.

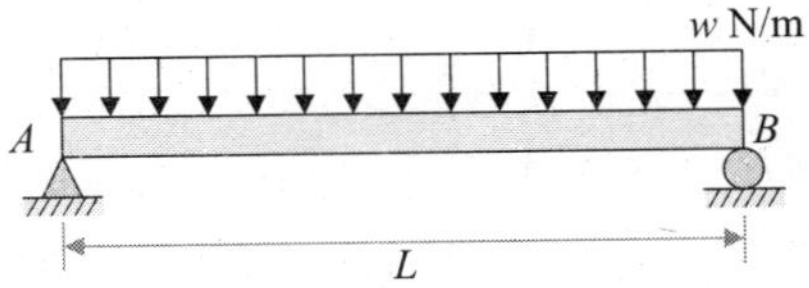

Fig. 11.28

Solution Keeping the end A of the beam fixed, let the end B be given a virtual displacement δy_1 in the vertical direction. Consider a strip of load at a distance x from the end A. Then the infinitesimally small force is wdx. The virtual displacement at this point is δy.

From the geometry of the triangle, we see that

$$\frac{\delta y}{\delta y_1} = \frac{x}{L} \quad \Rightarrow \quad \delta y = \frac{x}{L}\delta y_1$$

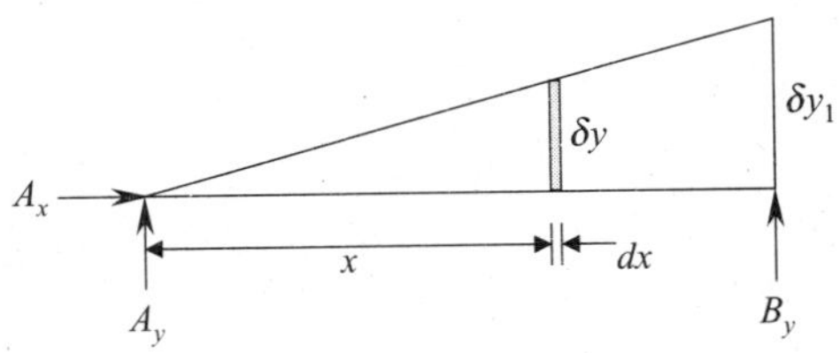

Fig. 11.28(a)

Therefore, virtual work done by the infinitesimally small force is

$$\delta U = -(w\,dx)\,\delta y$$

$$= -w\frac{x}{L}\,dx\,\delta y_1$$

Hence, total virtual work done by the entire load is

$$U = -\frac{w}{L}\delta y_1 \int_0^L x\,dx$$

$$= -\frac{w}{L}\delta y_1 \frac{L^2}{2}$$

$$= -\frac{w}{2}L\,\delta y_1$$

From the above expression, we can see that the total virtual work done by uniformly distributed load is w times the area of the triangle. Applying the principle of virtual work,

$$-\frac{w}{2}L\,\delta y_1 + B_y \delta y_1 = 0$$

$$\therefore \qquad B_y = \frac{wL}{2}$$

Similarly, $$A_y = \frac{wL}{2}$$

SUMMARY

Work Done on a Particle

When a force $\vec{F}$ acting on a particle causes a *displacement* $\vec{s}$ of the particle in the direction of the force then work done by the force is defined as the *product* of magnitudes of the *force* and *displacement*. Mathematically, this can be written as

$$W = Fs$$

In general, the displacement of a particle under the action of forces would not always occur in the direction of the force due to constraints involved. Hence, work done in general, can be defined as the product of the component of the force in the direction of motion and the displacement. Alternatively, it can also be defined as the product of the force and the component of displacement in the direction of the force. Mathematically, we can write this as

$$W = (F\cos\theta)s \quad \text{(or)} \quad W = (F)(s\cos\theta)$$

or written in vector form as

$$W = \vec{F}.\vec{s} = Fs\cos\theta$$

Work done is a *scalar* quantity that is having magnitude, but not direction. Hence, work done adds up algebraically.

(i) When $\theta = 0°$, i.e., when force and displacement are in the same direction, then $\cos\theta = 1$ and hence work done is $W = Fs$.

(ii) When $\theta = 180°$, i.e., when force and displacement are in the opposite direction, then $\cos\theta = -1$ and hence work done is $W = -Fs$. The force of friction always acts in the direction opposite to that of the motion; hence, work done by the force of friction is always *negative*.

Work done is **zero** in the following cases:

(i) When the displacement of the particle is zero

(ii) When the motion is at right angles to the direction of the force

(iii) When a particle is in static equilibrium

(iv) Work done by internal forces as in the case of interconnected members and mechanisms

Work Done in Stretching a Spring

A spring whose one end is fixed and the other end stretched is an example for a variable force constant in direction but varying in magnitude. When a spring is stretched by an amount x from its unstretched position, the restoring force acting on it is proportional to the displacement, i.e.,

$$F \propto -x$$

Introducing a constant of proportionality, we have

$$F = -kx$$

where the constant of proportionality k is called the *stiffness* of the spring or *spring constant.*

The work done by an external force P causing an elongation x_0 from its unstretched position is given as

$$W = \frac{1}{2}kx_o^2$$

Work Done on a Rigid Body

We know that a rigid body is subjected to moments in addition to the forces. Just as the forces cause linear displacements, moments cause angular displacements. If a moment M acting on a rigid body causes an angular displacement θ then work done by the moment is defined as product of moment and angular displacement, i.e.,

$$W = M\theta$$

Virtual Displacement

The displacement of a particle or a rigid body in equilibrium is not at all possible. However, we can assume an imaginary displacement to occur, particularly if the system is *partially* constrained. As this displacement is not actually occurring, but an imaginary one, we call it a *virtual displacement.* It is defined as any arbitrary infinitesimal change in position consistent with the constraints imposed on the motion.

Virtual Work

The total work done by the system of forces causing virtual displacement is termed *virtual work.* For the particle or rigid body to remain in equilibrium in the displaced position also, we know that the resultant force acting on it must be zero. Thus, we see that work done in causing this virtual displacement is also zero. This is known as the *principle of virtual work.*

For a system of concurrent forces acting on a particle, the virtual work principle states:

$$\delta U = \sum(F\,\delta r) = 0$$

Similarly, for a rigid body, we can write the principle of virtual work as:

$$\delta U = \sum F\,\delta r + \sum M\,\delta\theta = 0$$

EXERCISES

Objective-type Questions

1. In which of the following cases is work done by the force of gravity on the luggage? When the luggage is lying

(a) on the floor
(b) on a conveyor belt moving horizontally
(c) in a moving lift
(d) on a rotating turntable

2. Work done is zero when
   (a) the motion is at right angles to the direction of force
   (b) the body is in equilibrium
   (c) the displacement is zero
   (d) all of these
3. The restoring force in a spring is proportional to the
   (a) initial unstretched length of the spring
   (b) elongation of the spring
   (c) number of turns of coil in the spring
   (d) cross-sectional area of the spring
4. The work done in stretching a spring of spring constant k by a length Δ is
   (a) $k\Delta$ (b) $k\Delta^2$ (c) $k\Delta/2$ (d) $k\Delta^2/2$
5. Of the following forces, which force *always* does negative work?
   (a) Normal reaction
   (b) Force of friction
   (c) Weight
   (d) External applied force

Answers

1. (c) 2. (d) 3. (b) 4. (d) 5. (b)

Short-answer Questions

1. Define work done on a particle and a rigid body.
2. Discuss the cases when the work done becomes zero.
3. Derive the expression for work done in stretching a spring.
4. Define virtual displacement and virtual work.
5. Explain how virtual displacements are possible in partially and completely constrained bodies.
6. State the principle of virtual work.
7. Discuss the advantages of the method of virtual work over the method of equilibrium equations to determine the unknown forces in a system of forces acting on a body.

Numerical Problems

11.1 A uniform ladder *AB* of length l and weight W (refer Fig. E.11.1) rests against a smooth vertical wall and floor. To prevent it from sliding, the end B is connected to the wall by a string BC. By the method of virtual work, determine the tension in the string.

Ans. $\dfrac{W}{2\tan\theta}$

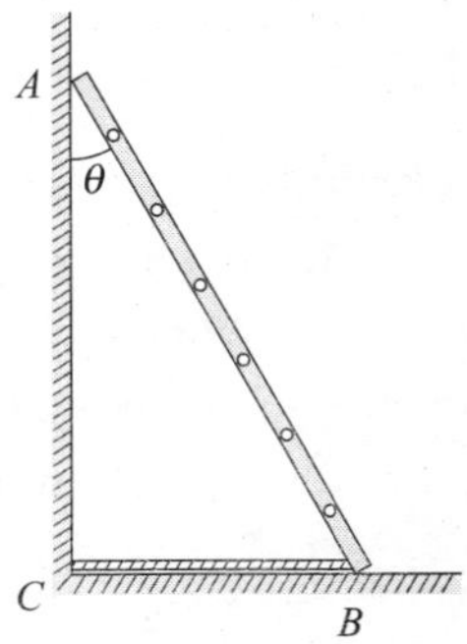

Fig. E.11.1

11.2 A uniform bar *AB* of length l and weight W rests against a smooth vertical wall and a smooth horizontal floor as shown in Fig. E.11.2. To prevent it from sliding, the end B is connected to the wall by a spring of stiffness k. If the bar is maintained in equilibrium in the position shown, by the method of virtual work, determine the unstretched length of the spring.

Ans. $l\sin\theta - (W\tan\theta)/2k$

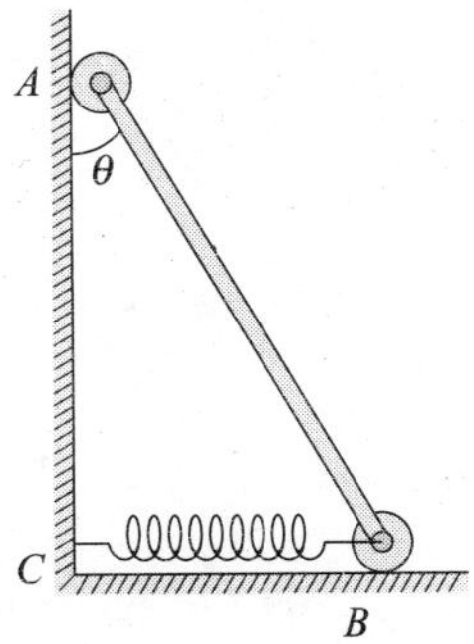

Fig. E.11.2

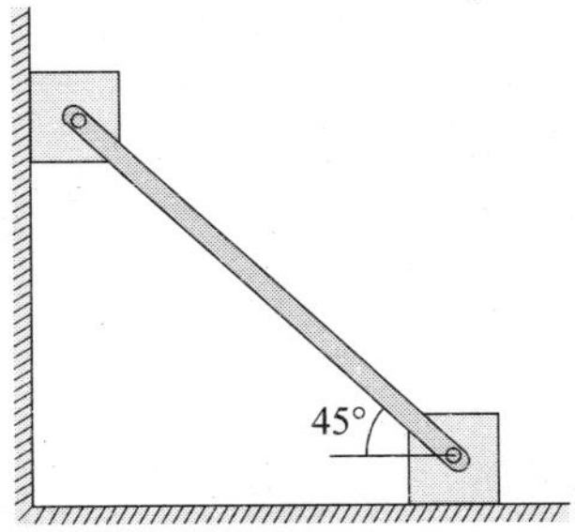

Fig. E.11.3

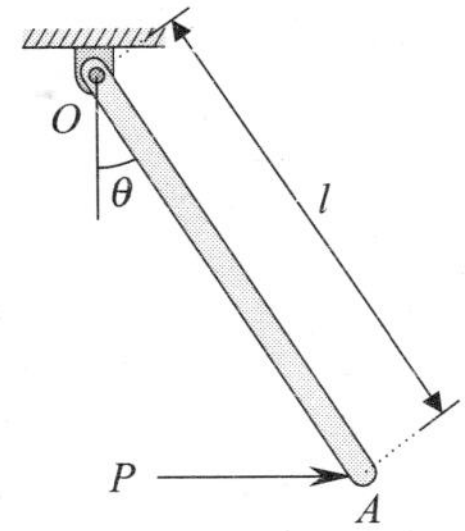

Fig. E.11.4

11.3 The system shown in Fig. E.11.3 consists of two equal masses connected by a massless bar hinged at its ends. Determine the coefficient of friction between the floor and block to maintain equilibrium in the position shown. The wall is assumed frictionless.

Ans. 1/2

11.4 By the method of virtual work, determine the horizontal force P required to be applied at the free end A of the rod OA hinged at O to maintain equilibrium. Refer Fig. E.11.4. The weight of the rod is W.

Ans. $W \tan \theta/2$

11.5 In Fig. E.11.5, by the method of virtual work, determine the effort P required to be applied to hold the block of weight W in equilibrium.

Ans. $W/4$

11.6 In Fig. E.11.6, using the principle of virtual work, determine the effort P required to be applied to hold the block of weight W in equilibrium.

Ans. $W/4$

11.7 Two equal spheres of weight W are suspended by strings as shown in Fig. E.11.7. They are also interconnected by a horizontal weightless rod. For a given applied force P, applied horizontally, determine the inclination of the strings with the vertical by the principle of virtual work.

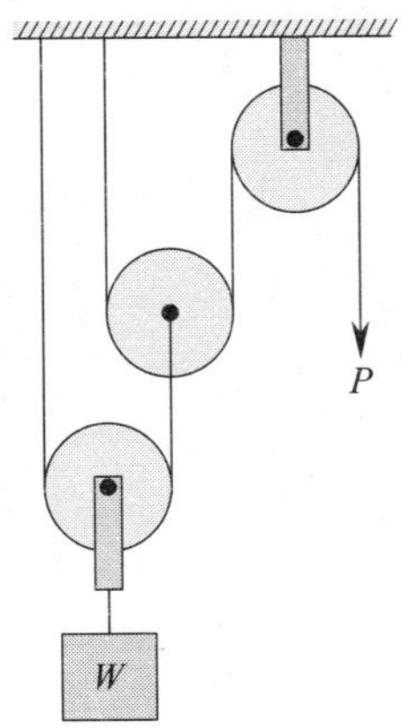

Fig. E.11.5

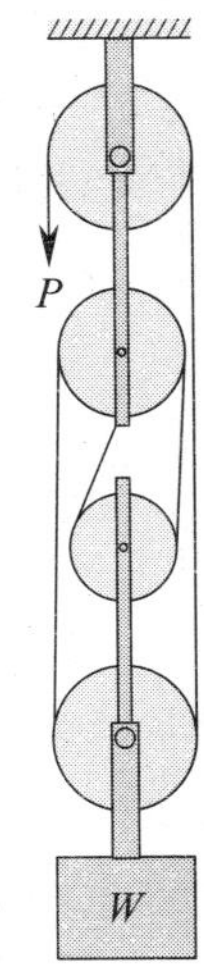

Fig. E.11.6

Ans. $\tan^{-1}\left[\dfrac{P}{2W}\right]$

11.8 What pressure p should be applied to the plunger of cross-sectional area A shown in Fig. E.11.8 to hold the mechanical system in equilibrium for a given value of moment M applied at the crank?

Ans. $$P = \frac{M}{Aa \sin\theta[1 + a\cos\theta/(b^2 - a^2\sin^2\theta)^{1/2}]}$$

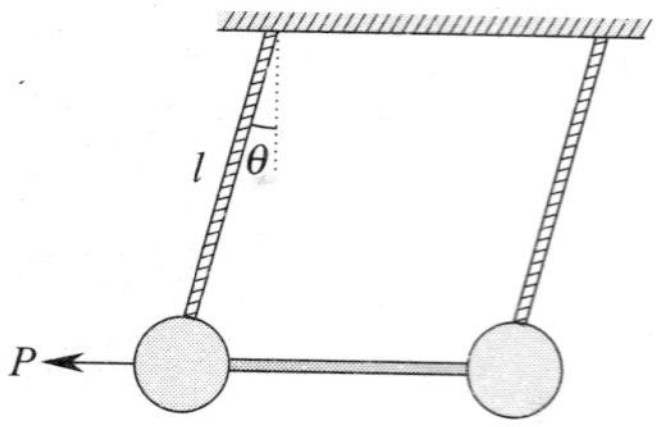

Fig. E.11.7

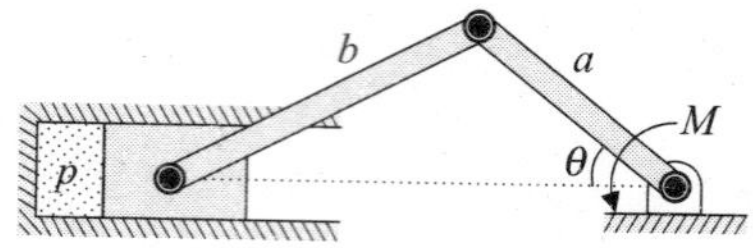

Fig. E.11.8

11.9 In Fig. E.11.9, by the method of virtual work, determine the force P required to be applied to maintain equilibrium. Each link is of length l and weight W.

Ans. W

11.10 In Fig. E.11.10, by the method of virtual work, determine the moment M required to be applied at the hinge A to maintain equilibrium of the mechanism shown. Assume that the members are weightless and their respective lengths are $\overline{AB} = \overline{BC} = \overline{CD} = \overline{DE} = \overline{BE} = \overline{BF} = l$.

Ans. $3Pl \cos\theta$

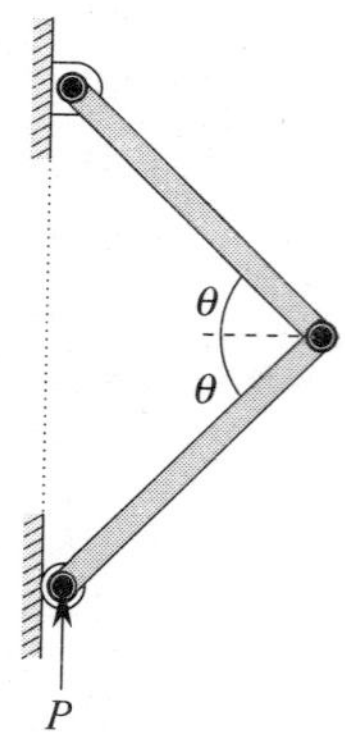

Fig. E.11.9

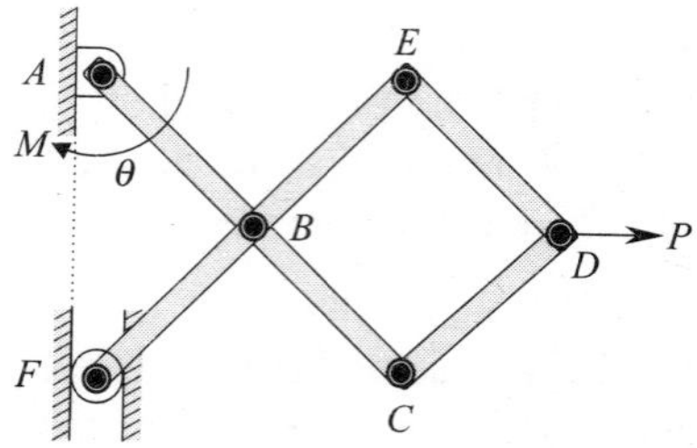

Fig. E.11.10

11.11 In the lazy tongs shown in Fig. E.11.11, using the principle of virtual work, determine the angle θ for which equilibrium is maintained for given values of forces P and Q applied. Length of the longer links is l and that of the shorter links is $l/2$.

Ans. $\tan^{-1}\left[\dfrac{3.5Q}{P}\right]$

11.12 In Fig. E.11.12, by the method of virtual work, determine the coefficient of friction between block of mass m_2 and horizontal plane to maintain equilibrium.

Ans. $m_1/2m_2$

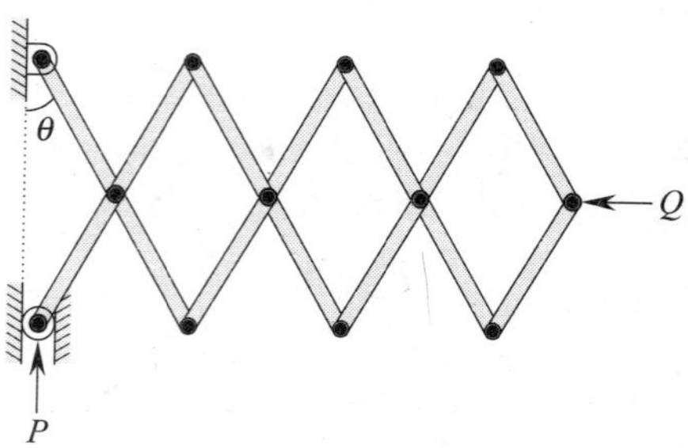

Fig. E.11.11

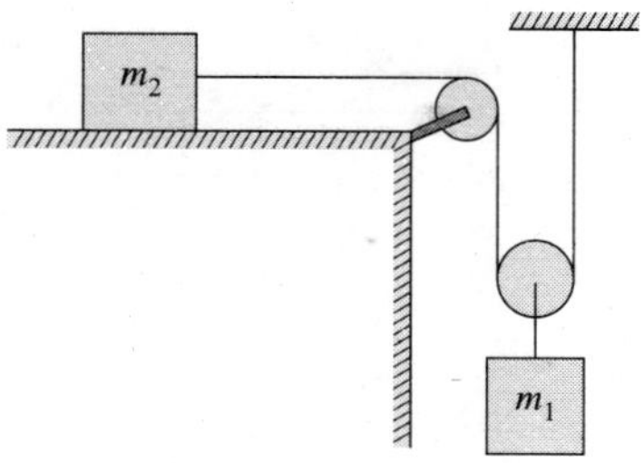

Fig. E.11.12

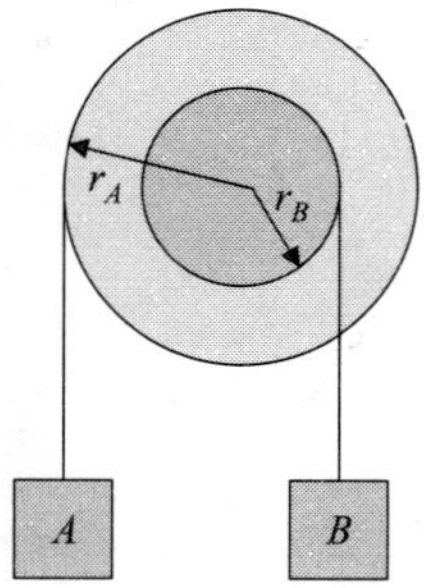

Fig. E.11.13

11.13 By the method of virtual work, determine the mass of block A in Fig. E.11.13, to maintain equilibrium if $m_B = 5$ kg, $r_A = 15$ cm and $r_B = 10$ cm.

Ans. 3.33 kg

11.14 By using the principle of virtual work, determine the force P required to hold the bar shown in Fig. E.11.14 under equilibrium in the position shown. The mass of the bar is 28.5 kg/m.

Ans. 790 N

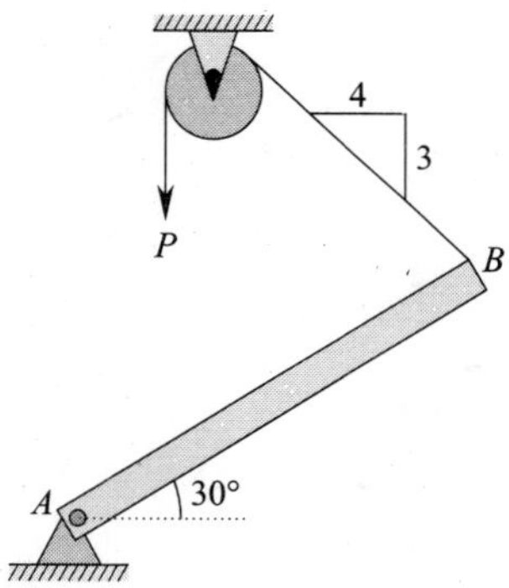

Fig. E.11.14

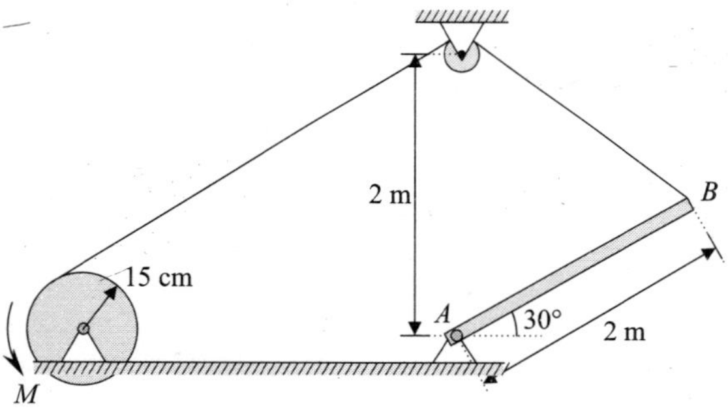

Fig. E.11.15

11.15 Using the principle of virtual work, determine the moment M required to be applied to maintain equilibrium of a uniform wooden bridge shown in Fig. E.11.15. The mass per unit length of the bridge is 100 kg/m. The size of the pulley at the top can be neglected.

Ans. 147.15 N.m

11.16 By the method of virtual work, determine the horizontal components of support reactions at A and B for the system shown in Fig. E.11.16. The length of each link is l. The weights of the links can be neglected.

Ans. $A_x = \dfrac{P\tan\theta}{2}$, $B_x = -\dfrac{P\tan\theta}{2}$

11.17 Using the method of virtual work, find the support reactions for the simply supported beam AB loaded as shown in Fig. E.11.17.

Ans. $A_y = 156$ N, $B_y = 124$ N

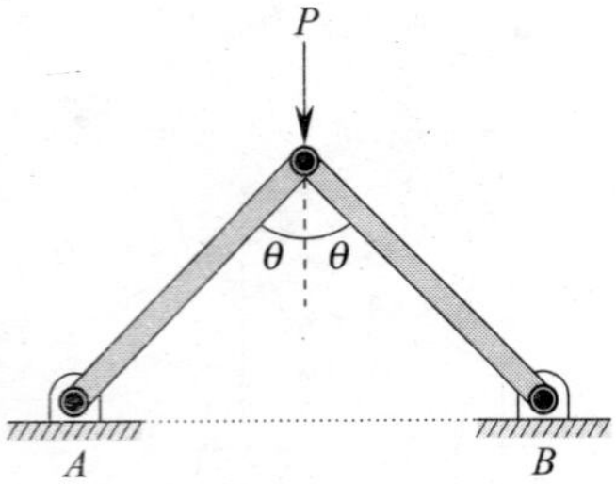

Fig. E.11.16

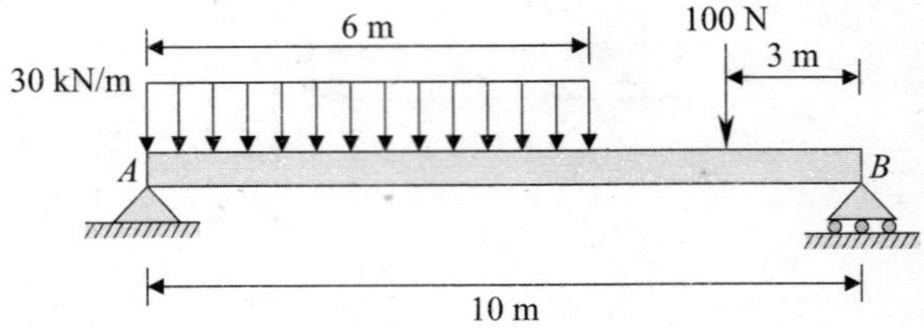

Fig. E.11.17

12 Kinematics of Particles (Rectilinear Motion)

12.1 INTRODUCTION

So far, in our study in the previous chapters, we have analyzed bodies at rest, which we called the statics part of mechanics. From this chapter onwards, we will analyze bodies under motion, which we call the dynamics part of mechanics. For a body to move from its state of rest or change its existing motion, it cannot do so by itself but must be acted on by some external force. This was stated by Newton in his famous laws of motion. As it is the *external force* acting on a body, which causes the change of motion of the body, we can understand that the resulting motion of the body is dependent upon the force or forces acting on the body.

In chapters 3 and 4, we saw that a system of forces acting on a body could be *replaced* by a single resultant force acting at a common point and a *moment*. This resultant force is equal to the summation of all individual forces and the moment is equal to the summation of all individual moments. When both the resultant force and the moment are **zero** then the body will be at rest or under static equilibrium. The study of bodies at rest is termed statics. This was covered in the previous chapters, where we treated bodies at rest and analyzed their equilibrium conditions. However, when the resultant force, moment, or both are non-zero then the body will be under motion. The study of bodies under motion is termed dynamics.

The resulting motion of a body under the action of a system of forces can be either translational rotational or a combination of both motions. The actual type of motion, whether it is translational, rotational or combination of both depends upon the nature of the resultant of the system of forces. When the resultant is a single force $\vec{R}$ acting at the centre of gravity, i.e., the moment being zero then the body is in pure translational motion in the direction of the resultant and there is *no* rotational motion.

Further, if the direction of the resultant force is constant then the translational motion is along a straight line, which we term as **rectilinear** or **one-dimensional** motion. For instance, a ball thrown vertically upwards and a car travelling on a straight road are examples of rectilinear motion. However, if the direction of the resultant force varies, then the motion will *not* be in a straight line and we term such a motion as **curvilinear** or **two-dimensional** motion. A golf ball hit from the ground and a motorist travelling on a curved road are examples of such curvilinear motion.

Examples of rectilinear motion

Examples of curvilinear motion

Fig. 12.1

When the resultant of a system of forces is a couple $\vec{M}$ then it tends to *rotate* the body. Further, if the body is *fixed* about a point then the motion is **pure rotational motion**. The motion of a pulley fixed at its centre of gravity is an example for fixed axis rotation.

When the resultant of a system of forces is a *centroidal force* $\vec{R}$ and a *couple* $\vec{M}$ then the body will have **general motion** which is neither pure translational nor pure rotational. However, such motions can be thought of as a **combination** of translational and rotational motions. A cylinder rolling down on an inclined plane is an example for such a motion. If the motion of the body lies in a plane then it is termed as planar motion, otherwise it is termed as general three-dimensional or spatial motion.

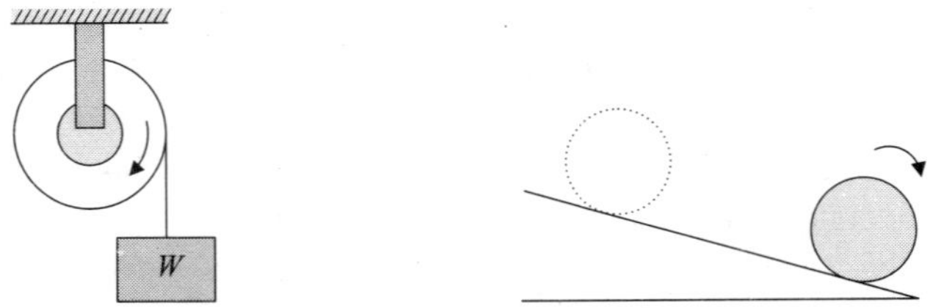

Example of rotational motion

Example of combined translational and rotational motion

Fig. 12.2

When a body is in pure translational motion, all the particles in the body will move in parallel paths with the *same* displacement, velocity and acceleration. Hence, instead of treating the body as a whole, we can analyze the motion of the whole body by idealizing it as a *particle*. However, if it is in pure rotational motion or a combination of both translational and rotational motions, the particles in the body move with different velocities and accelerations and hence, we can no more describe its motion by idealizing it as a particle but treating it as a rigid body itself. Thus, we can divide dynamics of bodies into **dynamics of particles** and **dynamics of rigid bodies**.

The dynamics of particles or rigid bodies can further be divided into two parts, namely, **kinematics** and **kinetics**. If we are interested only in the motion of particles or rigid bodies *without* considering the forces causing the motion, then that branch of dynamics is termed **kinematics**. It deals with the relationship between displacement, velocity and acceleration, and their variation with time. However, if we want to relate the motion of particles or rigid bodies with the forces causing the motion then it is termed **kinetics**.

The kinematics of particles is covered in this chapter and in the following chapter. In this chapter, we will discuss the rectilinear motion of particles and in the following chapter the curvilinear motion of particles. The kinetics of particles, i.e., relating the motion of particles with the forces causing the motion will be covered in Chapter 14.

As dynamics of rigid bodies involves rotational motion, it is studied separately from dynamics of particles. The kinematics of rigid bodies will be covered in Chapter 17, and kinetics of rigid bodies in Chapter 18.

12.2 MOTION OF A PARTICLE

Consider the motion of a body in the X–Y plane from A to B and from B to C as shown in Fig. 12.3(a), where X_1–Y_1 are local coordinate axes or axes fixed to the body. If throughout the motion, the X_1–Y_1 axes always remain *parallel* to the fixed reference axes X–Y then we see that the body is in *pure translational motion* and there is *no* rotational motion involved. In such a motion, all the particles in the body move in parallel paths with the same displacement, velocity and acceleration [refer Fig. 12.3(b)].

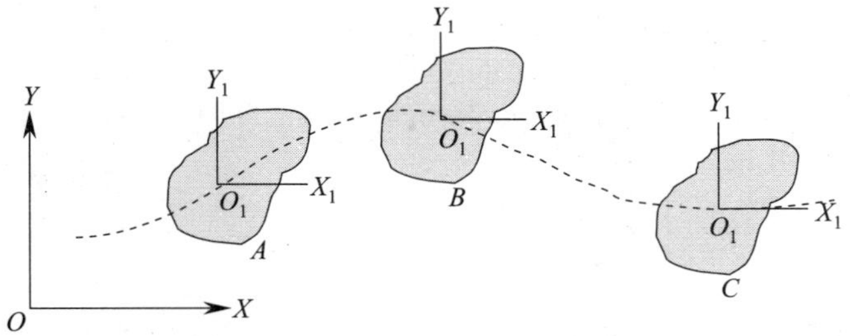

Fig. 12.3(a) Motion in X–Y plane

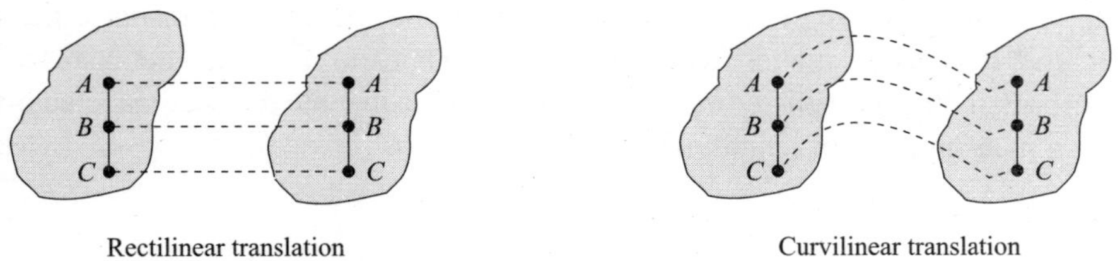

Fig. 12.3(b)

Hence, instead of analyzing the motion of the body as a whole, we can analyze the motion of a single particle in the body, which is a *representative* of the motion of the entire body. This is termed as *idealization* of the body as a **particle**, i.e., body without extent. Whenever we say particle, we should keep in mind that we are not dealing with minute bodies but rather gross bodies, which do not have rotational motion at all, or even if they have, they have been neglected. Mathematically, a particle is treated as a point and normally the centre of gravity of the body is chosen as this point.

To describe the motion of a particle, we must specify its position at any instant of time, and also its velocity and acceleration at that instant. Hence, we proceed to define each of these terms in the following section. We will define these terms considering a two-dimensional motion or plane motion, which can later on be extended to a more general three-dimensional or spatial motion.

12.3 DISPLACEMENT, VELOCITY AND ACCELERATION

12.3.1 Displacement

To describe the motion of a particle, we must specify its position at any instant of time with respect to a reference frame. As we are measuring the *motion*, the reference frame should be such that it is fixed, i.e., having no motion. For this reason, in astronomical studies, distant stars are chosen as non-inertial reference frames as they are considered to be fixed. However, for normal engineering analysis, we can choose a point on the earth's surface as the fixed origin for the reference frame, even though the earth is not fixed but rotating about itself and about the sun.

The position of a particle is then given by **position vector**, drawn from the origin of that reference frame to the particle. If at any instant of time t_1, say the particle is at A, whose position vector is $\vec{r}_1$ and at a later time t_2, it is at B, whose position vector is $\vec{r}_2$ then we say that the particle has displaced from the point A to the point B in time $(t_2 - t_1)$. Thus, we define **displacement vector** as the change in position of the particle during this interval of time. In Fig. 12.4, we see that

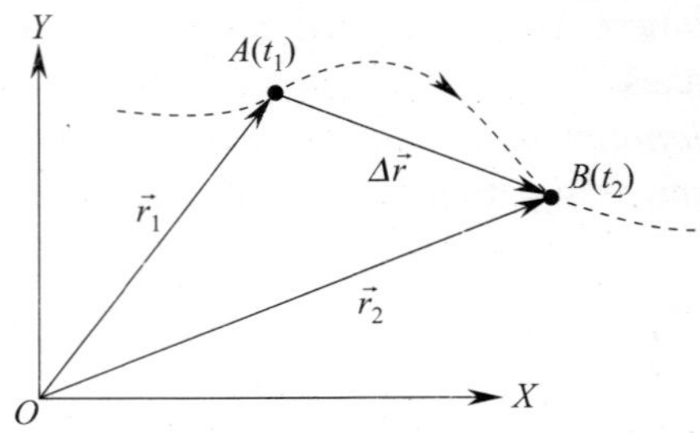

Fig. 12.4 Displacement vector

$$\overrightarrow{AB} = \overrightarrow{OB} - \overrightarrow{OA}$$

$$\therefore \quad \Delta\vec{r} = \vec{r}_2 - \vec{r}_1 \tag{12.1}$$

The displacement is a *vector* quantity and it is dependent only on the *initial* and *final* positions of the particle. It does not tell us anything about the *path* traced. It could be a straight line or curved. On the other hand, the **distance travelled** being a scalar quantity, is dependent on the actual path traced by the particle. In Fig. 12.4, the arc length AB (shown by dotted line) denotes the distance travelled by the particle in this time. If a particle moves from A to B and then back to A then the net displacement is zero as the particle is back to its initial position. However, we can see that the distance traveled is not zero. As displacement is a measurement of length, its SI unit is metre 'm'.

12.3.2 Velocity

To describe the motion of a particle at any instant of time, it would not suffice to define its position alone, but also the rate at which it gets displaced. For instance, a bullock cart will generally take a longer time to get displaced from A to B than a car. We define this change in displacement with respect to time as velocity.

Thus, **velocity** of a particle can be defined as the rate of change of displacement with time. In Fig. 12.4, as the particle moves from A to B, the **average velocity** during this time interval is then given as the ratio of the net displacement and elapsed time, i.e.,

$$\vec{v}_{\text{ave}} = \frac{\text{net displacement}}{\text{elapsed time}}$$

$$= \frac{\Delta\vec{r}}{\Delta t} = \frac{\vec{r}_2 - \vec{r}_1}{t_2 - t_1} \tag{12.2}$$

We can see that as the average velocity vector is a ratio of vector displacement and scalar time, its magnitude is equal to $|\Delta\vec{r}/\Delta t|$ and direction same as $\Delta\vec{r}$.

This average velocity is determined from the initial and final positions of the particle. Hence, it does not say anything about the velocity of the particle at intermediate points. If the average velocity measured between any two points along the path remains the same in magnitude and direction then the particle is said to move with **constant velocity**. It should be noted that as velocity is a vector quantity [being a ratio of vector displacement and scalar time], constant velocity can be maintained only when both magnitude and direction are constant. This is possible then only when the motion is along a straight line or rectilinear motion.

On the contrary, if the average velocity measured between any two points along the path does not remain constant then the particle is said to move with **variable velocity**. In such a case, we must specify the velocity of the particle at a particular instant of time, called the **instantaneous velocity**.

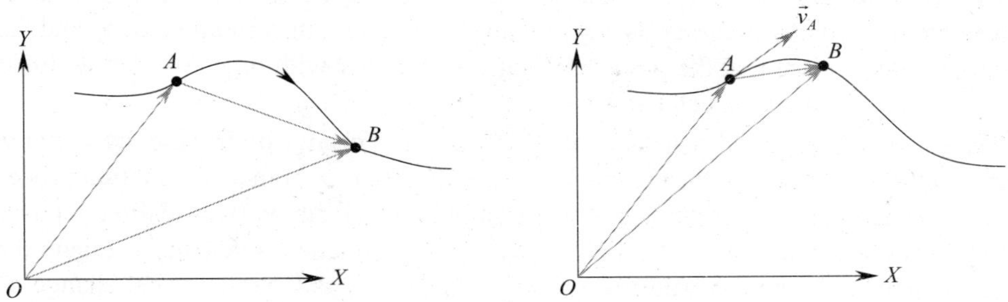

Fig. 12.5 Instantaneous velocity

To determine the instantaneous velocity, let us consider smaller time increments such that the point B approaches A. In the limiting case as $\Delta t \to 0$, the average velocity then very closely defines the instantaneous velocity at the point A. Mathematically,

$$\vec{v} = \lim_{\Delta t \to 0} \frac{\Delta \vec{r}}{\Delta t} \tag{12.3}$$

From calculus, we know that this can be expressed as

$$\vec{v} = \frac{d\vec{r}}{dt} \tag{12.4}$$

Also, in the limiting condition, we can see that the direction of $\Delta\vec{r}$ approaches that of the tangent to the path of the particle at A. Hence, the direction of instantaneous velocity is *always* **tangential** to the path of the particle and its magnitude is called the **speed** of the particle. The unit of velocity in SI units is **m/s**, but sometimes it is also expressed in **km/h**. The conversion factor for which is given as

$$1 \text{ km/h} = \frac{1000 \text{ m}}{3600 \text{ s}} = \frac{5}{18} \text{ m/s} \tag{12.5}$$

Similarly, the conversion factor of m/s to km/h is

$$1 \text{ m/s} = \frac{1/1000 \text{ km}}{1/3600 \text{ h}} = \frac{18}{5} \text{ km/h} \tag{12.6}$$

Corollary If the position vector of a particle is expressed in rectangular coordinates as

$$\vec{r} = x\vec{i} + y\vec{j} \tag{12.7}$$

then the velocity vector is obtained by differentiating the above expression with respect to time, i.e.,

$$\begin{aligned}\vec{v} &= \frac{d\vec{r}}{dt} = \frac{dx}{dt}\vec{i} + \frac{dy}{dt}\vec{j} \\ &= v_x\vec{i} + v_y\vec{j}\end{aligned} \tag{12.8}$$

where v_x and v_y are x and y components of velocity.

12.3.3 Acceleration

A particle may not always move with constant velocity throughout its motion. For instance, if it is starting from rest, it normally increases its velocity until it reaches a maximum velocity and then moves at this velocity. Thus, we see that the particle changes in velocity with time. We define this change in velocity with respect to time as **acceleration.**

Thus, *acceleration* of a particle can be defined as the rate of change of the velocity vector with time. The velocity vector may change either in magnitude, in direction or both as the motion proceeds. Suppose at time t_1, the particle is at the point A with instantaneous velocity v_1 (whose direction is tangential to the path at A) and at a later time t_2, it is at point B with instantaneous velocity v_2 (again direction is tangential to the path at B) then **average acceleration** is defined as the ratio of net change in velocity and time elapsed, i.e.,

$$\vec{a}_{\text{ave}} = \frac{\text{net change in velocity}}{\text{time interval}}$$

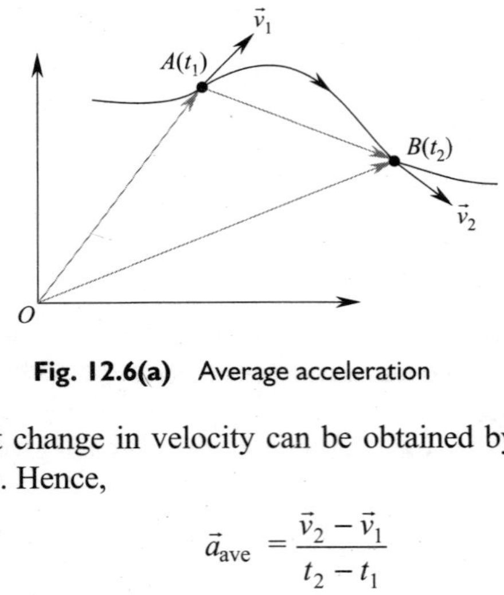

Fig. 12.6(a) Average acceleration

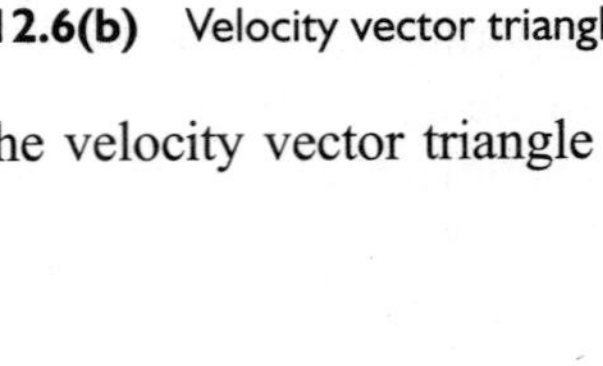

Fig. 12.6(b) Velocity vector triangle

The net change in velocity can be obtained by drawing the velocity vector triangle as shown in Fig. 12.6(b). Hence,

$$\begin{aligned}\vec{a}_{\text{ave}} &= \frac{\vec{v}_2 - \vec{v}_1}{t_2 - t_1} \\ &= \frac{\Delta\vec{v}}{\Delta t}\end{aligned} \tag{12.9}$$

Its direction is same as $\Delta\vec{v}$ and its magnitude is $|\Delta\vec{v}/\Delta t|$. Its unit as seen from the above equation is $\mathbf{m/s^2}$.

This average acceleration is based on the initial and final positions of the particle. Hence, it does not say anything about the acceleration of the particle at intermediate points. If the average acceleration measured between any two points along the path remains the same in magnitude and direction then the particle is said to move with **constant acceleration**. A body falling freely under gravity is an example of constant acceleration.

On the contrary, if average acceleration measured between any two points along the path does not remain constant then the particle is said to move with **variable acceleration**. In such a case, we must specify the acceleration of the particle at a particular instant of time, called the **instantaneous acceleration**. To determine the instantaneous acceleration, let us consider smaller time increments such that the point B approaches A. In the limiting case as $\Delta t \to 0$, the average acceleration very closely defines the instantaneous acceleration at the point A. Mathematically,

$$\vec{a} = \lim_{\Delta t \to 0} \frac{\Delta \vec{v}}{\Delta t}$$

and from calculus, we know that this can be expressed as

$$\vec{a} = \frac{d\vec{v}}{dt} \qquad (12.10)$$

The magnitude of instantaneous acceleration is $|d\vec{v}/dt|$ and its direction is the limiting direction of the change in velocity vector $\Delta\vec{v}$. If the velocity of the particle **decreases** as it moves from point A to B then the acceleration of the particle is **negative**, which we also call as **deceleration**.

Corollary If the position vector of a particle is expressed in rectangular coordinates as

$$\vec{r} = x\vec{i} + y\vec{j} \qquad (12.11)$$

then we know that the velocity vector is given as

$$\vec{v} = v_x\vec{i} + v_y\vec{j} \qquad (12.12)$$

Hence, acceleration is obtained by differentiating the above expression for velocity with respect to time, i.e.,

$$\vec{a} = \frac{d\vec{v}}{dt} = \frac{dv_x}{dt}\vec{i} + \frac{dv_y}{dt}\vec{j} \qquad (12.13)$$

or by differentiating twice the expression for displacement with respect to time, i.e.,

$$\vec{a} = \frac{d^2x}{dt^2}\vec{i} + \frac{d^2y}{dt^2}\vec{j} \qquad (12.14)$$

$$= a_x\vec{i} + a_y\vec{j} \qquad (12.15)$$

where a_x and a_y are x and y components of acceleration.

Note: From Fig. 12.6(b), we can see that even if the magnitude of the velocity vector is constant, i.e., $|\vec{v}_1| = |\vec{v}_2| = v$, a change in its direction results in a non-zero $\Delta\vec{v}$. Hence, the particle will be under acceleration, even if the velocity remains constant in magnitude, but changes in direction.

12.4 RECTILINEAR MOTION

The relationships between displacement, velocity and acceleration were derived in the previous section for a two-dimensional motion. Before analyzing problems related to two-dimensional motion, it is good to understand one-dimensional motion, which is treated as a special case of two-dimensional motion. The general two-dimensional motion will be covered in the next chapter.

When the motion of a particle is restricted along a *straight line*, the motion is said to be one-dimensional or **rectilinear motion**. We will consider motions either along a *horizontal* axis, i.e., *X*-axis or along a *vertical* axis, i.e., *Y*-axis. In Section 12.5, we will discuss rectilinear motion along the *X*-axis such as a car moving on a straight road or a sprinter running a race on a straight track, etc., and in Section 12.6, we will discuss rectilinear motion along the *Y*-axis such as a ball thrown vertically upwards or dropped from a building, etc.

12.5 RECTILINEAR MOTION ALONG THE X-AXIS

If the motion is along the *X*-axis then the position, velocity and acceleration can be obtained respectively by equating the *y*-components to zero in Eqs 12.11, 12.12 and 12.15.

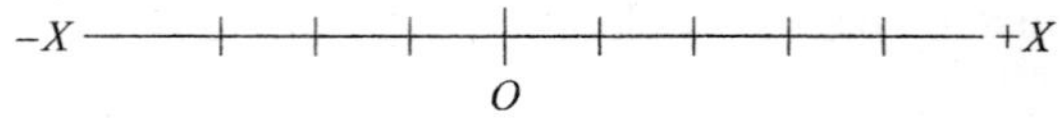

Fig. 12.7 Motion along the X-axis

Therefore, $\vec{r} = x\vec{i}$ or simply $r = \pm x$, where r is positive when measured towards right and negative when measured towards left from origin. Similarly, $\vec{v} = v_x\vec{i}$ or simply, $v = \pm v_x$, where velocity is positive when the velocity vector $\vec{v}$ points in the positive *X*-direction and negative when points in the negative *X*-direction and $a = \pm a_x$, where acceleration is positive when the acceleration vector $\vec{a}$ points in the positive *X*-direction and negative when it points in the negative *X*-direction. Hence, for one-dimensional motion, we can avoid the vector approach taking into consideration the above sign conventions. As the motion is only along the *X*-axis, we can also drop the subscript x. Hence, we can write

$$v = \frac{dx}{dt} \tag{12.16}$$

and

$$a = \frac{dv}{dt} = \frac{d^2x}{dt^2} \tag{12.17}$$

By separation of variables, we can also write acceleration as

$$a = \frac{dv}{dt} = \frac{dv}{dx}\frac{dx}{dt} = v\frac{dv}{dx} \tag{12.18}$$

The above three equations are known as **differential equations of motion** in a straight line. Hence, if the position of a particle is known as a function of time, i.e., $x = f(t)$ then velocity v and acceleration a can be determined by differentiating the position function with respect to time:

$$v = \frac{dx}{dt} = f'(t) \tag{12.19}$$

and

$$a = \frac{dv}{dt} = \frac{d^2x}{dt^2} = f''(t) \tag{12.20}$$

Example 12.1 The position of a particle in rectilinear motion is defined by the relation $x = t^3 - 2t^2 + 10t - 6$, where x is in metres and t is in seconds. Determine (i) the position, velocity and acceleration of the particle at time $t = 3$ s, (ii) the average velocity during $t = 2$ s and $t = 3$ s, and (iii) the average acceleration during the third second.

Solution *Given*

$$x = t^3 - 2t^2 + 10t - 6$$

The expressions for velocity and acceleration can be obtained by differentiating the above expression successively with respect to time:

$$v = \frac{dx}{dt} = 3t^2 - 4t + 10$$

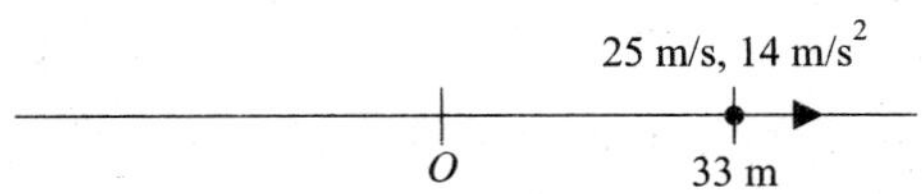

Fig. 12.8(a)

and
$$a = \frac{dv}{dt} = 6t - 4$$

(i) Position, velocity and acceleration of the particle at time t = 3 s

Substituting $t = 3$ s in the above expressions for position, velocity and acceleration, we get

$$x(3) = (3)^3 - 2(3)^2 + 10(3) - 6 = 33 \text{ m} \quad \text{(a)}$$
$$v(3) = 3(3)^2 - 4(3) + 10 = 25 \text{ m/s} \quad \text{(b)}$$
$$a(3) = 6(3) - 4 = 14 \text{ m/s}^2$$

Since velocity and acceleration are positive, both are pointing along the positive X-axis as shown in Fig. 12.8(a).

(ii) Average velocity during t = 2 s and t = 3 s

The displacements at these two time instants are

$$x(3) = 33 \text{ m [from the equation (a)]}$$
$$x(2) = (2)^3 - 2(2)^2 + 10(2) - 6$$
$$= 14 \text{ m}$$

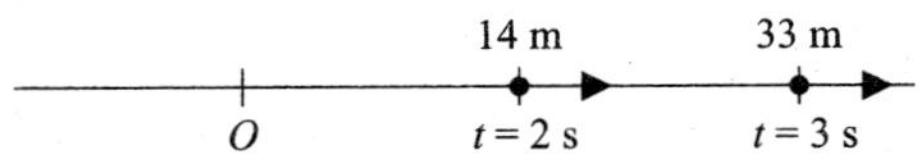

Fig. 12.8(b)

Therefore, average velocity during this time interval is given as

$$v_{ave} = \frac{\text{net displacement}}{\text{time elapsed}}$$
$$= \frac{33-14}{3-2} = 19 \text{ m/s}$$

(iii) Average acceleration during t = 2 s and t = 3 s

The instantaneous velocities at these two time instants are

$$v(2) = 3(2)^2 - 4(2) + 10 = 14 \text{ m/s}$$

and
$$v(3) = 25 \text{ m/s [from the equation (b)]}$$

Therefore, average acceleration during this time interval is given as

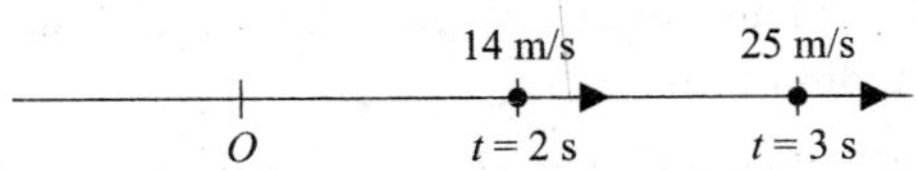

Fig. 12.8(c)

$$a_{ave} = \frac{\text{change in velocity}}{\text{time interval}}$$
$$= \frac{25-14}{3-2} = 11 \text{ m/s}^2$$

Example 12.2 The motion of a particle in rectilinear motion is defined by the relation $x = t^3 - 8t^2 + 16t - 5$, where x and t are expressed in metres and seconds respectively. Determine (i) the instants when velocity is zero, (ii) the position and acceleration at those instants of time, (iii) the instant when the acceleration is zero, and (iv) the position, the displacement, and the total distance travelled when the acceleration is zero.

Solution *Given*

$$x = t^3 - 8t^2 + 16t - 5 \tag{a}$$

The expressions for velocity and acceleration of the particle can be obtained by differentiating the above expression successively with respect to time:

$$v = \frac{dx}{dt} = 3t^2 - 16t + 16 \tag{b}$$

and

$$a = \frac{d^2x}{dt^2} = 6t - 16 \tag{c}$$

(i) Instants when the velocity is zero

In the equation (b), equating velocity to zero, we have

$$3t^2 - 16t + 16 = 0$$

Solving the above quadratic equation, we get

$$t = \frac{-(-16) \pm \sqrt{(-16)^2 - 4(3)(16)}}{2(3)}$$

$$\therefore \quad t = 1.33 \text{ (or) } 4 \text{ s} \tag{d}$$

(ii) Position and acceleration when the velocity is zero

$$x(1.33) = (1.33)^3 - 8(1.33)^2 + 16(1.33) - 5$$
$$= 4.48 \text{ m}$$
$$a(1.33) = 6(1.33) - 16$$
$$= -8.02 \text{ m/s}^2$$

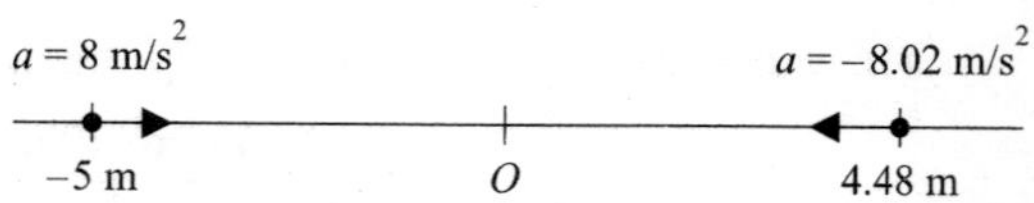

Fig. 12.9(a)

and

$$x(4) = (4)^3 - 8(4)^2 + 16(4) - 5$$
$$= -5 \text{ m}$$
$$a(4) = 6(4) - 16$$
$$= 8 \text{ m/s}^2$$

The negative acceleration at time $t = 1.33$ s indicates that it is pointing along the negative x-direction.

(iii) Instant when the acceleration is zero

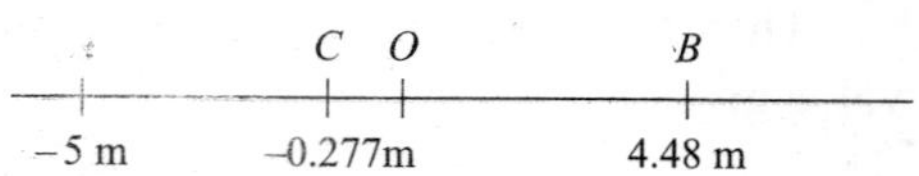

Fig. 12.9(b)

In the equation (c), equating acceleration to zero

$$\Rightarrow \quad 6t - 16 = 0$$

$$\therefore \quad t = 2.67 \text{ s}$$

(iv) Position, displacement and distance travelled when the acceleration is zero

Position, $x(2.67) = (2.67)^3 - 8(2.67)^2 + 16(2.67) - 5$

$= -0.277$ m

$x(0) = -5$m

Therefore, displacement from the point A to C is

$= |x(2.67) - x(0)|$

$= |-0.277 - (-5)| = 4.723$ m

From Fig. 12.9(b), we can see that the particle starts its motion from A and at time $t = 1.33$ s, it reaches the point B. At B, its velocity becomes zero as per the equation (d). Hence, it changes its direction of motion and reaches the point C at $t = 2.67$ s. As the direction of motion changes during the time $t = 0$ s to $t = 2.67$ s, the total distance travelled must be determined in two phases:

Total distance travelled $= |x(1.33) - x(0)| + |x(2.67) - x(1.33)|$

$= |4.48 - (-5)| + |-0.277 - 4.48|$

$= 9.48 + 4.757 = 14.237$ m

12.5.1 Integral Equations of Motion

In the previous section, we saw how if position of a particle is given as a function of *time*, its velocity and acceleration could be obtained by differentiating the position function successively with respect to time. Normally, position will not be specified for the motion, i.e., [x–t] relationship will not be given; instead, acceleration is specified for the motion, because physical phenomenon observed around us are such that acceleration is easy to specify. In such cases, v–t and x–t relationships are obtained through integration methods. Acceleration can be specified in terms of *time*, *position* or *velocity*. Accordingly, there are three different cases that we normally come across:

(i) Acceleration is given as a function of time, i.e., $a = f(t)$

We can write
$$a = \frac{dv}{dt} = f(t) \tag{12.21}$$

or
$$dv = f(t)\, dt \tag{12.22}$$

Upon integration between limits, we get the $v-t$ relationship:

$$\int_{v_o}^{v} dv = \int_{0}^{t} f(t)\, dt$$

$$\therefore \quad v = \int_{0}^{t} f(t)\, dt + v_o \tag{12.23}$$

We know $v = \dfrac{dx}{dt}$. Hence, the above equation can be written as

$$v = \frac{dx}{dt} = \int_{0}^{t} f(t)\, dt + v_o \tag{12.24}$$

Upon further integration between limits, we get the x–t relationship:

$$\therefore \quad \int_{x_o}^{x} dx = \int_{0}^{t} v\, dt \tag{12.25}$$

(ii) Acceleration is given as a function of velocity, i.e., $a = f(v)$

To find x–t and v–t relations, we proceed as follows:

Given
$$a = \frac{dv}{dt} = f(v) \tag{12.26}$$

or
$$\frac{dv}{f(v)} = dt \tag{12.27}$$

Upon integration, we get the v–t relationship, i.e., v is some function of t.

$$v = g(t) \tag{12.28}$$

From which the x–t relationship can be obtained by further integration as explained in the above case.

(iii) Acceleration is given as a function of position, i.e., $a = f(x)$

To find x–t and v–t relations, we proceed as follows:

We can write acceleration as

$$a = v\frac{dv}{dx} = f(x) \tag{12.29}$$

or
$$v\,dv = f(x)\,dx \tag{12.30}$$

Upon integration between limits,

$$\int_{v_o}^{v} v\,dv = \int_{x_o}^{x} f(x)\,dx \tag{12.31}$$

$\Rightarrow$
$$\frac{v^2}{2} - \frac{v_o^2}{2} = \int_{x_o}^{x} f(x)\,dx \tag{12.32}$$

Hence, velocity v can be expressed as a function of x, i.e., $v = g(x)$, then the $[x–t]$ relationship can be determined as follows:

Velocity can now be expressed as:

$$v = \frac{dx}{dt} = g(x) \tag{12.33}$$

or
$$\frac{dx}{g(x)} = dt \tag{12.34}$$

which can be integrated between the limits to get the x–t relationship. The above three cases are explained in the following examples.

Example 12.3 The acceleration of a particle in rectilinear motion is defined by the relation $a = kt$. The velocities of the particle at times $t = 1$ s and $t = 2$ s are respectively $v = 2$ m/s and $v = 3$ m/s. Write the equations of motion given that $x = 0$ at $t = 3$ s.

Solution Since acceleration is given as a function of time, upon integration with respect to time, we have

$$v = k\frac{t^2}{2} + v_o \tag{a}$$

where v_o is a constant of integration, which can be determined from the given conditions.

Given that at $t = 1$ s, $v = 2$ m/s

$$\Rightarrow \quad 2 = k\frac{1^2}{2} + v_o$$

$$2 = 0.5k + v_o \tag{b}$$

and at $t = 2$ s, $v = 3$ m/s,

$$\Rightarrow \quad 3 = k\frac{2^2}{2} + v_o$$

$$3 = 2k + v_o \tag{c}$$

From equations (b) and (c), solving for k and v_o, we get

$$k = 2/3 \quad \text{and} \quad v_o = 5/3$$

Therefore, equation (a) can be written as

$$v = \frac{2}{3}\frac{t^2}{2} + \frac{5}{3}$$

$$= \frac{1}{3}t^2 + \frac{5}{3} \tag{d}$$

Upon further integration of the above equation, we get

$$x = \frac{1}{3}\frac{t^3}{3} + \frac{5}{3}t + x_o \tag{e}$$

where x_o is a constant of integration. Given that at $t = 3$ s, $x = 0$,

$$\Rightarrow \quad x_o = -8 \text{ m}$$

Hence, the equation (e) can be written as:

$$x = \frac{1}{9}t^3 + \frac{5}{3}t - 8$$

Example 12.4 The acceleration of a particle in rectilinear motion is defined by the relation a = $3t^2 + 2$. Given that the initial velocity and displacement are respectively 2 m/s and 3 m, write the equations of motion. Also, determine the position, velocity and acceleration at $t = 2$s.

Solution Since acceleration is given as a function of time, upon integration with respect to time, we have

$$v = 3\frac{t^3}{3} + 2t + v_o \tag{a}$$

Given that at $t = 0$ s, $v = 2$ m/s,

$$\Rightarrow \quad v_o = 2 \text{ m/s}$$

Therefore, the equation (a) can be written as

$$v = t^3 + 2t + 2$$

Since, $v = \dfrac{dx}{dt}$, upon further integration,

$$x = \frac{t^4}{4} + 2\frac{t^2}{2} + 2t + x_o \tag{b}$$

Given that at $t = 0$ s, $x = 3$ m,

$$\Rightarrow \qquad x_o = 3 \text{ m}$$

Therefore, the equation (b) can be written as

$$x = \frac{t^4}{4} + t^2 + 2t + 3$$

Position, velocity and acceleration at t = 2 s

$$x(2) = \frac{(2)^4}{4} + (2)^2 + 2(2) + 3 = 15 \text{ m}$$
$$v(2) = (2)^3 + 2(2) + 2 = 14 \text{ m/s}$$
$$a(2) = 3(2)^2 + 2 = 14 \text{ m/s}^2$$

Example 12.5 The acceleration of a particle in rectilinear motion is defined by $a = k\sqrt{v}$, where a is in m/s^2, v is in m/s and k is a constant. Given that at times $t = 2$ s and $t = 3$ s, the velocities are respectively 4 m/s and 9 m/s, and the displacement at $t = 3$ s is 20 m, write the equations of motion.

Solution Since acceleration is given as a function of velocity, we can write

$$a = \frac{dv}{dt} = k\sqrt{v}$$

or

$$v^{-1/2}\, dv = k\, dt$$

Integrating on both sides,

$$\frac{v^{1/2}}{1/2} = kt + v_o \tag{a}$$

Given that at $t = 2$ s, $v = 4$ m/s,

$$\Rightarrow \qquad 2\sqrt{4} = k(2) + v_o$$

$$2(2) = k(2) + v_o \tag{b}$$

and at $t = 3$ s, $v = 9$ m/s

$$\Rightarrow \qquad 2\sqrt{9} = k(3) + v_o$$

$$2(3) = k(3) + v_o \tag{c}$$

From equations (b) and (c), solving for k and v_o, we get

$$k = 2 \quad \text{and} \quad v_o = 0$$

Therefore, the equation (a) can be written as

$$\frac{v^{1/2}}{1/2} = 2t$$

$$\sqrt{v} = t \quad \text{(or)} \quad v = t^2$$

Now,

$$v = \frac{dx}{dt} = t^2$$

Upon further integration,

$$x = \frac{t^3}{3} + x_o \qquad \text{(d)}$$

Given that at $t = 3$ s, $x = 20$ m,

$$20 = \frac{(3)^3}{3} + x_o$$

$\Rightarrow$ $$x_o = 11 \text{ m}$$

Therefore, the equation (d) can be written as

$$x = \frac{t^3}{3} + 11$$

Example 12.6 The acceleration of a particle in rectilinear motion is defined by $a \propto -v$, where a is in m/s^2, v is in m/s. Given that $x = 0$ and $v = v_o$ at $t = 0$, write the equations of motion.

Solution Since acceleration is proportional to velocity, introducing a constant of proportionality, we have

$$a = -kv$$

or $$a = \frac{dv}{dt} = -kv$$

Upon rearranging,

$$\frac{dv}{v} = -kdt$$

and integrating,

$$\ln v = -kt + c_1 \qquad \text{(a)}$$

Since $v = v_o$ at $t = 0$, we have $c_1 = \ln v_o$. Therefore, the equation (a) can be written as

$$\ln \frac{v}{v_o} = -kt$$

or $$\frac{v}{v_o} = e^{-kt}$$

$\therefore$ $$v = v_o\, e^{-kt} \qquad \text{(b)}$$

Further, we can write velocity as

$$v = \frac{dx}{dt} = v_o\, e^{-kt}$$

On rearranging,

$$dx = v_o\, e^{-kt}\, dt$$

and integrating,

$$x = -\frac{v_o}{k}\, e^{-kt} + c_2 \qquad \text{(c)}$$

Since $x = 0$ at $t = 0$, we have

$$c_2 = \frac{v_o}{k}$$

Therefore, the equation (c) can be written as

$$x = \frac{v_o}{k}[1 - e^{-kt}]$$

Example 12.7 The rectilinear motion of a particle is governed by $a = \frac{-16}{x^3}$, where a is in m/s^2 and x is in metres. Given that at time $t = 1$ s, $x = 2$ m and $v = 2$ m/s, (i) write the equations of motion; (ii) determine the position, velocity and acceleration at $t = 4$ s.

Solution

(i) Equations of motion

Since acceleration is given as a function of displacement, we can write

$$a = \frac{dv}{dt} = v\frac{dv}{dx} = -16x^{-3}$$

On rearranging,

$$v\,dv = -16x^{-3}\,dx$$

and integrating,

$$\frac{v^2}{2} = -16\frac{x^{-2}}{-2} + \frac{v_o^2}{2}$$

$$\therefore \quad v^2 = 16x^{-2} + v_o^2 \qquad \text{(a)}$$

Given that at $t = 1$ s, $x = 2$ m and $v = 2$ m/s,

$$(2)^2 = 16(2)^{-2} + v_o^2$$

$$\Rightarrow \quad v_o = 0$$

Therefore, the equation (a) can be written as

$$v^2 = 16x^{-2}$$

or

$$v = \frac{4}{x} \qquad \text{(b)}$$

Now,

$$v = \frac{dx}{dt} = \frac{4}{x}$$

Therefore,

$$x\,dx = 4\,dt$$

Upon further integration,

$$\frac{x^2}{2} = 4t + x_o \qquad \text{(c)}$$

At $t = 1$ s, $x = 2$ m

$$\therefore \quad \frac{2^2}{2} = 4(1) + x_o \Rightarrow x_o = -2 \text{ m}$$

Therefore, the equation (c) can be written as

$$\frac{x^2}{2} = 4t - 2 \quad \text{(or)} \quad x^2 = 8t - 4 \tag{d}$$

(ii) Position, velocity and acceleration at t = 4 s

$$x(4) = \sqrt{8(4) - 4} = 5.29 \text{ m}$$

$$v(4) = 4/x = 4/(5.29) = 0.756 \text{ m/s}$$

$$a = \frac{-16}{x^3} = \frac{-16}{(5.29)^3} = -0.108 \text{ m/s}^2$$

12.6 UNIFORMLY ACCELERATED MOTION

We saw in the previous section that acceleration could **vary** with time, position or velocity. In this section, we will consider a special case of rectilinear motion, in which the acceleration is **constant**. A number of real-life problems can very well be described by assuming the acceleration to be constant. The v–t and x–t relationships can be determined by the methods described in the previous section:

(i) [v–t] relationship

We can write acceleration as

$$\frac{dv}{dt} = a = \text{constant} \tag{12.35}$$

Upon integration between limits,

$$\int_{v_o}^{v} dv = a\int_{0}^{t} dt \tag{12.36}$$

Note that as a is a constant, it can be taken outside the integral sign. Hence,

$$v - v_o = at$$

or

$$v = v_o + at \tag{12.37}$$

(ii) [x–t] relationship

Since $v = \frac{dx}{dt}$, the Eq. 12.37 can be written as

$$\frac{dx}{dt} = v_o + at \tag{12.38}$$

Upon integration between limits,

$$x - x_o = v_o t + a\frac{t^2}{2} \tag{12.39}$$

Therefore,

$$x = x_o + v_o t + \frac{1}{2}at^2 \tag{12.40}$$

or

$$s = x - x_o = v_o t + \frac{1}{2}at^2 \tag{12.41}$$

Hence, if *initial velocity* v_o and *acceleration* a are known then the displacement in time t can be determined. Sometimes, instead of initial velocity, *final velocity* v may be specified; then in that case, we can rewrite the above equation by substituting the value of v_o from the Eq. 12.37, i.e.,

$$s = (v - at)t + \frac{1}{2}at^2$$

Therefore,
$$s = vt - \frac{1}{2}at^2 \tag{12.42}$$

(iii) [v–x] relationship

Also, we can express acceleration as

$$v\frac{dv}{dx} = a = \text{constant} \tag{12.43}$$

Upon integration between limits,

$$\int_{v_o}^{v} v\,dv = a\int_{x_o}^{x} dx$$

$$\frac{1}{2}(v^2 - v_o^2) = a(x - x_o)$$

$$v^2 = v_o^2 + 2a(x - x_o) \tag{12.44}$$

or
$$v^2 = v_o^2 + 2as \tag{12.45}$$

The student should note that as the above equations were derived assuming that the acceleration is constant, these equations can be used *only* when the acceleration of the particle is *constant*. If it is not constant then we must use the integration methods as discussed in the previous section.

Further, if acceleration is zero then the above equations reduce to

$$s = x - x_o = vt \qquad \text{and} \qquad v = v_o \tag{12.46}$$

We see that the velocity of the particle remains constant. Hence, such motion is called **uniform motion**, i.e., motion with *constant* velocity.

Example 12.8 An electric train starting from rest attains a maximum speed of 100 kmph in 20 seconds. Determine (i) its acceleration assuming it to be uniform, (ii) distance covered during this time period, and (iii) its velocity 15 seconds after staring from rest.

Solution As the train is starting from rest, its initial velocity is zero, i.e., $v_o = 0$. Its maximum speed given in kmph is converted to m/s as

$$v = 100 \text{ kmph} = 100 \times 5/18 = 27.78 \text{ m/s}$$

Time taken to reach the maximum speed is 20 s

(i) Acceleration of the train

As the acceleration is assumed to be uniform, we can apply the kinematic equations for constant acceleration. Since final velocity and time taken are known, we use the equation,

$$v = v_o + at$$

Substituting the corresponding values,

$$27.78 = 0 + a(20)$$

$$\Rightarrow \quad a = 1.389 \text{ m/s}^2$$

(ii) Distance covered during this time period

We know, $s = v_o t + (1/2)at^2$

$$= 0 + (1/2)(1.389)(20)^2 = 277.8 \text{ m}$$

(iii) *Velocity of the train 15 seconds after staring from rest*

Velocity at any instant of time is given as

$$v = v_o + at$$

$$= 0 + (1.389)(15) = 20.835 \text{ m/s (or) } 75 \text{ kmph}$$

Example 12.9 An airplane while taking off moves with a constant acceleration over a runway of 400 m in 8 seconds. Determine the velocity with which it takes off. Also, determine its constant acceleration.

Solution *Given data*

Initial speed of airplane, $v_o = 0$ m/s

Length of runway, $s = 400$ m

Time taken for take-off, $t = 8$ s

We know, $v = v_o + at$

$$= 0 + a(8)$$

$$= 8a \qquad \text{(a)}$$

and $v^2 = v_o^2 + 2as$

$$= 0 + 2a(400)$$

$$= 800a \qquad \text{(b)}$$

From equations (a) and (b), solving for v and a, we get

$$v = 100 \text{ m/s}$$

and $a = 12.5 \text{ m/s}^2$

Example 12.10 The speed of a truck moving at a constant speed of 30 m/s is reduced to 20 m/s in a distance of 200 m. Determine (i) the acceleration assuming it to be constant, and (ii) the time taken. Also, determine the distance in which the truck can be brought to a stop with the acceleration calculated in part (i).

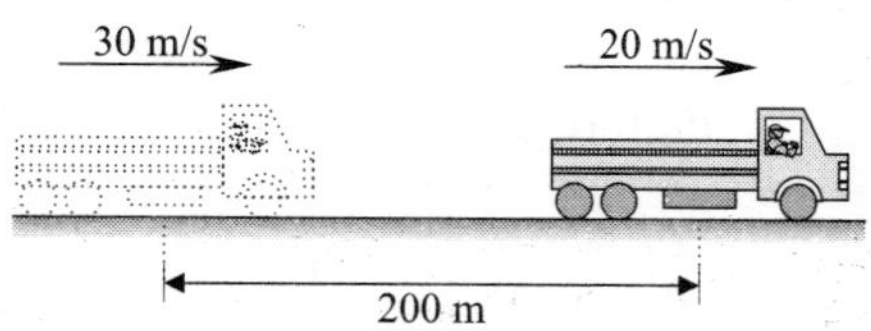

Fig. 12.10 (a)

Solution *Given data*

Initial speed of truck, $v_o = 30$ m/s

Final speed of truck, $v = 20$ m/s

Distance covered, $s = 200$ m

(i) Constant acceleration

Since the initial and final velocities are known, we choose the kinematic equation:

$$v^2 = v_o^2 + 2as$$

Hence, $$20^2 = 30^2 + 2a(200)$$

$\Rightarrow$ $$a = -1.25 \text{ m/s}^2$$

The negative sign indicates that the truck is under *deceleration*.

(ii) Time taken to reduce the speed from 30 m/s to 20 m/s

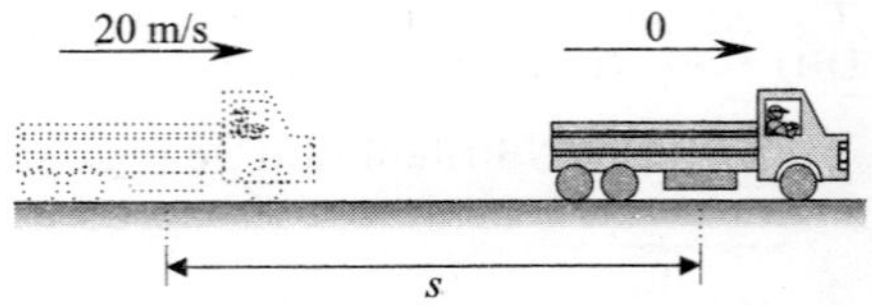

Fig. 12.10(b)

We know, $$v = v_o + at$$

Hence, $$20 = 30 + (-1.25)t$$

$\Rightarrow$ $$t = 8\text{s}$$

Distance travelled before the truck can be brought to a stop

For the remaining part of the motion, we consider

$$v_o = 20 \text{ m/s}, \quad v = 0 \text{ m/s} \quad \text{and} \quad a = -1.25 \text{ m/s}^2$$

Substituting the values in the equation of motion,

$$v^2 = v_o^2 + 2as$$

$$0 = 20^2 + 2(-1.25)s$$

$\Rightarrow$ $$s = 160 \text{ m}$$

Example 12.11 A car covers 100 m in 10 seconds, while accelerating uniformly at a rate of 1 m/s^2. Determine (i) initial and final velocities of the car, (ii) distance travelled before coming to this point assuming it started from rest, and (iii) its velocity after the next 10 seconds.

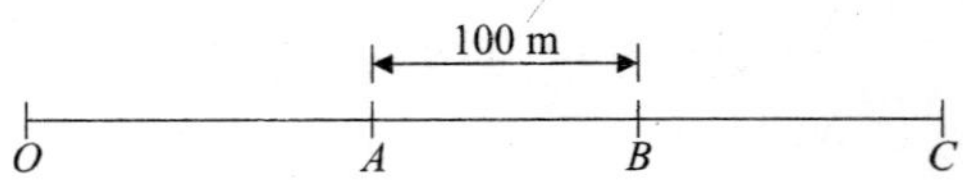

Fig. 12.11

Solution *Given data*

Distance covered, $s_{AB} = 100$ m

Time taken, $t = 10$ s

Acceleration, $a = 1$ m/s^2

(i) Initial and final velocities of the car

The equation of motion between A and B can be written as

$$s_{AB} = v_A t + \frac{1}{2} a t^2$$

$$100 = v_A(10) + \frac{1}{2}(1)(10)^2$$

$\Rightarrow$ $$v_A = 5 \text{ m/s} \quad \text{(or)} \quad 18 \text{ kmph}$$

Therefore, the velocity of the car at the point B is obtained as

$$v_B = v_A + at$$

$$= 5 + (1)(10) = 15 \text{ m/s (or) 54 kmph}$$

(ii) Distance travelled before coming to this point assuming it started from rest

The equation of motion between O and A can be written as

$$v_A^2 = v_O^2 + 2a(s_{OA})$$

$$5^2 = 0 + 2(1)\,(s_{OA})$$

$$\Rightarrow \quad s_{OA} = 12.5 \text{ m}$$

(iii) Velocity after the next 10 seconds

The equation of motion between B and C can be written as

$$v_C = v_B + at$$

$$= 15 + (1)\,(10) = 25 \text{ m/s (or) } 90 \text{ kmph}$$

Example 12.12 The driver of a car moving at a constant speed of 18 kmph realizes that if he moves at this speed, he will reach the office late by 10 seconds. Hence, he accelerates at a constant rate of 2 m/s^2 so that he reaches the office right in time. Determine the time taken to reach the office and the distance covered during this time.

Solution *Given data*

Initial speed of car, $v_o = 18 \text{ kmph} = 18 \times 5/18 = 5 \text{ m/s}$

Acceleration, $a = 2 \text{ m/s}^2$

Let t be the time taken to reach the office in right time. Then the time taken to reach it at constant speed is $(t + 10)$s.

Case I When moving under constant speed, the distance covered is obtained as

$$s_1 = v_o(t + 10)$$

$$= 5(t + 10) = 5t + 50 \qquad \text{(a)}$$

Case II When accelerating at the rate of 2 m/s^2, the distance covered is obtained as

$$s_2 = v_o t + \frac{1}{2}at^2$$

$$= 5t + \frac{1}{2}2t^2 = 5t + t^2 \qquad \text{(b)}$$

As the distance covered is same in both the cases, equating the two equations (a) and (b),

$$5t + 50 = 5t + t^2$$

$$\Rightarrow \quad t = \sqrt{50} = 7.07 \text{ s}$$

Then distance covered during this time is obtained by substituting the value of t in either of the equations (a) and (b).

$$s_2 = 5t + t^2$$

$$= 5(7.07) + (7.07)^2 = 85.33 \text{ m}$$

Example 12.13 A car travelling at a constant acceleration covers two points A and B, 200 m apart, in 10 s. If it is found to cross the point B with a speed of 90 kmph, determine (i) the speed with which it crossed the point A, (ii) its constant acceleration, (iii) the distance covered before coming to the point A if it started from rest.

Solution *Given data*

Speed of car while crossing the point B is

$$v_B = 90 \text{ kmph} = 90 \times 5/18 = 25 \text{ m/s}$$

Distance between points A and B is

$$s_{AB} = 200 \text{ m}$$

Time taken to travel from A to B is

$$t = 10 \text{ s}$$

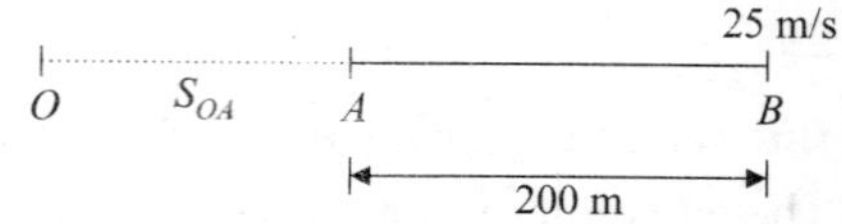

Fig. 12.12

Determination of constant acceleration

Since final velocity is known for the time interval, we use the expression

$$s_{AB} = v_B t - \frac{1}{2} a t^2$$

Therefore, $$200 = 25(10) - \frac{1}{2}(a)(10)^2$$

$\Rightarrow$ $$a = 1 \text{ m/s}^2$$

Determination of speed while crossing the point A

We know, $$v_B = v_A + at$$

$\Rightarrow$ $$v_A = v_B - at$$
$$= 25 - 1(10)$$
$$= 15 \text{ m/s (or) } 54 \text{ kmph}$$

Distance covered from origin to the point A

Let s_{OA} be the distance covered before reaching the point A. Then

$$v_A^2 = v_O^2 + 2as_{OA}$$
$$(15)^2 = 0 + 2(1)s_{OA}$$

$\Rightarrow$ $$s_{OA} = 225/2 = 112.5 \text{ m}$$

Example 12.14 The driver of a car travelling at a constant speed of 10 m/s sees the traffic signal ahead of him turning green, when he is at a distance of 200 m from the signal. If the signal remains green for 15 s, what should be his minimum acceleration in order to just cross the signal before the light turns orange? Also, determine the speed with which he crosses the signal.

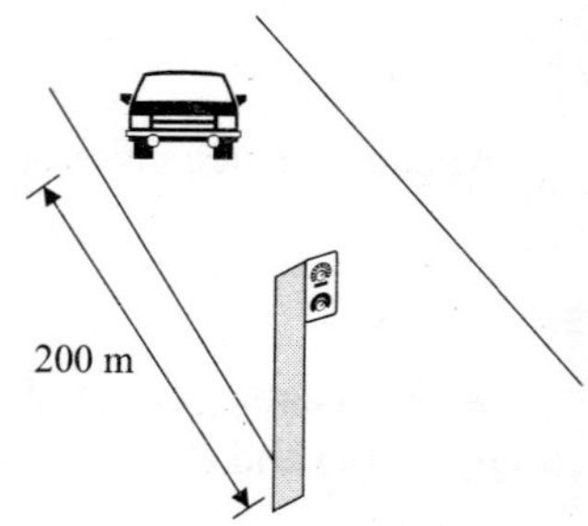

Fig. 12.13

Solution

Initial speed of car, $v_o = 10$ m/s

Distance covered, $s = 200$ m

If he continues to travel at constant speed, we can see that in 15 seconds, he can cover only 150 m. Hence, he has to accelerate. As he just crosses before the signal turns orange, time taken is same as the time for which the signal remains green, i.e.,

Time taken, $t = 15$ s

The kinematic equation of motion of the car is given as

$$s = v_o t + \frac{1}{2} at^2$$

Hence, $$200 = 10\,(15) + \frac{1}{2} a(15)^2$$

$\Rightarrow$ $$a = 100/225 = 0.44 \text{ m/s}^2$$

The speed of the car when it crosses the signal is given as

$$v = v_o + at$$
$$= 10 + (0.44)\,(15)$$
$$= 16.6 \text{ m/s}$$

Example 12.15 The driver of a car moving at a constant speed of 36 kmph sees the signal turning red when he is 50 m from the signal. The reaction time of the driver, i.e., the time interval between the perception of a signal to stop and the application of brakes is 0.7 s. If the car begins to decelerate at a constant rate upon the application of brakes, determine (i) the minimum deceleration of the car required to bring it to a halt just before the signal, and (ii) time taken to bring the car to a halt.

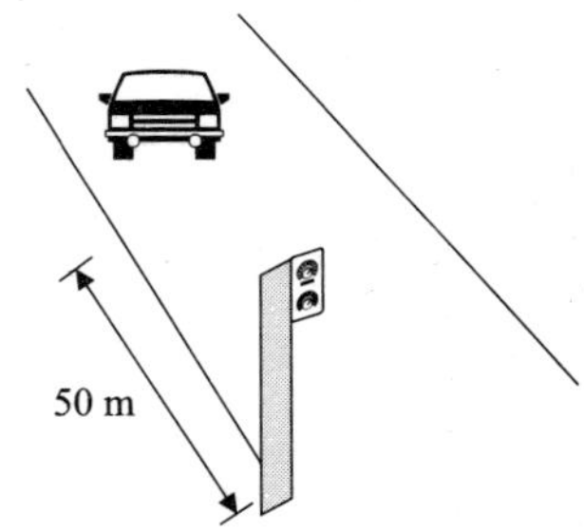

Fig. 12.14

Solution

Initial speed of the car, $v_o = 36$ kmph $= 36 \times 5/18 = 10$ m/s.

Distance travelled by the car before the brakes are applied (i.e., during the reaction time) is

$$s_1 = v_o t$$
$$= 10(0.7) = 7 \text{ m}$$

Hence, the distance travelled by the car from the instant that the brakes are applied to come to a halt just before the signal is

$$s_2 = 50 - s_1$$
$$= 50 - 7 = 43 \text{ m}$$

(i) Minimum deceleration of the car

The kinematic equation of motion of the car after the brakes are applied is given by

$$v^2 = v_o^2 + 2as_2$$

Hence, $$0 = 10^2 + 2a(43)$$

$\Rightarrow$ $$a = -1.16 \text{ m/s}^2$$

(ii) Time taken to bring the car to a halt

We know, $$v = v_o + at$$

Hence, $$0 = 10 + (-1.16)t$$

$\Rightarrow$ $$t = 8.62 \text{ s}$$

Therefore, the time taken to bring the car to a halt from the instant that the driver sees the signal turning red is 9.32 s (= 0.7 + 8.62).

Example 12.16 A man standing at a bus stand sees that a bus just leaves when he is about 20 m from the bus. If the bus accelerates at a constant rate of 1 m/s^2 then determine the acceleration with which the man must run to catch the bus within a distance of 30 m. Also, determine the speeds of the bus and the man at that instant.

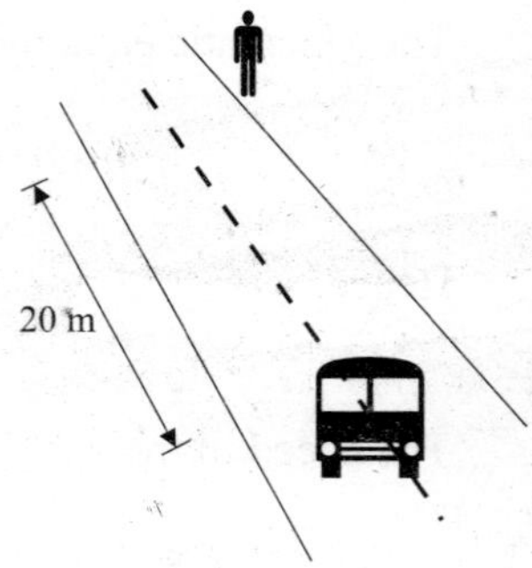

Fig. 12.15

Solution As the distance travelled by the man before catching the bus is 30 m, the distance travelled by the bus during this time is

$$30 - 20 = 10 \text{ m}$$

Hence, the kinematic equation of motion of the bus can be written as

$$s_b = v_{bo}t + (1/2)a_b t^2$$

$$10 = 0 + (1/2)(1)t^2$$

$$\Rightarrow \quad t = 4.47 \text{ s}$$

The equation of motion of the man can be written as

$$s_m = v_{mo}t + (1/2)a_m t^2$$

$$30 = 0 + (1/2)a_m(4.47)^2$$

$$\Rightarrow \quad a_m = 3 \text{ m/s}^2$$

Speed of the man when he catches the bus is given as

$$v_m = v_{mo} + a_m t$$

$$= 0 + 3(4.47) = 13.41 \text{ m/s}$$

Speed of the bus at that instant is given as

$$v_b = v_{bo} + a_b t$$

$$= 0 + (1)(4.47) = 4.47 \text{ m/s}$$

Example 12.17 At the intersection, when the traffic signal turns green, a car starts with a constant acceleration of 3 m/s^2. At the same instant, a bus travelling in the adjacent lane crosses the signal overtaking the car at a uniform speed of 36 kmph. Determine when and where the car will overtake the bus. Also, determine the speed of the car at that instant.

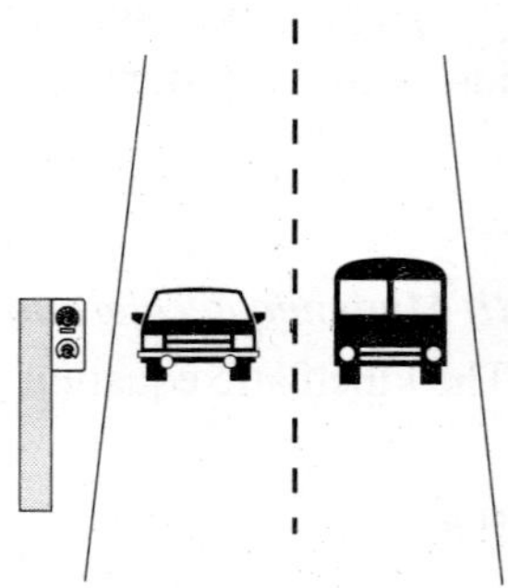

Fig. 12.16

Solution The uniform speed of the bus is 36 kmph (or) 36 × 5/18 = 10 m/s. Then kinematic equation of motion of the bus can be written as

$$s_b = v_b t = 10t$$

The kinematic equation of motion of the car can be written as

$$s_c = v_{co}t + \frac{1}{2}a_c t^2$$

$$= 0 + \frac{1}{2}(3)t^2$$

Note that as the bus crosses the car at the same time as the car starts, the time of motion is the same for both. At the instant when the car overtakes the bus, the distance travelled by both will be equal. Hence,

$$s_b = s_c$$

$\Rightarrow$ $$10t = \frac{1}{2}(3)t^2$$

$\therefore$ $$t = 0 \text{ or } 6.67 \text{ s}$$

Hence, the time at which the car overtakes the bus is 6.67 s. The distance travelled during this time is obtained by substituting this value of t in the expression for s_b or s_c

$$s_b = 10t = 10(6.67) = 66.7 \text{ m}$$

The speed of the car at this instant is obtained as

$$v_c = v_{co} + a_c t$$
$$= 0 + 3(6.67) = 20 \text{ m/s}$$

Example 12.18 In a car race, the car A starts and accelerates at a constant rate of 3 m/s^2. Car B starts 2 seconds later but accelerates at a constant rate of 4.6 m/s^2. Determine (a) when and where the car B will overtake A, and (b) speed of the two cars at that instant.

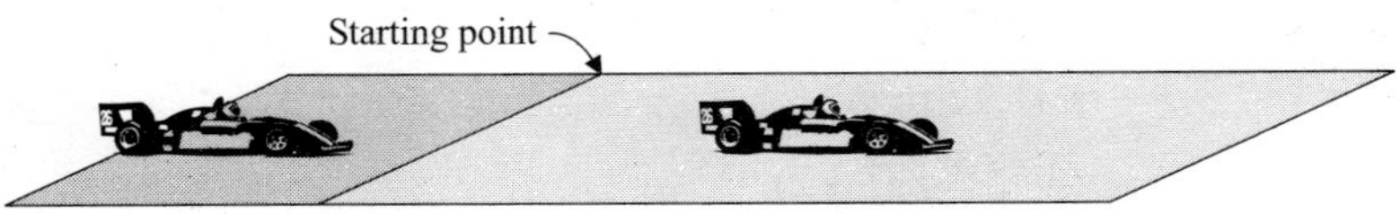

Fig. 12.17

Solution

(a) When and where the car B will overtake A

Let t s be the time taken for the car A from start to when the car B overtakes it. Then the time taken by the car B is $(t - 2)$ s. The kinematic equations of motion for the two cars can then be written as

$$s_A = v_{Ao}t + \frac{1}{2}a_A t^2$$
$$= 0 + \frac{1}{2}(3)t^2$$
$$= 1.5t^2 \quad \text{(a)}$$

$$s_B = v_{Bo}(t-2) + \frac{1}{2}a_B(t-2)^2$$
$$= 0 + \frac{1}{2}(4.6)(t-2)^2$$
$$= 2.3(t-2)^2 \quad \text{(b)}$$

Note that as both the cars start from rest, their initial velocities are zero. At the instant, that the car B overtakes the car A, the distance travelled by the two cars will be equal. Hence, equating the two equations (a) and (b), we have

$$1.5t^2 = 2.3(t-2)^2$$

or $$0.8t^2 - 9.2t + 9.2 = 0$$

On solving the quadratic equation, we get

$$t = 1.11 \text{ s or } 10.39 \text{ s}$$

Even though there are two roots for the above quadratic equation, we should not consider the time $t = 1.11$ s, as the car *B* starts only 2 s after the car *A* has left. Hence, considering the other value for time, i.e., $t = 10.39$ s, the distance travelled by the car *A* is obtained from equation (a) as:

$$s_A = 1.5t^2$$

$$= 1.5(10.39)^2 = 161.93 \text{ m}$$

Hence, the car *B* overtakes the car *A* 10.39 s after the car *A* has left and at a point 161.93 m from the starting point.

(b) Speeds of the two cars at that instant

The speeds of the two cars at that instant are obtained as

$$v_A = v_{Ao} + a_A t$$

$$= 0 + (3)\ 10.39$$

$$= 31.17 \text{ m/s}$$

$$v_B = v_{Bo} + a_B(t - 2)$$

$$= 0 + (4.6)(10.39 - 2) = 38.59 \text{ m/s}$$

Example 12.19 Two cars, *A* and *B*, start from rest at the same instant, with the car *A* initially trailing at some distance behind the car *B*. The uniform accelerations of the cars *A* and *B* are respectively 3 m/s^2 and 2 m/s^2. If the car *A* overtakes the car *B*, when *B* has moved 200 m, (i) determine the time taken to overtake, (ii) how far was the car *A* behind *B* initially, (iii) determine the speed of each car at that instant.

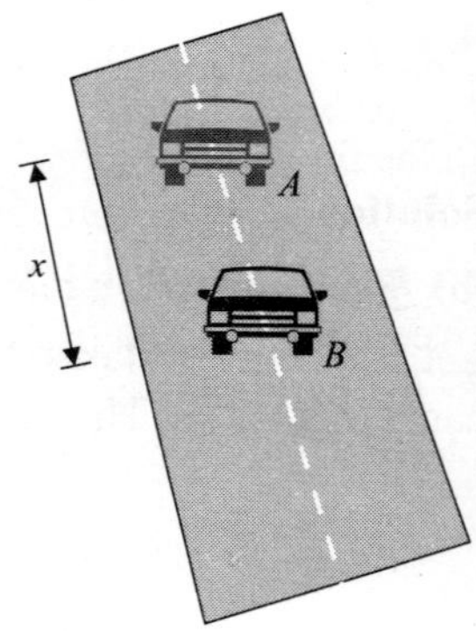

Fig. 12.18

Solution *Given*

As the two cars start from rest, their initial velocities are zero, i.e., $v_{Ao} = v_{Bo} = 0$.

$$a_A = 3 \text{ m/s}^2 \text{ and } a_B = 2 \text{ m/s}^2$$

(i) Time taken to overtake

As the car *B* moves through a distance of 200 m before the car *A* overtakes it, we can write the equation of motion of the car *B* as

$$s_B = v_{Bo}t + \frac{1}{2}a_B t^2$$

$$200 = 0 + \frac{1}{2}(2)t^2$$

$$\Rightarrow \quad t = \sqrt{200} = 14.14 \text{ s}$$

Thus, the car *A* overtakes the car *B* after 14.14 seconds from start.

(ii) Initial distance between the two cars

If x be the initial distance between cars A and B then the total distance travelled by the car A before overtaking the car B is

$$s_A = x + 200$$

Hence, the equation of motion of the car A can be written as

$$s_A = v_{Ao}t + \frac{1}{2} a_A t^2$$

$$x + 200 = 0 + \frac{1}{2}(3)(\sqrt{200})^2$$

$$\Rightarrow \quad x = 100 \text{ m}$$

(iii) Speed of the two cars at the instant of overtaking

The speeds of the two cars at the instant of overtaking can be determined as

$$v_A = v_{Ao} + a_A t$$

$$= 0 + (3)\sqrt{200} = 42.43 \text{ m/s}$$

$$v_B = v_{Bo} + a_B t$$

$$= 0 + (2)\sqrt{200} = 28.28 \text{ m/s}$$

12.7 MOTION CURVES

In the previous sections, we discussed the differential and integral equations of motion. These equations can also be represented graphically by plotting $x-t$, $v-t$ and $a-t$ graphs. Such graphical representations are termed as **motion curves**. Before drawing these curves, we must understand the relationship between them, so that if one of the curves is known then the other two can be easily obtained. If position of the particle is given as a function of t, i.e., $x = f(t)$ then the velocity and acceleration functions are obtained by differentiating the position function with respect to time, i.e.,

$$v = \frac{dx}{dt} = f'(t) = \text{slope of } x-t \text{ curve} \qquad (12.47)$$

and

$$a = \frac{dv}{dt} = f''(t) = \text{slope of } v-t \text{ curve} \qquad (12.48)$$

We know from calculus that differentiation of a function represents the slope of the function at a point. Thus, the slope of $x-t$ curve at any point gives the velocity of the particle at that instant and the slope of $v-t$ curve gives the acceleration of the particle at that instant. On the contrary, if acceleration of the particle is specified then velocity and position of the particle are obtained by integration. From calculus, we know that the integration [the reverse of differentiation] of a function represents the area of the function between limits. Thus, the area under $a-t$ curve between the limits gives the change in velocity of the particle and the area under $v-t$ curve between the limits gives the change in position of the particle. We will discuss below the motion curves for some of the common types of motions:

(i) Uniform motion Uniform motion is defined as that type of motion in which the acceleration is **zero** or in other words, the velocity is **uniform** or **constant**. The kinematic equations of motion are

$$v = v_o \quad \text{and} \quad x = x_o + v_o t \qquad (12.49)$$

Since velocity is constant at all times, the v–t graph is a horizontal straight line. Further, from the expression for position, we see that the x–t graph must also be a straight line as x is a linear function of time. It has a constant slope v_o and it intersects the X-axis at x_o, which is the initial position of the particle.

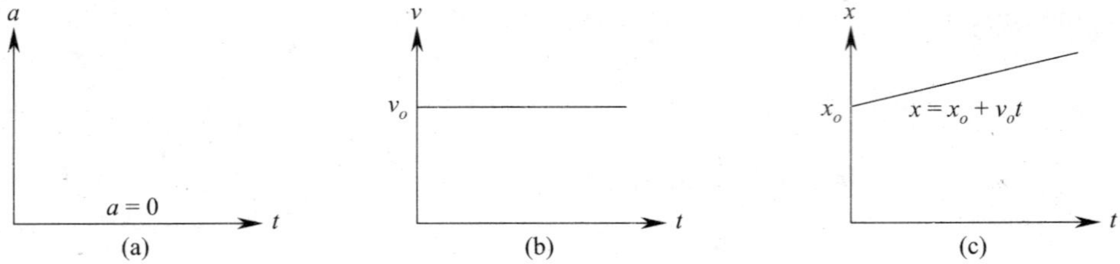

Fig. 12.19 Uniform motion

(ii) Uniformly accelerated motion Uniformly accelerated motion is that type of motion in which the acceleration is **constant** or in other words, the velocity is **uniformly varying**. The kinematic equations of motion are

$$a = \text{constant}; \quad v = v_o + at \quad \text{and} \quad x = x_o + v_o t + (1/2)at^2 \tag{12.50}$$

As acceleration is constant, the a–t graph is a horizontal straight line. Since velocity is a linear function of time, the v–t graph is also a straight line with a constant slope equal to a and it intersects the v-axis at v_o, which is the initial velocity. Further, since the position function contains *square* of time term, the x–t graph is parabolic in nature. Its slope at any point gives the velocity at that point.

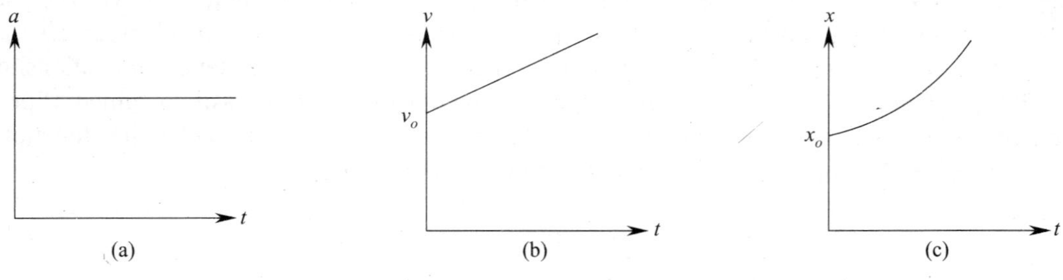

Fig. 12.20 Uniformly accelerated motion

(iii) Non-uniformly accelerated motion Non-uniformly accelerated motion is that type of motion in which the acceleration is **not constant** or in other words, the acceleration is **non-uniformly varying**.

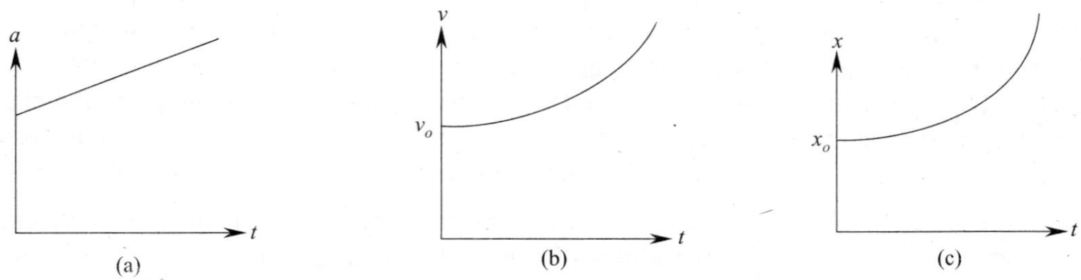

Fig. 12.21 Non-uniformly accelerated motion

If we assume that acceleration is uniformly varying, i.e., a linear function of time, then the $a-t$ graph is a straight line. Then the velocity obtained by integrating the linear acceleration function varies parabolically. Its slope at any point gives the acceleration of the particle at that point. Further, the position curve varies cubically and its slope at any point gives the velocity at that point.

12.8 DISTINCT PHASES OF MOTION (GRAPHICAL METHOD)

In all of the previous examples, we have seen so far, the particle moved either with constant velocity or with constant acceleration. However, the motion of a particle may have distinct *phases* of motion. For instance, a particle may move with *constant velocity* for some time and at some later times, it may move with *constant acceleration* or *deceleration,* and so on. In such cases, if we write the kinematic equations of motion for each phase separately and try to solve them, the procedure becomes cumbersome, as there will be more number of unknowns. In such type of problems, graphical methods provide easier solutions and hence, we explain it below.

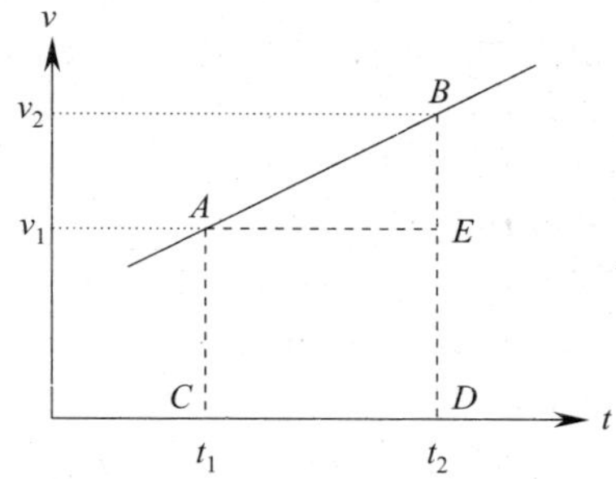

Fig. 12.22 $v-t$ graph for uniformly accelearted motion

When the motion of a particle has distinct phases, the graphical method, particularly the **$v-t$ graph** is very much useful in describing the motion. For a rectilinear motion with constant acceleration, the $v-t$ curve will have a constant slope as discussed above. Let the velocity of the particle be v_1 at a time t_1 and at a later time t_2 be v_2.

For this time interval, the equation of motion with constant acceleration can be written as

$$v_2 = v_1 + a(t_2 - t_1) \tag{12.51}$$

Then the acceleration a can be written as

$$a = \frac{(v_2 - v_1)}{(t_2 - t_1)} \tag{12.52}$$

From the above figure, we see that slope of line AB is

$$= \frac{BE}{AE} = \frac{(v_2 - v_1)}{(t_2 - t_1)} \tag{12.53}$$

which is same as the previous Eq. 12.52 obtained for the acceleration. Hence, we can conclude that *the acceleration of a particle is represented by the slope of the $v-t$ curve.* The slope is **positive** if the particle is **accelerated**; and it is **negative** if the particle is **decelerated**.

The distance travelled during this time interval is given by the kinematic equation,

$$s = v_1(t_2 - t_1) + \frac{1}{2} a(t_2 - t_1)^2 \tag{12.54}$$

Substituting the value of a from the Eq. 12.52 in the above equation,

$$s = v_1(t_2 - t_1) + \frac{1}{2}\left[\frac{v_2 - v_1}{t_2 - t_1}\right](t_2 - t_1)^2 \tag{12.55}$$

$$= v_1(t_2 - t_1) + \frac{1}{2}(v_2 - v_1)(t_2 - t_1) \tag{12.56}$$

From the graph, we can see that the first term in the above equation represents the area of the rectangle *ACDE* and the second term represents the area of the triangle *ABE*. Thus, their summation represents the area of the trapezium *ABDC*, which is nothing but the area of the $v-t$ curve between the time limits t_1 and t_2.

Hence, we can conclude that *the distance travelled by a particle during a certain interval of time is given by the area under the v–t curve between the time limits*. The following examples will illustrate the usefulness of motion curves in solving problems with distinct phases of motion.

Example 12.20 The motion curve of a particle moving in rectilinear path is shown in Fig. 12.23 Determine (i) the total distance travelled in 240 s, (ii) average velocity and average acceleration between 0–200 seconds.

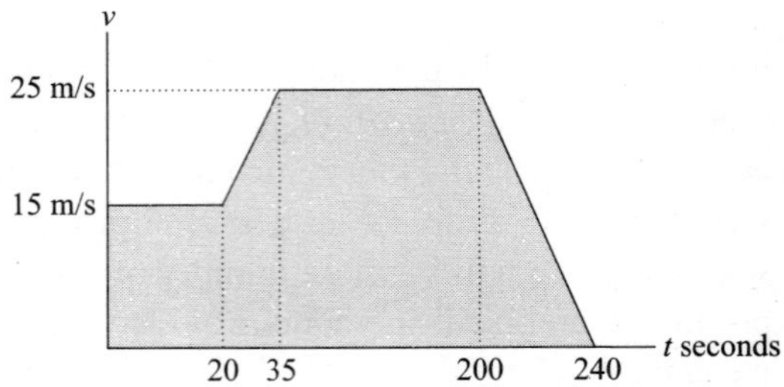

Fig. 12.23

Solution

(i) The total distance travelled in 240 s is given by the area under the curve between time limits $t = 0$ s and $t = 240$ s

$$\therefore \quad s = [15 \times 20] + [(1/2) \times 15(15 + 25)] + [165 \times 25] + [(1/2) \times 40 \times 25] = 5225 \text{ m}$$

(ii) The average velocity between 0–200 s is given by the ratio of net displacement and net time taken. Therefore,

$$v_{\text{ave}} = \frac{\text{net displacement}}{\text{time taken}}$$

$$= \frac{[15 \times 20] + [(1/2) \times 15(15 + 25)] + [165 \times 25]}{200} = 23.63 \text{ m/s}$$

The average acceleration is given by the ratio of change in velocity and time taken. Hence,

$$a_{\text{ave}} = \frac{\text{change in velocity}}{\text{time taken}}$$

$$= \frac{25 - 15}{200} = 0.05 \text{ m/s}^2$$

Example 12.21 An electric train starts from a station and accelerates at a constant rate of 1 m/s^2 for 20 s. It then runs at the maximum speed attained for the next 3 minutes and finally decelerates at a constant rate over a distance of 100 m until it comes to a stop at the next station. Find the total distance between the two stations and the time taken to travel this distance.

Solution As the motion has three distinct phases, accordingly we represent the v–t graph. In the first phase, the train is constantly accelerating to reach a maximum speed of v and then it runs at this maximum speed for some time. Finally, it decelerates at a constant rate to a come to a stop. Let t_1, t_2 and t_3 be the respective time intervals for the three phases.

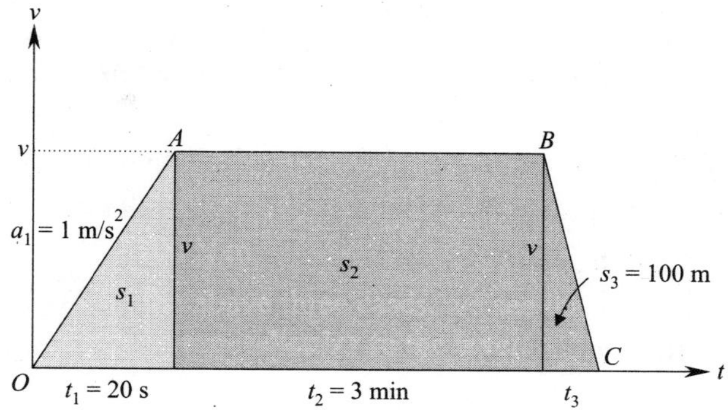

Fig. 12.24

Phase I Constant acceleration

We know that acceleration of the train is given by the slope of v–t curve. Hence,

$$a_1 = \text{slope of } OA = \frac{v}{t_1}$$

Therefore, maximum velocity of the particle is obtained as

$$v = a_1 t_1 = (1)(20) = 20 \text{ m/s}$$

Distance covered during the constant acceleration is given by the area under the v–t graph. Hence,

$$s_1 = \text{area under the curve } OA$$
$$= \frac{1}{2} v t_1 = \frac{1}{2}(20)\,(20) = 200 \text{ m}$$

Phase II Constant speed

Distance covered during constant speed is given by the area under the v–t graph between the time limits. Hence,

$$s_2 = \text{area under the curve } AB$$
$$= v t_2 = (20)(180) = 3600 \text{ m} \quad [\text{Note that } t_2 = 3 \text{ min} = 180 \text{ s}]$$

Phase III Constant deceleration

Distance covered during constant deceleration is given by the area under the v–t graph between the time limits. Hence,

$$s_3 = \text{area under the curve } BC$$
$$= \frac{1}{2} v t_3$$

$\Rightarrow$ $$t_3 = 2s_3/v = 2(100)/20 = 10 \text{ s}$$

Therefore, total distance between the two stations is obtained as

$$s = s_1 + s_2 + s_3$$
$$= 200 + 3600 + 100 = 3900 \text{ m (or) } 3.9 \text{ km}$$

The total time taken to cover this distance is obtained as

$$t = t_1 + t_2 + t_3$$
$$= 20 + 180 + 10 = 210 \text{ s}$$

Example 12.22 A car starts from rest and accelerates uniformly to reach a maximum speed of 72 kmph in 30 seconds. It then travels at this speed for 3 minutes and finally comes to rest in 45 seconds. Determine (i) the acceleration and deceleration of the car, (ii) the total distance travelled during this time, (iii) average velocity in this time, and (iv) average acceleration in this time.

Solution Maximum speed of car, $v_{max} = 72$ kmph $= 72 \times 5/18 = 20$ m/s.

(i) Acceleration and deceleration of the car

Since the time intervals for the first and last phases of motion are known, we can determine the acceleration and deceleration from the slopes of the v–t graph. Hence,

$$a_1 = \frac{v_{max} - 0}{t_1} = \frac{20 - 0}{30} = 0.67 \text{ m/s}^2$$

$$a_3 = \frac{0 - v_{max}}{t_3} = \frac{0 - 20}{45} = -0.44 \text{ m/s}^2$$

(ii) Total distance covered

The distances covered in each of the phases of motion are given by the areas under the v–t graph. Therefore,

$$s_1 = \frac{1}{2} v_{max} t_1$$

$$= \frac{1}{2} (20)(30) = 300 \text{ m}$$

$$s_2 = v_{max} t_2$$

$$= (20)(180) = 3600 \text{ m}$$

$$s_3 = \frac{1}{2} v_{max} t_3$$

$$= \frac{1}{2}(20)(45) = 450 \text{ m}$$

Therefore, total distance travelled is obtained as

$$s = s_1 + s_2 + s_3$$

$$= 300 + 3600 + 450 = 4350 \text{ m}$$

Total time taken is obtained as

$$t = t_1 + t_2 + t_3$$

$$= 30 + 180 + 45 = 255 \text{ s}$$

(iii) Average velocity during this time is given as

$$v_{ave} = \frac{\text{change in displacement}}{\text{total time elapsed}}$$

$$= \frac{4350 - 0}{255} = 17.06 \text{ m/s}$$

(iv) Similarly, average acceleration is given as

$$a_{ave} = \frac{\text{change in velocity}}{\text{total time elapsed}}$$

$$= \frac{0-0}{255} = 0$$

Note that change in velocity is zero as both initial and final velocities are zero.

Example 12.23 A train moving at 72 kmph, while crossing a bridge reduces its speed to 45 kmph at a constant rate of 0.2 m/s^2 due to speed limitations. It travels at the reduced speed for a distance of 300 m before entering into the bridge. While leaving the bridge at the other end, it travels a distance of 300 m before accelerating at a constant rate of 0.3 m/s^2 to its normal speed. If the total time elapsed from the instant that the train is decelerated to the instant that the train attains its normal speed is 6 minutes, determine the length of the bridge and the time taken to travel this length.

Solution

Initial speed of train: 72 kmph = 72 × 5/18 = 20 m/s

Speed of train while crossing the bridge: 45 kmph = 45 × 5/18 = 12.5 m/s

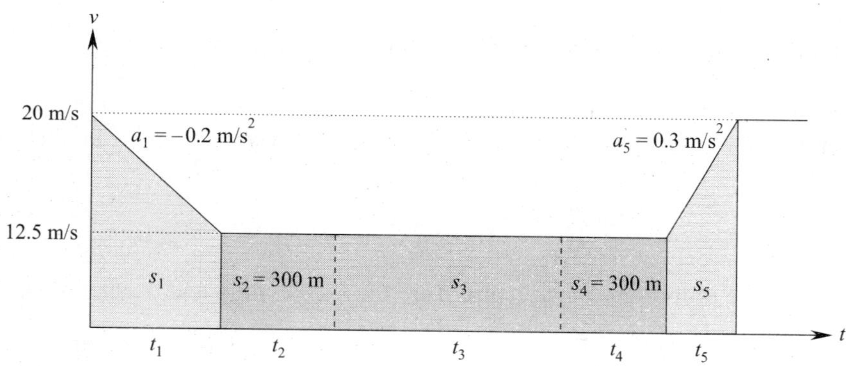

Fig. 12.25

Phase I

From the graph, we know that the constant deceleration is given by

$$a_1 = \frac{12.5-20}{t_1} = \frac{-7.5}{t_1}$$

$$\Rightarrow \quad t_1 = \frac{-7.5}{-0.2} = 37.5 \text{ s}$$

Phase II

$$s_2 = 12.5\, t_2$$

$$\Rightarrow \quad t_2 = \frac{300}{12.5} = 24 \text{ s}$$

Similarly, $t_4 = 24$ s

Phase V

The constant acceleration is given by

$$a_5 = \frac{20 - 12.5}{t_5} = \frac{7.5}{t_5}$$

$$\Rightarrow \qquad t_5 = \frac{7.5}{0.3} = 25 \text{ s}$$

The total time taken is given as

$$t = t_1 + t_2 + t_3 + t_4 + t_5$$

$$6 \times 60 = 37.5 + 24 + t_3 + 24 + 25$$

$$\Rightarrow \qquad t_3 = 249.5 \text{ s}$$

Therefore, the length of the bridge is obtained as

$$s_3 = 12.5\, t_3$$

$$= 12.5 \times 249.5 = 3118.75 \text{ m (or) } 3.12 \text{ km}$$

Example 12.24 A train starting from rest accelerates uniformly at the rate of α to reach a maximum speed of v. It then moves at this speed for some time and decelerates uniformly at the rate of β to come to rest. If the total distance covered is s, prove that the total time taken is $\frac{s}{v} + \frac{v}{2}\left[\frac{1}{\alpha} + \frac{1}{\beta}\right]$

Solution The motion curve for the entire journey is shown in Fig. 12.26. Let s_1, s_2 and s_3 be the distances covered in the respective phases of motion and similarly, t_1, t_2 and t_3 be the respective times. Then we can write

$$s = s_1 + s_2 + s_3$$

and $$t = t_1 + t_2 + t_3$$

where t is the total time of motion. Since α and β are the acceleration and deceleration respectively, we can also write

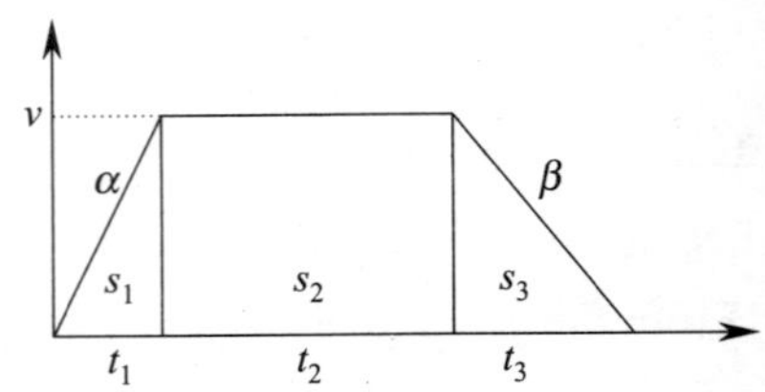

Fig. 12.26

$$t_1 = \frac{v}{\alpha} \quad \text{and} \quad t_3 = \frac{v}{\beta}$$

Therefore, $$s_1 = \frac{1}{2} v\, t_1 = \frac{1}{2}\frac{v^2}{\alpha}$$

and $$s_3 = \frac{1}{2} v\, t_3 = \frac{1}{2}\frac{v^2}{\beta}$$

Thus, $$s_2 = s - s_1 - s_3$$

$$= s - \frac{1}{2}\frac{v^2}{\alpha} - \frac{1}{2}\frac{v^2}{\beta}$$

Therefore, $$t_2 = \frac{s_2}{v} = \frac{s}{v} - \frac{v}{2\alpha} - \frac{v}{2\beta}$$

Hence, the total time taken is obtained as

$$t = \frac{v}{\alpha} + \left(\frac{s}{v} - \frac{v}{2\alpha} - \frac{v}{2\beta}\right) + \frac{v}{\beta}$$

$$= \frac{v}{2\alpha} + \frac{s}{v} + \frac{v}{2\beta}$$

$$= \frac{s}{v} + \frac{v}{2}\left(\frac{1}{\alpha} + \frac{1}{\beta}\right)$$

12.9 RECTILINEAR MOTION ALONG VERTICAL *Y*-AXIS

As rectilinear motion can be either along the *X*-axis or along the *Y*-axis, we will cover motion along the vertical *Y*-axis in this section. Bodies when thrown vertically upwards or dropped from a point above the ground level travel downwards due to pull of the earth. They will be subjected to an *acceleration* directed towards the centre of the earth. This acceleration is termed as **acceleration due to gravity** and it is denoted by *g*. In the absence of air resistance, it is observed that all bodies irrespective of their size and weight fall with the same acceleration due to gravity. This point was proved experimentally by Galileo.

The acceleration due to gravity at a particular location is dependent on the radius of the earth and hence it varies with latitude and altitude to a small extent. However, we can assume it to be *constant* near the earth's surface. Normally, an approximate value of *g* equal to 9.81 m/s^2 can be used for all calculation purposes. The idealized motion in which air resistance and change in acceleration with altitude are neglected is termed as **free fall**, although the term includes *rising* as well as *falling*.

Since the acceleration is constant, we can apply the equations for rectilinear motion with constant acceleration, by replacing *x* with *y* and *a* with *g*. As bodies can be thrown upwards or else thrown downwards, we try to differentiate the two cases and follow different sign conventions while writing the equations of motion as explained below:

12.9.1 Bodies Thrown Upwards

For bodies thrown upwards, it will be convenient to choose the ground as the fixed origin with the positive *Y*-axis directed upwards. Since acceleration is directed downwards, i.e., along the negative *Y*-axis, it is taken as negative. Hence, acceleration, $a = -g$. The equations of motion are then given by

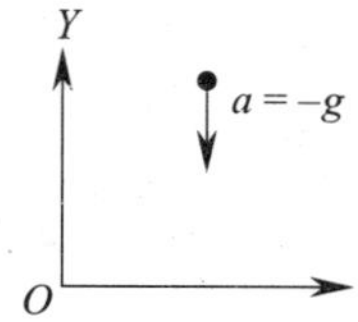

Fig. 12.27

$$v = v_o - gt \qquad (12.57)$$

$$v^2 = v_o^2 - 2g(y - y_o) \quad \text{(or)} \quad v^2 = v_o^2 - 2gs \qquad (12.58)$$

$$y = y_o + v_o t - \frac{1}{2}gt^2 \quad \text{(or)} \quad s = y - y_o = v_o t - \frac{1}{2}gt^2 \qquad (12.59)$$

where y_o is the initial displacement from the ground level. If the body is thrown from the ground level, then $y_o = 0$; if thrown from the top of a building then y_o = height of the building. It should be noted that the vertical path traced while ascent will not be strictly same as the vertical path traced during descent. However, to simplify the calculations, we assume these two paths to be the same.

At the highest point, the velocity of the body becomes *zero* and hence, the body changes its direction of motion and starts moving *downwards*. The maximum height reached by the body above the point of projection can be determined by equating $v = 0$ in the Eq. 12.58. Therefore,

$$0 = v_o^2 - 2gs$$

$$\Rightarrow s = h_{\max} = \frac{v_o^2}{2g} \tag{12.60}$$

The time taken to reach this maximum height is given by equating $v = 0$ in the Eq. 12.57. Hence,

$$0 = v_o - gt$$

$$\Rightarrow t = \frac{v_o}{g} \tag{12.61}$$

In the absence of air resistance, we know that the time of *ascent* is equal to the time of *descent*. Hence, the time taken to return to the point of projection is twice the time taken for ascent, i.e.,

$$T = \frac{2v_o}{g} \tag{12.62}$$

Also, it should be noted that the velocity of the body while crossing the point of projection would be same as the velocity of projection in magnitude but opposite in direction.

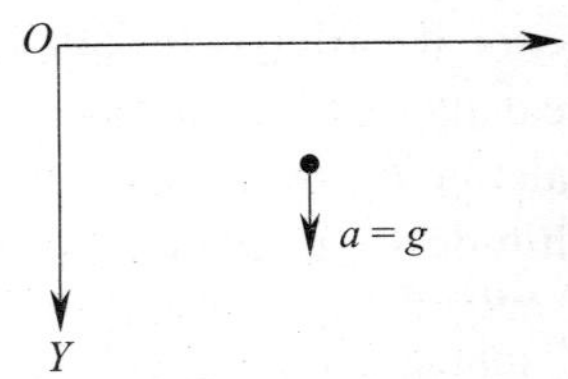

Fig. 12.28

12.9.2 Bodies Thrown Downwards or Dropped

For bodies thrown downwards or dropped from a point above the ground, the point of release is taken as the fixed origin and the positive Y-axis directed downwards. Hence, acceleration is taken as positive in this case. Therefore, the equations of motion can be written as

$$v = v_o + gt \tag{12.63}$$

$$v^2 = v_o^2 + 2gs \tag{12.64}$$

$$s = y = v_o t + \frac{1}{2}gt^2 \quad \text{[Note that } y_o = 0 \text{ in this case]} \tag{12.65}$$

For bodies dropped, the initial velocity is zero. Hence, the above equations can be simplified as

$$v = gt \tag{12.66}$$

$$v^2 = 2gs \tag{12.67}$$

$$s = y = \frac{1}{2}gt^2 \tag{12.68}$$

Example 12.25 A ball is thrown vertically upwards from the ground with an initial velocity of 20 m/s. Determine (i) the maximum height reached by the ball, (ii) the time taken to reach the maximum height, and (iii) the total time of flight.

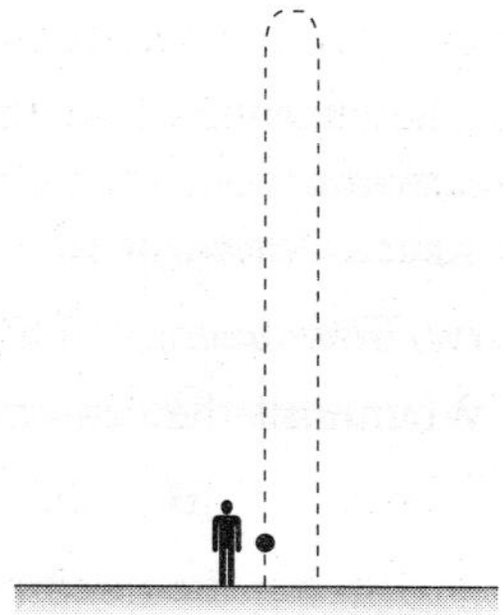

Fig. 12.29

Solution The maximum height reached by the ball is given by

$$h_{max} = \frac{v_o^2}{2g}$$

$$= \frac{(20)^2}{2 \times 9.81} = 20.39 \text{ m}$$

The time taken to reach this maximum height is given as

$$t = \frac{v_o}{g}$$

$$= \frac{20}{9.81} = 2.04 \text{ s}$$

The time of ascent is equal to the time of descent. Hence, the total time of flight is

$$T = 2 \times \text{time taken to reach the maximum height} = 4.08 \text{ s}$$

Example 12.26 A ball is thrown vertically upwards with an initial velocity of 20 m/s from the top of a building of 30 m height. Determine (i) the maximum height reached by the ball, (ii) the time taken to reach the maximum height, (iii) the velocity of the ball as it crosses the top of the building during its downward journey, (iv) the time taken to hit the ground and the corresponding velocity.

Solution As the ball is thrown from the top of a building, we take the ground as the origin and initial displacement as height of the building. The equation of motion can then be written as

$$y = y_o + v_o t - (1/2)\, g t^2$$

$$= 30 + 20t - 4.905t^2 \qquad \text{(a)}$$

(i) Maximum height reached

The maximum height reached by the ball from the top of the building is given as

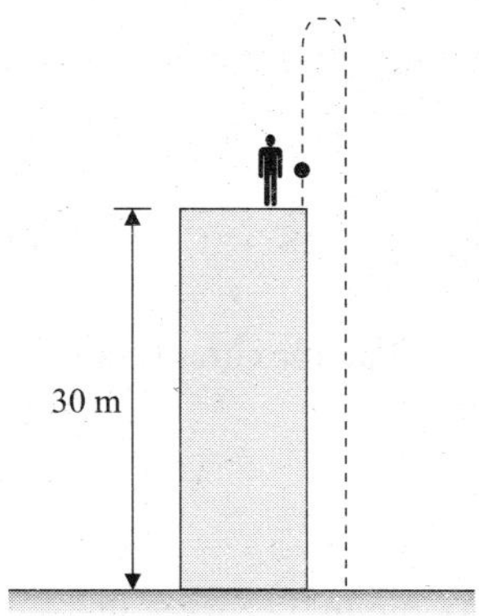

Fig. 12.30

$$h_{max} = \frac{v_o^2}{2g}$$

$$= \frac{(20)^2}{2 \times 9.81} = 20.39 \text{ m}$$

(ii) Time taken to reach the maximum height

$$t = \frac{v_o}{g} = \frac{20}{9.81} = 2.04 \text{ s}$$

(iii) Velocity as it crosses the top of the building

In the absence of air resistance, the velocity of the ball as it crosses the top of the building during its downward journey will be the same as the initial velocity in magnitude but opposite in direction, i.e., directed downwards.

(iv) Time taken to hit the ground and the corresponding velocity

When it hits the ground, we know that $y = 0$ in the equation (a). Hence,

$$30 + 20t - 4.905t^2 = 0$$

Solving the quadratic equation, we get

$$t = 5.24 \text{ s}$$

Note that as time cannot be negative, the other value of t satisfying the quadratic equation is not taken into account. The velocity of the ball at that instant is given as

$$v = v_o - gt$$
$$= 20 - (9.81)(5.24) = -31.4 \text{ m/s}$$

The negative sign indicates that the velocity is directed downwards.

Example 12.27 A ball is thrown vertically upwards from the ground. Two men standing at different heights on a building watch the ball pass by them at speeds of 20 m/s and 10 m/s respectively. Determine the height between their locations. How high does it rise beyond the second man?

Solution Let v_o be the initial speed of the ball when thrown from the ground. We know,

$$v^2 = v_o^2 - 2gs$$

When the ball crosses the first man, we can write it as

$$(20)^2 = v_o^2 - 2gs_1 \quad \text{(a)}$$

Similarly, when it crosses the second man, we can write

$$(10)^2 = v_o^2 - 2gs_2 \quad \text{(b)}$$

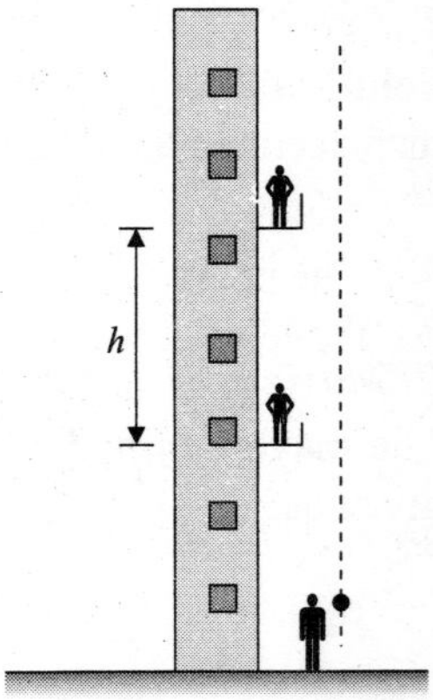

Fig. 12.31

(i) Height between their locations

Subtracting the equation (b) from the equation (a), we get

$$(20)^2 - (10)^2 = 2g(s_2 - s_1)$$

$$\Rightarrow \quad (s_2 - s_1) = h = \frac{(20)^2 - (10)^2}{2 \times 9.81} = 15.29 \text{ m}$$

The maximum height reached by the ball beyond the second man is given as

$$s_{max} = \frac{v^2}{2g}$$

As the ball crosses the second man with a speed of 10 m/s, we have

$$s_{max} = \frac{10^2}{2 \times 9.81} = 5.1 \text{ m}$$

Example 12.28 A ball is thrown upwards from the top of a 50 m high building with an initial velocity of 20 m/s. At the same instant, another ball is thrown upwards with an initial velocity of 30 m/s from the ground. Determine (i) when and where they will meet each other, and (ii) the velocity of each ball at that instant.

Solution

(i) Time and location where they meet each other

As both the balls are thrown upwards, we choose the ground as the fixed origin. Hence,

for the first ball, $v_{oA} = 20$ m/s and $y_{oA} = 50$ m and

for the second ball, $v_{oB} = 30$ m/s and $y_{oB} = 0$

The equations of motion for both the balls can be written as

$$y_A = y_{oA} + v_{oA}t - \frac{1}{2}gt^2$$

$$= 50 + 20t - \frac{1}{2}(9.81)t^2$$

$$y_B = y_{oB} + v_{oB}t - \frac{1}{2}gt^2$$

$$= 0 + 30t - \frac{1}{2}(9.81)t^2$$

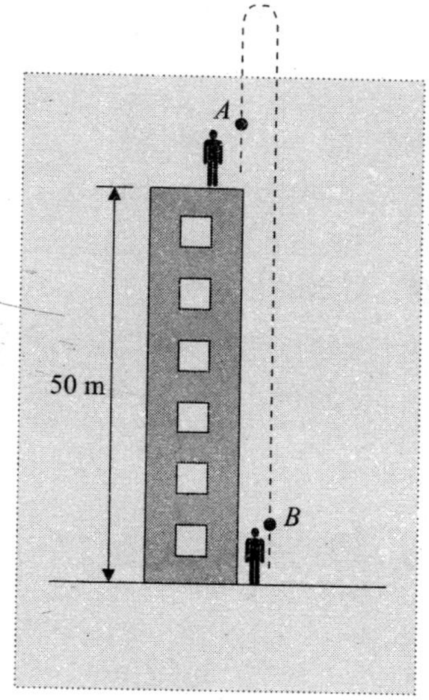

Fig. 12.32

When they meet each other, we know that $y_A = y_B$. Therefore,

$$50 + 20t - \frac{1}{2}(9.81)t^2 = 30t - \frac{1}{2}(9.81)t^2$$

$\Rightarrow$
$$t = 50/10 = 5 \text{ s}$$

The position with respect to the ground level when they meet each other is obtained by substituting the above value of time in the expression for either y_A or y_B:

$$y_B(5) = 30(5) - \frac{1}{2}(9.81)(5)^2 = 27.38 \text{ m}$$

(ii) Velocity of the two balls at that instant

$$v_A = v_{oA} - gt$$

$\therefore$
$$v_A(5) = 20 - 9.81(5) = -29.05 \text{ m/s}$$

Similarly,
$$v_B = v_{oB} - gt$$

$\therefore$
$$v_B(5) = 30 - 9.81(5) = -19.05 \text{ m/s}$$

The negative sign indicates that both the velocities are directed downwards. This implies that the two balls meet each other during the downward motion of the ball B thrown from the ground.

Example 12.29 A stone is dropped from the top of a building. Two seconds later, another stone is thrown downwards with an initial velocity of 30 m/s. If both the stones reach the ground at the same time, determine (i) the time taken by the first stone to reach the ground, and (ii) the height of the building.

Solution Since the displacements of both the stones are downwards, we can take the top of the building as the origin and the positive Y-axis as directed downwards. Let t s be the time taken by the first stone to reach the ground; then the time taken by the second stone to reach the ground is $(t-2)$ s. Hence, the equations of motion for the two stones can be written as

$$s_1 = v_{1o}t + (1/2)\,gt^2$$
$$= 0 + (1/2)gt^2 \qquad \text{(a)}$$
$$s_2 = v_{2o}(t-2) + (1/2)g(t-2)^2$$
$$= 30(t-2) + (1/2)g(t^2 + 4 - 4t) \qquad \text{(b)}$$

Since both the stones reach the ground at the same time, we can equate the two equations (a) and (b). Therefore,

$$(1/2)gt^2 = 30(t-2) + (1/2)g(t^2 + 4 - 4t)$$

On solving for t, we get

$$t = \frac{60-2g}{30-2g} = 3.89 \text{ s}$$

Therefore, height of the building can be determined by substituting the value of t in the equation (a),

$$h = s_1 = (1/2)gt^2$$
$$= (1/2)\,(9.81)(3.89)^2 = 74.22 \text{ m}$$

Example 12.30 The distance covered by a freely falling body in the last one second of its motion and that covered in the last but one second are in the ratio 5:4. Calculate the height from which the body was dropped and the velocity with which it strikes the ground.

Solution Let t be the time taken by the body to reach the ground. Then its equation of motion can be written as

$$y = y_o + v_o t + \frac{1}{2}gt^2$$

For a freely falling body, $y_o = 0$ and $v_o = 0$. Hence,

$$y = \frac{1}{2}gt^2 \qquad \text{(a)}$$

Therefore, distance covered in the last one second of its motion is given as

$$y_t - y_{t-1} = \frac{1}{2}gt^2 - \frac{1}{2}g(t-1)^2$$
$$= \frac{1}{2}g(2t-1) \qquad \text{(b)}$$

Similarly, the distance covered in the last but one second is given as

$$y_{t-1} - y_{t-2} = \frac{1}{2}g(t-1)^2 - \frac{1}{2}g(t-2)^2$$
$$= \frac{1}{2}g(2t-3) \qquad \text{(c)}$$

Given that these two distances are in the ratio 5:4, we have

$$\frac{5}{4} = \frac{\frac{1}{2}g(2t-1)}{\frac{1}{2}g(2t-3)} = \frac{(2t-1)}{(2t-3)}$$

$$10t - 15 = 8t - 4$$

$\therefore$ $$t = 11/2 = 5.5 \text{ s}$$

The height from which the body was dropped is determined from the equation (a),

$$y = \frac{1}{2}gt^2 = \frac{1}{2}(9.81)(5.5)^2$$
$$= 148.38 \text{ m}$$

The velocity with which it strikes the ground is given by

$$v = v_o + gt$$
$$= 0 + (9.81)(5.5)$$
$$= 53.96 \text{ m/s}$$

Example 12.31 $AB = BC = CD =$ and so on, and A, B, C, D being points in a vertical straight line; if a body falls from A, show that the times in covering $AB, BC, \ldots$ are as $1: \sqrt{2} - 1: \sqrt{3} - \sqrt{2} : \ldots$.

Solution Since the body is dropped from A, its initial velocity is zero, i.e., $v_0 = 0$. Let $B, C, D, \ldots$ be points in the path such that $AB = BC = CD = \ldots = s$. Let $t_1, t_2, t_3, \ldots$ be the respective times in covering AB, $AC, AD, \ldots$. Then the equations of motion for the different portions can be written as

$(A \to B)$ $$s = \frac{1}{2}gt_1^2 \Rightarrow t_1 = \sqrt{\frac{2s}{g}}$$

$(A \to C)$ $$2s = \frac{1}{2}gt_2^2 \Rightarrow t_2 = \sqrt{\frac{4s}{g}}$$

$(A \to D)$ $$3s = \frac{1}{2}gt_3^2 \Rightarrow t_3 = \sqrt{\frac{6s}{g}}$$

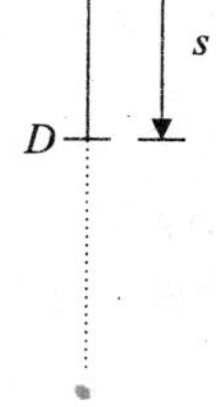

Fig. 12.33

Therefore, the time taken in covering the portion BC is

$$t_2 - t_1 = \sqrt{\frac{4s}{g}} - \sqrt{\frac{2s}{g}} = \sqrt{\frac{2s}{g}}[\sqrt{2} - 1]$$

Similarly, the time taken in covering the portion CD is

$$t_3 - t_2 = \sqrt{\frac{6s}{g}} - \sqrt{\frac{4s}{g}} = \sqrt{\frac{2s}{g}}[\sqrt{3} - \sqrt{2}]$$

Therefore, the ratio of times in covering $AB, BC, CD, \ldots$ are as

$$t_1 : (t_2 - t_1) : (t_3 - t_2) = 1 : (\sqrt{2} - 1) : (\sqrt{3} - \sqrt{2})$$

Example 12.32 A body that has fallen from rest is observed at one portion of its path to fall through h metres in n seconds; find the number of metres described in the next n seconds.

Solution Let v_A be the velocity of the body at the beginning of the considered portion AB of the path. Since $\overline{AB} = h$ and the time taken to cover this distance is n seconds, we can write the equation of motion for the portion AB as

$$h = v_A n + \frac{1}{2} g n^2 \qquad \text{(a)}$$

Let s be the distance covered in the next n seconds. Then the equation of motion considering both the portions AB and BC of the path can be written as

$$h + s = v_A(2n) + \frac{1}{2} g(2n)^2 \qquad \text{(b)}$$

Multiplying the equation (a) by 2 and subtracting from the above equation (b),

$$h + s - 2h = \frac{1}{2} g(4n^2 - 2n^2)$$

$$s - h = \frac{1}{2} g\, 2n^2$$

$$\therefore \quad s = h + gn^2$$

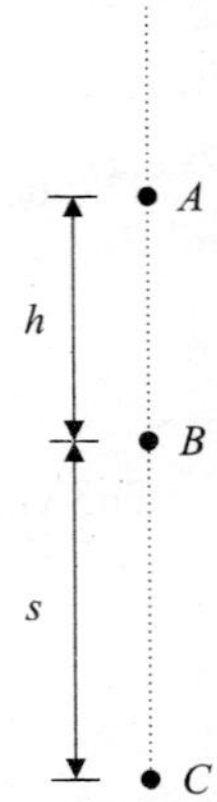

Fig. 12.34

Example 12.33 A stone is dropped into a well and the splash is heard after 3 s. If the speed of sound in air is 340 m/s, determine the depth of the well.

Solution Let t_1 be the time taken by the stone to strike the water and t_2 be the time taken by the sound of splash to travel in air. Then equation of motion of the stone can be written as

$$s_1 = v_{1o} t_1 + (1/2) g t_1^2$$

$$= 0 + (1/2) g t_1^2 \qquad \text{(a)}$$

and the equation of motion of sound is

$$s_2 = v_2 t_2 = 340 t_2 \qquad \text{(b)}$$

Given the total time taken to be

$$t = t_1 + t_2 = 3 \text{ s}$$

$$\Rightarrow \quad t_2 = 3 - t_1 \qquad \text{(c)}$$

As the distance travelled by the stone during its free fall and the distance travelled by sound in air are equal, we also have

$$s_1 = s_2$$

$$\Rightarrow \quad (1/2) g t_1^2 = 340 t_2$$

or $$(1/2)\, g t_1^2 = 340(3 - t_1)$$

$$4.905 t_1^2 + 340 t_1 - 1020 = 0$$

Solving the quadratic equation and disregarding the negative value of t, we have

$$t_1 = 2.88 \text{ s}$$

Therefore, $$t_2 = 3 - t_1 = 0.12 \text{ s}$$

Therefore, the depth of the well is given by substituting the value of t_2 in the equation (b):

$$s_2 = 340t_2$$
$$= 340 \times 0.12 = 40.8 \text{ m}$$

Example 12.34 An open platform elevator used at a construction site starts from ground level with an acceleration of 1.2 m/s^2 and after it has travelled for 10 s, a ball is released from the elevator. How much time it will take the ball to reach the ground? Also, determine the velocity with which it will strike the ground.

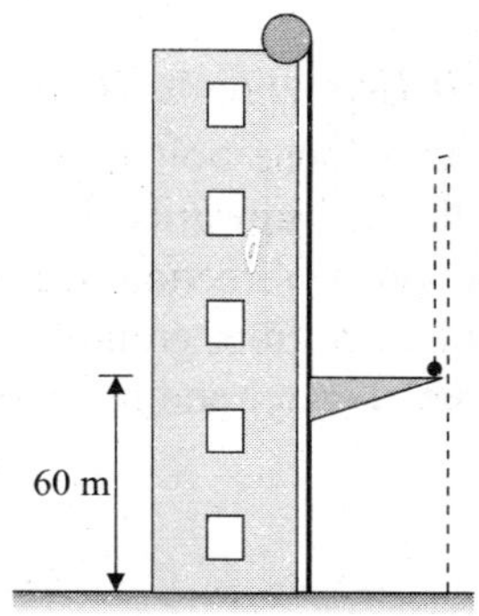

Fig. 12.35

Solution

Initial speed of elevator, $v_{eo} = 0$

Acceleration of elevator, $a_e = 1.2 \text{ m/s}^2$

Therefore, speed of elevator after 10 seconds is given as

$$v_e = v_{eo} + a_e t$$
$$= 0 + 1.2(10) = 12 \text{ m/s}$$

Distance travelled by the elevator during this time is given as

$$s_e = v_{eo}t + (1/2)a_e t^2$$
$$= 0 + (1/2)(1.2)10^2 = 60 \text{ m}$$

As a ball is released from the elevator after 10 seconds, its initial velocity will be the same as that of the elevator at that instant, i.e., its magnitude is equal to 12 m/s and directed upwards. Also, its initial position with respect to the ground level is 60 m. Hence, the equation of motion of the ball from the instant it is released can be written as

$$y_b = y_{bo} + v_{bo}t - \frac{1}{2}gt^2$$
$$= 60 + 12t - 4.905t^2$$

[Note that once the ball is released from the elevator, it is subjected to force of gravity or under free fall.] When it reaches the ground, we know $y_b = 0$. Hence,

$$4.905t^2 - 12t - 60 = 0$$

Solving the quadratic equation and neglecting the negative value of t, we have

$$t = 4.93 \text{ s}$$

The velocity with which it will strike the ground is determined from the kinematic equation,

$$v_b = v_{bo} - gt$$
$$= 12 - 9.81(4.93) = -36.36 \text{ m/s}$$

The negative sign indicates that it is directed downwards.

Example 12.35 An elevator ascends from rest with an upward acceleration of 1 m/s^2. After 2 seconds, a loose bolt drops from the ceiling of the elevator, whose height is 2.2 m. Determine (i) the time of flight of the bolt from the ceiling to the floor, (ii) distance travelled by the bolt relative to the elevator shaft.

Solution The velocity of the elevator at any instant is given as

$$v_e = v_{eo} + a_e t$$

$\therefore$ $$v_e(2) = 0 + 1(2) = 2 \text{ m/s}$$

The distance travelled by the elevator during this time is given as

$$s_e = v_e t + (1/2)a_e t^2$$
$$= 0 + (1/2)(1)2^2 = 2\text{m}$$

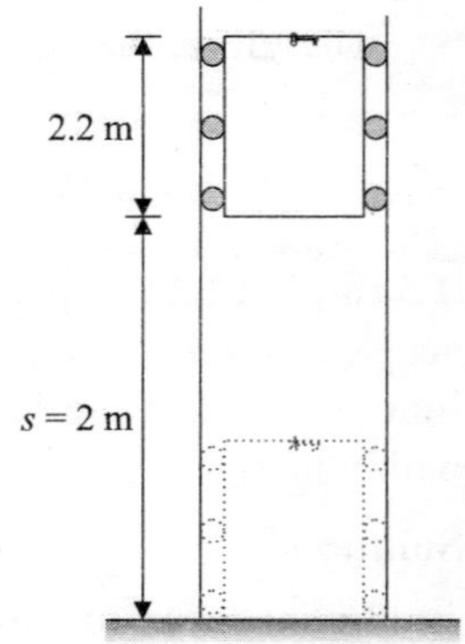

Fig. 12.36

(i) Time of flight of bolt

As a loose bolt falls from the ceiling at this instant, its velocity is 2 m/s directed upwards. Since the displacement of the bolt and elevator are in the opposite direction, we choose the position of the elevator *floor* at that instant, i.e., 2 m from ground level to be the origin. Hence, the equations of motion of the elevator and bolt can be written as

$$y_e = y_{eo} + v_e t + (1/2)a_e t^2$$
$$= 0 + 2t + (1/2)(1)t^2$$
$$= 2t + 0.5t^2 \quad \text{(a)}$$
$$y_b = y_{bo} + v_b t - (1/2)gt^2$$
$$= 2.2 + 2t - 4.905t^2 \quad \text{(b)}$$

Note that the height of the elevator or its shaft is 2.2 m. When the bolt hits the floor, we know that the displacement of the elevator and bolt are equal, i.e., $y_e = y_b$. Therefore,

$$2t + 0.5t^2 = 2.2 + 2t - 4.905t^2$$

$\Rightarrow$ $$5.405t^2 = 2.2$$

$\therefore$ $$t = 0.638 \text{ s}$$

(ii) Distance travelled relative to the shaft

Displacement of the bolt relative to the elevator shaft is given as

$$s = y_b - y_{bo} = 2t - (1/2)gt^2$$
$$= 2(0.638) - 4.905(0.638)^2 = -0.721 \text{ m}$$

The negative sign indicates that the bolt is below the elevator shaft by 0.721 m. Also, we know that the bolt goes upwards before coming down. Hence, the maximum height reached by the bolt with its initial velocity is

$$h_{max} = \frac{v_b^2}{g} = \frac{2^2}{9.81} = 0.408 \text{ m}$$

Note that the bolt moves down by the same distance as it moved up. Therefore, the total distance travelled by the bolt relative to the elevator shaft is obtained as

$$= 0.408 + 0.408 + 0.721 = 1.537 \text{ m}$$

SUMMARY

The study of bodies at rest is termed *statics* and the study of bodies under motion is termed *dynamics*. The motion of a body can be either *translational, rotational* or a *combination* of both. The actual type of motion, whether it is translational, rotational or a combination of both depends upon the nature of the *resultant* of the system of forces acting on the body.

Pure translational motion: The motion is translational when the resultant of a system of forces is a single force $\vec{R}$ acting at the centre of gravity of the body. There is no rotational motion involved.

Rectilinear or one-dimensional motion: The motion is rectilinear when it is along a straight line.

Curvilinear motion: When the translational motion is along a curve other than a straight line.

Pure rotational motion: When the resultant of a system of forces is a couple $\vec{M}$ then the body will rotate.

Combined translational and rotational motions: When the resultant of a force system is a centroidal force and a couple.

Kinematics: The study of motion of bodies without considering the forces causing the motion.

Kinetics: The study of motion of bodies together with the forces causing the motion.

Motion of a Particle

When a body is in pure translational motion, all the particles in the body move in *parallel* paths with the same displacement, velocity and acceleration. Hence, instead of analyzing the motion of the body as a whole, we can analyze the motion of a single particle in the body, which is a representative of the motion of the entire body. This is termed idealization of the body as a *particle*, i.e., a body without extent. Mathematically, this is treated as a point; and normally, the centre of gravity of the body is chosen as this point.

Position of a particle is defined with respect to a fixed reference frame.

Displacement of a particle is defined as the change in position of the particle with respect to time. It is dependent only on the *initial* and *final* positions of the particle, whereas the *distance* travelled is dependent on the actual path traced by the particle.

Velocity of a particle is defined as the rate of change of displacement with time.

Acceleration of a particle is defined as the rate of change of velocity with time.

Rectilinear Motion

When the motion of a particle is restricted along a *straight line*, the motion is said to be *one-dimensional* or *rectilinear motion*. If the position of the particle [x] is known as a function of time then velocity and acceleration can be determined by differentiation of the position function with respect to time, i.e.,

$$v = \frac{dx}{dt} \quad \text{and} \quad a = \frac{dv}{dt} = \frac{d^2x}{dt^2}$$

By separation of variables, we can also write acceleration as

$$a = \frac{dv}{dt} = \frac{dv}{dx}\frac{dx}{dt} = v\frac{dv}{dx}$$

On the other hand, if acceleration is known as a function of time, velocity or position then velocity and position functions are obtained by integration method.

Uniformly Accelerated Motion

A special type of rectilinear motion is that in which the acceleration is *constant*. The equations of motion in that case are the following

v–t relationship: $v = v_o + at$

x–t relationship: $s = x - x_o = v_o t + \frac{1}{2}at^2$

If final velocity is specified for the motion then

$$s = vt - \frac{1}{2}at^2$$

v–x relationship: $v^2 = v_o^2 + 2as$

Graphical Method

When the motion of a particle has distinct phases, the graphical method, particularly the *v–t* graph is very much useful in describing the motion. For a rectilinear motion with constant acceleration, the acceleration of a particle is represented by the slope of the *v–t* curve. The slope is positive if the particle is accelerated; it is negative if the particle is decelerated. The distance travelled by a particle during a certain interval of time is given by the area under the *v–t* curve.

Rectilinear Motion Along Y-Axis

The idealized motion in which air resistance and change in acceleration with altitude are neglected is termed as *free fall*, although the term includes *rising* as well as *falling*.

For *bodies thrown upwards*, the equations of motion are given as

$$v = v_o - gt$$

$$v^2 = v_o^2 - 2gs$$

$$y = y_o + v_o t - \frac{1}{2}gt^2 \quad \text{(or)} \quad s = y - y_o = v_o t - \frac{1}{2}gt^2$$

Maximum height reached is

$$h_{max} = \frac{v_o^2}{2g}$$

Time taken to reach this maximum height is

$$t = \frac{v_o}{g}$$

For *bodies thrown downwards or dropped*, the equations of motion are given as

$$v = v_o + gt$$

$$v^2 = v_o^2 + 2gs$$

$$s = y = v_o t + \frac{1}{2}gt^2$$

EXERCISES

Objective-type Questions

1. A rigid body can be idealized as a particle
   (a) only when its size is very minute
   (b) only when the body is at rest
   (c) when there is no translational motion involved
   (d) when there is no rotational motion involved
2. In rectilinear motion, all the particles in the body
   (a) have the same displacement (b) have the same velocity
   (c) have the same acceleration (d) all of these
3. Average velocity is defined as
   (a) average of initial and final velocities
   (b) ratio of change in displacement and elapsed time
   (c) ratio of distance travelled and elapsed time
   (d) average of initial and final speeds
4. When a car moves at a constant speed around a curved path, its velocity
   (a) is zero (b) is constant
   (c) changes in magnitude (d) changes in direction
5. A man walks from one town to another and then comes back. State which of the following statements is true concerning his journey:
   (a) his displacement is zero (b) distance traveled is zero
   (c) average speed is zero (d) time taken is zero
6. A man walks from one town to another at a constant speed of 15 kmph and then returns back at a constant speed of 10 kmph. His average speed for the journey is
   (a) 12.5 kmph (b) 12 kmph (c) 2.5 kmph (d) 25 kmph
7. A particle can move with constant velocity when motion is
   (a) rectilinear (b) curvilinear (c) rotational (d) general motion
8. Uniform motion implies that
   (a) acceleration is constant (b) velocity is constant
   (c) position is constant (d) time is constant
9. The area under an *a–t* curve represents
   (a) average acceleration (b) instantaneous acceleration
   (c) change in position of the particle (d) change in velocity of the particle
10. The area under a *v–t* curve represents
    (a) average velocity of the particle (b) instantaneous velocity of the particle
    (c) distance travelled by the particle (d) acceleration of the particle

11. The free fall of a body is an example for
    (a) uniform motion
    (b) uniformly accelerated motion
    (c) non-uniformly accelerated motion
    (d) curvilinear motion
12. When a ball is thrown vertically upwards with a velocity v_o, the total time taken to return to the point of projection is
    (a) $\frac{v_o^2}{g}$ (b) $\frac{v_o^2}{2g}$ (c) $\frac{v_o}{g}$ (d) $\frac{2v_o}{g}$
13. When a ball is thrown vertically upwards with a velocity v_o, the maximum height reached by it is
    (a) $\frac{v_o^2}{g}$ (b) $\frac{v_o^2}{2g}$ (c) $\frac{v_o}{g}$ (d) $\frac{2v_o}{g}$
14. Which of the following motion is not having a constant acceleration?
    (a) A car moving on a level road with uniformly varying velocity
    (b) A ball dropped from the top of a tower
    (c) A ball thrown vertically upwards
    (d) A bike moving on a curved path with constant speed

Answers

1. (d)	2. (d)	3. (b)	4. (d)	5. (a)	6. (b)	7. (a)	8. (b)
9. (d)	10. (c)	11. (b)	12. (d)	13. (b)	14. (d)		

Short-answer Questions

1. Distinguish between statics and dynamics.
2. Distinguish between kinematics and kinetics.
3. Distinguish between particle and rigid body.
4. Explain the types of motion with suitable examples.
5. Define position vector and displacement vector.
6. Distinguish between displacement vector and distance travelled.
7. Define velocity of a particle.
8. Define average velocity and instantaneous velocity.
9. Under what conditions is average velocity equal to instantaneous velocity?
10. Define average acceleration and instantaneous acceleration.
11. If a particle moves with constant speed but changes in direction, can there be acceleration?
12. Distinguish between rectilinear motion and curvilinear motion.
13. State the differential equations of motion.
14. Distinguish between uniform motion and uniformly accelerated motion.
15. Derive the x–t, v–t and a–t relationships for uniformly accelerated motion.
16. What are motion curves? What are they used for?
17. Define free fall.
18. What are the assumptions made in free fall?

Numerical Problems

12.1 The position of a particle moving along the X-axis is given by $x = A\cos 2\pi f t$, where A and f are constants. Find the expressions for velocity and acceleration of the particle at any time.

Ans. $v = -A(2\pi f)\sin 2\pi f t,\ a = -A(2\pi f)^2 \cos 2\pi f t$

12.2 The position of a particle in rectilinear motion is defined by the relation $x = 2t^3 - 8t^2 + 4t - 2$, where x is in metres and t is in seconds. Determine its position, velocity and acceleration at time $t = 3$ s.

Ans. -8 m, 10 m/s, 20 m/s^2

12.3 The position of a particle in rectilinear motion is defined by the relation $x = t^3 + t^2 + t + 1$, where x is in metres and t is in seconds. Determine its position, velocity and acceleration at time $t = 2$ s.

Ans. 15 m, 17 m/s, 14 m/s^2

12.4 The position of a particle in rectilinear motion is defined by the relation $x = t^3 + 3t + 5$, where x is in metres and t is in seconds. Determine its position, velocity and acceleration at $t = 0$ and $t = 2$ s.

Ans. (i) 5 m, 3 m/s, 0 m/s^2; (ii) 19 m, 15 m/s, 12 m/s^2

12.5 The motion of a particle in rectilinear motion is defined by the relation $x = t^4 - 3t^3 + t^2 + 6$, where x and t are expressed in metres and seconds respectively. Determine (i) the position, velocity and acceleration of the particle at time $t = 2$ s, (ii) the average velocity during $t = 2$ s and $t = 3$ s, (iii) the average acceleration during the third second, and (iv) the instants when the acceleration is zero.

Ans. (i) 2 m, 0 m/s, 14 m/s^2; (ii) 13 m/s; (iii) 33 m/s^2; (iv) 0.12 s, 1.38 s

12.6 The position of a particle in rectilinear motion is defined by the relation $x = \dfrac{t^4}{12} - \dfrac{2t^3}{3} - \dfrac{5t^2}{2} + 3t + 4$, where x is in metres and t is in seconds. Determine its position and velocity when the acceleration is zero.

Ans. -74.75 m, -30.33 m/s

12.7 The position of a particle in rectilinear motion is defined by the relation $x = t^3 - 5.5t^2 + 8t - 6$, where x is in metres and t is in seconds. Determine the time at which the particle reaches a momentary rest. Also, determine the position, velocity and total distance travelled when the acceleration is zero.

Ans. 1 s and 2.67 s; -3.65 m, -2.08 m/s, 4.65 m

12.8 The position of a particle in rectilinear motion is given by the relation $x = \dfrac{t^3}{3} - 1.5t^2 - 4t + 5$, where x is in metres and t is in seconds. Determine the position and acceleration when the velocity is zero.

Ans. -13.67 m, 5 m/s^2

12.9 The acceleration of a particle in rectilinear motion is defined by the relation $a = 3t$. Given that at time $t = 0$, $v = 0$ m/s and $x = 2$ m, determine its position, velocity and acceleration at $t = 4$ s.

Ans. 34 m, 24 m/s, 12 m/s^2

12.10 The acceleration of a particle in rectilinear motion is defined by the relation $a = 3t^2 - 4t^3$. Given that at $t = 0$ s, $v = 0$ m/s and at $t = 1$ s, $x = 0$, write the equations of motion.

Ans. $v = t^3 - t^4,\ x = \dfrac{t^4}{4} - \dfrac{t^5}{5} - \dfrac{1}{20}$

12.11 The acceleration of a particle in rectilinear motion is defined by the relation $a = kt - 1$. Given that at $t = 0$ s, $v = 0$ m/s, $x = 0$ and at $t = 2$ s, $v = 2$ m/s, determine its position, velocity and acceleration at $t = 3$ s.

Ans. 4.5 m, 6 m/s, 5 m/s^2

12.12 The acceleration of a particle in rectilinear motion is defined by the relation $a = kt^2 - 3$. Given that at $t = 0$ s, $v = 0$ m/s, $x = 0$ and at $t = 3$ s, $v = 9$ m/s, determine its position, velocity and acceleration at $t = 2$ s.

Ans. –3.33 m, – 0.67 m/s, 5 m/s^2

12.13 The acceleration of a particle in rectilinear motion is defined by the relation $a = 1/v$, where a is in m/s^2 and v is in m/s. Given that at time $t = 2$ s, $v = 2$ m/s and at time $t = 0$, the displacement $x = 0$, write the equations of motion.

Ans. $v = \sqrt{2t},\ x = \dfrac{2\sqrt{2}}{3} t^{3/2}$

12.14 The acceleration of a particle in rectilinear motion is defined by the relationship $a = kv$. Given that at time $t = 0$, $v = v_o$ and $x = 0$, write the equations of motion.

Ans. $v = v_o e^{kt},\ x = \dfrac{v_o}{k}[e^{kt} - 1]$

12.15 A body is freely falling and the resistance of the air is proportional to velocity given by the relation: $a = g - kv$. Given that at $t = 0$, velocity and displacement are zero, derive the expressions for velocity and position as functions of time.

Ans. $v = \dfrac{g}{k}[1 - e^{-kt}],\ x = \dfrac{g}{k}t - \dfrac{g}{k^2}[1 - e^{-kt}]$

12.16 The acceleration of a particle in rectilinear motion is defined by the relation $a = -1/x^2$. Given that at $t = 4/3$ s, $x = 2$ m, and $v = 1$ m/s, write the equations of motion.

Ans. $v = \sqrt{2}\, x^{-1/2},\ x^{3/2} = \dfrac{3}{\sqrt{2}} t$

12.17 An airplane while landing touches the runway with a speed of 150 kmph and decelerates at a constant rate of 2.5 m/s^2 to 1 kmph. Determine (i) the time taken to decelerate, and (ii) the distance travelled during this time.

Ans. 16.6 s, 347.3 m

12.18 In a 100 m dash, a sprinter starts with a constant acceleration of 2.2 m/s^2 wishing to break the world record of 9.6 s. Determine whether he will achieve this or not.

Ans. yes, $t = 9.53$ s

12.19 A car moving at 54 kmph accelerates uniformly at the rate of 1 m/s^2 for the next 10 seconds. Determine the velocity of the car after 10 seconds.

Ans. 90 kmph

12.20 An electric train enters a station with a speed of 60 kmph. If the length of the platform is 150 m, determine the constant deceleration, which the driver must apply such that the front part just stops at the end of the station.

Ans. 0.926 m/s^2

12.21 A truck moving at a certain velocity accelerates at a rate of 1 m/s^2 to reach a speed of 72 kmph in 10 seconds. Determine (i) the distance travelled during this time, and (ii) its initial velocity.

Ans. (i) 150 m, (ii) 36 kmph

12.22 A car covers 200 m in 10 seconds, while being accelerated at a rate of 1 m/s^2. Determine its initial and final velocities.

Ans. v_o = 15 m/s, v = 25 m/s

12.23 An airplane lands with a speed of 80 m/s. If it is brought to a stop in a distance of 1.2 km, determine (i) its uniform deceleration, and (ii) its velocity 5 seconds after landing.

Ans. 2.67 m/s^2, 66.65 m/s

12.24 A train moving at a constant speed of 80 kmph crosses a station without stopping. If it takes 20 s for the train to totally cross the station of length 200 m, determine the length of the train.

Ans. 244.4 m

12.25 A cyclist moving at a constant speed of 1 m/s realizes that he will be late to the post office by 30 seconds. Hence, he accelerates at a constant rate of 0.3 m/s^2 so that he reaches just in time. Determine the time taken and the distance travelled.

Ans. 14.14 s, 44.14 m

12.26 A truck is found to cross two points, 200 m apart on a straight road, with speeds 60 kmph and 80 kmph respectively. Determine (a) its acceleration assuming it to be constant, (b) time taken to cover this distance, (c) how far from the first point did it start its motion assuming constant acceleration determined in part – a, and (d) time taken to reach the first point.

Ans. (a) 0.54 m/s^2, (b) 10.3 s, (c) 257.3 m, (d) 30.9 s

12.27 A car starting from rest accelerates uniformly and crosses points *A* and *B*, 1 km apart, with velocities 45 kmph and 60 kmph respectively. Determine (i) its uniform acceleration, (ii) time taken to cover the distance from the point *A* to the point *B* and (iii) distance between the point *A* and the starting point assuming the same constant acceleration throughout the motion.

Ans. (i) 0.061 m/s^2; (ii) 68.4 s; (iii) 1.28 km

12.28 A car starting from rest accelerates uniformly and crosses the point *A* with a velocity of 54 kmph and the point *B* with a velocity of 90 kmph. If the time taken to cross from *A* to *B* is 8 seconds, determine the distance between the points *A* and *B*.

Ans. 160 m

12.29 The velocity of a particle moving in rectilinear motion is 10 m/s at 5 m from the origin and 15 m/s at 15 m from the origin. Determine its acceleration assuming it to be uniform and its initial velocity, i.e., at the origin.

Ans. 6.25 m/s^2, 6.12 m/s

12.30 A train moving with constant acceleration takes t seconds to cross a point. If its front end crosses the point with velocity u and the rear end with velocity v in time t, determine (i) the uniform acceleration, (ii) length of the train, and (iii) velocity with which the midpoint of the train crosses the point.

Ans. (i) $\frac{v-u}{t}$, (ii) $\frac{v+u}{2}t$, (iii) $\sqrt{\frac{v^2+u^2}{2}}$

12.31 Car *A* starts from rest and accelerates uniformly at the rate of 1 m/s^2. Four seconds later, the car *B* starts from the same point and accelerates uniformly at the rate of 1.2 m/s^2. Determine when and where the car *B* will overtake the car *A*.

Ans. 45.9 s after the car *A* started, 1.05 km

12.32 A truck moving at a constant speed of 60 kmph passes a car at rest on the way. If the car starts 3 s after the truck crosses and accelerates at a constant rate of 2 m/s^2, determine when and where it will overtake the truck.

Ans. 22.27 s after the truck has crossed, 371.2 m

12.33 A truck moving at a constant speed of 54 kmph crosses a point *A*. Two seconds later, a car moving at a constant speed of 72 kmph crosses the same point. Determine when and where the car will overtake the truck.

Ans. 6 s after the car crosses the point *A*, 120 m

12.34 A car starts from rest and accelerates uniformly. At the same instant that car starts, a truck crosses with a uniform speed of 45 kmph in the same direction as the car moves. If the car overtakes the truck after 15 seconds, determine (i) the uniform acceleration of the car, (ii) distance covered by it during this time, and (iii) its velocity at the instant it crosses the truck.

Ans. (i) 1.67 m/s^2; (ii) 187.5 m; (iii) 90 kmph

12.35 Two trains travel on the same track in the opposite direction with constant speeds of 45 kmph and 60 kmph. The drivers of the two trains realize that they are on the same track when they are 300 m apart and apply the brakes. The reaction time for both the drivers is 0.7 s. If the deceleration of the slower train is 0.5 m/s^2, determine the minimum deceleration of the faster train to just avoid collision.

Ans. 1.13 m/s^2

12.36 A truck moves at a constant speed of 15 m/s. A car following the truck in the adjacent lane moves with a constant speed of 10 m/s. When they are 100 m apart, the driver of the car accelerates at a constant rate of 0.5 m/s^2. Determine when and where the car will overtake the truck. Also, determine the speed of the car at that instant.

Ans. 32.36 s, 585.4 m, 26.2 m/s

12.37 The driver of a car travelling at a constant speed of 72 kmph, seeing a procession ahead of him moving in the same direction, decelerates at a constant rate of 1 m/s^2 until his speed is reduced to 3 kmph. He moves at this speed for 2 minutes until the procession takes a left turn. He then accelerates at a constant rate of 0.5 m/s^2 to his normal speed. Determine the total distance travelled during this time and the time lost.

Ans. 699 m, 142.6 s

12.38 A car starts from rest and accelerates uniformly at the rate of 1 m/s^2 to reach a maximum speed of 60 kmph. It then travels at this speed for some time and finally decelerates at a uniform rate of 0.5 m/s^2 to come to rest. If the total distance travelled is 5 km, determine the total time taken for the car to cover this distance.

Ans. 325 s

12.39 A lift starts with a uniform acceleration of 1 m/s^2 for 2 seconds and moves with the speed attained for another 10 seconds and retards at the rate of 0.5 m/s^2 to come to rest. Determine the maximum velocity of the lift, height up to which it reaches and the total time taken.

Ans. 2 m/s, 26 m, 16 s

12.40 A train crosses a station without stopping at 54 kmph. After 10 seconds, it increases its speed to 72 kmph in 15 seconds and travels at this speed for another 4 minutes. It then retards at the rate of 0.5 m/s^2 to come to rest at the next station. Determine the distance between the stations and the time taken to cover the distance.

Ans. 5.6 km, 305 s

12.41 Two trains start at the same time from stations *A* and *B* (6 km apart) respectively in opposite tracks. Train *A* accelerates uniformly at the rate of 0.5 m/s^2 until it reaches a speed of 54 kmph, while Train *B* accelerates uniformly at the rate of 0.6 m/s^2 until it reaches a speed of 72 kmph and then travels at this speed. Determine when and where both will cross each other.

Ans. 187.4 s, 2.59 km from the station *A*

12.42 A ball is thrown upwards with a velocity of 20 m/s. Determine (i) its velocity after 2 s and after 3 s, (ii) the maximum height reached, (iii) time taken to reach the maximum height, and (iv) total time of flight.

Ans. (i) 0.38 m/s (directed upwards), 9.43 m/s (directed downwards); (ii) 20.39 m; (iii) 2.04 s; (iv) 4.08 s

12.43 A ball is thrown upwards with a velocity of 25 m/s. Determine (i) the maximum height reached by the ball, (ii) the time taken to reach it, and (iii) the time when its velocity is 10 m/s pointing downward.

Ans. (i) 31.86 m, (ii) 2.55 s, (iii) 3. 57 s

12.44 A ball thrown upwards from the ground crosses the top of a building with a velocity of 5 m/s. If the total time of flight is 6 s, determine (i) the height of the building, (ii) initial velocity of the ball, and (iii) height reached by the ball above the top of the building.

Ans. (i) 42.9 m, (ii) 29.4 m/s, (iii) 1.27 m

12.45 A stone is thrown vertically upwards from the top of a building with a velocity of 20 m/s. If it reaches the ground after 5 seconds, determine the height of the building.

Ans. 22.63 m

12.46 A helicopter used for relief activities drops food packets to the relief camp. If the food packets reach the ground after 3 seconds, determine the height from which they were dropped and the velocity with which they strike the ground.

Ans. 44.2 m, 29.4 m/s

12.47 A ball is thrown upwards from the ground with an initial velocity of 20 m/s. Two seconds later another ball is thrown upwards with an initial velocity of 30 m/s. Determine when and where they will meet.

Ans. 2.7 s, 18.2 m

12.48 A ball is thrown upwards from a 40 m high tower with an initial velocity of 10 m/s. At the same instant, another ball is thrown upwards from ground with an initial velocity of 20 m/s. Determine (i) when and where they will meet each other, and (ii) the velocity of each ball at that instant.

Ans. (i) 4 s after the first ball is thrown upwards, 1.52 m up from the ground, (ii) 29.24 m/s and 19.24 m/s (both the velocities are directed downwards)

12.49 A ball is thrown upwards from the top of a tower with an initial velocity of 15 m/s. Two seconds later, another ball is dropped from the tower. Determine (i) when and where they will meet, (ii) their respective velocities at that instant.

Ans. (i) 4.25 s, 24.8 m; (ii) 26.7 m/s, 22.07 m/s both directed downwards

12.50 A man throws a ball from the foot of a building with a velocity of 20 m/s. Another man standing at 10 m above the ground level tries to catch the ball. Determine the velocity of the ball at that instant and the time taken if he catches the ball (i) when it moves up, (ii) when it moves down.

Ans. (i) 14.3 m/s (↑), 0.58 s; (ii) 14.3 m/s (↓), 3.5 s

12.51 An open platform elevator used at a construction site starts from the ground level with an acceleration of 1.2 m/s^2 and after it has travelled a distance of 30 m, a ball is released from the elevator. How much time it will take the ball to reach the ground? With what velocity will it strike the ground?

Ans. 3.5 s, 25.9 m/s (directed downwards)

12.52 A balloon is ascending from the ground at a constant acceleration of 0.5 m/s^2. After 30 seconds from the start, a ball is released from the balloon. Determine the velocity with which it will strike the ground and the time taken to reach the ground.

Ans. 68.1 m/s vertically downwards, 8.47 s

13 Kinematics of Particles (Curvilinear Motion)

13.1 INTRODUCTION

In the previous chapter, we analyzed rectilinear motion, treating it as a special case of translational motion in a plane. In this chapter, we return to the translational motion in a plane, which we also called *curvilinear motion*. When a particle undergoes translational motion along a curved path other than a straight line, the motion is said to be **curvilinear**. This type of motion can occur along a vertical plane such as a missile fired from a cannon or along a horizontal plane such as a car or a bike travelling on a curved road.

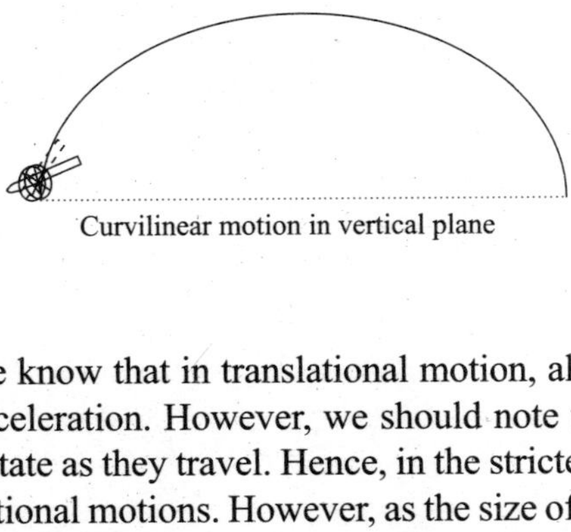

Curvilinear motion in vertical plane

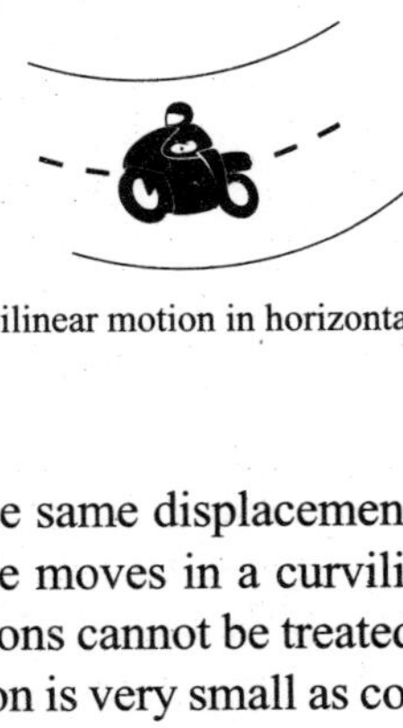

Curvilinear motion in horizontal plane

Fig. 13.1

We know that in translational motion, all the particles move with the same displacement, velocity and acceleration. However, we should note that when a ball or a missile moves in a curvilinear path, they rotate as they travel. Hence, in the strictest sense, such types of motions cannot be treated as *purely* translational motions. However, as the size of the body under consideration is very small as compared to the trajectory or path, we can neglect such rotational effect without appreciable error. Thus, we assume that all the particles in the body move with the *same* displacement, velocity and acceleration. As a result, the body can be idealized as a *particle* as before and its motion can be analyzed.

The motion of the particle can be described by stating its position, velocity and acceleration at different instants of time. The relationship between displacement, velocity and acceleration vectors and their variation with time derived in the previous chapter for curvilinear motion are restated below:

$$\vec{v} = \frac{d\vec{r}}{dt} \quad \text{and} \quad \vec{a} = \frac{d\vec{v}}{dt} = \frac{d^2\vec{r}}{dt^2} \tag{13.1}$$

Hence, velocity and acceleration vectors are obtained by differentiating the position vector successively with respect to time. We also saw that the direction of instantaneous velocity is always tangential to the path of the particle at that instant, while the acceleration of the particle does not bear any relationship to the direction of motion. However, to understand the motion better, we can resolve such motion into perpendicular components along the reference axes. Various reference axes are chosen for this purpose and each one is useful in studying a particular type of curvilinear motion.

For instance, in the motion of a missile, as it is subjected to the force of gravity, the acceleration of the particle always acts vertically downwards. In this case, it will be convenient to represent the motion using *rectangular* coordinates. This is covered in the following few Sections 13.2–13.4. When a car or a bike moves along a curved path, its acceleration varies in direction apart from varying in magnitude. In such cases, it will be convenient to resolve the motion along the *tangent* to the path and *normal* to the path. This is covered in Section 13.5. We also come across curvilinear motions in some of the mechanisms, which are also of interest to us. Here the particles may rotate about an axis and at the same time, move radially. Hence, it will be convenient in such cases to represent the motion using *polar* coordinates. This will be discussed in Section 13.6.

13.2 RECTANGULAR COORDINATES

Consider a particle moving in the *XY*-plane. Let its position at an instant of time be A, whose position vector is $\vec{r}$. If x and y be the rectangular coordinates of the point A then its position vector $\vec{r}$ can be expressed as

$$\vec{r} = x\vec{i} + y\vec{j} \tag{13.2}$$

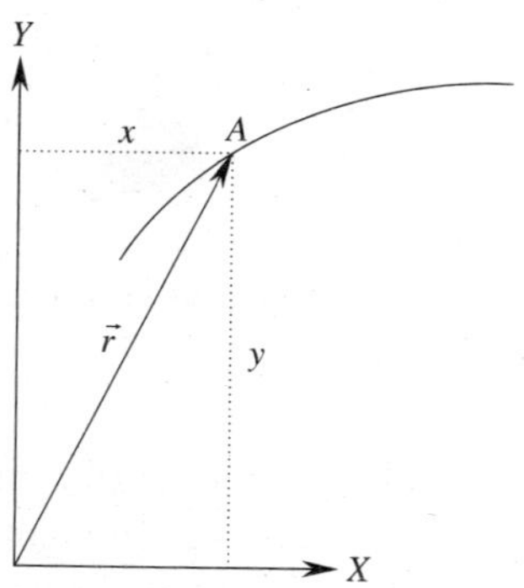

Fig. 13.2

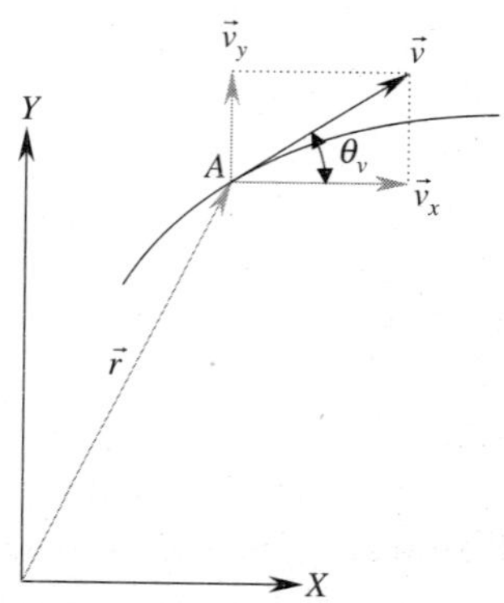

Fig. 13.3

Then velocity vector can be obtained by differentiating the above expression with respect to time, i.e.,

$$\vec{v} = \frac{d\vec{r}}{dt} = \frac{dx}{dt}\vec{i} + \frac{dy}{dt}\vec{j}$$

$$= v_x\vec{i} + v_y\vec{j} \tag{13.3}$$

where v_x and v_y are x and y components of velocity (refer Fig. 13.3). The magnitude and direction of instantaneous velocity can be expressed in terms of its components as

$$v = \sqrt{v_x^2 + v_y^2} \quad \text{and} \quad \theta_v = \tan^{-1}\left[\frac{v_y}{v_x}\right] \tag{13.4}$$

The direction of this instantaneous velocity is tangential to the path of the particle at that instant. If the equation of path of the particle is known in the form, $y = f(x)$ then it can be proved that the direction of velocity vector coincides with the slope of the curve or tangent to the curve at that point.

Similarly, the acceleration vector can be obtained by differentiating the expression for velocity vector (Eq. 13.3) with respect to time, i.e.,

$$\vec{a} = \frac{d\vec{v}}{dt} = \frac{dv_x}{dt}\vec{i} + \frac{dv_y}{dt}\vec{j}$$

$$= \frac{d^2x}{dt^2}\vec{i} + \frac{d^2y}{dt^2}\vec{j}$$

$$= a_x\vec{i} + a_y\vec{j} \qquad (13.5)$$

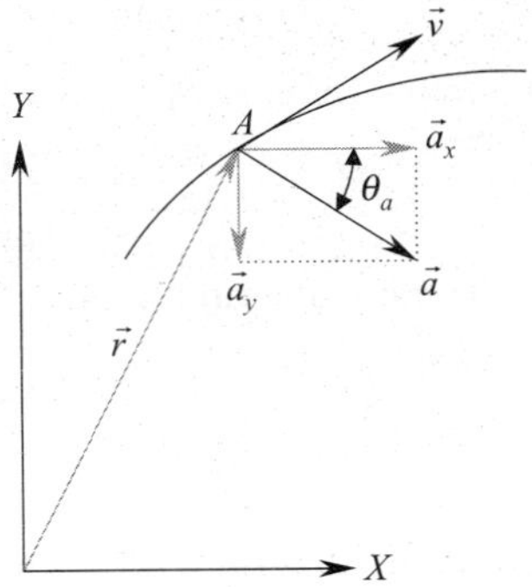

Fig. 13.4

where a_x and a_y are x and y components of acceleration (refer Fig. 13.4).

The magnitude and direction of instantaneous acceleration in terms of its components are

$$a = \sqrt{a_x^2 + a_y^2} \quad \text{and} \quad \theta_a = \tan^{-1}\left[\frac{a_y}{a_x}\right] \qquad (13.6)$$

It should be noted that though velocity is tangential to the path of the particle, the acceleration does not bear any relationship with the direction of motion.

Example 13.1 The x and y coordinates of the position of a particle moving in curvilinear motion are defined by $x = 2 + 3t^2$ and $y = 3 + t^3$. Determine the particle's position, velocity and acceleration at $t = 3$ s.

Solution *Given* $x = 2 + 3t^2$ and $y = 3 + t^3$

Therefore, the x and y components of velocity and acceleration can be obtained by differentiating successively the above expressions with respect to time.

$$v_x = \frac{dx}{dt} = 6t, \quad v_y = \frac{dy}{dt} = 3t^2$$

and

$$a_x = \frac{d^2x}{dt^2} = 6, \quad a_y = \frac{d^2y}{dt^2} = 6t$$

Particle's position at $t = 3$ s

$$x(3) = 2 + 3(3)^2 = 29 \text{ m}$$

$$y(3) = 3 + (3)^3 = 30 \text{ m}$$

Therefore, magnitude and direction of position vector at $t = 3$ s are

$$r = \sqrt{x^2 + y^2}$$

$$= \sqrt{29^2 + 30^2} = 41.73 \text{ m}$$

and
$$\theta_r = \tan^{-1}\left[\frac{y}{x}\right]$$

$$= \tan^{-1}\left[\frac{30}{29}\right] = 45.97°$$

Particle's velocity at t = 3 s

$$v_x(3) = 6(3) = 18 \text{ m/s}$$
$$v_y(3) = 3(3)^2 = 27 \text{ m/s}$$

Hence, magnitude of velocity at time $t = 3$ s is given as

$$v = \sqrt{v_x^2 + v_y^2}$$

$$= \sqrt{18^2 + 27^2} = 32.45 \text{ m/s}$$

and its inclination with respect to the X-axis is obtained as

$$\theta_v = \tan^{-1}\left[\frac{v_y}{v_x}\right]$$

$$= \tan^{-1}\left[\frac{27}{18}\right] = 56.31°$$

Particle's acceleration at t = 3 s

$$a_x(3) = 6 \text{ m/s}^2$$
$$a_y(3) = 6(3) = 18 \text{ m/s}^2$$

Therefore, magnitude of acceleration at $t = 3$ s is obtained as

$$a = \sqrt{a_x^2 + a_y^2}$$

$$= \sqrt{6^2 + 18^2} = 18.97 \text{ m/s}^2$$

and its inclination with respect to the X-axis is obtained as

$$\theta_a = \tan^{-1}\left[\frac{a_y}{a_x}\right]$$

$$= \tan^{-1}\left[\frac{18}{6}\right] = 71.57°$$

Example 13.2 A particle moves along the path $y^2 = 4x$, where x and y are in metres. The x coordinate of the particle at any time is given by $x = t^2$. Determine the y component of velocity and acceleration at $x = 4$ m.

Solution *Given* $x = t^2$ (a)

Upon differentiating it twice with respect to time, we get

$$\frac{dx}{dt} = 2t \quad \text{and} \quad \frac{d^2x}{dt^2} = 2$$

Given $$y^2 = 4x \tag{b}$$

Upon differentiating it with respect to time, we get

$$2y\frac{dy}{dt} = 4\frac{dx}{dt} \tag{c}$$

Upon further differentiation with respect to time,

$$2y\frac{d^2y}{dt^2} + 2\left[\frac{dy}{dt}\right]^2 = 4\frac{d^2x}{dt^2} \tag{d}$$

At $x = 4$ m, we have from the equation (a), $t = 2$ s. Hence,

$$\frac{dx}{dt} = 2(2) = 4 \text{ m/s}$$

Also, when $x = 4$ m, from the equation (b)

$$y = \sqrt{4x} = 4 \text{ m}$$

y-component of velocity at x = 4 m

Substituting the values in the equation (c), we have,

$$2(4)\frac{dy}{dt} = 4(4)$$

$$\Rightarrow \quad \frac{dy}{dt} = 2 \text{ m/s}$$

y-component of acceleration at x = 4 m

Substituting all the values in the equation (d), we have

$$2(4)\frac{d^2y}{dt^2} + 2(2)^2 = 4(2)$$

$$\Rightarrow \quad \frac{d^2y}{dt^2} = 0$$

Alternative method

The y-coordinate of the particle can be expressed as a function of time as

$$y^2 = 4x = 4(t^2) \quad \Rightarrow \quad y = 2t$$

Therefore, differentiating it twice with respect to time,

$$\frac{dy}{dt} = 2 \quad \text{and} \quad \frac{d^2y}{dt^2} = 0$$

Therefore, at $t = 2$ s,

$$\frac{dy}{dt} = 2 \text{ m/s} \quad \text{and} \quad \frac{d^2y}{dt^2} = 0 \text{ m/s}^2$$

Example 13.3 A particle moves along the path $y^2 = x^3$, where x and y are in metres. The x coordinate of the particle at any time is given by $x = t^2/5$. Determine the y component of velocity and acceleration at $x = 5$ m.

Solution *Given* $x = t^2/5$ (a)

Upon differentiating it twice with respect to time, we get

$$\frac{dx}{dt} = 2t/5 \quad \text{and} \quad \frac{d^2x}{dt^2} = 2/5$$

Given $y^2 = x^3$ (b)

Upon differentiating it with respect to time, we get

$$2y\frac{dy}{dt} = 3x^2\frac{dx}{dt} \tag{c}$$

Upon further differentiation with respect to time,

$$2y\frac{d^2y}{dt^2} + 2\left[\frac{dy}{dt}\right]^2 = 3x^2\frac{d^2x}{dt^2} + 6x\left[\frac{dx}{dt}\right]^2 \tag{d}$$

At $x = 5$ m, we have from the equation (a), $t = 5$ s. Hence,

$$\frac{dx}{dt} = 2(5)/5 = 2 \text{ m/s}$$

Also, when $x = 5$ m, from the equation (b)

$$y = \sqrt{x^3} = 11.18 \text{ m}$$

y-component of velocity at x = 5 m

Substituting the values in the equation (c), we have,

$$2(11.18)\frac{dy}{dt} = 3(5)^2\,(2)$$

$$\Rightarrow \quad \frac{dy}{dt} = 6.71 \text{ m/s}$$

y-component of acceleration at x = 5 m

Substituting all the values in the equation (d), we have

$$2(11.18)\frac{d^2y}{dt^2} + 2(6.71)^2 = 3(5)^2\,(2/5) + 6(5)(2)^2$$

$$\Rightarrow \quad \frac{d^2y}{dt^2} = 2.68 \text{ m/s}^2$$

Alternative method

The y-coordinate of the particle can be expressed as a function of time as

$$y = x^{3/2} = \left(\frac{t^2}{5}\right)^{3/2} = \left(\frac{t^3}{\sqrt{125}}\right)$$

Therefore, differentiating it twice with respect to time,

$$\frac{dy}{dt} = \frac{3t^2}{\sqrt{125}} \quad \text{and} \quad \frac{d^2y}{dt^2} = \frac{6t}{\sqrt{125}}$$

Substituting $t = 5$ s,

$$\frac{dy}{dt} = \frac{3(5)^2}{\sqrt{125}} = 6.71 \text{ m/s}$$

and

$$\frac{d^2y}{dt^2} = \frac{6(5)}{\sqrt{125}} = 2.68 \text{ m/s}^2$$

13.3 PROJECTILE MOTION

We will discuss a special case of curvilinear motion in this section, namely, motion with **constant acceleration**. We saw in the previous section that the acceleration vector in rectangular coordinates could be expressed as

$$\vec{a} = a_x\vec{i} + a_y\vec{j} \qquad (13.5^*)$$

Its magnitude and direction in terms of the components are

$$a = |\vec{a}| = \sqrt{a_x^2 + a_y^2}, \quad \theta = \tan^{-1}\left[\frac{a_y}{a_x}\right] \qquad (13.6^*)$$

As acceleration is a vector, it can be **constant** only when **both** its magnitude $|\vec{a}|$ and direction θ are constant. Then from the above expressions for magnitude and direction, we can see that the *components* of acceleration a_x and a_y must also be constants. Further, the components are independent of each other. Thus, curvilinear motion with *constant acceleration* can be considered to be a **combined motion** of **two** rectilinear motions occurring simultaneously along two mutually perpendicular x and y-directions.

An example of such a motion is the motion of a missile or a ball hit in air. In the absence of *air resistance* and wind *velocity*, the motion is subjected to constant acceleration in the vertically downward direction while the acceleration along the horizontal direction is zero. Further, if we also neglect the rotation of the earth, the motion will occur only along a single vertical plane.

Consider a particle projected with an initial velocity v_o at an angle of inclination α to the horizontal. Then the initial velocity can be resolved into rectangular components along the x and y directions respectively as

$$v_o \cos\alpha \quad \text{and} \quad v_o \sin\alpha \qquad (13.7)$$

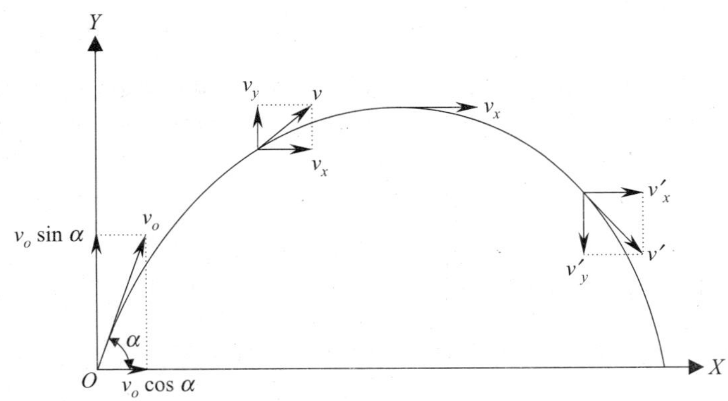

Fig. 13.5 Projectile motion

As there is no acceleration along the horizontal direction, the **horizontal component** of velocity v_x remains **constant** throughout the motion, i.e., $v_x = v_o \cos\alpha$ at any instant of time during the motion. However, the **vertical component** of velocity v_y **changes** with time due to acceleration due to gravity in the vertically downward direction. Thus, the equation of motion in the horizontal direction can be considered to be that of rectilinear motion with constant velocity and the equation of motion in the vertical direction to be that of free fall.

Choice of Origin and Reference Axes The point of projection is always chosen as the origin unlike in rectilinear motion in the vertical direction. The positive Y-axis is taken as directed upwards. Hence, the acceleration due to gravity acting vertically downwards is taken as negative, i.e., $-g$. Some of the cases that we normally encounter and the corresponding points of projections and the choice of axes are shown in the following figures.

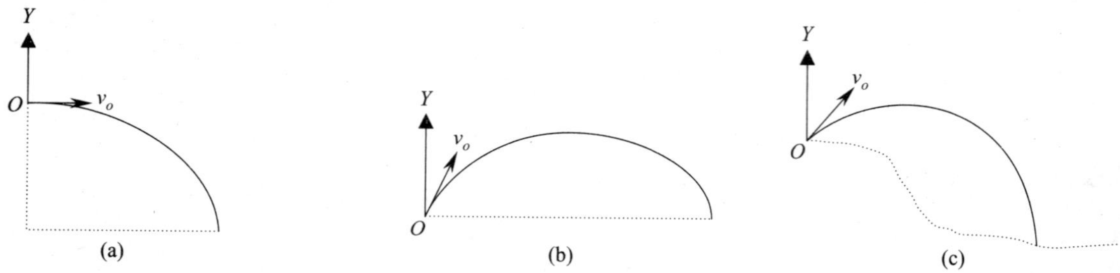

Fig. 13.6 Choice of origin and reference axes

As the projectile motion can be resolved into independent motions along the x and y directions, the equations of motion can be written separately along these two directions and they are summarized below:

MOTION ALONG THE X-DIRECTION

(Uniform motion)

$$a_x = 0 \qquad (13.8\text{ a})$$

$$v_x = v_o \cos\alpha \qquad (\text{b})$$

$$x = (v_o \cos\alpha)t \qquad (\text{c})$$

MOTION ALONG THE Y-DIRECTION

(Uniformly accelerated motion)

$$a_y = -g \qquad (13.9\text{ a})$$

$$v_y = v_o \sin\alpha - gt \qquad (\text{b})$$

$$v_y^2 = (v_o \sin\alpha)^2 - 2gy \qquad (\text{c})$$

$$y = (v_o \sin\alpha)t - \frac{1}{2}gt^2 \qquad (\text{d})$$

It should be noted that as the point of projection is chosen as the origin, y_o is zero. The total velocity at any instant of time is obtained by the vector addition of the components of velocity at that instant. Its magnitude and direction are

$$v = \sqrt{v_x^2 + v_y^2} \qquad (13.10)$$

and

$$\theta = \tan^{-1}\left[\frac{v_y}{v_x}\right] \qquad (13.11)$$

Eliminating time t from the two Eqs 13.8 c and 13.9 d, we get

$$y = x\tan\alpha - \frac{1}{2}g\frac{x^2}{(v_o \cos\alpha)^2} \qquad (13.12)$$

which is of the form $y = ax - bx^2$, i.e., the equation of a parabola. Hence, we can understand that the trajectory of projectile is a **parabola**.

Time taken to reach maximum height and time of flight When the particle reaches the maximum height, we know that the vertical component of velocity, i.e., v_y is zero. Therefore, from the Eq. 13.9 b,

$$0 = v_o \sin\alpha - gt$$

Hence, the time taken to reach the maximum height is

$$t = \frac{v_o \sin\alpha}{g} \quad (13.13)$$

Since the time of ascent is equal to the time of descent, the total time taken for the projectile to return to the same level of projection is

$$T = \frac{2v_o \sin\alpha}{g} \quad (13.14)$$

Maximum height reached Substituting the value of time of ascent in the Eq. 13.9 d, we get

$$y = v_o \sin\alpha\left(\frac{v_o \sin\alpha}{g}\right) - \frac{1}{2}g\left(\frac{v_o \sin\alpha}{g}\right)^2$$

$$= \frac{v_o^2 \sin^2\alpha}{g} - \frac{1}{2}g\left(\frac{v_o^2 \sin^2\alpha}{g^2}\right)$$

$$= \frac{v_o^2}{2g}\sin^2\alpha$$

Hence, the maximum height reached is

$$h_{\max} = \frac{v_o^2 \sin^2\alpha}{2g} \quad (13.15)$$

We notice that when the angle of projection is vertically upward, i.e., $\alpha = 90°$, the maximum height reached is $\frac{v_o^2}{2g}$, which is same as that obtained for the rectilinear motion along vertical direction in the previous chapter.

Range The horizontal distance between the point of projection and point of return of projectile to the *same level of projection* is termed **range**. Hence, range is obtained by substituting the value of total time of flight in the Eq. 13.8 c,

$$R = (v_o \cos\alpha)T$$

$$= (v_o \cos\alpha)\left[\frac{2v_o \sin\alpha}{g}\right]$$

Since $\sin 2\alpha = 2 \sin\alpha \cos\alpha$, we can write,

$$R = \frac{v_o^2 \sin 2\alpha}{g} \tag{13.16}$$

Corollary While calculating range, we must bear in mind that it is always the horizontal distance between the point of projection and the point of fall at the **same level** as the projection and otherwise not. Thus, in Fig. 13.6(c), the horizontal distance in the path shown cannot be calculated by the formula for range given above, but by using the equation of path (13.12).

Further, we can see from the above expression that the range is *maximum* when $\sin 2\alpha = 1$, or, in other words, $\alpha = 45°$. Hence, the maximum range is

$$R_{max} = \frac{v_o^2}{g} \tag{13.17}$$

Example 13.4 A ball is thrown from the ground with a velocity of 20 m/s at an angle of 30° to the horizontal. Determine (i) the velocity of the ball at $t = 0.5$ s and $t = 1.5$ s, (ii) total time of flight of the ball, (iii) maximum height reached, (iv) range of the ball, and (v) maximum range.

Solution The initial velocity of the ball can be resolved into horizontal and vertical components as

$$v_{ox} = v_o \cos\alpha = 20 \cos 30° = 17.32 \text{ m/s}$$

and

$$v_{oy} = v_o \sin\alpha = 20 \sin 30° = 10 \text{ m/s}$$

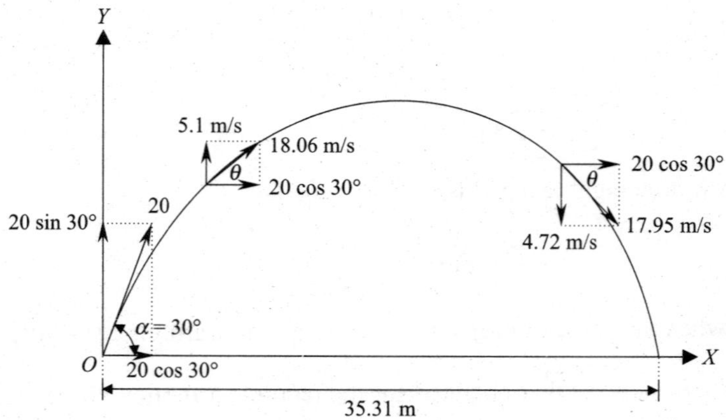

Fig. 13.7

(i) Velocity of the ball at t = 0.5 s

We know that the horizontal component of velocity always remains constant and only the vertical component of velocity varies with time. Thus,

$$v_y(0.5) = v_o \sin\alpha - gt$$
$$= 10 - 9.81\,(0.5) = 5.1 \text{ m/s}$$

Therefore, the total velocity at that instant is obtained as

$$v = \sqrt{v_x^2 + v_y^2}$$

$$= \sqrt{(17.32)^2 + (5.1)^2} = 18.06 \text{ m/s}$$

and its inclination with respect to the X-axis is obtained as

$$\theta = \tan^{-1}\left[\frac{|v_y|}{|v_x|}\right]$$

$$= \tan^{-1}\left[\frac{5.1}{17.32}\right] = 16.41°$$

As both the components are positive, the instantaneous velocity is directed upwards (refer Fig. 13.7).

Velocity of the ball at t =1.5 s

Similarly, $$v_y(1.5) = 10 - 9.81(1.5) = -4.72 \text{ m/s}$$

Therefore, $$v = \sqrt{v_x^2 + v_y^2}$$

$$= \sqrt{(17.32)^2 + (-4.72)^2} = 17.95 \text{ m/s}$$

and $$\theta = \tan^{-1}\left[\frac{|v_y|}{|v_x|}\right]$$

$$= \tan^{-1}\left[\frac{4.72}{17.32}\right] = 15.24°$$

As the y-component is negative, the instantaneous velocity is directed downwards (refer Fig. 13.7).

(ii) Time of flight

We know that total time of flight of the ball is given as

$$T = \frac{2v_o \sin\alpha}{g} = \frac{2(10)}{9.81} = 2.04 \text{ s}$$

(iii) Maximum height reached

Maximum height reached by the ball is given as

$$h = \frac{v_o^2 \sin^2\alpha}{2g} = \frac{(10)^2}{2 \times 9.81} = 5.1 \text{ m}$$

(iv) Range

Range of the projectile is given as

$$R = \frac{v_o^2 \sin 2\alpha}{g}$$

$$= \frac{(20)^2 \sin 60°}{9.81} = 35.31 \text{ m}$$

(v) Maximum range

$$R_{\max} = \frac{v_o^2}{g} = \frac{(20)^2}{9.81} = 40.77 \text{ m}$$

Example 13.5 A ball is thrown from the top of a building of 20 m height with a velocity of 30 m/s at an angle of 45° to the horizontal. Determine its velocity at $t = 2$ s. How high does it rise? Determine the horizontal distance it will travel before striking the ground. Also, determine its velocity at that instant.

Solution The horizontal and vertical components of initial velocity are

$$v_{ox} = 30 \cos 45° = 21.21 \text{ m/s}$$

and

$$v_{oy} = 30 \sin 45° = 21.21 \text{ m/s}$$

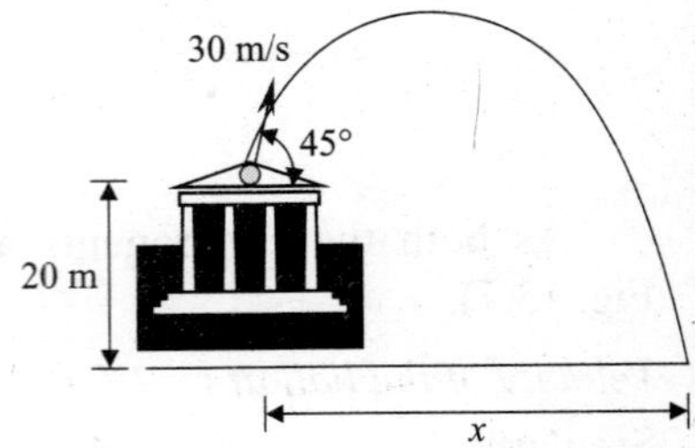

Fig. 13.8(a)

Velocity at time t = 2 s

The x-component of velocity remains constant, while the y-component of velocity varies with time. At time $t = 2$ s, the y-component of velocity is given as

$$v_y = v_{oy} - gt$$

$$= 21.21 - 9.81(2) = 1.59 \text{ m/s}$$

Therefore, total velocity at time $t = 2$ s is given as

$$v = \sqrt{v_x^2 + v_y^2}$$

$$= \sqrt{(21.21)^2 + (1.59)^2} = 21.27 \text{ m/s}$$

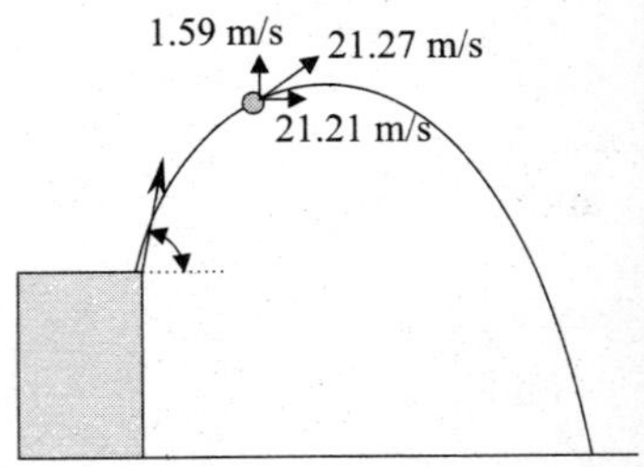

Fig. 13.8(b)

Its inclination with respect to the X-axis is

$$\theta = \tan^{-1}\left[\frac{|v_y|}{|v_x|}\right] = 4.29° \text{ [directed upwards]}$$

Maximum height reached

$$h_{\max} = \frac{v_o^2 \sin^2 \alpha}{2g}$$

$$= \frac{(21.21)^2}{2 \times 9.81} = 22.93 \text{ m above the top of the building}$$

Horizontal distance travelled before striking the ground

Since the point of projection is not at the ground level, we cannot determine the horizontal distance travelled using the expression for range, which is valid only when the two points are at the same level. Instead, we use the equation of motion:

$$y = x \tan \alpha - \frac{1}{2} g \frac{x^2}{(v_o \cos \alpha)^2}$$

Taking into consideration the fact that the positive Y-axis points upwards,

$$-20 = x \tan (45)° - \frac{1}{2}(9.81)\frac{x^2}{(21.21)^2}$$

$$0.0109x^2 - x - 20 = 0$$

$$\Rightarrow \quad x = 108.63 \text{ m} \quad \text{[neglecting the other trivial root with a negative sign]}$$

Velocity with which it strikes the ground

We know that the horizontal distance travelled by the ball is given as

$$x = v_o \cos \alpha t$$

Therefore, time taken to travel this distance is given as

$$t = \frac{x}{v_o \cos \alpha}$$

$$= \frac{108.63}{21.21} = 5.12 \text{ s}$$

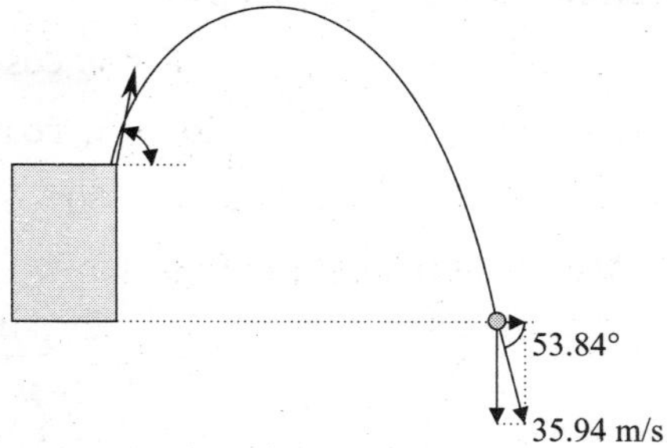

Fig. 13.8(c)

Therefore, $v_y = v_{oy} - gt$

$$= 21.21 - 9.81(5.12) = -29.02 \text{ m/s}$$

Hence, the total velocity with which it will strike the ground is obtained as

$$v = \sqrt{v_x^2 + v_y^2}$$

$$= \sqrt{(21.21)^2 + (-29.02)^2} = 35.94 \text{ m/s}$$

and $\theta = \tan^{-1}\left[\frac{|v_y|}{|v_x|}\right] = 53.84°$ directed downwards

Example 13.6 In a shooting practice, a bullet is to be fired at a target, 150 m away at the same level. If the release velocity of the bullet is 60 m/s, determine what should be the angle of inclination of the gun with respect to the horizontal.

Solution We know that the range is given as

$$\text{Range} = \frac{v_o^2 \sin 2\alpha}{g}$$

$$150 = \frac{60^2 \sin 2\alpha}{9.81}$$

$\therefore$ $\sin 2\alpha = 0.409$

Since $\sin\alpha = \sin(180° - \alpha)$, there are two values of the angle, which satisfy the above equation. Hence,

$$2\alpha = 24.14° \text{ (or) } 155.86°$$

$\Rightarrow$ $\alpha = 12.07° \text{ (or) } 77.93°$

Though there are two values of α which will give the same range, we consider the smaller value, as in shooting practice, normally, the angle of inclination is smaller.

Example 13.7 Find the velocity of projection of a missile, which has a horizontal range of 200 m, if its time of flight for that range is 4 s. Also, determine the angle of projection.

Solution Range R is given as

$$R = v_o \cos\alpha . t$$

Therefore, $$200 = v_o \cos\alpha(4)$$

$$\Rightarrow \quad v_o \cos\alpha = 50 \qquad \text{(a)}$$

Also, total time of flight is

$$t = \frac{2v_o \sin\alpha}{g}$$

$$\Rightarrow \quad v_o \sin\alpha = 4g/2 = 2g \qquad \text{(b)}$$

From equations (a) and (b), we get

$$\tan\alpha = \frac{2g}{50}$$

$$\Rightarrow \quad \alpha = 21.43°$$

Substituting the value of α in the equation (a), we get

$$v_o = \frac{50}{\cos(21.43)°} = 53.71 \text{ m/s}$$

Example 13.8 In shot-put game, a man throws an iron ball with an initial velocity of 10 m/s. What should be the angle of projection if he has to reach a record of 10 m? Take the height at which the ball is released as 1.2 m.

Solution The equation of motion of the iron ball can be written as

$$y = x\tan\alpha - \frac{1}{2}\frac{gx^2}{v_o^2\cos^2\alpha}$$

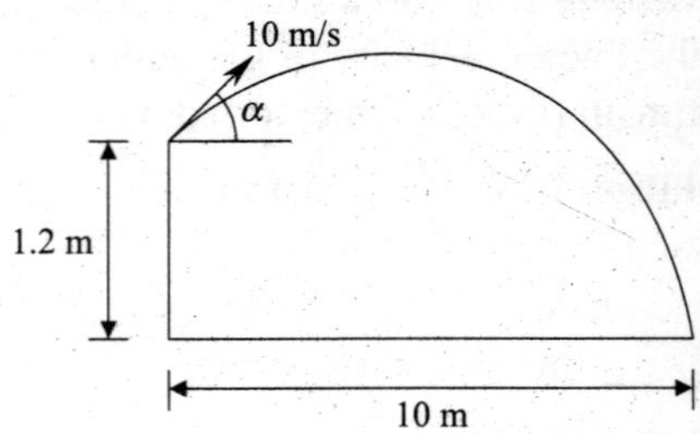

Fig. 13.9

Since it is released at a height of 1.2 m, $y = -1.2$. Hence,

$$-1.2 = 10\tan\alpha - \frac{1}{2}\frac{9.81(10)^2}{(10)^2\cos^2\alpha}$$

$$-1.2 = 10\tan\alpha - 4.905\sec^2\alpha$$

Since $$\sec^2\alpha = 1 + \tan^2\alpha,$$

$$-1.2 = 10\tan\alpha - 4.905\,[1 + \tan^2\alpha]$$

$$4.905\tan^2\alpha - 10\tan\alpha + 3.705 = 0$$

Solving the quadratic equation, we get

$$\tan\alpha = 0.486 \text{ (or) } 1.552$$

$$\therefore \quad \alpha = 25.92° \text{ (or) } 57.21°$$

Example 13.9 A basketball player who is 2 m tall jumps 0.25 m above the ground and shoots the ball into the basket from a point that is 4 m from the basket as shown in Fig. 13.10. Determine the velocity with which he must throw the ball so that the ball enters the basket.

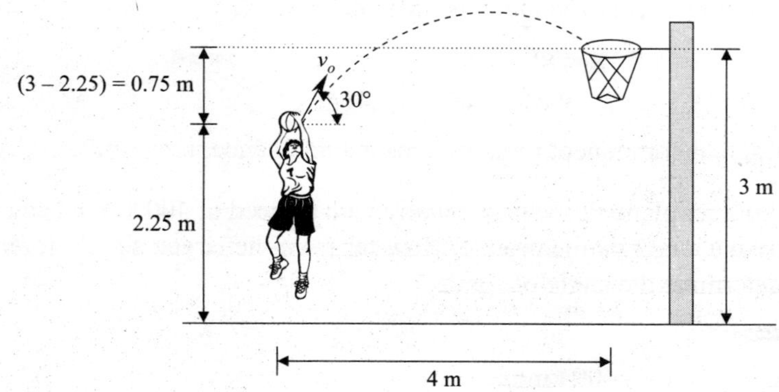

Fig. 13.10

Solution The equation of motion of the ball can be written as

$$y = x(\tan\alpha) - \frac{1}{2}\frac{g}{(v_o\cos\alpha)^2}x^2$$

$$0.75 = 4\tan 30° - \frac{1}{2}\frac{9.81}{(v_o\cos 30°)^2}(4)^2$$

$$\Rightarrow \qquad v_o = 8.19 \text{ m/s}$$

Example 13.10 A block slides off a horizontal 1 m high tabletop with a speed of 3 m/s. Find (a) the horizontal distance from the edge of the table at which the block strikes the floor, and (b) the horizontal and vertical components of velocity when it reaches the floor.

Solution As the block leaves the table tangentially, it should be noted that the vertical component of initial velocity v_{oy} is zero. The horizontal component of velocity does not change with time. Hence, it is equal to 3 m/s.

(a) Horizontal distance from the table at which the block strikes the floor

The equation of motion of the block can be written as

$$y = x(\tan\alpha) - \frac{1}{2}\frac{g}{(v_o\cos\alpha)^2}x^2$$

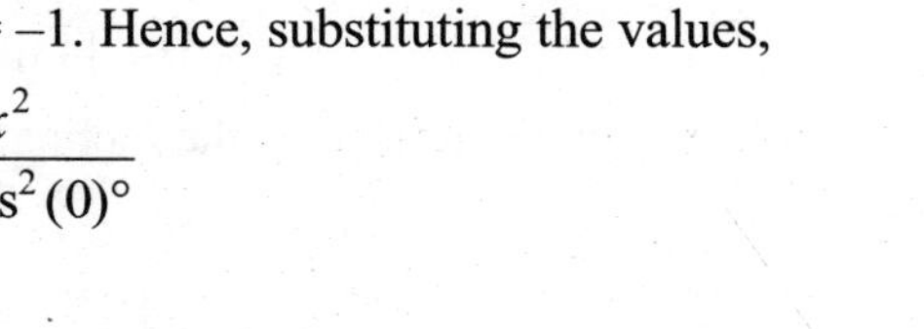

Fig. 13.11

Since the point of projection is chosen as the origin, $y = -1$. Hence, substituting the values,

$$-1 = x\tan(0)° - \frac{1}{2}(9.81)\frac{x^2}{3^2\cos^2(0)°}$$

$$\Rightarrow \qquad x = 1.35 \text{ m}$$

(b) Horizontal and vertical components of velocity when it strikes the floor

We know, $\qquad x = (v_o\cos\alpha)t$

$$\Rightarrow \qquad t = \frac{1.35}{3\cos(0)°} = 0.45 \text{ s}$$

Therefore, vertical component of velocity is obtained as

$$v_y = v_{oy} - gt$$
$$= 0 - 9.81\ (0.45) = -4.41\text{m/s}$$

Note that the horizontal component remains constant and hence, it is 3 m/s.

Example 13.11 A bomber plane flying horizontally with a speed of 400 kmph at an altitude of 150 m releases a bomb to strike a target on the ground. How far from the target, should it release the bomb to hit the target? Also, determine the angle of sight.

Solution *Given data*

Speed of plane, $v_o = 400$ kmph

$= 400 \times 5/18 = 111.11$ m/s

Altitude = 150 m

Fig. 13.12(a)

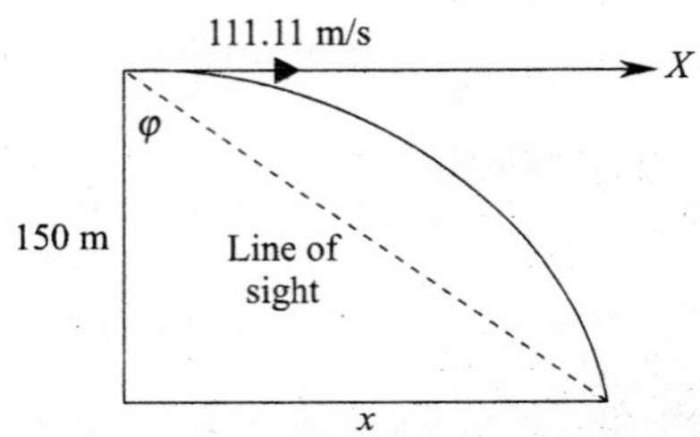

Fig. 13.12(b)

When the bomb is released, we know that its speed is the same as that of the plane and the direction is horizontal. Hence, the angle of projection is taken as 0°. Therefore, the equation of motion of the bomb can be written as

$$y = x\,(\tan\alpha) - \frac{1}{2}\frac{g}{(v_o\cos\alpha)^2}x^2$$

Substituting the values,

$$-150 = x\tan(0)° - \frac{1}{2}9.81\frac{x^2}{[111.11\cos(0)°]^2}$$

$\Rightarrow$ $x = 614.44$ m

Angle of sight is defined as the inclination of line of sight with the vertical. Hence,

$$\tan\varphi = \frac{|x|}{|y|}$$
$$= \frac{614.44}{150} = 4.096$$

$\Rightarrow$ $\phi = 76.28°$

Example 13.12 A bomber flying horizontally at a speed of 300 kmph at an altitude of 150 m releases a bomb targeting a ship moving in the same direction as the plane at a constant speed of 10 m/s. How far from the ship, should it release the bomb to hit it? Also, determine the angle of sight.

Solution *Given data*

Speed of plane, $v_o = 300$ kmph

$= 300 \times 5/18 = 83.33$ m/s

Altitude = 150 m

Fig. 13.13(a)

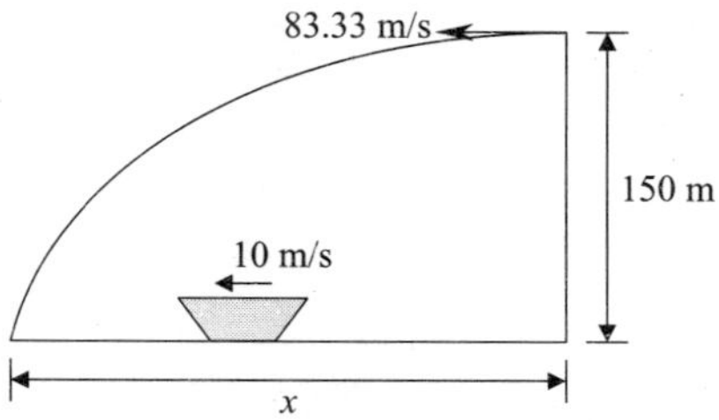

Fig. 13.13(b)

When the bomb is released, we know that its speed is the same as that of the plane and the direction is horizontal. Hence, the angle of projection is taken as 0°. Therefore, the equation of motion of the bomb can be written as

$$y = x(\tan\alpha) - \frac{1}{2}\frac{g}{(v_o \cos\alpha)^2}x^2$$

Substituting the values,

$$-150 = 0 - \frac{1}{2}\frac{9.81}{(83.33)^2}x^2$$

$\Rightarrow$ $$x = 460.82 \text{ m}$$

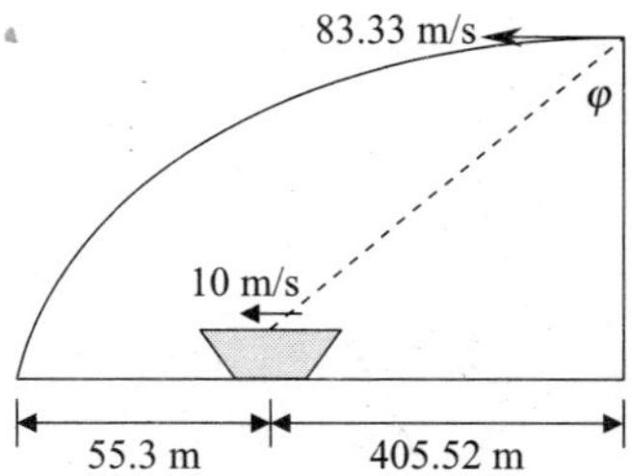

Fig. 13.13(c)

The time taken by the bomb to hit the target is obtained as

$$t = \frac{x}{v_o \cos\alpha}$$

$$= \frac{460.82}{83.33} = 5.53 \text{ s}$$

Hence, the distance travelled by the ship during this time is obtained as

$$s_s = v_s t$$

$$= 10 \times 5.53 = 55.3 \text{ m}$$

Hence, the distance from which the bomb must be released in order to hit the ship is

$$= x - s_s$$

$$= 460.82 - 55.3 = 405.52 \text{ m}$$

Angle of sight is defined as the inclination of line of sight with the vertical. Hence,

$$\tan\varphi = \frac{|x - s_s|}{|y|}$$

$$= \frac{405.52}{150} = 2.703$$

$\Rightarrow \qquad \varphi = 69.7°$

Example 13.13 Find the angle of projection at which the horizontal range and the maximum height of a projectile are equal.

Solution We know,

$$\text{maximum height} = \frac{v_o^2 \sin^2\alpha}{2g} \quad \text{and}$$

$$\text{horizontal range} = \frac{v_o^2 \sin 2\alpha}{g}$$

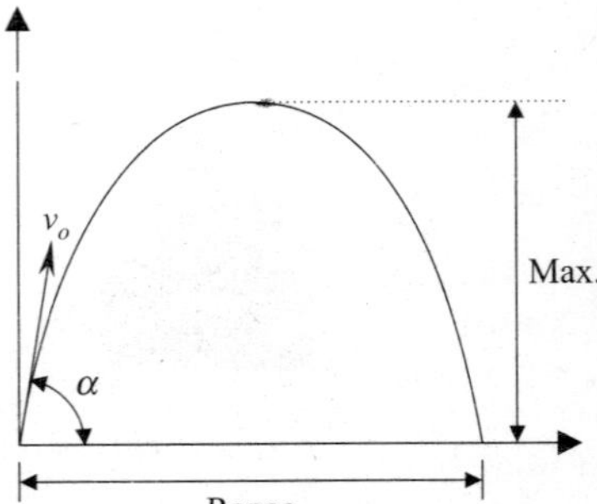

Fig. 13.14

When the horizontal range and maximum height are equal,

$$\frac{v_o^2 \sin^2\alpha}{2g} = \frac{v_o^2 \sin 2\alpha}{g}$$

$$\frac{v_o^2 \sin^2\alpha}{2g} = \frac{v_o^2 2\sin\alpha\cos\alpha}{g} \qquad [\text{since } \sin 2\alpha = 2\sin\alpha\cos\alpha]$$

$\Rightarrow \qquad \tan\alpha = 4$

$\therefore \qquad \alpha = 75.96°$

Example 13.14 A particle is projected at an angle of inclination α to the horizontal and after time t seconds, it is found that the velocity is inclined at an angle β to the horizontal. Show that the initial velocity of projection was $\dfrac{gt\cos\beta}{\sin(\alpha-\beta)}$

Solution If v_o be the initial velocity of projection then the components of initial velocity are

$$(v_o)_x = v_o\cos\alpha \quad \text{and} \quad (v_o)_y = v_o\sin\alpha$$

The horizontal component of velocity remains the same at all times. However, the vertical component of velocity changes with time and it is given as

$$v_y = (v_o)_y - gt$$

$$= v_o\sin\alpha - gt$$

and $\qquad v_x = v_o\cos\alpha$

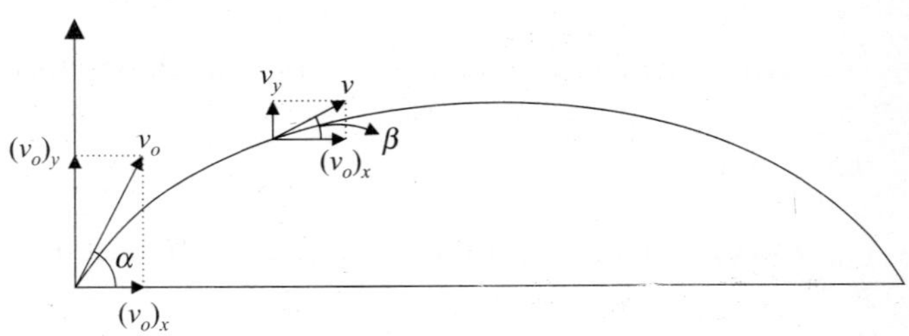

Fig. 13.15

Since the inclination of the instantaneous velocity after time t seconds is β, w can write

$$\tan\beta = \left[\frac{v_y}{v_x}\right] = \frac{v_o \sin\alpha - gt}{v_o \cos\alpha}$$

Upon rearranging,

$$v_o \cos\alpha \tan\beta = v_o \sin\alpha - gt$$

$$v_o[\sin\alpha - \cos\alpha \tan\beta] = gt$$

$$v_o[\sin\alpha \cos\beta - \cos\alpha \sin\beta] = gt\cos\beta \quad [\text{since } \tan\beta = \sin\beta/\cos\beta]$$

Since $\sin(\alpha - \beta) = \sin\alpha\cos\beta - \cos\alpha\sin\beta$, we can write

$$v_o[\sin(\alpha - \beta)] = gt\cos\beta$$

$$\Rightarrow \quad v_o = \frac{gt\cos\beta}{\sin(\alpha - \beta)}$$

Example 13.15 A particle is projected from the ground level. If t_1 be the time taken by the particle to reach a point A on its path and t_2 be the time taken from A to the ground level then show that the height of point A above the ground level is $\frac{1}{2}gt_1t_2$.

Solution Since t_1 is the time taken to reach point A and t_2 is the time taken from A to the ground level, the total time of flight is $t_1 + t_2$. Hence, we can write

$$t_1 + t_2 = \frac{2v_o \sin\alpha}{g}$$

$$\Rightarrow \quad v_o \sin\alpha = \frac{(t_1 + t_2)g}{2}$$

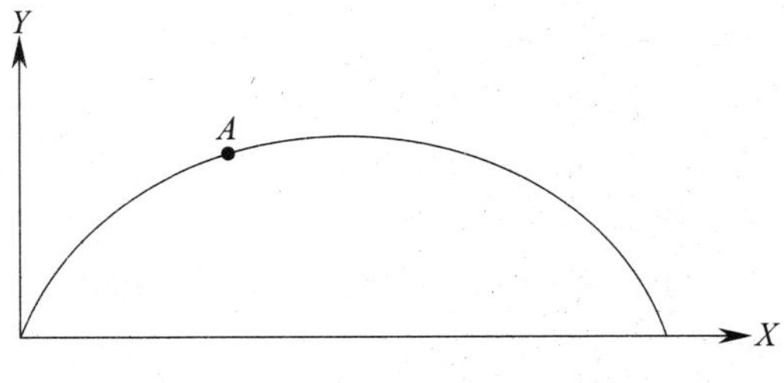

Fig. 13.16

If y be the height of point A above ground level,

$$y = v_o \sin\alpha t_1 - \frac{1}{2}gt_1^2$$

$$= \frac{(t_1 + t_2)g}{2}t_1 - \frac{1}{2}gt_1^2$$

$$= \frac{1}{2}g[t_1^2 + t_1t_2 - t_1^2]$$

$$= \frac{1}{2}gt_1t_2$$

Example 13.16 A football player hits the ball to impart an initial velocity of 30 m/s at an angle of 15° with the horizontal from a point that is 15 m from the goal post (of 2.2 m height). Determine whether the ball will enter into the goal post or not.

Solution

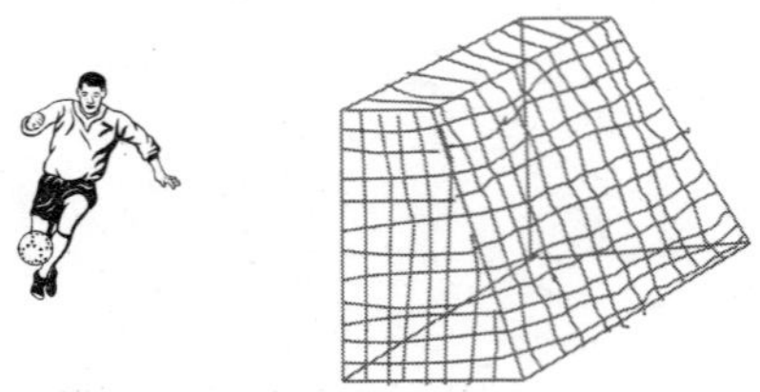

Fig.13.17(a)

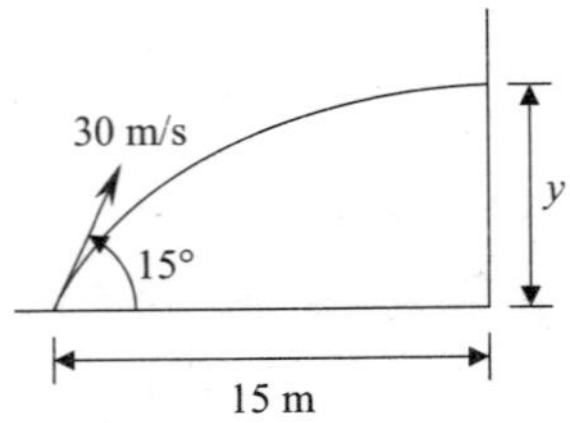

Fig.13.17(b)

The equation of motion of the ball can be written as

$$y = x\tan\alpha - \frac{1}{2}g\frac{x^2}{(v_o\cos\alpha)^2}$$

Hence, at $x = 15$ m

$$y = 15\tan 15° - \frac{1}{2}\times 9.81\times\frac{15^2}{(30\cos 15°)^2} = 2.7 \text{ m}$$

Since the value of y is greater than the height of the goal post (2.2 m), the ball will go over the post.

Example 13.17 A cricket batsman hits the ball with a velocity of 40 m/s almost from the ground level at an angle of 20° with the horizontal. A fielder is standing in the plane of motion of the ball, just before the boundary line at a distance of 98 m from the batsman. If the height of the fielder is 2 m, how high must he jump in order to catch the ball before crossing the boundary?

Solution *Given data*

Initial velocity of the ball, $v_o = 40$ m/s

Angle of projection, $\alpha = 20°$

Horizontal distance, $x = 98$ m

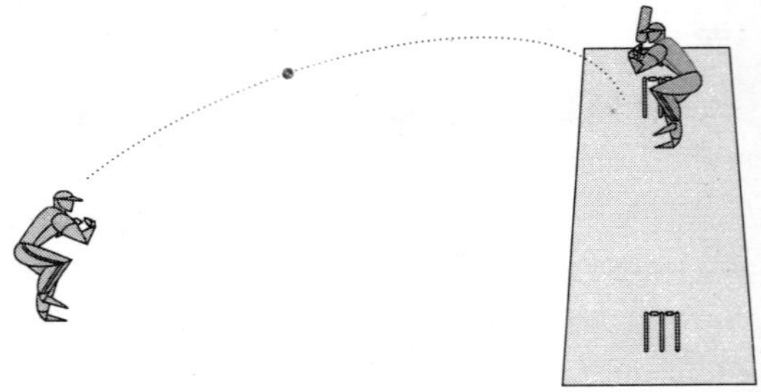

Fig. 13.18

The equation of motion of the ball can be written as

$$y = x\tan\alpha - \frac{1}{2}g\frac{x^2}{(v_o\cos\alpha)^2}$$

Substituting the values,

$$y = 98\tan 20° - \frac{1}{2}\times 9.81\times\frac{98^2}{(40\cos 20°)^2} = 2.33 \text{ m}$$

As the y coordinate of the trajectory of the ball is 2.33 m when x is 98 m, the height by which he should jump to catch the ball is

$$= 2.33 - 2 = 0.33 \text{ m}$$

Example 13.18 A motorist in a circus tries to jump from one incline to the other as shown in the figure. Determine the speed with which he must jump in order to just cross the boards and the time taken to jump.

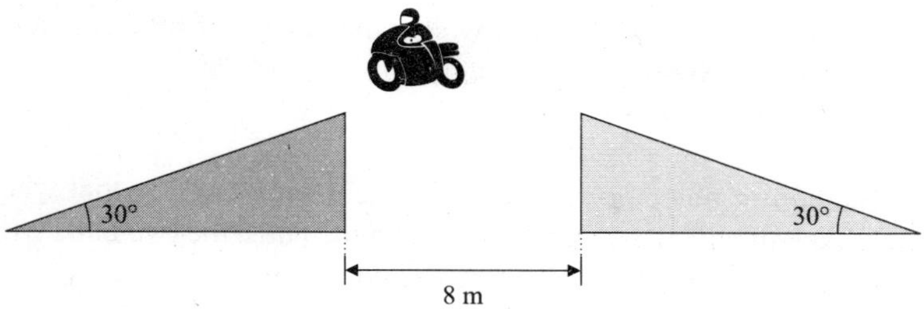

Fig. 13.19

Solution The launch angle of the motorist is the same as the angle of the incline, i.e., $\alpha = 30°$. The equation of motion of the motorist is

$$y = x\tan\alpha - \frac{1}{2}g\frac{x^2}{(v_o\cos\alpha)^2}$$

Speed with which he must jump

Since the height of the two inclines are same, the vertical displacement, $y = 0$. Hence,

$$0 = 8\tan 30° - \frac{1}{2}\times 9.81\times\frac{8^2}{v_o^2\cos^2 30°}$$

$\Rightarrow$ $$v_o = 9.52 \text{ m/s}$$

Time taken to jump

We know, $$x = v_o\cos\alpha t$$

$\Rightarrow$ $$t = \frac{8}{9.52\cos 30°} = 0.97 \text{ s}$$

Example 13.19 A ball is projected such that its initial velocity is directed towards a second ball. The second ball is released from rest at the instant the first is projected. Prove that the two balls collide regardless of the value of the initial velocity.

Solution Let t be the time taken by the second ball to reach the point B lying vertically below it and on the trajectory of the first ball. Then the height AB through which it has fallen is obtained from the equation of motion for free fall:

$$s = AB = (1/2)gt^2$$

From the above figure, we see that

$$\tan\alpha = \frac{(1/2)gt^2 + y}{x}$$

$\Rightarrow$ $$y = x\tan\alpha - (1/2)gt^2 \quad \text{(a)}$$

The horizontal distance x travelled by the first ball in the same time t is

$$x = v_o\cos\alpha t$$

$\Rightarrow$ $$t = \frac{x}{v_o\cos\alpha}$$

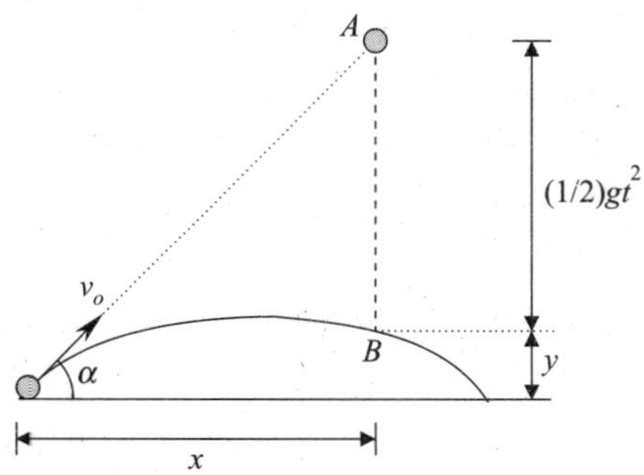

Fig. 13.20

Substituting this value of t in the equation (a), we get the vertical displacement of the first ball as

$$y = x\tan\alpha - \frac{1}{2}g\frac{x^2}{(v_o\cos\alpha)^2}$$

which is same as the equation of trajectory of the projectile. Hence, we see that when the second ball reaches the point B, the first ball will also reach the same point. Thus, the two balls will collide irrespective of the initial speed with which the first ball is projected.

Example 13.20 A football player kicks a ball at an angle of 37° with the horizontal and with an initial velocity of 25 m/s. The goalkeeper standing at a distance of 70 m in the direction of the kick starts running to meet the ball at the instant it is kicked. What must be his rate of acceleration in order to catch the ball just before it hits the ground?

Solution The horizontal range of the ball is given as

$$R = \frac{v_o^2\sin 2\alpha}{g}$$

$$= \frac{(25)^2\sin(2\times 37)^\circ}{9.81} = 61.24\text{ m}$$

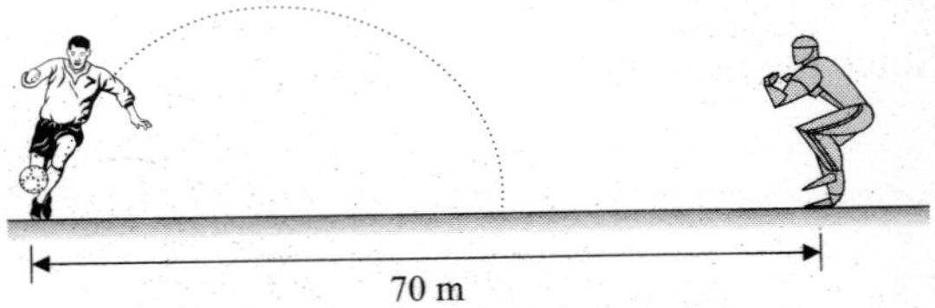

Fig. 13.21

Time taken to travel this distance is

$$t = 2\frac{v_o\sin\alpha}{g}$$

$$= 2\times\frac{(25)\sin 37^\circ}{9.81} = 3.07\text{ s}$$

Distance to be travelled by the goalkeeper to meet the ball before it hits the ground is

$$s = 70 - 61.24 = 8.76\text{ m}$$

Therefore, the acceleration of the goalkeeper is determined as follows:

The equation of motion of the goalkeeper considering the same time t as the ball is

$$s = v_{go}t + (1/2)a_g t^2$$

$$8.76 = 0 + (1/2)a_g(3.07)^2$$ [***Note:*** Initial velocity of the goalkeeper is zero.]

$$\Rightarrow \quad a_g = 1.86\text{ m/s}^2$$

Example 13.21 A ball is thrown as shown with an initial velocity at an angle of 30° above the horizontal, from a point that is 100 m from the edge of a 100 m high vertical cliff. The ball just misses the edge of the cliff. Find (a) the initial velocity of the ball, and (b) the distance beyond the foot of the cliff where the ball strikes the ground.

Solution Since points A and B are at the same elevation, the horizontal distance $\overline{AB}$ is nothing but the range. Hence,

$$\text{range} = \frac{v_o^2\sin 2\alpha}{g}$$

$$100 = \frac{v_o^2 \sin(2 \times 30)°}{9.81}$$

$\Rightarrow$ $$v_o = 33.66 \text{ m/s}$$

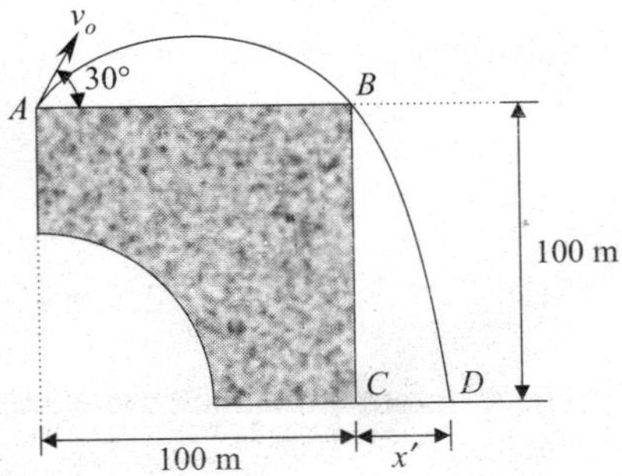

Fig. 13.22

The equation of trajectory of the ball is

$$y = x \tan \alpha - \frac{1}{2} g \frac{x^2}{(v_o \cos \alpha)^2}$$

Substituting the values,

$$-100 = x \tan 30° - \frac{1}{2}(9.81)\frac{x^2}{(33.66 \cos 30)^2}$$

[Note that as the point D is below point A, we see that $y = -100$]

$$5.772 \times 10^{-3} x^2 - 0.5774x - 100 = 0$$

Upon solving the quadratic equation and neglecting the other trivial root with negative value,

$$x = 190.82 \text{ m}$$

Therefore, the distance beyond the foot of the cliff where the ball strikes the ground is:

$$x' = CD = x - 100$$

$$= 190.82 - 100 = 90.82 \text{ m}$$

Example 13.22 A boy while playing with a ball throws it against a wall as shown in Fig. 13.23. What should be the initial velocity of the ball and its angle of projection so that it hits the wall at right angles?

Solution The components of initial velocity are

$$v_{ox} = v_o \cos \alpha \quad \text{and} \quad v_{oy} = v_o \sin \alpha$$

Since the ball strikes the wall at right angles, its velocity along the vertical or y-direction at that instant is *zero*. Hence, its total velocity is only along the horizontal or x-direction. We know that the horizontal component always remains constant; hence, its instantaneous velocity when it strikes the wall is $v_o \cos \alpha$.

The equation of motion along the y-direction can be written as

$$v_y = v_{oy} - gt$$

When it strikes the wall, $v_y = 0$. Therefore,

$$0 = v_o \sin \alpha - gt$$

$\Rightarrow$ $$t = \frac{v_o \sin \alpha}{g}$$

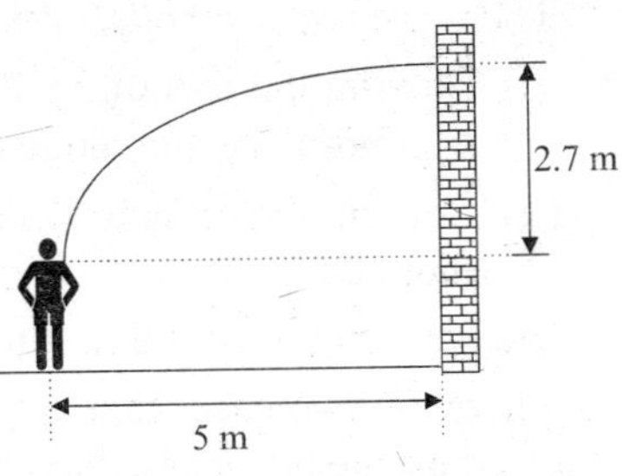

Fig. 13.23

The vertical displacement of the ball is given as

$$y = (v_o \sin \alpha)t - \frac{1}{2} gt^2$$

Substituting the value of t

$$y = \frac{v_o^2 \sin^2 \alpha}{2g}$$

$$2.7 = \frac{v_o^2 \sin^2 \alpha}{2g} \qquad \text{(a)}$$

The horizontal distance travelled by the ball is

$$x = v_o \cos \alpha\, t$$

Substituting the value of t

$$x = \frac{v_o^2 \cos\alpha \sin\alpha}{g}$$

$$5 = \frac{v_o^2 \cos\alpha \sin\alpha}{g} \qquad \text{(b)}$$

From equations (a) and (b), we get

$$\frac{2.7}{5} = \frac{\tan\alpha}{2}$$

$$\therefore \qquad \alpha = \tan^{-1}\left[\frac{5.4}{5}\right] = 47.2°$$

Substituting this value of α in the equation (a), we get

$$v_o = 9.92 \text{ m/s}$$

13.4 PROJECTION ON AN INCLINED PLANE

In the previous section, we discussed projectile motion along level ground as when a body is projected from the ground level or from the top of a building, etc., where we determined the horizontal distance travelled, maximum height reached and time of flight. In this section, we will discuss bodies projected on an inclined plane, where we have to determine the range on an inclined plane and time of flight.

There are two methods, which can be adopted to solve these types of problems:

(i) Solving the motion by the normal method described in the previous section and then by using trigonometry, the range on the inclined plane can be determined.

(ii) Solving the motion by resolving it into components along the inclined plane and normal to the inclined plane.

We have followed the first method in all of the derivations and examples described below.

There are two cases that we come across in motion on an inclined plane, one in which the body is projected **up** the plane and the other in which the body is projected **down** the plane. These are dealt with separately in the following sections.

13.4.1 Body Projected Up an Incline

Consider a body projected up an incline with an initial velocity v_o at an angle of inclination α to the horizontal. Let the inclination of the plane with the horizontal be β and C be the point at which the body strikes the plane.

Time of flight If t be the time of flight of the projectile then the horizontal displacement in time t is

$$AB = x = (v_o \cos\alpha)\, t \qquad (13.18)$$

Similarly, the vertical displacement in the same time t is

$$BC = y = (v_o \sin\alpha)\, t - \frac{1}{2}gt^2 \qquad (13.19)$$

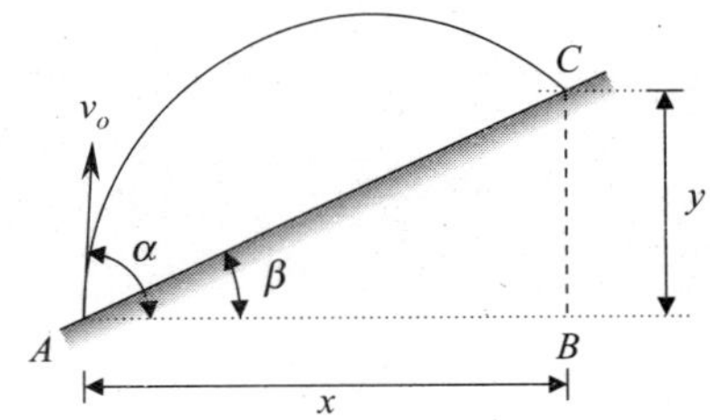

Fig. 13.24 Body projected up the plane

From the Fig. 13.24, we see that

$$\tan\beta = \frac{y}{x}$$

$$= \frac{(v_o \sin\alpha)t - \frac{1}{2}gt^2}{(v_o \cos\alpha)t}$$

$$= \frac{(v_o \sin\alpha) - \frac{1}{2}gt}{(v_o \cos\alpha)}$$

$$\Rightarrow \qquad v_o \cos\alpha \tan\beta = v_o \sin\alpha - \frac{1}{2}gt$$

$$\therefore \qquad t = \frac{2v_o}{g}[\sin\alpha - \cos\alpha \tan\beta]$$

$$= \frac{2v_o}{g\cos\beta}[\sin\alpha\cos\beta - \cos\alpha\sin\beta]$$

As $\sin(A - B) = \sin A \cos B - \cos A \sin B$, the above expression can be written as

$$t = \frac{2v_o}{g\cos\beta}[\sin(\alpha - \beta)] \qquad (13.20)$$

Range along the inclined plane Substituting this value of t in the Eq. 13.18, we have

$$x = (v_o \cos\alpha)\, t$$

$$= (v_o\cos\alpha)\frac{2v_o}{g\cos\beta}[\sin(\alpha - \beta)]$$

$$= \frac{2v_o^2}{g\cos\beta}\sin(\alpha - \beta)\cos\alpha \qquad (13.21)$$

Therefore, range along the incline is

$$R = AC = \frac{x}{\cos\beta} = \frac{2v_o^2}{g\cos^2\beta}\sin(\alpha - \beta)\cos\alpha \qquad (13.22)$$

Since $2\cos A \sin B = \sin(A+B) - \sin(A-B)$,

$$R = \frac{v_o^2}{g\cos^2\beta}[\sin(2\alpha - \beta) - \sin(\beta)] \tag{13.23}$$

Maximum range and angle of projection Since the angle of the incline β is fixed, we see that range is *maximum* when

$$\sin(2\alpha - \beta) = 1$$

or

$$2\alpha - \beta = 90°$$

$\Rightarrow$

$$\alpha = \frac{90° + \beta}{2} = 45° + \frac{\beta}{2} \tag{13.24}$$

Hence, maximum range is

$$R = \frac{v_o^2}{g\cos^2\beta}[\sin 90° - \sin\beta]$$

$$= \frac{v_o^2}{g}\frac{(1 - \sin\beta)}{(1 - \sin^2\beta)} \quad [\text{since } \cos^2\beta = 1 - \sin^2\beta]$$

$$R = \frac{v_o^2}{g(1 + \sin\beta)} \tag{13.25}$$

13.4.2. Body Projected Down an Incline

Consider a body projected down an incline with an initial velocity v_o at an angle of inclination α with the horizontal. Let the inclination of the plane with the horizontal be β and C be the point at which the body strikes the plane.

Time of flight If t be the time of flight of the projectile, then the horizontal displacement in time t is

$$x = (v_o \cos\alpha)\, t \tag{13.26}$$

Similarly, the vertical displacement in the same time t is

$$-y = (v_o \sin\alpha)\, t - \frac{1}{2}gt^2 \quad \text{[Note that } y \text{ is negative downwards.]} \tag{13.27}$$

From Fig. 13.25, we see that

$$\tan\beta = \frac{AB}{BC}$$

$$= \frac{-(v_o \sin\alpha)t + \frac{1}{2}gt^2}{(v_o \cos\alpha)t}$$

$$= \frac{-(v_o \sin\alpha) + \frac{1}{2}gt}{(v_o \cos\alpha)}$$

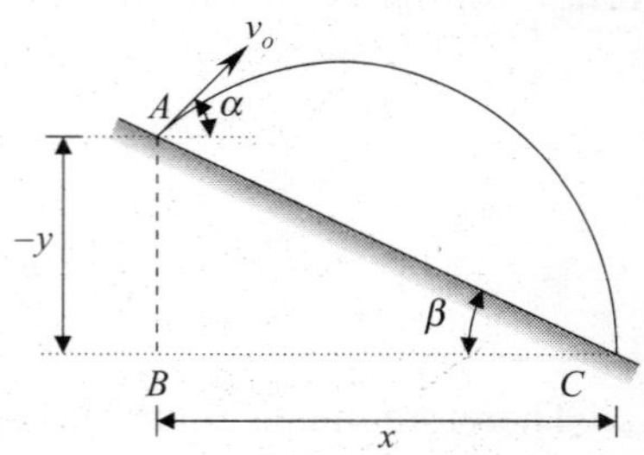

Fig. 13.25 Body projected down the plane

$$\Rightarrow \qquad v_o \cos\alpha \tan\beta = -v_o \sin\alpha + \frac{1}{2}gt$$

$$\therefore \qquad t = \frac{2v_o}{g\cos\beta}[\sin\alpha\cos\beta + \cos\alpha\sin\beta]$$

$$= \frac{2v_o}{g\cos\beta}[\sin(\alpha + \beta)] \qquad (13.28)$$

Range along the inclined plane From this value of t, we can get the range along the inclined plane in a similar manner as

$$R = \frac{v_o^2}{g\cos^2\beta}[\sin(2\alpha + \beta) + \sin(\beta)] \qquad (13.29)$$

Maximum range and angle of projection From the above expression, we can see that range is *maximum* when

$$\sin(2\alpha + \beta) = 1$$

or

$$2\alpha + \beta = 90°$$

Hence, the condition for maximum range is

$$\alpha = \frac{90° - \beta}{2} = 45° - \frac{\beta}{2}$$

and the maximum range is

$$R_{max} = \frac{v_o^2}{g(1 - \sin\beta)} \qquad (13.30)$$

The above results for the two types of motions are summarized below:

Table 13.1

	Projection up the plane	*Projection down the plane*
Time of flight, t	$\frac{2v_o}{g\cos\beta}[\sin(\alpha - \beta)]$	$\frac{2v_o}{g\cos\beta}[\sin(\alpha + \beta)]$
Range	$\frac{v_o^2}{g\cos^2\beta}[\sin(2\alpha - \beta) - \sin(\beta)]$	$\frac{v_o^2}{g\cos^2\beta}[\sin(2\alpha + \beta) + \sin(\beta)]$
Condition for maximum range	$\alpha = 45° + \frac{\beta}{2}$	$\alpha = 45° - \frac{\beta}{2}$
Maximum range	$\frac{v_o^2}{g(1 + \sin\beta)}$	$\frac{v_o^2}{g(1 - \sin\beta)}$

Example 13.23 A boy and a girl of same height stand on an incline of 30° to the horizontal as shown in Fig. 13.26. If the girl throws the ball with an initial velocity of 30 m/s at an angle of 50° to the

horizontal to the boy standing uphill, determine the time of flight of the ball and the position of the boy on the incline to catch the ball. Also, determine the maximum range and the angle of projection required for maximum range.

Solution *Given data*

$$v_o = 30 \text{ m/s}$$

$$\alpha = 50° \text{ and } \beta = 30°$$

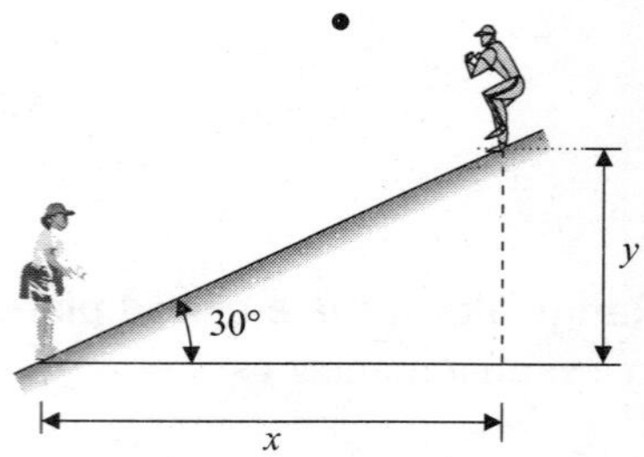

Fig. 13.26

Time of flight

The time of flight is given by

$$t = \frac{2v_o}{g\cos\beta}[\sin(\alpha - \beta)]$$

$$= \frac{2 \times 30}{9.81 \times \cos(30)°}[\sin(50 - 30)°] = 2.42 \text{ s}$$

Range

The position of the boy to catch the ball is given by the range along the incline. Hence,

$$R = \frac{v_o^2}{g\cos^2\beta}[\sin(2\alpha - \beta) - \sin(\beta)]$$

$$= \frac{(30)^2}{9.81 \times \cos^2(30)°}[\sin\{2(50) - 30\}° - \sin\{30\}°] = 53.79 \text{ m}$$

Maximum Range

$$R = \frac{v_o^2}{g(1 + \sin\beta)}$$

$$= \frac{(30)^2}{9.81[1 + \sin(30)°]} = 61.16 \text{ m}$$

The angle of projection for maximum range is

$$\alpha = 45° + \frac{\beta}{2}$$

$$= 45° + (30°/2) = 60°$$

Example 13.24 In the above problem, if the boy throws the ball with an initial velocity of 20 m/s at an angle of 20° to the horizontal to the girl standing downhill, determine the time of flight of the ball and the position of the girl down the incline to catch the ball. Also, determine the maximum range and the angle of projection required for maximum range.

Solution *Given data*

$$v_o = 20 \text{ m/s}$$

$$\alpha = 20° \text{ and } \beta = 30°$$

Time of flight

The time of flight is given by

$$t = \frac{2v_o}{g\cos\beta}[\sin(\alpha + \beta)]$$

$$= \frac{2 \times 20}{9.81 \times \cos(30)^\circ}[\sin(20 + 30)^\circ] = 3.61 \text{ s}$$

Range

The range along the incline is given as

$$R = \frac{v_o^2}{g\cos^2\beta}[\sin(2\alpha + \beta) + \sin(\beta)]$$

$$= \frac{(20)^2}{9.81 \times \cos^2(30)^\circ}[\sin\{2(20) + 30\}^\circ + \sin\{30\}^\circ] = 78.27 \text{ m}$$

Maximum Range

$$R = \frac{v_o^2}{g(1 - \sin\beta)}$$

$$= \frac{(20)^2}{9.81[1 - \sin(30)^\circ]} = 81.55 \text{ m}$$

The angle of projection for maximum range is

$$\alpha = 45^\circ - \frac{\beta}{2}$$

$$= 45^\circ - (30^\circ/2) = 30^\circ$$

13.5 TANGENTIAL AND NORMAL COMPONENTS OF ACCELERATION

In the previous sections, we expressed the acceleration of the particle in rectangular coordinates. This is particularly useful, when the position of the particle is given explicitly in x and y as functions of time, i.e., $x = f(t)$ and $y = g(t)$ respectively. However, if the path of the particle is specified in the form, $y = f(x)$ or if we want to relate the motion of the particle with its path then it will be found convenient to express the acceleration of the particle in terms of *tangential* and *normal* components.

Consider the motion of a particle in curvilinear path. Let its position at a particular instant of time be A and the corresponding instantaneous velocity at that point be $\vec{v}$. We know that the direction of velocity is tangential to the path of the particle. However, it should be noted that the velocity vector is *not* normal to the position vector $\overrightarrow{OA}$ of the particle, which is true only in the case of circular motion.

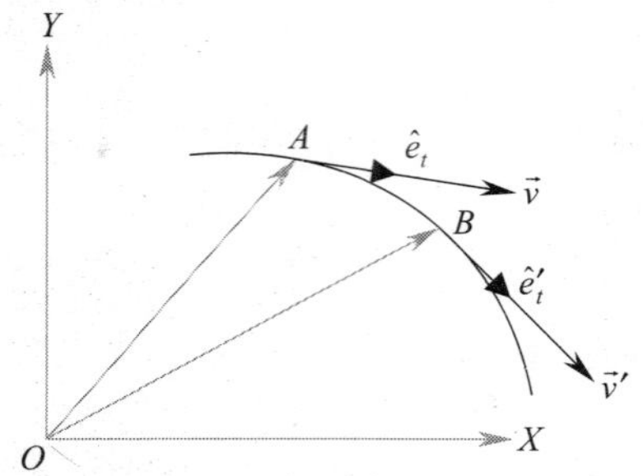

Fig. 13.27(a)

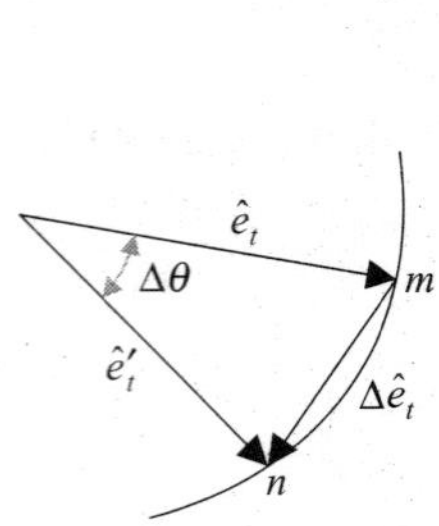

Fig. 13.27(b)

Let $\hat{e}_t$ be the unit vector along the tangential direction at A. Then the velocity vector at A can be written as product of speed of the particle at that instant and the unit vector along the tangential direction: $\vec{v} = v\hat{e}_t$. Unlike in rectangular coordinates, where unit vectors are constant in direction, here the direction of unit vector keeps on changing with time. Hence, at a later instant of time, suppose the particle be at point B, then $\hat{e}'_t$ is the unit vector along the tangential direction to the path of the particle.

Differentiating the velocity vector with respect to time, we get acceleration:

$$\vec{a} = \frac{d\vec{v}}{dt} = \frac{d(v\hat{e}_t)}{dt} = \frac{dv}{dt}\hat{e}_t + v\frac{d\hat{e}_t}{dt}$$

It should be noted in the above expression that both magnitude [v] and direction [$\hat{e}_t$] of the velocity vector change with time. Writing the rate of change of tangential velocity as a_t and expressing $\frac{d\hat{e}_t}{dt}$ by chain rule of differentiation, we have

$$\vec{a} = a_t\hat{e}_t + v\frac{d\hat{e}_t}{d\theta}\frac{d\theta}{ds}\frac{ds}{dt} \tag{13.31}$$

We also know that $\frac{ds}{dt} = v$. Hence, the above equation reduces to

$$\vec{a} = a_t\hat{e}_t + v^2\frac{d\hat{e}_t}{d\theta}\frac{d\theta}{ds} \tag{13.32}$$

To determine the quantities $\frac{d\hat{e}_t}{d\theta}$ and $\frac{d\theta}{ds}$ in the second term on the right-hand side, we proceed as follows:

Redrawing the unit vectors, $\hat{e}_t$ and $\hat{e}'_t$, with a common origin as shown in Fig. 13.27(b), we see that the change in unit vector,

$$\Delta\hat{e}_t = \hat{e}'_t - \hat{e}_t \tag{13.33}$$

As we let B approach A, that is, as the time interval $\Delta t \to 0$, then the angle $\Delta\theta \to 0$. Hence, the arc length mn becomes equal to the chord length mn and the direction of $\Delta\hat{e}_t$ becomes tangential to the direction of $\hat{e}_t$, or in other words, it becomes normal to the path of the particle at A. Let the unit vector along the normal to the path of the particle at A be denoted as $\hat{e}_n$. Hence, in the limiting case, we can write

$$\lim_{\Delta\theta \to 0} \frac{\Delta \hat{e}_t}{\Delta \theta} = \frac{d\hat{e}_t}{d\theta} = (1)\hat{e}_n \quad (13.34)$$

It should be noted that the magnitude of $\frac{d\hat{e}_t}{d\theta}$ is unity because $|\hat{e}_t| = |\hat{e}'_t| = 1$ and $|\Delta \hat{e}'_t| = |\hat{e}_t|\Delta\theta = \Delta\theta$.

From points A and B, if we draw normal to the path of the particle as shown in Fig. 13.28, they will meet at a point say C, called the **centre of curvature** of the path. The distance from the centre of curvature to the point considered is called the **radius of curvature** [ρ].

From the figure, we see that in the limiting case,

$$ds = \rho d\theta \quad (13.35)$$

$$\Rightarrow \quad \kappa = \frac{1}{\rho} = \frac{d\theta}{ds} \quad (13.36)$$

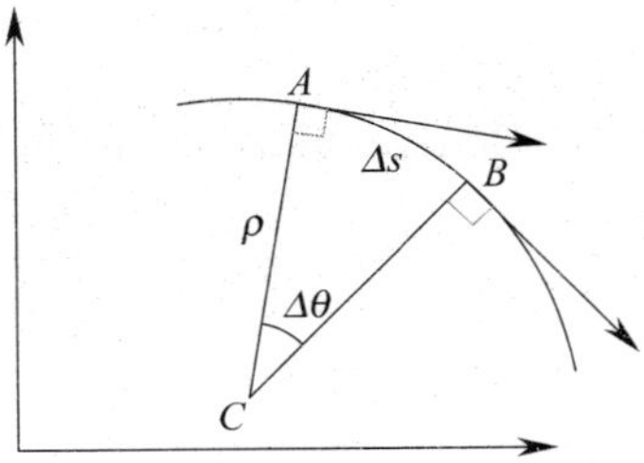

Fig. 13.28 Centre of curvature

where κ, the inverse of radius of curvature is termed the **curvature**. If the path of the particle is known in the form $y = f(x)$ then the curvature is given from calculus as

$$\kappa = \frac{1}{\rho} = \frac{[d^2y/dx^2]}{[1 + (dy/dx)^2]^{3/2}} \quad (13.37)$$

Substituting the values of $\frac{d\hat{e}_t}{d\theta}$ and $\frac{d\theta}{ds}$ respectively from Eqs 13.34 and 13.36 in Eq. 13.32, we have

$$\vec{a} = a_t\hat{e}_t + v^2\hat{e}_n\frac{1}{\rho}$$

$$\vec{a} = a_t\hat{e}_t + \frac{v^2}{\rho}\hat{e}_n \quad (13.38)$$

Hence, the total acceleration can be written as the **sum** of **tangential** and **normal** components. The direction of tangential component is *same* as the direction of velocity vector when the particle is *accelerating* and *opposite* to the direction of velocity vector when the particle is *decelerating*. The direction of normal component of acceleration is **always inward** toward the centre of curvature.

Physical significance of tangential and normal components of acceleration The first term on the right-hand side of the Eq. 13.38, the tangential component of acceleration, arises due to a change in *magnitude* of the velocity vector $\vec{v}$, while the second term, the normal component of acceleration, arises due to a change in *direction* of the velocity vector. Suppose the speed of the particle moving in a curvilinear path is *constant* then the particle will have acceleration even though the tangential component is zero. This is because it has non-zero normal component. The normal component of acceleration can be *zero* only when the radius of curvature of the path tends to infinity or in other words, the particle moves along a straight line, i.e., rectilinear motion.

Corollary If the path of the particle is known in the form $y = f(x)$ then radius of curvature is obtained from the Eq. 13.37. However, if x and y are given explicitly as functions of time, then the radius of curvature of the particle can be expressed in an alternate form as discussed below:

By chain rule of differentiation, we can express

$$\frac{dy}{dx} = \frac{dy}{dt}\frac{dt}{dx} = \frac{\dot{y}}{\dot{x}} \tag{13.39}$$

and

$$\frac{d^2y}{dx^2} = \frac{d}{dx}\left[\frac{dy}{dx}\right] = \frac{d}{dt}\left[\frac{dy}{dx}\right]\frac{dt}{dx}$$

$$= \frac{d}{dt}\left[\frac{\dot{y}}{\dot{x}}\right]\frac{1}{\dot{x}}$$

$$= \frac{\dot{x}\ddot{y} - \ddot{x}\dot{y}}{\dot{x}^2}\frac{1}{\dot{x}} \tag{13.40}$$

Hence, radius of curvature can also be expressed as

$$\kappa = \frac{1}{\rho} = \frac{[d^2y/dx^2]}{[1+(dy/dx)^2]^{3/2}}$$

$$\kappa = \frac{1}{\rho} = \frac{\dot{x}\ddot{y} - \ddot{x}\dot{y}}{[\dot{x}^2 + \dot{y}^2]^{3/2}} \tag{13.41}$$

Example 13.25 A car moving at a constant speed of 60 kmph enters a curved path of radius of curvature measuring 100 m. Determine its total acceleration.

Solution *Given data*

Speed of car, $v = 60 \text{ kmph} = 60 \times 5/18 = 16.67 \text{ m/s}$

Radius of curvature, $\rho = 100 \text{ m}$

As the car moves with constant speed, its tangential component of acceleration is zero, i.e.,

$$a_t = 0$$

However, the normal component of acceleration will not be zero as it passes a curved path. The normal component of acceleration is given as

$$a_n = \frac{v^2}{\rho} = \frac{(16.67)^2}{100} = 2.78 \text{ m/s}^2$$

Hence, the total acceleration is given as

$$a = \sqrt{a_t^2 + a_n^2} = 2.78 \text{ m/s}^2$$

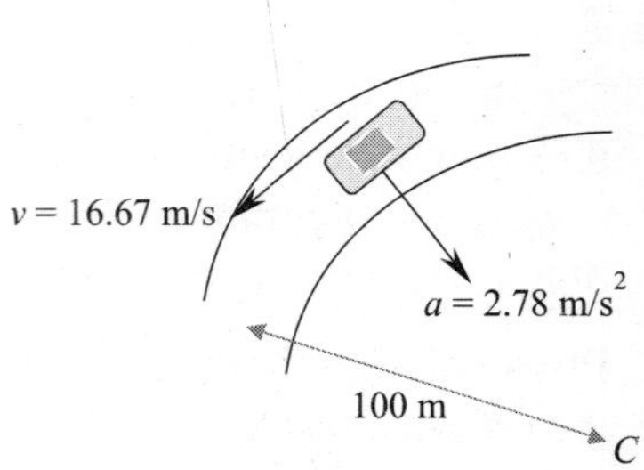

Fig. 13.29

Example 13.26 A truck moving along a curved road with a speed of 36 kmph begins to accelerate at a constant rate to reach a speed of 54 kmph in 5 seconds. Determine the total acceleration (i) at the instant it is accelerated, and (ii) after 3 seconds. Radius of curvature of the road is 90 m.

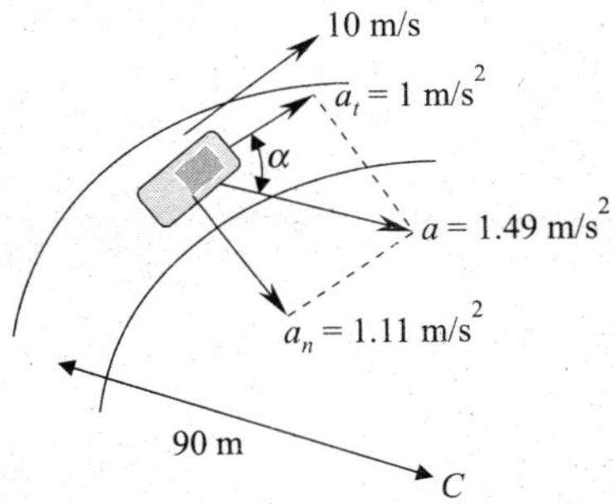

Fig. 13.30(a)

Solution *Given data*

Initial speed of truck, $v_o = 36$ kmph $= 36 \times 5/18 = 10$ m/s

Final speed of truck, $v = 54$ kmph $= 54 \times 5/18 = 15$ m/s

Radius of curvature, $\rho = 90$ m

(i) Total acceleration at the instant it is accelerated

We know, $v = v_o + a_t\, t$

Hence, its tangential component of acceleration is given as

$$a_t = \frac{15-10}{5} = 1 \text{ m/s}^2$$

At the instant it is accelerated, its velocity is 10 m/s. Hence, its normal component of acceleration at that instant is given as

$$a_n = \frac{v^2}{\rho} = \frac{10^2}{90} = 1.11 \text{ m/s}^2$$

Hence, the total acceleration is given as

$$a = \sqrt{a_t^2 + a_n^2}$$

$$= \sqrt{(1)^2 + (1.11)^2} = 1.49 \text{ m/s}^2$$

Inclination of total acceleration with respect to tangential direction is given as

$$\alpha = \tan^{-1}\frac{|a_n|}{|a_t|} = \tan^{-1}\left[\frac{1.11}{1}\right] = 47.98°$$

(ii) Acceleration after 3 seconds

As the truck is constantly accelerated, its tangential component of acceleration after 3 seconds remains the same, i.e.,

$$a_t = 1 \text{ m/s}^2$$

However, its velocity changes and after 3 s, it is given as

$$v = v_o + at$$

$$= 10 + (1)3 = 13 \text{ m/s}$$

Hence, the normal component of acceleration after 3 s is given as

$$a_n = \frac{v^2}{\rho}$$

$$= \frac{13^2}{90} = 1.88 \text{ m/s}^2$$

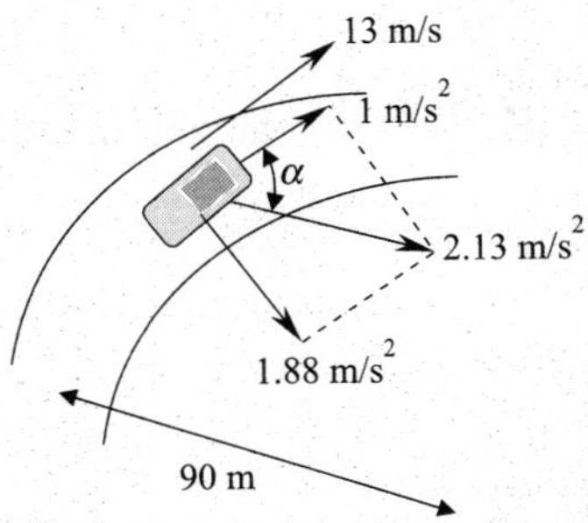

Fig. 13.30(b)

Hence, the total acceleration is given as

$$a = \sqrt{a_t^2 + a_n^2}$$

$$= \sqrt{(1)^2 + (1.88)^2}$$
$$= 2.13 \text{ m/s}^2$$

Inclination of total acceleration with respect to tangential direction is given as

$$\alpha = \tan^{-1}\frac{|a_n|}{|a_t|}$$
$$= \tan^{-1}\left[\frac{1.88}{1}\right] = 62°$$

Example 13.27 A bus moving along a curved road with a constant speed of 45 kmph decelerates at a constant rate to a halt in 10 seconds. Determine the total acceleration at the instant the brake is applied. Radius of curvature is 100 m.

Solution *Given data*

Initial speed of bus, $v_o = 45 \text{ kmph} = 45 \times 5/18 = 12.5 \text{ m/s}$

Final speed of bus, $v = 0 \text{ m/s}$

We know, $v = v_o + a_t t$

Hence, its tangential component of acceleration is given as

$$a_t = \frac{0 - 12.5}{10} = -1.25 \text{ m/s}^2$$

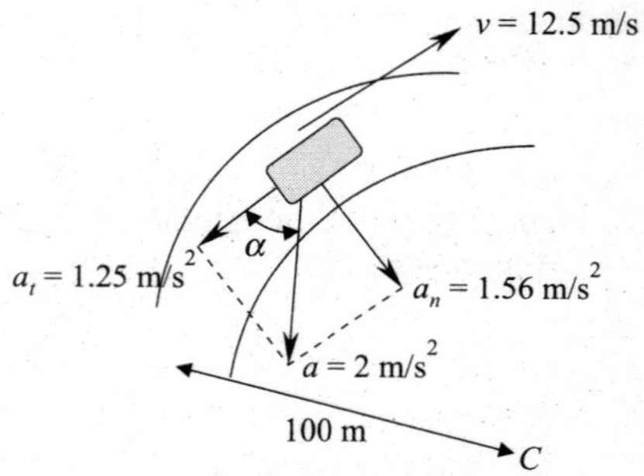

Fig. 13.31

The negative sign indicates that the direction of tangential acceleration is opposite to that of the velocity as shown in Fig. 13.31. Since the velocity at the instant the brake is applied, is 12.5 m/s, the normal component of acceleration at that instant is given as

$$a_n = \frac{v^2}{\rho}$$
$$= \frac{12.5^2}{100} = 1.56 \text{ m/s}^2$$

Hence, the total acceleration is given as

$$a = \sqrt{a_t^2 + a_n^2}$$
$$= \sqrt{(-1.25)^2 + (1.56)^2} = 2 \text{ m/s}^2$$

Inclination of total acceleration with respect to tangential direction is given as

$$\alpha = \tan^{-1}\frac{|a_n|}{|a_t|}$$
$$= \tan^{-1}\left[\frac{1.56}{1.25}\right] = 51.3°$$

Example 13.28 A particle moves along a curvilinear path defined by $y = ax^2$ where x and y are in metres. The velocity and acceleration of the particle at a point (5 m, 2.5 m) are respectively 5 m/s and 2 m/s^2. Determine the total acceleration of the particle at that point.

Solution The equation of path of the particle is given as

$$y = ax^2$$

Given that when $x = 5$ m, $y = 2.5$ m. Hence,

$$a = y/x^2 = 0.1$$

Therefore, the given equation can be written as

$$y = 0.1x^2$$

Differentiating the equation of the curve with respect to x,

$$\frac{dy}{dx} = (0.1)\,2x = 0.2x$$

and

$$\frac{d^2y}{dx^2} = 0.2$$

Hence, at the given point (5 m, 2.5 m),

$$\frac{dy}{dx} = 1 \text{ and } \frac{d^2y}{dx^2} = 0.2$$

Therefore, radius of curvature of the path at the given point is given as

$$\rho = \frac{[1 + (dy/dx)^2]^{3/2}}{d^2y/dx^2}$$

$$= \frac{[1 + (1)^2]^{3/2}}{0.2} = 14.14 \text{ m}$$

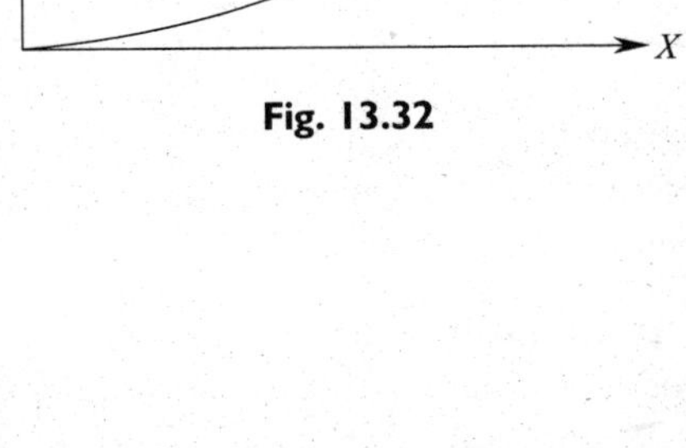

Fig. 13.32

Hence, normal component of acceleration is given as

$$a_n = \frac{v^2}{\rho}$$

$$= \frac{5^2}{14.14} = 1.77 \text{ m/s}^2$$

Therefore, total acceleration is given as

$$a = \sqrt{a_t^2 + a_n^2}$$

$$= \sqrt{(2)^2 + (1.77)^2} = 2.67 \text{ m/s}^2$$

13.6 RADIAL AND TRANSVERSE COMPONENTS OF ACCELERATION

For certain motions, particularly in some of the mechanisms, it is convenient to express the acceleration in terms of *radial* and *transverse* components. As we require two independent coordinates to represent the motion of a particle in a plane, we choose one as the radial position $[\vec{r}]$ of the particle with respect to the origin and the other as the angle of inclination $[\theta]$ of the position vector with respect to the reference axis, namely, X-axis. Such a coordinate system is known as **polar coordinate** system.

Consider a particle moving in a curvilinear path as shown in Fig. 13.33(a). Let it be at a point A at a particular instant of time. Its position is then specified by the radial vector $\vec{r}$ and inclination of $\vec{r}$ with respect to X-axis, i.e., θ. The instantaneous velocity $\vec{v}$ of the particle is tangential to the path at that instant.

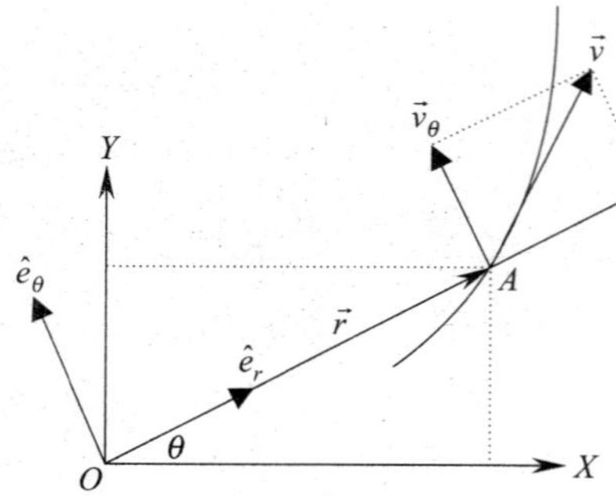

Fig. 13.33(a)

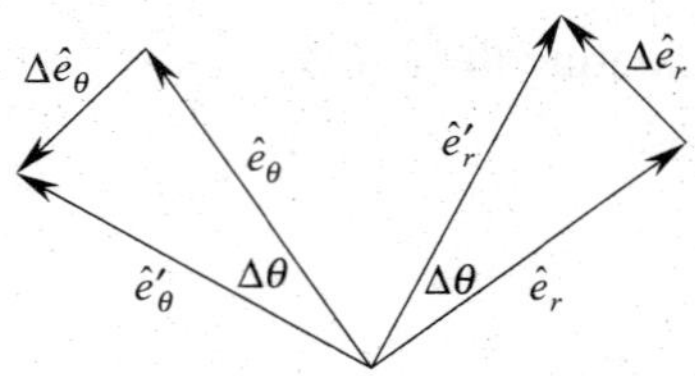

Fig. 13.33(b)

This tangential velocity can be resolved into orthogonal components along the radial and transverse directions. For this, let us consider unit vectors $\hat{e}_r$ and $\hat{e}_\theta$ along the radial and transverse directions respectively as shown in Fig. 13.33(a). It should be noted that they are considered to be positive when acting along increasing directions of r and θ. As the particle moves from point A to another point in a small interval of time, we can see that the directions of unit vectors also change. To determine this change in unit vectors, we proceed as follows.

Draw the unit vectors with a common origin as shown in Fig. 13.33(b). Le the unit vector along the direction of radial vector at a later instant of time be $\hat{e}'_r$ and along the transverse direction be $\hat{e}'_\theta$. As we let the time interval $\Delta t \to 0$ then the angle $\Delta\theta \to 0$. In the limiting case, we have

$$\lim_{\Delta\theta\to 0}\frac{\Delta\hat{e}_r}{\Delta\theta} = \frac{d\hat{e}_r}{d\theta} = \hat{e}_\theta \quad (13.42)$$

and

$$\lim_{\Delta\theta\to 0}\frac{\Delta\hat{e}_\theta}{\Delta\theta} = \frac{d\hat{e}_\theta}{d\theta} = -\hat{e}_r \quad (13.43)$$

That is, in the limiting case, the change in radial unit vector points in the direction of angular unit vector and the change in angular unit vector points in the direction opposite to that of the radial unit vector.

Determination of velocity and acceleration vectors The radius vector can be expressed as a product of the radial distance and the unit vector along that direction, i.e.,

$$\vec{r} = r\hat{e}_r \quad (13.44)$$

Differentiating it with respect to time, we can get the expression for velocity:

$$\vec{v} = \frac{d\vec{r}}{dt}$$

$$= \frac{dr}{dt}\hat{e}_r + r\frac{d\hat{e}_r}{dt}$$

$$= \frac{dr}{dt}\hat{e}_r + r\frac{d\hat{e}_r}{d\theta}\frac{d\theta}{dt}$$

$$= \frac{dr}{dt}\hat{e}_r + r\frac{d\theta}{dt}\hat{e}_\theta$$

$$\vec{v} = \dot{r}\hat{e}_r + r\dot{\theta}\hat{e}_\theta \tag{13.45}$$

Differentiating the above expression with respect to time, we get the expression for acceleration:

$$\vec{a} = \ddot{r}\hat{e}_r + \dot{r}\frac{d\hat{e}_r}{dt} + \dot{r}\dot{\theta}\hat{e}_\theta + r\ddot{\theta}\hat{e}_\theta + r\dot{\theta}\frac{d\hat{e}_\theta}{dt}$$

$$= \ddot{r}\hat{e}_r + \dot{r}\frac{d\hat{e}_r}{d\theta}\frac{d\theta}{dt} + \dot{r}\dot{\theta}\hat{e}_\theta + r\ddot{\theta}\hat{e}_\theta + r\dot{\theta}\frac{d\hat{e}_\theta}{d\theta}\frac{d\theta}{dt}$$

$$= \ddot{r}\hat{e}_r + \dot{r}\dot{\theta}\hat{e}_\theta + \dot{r}\dot{\theta}\hat{e}_\theta + r\ddot{\theta}\hat{e}_\theta - r(\dot{\theta})^2\hat{e}_r$$

$$\vec{a} = [\ddot{r} - r(\dot{\theta})^2]\,\hat{e}_r + [r\ddot{\theta} + 2\dot{r}\dot{\theta}]\hat{e}_\theta \tag{13.46}$$

Example 13.29 The motion of a particle is defined as $r = 2t^2$ and $\theta = t$, where r is in metres, θ is in radians and t is in seconds. Determine the velocity and acceleration of the particle at $t = 2$ s.

Solution Differentiating the radial and angular displacement functions, we have

$$\dot{r} = 4t \qquad \dot{\theta} = 1$$

$$\ddot{r} = 4 \qquad \ddot{\theta} = 0$$

We know velocity vector is given as

$$\vec{v} = \dot{r}\hat{e}_r + r\dot{\theta}\hat{e}_\theta$$

Substituting the values, we have

$$\vec{v} = (4t)\hat{e}_r + (2t^2)\,(1)\hat{e}_\theta$$

Hence, the velocity at $t = 2$ s is obtained as

$$\vec{v} = (4)\,(2)\hat{e}_r + [2(2)^2]\,(1)\hat{e}_\theta$$

$$= 8\hat{e}_r + 8\hat{e}_\theta$$

whose magnitude is 11.31 m/s.

The acceleration vector is given as

$$\vec{a} = [\ddot{r} - r(\dot{\theta})^2]\hat{e}_r + [r\ddot{\theta} + 2\dot{r}\dot{\theta}]\hat{e}_\theta$$

$$= [\,(4) - 2t^2(1)^2]\hat{e}_r + [2t^2(0) + 2(4t)\,(1)]\hat{e}_\theta$$

Hence, the acceleration at $t = 2$ s is obtained as

$$\vec{a} = [(4) - 2(2)^2]\hat{e}_r + [(8)(2)]\hat{e}_\theta$$
$$= -4\hat{e}_r + 16\hat{e}_\theta$$

whose magnitude is 16.49 m/s^2.

Example 13.30 In a telescopic mechanism, the inner cylinder slides within the outer cylinder, which rotates about a fixed axis. If the position of the point *A* on the inner cylinder with respect to time is given as $r = 2 + \sqrt{t}$ and the angular displacement of outer cylinder with respect to time is $\theta = \frac{t^2}{2} + 2$ then determine the total velocity and acceleration of the point *A* at time $t = 2$ s.

Solution *Given* $r = 2 + \sqrt{t}$ and $\theta = \frac{t^2}{2} + 2$

Differentiating the radial position function and angular displacement function with respect to time, we have

$$\dot{r} = \frac{1}{2}t^{-1/2} \qquad \dot{\theta} = t$$

$$\ddot{r} = \frac{-1}{4}t^{-3/2} \qquad \ddot{\theta} = 1$$

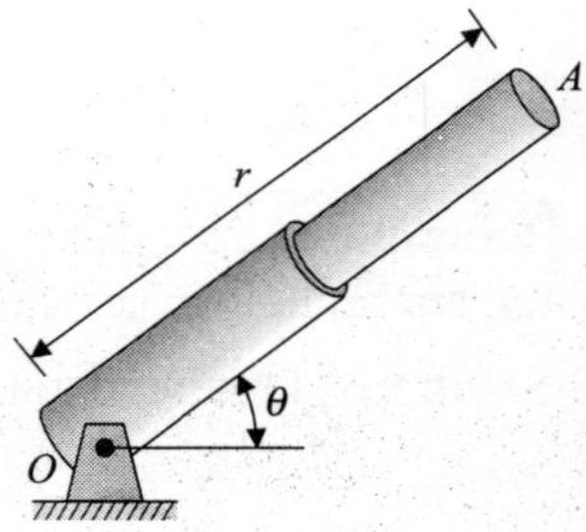

Fig. 13.34

The total velocity of the point *A* is given as

$$\vec{v} = \dot{r}\hat{e}_r + r\dot{\theta}\hat{e}_\theta$$
$$= \frac{1}{2}(t)^{-1/2}\hat{e}_r + (2 + \sqrt{t})t\hat{e}_\theta$$

Hence, the velocity of the point *A* at time $t = 2$ s is obtained as

$$\vec{v}(2) = \frac{1}{2}(2)^{-1/2}\hat{e}_r + (2 + \sqrt{2})2\hat{e}_\theta$$
$$= 0.354\hat{e}_r + 6.828\hat{e}_\theta$$

Similarly, the acceleration of the point *A* is given as

$$\vec{a} = [\ddot{r} - r(\dot{\theta})^2]\hat{e}_r + [r(\ddot{\theta}) + 2\dot{r}\dot{\theta}]\hat{e}_\theta$$
$$= \left[\frac{-1}{4}t^{-3/2} - (2 + \sqrt{t})t^2\right]\hat{e}_r + \left[(2 + \sqrt{t})1 + 2\left(\frac{1}{2}t^{-1/2}\right)t\right]\hat{e}_\theta$$

Hence, the acceleration of the point *A* at time $t = 2$ s is obtained as

$$\vec{a}(2) = \left[\frac{-1}{4}(2)^{-3/2} - (2 + \sqrt{2})2^2\right]\hat{e}_r + \left[(2 + \sqrt{2})1 + 2\left(\frac{1}{2}2^{-1/2}\right)2\right]\hat{e}_\theta$$
$$= -13.745\hat{e}_r + 4.828\hat{e}_\theta$$

Example 13.31 The motion of a rocket fired vertically upwards is tracked by a radar as shown in Fig. 13.35. At a particular instant, it is observed that r = 15 km, θ = 70°, $\dot{\theta}$ = 0.02 rad/s and $\ddot{\theta}$ = 0.001 rad/s^2, determine the velocity and acceleration of the rocket at this instant.

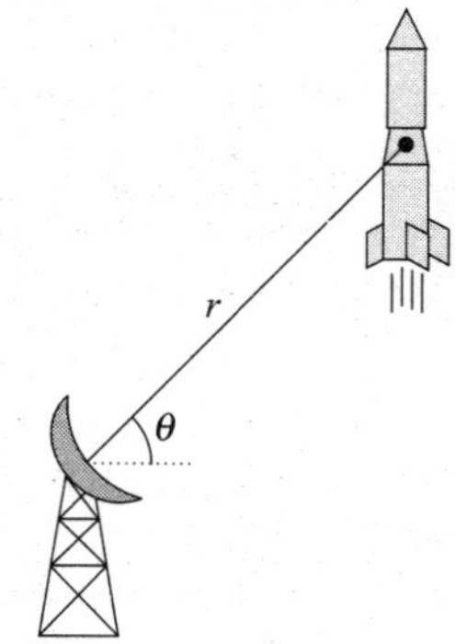

Fig. 13.35

Solution We know that the velocity vector is given as

$$\vec{v} = v_r\hat{e}_r + v_\theta\hat{e}_\theta = \dot{r}\hat{e}_r + r\dot{\theta}\hat{e}_\theta$$

We see that the component of velocity along the radial direction is $v_r = \dot{r}$ and the component along the transverse direction is $v_\theta = r\dot{\theta}$. As the rocket moves vertically upwards, the direction of its velocity is also vertically upwards. Thus, from the velocity triangle, we have

$$\frac{v_\theta}{v} = \cos 70°$$

where $v_\theta = r\dot{\theta}$ = 15 000(0.02) = 300 m/s

Therefore, $$v = \frac{v_\theta}{\cos 70°} = 877.14 \text{ m/s}$$

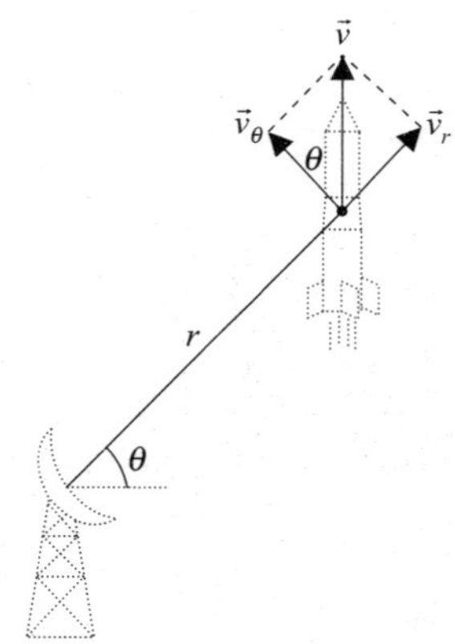

Fig. 13.35(a)

Also, from the velocity triangle, we have

$$\frac{v_r}{v} = \sin 70°$$

Therefore, $$v_r = \dot{r} = v \sin 70° = 824.24 \text{ m/s}$$

We know that the acceleration vector is given as

$$\vec{a} = [\ddot{r} - r(\dot{\theta})^2]\hat{e}_r + [r\ddot{\theta} + 2\dot{r}\dot{\theta}]\hat{e}_\theta$$

whose direction is vertically upwards, as the rocket moves vertically upwards. We can see that the component of acceleration along the transverse direction is

$$a_\theta = r\ddot{\theta} + 2\dot{r}\dot{\theta}$$
$$= (15\ 000)(0.001) + 2(824.24)(0.02) = 47.97 \text{ m/s}^2$$

As the acceleration triangle is similar to the velocity triangle, we have

$$\frac{a_\theta}{a} = \cos 70°$$

$$\Rightarrow \quad a = \frac{a_\theta}{\cos 70°} = 140.25 \text{ m/s}^2$$

Example 13.32 A shaper quick return mechanism shown below rotates at a constant angular speed of 60 r.p.m. in the counterclockwise direction. Determine the angular acceleration of the connecting rod BA (i) when the connecting rod BA makes the maximum oblique angle and (ii) when the crank OA is horizontal.

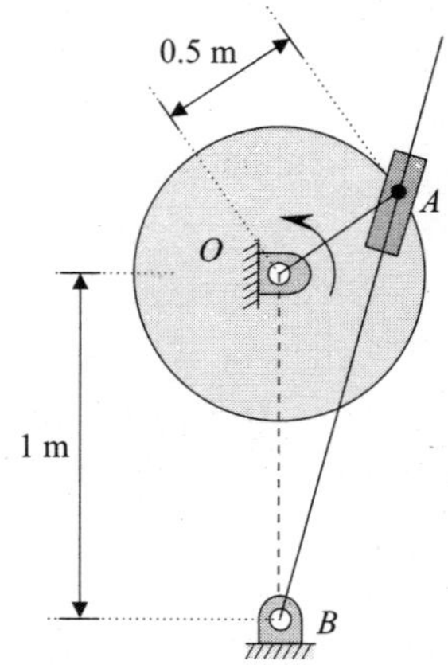

Fig. 13.36

Solution Angular velocity of the crank OA is

$$\omega = \frac{2\pi N}{60} = \frac{2\pi 60}{60} = 6.28 \text{ rad/s}$$

Therefore, tangential velocity and radial acceleration of the point A of the crank OA are

$$v_A = r_{OA}\omega$$

$$= 0.5 \times 6.28 = 3.14 \text{ m/s} \qquad \text{[tangential to the circular path]} \qquad \text{(a)}$$

$$a_A = r_{OA}\omega^2$$

$$= 0.5 \times (6.28)^2 = 19.72 \text{ m/s}^2 \qquad \text{[directed radially inwards]} \qquad \text{(b)}$$

(i) When the connecting rod BA makes maximum oblique angle

We can see from the figure that the connecting rod BA will make maximum oblique angle when BA is tangential to the circular path. Hence, OA and BA must be perpendicular to each other. Therefore,

$$r = AB = \sqrt{1^2 - (0.5)^2} = 0.866 \text{ m}$$

As the motion of the connecting rod can be represented by the polar coordinates, we know that the velocity of the point A on BA is given as

$$\vec{v} = \dot{r}\hat{e}_r + r\dot{\theta}\hat{e}_\theta \qquad \text{(c)}$$

where $\vec{v}_r$ is directed tangential to the path and $\vec{v}_\theta$ is directed radially inwards in the given position. Since the point A is a common point on both OA and BA, the velocity of the end A on the crank OA and velocity of point A on BA must be equal. Comparing equations (a) and (c) then, we have

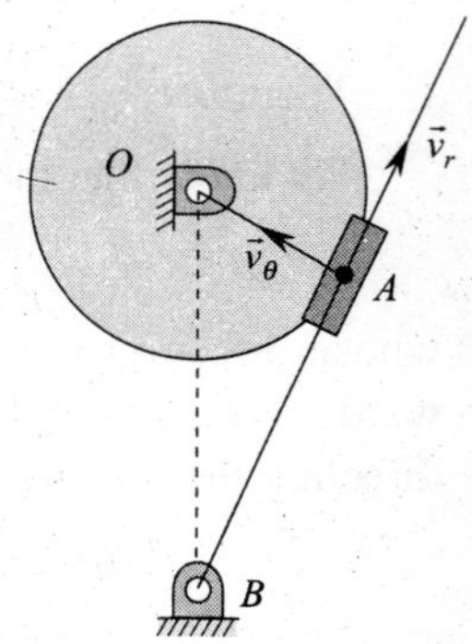

Fig. 13.36(a)

$$v_r = \dot{r} = 3.14 \text{ m/s}$$

and

$$v_\theta = r\dot{\theta} = 0$$

$$\Rightarrow \quad \dot{\theta} = 0$$

Similarly, the acceleration of the point A on BA is given as

$$\vec{a} = [\ddot{r} - r(\dot{\theta})^2]\hat{e}_r + [r\ddot{\theta} + 2\dot{r}\dot{\theta}]\hat{e}_\theta \qquad \text{(d)}$$

Comparing equations (b) and (d), we get angular acceleration of the connecting rod as follows:

$$a_r = \ddot{r} - r(\dot{\theta})^2 = 0$$

and

$$a_\theta = r(\ddot{\theta}) + 2\dot{r}\dot{\theta} = 19.72 \text{ m/s}^2$$

$$(0.866)\ddot{\theta} + 2(3.14)(0) = 19.72$$

$$\Rightarrow \quad \ddot{\theta} = 22.77 \text{ rad/s}^2$$

(ii) When the crank OA is horizontal

When the crank OA is horizontal then

$$\theta = \tan^{-1}\left[\frac{0.5}{1}\right] = 26.57°$$

and
$$r = AB = \sqrt{1^2 + (0.5)^2} = 1.12 \text{ m}$$

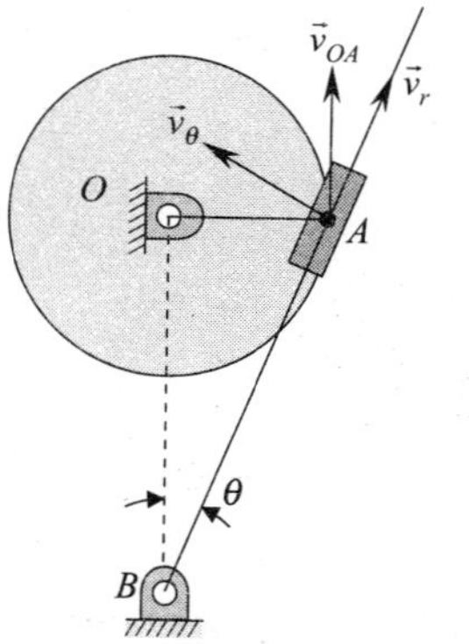

Fig. 13.36(b)

The components of velocity of the end A on OA along radial and transverse directions to BA are

$$v_r = v_{OA} \cos\theta = 2.81 \text{ m/s}$$

$$v_\theta = v_{OA} \sin\theta = 1.4 \text{ m/s}$$

Comparing the values with the equation for velocity, $\vec{v} = \dot{r}\hat{e}_r + r\dot{\theta}\hat{e}_\theta$, we see that

$$v_r = \dot{r} = 2.81 \text{ m/s}$$

$$v_\theta = 1.4 = r\dot{\theta}$$

$\Rightarrow$
$$\dot{\theta} = 1.4/1.12 = 1.25 \text{ rad/s}$$

The components of the acceleration of the end A on OA along radial and transverse directions to BA are

$$a_r = -a_{OA} \sin\theta = -8.82 \text{ m/s}^2$$

[The negative sign indicates that the direction is opposite to that of $\hat{e}_r$].

$$a_\theta = a_{OA} \cos\theta = 17.64 \text{ m/s}^2$$

Comparing the values with the equation for acceleration, $\vec{a} = [\ddot{r} - r(\dot{\theta})^2]\hat{e}_r + [r(\ddot{\theta}) + 2\dot{r}\dot{\theta}]\hat{e}_\theta$, we see that

$$a_\theta = r(\ddot{\theta}) + 2\dot{r}\dot{\theta}$$

$$17.64 = (1.12)\ddot{\theta} + 2(2.81)(1.25)$$

$\Rightarrow$
$$\ddot{\theta} = 9.48 \text{ rad/s}^2$$

SUMMARY

When a particle moves along a curved path other than a straight line, the motion is said to be *curvilinear*. The velocity vector for such a motion is tangential to the path of the particle, while the acceleration is not so. The velocity and acceleration can be represented in rectangular, path (intrinsic) or polar coordinates.

Rectangular Coordinates

The position vector, velocity vector and acceleration vector can be expressed in rectangular coordinates as

$$\vec{r} = x\vec{i} + y\vec{j}$$

$$\vec{v} = \frac{d\vec{r}}{dt} = \frac{dx}{dt}\vec{i} + \frac{dy}{dt}\vec{j}$$
$$= v_x\vec{i} + v_y\vec{j}$$
$$\vec{a} = \frac{d\vec{v}}{dt} = \frac{dv_x}{dt}\vec{i} + \frac{dv_y}{dt}\vec{j}$$
$$= \frac{d^2x}{dt^2}\vec{i} + \frac{d^2y}{dt^2}\vec{j}$$
$$= a_x\vec{i} + a_y\vec{j}$$

where v_x and v_y are respectively x and y components of velocity and a_x and a_y are respectively x and y components of acceleration. The magnitude and direction of instantaneous velocity can be expressed in terms of its components as

$$v = \sqrt{v_x^2 + v_y^2} \quad \text{and} \quad \theta_v = \tan^{-1}\left[\frac{v_y}{v_x}\right]$$

Similarly, the magnitude and direction of instantaneous acceleration in terms of its components are

$$a = \sqrt{a_x^2 + a_y^2} \quad \text{and} \quad \theta_a = \tan^{-1}\left[\frac{a_y}{a_x}\right]$$

Projectile Motion

Projectile motion is a special case of curvilinear motion with constant acceleration occurring in the vertical plane. An example of such a motion is the motion of a missile or a ball hit in air. In the absence of air resistance, wind velocity and rotation of the earth, the motion can be considered to occur in a plane with constant acceleration in the vertically downward direction and with zero acceleration along the horizontal direction.

Such a motion can be considered to be a *combined* motion of two rectilinear motions occurring simultaneously along mutually perpendicular x and y-directions. The equations of motion along the two directions are

MOTION ALONG THE X-DIRECTION	*MOTION ALONG THE Y-DIRECTION*
$a_x = 0$	$a_y = -g$
$v_x = v_o \cos\alpha$	$v_y = v_o \sin\alpha - gt$
	$v_y^2 = (v_o \sin\alpha)^2 - 2gy$
$x = (v_o \cos\alpha)t$	$y = (v_o \sin\alpha)t - \frac{1}{2}gt^2$

The total velocity at any instant of time is obtained by the vector addition of the components at that instant. Its magnitude and direction are

$$v = \sqrt{v_x^2 + v_y^2} \quad \text{and} \quad \theta = \tan^{-1}\left[\frac{v_y}{v_x}\right]$$

The trajectory of the projectile is of the same form as that of a parabola:

$$y = x\tan\alpha - \frac{1}{2}g\frac{x^2}{(v_o\cos\alpha)^2}$$

The time taken to reach the maximum height is

$$t = \frac{v_o\sin\alpha}{g}$$

Since the time of ascent is equal to the time of descent, the total time taken for the projectile to return to the same level of projection is

$$T = \frac{2v_o\sin\alpha}{g}$$

The maximum height reached is

$$h_{max} = \frac{v_o^2\sin^2\alpha}{2g}$$

The horizontal distance between the point of projection and point of return of projectile to the same level of projection is termed the **range**.

$$R = \frac{v_o^2\sin 2\alpha}{g}$$

The condition for maximum range is angle of projection, $\alpha = 45°$ and the maximum range is

$$R_{max} = \frac{v_o^2}{g}$$

Projection on an Inclined Plane

The time of flight, range, maximum range and the condition for maximum range for motion *up* the inclined plane and *down* the inclined plane are as follows:

	Projection up the plane	*Projection down the plane*
Time of flight, t	$\frac{2v_o}{g\cos\beta}[\sin(\alpha-\beta)]$	$\frac{2v_o}{g\cos\beta}[\sin(\alpha+\beta)]$
Range	$\frac{v_o^2}{g\cos^2\beta}[\sin(2\alpha-\beta)-\sin(\beta)]$	$\frac{v_o^2}{g\cos^2\beta}[\sin(2\alpha+\beta)+\sin(\beta)]$
Condition for maximum range	$\alpha = 45° + \frac{\beta}{2}$	$\alpha = 45° - \frac{\beta}{2}$
Maximum range	$\frac{v_o^2}{g(1+\sin\beta)}$	$\frac{v_o^2}{g(1-\sin\beta)}$

Tangential and Normal Components of Acceleration

Acceleration vector can be resolved into orthogonal components along tangential and normal directions as they have a physical significance:

$$\vec{a} = a_t\hat{e}_t + \frac{v^2}{\rho}\hat{e}_n$$

where $\hat{e}_t$ is the unit vector along the tangential direction while $\hat{e}_n$ is the unit vector along the normal to the path of the particle. ρ is the radius of curvature of the path and v is the speed of the particle.

If the path of the particle is known in the form $y = f(x)$ then curvature is given from calculus as

$$\kappa = \frac{1}{\rho} = \frac{d^2y/dx^2}{[1+(dy/dx)^2]^{3/2}}$$

However, if x and y are given explicitly as functions of time then the curvature of the particle can be expressed in an alternative form as

$$\kappa = \frac{1}{\rho} = \frac{\dot{x}\ddot{y} - \ddot{x}\dot{y}}{[\dot{x}^2 + \dot{y}^2]^{3/2}}$$

Radial and Transverse Components of Acceleration

For certain motions, particularly in some of the mechanisms, it is convenient to express the velocity and acceleration in terms of radial and transverse components, also known as **polar coordinate** system.

If radius vector can be expressed as

$$\vec{r} = r\hat{e}_r$$

then velocity and acceleration vectors can be obtained as

$$\vec{v} = \dot{r}\,\hat{e}_r + r\dot{\theta}\hat{e}_\theta$$

$$\vec{a} = [\ddot{r} - r(\dot{\theta})^2]\hat{e}_r + [r\ddot{\theta} + 2\dot{r}\dot{\theta}]\,\hat{e}_\theta$$

EXERCISES

Objective-type Questions

1. A man standing on an open truck moving at a constant speed throws a ball vertically upwards. The ball will fall

(a) behind the truck (b) into his hands
(c) ahead of the truck (d) on the truck, but not into his hands

2. A man standing at the rear end on an open truck moving with uniform acceleration throws a ball vertically upwards. The ball will fall

(a) behind the truck (b) into his hands
(c) ahead of the truck (d) on the truck, but not into his hands

3. A man standing on an open truck moving with uniform retardation throws a ball vertically upwards. The ball will fall

(a) behind the truck (b) into his hands
(c) ahead of the truck (d) none of these

4. The maximum height reached by a ball thrown upwards with an initial velocity v_o at an angle α to the horizontal is

(a) $\frac{v_o^2 \sin 2\alpha}{g}$ (b) $\frac{v_o^2 \sin 2\alpha}{2g}$ (c) $\frac{v_o^2 \sin^2 \alpha}{g}$ (d) $\frac{v_o^2 \sin^2 \alpha}{2g}$

5. The horizontal range of a ball thrown upwards with an initial velocity v_o at an angle α to the horizontal is

(a) $\frac{v_o^2 \sin 2\alpha}{g}$ (b) $\frac{v_o^2 \sin 2\alpha}{2g}$ (c) $\frac{v_o^2 \sin^2 \alpha}{g}$ (d) $\frac{v_o^2 \sin^2 \alpha}{2g}$

6. Horizontal range is defined as the horizontal distance between
(a) point of projection and point of fall on the ground
(b) point of projection and point of fall at the same level
(c) point of projection and point of maximum height
(d) point of maximum height and point of fall on the ground

7. A missile is fired so as to reach maximum range then the maximum height reached is
(a) same as range (b) three-fourth of range
(c) half of range (d) one-fourth of range

8. State which of the following statements is correct:
When a ball is at the highest point of its projectile motion,
(a) its acceleration is zero (b) its velocity is zero
(c) its velocity is directed downward (d) its velocity is directed forward

9. A missile fired at an angle α to the horizontal hits a target. What should be the other angle of projection to hit the same target, initial velocity remaining the same?
(a) 2α (b) $90° + \alpha$ (c) $90° - \alpha$ (d) $45° + \alpha$

10. The maximum range in a projectile motion is obtained when angle of inclination is
(a) 90° (b) 60° (c) 45° (d) 30°

11. The velocity of a particle in projectile motion at the top point of its path is
(a) zero
(b) equal to the initial velocity of projection
(c) equal to the vertical component of initial velocity of projection
(d) equal to the horizontal component of initial velocity of projection

12. When a particle projected from the top of a building strikes the ground away from the building then its horizontal distance is
(a) same as range (b) less than range
(c) greater than range (d) zero

13. The condition for maximum range on an inclined plane (angle of inclination being β), when projected up the plane is
(a) $45° + \beta/2$ (b) $45° - \beta/2$ (c) $45° + \beta$ (d) $45° - \beta$

14. The condition for maximum range on an inclined plane (angle of inclination being β), when projected down the plane is
(a) $45° + \beta/2$ (b) $45° - \beta/2$ (c) $45° + \beta$ (d) $45° - \beta$

15. The normal component of acceleration is zero when motion is
(a) rectilinear
(b) curvilinear
(c) combined translational and rotational
(d) general plane motion

Answers

1. (b)	**2.** (a)	**3.** (c)	**4.** (d)	**5.** (a)	**6.** (b)	**7.** (d)	**8.** (d)
9. (c)	**10.** (c)	**11.** (d)	**12.** (c)	**13.** (a)	**14.** (b)	**15.** (a)	

Short-answer Questions

1. Define curvilinear motion with suitable examples.
2. Express velocity and acceleration vectors in terms of rectangular components.
3. Define projectile motion and state how such a motion can be considered as a combination of two independent motions occurring simultaneously along perpendicular directions.
4. What are the assumptions made in projectile motion?
5. Derive the equation of path of projectile motion.
6. Define range of projectile and the condition for maximum range.
7. Derive the expressions for (i) time of flight, (ii) range when a particle is projected on an inclined plane.
8. Derive the condition for maximum range when a particle is projected on an inclined plane and determine the maximum range.
9. Express the acceleration of a particle in tangential and normal components.
10. Define curvature, centre of curvature and radius of curvature in a curvilinear path.
11. Express radius of curvature in mathematical form.
12. Derive the expressions for velocity and acceleration vectors in radial and transverse components.

Numerical Problems

13.1 The position of a particle moving in curvilinear motion is defined by $y = x^2 + x - 4$, where x and y are in metres. Determine the displacement vector from A to B, where $x_A = 2$ m and $x_B = 3$ m.

Ans. 6.08 m at 80.5° to the horizontal

13.2 The position of a particle moving in curvilinear motion is defined by $y = 25 - x^2/2$, where x and y are in metres. Determine the displacement vector from A to B, where $x_A = 0$ and $x_B = 2$ m.

Ans. 2.83 m at 45° to the horizontal directed downwards.

13.3 The position coordinates of a particle moving in curvilinear motion is defined as $x = t^2$, $y = 2t + 3$. Determine the position, velocity and acceleration of the particle at $t = 2$ s.

Ans. (i) $x = 4$ m, $y = 7$ m; (ii) 4.47 m/s at 26.6° to the horizontal; (iii) 2 m/s^2 along the X-axis

13.4 The velocity of a particle moving in curvilinear motion is defined as $v_x = 4t^2 + 3t + 2$, $v_y = 2t^2 - 3$. Determine the position, velocity and acceleration of the particle at $t = 3$ s if it started from the origin.

Ans. (i) 56.2 m at 9.2° to the X-axis; (ii) 49.3 m/s at 17.7° to the X-axis (iii) 29.5 m/s^2 at 24° to the X-axis

13.5 A missile is fired with an initial velocity of 60 m/s to hit a target that is 300 m away on a level ground. What should be the angle of projection?

Ans. 27.4° (or) 62.6°

13.6 Find the velocity of projection of a ball, which has a horizontal range of 60 m if its time of flight for this range is 3 s.

Ans. 24.8 m/s at an angle of 36.3° to the horizontal

13.7 A passenger travelling in a bus moving at 54 kmph drops a stone through a window from a height of 2 m. Determine when and where it will fall on the ground. What will be the distance between the stone and the bus at the instant it reaches the ground.

Ans. 0.64 s, 9.6 m, 0 m

13.8 A ball is thrown from the top of a building of 30 m height with an initial velocity of 30 m/s at an angle of 30° downwards to the horizontal. Determine the time of flight and the distance from the foot of the building to where it strikes the ground.

Ans. 1.38 s, 35.9 m

13.9 Solve the above problem, if the ball is thrown horizontally.

Ans. 2.47 s, 74.1 m

13.10 Solve the above problem, if the ball is thrown at an angle of 30° upwards to the horizontal.

Ans. 4.44 s, 115.35 m

13.11 The velocity of a particle at the highest point of projection is 10 m/s and the time of flight is 3.5 s. Determine the (i) initial velocity of the particle, (ii) angle of projection, and (iii) maximum height reached.

Ans. (i) 19.9 m/s; (ii) 59.8°; (iii) 15.1 m.

13.12 A ball thrown vertically upwards reaches a maximum height of 15 m. If it is thrown at an angle of 60° to the horizontal with the same initial velocity, (i) determine the maximum height reached, (ii) determine the initial velocity of the ball, if it has to reach the same height as when thrown vertically upwards.

Ans. (i) 11.3 m; (ii) 14.9 m/s

13.13 A ball is thrown upwards from the top of a building with an initial velocity of 20 m/s and at an angle of 30° with the horizontal. The height of the building from the ground level is 25 m. Determine (i) where and when it will strike the ground, ii) velocity with which it strikes the ground, (iii) maximum height reached by the ball above the ground level.

Ans. (i) 3.5 s, 60.6 m form the foot of the building; (ii) 29.9 m/s downwards at an angle of 54.6° to the horizontal; (iii) 30.1 m

13.14 A ball is thrown downwards from the top of a building with an initial velocity of 20 m/s and at angle of 20° with the horizontal. The height of the building from the ground level is 30 m. Determine (i) where and when it will strike the ground, (ii) velocity with which it strikes the ground.

Ans. (i) 1.87 s, 35.2 m from the foot of the building; (ii) 31.4 m/s downwards at an angle of 53.3° to the horizontal

13.15 A missile is fired with an initial velocity of 60 m/s at angle of 40° to the horizontal. What is the position of the particle with respect to the initial point when its total velocity is 50 m/s while ascending?

Ans. $x = 88.2$ m, $y = 56$ m

13.16 If a body is projected at an angle α to the horizon so as just to clear two walls of equal height a and at a distance $2a$ from one another, show that the range is $2a \cot \dfrac{\alpha}{2}$

13.17 A fielder throws a ball with an initial velocity of 30 m/s at an angle of 45° to the horizontal towards the stumps. The ball is caught by the wicketkeeper. The height at which the fielder throws the ball is 1.6 m and the height at which the keeper receives it is 1m. Determine the speed and the angle at which he will receive the ball. Also, determine the distance between them.

Ans. 30.2 m/s at 45.3° downwards, $x = 92.3$ m

13.18 A man standing on a truck at rest throws a ball with a velocity of 20 m/s at an angle of 60° to the horizontal. Determine (i) the horizontal distance between the truck and the ball when it hits the ground, and (ii) the velocity with which it strikes the ground. Neglect the height of the truck.

Ans. (i) 35.3 m, (ii) 20 m/s at 60° downwards

13.19 Solve the above problem if the ball is thrown when the truck moves with a constant velocity of 5 m/s in the direction of motion of the ball. Also, determine how far away from the previous point does it hit the ground.

Ans. (i) 35.3 m; (ii) 22.9 m/s at 49.1° downwards; (iii) 17.65 m

13.20 A missile is fired with a velocity of 60 m/s at an angle of 50° to the horizontal so as to hit a target on the ground. If it is moved 50 m towards the target and then fired, what should be the angle of projection so as to hit the same target?

Ans. 29.03° (or) 60.97°

13.21 A missile is fired with a velocity of 60 m/s at an angle of 50° to the horizontal so as to hit a target on a mountain that is 100 m above the ground level and the missile hits the target during upward motion. Determine the horizontal distance of the target from the point of projection.

Ans. 132.4 m

13.22 A bomb is released from an airplane moving horizontally at a constant speed of 400 kmph, when it is at an altitude of 800 m from the ground level. Determine how far from the target must the bomb be released.

Ans. 1.42 km

13.23 A missile is fired with a velocity of 60 m/s at an angle of 45° to the horizontal so as to hit a high-rise building at the highest point of its trajectory. If it has to be fired at 60° to the horizontal to hit the same point, determine how much further must it move toward the building.

Ans. 116.3 m

13.24 A boy and a girl of same height stand on an incline of 30° to the horizontal. If the girl throws the ball with an initial velocity of 15 m/s at an angle of 50° to the horizontal to the boy standing uphill, determine the time of flight of the ball and the position of the boy on the incline to catch the ball. Also, determine the maximum range and the angle of projection required for maximum range.

Ans. 1.21 s, 13.45 m, 15.3 m and 60°

13.25 In the above problem, if the boy throws the ball with an initial velocity of 15 m/s at an angle of 20° to the horizontal to the girl standing downhill, determine the time of flight of the ball and the position of the girl down the incline to catch the ball. Also, determine the maximum range and the angle of projection required for maximum range.

Ans. 2.71 s, 44 m, 45.9 m and 30°

13.26 A particle moving in curvilinear motion describes the path $y = 3x^2$, where x and y are in metres. Determine the radius of curvature of the path at $x = 2$ m. If the particle is moving at a constant speed of 15 m/s, determine acceleration at that point.

Ans. 291 m, 0.773 m/s^2 normal to the path of the particle at that instant.

13.27 A particle moving in curvilinear path has positions with respect to time as $x = 2t + 1$ and $y = t^2$, determine the radius of curvature of the path at $t = 2$ s.

Ans. 22.4 m

13.28 A ball is projected upwards with an initial velocity of 20 m/s at an angle of 30° to the horizontal. Determine the radius of curvature of the path of the ball at the highest point.

Ans. 30.6 m

13.29 In the above problem, determine the radius of curvature of the path of the ball at the point where (i) $x = 10$ m, and at (ii) $x = 30$ m.

Ans. (i) 33.5 m, (ii) 38.4 m

13.30 A particle moving in curvilinear motion describes the path, $y^2 = 4ax$. Determine the radius of curvature of the path (i) at $x = a$ and (ii) at $x = 2a$.

Ans. (i) $5.66a$; (ii) $10.4a$

13.31 A particle moving in curvilinear motion describes the path, $y = e^x$, where x and y are in metres. Determine the radius of curvature of the path at $x = 2$ m.

Ans. 56.1 m

13.32 A particle moving in curvilinear motion describes the path, $y = x^3$, where x and y are in metres. Determine the radius of curvature of the path at $x = 1$ m and $x = 2$ m.

Ans. 5.27 m, 145.5 m

13.33 A motorist enters a curved road with a 90 m radius of curvature with a velocity of 90 kmph and reduces the speed uniformly to 60 kmph in 10 seconds. Determine the total acceleration at the end of 5 seconds.

Ans. 4.9 m/s^2 at 80.2° to the opposite direction of velocity

13.34 A motorist waiting at a traffic signal takes a right turn when the light turns green. If he accelerates at a constant rate of 1 m/s^2, determine his acceleration when he exits the quarter circular path *AB* of 25 m radius at *B*.

Ans. 3.3 m/s^2

13.35 At a traffic intersection, a motorist moving with a constant speed of 30 kmph takes a free left turn. What is the total acceleration at any point between A and B, the end points of quarter circular path? Take radius of curvature as 15 m.

Ans. 4.63 m/s^2

13.36 A particle moving in curvilinear motion at a constant speed of 10 m/s describes the path, $y = x^2$, where x and y are in metres. Determine (i) the radius of curvature of the path at $x = 1/3$ m, (ii) components of velocity at the point, and (iii) acceleration at that point.

Ans. (i) 0.868 m; (ii) $v_x = 8.32$ m/s, $v_y = 5.55$ m/s; (iii) $a_t = 0$, $a_n = 115.2$ m/s^2.

13.37 The shaper quick return mechanism is shown in Fig. E.13.37. The crank O_2A rotates in the anti-clockwise direction with a constant angular velocity of 30 rpm. Determine the angular velocity and angular acceleration of the connecting rod O_1B at the position shown. $\overline{O_1B} = 8.5$ cm.

Ans. 0.9 rad/s, 1.04 rad/s^2

13.38 In the slider crank mechanism shown in Fig. E.13.38, the slider B moves upwards along the crank OA at a constant speed of 3 m/s, while the crank rotates at constant speed of 30 rpm in the clockwise direction. Determine the velocity and acceleration of the slider, if the radial position of the slider at that instant is 1.5 m.

Ans. 5.6 m/s at 57.5° to the crank; 24 m/s^2 at 51.9° to the crank

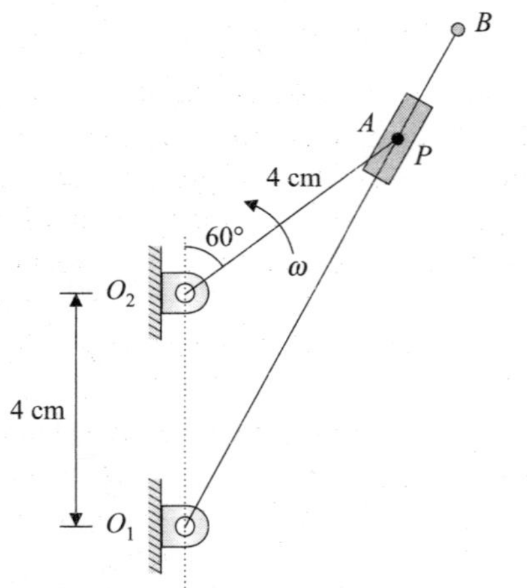

Fig. E.13.37

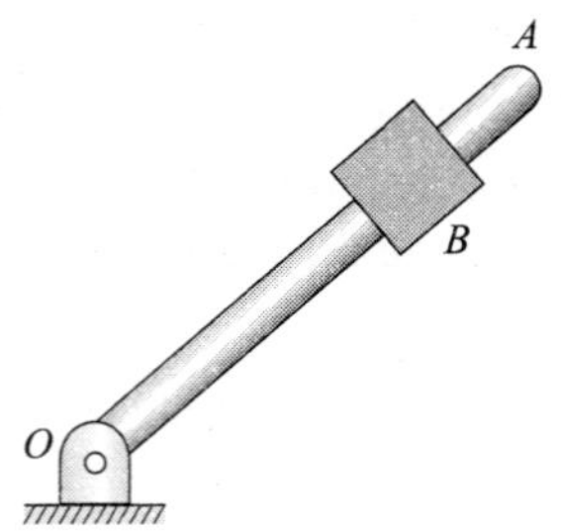

Fig. E. 13.38, E.13.39

13.39 In the above problem, the radial position of the slider at a particular instant is 1.5 m; its velocity is 2 m/s and acceleration 1 m/s^2 both radially outward. The angular velocity and angular acceleration of the crank are respectively 2 rad/s and 1.5 rad/s^2 in the anti-clockwise direction. Determine the total velocity and acceleration of the slider. Refer Fig. E.13.39.

Ans. 3.61 m/s at 56.3° to the crank; 11.4 m/s^2 at 64° to the crank

13.40 The crank O_2A of an oscillating arm quick-return mechanism shown in Fig. E.13.40 rotates at a constant angular velocity of 10 rad/s in the clockwise direction. Determine (i) the angular velocity, and (ii) angular acceleration of the arm O_1B, when O_2A is vertical.

Ans. (i) 1 rad/s; (ii) 24 rad/s^2

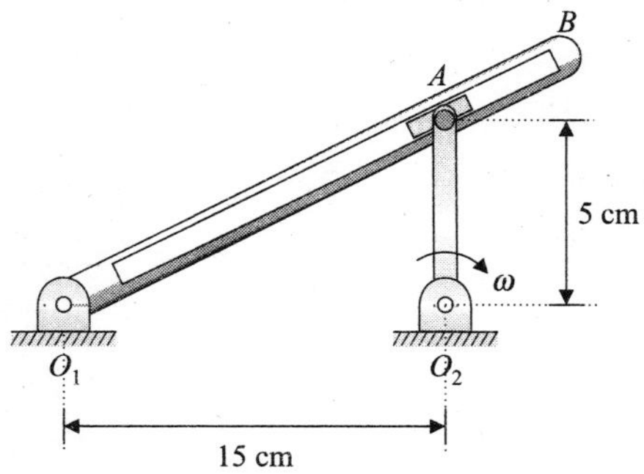

Fig. E.13.40

14 Kinetics of Particles

14.1 INTRODUCTION

In the previous two chapters, we discussed the *kinematics* of particles, where we described the motion of the particle by stating its position, velocity and acceleration at different instants of time. In this chapter, we will discuss what actually *causes* such types of motions and *relate* the cause of motion with the resulting motion. The branch of dynamics, which deals with the motion of particles *without considering* the cause of motion, i.e., just the geometry of motion is termed **kinematics**; whereas the branch of dynamics which deals with the motion of particles *considering* together the cause of motion is termed **kinetics**.

In the following section, we will discuss the laws, which govern the motion of particles. These laws were put forth by Sir Isaac Newton in the sixteenth century. They can also be expressed mathematically and they are known as **kinetic equations of motion** just like the *equations of equilibrium* seen under statics. In Sections 14.3–14.4, we will discuss the kinetic equations of motion in rectangular coordinates for rectilinear motion with *constant* acceleration and in Section 14.6 for rectilinear motion with *variable* acceleration. In Section 14.5, we will discuss an alternative way of expressing the equation of motion as given by D'Alembert and it is known as D'Alembert's principle. It solves dynamics problems by applying the equilibrium condition. In Section 14.7, we will discuss the forces causing the acceleration in curvilinear motion resolving the motion along tangential and normal directions and establish the corresponding kinetic equations of motion.

14.2 LAWS OF MOTION

The early contributions to the development of statics could be dated back to the time before Christ. However, the contributions to the development of dynamics can be considered to have started only from the time of Galileo (1564–1642). The reason for this delay was due to lack of proper time measuring devices such as pendulum clock and balance wheel watch, which were not developed at that time.

Galileo's experimental studies on motion of bodies, namely, the one at the leaning tower of Pisa, blocks sliding down an inclined plane, etc., revealed the fallacies in the beliefs held till then and later paved the way for Sir Isaac Newton (1642–1727) to put forth his famous laws of motion. The branch of dynamics was also further refined by the contributions of mathematicians like Sir. W. Hamilton, J.L. Lagrange and Pierre S. De Laplace, who developed formulas that described motion in dynamic systems through time.

Before Galileo's time, most philosophers, including Aristotle held the view that some *force* was *always* needed to keep a body *moving*; otherwise, it would come to *rest*. This they believed based on the experiment on motion of a block on a horizontal plane. Suppose we push a block, which is at rest on a horizontal plane, it causes the motion of the block. However, when the force is removed, which in this case is the hand, it can be observed that the block slows down and comes to rest. Hence, they believed that bodies at rest remained at rest unless a force acted on them, but that bodies in motion did not continue in that motion unless the force continued to act on it. Hence, they concluded that a **force** must always **continue** to act on a body to keep it in **motion**; otherwise, it would come to **rest**.

The first part of the conclusion, that is, a body at rest must be acted on by an external force to make it move could be easily accepted. However, a body in motion also requires an external force to keep it moving was hard to accept for Galileo. He realized that the analysis of Aristotle was incorrect, because it failed to account properly for a hidden force, namely, the **frictional force** between the surface and the body. He realized that the reduction in velocity is not due to the removal of the applied force but due to the action of a hidden force, namely, the frictional force and air resistance coming into play.

He conducted experiments with a smoother block placed on a smoother plane and a lubricant placed in between the sliding surfaces, and found that the block moved a further distance before coming to rest. With still smoother contact surfaces, he found that the block moved still further before it came to rest. In this way, he argued that if the total friction could somehow be eliminated, the body once set in motion would continue to do so *forever*, even if the force, which caused the motion, is removed. Thus, he concluded that force is always needed to change the state of rest of a body, but once motion is set, force is not needed to keep it in motion.

Based on this observation, he introduced the concept of **inertia** stating that an object in a state of rest or of motion possesses an "inertia" that causes it to remain in that state of rest or of motion unless an external force acts on it. Stated in simple words, if a body is at rest, it cannot of itself change that state of rest. Similarly, if a body is moving with a uniform velocity along a straight line, the body is incapable of altering the magnitude or the direction of its motion unless an external force acts on it. Thus, inertia can be defined *as the tendency of a body to resist any attempt to change the state of rest or of motion.*

To make this point clear, let us consider for instance a railway coach at rest. It would remain at rest forever unless it is pulled by a locomotive. Once set in motion, it would continue to do so, even if it is not pulled by the locomotive. However, we observe that this does not happen and the coach comes to rest after some interval of time. This is because the motion is retarded by an external force, namely, the frictional force between the wheel and the rail and also by air resistance.

First-Law of Motion Sir Isaac Newton (1642–1727), who was born the same year that Galileo died, adopted the ideas of Galileo and had put forth his three laws of motion and the universal law of gravitation. Based on the above observation of Galileo, Newton stated his first law of motion as follows: *Every body continues in its state of rest or of uniform motion in a straight line unless it is compelled to change that state by forces acting on it.*

Stated in an alternate way, if no external force acts on a body, it will not have *acceleration*, i.e., its velocity will remain *constant*. If a body were initially at rest, it would continue to remain at rest until a net external force acts on it. Similarly, if it were initially moving with constant velocity along a straight line, it would continue to do so until a net external force acts on it. Hence, to cause a change in velocity or in other words, to accelerate a body, an external force must act on it.

Second-Law of Motion If a resultant force acts on a body at rest or moving with constant velocity along a straight line, it will always cause a change in velocity or in other words **accelerate** the body. Depending upon the direction of the resultant force acting on it, the body either *accelerates* or *decelerates* or *changes its direction* of motion. If the force acts on a moving body in the same direction as that of the motion then the body will accelerate. If it acts in the direction opposite to that of the direction of motion, it will cause deceleration. If it acts in a direction other than that of the direction of motion, it will cause a change in direction of motion.

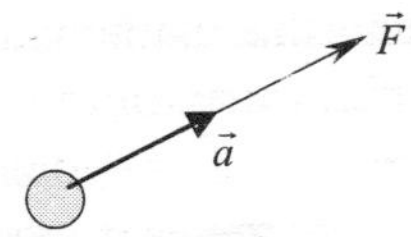

Fig. 14.1 Acceleration in the direction of force

The relationship between the external force acting on a particle and the resulting motion was stated by Newton in his second law of motion. It relates the force causing the change in motion with the resulting acceleration and it can be stated as follows: *If the resultant force acting on a particle is not zero, the particle will have an acceleration proportional to the magnitude of the resultant and in the direction of this resultant force.*

The mathematical expression for this law can be derived based on experimental studies. The following observations could be made based on the experimental studies on the motion of a block acted upon by a force:

(i) When a force of *constant* magnitude and direction is applied on a block resting on a frictionless surface, it can be found that its acceleration is also of *constant* magnitude and that it is in the *same* direction as that of the applied force.

(ii) Suppose the magnitude of the force is doubled or tripled. Then the acceleration also **increases proportionately** and the motion of the block is in the same **direction** as that of the applied force.

(iii) Further, if a number of forces act on a block then it can be found that each force produces a proportional acceleration as if it is acting alone. The resultant acceleration is proportional to the magnitude of the resultant force and the motion is in the direction of the resultant force.

(iv) If the same force is applied on different bodies, it can be seen that it produces different accelerations on different bodies. This is because different bodies offer different *resistances* to change in motion. This property of the body, offering resistance to change in motion as discussed above is the *inertia*. The physical quantity, which measures this property, is defined as **mass** of the body. If the mass of the body is greater then it offers greater resistance to change in motion and thus, its acceleration is lesser and vice versa. Thus, we can see that for a given force, acceleration is **inversely** proportional to the **mass** of the body.

Combining all of the above observations, we can see that acceleration is proportional to the applied force and is in the direction of the force, and inversely proportional to the mass of the particle and this can be mathematically stated as follows:

$$\vec{a} \propto \vec{F} \quad \text{and} \quad \vec{a} \propto \frac{1}{m} \tag{14.1}$$

Hence, we can write

$$\vec{a} \propto \frac{\vec{F}}{m} \tag{14.2}$$

Introducing a constant of proportionality c, we have

$$\vec{a} = c\frac{\vec{F}}{m} \tag{14.3}$$

By choosing a proper system of units such that unit force acting on unit mass produces unit acceleration, we make the constant of proportionality c equal to 1. Hence, we can also write

$$\vec{F} = m\vec{a} \tag{14.4}$$

which is the mathematical expression of Newton's second law of motion, where $\vec{F}$ represents the resultant of a system of forces acting on a particle. A special case of the second law is that if the resultant force is zero, i.e., $\vec{F} = \vec{O}$ then acceleration is also zero, $\vec{a} = \vec{o}$, which is same as the first law. Hence, first law can be treated as a special case of second law.

Alternatively, we can also write the Eq. 14.4 as

$$\vec{F} = m\,\vec{a} = m\frac{d\vec{v}}{dt} \tag{14.5}$$

Since the mass m of the particle is constant, which is true in most of the situations we encounter except in those cases such as motion of rockets, we can take it inside the differential sign and hence we can write the above expression as

$$\vec{F} = \frac{d(m\vec{v})}{dt} \tag{14.6}$$

The product of mass and velocity vector is defined as **linear momentum** of the particle. Thus, Newton's second law of motion can also be stated as that *the rate of change of momentum of a particle is proportional to the resultant force acting on the particle and is in the same direction.* Actually, Newton stated his second law of motion in the above form.

It is worthwhile mentioning here that Newton's law has been found to hold good for motion of gross bodies. This is the size of bodies that we normally deal in the study of mechanics. However, it fails in the motion of minute particles such as electrons and particles reaching the speed of light.

14.3 MOTION OF BODIES IN RECTANGULAR COORDINATES

If force and acceleration vectors can be resolved into rectangular components along x and y directions for plane motion then Newton's second law can also be expressed in scalar forms as:

$$\sum F_x = ma_x \tag{14.7}$$

and

$$\sum F_y = ma_y \tag{14.8}$$

where F_x and F_y are respectively x and y components of the individual forces in a system of forces. If the resultant force is constant in magnitude and direction then the body is under constantly accelerated rectilinear motion in the direction of the resultant. The motion of a body in free fall is an example for such type of motion. Here the body is subjected to constant force of gravity, resulting in constant acceleration due to gravity vertically downwards. If the resultant force is varying in magnitude, but constant in direction then the body is under non-uniformly accelerated rectilinear motion in the direction of the resultant. If the resultant force is varying in direction and magnitude either constant or varying then it has a curvilinear motion.

While solving problems in statics, we initially chose some axes of reference and accordingly the sign of the forces were considered, i.e., if they pointed along positive axes, forces were taken as positive; and negative if they pointed along negative direction of axes. However, here in dynamics, it will be found convenient to choose the initial direction of motion of the bodies as positive, irrespective of whether they point along positive or negative direction of the axes. Hence, throughout this book, we have followed this sign convention.

Example 14.1 The displacement of a body of 5 kg mass with respect to time is given as $x = 3t^2 + 2$, where x is in metres and t is in seconds. Determine the force acting on it causing the motion.

Solution Given that the displacement of the body with respect to time is $x = 3t^2 + 2$, its velocity and acceleration could be determined by differentiating successively the above expression with respect to time.

$$v = \frac{dx}{dt} = 6t$$

and
$$a = \frac{d^2x}{dt^2} = 6$$

We observe that the acceleration is of constant magnitude and hence the force causing this change in velocity or acceleration is also of constant magnitude. We know that these two are related by the Newton's second law of motion. Hence, the force acting on the body is obtained as

$$F = ma$$
$$= (5)(6) = 30 \text{ N}$$

Example 14.2 A block of 150 N weight is resting on a rough horizontal table. What horizontal force P is required to move the block with an acceleration of 1.5 m/s^2? The coefficient of kinetic friction between the contact surfaces is 0.2.

Solution The free-body diagram of the block is shown in Fig. 14.2(a). The forces acting on the block are its weight mg, normal reaction N, frictional force $\mu_k N$ acting in the direction opposite to the direction of motion and the applied force P. Writing the equations of motion along the X and Y directions, taking the initial direction of motion as positive:

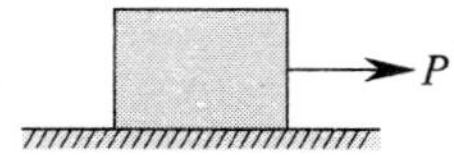

Fig. 14.2

$$\sum F_y = ma_y \Rightarrow$$
$$N - mg = 0$$
$$\therefore \quad N = mg$$

[Note that as the block moves horizontally, its acceleration along the Y-direction, i.e., a_y is zero.]

$$\sum F_x = ma_x \Rightarrow$$
$$P - \mu_k N = ma$$
$$P - \mu_k mg = ma$$
$$\Rightarrow \quad P = \mu_k mg + ma$$

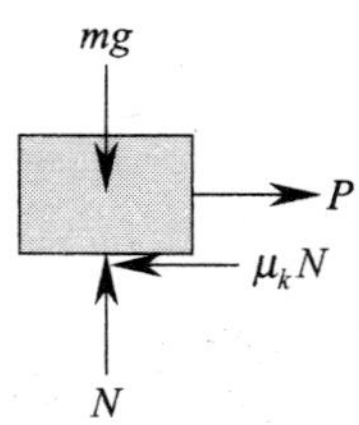

Fig. 14.2(a)

Substituting the values, we get

$$P = (0.2)(150) + \frac{150}{9.81}(1.5) = 52.94 \text{ N}$$

Example 14.3 A block of 100 N weight is resting on a rough horizontal table. What force P inclined at 30° to the horizontal is required to move the block horizontally with an acceleration of 2 m/s^2? The coefficient of kinetic friction between the contact surfaces is 0.2.

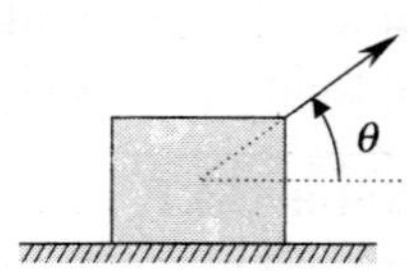

Fig. 14.3

Solution The free-body diagram of the block is shown in Fig. 14.3(a). Writing the equations of motion along the X and Y directions, taking the initial direction of motion as positive:

$$\sum F_y = ma_y \Rightarrow$$

$$N + P\sin\theta - mg = 0$$

$$\therefore \quad N = mg - P\sin\theta$$

[Note that as the block moves horizontally, its acceleration along the Y-direction is zero.]

$$\sum F_x = ma_x \Rightarrow$$

$$P\cos\theta - \mu_k N = ma$$

$$P\cos\theta - \mu_k(mg - P\sin\theta) = ma$$

$$\Rightarrow \quad P = \frac{\mu_k mg + ma}{\mu_k \sin\theta + \cos\theta}$$

$$= \frac{(0.2 \times 100) + (100/9.81)(2)}{(0.2)\sin 30° + \cos 30°} = 41.81 \text{ N}$$

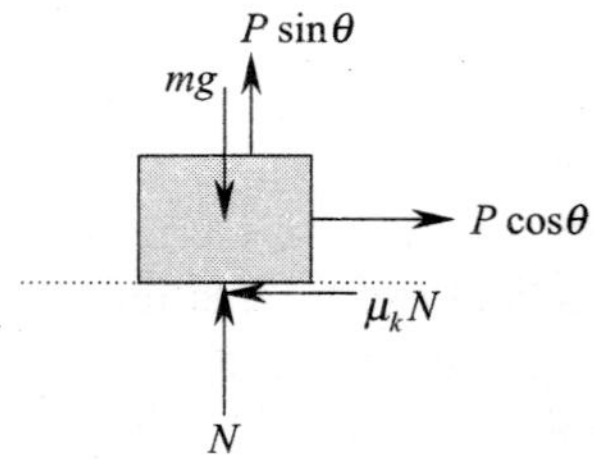

Fig. 14.3(a)

Example 14.4 A block of 2 kg mass rests on a rough horizontal surface, whose coefficient of kinetic friction is 0.2. It is acted on by a horizontal force of 10 N for 5 seconds and then it is removed. Determine how far it would travel before coming to rest, assuming the frictional resistance to be uniform. Also, determine the total distance travelled starting from rest.

Solution The free-body diagram of the block is shown in Fig. 14.4(a). Writing the kinetic equations of motion along the X and Y directions, taking the initial direction of motion as positive:

$$\sum F_y = ma_y \Rightarrow$$

$$N - mg = 0$$

$$\therefore \quad N = mg$$

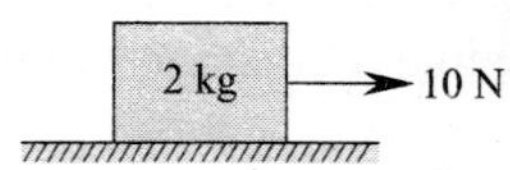

Fig. 14.4

$$\sum F_x = ma_x \Rightarrow$$

$$P - \mu_k N = ma$$

$$10 - (0.2)(2)(g) = 2a$$

$$\Rightarrow \quad a = 3.04 \text{ m/s}^2$$

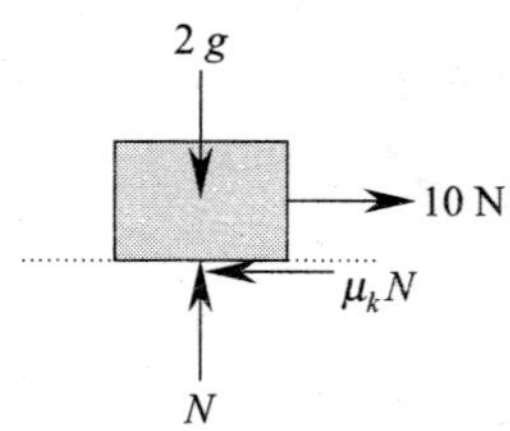

Fig. 14.4(a)

Applying the kinematic equation of motion of the block,

$$s = v_o t + \frac{1}{2}at^2$$

Therefore, the distance travelled until the force is applied is given as

$$s(5) = 0 + \frac{1}{2}(3.04)(5)^2 = 38 \text{ m}$$

and its velocity at the instant the force is removed is given as

$$v = v_o + at$$

$$v(5) = 0 + (3.04)(5) = 15.2 \text{ m/s}$$

Motion after the force is removed

Once the applied force is removed, the frictional force acting against the direction of motion brings the block to a halt. Hence, the kinetic equation of motion can be written as

$$-f = ma'$$

$$\Rightarrow \quad a' = \frac{-f}{m} = \frac{-\mu_k mg}{m} = -1.962 \text{ m/s}^2$$

Hence, using the kinematic equation of motion,

$$v^2 = v_o^2 + 2a's'$$

we get $$s' = \frac{v^2 - v_o^2}{2a'} = \frac{0-(15.2)^2}{2(-1.962)} = 58.88 \text{ m}$$

Therefore, the total distance travelled is 38 + 58.88 = 96.88 m.

Example 14.5 A block of 25 kg mass at rest on an inclined plane is pulled up by a force of 175 N magnitude acting parallel to the inclined plane. Determine the acceleration of the block, if the coefficient of kinetic friction between the block and the plane is 0.3.

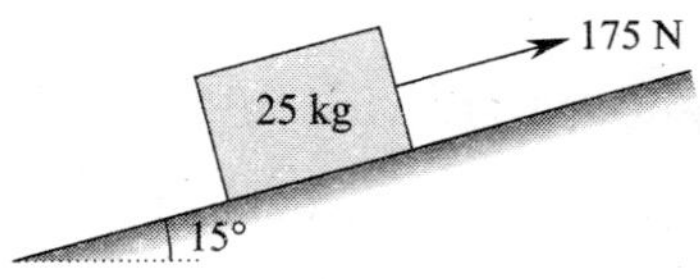

Fig. 14.5

Solution The free-body diagram of the block is shown in Fig. 14.5(a). Writing the equations of motion along the X and Y directions, taking the initial direction of motion as positive:

$$\sum F_y = ma_y \quad \Rightarrow$$

$$N - mg\cos 15° = 0$$

$$\therefore \quad N = mg\cos 15°$$

[Note that as the block moves along the plane, its acceleration normal to the plane is zero.]

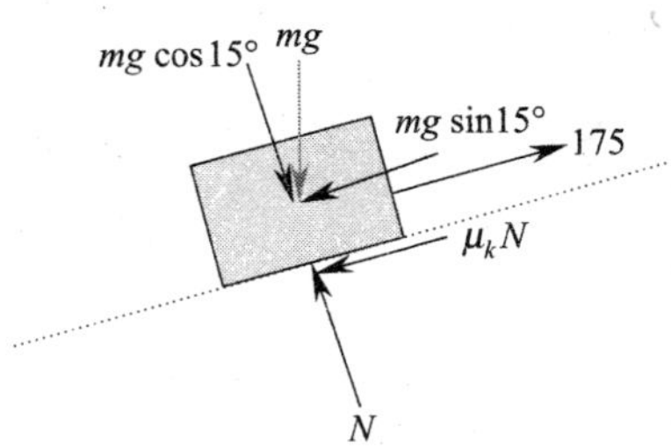

Fig. 14.5(a)

$$\sum F_x = ma_x \Rightarrow$$

$$175 - mg\sin 15° - \mu_k N = ma$$

$$175 - mg\sin 15° - \mu_k\, mg\cos 15° = ma$$

$$175 - 25 \times g\sin 15° - (0.3)\, 25 \times g\cos 15° = 25a$$

$$\Rightarrow \quad a = 1.618 \text{ m/s}^2$$

Example 14.6 A 20 kg block rests on a rough horizontal surface over which another block of 10 kg mass is placed. A horizontal force P is applied to the lower block. Determine the maximum force that can be applied without allowing the upper block to slide backward over the lower block. Assume coefficient of static and dynamic friction to be respectively 0.25 and 0.2 for all contact surfaces.

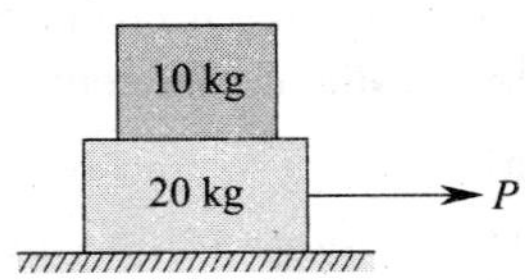

Fig. 14.6

Solution The free-body diagrams of the two blocks are shown in the adjacent figures. The upper block is also pulled along with the lower block due to friction between the blocks. This friction is a sliding friction, while the friction between the lower block and the plane is kinetic in nature due to motion. Let a be the acceleration of the blocks.

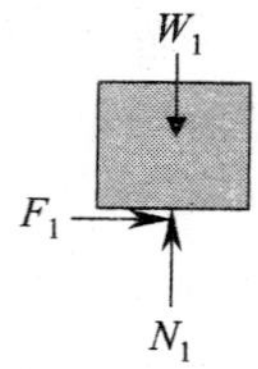

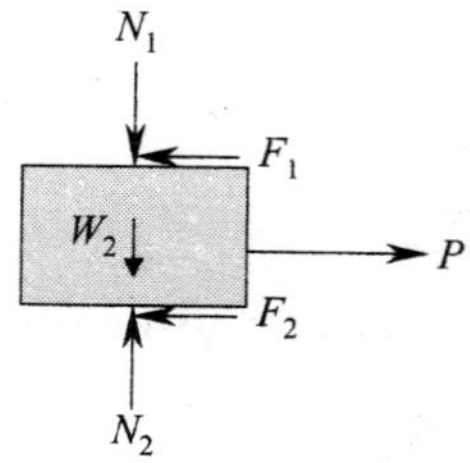

Fig. 14.6(a)

Upper block

Applying the equations of motion for the upper block,

$$\sum F_y = ma_y \Rightarrow$$

$$N_1 - W_1 = 0$$

$$\therefore \quad N_1 = W_1$$

$$\sum F_x = ma_x \Rightarrow$$

$$F_1 = m_1 a$$

We know that as the upper block is about to slide, the frictional force reaches the maximum value and it is equal to the limiting static friction, i.e., $F_1 = \mu_s N_1 = \mu_s W_1$. Hence, the above equation can be written as

$$\mu_s W_1 = \mu_s m_1 g = m_1 a$$

$$\Rightarrow \quad a = \mu_s g$$

$$= (0.25)9.81 = 2.45 \text{ m/s}^2$$

Lower block

Applying the equations of motion for the lower block,

$$\sum F_y = ma_y \Rightarrow$$

$$N_2 - N_1 - W_2 = 0$$

$$\therefore \quad N_2 = N_1 + W_2 = W_1 + W_2$$

$$\sum F_x = ma_x \Rightarrow$$

$$P - F_1 - F_2 = m_2 a$$

$$\Rightarrow \quad P = F_1 + F_2 + m_2 a$$

$$= \mu_s W_1 + \mu_k [W_1 + W_2] + m_2 a$$

$$= (0.25)(10)(9.81) + (0.2)(10 + 20)(9.81) + 20(2.45) = 132.39 \text{ N}$$

Example 14.7 A block of mass m slides down an inclined plane, having an inclination θ with respect to the horizontal. Determine the acceleration of the block if the coefficient of kinetic friction between the contact surfaces is μ.

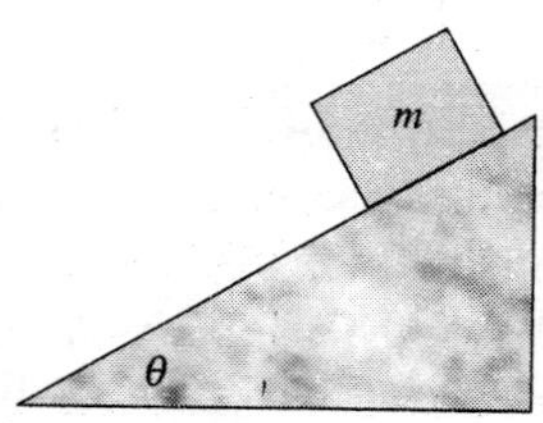

Fig. 14.7

Solution The free-body diagram of the block is shown in Fig. 14.7(a). The forces acting on the block are its weight resolved into components $mg \sin\theta$ along the plane and $mg \cos\theta$ normal to the plane, and normal reaction N and frictional force μN acting up the plane. Applying the equations of motion along the plane and normal to the plane, we get

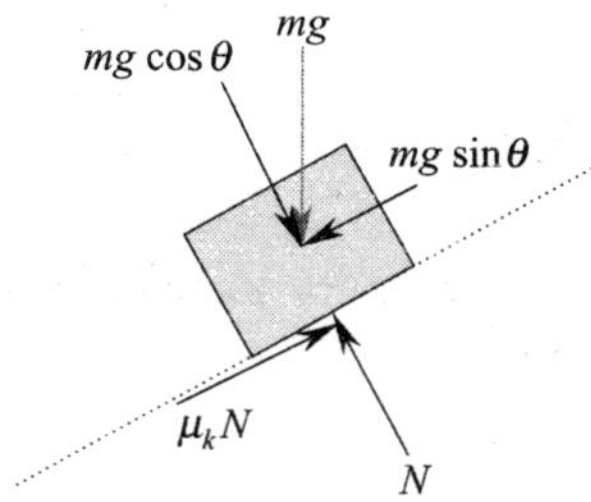

Fig. 14.7(a)

$$\sum F_y = ma_y \Rightarrow$$

$$N - mg\cos\theta = 0$$

$$\therefore \quad N = mg\cos\theta$$

$$\sum F_x = ma_x \Rightarrow$$

$$mg\sin\theta - \mu_k N = ma$$

$$\therefore \quad a = g(\sin\theta - \mu_k\cos\theta)$$

Example 14.8 A block of 1 kg mass slides down a 30° incline through a distance of 10 m in 3 seconds. Using the result of the previous problem, determine the coefficient of kinetic friction between the contact surfaces.

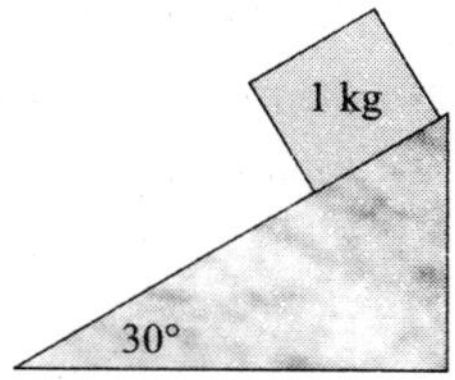

Fig. 14.8

Solution The free-body diagram of the block is shown in Fig. 14.8(a).

Acceleration of the block

As the block slides down 10 m in 3 seconds, its acceleration can be determined from the kinematic equation for constant acceleration as follows:

$$s = v_0 t + \frac{1}{2}at^2$$

$$10 = 0 + \frac{1}{2}(a)(3)^2$$

$$\Rightarrow \quad a = 2.22 \text{ m/s}^2$$

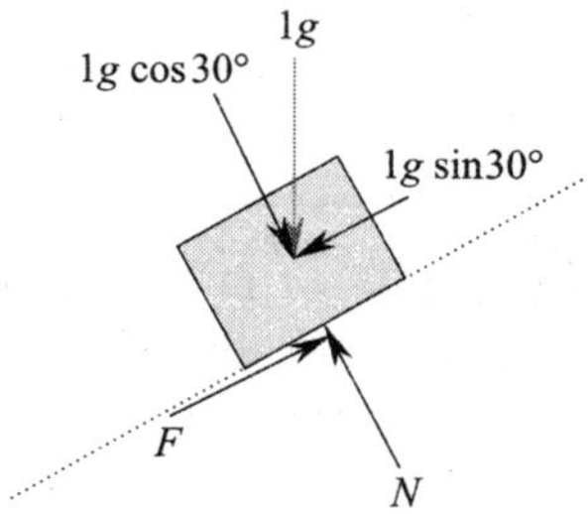

Fig. 14.8(a)

[Note that the initial velocity of the block is zero.]

Coefficient of kinetic friction between the contact surfaces

We notice that the acceleration of the block is less than acceleration due to gravity g. This is because of the frictional force acting between the contact surfaces.

From the result of the previous problem, we have

$$a = g[\sin 30° - \mu_k\cos 30°]$$

$$2.22 = 9.81\,[\sin 30° - \mu_k\cos 30°]$$

$$\Rightarrow \quad \mu_k = 0.32$$

Example 14.9 In a game of carrom, a player strikes a coin into the pocket, which is 0.8 m from the coin. If the coefficient of kinetic friction is 0.05, what minimum initial velocity must he impart to the coin so as to just get it into the pocket?

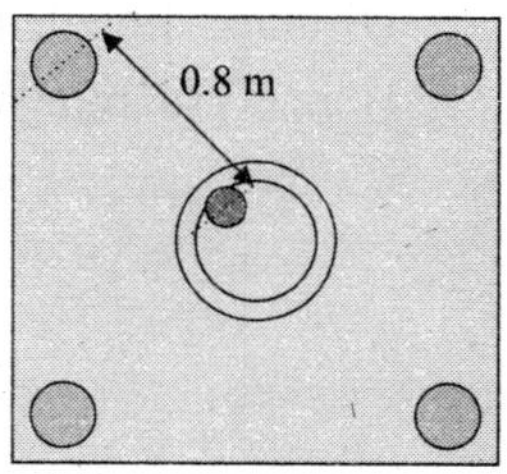

Fig. 14.9

Solution In the limiting case, as the coin just gets into the pocket, its final velocity is zero. The force acting on the coin opposing the motion is the frictional force acting in the opposite direction. Assuming constant deceleration, we have

$$-\mu mg = ma$$

$$\Rightarrow \quad a = -\mu g$$

$$= -0.05 \times 9.81 = -0.491 \text{ m/s}^2$$

Distance to be covered, $s = 0.8$ m. Therefore, writing the kinematic equation for constant acceleration,

$$v^2 = v_o^2 + 2as$$

$$0 = v_o^2 + 2(-0.491)(0.8)$$

$$\Rightarrow \quad v_o = 0.886 \text{ m/s}$$

Example 14.10 A cannon can fire a bomb with a release velocity of 75 m/s. If the length of the barrel is 1 m and the mass of the bomb is 1200 g then determine the force accelerating the bomb.

Solution The distance travelled by the bomb within the barrel is equal to the length of the barrel and if we assume that its acceleration is uniform then we can write the kinematic equation as

$$v^2 = v_o^2 + 2as$$

$$75^2 = 0 + 2a(1)$$

$$\Rightarrow \quad a = 2812.5 \text{ m/s}^2$$

Therefore, the force accelerating the bomb is obtained as

$$F = ma$$

$$= 1.2 \times 2812.5 = 3.38 \text{ kN}$$

Example 14.11 A car of 2 ton mass moving at a speed of 54 kmph is to be brought to a halt in a distance of 40 m. What should be the braking force applied assuming it to be uniform?

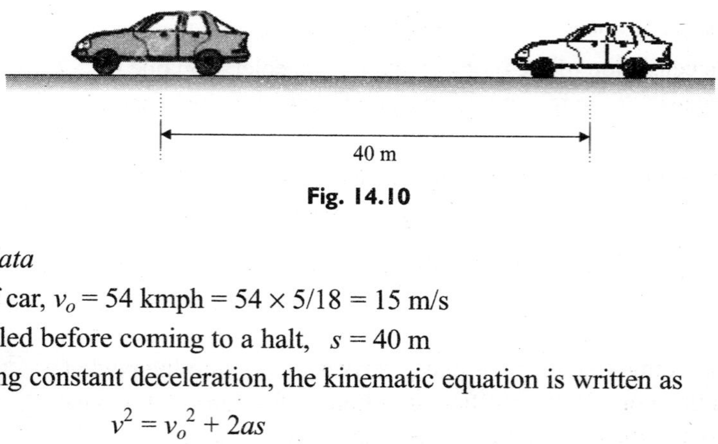

Fig. 14.10

Solution *Given data*

Initial speed of car, $v_o = 54$ kmph $= 54 \times 5/18 = 15$ m/s

Distance travelled before coming to a halt, $s = 40$ m

Hence, assuming constant deceleration, the kinematic equation is written as

$$v^2 = v_o^2 + 2as$$

$$0 = 15^2 + 2a(40)$$

$$\Rightarrow \quad a = -\frac{225}{80} = -2.81 \text{ m/s}^2$$

Therefore, braking force is

$$F = ma = (2000) \times (-2.81) = -5.62 \text{ kN} \qquad \text{[Note: 1 ton = 1000 kg]}$$

Example 14.12 A short commuter train consists of three coaches, each of 6-ton mass. If the frictional resistance is 0.5 kN/ton mass, determine the tractive force of the train, if it has to attain a speed of 60 kmph in 10 s. Also, determine the tension in each of the coupling between the coaches.

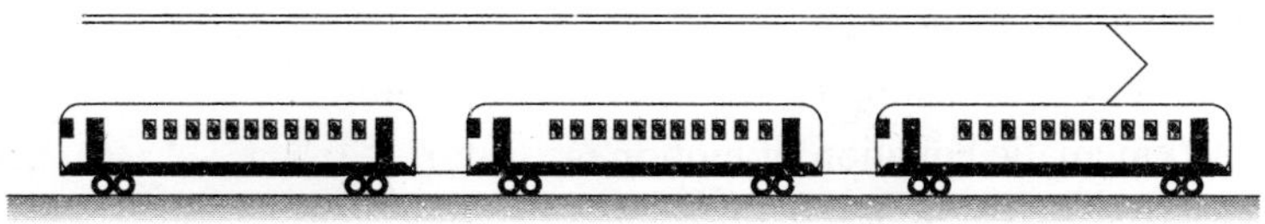

Fig. 14.11

Solution *Given data*

Initial speed of train, $v_o = 0$

Final speed of train, $v = 60 \times \dfrac{5}{18} = 16.67$ m/s

Time taken, $t = 10$ s

Total mass of train, $m = 3(6) = 18$ tons

Acceleration of the train

We know that the kinematic equation of motion is

$$v = v_o + at$$

$$\therefore \qquad a = \frac{v - v_o}{t} = 1.67 \text{ m/s}^2$$

Tractive force of the train

If P be the tractive force and F be the frictional force then we can write the equation of motion as

$$P - F = ma$$

$$P - (0.5 \times 10^3)(18) = (18 \times 10^3) \times 1.67$$

$$\therefore \qquad P = 18 \times 10^3\,[0.5 + 1.67] = 39.06 \text{ kN}$$

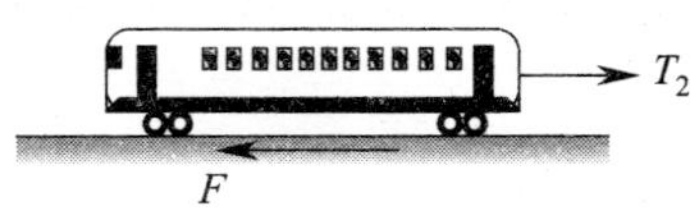

Fig. 14.11(a)

Tension in the couplings

Let T_1 and T_2 be the respective tensions in the couplings between the first coach and the second and that between the second and the last. Considering the free-body diagram of the last coach, its equation of motion can be written as

$$T_2 - 0.5 \times 10^3(6) = (6 \times 10^3)1.67$$

$$\Rightarrow \qquad T_2 = 6 \times 10^3\,(0.5 + 1.67) = 13.02 \text{ kN}$$

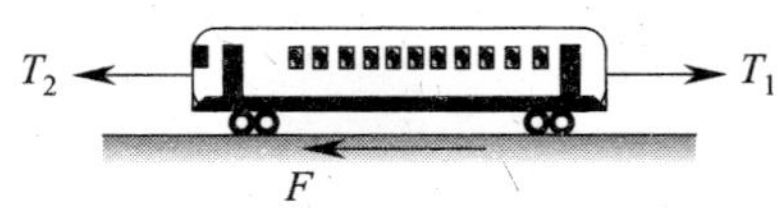

Fig. 14.11(b)

Considering the free–body diagram of the middle coach, its equation of motion can be written as

$$T_1 - T_2 - 0.5 \times 10^3(6) = (6 \times 10^3)1.67$$

$$\Rightarrow \quad T_1 = T_2 + 6 \times 10^3(0.5 + 1.67) = 26.04 \text{ kN}$$

Example 14.13 The tractive force of a locomotive of 15 ton mass is 200 kN. The mass of each bogie is 8 tons. How many bogies can it haul if it has to attain a speed of 60 kmph in a distance of 150 m? The frictional force is 1 kN/ton.

Solution *Acceleration of the train*

Final speed of the train, $v = 60$ kmph $= 60 \times 5/18 = 16.67$ m/s

We know that the kinematic equation of motion is

$$v^2 = v_o^2 + 2as$$

$$\therefore \quad a = \frac{v^2 - v_o^2}{2s} = \frac{16.67^2}{2 \times 150} = 0.93 \text{ m/s}^2$$

If we consider n number of bogies then the mass of the train is $(15 + 8n)$ tons. Hence, the kinetic equation of motion can be written as

$$P - F = ma$$

$$200 \times 10^3 - (1 \times 10^3)(15 + 8n) = [(15 + 8n) \times 10^3] \times 0.93$$

$$\Rightarrow \quad n = \frac{1}{8}\left[\frac{200}{1.93} - 15\right] = 11.08$$

The number of bogies must be a whole number. Hence, n should be either 11 or 12. If we take $n = 12$ then the tractive force required will be more than the given value. Thus, we conclude that the locomotive can haul a maximum of 11 bogies.

Example 14.14 When an airplane is moving on a runway during take off, the pilot realizes a fault in the engine and suddenly applies the brakes. If the speed of the plane when the brakes were applied was 200 kmph and the braking force is 2000 kN, determine the time in which the plane would be brought to a stop and the distance covered in this time. Mass of the plane is 200 tons.

Solution As the frictional force acts in the direction opposite to that of the motion, we can write the equation of motion of the plane as

$$-F = ma$$

$$\Rightarrow \quad a = -\frac{F}{m} = -\frac{2000 \times 10^3}{200 \times 10^3} = -10 \text{ m/s}^2$$

Initial speed of plane, $v_o = 200$ kmph $= 200 \times 5/18 = 55.56$ m/s

Final speed of plane, $v = 0$

Hence, using the kinematic equation,

$$v = v_o + at$$

$$\Rightarrow \quad t = \frac{v - v_o}{a} = 5.56 \text{ s}$$

and

$$v^2 = v_o^2 + 2as$$

$$\Rightarrow \quad s = \frac{v^2 - v_o^2}{2a} = 154.35 \text{ m}$$

Example 14.15 An elevator together with passengers weighing 3 tons is supported by a cable. Find the tension in the cable when the elevator is (i) moving upward with an acceleration of 1 m/s^2, (ii) moving upward with a deceleration of 0.5 m/s^2, (iii) moving downward with an acceleration of 1 m/s^2, and (iv) moving downward with a deceleration of 0.6 m/s^2 .

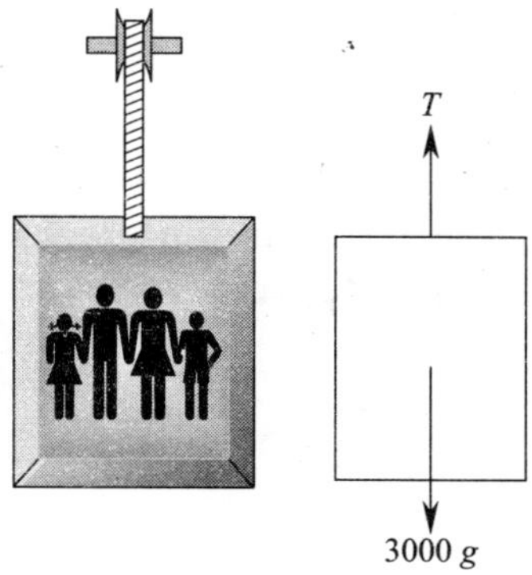

Fig. 14.12

Solution When the elevator is moving upwards, considering the initial direction of motion as positive, the equation of motion can be written as

$$T - 3000g = 3000a$$

$$\Rightarrow \quad T = 3000(g + a)$$

(i) When the elevator is moving upwards with an *acceleration* of 1 m/s^2, the tension in the cable is obtained as

$$T = 3000 \times (9.81 + 1) = 32.43 \text{ kN}$$

(ii) Similarly, when the elevator is moving upwards with a *deceleration* of 0.5 m/s^2, the tension in the cable is obtained as

$$T = 3000 \times [9.81 + (-0.5)] = 27.93 \text{ kN}$$

When the elevator is moving downwards, considering the initial direction of motion as positive, its equation of motion can be written as

$$3000g - T = 3000a$$

$$\Rightarrow \quad T = 3000(g - a)$$

(iii) When the elevator is moving downwards with an *acceleration* of 1 m/s^2, the tension in the cable is obtained as

$$T = 3000 \times (9.81 - 1) = 26.43 \text{ kN}$$

(iv) Similarly, when the elevator is moving downwards with a *deceleration* of 0.6 m/s^2, the tension in the cable is obtained as

$$T = 3000 \times [9.81 - (-0.6)] = 31.23 \text{ kN}$$

Example 14.16 Determine the force exerted by the floor of the lift on a passenger of 75 kg mass, when the lift is (i) accelerating upwards at 1 m/s^2, and (ii) accelerating downwards at 0.9 m/s^2.

Solution

(i) When the lift is accelerating upwards at 1 m/s^2

As the lift is moving upwards, we consider the upward direction as positive and hence, we can write the equation of motion as

$$R - mg = ma$$

where R is the force exerted by the floor of the lift on the man. Therefore,

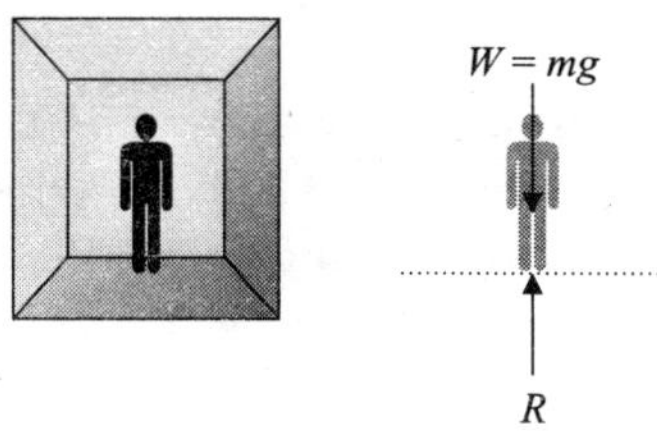

Fig. 14.13

$$R = m(g + a)$$
$$= 75(9.81 + 1) = 810.75 \text{ N}$$

(ii) When the lift is accelerating downwards at 0.9 m/s²

As the lift is moving downwards, we consider the downward direction as positive and hence, we can write the equation of motion as

$$mg - R = ma$$

Therefore,

$$R = m(g - a)$$
$$= 75(9.81 - 0.9) = 668.25 \text{ N}$$

Example 14.17 A body of 5 kg mass is hung from a spring balance suspended from the ceiling of an elevator. What is the reading in the balance (i) if the elevator is accelerating upwards at 2 m/s², and (ii) if the elevator is accelerating downwards at 1.5 m/s².

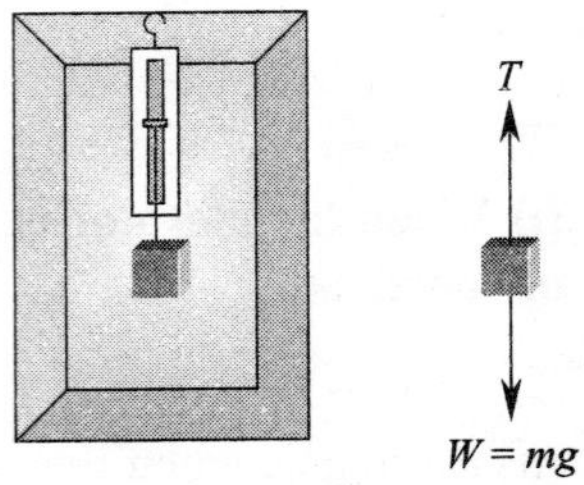

Fig. 14.14

Solution If we can determine the tension in the string by which the body is suspended to the spring balance then that gives the reading of the balance. Hence, we proceed as follows:

(i) When the elevator is accelerating upwards at 2 m/s²

As the elevator is moving upwards, we consider the upward direction as positive and hence, we can write the equation of motion as

$$T - mg = ma$$

$$\Rightarrow \quad T = m(g + a)$$
$$= 5(9.81 + 2) = 59.05 \text{ N}$$

(ii) When the elevator is accelerating downwards at 1.5 m/s²

As the elevator is moving downwards, we consider the downward direction as positive and hence, we can write the equation of motion as

$$mg - T = ma$$

$$\Rightarrow \quad T = m(g - a)$$
$$= 5(9.81 - 1.5) = 41.55 \text{ N}$$

Example 14.18 A 75 kg man stands on a scale in an elevator. Determine the acceleration of the elevator when the scale reads (i) 100 kg, (ii) 60 kg, and (iii) zero.

Solution We know that the reading on the scale is the force exerted by the man on the scale. By Newton's third law of motion, we know that this force is equal and opposite to the force exerted by the floor of the elevator on the man. Hence, we can write the equation of motion of the man considering upward direction as positive:

$$R - mg = ma$$

$$\Rightarrow \quad a = \frac{R - mg}{m}$$

(i) Acceleration of the elevator when the scale reads 100 kg

$$a = \frac{100g - 75g}{75}$$
$$= 3.27 \text{ m/s}^2$$

(ii) Acceleration of the elevator when the scale reads 60 kg

$$a = \frac{60g - 75g}{75}$$
$$= -1.96 \text{ m/s}^2$$

From the negative value of acceleration, we can conclude that the elevator is either accelerating downwards or decelerating upwards.

(iii) Acceleration of the elevator when the scale reads zero:

$$a = \frac{0 - 75g}{75}$$
$$= -g$$

Here also as the acceleration is a negative value, we conclude that it is either accelerating downwards or decelerating upwards. However, as the magnitude is equal to g, which being a very high value, the only possibility is that it is accelerating downwards with a value equal to g or in other words, it is freely falling. Since the man inside the elevator is also freely falling with the same acceleration due to gravity, he looses contact with the floor of the elevator and hence the scale reads zero.

Example 14.19 A lift can operate under a maximum of 8 persons. Mass of the lift is 800 kg. Determine the limits of tension if the lift accelerates at a constant rate of 1 m/s^2 either upwards or downwards. Take average weight of a person = 750 N.

Solution *Given data*

Mass of lift, $M = 800$ kg

Total mass of passengers, $m = 8 \times \frac{750}{9.81} = 611.62$ kg

When the lift is accelerating upwards, we can write the equation of motion as

$$T - [M + m]g = [M + m]a$$

$$\Rightarrow \quad T = [M + m][g + a]$$
$$= [800 + 611.62][9.81 + 1] = 15.26 \text{ kN}$$

Similarly, when the lift is accelerating downwards, we can write the equation of motion as

$$[M + m]g - T = [M + m]a$$

$$\Rightarrow \quad T = [M + m][g - a]$$
$$= [800 + 611.62][9.81 - 1] = 12.44 \text{ kN}$$

Thus, the limits of tension are $12.44 \text{ kN} \le T \le 15.26 \text{ kN}$

14.4 MOTION OF CONNECTED BODIES

In the previous section, we considered motion of isolated bodies subjected to an externally applied force. In this section, we will discuss the motion of connected bodies such as a block-and-pulley system. Here we must draw separate free-body diagrams for each body or sub-systems forming the system and apply the equations of motion to each body and solve for the unknowns. Here we assume that the strings connecting the bodies are inextensible such that the motions of the bodies are all related. Further, the pulleys over which a string passes are assumed to be massless such that their moments of inertia can be neglected and frictionless such that the tensions on both the ends of the string are equal.

Example 14.20 Two blocks of masses m_1 and m_2 are suspended by an inextensible string passing over a massless and frictionless pulley as shown. When released from rest, determine the expressions for the acceleration of the system and the tension in the string. Take $m_1 = 2$ kg and $m_2 = 3$ kg.

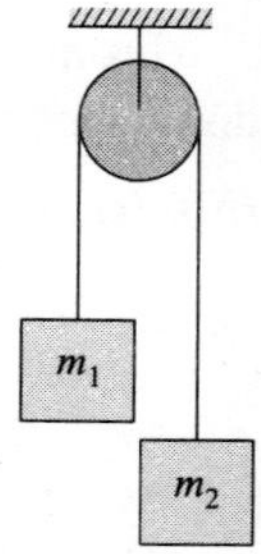

Fig. 14.15

Solution The free-body diagrams of the blocks and the pulley are shown below. Since the mass of the block II is greater than the mass of the block I, the block I will move up and the block II will move down. Since the string is inextensible, the magnitude of acceleration of the two blocks will be same and let it be a. Also, as the pulley is massless and frictionless, its moment of inertia is zero and the tension on both the ends of the string are equal.

Applying the equations of motion along the y-direction for both the blocks taking the initial direction of motion as positive,

Block I $\quad T - m_1 g = m_1 a \quad$ (a)

Block II $\quad m_2 g - T = m_2 a \quad$ (b)

Solving for a and T from the above two equations, we get

$$a = \frac{m_2 - m_1}{m_2 + m_1} g$$

and

$$T = \frac{2m_1 m_2}{m_1 + m_2} g$$

Acceleration and tension for the given data

$$a = \left[\frac{3-2}{3+2}\right] 9.81 = 1.96 \text{ m/s}^2$$

$$T = \left[\frac{2(2)(3)}{2+3}\right] 9.81 = 23.54 \text{ N}$$

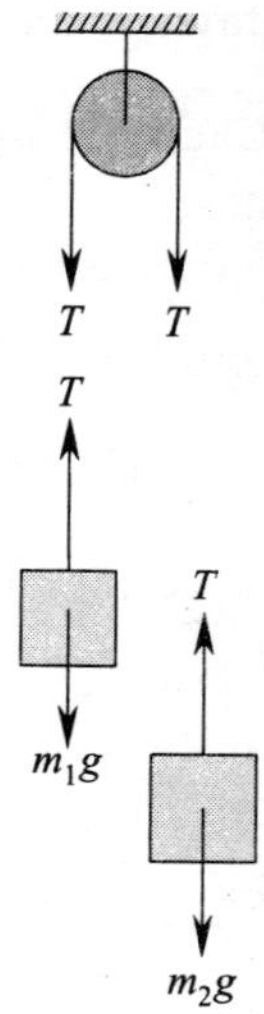

Fig. 14.15(a)

Example 14.21 For the system shown below, write the kinetic equations of motion for various bodies in the system. Assume the pulleys are frictionless and massless. Assume the block II moves down and the block I moves up.

Solution As the string is inextensible, its total length is constant, i.e.,

$$AB + \text{arc } BC + CD + \text{arc } DE + EF = \text{constant}$$

We see from the figure that the lengths of the string embracing the pulleys, i.e., arcs BC and DE are always constants. Therefore,

$$AB + CD + EF = \text{constant}$$

Since the lengths y_1' and y_2' are constants, we can also write the above expression as

$$(AB + y_1') + (CD + y_1') + EF = \text{constant}$$

$$(AB + y_1') + 2EF = \text{constant}$$

$$(AB + y_1') + 2(EF + y_2') = \text{constant}$$

$$y_1 + 2y_2 = \text{constant}$$

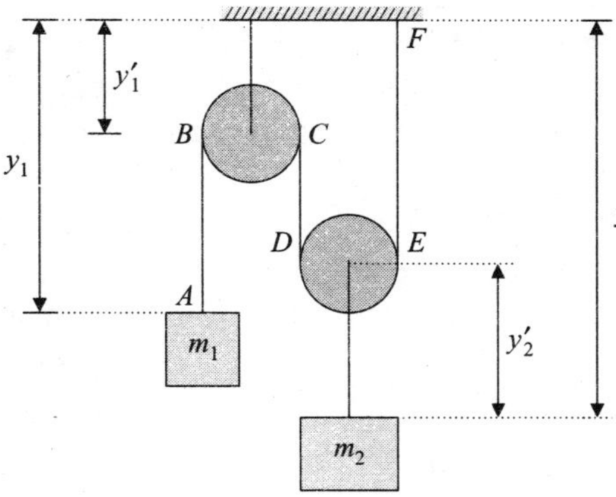

Fig. 14.16

As y_1 and y_2 vary with time, differentiating them with respect to time, we have

$$v_1 + 2v_2 = 0$$

and $$a_1 + 2a_2 = 0$$

Hence, if the acceleration of the block I is a, then the acceleration of the block II is $-a/2$. The free-body diagram of the system is shown in Fig. 14.16(a). As the pulleys are frictionless, the tension at every point of the string is equal to T_1.

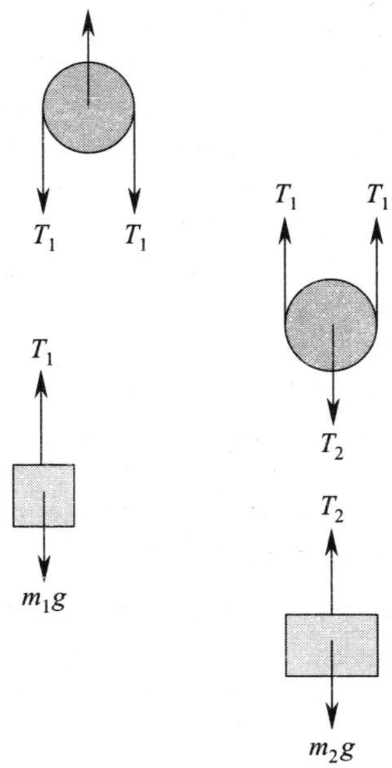

Fig. 14.16(a)

Motion of the block I

Applying the equations of motion for the block I taking the initial direction of motion as positive,

$$\sum F_y = ma_y \Rightarrow$$

$$T_1 - m_1 g = m_1 a_1 \qquad \text{(a)}$$

Motion of the block II

Similarly, applying the equations of motion for the block II taking the initial direction of motion as positive,

$$\sum F_y = ma_y \Rightarrow$$

$$m_2 g - T_2 = m_2 a_2 \qquad \text{(b)}$$

Motion of right pulley

$$\sum F_y = ma_y \Rightarrow$$

$$T_2 - 2T_1 = 0 \qquad \text{(c)}$$

Note that because the pulley is massless, the right-hand side of the above equation is zero. From these three equations (a), (b) and (c), we can solve for the unknowns.

Example 14.22 A block of mass m_1 resting on a rough horizontal plane is pulled by an inextensible string, whose other end is attached to a block of mass m_2 and passing over a pulley as shown. Assume the pulley to be frictionless and massless. If the coefficient of kinetic friction between the plane and the block is μ, derive the expressions for acceleration of the system and tension in the string. If $m_1 = 3$ kg, $m_2 = 2$ kg and $\mu = 0.2$ then determine the acceleration of the system and tension in the string.

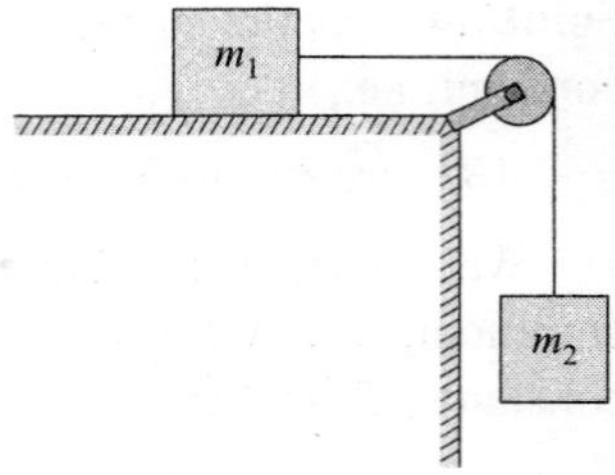

Fig. 14.17

Solution The free-body diagrams of the two blocks and the pulley are shown below. As the pulley is frictionless, the tensions on both ends of the string are equal.

Motion of the block 1

As the block on the horizontal plane is restricted from moving in the vertical direction, its acceleration along the *Y*-direction is zero, i.e., $a_{1y} = 0$. Hence, writing the equations of motion along the *X* and *Y* directions, taking the initial direction of motion as positive:

$$\Sigma F_y = ma_y \Rightarrow$$

$$N - m_1 g = m_1 a_{1y} = 0$$

$$\Rightarrow \quad N = m_1 g$$

$$\Sigma F_x = ma_x \Rightarrow$$

$$T - \mu N = m_1 a_{1x}$$

$$T - \mu m_1 g = m_1 a_{1x} \qquad \text{(a)}$$

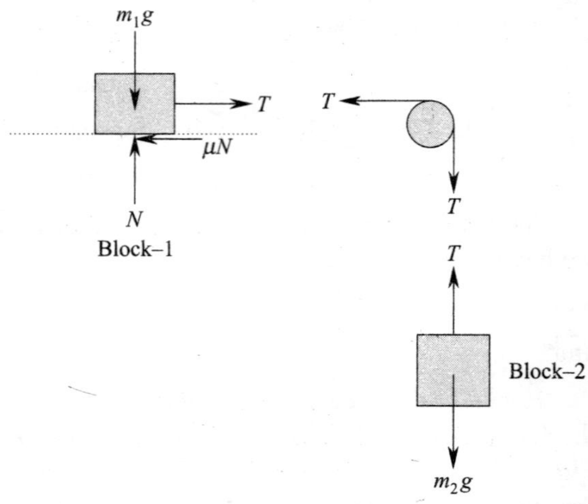

Fig. 14.17(a)

Motion of the block 2

As the net force acts along the *y*-direction, it has only vertical acceleration in the downward direction. Hence, writing the equation of motion along the *Y*-direction,

$$\Sigma F_y = ma_y \Rightarrow$$

$$m_2 g - T = m_2 a_{2y} \qquad \text{(b)}$$

As the string is inextensible, the acceleration of the two blocks will be equal, i.e., $a_{1x} = a_{2y} = a$. Hence, equations (a) and (b) can be written as

$$T - \mu m_1 g = m_1 a \qquad \text{(c)}$$

and $$m_2 g - T = m_2 a \qquad \text{(d)}$$

From the above two equations, solving for T and a, we have

$$a = \left[\frac{m_2 - \mu m_1}{m_1 + m_2}\right] g$$

and $$T = \frac{m_1 m_2 g}{m_1 + m_2}[1 + \mu]$$

Acceleration and tension in the string for the given data

$$a = \left[\frac{2-(0.2)(3)}{3+2}\right]9.81 = 2.75 \text{ m/s}^2$$

$$T = \frac{(3)(2)(9.81)}{3+2}[1+0.2] = 14.13 \text{ N}$$

Example 14.23 Find the expressions for the acceleration of the system shown in Fig. 14.18 and the tension in the string. If $m_1 = 2$ kg, $m_2 = 1$ kg, $\theta = 30°$ and $\mu = 0.2$ for all contact surfaces, determine the tension in the string and the acceleration of the system. Assume the pulleys are massless and frictionless and the string is inextensible.

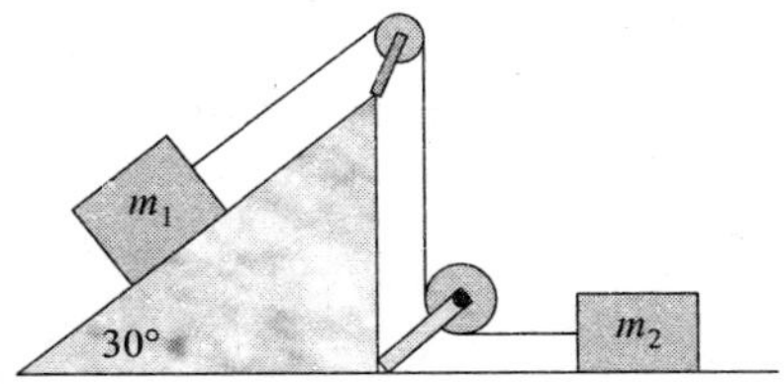

Fig. 14.18

Solution The free-body diagrams of the two blocks and the pulleys are shown below. As the pulleys are frictionless, the tensions in different portions of the string are equal. In addition, as the string is inextensible, the acceleration of the block I down the plane is same as the acceleration of the block II towards its left.

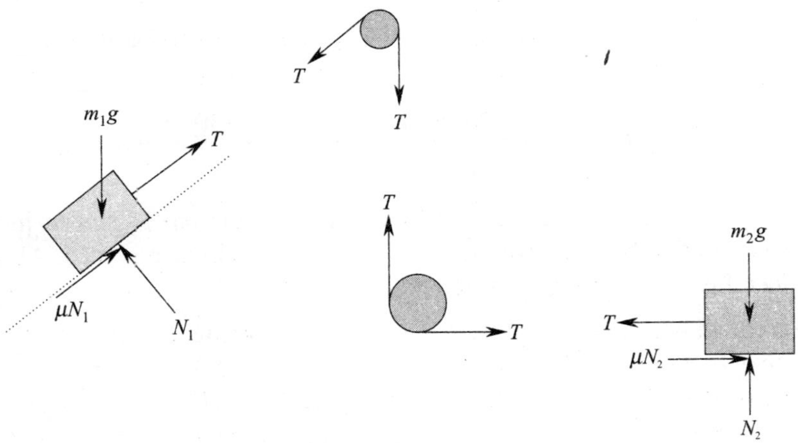

Fig. 14.18(a)

Motion of the block I

Writing the equations of motion normal to the inclined plane and along the inclined plane,

$\Sigma F_y = ma_y \Rightarrow$

$$N_1 - m_1 g \cos\theta = 0$$

$$\therefore \quad N_1 = m_1 g \cos\theta$$

$\Sigma F_x = ma_x \Rightarrow$

$$m_1 g \sin\theta - \mu N_1 - T = m_1 a$$

$$m_1 g \sin\theta - \mu m_1 g \cos\theta - T = m_1 a \qquad \text{(a)}$$

Motion of the block II

Applying the equations of motion for the block II taking the initial direction of motion as positive,

$\sum F_y = ma_y \Rightarrow$

$$N_2 - m_2 g = 0$$

$$\therefore \qquad N_2 = m_2 g$$

$\sum F_x = ma_x \Rightarrow$

$$T - \mu N_2 = m_2 a$$

$$T - \mu m_2 g = m_2 a \qquad \text{(b)}$$

From equations (a) and (b), solving for a and T, we have

$$\Rightarrow \qquad a = \frac{m_1 g(\sin\theta - \mu\cos\theta) - \mu m_2 g}{m_1 + m_2}$$

Also,

$$T = \frac{m_1 m_2 g[\sin\theta + \mu(1 - \cos\theta)]}{m_1 + m_2}$$

For the given data,

$$a = \frac{2 \times 9.81[\sin 30° - (0.2) \times \cos 30°] - (0.2)(1)(9.81)}{2 + 1} = 1.48 \text{ m/s}^2$$

and

$$T = \frac{2 \times 1 \times 9.81[\sin 30° + 0.2\,(1 - \cos 30°)]}{2 + 1} = 3.45 \text{ N}$$

Example 14.24 Determine the acceleration of the system of the blocks shown in Fig. 14.19. The coefficient of kinetic friction between the block 1 and the inclined plane is 0.15. Also, determine the tension in the string. Take $m_1 = 75$ kg, $m_2 = 50$ kg.

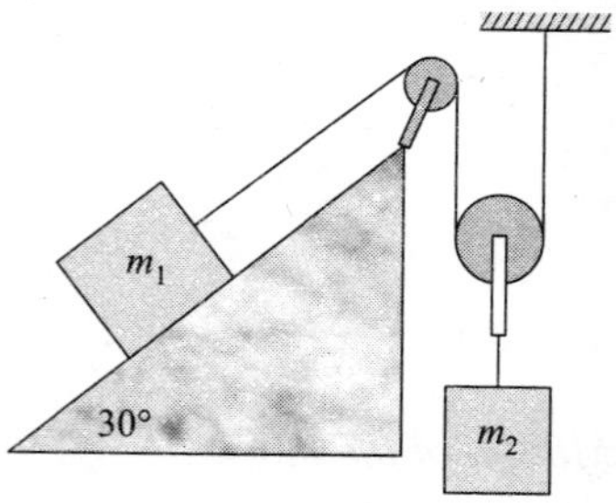

Fig. 14.19

Solution This problem is quite different from the previous problems in that the direction of motion of the blocks is not readily known. Block I may move up or down the plane. Hence, before drawing the free-body diagrams, we must determine the direction of motion, as the direction of friction is dependent upon the direction of motion. Suppose the plane is frictionless. Then the direction could be determined by calculating the tension in different portions of the string and the one with higher tension would be the determining factor. In Fig. 14.19(a), for equilibrium of the block II, we see that the tension $T_2 = m_2 g = 490.5$ N. Hence, $T_1 = T_2/2 = 245.25$ N. For equilibrium of the block I, the tension in the string is $T_1 = m_1 g \sin 30°$ = 367.88 N. As this tension is greater than 245.25 N, we can readily see that the block I would move down the plane. However, in the current situation the friction also comes into play. If the difference in tensions, i.e., 367.88 – 245.25 = 122.63 N is greater than the limiting force of friction, then the block I would move down the plane. Otherwise, there would be no motion at all. We see that the limiting force of friction is $\mu m_1 g \cos 30° = 95.58$ N. Thus, we conclude that the block I would move down the plane and block II would move up. Based on this discussion, we draw the free-body diagrams showing the force of friction in the direction opposite to the direction of motion.

We can see from the figure that if acceleration of the block I is a, then acceleration of the block II is $a/2$.

Motion of the block I

Applying the equations of motion for the block I taking the initial direction of motion as positive,

$$\Sigma F_y = ma_y \Rightarrow$$

$$N_1 - m_1 g \cos\theta = 0$$

$$\therefore \quad N_1 = m_1 g \cos\theta$$

$$\Sigma F_x = ma_x \Rightarrow$$

$$m_1 g \sin\theta - \mu N_1 - T_1 = m_1 a$$

$$m_1 g \sin\theta - \mu m_1 g \cos\theta - T_1 = m_1 a \qquad \text{(a)}$$

Motion of the block II

Similarly, applying the equation of motion for the block II taking the initial direction of motion as positive,

$$\Sigma F_y = ma_y \Rightarrow$$

$$T_2 - m_2 g = m_2 a/2 \qquad \text{(b)}$$

Motion of the pulley

$$\Sigma F_y = ma_y \Rightarrow$$

$$2T_1 - T_2 = 0 \qquad \text{(c)}$$

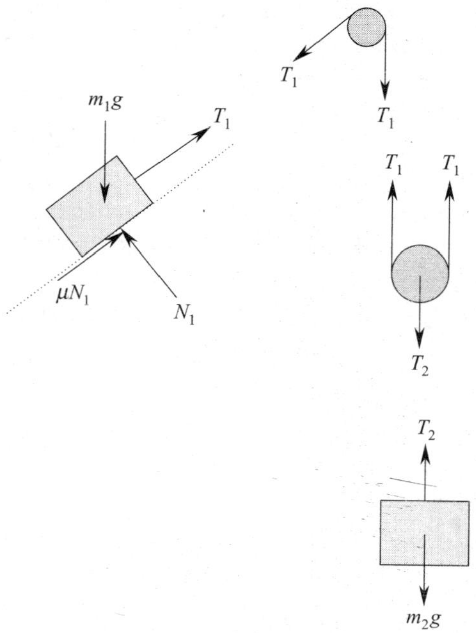

Fig. 14.19(a)

Note that because the pulley is massless, the right-hand side of the above equation is zero. From equations (a), (b) and (c), solving for a, T_1 and T_2, we have

$$a_1 = 0.309 \text{ m/s}^2$$

$$a_2 = 0.155 \text{ m/s}^2$$

$$T_1 = 249.11 \text{ N}$$

and $$T_2 = 498.22 \text{ N}$$

14.5 D'ALEMBERT'S PRINCIPLE

Newton's second law of motion can also be written as

$$\Sigma \vec{F} - m\vec{a} = \vec{O} \qquad (14.9)$$

We note that the above equation represents the condition for *equilibrium*. Thus, by adding a **fictitious** force, $-m\vec{a}$ to the system of forces, we can bring the particle into equilibrium. Since the particle is not at rest in this condition, but actually moving, we term it **dynamic equilibrium**. The force $-m\vec{a}$ is a fictitious force called an **inertial force** and it is equal to the product of mass of the particle and its acceleration, directed opposite to that of the acceleration.

This alternative form of Newton's second law was stated by Jean le Rond d' Alembert, a French mathematician, in 1742 and it is known D'Alembert's principle. In effect, the principle reduces a problem in dynamics to a problem in statics. D'Alembert showed that one could transform an accelerating system into an equivalent static system by adding the so-called "inertial force."

Example 14.25 A bob of mass m is hung by an inextensible string attached to the roof of a moving vehicle. If the inclination of the string with the vertical is θ then derive an expression for the acceleration of the vehicle.

Solution Initially when the vehicle is at rest, the bob is vertical and the forces acting on it are (i) its own weight acting vertically downwards, and (ii) the tension in the string acting along the string and away from the bob. When the vehicle is moving with constant acceleration, the bob will swing backwards due to the accelerating force. Hence, for its equilibrium, the resultant of the forces along the X and Y directions must be zero.

Therefore, we have,

$\Sigma F_y = 0 \Rightarrow$

$$T\cos\theta - mg = 0 \qquad \text{(a)}$$

$\Sigma F_x = 0 \Rightarrow$

$$T\sin\theta - ma = 0 \qquad \text{(b)}$$

From the above two equations (a) and (b), we have

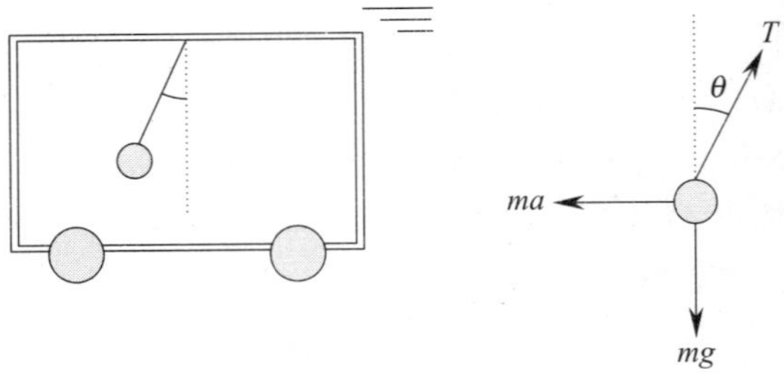

Fig. 14.20

$$\tan\theta = \frac{a}{g}$$

or $$a = g\tan\theta$$

Thus, we can see that by measuring the inclination of the string with the vertical, the acceleration of the vehicle can be determined.

14.6 VARIABLE ACCELERATION

In the previous sections, we considered particles subjected to constant forces, i.e., constant in magnitude and direction. As the force was constant, the acceleration was also constant and hence, we could use the kinematic equations for constant acceleration to determine the motion of particles. In this section, we will consider particles, which are subjected to variable forces, and as a result, the accelerations will also vary. To determine the motion, we use the method of integration as discussed in the previous chapters.

Example 14.26 A block of 10 kg mass resting on a smooth horizontal plane is acted on by a horizontal force F that varies with time as shown in Fig. 14.21. Determine the velocity and displacement of the block just after 10 seconds.

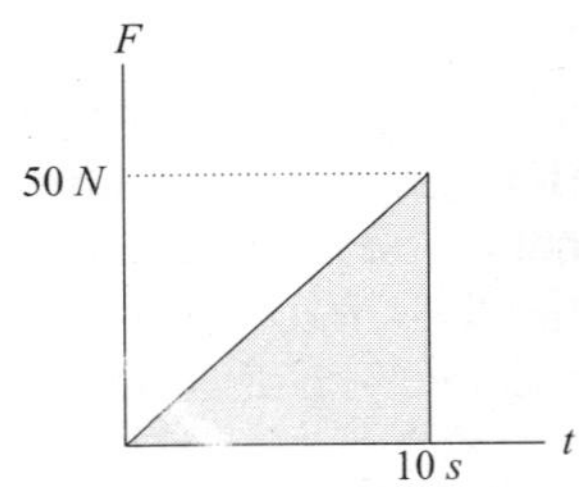

Fig. 14.21

Solution From the figure, we can write the variation of the force with respect to time as

$$F = \frac{50}{10}t = 5t$$

Since mass of the block is 10 kg, the variation of acceleration with respect to time can be obtained as

$$a = \frac{F}{m} = \frac{5t}{10} = \frac{t}{2}$$

Therefore, velocity and displacement are obtained by integrating the acceleration function with respect to time,

$$a = \frac{dv}{dt} = \frac{t}{2}$$

$$\Rightarrow \quad v = \frac{t^2}{4} + v_o$$

Note that as the block starts from rest, its initial velocity, v_o is zero. Hence,

$$v = \frac{dx}{dt} = \frac{t^2}{4}$$

Upon further integration, we get

$$x = \frac{t^3}{12} + x_o$$

If we take initial position of the block as zero then we have

$$x = \frac{t^3}{12}$$

Therefore, velocity and displacement just after 10 seconds are

$$v(10) = \frac{10^2}{4} = 25 \text{ m/s}$$

$$x(10) = \frac{10^3}{12} = 83.33 \text{ m}$$

Example 14.27 In the above problem, the same force acts on a block of 5 kg mass resting on a rough horizontal surface whose coefficient of kinetic friction is 0.2. Determine the velocity and displacement of the block just after 10 seconds.

Solution The variation of the force with respect to time as before is

$$F = \frac{50}{10} t = 5t$$

Since mass of the block is 5 kg, the variation of acceleration with respect to time can be obtained as

$$F - \mu mg = ma$$

$$\Rightarrow \quad a = \frac{F}{m} - \mu g$$

$$= t - 1.962$$

Therefore, velocity and displacement are obtained by integrating the acceleration function with respect to time:

$$a = \frac{dv}{dt} = t - 1.962$$

$$v = \frac{t^2}{2} - 1.962t + v_o$$

Note that as the block starts from rest, its initial velocity, v_o is zero. Hence,

$$v = \frac{dx}{dt} = \frac{t^2}{2} - 1.962t$$

Therefore,

$$x = \frac{t^3}{6} - 1.962\frac{t^2}{2} + x_o$$

If we take initial position of the block as zero then we have

$$x = \frac{t^3}{6} - 0.981t^2$$

Therefore, velocity and displacement just after 10 seconds are

$$v(10) = \frac{10^2}{2} - 1.962(10) = 30.38 \text{ m/s}$$

$$x(10) = \frac{10^3}{6} - 0.981(10)^2 = 68.57 \text{ m}$$

14.7 TANGENTIAL AND NORMAL COMPONENTS OF ACCELERATION

We saw in the previous chapter, that acceleration of a particle in curvilinear motion could also be represented in tangential and normal components. Hence, the vector equation of motion,

$$\sum \vec{F} = m\vec{a} \tag{14.10}$$

can be expressed in component forms along tangential and normal directions as

$$\sum F_t = ma_t \tag{14.11}$$

and

$$\sum F_n = ma_n \tag{14.12}$$

A special case of curvilinear motion is *circular* motion. Suppose a particle moves with constant speed in a circular path. Then we know that its tangential component of acceleration is zero. However, its normal component is non-zero, as the particle continuously changes its direction of motion. Hence,

$$\sum F_t = 0 \tag{14.13}$$

and

$$\sum F_n = ma_n = \frac{mv^2}{r}$$

where

$$a_n = \frac{v^2}{r} = r\omega^2, \text{ acting towards the centre of the circular path} \tag{14.14}$$

From Newton's first law of motion, we can understand that an external force must act on the particle to impart this normal acceleration (even though it is moving with constant speed). It is this external force, which causes change in direction of motion and thus, keeps the particle in circular path.

For instance, consider a stone tied by a string and the other end held in hand; if it is whirled with initial velocity then the *tension* in the string provides this normal acceleration. Thus, it follows the circular path. Suppose the string snaps then the tension in the string ceases to act. As a result, its normal acceleration becomes zero and hence the particle no more follows the circular path. The particle is then subjected purely to the force of gravity and it follows a parabolic path and falls to the ground. Similarly, when a vehicle moves on a curved road, the *friction* between the tires and the road provide the normal acceleration. Suppose the surface is smooth; then the vehicle would skid.

Example 14.28 A block of mass m resting on a smooth table is tied by a string, whose other end passes through a hole in the table at its centre and supports another block of mass M. If the block on the table is rotated at a constant angular velocity, determine the mass M required to hold it in equilibrium. For the given data, $m = 3$ kg, $r = 0.5$ m and $v = 3.141$ m/s, determine M.

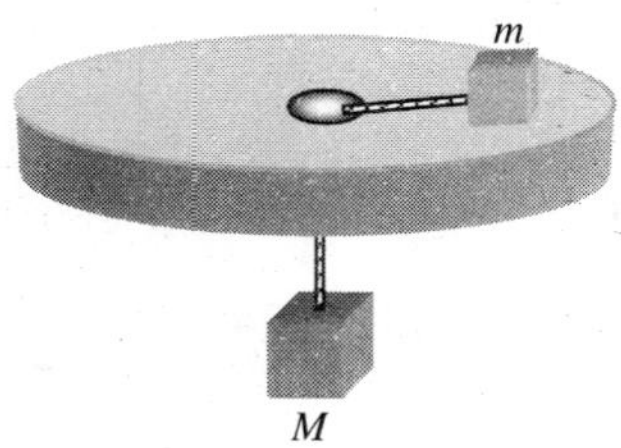

Fig. 14.22

Solution The force required to keep the block on the table in a circular path is provided by the tension in the string. Note that as the table is smooth, the frictional force is zero. Hence, the equation of motion in the direction towards the centre of the table can be written as

$$T = \frac{mv^2}{r}$$

If the other end of the string were free, the block would be thrown away. However, as it is held in equilibrium by the mass suspended about the centre of the circular path, the tension in the string must be equal to the weight of the suspended mass, i.e.,

$$Mg = \frac{mv^2}{r}$$

$$\therefore \quad M = \frac{mv^2}{gr}$$

For the given data,

$$M = \frac{(3)(3.141)^2}{(9.81)(0.5)} = 6.03 \text{ kg}$$

Example 14.29 A string of 1 m length can support a maximum load of 6 kg. If a sphere of 2 kg mass is suspended at the bottom end of the string and rotated in a horizontal circle, determine the maximum number of revolutions per minute that can be made by the string without breaking.

Solution Since the string can support a maximum load of 6 kg, the maximum tension in the string is given as

$$T_{max} = mg = 6 \times 9.81 = 58.86 \text{ N}$$

When a sphere of 2 kg mass is suspended at the end of the same string of length $l = 1$ m and rotated with an angular velocity ω in a circle of radius $r = l = 1$ m then the equation of motion in the radial direction can be written as

$$T = mr\omega^2$$

To determine the maximum angular velocity, we equate the tension in the string to the maximum value and hence,

$$58.86 = (2)(1)\omega^2$$

$$\Rightarrow \quad \omega = 5.42 \text{ rad/s} \quad \text{(or)} \quad \frac{5.42 \times 60}{2\pi} = 51.8 \text{ rpm}$$

Example 14.30 A body of 3 kg mass is suspended by an inextensible string of 1 m length. It is rotated in a circular path of 0.5 m radius as shown in Fig. 14.23. Determine the tension in the string and the constant speed of the body.

Solution From the figure, we see that

$$\sin\theta = 0.5/1 = 0.5 \quad \Rightarrow \quad \theta = 30°$$

As there is no motion in the vertical direction, the vertical component of tension must balance the weight of the body, i.e.,

$$T\cos\theta = mg \qquad \text{(a)}$$

The normal acceleration, $a_n = \dfrac{v^2}{r}$ required to keep the body in a circular path is provided by the horizontal component of the tension in the string. Therefore,

$$T\sin\theta = \frac{mv^2}{r} \qquad \text{(b)}$$

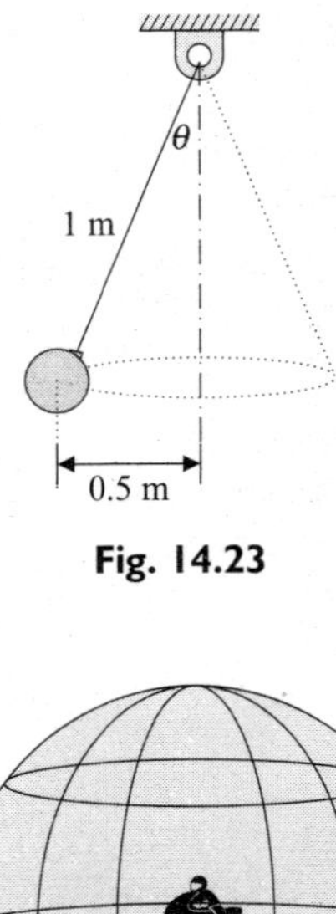

Fig. 14.23

Dividing equation (b) by equation (a), we get

$$\tan\theta = \frac{v^2}{gr}$$

or

$$v^2 = gr\tan\theta$$
$$= (9.81)(0.5)\tan 30°$$

$$\therefore \quad v = 1.683 \text{ m/s}$$

From the equation (a), we get the tension in the string as

$$T = \frac{mg}{\cos\theta} = \frac{3 \times 9.81}{\cos 30°} = 33.98 \text{ N}$$

Example 14.31 In a circus, the motorist moves in a cage of radius r in a horizontal circle ABA at a constant speed. If coefficient of friction between the tires and the cage is μ, determine the minimum speed with which he must move around without sliding down.

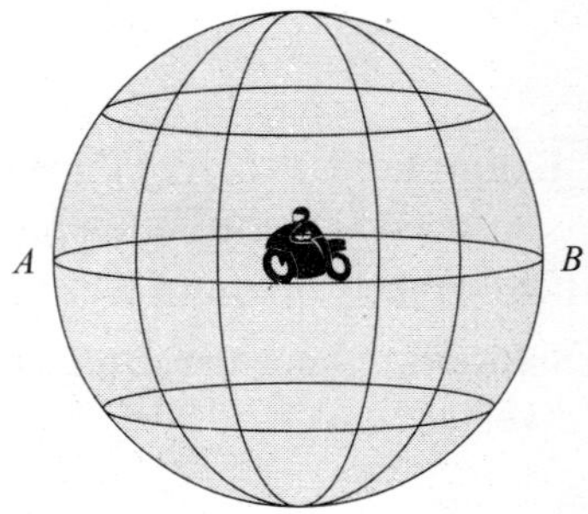

Fig. 14.24

Solution When the bike is moving in a horizontal circle, the forces acting on it are its weight mg, force of friction F and normal reaction R, which provides the normal acceleration towards the centre of the circular path.

As the bike is at the point of sliding, it reaches the limiting friction, i.e., $F = \mu R$. As there is no motion in the vertical direction, we also know that

$$F = mg$$

$$\therefore \quad \mu R = mg \qquad \text{(a)}$$

Since the normal acceleration is provided by the normal reaction R, we also have

$$R = m\frac{v^2}{r} \qquad \text{(b)}$$

From the two equations (a) and (b), we get

$$\frac{mg}{\mu} = m\frac{v^2}{r}$$

$$\Rightarrow \qquad v = \sqrt{\frac{gr}{\mu}}$$

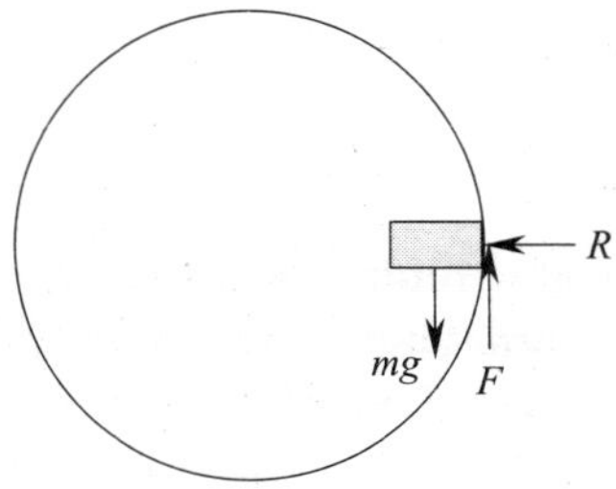

Fig. 14.24(a)

Example 14.32 In the above problem, if the motorist moves in the cage in a vertical circle at a constant speed, determine the reaction (i) at the lowest point A, (ii) at the highest point B, (iii) at the extreme point C, and (iv) the minimum speed required to maintain the circular path.

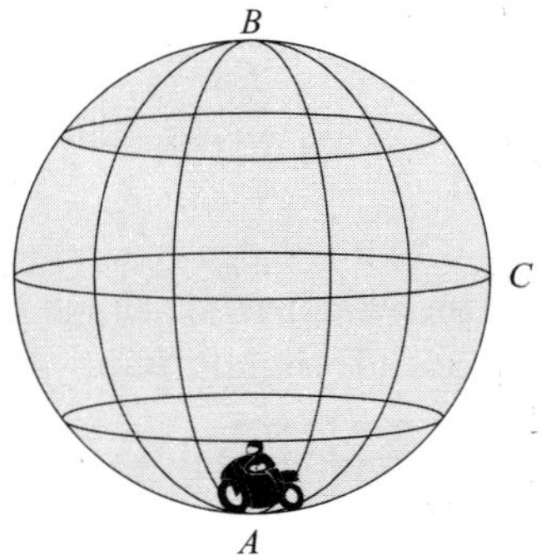

Fig. 14.25

Solution As the bike is moving with constant speed, its tangential acceleration is zero. However, as the direction is constantly changing, its normal acceleration at any instant is non-zero and equal to a constant value, $\frac{v^2}{r}$. This acceleration must be provided to the bike by a force, which in this case is the normal reaction exerted by the cage on the bike. Its magnitude is different at different points along the vertical circular path.

At the lowest point A, the forces acting on the bike are its weight mg acting vertically downwards and the reaction R acting vertically upwards. Hence, applying the equation of motion in the vertical direction, we have

$$R - mg = ma_n$$

$$= \frac{mv^2}{r}$$

Therefore,
$$R = mg + \frac{mv^2}{r}$$

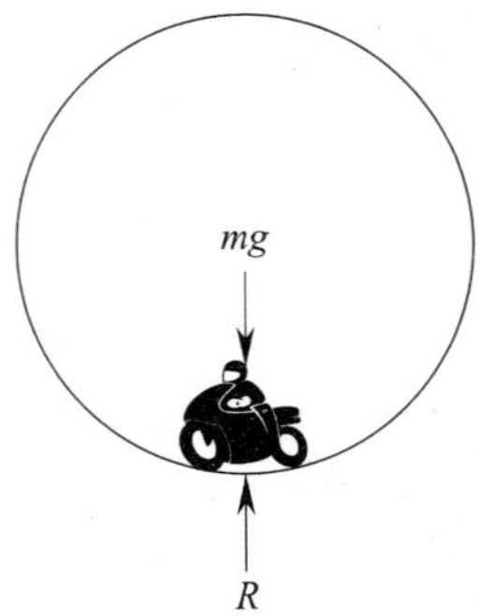

Fig. 14.25(a)

When the bike is at the highest point B, the forces acting on it are its weight mg and the reaction R exerted by the cage on the bike, both acting vertically downwards. Hence, writing the equation of motion in the vertical direction,

$$R + mg = \frac{mv^2}{r}$$

$$\therefore \qquad R = \frac{mv^2}{r} - mg$$

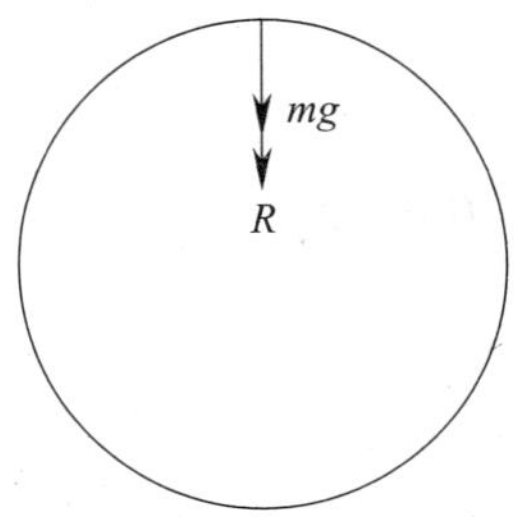

Fig. 14.25(b)

When the bike is at the extreme right point C, the forces acting on it are the weight mg acting vertically downwards, the frictional force preventing it from sliding downwards and the reaction R acting horizontally towards the centre of the cage. Hence, writing the equation of motion in the horizontal direction,

$$R = \frac{mv^2}{r}$$

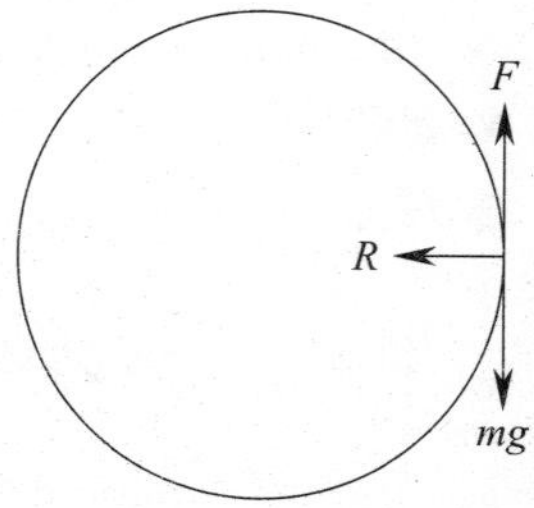

Fig. 14.25 (c)

From the above expressions for the reaction at the lowest, extreme right and highest points, we see that the reaction is the least at the highest point. Hence, the minimum speed with which the bike must travel must be such that it maintains contact with the cage is determined by making the normal reaction R at the highest point equal to zero as the limiting case, i.e.,

$$0 = \frac{mv^2}{r} - mg$$

$$\Rightarrow \quad v^2 = gr$$

$$\text{(or)} \quad v = \sqrt{gr}.$$

As it is the normal reaction which is keeping the bike to stay in the circular path, if it becomes zero [which happens when the velocity is less than the above value] then the bike will lose contact with the cage and will not follow the circular path and eventually fall down.

Example 14.33 A stone of mass m is tied to one end of an inextensible string and the other end of the string is held in hand. If the stone is rotated in a vertical circle of radius r at a constant angular speed, determine (i) the tension at the lowest point, (ii) the tension at the highest point, (iii) the minimum velocity with which the stone must be whirled to keep it in circular path, and (iv) the maximum velocity with which the stone must be whirled, if the allowable tension in the string is $4mg$.

Solution

(i) When the stone is at the lowest point

$$T - mg = \frac{mv^2}{r}$$

$$\Rightarrow \quad T = mg + \frac{mv^2}{r} \quad \text{(a)}$$

(ii) When the stone is at the highest point

$$T + mg = \frac{mv^2}{r}$$

$$\Rightarrow \quad T = \frac{mv^2}{r} - mg \quad \text{(b)}$$

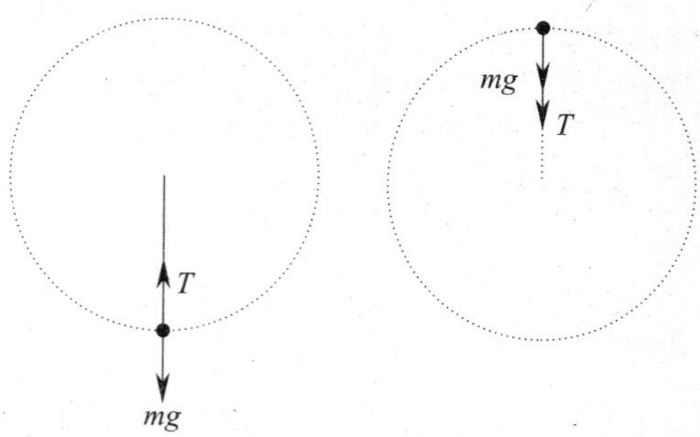

Fig. 14.26

(iii) Minimum velocity to maintain circular path

As the tension is the lowest at the highest point, the minimum velocity required to maintain circular path is obtained by equating T to zero in the above equation (b),

$$T = 0$$

$$\Rightarrow \quad v = \sqrt{gr}$$

(iv) Maximum velocity with which the stone must be whirled

As the tension at the lowest point is the maximum, we obtain the maximum velocity with which the stone must be whirled by equating the tension to the allowable tension. Hence,

$$4mg = mg + \frac{mv^2}{r}$$

$$\Rightarrow \quad 3mg = \frac{mv^2}{r}$$

$$\Rightarrow \quad v = \sqrt{3gr}$$

Example 14.34 A block of mass m is at rest at the topmost point of a hemispherical shell. If the block begins to slide over the hemispherical shell, determine the position of a point on the hemisphere at which the block loses contact with the shell.

Solution When the block begins to slide, it undergoes circular motion in vertical plane. The forces acting on the block are its weight mg, normal reaction R exerted by the shell on the block. Resolving the motion along normal direction, we get

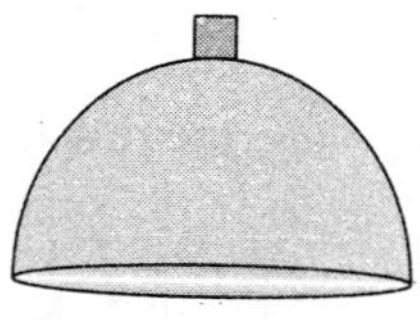

Fig. 14.27

$$mg\cos\theta - R = \frac{mv^2}{r}$$

or

$$R = mg\cos\theta - \frac{mv^2}{r}$$

We see from the figure that $\cos\theta = (r-h)/r$ and as the block slides from rest, its velocity after falling though a height h is $v = \sqrt{2gh}$. Therefore,

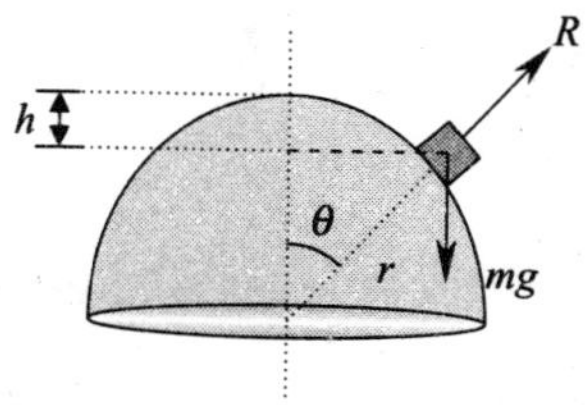

Fig. 14.27(a)

$$R = m\left[g\left(\frac{r-h}{r}\right) - \frac{2gh}{r}\right]$$

$$= mg\left[\left(\frac{r-h}{r}\right) - \frac{2h}{r}\right]$$

$$= \frac{mg}{r}[r-h-2h]$$

$$= \frac{mg}{r}[r-3h]$$

We see that when the block loses contact with the shell, its reaction R is zero. Therefore,

$$0 = \frac{mg}{r}[r-3h]$$

$$\Rightarrow \quad h = \frac{r}{3}$$

Example 14.35 A motorcyclist travels along a level curved track with a radius of curvature of 90 m. If the coefficient of friction between the wheels and the road is 0.25, determine the maximum speed with which he can travel without skidding.

Solution When the motorcycle is about to slide, the force of friction reaches limiting friction. Hence,

$$F = \mu N = \mu mg$$

This frictional force provides the normal acceleration of the vehicle. Hence, writing the equation of motion,

$$\mu mg = \frac{mv^2}{r}$$

$$\Rightarrow \qquad v = \sqrt{\mu gr}$$

$$= \sqrt{(0.25)(9.81)(90)} = 14.86 \text{ m/s (or) } 53.5 \text{ kmph}$$

Example 14.36 A particle is projected with an initial velocity v_o at an angle of inclination α to the horizontal. Determine the equation of motion considering air resistance of k times the velocity.

Solution As the initial velocity is v_o, the initial components of velocity in the x and y directions are respectively

$$v_{ox} = v_o \cos\alpha \quad \text{and} \quad v_{oy} = v_o \sin\alpha$$

As the resistance is k times the velocity, the acceleration along the x and y directions are given as

$$a_x = -kv_x \quad \text{and} \quad a_y = -g - kv_y$$

Note that as the air medium offers resistance to the motion, the acceleration has a negative sign.

We can see that the accelerations in both the directions are not constant but vary with time. Hence, we cannot apply the equations of motion with constant acceleration. Instead, we use the method of integration to determine the motion of the particle.

Motion along the y-direction

Considering the motion along the y-direction,

$$a_y = \frac{dv_y}{dt} = -g - kv_y$$

By separation of variables, we can write,

$$\frac{dv_y}{g + kv_y} = -dt$$

Upon integration,

$$\frac{1}{k}\ln[g + kv_y] = -t + c_1 \qquad \text{(a)}$$

where c_1 is the constant of integration. We know that at $t = 0$, $v_y = v_o \sin\alpha$. Therefore,

$$c_1 = \frac{1}{k}\ln[g + kv_o \sin\alpha] \qquad \text{(b)}$$

Substituting this value in the equation (a),

$$\ln\left[\frac{g + kv_y}{g + kv_o \sin\alpha}\right] = -kt$$

or

$$g + kv_y = [g + kv_o \sin\alpha]e^{-kt}$$

$$\therefore \qquad v_y = \left[\frac{g}{k} + v_o \sin\alpha\right] e^{-kt} - \frac{g}{k} \qquad \text{(c)}$$

Thus, we obtain the velocity along the *y*-direction as a function of time. Similarly, we can obtain displacement as a function of time as follows:

Equation (c) can be written as:

$$\frac{dy}{dt} = \left[\frac{g}{k} + v_o \sin\alpha\right] e^{-kt} - \frac{g}{k}$$

Upon integration,

$$y = \left[\frac{g}{k} + v_o \sin\alpha\right] \frac{e^{-kt}}{-k} - \frac{g}{k} t + c_2$$

We know that at $t = 0$, $y = 0$. Therefore,

$$c_2 = \frac{\left[\frac{g}{k} + v_o \sin\alpha\right]}{k} \qquad \text{(d)}$$

Hence,

$$y = \frac{1}{k}\left[\frac{g}{k} + v_o \sin\alpha\right][1 - e^{-kt}] - \frac{g}{k} t \qquad \text{(e)}$$

Motion along the x-direction

Considering the motion along the *x*-direction,

$$a_x = \frac{dv_x}{dt} = -kv_x$$

By separation of variables, we can write,

$$\frac{dv_x}{v_x} = -kdt$$

Upon integration,

$$\ln v_x = -kt + c_3 \qquad \text{(f)}$$

where c_3 is the constant of integration. We know that at $t = 0$, $v_x = v_o \cos\alpha$. Therefore,

$$c_3 = \ln (v_o \cos\alpha) \qquad \text{(g)}$$

Substituting this value in the equation (f),

$$\ln\left[\frac{v_x}{v_o \cos\alpha}\right] = -kt$$

or

$$v_x = v_o \cos\alpha\, e^{-kt} \qquad \text{(h)}$$

Equation (h) can be written as

$$\frac{dx}{dt} = v_o \cos\alpha\, e^{-kt}$$

Upon integration,

$$x = \frac{v_o \cos\alpha\, e^{-kt}}{-k} + c_4$$

At $t = 0$, $x = 0$. Therefore,

$$c_4 = \frac{v_o \cos\alpha}{k} \tag{i}$$

Hence,

$$x = \frac{v_o \cos\alpha}{k}[1 - e^{-kt}] \tag{j}$$

Example 14.37 In the previous problem, what is the total acceleration at any instant of time and its direction with respect to the horizontal? What do you infer from the result? At what time does it reach the maximum height? Also, determine the maximum height.

Solution From the equation (c), we have

$$v_y = \left[\frac{g}{k} + v_o \sin\alpha\right]e^{-kt} - \frac{g}{k}$$

Upon differentiating it with respect to time,

$$a_y = \frac{dv_y}{dt} = \left[\frac{g}{k} + v_o \sin\alpha\right](-k)e^{-kt}$$

$$= -[g + kv_o \sin\alpha]\, e^{-kt}$$

From the equation (h), we have

$$v_x = v_o \cos\alpha\, e^{-kt}$$

Upon differentiating it with respect to time,

$$a_x = \frac{dv_x}{dt} = -kv_o \cos\alpha\, e^{-kt}$$

Therefore, the magnitude of total acceleration at any instant of time is given as

$$a = \sqrt{a_x^2 + a_y^2} = e^{-kt}\sqrt{(-kv_o \cos\alpha)^2 + (-g - kv_o \sin\alpha)^2}$$

$$= e^{-kt}\sqrt{k^2 v_o^2 + g^2 + 2kv_o g \sin\alpha}$$

The inclination of the acceleration at any instant of time is given as

$$\theta = \tan^{-1}\left[\frac{a_y}{a_x}\right] = \tan^{-1}\left[\frac{-(g + kv_o \sin\alpha)e^{-kt}}{-kv_o \cos\alpha\, e^{-kt}}\right]$$

$$= \tan^{-1}\left[\frac{g + kv_o \sin\alpha}{kv_o \cos\alpha}\right] = \text{constant}$$

Thus, we see that the total acceleration at any instant of time is inclined at a constant angle to the horizontal.

Time taken to reach the maximum height

At the maximum height, the y-component of velocity is zero. Hence,

$$v_y = \left[\frac{g}{k} + v_o \sin\alpha\right] e^{-kt} - \frac{g}{k} = 0$$

$$\Rightarrow \qquad e^{-kt} = \frac{g}{g + kv_o \sin\alpha}$$

Therefore,

$$t = -\frac{1}{k} \ln\left[\frac{g}{g + kv_o \sin\alpha}\right]$$

Maximum height reached

Substituting the value of t in the equation (e), we get the maximum height reached,

$$h_{\max} = \frac{v_o \sin\alpha}{k} + \frac{g}{k^2} \ln\left[\frac{g}{g + kv_o \sin\alpha}\right]$$

SUMMARY

The branch of dynamics, which deals with the motion of bodies without considering the cause of motion, i.e., just the geometry of the motion is termed *kinematics*; whereas the branch of dynamics which deals with the motion of bodies considering together the cause of motion is termed *kinetics*.

First Law of Motion

Every body continues in its state of rest or of uniform motion in a straight line unless it is compelled to change that state by forces acting on it.

Stated in an alternate way, if no external force acts on a body, it will not have acceleration, i.e., its velocity will remain constant. If it were initially at rest, it would continue to remain at rest and if it were initially moving with constant velocity along a straight line, it would continue to do so. Hence, to cause a change in velocity or in other words, to accelerate a body, an external force must act on it.

Second Law of Motion

It relates the force causing the change in motion with the resulting acceleration and it can be stated as follows: *If the resultant force acting on a particle is not zero, the particle will have acceleration proportional to the magnitude of the resultant and in the direction of this resultant force.* Mathematically, Newton's second law of motion can be stated as

$$\vec{F} = m\vec{a}$$

where $\vec{F}$ represents the resultant of a system of forces acting on a particle.

Motion of Bodies in Rectangular Coordinates

If force and acceleration vectors can be resolved into rectangular components along x and y directions for plane motion, then Newton's second law can also be expressed in scalar forms as

$$\sum F_x = ma_x$$

and

$$\sum F_y = ma_y$$

where F_x and F_y are respectively x and y components of the individual forces in a system of forces. While solving problems in dynamics, it will be found convenient to choose the initial direction of motion of the bodies as positive, irrespective of whether they point along positive or negative direction of the axes.

Variable Acceleration

When a particle is subjected to a variable force, the resulting acceleration will also vary. To determine such types of motion, we use the method of integration.

Tangential and Normal Components of Acceleration

When a particle moves in a curvilinear path, the vector equation of motion $\Sigma\vec{F} = m\,\vec{a}$ can be represented in tangential and normal components as

$$\Sigma F_t = ma_t$$

$$\Sigma F_n = ma_n$$

A special case of curvilinear motion is *circular motion*. Suppose a particle moves with constant speed in a circular path. Then we know that its tangential component of acceleration is zero. However, its normal component is non-zero, as the particle continuously changes its direction of motion. Hence,

$$\Sigma F_t = 0 \quad \text{and} \quad \Sigma F_n = ma_n = mv^2/r$$

From Newton's first law of motion, we can understand that an external force must act on the particle to impart this normal acceleration (even though it is moving with constant speed). It is this external force, which causes change in direction of motion and thus, keeps the particle in a circular path.

EXERCISES

Objective-type Questions

1. State which of the following statements is false:

(a) A force is required to move a body at rest

(b) A force is required to keep a body moving with uniform velocity

(c) A force is required to change the direction of motion of a body moving with uniform velocity

(d) A force is required to accelerate a body moving with uniform velocity

2. The physical quantity, which is a measure of inertia, is

(a) mass (b) velocity (c) acceleration (d) force

3. What is the inclination of a smooth inclined plane such that the accelerating force acting on a block sliding down it may be $g/2$?

(a) 30° (b) 45° (c) 60° (d) 75°

4. The acceleration of a block sliding down an inclined plane is

(a) same as acceleration due to gravity

(b) less than acceleration due to gravity

(c) greater than acceleration due to gravity

(d) uniformly increasing

5. In the system of pulleys shown, the ratio of velocities of blocks *A* and *B* are

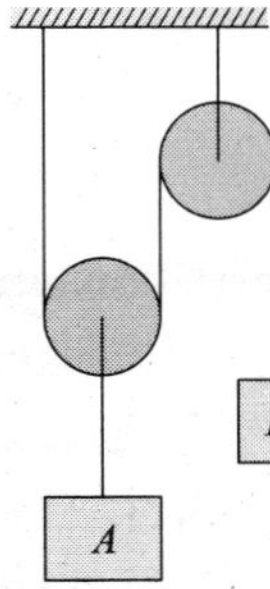

(a) $v_A = v_B$ (b) $2v_A = v_B$
(c) $3v_A = v_B$ (d) $v_A = 2v_B$

6. Two blocks *A* and *B* such that $W_A > W_B$ slide down an inclined plane at the same time. If the coefficient of friction is same for both then which of the following statements is true?
(a) Acceleration of *A* is greater than acceleration of *B*.
(b) Acceleration of *A* is lesser than acceleration of *B*.
(c) Both the accelerations are equal.
(d) Their accelerations are equal to acceleration due to gravity.

7. When a lift is accelerating up, the weight of a man standing on the floor of the lift is
(a) same as that when on ground (b) greater than that on ground
(c) less than that on ground (d) zero

8. When a lift is decelerating when moving upwards, the weight of a man standing on the floor of the lift is
(a) same as that when on ground (b) greater than that on ground
(c) less than that on ground (d) zero

9. When a lift is accelerating downwards, the weight of a man standing on the floor of the lift is
(a) same as that when on ground (b) greater than that on ground
(c) less than that on ground (d) zero

10. If a lift is decelerating while moving downwards then the weight of a man standing on the floor is
(a) same as that when on ground (b) greater than that on ground
(c) less than that on ground (d) zero

11. When a lift is freely falling, the reaction exerted by the floor of the lift on a man standing on the floor is
(a) same as that when on ground (b) greater than that on ground
(c) less than that on ground (d) zero

12. When a car moves over a bump, the pressure exerted by the wheels on the road is
(a) same as that on level road (b) greater than that on level road
(c) less than that on level road (d) zero

13. When a stone tied to one end of a string is whirled in a vertical circle, the tension in the string is maximum at
(a) the lowest point (b) the highest point (c) mid-height (d) 45° to the vertical

14. When a stone tied to one end of a string is whirled in a vertical circle, the tension in the string is the least at
(a) the lowest point (b) the highest point (c) mid-height (d) 45° to the vertical

Answers

1. (b)	**2.** (a)	**3.** (a)	**4.** (b)	**5** (b)	**6.** (c)	**7.** (b)	**8.** (c)
9. (c)	**10.** (b)	**11.** (d)	**12.** (c)	**13.** (a)	**14.** (b)		

Short-answer Questions

1. Distinguish between kinematics and kinetics.
2. Discuss on the experiments of Galileo and his conclusions.
3. Define inertia and how can it be measured.
4. State Newton's first and second-laws of motion.
5. Derive the mathematical expression for Newton's second law of motion.
6. Express the scalar forms of equations of motion.
7. State D'Alembert's principle.
8. Discuss the forces providing the normal acceleration in circular motions considering various examples.

Numerical Problems

14.1 A block of 5 kg mass rests on a frictionless horizontal table. It is acted upon by a system of forces lying on a plane parallel to the table as shown in Fig. E.14.1. Determine the acceleration of the block and the direction of motion.

Ans. 0.9 m/s^2, 28.4° to X-axis

14.2 A block of 10 kg mass rests on a rough horizontal surface, whose coefficient of kinetic friction is 0.2. It is being pulled by a constant force of 50 N magnitude inclined at 30° to the horizontal. Determine the velocity and the distance travelled by the block after 5 seconds. Refer Fig. E.14.2.

Ans. 14.35 m/s; 35.88 m

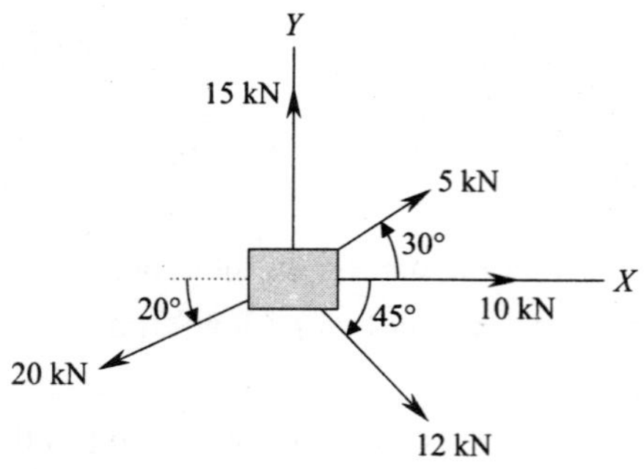

Fig. E.14.1

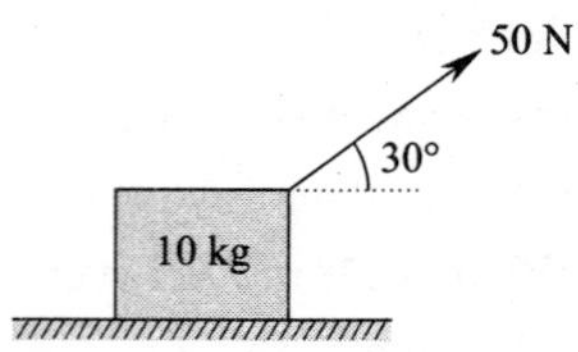

Fig. E.14.2

14.3 A block of 200 kg mass is pushed up a 25° incline as shown in Fig. E.14.3. Determine the horizontal force exerted by the truck on the block if (i) the block moves at a constant speed, and (ii) at an acceleration of 0.5 m/s^2. The coefficient of kinetic friction between the block and the incline is 0.25.

Ans. (i) 1273.7 N; (ii) 1373.7 N

14.4 A body of 2 kg mass is acted on by a force $\vec{F} = 5t\vec{i} + 2t^2\vec{j}$. Determine the position, velocity and acceleration of the body at $t = 2$ s, if the body started from rest at the origin.

Ans. 3.59 m at 21.8° to x-axis; 5.67 m/s at 28.1° to x-axis, 6.4 m/s^2 at 38.7° to x-axis

14.5 A constant force of 30 N magnitude acts horizontally on a body of 10 kg mass moving at 5 m/s. Determine its displacement and velocity after 3 seconds.

Ans. 28.5 m; 14 m/s

14.6 A block of 5 kg mass is at rest on a rough horizontal surface, whose coefficient of friction is 0.25. A constant force of 20 N magnitude acts on the block parallel to the plane for a duration of 3 seconds [refer Fig. E.14.6]. Determine (i) the velocity of the block after 3 s, and (ii) distance travelled in that time. Also, determine what further distance will the block move after the force is removed and the time taken.

Ans. (i) 4.65 m/s, (ii) 6.98 m; 4.41 m, 1.9 s

Fig. E.14.3

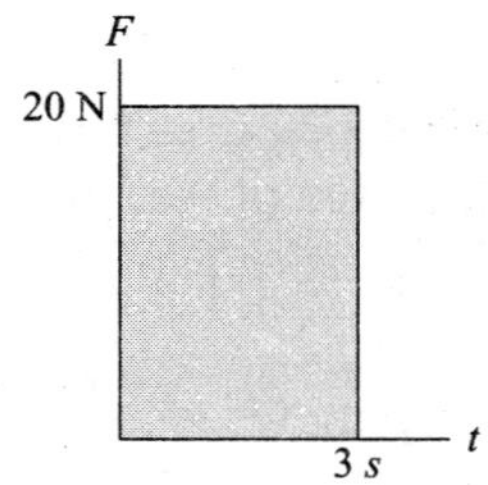

Fig. E.14.6

14.7 A car of 3 ton mass is moving at a constant speed of 60 kmph. What is the braking force required to bring the car to a stop in a distance of 150 m? Also, determine the time taken to bring the car to a stop.

Ans. 2.8 kN, 18 s

14.8 A truck of 10 ton mass must reach a velocity of 54 kmph in 10 seconds. If the frictional resistance is 1 kN/ton, what should be the tractive force of the truck?

Ans. 20 kN

14.9 An airplane of 20 ton mass has to take off with a velocity of 200 kmph on a runway of 300 metres. If the frictional resistance is 1 kN/ton mass, what is the minimum tractive force required for take off?

Ans. 122.8 kN

14.10 A car of 2 ton mass is travelling at a constant speed of 60 kmph. The brakes are applied and it comes to rest in 5 seconds. Determine the braking force.

Ans. 6.67 kN

14.11 A car of 2 ton mass starts from rest and attains a speed of 60 kmph in 15 seconds. If the frictional resistance is 0.5 kN/ton mass, determine the tractive force. If the engine is switched off when moving with a constant speed of 60 kmph, in how much time will it come to a stop and what will be the distance travelled during this time? Assume the same resistance as before.

Ans. 3.22 kN, 33.34 s, 277.9 m

14.12 A long-bed truck with a heavy mass of 500 kg resting on it is travelling at a uniform speed of 45 kmph. On seeing another vehicle just stopping ahead, the driver of the truck applies the brake to bring it to a halt in 6 seconds. Determine (i) the minimum coefficient of friction between the mass and the truck bed to prevent sliding of the mass, and (ii) the force of friction.

Ans. (i) 0.212; (ii) 1.04 kN

14.13 A body of 5 kg mass is lifted up by a rope tied to it. What is the maximum acceleration with which it can be raised, if the allowable tension in the string is 60 N?

Ans. 2.19 m/s^2

14.14 An elevator together with the passengers weighing 2 tons is supported by a cable. Determine the acceleration of the lift when the tension in the cable is (i) 23 kN when the lift is moving upwards, (ii) 18 kN when it is moving upwards, (iii) 16 kN when the lift is moving downwards, and (iv) 21 kN when the lift is moving downwards.

Ans. (i) 1.69 m/s^2 accelerating upwards, (ii) 0.81 m/s^2 decelerating upwards (iii) 1.81 m/s^2 accelerating downwards, (iv) 0.69 m/s^2 decelerating downwards

14.15 A lift starting from rest attains a maximum speed of 3 m/s in 2 seconds. The allowable tension in the cable of the lift is 12 kN and the mass of the lift is 600 kg. Determine the maximum carrying capacity of the lift. [Take average weight of one person to be 60 kg].

Ans. 7 persons

14.16 An elevator weighing 6 tons together with the passengers descends with a speed of 4 m/s. If the tension in the cable must not exceed 50 kN, what is the shortest distance in which the elevator can be stopped?

Ans. 5.4 m

14.17 A 60 kg man stands on a scale in an elevator. Determine the acceleration of the lift when the scale reads (i) 80 kg, (ii) 50 kg, and (iii) zero.

Ans (i) 3.27 m/s^2 accelerating upwards or decelerating downwards; (ii) 1.64 m/s^2 decelerating upwards or accelerating downwards; (iii) 9.81 m/s^2 (freely falling)

14.18 A body attached to a spring balance suspended from the ceiling of a moving lift weighed 50 N. If the lift moved up at 1 m/s^2 at this instant, determine its true weight.

Ans. 45.4 N

14.19 In Fig. E.14.19, a force of 30 N is applied on the lower block of 5 kg mass, over which another block of 3 kg mass rests. Determine the acceleration of the blocks and the tension in the string assuming it to be inextensible. The coefficient of kinetic friction for all contact surfaces is 0.15.

Ans. 1.18 m/s^2; 7.95 N

14.20 A block of mass m_2 = 8 kg resting on a rough horizontal plane is pulled by an inextensible string, whose other end is attached to a block of mass m_1 = 5 kg and passing over a rough surface as shown in Fig. E.14.20. Determine the acceleration of the system and the tension in each portion of the string. The coefficient of friction at all contact surfaces is 0.2.

Ans. 1.73 m/s^2, 40.4 N in the vertical portion of the string; 29.5 N in the horizontal portion of the string

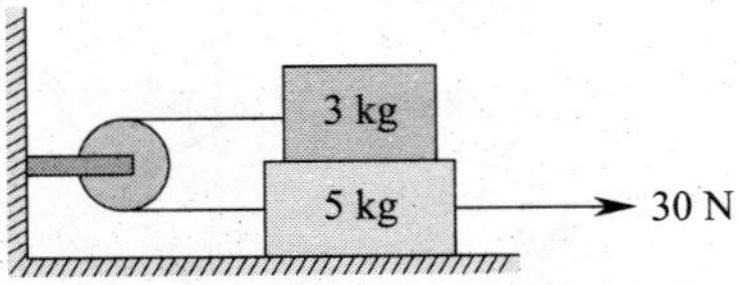

Fig. E.14.19

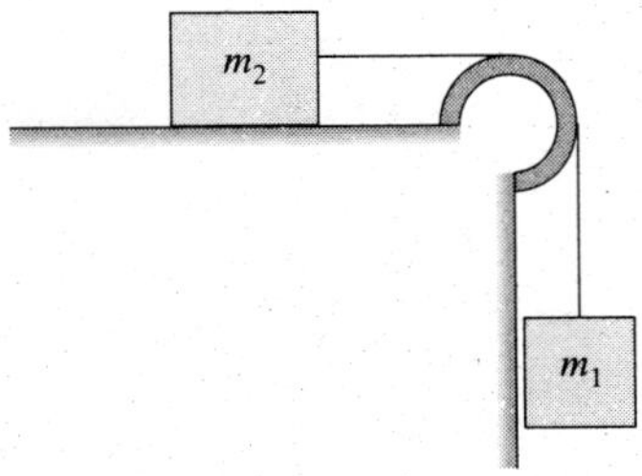

Fig. E.14.20

14.21 Determine the acceleration of the system of blocks shown in Fig. E.14.21. Assume the pulleys to be massless and frictionless. Take $m_1 = 8$ kg, $m_2 = 2$ kg and $m_3 = 3$ kg. Also, determine the tension in the strings connecting the blocks.

Ans. $a_1 = g/4$ m/s^2, $a_2 = g/2$ m/s^2, $a_3 = 0$; $T_1 = 6\,g$ N, $T_2 = T_3 = 3\,g$ N

14.22 A block of mass $m_2 = 6$ kg resting on a rough horizontal plane is pulled by an inextensible string passing over a pulley, which supports a block of mass $m_1 = 5$ kg as shown in Fig. E.14.22. Determine the acceleration of the system and the tensions in each portion of the string connecting the blocks. The coefficient of friction between the block II and the plane is 0.2.

Ans. $a_1 = 0.88$ m/s^2; $a_2 = 1.76$ m/s^2; 44.65 N in the string supporting mass m_1, 22.33 N in the string connected to mass m_2

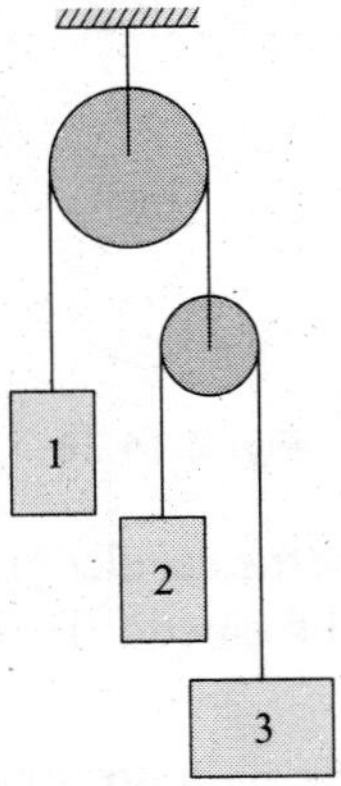

Fig. E.14.21

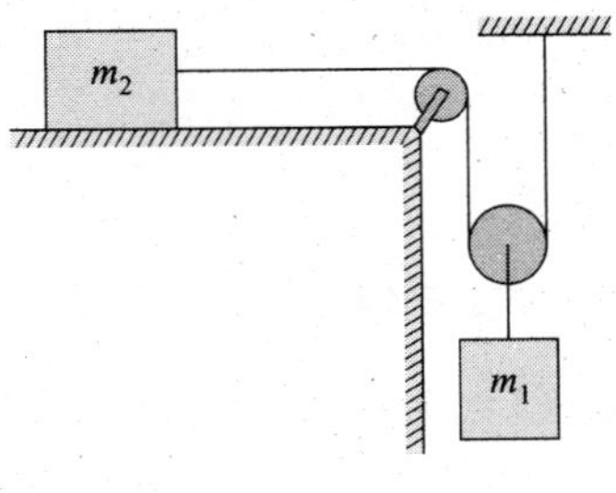

Fig. E.14.22

14.23 Determine the acceleration of the system of blocks shown in Fig. E.14.23. The coefficient of kinetic friction between the block 2 and the inclined plane is 0.2. Also, determine the tension in the string. Take $m_1 = 6$ kg, $m_2 = 8$ kg. Assume the pulley to be massless and frictionless.

Ans. 0.43 m/s^2; 56.3 N

14.24 Two blocks placed on two inclined planes are connected by a string passing over a pulley at the vertex as shown in Fig. E.14.24. If released from rest, determine the acceleration of the system and the tension in the string. Take $m_1 = 30$ kg and $m_2 = 25$ kg. Assume the pulley to be massless and frictionless.

Ans. 2.404 m/s^2, 182.7 N

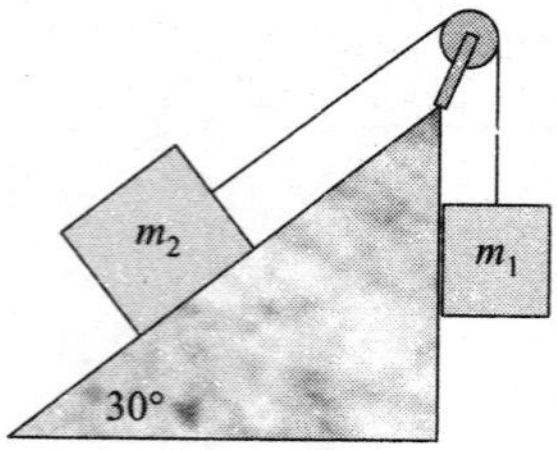

Fig. E.14.23

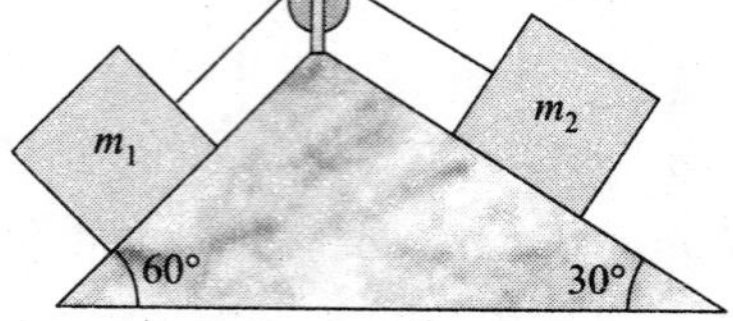

Fig. E.14.24

14.25 Determine the acceleration of the system of blocks shown in Fig. E.14.25. The coefficient of kinetic friction between all contact surfaces is 0.2. Also, determine the tension in the string. Assume the pulley to be massless and frictionless.

Ans. 0.33 m/s^2; 11.5 N

14.26 Determine the acceleration of the system of blocks shown in Fig. E.14.26. The coefficient of kinetic friction between block of mass M and the surface is 0.2. Also, determine the tension in each portion of the string. Take $m_1 = 3$ kg, $m_2 = 5$ kg and $M = 3$ kg. Assume the string to be inextensible, the pulleys to be frictionless and massless.

Ans. 1.25 m/s^2; 33.2 N on the left portion of the string, 42.8 N on the right portion

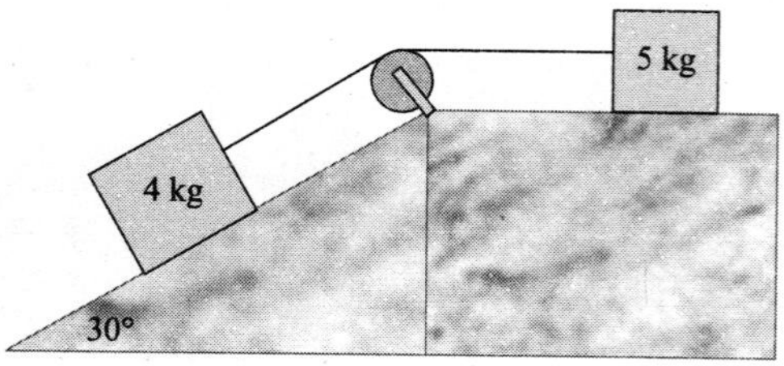

Fig. E.14.25

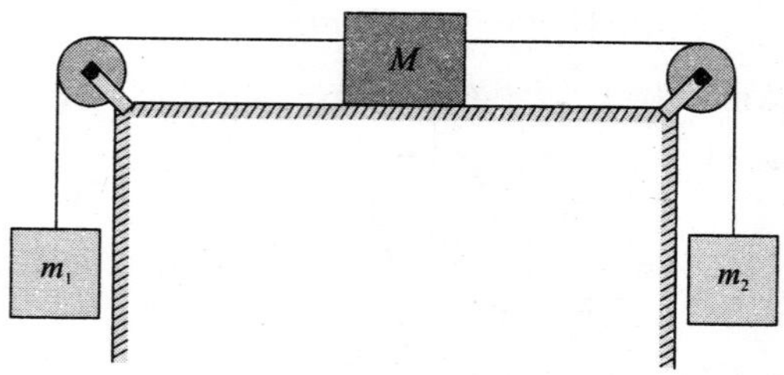

Fig. E.14.26

14.27 In the Fig. E.14.27, if the blocks are released from rest, determine the acceleration of the blocks and the tension in each portion of the string. The coefficient of kinetic friction between all contact surfaces is 0.2.

Ans. 1.4 m/s^2; 22.42 N on the left portion of the string; 42.05 N on the right portion of the string

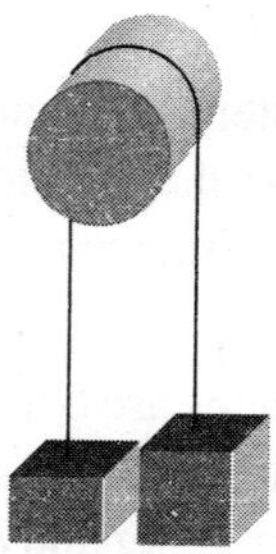

Fig. E.14.27

14.28 A block of mass m rests on a rough turntable at a distance r from the centre. If the coefficient of friction between the block and the table is μ, determine the maximum speed at which the table can be rotated without the block sliding.

Ans. $\sqrt{\mu g r}$

14.29 Two blocks of equal mass resting on a smooth table are tied by a string such that one is at a distance r from the centre of rotation and the other at a distance of $2r$ from the centre of rotation. When the blocks rotate in a horizontal circle such that both are on the same radial line, determine the tension in each part of the string.

Ans. 3 m $r\omega^2$, 2 m $r\omega^2$

14.30 A string can support a maximum load of 2 kg. If a stone of 500 g mass is tied to it and whirled in a horizontal circle of 1 m radius, determine the maximum speed with which it can be whirled.

Ans. 6.26 m/s

14.31 A bob of 1 kg mass is attached to a vertical shaft as shown in Fig. E.14.31 by a string of 1 m length. If the shaft rotates at a constant angular speed of 60 rpm, determine (i) the inclination of the string with the shaft, and (ii) the tension in the string.

Ans. (i) 75.6°, (ii) 39.5 N

14.32 In the above problem, if the allowable tension in the string is 50 N, determine (i) the maximum allowable angular speed of the shaft, and (ii) the corresponding inclination of the string with the shaft.

Ans. (i) 67.5 rpm, (ii) 78.7°

14.33 A smooth steel ball of 1 g mass and 2 cm diameter is placed in a smooth hemispherical cup with an inner radius of 15 cm. If the cup is rotated at a constant angular speed of 100 rpm about its vertical axis, determine the equilibrium position of the ball with respect to the vertical axis. Refer Fig. E.14.33.

Ans. 50.3°

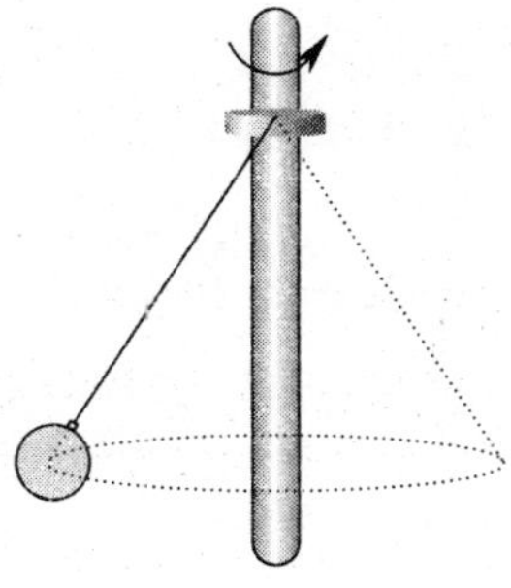

Fig. E.14.31, Fig. E.14.32

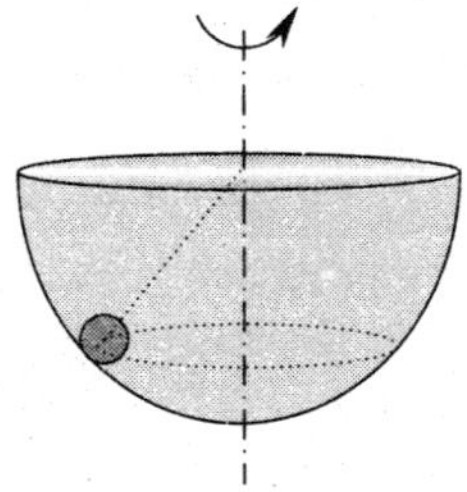

Fig. E.14.33

15 Work and Energy

15.1 INTRODUCTION

In the preceding chapter, we solved kinetic problems using Newton's laws of motion. From the second law of motion, we determined the *acceleration* of a body or a system of bodies. Once acceleration is known, we could describe the kinematics of the system, i.e., its *displacement* and *velocity* as functions of time. In this chapter, we will introduce an alternative approach, called **work–energy** method to solve the same type of kinetic problems.

Unlike Newton's second law of motion, which relates force and acceleration, the work–energy method relates *force*, *velocity* and *displacement*. The work–energy method has certain advantages over Newton's method for the following reasons: Firstly, work and energy are *scalar* quantities and hence, they add up algebraically; thus, avoiding the need to deal with directional aspects of force vectors under Newton's method. Secondly, the kinematics of the problem, i.e., displacement and velocity could be determined directly without knowing the acceleration of the system. Displacement and velocity are more real to understand than acceleration, which seems to be an *abstract* quantity. Thirdly, motion of inter connected bodies could be solved without drawing separate free-body diagrams for each body in the system.

In Sections 15.2–15.4, we will define the concept of work in general and works done by a constant force and a variable force. In Section 15.5, we will define power of the driving force, since the rate at which work is done is more important than the work itself. In Sections 15.6 and 15.8, we will define energy, various forms of energy, kinetic and potential energies. In Section 15.7, we will discuss applying work–energy method to solve motion of interconnected bodies.

15.2 WORK DONE BY A FORCE

When a force acting on a particle causes a *displacement* of the particle, the force is then said to have done **work** on the particle. This definition of work is quite different from our daily usage of the word 'work,' which we refer to any activity involving muscular or mental effort. Consider a force $\vec{F}$ acting on a particle at A causing a displacement $\vec{s}$ (from the point A to B) in the direction of the force. We then define work done on the particle as a **product** of magnitudes of **force** and **displacement**. Mathematically, we can write this as

$$W = F s \tag{15.1}$$

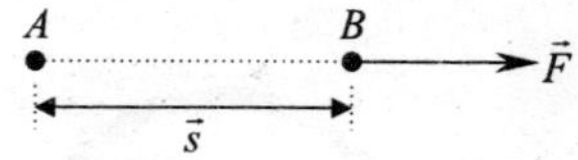

Fig. 15.1 Work done by a force displacing a particle in the direction of the force

A ball freely falling under gravity or a block being pulled on a smooth plane by a horizontal force (refer Fig. 15.2) are examples in which the motion of the body is in the direction of the applied force. (It should be noted that as the line of action of the force passes through the centre of gravity of each body, we could idealize each body as a particle). However, we should note that due to *constraints* involved, the displacement of a particle under the action of forces would not always occur in the direction of the force. Consider for instance, a block sliding down a smooth inclined plane due to pull of gravity or a block being pulled on a smooth plane by an inclined force (refer Fig. 15.3). In these two cases, we observe that the direction of motion is *different* to the direction of the force acting, i.e., inclined at an angle θ to the direction of the force.

Fig. 15.2

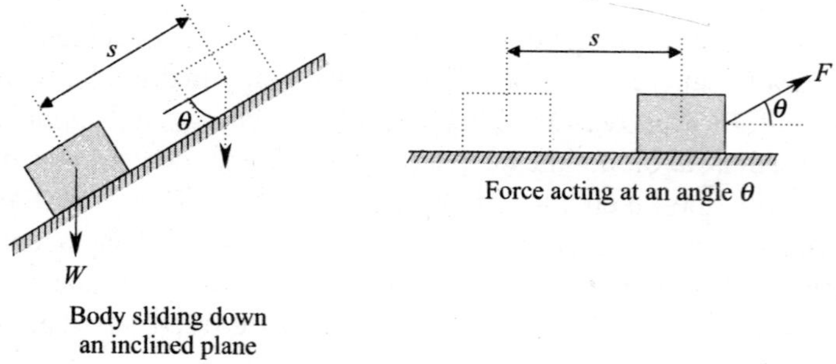

Fig. 15.3

Hence, we can define work done in general, as a product of the **component** of the force in the direction of motion and the displacement. Mathematically, we can write this as

$$W = (F\cos\theta)s \tag{15.2}$$

As the above expression can also be written as

$$W = (F)\,(s\cos\theta) \tag{15.3}$$

work done can also be defined as a product of the force and *component* of displacement in the direction of the force.

The unit of work done is dependent on the units of force and displacement. Hence, its S.I. unit is N.m. This is also called Joule, in honour of the British scientist James P. Joule, who worked on the relationship between heat and work. It is defined as the work done by a unit force (1 newton) when it causes a unit displacement (1 metre) along the direction of the force.

For a system of forces acting on a particle, the work done by the system of forces is given by the *algebraic sum* of works done by individual forces. Since work done adds up algebraically, it is a *scalar* quantity, that is, having magnitude, but not direction. However, we see that force and displacement are vectors. The product of two vectors resulting in a scalar quantity can best be represented by the dot product of two vectors. Thus, work done can be written in vector form as

$$W = \vec{F}.\vec{s} = Fs\cos\theta \tag{15.4}$$

If force and displacement are in the same direction, i.e., $\theta = 0°$, then $\cos\theta = 1$ and hence work done is $W = Fs$, which is the same as the Eq. 15.1. If force and displacement are in the opposite direction, i.e., $\theta = 180°$, then $\cos\theta = -1$ and hence work done is $W = -Fs$. We saw in Chapter 6 that the force of friction always acts in the direction *opposite* to that of the motion; hence, work done by the force of friction is always **negative**. In general, if the component of force is in the *direction* of displacement then θ is an *acute* angle and $\cos\theta$ is *positive*. Hence, work done in such a case is *positive*. If the component of force is in the *direction opposite* to that of displacement then θ is an *obtuse* angle and $\cos\theta$ is *negative*. Hence, work done in such a case is *negative*.

We also come across situations in which the work done by a force or a system of forces is **zero**. These are discussed below in detail.

(i) When the displacement [s] is zero Even though forces may act on a particle, if there is **no** displacement of the particle then *no* work is done on the particle. Consider a block resting on a table. In its free-body diagram, we see that even though its weight W and normal reaction R are acting, they do *no* work on the block, as there is *no* displacement of the block. Similarly, in the case of a ball suspended by a string and a beam supported at one end by a hinge support, no work is done by the forces acting, as there is no displacement involved.

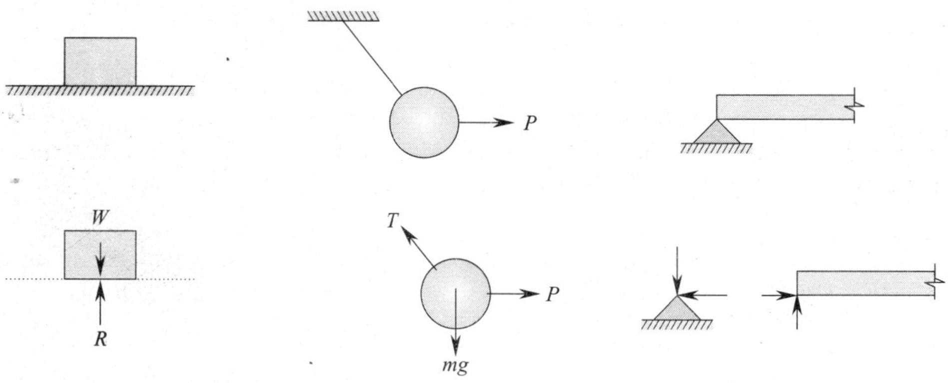

Fig. 15.4 No work is done by forces as there is no displacement

(ii) When the motion is at right angles to the direction of the force When the motion is at right angles to the direction of the force, we see that $\theta = 90°$ and hence, $\cos\theta = 0$. Thus, work done is **zero** as per the Eq. 15.4. Consider a block moving along a horizontal plane as shown in Fig. 15.5. Since the displacement is at right angles to the direction of the forces, namely, its weight and normal reaction, the two forces do *no* work on the block. Similarly, when a body is moving in a circle, the centripetal force does *no* work on the body.

(iii) Total work done is zero When a particle is in *static equilibrium* then the resultant force acting on it is **zero**. Hence, the **total work done** by the system of forces is also **zero**. Consider a system of concurrent forces $\vec{F}_1, \vec{F}_2, \ldots \vec{F}_n$ acting on a particle causing a displacement $\Delta\vec{s}$. Then we can write the total work done on the particle as

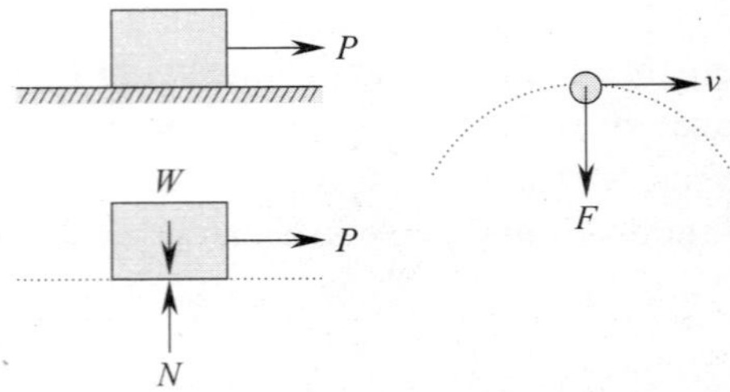

Fig. 15.5 No work is done by forces acting at right angles to the direction of motion

$$W = \vec{F}_1.\Delta\vec{s} + \vec{F}_2.\Delta\vec{s} + \ldots + \vec{F}_n.\Delta\vec{s}$$
$$= [\vec{F}_1 + \vec{F}_2 + \ldots + \vec{F}_n].\Delta\vec{s}$$
$$= \Sigma\vec{F}.\Delta\vec{s}$$
$$= 0 \text{ [since the resultant of the forces } \Sigma\vec{F} \text{ is zero]} \quad (15.5)$$

Corollary The way we have defined work done is quite different from our colloquial usage of the word "work." For instance, a passenger carrying a heavy luggage and waiting for a train is said to do *no* work, even though he has to exert effort to hold the luggage. This is because his effort is causing *no* displacement of the luggage. However, when the train arrives and he lifts the luggage, boards the train and places the luggage on the rack, he is said to do work as it causes displacement of the luggage.

15.3 WORK DONE BY A VARIABLE FORCE

In the previous section, we considered the force as constant, i.e., constant in magnitude and direction and thus we defined work done by the force. In this section, we will discuss work done by a *varying* force acting on a particle.

Consider a variable force $\vec{F}$ acting on a particle as shown in Fig. 15.6. As force is a vector, it can vary in magnitude, direction or both. In such a case, the displacement in general will be along a **curvilinear** path. As shown in the figure, we see that the force vector changes in magnitude as well as in the direction as it moves from the point A to the point B along the path of motion.

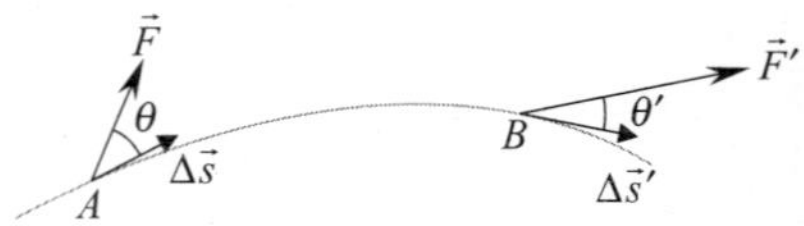

Fig. 15.6 Work done by a variable force

As we let B approach A, the displacement becomes *infinitesimally* small, that is $d\vec{s}$. Then over this infinitesimally small displacement, we can assume the force to remain *constant* in magnitude and direction. Hence, we can write work done over this infinitesimally small displacement $d\vec{s}$ as

$$dW = \vec{F}.d\vec{s} \quad (15.6)$$

Therefore, work done over the entire path is obtained by integrating the above expression between limits, i.e.,

$$W_{AB} = \int_A^B dW = \int_A^B \vec{F}.d\vec{s}$$
$$= \int_A^B F\cos\theta\, ds \quad (15.7)$$

The above integral is difficult to evaluate as both F and θ vary from point to point along the path. If we express $\vec{F}$ and $d\vec{s}$ in terms of the rectangular components, i.e.,

$$\vec{F} = F_x\vec{i} + F_y\vec{j} + F_z\vec{k} \tag{15.8}$$

and

$$d\vec{s} = dx\vec{i} + dy\vec{j} + dz\vec{k} \tag{15.9}$$

then work done is given as

$$W_{AB} = \int_A^B [F_x\vec{i} + F_y\vec{j} + F_z\vec{k}].\,[dx\vec{i} + dy\vec{j} + dz\vec{k}]$$

$$= \int_A^B [F_x dx + F_y dy + F_z dz] \tag{15.10}$$

The above integral, which is evaluated over the curvilinear path of motion of the particle, is termed as **line integral**.

15.4 WORK DONE IN STRETCHING A SPRING

As a special case of varying force, consider a force varying in magnitude but constant in direction, acting on a particle. Then the resulting displacement is along a *rectilinear* path. A spring, whose one end is fixed and the other end stretched is an example for a variable force constant in direction but varying in magnitude.

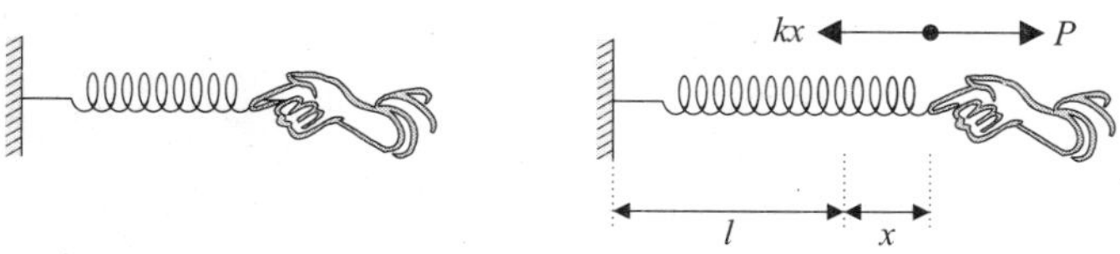

Fig. 15.7 Work done by a variable force

Consider a force P applied to an unstretched spring of length l. Let it cause a displacement x of the free end of the spring, where x is measured from the unstretched position. Due to elastic nature of the spring, a restoring force is developed in the spring, which tries to regain its original unstretched position. Within elastic limit, Hooke's law says that this restoring force is proportional to the elongation and opposite to the direction of displacement. Mathematically,

$$F \propto -x \tag{15.11}$$

Introducing a constant of proportionality, we have

$$F = -kx \tag{15.12}$$

where the constant of proportionality k is called **stiffness** of spring or **spring constant**.

If the spring is stretched slowly, such that it is not accelerated then the restoring force must be equal and opposite to the applied force for equilibrium to be maintained, i.e.,

$$P = kx \tag{15.13}$$

The relationship between the applied force P and elongation x can be represented graphically as shown in Fig. 15.8.

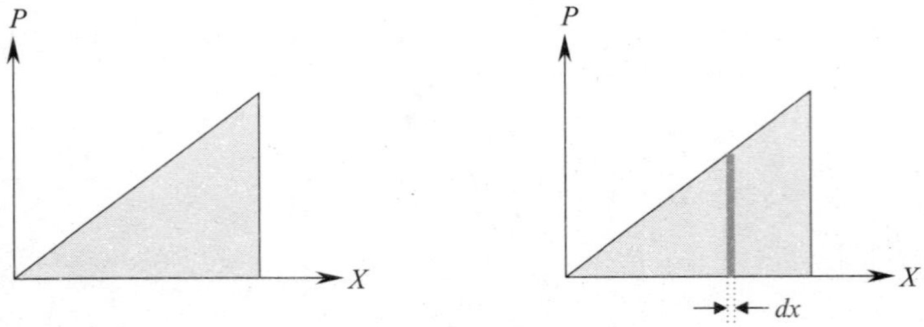

Fig. 15.8 Graphical representation of applied force and elongation

We see that the force and the elongation vary *linearly*. Suppose we consider an infinitesimally small elongation dx then over this infinitesimally small elongation, we can assume the force acting on the spring to be *constant*. Hence, work done over this infinitesimally small displacement is given as

$$dW = P\,dx \tag{15.14}$$

Therefore, work done in stretching the spring to an elongation of x_o from its unstretched position is obtained by integrating the above expression between limits.

$$W = \int_0^{x_o} dW = \int_0^{x_o} P\,dx$$

$$= \int_0^{x_o} kx\,dx$$

$$= \frac{1}{2}kx_o^2 \tag{15.15}$$

As the term on the right-hand side represents the area of the triangle in Fig. 15.8, we see that the work done in stretching a spring is given by the area under the P–x curve.

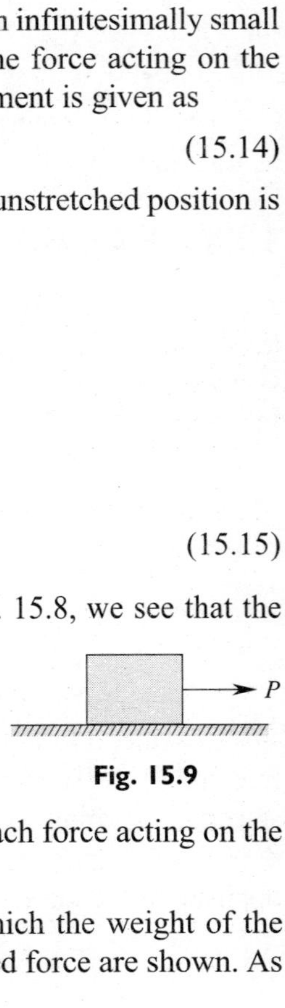

Fig. 15.9

Example 15.1 A block of 5 kg mass is resting on a rough horizontal plane having coefficient of kinetic friction 0.15. It is pulled by a horizontal force P at a constant velocity over a distance of 5 m. Sketch the free-body diagram of the block showing all the forces acting on it. Also, determine (i) the work done by each force acting on the free body, and (ii) the total work done on the body.

Solution The free-body diagram of the block is shown in Fig. 15.9(a), in which the weight of the block, the normal reaction and frictional force together with the externally applied force are shown. As there is no motion along the Y-direction, we know that $\Sigma F_y = 0$. Therefore,

$$N - 5g = 0$$

$$\therefore \quad N = 5g \tag{a}$$

Since the block is moving with constant velocity along the horizontal direction, its acceleration in that direction is zero. Hence, we can write

$$\Sigma F_x = 0 \Rightarrow$$

$$P - F = 0 \tag{b}$$

Since $F = \mu_k N$, , the above equation can be written as

$$P - \mu_k N = 0$$

$$P - \mu_k (5g) = 0$$

$$\Rightarrow \quad P = \mu_k(5g)$$

$$= 0.15\,(5 \times 9.81) = 7.36 \text{ N}$$

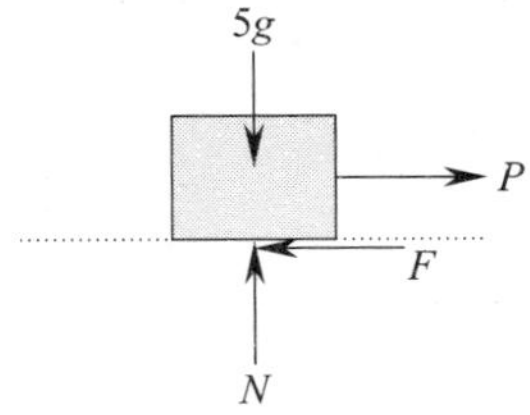

Fig. 15.9(a)

From the equation (b), the force of friction is

$$F = P = 7.36 \text{ N}$$

(i) Work done by each force

Since the displacement of the block is 5 m, work done by the applied force P is

$$W_p = 7.36 \times 5 = 36.8 \text{ J}$$

and work done by the frictional force is

$$W_F = -7.36 \times 5 = -36.8 \text{ J}$$

Note that as the frictional force acts opposite to the direction of displacement of the block, the work done is *negative*. Since the other forces acting on the block, mg and N, are all perpendicular to the direction of displacement of the block, the work done by each of them is zero.

(ii) Hence, the total work done on the block is the algebraic sum of the works done by the forces acting on the block.

$$W = 36.8 - 36.8 = 0$$

Alternatively, we could say that as the block is moving with constant velocity, the resultant force acting on it is zero; hence, the work done on the block is zero.

Example 15.2 A block of 10 kg mass resting on a rough horizontal plane is pulled by an inclined force P as shown in Fig. 15.10, at a constant velocity over a distance of 5 m. The coefficient of kinetic friction between the contact planes is 0.2. Sketch the free-body diagram of the block showing all the forces acting on it. Also, determine (i) the work done by each force acting on the free body, and (ii) the total work done on the block.

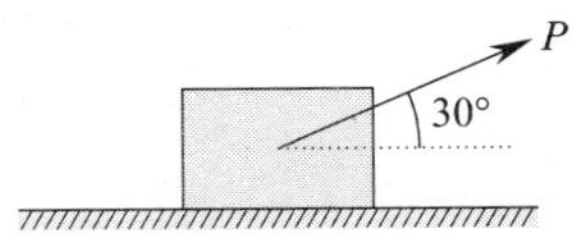

Fig. 15.10

Solution The free-body diagram of the block is shown in Fig. 15.10(a) below. As there is no motion along the Y-direction, we know that $\Sigma F_y = 0$. Therefore,

$$N + P \sin 30° - 10\,g = 0$$

$$\therefore \quad N = 10\,g - P \sin 30° \quad \text{(a)}$$

Since the block is moving with constant velocity along the horizontal direction, its acceleration in that direction is zero. Hence, we can write

$$\Sigma F_x = 0 \Rightarrow$$

$$P \cos 30° - F = 0 \quad \text{(b)}$$

Since $F = \mu_k N$, the above equation can be written as

$$P \cos 30° - \mu_k N = 0$$

Substituting equation (a) in the above equation,

$P\cos 30° - \mu_k(10g - P\sin 30°) = 0$

$$\Rightarrow \quad P = \frac{10\,\mu_k g}{[\cos 30° + \mu_k \sin 30°]}$$

$$= \frac{10(0.2)(9.81)}{[\cos 30° + (0.2)\sin 30°]}$$

$$= 20.31 \text{ N}$$

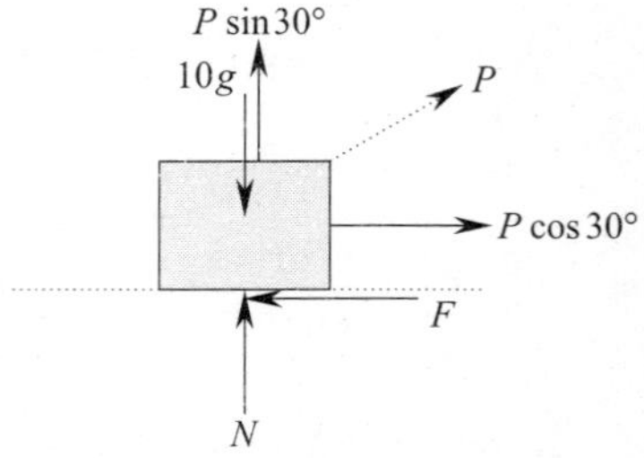

Fig.15.10(a)

Also, force of friction is given as

$$F = P\cos 30°$$

$$= 20.31 \cos 30° = 17.59 \text{ N}$$

(i) Work done by each force

Work done by the horizontal component of P, i.e., $P\cos\theta$ is

$$W_{P\cos\theta} = 20.31 \cos 30° \times 5 = 87.9 \text{ J}$$

and work done by the frictional force is

$$W_F = -17.59 \times 5 = -87.9 \text{ J}$$

Since the other forces acting on the block, $P\sin\theta$, mg and N are all perpendicular to the direction of displacement of the block, the work done by each of them is zero.

(ii) Hence, the total work done on the block is the algebraic sum of works done by each of the forces acting on the block.

$$W = 87.9 - 87.9 = 0$$

Alternatively, we could say that as the block is moving with constant velocity, the resultant force acting on it is zero; hence, the work done on the block is zero.

Example 15.3 Solve the above problem, if the externally applied force P pulls the block at a constant acceleration of 1 m/s^2.

Solution The free body diagram of the block is shown in Fig. 15.11. As before

$$N = 10\,g - P\sin 30° \qquad \text{(a)}$$

Since the block is moving with constant acceleration along the X-axis,

$\sum F_x = ma_x \Rightarrow$

$$P\cos 30° - F = 10a_x$$

$$P\cos 30° - \mu_k N = 10a_x$$

Substituting equation (a) in the above equation,

$P\cos 30° - \mu_k(10g - P\sin 30°) = 10a_x$

$$\Rightarrow \quad P = \frac{10(a_x + \mu_k g)}{[\cos 30° + \mu_k \sin 30°]}$$

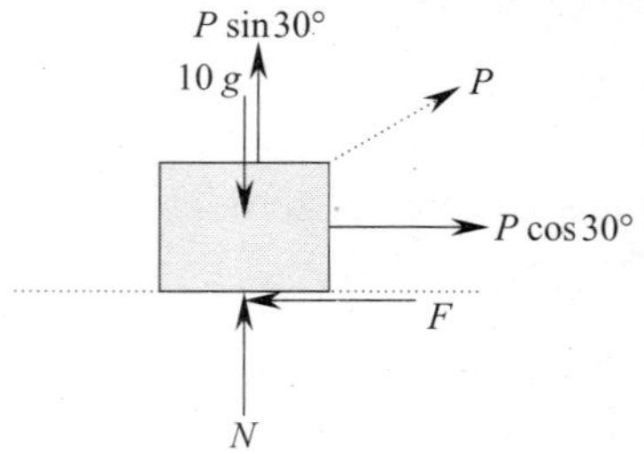

Fig. 15.11

$$= \frac{10[1 + (0.2)(9.81)]}{[\cos 30° + (0.2)\sin 30°]}$$

$$= 30.66 \text{ N}$$

Also, force of friction is given as

$$F = P \cos 30° - 10a_x$$

$$= 30.66 \cos 30° - 10(1) = 16.55 \text{ N}$$

(i) Work done by each force

Work done by the horizontal component of P, i.e., $P \cos\theta$ is

$$W_{P\cos\theta} = 30.66 \cos 30° \times 5 = 132.76 \text{ J}$$

and work done by frictional force is

$$W_F = -16.55 \times 5 = -82.75 \text{ J}$$

Since the other forces acting on the body, $P \sin\theta$, mg and N are all perpendicular to the direction of displacement of the body, the work done by each of them is zero.

(ii) Hence, the total work done on the body is the algebraic sum of the works done by each of the forces acting on the body.

$$W = 132.76 - 82.75 \approx 50 \text{ J}$$

Alternatively, we could say that as the body is moving with constant acceleration, the resultant force acting on it is

$$R = ma_x = 10(1) = 10 \text{ N}$$

and the work done by this force is

$$W_R = 10(5) = 50 \text{ J}$$

Hence, we see that the total work done on the body is same as the work done by the resultant force acting on the body as expected.

Example 15.4 A horse pulls a chariot of 200 kg mass at a constant speed. Determine the work done by the horse in pulling the chariot through a distance of 20 m, if the total frictional resistance is 400 N. Also, determine the work done if the horse pulls the chariot at a constant acceleration of 0.5 m/s^2.

Solution *When the horse pulls the chariot at constant speed*

When the horse pulls the chariot at constant speed, the force exerted by the horse on the chariot is equal to the frictional resistance. Hence,

$$F = 400 \text{ N}$$

Fig. 15.12

Therefore, work done by the horse in pulling the chariot through a distance of 20 m is

$$W = (400)(20) = 8 \text{ kJ}$$

When the horse pulls the chariot at constant acceleration

When the horse pulls the chariot at a constant acceleration, the total pulling force exerted by the horse is given as

$$F - \text{force of friction} = ma$$

$$\therefore \quad F = \text{force of friction} + ma$$

$$= 400 + 200(0.5) = 500 \text{ N}$$

Therefore, work done by the horse in pulling the chariot through a distance of 20 m is

$$W = (500)(20) = 10 \text{ kJ}$$

Example 15.5 Find the work done by the force of gravity on a body of 5 kg mass as (i) it falls vertically downwards through a distance of 3 m, and (ii) as it slides down an inclined plane with a slope of 0.75. What do you infer from the result?

Solution

(i) When it falls vertically downwards over a distance of 3 m

Since the force of gravity and the displacement are in the same direction, the work done by the force of gravity on the body as it descends by a distance of 3 m is given as

$$W_{mg} = (mg)(s)$$

$$= 5 \times 9.81 \times 3 = 147.15 \text{ J}$$

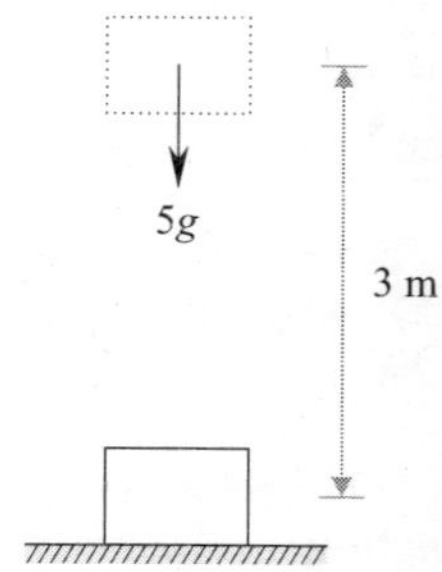

Fig. 15.13(a)

(ii) When it slides down an inclined plane with a slope of 0.75

Since the slope of the plane is 0.75, the sides of the triangle are as indicated in Fig. 15.13(b). The force of gravity can be resolved into two components '$mg \cos \theta$' normal to the plane and '$mg \sin \theta$' along the plane. As the displacement is along the plane, the normal component does not contribute to the work done. Hence,

$$W = (mg \sin \theta)(5)$$

which can also be written as

$$W = (mg)(5 \sin \theta)$$

$$= (mg)(3) = 147.15 \text{ J}$$

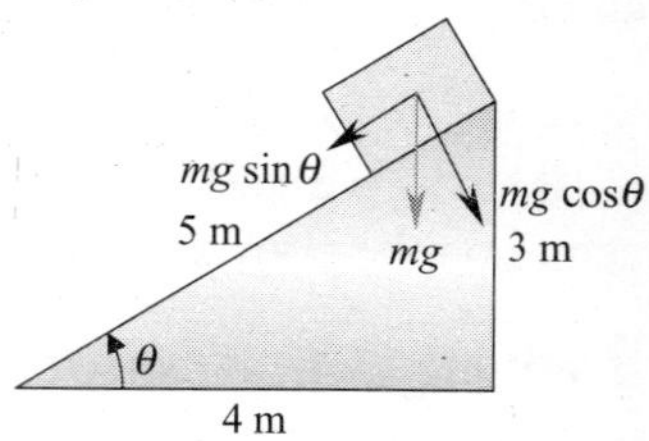

Fig. 15.13(b)

We see that the work done is same as the previous case.

Thus, we see that the work done by the force of gravity is independent of the path traced but only on *initial* and *final* positions of the block in the vertical direction. Such a kind of force is termed *conservative* force.

Example 15.6 A boy lifts water from a 20 m deep well using a bucket-and-pulley arrangement. The mass of the bucket with water is 5 kg. Determine the work done by him (i) if he lifts it at a constant speed, (ii) if he lifts it with a constant acceleration such that the velocity of the bucket at mid-depth is 1.5 m/s.

Solution

(i) When the boy lifts water at constant speed

When he lifts water at a constant speed, the force he exerts is equal to the weight of the bucket with water, i.e.,

$$F = 5(9.81) = 49.05 \text{ N}$$

As the displacement is in the same direction as the force exerted, work done by the boy is

$$W = 49.05(20) = 981 \text{ J}$$

(ii) When the boy lifts water at constant acceleration

When he lifts water at a constant acceleration, the force he exerts is equal to the weight of the bucket with water plus the inertia force, i.e.,

$$F = 49.05 + ma \qquad \text{(a)}$$

Using the kinematic equation of motion of the bucket,

$$v^2 = v_o^2 + 2as$$

$$\Rightarrow \qquad a = \frac{v^2 - v_o^2}{2s}$$

$$= \frac{(1.5)^2 - 0}{2(10)} = 0.1125 \text{ m/s}^2$$

Substituting this value of acceleration in the equation (a), we have

$$F = 49.05 + 5(0.1125) = 49.61 \text{ N}$$

As the displacement is in the same direction as the force exerted, work done by the boy is

$$W = 49.61(20) = 992.2 \text{ J}$$

Example 15.7 A body of 5 kg mass is tied to an inextensible string. Determine the work done by the external agent on the body, if (i) it is lowered down at a constant speed through a distance of 3 m, (ii) if it is lowered down at a constant acceleration of 1 m/s^2 through the same distance, (iii) if it is lifted up at a constant velocity by a distance of 3 m, and (iv) if it is lifted up at a constant acceleration of 1 m/s^2 by the same distance.

Solution *(i) When the body is lowered down at a constant velocity through a distance of 3 m*

Since the body is lowered down at a constant velocity, we know $\Sigma F_y = 0$. Therefore,

$$5g - T = 0$$

$$\Rightarrow \qquad T = 5g = 5(9.81) = 49.05 \text{ N}$$

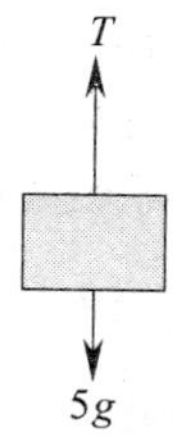

Fig. 15.14

As this force T acts in the direction opposite to that of the displacement, the work done by the external agent on the body is given as

$$W_T = -(T)(s)$$

$$= -49.05 \times 3 = -147.15 \text{ J}$$

(ii) When the body is lowered down at a constant acceleration through a distance of 3 m

Since the body is lowered down at a constant acceleration of 1 m/s^2, we know

$$\Sigma F_y = ma_y \Rightarrow$$

$$5g - T = 5(1)$$

$$\therefore \qquad T = 5(g - 1) = 5(8.81) = 44.05 \text{ N}$$

As this force T acts in the direction opposite to that of the displacement, the work done by the external agent on the body is given as

$$W_T = -(T)(s)$$
$$= -44.05 \times 3 = -132.15 \text{ J}$$

(iii) *When the body is lifted up at a constant velocity through a distance of 3 m*

Since the body is lifted up at a constant velocity, we know $\Sigma F_y = 0$. Therefore,

$$T - 5g = 0$$

$$\Rightarrow \quad T = 5g = 5(9.81) = 49.05 \text{ N}$$

As this force T acts in the same direction as that of the displacement, the work done by the external agent on the body is given as

$$W_T = (T)(s)$$
$$= 49.05 \times 3 = 147.15 \text{ J}$$

(iv) *When the body is lifted up at a constant acceleration through a distance of 3 m*

Since the body is lifted up at a constant acceleration of 1 m/s^2, we know,

$$\sum F_y = ma_y \Rightarrow$$

$$T - 5g = 5(1)$$

$$\therefore \quad T = 5(g+1) = 5(10.81) = 54.05 \text{ N}$$

As this force T acts in the same direction as that of the displacement, the work done by the external agent on the body is given as

$$W_T = (T)(s)$$
$$= 54.05 \times 3 = 162.15 \text{ J}$$

Example 15.8 The motion of an automobile of 3 ton mass is represented by v–t graph as shown in Fig. 15.15. If the frictional resistance to the motion is 100 N/ton, determine the work done by the driving force of the automobile in the first 10 seconds and in the next 2 minutes.

Solution From the graph, we see that the acceleration of the automobile is $a_1 = 15/10 = 1.5$ m/s^2 and the distance travelled in the first 10 seconds is $s_1 = \frac{1}{2}(15)(10) = 75$ m. The driving force of the automobile during acceleration is given as

$$F_1 - f = ma_1$$

$$\Rightarrow \quad F_1 = f + ma_1$$
$$= (100 \times 3) + (3000 \times 1.5) = 4.8 \text{ kN}$$

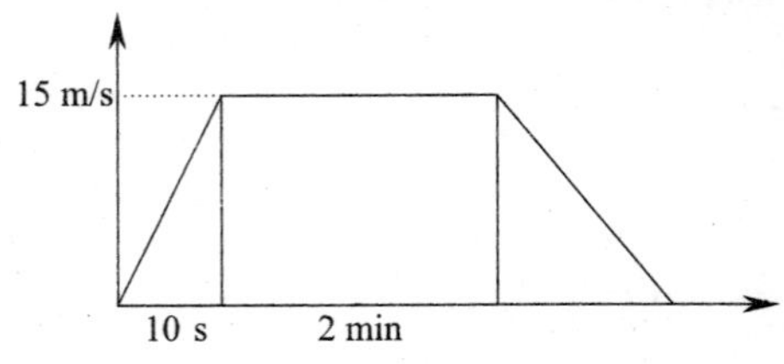

Fig. 15.15

Hence, work done by the driving force is given as

$$W = F_1 s_1 = 4800 \times 75 = 360 \text{ kJ}$$

During the next 2 minutes, it moves at a constant speed. Hence, the driving force is equal to the force of friction, i.e., $F_2 = f = 300$ N. The total distance travelled during this time is $s_2 = (15)(120) = 1800$ m. Hence, the work done by the driving force is given as

$$W = F_2 s_2 = 300 \times 1800 = 540 \text{ kJ}$$

15.5 POWER

In the previous section, we saw work done by a force on a particle. In this section, we will see the *rate* at which this work is done, as it is more important than the work itself. Two different agents may perform the *same* amount of work. However, the one that performs the work faster is of concern to us. For instance, to draw water from a well, if human effort is used, the time taken will be more as compared to that when an electric pump is used. Similarly, a wagon pulled by a horse is slower as compared to that pulled by an engine. Hence, the one doing work faster than the other is said to be more **powerful**.

Thus, **power** is defined as the rate at which work is done. The capacity of an engine or a machine used to do work is normally expressed as its *rated power*. If W is the total work done in a time interval t, then *average* power is given as

$$P_{\text{ave}} = \frac{\text{total work done}}{\text{time taken}} = \frac{W}{t} \tag{15.16}$$

The *instantaneous* power, i.e., power at a particular instant of time is given as

$$P = \frac{dW}{dt} = \frac{d(Fs)}{dt} \tag{15.17}$$

The force can be assumed to be constant over this infinitesimally small time interval dt. Hence, we can write the above expression as

$$P = \frac{Fds}{dt} = Fv \tag{15.18}$$

Thus, power can be defined as product of applied force and velocity of point of application of the force. In S.I system of units, the unit of power is Joule per second (J/s), also called Watt (W), in honor of James Watt, the inventor of steam engine. He also expressed power in terms of power of a horse. Though S.I. system of units is widely used, some of the measuring equipments, engines and machines still measure quantities in older units. For instance, the speedometer records the speed in kilometres per hour (kmph), though the S.I. unit is m/s. Similarly, electrical and mechanical machines still express power in **horsepower**. Hence, it is worthwhile mentioning the relationship between watt and horsepower.

The term horsepower was invented by the engineer James Watt in 1782. He found that a horse could turn a mill wheel (of 12 ft radius) 144 times in an hour or 2.4 times in a minute. Also, a horse could pull with a force of 180 lb. Hence, he defined horsepower as

$$1\text{ hp} = \frac{\text{work done}}{\text{time}} = \frac{(180\text{ lb})(2.4 \times 2\pi \times 12\text{ ft})}{\text{min}} = 32\,572\ \frac{\text{ft.lb}}{\text{min}} \tag{15.19}$$

which is approximated to 33 000 ft.lb/min (or) 745.69987158227022 W. Thus, in the British system of units, one horsepower is given as

$$1\text{ hp} \approx 746\text{ W} \tag{15.20}$$

There is also another unit followed for power, namely, *metric horsepower*. It began in Germany in the nineteenth century and became popular in Europe and Asia. Metric horsepower is defined as roughly 98.6% of mechanical horsepower, i.e., 735.49875 W. Hence, in metric system of units,

$$1\text{ hp} \approx 736\text{ W} \tag{15.21}$$

Example 15.9 Water is to be pumped from a 100 m deep well. If the pump used discharges water at a rate of 0.05 m^3/s, determine metric horsepower of the pump. Weight density of water is 9810 N/m^3.

Solution Discharge of water is defined as the volume of water discharged in unit time. Hence,

$$Q = \frac{\text{volume of water}}{\text{time}} = 0.05 \text{ m}^3/\text{s}$$

Hence, mass of water discharged per unit time is given as

$$m = \frac{\text{density} \times \text{volume of water}}{\text{time}}$$
$$= \text{density} \times Q = \rho Q$$

Therefore, power of the pump is obtained as

$$P = \frac{\text{work done}}{\text{time}}$$
$$= \frac{\text{Force} \times \text{displacement}}{\text{time}}$$
$$= \frac{(\text{mass} \times \text{g}) \times \text{height}}{\text{time}}$$
$$= \frac{\text{mass}}{\text{time}} \times g \times \text{height}$$
$$= \rho Q\, g(\text{height})$$
$$= \gamma Q\,(\text{height}) \qquad [\text{since weight density, } \gamma = \rho g]$$
$$= (9810)(0.05)(100)$$
$$= 49\,050 \text{ W}$$
$$= (49\,050)/736 = 66.64 \text{ metric hp}$$

Example 15.10 A car of 2 ton mass starts from rest and accelerates at a uniform rate to reach a speed of 60 kmph in 20 seconds. If the frictional resistance is 600 N/ton, determine the driving power of the engine when it reaches a speed of 60 kmph.

Solution

Initial speed of car, $v_o = 0$

Final speed of car, $v = 60 \text{ kmph} = 60 \times \frac{5}{18} = 16.67 \text{ m/s}$

For uniform acceleration, we know that the kinematic equation of motion of the car is

$$v = v_o + at$$

Therefore, acceleration of the car is given as

$$a = \frac{v - v_o}{t} = \frac{16.67 - 0}{20} = 0.8335 \text{ m/s}^2$$

The kinetic equation of motion of the car is given as

$$F - f = ma$$

where F is the driving force and f is the force of friction. Therefore,

$$F = f + ma$$

$$= (600)(2) + (2 \times 10^3)(0.8335) = 2867 \text{ N}$$

Therefore, driving power of the engine when the car is moving at 60 kmph is given as

$$P = Fv$$

$$= (2867)(16.67) = 47\,792.89 \text{ W (or) } 47.8 \text{ kW}$$

Example 15.11 A train of 400 ton mass is pulled by a locomotive of 20 tons along level rails at a constant speed of 80 kmph. If the force of friction is 100 N/ton, determine the driving power of the locomotive.

Solution

Speed of train, $v = 80 \text{ kmph} = 22.22 \text{ m/s}$

Total force of friction is

$$f = (100)(400 + 20) = 42\,000 \text{ N (or) } 42 \text{ kN}$$

As the train runs at constant speed, the pulling force of the locomotive is equal to the force of friction resisting the motion. Hence,

$$F = f = 42 \text{ kN}$$

Therefore, driving power of the locomotive is given as

$$P = Fv$$

$$= (42 \times 10^3)(22.22)$$

$$= 933\,240 \text{ W (or) } 933.24 \text{ kW}$$

Example 15.12 In the previous problem, determine the maximum speed the train can attain with the same locomotive when it moves on an incline 1 in 100; the frictional resistance being the same. Also, determine the required power of the engine to maintain the same speed of 80 kmph on incline as on level track.

Solution On the incline, the pulling force of the locomotive should be such as to overcome the force of friction f and the component of weight of the train '$mg \sin\theta$' along the incline. Therefore,

$$F = f + mg \sin\theta$$

$$= (42 \times 10^3) + (420 \times 10^3) \times 9.81 \times \frac{1}{100} = 83.2 \text{ kN}$$

Maximum speed attainable

We know power is given as

$$P = Fv$$

$$\therefore \quad v = \frac{P}{F} = \frac{933.24 \times 10^3}{83.2 \times 10^3} = 11.22 \text{ m/s} = 40.4 \text{ kmph}$$

Required power of the engine to maintain the same speed of 80 kmph on incline

If the train has to run at the same speed as on terrain then required power of the locomotive is given as

$$P = Fv$$
$$= 83.2 \times 10^3 \times 22.22$$
$$= 1.85 \text{ MW}$$

Example 15.13 A train of total 300 ton mass descends an incline 1 in 120 with a uniform velocity of 8 m/s. If the frictional resistance is 250 N/ton, determine the power of the engine.

Solution Down the incline, the pulling force of the locomotive and the component of weight of the train down the incline should be such as to overcome the force of friction. Therefore,

$$F + mg \sin\theta - f = 0$$

$$\therefore \quad F = f - mg \sin\theta$$
$$= (250 \text{ N/ton})(300 \text{ ton}) - (300 \times 10^3) \times 9.81 \times \frac{1}{120}$$
$$= 50.5 \text{ kN}$$

Therefore, driving power of the engine is given as

$$P = Fv$$
$$= (50.5 \times 10^3) \times 8$$
$$= 404\,000 \text{ W (or) } 404 \text{ kW}$$

Example 15.14 A block of mass m resting on a rough inclined plane is pulled by a force F acting parallel to the incline. If the block starts from rest and attains a speed of v in t seconds, determine the maximum power exerted on the block. The coefficient of friction between the contact surfaces is μ.

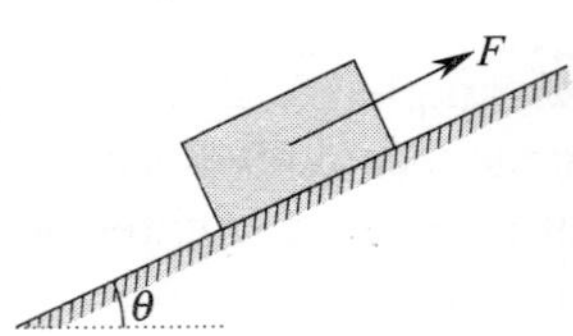

Fig. 15.16

Solution The free-body diagram of the block is shown in Fig. 15.16(a). The kinetic equation of motion of the block along the X and Y directions can be written as

$$\Sigma F_y = ma_y \Rightarrow$$

$$N - mg \cos\theta = 0$$

$$\therefore \quad N = mg \cos\theta$$

Hence, the force of friction is

$$f = \mu N = \mu mg \cos\theta$$

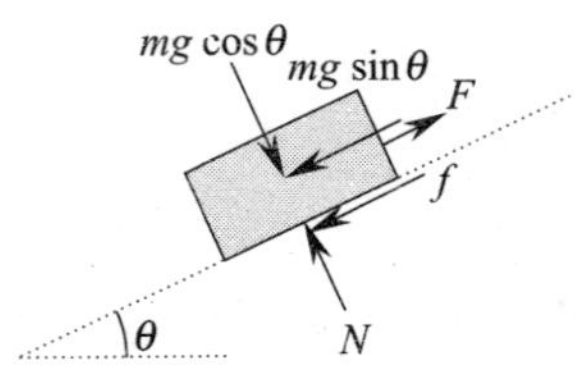

Fig. 15.16(a)

$$\Sigma F_x = ma_x \quad \Rightarrow$$

$$F - mg \sin\theta - f = ma$$

$$\therefore \quad F = ma + mg\sin\theta + f$$

$$= ma + mg\sin\theta + \mu mg\cos\theta \qquad \text{(a)}$$

We know that the kinematic equation of motion of the block is

$$v = v_o + at$$

Since the initial velocity v_o is zero, the acceleration of the block is given as

$$a = v/t$$

Substituting this in the kinetic equation (a), we get

$$F = m\frac{v}{t} + mg\sin\theta + \mu mg\cos\theta$$

$$= mg\left[\frac{v}{gt} + \sin\theta + \mu\cos\theta\right]$$

Therefore, the power exerted on the block is given as

$$P = Fv = mgv\left[\frac{v}{gt} + \sin\theta + \mu\cos\theta\right]$$

Example 15.15 A locomotive exerts its full power to draw a train uphill of slope 1 in n at a constant speed of v. The frictional resistance is 1/mth of the weight of the train. If the same train runs on a level track, prove that the maximum speed that can be attained against the same friction is $v\left[1 + \frac{m}{n}\right]$.

Solution Let W be the total weight of train including the weight of locomotive. When the train is moving uphill at a constant speed v, the driving force of locomotive can be determined as

$$F - f - mg\sin\theta = 0$$

$$\Rightarrow \quad F = f + mg\sin\theta$$

$$= W\left[\frac{1}{m}\right] + W\left[\frac{1}{n}\right]$$

$$= W\left[\frac{m+n}{mn}\right]$$

Therefore, power of the locomotive is

$$P = Fv = Wv\left[\frac{m+n}{mn}\right]$$

Similarly, when the train is running on level track,

$$F' - f = 0$$

$$\Rightarrow \quad F' = f = \frac{W}{m}$$

If v' be the maximum speed of the train on level track then

$$P = F'v'$$

$$Wv\left[\frac{m+n}{mn}\right] = \frac{W}{m}v'$$

$$\Rightarrow \qquad v' = v\left[\frac{m+n}{n}\right] = v\left[1+\frac{m}{n}\right]$$

Example 15.16 The power output of an electric motor is measured using a brake drum arrangement as shown in Fig. 15.17. The brake drum is coupled to the motor and load is applied on the drum by a belt wound over the drum. If the difference in the two spring balance readings is 2.5 kgf, radius of the drum is 0.15 m and the speed of the motor is 1000 rpm, then determine the power output of the motor.

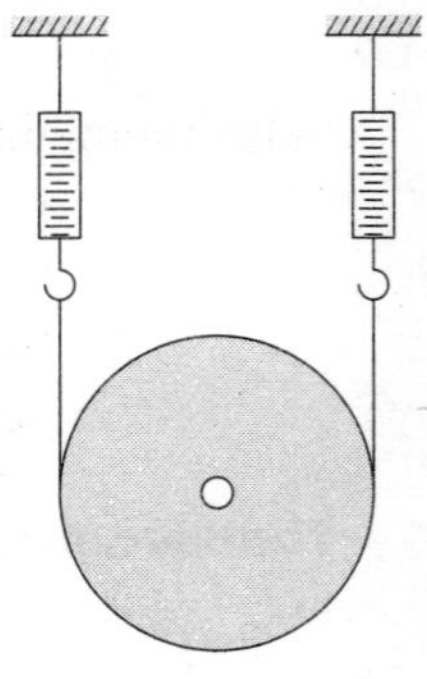

Fig. 15.17

Solution Since the application of tension in the belt retards the motion of the drum coupled with the motor, the retarding torque is given as

$$\tau = \text{(net force acting on the drum)} \times \text{(radius of the drum)}$$

Just like the power for linear motion is given as a product of force and linear velocity, the power for angular motion is given as a product of torque and angular velocity, i.e.,

$$P = \tau\omega$$
$$= \text{(net force)} \times \text{(radius of the drum)}\,\omega$$

Note that the product of radius of the drum and angular velocity is equal to the linear velocity of a point on the rim of the drum and thus we see that the power has the same form as that for linear motion.

Net force acting on the drum is the difference in the two spring balance readings. Hence,

$$F = 2.5 \text{ kgf}$$

The conversion factor for force is

$$1 \text{ N force} = (1 \text{ kgf})(9.81 \text{ m/s}^2)$$

Therefore, $$F = 2.5 \times 9.81 = 24.53 \text{ N}$$

The angular velocity of the motor is given as

$$\omega = \frac{2\pi N}{60} = \frac{2\pi(1000)}{60} = 104.72 \text{ rad/s}$$

Hence, the power output of the motor is obtained as

$$P = Fr\omega$$
$$= (24.53)(0.15)(104.72)$$
$$= 385.3 \text{ W}$$

Since metric horsepower = watt power/736, the power output of the motor can also be expressed as

$$P = 0.52 \text{ metric hp}$$

15.6 ENERGY

Energy is defined as that property of a body by virtue of which work can be performed or in other words, it is the *capacity* of a body to *do work*. Energy exists in *different forms* such as chemical, electrical, thermal, elastic, nuclear, mechanical, etc. We know from the law of conservation of energy that energy can neither be created nor destroyed, but can be transformed from one form to another. For instance, in the case of firing a bomb, chemical energy is converted into mechanical energy, which makes the bomb move forward and hit the target. Here we see that the chemical composition of the bomb has the energy to do work on the bomb, i.e., to make it move forward. In most of the cases, all the available energy is not converted into useful work, but is lost or dissipated in the form of *heat* energy. Since energy is the capacity of a body to do work, its unit is the same as that of work done, i.e., N.m (or) Joule.

In our study, we are mainly concerned with *mechanical* energy. Mechanical energy is further divided into two types, namely, **kinetic energy** and **potential energy**. Kinetic energy of a body is defined as the energy possessed by virtue of its **motion** and potential energy of a body is defined as the energy possessed by virtue of its **position**. In the following section, we will discuss kinetic energy and in Section 15.8, we will discuss potential energy.

15.6.1 Kinetic Energy

The word 'kinetic' is derived from the Greek word 'kinesis' meaning to *move* and the word 'energy' is the ability to move. A body under motion is said to possess energy called kinetic energy because it is able to do work on any body that comes across its path. This is the reason we do not step in front of a moving vehicle as it has the capacity to do work on us.

Consider an automobile moving with a certain velocity. If the engine is switched off, the automobile will continue to move with the same velocity if there is *no* resistance to its motion. Hence, no work is done by the body as it causes its own displacement. However, we know that the motion is always retarded by frictional resistance, and hence the speed goes on decreasing until the automobile comes to a stop. Here, the body does work against the force of friction in moving forward. Thus, kinetic energy possessed by a body can be measured by the amount of work the moving body will do if brought to rest or by the amount of work originally needed to impart the velocity to it.

15.6.2 Mathematical Expression of Kinetic Energy

When the resultant force acting on a particle is *non-zero*, it causes *acceleration* of the particle. By Newton's second law of motion, the force and the resulting acceleration are related as

$$\vec{F} = m\vec{a} = m\frac{d\vec{v}}{dt} \tag{15.22}$$

Then work done by the force in causing an infinitesimally small displacement $d\vec{r}$ is given as $\vec{F}.d\vec{r}$. Therefore, the total work done in displacing from $\vec{r}_1$ to $\vec{r}_2$ is

$$W = \int_{r_1}^{r_2} \vec{F}.d\vec{r} = \int_{r_1}^{r_2} m\frac{d\vec{v}}{dt}.d\vec{r} \tag{15.23}$$

Multiplying numerator and denominator by *dt*, we change the variable of integration from *dr* to *dt*. In addition, if the particle also started from rest, then

$$W = m\int_0^t \frac{d\vec{v}}{dt}\cdot\frac{d\vec{r}}{dt}\,dt$$

$$= m\int_0^t \frac{d\vec{v}}{dt}\cdot\vec{v}\,dt$$

$$= \frac{1}{2}m\int_0^t \frac{d}{dt}(\vec{v}\cdot\vec{v})dt$$

$$= \frac{1}{2}m\int_0^t \frac{d}{dt}(v^2)dt$$

$$= \frac{1}{2}m\int_0^v d(v)^2$$

$$= \frac{1}{2}mv^2 \tag{15.24}$$

Just as we derived the work done in imparting a velocity to a particle at rest, we can also determine in a similar manner the work done in bringing a moving particle to rest. Since kinetic energy is defined as the amount of work done in bringing to rest a moving particle or the amount of work originally needed to impart the velocity to it, we see that kinetic energy can be represented mathematically as

$$\text{K.E} = W = \frac{1}{2}mv^2 \tag{15.25}$$

We see that the kinetic energy is proportional to *square* of the speed. Thus, a twofold increase in speed will increase the kinetic energy by fourfold, and so on.

15.6.3 Work–Energy Principle

Suppose the net force acting on a particle changes its velocity from v_o to v, then we can derive the work done as before:

$$W = \frac{1}{2}m(v^2 - v_o^2) = \frac{1}{2}mv^2 - \frac{1}{2}mv_o^2$$

$$= (\text{K.E})_{\text{final}} - (\text{K.E})_{\text{initial}}$$

$$= \text{change in K.E} \tag{15.26}$$

Thus, we see that the change in kinetic energy of a particle during any displacement is equal to the work done by the net force acting on it. This is known as **work–energy principle**.

Example 15.17 A block is projected up an inclined plane with an initial velocity of 10 m/s. If coefficients of static and kinetic friction between the contact surfaces are respectively 0.25 and 0.2, determine how far up the plane will the block move before coming to rest.

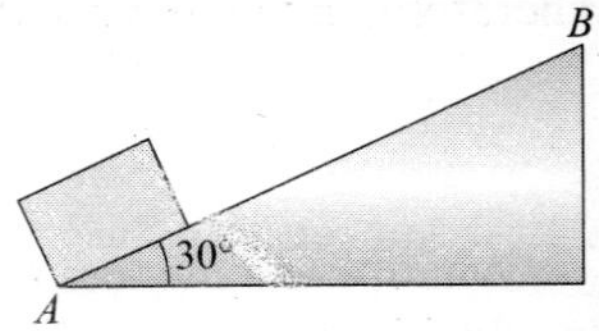

Fig. 15.18

Solution The forces acting on the block are shown in the free-body diagram below: the component of weight '$mg \sin\theta$' acting along the inclined plane and frictional force F acting in the direction opposite to that of the displacement. As there is no displacement along the Y-direction, we know

$$N - mg\cos\theta = 0$$

$$\Rightarrow \quad N = mg\cos\theta$$

Hence, force of friction is

$$F = \mu_k N = \mu_k mg\cos\theta$$

Let s be the displacement of the block along the plane before coming to rest. Then applying the work–energy equation, we have

$$(\text{K.E})_f - (\text{K.E})_i = \text{work done}$$

$$0 - \frac{1}{2}mv^2 = -(mg\sin\theta)s - (F)s$$

Note that the final kinetic energy is zero as the block comes to rest. In addition, the works done by the forces are negative as they act in the direction opposite to that of the displacement. Therefore,

$$\frac{1}{2}mv^2 = (mg\sin\theta + \mu_k mg\cos\theta)s$$

$$\therefore \quad s = \frac{v^2}{2g(\sin\theta + \mu_k\cos\theta)}$$

Substituting the values,

$$s = \frac{(10)^2}{2(9.81)(\sin 30° + (0.2)(\cos 30°))} = 7.57 \text{ m}$$

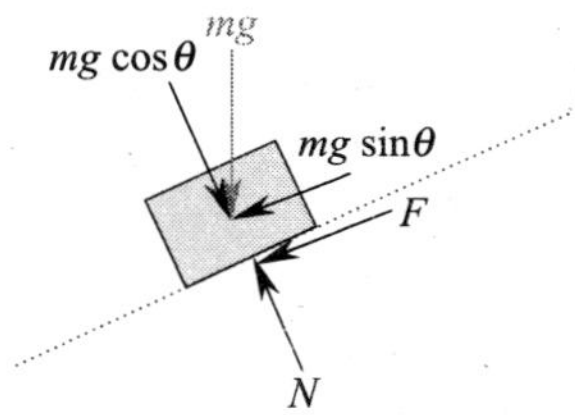

Fig. 15.18(a)

Example 15.18 In the previous problem, state whether the block after coming to rest will slide down under its own weight or not. If so, what will be the velocity of the block as it slides down from the point B and reaches the point A, the bottom of the incline?

Solution After coming to rest, the body will begin to slide only if $\tan\theta > \mu_s$. For given values of 'θ' and 'μ_s,' we see that the condition is satisfied and hence, the block will begin to slide downwards under its own weight. From the free-body diagram of the block as it slides down (shown below), we see that the component of weight $mg\sin\theta$ acts in the direction of displacement and force of friction F acts in the direction opposite to that of displacement. Hence, writing the work–energy equation,

$$[mg\sin\theta - F]s = \frac{1}{2}mv^2 - 0$$

Taking $F = \mu_k N = \mu_k mg\cos\theta$ as before,

$$[mg\sin\theta - \mu_k mg\cos\theta]s = \frac{1}{2}mv^2$$

$$\therefore \quad v = \sqrt{2gs[\sin\theta - \mu_k\cos\theta]}$$

Substituting the values,

$$v = \sqrt{2(9.81)(7.57)[\sin 30° - (0.2)\cos 30°]} = 6.97 \text{ m/s}$$

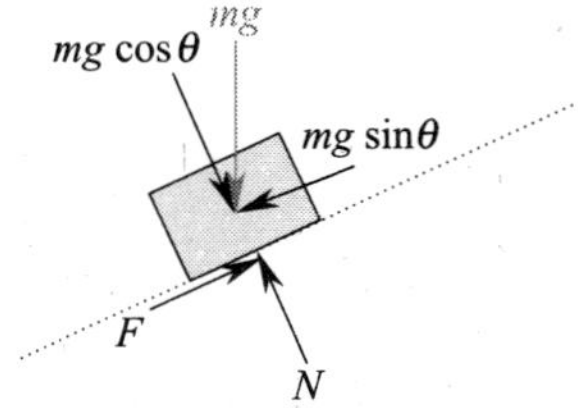

Fig. 15.19

Example 15.19 A block of 5 kg mass slides down an inclined plane from rest. How far along the horizontal plane, will it reach before coming to rest? The coefficient of kinetic friction between the block and the inclined plane is 0.15 and that between the block and the horizontal plane is 0.2.

Solution This problem can be solved in two ways as explained below:

Method I

We consider two different phases, namely, the motion of the block from *A* to *B* and from *B* to *C*. The free-body diagram of the block in the two phases of motion is shown in Fig. 15.20(a). Let v be the velocity at *B*.

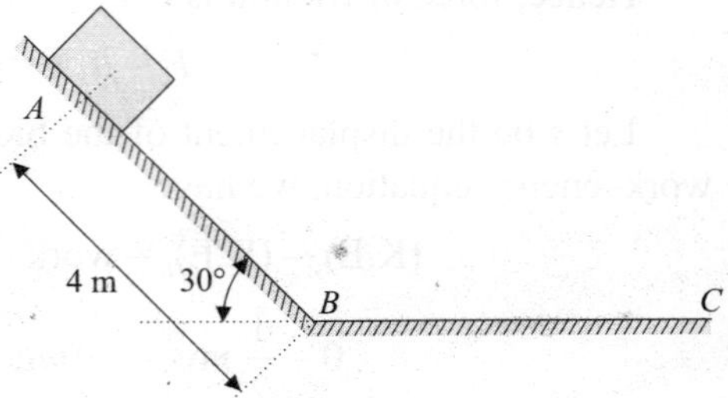

Fig. 15.20

Motion from A to B

Writing the work energy equation for the motion from A to B,

$$\frac{1}{2}mv^2 - 0 = [5g\sin 30° - F_1]4$$

$$\frac{1}{2}5v^2 - 0 = [5g\sin 30° - 0.15(5g\cos 30°)]4 \quad [\text{Since } F_1 = = \mu_{k1} N_1]$$

$$\Rightarrow \quad v = 5.39 \text{ m/s}$$

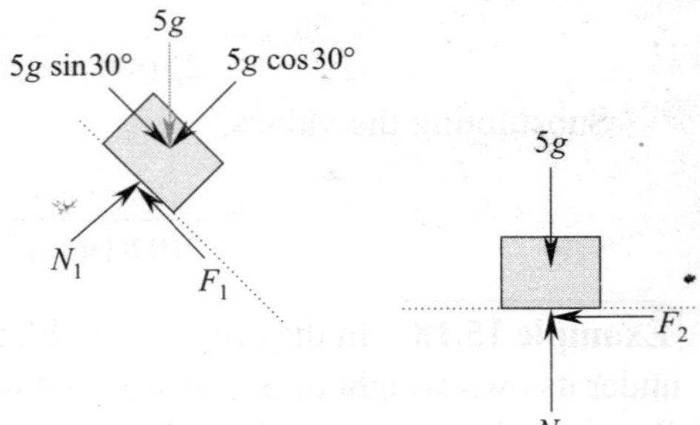

Fig. 15.20(a)

Motion from B to C

In this portion of motion, the initial velocity of the block is the velocity at *B*, i.e., v calculated as above. Writing the work energy equation for the motion from *B* to *C*,

$$0 - \frac{1}{2}mv^2 = -F_2(s)$$

$$-\frac{1}{2}5(5.39)^2 = -0.2(5g)(s) \quad [\text{Since } F_2 = \mu_{k_2} N_2]$$

$$\Rightarrow \quad s = 7.4 \text{ m}$$

Method II

Instead of considering two different phases, we can also consider the motion directly from *A* to *C*. Noting that the initial and final velocities are zero, we can write the work energy equation as

$$0 = (5g\sin 30° - F_1)4 - F_2(s)$$

$$0 = [5g\sin 30° - 0.15(5g\cos 30°)]4 - 0.2(5g)(s)$$

$$\Rightarrow \quad s = 7.4 \text{ m}$$

Thus, we see that the *same* result is obtained using both the methods.

Example 15.20 A truck of 6 ton mass is moving on a level terrain at a speed of 80 kmph, when brakes are applied. If the braking force is of constant magnitude 4 kN/ton, determine the distance it travels before coming to a stop.

Solution *Given data*

Speed of truck, $v = 80 \text{ kmph} = 80 \times 5/18 = 22.22 \text{ m/s}$

Total braking force, $F = (4 \times 10^3 \text{ N/ton})(6 \text{ tons}) = 24 \times 10^3 \text{ N}$

The initial kinetic energy of the truck is $(\text{K.E})_i = (1/2)\, mv^2$. As the braking force brings the truck to a stop, its final kinetic energy is zero. It should be noted that the displacement of the truck is opposite to that of the braking force; hence, work done by the braking force is negative. Applying the work–energy equation, we have

$$(\text{K.E})_f - (\text{K.E})_i = \text{work done}$$

$$0 - \frac{1}{2}mv^2 = -(F)(s)$$

$$0 - \frac{1}{2}(6 \times 10^3)(22.22)^2 = -(24 \times 10^3)(s)$$

$$\Rightarrow \quad s = 61.72 \text{ m}$$

Example 15.21 A block of mass m attached to a spring of stiffness k, is pushed to the right with a velocity v when the spring is in the unstretched position. Determine the maximum extension of the spring assuming the contact surfaces to be smooth.

Solution We know that when the block reaches the extreme position, its velocity is zero. Let s be distance of the extreme position with respect to the unstretched position of the spring. Then the restoring force acting on the block is ks. Then applying the principle of work–energy, we have

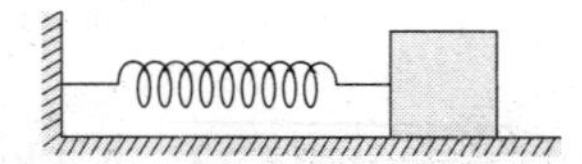

Fig. 15.21

$$(\text{K.E})_f - (\text{K.E})_i = \text{average work done}$$

$$0 - \frac{1}{2}mv^2 = -\left(\frac{1}{2}ks\right)s$$

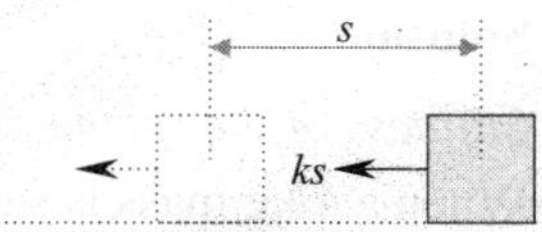

Fig. 15.21(a)

It should be noted that the weight and normal reaction do no work, as the displacement is perpendicular to the direction of these two forces. Therefore,

$$s = \sqrt{\frac{m}{k}}\, v$$

Example 15.22 In the previous problem, suppose the contact surfaces are rough having coefficient of friction μ then determine the maximum extension of the spring.

Solution The free-body diagram of the block is shown in Fig. 15.22. The frictional force acts opposite to that of the displacement. As before applying the principle of work–energy, we have

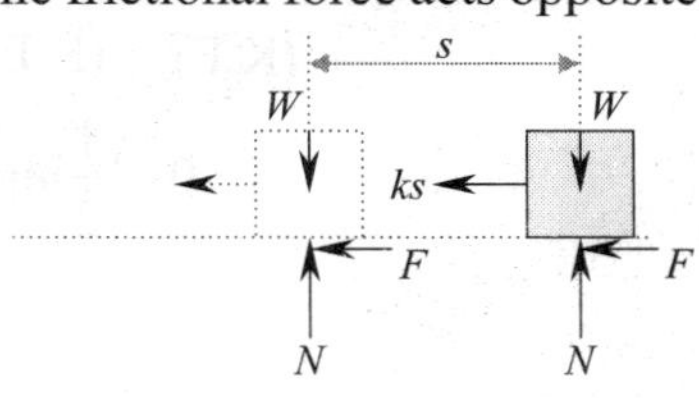

Fig. 15.22

$$(\text{K.E})_f - (\text{K.E})_i = \text{work done}$$

$$0 - \frac{1}{2}mv^2 = -\frac{1}{2}ks^2 - Fs$$

$$mv^2 = ks^2 + 2Fs$$

We know that the frictional force is

$$F = \mu N = \mu mg$$

Therefore, $ks^2 + 2\mu\, mgs - mv^2 = 0$

On solving the quadratic equation, the roots are

$$s = \frac{-2\mu mg \pm \sqrt{4\mu^2 m^2 g^2 + 4kmv^2}}{2k}$$

$$= \frac{-\mu mg \pm \sqrt{\mu^2 m^2 g^2 + kmv^2}}{k}$$

Dividing both numerator and denominator by m, we have

$$s = \frac{-\mu g \pm \sqrt{\mu^2 g^2 + (k/m)v^2}}{(k/m)}$$

As we know displacement has to be positive, we take only positive root, i.e.,

$$s = \frac{-\mu g + \sqrt{\mu^2 g^2 + (k/m)v^2}}{(k/m)}$$

Example 15.23 A 4 kg mass when suspended from a spring extends it by 10 cm. If the same spring is used to stop a block moving horizontally at a speed of 4 m/s, determine the compression of the spring (i) assuming the contact surfaces to be smooth, and (ii) the frictional resistance to motion as 12 N.

Solution

Stiffness of the spring

When a 4 kg mass is suspended from the spring, it extends by 10 cm. Therefore, stiffness of the spring is given as

$$k = \frac{F}{\Delta} = \frac{mg}{\Delta}$$

$$= \frac{(4)(9.81)}{(0.1)} = 392.4 \text{ N/m}$$

Compression of the spring when the contact surfaces are smooth

Initial velocity of block, $v_i = 4$ m/s. Final velocity, v_f of the block after compressing the spring is zero. If x be the compression of the spring, then applying the work–energy equation,

$$(\text{K.E})_f - (\text{K.E})_i = \text{average work done}$$

$$0 - \frac{1}{2}mv_i^2 = -\frac{1}{2}kx^2$$

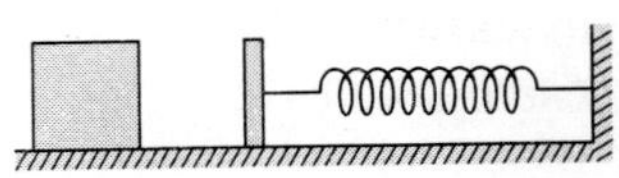

Fig. 15.23

$$\Rightarrow \quad x = \sqrt{\frac{m}{k}}\, v_i$$

$$= \sqrt{\frac{4}{392.4}} \times 4 = 403.9 \text{ mm}$$

Compression of the spring when the contact surfaces are rough

As the force of friction acts opposite to the direction of motion of the block, it does negative work. Hence, applying the work–energy equation as before,

$$0 - \frac{1}{2}mv_i^2 = -fx - \frac{1}{2}kx^2$$

$$-\frac{1}{2}(4)(4)^2 = -12x - \frac{1}{2}(392.4)x^2$$

$$\Rightarrow \qquad 196.2x^2 + 12x - 32 = 0$$

Solving the above quadratic equation for the roots, we get

$$x = \frac{-12 \pm \sqrt{(12)^2 + 4(196.2)(32)}}{392.4}$$

Neglecting the negative value, we have $x = 374.4$ mm.

Example 15.24 A block of mass m attached to a spring of stiffness k is resting on an inclined plane. If the block is released when the spring is in the unstretched position, derive the expression for maximum extension of the spring. The coefficient of friction between contact surfaces is μ.

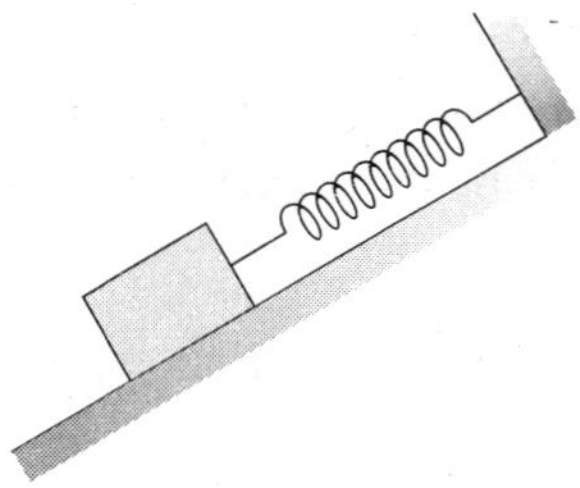

Fig. 15.24

Solution Since the block is released from rest, its initial velocity is zero. In addition, as the block stretches the spring to the maximum extent, its final velocity is zero. The normal reaction N and component $mg \cos\theta$ of the weight do no work, as the displacement is perpendicular to the forces. The only forces contributing to the work done are component $mg \sin\theta$ of the weight, force of friction $F = \mu N = \mu mg \cos\theta$ and restoring force ks of the spring. Hence, applying the principle of work–energy, we have

$$(K.E)_f - (K.E)_i = \text{work done}$$

$$0 = [mg \sin\theta]s - [\mu\, mg \cos\theta]s - \frac{1}{2}ks^2$$

$$mg[\sin\theta - \mu\cos\theta]s = \frac{1}{2}ks^2$$

Since $s \neq 0$, we have

$$s = \frac{2mg[\sin\theta - \mu\cos\theta]}{k}$$

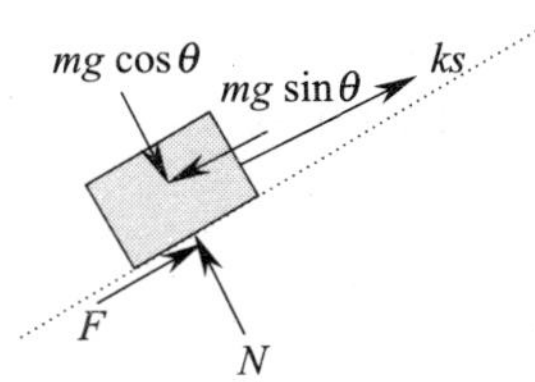

Fig. 15.24(a)

Example 15.25 A block of 10 kg mass slides down an inclined plane with a slope angle of 35°. It is stopped by a spring of stiffness 1 kN/m. If the block slides down 5 m before hitting the spring then determine the maximum compression of the spring. The coefficient of friction between the block and the inclined plane is 0.15.

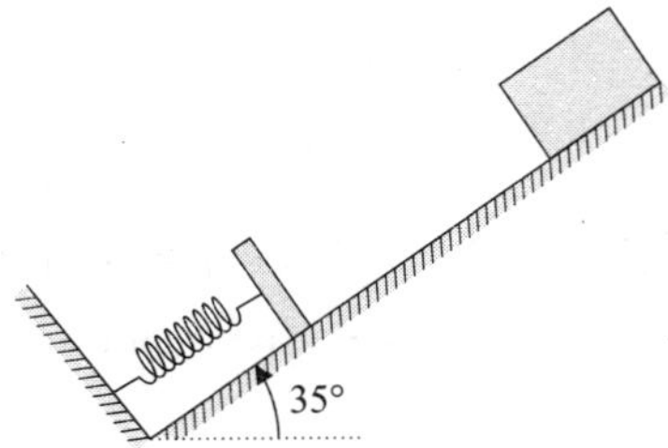

Fig. 15.25

Solution The free-body diagram of the block is shown in Fig. 15.25(a). Since the block is released from rest, its initial velocity is zero. In addition, as the block compresses the spring to the maximum extent, its final velocity is zero. Let s be the compression of the spring from the unstretched position. Then the total displacement of the block before coming to rest is $(5 + s)$ m. Applying the principle of work–energy, we have

$$[10g \sin 35° - \mu\, 10g \cos 35°][5 + s] - \frac{1}{2}ks^2 = 0$$

$$10g[\sin 35° - (0.15)\cos 35°]\,[5 + s] - \frac{1}{2}(10^3)(s)^2 = 0$$

$$(44.21)(5 + s) - 500s^2 = 0$$

$$500s^2 - 44.21s - 221.05 = 0$$

Upon solving the quadratic equation, we get

$$s = 0.711 \text{ m (or) } 71.1 \text{ cm}$$

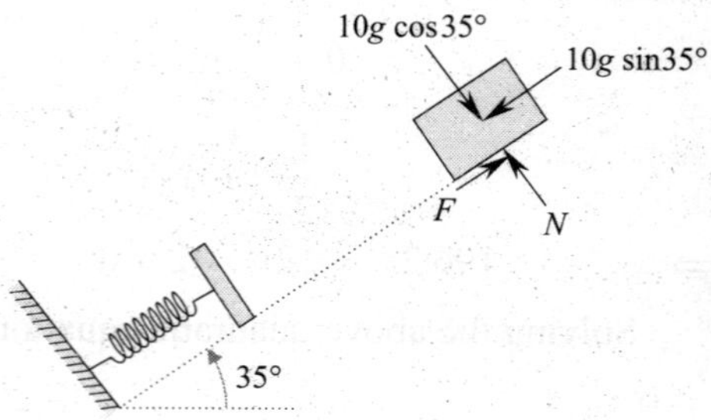

Fig. 15.25(a)

Example 15.26 A car of 2 ton mass powered by an engine of 40 kW capacity, starts from rest and attains maximum speed in 30 seconds. If the frictional resistance to motion is 0.75 kN/ton, determine the maximum speed it can attain. If after attaining the maximum speed, the engine is switched off, determine the distance it would travel before coming to rest.

Solution

Maximum speed attained

If v be the maximum speed attained by the car in 30 seconds, its uniform acceleration is

$$a = v/t = v/30$$

The total frictional resistance to the motion of the car is

$$f = (0.75 \times 10^3 \text{ N/ton})(2 \text{ tons}) = 1500 \text{ N}$$

Since the car is uniformly accelerating, the kinetic equation of motion is given as

$$F - f = ma$$

Therefore, the driving force of the car is obtained as

$$F = f + ma$$

$$= (1500) + (2 \times 10^3)\left[\frac{v}{30}\right]$$

Therefore, maximum power of the engine is given as

$$P = Fv = \left[1500 + \frac{2000}{30}v\right]v$$

$$40 \times 10^3 = 1500v + 66.67v^2$$

$$66.67v^2 + 1500v - 40 \times 10^3 = 0$$

Upon solving the quadratic equation, we get

$$v = \frac{-1500 \pm \sqrt{(1500)^2 + 4(40 \times 10^3)(66.67)}}{2 \times 66.67}$$

$$= 15.7 \text{ m/s (or) } 56.5 \text{ kmph}$$ [neglecting the negative root]

Distance travelled before coming to rest

If after attaining the maximum speed, the engine is switched off, then the final velocity is zero. Hence, applying the work–energy equation,

$$(K.E)_f - (K.E)_i = \text{work done}$$

$$0 - \frac{1}{2}mv^2 = -fs$$

$$\Rightarrow \quad s = \frac{mv^2}{2f}$$

$$= \frac{(2000)(15.7)^2}{2(1500)} = 164.33 \text{ m}$$

Example 15.27 A bullet of 20 g mass moving at 300 m/s pierces a fixed 5 cm thick metal plate and emerges out with a velocity of 50 m/s. Determine the resistance offered by the plate assuming it to be uniform.

Solution

Initial speed of bullet, $v_i = 300$ m/s

Final speed of bullet, $v_f = 50$ m/s

Applying the work–energy equation,

$$(K.E)_f - (K.E)_i = \text{work done}$$

$$\frac{1}{2}mv_f^2 - \frac{1}{2}mv_i^2 = -F(s)$$

$$\frac{1}{2}(20 \times 10^{-3})[50^2 - 300^2] = -F[5 \times 10^{-2}]$$

$$\Rightarrow \quad F = 17.5 \text{ kN}$$

15.7 WORK DONE BY INTERNAL FORCES

In this section, we will discuss work done by internal forces such as tension in the string in interconnected bodies. Consider a system of bodies as shown in Fig. 15.26. The internal force, namely, the tension in the string connecting the blocks, acts on the blocks in the opposite directions and hence the net work done for any displacement is *zero*. Thus, this method is advantageous over the Newton's method as separate free-body diagrams need not be drawn for individual members.

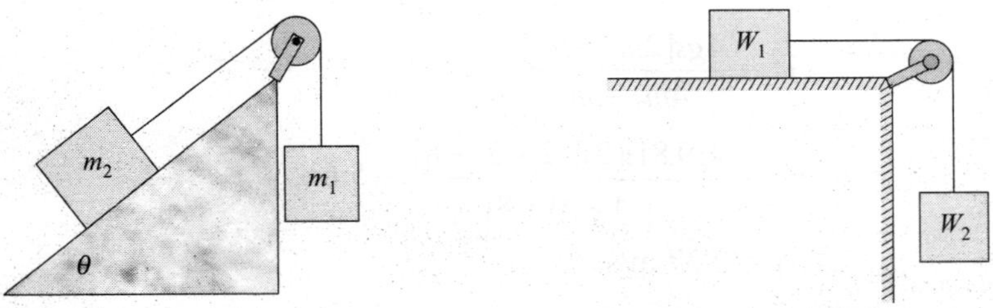

Fig. 15.26 Net work done by internal forces is zero

For illustration, consider the system of interconnected bodies shown on the left in Fig. 15.26. Let us isolate each body and analyze their motion using work–energy method assuming that the block-I moves downwards. Since the string is inextensible, the displacement of both the blocks will be same and let it be 's'. Applying the work–energy equation for the block 1,

$$(m_1 g - T)s = \frac{1}{2} m_1 v^2 - 0 \qquad (15.27)$$

Similarly, applying the work–energy equation for the block 2,

$$(T - \mu_2 N_2 - m_2 g \sin\theta)s = \frac{1}{2} m_2 v^2 - 0 \qquad (15.28)$$

Adding the two equations,

$$(m_1 g - \mu_2 N_2 - m_2 g \sin\theta)s = \frac{1}{2}(m_1 + m_2)v^2 \qquad (15.29)$$

We see that the summation of works done by forces acting on both the bodies is equal to summation of kinetic energies of individual bodies. In addition, we see that the work done by the internal force, i.e., tension T is eliminated. Thus, it need not be considered at all in the work–energy equation. Moreover, the kinetic energy of the system is given by the summation of kinetic energies of the individual bodies in the system. This is the advantage of using the work–energy method over Newton's method. The system can be considered as such and there is no need to draw individual free-body diagrams.

Example 15.28 In the system of blocks shown, if m_1 = 8 kg and m_2 = 5 kg, determine the velocities of the blocks after the block of mass m_2 displaces by 2 m.

Solution Since m_1 = 8 kg, we can see that to maintain its equilibrium, the tension in the string passing over the pulley must be $T = 8g/2 = 4g$. Since this value is less than the weight of the second block, i.e., $5g$, we can conclude that the second block moves down while the first block moves up. Hence, we can write the work–energy equation for the system as

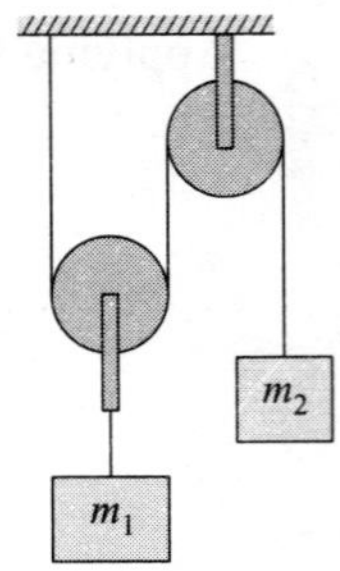

Fig. 15.27

$$m_2 g s_2 - m_1 g s_1 = \frac{1}{2} m_2 v_2^{\,2} + \frac{1}{2} m_1 v_1^{\,2}$$

Note that as the displacement of the block I is opposite to the direction of force of gravity, the work done by the force of gravity is negative. We also know that the displacement of the block I is half that of the displacement of the block II and the velocity of the block I is half that of the velocity of the block II. Hence, if $s_2 = s$ and $v_2 = v$, then $s_1 = s/2$ and $v_1 = v/2$. Therefore, we can write the above expression as

$$m_2 gs - m_1 g\frac{s}{2} = \frac{1}{2}\left[m_2 + \frac{m_1}{4}\right]v^2$$

$$\Rightarrow \qquad v^2 = \frac{4gs[2m_2 - m_1]}{4m_2 + m_1}$$

$$= \frac{4(9.81)(2)[(2\times 5) - 8]}{[(4\times 5) + 8]} = 5.61$$

$$v = 2.37 \text{ m/s}$$

Therefore, the velocities of the two blocks are

$$v_2 = 2.37 \text{ m/s} \quad \text{and} \quad v_1 = 1.19 \text{ m/s}$$

Example 15.29 In the system of blocks shown, if $m_1 = 3$ kg and $m_2 = 5$ kg, determine the velocities of the blocks after the block of mass m_2 displaces by 2 m. Take $\mu = 0.15$.

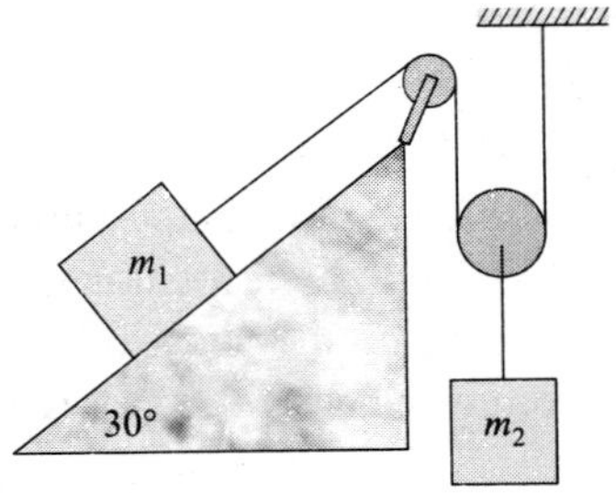

Fig. 15.28

Solution For equilibrium of the block II, we see that the tension in the string passing over the pulley must be such that $2T = m_2 g$ (or) $T = m_2 g/2 = 2.5g$. For equilibrium of the block I, the tension in the string must be such that $T = m_1 g \sin\theta + \mu m_1 g \cos\theta = 1.89g$. Since this value is less than $2.5g$, we conclude that the block II moves down while the block I moves up as shown in Fig. 15.28(a)

Writing the work–energy equation for the system, we have

$$W_2 s_2 - [W_1 \sin\theta + \mu W_1 \cos\theta] s_1 = \frac{1}{2}\frac{W_2}{g} v_2^2 + \frac{1}{2}\frac{W_1}{g} v_1^2$$

We know that the displacement of the block I is twice that of the displacement of the block II and the velocity of the block I is twice that of the velocity of the block II. If $s_2 = s$ and $v_2 = v$, then $s_1 = 2s$ and $v_1 = 2v$. Hence, we can write the above expression as

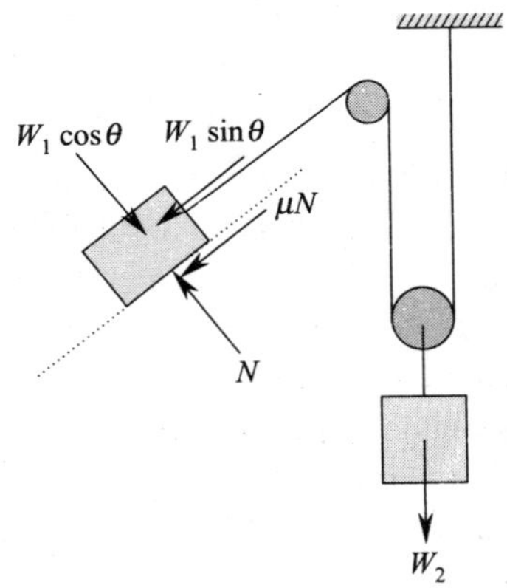

Fig. 15.28(a)

$$W_2 s - [W_1 \sin\theta + \mu W_1 \cos\theta]\, 2s = \frac{1}{2}\frac{v^2}{g}[W_2 + 4W_1]$$

$$\Rightarrow \qquad v^2 = \frac{2gs[W_2 - 2W_1(\sin\theta + \mu\cos\theta)]}{4W_1 + W_2}$$

$$= \frac{2gs[m_2 - 2m_1(\sin\theta + \mu\cos\theta)]}{4m_1 + m_2}$$

$$= \frac{2(9.81)(2)\{5 - 2(3)[\sin 30° + (0.15)(\cos 30°)]\}}{4(3) + 5} = 2.82$$

$$\therefore \qquad v = 1.68 \text{ m/s}$$

Therefore, the velocities of the two blocks are

$$v_1 = 3.36 \text{ m/s} \quad \text{and} \quad v_2 = 1.68 \text{ m/s}$$

15.8 POTENTIAL ENERGY

Potential energy of a body is defined as the capacity to do work by virtue of its *position*. There are many types of potential energies such as gravitational, electrical, elastic, etc. In our study, we are mainly concerned with *gravitational* potential energy and *elastic* potential energy.

When a body is placed in a uniform gravitational field, it is attracted by the earth. Thus, work is always required to lift a body **up** against the force of gravity. After lifting, if the force is removed, it is said to posses the capacity to do work. The body allowed to fall down freely would do work on any body that comes across its path. This principle is used in a pile hammer to drive piles. It is also used in a roller coaster, where the train is initially lifted to a high hill to increase its potential energy. When released from that height, it has more potential and hence the train moves over the coaster with high speed. Though the acceleration due to gravity remains constant at elevations near the earth's surface, the velocity of the body with which it reaches the ground is dependent on the height from which it falls. Hence, we could say that the body has more potential to work at a higher elevation than it has when at the ground level and this is termed potential energy.

To determine the potential energy, a zero height position must be chosen. Typically, the ground is considered to be a position of zero height. Potential energy is measured by the amount of work that has to be done on the body to lift it from ground level to a higher elevation. Thus, to lift a body to a height h from ground level, work required to be done against gravity is

$$W = mgh \tag{15.30}$$

$$\therefore \quad \text{P.E} = W = mgh \tag{15.31}$$

The student should recall the discussion made in Example 15.5 that the work done in a gravitational field is independent of the path followed and thus the force of gravity is termed **conservative force**. In a conservative force field, the total mechanical energy remains constant.

Principle of conservation of mechanical energy If a body is subject to a conservative system of forces, (say gravitational force) then its mechanical energy remains constant for any position in the force field. Consider a body either sliding down a *smooth* incline or freely falling. Since it is initially at rest, all of its energy is potential energy. As it accelerates downwards, some of its potential energy is converted into kinetic energy. At the bottom of the incline or at the ground level, the energy will be purely kinetic, assuming the bottom of the slope or the ground level as the datum for potential energy. By the principle of conservation of energy, we see that the loss in potential energy is equal to the gain in kinetic energy. Mathematically,

$$(\text{P.E})_i - (\text{P.E})_f = (\text{K.E})_f - (\text{K.E})_i \tag{15.32}$$

On rearranging, we have

$$(\text{P.E})_i + (\text{K.E})_i = (\text{P.E})_f + (\text{K.E})_f \tag{15.33}$$

$$\Rightarrow \quad (\text{P.E}) + (\text{K.E}) = \text{constant} \tag{15.34}$$

Thus, we see that the total mechanical energy, i.e., sum of potential and kinetic energies remain constant. This is known as principle of conservation of mechanical energy.

However, we must understand this is true only in the case of conservative force field such as gravitational force. For instance, the force of friction, which is a non-conservative force, when acting on a body, the mechanical energy will not be conserved. Actually, the energy to overcome friction is converted into heat energy.

Elastic potential energy is the energy stored in elastic materials as a result of their stretching or compressing. We already saw in Section 15.4 that work done in stretching in a spring of stiffness k to an elongation x_o is

$$W = \frac{1}{2} k x_o^2$$

This is also true if the spring is compressed. By conservation of work–energy, this work done on the spring is stored in it as what is known as potential energy, because a compressed or stretched spring has potential to do work. Since this energy is dependent upon the elongation or in other words, its position with respect to its unstretched position, it is known as potential energy.

Example 15.30 A ball is dropped from the top of a tower. If it reaches the ground with a velocity of 30 m/s, determine the height of the tower by the conservation of energy method.

Solution By the principle of conservation of energy, we know that the total mechanical energy remains constant. Hence, the total energy at the top of the tower must be equal to that at the base of the tower, i.e.,

$$(\text{K.E} + \text{P.E})_{top} = (\text{K.E} + \text{P.E})_{base}$$

Since the ground surface is taken as the datum, the potential energy at the top is mgh [where h is height of the tower] and that at the bottom is zero. If v is the velocity of the ball at the base, we can write

$$0 + (mg)(h) = \frac{1}{2}mv^2 + 0$$

$$\Rightarrow \quad h = \frac{v^2}{2g}$$

$$= \frac{(30)^2}{2(9.81)} = 45.87 \text{ m}$$

Fig. 15.29

Example 15.31 A small bob suspended by a string of length l oscillates about its mean position. Determine the velocity of the bob at its lowest position if the maximum angle of inclination of the bob with respect to the vertical is θ.

Solution The velocity of the bob at the extreme position is zero and hence, it swings back to the mean position. At the mean position, its velocity is maximum. If we take the lowest position of the bob as the datum then potential energy is maximum at the extreme position. Hence, applying the principle of conservation of energy, we have

$$(\text{K.E} + \text{P.E})_{extreme} = (\text{K.E} + \text{P.E})_{lowest}$$

$$0 + mgh = \frac{1}{2}mv^2 + 0$$

$$\Rightarrow \quad v^2 = 2gh$$

$$= 2g(l - l\cos\theta)$$

$$= 2gl(1 - \cos\theta)$$

$$\therefore \quad v = \sqrt{2gl(1 - \cos\theta)}$$

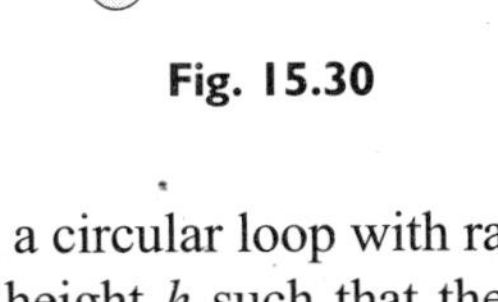

Fig. 15.30

Example 15.32 *ABCDE* is a channel in the vertical plane, *BCDE* being a circular loop with radius r. If a block is released from rest at the point A, determine the minimum height h such that the block completes the loop.

Solution The difference in height between points A and D is $(h - 2r)$. Hence, the loss in potential energy as the block slides from A to D is $mg(h - 2r)$. By the principle of conservation of energy, we know that the reduction in potential energy is equal to the increase in kinetic energy. Applying the conservation of energy,

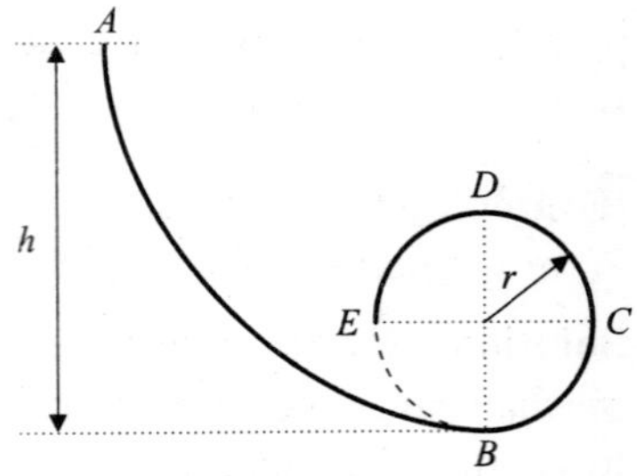

Fig. 15.31

$$P.E + K.E = \text{const}$$

$$mgh + 0 = mg(2r) + \frac{1}{2}mv^2$$

$$\Rightarrow \quad mg(h - 2r) = \frac{1}{2}mv^2$$

$$\therefore \quad v^2 = 2g(h - 2r)$$

From Chapter 14, we know that at the highest point *D*, the kinetic equation of motion can be written as

$$R + mg = \frac{mv^2}{r}$$

where *R* is the normal reaction exerted by the loop on the block. In addition, the normal reaction *R* is the *least* at the *highest* point. Hence, to determine the minimum height *h* for the block to stay in the loop, we equate *R* to zero. Therefore,

$$mg = \frac{mv^2}{r}$$

$$g = \frac{v^2}{r}$$

$$gr = 2g(h - 2r)$$

$$r = 2(h - 2r)$$

$$r = 2h - 4r \quad \Rightarrow \quad h = \frac{5r}{2}$$

Example 15.33 A ball is projected with an initial velocity v_o at an angle of α to the horizontal. Using the principle of conservation of energy, determine the maximum height reached by the ball.

Solution In a projectile motion, we know that the horizontal component of velocity always remains constant and it is equal to $v_o \cos\alpha$. At the highest point, the vertical component of velocity is zero. Applying the principle of conservation of energy, we have

$$(K.E + P.E)_{\text{point of projection}} = (K.E + P.E)_{\text{highest point}}$$

$$\frac{1}{2}mv_o^2 + 0 = \frac{1}{2}m(v_o \cos\alpha)^2 + mgh$$

$$mgh = \frac{1}{2}mv_o^2[1 - \cos^2\alpha] = \frac{1}{2}mv_o^2 \sin^2\alpha$$

$$\Rightarrow \quad h = \frac{v_o^2 \sin^2\alpha}{2g}$$

Example 15.34 A pile hammer of 40 kg is lifted to a height of 3 m and dropped onto a pile. If the resistance offered by the soil is uniform equal to 3 kN, determine by how much the pile would advance into the soil.

Solution When the pile is raised to a height of 3 m, its potential energy increases by an amount of

$$\Delta \text{P.E} = mgh = 40(9.81)(3) = 1177.2 \text{ J}$$

When it is dropped, the potential energy is converted into kinetic energy due to conservation of energy. Hence, the kinetic energy of the hammer at the instant that it hits the pile is equal to the loss in potential energy, i.e., K.E = 1177.2 J.

To determine by how much the pile would advance into the soil, we apply the work–energy equation. Hence,

$$\Delta \text{K.E} = -fs$$

where f is the average resistance offered by the soil. Hence,

$$0 - 1177.2 = -(3 \times 10^3)s$$

$$\Rightarrow \qquad s = 0.3924 \text{ m (or) } 392.4 \text{ mm}$$

Example 15.35 A 5 kg block is dropped from a height of 1 m onto a spring of stiffness 3 kN/m as shown. Determine the maximum compression of the spring.

Solution Let s be the compression of the spring. Then the total distance over which the block falls is (1 + s) metre. The change in potential energy of the block is stored in the spring as its potential energy. Therefore,

$$5(9.81)[1 + s] = \frac{1}{2} ks^2$$

$$1500s^2 - 49.05s - 49.05 = 0$$

On solving the quadratic equation for roots, we get

$$s = \frac{49.05 \pm \sqrt{(49.05)^2 + 4(1500)(49.05)}}{3000} = 197.9 \text{ mm}$$

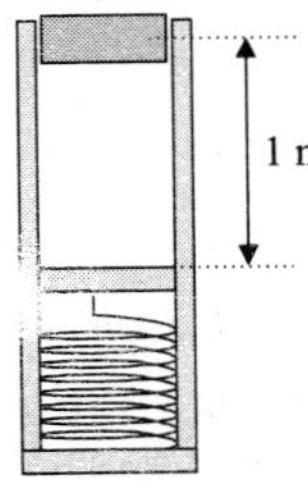

Fig. 15.32

SUMMARY

Work–Energy Method

Work–energy method is an alternative approach to solve kinetic problems. The advantages of this method over Newton's method are that work and energy are scalars; acceleration need not be determined to know the kinematics of the problem; separate free-body diagrams need not be drawn for interconnected bodies.

Work Done by a Constant Force

When a force acting on a particle causes a *displacement* of the particle, the force is then said to have done *work* on the particle. If the displacement $\vec{s}$ is in the direction of the force $\vec{F}$, then work done on the particle is defined as *product* of magnitudes of *force* and *displacement*. Mathematically, this can be written as

$$W = Fs$$

In general, the displacement of a particle under the action of forces would not always occur in the direction of the force due to constraints involved. Hence, work done in general, can be defined as product of component of the force in the direction of motion and the displacement. Alternatively, it can also be defined as product of force and component of displacement in the direction of the force. Mathematically, we can write this as

$$W = (F\cos\theta)s \quad \text{(or)} \quad W = (F)(s\cos\theta)$$

or written in vector form as

$$W = \vec{F}.\vec{s} = Fs\cos\theta$$

Work done is a *scalar* quantity that is having magnitude, but not direction. Hence, work done adds up algebraically. The S.I. unit of work done is N.m (or) Joule.

(i) When $\theta = 0°$, i.e., when force and displacement are in the same direction then $\cos\theta = 1$ and hence work done is $W = Fs$.

(ii) When $\theta = 180°$, i.e., when force and displacement are in the opposite direction then $\cos\theta = -1$ and hence work done is $W = -Fs$. The force of *friction* always acts in the direction opposite to that of the motion; hence, work done by the force of friction is always *negative*.

Work done is *zero* in the following cases:

(i) When the displacement of the particle is zero

(ii) When the motion is at right angles to the direction of the force

(iii) When a particle is in static equilibrium

(iv) Work done by internal forces as in the case of interconnected members

Work Done by a Variable Force

When a variable force acts on a particle, the resulting displacement in general will be along a *curvilinear* path. The work done in causing a displacement from the point A to B is given as

$$W_{AB} = \int_A^B dW = \int_A^B \vec{F}.d\vec{s} = \int_A^B F\cos\theta\, ds$$

Work Done in Stretching a Spring

A spring, whose one end is fixed and the other end stretched is an example for a variable force constant in direction but varying in magnitude. Work done in stretching a spring of stiffness k to an elongation of x_o from its unstretched position is given as

$$W = \int_0^{x_o} dW = \int_0^{x_o} kx\, dx = \frac{1}{2}kx_o^2$$

Power

Power is defined as the rate at which work is done. The capacity of an engine or a machine used to do work is normally expressed as its *rated power*. The one doing work faster than the other is said to be more *powerful*. If W is the total work done for a time interval t, then average power is given as

$$P_{\text{ave}} = \frac{\text{total work done}}{\text{time taken}} = \frac{W}{t}$$

The instantaneous power is given as

$$P = Fv$$

Its S.I unit is Joule per second (J/s), also called Watt (W) and the non-S.I unit is horsepower.

Energy

Energy is defined as that property of a body by virtue of which work can be performed or in other words, it is the *capacity* of a body to *do work*. Since energy is the capacity of a body to do work, its unit is same as that of work done, i.e., N.m (or) Joule.

Kinetic Energy

Kinetic energy of a body is defined as the energy possessed by virtue of its motion. It is measured by the amount of work a moving body will do if brought to rest or by the amount of work originally needed to impart the velocity to it. Mathematically,

$$\text{K.E} = \frac{1}{2} mv^2$$

Work–Energy Principle

The change in kinetic energy of a particle during any displacement is equal to the work done by the net force acting on it. Mathematically,

$$W = \frac{1}{2} mv^2 - \frac{1}{2} mv_o^2 = \text{change in K.E}$$

Potential Energy

Potential energy of a body is defined as the capacity to do work by virtue of its position. It is measured by the amount of work that has to be done on the body to lift it from ground level to a higher elevation. Thus, to lift a body to a height h from ground level, work required to be done against gravity is

$$\text{P.E} = W = mg\,h$$

It should be noted that the work done is independent of the path followed, thus the force of gravity is termed as *conservative* force. In a conservative force field, the total mechanical energy remains constant.

Principle of conservation of mechanical energy

If a body is subject to conservative system of forces, (say gravitational force) then its mechanical energy remains constant.

$$(\text{K.E} + \text{P.E})_1 = (\text{K.E} + \text{P.E})_2$$

EXERCISES

Objective-type Questions

1. Work energy method relates

(a) force, acceleration and time

(b) force, velocity and time

(c) force, velocity and displacement

(d) force, displacement and time

2. When a body is lifted up, the work done by force of gravity is
   (a) positive (b) negative (c) zero
3. When a body is freely falling, the work done by force of gravity is
   (a) positive (b) negative (c) zero
4. When a body displaces normal to the force of gravity, the work done by force of gravity is
   (a) positive (b) negative (c) zero
5. The work done on a body is zero when
   (a) there is no displacement of the body
   (b) the resultant of forces acting on it is zero
   (c) the displacement is perpendicular to the force
   (d) all of these
6. The restoring force in a spring is proportional to
   (a) the initial unstretched length of the spring
   (b) the elongation of the spring
   (c) the number of turns of coil in the spring
   (d) the cross-sectional area of the spring
7. The work done in stretching a spring of spring constant k by a length Δ is
   (a) $k\Delta$ (b) $k\Delta^2$ (c) $k\Delta/2$ (d) $k\Delta^2/2$
8. Two agents A and B do the same amount of work on a body, in which agent A causes a displacement less than that of B. Then state which of the two agents is powerful.
   (a) Both the agents are equally powerful.
   (b) A is more powerful than B.
   (c) B is more powerful than A.
   (d) Cannot be determined from the given conditions
9. One metric horsepower is equal to
   (a) 1 watt (b) 736 watts (c) 746 watts (d) 1000 watts
10. When the speed of a particle is doubled, its kinetic energy
    (a) remains the same (b) increases twofold
    (c) increases threefold (d) increases fourfold
11. In a conservative force field,
    (a) work done is zero (b) kinetic energy is constant
    (c) potential energy is constant (d) total mechanical energy is constant

Answers

1. (c) **2.** (b) **3.** (a) **4.** (c) **5.** (d) **6.** (b) **7.** (d) **8.** (b)
9. (b) **10.** (d) **11.** (d)

Short-answer Questions

1. Define work done on a body (a) by a constant force, and (b) by a varying force.
2. When is the work done upon a body positive and when is it negative?

3. Under what conditions does the work done upon a body become zero?
4. The work done upon a body by a system of forces causing uniform velocity is zero. Discuss.
5. Derive the expression for work done upon stretching a spring without accelerating it.
6. Define power.
7. What is the relationship between Watt power and horsepower?
8. Define energy. What are the various forms of energy?
9. Differentiate between kinetic energy and potential energy.
10. State the work–energy principle.
11. Explain the work done by internal forces in a connected system.
12. Show that the energy of a freely falling body is constant.

Numerical Problems

15.1 A block of 5 kg mass is resting on a rough horizontal plane having a coefficient of kinetic friction of 0.15. It is pulled by a horizontal force P at a constant acceleration of 1 m/s^2 over a distance of 5 m. Sketch the free-body diagram of the block showing all the forces acting on it. Determine the works done by the externally applied force and force of friction.

Fig. E.15.1

Ans. 61.8 J, – 36.8 J

15.2 In the Guinness world record, a man pulled a Boeing 747-400 weighing 187 tons, a distance of 91 m in 1 min 27.7 s. If the force of friction is 1 kN/ton then determine the work done by the man and power exerted by him, if he pulled it at a constant speed.

Ans. 17.02 MJ, 194.1 kW

15.3 Water is to be lifted from a well of 80 m depth. If the power of the motor available is 10 kW, determine the discharge of water. The weight density of water is 9810 N/m^3.

Ans. 0.013 m^3/s

15.4 Five men push a bus, which had a breakdown. The mass of the bus is 7 tons and the frictional resistance is 0.25 kN/ton. Determine the work done by the men in pushing the bus at a constant speed over a distance of 15 m.

Ans. 26.25 kJ

15.5 A mass of 5 kg when attached to a spring extends it vertically by 10 cm. Determine the work done in stretching the same spring horizontally by 5 cm.

Ans. 0.613 J

15.6 A man lifts a weight of 500 N through a height of 6 m as shown in Fig. E.15.6. Determine the work done by him if the coefficient of kinetic friction between rope and contact surface is 0.2.

Ans. 2.2 kJ

15.7 In a workshop, a crane lifts a load of 1 ton from the floor to a height of 2 m and it moves horizontally a distance of 5 m and finally drops it onto a platform that is 1 m from the floor level. Determine the total work done by the crane.

Ans. 9810 J

15.8 Compare the works done by external agents in each case: (i) when a body of 20 kg mass is tied to an inextensible string and lifted up by a distance of 3 m at a constant velocity; (ii) when the same body is pushed up at constant velocity over a smooth inclined plane with a slope of 0.75 to reach the same height. What do you infer from the result?

Ans. (i) 588.6 J; (ii) 588.6 J; work done is same

15.9 A heavy block of 500 kg mass is to be loaded onto a truck. A man pulls it by a rope attached to the block and parallel to the plank as shown in Fig. E.15.9. Determine the work done by him if he pulls it up at a constant speed; the length of plank being 4 m and coefficient of friction between plank and block is 0.2.

Ans. 8.87 kJ

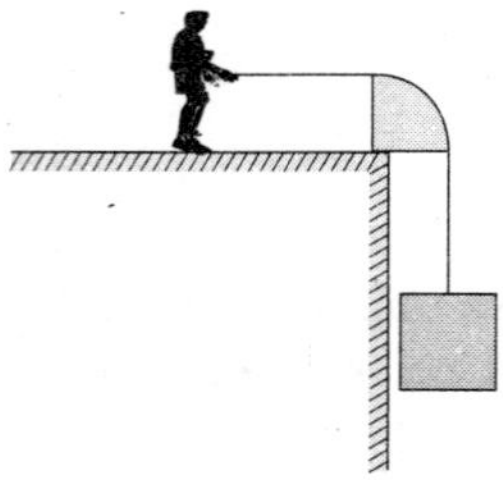

Fig. E.15.6

Fig. E.15.9

15.10 A high-speed train of 40 ton mass attains a maximum speed of 270 kmph in 90 seconds. If the frictional resistance is 0.5 kN/ton, determine the power of the engine.

Ans. 4 MW

15.11 A train of 300 ton mass is pulled by a locomotive at a constant speed of 100 kmph. If the resistance due to friction is 2 kN/ton, determine (i) the power of the engine, and (ii) work done by the engine in travelling a distance of 1/2 km.

Ans. 16.67 MW, 300 MJ

15.12 A car of mass m ascends a smooth incline of slope 1 in n at a constant acceleration of a and attains a maximum speed v. Determine the power of the engine.

Ans. $mv[(g/n) + a]$

15.13 A bus of 8 ton mass starts from rest and moves up an incline of 1 in 100. If it attains a maximum speed of 30 kmph in 30 seconds, determine the power of the engine; the frictional resistance being 1.5 kN/ton.

Ans. 125 kW

15.14 A car of 2 ton mass travelling at 45 kmph approaches a traffic junction. When the car is 100 m before the signal, the driver realizes that the green light is about to turn into red in 6 seconds and hence, he accelerates the car uniformly and crosses just before the light turns red. If the frictional resistance is 0.6 kN/ton, determine the power imparted by the engine at that instant.

Ans. 82.9 kW

15.15 A car of 2 ton mass attains a maximum speed of 80 kmph and moves at this speed on a level terrain. If the frictional resistance is 0.8 kN/ton, determine the power of the engine. If the same car moves up an incline of 1 in 110, determine the maximum speed it can attain.

Ans. 35.6 kW, 72.1 kmph

15.16 A cable train used in mountains has 2 bogies, each of 6 ton mass. The frictional resistance to motion is 0.4 kN/ton. Determine the driving power of the engine if the train has to move at 1 m/s (i) up the incline of 1 in 40, and (ii) down the same incline.

Ans. (i) 7.74 kW, (ii) 1.86 kW

15.17 A car of 2.5 ton mass going uphill of grade 1 in 100 can attain a maximum speed of 45 kmph. If the frictional resistance is 0.5 kN/ton, determine the power of the engine. If the same car moves on level terrain, determine the maximum velocity it can attain if the frictional resistance remains the same.

Ans. 18.69 kN, 53.82 kmph

15.18 A bus of 5 ton mass, moving at 60 kmph is stopped by applying brakes in a distance of 40 m. Determine the braking force, assuming it to be constant.

Ans. 17.4 kN

15.19 A bullet of 20 g mass moving at 300 m/s pierces a 3 cm thick metal plate and emerges out with a velocity of 200 m/s. Determine the resistance offered by the plate assuming it to be uniform. Also, determine the minimum number of such plates, each of 3 cm thickness, to be placed together to stop the bullet. Assume the same frictional force to be acting.

Ans. 16.67 kN, 2 plates

15.20 In a toy as shown in Fig. E.15.20, a boy compresses the spring of stiffness 200 N/m by 10 cm and places a ball of 200 g mass over it and releases it. Determine the height to which the ball would rise against gravity.

Ans. 510 mm

15.21 A car of 2 ton mass and moving with a speed of 60 kmph collides with another car of the same mass and at rest. Determine how far both will move together, if the frictional resistance is 0.75 kN/ton. Refer Fig. E.15.21.

Ans. 92.6 m

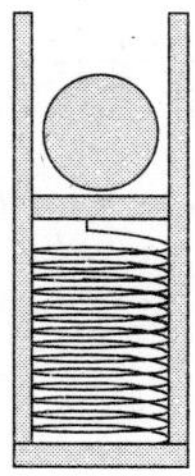

Fig. E.15.20

Fig. E.15.21

15.22 An aeroplane of 200 ton mass lands with a speed of 50 m/s. If it has to be brought to a halt within a distance of 500 m, determine (i) the average braking force, and (ii) constant deceleration.

Ans. 500 kN, 2.5 m/s^2

15.23 The driver of a car wishing to save the fuel takes his leg off the accelerator when the car is moving at a maximum speed of 80 kmph. Determine the speed of the car after travelling a distance of 100 m. The mass of car is 2 tons and the force of friction is 0.6 kN/ton.

Ans. 70 kmph

15.24 A lift starts upwards with a uniform acceleration and reaches a speed of 2.4 m/s in 2 seconds. Determine the power of the motor, if the total weight of the lift including the maximum number of occupants is 1 ton. The frictional resistance in the guide rollers is 1 kN.

Ans. 28.8 kW

15.25 In the above problem, if the lift descends down with the same acceleration, determine the power of the motor.

Ans. 18.3 kW

15.26. Using the method of work–energy, determine the acceleration of the system shown in Fig. E.15.26 and tension in the string. Assume the contact surfaces to be smooth.

Ans. $\dfrac{W_2}{W_1 + W_2} g, \; \dfrac{W_1 W_2}{W_1 + W_2}$

15.27 A skier of 80 kg mass starts sliding down an incline of slope angle 10° as shown in Fig. E.15.27. Determine the speed with which she reaches the bottom of the incline, if the inclined length is 60 m. Assume no resistance is offered to the motion. On reaching the horizontal plain, determine what uniform retarding force she has to exert to bring herself to a halt in a distance of 75 m.

Ans. 14.3 m/s, 109.1 N

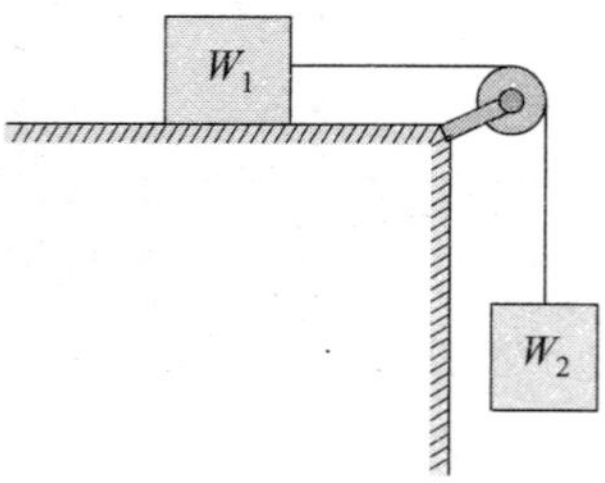

Fig. E.15.26

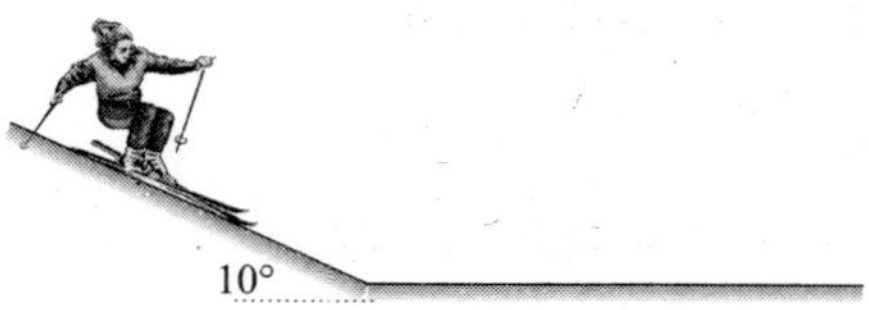

Fig. E.15.27

15.28 A block of 3 kg mass slides down a roller coaster as shown in Fig. E.15.28. If the block starts from rest at the point A, determine its velocity when it reaches the (i) point B, and (ii) point C.

Ans. (i) 12.53 m/s, (ii) 8.86 m/s

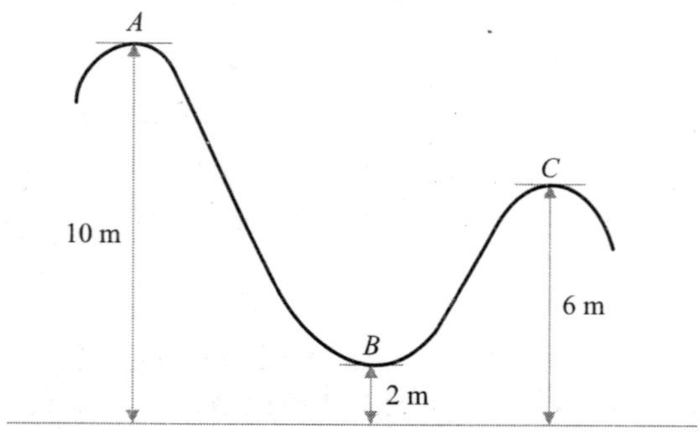

Fig. E.15.28

15.29 A block of 3 kg mass slides down a frictionless loop of 3 m radius and enters a rough horizontal plane as shown in Fig. E.15.29. Determine the distance it travels on the horizontal plane before coming to rest; the coefficient of friction between the block and plane being 0.25.

Ans. 12 m

15.30 A block of 3 kg mass slides down a frictionless loop of 3 m radius and enters a rough horizontal plane and compresses a spring of stiffness 250 N/m as shown in Fig. E.15.30. Determine the compression of the spring; the coefficient of friction between the block and plane being 0.25.

Ans. 456.7 mm

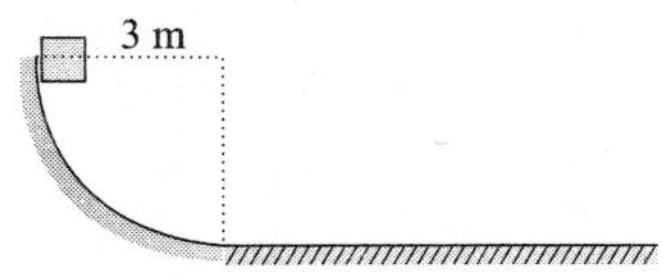

Fig. E.15.29

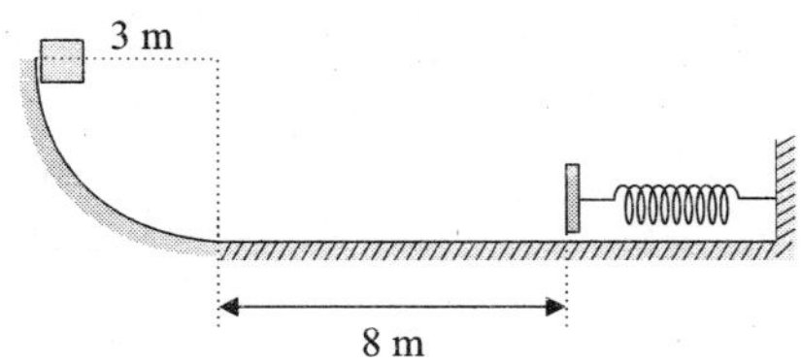

Fig. E.15.30

15.31 A block of 5 kg mass resting on a smooth horizontal plane is attached to a spring as shown in Fig. E.15.31. It is pulled by a distance of 50 cm from the unstretched position of the spring and released; determine the velocity with which the block crosses the unstretched position. The spring extends by 20 cm when a 4 kg mass is suspended from it.

Ans. 3.13 m/s

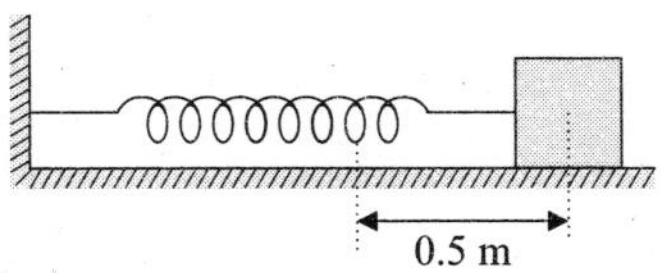

Fig. E.15.31, E.15.32

15.32 In the previous problem, determine the velocity with which the block crosses the unstretched position considering the surface to be rough with coefficient of kinetic friction of 0.2.

Ans. 2.8 m/s

16 Impulse and Momentum

16.1 INTRODUCTION

In this chapter, we will introduce an alternative approach, called **impulse momentum** method to solve kinetics problems. This method is based on integration of equation of motion with respect to time. It relates *force*, *velocity* and *time*.

Just like the work–energy method, this method avoids the need to determine the *acceleration* of the body for knowing the kinematics of problems. However, unlike the work–energy method, the quantities involved, namely, *impulse* and *momentum* are vectors; thus, necessitating the need to take care of directional aspects. In addition, the effects of internal forces do not always directly cancel out as seen in work–energy method and hence, separate free-body diagrams need to be drawn to solve kinetics of connected bodies.

In spite of these drawbacks as compared to the work–energy method, this method is particularly useful in the case of *impulsive* forces, which are large forces acting for a very short duration and causing a sudden change in velocity of the body, and also in *impact* problems. This method can also be used to solve any kinetic problem, where *duration* of motion needs to be determined.

16.2 IMPULSIVE FORCES

When a moving body strikes a *fixed* object, as in the case of a ball impinging upon a floor or a wall, or when two moving bodies collide with each other, as in the case of a cricketer striking a moving ball with a bat or when two vehicles make a head-on collision, a *large* force of impact acts on each body at the point of contact for a *short* duration. Such types of forces are known as **impulsive forces**. We say short duration, because it is very small as compared to the time of observation of motion of the bodies.

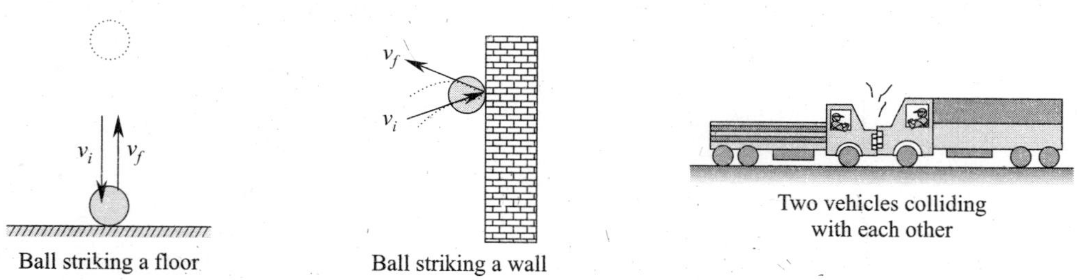

Fig. 16.1

As we know that whenever forces act on a body, they cause acceleration or in other words, change in velocity of the body. Hence, in these cases too, we observe that there is change in velocity, but they are found to occur *abruptly* unlike the normal cases of motion. As a result, a moving body after collision with a fixed or moving body experiences an abrupt change in velocity (both in magnitude and in direction).

16.3. IMPULSE AND MOMENTUM

Since impulsive forces are very *large* in magnitude and vary over the *short* duration during which they act, it is difficult to exactly determine their intensity and the time during which they act. However, the **effect** of forces could be understood by knowing the change in velocities they produce. Newton's second law of motion can be written as

$$\vec{F} = m\vec{a} = m\frac{d\vec{v}}{dt} \tag{16.1}$$

Multiplying both sides by dt

$$\vec{F}\,dt = m\,d\vec{v} \tag{16.2}$$

and integrating between the limits, we have

$$\int_{t_1}^{t_2} \vec{F}\,dt = m[\vec{v} - \vec{v}_o] \tag{16.3}$$

where t_1 and t_2 are initial and final times of impact, and $\vec{v}_o$ and $\vec{v}$ are initial and final velocities of the body respectively just before impact and just after impact.

The integral on the left-hand side, i.e., $\int_{t_1}^{t_2} \vec{F}\,dt$ is defined as **impulse** of the force and the product $m[\vec{v} - \vec{v}_o]$ on the right-hand side is defined as the **change in momentum** of the body. (It should be noted that the product of mass of a body and its velocity is termed the *momentum* of the body). Thus, impulse of a force is equal to change in momentum of the body. Hence, if we know the *initial* and *final* velocities of the body then the effect of impulsive forces measured by the quantity, *impulse of the force*, can be determined.

Force being a vector and time being a scalar, their product impulse is a vector. Similarly, mass being a scalar and velocity being a vector, their product defined as change in momentum is a vector. Thus, we see that unlike the work–energy method, both the quantities involved in the impulse–momentum method are vector quantities and hence, we should take care of the direction in solving the kinetic equations. If we express force and velocity vectors in rectangular components, then the above Eq. 16.3 can be written in scalar forms as

$$\int_{t_1}^{t_2} F_x\,dt = m[v_x - (v_o)_x] \tag{16.4}$$

and

$$\int_{t_1}^{t_2} F_y\,dt = m[v_y - (v_o)_y] \tag{16.5}$$

where F_x and F_y are components of impulsive force in x and y directions respectively, and v_{ox}, v_{oy}, v_x and v_y are x and y components of initial and final velocities respectively. In solving numerical problems, generally we take velocities pointing along positive axes as positive. However, in some cases, we may take initial direction of velocity of a body as positive too.

16.3.1 Graphical Representation of Impulse and Momentum

If we could represent the variation of component of impulsive force (say F_x or F_y) with respect to time graphically as shown in the figures below then we could determine the impulse of the force as follows. Consider an infinitesimally small time interval dt. Then the force acting over this time interval can be assumed to be *constant*. Hence, Fdt represents the area of the shaded strip shown in Fig. 16.3. Then integrating over the time intervals, $\int_{t_1}^{t_2} Fdt$ represents the area under the graph between the time intervals. Thus, impulse of the force is obtained by the *area* under the F–t graph.

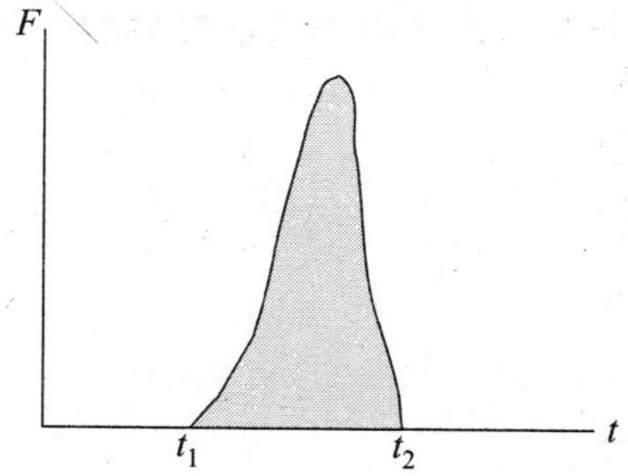

Fig. 16.2 F–t graph

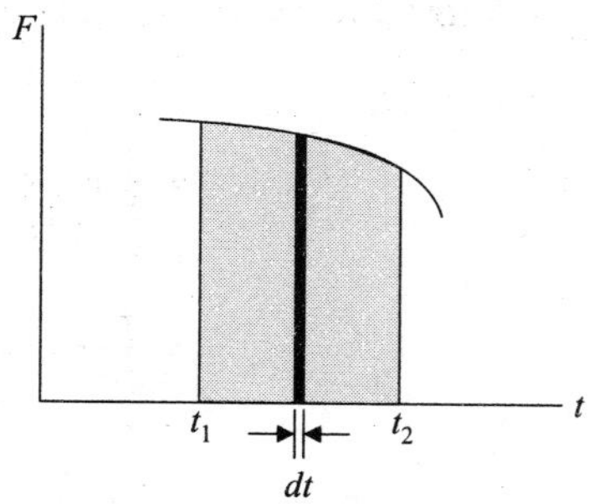

Fig. 16.3 Graphical representation of impulse

If, for instance, the force is constant, i.e., constant in magnitude and direction then it can be taken outside the integral sign. The impulse of the force then reduces to

$$I = \int_{t_1}^{t_2} F dt = F \int_{t_1}^{t_2} dt = F(t_2 - t_1) \tag{16.6}$$

that is, impulse is a product of impulsive force and the time duration for which it acts. Hence, we can write

$$F(t_2 - t_1) = m(v - v_o) \tag{16.7}$$

Thus, for a constant force acting for a very short duration t, the impulse of the force is defined as a **product** of **force** and **time** and it is equal to **change in momentum**. Even for a varying force of impulse, we can replace it by an **average** force of impulse such that the area under the graph between the time limits for both are maintained equal, i.e., $\int_{t_1}^{t_2} Fdt = F_{ave}(t_2 - t_1)$. Graphically, it is represented as shown in Fig. 16.4, where the shaded area represents the impulse of average force.

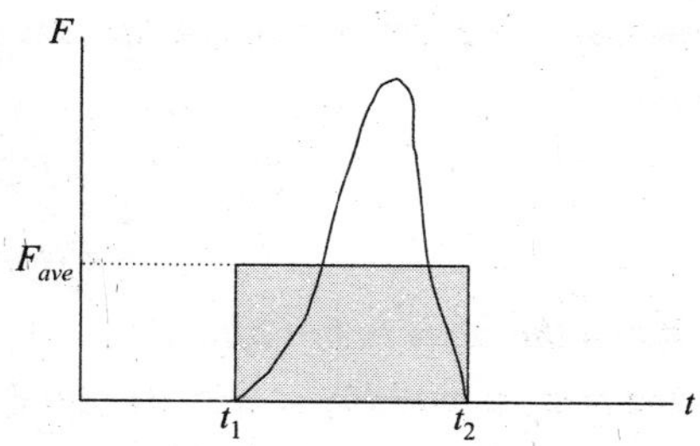

Fig. 16.4 Impulse of average force

Since impulse is the product of force and time, its S.I. unit is **N.s**. As newton can also be represented in kg.m/s², we see that unit of impulse can also be given as

$$[\text{kg.m/s}^2][\text{s}] = \text{kg.m/s}$$

We already saw that impulse is equal to change in momentum. Since momentum is product of mass and velocity, its S.I. unit is kg.m/s, which is same as the unit of impulse. Thus, we see that the impulse–momentum equation is *dimensionally homogeneous*.

If somehow using a snapshot, we could estimate the time duration for which the impulsive force acts then we can also determine the magnitude of the average force of impulse, i.e.,

$$F_{\text{ave}} = \frac{m(v - v_o)}{(t_2 - t_1)} = \frac{m\Delta v}{\Delta t} = \frac{\text{change in momentum}}{\text{change in time}} \qquad (16.8)$$

Hence, **force of impulse** is equal to the **rate of change** of **momentum**.

Example 16.1 A varying force $\vec{F} = t^2\vec{i} + 2t\vec{j}$ is applied on a particle from time $t = 0$ to 3s. Determine the impulse of the force in 3 seconds.

Solution We know that impulse of a force is obtained by integration of $\vec{F}$ with respect to time between limits. Thus,

$$I = \int_0^3 \vec{F}\, dt$$

$$= \int_0^3 (t^2\vec{i} + 2t\vec{j})dt$$

$$= \left[\frac{t^3}{3}\right]_0^3 \vec{i} + \left[2\frac{t^2}{2}\right]_0^3 \vec{j}$$

$$= 9\vec{i} + 9\vec{j}$$

Therefore, magnitude of the impulse is given as

$$\sqrt{(9)^2 + (9)^2} = 12.73 \text{ N.s}$$

Example 16.2 A force acting on a body of 2 kg mass for a short duration varies with time as shown in Fig. 16.5. Determine the final velocity of the body after 3 seconds, if the body is initially (i) at rest, (ii) moving with a velocity of 5 m/s in the positive x direction, and (iii) moving with a velocity of 5 m/s in the negative x direction.

Solution The impulse of the force is given by the area under the F–t graph, i.e.,

$$I = \int F\, dt$$

$$= F(t_2 - t_1) = 10(3 - 1) = 20 \text{ N.s}$$

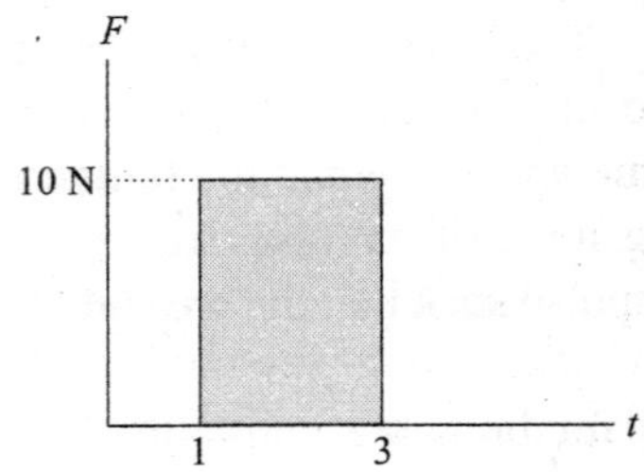

Fig. 16.5

(i) *When the body is initially at rest*

As the body is initially at rest, its initial velocity, $v_i = 0$. If v_f be its final velocity then applying the impulse momentum equation, we have

$$I = F\,\Delta t = m[v_f - v_i]$$

$$20 = 2[v_f - 0]$$

$$\Rightarrow \quad v_f = 10 \text{ m/s}$$

(ii) *When the body is initially moving at 5 m/s in the +x direction*

$$I = m[v_f - v_i]$$

$$20 = 2[v_f - 5]$$

$$\Rightarrow \quad v_f = \frac{20}{2} + 5 = 15 \text{ m/s}$$

The positive value of v_f indicates that the motion is along the $+x$ direction.

(iii) *When the body is initially moving at 5 m/s in the –x direction*

$$I = m[v_f - v_i]$$

$$20 = 2[v_f - (-5)]$$

Note that as the initial velocity is pointing along the $-x$ direction, it is considered as negative.

$$\therefore \quad v_f = \frac{20}{2} - 5 = 5 \text{ m/s}$$

The positive value of v_f indicates that the motion is along the $+x$ direction.

Example 16.3 A ball of 50 g mass is dropped from a height of 10 m, and after striking the floor, it rebounds to a height of 7 m. Determine (i) the impulse of the force, and (ii) the average force exerted by the floor on the ball, if the force acts for a fraction 1/60th of a second.

Solution

(i) Impulse of the force

As the ball is dropped from a height of 10 m, its velocity as it strikes the floor is given as

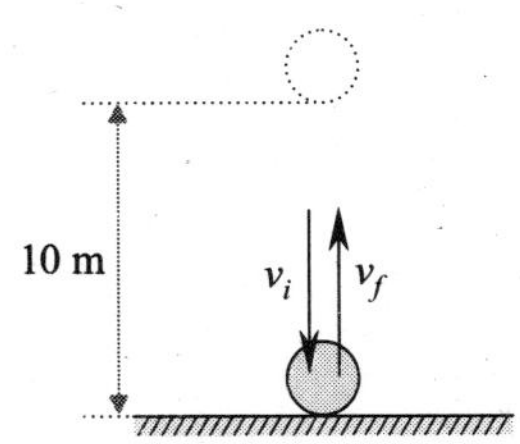

Fig. 16.6

$$v_i = \sqrt{2gh_i} = \sqrt{2 \times 9.81 \times 10} = 14.01 \text{ m/s}$$

[Note that $v_i^2 = v_o^2 + 2gh_i = 0 + 2gh_i$]. Similarly, as the ball rises to a height of 7 m after the impact, the velocity of the ball immediately after impact is given as

$$v_f = \sqrt{2 \times 9.81 \times 7} = 11.72 \text{ m/s}$$

Therefore, impulse of the force is obtained from the change in momentum of the ball due to impact. Thus,

$$I = m[-v_f - v_i]$$

$$= -50 \times 10^{-3}[11.72 + 14.01] = -1.29 \text{ N.s}$$

Note that we have taken the initial velocity of the ball as positive; and since the final velocity is opposite to the direction of initial velocity, it is taken as negative. The negative sign in impulse indicates that the impulse is acting in the direction opposite to the direction of the initial velocity of the ball.

(ii) Average force exerted by the floor on the ball

The net force acting upwards on the ball during its impact with the floor is

$$F = R - mg$$

Therefore,

$$\text{Impluse} = Ft$$
$$= (R - mg)t$$
$$1.29 = [R - (0.05)(9.81)](1/60)$$
$$\Rightarrow \quad R = 77.89 \text{ N}$$

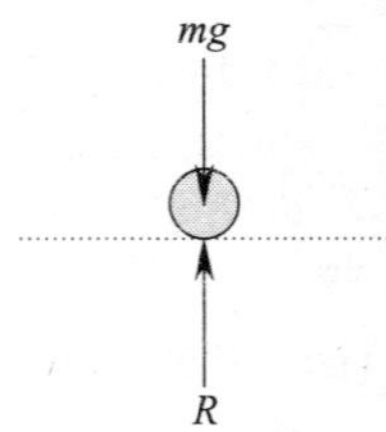

Fig. 16.6(a)

Example 16.4 During a free kick, a football player kicks a football of 250 g mass, which is at rest; and it leaves his foot with a velocity of 25 m/s at an angle of 25° with respect to the ground level. Determine the force exerted by the player, if the duration of the strike is 1/60th of a second.

Solution As the football is initially at rest, its initial velocity is zero. The *x*-component of the final velocity is 25 cos 25° m/s [refer Fig. 16.7(a)]. Let F_x and F_y be the *x* and *y* components of force exerted by the player on the ball. Then writing the impulse–momentum equation along the *X* and *Y* directions, we have

Fig. 16.7

$$F_x t = m[25 \cos 25° - 0]$$
$$\Rightarrow \quad F_x = \frac{m(25 \cos 25°)}{t}$$
$$= \frac{0.25(25 \cos 25°)}{1/60} = 339.87 \text{ N}$$
$$F_y t = m[25 \sin 25° - 0]$$
$$\Rightarrow \quad F_y = \frac{m(25 \sin 25°)}{t}$$
$$= \frac{0.25(25 \sin 25°)}{1/50} = 158.48 \text{ N}$$

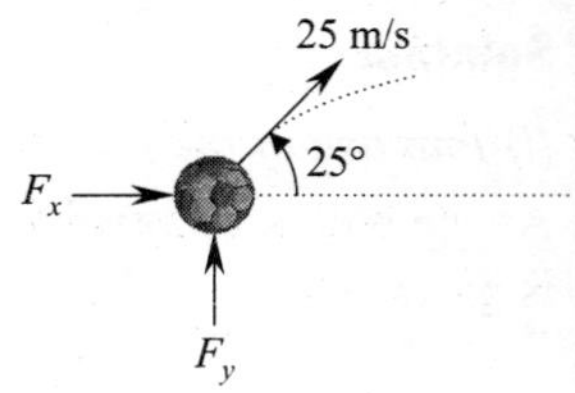

Fig. 16.7(a)

Therefore, the average force exerted by the player on the ball is obtained as

$$F = \sqrt{F_x^2 + F_y^2}$$
$$= \sqrt{(339.87)^2 + (158.48)^2} = 375 \text{ N}$$

Example 16.5 A cricket ball of 150 g mass moving at 25 m/s is hit by a cricketer and the ball leaves the bat with a velocity of 35 m/s at an angle of 30° to the initial direction. Determine the average force of impulse exerted by the bat on the ball if the contact duration is 1/100th of a second.

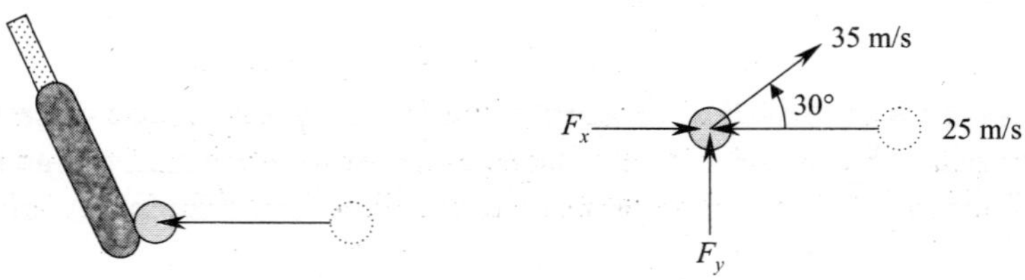

Fig. 16.8

Solution The initial direction of motion of the ball is taken as positive and thus the final direction of motion of the ball, being opposite to the initial direction, is taken as negative. Then, writing the impulse–momentum equation along the *X* and *Y* directions, we have

$$F_x t = m[-35 \cos 30° - 25]$$

$$\Rightarrow \qquad F_x = \frac{(0.15)[-35 \cos 30° - 25]}{1/100}$$

$$= -\,829.66 \text{ N}$$

The negative sign indicates that the *x*-component of the impulse is acting in the direction opposite to the direction of initial velocity of the ball.

Similarly, $$F_y t = m[35 \sin 30° - 0]$$

$$\Rightarrow \qquad F_y = 262.5 \text{ N}$$

Therefore, the force exerted by the bat on the ball is given as

$$F = \sqrt{F_x^{\,2} + F_y^{\,2}}$$

$$= \sqrt{(829.66)^2 + (262.5)^2} = 870.2 \text{ N}$$

16.4 NON-IMPULSIVE FORCES

In this section, we will discuss some of the kinetics problems, which do not involve impulsive forces, i.e., forces acting for a short duration. The examples for such non-impulsive forces are force of gravity, force of friction, etc. If we are interested in knowing only the time of motion in such type of problems, we need not determine the acceleration of the motion; hence, we can directly determine the time of motion by applying the impulse–momentum equation.

Example 16.6 The driver of a car of 2 ton mass moving at 60 kmph applies sudden brakes to bring the car to a stop in 2 seconds. Determine the average braking force.

Solution The initial speed of the car, $v_i = 60$ kmph $= 60 \times 5/18 = 16.67$ m/s; the final speed of the car is zero. As the braking force acts in the direction opposite to the direction of motion, we can write the impulse–momentum equation as

Fig.16.9

$$-Ft = m[v_f - v_i]$$

$$-F(2) = [2000][0 - 16.67]$$

$$\Rightarrow \qquad F = \frac{2000 \times 16.67}{2} = 16.67 \text{ kN}$$

Example 16.7 In the previous problem, the same car moves down an incline of 1 in 100 with the same speed. When a sudden brake is applied, it is brought to a stop in 3.5 seconds. Determine the average braking force.

Fig. 16.10

Solution The initial speed of the car, $v_o = 60$ kmph $= 60 \times 5/18 = 16.67$ m/s; its final speed is zero. The forces acting on the car are the component of its weight $mg \sin\theta$ acting down the incline and force of friction acting up the plane. Hence, we can write the impulse–momentum equation taking the direction of motion as positive

$$[mg \sin\theta - F]\, t = m[0 - v_o]$$

$$\Rightarrow \quad F = mg \sin\theta + \frac{mv_o}{t}$$

$$= 2000(9.81)\frac{1}{100} + \frac{(2000)(16.67)}{3.5} = 9.72 \text{ kN}$$

Example 16.8 A ball is thrown upwards with an initial velocity of v_o. Determine the time taken for the ball to reach the maximum height. Use the impulse—momentum method.

Solution The force acting on the ball is the force of gravity, acting against the motion of the ball. The velocity of the ball as it reaches the maximum height is zero. Writing the impulse–momentum equation along the vertical direction,

$$-mgt = m[0 - v_o]$$

$$\Rightarrow \quad t = \frac{v_o}{g}$$

Example 16.9 A block is pushed with a velocity of 10 m/s along a rough horizontal plane, whose coefficient of kinetic friction is 0.25 and that of static friction is 0.3. Determine the time taken for the block to come to a stop.

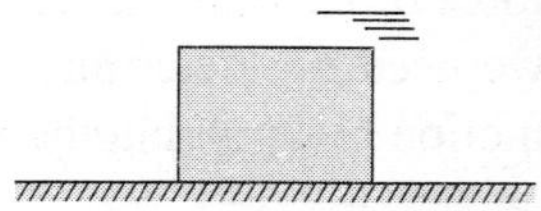

Fig. 16.11

Solution The free-body diagram of the block is shown in Fig. 16.11(a) below. The only tangential force acting on the body is the force of friction opposing the motion of the body. As there is no motion along the normal direction, we know that $N = mg$. Therefore, the force of friction is given as

$$F_k = \mu_k N = \mu_k mg$$

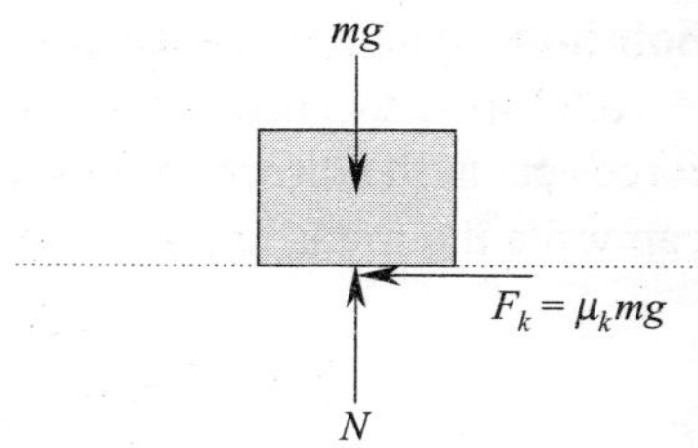

Fig. 16.11(a)

Writing the impulse–momentum equation along the horizontal plane, we have

$$-[\mu_k mg]\, t = m[0 - v]$$

Note that as the force of friction acts opposite to the direction of motion, the negative sign is used. From the above equation, we have

$$t = \frac{v}{\mu_k g}$$

$$= \frac{10}{(0.25)(9.81)} = 4.08 \text{ s}$$

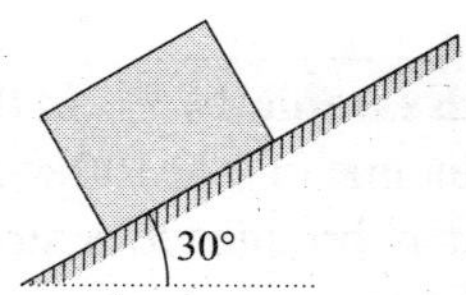

Fig. 16.12

Example 16.10 If the same block in the above problem is pushed up an incline of 30° with the same initial velocity, determine the time taken for its velocity to become zero. The coefficients of friction remain the same.

Solution The free-body diagram of the block is shown in Fig. 16.12(a). The forces acting on the block along the incline are the component of its weight $mg \sin\theta$ and force of friction F, both acting down the incline. As there is no motion along the normal direction, we know that $N = mg \cos\theta$ and hence,

$$F = \mu_k N = \mu_k mg \cos\theta$$

Writing the impulse–momentum equation along the incline,

$$[-\mu_k mg \cos\theta - mg \sin\theta]\, t = m[0 - v]$$

Note that as both the forces act down the plane, negative sign is used.

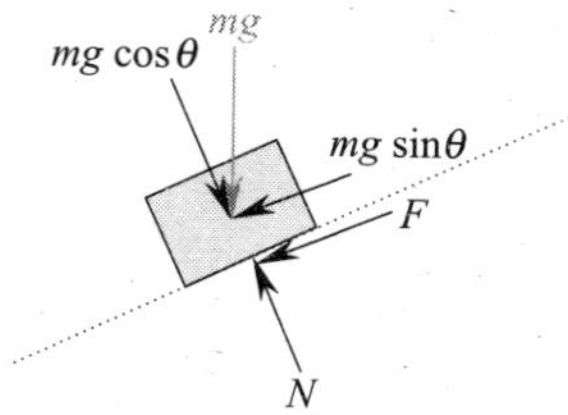

Fig. 16.12(a)

$$\therefore \quad t = \frac{v}{g[\sin\theta + \mu_k \cos\theta]}$$

$$= \frac{10}{9.81[\sin 30° + (0.25)\cos 30°]} = 1.42 \text{ s}$$

Example 16.11 In the above problem, after the velocity of the block becoming zero, will the block slide down by itself? If so, what is the time taken for it to reach the initial velocity of 10 m/s?

Solution We see that tangent of the angle of inclination of the plane is greater than the coefficient of static friction , i.e., tan 30° > μ_s. Hence, the block after reaching a momentary rest, will begin to slide down by its own weight. The free-body diagram of the block sliding down the plane is shown below. The forces acting on the block are $mg \sin\theta$ acting down the plane and force of friction acting up the plane.

Hence, writing the impulse–momentum equation along the incline,

$$[mg \sin\theta - F]\, t = m[v - 0]$$

$$[mg \sin\theta - \mu_k\, mg \cos\theta]t = m[v - 0]$$

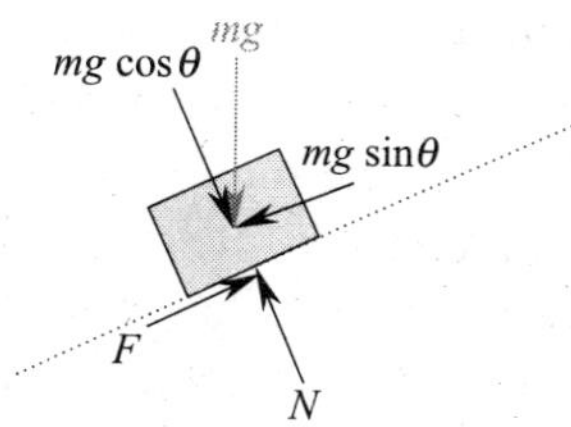

Fig. 16.13

$$\Rightarrow \quad t = \frac{v}{g[\sin\theta - \mu_k \cos\theta]}$$

$$= \frac{10}{(9.81)[\sin 30° - (0.25)\cos 30°]} = 3.6 \text{ s}$$

Example 16.12 The block-and-pulley arrangement shown is released from rest. Determine the time taken for the block of mass m_2 to reach a velocity of 1 m/s. Neglect the mass of the pulleys and assume that they are frictionless. Take m_1 = 20 kg and m_2 = 15 kg.

Solution The free-body diagrams of the two blocks are shown in Fig. 16.14(a). Since the pulleys are frictionless, the tensions on both the ends of the string in the free-body diagrams are equal. For the equilibrium of the first block, we see that $2T = m_1 g = 20g$ and hence, $T = 10g$. Since $W_2 = 15g$ is greater than this tension, we can see that the second block will move downwards, while the first block will move upwards. Since the length of the string remains constant, we see that displacement of the first block is *half* of the displacement of second block and hence the velocity of first block is *half* of the velocity of second block.

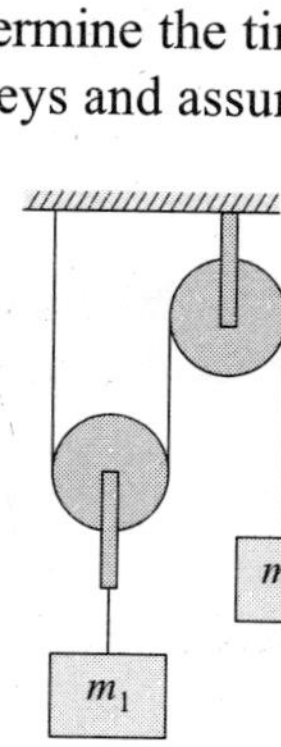

Fig. 16.14

Applying the impulse–momentum equation to the two blocks respectively, we have

$$[2T - 20g]t = 20\left[\frac{v}{2} - 0\right] \quad \text{(a)}$$

$$[15g - T]t = 15[v - 0] \quad \text{(b)}$$

Multiplying the equation (b) by 2 and adding it with the equation (a), we have

$$(10g)t = 40v$$

$$\Rightarrow \quad t = \frac{4v}{g}$$

$$= \frac{4(1)}{9.81} = 0.408 \text{ s}$$

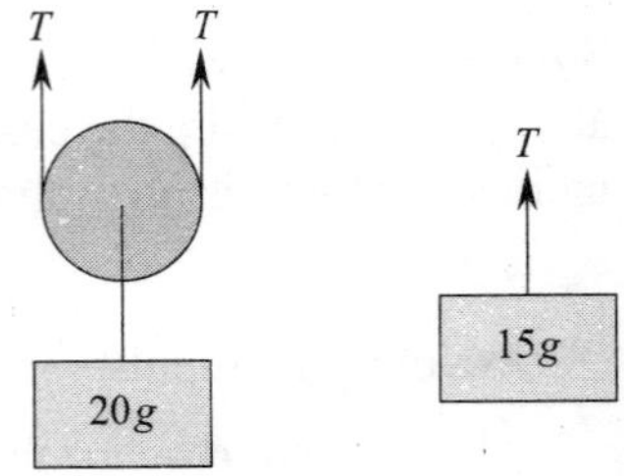

Fig. 16.14(a)

16.5 IMPACT OF JET ON PLATES (OR) VANES

Consider a jet of water issued from a nozzle of a certain cross-sectional area. If an obstruction such as a *smooth* flat plate is placed across its path, it causes a change in momentum of the flowing water. This is due to the impulsive force exerted by the plate on the jet of water.

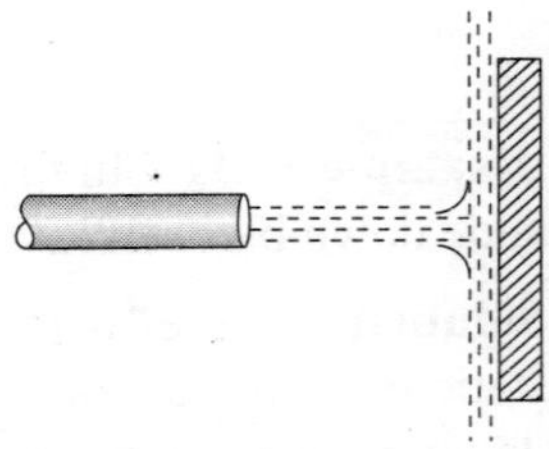
Fig. 16.15 Impact of jet of water on a plate

However, we are interested in knowing the force exerted by the moving jet of water on the plate, as it finds application in the design of *hydraulic turbines*. By Newton's third law of motion, we know that the force exerted by the jet of water on the plate is equal and opposite to the force exerted by the plate on the jet of water. Hence, by determining the force exerted by the plate on the mass of water, we can know the force exerted by the jet of water on the plate.

To determine this force, we can proceed as in the previous section by applying the impulse–momentum equation. However, we are faced with an important question as to what mass of liquid should we consider as striking the plate at a particular instant of time, as the water is continuously flowing out of the nozzle. We address this problem in the following manner. Consider the impulse momentum equations in rectangular components,

$$\int_{t_1}^{t_2} F_x\, dt = m[v_x - (v_o)_x] \tag{16.9}$$

$$\int_{t_1}^{t_2} F_y\, dt = m[v_y - (v_o)_y] \tag{16.10}$$

If the impulsive force is assumed to be constant then we can take the components of the force outside the integral sign and hence we can write them as

$$F_x t = m[v_x - (v_o)_x] \tag{16.11}$$

$$F_y t = m[v_y - (v_o)_y] \tag{16.12}$$

where $t = t_2 - t_1$. Dividing both sides by t, we have

$$F_x = \left[\frac{m}{t}\right][v_x - (v_o)_x]$$

$$= [\text{mass/unit time}][v_x - (v_o)_x] \tag{16.13}$$

$$F_y = \left[\frac{m}{t}\right][v_y - (v_o)_y]$$

$$= [\text{mass/unit time}][v_y - (v_o)_y] \tag{16.14}$$

Thus, the force of impulse is represented as a product of **mass per unit time** and **change in velocity**. As water is continuously striking the plate, we consider the mass of the liquid striking *per unit time* as the free body and solve as before.

Suppose the nozzle is discharging water at a rate of Q m^3/s. Then the mass of water discharged per unit time is given as

$$\rho Q \tag{16.15}$$

Its S.I. unit is obtained from the product of units of density and discharge, i.e., $\frac{kg}{m^3}\cdot\frac{m^3}{s}$ = kg/s and we see that it agrees with the definition of mass per unit time. In addition, we know from principles of fluid mechanics that discharge, $Q = av$, where a is the cross-sectional area of jet and v is the velocity of jet. Therefore,

$$\text{mass/unit time} = \rho Q = \rho av \tag{16.16}$$

The density of water is taken as 1000 kg/m^3. The following illustrative examples will explain the force exerted by the jet of water on various obstructions placed across its path under various kinematical conditions.

Example 16.13 A jet of water issued from a nozzle strikes normally a smooth flat fixed plate. The water after striking the plate leaves parallel to the plate. Derive an expression for the force exerted by the jet of water on the plate. If the jet diameter is 6 mm and water moves at 15 m/s, determine the force exerted by the jet of water on the plate.

Solution Consider a jet of water issued from a nozzle of cross-sectional area a. After striking the plate normally, the water leaves parallel to the plate. Since the plate is *smooth*, we can assume that the *exit* velocity of the jet to be the same as the *entry* velocity. The free-body diagrams of both the mass of water striking the plate and that of the plate are shown below. Though we are interested in the force exerted by the jet of water on the plate (shown in the free-body diagram of the plate), we solve the problem considering the free-body diagram of the mass of water. This is because as stated before at the beginning of this section, the force exerted by the jet of water on the plate is equal and opposite to that exerted by the plate on the jet of water.

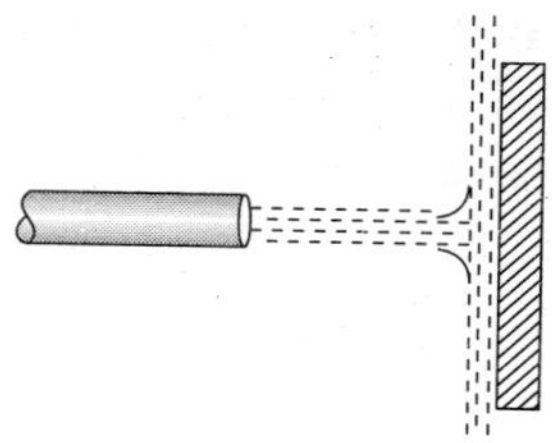

Fig. 16.16

Considering the free-body diagram of the mass of water, we see that the component of exit velocity along the direction of initial velocity is zero. Hence, applying the impulse momentum equation along the X direction, we have

$$F_x = \rho Q[0 - v]$$

$$= -\rho Qv$$

$$= -\rho av^2 \quad [\text{since } Q = av]$$

The negative sign indicates that this force acts in the direction opposite to that of the initial velocity. As the plate is *smooth*, no resistance is offered to the motion of water parallel to the plate. Hence, we can note that $F_y = 0$. Therefore, the total force exerted by the plate on the jet is $F = F_x = -\rho a v^2$. Hence, the force exerted by the jet of water on the plate is equal and opposite to that exerted by the plate on the jet, i.e., $\rho a v^2$.

If the jet diameter is 6 mm and water is issued at 15 m/s then the force exerted by the jet of water on the plate is obtained as

$$F = \rho a v^2 = (1000)[(\pi/4)(0.006)^2](15)^2 = 6.36 \text{ N}$$

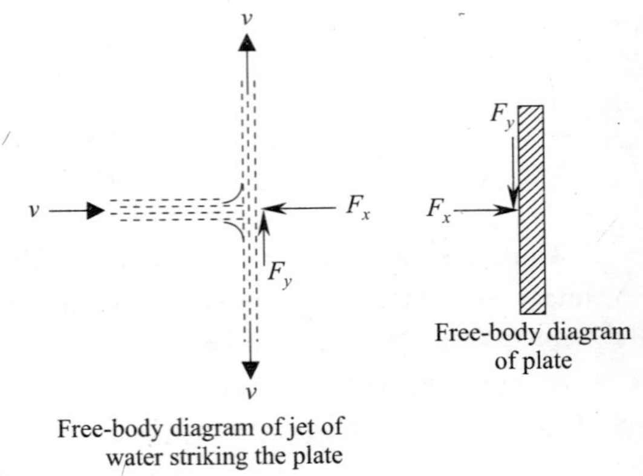

Fig. 16.16(a)

Example 16.14 In the previous problem, if the plate is free to move in the direction of the jet, derive the expression for the force of impact. If the speed of the plate is 5 m/s in the direction of the jet, determine the force exerted by the jet on the plate; the other data remaining the same.

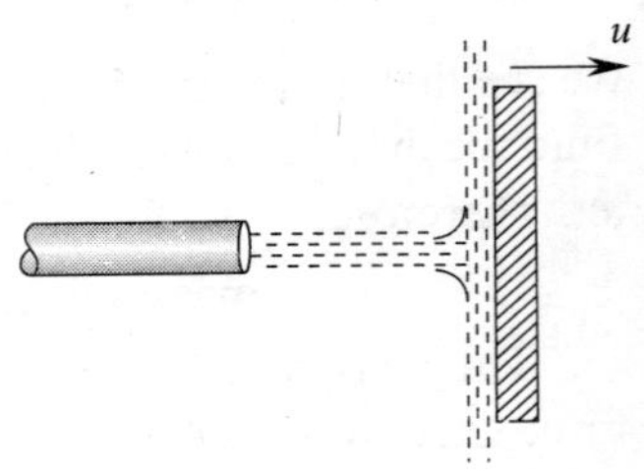

Fig. 16.17

Solution Let u be the speed of the plate moving in the direction of the jet. In this case, the mass of the liquid striking the plate per unit time will not be the same as before, because as the liquid is about to strike the plate, the plate would have moved farther away from the jet. Thus, the effective mass of water striking the plate per unit time is dependent upon the relative velocity of the jet. Therefore,

$$\text{mass/unit time} = \rho a(v - u)$$

The velocities of the jet striking and leaving parallel to the plate are equal to $v - u$. Applying the impulse–momentum equation along the X-direction, we have

$$F_x = \rho a(v - u)[0 - (v - u)]$$
$$= -\rho a(v - u)^2$$

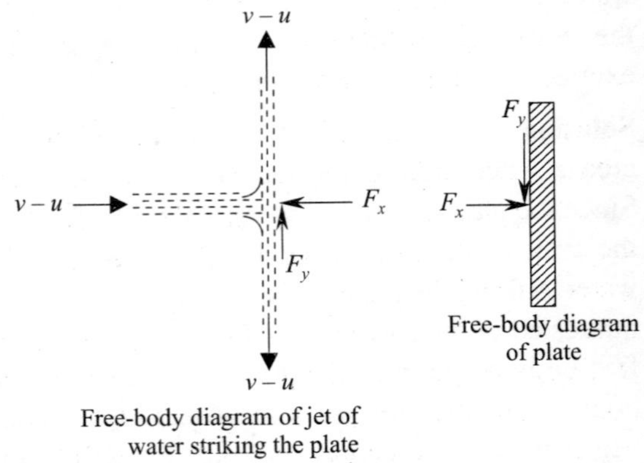

Fig. 16.17(a)

As the plate is smooth, no resistance is offered to the motion of water parallel to the plate. Thus, we can note as before that $F_y = 0$. Therefore, the total force exerted by the plate on the jet of water is $F = F_x = -\rho a(v - u)^2$ and hence, the force exerted by the jet on the plate is $\rho a(v - u)^2$.

If the plate is moving at 5 m/s in the direction of jet, then the force exerted by the jet of water on the plate is given as

$$F = \rho a(v - u)^2 = (1000)[(\pi/4)(0.006)^2](15 - 5)^2 = 2.83 \text{ N}$$

Though we have derived the expression for force exerted by the jet of water on a moving flat plate, it should be noted that practically this condition is not possible. This is because as the plate is moving away from the jet, the jet also must go along with it to make the jet strike on the plate. In actual practice, a series of plates are arranged on the periphery of a wheel, such that as one plate moves away from the jet, the next plate comes to the vertical position to intercept the jet. Thus, there is always one plate to intercept the jet in the normal direction. Hence, the wheel rotates at a constant angular speed. The mechanical energy thus generated is utilized in driving the generator for power generation.

Example 16.15 A jet of water issued from a nozzle strikes a smooth inclined plate kept fixed. The water after striking the plate leaves parallel to the plate. Derive an expression for the force exerted by the jet of water on the plate. If the jet diameter is 4 mm and water moves at 10 m/s impinging on an inclined plate inclined at 60° to the direction of the jet, determine the force exerted by the jet of water on the plate.

Solution Let the jet of water strike an inclined plate inclined at an angle of 'θ' to the direction of the jet. After striking the plate, the water leaves parallel to the plate. Since the plate is smooth, we can assume that the exit velocity of the jet to be same as the entry velocity. The free-body diagrams of the mass of water striking the plate and that of the plate are shown in Fig. 16.18(a). For convenience, we have resolved the force exerted by the plate on the mass of water into components F_n and F_t along the normal and tangential directions respectively.

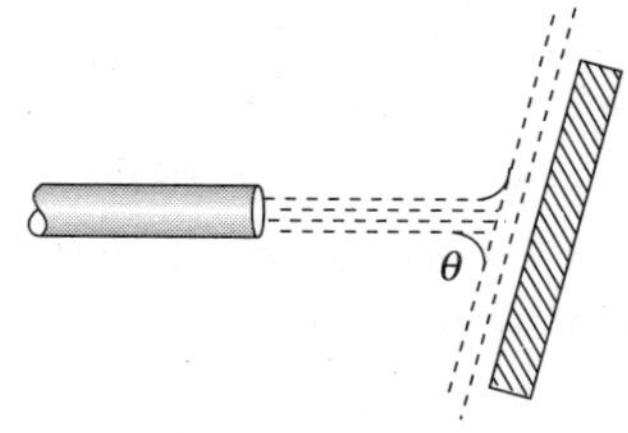

Fig. 16.18

Accordingly, we apply the impulse–momentum equation along the normal and tangential directions. The component of velocity of entry normal to the plate is '$v \sin\theta$' and the component of exit velocity along the normal direction is zero. Therefore,

$$F_n = \rho Q[0 - v \sin\theta] = -\rho a v^2 \sin\theta$$

Since the plate is smooth, as before, the force exerted by the plate on the mass of water parallel to the plate is zero, i.e., $F_t = 0$. Therefore, the total force exerted by the plate on the jet is

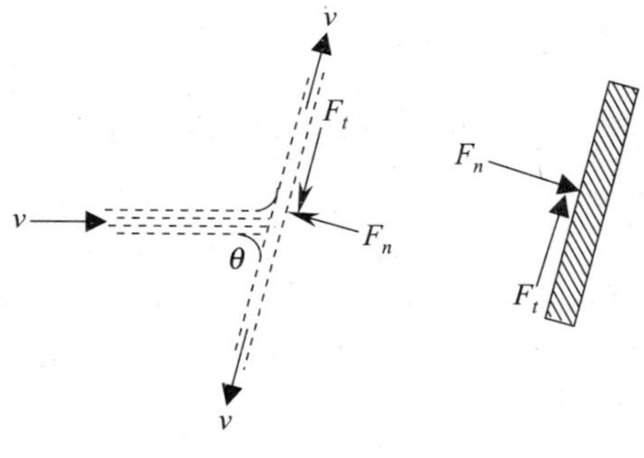

Fig. 16.18(a)

$$F = F_n = -\rho a v^2 \sin\theta$$

Thus, the force exerted by the jet of water on the plate is $\rho a v^2 \sin\theta$ acting normal to the plate. This total force can also be resolved along the x and y directions as

$$F_x = [\rho a v^2 \sin\theta]\sin\theta = \rho a v^2 \sin^2\theta$$

and $$F_y = [\rho a v^2 \sin\theta]\cos\theta = \rho a v^2 \sin\theta\cos\theta$$

For the given data

Velocity of jet, $v = 10$ m/s

Area of jet, $a = \dfrac{\pi}{4}(0.004)^2 = 1.26 \times 10^{-5}$ m^2

Angle of inclination, $\theta = 60°$

Therefore, the force exerted by the jet of water on the plate is given as

$$F = \rho a v^2 \sin\theta = (1000)(1.26 \times 10^{-5})(10)^2(\sin 60°)$$
$$= 1.09 \text{ N acting normal to the plate}$$

Example 16.16 A jet of water issued from a nozzle strikes at the centre of a smooth curved vane. The water after striking the vane leaves tangential to the vane at the exit. Derive an expression for the force exerted by the jet of water on the vane. If the jet diameter is 4 mm, the velocity of the jet is 10 m/s and after striking, it gets deflected through 120°, determine the force exerted by the jet of water on the vane.

Solution Let the tangent to the vane at the exit make an angle 'θ' with the axis of jet. Then we see that the jet gets deflected through an angle of (180° – θ). As the vane is smooth, we can assume that the velocity of water at exit is same as that at entry. The component of velocity at exit in the direction of velocity at entry is

$$-v \cos\theta$$

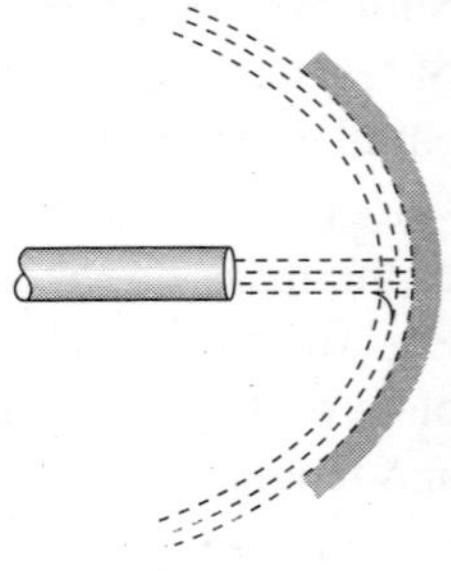

Fig. 16.19

The negative sign indicates that its direction is opposite to that of the initial direction.

As the jet strikes the vane at the centre, we can consider that respectively half of the discharge leaves through the top and bottom exits. Thus, applying the impulse–momentum equation in the direction of the jet, we have

$$F_x = \left\{\rho\frac{Q}{2}[-v\cos\theta] + \rho\frac{Q}{2}[-v\cos\theta]\right\} - \rho Q v$$

$$= -\rho Q v[1 + \cos\theta]$$
$$= -\rho a v^2[1 + \cos\theta]$$

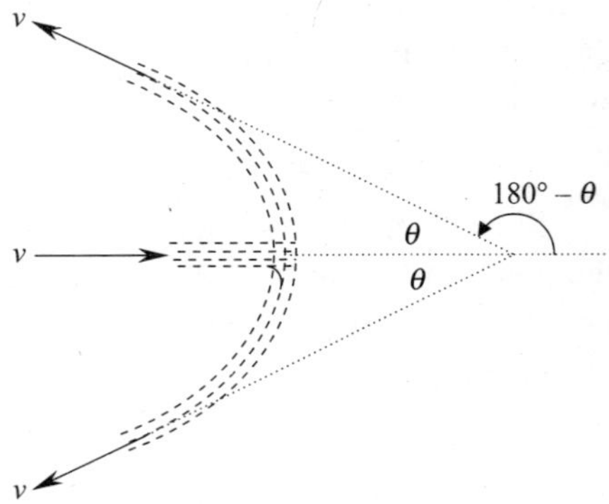

Fig. 16.19(a)

As the vane is smooth, we know that no resistance is offered to the motion of water in the vertical direction and therefore, $F_y = 0$. Hence, the total force of impact exerted by the vane on the jet of water is

$$F = F_x = -\rho a v^2 [1 + \cos\theta]$$

Therefore, the force exerted by the jet of water on the vane is $\rho a v^2[1 + \cos\theta]$.

It is given that the jet gets deflected through 120°, i.e., 180° – θ = 120° and hence θ = 60°. Therefore, for the given data, the force exerted by the jet of water on the vane is obtained as

$$F = \rho a v^2[1 + \cos\theta]$$
$$= (1000)[(\pi/4)(0.004)^2](10)^2[1 + \cos 60°] = 1.88 \text{ N}$$

Example 16.17 A nozzle is fitted at the exit of a water pipe. If the pipe diameter is 5 cm, nozzle diameter is 1 cm and the discharge is 0.005 m^3/s, determine the force required to hold the nozzle in position.

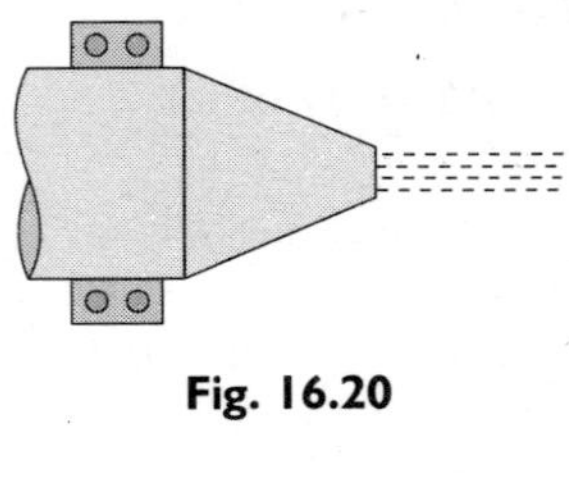

Fig. 16.20

Solution A nozzle is an exit of a pipe in which the cross-sectional area gradually decreases to increase the kinetic energy of the flowing water. Let

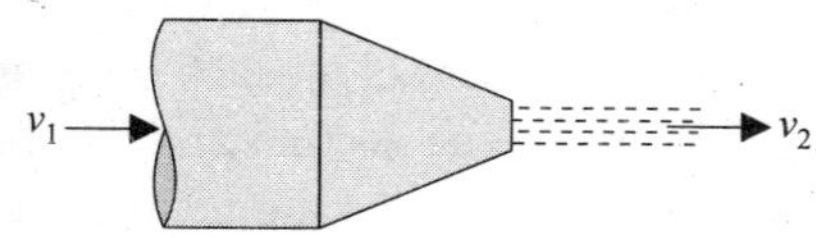

Fig. 16.20(a)

v_1 and v_2 be the velocities of the water at the base and exit of the nozzle. From fluid mechanics, we know that the velocities are inversely proportional to the cross-sectional areas of the pipe. It should be noted that the nozzle does not change the direction of velocity, but its magnitude. We know that whenever there is a change in velocity, i.e., change in momentum, a force of impulse acts on the flowing water as well as on the nozzle, which are equal and opposite. As a result, the nozzle does not stay in position. To keep it in position, clamps are used such that they resist this force of impulse.

Applying the impulse–momentum equation along the direction of flow, we have

$$F_x = \rho Q[v_2 - v_1]$$

Note that as both the initial and final velocities are in the same direction, they are taken as positive. Since there is no change in momentum in the vertical direction, $F_y = 0$. Therefore, total force of impulse is

$$F = F_x = \rho Q[v_2 - v_1]$$

Since the discharge is constant,

$$Q = a_1 v_1 = a_2 v_2$$

Therefore,
$$v_1 = \frac{Q}{a_1} = \frac{0.005}{(\pi/4)(0.05)^2} = 2.55 \text{ m/s}$$

and
$$v_2 = \frac{Q}{a_2} = \frac{0.005}{(\pi/4)(0.01)^2} = 63.66 \text{ m/s}$$

Therefore,
$$F = \rho Q[v_2 - v_1]$$
$$= (1000)(0.005)[63.66 - 2.55] = 305.55 \text{ N}$$

16.6 CONSERVATION OF MOMENTUM

From Newton's first law of motion, we know that when a body is moving with *uniform* velocity, it will continue to move with the *same* velocity, if *no* external force acts on it. Expressing the motion of the body using Newton's second law of motion, we have

$$\vec{F} = m\vec{a} = m\frac{d\vec{v}}{dt} = \frac{d}{dt}[m\vec{v}] = \text{rate of change of momentum} \qquad (16.17)$$

Thus, we see that the momentum of the body remains **constant** if **no** external force acts on it. In the same way, consider two bodies [of masses m_1 and m_2] each moving with different velocities u_1 and u_2 [one of them could be zero too], we can consider them to form a system. By the same explanation as that for a single body, the momentum of the system remains constant provided no external force acts on it.

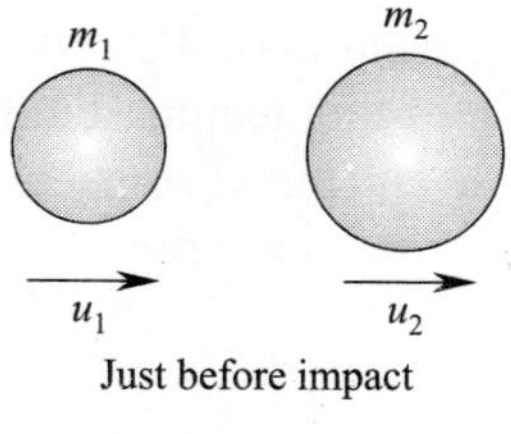

Fig. 16.21(a)

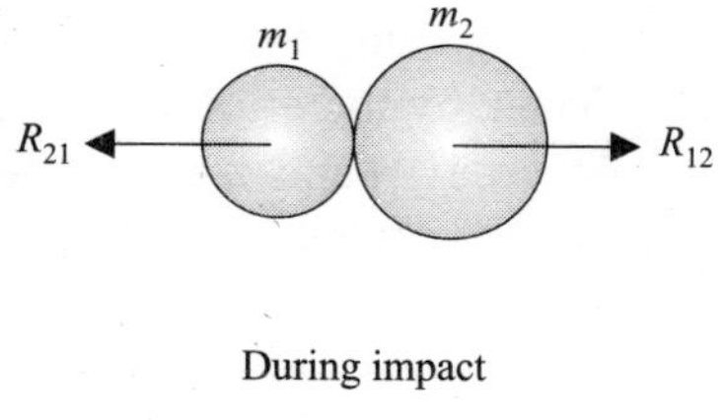

Fig. 16.21(b)

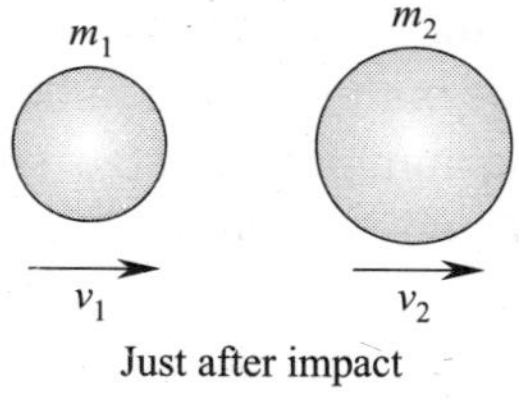

Fig. 16.21(c)

Suppose during the course of their motion, they collide with one another. Then as already seen, impulsive forces act on each body as exerted by the other. Let R_{12} be the impulsive force exerted by the first body on the second and R_{21} be the force exerted by the second body on the first. Due to these impulsive forces, the momentum of the individual bodies change, i.e.,

$$\int \vec{R}_{21}\, dt = m_1[\vec{v}_1 - \vec{u}_1] \tag{16.18}$$

Similarly,
$$\int \vec{R}_{12}\, dt = m_2[\vec{v}_2 - \vec{u}_2] \tag{16.19}$$

where $\vec{v}_1$ and $\vec{v}_2$ are respectively the final velocities of the two bodies after collision. Adding the above two equations, we get

$$\int \vec{R}_{21}\, dt + \int \vec{R}_{12}\, dt = m_1[\vec{v}_1 - \vec{u}_1] + m_2[\vec{v}_2 - \vec{u}_2] \tag{16.20}$$

By Newton's third law of motion, we know that $\vec{R}_{12} = -\vec{R}_{21}$. Therefore, in the above equation, the left-hand term reduces to zero and hence on rearranging, we can write

$$m_1\vec{u}_1 + m_2\vec{u}_2 = m_1\vec{v}_1 + m_2\vec{v}_2 \tag{16.21}$$

i.e., [initial momentum of the system] = [final momentum of the system]

Thus, we see that though the momentum of the individual bodies does change, the total momentum of the system remains constant. This principle is known as **conservation of momentum**.

We must understand that the motion of bodies just before impact and just after impact are affected by the external forces such as force of gravity and force of friction. These forces are termed **non-impulsive forces** as they act for a *longer* duration of time of motion of the bodies. However, during the very short duration of time during which the bodies collide with each other, the impulsive forces acting on each body as exerted by the other are very *large* in magnitude as compared to these external non-impulsive forces. Hence, during this small time interval, we can as well assume that there are **no external forces** acting on the bodies, except the internal impulsive forces. This assumption is justified because the duration of impact is very small, normally a small fraction of a second, say $1/60^{th}$ or $1/100^{th}$ of a second. Thus, we can neglect the effect of external forces in causing the change in momentum of bodies. Hence, we can state the law of conservation of momentum as follows:

When no external forces act on bodies forming a system (which is true for the small duration of collision), the momentum of the system is conserved, i.e., the initial momentum of the system is equal to the final momentum of the system.

The following illustrative examples will explain the application of conservation of momentum principle in solving some of the numerical problems.

Example 16.18 A rifle of 5 kg mass fires a bullet of 10 g mass at a velocity of 300 m/s. Determine the velocity with which the rifle recoils.

Solution We consider the rifle and bullet as a single system. Initially the momentum of the system is zero as both rifle and bullet are at rest. Thus, by the principle of conservation of momentum, the final momentum of the system should also be zero. If M and m are masses of the rifle and bullet respectively, and V and v are their respective velocities immediately after bullet is fired, then we can write

$$MV + mv = 0$$

$$\Rightarrow \qquad V = -\frac{m}{M}v = -\frac{0.01}{5}(300) = -0.6 \text{ m/s}$$

The negative sign indicates that the direction of recoil of the rifle is in the direction opposite to the direction of velocity of the bullet.

Example 16.19 A truck of 6 ton mass moving at 60 kmph collides with a car of 2 ton mass moving at 45 kmph in the same direction as that of the truck. If after collision, they coalesce as one body, determine their common velocity. Determine the loss in kinetic energy, considering the two vehicles to be a single system.

Solution

Fig. 16.22

Initial velocity of truck, $v_{o1} = 60$ kmph = 16.67 m/s

Initial velocity of car, $v_{o2} = 45$ kmph = 12.5 m/s

Applying the principle of conservation of momentum,

$$m_1 v_{o1} + m_2 v_{o2} = [m_1 + m_2]v$$

$$[6000 \times 16.67] + [2000 \times 12.5] = [6000 + 2000] \times v$$

$$\Rightarrow \qquad v = 15.63 \text{ m/s}$$

Kinetic energy of truck and car system

Initial kinetic energy of the system,

$$(\text{K.E})_i = \frac{1}{2}m_1 v_{o1}^2 + \frac{1}{2}m_2 v_{o2}^2$$

$$= \frac{1}{2}(6 \times 10^3)(16.67)^2 + \frac{1}{2}(2 \times 10^3)(12.5)^2 = 989.92 \text{ kJ}$$

Final kinetic energy of the system,

$$(\text{K.E})_f = \frac{1}{2}[m_1 + m_2]v^2$$

$$= \frac{1}{2}[(6 + 2) \times 10^3][15.63]^2 = 977.19 \text{ kJ}$$

Therefore, loss of kinetic energy is given as

$$\Delta\text{K.E} = (\text{K.E})_i - (\text{K.E})_f$$

$$= 989.92 - 977.19 = 12.73 \text{ kJ}$$

Example 16.20 Solve the previous problem if the car is moving in the direction opposite to that of the truck.

Solution As the car is moving in the direction opposite to that of the truck, its initial velocity is taken as negative. Applying the principle of conservation of momentum, we have

$$m_1 v_{o1} - m_2 v_{o2} = [m_1 + m_2]v$$

$$(6 \times 10^3)(16.67) - (2 \times 10^3)(12.5) = [(6 + 2) \times 10^3]\, v$$

$$\Rightarrow \qquad v = 9.38 \text{ m/s}$$

$$(\text{K.E})_i = \frac{1}{2}m_1 v_{o1}^2 + \frac{1}{2}m_2 v_{o2}^2 = 989.92 \text{ kJ}$$

$$(\text{K.E})_f = \frac{1}{2}[m_1 + m_2]v^2 = 351.94 \text{ kJ}$$

Therefore, loss of kinetic energy is

$$= 989.92 - 351.94 = 637.98 \text{ kJ}$$

Example 16.21 A bullet of 20 g mass moving with a speed of 300 m/s, strikes sand bags loaded on a cart having a total mass of 500 kg. If the bullet gets embedded in the sand, determine the speed of the cart after impact. Assume that there is no friction to the movement of the cart. Also, determine the loss of kinetic energy due to the impact.

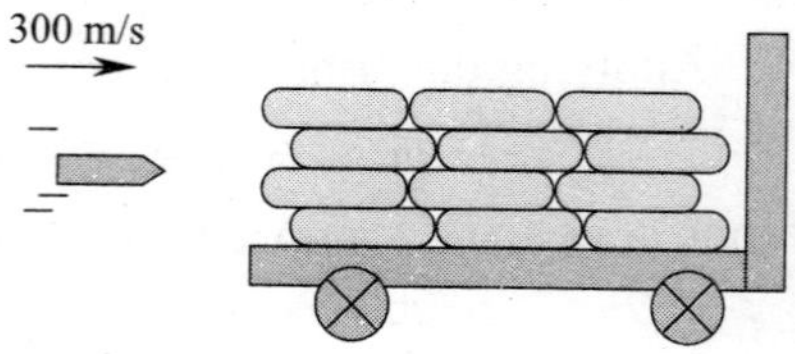

Fig. 16.23

Solution Initial speed of bullet, v = 300 m/s and let V be the speed of bullet and cart together after the bullet gets embedded in the sand bags. Then applying the principle of conservation of impulse and momentum, we have

$$mv + M(0) = [M + m]V$$

$$\Rightarrow \qquad V = \frac{0.02 \times 300}{[500 + 0.02]} = 0.012 \text{ m/s}$$

Initial kinetic energy of the system,

$$(\text{K.E})_i = \frac{1}{2}mv^2$$

$$= \frac{1}{2}(0.02)(300)^2 = 900 \text{ J}$$

Final kinetic energy of the system,

$$(\text{K.E})_f = \frac{1}{2}[m + M]V^2$$

$$= \frac{1}{2}[0.02 + 500][0.012]^2 = 0.036 \text{ J}$$

Therefore, loss of kinetic energy is obtained as

$$= (\text{K.E})_i - (\text{K.E})_f$$

$$= 900 - 0.036 = 899.964 \text{ J}$$

Example 16.22 A woman of mass m stands in a boat of mass M. If she jumps horizontally with a velocity of v relative to the boat, determine the velocity of the boat immediately after she jumps off the boat.

Fig. 16.24

Solution Let V be the velocity of the boat after the woman jumps off the boat. Since the velocity of the woman relative to the boat is v, her absolute velocity is $[V + v]$. Applying the conservation of momentum to the system made up of boat and the woman,

$$[\text{momentum}]_{\text{initial}} = [\text{momentum}]_{\text{final}}$$

$$0 = m[V + v] + MV$$

$$\Rightarrow \quad V = -\frac{mv}{M + m}$$

Example 16.23 A man of 60 kg mass standing on a bridge jumps on to a cart below him such that he lands with a velocity of 5 m/s at an angle of 30° to the horizontal direction. If the cart is free to move, determine its velocity after he has jumped in for the following cases: the cart is initially (i) at rest, (ii) moving with a velocity of 1 m/s away from the bridge, and (iii) moving with a velocity of 1 m/s towards the bridge. Take the mass of the cart as 125 kg. Also, determine the loss in kinetic energy in each case.

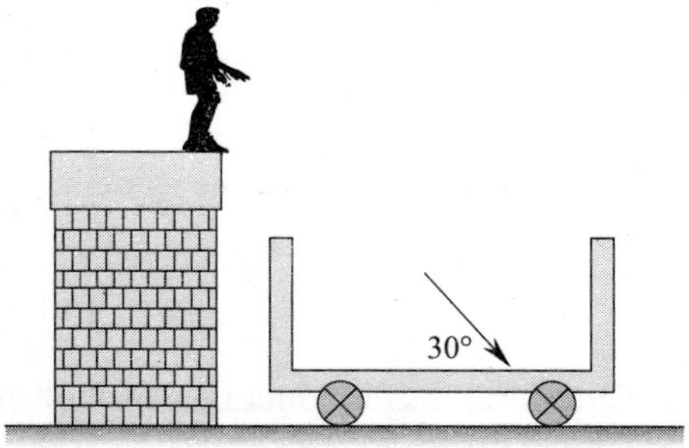

Fig. 16.25

Solution

(i) Cart initially at rest

The components of velocity of the man in the horizontal and vertical directions are respectively 5 cos 30° and 5 sin 30°. Then applying the principle of conservation of momentum along the horizontal direction, we have

$$m_1 v_{1x} + m_2 v_{2x} = [m_1 + m_2] V_x$$

$$60\,(5\cos 30°) + 125(0) = [60 + 125] V_x$$

$$\Rightarrow \quad V_x = 1.404 \text{ m/s}$$

Just like horizontal direction, we can also apply the conservation principle in the vertical direction to deduce the final velocity of the system in the vertical direction after the man lands in. As the system has to go against the gravity in the vertical direction, it just results in up and down movements at the time of landing.

Initial kinetic energy of the system:

$$(\text{K.E})_i = \frac{1}{2} m_1 v_1^2 + \frac{1}{2} m_2 v_2^2$$

$$= \frac{1}{2}(60)(5)^2 + \frac{1}{2}(125)(0)^2 = 750 \text{ J}$$

Final kinetic energy of the system:

$$(\text{K.E})_f = \frac{1}{2}[m_1 + m_2] V^2$$

$$= \frac{1}{2}[60 + 125][1.404]^2 = 182.34 \text{ J}$$

Therefore, loss in kinetic energy of the system is 750 – 182.34 = 567.66 J

(ii) Cart pushed with a velocity of 1 m/s away from the bridge

Similarly, applying the principle of conservation of momentum along the horizontal direction, we have

$$60(5\cos 30°) + 125(1) = (60 + 125)[V]$$

$$\Rightarrow \quad V = 2.08 \text{ m/s}$$

Initial kinetic energy of the system: $\frac{1}{2}m_1 v_1^2 + \frac{1}{2}m_2 v_2^2$

$$= \frac{1}{2}(60)(5)^2 + \frac{1}{2}(125)(1)^2 = 812.5 \text{ J}$$

Final kinetic energy of the system: $\frac{1}{2}[m_1 + m_2]V^2$

$$= \frac{1}{2}[60 + 125][2.08]^2 = 400.19 \text{ J}$$

Therefore, loss in kinetic energy of the system is 812.5 – 400.19 = 412.31 J

(iii) Cart pushed with a velocity of 1 m/s towards the bridge

Similarly, applying the principle of conservation of momentum along the horizontal direction, we have

$$60(5\cos 30°) - 125(1) = (60 + 125)[V]$$

$$\Rightarrow \quad V = 0.73 \text{ m/s}$$

It should be noted that as the cart moves in a direction opposite to that of the man, its velocity is considered as negative.

Initial kinetic energy of the system:

$$(\text{K.E})_i = \frac{1}{2}m_1 v_1^2 + \frac{1}{2}m_2 v_2^2$$

$$= \frac{1}{2}(60)(5)^2 + \frac{1}{2}(125)(-1)^2 = 812.5 \text{ J}$$

Final kinetic energy of the system:

$$(\text{K.E})_f = \frac{1}{2}[m_1 + m_2]V^2$$

$$= \frac{1}{2}[60 + 125][0.73]^2 = 49.29 \text{ J}$$

Therefore, loss in kinetic energy of the system is 812.5 – 49.29 = 763.21 J

Example 16.24 The block-and-pulley arrangement shown is released from rest, when the block *A* is 1 m from the platform *CD*. When the block *A* reaches the platform, it strikes a 3 kg block resting on the platform. If the 3 kg block adheres to the block *A*, determine the height the system will reach before instantaneous rest. Mass of blocks *A* and *B* are respectively 3 kg and 5 kg. Assume the pulley is frictionless and massless.

Solution Since the mass of the block B is greater than the mass of the block A, the block A will move upwards, while the block B moves downwards. Drawing separate free-body diagrams for both, and applying the law of motion, we have

$$5g - T = 5a \quad \text{(a)}$$

$$T - 3g = 3a \quad \text{(b)}$$

$$\Rightarrow \quad a = \frac{2g}{8} = \frac{g}{4}$$

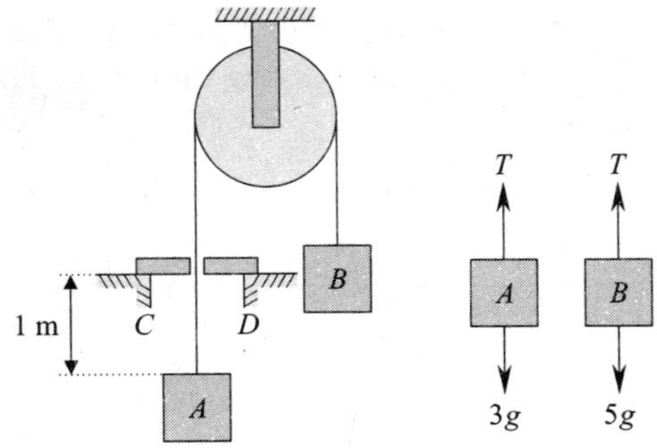

Fig. 16.26

The velocity of the block A as it reaches the platform is obtained from kinematics as follows:

$$v^2 = v_o^2 + 2as$$

$$v^2 = 0 + 2\frac{g}{4}(1)$$

$$\therefore \quad v^2 = \frac{g}{2} \quad \Rightarrow \quad v = \sqrt{\frac{g}{2}} \text{ m/s}$$

Before impact, the initial momentum of the system consisting of blocks A and B, whose respective masses are 3 kg and 5 kg is given as

$$(\text{momentum})_i = [3 + 5]\, v$$

After impact, the final momentum of the system consisting of blocks A, B and the 3 kg block resting on the platform is given as

$$(\text{momentum})_f = [3 + 3 + 5]\, V$$

Therefore, applying the principle of conservation of momentum, we get

$$[3 + 5]\, v = [3 + 3 + 5]\, V$$

$$\Rightarrow \quad V = \frac{8}{11} v = \frac{8}{11}\sqrt{\frac{g}{2}} \text{ m/s}$$

Now we can see that the blocks on the left side together have mass greater than the mass of 5 kg on the right side. Hence, the blocks on the left side try to go down. However, due to the non-zero velocity of the block A as it approaches the platform, it still continues to move upward with the 3 kg block till they reach momentary rest. Thus, again applying the laws of motion,

$$T - 6g = 6a \quad \text{(c)}$$

$$5g - T = 5a \quad \text{(d)}$$

$$\Rightarrow \quad a = -\frac{g}{11} \text{ m/s}^2$$

The negative sign in the above value for acceleration indicates that after impact, the blocks begin to decelerate. Again applying the kinematics equation, we get

$$v^2 = v_0^2 + 2as$$

When the blocks reach a momentary rest, their velocities are zero and hence,

$$0 = \left(\frac{8}{11}\right)^2 \frac{g}{2} - 2\frac{g}{11}s$$

$$\Rightarrow \qquad s = \frac{8^2}{(11)(4)} = 1.45 \text{ m}$$

16.7 COLLISIONS

In this section, we will analyse the motion of colliding bodies immediately after *collision*. Consider two bodies of known masses [m_1 and m_2] and known initial velocities [u_1 and u_2] colliding with each other. In the previous section, we assumed the colliding bodies to *coalesce* or *lock* together immediately after collision and move with the *same* velocity. Hence, we could determine the resulting common velocity of the bodies using the conservation of momentum equation. However, this is an ideal condition as we always observe in practice that bodies separate immediately after collision and move with *different* velocities. This is due to the *elastic* nature of material of the colliding bodies.

The impulsive forces acting on the bodies during collision are large in magnitude when compared to the external non-impulsive forces and the time during which they act also being very small, we can assume as before that the momentum is conserved during collision. Hence, we can write

$$m_1u_1 + m_2u_2 = m_1v_1 + m_2v_2 \qquad (16.22)$$

where v_1 and v_2 are respective velocities of the two bodies after collision. There being two unknown velocities, namely, v_1 and v_2, they cannot be determined from a single equation. Hence, to solve these unknowns, the conservation of momentum equation must be supplemented by another equation relating to elasticity of material of the bodies.

In our study throughout this book, we have idealized the bodies to be rigid, i.e., they do not deform under the action of loads. However, we should understand this is just an idealized condition and that materials do deform. Hence, when two bodies collide with each other, the impulsive forces exerted by each body on the other cause the other body to deform. The extents to which they deform depend upon the elasticity of material of the two bodies.

Suppose we drop a rubber ball onto a hard floor. It bounces back almost to its original height, while a glass ball dropped likewise, does not rise to the same height. The height from which a body falls, determines the initial velocity of impact and the height to which a ball rises determines the final velocity after collision. Thus, we see that both the rubber and glass balls, though having the same initial velocity (as the initial velocity is just dependent upon the height from which the body was dropped), they rebound to different heights indicating different velocities after collision. This difference in velocities is due to the elastic nature of the material of the bodies.

Due to this elastic nature of the material of the bodies, they tend to regain their original shape during collision. In doing so, they rebound from the point of contact of collision and move with different velocities. Thus, if a body is made up of a material, which exhibits higher elasticity, it will have more velocity after collision than a body made of material, which has lower elasticity. If the final velocity is totally restored then the collision is said to be perfectly **elastic**. Otherwise, it is said to be **inelastic**. This property of the colliding bodies is expressed by what is known as Newton's experimental law. The experimental evidence suggests that the relative velocities of the colliding bodies before impact bears a constant relationship with their relative velocities after impact and is in opposite direction. Stated mathematically,

$$(v_1 - v_2)\ \alpha\,[-(u_1 - u_2)] \tag{16.23}$$

$$(v_1 - v_2) = -e(u_1 - u_2)$$

$$\Rightarrow \quad -e = \frac{v_1 - v_2}{u_1 - u_2} \tag{16.24}$$

The constant of proportionality, e is called **coefficient of restitution**. It depends upon the material of the bodies, but independent of their masses and their velocities before impact. The value of e is found to vary from 0 to 1.

When the direction of motion of each body is along the line joining their centres then the impact is said to be **direct**. When the direction of motion of either or both is inclined to the line joining their centres then the impact is said to be **oblique**.

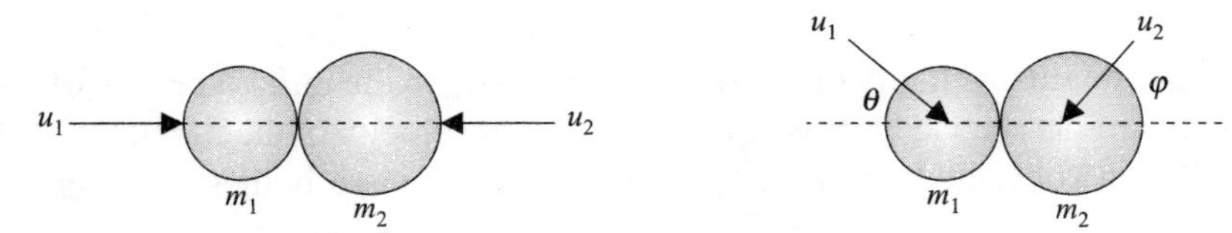

Direct impact   Oblique impact

Fig. 16.27

16.7.1 Direct Impact

Consider two smooth spheres of masses m_1 and m_2, moving with initial velocities u_1 and u_2 respectively. Let them collide with each other along the line joining their centres and let v_1 and v_2 be their respective velocities after collision.

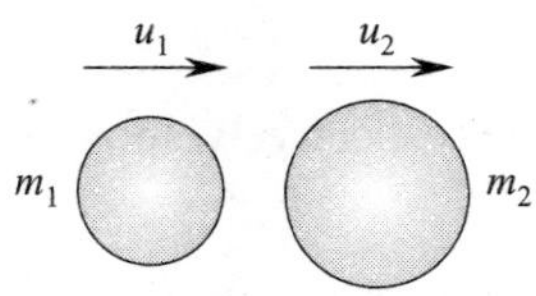

Before impact

Fig. 16.28(a)

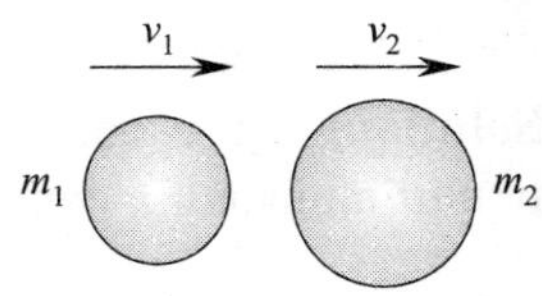

After impact

Fig. 16.28(b)

As the impulsive force exerted by each body on the other during the collision is equal and opposite, we know that the total momentum of the system is conserved. Thus, we can write,

$$m_1u_1 + m_2u_2 = m_1v_1 + m_2v_2 \tag{16.25}$$

From Newton's experimental law, we can write

$$-e = \frac{v_1 - v_2}{u_1 - u_2} \tag{16.26}$$

Solving for v_1 and v_2 from the above two equations, we have

$$v_1 = \frac{m_1u_1 + m_2u_2 - m_2e(u_1 - u_2)}{m_1 + m_2} \tag{16.27}$$

and
$$v_2 = \frac{m_1u_1 + m_2u_2 + m_1e(u_1 - u_2)}{m_1 + m_2} \tag{16.28}$$

We can get some important conclusions from the above two expressions derived for final velocities of the colliding bodies immediately after collision. If we assume that the collision is *inelastic* then substituting the value of the coefficient of restitution '$e = 0$' in the above Eqs 16.27 and 16.28, we get

$$v_1 = v_2 = \frac{m_1u_1 + m_2u_2}{m_1 + m_2} \tag{16.29}$$

Thus, we see that if the collision is inelastic then after impact, the two bodies *coalesce* as one body and move with the *same* velocity.

If we assume that the collision is *elastic* then substituting the value of the coefficient of restitution '$e = 1$' in the above Eqs 16.27 and 16.28, we get

$$v_1 = \frac{(m_1 - m_2)u_1 + 2m_2u_2}{m_1 + m_2}, \quad v_2 = \frac{2m_1u_1 + u_2(m_2 - m_1)}{m_1 + m_2} \tag{16.30}$$

Further, if the masses of the two colliding bodies are equal, i.e., $m_1 = m_2$, then we get

$$v_1 = u_2 \quad \text{and} \quad v_2 = u_1 \tag{16.31}$$

Stated in words, the final velocity of the first body is equal to the initial velocity of the second body and the final velocity of the second body is equal to the initial velocity of the first body. Thus, we can see that when the collision is elastic between two equal masses, the two bodies exchange their velocities after impact.

Example 16.25 Two spheres of masses m_1 amd m_2 respectively approach each other with respective velocities u_1 and u_2. Determine the expressions for velocities of spheres after impact, if the coefficient of restitution between the contact surfaces is e. Also, determine the velocities, if collision is (i) plastic, and (ii) elastic.

Solution Let v_1 and v_2 be the respective velocities of the spheres after impact. Applying the conservation of momentum equation,

$$m_1u_1 - m_2u_2 = m_1v_1 + m_2v_2 \tag{a}$$

Note that as the motion of the second sphere is in the direction opposite to that of the first, the negative sign is used for its initial velocity.

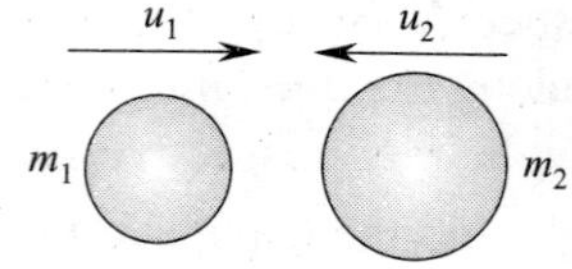

Before impact

Fig. 16.29

In addition, coefficient of restitution is given as

$$-e = \frac{v_1 - v_2}{u_1 - (-u_2)}$$

$$\Rightarrow \quad -e(u_1 + u_2) = (v_1 - v_2) \tag{b}$$

Multiplying the equation (b) by m_1 and subtracting it from the equation (a), we can eliminate v_1. Thus, we can get the expression for v_2,

$$v_2 = \frac{m_1u_1 - m_2u_2 + m_1e(u_1 + u_2)}{(m_1 + m_2)}$$

Similarly, multiplying the equation (b) by m_2 and adding it with the equation (a), we can eliminate v_2. Hence, we can get the expression for v_1,

$$v_1 = \frac{m_1u_1 - m_2u_2 - m_2e(u_1 + u_2)}{(m_1 + m_2)}$$

(i) When the collision is inelastic, i.e., e = 0

$$v_1 = v_2 = \frac{m_1u_1 - m_2u_2}{(m_1 + m_2)}$$

Thus, when the coefficient of restitution is zero, i.e., if the collision is inelastic, they coalesce as one body and move with the same velocity.

(ii) When the collision is elastic, i.e., e = 1

$$v_1 = \frac{u_1(m_1 - m_2) - 2m_2u_2}{(m_1 + m_2)}$$

and

$$v_2 = \frac{2m_1u_1 + u_2(m_1 - m_2)}{(m_1 + m_2)}$$

Further, if $m_1 = m_2$, then

$$v_1 = -u_2 \quad \text{and} \quad v_2 = u_1$$

Thus, we can see that when the collision is elastic between two equal masses, the two bodies exchange their velocities after impact.

Example 16.26 Two balls of equal masses approach each other with velocities u and v respectively. If after impact, the ball moving with velocity u is brought to rest, show that

$$\frac{u}{v} = \frac{1+e}{1-e}$$

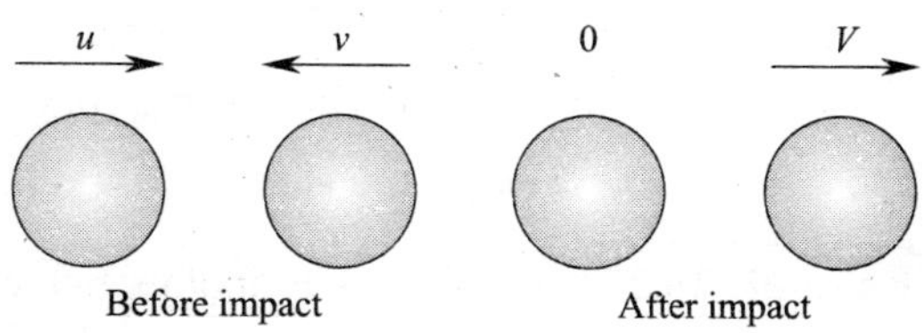

Fig. 16.30

Solution It is stated that the two equal balls approach each other with velocities u and v respectively and that the ball moving with velocity u is brought to rest after impact. Let V be the final velocity after impact of the ball moving with initial velocity v. Applying the conservation of momentum equation, we have

$$mu - mv = m(0) + mV$$

$$\Rightarrow \quad u - v = V \qquad \text{(a)}$$

We know that coefficient of restitution is given as

$$-e = \frac{0 - V}{u - (-v)} \qquad \text{(b)}$$

Note that as the initial velocity of the second ball is towards left, it is taken as negative. Hence,

$$V = e(u + v) \tag{c}$$

Substituting this value of V in the equation (a), we have

$$u - v = e(u + v)$$

On rearranging, we get

$$\frac{u}{v} = \frac{1+e}{1-e}$$

Example 16.27 If a ball overtakes a ball of twice its mass moving with 1/7th of its velocity and if the coefficient of restitution between them is 3/4, show that the first ball after striking the second ball will remain at rest.

Solution It is stated that the velocity of the second ball is 1/7th of the velocity of the first ball. Hence, applying the conservation of momentum equation,

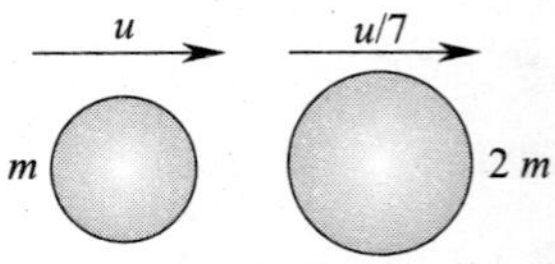

Fig. 16.31

$$mu + 2m\frac{u}{7} = mv_1 + 2mv_2$$

$$\Rightarrow \quad v_1 + 2v_2 = \frac{9u}{7} \tag{a}$$

In addition, coefficient of restitution is given as

$$-e = \frac{v_1 - v_2}{u - u/7}$$

$$\Rightarrow \quad v_1 - v_2 = -\frac{6}{7}eu$$

$$= -\frac{6}{7}\left[\frac{3}{4}\right]u = -\frac{9}{14}u \tag{b}$$

From equations (a) and (b), solving for v_1, we get

$$v_1 = 0$$

Example 16.28 A ball strikes another ball of twice its mass but moving at half its speed in the opposite direction; determine the expressions for final velocities of the two balls after impact.

Solution Let v_1 and v_2 be the respective velocities of the balls after impact. Applying the conservation of momentum equation,

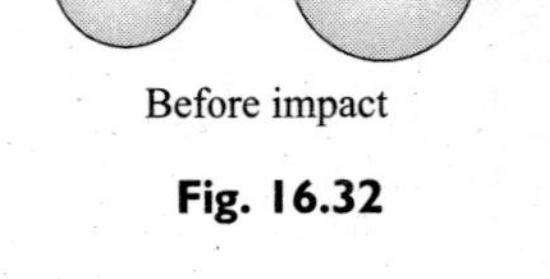

Fig. 16.32

$$m(2u) - 2mu = mv_1 + 2mv_2$$

$$\Rightarrow \quad v_1 + 2v_2 = 0 \tag{a}$$

In addition, coefficient of restitution is given as

$$-e = \frac{v_1 - v_2}{2u - (-u)} \Rightarrow \frac{v_1 - v_2}{3u}$$

$$v_1 - v_2 = -3eu \tag{b}$$

From equations (a) and (b), solving for v_1 and v_2, we get

$$3v_1 = -6eu \Rightarrow \quad v_1 = -2eu \tag{c}$$

and

$$3v_2 = 3eu \Rightarrow \quad v_2 = eu \tag{d}$$

Thus, we see that each ball rebounds with a velocity, which is e times the original velocity.

Example 16.29 A ball overtakes another ball of twice its mass but moving at half its speed in the same direction; determine the expressions for final velocities of the two balls after impact. Also, determine the final velocities if coefficient of restitution is (i) $e = 0$, (ii) $e = 1/2$, (iii) $e = 1$.

Solution Let v_1 and v_2 be the respective velocities of the balls after impact. Applying the conservation of momentum equation,

$$m2u + 2mu = mv_1 + 2mv_2$$

$$v_1 + 2v_2 = 4u \tag{a}$$

In addition, the coefficient of restitution is given as

$$-e = \frac{v_1 - v_2}{2u - u}$$

$$\Rightarrow \quad (v_1 - v_2) = -eu \tag{b}$$

From equations (a) and (b), solving for v_1 and v_2, we get

$$v_1 = \frac{u}{3}[4 - 2e] \quad \text{and} \quad v_2 = \frac{u}{3}[4 + e]$$

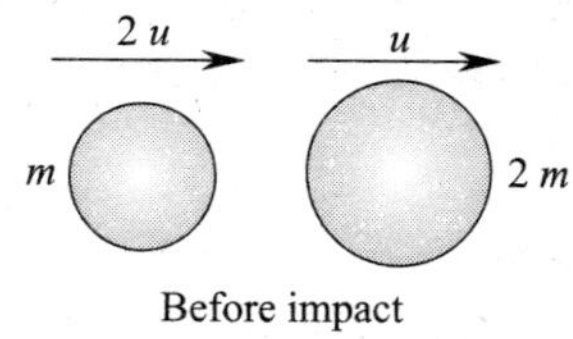

Fig. 16.33

(i) When $e = 0$,

$$v_1 = v_2 = \frac{4u}{3}$$

(ii) When $e = 1/2$,

$$v_1 = u, \; v_2 = \frac{3u}{2}$$

(iii) When $e = 1$,

$$v_1 = \frac{2u}{3}, \; v_2 = \frac{5u}{3}$$

16.7.2 Oblique Impact

Consider two smooth spheres of masses m_1 and m_2 approaching each other with velocities u_1 and u_2 such that their directions are inclined to the line joining their centres at the instant of impact at θ and φ respectively. Let v_1 and v_2 be the respective velocities immediately after impact and their directions be inclined to the line joining centres at α and β respectively.

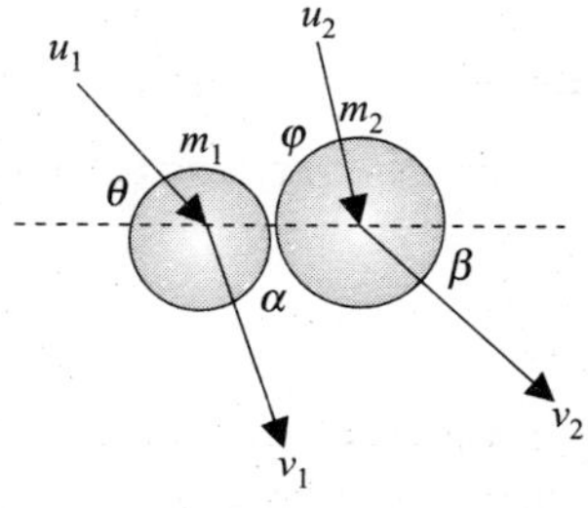

Fig. 16.34 Oblique impact

As the spheres are smooth, there is no impulsive force acting on each body along their common tangential plane during their time of collision. Thus, there is no change in momentum of individual bodies in that direction. Hence, we can write,

$$v_1 \sin\alpha = u_1 \sin\theta \tag{16.32}$$

and
$$v_2 \sin\beta = u_2 \sin\varphi \tag{16.33}$$

As the impulsive force exerted by each sphere on the other in the direction of line joining their centres is equal and opposite, the momentum of the system is conserved. Thus, we can write,

$$m_1(u_1 \cos\theta) + m_2(u_2 \cos\varphi) = m_1(v_1 \cos\alpha) + m_2(v_2 \cos\beta) \tag{16.34}$$

Applying Newton's experimental law,

$$-e = \frac{v_1 \cos\alpha - v_2 \cos\beta}{u_1 \cos\theta - u_2 \cos\varphi} \tag{16.35}$$

The following example will explain the oblique impact of bodies.

Example 16.30 A smooth sphere moving at 10 m/s in the direction shown collides with another smooth sphere of double its mass and moving with 5 m/s in the direction shown. If the coefficient of restitution is 2/3, determine their velocities after collision.

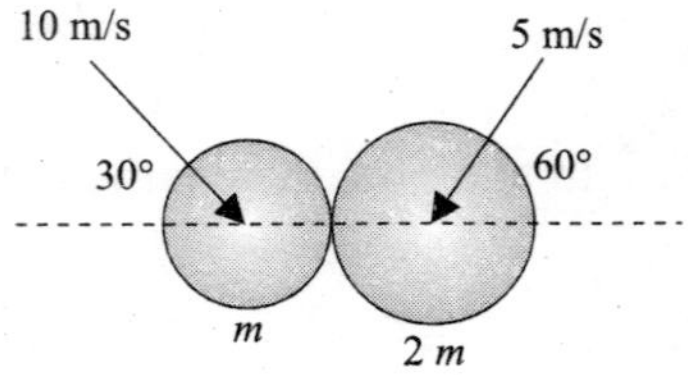

Fig. 16.35

Solution Let v_1 and v_2 be the velocities of the spheres after impact and let their directions with respect to the line joining their centres at the instant of impact be α and β respectively. As the spheres are smooth, there is no impulsive force acting on each body along their common tangential plane during their time of collision. Thus, there is no change in momentum of individual bodies in that direction. Hence, we can write,

$$v_1 \sin\alpha = 10\sin 30° \tag{a}$$

and
$$v_2 \sin\beta = 5\sin 60° \tag{b}$$

As the impulsive force exerted by each sphere on the other in the direction of the line joining their centres is equal and opposite, the momentum of the system is conserved. Thus, we can write

$$m(10\cos 30°) - 2m(5\cos 60°) = m(v_1 \cos\alpha) + 2m(v_2 \cos\beta)$$

$$\Rightarrow \quad v_1 \cos\alpha + 2v_2 \cos\beta = 3.66 \tag{c}$$

Applying Newton's experimental law, we have

$$-e = \frac{v_1 \cos\alpha - v_2 \cos\beta}{10\cos 30° - (-5\cos 60°)}$$

$$\Rightarrow \quad v_1 \cos\alpha - v_2 \cos\beta = -7.44 \tag{d}$$

From equations (c) and (d), solving for $v_1 \cos\alpha$ and $v_2 \cos\beta$, we get

$$v_1 \cos\alpha = -3.74 \text{ m/s} \tag{e}$$

and
$$v_2 \cos\beta = 3.7 \text{ m/s} \tag{f}$$

From equations (a) and (e), we get $v_1 = 6.24$ m/s in the direction opposite to that of the initial velocity at an angle of $\alpha = 53.2°$ to the line joining their centres. Similarly, from equations (b) and (f), we get $v_2 = 5.7$ m/s at an angle of $\beta = 49.49°$ to the line joining their centres.

16.7.3 Impact Against a Fixed Plane

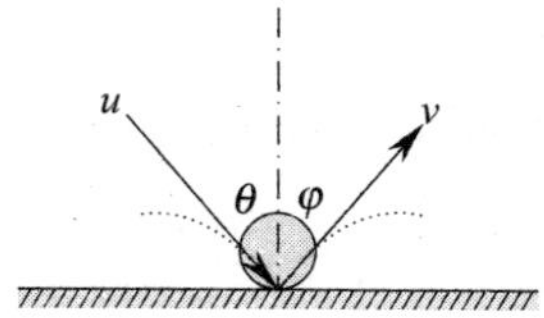

Fig. 16.36

Consider a *smooth* ball or sphere impinging upon a *smooth* fixed plane. Let u be the initial velocity of the ball before striking the plane, making an angle of 'θ' with respect to the normal to the plane. Let v be its final velocity after impact, making an angle of φ with respect to the normal to the plane.

Since the plane is *smooth*, we know that there is *no* force of impulse acting on the ball in the direction parallel to the plane. Hence, the momentum of the ball in that direction remains constant or in other words, its velocity is unaltered. Thus, we have

$$u\sin\theta = v\sin\varphi \tag{16.36}$$

Applying Newton's experimental law for the components of relative velocity along the normal to the plane,

$$-e = \frac{-v\cos\varphi}{u\cos\theta}$$

$$e\,u\cos\theta = v\cos\varphi \tag{16.37}$$

Note that as the component of final velocity is in the direction opposite to that of the initial velocity, it is considered as negative. From Eqs 16.36 and 16.37, solving for v and φ, we have

$$v^2 = u^2[\sin^2\theta + e^2\cos^2\theta] \tag{16.38}$$

$$\tan\varphi = \frac{1}{e}\tan\theta \tag{16.39}$$

Example 16.31 A ball of 100 g mass is projected up with a velocity of 20 m/s. It hits a ceiling that is 10 m above the point of projection. If $e = 3/4$, determine the speed of the ball as it descends to the point of projection. If the impact duration is $1/150^{\text{th}}$ of a second, determine the impulsive force.

Solution

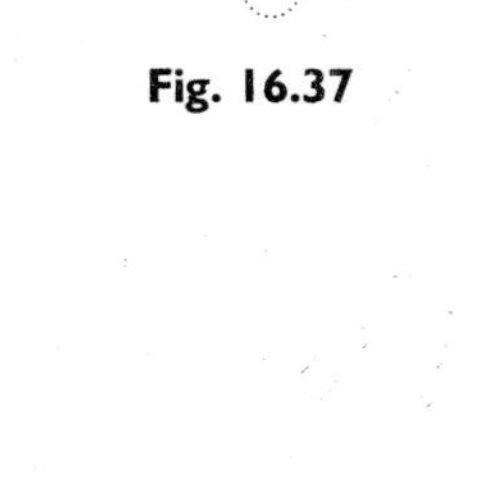

Fig. 16.37

Motion before impact

Initial velocity of the ball is $u_o = 20$ m/s

Therefore, velocity of the ball just before impact is obtained as

$$u^2 = u_o^2 - 2gs$$
$$= (20)^2 - 2(9.81)(10)$$
$$= 203.8$$

$$\Rightarrow \quad u = 14.28 \text{ m/s}$$

During impact

During impact, we can write

$$-e = \frac{-v_o - 0}{u - 0}$$

where v_o is the final velocity after impact and it is considered as negative as it is in the direction opposite to that of the initial velocity of impact. Hence,

$$v_o = eu$$

Motion after impact

Considering the motion after impact,

$$v^2 = v_o^2 + 2gs$$

$$= e^2u^2 + 2gs$$

$$= \left(\frac{3}{4}\right)^2 (203.8) + 2(9.81)(10)$$

$\Rightarrow$ $$v = 17.63 \text{ m/s}$$

Impulsive force

The impulsive force can be determined as follows:

$$Ft = m[-v_o - u]$$

$\therefore$ $$F = \frac{-m[eu + u]}{t} = -\frac{mu}{t}[1 + e]$$

$$= -\frac{(0.1)\,(14.28)}{1/150}[1 + (3/4)] = -374.9 \text{ N}$$

Example 16.32 A tennis ball hits the court with a velocity u at such an angle that the ball after bouncing leaves at right angles to the initial direction with a velocity of 0.9 u. Determine the angle at which the ball strikes the court and the coefficient of restitution. Assume the court to be smooth.

Solution Let θ be the angle made by the initial velocity with respect to the normal to the court. Then from the given condition, we know that the angle made by the velocity after impact, i.e., $\alpha = 90° - \theta$. Since the court is smooth, there is no impulsive force acting on the ball in the direction parallel to the court. Hence, the velocity of the ball parallel to the court is altered. Thus, we can write

$$u\sin\theta = v\sin\alpha = v\sin(90° - \theta) = v\cos\theta$$

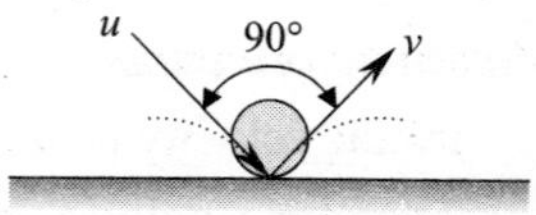

Fig. 16.38

$\Rightarrow$ $$\tan\theta = \frac{v}{u}$$

$$= \frac{0.9u}{u} = 0.9$$

Therefore, $\theta = \tan^{-1}(0.9) = 41.99°$

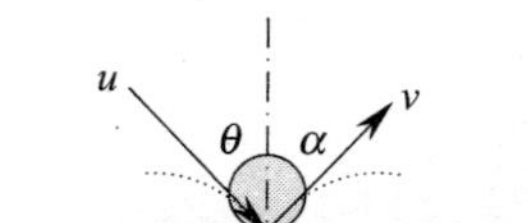

Fig. 16.38(a)

From Newton's experimental law, we have

$$-e = \frac{-v\cos\alpha}{u\cos\theta} = \frac{-v\cos(90° - \theta)}{u\cos\theta} = \frac{-v\sin\theta}{u\cos\theta}$$

$$e = \frac{v}{u}\tan\theta$$

$$= \left[\frac{v}{u}\right]^2 = \left[\frac{0.9u}{u}\right]^2 = 0.81$$

Example 16.33 A ball is projected from a point in a horizontal plane, and makes one rebound. Show that if the second range is equal to the greatest height which the ball attains

$$\tan \alpha = 4e$$

'α' being the angle of projection and e, the measure of elasticity.

Solution Let u be the initial velocity of projection at an angle α to the horizontal. Then the horizontal and vertical components of the initial velocity are respectively $u \cos \alpha$ and $u \sin \alpha$. Since it is projected on level ground, the components of velocity when it strikes the ground will be same as the components of initial velocity. Let v be the velocity of the ball after impact at an angle β to the horizontal. Then during impact, we can write Newton's experimental law along the direction normal to the plane as

$$-e = \frac{-v \sin \beta - 0}{u \sin \alpha - 0}$$

$$\Rightarrow \qquad e = \frac{v \sin \beta}{u \sin \alpha} \tag{a}$$

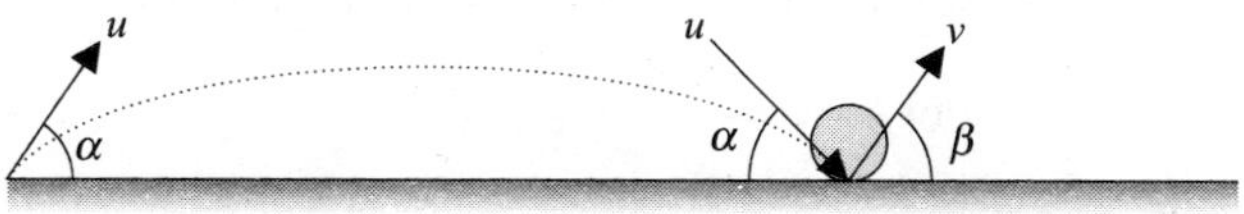

Fig. 16.39

Since the impact is not elastic, we can know that the height reached by the ball is maximum during the first flight. Hence,

$$h_{\text{max}} = \frac{u^2 \sin^2 \alpha}{2g} \tag{b}$$

The second range of the ball is given as

$$R_2 = \frac{v^2 \sin 2\beta}{g} \tag{c}$$

Since it is given that the second range is equal to the greatest height, which the ball attains, we have

$$h_{\text{max}} = R_2$$

Therefore,

$$\frac{u^2 \sin^2 \alpha}{2g} = \frac{v^2 \sin 2\beta}{g}$$

$$\frac{u^2 \sin^2 \alpha}{2g} = \frac{2v^2 \sin \beta \cos \beta}{g}$$

Upon rearranging, we have

$$\frac{1}{4} = \frac{v^2 \sin\beta \cos\beta}{u^2 \sin^2\alpha}$$

$$= \left[\frac{v^2 \sin^2\beta}{u^2 \sin^2\alpha}\right] \frac{\cos\beta}{\sin\beta} = e^2 \cot\beta \quad \text{(d)}$$

If we assume that the surface of the plane is smooth, we know that the velocity remains unaltered in the tangential direction. Hence,

$$u\cos\alpha = v\cos\beta \quad \text{(e)}$$

Substituting this in equation (a), we have

$$e = \frac{v \sin\beta}{u \sin\alpha} = \frac{u \cos\alpha \sin\beta}{u \sin\alpha \cos\beta}$$

$$\Rightarrow \quad \cot\alpha = e \cot\beta \quad \text{(f)}$$

Substituting the value of $\cot\beta$ from the above equation in the equation (d), we have

$$\frac{1}{4} = e^2 \cot\beta = e^2 \frac{\cot\alpha}{e} = e \cot\alpha$$

$$\Rightarrow \quad \tan\alpha = 4e$$

SUMMARY

Impulse momentum method is an alternative approach to solve kinetics problems. This method is based on integration of equation of motion with respect to time and it relates *force*, *velocity* and *time*. Just like work–energy method, this method avoids the need to determine the acceleration of the body for knowing the kinematics of problems. This method is particularly useful in the case of impulsive forces, which are large forces acting for a very short duration, and cause a sudden change in velocity of the body, and in impact problems.

Impulsive Forces

Impulsive forces are very *large* forces acting for a very *short* duration, arising whenever a moving body strikes a fixed object or when two moving bodies collide with each other. These impulsive forces cause sudden changes in velocities of the bodies. Their intensity and the time during which they act cannot be determined exactly. However, their effect on bodies can be understood by knowing the change in velocities they produce.

Integrating Newton's law of motion with respect to time, we get

$$\int_{t_1}^{t_2} \vec{F}\, dt = m[\vec{v} - \vec{v}_o]$$

The integral on the left-hand side is defined as the *impulse* of the force, and the product $m[\vec{v} - \vec{v}_o]$ on the right-hand side is defined as the *change in momentum* of the body. Thus, impulse of a force is equal to the change in momentum of the body. Expressed in rectangular components, we have

$$\int_{t_1}^{t_2} F_x\, dt = m[v_x - (v_0)_x]$$

and
$$\int_{t_1}^{t_2} F_y\, dt = m[v_y - (v_0)_y]$$

If we could represent the variation of impulsive force with respect to time as F–t graph then the area under the graph between the time limits gives the *impulse* of the force. The unit of impulse in S.I. units is N.s.

Impact of Jet on Plates (or) Vanes

When a jet of water strikes a *fixed* or *moving* obstruction placed across its path, it exerts a *force of impact* on the obstruction. The determination of this force of impact finds application in the design of hydraulic turbines. By Newton's third law of motion, this force of impact is equal and opposite to that of the force exerted by the plate on the jet. Thus, we can evaluate this force of impact by considering the free-body diagram of the mass of water striking the plate per unit time. The mass of water striking the plate or vane per unit time is defined as the product of density and discharge of water through the pipeline. Mathematically,

$$\text{mass/unit time} = \rho Q = \rho a v$$

where ρ is density of water, a is area of cross section of the jet and v is the velocity of the jet.

Force exerted by a jet of water:

on a smooth flat fixed plate : $F = F_n = \rho a v^2$

on a smooth flat moving plate : $F = F_n = \rho a(v - u)^2$

on a smooth inclined plate : $F = F_n = \rho a v^2 \sin\theta$

on a smooth curved vane : $F = F_n = \rho a v^2[1 + \cos\theta]$

Conservation of Momentum

When two or more bodies forming a system of bodies move with different velocities and in their course of motion if they collide with one another, impulsive forces act on each body exerted by the other. These impulsive forces cause change in momentum of the individual bodies. However, they being equal and opposite cancel out and the momentum of the entire system remains constant, i.e.,

$$m_1\vec{u}_1 + m_2\vec{u}_2 = m_1\vec{v}_1 + m_2\vec{v}_2$$

initial momentum of the system = final momentum of the system

Thus, we see that though the momentum of the individual bodies does change, the total momentum of the system remains constant. This principle is known as *conservation of momentum*.

When no external forces act on bodies forming a system, the momentum of the system is conserved, i.e., the initial momentum of the system is equal to the final momentum of the system.

Collisions

When two bodies collide with each other, the impulsive forces exerted by each body on the other cause the other body to deform. The extents to which they deform depend upon the elasticity of material of the two bodies. If the final velocity is totally restored then the collision is said to be perfectly *elastic*.

Otherwise, it is said to be *inelastic*. When the direction of motion of each body is along the line joining their centres then the impact is said to be *direct*. When the direction of motion of either or both is inclined to the line joining their centres then the impact is said to be *oblique*.

During collision, we can apply the conservation of momentum as

$$m_1u_1 + m_2u_2 = m_1v_1 + m_2v_2$$

This equation must be supplemented by another equation relating to elasticity of material of the bodies, namely, *Newton's experimental law*. The experimental evidence suggests that the relative velocities of the colliding bodies before impact bears a constant relationship with their relative velocities after impact and is in opposite direction. Stated mathematically,

$$-e = \frac{v_1 - v_2}{u_1 - u_2}$$

The constant of proportionality e is called *coefficient of restitution*. It depends upon the material of the bodies, but independent of their masses and their velocities before impact. The value of 'e' is found to vary from 0 to 1.

EXERCISES

Objective-type Questions

1. Impulse of a force acting on a body is equal to
   (a) momentum of the body (b) change in momentum of the body
   (c) rate of change in momentum of the body (d) product of momentum and time
2. Unit of impulse of a force is
   (a) N (b) N.m (c) N/s (d) N.s
3. Force of impulse is equal to
   (a) momentum (b) change in momentum
   (c) rate of change in momentum (d) product of momentum and time
4. Identify the impulsive force from the following:
   (a) Force of gravity (b) Force of friction
   (c) Normal reaction force (d) Collision force
5. Impulse momentum equation relates
   (a) force, velocity and displacement b) force, velocity and time
   (c) force, displacement and time d) force and acceleration
6. Impulse of a force between time limits is equal to area under ____________ graph between the time limits.
   (a) force–time (b) acceleration–time (c) velocity–time (d) displacement–time
7. Impulse of an average force is equal to the product of
   (a) force and time (b) force and displacement
   (c) force and velocity (d) force and acceleration

8. If a jet of water of cross sectional area a moving with velocity v strikes a smooth plate or vane, the mass of water striking per unit time is

(a) ρav (b) ρav^2 (c) $\frac{\rho a}{v}$ (d) $\frac{\rho}{av}$

9. The force exerted by a jet of water of cross-sectional area a and moving at v on a smooth fixed flat plate is

(a) ρav (b) ρav^2 (c) $\frac{\rho a}{v}$ (d) $\frac{\rho}{av}$

10. When two bodies collide with each other,
(a) their individual momenta are conserved (b) their overall momentum is conserved
(c) their individual energies are conserved d) their overall energies are conserved

11. If the kinetic energies of a lighter mass and a heavier mass are equal then
(a) the lighter mass has greater momentum (b) the heavier mass has greater momentum
(c) both have the same momentum (d) their velocities are equal

12. If u_1 and u_2 are the initial velocities of two bodies making direct collision and if v_1 and v_2 are their respective velocities after collision then the coefficient of restitution is

(a) $\frac{v_1 - v_2}{u_1 - u_2}$ (b) $\frac{v_1 + v_2}{u_1 + u_2}$ (c) $\frac{v_1 - v_2}{u_2 - u_1}$ (d) $\frac{u_1 - u_2}{v_1 - v_2}$

13. In a perfectly elastic collision,
(a) momentum is conserved
(b) kinetic energy is conserved
(c) both momentum and kinetic energy are conserved
(d) neither momentum nor energy is conserved

14. In an elastic collision of two bodies of equal masses, the two bodies after impact
(a) coalesce together and move with a common velocity
(b) rebound with their initial speeds
(c) exchange their velocities
(d) are brought to rest

Answers

1. (b) 2. (d) 3. (c) 4. (d) 5. (b) 6. (a) 7. (a) 8. (a)
9. (b) 10. (b) 11. (b) 12. (c) 13. (c) 14. (c)

Short-answer Questions

1. What are impulsive forces? Give examples.
2. What is the effect of impulsive forces upon the motion of bodies?
3. Define impulse of a force.
4. Derive the mathematical expression for the impulse–momentum equation.
5. Describe how to estimate the impulsive force.

6. Differentiate between impulsive force and impulse of a force.
7. What are non-impulsive forces? Give examples.
8. Discuss the effect of an impact of jet of water on plates or vanes and where they find application.
9. Derive the expression for a mass of water striking an obstruction.
10. What is the practical difficulty involved when a jet of water strikes a moving plate or vane and how is it overcome?
11. State the principle of conservation of momentum. Give some examples where this principle is applied.
12. Differentiate between the work–energy and impulse–momentum methods.
13. Distinguish between direct and oblique collisions.
14. Distinguish between elastic and inelastic collisions.
15. State Newton's experimental law.

Numerical Problems

16.1 A varying force $\vec{F} = t^3\vec{i} + 4t\vec{j}$, where F is in newtons and t in seconds, is applied on a body of 5 kg mass from time $t = 0$ to 4 s. Determine the impulse of the force in 4 seconds. If the body is initially at rest, determine its velocity at the end of 4 seconds.

Ans. 71.55 N.s, 14.31 m/s

16.2 A body of 2 kg mass, which is at rest, is acted on by a force varying with time as $\vec{F} = 5t\vec{i} + t^2\vec{j}$, where F is in newtons and t in seconds. Determine the velocity of the body at $t = 3$ s.

Ans. 12.12 m/s

16.3 A body of 2 kg mass moving with velocity $\vec{v}_i = 5\vec{i} + 3\vec{j}$ impacts with another body and its velocity after impact is changed to $\vec{v}_f = -2\vec{i} + 4\vec{j}$. If the impact duration is 1/100th of a second, determine the force of impulse.

Ans. $200[-7\vec{i} + \vec{j}]$

16.4 A varying force acts on a body of 5 kg mass, which is at rest. If the variation of force with respect to time is as shown in Fig. E.16.4, determine the final velocity of the body after 4 seconds in each case.

Ans. (i) 6 m/s, (ii) 9 m/s

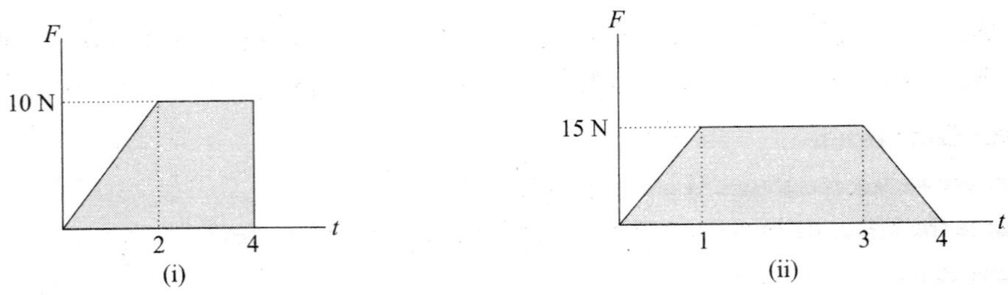

Fig. E.16.4

16.5 A varying force acts on a body of 5 kg mass, which is initially moving at 2 m/s. If the variation of force with respect to time is as shown in Fig. E.16.5, determine the final velocity of the body (i) after 2 seconds [refer Fig. (a)], (ii) after 4seconds [refer Fig. (b)], and (iii) after 3 seconds [refer Fig. (c)].

Ans. (i) 6 m/s, (ii) 14 m/s, (iii) 9.5 m/s

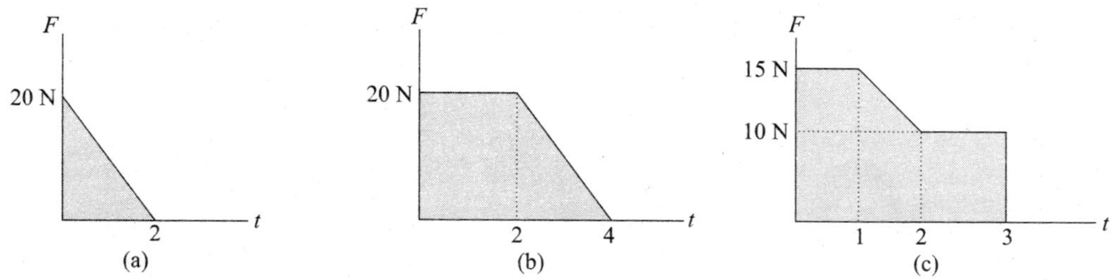

Fig. E.16.5

16.6 A boy swings his right foot such that it hits a wall normally with a velocity of 3 m/s and is brought to rest in 1/100th of a second. Determine (i) the impulse, (ii) the average force exerted by the wall on the foot. Assume mass of the right leg to be 10 kg swinging about the hip joint.

Ans. 30 N.s opposite to the direction of initial velocity, 3 kN

16.7 A girl of 20 kg mass slips from the top of a 10 m high building (refer Fig. E.16.7). If she can sustain a maximum impact force of 20 kN, state whether she will survive or not; the impact duration being 1/60th of a second.

Ans. Yes, she will survive; force of impact is 16.81 kN, which is less than 20 kN

16.8 The driver of a car moving at 60 kmph applies brakes to bring it to a stop. If coefficient of friction between the wheels and the road is 0.5, determine the time required to bring it to a stop.

Ans. 3.4 s

Fig. E.16.7

16.9 A body of 10 kg mass falling freely from a height of 20 m is brought to rest by penetrating into sand. Determine the average resistance offered by the sand if it is brought to rest in 1 s.

Ans. 198.1 N

16.10 A car of 2 ton mass moving with a speed of 45 kmph hits a road divider made of concrete and brought to rest in 1/6th of a second. Determine the average force exerted by the divider.

Ans. 150 kN

16.11 A heavy cargo block of 1.5 ton mass, while being lifted by a crane slips and drops from a height of 6 m onto a concrete floor (refer Fig. E.16.11). Determine the force exerted by the floor on the block; the impact duration being 1/60th of a second.

Ans. 976.5 kN

Fig. E.16.11

16.12 An archer strikes an arrow of 100 g mass, which moves at 30 m/s. If it hits the target in 1/60th of a second and comes to rest, determine the average resistance offered by the target. Refer Fig. E.16.12.

Ans. 180 N

16.13 A block of 10 kg mass is pulled up an inclined plane by a force of 100 N acting parallel to the plane as shown in Fig. E.16.13. If the block is initially at rest, determine the time taken for the body to reach a velocity of 10 m/s; the coefficient of kinetic friction being 0.2.

Ans. 2.94 s

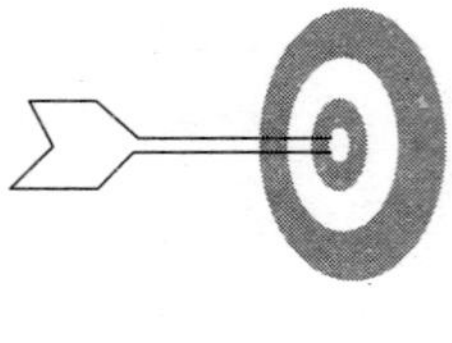

Fig. E.16.12

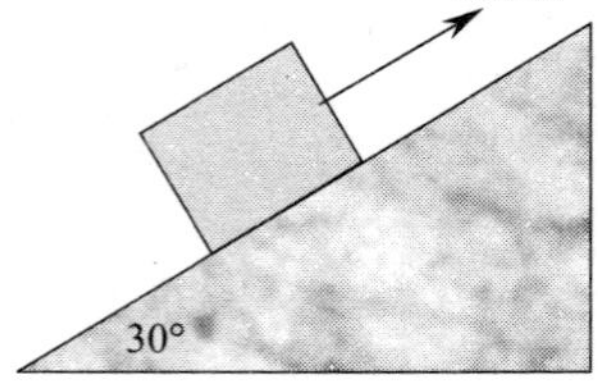

Fig. E.16.13

16.14 A block of mass m_1 resting on a rough inclined plane is attached by a string, whose other end is attached to a block of mass m_2 hanging vertically as shown in Fig. E.16.14. If released from rest, determine the velocity of blocks after 3 s. Take $m_1 = 7$ kg, $m_2 = 5$ kg, $\theta = 30°$ and $\mu_k = 0.2$.

Ans. 0.71 m/s

16.15 The block-and-pulley arrangement shown in Fig. E.16.15 is released from rest. If the 6 kg block comes to rest in 1/100th of a second after striking the floor, determine the force of impact and height to which the lighter block would rise?

Ans. 2376 N, 0.8 m

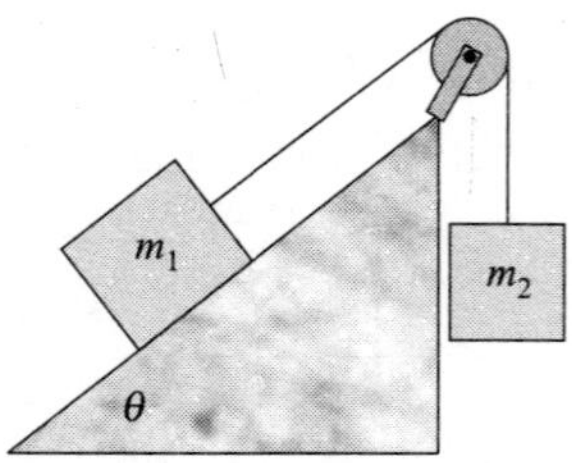

Fig. E.16.14

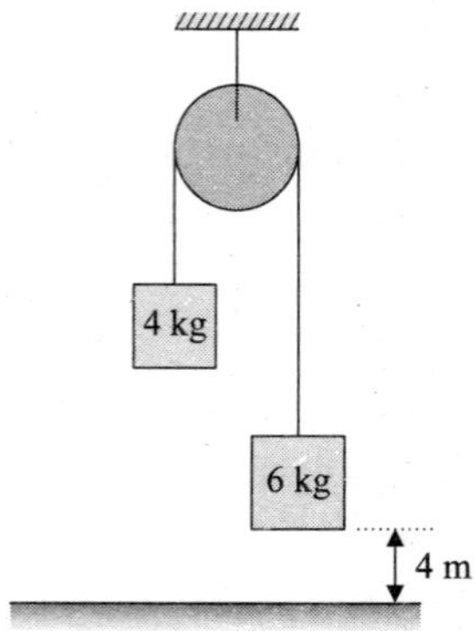

Fig. E.16.15

16.16 A block of 5 kg mass rests on a rough horizontal plane, whose coefficient of static friction is 0.2. If it is hit by a horizontal jet of water (the diameter of jet being 2 cm), determine the minimum speed of the jet of water to just move the block. Take density of water as 1000 kg/m^3.

Assume the face of the block on which water is striking to be smooth and that the water leaves parallel to this face after striking. Refer Fig. E.16.16.

Ans. 5.6 m/s

16.17 A jet of water of 1 cm diameter and moving at 10 m/s strikes a cart that is free to move on a horizontal plane as shown in Fig. E.16.17. If the force of impact is 5 N, determine the velocity of the cart.

Ans. 2.02 m/s

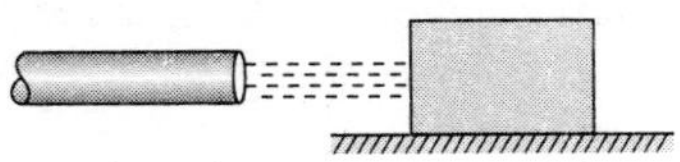

Fig. E.16.16

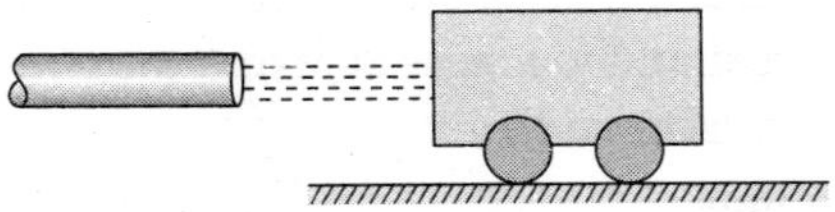

Fig. E.16.17

16.18 A jet of water of 1 cm diameter and discharging 10^{-3} m^3/s of water strikes normally a flat, fixed plate. If the water after striking the plate leaves parallel to the plate, determine the force of impact of the jet on the plate.

Ans. 12.73 N

16.19 A jet of water issued from a nozzle strikes a smooth inclined plate, which is free to move in the direction of the jet as shown in Fig. E.16.19. The water after striking the plate leaves parallel to the plate. Derive an expression for the force exerted by the jet of water on the plate. The jet diameter is 6 mm and water moves at 15 m/s, while the plate moves at 6 m/s, the inclination of the plate is 60° to the direction of jet. Determine the force exerted by the jet of water on the plate.

Ans. $F_n = \rho a(v - u)^2 \sin\theta$, 1.98 N

16.20 A jet of water issued from a nozzle of 6-mm diameter and moving at 15 m/s strikes at the centre of a smooth curved vane. The water after striking the vane gets deflected through 105° and leaves tangential to the vane at the exit. Determine the force exerted by the jet of water on the vane.

Ans. 8 N

16.21 Water flows through a pipe bend of uniform cross-sectional diameter of 5 cm as shown in Fig. E.16.21. If the discharge through the pipe is at a rate of 0.005 m^3/s, determine the force required to hold the pipe in position. [*Note:* Though the cross section of the pipe is constant, the direction of velocity changes, thus, causing change of momentum or impulse to act on the pipe body].

Ans. 12.74 N at 60° to initial direction of flow

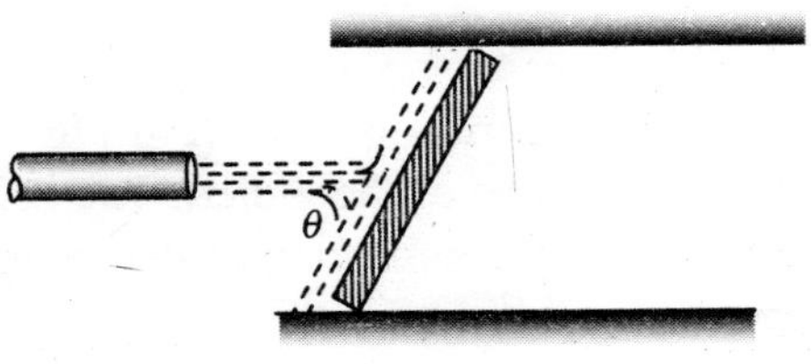

Fig. E.16.19

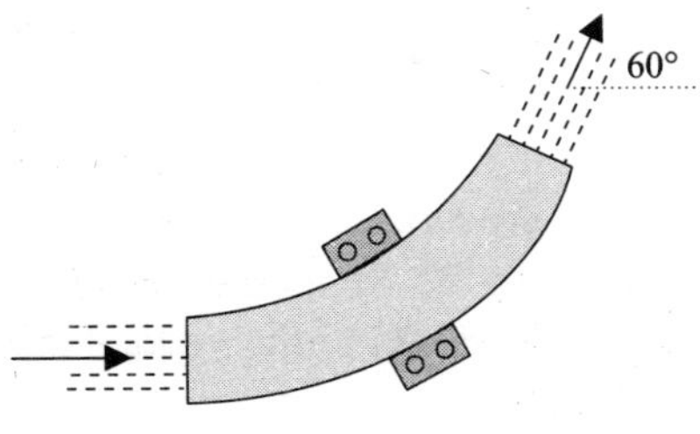

Fig. E.16.21

16.22 A bullet of 10 g mass is fired from a rifle of 4 kg mass. If the rifle recoils back at 0.5 m/s, determine the speed of the bullet.

Ans. $v = 200$ m/s

16.23 A railway wagon of 10 ton mass is at rest on a level track. It is hit by a locomotive of 15 tons moving at 20 kmph. If after collision, they lock and move together, determine their common velocity.

Ans. 12 kmph

16.24 A car of 2 ton mass moving at 30 kmph has brake failure. To stop the car, the driver decides to hit a truck of 6 ton mass at rest. If the two vehicles lock after collision, determine their common velocity after collision.

Ans. 7.5 kmph

16.25 Two 10 kg boxes are to be unloaded from a cart, which is free to move. A man standing on the cart, as shown in Fig. E.16.25, pushes each one horizontally with a velocity of 1 m/s. After pushing the second block, determine the velocity of the cart. Take mass of the cart together with the man as 100 kg.

Ans. 0.2 m/s

16.26 A cannon fires a bomb at an angle of 45° to the horizontal ground with a release velocity of 200 m/s. At the highest point of its trajectory, the bomb explodes into two pieces of equal mass. If one of the pieces just drops freely, determine the distance from the cannon at which the other piece would strike the ground.

Ans. 6.12 km from the point of projection

16.27 A ball of 1 kg mass is released from the position shown in Fig. E.16.27, when the string is horizontal. At the lowest point of its path, it strikes a block of 2 kg mass at rest on a frictionless surface. Determine the speed of the ball and block immediately after collision assuming the collision is elastic.

Ans. $V_{\text{ball}} = 1.48$ m/s in the direction opposite to the initial direction of motion of the ball,
$V_{\text{block}} = 2.95$ m/s in the same direction as the initial direction of motion of the ball

Fig. E. 16.25

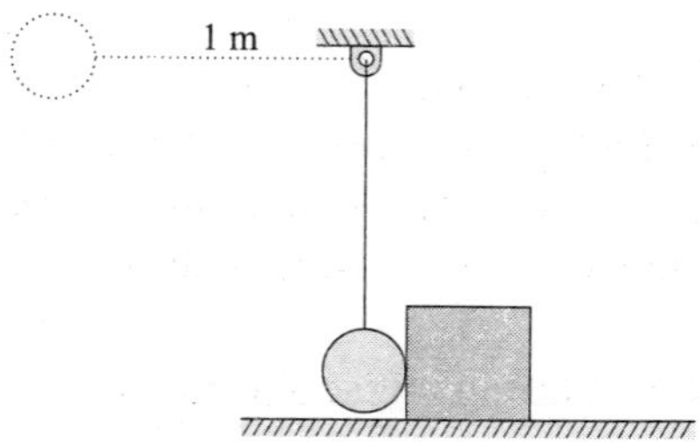

Fig. E.16.27

16.28 A bullet of 20 g mass strikes a ballistic pendulum of 2 kg mass as shown in Fig. E.16.28. [A ballistic pendulum is a device used to determine the speed of the bullet]. The bullet gets embedded in the wooden block after impact. If the maximum inclination of the string of 2 m length with respect to the vertical is 30°, determine the speed of the bullet.

Ans. 231.6 m/s

16.29 A bullet of 20 g mass is fired into a 5 kg wooden block resting on a rough horizontal surface as shown in Fig. E.16.29. The bullet gets embedded in the block and both move further by 0.5 m. If the coefficient of friction is 0.15, determine the initial speed of the bullet.

Ans. 304.5 m/s

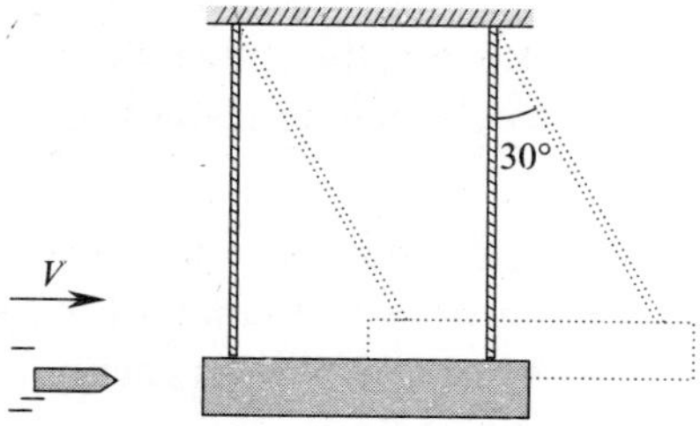

Fig. E.16.28

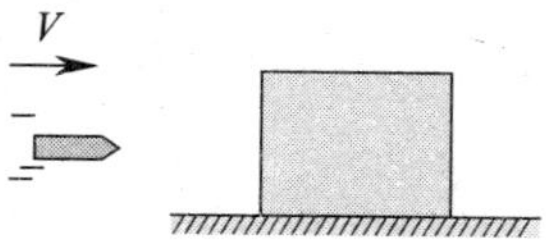

Fig. E.16.29

16.30 Two elastic spheres impinge directly with equal and opposite velocities; find the ratio of their masses so that one of them may be reduced to rest by the impact, the coefficient of elasticity being e.

Ans. $1 + 2e$

16.31 A sphere of mass m moving with a speed of u strikes a sphere of same mass at rest. If after striking, the two spheres exchange their velocities then determine the coefficient of restitution.

Ans. $e = 1$

16.32 A sphere moving with a velocity u strikes another sphere of equal mass, but at rest. If e be the coefficient of restitution, prove that the ratio of their final velocities is $\frac{1-e}{1+e}$. If $e = 3/4$, what are their respective final velocities.

Ans. $\frac{u}{8}, \frac{7u}{8}$

16.33 A car of 2 ton mass moving at 60 kmph collides with another car of 3 ton mass at rest. If they coalesce after collision, determine their common velocity.

Ans. 24 kmph

16.34 A body of elasticity e is projected from a point in a horizontal plane. If the distance of the point of nth impact be equal to four times the sum of the vertical spaces described, the tangent of the angle of projection is $\frac{1-e}{1 \mp e^n}$.

16.35 Two balls of equal masses approach each other with velocities u and v respectively. If after impact, the ball moving with velocity u is brought to rest, show that $\frac{u}{v} = \frac{1+e}{1-e}$.

16.36 If a ball of mass m, strikes another of mass nm at rest, the inclination of its direction to the line joining the centres at the time of impact being 30°, find n in order that m may go off perpendicular to its original.

Ans. 2

16.37 A ball moving with a speed of u strikes another ball of twice its mass but at rest; determine the expressions for final velocities of the two balls after impact. Also, determine the final velocities, if coefficient of restitution is (i) $e = 0$, (ii) $e = 1/4$ (iii) $e = 1/2$ (iv) $e = 3/4$ and (v) $e = 1$.

Ans. $v_1 = \frac{u}{3}(1 - 2e)$, $v_2 = \frac{u}{3}(1 + e)$; (i) $\frac{u}{3}, \frac{u}{3}$; (ii) $\frac{u}{6}, \frac{5u}{12}$; (iii) 0, $\frac{u}{2}$; (iv) $-\frac{u}{6}, \frac{7u}{12}$; (v) $-\frac{u}{3}, \frac{2u}{3}$

16.38 A body of 1 kg mass strikes another body at rest. If after collision, the body rebounds with half its initial velocity, determine the mass of the other body. Coefficient of restitution is 2/3.

Ans. 9 kg

16.39 While a car of 2 ton mass is moving in reverse gear at 2 m/s, a truck of 6 ton mass moving at 30 kmph comes and hits the car at the back. If the coefficient of restitution of buffers is 0.5, describe their motion after collision. Also, determine the amount of kinetic energy lost.

Ans. $v_{car} = 9.63$ m/s, $v_{truck} = 4.46$ m/s both in the forward direction; 59.8 J

16.40 In goods-shed, during shunting, a bogie of 15 ton mass moving on a level track at 9 kmph hits another bogie of 10 ton mass at rest. If the coefficient of restitution of the buffers is 3/4, determine their velocities immediately after impact.

Ans. 0.75 m/s, 2.625 m/s

16.41 A bus of 6 ton mass moving at 60 kmph collides with a car of 2 ton mass (i) which is at rest, (ii) moving at 45 kmph in the same direction, and (iii) moving at 45 kmph in the opposite direction. Determine their velocities immediately after collision if the coefficient of restitution of bumpers is 2/3.

Ans. (i) 35 kmph, 75 kmph, (ii) 53.75 kmph, 63.75 kmph, (iii) 16.25 kmph, 86.25 kmph

16.42 Three spheres of masses $3m$, $2m$ and m are placed in a line along a horizontal plane as shown in Fig. E.16.42. If the sphere of mass $3m$ is given an initial velocity of u, determine the velocity of the sphere of mass m.

Ans. $\frac{2}{5}(1 + e)^2 u$

16.43 In the previous problem, if the spheres are arranged as m, $2m$ and $3m$ as shown in Fig. E.16.43 and if the sphere of mass m is given an initial velocity of u, determine the velocity of the sphere of mass $3m$.

Ans. $\frac{2}{15}(1 + e)^2 u$

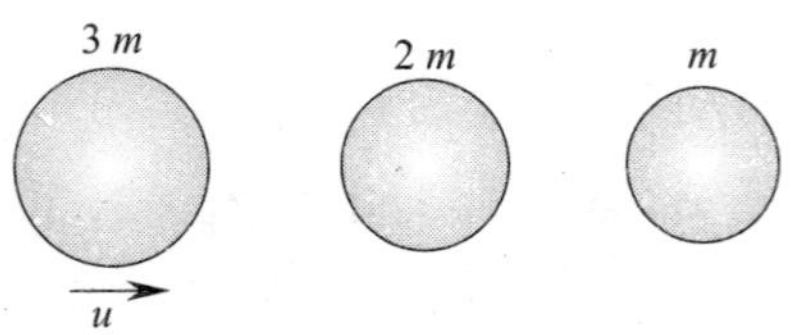

Fig. E.16.42

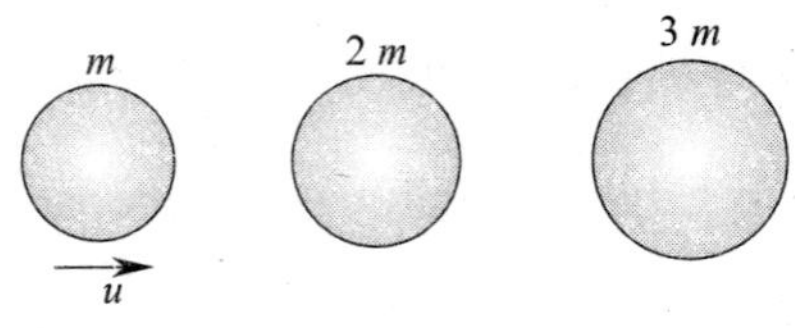

Fig. E. 16.43

16.44 A ball dropped from a height of 10 m onto a floor rebounds to a height of 8 m. Determine the coefficient of restitution of the floor.

Ans. 0.89

16.45 A ball is dropped from a height of 5 m onto a horizontal plane. If the coefficient of restitution is 3/4, determine the height it would rise after the 4th rebound.

Ans. 0.5 m

16.46 A ball is dropped from a height of 15 m onto a hard floor. Determine the coefficient of restitution, if it rebounds to a height of two-third of the height from which it was dropped. Suppose another ball of same material is dropped onto the same floor from a height of 10 m, determine the height to which it would rise.

Ans. 0.816, 6.67 m

16.47 A football moving at 20 m/s hits the ground at an angle of 45° to the ground level as shown in Fig. E.16.47. If the coefficient of restitution is 2/3, determine the magnitude and direction of velocity of the ball after impact.

Ans. 17 m/s at 33.7° to the ground level

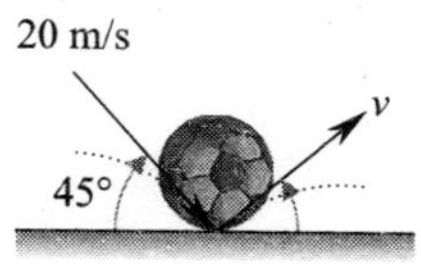

Fig. E.16.47

17 Kinematics of Rigid Bodies

17.1 INTRODUCTION

So far, in dynamics, we have analyzed the motion of *particles*. In this chapter and in the following, we will analyze the motion of **rigid bodies**. In this chapter, we will discuss the **kinematics** of rigid bodies, where we will describe the motion of rigid bodies by specifying displacement, velocity and acceleration, and their variation with time. In the next chapter, we will discuss the **kinetics** of rigid bodies, where we will relate the motion with the forces causing the motion.

In the previous chapters on dynamics, we saw that under **pure translational motion**, either *rectilinear* or *curvilinear* motion, the bodies moved in such a way that the local coordinate axes or axes fixed to the bodies always remained *parallel* to the fixed reference axes (refer Fig. 17.1). In such types of motions, there is **no** rotational motion at all. Every particle in the body moves in *parallel* paths in the same direction as shown in Fig. 17.2.

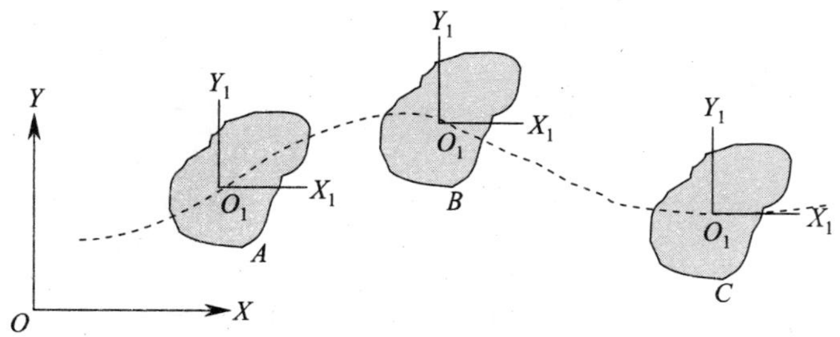

Fig. 17.1 Pure translational motion

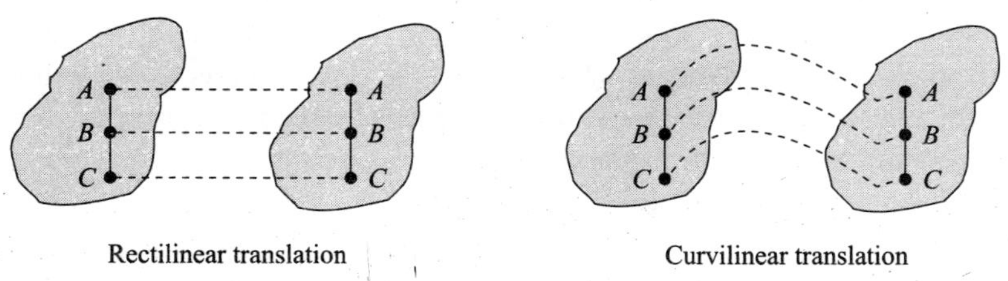

Fig. 17.2

As the particles move in parallel paths, the *displacement*, *velocity* and *acceleration* of each particle is the *same* as that of the other. Thus, instead of considering the body as a whole, we considered a single particle in the body (generally, the centre of mass of the body) and by describing the motion of that particle, we described the motion of the entire body. This we termed as idealization of the rigid body as a **particle**.

The student should recall that even during such pure translational motion, a body could **rotate** as in the case of a cricket ball spinning or rotating about itself while moving in its trajectory. However, we have neglected such spinning or rotation as the size of the ball was very small as compared to its trajectory. Thus, we concluded and said that whenever a body moves in pure translation, there is no rotational motion involved or even if rotational motion is involved, it can be neglected for the reason stated above. Thus, we could idealize the rigid body as a particle.

However, there are instances in which the rotational motion is considerable and cannot be neglected. In such type of motions, the particles in the body undergo *different* displacements, velocities and accelerations. As a result, we can no more describe the motion of the entire body by describing the motion of a *single* particle and thus we have to treat the body as a whole, i.e., as the rigid body itself. The motion of a rigid body is thus **translational** motion together with **rotational** motion.

To understand the **general motion** of a rigid body, we have taken a body of regular shape instead of an arbitrary shape as shown in Fig. 17.3. We see that the body not only translates but also rotates in an arbitrary manner. To represent the translational motion of a rigid body in space, we require **three** independent coordinates, namely, x, y and z, and to represent its rotational motion, we require **three** more coordinates, namely, angular coordinates θ_x, θ_y and θ_z. Hence, six *independent* coordinates are required to represent the motion of a rigid body. Thus, we say that a rigid body has **six degrees of freedom**.

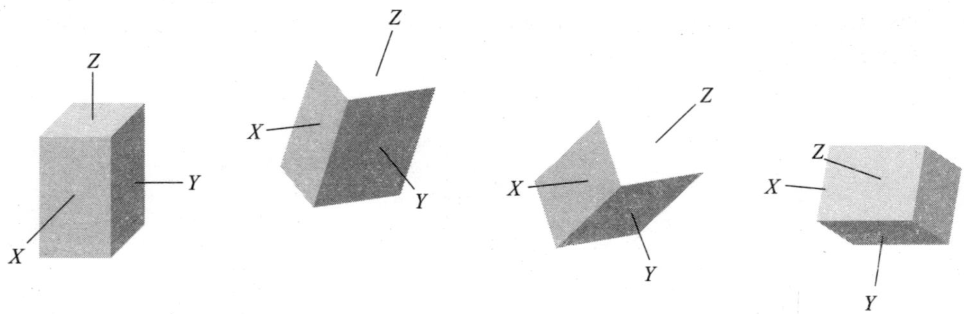

Fig. 17.3 General motion of a rigid body

If we constrain or restrict the motion of the rigid body to a single plane, say x–y plane as shown in Fig. 17.4 then we call that the **general plane motion** because each particle in the body is then restricted to move in **parallel planes**. The translational motion of the body occurs in the plane and the rotation is about an axis perpendicular to the plane; in our case, it is the Z-axis. Thus, considering the motion of a plane taken perpendicular to the axis of rotation, we can describe the motion of the entire body and hence the problem reduces to that of *two-dimensional* case.

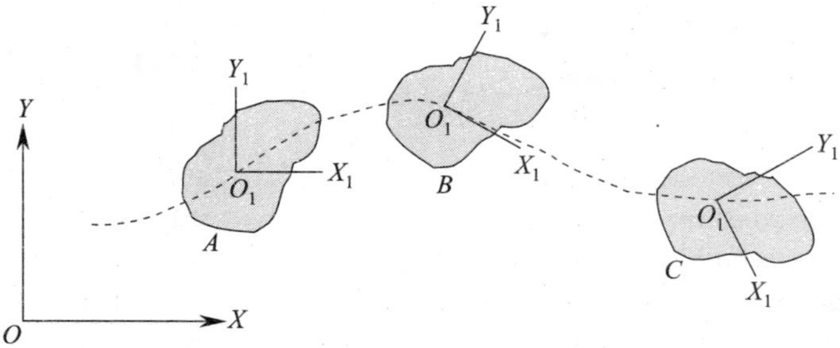

Fig. 17.4 General plane motion

To describe the general plane motion of a rigid body, we require *two* independent coordinates, x and y, to represent its translational motion along the x–y plane and *one* angular coordinate θ_z to represent rotation about the z-axis, i.e., perpendicular to the plane of the figure. Thus, we say that a body in general plane motion has **three** degrees of freedom. A wheel rolling without slipping on a horizontal floor, a ladder sliding down with its ends in contact with a wall and a floor are examples of such types of general plane motion. This type of motion will be discussed later in Section 17.6.

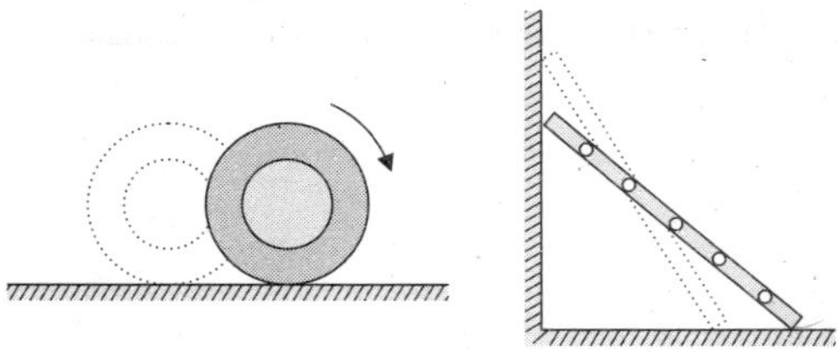

Fig. 17.5 Example of general plane motion

If we further constrain the *translational* motion of the rigid body by fixing the body about an axis (in our case, the Z-axis) then the resulting motion is **pure rotation** about a fixed axis. Such types of motions are very common in all man-made machines such as electric motor, turntable, turbines, etc., In such a motion, we see that all the particles in the body move in parallel planes (planes which are normal to the axis of rotation) in **circular** paths about the axis of rotation. As the paths of all the particles are concentric circles, their distances from the axis of rotation remain **constant**. Thus, to describe the motion of particles, we require only one coordinate, namely, the angular coordinate in the Z-direction. Hence, the body has *one degree of freedom*. We must note that the particles on the axis of rotation have *no* motion at all.

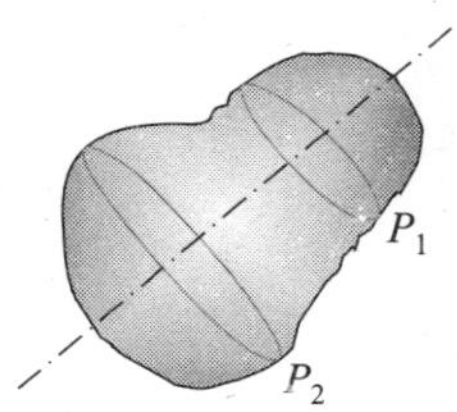

Fig. 17.6 Fixed axis rotation

The rotation about a fixed axis will be discussed in the following section, where we will describe angular displacement, velocity and acceleration, and their variation with time. In Section 17.5, we will establish a relationship between angular motion about a fixed axis and linear motion.

17.2 ROTATIONAL MOTION ABOUT A FIXED AXIS

As discussed above, when a body rotates about a fixed axis, all the particles in the body move in parallel planes and thus by describing the motion of particles in any given plane, we can describe the motion of the entire body. Consider one such plane as seen from the top of the fixed axis. Then we see that all particles in this plane move in *concentric circles* as shown in Fig. 17.7

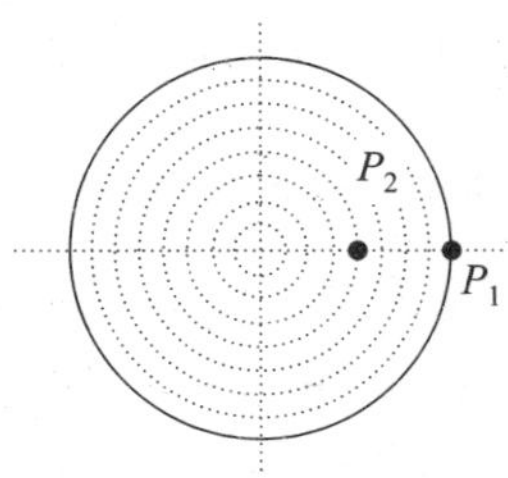

Fig. 17.7 Representative plane for rotational motion about fixed axis

about the fixed axis of rotation, i.e., the axis perpendicular to the plane of the figure. We will describe the angular displacement, velocity and acceleration of a particle in such a representative plane.

17.2.1 Angular Displacement

Consider a particle P at a distance r from the axis of rotation. As it moves in circular path, its distance from the axis is always *constant*. Hence, to define its position at any instant of time, we need only one independent coordinate, which describes its angular position with respect to a reference. Let us choose the OX axis to be that reference. At time, $t = t_1$, let the particle be at the position A, where OA makes an angle of θ_1 with respect to OX and at a later time, $t = t_2$, let it be at B at an angular position of θ_2 with respect to OX. Then we say that the particle has *displaced* by $\theta_2 - \theta_1$ in time $t_2 - t_1$. This displacement being angular, we call it **angular displacement**, to differentiate it from linear displacement, which we have already seen in particle kinematics.

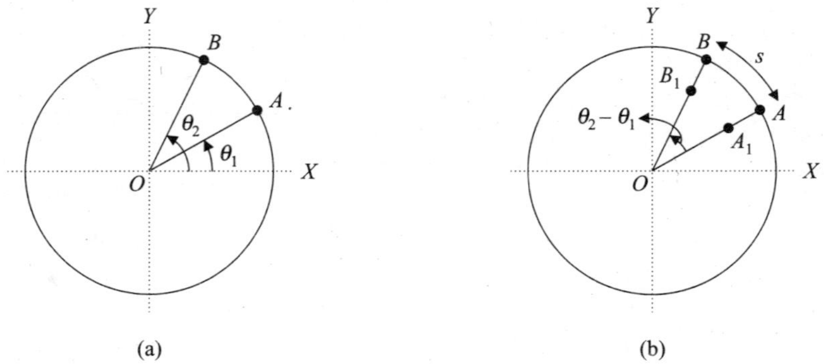

Fig. 17.8 Angular displacement

By convention, we choose the **anticlockwise** displacement as **positive** and **clockwise** displacement as **negative**. It is convenient to express angular coordinate in **radians** rather than in degrees. The conversion factor between them is

$$2\pi \,\text{rad} = 360° \tag{17.1}$$

At times, it is also expressed in revolutions, where

$$1 \text{ rev} = 2\pi \,\text{rad} = 360° \tag{17.2}$$

We can also note that when the particle moves from A to B, other particles (say A_1) on the same radial line also get displaced (from A_1 to B_1) through the *same* angle. Similarly, other radial lines also displace through the same angle. Thus, we see that the angular displacement of every particle in fixed rotation remains the same.

17.2.2 Angular Velocity

Just as in particle linear motion, it would not suffice to describe the angular displacement alone to describe the motion of a rigid body about a fixed axis; we must also specify the rate at which it gets displaced, which we define as **angular velocity**.

The **average angular velocity** is defined as the ratio of angular displacement to time elapsed. Mathematically,

$$\omega_{\text{ave}} = \frac{\theta_2 - \theta_1}{t_2 - t_1} = \frac{\Delta\theta}{\Delta t} \tag{17.3}$$

As we let Δt approach zero, i.e., as we let B approach A, we define instantaneous angular speed as

$$\omega = \lim_{\Delta t \to 0} \frac{\Delta\theta}{\Delta t} = \frac{d\theta}{dt} \tag{17.4}$$

As we saw that all the particles undergo the same displacement in a given time interval, their angular velocity, being the ratio of angular displacement and time, is also the same. Its unit is rad/s (or) rpm. It is a common engineering practice to represent angular velocity as revolutions per minute (abbreviated as rpm). If N revolutions are made in one minute then we express it in radians per second as

$$\omega = \frac{2\pi N}{60}\text{rad/s} \tag{17.5}$$

Although angular displacement may seem to be a vector, it is not, because though it has magnitude and direction, it does not obey the *parallelogram* law of vector addition. For example, consider a book in the position shown on the left end of the figure below. If it is given 90° displacements succesively in two perpendicular directions in two ways as shown, we see that the angular displacements do not follow the *commutative law of addition*, i.e., $\theta_1 + \theta_2 \neq \theta_2 + \theta_1$. Hence, angular displacement, though having magnitude and direction cannot be treated as a vector.

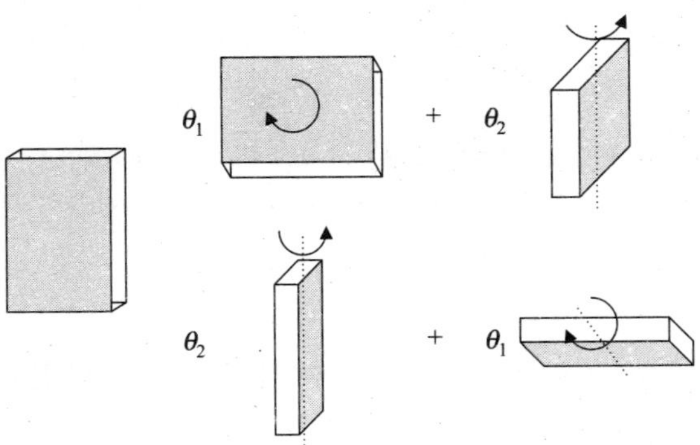

Fig. 17.9 Angular displacements are not commutative

However, if we reduce the displacements to say 2° or even further, the angular displacement tends to obey the commutative law (which the student can experimentally check). Thus, we can conclude that **finite** angular displacements are **scalars**, while **infinitesimally small** angular displacements are **vectors**. Though angular displacement is not a vector, angular velocity, which is the ratio of *infinitesimally small* angular displacement and time, is a vector.

Angular velocity being a vector, its direction is perpendicular to the plane of motion, either into the plane or out of the plane. Since the positive Z-axis points out of the plane, angular velocity pointing out of the plane is considered as positive. When we curl the fingers of our right hand in the anticlockwise direction, the thumb points along the positive Z-axis. This is the reason why we chose anticlockwise angular displacements as positive.

17.2.3 Angular Acceleration

Angular acceleration is defined as the rate at which angular velocity changes. Let ω_1 and ω_2 be the instantaneous angular speeds at times t_1 and t_2 respectively. Then average angular acceleration is defined as

$$\alpha_{\text{ave}} = \frac{\omega_2 - \omega_1}{t_2 - t_1} = \frac{\Delta\omega}{\Delta t} \tag{17.6}$$

The instantaneous angular acceleration is defined as

$$\alpha = \lim_{\Delta t \to 0} \frac{\Delta\omega}{\Delta t} = \frac{d\omega}{dt} \tag{17.7}$$

Just like angular velocity, angular acceleration is also same for all the particles in the rigid body. Its unit is rad/s^2. Since angular velocity is a vector, angular acceleration is also a vector.

Thus, we see that in fixed axis rotation, the angular displacement, velocity and acceleration are *same* for *every* particle in the body. Hence, by describing the motion of one particle in the body, we can describe the motion of the entire body.

If the angular position of a particle is given as a function of time then we can determine the angular velocity and acceleration by differentiating the displacement function with respect to time. In general, instead of angular position of a particle, its angular acceleration is specified. In which case, we derive the kinematic equations of motion by integration methods as described for particle motion. The following examples will explain these.

Example 17.1 The angular motion of a disc is defined by the relation $\theta = 2t + t^3$, where θ is in radians and t is in seconds. Determine the angular position, velocity and acceleration at $t = 2$ s.

Solution Given that angular displacement is

$$\theta = 2t + t^3$$

the angular velocity and angular acceleration are obtained by differentiating it successively with respect to time, i.e.,

$$\omega = \frac{d\theta}{dt} = 2 + 3t^2$$

and
$$\alpha = \frac{d^2\theta}{dt^2} = 6t$$

Hence,the angular position, velocity and acceleration at $t = 2$ s are obtained as

$$\theta(2) = 2(2) + (2)^3 = 12 \text{ rad}$$

$$\omega(2) = 2 + 3(2)^2 = 14 \text{ rad/s}$$

$$\alpha(2) = 6(2) = 12 \text{ rad/s}^2$$

Example 17.2 The angular motion of a rigid body is defined by the relation $\theta = 3t + 2t^2 - t^3$, where θ is in radians and t is in seconds. Determine the angular position, velocity and acceleration at $t = 2$ s. Also, determine the angular acceleration when angular velocity is zero.

Solution Given that angular displacement is

$$\theta = 3t + 2t^2 - t^3 \tag{a}$$

the angular velocity and angular acceleration are obtained by differentiating it successively with respect to time.

$$\omega = \frac{d\theta}{dt} = 3 + 4t - 3t^2 \tag{b}$$

$$\alpha = \frac{d\omega}{dt} = 4 - 6t \tag{c}$$

Therefore, the angular displacement, velocity and acceleration at $t = 2$ s are obtained as

$$\theta(2) = 3(2) + 2(2)^2 - (2)^3 = 6 \text{ rad}$$

$$\omega(2) = 3 + 4(2) - 3(2)^2 = -1 \text{ rad/s}$$

$$\alpha(2) = 4 - 6(2) = -8 \text{ rad/s}^2$$

To determine the angular acceleration when angular velocity is zero, we equate $\omega = 0$ in equation (b),

$$0 = 3 + 4t - 3t^2$$

Solving the quadratic equation and neglecting the negative root, we have

$$t = 1.87 \text{ s}$$

Therefore, the angular acceleration is given as

$$\alpha(1.87) = 4 - 6(1.87) = -7.22 \text{ rad/s}^2$$

Example 17.3 A motor running freely at 1200 rpm is switched off. The deceleration due to bearings varies with time as $\alpha = 3t - 4t^2$. Determine angular velocity and displacement at $t = 3$ s. Also, determine the time taken to come to a stop.

Solution Given that the angular deceleration of the motor varies as

$$\alpha = \frac{d\omega}{dt} = 3t - 4t^2$$

the angular velocity is obtained by integrating the above expression with respect to time.

$$\omega = 3\frac{t^2}{2} - 4\frac{t^3}{3} + \omega_o$$

Since the motor is initially running at 1200 rpm, we know that $\omega_o = 1200$ rpm $= 1200 \times (2\pi/60) =$ 125.66 rad/s. Hence,

$$\omega = 3\frac{t^2}{2} - 4\frac{t^3}{3} + 125.66 \qquad \text{(a)}$$

We can also write the above equation as

$$\omega = \frac{d\theta}{dt} = 3\frac{t^2}{2} - 4\frac{t^3}{3} + 125.66$$

Upon further integration with respect to time, we get

$$\theta = \frac{t^3}{2} - \frac{t^4}{3} + 125.66t + \theta_0$$

Taking the initial displacement when the motor is switched off as zero, we get

$$\theta = \frac{t^3}{2} - \frac{t^4}{3} + 125.66t \qquad \text{(b)}$$

Substituting $t = 3$ s in the two equations (a) and (b), we get

$$\omega(3) = 3\frac{(3)^2}{2} - 4\frac{(3)^3}{3} + 125.66 = 103.16 \text{ rad/s}$$

$$\theta(3) = \frac{(3)^3}{2} - \frac{(3)^4}{3} + 125.66(3) = 363.48 \text{ rad}$$

To determine the time taken by the motor to come to a stop, we make $\omega = 0$ in equation (a), i.e.,

$$0 = 3\frac{t^2}{2} - 4\frac{t^3}{3} + 125.66$$

By trial and error method, we can get the value of time t as 4.96 s.

Example 17.4 A circular disc starts from rest at $\theta = 0$ and accelerates at a rate given by the relation $\alpha = 3t - t^2$. Determine the time at which the body comes to a momentary rest before changing the direction of motion. Also, determine the angular displacement in this time.

Solution By integrating the given expression for acceleration with respect to time, we get the expression for angular velocity.

$$\omega = 3\frac{t^2}{2} - \frac{t^3}{3} + \omega_o$$

Since it starts from rest, we know that $\omega_o = 0$. Hence,

$$\omega = 3\frac{t^2}{2} - \frac{t^3}{3}$$

Integrating it further with respect to time, we get

$$\theta = \frac{t^3}{2} - \frac{t^4}{12} + \theta_o$$

$$= \frac{t^3}{2} - \frac{t^4}{12} \quad \text{[since it is given that } \theta_o = 0\text{]}$$

When the circular disc comes to a momentary rest, we know that its angular velocity is zero. Thus equating the expression for ω to zero,

$$0 = 3\,\frac{t^2}{2} - \frac{t^3}{3} = t^2\left[\frac{3}{2} - \frac{t}{3}\right]$$

$\Rightarrow$ $$t = 9/2 = 4.5 \text{ s}$$

Therefore, the angular displacement in this time is given as

$$\theta(4.5) = \frac{(4.5)^3}{2} - \frac{(4.5)^4}{12} = 11.39 \text{ rad}$$

17.3 ROTATIONAL MOTION WITH CONSTANT ANGULAR ACCELERATION

In this section, we will discuss a special case of fixed-axis rotation in which the angular acceleration is *constant*.

$$\alpha = \frac{d\omega}{dt} = \text{constant} \tag{17.8}$$

On multiplying both sides by dt, we have

$$d\omega = \alpha dt \tag{17.9}$$

Upon integration between limits,

$$\omega - \omega_o = \alpha t$$

Therefore, $$\omega = \omega_o + \alpha t \tag{17.10}$$

where ω_o is initial angular velocity, i.e., at time $t = 0$. The above equation can also be written as

$$\omega = \frac{d\theta}{dt} = \omega_o + \alpha t \tag{17.11}$$

Upon further integration between limits, we get

$$\theta - \theta_o = \omega_o t + \frac{\alpha t^2}{2}$$

Therefore, $$\theta = \theta_o + \omega_o t + \frac{\alpha t^2}{2}$$

where θ_o is initial angular displacement. If it is zero then the above equation reduces to

$$\theta = \omega_o t + \frac{\alpha t^2}{2} \tag{17.12}$$

The angular acceleration can also be written as

$$\alpha = \frac{d\omega}{dt} = \frac{d\omega}{d\theta}\frac{d\theta}{dt} = \omega\frac{d\omega}{d\theta} = \text{constant} \tag{17.13}$$

Upon rearranging,

$$\omega d\omega = \alpha d\theta$$

and integrating between limits,

$$\frac{\omega^2}{2} - \frac{\omega_o^2}{2} = \alpha\theta$$

$$\omega^2 = \omega_o^2 + 2\alpha\theta \tag{17.14}$$

We can see that there is a complete analogy between the kinematic equations of linear motion and rotational motion about fixed axis. Equations (17.10), (17.12) and (17.14) can also be obtained directly by substituting θ for s, ω for v and α for a in the linear kinematic equations of motion.

17.4 ROTATIONAL MOTION WITH CONSTANT ANGULAR VELOCITY

If the angular velocity of the body remains *constant* then its angular acceleration is *zero*. Then the kinematic equations of rotational motion reduce to

$$\omega = \omega_o \tag{17.15}$$

and

$$\theta = \omega_o t \tag{17.16}$$

This type of motion is also known as **uniform angular motion**.

Example 17.5 Determine the angular speed of the second hand and the minute hand in a clock.

Solution We know that the second hand completes one revolution in 1 minute. Hence, its angular speed is given as

$$\omega = \frac{1 \text{ rev}}{1 \text{ min}} = 1 \text{ rpm} = \frac{2\pi}{60} \text{ rad/s (or) } \frac{\pi}{30} \text{ rad/s}$$

The minute hand completes one revolution in 1 hour. Hence, its angular speed is given as

$$\omega = \frac{1 \text{ rev}}{1 \text{ hour}} = \frac{1}{60} \text{ rpm} = \frac{2\pi}{3600} \text{ rad/s (or) } \frac{\pi}{1800} \text{ rad/s}$$

Example 17.6 A ceiling fan when switched on attains a maximum angular speed of 240 rpm in 10 seconds. Determine (i) the constant angular acceleration, and (ii) the number of revolutions made in 10 seconds. The regulator of the fan is then rotated so that its speed is reduced from 240 rpm to 180 rpm in 5 seconds. Determine the uniform retardation.

Solution Initial angular speed of the fan, $\omega_o = 0$

Final angular speed of the fan, $\omega = 240$ rpm $= 240 \times (2\pi/60) = 8\pi$ rad/s

(i) We know that kinematic equation for angular motion with constant angular acceleration is given as

$$\omega = \omega_o + \alpha t$$

$\therefore$ $$8\pi = 0 + \alpha[10]$$

$\Rightarrow$ $$\alpha = 0.8\pi \text{ rad/s}^2$$

(ii) The number of revolutions made in 10 seconds is given as

$$\theta = \omega_o t + (1/2)\alpha t^2$$

$$= 0 + \frac{1}{2}[0.8\pi][10]^2$$

$$= 40\pi \text{ rad} = 20 \text{ rev} \quad [\text{since } 1 \text{ rev} = 2\pi \text{ rad}]$$

(iii) When the regulator is turned, the final angular speed of the fan after 5 seconds is

$$\omega = 180 \text{ rpm} = 180 \times (2\pi/60) = 6\pi \text{ rad/s}$$

Therefore, its constant retardation is determined as follows:

$$\omega = \omega_o + \alpha t$$

$$6\pi = 8\pi + \alpha(5)$$

$\Rightarrow$ $$\alpha = -0.4\pi \text{ rad/s}^2$$

Example 17.7 An electric motor when switched on rotates through 600 radians in 5 seconds. Determine its angular velocity after 5 seconds, and its angular acceleration assuming it to be uniform.

Solution The angular displacement of the motor is given as $\theta = 600$ rad. Therefore, its angular acceleration is obtained by substituting the values in the kinematic equation

$$\theta = \omega_o t + (1/2)\alpha t^2$$

$$600 = 0 + (1/2)\,\alpha(5)^2$$

$\Rightarrow$ $$\alpha = 1200/25 = 48 \text{ rad/s}^2$$

The angular velocity after 5 seconds is obtained as

$$\omega = \omega_o + \alpha t$$

$$= 0 + (48)(5) = 240 \text{ rad/s}$$

Example 17.8 An electric motor when switched on increases its speed at a rate of π rad/s^2 to reach a speed of 200 rpm. Determine the time taken to come to this speed and the number of revolutions made in that time.

Solution Initial angular speed, $\omega_o = 0$

Final angular speed, $\omega = 200 \text{ rpm} = 200 \times (2\pi/60) = \dfrac{20\pi}{3}$ rad/s

The time taken to reach this speed is obtained by substituting the given values in the kinematic equation,

$$\omega = \omega_o + \alpha t$$

$$\frac{20\pi}{3} = 0 + (\pi)t$$

$\Rightarrow$ $$t = 20/3 = 6.67 \text{ s}$$

The angular displacement during this time is obtained as

$$\theta = \omega_o t + \frac{1}{2}\alpha t^2$$

$$= 0 + \frac{1}{2}(\pi)(20/3)^2 = 69.81 \text{ rad}$$

Therefore, the number of revolutions made is given as

$$\theta = 69.81/(2\pi) = 11.11 \text{ rev}$$

Example 17.9 A flywheel increases its speed from 200 rpm to 300 rpm in 5 seconds. Determine the (i) acceleration assuming it to be uniform, and (ii) angular displacement made during this time.

Solution The initial angular speed of the flywheel is 200 rpm (or) 6.67 π rad/s and its final angular speed is 300 rpm (or) 10 π rad/s. Therefore, substituting the values in the kinematic equation,

$$\omega = \omega_o + \alpha t$$

$$10\pi = 6.67\pi + \alpha(5)$$

$\Rightarrow$ $$\alpha = 2.09 \text{ rad/s}^2$$

The angular displacement during this time is obtained as

$$\theta = \omega_o t + (1/2)\alpha t^2$$

$$= (6.67\pi)(5) + (1/2)(2.09)(5)^2 = 130.9 \text{ rad}$$

Example 17.10 A flywheel that is rotating at 300 rpm attains a rotational rate of 180 rpm after 20 seconds. Determine the angular retardation of the flywheel assuming it to be uniform. Also, determine the time taken to come to rest from a speed of 300 rpm if the retardation rate remains the same and the number of revolutions made during this time.

Solution *Motion during the first 20 seconds*

Initial angular speed, $\omega_o = 300 \text{ rpm} = 300 \times (2\pi/60) = 10\pi$ rad/s

Final angular speed, $\omega = 180 \text{ rpm} = 180 \times (2\pi/60) = 6\pi$ rad/s

Therefore, substituting the values in the kinematic equation,

$$\omega = \omega_o + \alpha t$$

$$6\pi = 10\pi + \alpha(20)$$

$\Rightarrow$ $$\alpha = -0.2\,\pi \text{ rad/s}^2$$

Time taken to come to rest from a speed of 300 rpm

Initial angular speed, $\omega_o = 300$ rpm $= 10\pi$ rad/s and final angular speed, $\omega = 0$. Hence, substituting the values in the kinematic equation, assuming the same retardation as above,

$$\omega = \omega_o + \alpha t$$

$$0 = 10\pi + (-0.2\pi)(t)$$

$$\Rightarrow \quad t = 50 \text{ s}$$

Number of revolutions made before coming to a stop

The number of revolutions made before coming to a stop is determined from the equation

$$\theta = \omega_o t + (1/2)\alpha t^2$$

$$= (10\pi)(50) + (1/2)(-0.2\pi)(50)^2 = 785.4 \text{ rad (or) } 125 \text{ rev}$$

Example 17.11 A flywheel rotating at 300 rpm reduces its speed to 240 rpm while making 10 complete revolutions. Determine its angular retardation assuming it to be uniform. What is its speed after 3 seconds assuming the same retardation? Also, determine how much time is taken to come to a stop from a speed of 300 rpm.

Solution *Motion during the first 10 complete revolutions*

Initial angular speed, $\omega_o = 300$ rpm $= 300 \times (2\pi/60) = 10\pi$ rad/s

Final angular speed, $\omega = 240$ rpm $= 240 \times (2\pi/60) = 8\pi$ rad/s

Angular displacement, $\theta = 10$ rev $= 10 \times 2\pi = 20\pi$ rad

Substituting the values in the kinematic equation for rotational motion, we have

$$\omega^2 = \omega_o^2 + 2\alpha\theta$$

$$(8\pi)^2 = (10\pi)^2 + 2\alpha(20\pi)$$

$$\Rightarrow \quad \alpha = -0.9\pi \text{ rad/s}^2$$

Motion during the next 3 seconds

Initial angular velocity is equal to the final angular velocity in the first phase of motion, i.e., $\omega_o = 8\pi$ rad/s. Hence, the final angular velocity can be determined by taking the same retardation as

$$\omega = \omega_o + \alpha t$$

$$= 8\pi + (-0.9\pi)(3) = 5.3\pi \text{ rad/s}$$

Time taken to come to a stop from a speed of 300 rpm

$$\omega = \omega_o + \alpha t$$

$$0 = 10\pi + (-0.9\pi)t$$

$$\Rightarrow \quad t = 11.11 \text{ s}$$

Example 17.12 When a Pelton wheel turbine is rotating at a constant angular speed of 1200 rpm, the water jet is shut off. If the frictional resistance in the bearings reduces the speed at a rate of 2π rad/s^2, determine the time taken to come to a stop and the number of revolutions made in that time.

Solution Initial speed of turbine, $\omega_o = 1200 \text{ rpm} = 1200 \times (2\pi/60) = 40\pi \text{ rad/s}$

Final speed of turbine, $\omega = 0$

(i) *Time taken to come to a stop*

$$\omega = \omega_o + \alpha t$$

$$0 = 40\pi + (-2\pi)t$$

$$\Rightarrow \quad t = 20 \text{ s}$$

(ii) *Number of revolutions made before coming to a stop*

$$\theta = \omega_o t + \frac{1}{2}\alpha t^2$$

$$= 40\pi[20] + \frac{1}{2}[-2\pi][20]^2$$

$$= 800\pi - 400\pi = 400\pi \text{ rad (or) } 200 \text{ rev}$$

17.5 RELATIONSHIP BETWEEN ANGULAR AND LINEAR MOTIONS

In the previous sections, we saw that in fixed-axis rotation, the particles move in *circular* paths. In Chapter 13, we saw that a curvilinear motion (circular motion being a special case of curvilinear motion) can be represented by *tangential* and *normal* components. Hence, we can also represent the fixed-axis rotation using linear quantities such as linear displacement, velocity and acceleration. There exists a relationship between the linear variables and angular variables which we will establish below. This way the two variables can be used interchangeably.

In Fig. 17.8(b), we note that when the particle P moves from A to B, for the **angular displacement** θ, there is a corresponding **linear displacement** s along the curve. Since θ is expressed in radians, we can then relate them as

$$s = r\theta \tag{17.17}$$

where r is the radial distance of the particle from the axis of rotation. Upon differentiating both sides of the above expression with respect to time, we get

$$\frac{ds}{dt} = r\frac{d\theta}{dt} \tag{17.18}$$

Note that the radius of the particle is constant for circular motion. Since ds/dt is the linear velocity v of the particle and $d\theta/dt$ is the angular velocity ω of the particle, we can write the above expression as:

$$v = r\omega \tag{17.19}$$

We know that the linear velocity is a vector pointing along the tangent to the path of the particle. Differentiating the above equation further with respect to time and noting that velocity is a vector, it results in two components of acceleration, namely, tangential and normal components. As seen in Chapter 13, the tangential component is given as

$$a_t = \frac{dv}{dt} = r\frac{d\omega}{dt} = r\alpha \tag{17.20}$$

and the normal component is given as

$$a_n = \frac{v^2}{r} = \frac{(r\omega)^2}{r} = r\omega^2 \quad (17.21)$$

From the relationships established above, we can readily see that though the angular quantities are same for all the particles of the rigid body in fixed-axis rotation, the linear quantities are not the same. They are dependent upon the position of the particle r from the axis of rotation and hence, they vary.

Example 17.13 A disc of 2 m diameter is rotating at a constant angular speed of $\pi/8$ rad/s about a vertical axis passing through its centre. After 2 seconds, determine the angular position, and linear velocities of points A and B, assuming their initial positions were along the X-axis. Point A is on the rim of the disc and the point B is at half the distance from the centre.

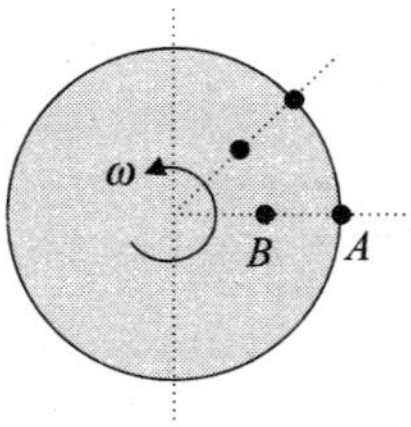

Fig. 17.10

Solution Since the disc is rotating at a constant angular speed, its angular displacement after 2 seconds is given as

$$\theta = \omega t = \left[\frac{\pi}{8}\right]2 = \frac{\pi}{4} \text{ rad}$$

Since 2π rad = 360°, θ = 45°. Since the point A is on the rim of the disc, its radial distance from the axis of rotation is 1 m and that of the point B is 0.5 m. Hence, their linear velocities are given as

$$v_A = r_A\omega = 1\frac{\pi}{8} = \frac{\pi}{8} \text{ m/s}$$

and

$$v_B = r_B\omega = \frac{1}{2}\frac{\pi}{8} = \frac{\pi}{16} \text{ m/s}$$

Both the velocities are directed tangential to the respective circular paths at 45° to the horizontal direction.

Example 17.14 In the above problem, if the disc is rotating at a uniform acceleration of π rad/s^2 starting from rest, determine (i) its angular velocity and displacement after 1 second, (ii) linear velocities of points A and B, and (iii) acceleration of points A and B.

Solution Since the disc is rotating at a uniform acceleration, we can apply the kinematic equations for angular motion to determine the angular velocity and displacement after 1 second as

$$\omega = \omega_o + \alpha t$$

$$= 0 + \pi(1) = \pi \text{ rad/s}$$

and

$$\theta = \omega_o t + (1/2)\alpha t^2$$

$$= 0 + (1/2)[\pi][1]^2 = \frac{\pi}{2} \text{ rad}$$

(ii) Therefore, linear velocities of points A and B are given as

$$v_A = (1)(\pi) = \pi \text{ m/s}$$

$$v_B = (1/2)\pi = \frac{\pi}{2} \text{ m/s}$$

(iii) The tangential accelerations of points A and B are given as

$$(a_A)_t = r_A\alpha = (1)\pi = \pi \text{ m/s}^2$$

$$(a_B)_t = r_B\alpha = (0.5)\pi = 0.5\pi \text{ m/s}^2$$

and the radial accelerations of points A and B are given as

$$(a_A)_r = r_A\omega^2 = (1)\pi^2 = \pi^2 \text{ m/s}^2$$

$$(a_B)_r = r_B\omega^2 = (0.5)\pi^2 = 0.5\pi^2 \text{ m/s}^2$$

Example 17.15 Two points A and B located 10 cm apart on a rotating disc have velocities respectively 10 m/s and 15 m/s. Determine (a) the angular velocity of the disc, and (b) the radial distances of points A and B.

Solution If ω be the angular velocity of the disc then the linear velocities of points A and B are given as

$$v_A = r_A\omega \tag{a}$$

and

$$v_B = r_B\omega \tag{b}$$

(a) *Angular velocity of the disc*

Since the velocity of the point B is greater than that of the point A, subtracting the equation (a) from the equation (b), we have

$$v_B - v_A = (r_B - r_A)\omega$$

Substituting the values, we have

$$15 - 10 = (0.1)\omega$$

$$\Rightarrow \quad \omega = 50 \text{ rad/s}$$

(b) *Radial distances of points A and B*

Substituting the value of ω in equations (a) and (b), we get

$$r_A = v_A/\omega$$

$$= 10/50 = 0.2 \text{ m (or) } 20 \text{ cm}$$

and

$$r_B = v_B/\omega$$

$$= 15/50 = 0.3 \text{ m (or) } 30 \text{ cm}$$

Example 17.16 Determine the angular velocity of the earth assuming it to be a perfect sphere revolving about the north and south poles. If the radius of the earth is 6370 km, determine the velocity and acceleration of a particle (i) at the equator, (ii) at 45° latitude, and (iii) at the poles.

Solution Since the earth makes one complete revolution in a day, its angular velocity is given as

$$\omega = \frac{1 \text{ rev}}{24 \text{ h}} = \frac{1 \text{ rev}}{24 \times 3600 \text{ s}} = \frac{1}{86400} \text{ rev/s}$$

It can also be expressed as

$$\omega = \frac{2\pi \text{ rad}}{24 \text{ h}} = \frac{2\pi \text{ rad}}{24 \times 3600 \text{ s}} = 7.27 \times 10^{-5} \text{ rad/s}$$

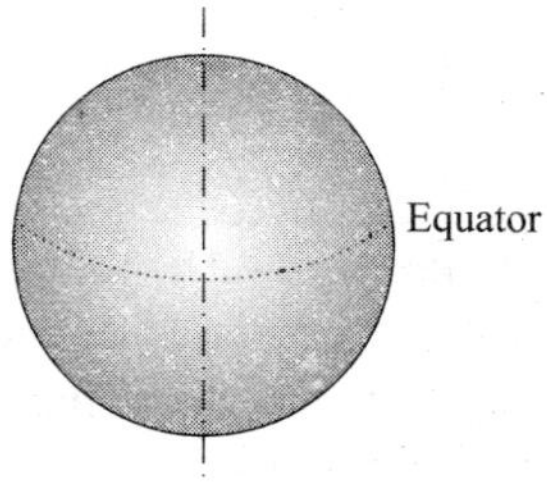

Fig. 17.11

The velocity and acceleration of a particle at the equator are given as

$$v_E = R\omega$$
$$= (6370 \times 10^3)(7.27 \times 10^{-5}) = 463.1 \text{ m/s}$$
$$a_E = R\omega^2$$
$$= (6370 \times 10^3)(7.27 \times 10^{-5})^2 = 0.0337 \text{ m/s}^2$$

The normal component of acceleration a_E is directed radially inwards. Since the earth is rotating at a constant angular velocity, its angular acceleration is zero. Hence, the tangential component of the linear acceleration is also zero.

At 45° latitude radial distance of the particle from the axis of the earth is $R_l = R \cos 45°$. Hence, the angular velocity and acceleration of the particle are

$$v_l = R_l\,\omega$$
$$= [(6370 \times 10^3)\cos 45°](7.27 \times 10^{-5}) = 327.5 \text{ m/s}$$
$$a_l = R_l\,\omega^2$$
$$= [(6370 \times 10^3)\cos 45°](7.27 \times 10^{-5})^2 = 0.0238 \text{ m/s}^2$$

At the poles, the radial distance from the axis of the earth is zero, and hence, both the angular velocity and angular acceleration are zero.

Example 17.17 Two hard rubber rollers are in contact with each other as shown. The diameter of the smaller roller is 2 cm and that of the bigger one is 4 cm. If the smaller roller is rotating at a constant angular speed of 30 rpm in the direction shown, determine the magnitude and direction of the speed of the bigger roller assuming there is no slip between the rollers.

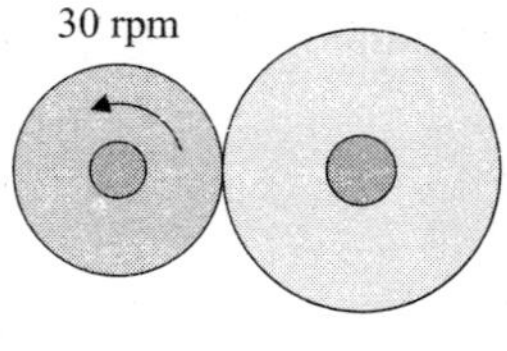

Fig. 17.12

Solution *Given data*

Radius of smaller roller, $r_1 = 1$ cm

Radius of bigger roller, $r_2 = 2$ cm

Angular speed of smaller roller, $\omega_1 = 30$ rpm (anticlockwise)

As there is no slip between the rollers, the tangential velocity at the contact point for both the rollers must be equal, i.e.,

$$v_1 = v_2$$

$$\Rightarrow \quad r_1\omega_1 = r_2\omega_2$$

Therefore, the angular speed of the bigger roller is given as

$$\omega_2 = \frac{r_1\omega_1}{r_2} = \frac{(1)(30)}{2} = 15 \text{ rpm (clockwise)}$$

Example 17.18 A seesaw *AB* is at rest as shown. If the end *B* is given a vertically downward velocity of 2 m/s, determine (i) the angular velocity of the seesaw at that instant, and (ii) the velocity of the midpoint of *OB*.

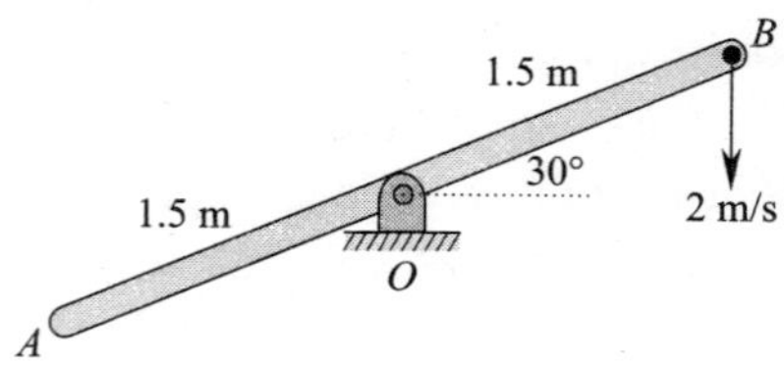

Fig. 17.13

Solution The vertical velocity of the end *B* can be resolved into components along the axis of the seesaw and along normal directions as 2 sin 30° and 2 cos 30° respectively. The angular velocity of seesaw at that instant is given as

$$\omega = \frac{v_{\text{normal}}}{r}$$

$$= \frac{2\cos 30°}{1.5} = 1.15 \text{ rad/s (clockwise)}$$

Therefore, the tangential velocity of the mid-point of *OB* is obtained as

$$v = r_{\text{midpoint}}\,\omega$$

$$= (1.5/2)(1.15) = 0.8625 \text{ m/s directed perpendicular to } OB$$

Example 17.19 In the stepped pulley arrangement shown in Fig. 17.14 if the block *B* moves down with a velocity of 1 m/s, determine the velocity of the block *A*.

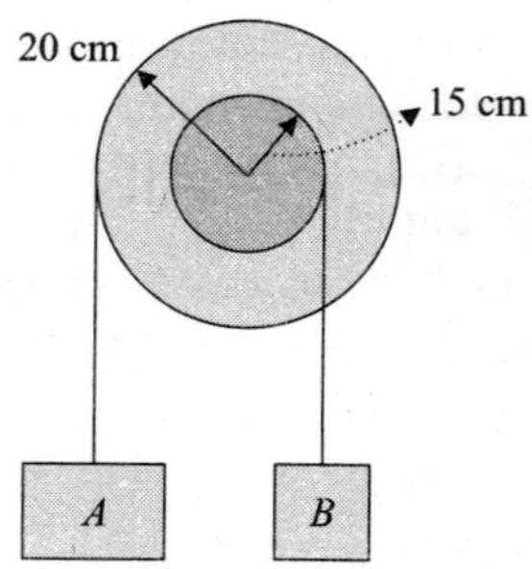

Fig. 17.14

Solution Since the string supporting the block *B* passes over the smaller-diameter pulley, the tangential velocity of a point on the rim of the smaller-diameter pulley is the same as the velocity of the block *B*, i.e., 1 m/s. Hence, the angular velocity of the smaller-diameter pulley is given as

$$\omega = \frac{v_B}{r_B} = \frac{1}{0.15} = 6.67 \text{ m/s (clockwise direction)}$$

We know that in fixed-axis rotation, the angular velocity is the same for every particle on the rigid body. Hence, the angular velocity of the larger-diameter pulley is also the same as that of the smaller-diameter pulley. The linear velocity of the block *A* is the same as the tangential velocity of a point on the rim of larger-diameter pulley and it is given as

$$v_A = r_A\omega$$

$$= (0.2)(6.67) = 1.334 \text{ m/s (directed upwards)}$$

17.6 GENERAL PLANE MOTION

When the motion of a rigid body is constrained along a plane then the motion is said to be a **general plane motion**. Such a motion is neither translational nor rotational motion. However, to analyze its motion, we can consider it to be a combination of **translational** and **rotational motions** occurring in *succession*.

For example, consider a wheel or a cylinder rolling *without slipping* on a horizontal plane. If we consider points A and B in the body (the imaginary line joining them initially a vertical straight line) then in a certain interval of time, they would have moved to positions A_1 and B_1. This motion can be considered to be made up of a translational motion, in which points A and B move in parallel paths to positions A_1 and B', followed by a fixed-axis rotation about the point A_1, in which the point B' moves to B_1. Though we have considered the motion to be first a translational motion followed by a fixed-axis rotation, it should be noted that the same motion would have resulted had we considered the fixed-axis rotation first, followed by translation. Thus, it is immaterial in which order we take the combination of motions. We can see that although the combination of translational and rotational motions does not represent the actual general motion taking place, the simultaneous occurrence of the two motions does represent the actual motion.

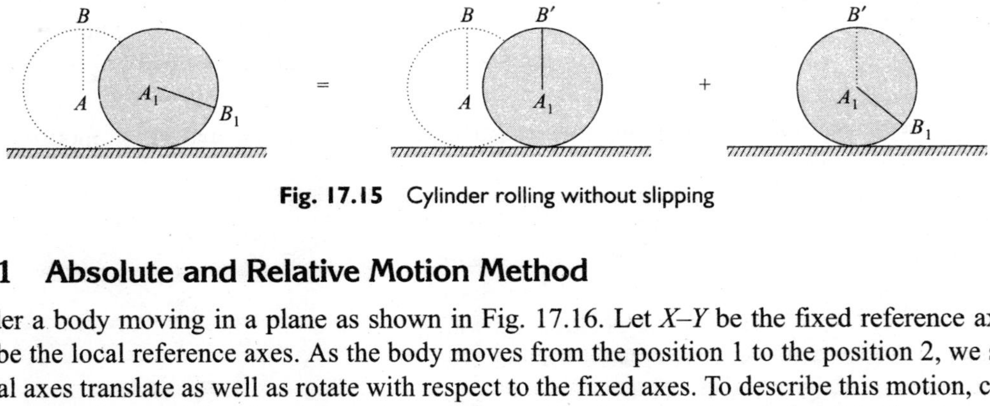

Fig. 17.15 Cylinder rolling without slipping

17.6.1 Absolute and Relative Motion Method

Consider a body moving in a plane as shown in Fig. 17.16. Let X–Y be the fixed reference axes and X_1–Y_1 be the local reference axes. As the body moves from the position 1 to the position 2, we see that the local axes translate as well as rotate with respect to the fixed axes. To describe this motion, consider two representative particles A and B as shown, where A is chosen as the origin of the local coordinate axes.

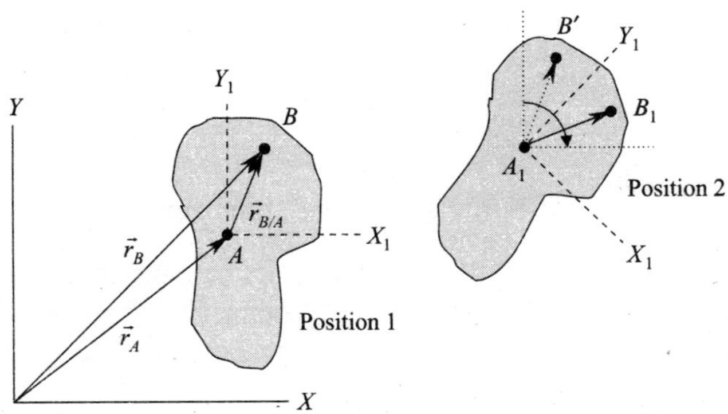

Fig. 17.16 Absolute and relative motions

The position vectors of A and B with respect to the fixed reference axes are respectively $\vec{r}_A$ and $\vec{r}_B$. The position vector of the point B with respect to the local reference axes with A as the origin is represented as $\vec{r}_{B/A}$. Then, we can see that the position vector of the point B can be written as

$$\vec{r}_B = \vec{r}_A + \vec{r}_{B/A} \tag{17.22}$$

It should be noted that as the body is rigid, the distance between any two particles remains **constant**. Thus, the magnitude of vector $\vec{r}_{B/A}$ always remains constant. As the particle moves to the position 2 in a certain interval of time t, we see that the particle A moves to the position A_1 and the particle B to the position B_1. Thus, during this time we see that there is a change in position of the particles with respect to the fixed reference frame. This change in position with respect to time as the time interval tends to zero is termed as the velocity of the particles. Mathematically, it is expressed by differentiating the Eq. (17.22) with respect to time, and hence we have

$$\vec{v}_B = \vec{v}_A + \vec{v}_{B/A} \tag{17.23}$$

The first vector on the right-hand side represents the velocity of the point A. It should be noted that as the body is rigid, the length AB remains constant, but the position vector of B with respect to A changes in *direction* and hence $\vec{v}_{B/A}$ is *non-zero*. This velocity refers to the linear velocity of the particle B due to the fixed axis rotation about A and its direction is tangential to the circular path traced by B as it rotates about A. Thus, we can see that the absolute velocity of the point B is equal to the vector sum of translational velocity of an arbitrary reference point A and tangential velocity of B due to fixed-axis rotation about A.

The following examples will explain how this method is useful in the analysis of motion of many pin-jointed mechanisms.

Example 17.20 The top view of a door in a luxury bus is shown in Fig. 17.17, where the point A is hinged and the point C is free to slide along the groove. If the velocity of the pin-joint B is 1 m/s for the position shown, determine (i) the angular velocities of panels AB and BC, and (ii) linear velocity of C, given that $AB = BC = 40$ cm.

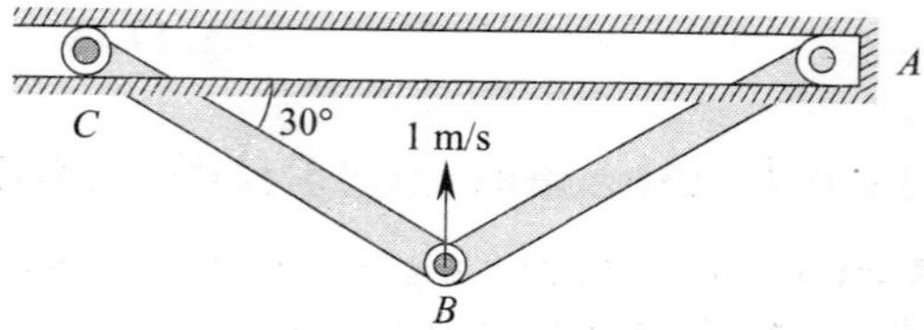

Fig. 17.17

Solution Since the velocity of pin joint B is in the direction shown, the component of velocity perpendicular to the panel AB is

$$1 \cos 30° \text{ (or) } 1 \sin 60°$$

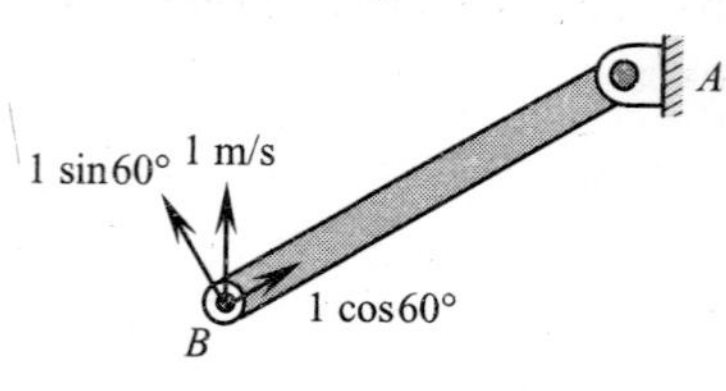

Fig. 17.17(a)

The angular velocity of AB is then obtained as

$$v_{\text{perpendicular}} = r_{AB}\,\omega_{AB}$$

$$\Rightarrow \qquad \omega_{AB} = \frac{1 \cos 30°}{0.4} = 2.17 \text{ rad/s (clockwise direction)}$$

The motion of the panel BC, represented as a combination of translational and rotational motions is shown in Fig. 17.17(b). Note that as the motion of the point C is constrained along the groove, its velocity is directed to the left. Thus, we can construct the velocity triangle for the motion of the end C and the included angles are as shown in Fig. 17.17(c).

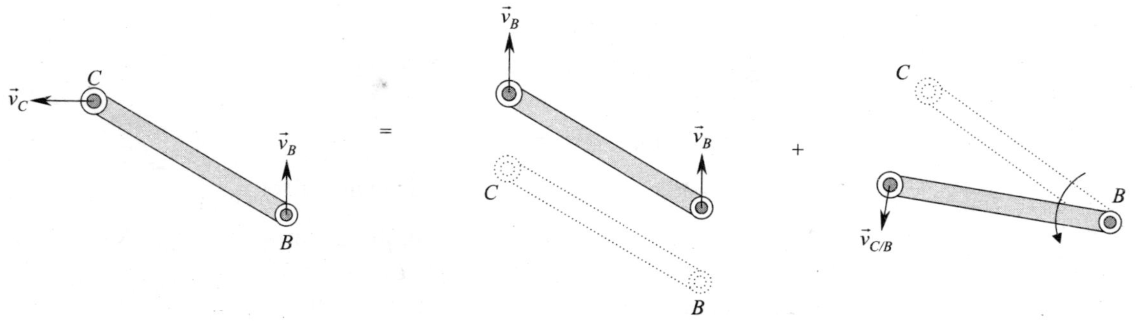

Plane motion of *BC* Translational motion of *BC* Fixed-axis rotation of *BC*

Fig. 17.17(b)

The absolute velocity of the point *C* can be written as

$$\vec{v}_C = \vec{v}_B + \vec{v}_{C/B}$$

From the velocity triangle, we have

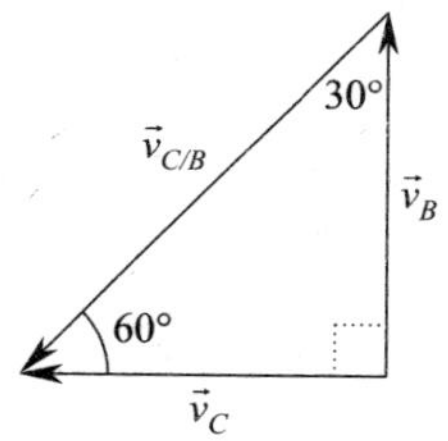

Fig. 17.17(c)

$$\frac{v_B}{\sin 60°} = \frac{v_C}{\sin 30°} = \frac{v_{C/B}}{\sin 90°}$$

$\Rightarrow$
$$v_{C/B} = \frac{\sin 90°}{\sin 60°} v_B = 1.155 \text{ m/s}$$

Therefore, angular velocity of the panel *BC* is given as

$$\omega_{CB} = \frac{v_{C/B}}{r_{CB}}$$

$$= \frac{1.155}{0.4} = 2.89 \text{ rad/s (anticlockwise direction)}$$

In addition, from the velocity triangle, we have

$$v_C = \frac{\sin 30°}{\sin 60°} v_B = 0.577 \text{ m/s}$$

Example 17.21 A reciprocating engine mechanism is shown in Fig. 17.18, in which the crank *OA* rotates at a constant angular velocity of 1200 rpm in the anticlockwise direction. For the position shown, determine (i) the angular velocity of the connecting rod *AB*, and (ii) the velocity of piston in the engine. Take *OA* = 10 cm, *AB* = 25 cm.

Solution The angle φ made by the connecting rod *AB* with the horizontal line is calculated as follows

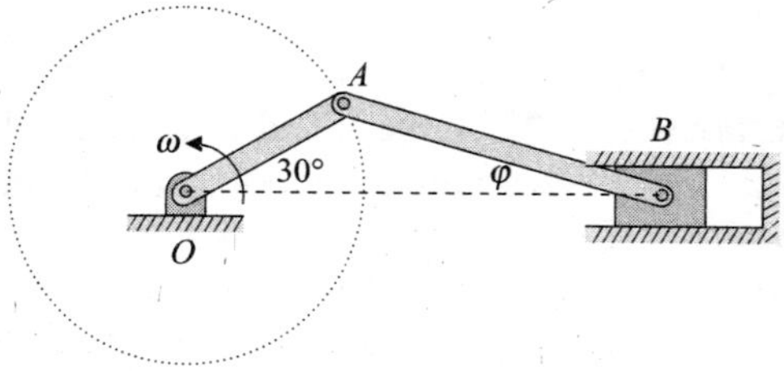

Fig. 17.18

$$OA \sin 30° = AB \sin \varphi$$

$$10 \sin 30° = 25 \sin \varphi$$

$\Rightarrow$
$$\varphi = 11.54°$$

Given that the angular velocity of the crank is

$$\omega = 1200 \times \frac{2\pi}{60} = 40\pi \text{ rad/s}$$

the linear velocity of the end A of the crank is given as

$$v_A = r_{OA}\omega = 0.1 \times 40\pi = 12.57 \text{ m/s (directed tangentially)}$$

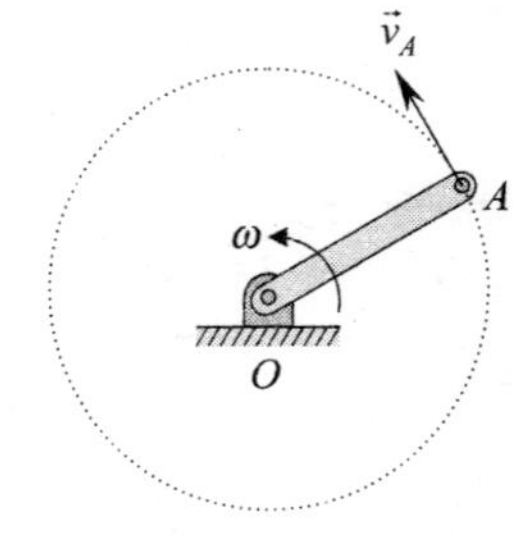

Fig. 17.18(a)

The motion of the connecting rod can be considered as a combination of translational and rotational motions. Figure17.18(b) shows the two motions of the connecting rod, where in the translational motion, the end B moves with the velocity of the end A and in the fixed-axis rotation, the end B moves with an angular velocity. However, we know that as the piston moves horizontally, the total velocity of the end B should be horizontal to the left. Thus, we can construct the velocity triangle for the motion of the end B and the included angles are as shown below.

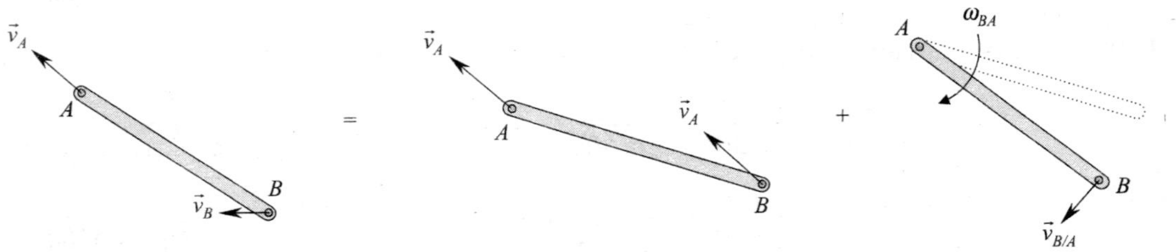

Fig. 17.18(b)

From the velocity triangle, we see that

$$\frac{v_A}{\sin 78.46°} = \frac{v_B}{\sin 41.54°} = \frac{v_{B/A}}{\sin 60°}$$

$$\Rightarrow \qquad v_{B/A} = \frac{\sin 60°}{\sin 78.46°} v_A = 11.11 \text{ m/s}$$

and

$$v_B = \frac{\sin 41.54°}{\sin 78.46°} v_A = 8.51 \text{ m/s}$$

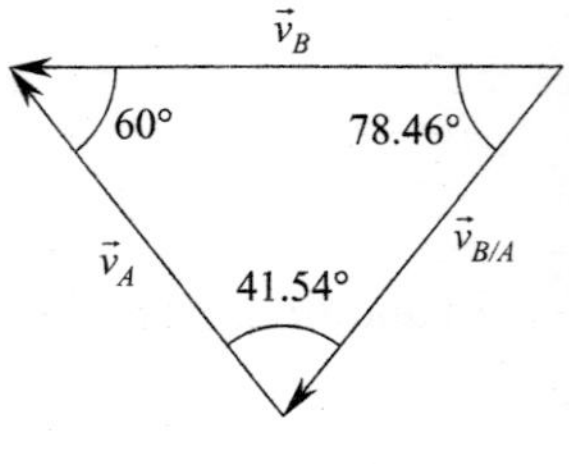

Fig. 17.18(c)

Therefore, angular velocity of the connecting rod AB is given as

$$\omega_{BA} = v_{B/A}/r_{AB}$$

$$= 11.11/0.25 = 44.44 \text{ rad/s (clockwise direction)}$$

Example 17.22 One of the wheels of an automobile is shown in Fig. 17.19. The automobile moves at a constant speed of 60 kmph. If the wheel of 60 cm diameter rolls without slipping, determine the velocities of points A, B, C, D and E on the rim of the wheel. [Note that points A and C are respectively the lowest and highest points, B is the extreme right point, and D and E are at 45° to the horizontal diameter].

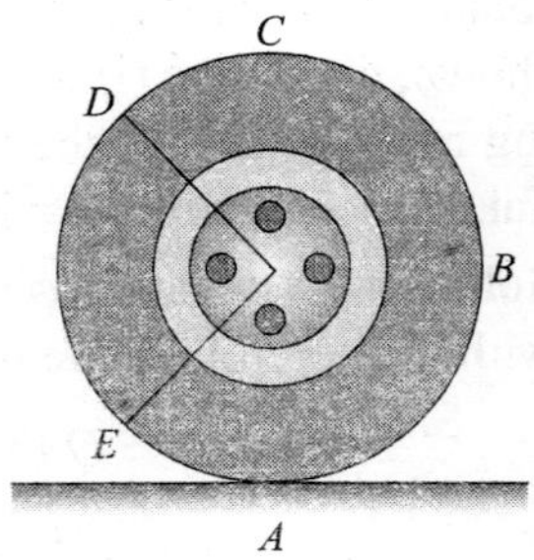

Fig. 17.19

Solution The velocity of the automobile is $v = 60$ kmph $= 60 \times (5/18) = 16.67$ m/s. Since the centre O of the wheel does not have any rotational motion as it lies on the axle, its velocity is same as the linear velocity of the automobile, i.e., $v_O = 16.67$ m/s.

The absolute velocity of the bottommost point A on the wheel is then given as

$$\vec{v}_A = \vec{v}_O + \vec{v}_{AO}$$

where $\vec{v}_{AO}$ is the tangentail velocity due to rotation about the centre O. Since the wheel is rolling without slipping, the point A is momentarily at rest or its velocity is zero. Therefore, we can write the above equation as

$$0 = \vec{v}_O + \vec{v}_{AO} \Rightarrow \vec{v}_{AO} = -\vec{v}_O$$

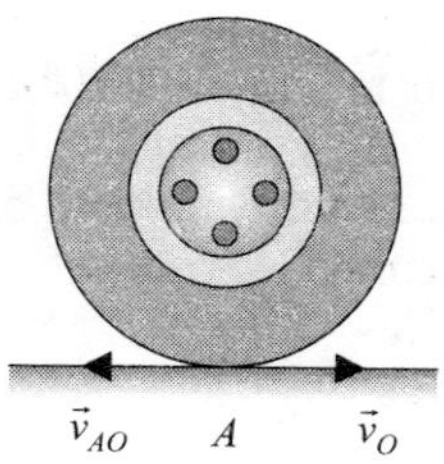

Fig. 17.19(a)

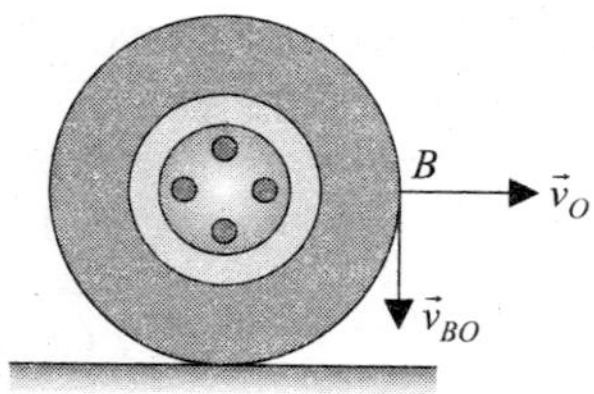

Fig. 17.19(b)

The negative sign indicates that the tangential velocity of the wheel at A is in the direction opposite to the direction of motion of the wheel. Therefore, the angular velocity of the wheel is given as

$$\therefore \quad \omega = \frac{v_{AO}}{r} = \frac{v_O}{r} = \frac{16.67}{0.3} = 55.57 \text{ rad/s (clockwise direction)}$$

Since the automobile is moving at a constant speed, the angular velocity is also constant. Further, the tangential velocity of any particle on the rim of the wheel remains the same and equal to $v_{AO} = v_O$.

The velocity of the point B can be written as

$$\vec{v}_B = \vec{v}_O + \vec{v}_{BO}$$

Since the two velocities $\vec{v}_O$ and $\vec{v}_{BO}$ are directed at right angles to each other [refer Fig. 17.19(b)], the magnitude of velocity of the point B is given as

$$v_B = \sqrt{(v_O)^2 + (v_{BO})^2}$$

$$= \sqrt{(v_O)^2 + (v_O)^2} = 23.6 \text{ m/s}$$

Similarly, velocity of the point C can be written as

$$\vec{v}_C = \vec{v}_O + \vec{v}_{CO}$$

Since the two velocities $\vec{v}_O$ and $\vec{v}_{CO}$ are pointing in the same direction [refer Fig. 17.19(c)], the magnitude of velocity of the point C is given as the algebraic sum of the two velocities. Hence,

$$v_C = v_O + v_{CO}$$

$$= 2v_O = 33.34 \text{ m/s}$$

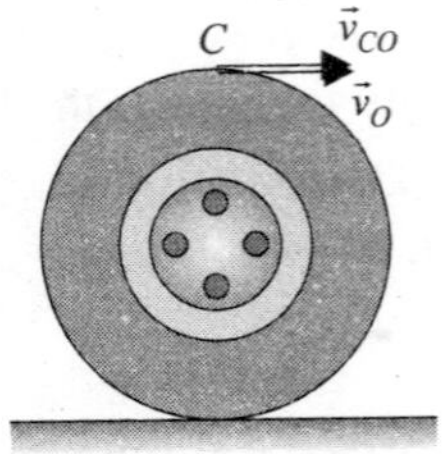

Fig. 17.19(c)

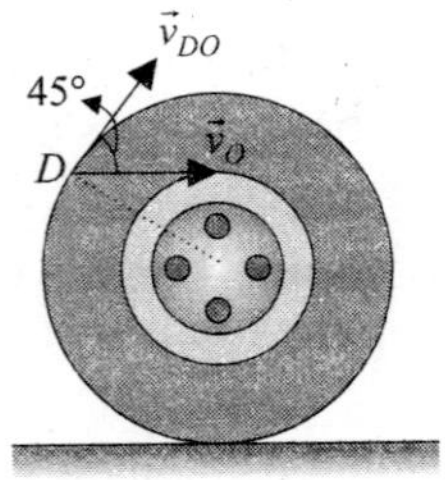

Fig. 17.19(d)

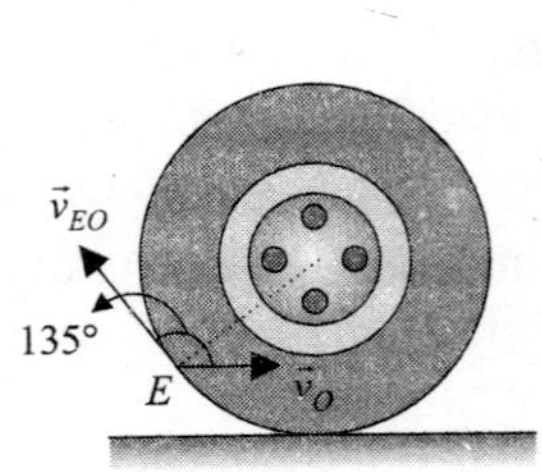

Fig. 17.19(e)

The velocity of the point D can be written as

$$\vec{v}_D = \vec{v}_O + \vec{v}_{DO}$$

Since the angle between the two velocities is 45° as shown in Fig. 17.19(d), the magnitude of velocity of the point D is given as

$$v_D = \sqrt{(v_O)^2 + (v_{DO})^2 + 2(v_O)(v_{DO})\cos 45°}$$

$$= \sqrt{2\,(v_O)^2(1+\cos 45°)} = 30.8 \text{ m/s}$$

The velocity of the point E can be written as:

$$\vec{v}_E = \vec{v}_O + \vec{v}_{EO}$$

Since the angle between the two velocities is 135° as shown in Fig. 17.19(e), the magnitude of velocity of the point E is given as

$$v_E = \sqrt{(v_O)^2 + (v_{EO})^2 + 2(v_O)(v_{EO})\cos 135°}$$

$$= \sqrt{2(v_O)^2(1+\cos 135°)} = 12.8 \text{ m/s}$$

Example 17.23 A car is moving at 54 kmph. The wheels are of 50 cm in diameter. Determine the angular velocity of the wheels about the axle. The car is brought to a stop by the application of brakes. If it covers 60 m during this time, determine the number of revolutions made by the wheel and its angular deceleration assuming it to be constant. Assume there is no slip between the wheels and the road.

Solution The speed of the car is $v = 54$ kmph $= 15$ m/s. Since the centre point O on the axle of the wheel does not rotate, its absolute velocity is the same as that of the linear velocity, i.e., $v_O = 15$ m/s. Considering a point A on the periphery of the wheel in contact with the road, its absolute velocity is given as

$$\vec{v}_A = \vec{v}_O + \vec{v}_{AO}$$

Since the wheel is rolling without slipping, the point A is momentarily at rest. Hence, its velocity is zero. Therefore, we can write the above equation as

$$0 = \vec{v}_O + \vec{v}_{AO} \quad \Rightarrow \quad \vec{v}_{AO} = -\vec{v}_O$$

Since $v_O = v_{AO} = r\omega$, we get the angular velocity of the wheel as

$$\omega = \frac{v_O}{r} = \frac{15}{0.25} = 60 \text{ rad/s}$$

Using the kinematic equation for rectilinear motion of the car as a whole,

$$v^2 = v_o^2 + 2as$$

$$0 = (15)^2 + 2a(60)$$

$$\Rightarrow \quad a = -1.875 \text{ m/s}^2$$

Therefore, the angular retardation is given as

$$\alpha = a/r = -1.875/(0.25) = -7.5 \text{ rad/s}^2$$

Since there is no slip between the wheels and the road, the angular displacement of the car can be obtained as

$$s = r\theta$$

$$\Rightarrow \quad \theta = \frac{s}{r} = \frac{60}{0.25} = 240 \text{ rad (or) } 38.2 \text{ rev}$$

Example 17.24 A spool rolls without slipping as shown in Fig. 17.20. Determine the velocity with which it moves if the string is pulled at a constant velocity of 1 m/s.

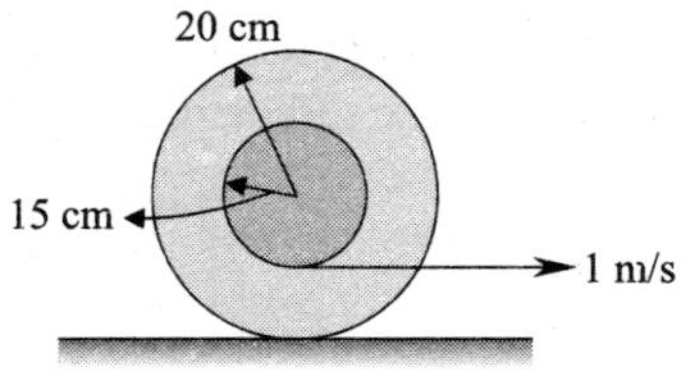

Fig. 17.20

Solution Since the string is pulled at a constant linear velocity of 1 m/s, the angular velocity of the spool is given as

$$\omega = \frac{v}{r_i}$$

where r_i is the radius of the smaller cylinder. The contact point of the spool with the plane is momentarily at rest. Its velocity is zero and hence we can write

$$v_O = v_{AO} = r_o\omega$$

$$= \frac{r_o}{r_i} v = \frac{20}{15}(1) = 1.33 \text{ m/s (towards left)}$$

Example 17.25 Arm OC of 5 cm length attached to a cylinder of 2 cm diameter rotates within a fixed outer rim without slipping at a constant angular velocity of $\pi/4$ rad/s. Determine (i) the angular velocity of the inner cylinder, and (ii) velocities of points B, C and D on the cylinder.

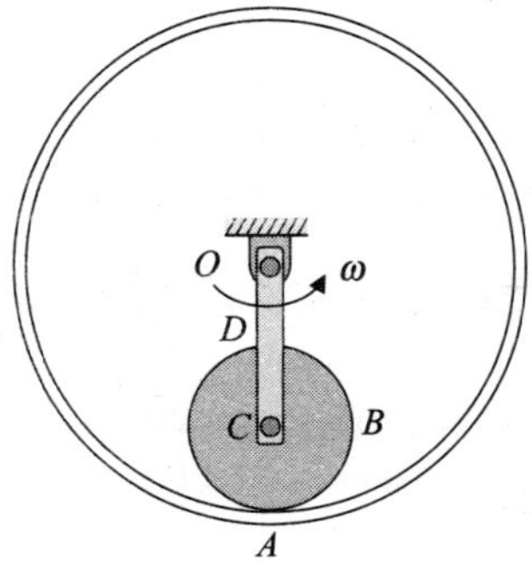

Fig. 17.21

Solution The linear velocity of the centre C of the inner cylinder is given as

$$v_C = r_{OC}\omega = 0.05\left(\frac{\pi}{4}\right) = 0.039 \text{ m/s}$$

The absolute velocity of the contact point A is given as

$$\vec{v}_A = \vec{v}_C + \vec{v}_{A/C}$$

Since the cylinder is rolling without slipping, we know that the instantaneous velocity of the point A is zero. Hence,

$$0 = \vec{v}_C + \vec{v}_{A/C}$$

$$\Rightarrow \quad \vec{v}_{A/C} = -\vec{v}_C$$

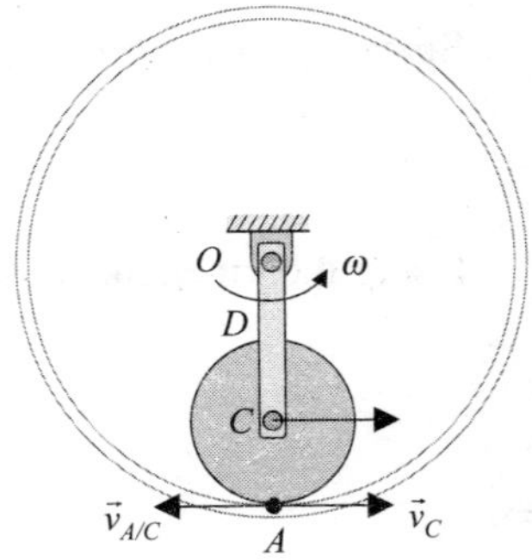

Fig. 17.21(a)

The negative sign indicates that it points in the direction opposite to that of the linear velocity of the point C as shown in Fig. 17.21(a). Therefore,

$$v_{A/C} = v_C = r_{AC}\,\omega_{AC}$$

$$= (0.01)\,\omega_{AC}$$

$$\Rightarrow \quad \omega_{AC} = \frac{v_C}{0.01} = 3.9 \text{ rad/s}$$

The velocity of the point B can be written as:

$$\vec{v}_B = \vec{v}_C + \vec{v}_{B/C}$$

We know that the velocity $\vec{v}_C$ points towards the right, while the velocity $\vec{v}_{B/C}$ points downwards. Since they are directed at right angles, we can write

$$v_B = \sqrt{(v_C)^2 + (v_{B/C})^2}$$

$$= \sqrt{(v_C)^2 + (r_{BC}\,\omega_{AC})^2}$$

$$= \sqrt{(0.039)^2 + (0.01 \times 3.9)^2} = 0.055 \text{ m/s.}$$

The velocity of the point D is given as

$$\vec{v}_D = \vec{v}_C + \vec{v}_{D/C}$$

We can see that both the velocities on the right-hand side are directed towards right and hence the velocity of the point D is given as the algebraic sum of the two velocities. Thus,

$$v_D = v_C + v_{D/C}$$

$$= v_C + r_{DC}\,(\omega_{AC})$$

$$= 0.039 + 0.01\,(3.9) = 0.078 \text{ m/s}$$

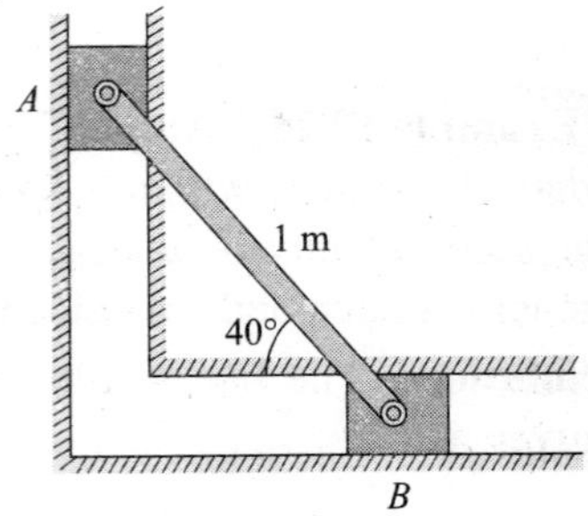

Fig. 17.22

Example 17.26 A bar AB of 1 m length attached to blocks at its ends is constrained to move as shown. If the block attached to the end B moves with a velocity of 2 m/s to the right, determine the velocity of the block attached at the end A and the angular velocity of the bar for the position shown.

Solution Since the bar is in general plane motion, we can consider the motion to be a combination of translational and rotational motions as shown below, in which the bar undergoes initially translational motion with velocity v_B followed by rotation about the end B.

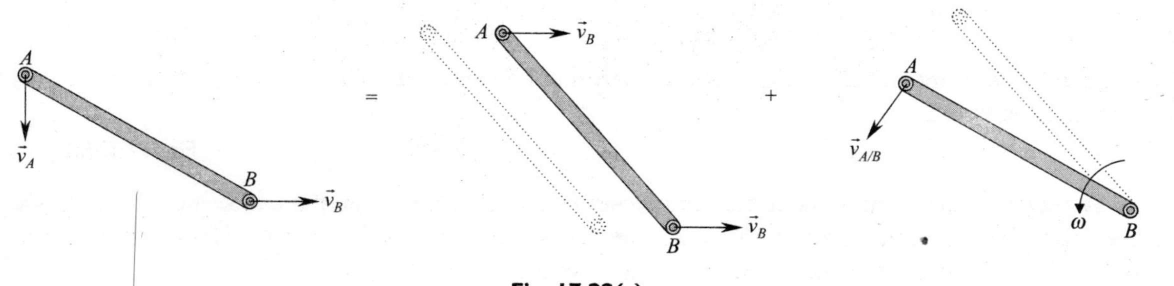

Fig. 17.22(a)

The absolute velocity of the point A is then given as

$$\vec{v}_A = \vec{v}_B + \vec{v}_{A/B}$$

We can construct the velocity triangle with the included angles as shown in Fig. 17.22(b) and applying the sine law to the triangle, we have

$$\frac{v_B}{\sin 40^\circ} = \frac{v_{A/B}}{\sin 90^\circ} = \frac{v_A}{\sin 50^\circ}$$

$$\Rightarrow \qquad v_{A/B} = \frac{\sin 90^\circ}{\sin 40^\circ} v_B = 3.11 \text{ m/s}$$

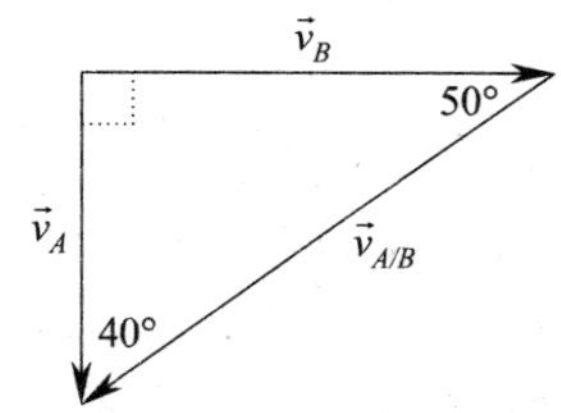

Fig. 17.22(b)

Therefore, angular velocity of the bar AB is given as

$$\omega_{AB} = \frac{v_{A/B}}{AB}$$

$$= \frac{3.11}{1} = 3.11 \text{ rad/s (anticlockwise direction)}$$

In addition, from the velocity triangle, we have

$$v_A = \frac{\sin 50^\circ}{\sin 40^\circ} v_B = 2.38 \text{ m/s (directed downwards)}$$

Example 17.27 The shaper quick return mechanism is shown in Fig. 17.23. The crank O_2A rotates in the anticlockwise direction with an angular velocity of 200 rpm. Determine the angular velocity of the connecting rod and the velocity of the end B of the connecting rod.

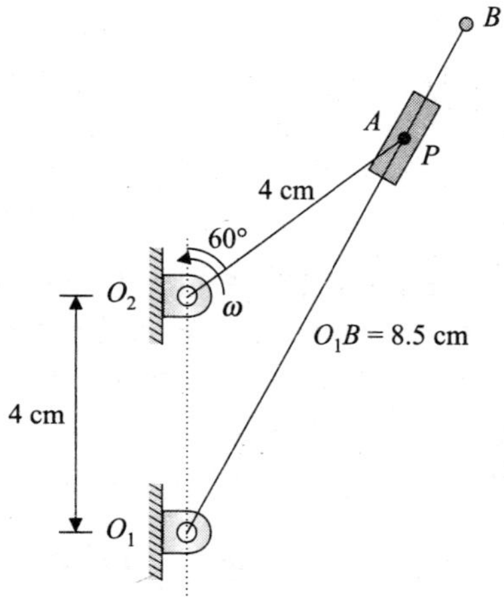

Fig. 17.23

Solution Given that the angular velocity of the crank is

$$\omega = 200 \text{ rpm} = 200 \times \frac{2\pi}{60} = 20.94 \text{ rad/s}$$

the linear velocity of the end A of the crank is given as

$$v_A = r_{O_2A}\,\omega$$

$$= (0.04)(20.94)$$

$$= 0.8376 \text{ m/s}$$

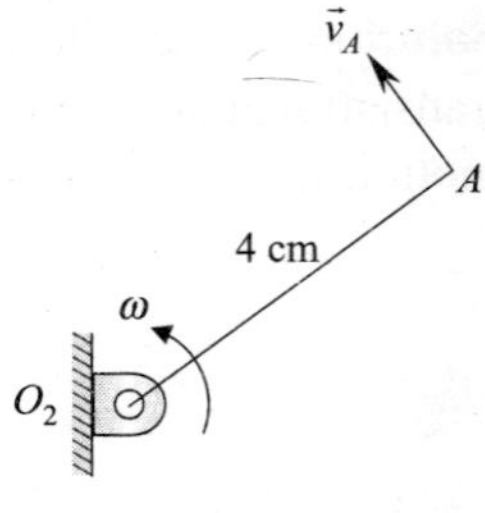

Fig. 17.23(a)

Let P be the point on the slider just below the point A on the crank. Then its absolute velocity is given as:

$$\vec{v}_p + \vec{v}_{PO_1}$$

As points A and P are lying at the same position but respectively on the crank and the slider, we know that their velocities should be equal. Hence, we construct the velocity triangle as shown in Fig. 17.23(c). Applying the sine law to the velocity triangle, we have

$$\frac{v_A}{\sin 90°} = \frac{v_P}{\sin 60°} = \frac{v_{PO_1}}{\sin 30°}$$

$$\Rightarrow \qquad v_P = v_A \frac{\sin 60°}{\sin 90°}$$

$$= 0.725 \text{ m/s}$$

and $$v_{PO_1} = v_A \frac{\sin 30°}{\sin 90°} = 0.4188 \text{ m/s}$$

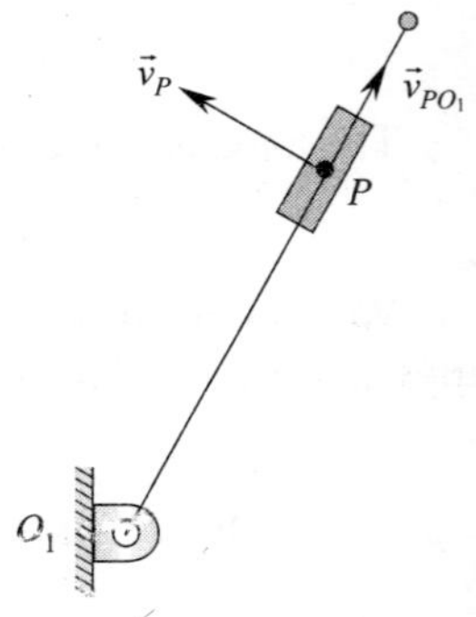

Fig. 17.23(b)

The length O_1P is obtained from the geometry of the figure as

$$O_1P = [4 \cos 60° + 4]/\cos 30° = 6.93 \text{ cm}$$

Therefore, angular velocity of the link O_1P or O_1B is given as

$$\omega_{PO_1} = \frac{v_P}{O_1P}$$

$$= \frac{0.725}{0.0693} = 10.46 \text{ rad/s (or) } 99.9 \text{ rpm}$$

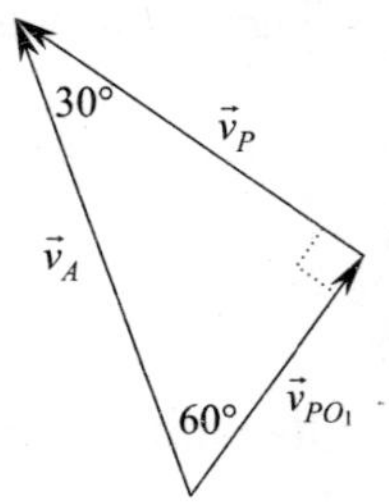

Fig. 17.23(c)

Therefore, velocity of the end B is obtained as

$$v_B = r_{O_1B}\,\omega_{PO_1}$$

$$= (0.085)(10.46) = 0.889 \text{ m/s}$$

Example 17.28 Figure 17.24 shows a four-bar mechanism. If the crank O_1A rotates with an angular velocity of 150 rpm in the clockwise direction, determine the angular velocities of links AB and O_2B for the postion shown.

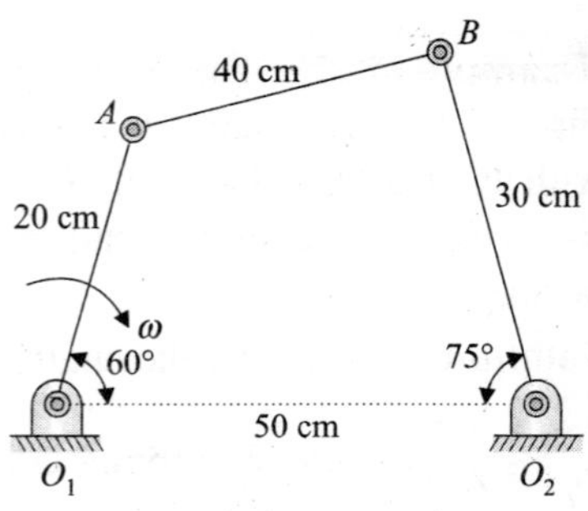

Fig. 17.24

Solution Given that the angular velocity of the crank O_1A is

$$\omega_{O_1A} = 150 \text{ rpm} = 150 \times \frac{2\pi}{60} = 15.71 \text{ rad/s}$$

the velocity of the end A of the crank O_1A is given as

$$v_A = r_{O_1A}\,\omega_{O_1A}$$
$$= (0.2)(15.71) = 3.142 \text{ m/s}$$

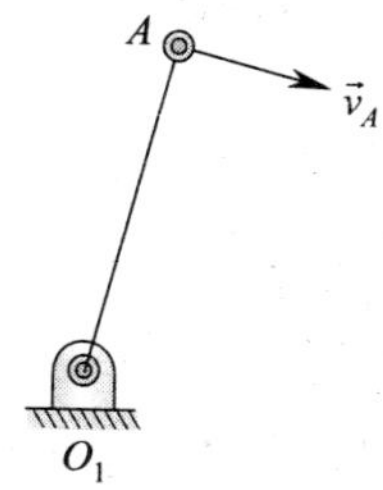

Fig. 17.24(a)

Since link AB undergoes general plane motion, the absolute velocity of the end B is given as:

$$\vec{v}_B = \vec{v}_A + \vec{v}_{B/A}$$

The motion of the link AB is shown in Fig. 17.24(b). If in case the end B is not linked to any other member then we see that velocity of the end B can be in any direction and hence determination of its magnitude is quite difficult. However, as it is connected to the link O_2B, we can readily understand that the velocity of B must be constrained to lie in a direction perpendicular to the link O_2B as shown in Fig. 17.24(b).

From the geometry of the mechanism, we can determine the inclination of the link AB with respect to the horizontal as shown in Fig. 17.24(c).

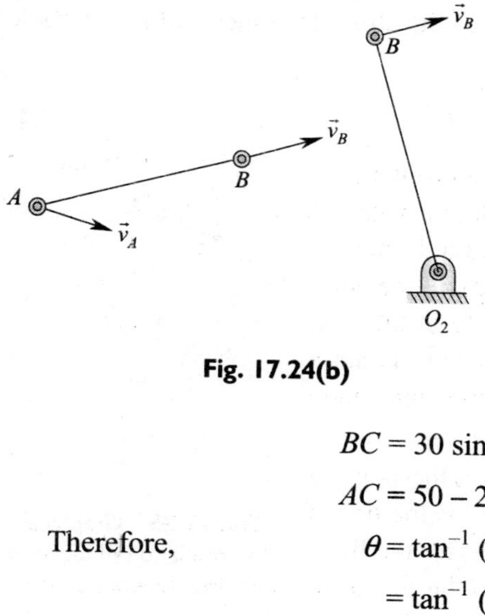

Fig. 17.24(b)

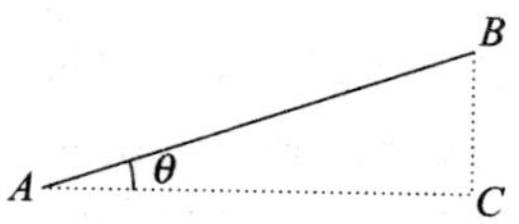

Fig. 17.24(c)

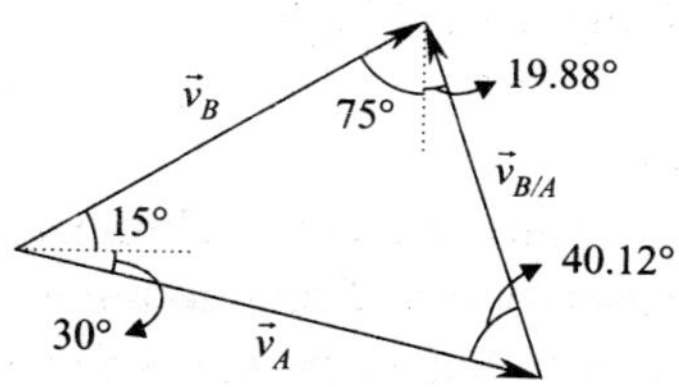

Fig. 17.24(d)

$$BC = 30 \sin 75° - 20 \sin 60° = 11.66 \text{ m}$$

$$AC = 50 - 20 \cos 60° - 30 \cos 75° = 32.24 \text{ m}$$

Therefore,
$$\theta = \tan^{-1}(BC/AC)$$
$$= \tan^{-1}(11.66/32.24) = 19.88°$$

Hence, we can construct the velocity triangle for the link AB as shown in Fig. 17.24(d). Applying sine rule to the triangle, we get

$$\frac{v_A}{\sin 94.88°} = \frac{v_B}{\sin 40.12°} = \frac{v_{B/A}}{\sin 45°}$$

Therefore,
$$v_B = \frac{\sin 40.12°}{\sin 94.88°} v_A = 2.03 \text{ m/s}$$

and
$$v_{B/A} = \frac{\sin 45°}{\sin 94.88°} v_A = 2.23 \text{ m/s}$$

Hence, angular velocities of links AB and O_2B are given respectively as

$$\omega_{AB} = \frac{v_{B/A}}{AB} = \frac{2.23}{0.4} = 5.58 \text{ rad/s}$$

and
$$\omega_{O_2B} = \frac{\vec{v}_B}{O_2B} = \frac{2.03}{0.3} = 6.77 \text{ rad/s}$$

17.6.2 Instantaneous Centre of Rotation

In the previous section, we saw how to analyze plane motion of rigid bodies by absolute and relative motion method. In this section, we will introduce another simpler method, called **instantaneous centre of rotation** method to solve the same type of problems. In the previous section, we saw that the general plane motion could be considered as a combination of translational motion and rotational motion. Thus, the absolute velocity of any particle on the rigid body is equal to the vector sum of translational velocity of an arbitrary reference point A and tangential velocity of the particle due to fixed axis rotation about A.

It is possible at any instant of time to locate a point C about which all the particles in the body rotate. Since all the particles rotate about this point at that instant, it is momentarily at *rest* or in other words, its translational velocity at that instant is *zero*. Hence, we can express the absolute velocity of any particle B on the rigid body as

$$\vec{v}_B = \vec{v}_C + \vec{v}_{B/C} = \vec{v}_{B/C} \quad [\text{since } \vec{v}_C = \vec{O}] \tag{17.24}$$

Thus, we could eliminate the translational motion and the problem reduces to that of pure rotational motion about a fixed axis passing through the point C. Such a point is known as **instantaneous centre** of rotation. It may lie within the rigid body or outside the rigid body. It should be noted that the point C is momentarily at rest and that it may have a different velocity in the next infinitesimally small interval of time. Thus, the instantaneous centre is not a general point but keeps on changing for various positions of the rigid body during its plane motion.

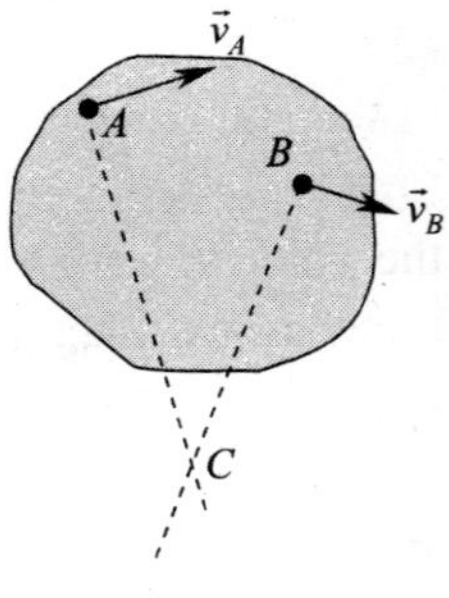

Fig. 17.25 Graphical determination of location of instantaneous centre

Suppose the velocity of the reference point A is zero. Then the point A happens to be the instantaneous centre. Suppose the directions of velocities of any two particles in the rigid body are known. Then we know that for the fixed axis rotation about C, these two velocities must act tangential to the line joining the particle with the axis of rotation. Hence, we draw lines perpendicular to the direction of velocities. The point of intersection of these two lines drawn from the two particles then defines the location of the instantaneous centre.

Example 17.29 A ladder AB leaning against a wall and resting on a floor, slides down as shown in Fig. 17.26. If the velocity of the end A is 2 m/s towards the right, for the position shown, determine (i) the angular velocity of the ladder AB, and (ii) the velocity of the end B.

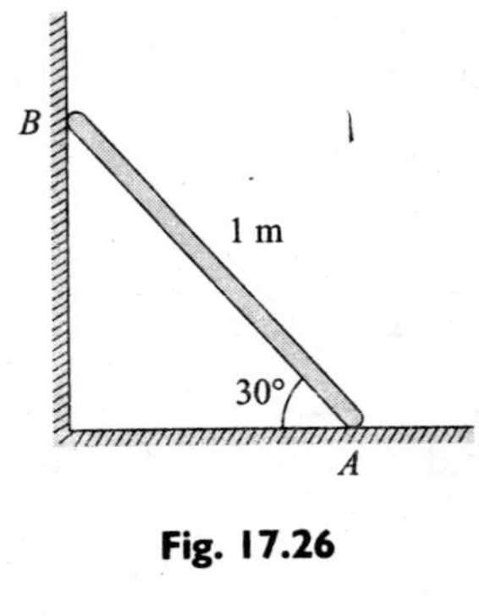

Fig. 17.26

Solution The ladder is constrained to move such that the velocity of the end A is horizontal and that of the end B is vertical. Thus, the instantaneous centre of rotation must lie on the intersection of lines AC and BC, which are drawn, respectively perpendicular to the direction of velocities of ends A and B.

Thus, we can write

$$\omega = \frac{v_A}{AC} = \frac{v_B}{BC}$$

$$= \frac{v_A}{1 \sin 30°} = \frac{v_B}{1 \cos 30°}$$

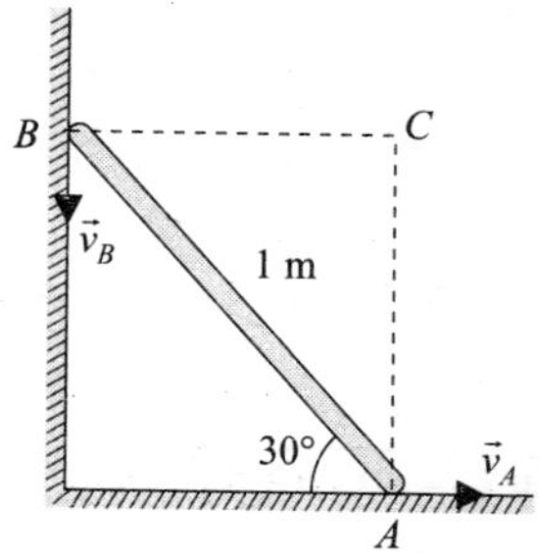

Fig. 17.26(a)

Therefore, angular velocity of the bar at the instant shown is

$$\omega = \frac{2}{1 \sin 30°} = 4 \text{ rad/s}$$

The velocity of the end B is obtained as

$$v_B = v_A \frac{1 \cos 30°}{1 \sin 30°}$$

$$= \frac{v_A}{\tan 30°} = 3.46 \text{ m/s}$$

Example 17.30 A prismatic bar AB is constrained to move as shown in Fig. 17.27. If the end A moves with a velocity of 2 m/s, determine (i) the angular velocity of the bar at the position shown, and (ii) velocity of the end B.

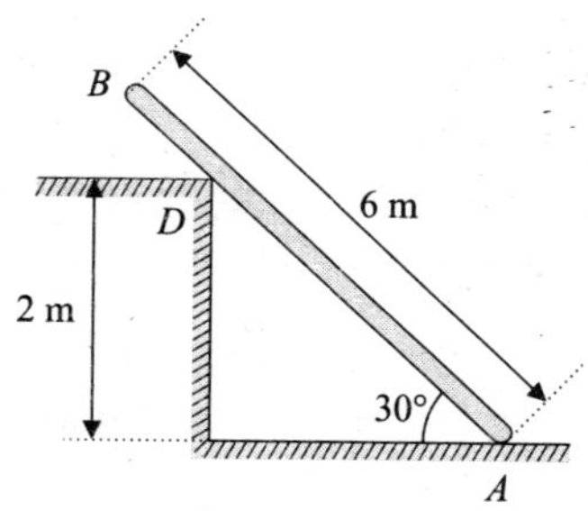

Fig. 17.27

Solution Since the bar is constrained to move as shown, we know that the velocity of the end A is tangential to the horizontal plane and directed towards right and the velocity of the contact point D is tangential to the bar at D and directed downwards. Hence, we can determine the instantaneous centre by drawing perpendiculars to the two velocities and finding the point of intersection C as shown in Fig. 17.27(a).

From the figure, we see that

$$\overline{AD} = 2/\sin 30° = 4 \text{ m}$$

Therefore,

$$\overline{BD} = \overline{AB} - \overline{AD}$$

$$= 6 - 4 = 2 \text{ m}$$

and

$$\overline{AC} = \overline{AD}/\cos 60° = 8 \text{ m}$$

Hence, the angular velocity of the bar is given as

$$\omega = \frac{v_A}{AC} = \frac{2}{8} = 0.25 \text{ rad/s}$$

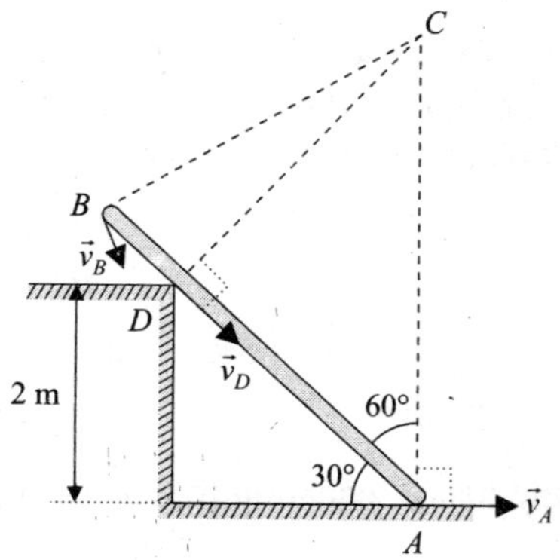

Fig. 17.27(a)

We can also see from the figure that

$$\overline{DC} = \overline{AD}\ \tan 60° = 6.93 \text{ m}$$

Therefore,

$$\overline{BC} = \sqrt{(BD)^2 + (DC)^2}$$

$$= \sqrt{(2)^2 + (6.93)^2} = 7.21 \text{ m}$$

Hence, we can determine the velocity of the end B as

$$v_B = r_{BC}\,\omega = (7.21)(0.25) = 1.8 \text{ m/s}$$

Example 17.31 Solve Example 17.21 by the method of instantaneous centre.

Solution When the crank OA rotates in the anticlockwise direction with the angular velocity ω then the linear velocity of the end A is tangential to the circular path at A. Hence, we can see that the instantaneous centre of rotation must lie perpendicular to the tangential direction, i.e., along the extended line OA. Similarly, as the piston is constrained to move in the cylinder, the end B of the connecting rod moves in the horizontal direction and thus the instantaneous centre must lie perpendicular to this direction. Thus, we can arrive at the intersection point C, which is the instantaneous centre of rotation for the position given.

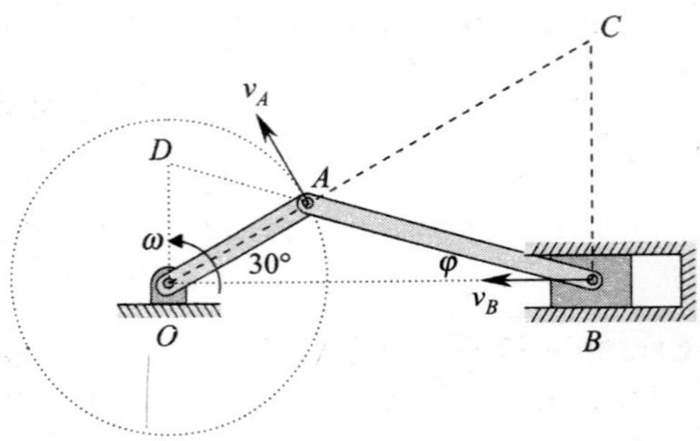

Fig. 17.28

From the figure, we see that with the instantaneous centre of rotation, the angular velocity is given as

$$\omega = \frac{v_A}{AC} = \frac{v_B}{BC} \tag{a}$$

Hence,

$$v_B = v_A \frac{BC}{AC} \tag{b}$$

To simplify the calculations for determination of the lengths BC and AC, we apply the geometrical conditions as follows. Draw OD parallel to BC and extend BA to meet OD at D. As the two triangles, OAD and BAC are similar, we can write as

$$\frac{BC}{AC} = \frac{OD}{OA}$$

In addition, we know that

$$v_A = \omega r_{OA} = \omega OA$$

and hence, equation (b) can be written as

$$v_B = \omega OA \frac{OD}{OA} = \omega OD$$

From the figure, we see that

$$OD = OB \tan\varphi$$

Since OB can be written as

$$OB = [OA(\cos 30°) + BA(\cos \varphi)]$$

we have

$$OD = [OA(\cos 30°) \tan\varphi + BA(\sin \varphi)]$$

$$= [0.1(\cos 30°) \tan 11.54° + 0.25 (\sin 11.54°)] = 0.0677 \text{ m}$$

Therefore, $$v_B = \omega\, OD = 40\pi (0.0677) = 8.51 \text{m/s}$$

Example 17.32 Solve Example 17.22 by the method of instantaneous centre.

Solution We saw in Example 17.22 that the lowest point A is momentarily at rest and hence all the particles in the wheel rotate about this point. Thus, the point A happens to be the instantaneous centre of rotation. The velocities of all the points are perpendicular to the line joining them with the point A as shown in Fig. 17.29.

As the centre point O of the wheel rotates about the instantaneous centre A, the angular velocity of the wheel about the instantaneous centre is given as

$$\omega = \frac{v_O}{r}$$

$$= \frac{16.67}{0.3} = 55.57 \text{ rad/s}$$

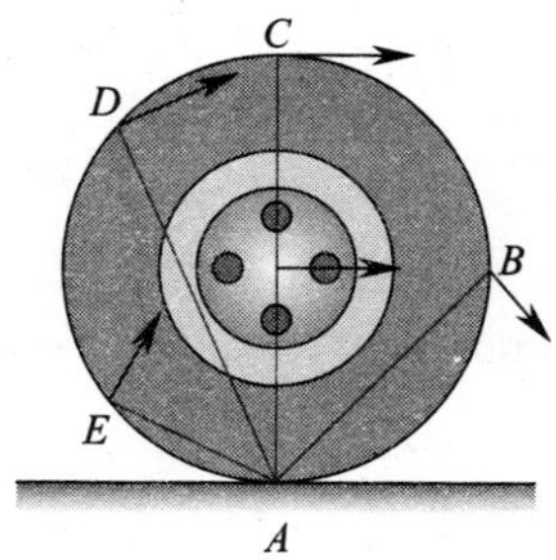

Fig. 17.29

Point B The velocity of the point B is given as

$$v_B = \omega(AB)$$

Since B is the extreme right point, we know that

$$AB = \sqrt{r^2 + r^2} = \sqrt{2}r = \sqrt{2}(0.3) \text{ m}$$

Therefore, $$v_B = 55.57(\sqrt{2})(0.3) = 23.6 \text{ m/s}$$

Point C The velocity of the point C is given as

$$v_C = \omega(AC)$$

Since C is the highest point, $AC = 2r = 2(0.3)$ m

Hence, $$v_C = 55.57\,(2)(0.3) = 33.34 \text{ m/s}$$

Point D The velocity of the point D is given as:

$$v_D = \omega\,(AD)$$

Since OD makes an angle of 45° to the vertical or horizontal, from the geometry, we can get

$$AD = AC \cos 22.5°$$

Therefore, $$v_D = 55.57\,(0.6)(\cos 22.5°) = 30.8 \text{ m/s}$$

Point E The velocity of the point E is given as

$$v_E = \omega(AE)$$

Just like AD, we can get the value of AE as

$$AE = AC \sin 22.5°$$

Therefore, $$v_E = 55.57(0.6)(\sin 22.5°) = 12.8 \text{ m/s}$$

Example 17.33 Solve Example 17.28 by the method of instantaneous centre.

Solution When the crank O_1A rotates in the clockwise direction with angular velocity ω then the linear velocity of the end A is tangential to the crank at A. Hence, we see that the instantaneous centre of rotation must lie on the extended line O_1A. Similarly, as the link O_2B rotates in the clockwise direction about hinge point O_2, the velocity of the end B of the link is tangential to the link at B and hence the instantaneous centre must lie perpendicular to this direction. Thus, we arrive at the intersection point C, which is the instantaneous centre of rotation for the position given.

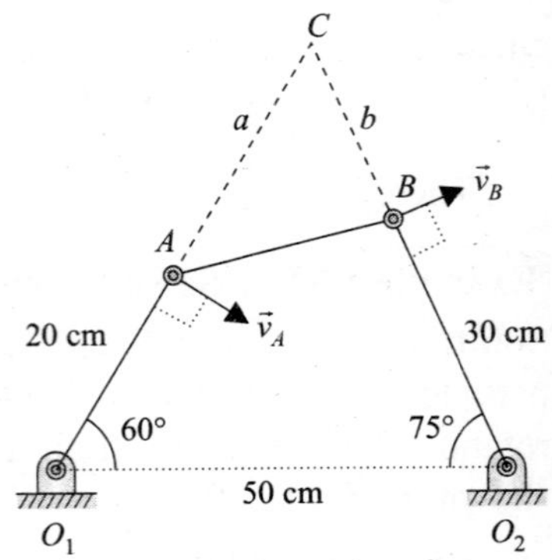

Fig. 17.30

Let $AC = a$ and $BC = b$. Then from the geometry of the figure, we see that

$$(20 + a) \sin 60° = (30 + b) \sin 75°$$

$$\Rightarrow \quad 0.897a - 12.068 = b \qquad \text{(i)}$$

and $$(20 + a) \cos 60° + (30 + b) \cos 75° = 50$$

$$\Rightarrow \quad b = 124.548 - 1.932a \qquad \text{(ii)}$$

From the two equations, solving for a and b we get

$$a = 48.29 \text{ cm and } b = 31.25 \text{ cm}$$

Hence, angular velocity of the link AB about C is given as

$$\omega = \frac{v_A}{AC} = \frac{v_B}{BC}$$

$$\Rightarrow \quad \omega = \frac{3.142}{0.4829} = 6.51 \text{ rad/s}$$

Therefore, velocity of the end B is given as:

$$v_B = \omega(BC)$$

$$= (6.51)(0.3125) = 2.03 \text{ m/s}$$

Hence, angular velocity of the link O_2B is given as

$$\omega_{O_2B} = \frac{v_B}{O_2B} = \frac{2.03}{0.3} = 6.77 \text{ rad/s}$$

SUMMARY

When a body is in *pure translational motion,* either rectilinear or curvilinear, every particle in the body moves in *parallel* paths in the *same* direction. The displacement, velocity and acceleration are *same* for all the particles in the body. In such type of motions, there is *no* rotational motion at all. Thus, instead of considering the body as a whole, we consider a single particle in the body (generally, the centre of mass of the body) and by describing the motion of that particle, we can describe the motion of the entire body. This we term as idealization of the rigid body as a *particle*.

It should be noted that even during such pure translational motion, a body could *rotate* as in the case of a cricket ball spinning or rotating about itself while moving in its trajectory. Such rotational motion can be neglected when the size of the body is very small as compared to its trajectory. However, there are instances in which the rotational motion is considerable and cannot be neglected. In such type of motions, the particles in the body undergo *different* displacements, velocities and accelerations. As a result, we can no more describe the motion of the entire body by describing the motion of a single particle and thus we have to treat the body as a whole, i.e., as the rigid body itself. The motion of a rigid body is thus *translational* motion together with *rotational* motion.

General Motion of a Rigid Body

It is the arbitrary of motion of a rigid body in space. To describe this motion, we require *six* independent coordinates, namely x, y and z to represent the translational motion and angular coordinates θ_x, θ_y, θ_z to represent the rotational motion. Hence, we say that a rigid body has *six* degrees of freedom.

General Plane Motion

When the general motion of a rigid body is constrained to a plane then it is said to be *general plane motion*. The translational motion of the body occurs in the plane and the rotation is about an axis perpendicular to the plane. A body in general plane motion has *three* degrees of freedom. A wheel rolling without slipping on a horizontal floor, a ladder sliding down with its ends in contact with a wall and a floor are examples of such type of general plane motion.

Such a motion is neither translational nor rotational motion. However, to analyze its motion, we can consider it to be a combination of *translational* and *rotational* motions occurring in *succession*. There are two methods to analyze such type of motions.

Absolute and Relative Motion Method

The absolute velocity of any point on the body is equal to the vector sum of translational velocity of an arbitrary reference point A and tangential velocity of the point due to fixed-axis rotation about A.

Instantaneous Centre of Rotation Method

It is possible at any instant of time to locate a point C, either on the rigid body or outside the rigid body about which all the particles in the body rotate. Since all the particles rotate about this point at that instant, it is momentarily at rest or in other words, its translational velocity at that instant is zero. Such a point is known as **instantaneous centre** of rotation.

Rotational Motion About a Fixed Axis

If the translational motion of a rigid body in general plane motion is constrained then the motion is said to be *pure rotation* about a fixed axis. All the particles in the body move in parallel planes (planes which are normal to the axis of rotation) in *circular* paths about the axis of rotation. To describe the motion of particles, we require only one coordinate, namely, the angular coordinate in the Z-direction. Hence, the body has one degree of freedom.

Angular Displacement

When the radial line from the axis of rotation to the particle sweeps through an angle with respect to a reference axis then we call that angle as angular displacement. By convention, *anticlockwise* displacements are chosen as *positive* and *clockwise* displacements are chosen as *negative*. It is expressed in radians or revolutions, where

$$1 \text{ rev} = 2\pi \text{ rad} = 360^\circ$$

Angular Velocity

The *average angular velocity* is defined as the ratio of angular displacement to time elapsed. Mathematically,

$$\omega_{\text{ave}} = \frac{\theta_2 - \theta_1}{t_2 - t_1} = \frac{\Delta\theta}{\Delta t}$$

and instantaneous angular velocity is defined as

$$\omega = \lim_{\Delta t \to 0} \frac{\Delta\theta}{\Delta t} = \frac{d\theta}{dt}$$

Its unit is rad/s (or) rpm.

Angular Acceleration

Angular acceleration is defined as the rate at which angular velocity changes. Then average angular acceleration is defined as

$$\alpha_{\text{ave}} = \frac{\omega_2 - \omega_1}{t_2 - t_1} = \frac{\Delta\omega}{\Delta t}$$

and the instantaneous angular acceleration is defined as:

$$\alpha = \lim_{\Delta t \to 0} \frac{\Delta\omega}{\Delta t} = \frac{d\omega}{dt}$$

In fixed-axis rotation, the angular displacement, velocity and acceleration are *same* for *every* particle in the body. Hence, by describing the motion of one particle in the body we can describe the motion of the entire body.

Rotational Motion with Constant Angular Acceleration

The rotational motion with constant acceleration is very much analogous to translational motion with constant acceleration. The equations of motion can be obtained by replacing s with θ, v with ω and a with α in the linear equations of motion. Thus, we can write

$$\omega = \omega_o + \alpha t$$

$$\theta = \omega_o t + \frac{\alpha t^2}{2}$$

$$\omega^2 = \omega_o^2 + 2\alpha\theta$$

If the angular acceleration is zero then it is said to be uniform angular motion. The above equations then reduce to

$$\omega = \omega_o \text{ and } \theta = \omega_o t$$

Relationship between Angular and Linear Motions

The fixed axis rotation can be represented by either angular variables or linear variables and their relationships are

$$s = r\theta \qquad v = r\omega$$

$$a_t = r\alpha \qquad a_n = r\omega^2$$

EXERCISES

Objective-type Questions

1. General motion of a rigid body in space has ________ degree(s) of freedom.

(a) one (b) two (c) three (d) six

2. A general plane motion has ________ degree(s) of freedom

(a) one (b) two (c) three (d) six

3. A fixed-axis rotation has ________ degree(s) of freedom

(a) one (b) two (c) three (d) six

4. Which of the following statements is true?

(a) In translational motion, the particles have different displacement, velocity and acceleration.

(b) In fixed-axis rotation, the particles have same angular displacement, velocity and acceleration.

(c) In fixed-axis rotation, the particles have same linear displacement, velocity and acceleration.

(d) In general plane motion, the particles have same displacement, velocity and acceleration.

5. Which of the following is not a vector?

(a) Angular displacement (b) Angular velocity

(c) Angular acceleration (d) Linear velocity

6. A disc is rotating at a constant angular speed about an axis passing through the centre and normal to it. State which of the following statements is true.

(a) Tangential velocity is zero. (b) Tangential acceleration is zero.

(c) Normal acceleration is zero. (d) Angular displacement is zero.

7. As we move away from the axis in a fixed axis rotation, the angular velocity

(a) remains constant (b) increases linearly

(c) increases parabolically (d) decreases

8. Instantaneous centre of rotation at that instant has
   (a) zero linear velocity
   (b) zero angular velocity
   (c) zero linear velocity and zero angular velocity
   (d) non-zero linear velocity
9. In fixed-axis rotation, all the particles have
   (a) the same angular displacement
   (b) the same angular velocity
   (c) the same angular acceleration
   (d) all of these
10. In fixed-axis rotation, the particles move in
    (a) parallel paths
    (b) arbitrary paths
    (c) concentric circular paths
    (d) parabolic paths
11. A particle on the axis in fixed-axis rotation has
    (a) zero angular displacement
    (b) zero angular velocity
    (c) zero angular acceleration
    (d) all of these
12. Which of the following statements is true concerning fixed-axis rotation?
    (a) All the particles have the same angular quantities.
    (b) All the particles have the same linear quantities.
    (c) Linear quantities are inversely proportional to the radial distance.
    (d) Fixed-axis rotation is a general plane motion.
13. In the motion of a wheel rolling without slipping, the instantaneous centre of rotation
    (a) is the centre of the wheel
    (b) is the point of contact of wheel with the ground
    (c) is the highest point of the wheel
    (d) lies outside the wheel
14. Which of the following point is always at rest?
    (a) Instantaneous centre in general plane motion
    (b) Particle on the axis in fixed axis rotation
    (c) Centre of mass of the body
    (d) Centre of gravity of the body

Answers

1. (d)	**2.** (c)	**3.** (a)	**4.** (b)	**5.** (a)	**6.** (b)	**7.** (a)	**8.** (c)
9. (d)	**10.** (c)	**11.** (d)	**12.** (a)	**13.** (b)	**14.** (b)		

Short-answer Questions

1. Differentiate between particle and rigid-body dynamics.
2. What are the different types of rigid-body motions?

3. Define general plane motion and give examples.
4. Define fixed-axis rotation and give examples.
5. Define angular displacement, velocity and acceleration.
6. Finite angular displacement is a scalar, while infinitesimally small angular displacement is a vector. Discuss.
7. Discuss on the various units of angular displacement, velocity and acceleration.
8. Derive the kinematic equations of angular motion with constant angular acceleration.
9. Define uniformly accelerated angular motion and uniform angular motion.
10. Establish the relationships between angular motion and linear motion.
11. What are the various methods used to analyze general plane motion. Discuss them.
12. Define instantaneous centre of rotation.
13. Explain how to locate instantaneous centre of rotation in general plane motion.
14. Under what conditions can we neglect the rotational motion of a body?

Numerical Problems

17.1 The angular motion of a disc is given by the relation $\theta = 4t^3 - 3t^2 + 2t$. Determine the angular displacement, velocity and acceleration at time $t = 2$ s.

Ans. 24 rad, 38 rad/s, 42 rad/s^2

17.2 The angular motion of a rigid body rotating about a fixed axis is given by the relation $\theta = t^2 + 2t$. Determine the angular displacement, velocity and acceleration at time $t = 3$ s.

Ans. 15 rad, 8 rad/s, 2 rad /s^2

17.3 The angular motion of a rigid body rotating about a fixed axis is given by the relation $\theta = \frac{\pi}{8}t^2 + \frac{\pi}{4}t$. Determine the time at which the angular velocity is 3 rad/s. Also, determine the angular displacement and angular acceleration at that instant.

Ans. 2.82 s, 5.34 rad, $\pi/4$ rad/s^2

17.4 A circular disc moving at ω_o rad/s accelerates at a rate given by the relation $\alpha = kt$. Derive the expressions for angular velocity and angular displacement. The initial angular displacement may be taken as θ_o rad.

Ans. $\omega = (kt^2/2) + \omega_o$, $\theta = \theta_o + \omega_o t + (1/6)\,kt^3$

17.5 The angular acceleration of a disc is defined by the relation $\alpha = 3t^2 - 2t$. Determine the expressions for angular velocity and displacement given that the disc is initially at rest at $\theta = 0$.

Ans. $\omega = t^3 - t^2$; $\theta = t^4/4 - t^3/3$

17.6 The acceleration of a circular disc due to resistance to its motion is given as $\alpha = 4t^2 - 3t$. Determine the angular velocity and angular displacement in 3 seconds, if its initial angular velocity is 2 rad/s.

Ans. $\omega = 24.5$ rad/s, $\alpha = 19.5$ rad/s^2

17.7 An electric motor runs at an angular speed of 2400 rpm. Express it in rad/s.

Ans. 80π rad/s

17.8 A turntable revolves at an angular speed of 4π rad/s. Express it in rpm.

Ans. 120 rpm

17.9 An electric motor makes 32.25 revolutions. Express this in radians.

Ans. 64.5 π rad

17.10 When power is switched on, a flywheel attains an angular speed of 300 rpm in 10 seconds. Determine its (i) uniform acceleration, (ii) angular displacement and velocity in 5 seconds.

Ans. (i) π rad/s^2, (ii) 6.25 rev, 5π rad/s

17.11 In the above problem, when the power is switched off, it comes to rest in 15 seconds. Determine the uniform angular retardation. Also, determine the angular displacement.

Ans. $\frac{-2\pi}{3}$ rad/s^2, 75 π rad

17.12 The hard disk of a *PC* attains a maximum speed of 7200 rpm in 6 seconds. Determine the angular acceleration assuming it to be uniform. Also, determine the number of complete revolutions made during this time. When switched off, it makes 100 revolutions before coming to rest. Determine the angular retardation and the time taken to come to a stop.

Ans. 40π rad/s^2, 360 rev, -144 π rad/ s^2, 1.67 s

17.13 The motor of a mixie used in kitchen increases its rate from 6000 rpm to 10,000 rpm when switched from medium speed to high speed in 2.5 seconds. Determine the angular acceleration assuming it to be uniform. Also, determine the number of complete revolutions made during this time.

Ans. 167.55 rad /s^2, 333 rev

17.14 A turntable covers $\pi/4$ rad in 3 seconds. If it started from rest, determine its angular velocity after 3 seconds.

Ans. $\pi/6$ rad /s

17.15 A turntable is rotating about a vertical axis as shown in Fig. E.17.15 at a constant angular speed of 45 rpm. Determine the velocity and acceleration of points *A* and *B*, which are respectively at distances 1 m and 0.4 m from the axis of rotation.

Ans. *A*: 4.71 m/s, 22.2 m/s^2 radially inward; *B*: 1.88 m/s, 8.87 m/s^2 radially inward

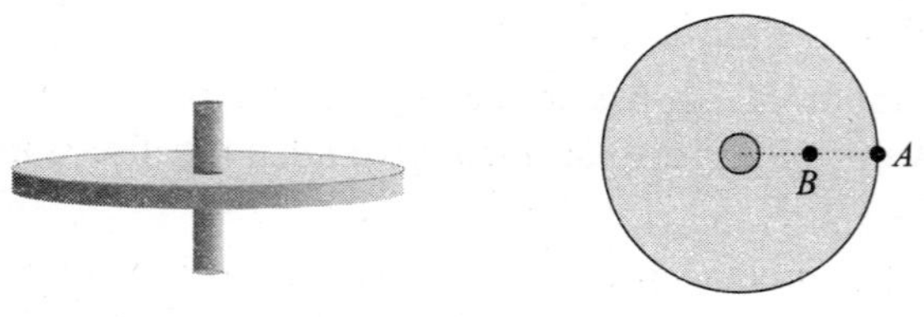

Fig. E.17.15, Fig. E.17.16

17.16 In the previous problem, if the turntable starts from rest and accelerates at a constant angular acceleration of $\pi/4$ rad/s^2, determine the velocity and acceleration of points *A* and *B* after 4 seconds from the start.

Ans. *A*: 3.14 m/s, $a_t = 0.785$ m/s^2, $a_r = 9.87$ m/s^2

B: 1.26 m/s, $a_t = 0.314$ m/s^2, $a_r = 3.95$ m/s^2

17.17 A block is placed on a disc, which is at rest at a radial distance of 20 cm. When the disk begins to rotate at a uniform acceleration of $\pi/4$ rad /s^2, the block is thrown off the disc after 2 seconds. Determine the velocity with which the block is thrown off, the tangential and radial components of acceleration of the block just before it is thrown off.

Ans. 0.314 m/s; 0.16 m/s^2; 0.49 m/s^2

17.18 A turntable of 60 cm diameter starts from rest and accelerates uniformly at a rate of $\pi/2$ rad/s^2 to reach a maximum speed of 150 rpm. Determine the velocity and acceleration of a point on the rim of the table at (i) $t = 4$ s, (ii) $t = 8$ s and (iii) $t = 12$ s.

Ans. (i) 1.88 m/s, 0.47 m/s^2, 11.84 m/s^2; (ii) 3.77 m/s, 0.47 m/s^2, 47.4 m/s^2; (iii) 4.71 m/s, 0.74 m/s^2

17.19 In an audio cassette, at a particular instant of time, the diameter of the roller on the left side together with tape is 2 cm and that on the right side is 5 cm. If the tape moves forward without slipping in the spool, determine the speed of the driven roller if the driver roller is running at 30 rpm in the direction shown in Fig. E.17.19.

Ans. 12 rpm

17.20 A square plate *ABCD* of side *a* is rotated in a horizontal plane at a constant angular velocity ω as shown in Fig. E.17.20 about a vertical axis passing through the corner *A*. Determine the linear velocity of corners *B*, *C* and its centre of gravity.

Ans. $v_B = a\omega$; $v_C = \sqrt{2}\,a\omega$; $v_{c.g.} = \dfrac{a}{\sqrt{2}}\omega$

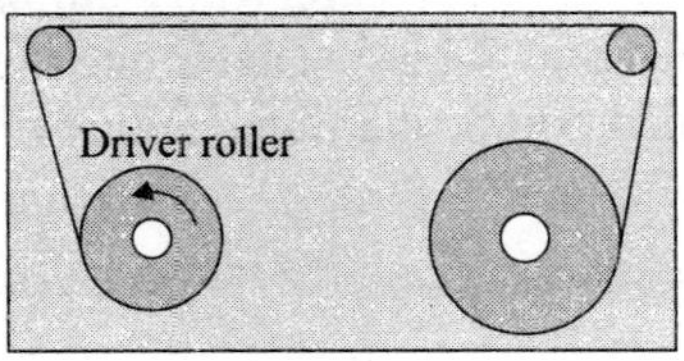

Fig. E.17.19

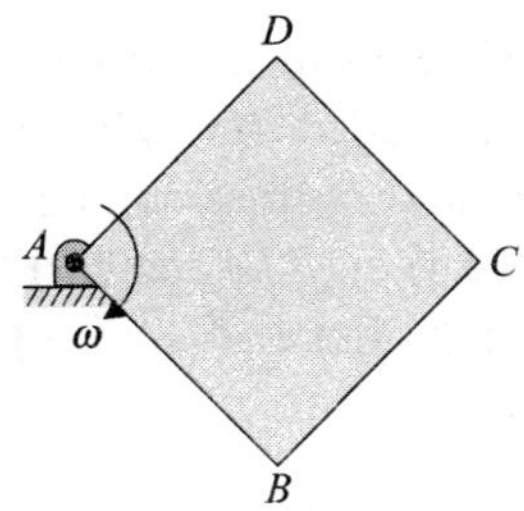

Fig. E.17.20

17.21 Two cylinders are in direct contact as shown in Fig. E.17.21. The diameter of the smaller cylinder is 10 cm and that of the bigger one is 20 cm. If the smaller cylinder is rotating at a constant speed of 50 rpm in the direction shown, determine the magnitude and direction of the speed of the bigger cylinder assuming there is no slip between the cylinders.

Ans. 25 rpm in clockwise direction

17.22 An escalator moves at a constant velocity of 0.25 m/s. If the diameter of the drums over which the escalator is rotating is 1 m, determine the angular velocity and angular acceleration of the drums assuming there is no slip. Refer Fig. E.17.22.

Ans. 0.5 rad/s, 0.125 rad/s^2 radially inward

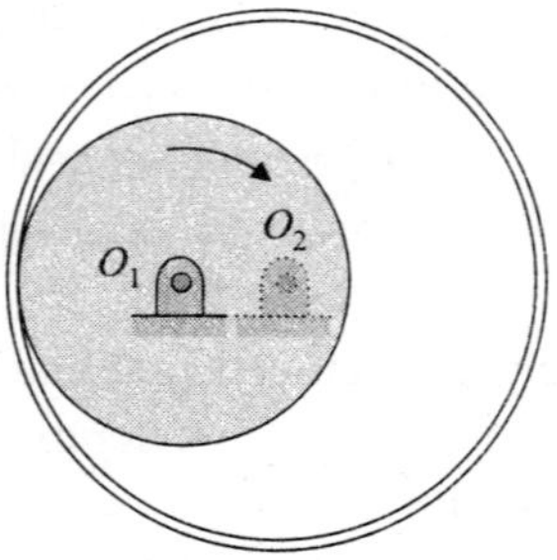

Fig. E.17.21

Fig. E. 17.22, E.17.23

17.23 In the above problem, if the escalator started from rest and moves at a constant acceleration of 0.1 m/s^2, determine the angular velocity and acceleration of the drums after 2 seconds, assuming there is no slip. Also, determine the normal acceleration of a point on the escalator as it moves over the drum.

Ans. 0.4 rad/s, α = 0.2 rad/s^2, a_r = 0.08 m/s^2

17.24 A block attached by a string, whose other end is wound over a cylindrical drum free to rotate about a horizontal axis, is released from rest. If the block has a velocity of 10 m/s downwards after 3 seconds, determine the angular velocity and acceleration of the drum after 3 seconds. Refer Fig. E.17.24.

Ans. 40 rad/s, 40/3 rad/s^2

17.25 A giant wheel in an amusement park (refer Fig. E.17.25) starts with a uniform angular acceleration of π/50 rad/s^2 to reach a maximum speed of π/4 rad/s. Determine the tangential velocity, tangential and normal components of acceleration at the outer radius of the wheel at times (i) t = 5 s, (ii) t = 10 s and (iii) t = 15 s. The diameter of the wheel is 10 m.

Ans. (i) π m/s, 0.2π rad/s^2, 0.1 π^2 rad/s^2; (ii) 2π m/s, 0.2 π rad/s^2, 0.4π^2 rad/s^2; (iii) 2.5π m/s, 0 rad/s^2, 0.625π^2 rad/s^2

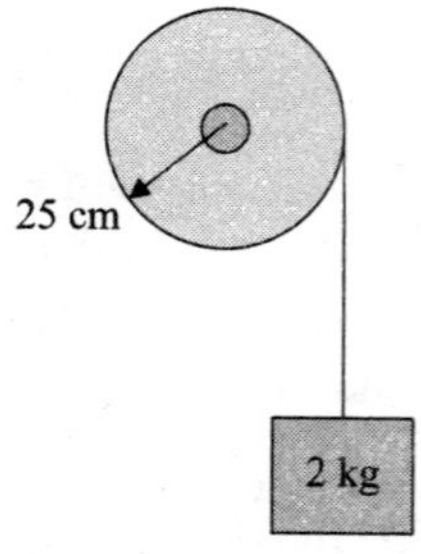

Fig. E.17.24

Fig. E.17.25

17.26 A cyclist is pedaling (refer Fig. E.17.26) at a constant angular velocity of 2π rad/s, determine the linear speed of the cycle. The diameter of gear wheel at the pedal is 15 cm and that at the rear wheel is 6 cm; the diameter of the wheels is 60 cm. Assume there is no slippage at contact points.

Ans. 4.7 m/s

17.27 A cylinder of 50 cm diameter rolls without slipping on a horizontal plane as shown in Fig. E.17.27. If the centre O of the cylinder moves at a constant speed of 5 m/s, determine (i) the angular velocity of the cylinder, (ii) the velocities of the points B, C and D on the rim of the cylinder. (B is the extreme right point, C is the highest point and radial line OD makes an angle of 45° to the horizontal).

Ans. 20 rad/s, $V_B = 7.07$ m/s directed downwards at 45° to the horizontal
$V_C = 10$ m/s directed towards right
$V_D = 3.83$ m/s directed upwards at 67.5° to the horizontal

Fig. E.17.26

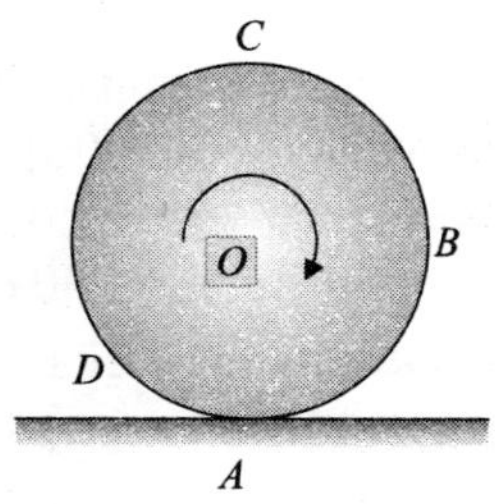

Fig. E.17.27

17.28 A bar AB of 1.5 m length slides down with its ends in contact with the floor and vertical wall as shown in Fig. E.17.28. If the end A moves with a constant velocity of 2 m/s away from the wall, determine (i) the angular velocity of the bar, and (ii) the velocity of the end B.

Ans. (i) 2.07 rad/s anticlockwise, (ii) 2.38 m/s vertically downwards

17.29 In Fig. E.17.29, if the end A is pushed with a constant velocity of 1 m/s, determine (i) the angular velocity of the bar, and (ii) the velocity of the end B. At this instant, the bar is inclined at 60° to the vertical.

Ans. (i) 1.33 rad/s anticlockwise, (ii) 1.732 m/s vertically upwards

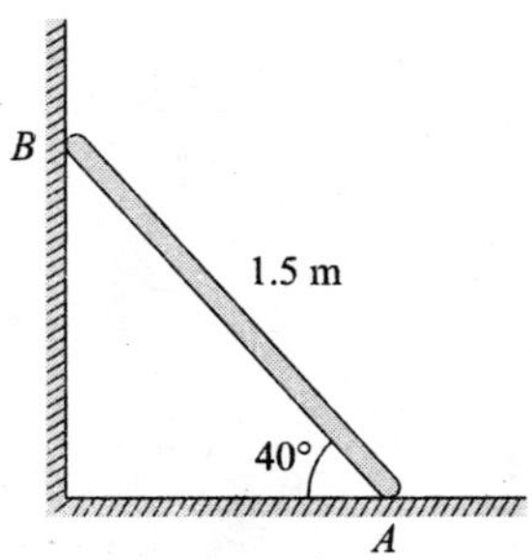

Fig. E.17.28

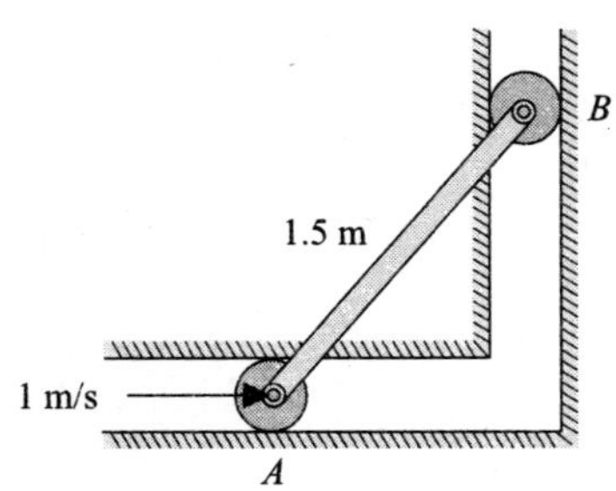

Fig. E.17.29

17.30 A bar AB of 2 m length slides down with its ends in contact with the horizontal floor and inclined surface as shown in Fig. E.17.30. If the end A moves with a constant velocity of 2 m/s towards right, determine (i) the angular velocity of the bar, and (ii) the velocity of the end B.

Ans. (i) 1 rad/s anticlockwise, (ii) 2 m/s downwards along the incline

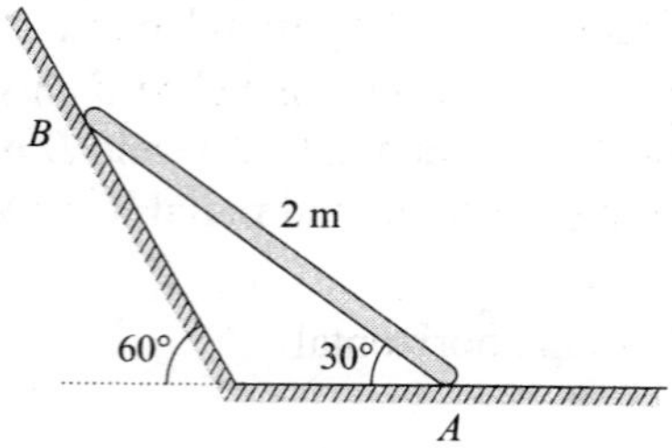

Fig. E.17.30

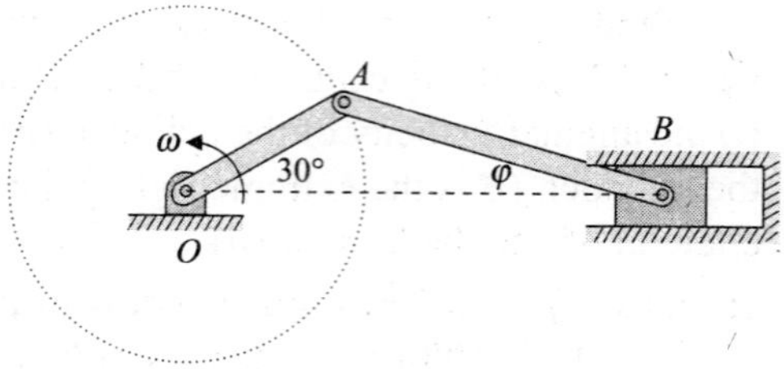

Fig. E.17.31

17.31 A reciprocating engine mechanism is shown in Fig. E.17.31, in which the crank OA rotates at a constant angular velocity of 200 rpm in the anticlockwise direction. For the position shown, determine (i) the angular velocity of the connecting rod AB and (ii) the velocity of piston in the engine. Take $OA = 10$ cm, $AB = 40$ cm.

Ans. (i) 4.55 rad/s clockwise, (ii) 1.27 m/s to the left

17.32 In a slider crank mechanism shown in Fig. E.17.32, the crank OA rotates with an angular velocity 'ω' in the anticlockwise direction. Determine (i) the angular velocity of the connecting rod AB and (ii) the velocity of the slider. $OA = a$ and $AB = b$.

Ans. (i) $\omega_{AB} = \omega\left(\tan\varphi + \frac{c}{b}\sec\varphi\right)\cot\theta$, (ii) $v_B = \omega(b\sin\varphi + c)(1 + \cot\theta\tan\varphi)$

17.33 In the slider crank mechanism shown in Fig. E.17.33, the crank OA rotates with an angular velocity of 100 rpm in the anticlockwise direction. Determine (i) the angular velocity of the connecting rod AB and (ii) the velocity of the slider B.

Ans. (i) 2.17 rad/s (clockwise direction), (ii) 0.495 m/s to the left

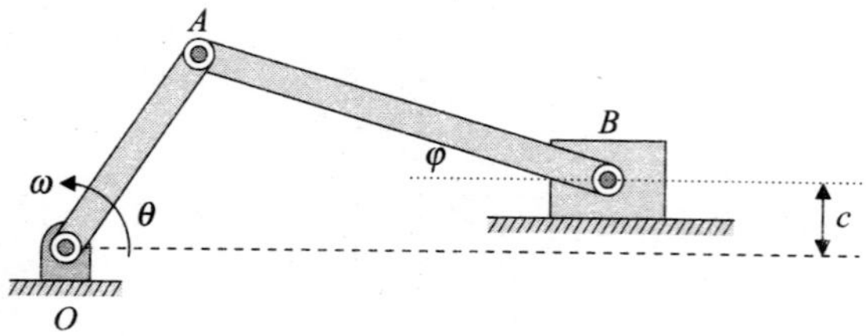

Fig. E.17.32

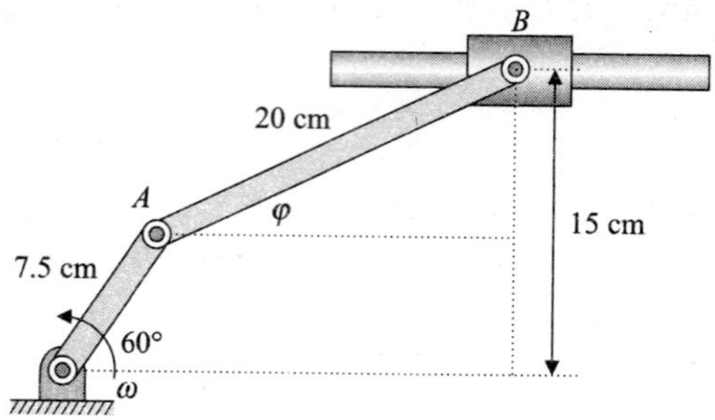

Fig. E.17.33

17.34. In the four-bar mechanism shown in Fig. E.17.34, the link O_1A moves in the anti-clockwise direction with an angular velocity of 60 rpm. Determine (i) the angular velocity of links AB and O_2B, and (ii) the velocity of the point B. The inclinations of links O_1A and O_2B with respect to the horizontal are respectively 60° and 75°.

Ans. (i) 1.38 rad/s, 3.21 rad/s (ii) 0.385 m/s

17.35 In the mechanism shown in Fig. E.17.35, if the block attached to the end C moves to the left at 1 m/s, determine the angular velocities of the links AB and BC.

Ans. ω_{AB} = 1.414 rad/s anticlockwise, ω_{BC} = 1.414 rad/s clockwise

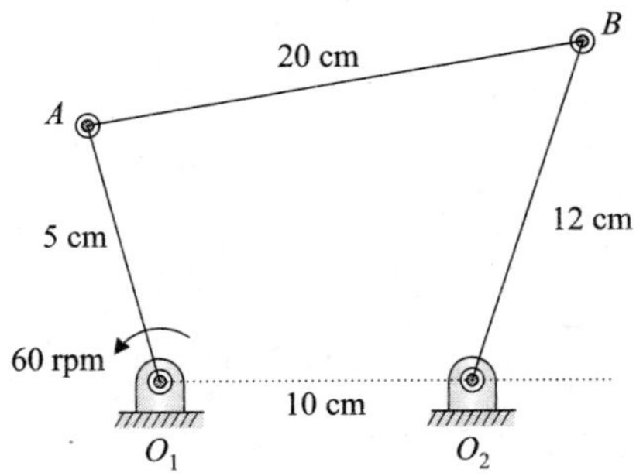

Fig. E.17.34

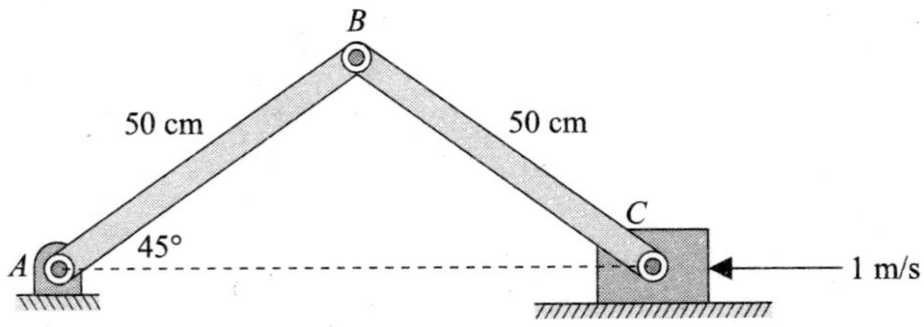

Fig. E.17.35

18 Kinetics of Rigid Bodies

18.1 INTRODUCTION

In the previous chapter, we discussed the *kinematics*, that is, the geometry of motion of rigid bodies, particularly, the fixed-axis rotation and the general plane motion. In this chapter, we will discuss the *kinetics* of such motions, that is, what actually *causes* such types of motions and *relate* the cause of motion with the resulting motion. Before proceeding to describe the equations of motion of a rigid body, we analyze the equations of motion of a *system of particles*. The results obtained for a system of particles can be extended to rigid body motion, as a rigid body can be assumed to be made up of a *large* number of particles.

18.2 SYSTEM OF PARTICLES

Consider a system of n particles as shown in Fig. 18.1. The particles may remain *fixed* (rigid body) or may *move* relative to one another. The forces acting on the ith particle are shown in the Fig. 18.1. They are the external and the internal forces. The net force acting on each particle is equal to the sum of the external and the internal forces. The internal forces refer to the forces exerted by all the other particles in the system upon the particle under consideration. If $\vec{f}_{ij}$ is the force exerted on the ith particle by the jth particle then the total internal forces acting on the ith particle is given as

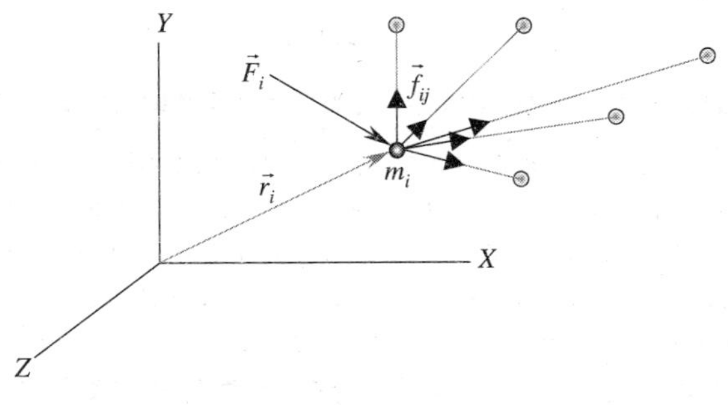

Fig. 18.1 Forces acting on the ith particle

$$\sum_{j=1}^{n} \vec{f}_{ij} \quad [\text{where } \vec{f}_{ij} = \vec{O}, \text{ when } i = j] \tag{18.1}$$

In the following section, we will derive the equations of translational motion of a system of particles and in Section 18.4, the rotational motion of a system of particles.

18.3 TRANSLATIONAL MOTION OF A SYSTEM OF PARTICLES

If m_i is the mass of the ith particle and $\vec{a}_i$ is its acceleration then we can write its equation of motion by equating the forces acting on the particle to the product of its mass and acceleration. From Newton's second law of motion, we have for the ith particle:

$$\vec{F}_i + \sum_{j=1}^{n} \vec{f}_{ij} = m_i \vec{a}_i \qquad (18.2)$$

Similarly, writing for each particle in the system and summing up, we have

$$\sum_{i=1}^{n} \vec{F}_i + \sum_{i=1}^{n} \sum_{j=1}^{n} \vec{f}_{ij} = \sum_{i=1}^{n} m_i \vec{a}_i \qquad (18.3)$$

To evaluate the second term on the left-hand side, let us consider a system of only three particles for simplicity as shown in Fig. 18.2.

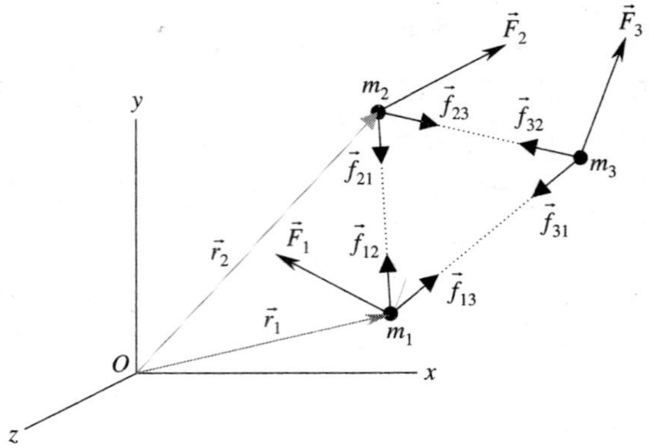

Fig. 18.2 External and internal forces acting on a system of three particles

According to Newton's law of gravitation, the forces $\vec{f}_{ij}$ and $\vec{f}_{ji}$ acting on each particle as exerted by the other are *equal* and *opposite* having the *same* line of action. Hence, the net internal forces acting between the particles cancel out. Thus, $\vec{F}_i$ represents purely the external system of forces acting on the particles and hence we can write

$$\sum_{i=1}^{n} \vec{F}_{\text{ext}} = \sum_{i=1}^{n} m_i \vec{a}_i \qquad (18.4)$$

or simply

$$\sum \vec{F}_{\text{ext}} = \sum m_i \vec{a}_i \qquad (18.5)$$

18.3.1 Motion of Centre of Mass of a System of Particles

The above equation can further be simplified if we consider the motion of *centre of mass* of the system of particles. This point moves in a *unique* way in which no other particle in the system moves. When we throw a solid body of any arbitrary shape at an angle to the horizontal, we know that the body would not only *translate* but also *rotate*. As a result, the path and velocity of each particle in the body is different. However, it is interesting to notice that its centre of mass moves in such a way that the entire mass of the body is concentrated at that point and all the forces act at that point. As a result, only the motion of the centre of mass of the body follows a *parabolic* path and no other point in the body follows such a path. Thus, the motion of the centre of mass represents the *translational* motion of the body.

Recalling what we studied in Chapter 8, we can write the coordinates of the centre of mass of a system of particles as

$$\bar{x} = \frac{\sum m_i x_i}{\sum m_i} \qquad \bar{y} = \frac{\sum m_i y_i}{\sum m_i} \quad \text{and} \quad \bar{z} = \frac{\sum m_i z_i}{\sum m_i} \qquad (18.6)$$

It should be noted that as the number of particles is *finite*, we have taken the *summation* sign instead of integral sign. Using the vector approach, the position vector of the centre of mass is given as

$$\vec{r}_{\text{cm}} = \bar{x}\vec{i} + \bar{y}\vec{j} + \bar{z}\vec{k}$$

$$= \frac{\sum[m_i(x_i\vec{i} + y_i\vec{j} + z_i\vec{k})]}{\sum m_i}$$

$$= \frac{\sum m_i\vec{r}_i}{\sum m_i} \tag{18.7}$$

Since $\sum m_i$ is equal to 'm,' the total mass of the system of particles, we can write

$$m\vec{r}_{\text{cm}} = \sum m_i\vec{r}_i \tag{18.8}$$

Differentiating the above equation with respect to time t, we have

$$m\frac{d\vec{r}_{\text{cm}}}{dt} = \sum m_i \frac{d\vec{r}_i}{dt}$$

or

$$m\vec{v}_{\text{cm}} = \sum m_i\vec{v}_i \tag{18.9}$$

Differentiating it further with respect to time t, we have

$$m\vec{a}_{\text{cm}} = \sum m_i\vec{a}_i \tag{18.10}$$

From the above result, we can write the Eq. 18.5 as

$$\sum \vec{F}_{\text{ext}} = m\vec{a}_{\text{cm}} \tag{18.11}$$

Comparing the above equation with Newton's second law of motion in general form, i.e., $\vec{F} = m\vec{a}$, we can see that the centre of mass of a system of particles moves in such a way as if the total mass of the system were concentrated at that point and all the external forces were acting at that point.

18.4 ROTATIONAL MOTION OF A SYSTEM OF PARTICLES

We saw in the previous chapter that there exists an analogy between the kinematics of rotational motion about a fixed axis and the kinematics of linear motion. In a similar manner, there exists an analogy between the kinetics of these two motions.

To derive the rotational effect of forces acting on particles, let us take moments of all the terms in the Eq. 18.2 about the origin O of the fixed reference frame.

$$\vec{r}_i \times \vec{F}_i + \sum_{j=1}^{n}(\vec{r}_i \times \vec{f}_{ij}) = \vec{r}_i \times m_i\vec{a}_i \tag{18.12}$$

where $\vec{r}_i$ is the position vector of the ith particle. Similarly, we can write for each particle in the system and summing up, we have

$$\sum_{i=1}^{n}(\vec{r}_i \times \vec{F}_i) + \sum_{i=1}^{n}\sum_{j=1}^{n}(\vec{r}_i \times \vec{f}_{ij}) = \sum_{i=1}^{n}(\vec{r}_i \times m_i\vec{a}_i) \tag{18.13}$$

Just as we saw that the sum of the internal forces for a system of particles became zero, it can be shown that the sum of moments of these internal forces about the origin is also zero. Consider the pair of internal forces $\vec{f}_{ij}$ and $\vec{f}_{ji}$ (refer Fig. 18.2) and taking sum of moments of these forces about the origin O, we can write

$$\vec{r}_i \times \vec{f}_{ij} + \vec{r}_j \times \vec{f}_{ji} = \vec{r}_i \times (\vec{f}_{ij} + \vec{f}_{ji}) + (\vec{r}_j - \vec{r}_i) \times \vec{f}_{ji} = 0 \tag{18.14}$$

Note that the sum of the forces $\vec{f}_{ij}$ and $\vec{f}_{ji}$ is zero and that the vectors $\vec{r}_j - \vec{r}_i$ and $\vec{f}_{ji}$ are collinear and hence their cross product is zero. Thus, we can see that the sum of moments of all the internal forces is zero and hence the Eq. 18.13 reduces to

$$\sum_{i=1}^{n} (\vec{r}_i \times \vec{F}_i) = \sum_{i=1}^{n} (\vec{r}_i \times m_i \vec{a}_i)$$

or simply,
$$\sum (\vec{r}_i \times \vec{F}_i) = \sum (\vec{r}_i \times m_i \vec{a}_i) \tag{18.15}$$

Since the product of the position vector of the point of the application of the force and the force vector is the moment of the force about the origin O, we can also write

$$\sum \vec{M}_O = \sum (\vec{r}_i \times m_i \vec{a}_i) \tag{18.16}$$

where $\sum \vec{M}_O$ is the resultant moment of the system of forces about O.

To describe a general plane motion, we saw in the previous chapter that kinematics of rotation is specified with respect to an *arbitrary* reference point. In the previous section, as we described the *translational* motion of a system of particles in terms of the motion of its *centre of mass*, it will be convenient to describe the rotational motion of the system of particles about the centre of mass instead of the origin O of the reference frame. Therefore, taking moment of the various terms on both sides of the Eq. 18.2 about the centre of mass G,

$$\vec{r}_i' \times \vec{F}_i + \sum_{j=1}^{n} (\vec{r}_i' \times \vec{f}_{ij}) = \vec{r}_i' \times m_i \vec{a}_i' \tag{18.17}$$

where $\vec{r}_i'$ is the position vector of the particle with respect to the centroidal reference frame. In a similar manner, writing for each particle in the system and summing up, we have

$$\sum_{i=1}^{n} (\vec{r}_i' \times \vec{F}_i) + \sum_{i=1}^{n} \sum_{j=1}^{n} (\vec{r}_i' \times \vec{f}_{ij}) = \sum_{i=1}^{n} (\vec{r}_i' \times m_i \vec{a}_i') \tag{18.18}$$

The sum of moments of internal forces becomes zero as before and hence we can write

$$\sum (\vec{r}_i' \times \vec{F}_i) = \sum (\vec{r}_i' \times m_i \vec{a}_i')$$

or
$$\sum \vec{M}_G = \sum (\vec{r}_i' \times m_i \vec{a}_i') \tag{18.19}$$

18.5 KINETIC EQUATIONS OF MOTION FOR A RIGID BODY

The results derived in the previous sections for a system of particles can be applied to a rigid body as a rigid body can be assumed to be made up of a *large* number of particles. The only difference being that the particles in the rigid body are *closely spaced* and the relative positions of the particles remain *fixed*.

18.5.1 General Plane Motion of a Rigid Body

Consider a system of forces $\vec{F}_1$, $\vec{F}_2$, ..., $\vec{F}_n$ acting on a rigid body in its plane. A rigid body can be considered to be made up of a large number n of particles of mass Δm_i.

As these forces are non-concurrent, we can replace them by a resultant force $\Sigma \vec{F}_i$ acting at the centre of mass together with the resultant moment $\Sigma \vec{M}_G$. The resultant force causes the translational motion of the rigid body and hence we can write

$$\Sigma \vec{F}_i = \Sigma \vec{F}_{\text{ext}} = \Sigma \Delta m_i \vec{a}_i$$

$$= m\vec{a}_{\text{cm}} \qquad (18.20)$$

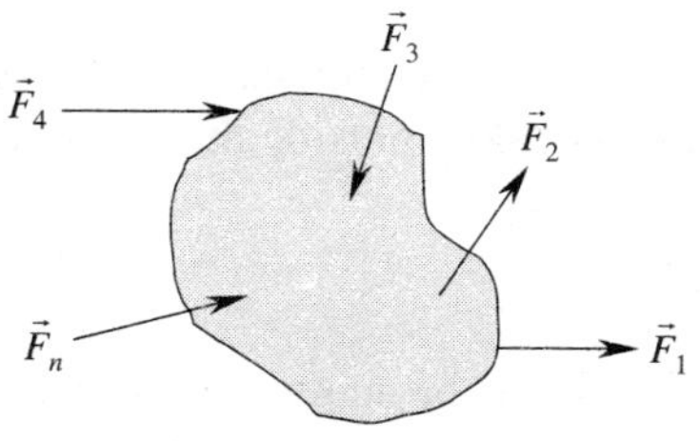

Fig. 18.3 System of forces acting on a rigid body in plane motion

It can also be expressed in scalar forms as

$$\Sigma F_x = m(a_{\text{cm}})_x \qquad (18.21)$$

and

$$\Sigma F_y = m(a_{\text{cm}})_y \qquad (18.22)$$

Similarly, the resultant moment causes the rotational motion of the rigid body and hence we can write

$$\Sigma \vec{M}_G = \Sigma(\vec{r}_i' \times \vec{F}_i) = \Sigma(\vec{r}_i' \times \Delta m_i \vec{a}_i') \qquad (18.23)$$

We know that the *linear momentum* of a particle is $m_i\vec{v}_i$. In a like manner, the **angular momentum** of the particle can be defined as

$$\Delta \vec{H}_G = \vec{r}_i' \times \Delta m_i \vec{v}_i' \qquad (18.24)$$

Thus, the angular momentum of the entire system is equal to the sum of angular momenta of individual particles and hence we can write

$$\vec{H}_G = \sum_{i=1}^{n} \vec{r}_i' \times \Delta m_i \vec{v}_i' \qquad (18.25)$$

Differentiating the above equation with respect to time,

$$\frac{d\vec{H}_G}{dt} = \sum_{i=1}^{n} \frac{d\vec{r}_i'}{dt} \times \Delta m_i \vec{v}_i' + \sum_{i=1}^{n} \vec{r}_i' \times \Delta m_i \frac{d\vec{v}_i'}{dt}$$

$$= \sum_{i=1}^{n} \vec{v}_i' \times \Delta m_i \vec{v}_i' + \sum_{i=1}^{n} \vec{r}_i' \times \Delta m_i \vec{a}_i'$$

$$= \sum_{i=1}^{n} \vec{r}_i' \times \Delta m_i \vec{a}_i' \qquad (18.26)$$

Note that the first term on the right side vanishes, as cross product of collinear vectors is zero. From the above result, we can write the Eq. 18.23 as

$$\Sigma \vec{M}_G = \frac{d\vec{H}_G}{dt} \qquad (18.27)$$

That is, *the moment resultant of the forces is equal to the rate of change of angular momentum.*

In the Eq. 18.25, since $\vec{v}_i'$ and $\vec{r}_i'$ lie on the same plane, the magnitude of their cross product is equal to the product of their magnitudes and direction perpendicular to the plane. Hence, we can eliminate the vector notation and write simply as

$$H_G = \sum_{i=1}^{n} \Delta m_i r_i' v_i' \qquad (18.28)$$

Since for a fixed axis rotation, $v_i' = r_i' \omega$, we have

$$H_G = \sum_{i=1}^{n} \Delta m_i r_i'^2 \omega \tag{18.29}$$

Since the angular velocity ω is the *same* for every particle in the body for a fixed axis rotation, it can be taken outside the summation sign and hence,

$$H_G = \omega \sum_{i=1}^{n} \Delta m_i r_i'^2 \tag{18.30}$$

Since the summation represents the mass moment of inertia about the centroidal axis perpendicular to the plane, we can write

$$H_G = \bar{I}\omega \tag{18.31}$$

Differentiating it with respect to time, we have

$$\frac{dH_G}{dt} = \bar{I}\frac{d\omega}{dt} = \bar{I}\alpha \tag{18.32}$$

Comparing the magnitudes of Eqs 18.27 and 18.32, we can write

$$\sum M_G = \bar{I}\alpha \tag{18.33}$$

This is the kinetic equation of rotational motion of a rigid body about its mass centre. It can be seen that it is analogous to the kinetic equation of translational motion, i.e., $F = ma$.

18.5.2 Centroidal Rotation of a Rigid Body

If the rigid body is in pure rotation about a perpendicular axis passing through the centre of mass, the translational motion of the centre of mass is zero. Hence, the kinetic equation of motion for fixed-axis rotation reduces to

$$\sum M_G = \bar{I}\alpha \tag{18.33'}$$

Taking moment of all the external forces about the axis of rotation and applying the above equation of motion, we can determine the kinematics of motion.

Example 18.1 A Pelton wheel turbine used in a lab runs at 1200 rpm, when the water jet is suddenly shut off. If it comes to rest in 30 seconds, determine the retarding torque due to friction in the bearings assuming it to be uniform. The mass of the wheel is 20 kg and the radius of gyration is 20 cm.

Solution The initial speed of the turbine is ω_o = 1200 rpm = 1200 × $(2\pi/60)$ = 40π rad/s and its final speed is $\omega = 0$. Therefore, the uniform angular retardation of the wheel is given as

$$\alpha = \frac{\omega - \omega_o}{t} = \frac{0 - 40\pi}{30} = -4.19 \text{ rad/s}^2$$

Since the mass of the wheel and its radius of gyration are given, its mass moment of inertia about the axis of rotation is then

$$\bar{I} = mk^2$$
$$= (20)(0.2)^2 = 0.8 \text{ kg.m}^2$$

Since the wheel undergoes fixed axis rotation about its centroid, its kinetic equation of motion can be written as

$$M = \bar{I}\alpha$$
$$= (0.8)(-4.19) = -3.35 \text{ N.m}$$

The negative sign indicates that the torque acting on the wheel is retarding.

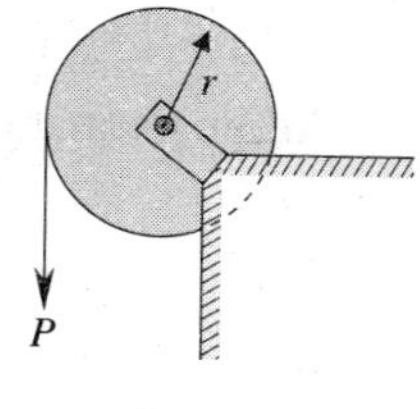

Fig. 18.4

Example 18.2 A light inextensible string is wrapped several times around a solid cylinder of mass m and radius r, which is free to rotate about a horizontal axis as shown in Fig. 18.4. If a steady force P is applied at the free end of the string, derive the expressions for angular acceleration of the cylinder and tangential acceleration of a point on the rim of the cylinder. Solve for $m = 20$ kg, $r = 15$ cm and $P = 30$ N.

Solution The steady force applied at the free and of the string causes the cylinder to rotate about its centroidal axis. We can relate the force and the resulting motion by the kinetic equation of motion for fixed axis rotation. The external applied moment on the cylinder is equal to the moment of the applied force about the axis of the cylinder, i.e., $M = Pr$. Applying the kinetic equation of motion of the cylinder,

$$\sum M = \bar{I}\alpha \Rightarrow$$

$$Pr = \frac{mr^2}{2}\alpha$$

$$\Rightarrow \qquad \alpha = \frac{2P}{mr}$$

Therefore, the tangential acceleration of a point on the rim of the cylinder is given as

$$a = r\alpha = \frac{2P}{m}$$

For the given data, we have

$$\alpha = \frac{2(30)}{(20)(0.15)} = 20 \text{ rad/s}^2$$

and

$$a = \frac{2(30)}{(20)} = 3 \text{ m/s}^2$$

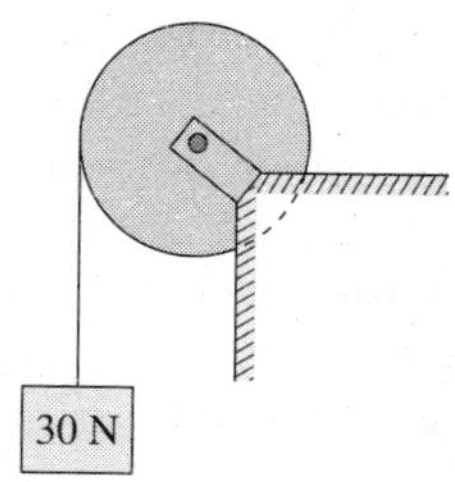

Fig. 18.5

Example 18.3 In the above problem, if the free end of the string supports a block of 30 N weight, determine the acceleration of the system when released from rest and the tension in the string. In addition, determine the velocity of the block after it has fallen through 2 m.

Solution The block moves down due to force of gravity and the tension in the string causes the cylinder to rotate about its centroidal axis. The free-body diagrams of the block and the cylinder are shown in Fig. 18.5(a). If we take m_1 as the mass of the block then we can write the kinetic equation of motion for the block as

$$m_1 g - T = m_1 a \qquad \text{(a)}$$

Similarly, for the cylinder, we can write

$$Tr = \bar{I}\alpha = \frac{mr^2}{2}\alpha$$

$$\Rightarrow \quad T = \frac{mr}{2}\alpha = \frac{ma}{2} \quad [\text{since } a = r\alpha] \qquad \text{(b)}$$

Adding equations (a) and (b), we can eliminate the tension T and hence we have

$$m_1 g = m_1 a + \frac{m}{2}a$$

$$\Rightarrow \quad a = \frac{2m_1 g}{2m_1 + m}$$

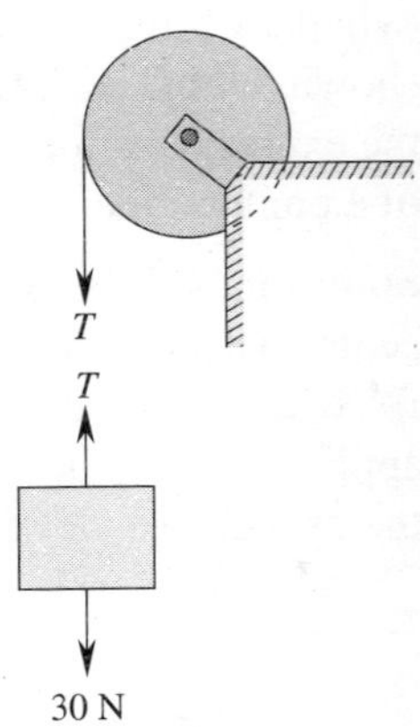

Fig. 18.5(a)

For the given data,

$$a = \frac{2(30)}{2(30/9.81) + 20} = 2.3 \text{ m/s}^2$$

Therefore, the angular acceleration of the cylinder is given as

$$\alpha = \frac{a}{r} = \frac{2.3}{0.15} = 15.33 \text{ rad/s}^2$$

From the equation (a), we get the tension in the string as

$$T = m_1 g - m_1 a$$
$$= 30 - (30/9.81)(2.3) = 23 \text{ N}$$

The velocity of the block can be determined using the kinematic equation of motion,

$$v^2 = v_o^2 + 2as$$
$$= 0 + 2(2.3)(2) = 9.2$$

$$\therefore \quad v = 3.03 \text{ m/s}$$

We see that the angular acceleration of the cylinder is lesser though the same force of 30 N is acting. This is because the tension in the string acting on the cylinder is lesser than the previous case, i.e., 30 N, to provide the acceleration for the block.

Example 18.4 The block-and-pulley arrangement shown in Fig. 18.6, when released from rest, determine the acceleration of the system and the tension in each portion of the string. Assume the pulley to be a solid cylinder of 20 kg mass and 15 cm radius.

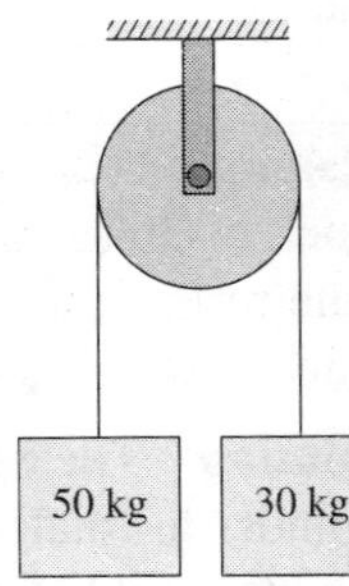

Fig. 18.6

Solution The free-body diagrams of the two blocks and the pulley are shown in Fig. 18.6(a). It should be noted that the tensions on the two sides of the string passing over the pulley would not be equal as to cause a net driving moment acting on the pulley. Writing the kinetic equations of motion for the two blocks taking the initial directions of motion as positive,

$\sum F_y = ma_y \Rightarrow$

$$50g - T_1 = 50a \qquad \text{(a)}$$

$$T_2 - 30g = 30a \qquad \text{(b)}$$

Similarly, writing the kinetic equation of motion for the pulley,

$\sum M = \bar{I}\alpha \Rightarrow$

$$(T_1 - T_2)r = \frac{mr^2}{2}\alpha$$

Since the tangential acceleration of a point on the rim of the pulley is same as the acceleration of the blocks,

$$T_1 - T_2 = \frac{m}{2}(r\alpha) = \frac{m}{2}a$$

$$= \frac{20}{2}a = 10a \qquad \text{(c)}$$

Adding the equations (a), (b) and (c), we can eliminate the tensions T_1 and T_2 and hence we have

$$50g - 30g = 50a + 30a + 10a$$

$$\Rightarrow \qquad a = \frac{20}{90}g = \frac{20}{90}(9.81) = 2.18 \text{ m/s}^2$$

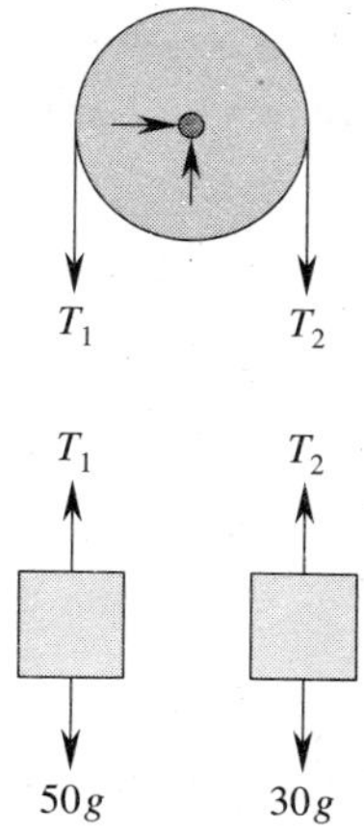

Fig. 18.6(a)

Therefore, the angular acceleration of the pulley is given as

$$\alpha = \frac{a}{r} = \frac{2.18}{0.15} = 14.53 \text{ rad/s}^2$$

From the equation (a), we get

$$T_1 = 50[g - a]$$

$$= 50[9.81 - 2.18] = 381.5 \text{ N}$$

From the equation (b), we get

$$T_2 = 30[g + a]$$

$$= 30[9.81 + 2.18] = 359.7 \text{ N}$$

Example 18.5 The stepped pulley system shown in Fig. 18.7, when released from rest, determine the acceleration of the blocks, angular acceleration of the pulley and tension in the strings connecting the blocks. The mass of the pulley is 20 kg and its radius of gyration is 20 cm.

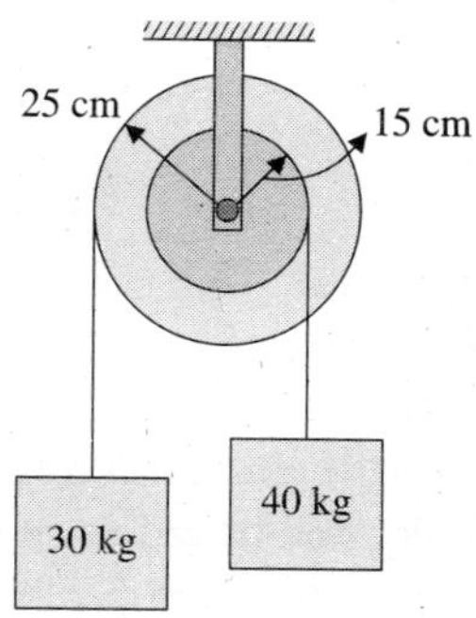

Fig. 18.7

Solution The free-body diagrams of the blocks and the pulley are shown in Fig. 18.7(a). For equilibrium of the 30 kg block, the tension T_1 must be $30g$ and for equilibrium of the 40 kg block, the tension T_2 must be $40g$. However, we see that the moment created by the 30 kg block about the axis of the pulley is greater than that created by the 40 kg block. Hence, the pulley will rotate in the anticlockwise direction. We should note that the angular accelerations of the pulleys are same, but the tangential accelerations of the blocks, being dependent upon the radial distances are not the same. Writing the kinetic equations of motion in the vertical direction for the two blocks,

$\sum F_y = ma_y \Rightarrow$

$$30g - T_1 = 30a_1 = 30r_1\alpha$$
$$= 30(0.25)\alpha = 7.5\alpha \quad \text{(a)}$$
$$T_2 - 40g = 40a_2 = 40r_2\alpha$$
$$= 40(0.15)\alpha = 6\alpha \quad \text{(b)}$$

Similarly, applying the kinetic equation of motion for the stepped pulley,

$\sum M = \bar{I}\alpha \Rightarrow$

$$T_1 r_1 - T_2 r_2 = mk^2\alpha$$
$$\Rightarrow \quad (0.25)T_1 - (0.15)T_2 = 20(0.2)^2\alpha$$
$$= 0.8\alpha \quad \text{(c)}$$

Multiplying the equation (a) by 0.25 and the equation (b) by 0.15 and adding them with the equation (c), we can eliminate the tensions T_1 and T_2 and hence we have

$$1.5g = 3.575\alpha$$
$$\Rightarrow \quad \alpha = \frac{1.5}{3.575}g$$
$$= \frac{1.5}{3.575}(9.81) = 4.12 \text{ rad/s}^2$$

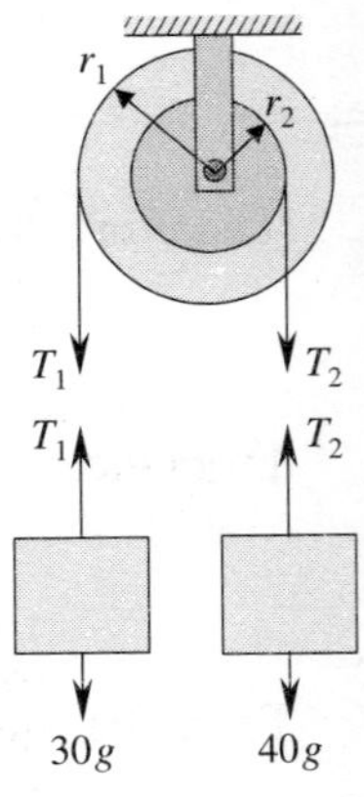

Fig. 18.7(a)

Therefore, the accelerations of the two blocks are given as

$$a_1 = r_1\alpha = (0.25)(4.12) = 1.03 \text{ m/s}^2$$

and
$$a_2 = r_2\alpha = (0.15)(4.12) = 0.62 \text{ m/s}^2$$

From the equation (a), we get

$$T_1 = 30g - 7.5\alpha$$
$$= 30(9.81) - 7.5(4.12) = 263.4 \text{ N}$$

From the equation (b), we get

$$T_2 = 40g + 6\alpha = 40(9.81) + 6(4.12) = 417.12 \text{ N}$$

Example 18.6 The stepped pulley arrangement shown in Fig. 18.8, when released from rest, determine the acceleration of the blocks, angular acceleration of the pulley and tension in the strings connecting the blocks. The mass of the pulley is 50 kg and its radius of gyration is 18 cm, and the coeffi..ient of friction between the horizontal plane and the block resting on it is 0.2.

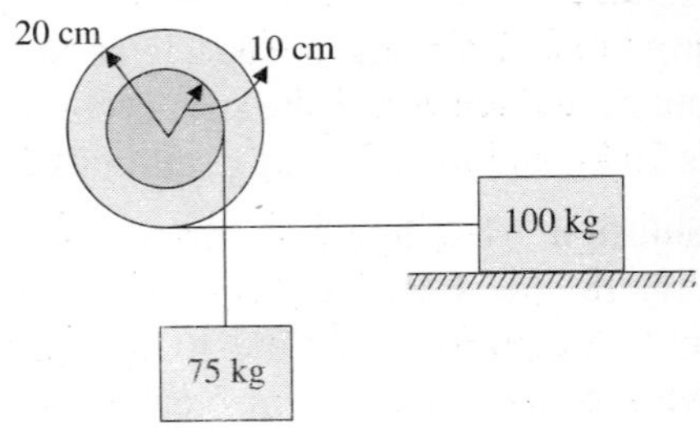

Fig. 18.8

Solution The free-body diagrams of the two blocks and the pulley are shown in Fig. 18.8(a). We should note that the angular accelerations of the pulleys are same, but the tangential accelerations of the blocks, being dependent upon the radial distances are not the same.

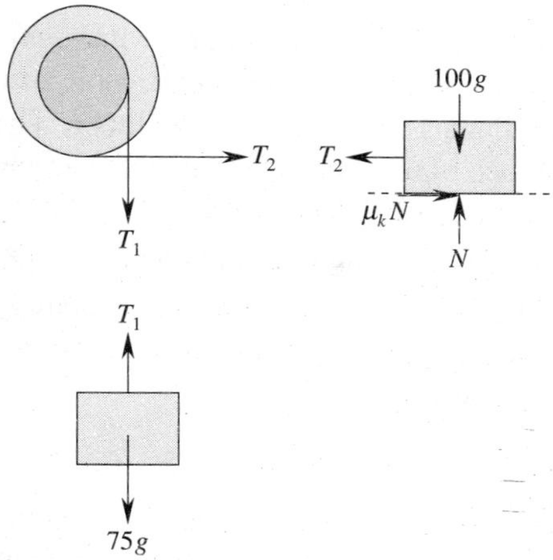

Fig. 18.8(a)

Writing the kinetic equation of motion for the hanging block in the vertical direction,

$$\sum F_y = ma_y \Rightarrow$$

$$75g - T_1 = 75a_1 = 75r_1\alpha$$

$$\therefore \quad 75g - T_1 = 75(0.1)\alpha = 7.5\alpha \quad \text{(a)}$$

Similarly, writing the kinetic equation of motion along the horizontal direction for the block lying on the plane,

$$\sum F_x = ma_x \Rightarrow$$

$$T_2 - \mu_k N = 100a_2 = 100r_2\alpha$$

$$T_2 - (0.2)(100g) = 100(0.2)\alpha$$

$$\therefore \quad T_2 - 20g = 20\alpha \quad \text{(b)}$$

Writing the kinetic equation of motion for the stepped pulley,

$$\sum M = \bar{I}\alpha \Rightarrow$$

$$T_1 r_1 - T_2 r_2 = mk^2\alpha$$

$$\therefore \quad (0.1)T_1 - (0.2)T_2 = 50(0.18)^2\alpha = 1.62\alpha \quad \text{(c)}$$

Multiplying the equation (a) by 0.1 and the equation (b) by 0.2 and adding them with the equation (c), we can eliminate the tensions T_1 and T_2 and hence we have,

$$7.5g - 4g = 0.75\alpha + 4\alpha + 1.62\alpha$$

$$\Rightarrow \quad \alpha = \frac{3.5g}{6.37} = 5.39 \text{ rad/s}^2$$

Therefore, the accelerations of the two blocks are given as

$$a_1 = r_1\alpha = (0.1)(5.39) = 0.54 \text{ m/s}^2$$

and

$$a_2 = r_2\alpha = (0.2)(5.39) = 1.08 \text{ m/s}^2$$

From the equation (a), we get

$$T_1 = 75g - 7.5\alpha$$

$$= 75(9.81) - 7.5(5.39) = 695.33 \text{ N}$$

From the equation (b), we get

$$T_2 = 20g + 20\alpha$$

$$= 20(9.81 + 5.39) = 304 \text{ N}$$

Example 18.7 A block of mass m_1 resting on a frictionless horizontal plane is attached to an inextensible string, which passes over a pulley (solid cylinder) of mass m and radius r. The other end of the string supports a block of mass m_2. Determine the acceleration of the system and the tension in each part of the string.

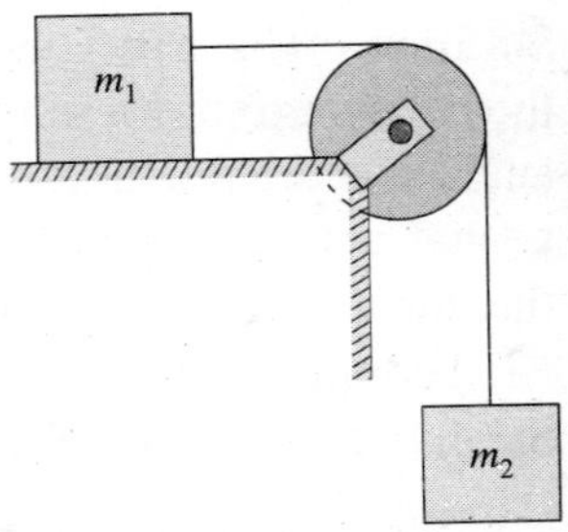

Fig. 18.9

Solution The free-body diagrams of the blocks and the pulley are shown in Fig. 18.9(a). Writing the kinetic equation of motion for the hanging block in the vertical direction,

$$\sum F_y = ma_y \Rightarrow$$

$$m_2 g - T_2 = m_2 a \qquad \text{(a)}$$

Similarly, writing the kinetic equation of motion for the block lying on the plane in the horizontal direction,

$$\sum F_x = ma_x \Rightarrow$$

$$T_1 = m_1 a \qquad \text{(b)}$$

Note that as the string is inextensible, the accelerations of the two blocks are equal. Writing the kinetic equation of motion for the pulley,

$$\sum M = \bar{I}\alpha \Rightarrow$$

$$T_2 r - T_1 r = \frac{mr^2}{2}\alpha = \frac{mr}{2}(r\alpha)$$

$$\therefore \qquad T_2 - T_1 = \frac{m}{2}a \qquad \text{(c)}$$

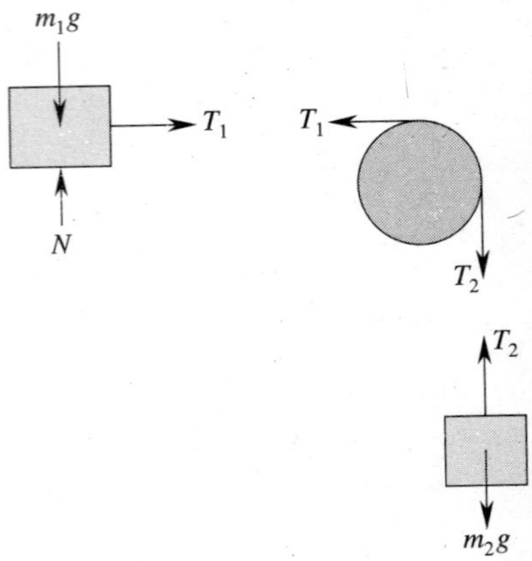

Fig. 18.9(a)

Adding equations (a), (b) and (c), we can eliminate the tensions T_1 and T_2 and hence we have

$$m_2 g = m_1 a + m_2 a + \frac{m}{2}a$$

$$\Rightarrow \qquad a = \frac{2m_2 g}{[2m_1 + 2m_2 + m]}$$

From the equation (b), we get

$$T_1 = m_1 a = \frac{2m_1 m_2 g}{[2m_1 + 2m_2 + m]}$$

Similarly, from the equation (a), we get

$$T_2 = m_2 g\left[1 - \frac{2m_2}{2m_1 + 2m_2 + m}\right]$$

$$= \frac{[2m_1 + m]m_2 g}{[2m_1 + 2m_2 + m]}$$

Example 18.8 Two circular disks mounted on frictionless bearings are in contact with each as shown in Fig. 18.10. The mass and radius of the larger disk are respectively 4 kg and 10 cm and that of the smaller disk are respectively 2 kg and 6 cm. If a clockwise moment of 1 N.m is applied to the larger disk, determine (i) the frictional force at the contact point, and (ii) the angular accelerations of each disk. Assume there is no slip between the disks.

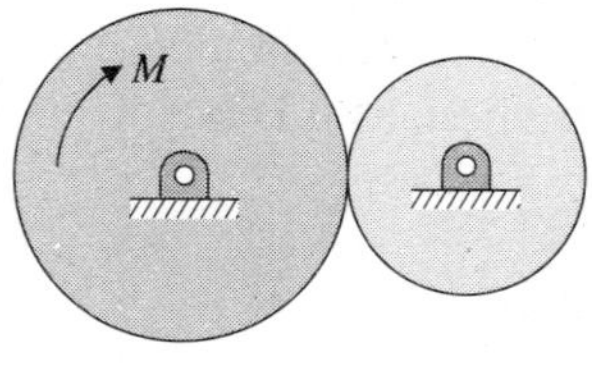

Fig. 18.10

Solution The free-body diagrams of the two disks are shown in Fig. 18.10(a). The force opposing the motion of the larger disk is the force of friction developed at the contact point and the driving torque for the smaller disk is provided by the force of friction developed at the contact point. They act in the opposite directions on each of the disks.

Applying the kinetic equation of motion for the larger disk then we have

$\sum M = \bar{I}\alpha \Rightarrow$

$$M - fr_A = \frac{m_A r_A^2}{2}\alpha_A \qquad \text{(a)}$$

and for the smaller disk,

$$fr_B = \frac{m_B r_B^2}{2}\alpha_B \qquad \text{(b)}$$

Since there is no slip, we know that at the contact point, the tangential acceleration must be the same for both the disks and hence,

$$r_A \alpha_A = r_B \alpha_B \qquad \text{(c)}$$

Solving for α_A and α_B from the three equations (a), (b) and (c), we get

$$\alpha_A = \frac{2M}{(m_A + m_B)r_A^2} = \frac{2(1)}{(4+2)(0.1)^2} = 33.33 \text{ rad/s}^2$$

and

$$\alpha_B = \frac{r_A \alpha_A}{r_B} = \frac{(0.1)(33.33)}{(0.06)} = 55.55 \text{ rad/s}^2$$

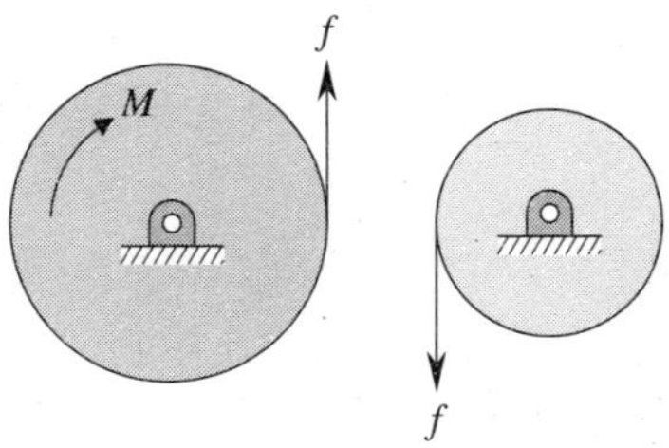

Fig. 18.10(a)

18.5.3 Non-Centroidal Rotation of a Rigid Body

We also come across situations, in which the body rotates about a perpendicular axis passing through a point O other than the centre of mass, in which case we write the kinetic equations of motion as

$$\sum M_O = I_O \alpha = (\bar{I} + md^2)\alpha \qquad (18.34)$$

Here it should be noted that unlike centroidal rotation, the centre of mass undergoes translational motion, as it is not fixed. Hence, we also include the equations for translational motion.

$$\sum F_x = m(a_{\text{cm}})_x \tag{18.21'}$$

and

$$\sum F_y = m(a_{\text{cm}})_y \tag{18.22'}$$

These two equations are used to determine the reactions developed at the point of support. The following examples will clarify this.

Example 18.9 A boy kicks open a door by exerting a force of 100 N in the horizontal direction. The plan view is shown in Fig. 18.11. Determine the angular acceleration of the door at that instant, if the mass of the door is 10 kg. Assume the door to be a perfect rectangular panel.

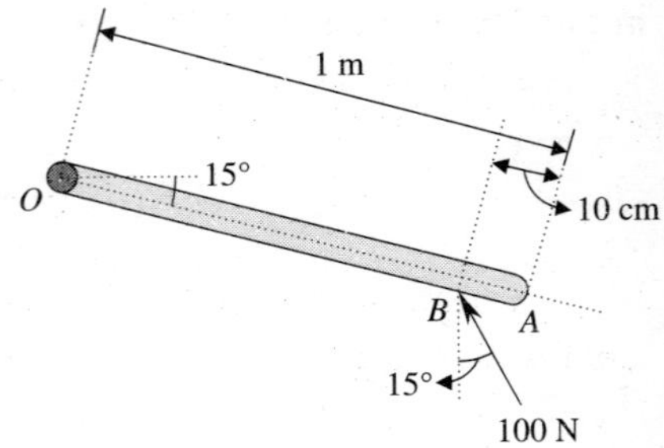

Fig. 18.11

Solution From the figure, we can see that the inclination of the applied force with respect to the axis along the door panel is 60°. Hence, the component of the applied force perpendicular to the door panel effective in causing moment is 100 sin 60°. Therefore, taking moment of the applied force about the hinge O, we have

$$M_O = (100 \sin 60°)(0.9)$$

$$= 77.94 \text{ N.m}$$

Since the door can be considered to be a perfect rectangular panel, we know that the mass moment of inertia of the door about a centroidal axis lying in the plane of the door and parallel to its hinge is

$$\bar{I} = \frac{ml^2}{12}$$

Therefore, mass moment of inertia about the hinge is given as

$$I_{\text{hinge}} = \frac{ml^2}{12} + m(l/2)^2$$

$$= \frac{ml^2}{3} = (10)(1)^2/3 = 3.33 \text{ kg.m}^2$$

Therefore, writing the kinetic equation for angular motion,

$$M_O = I_{\text{hinge}}\,\alpha$$

$$\Rightarrow \quad \alpha = M_O/I_{\text{hinge}}$$

$$= (77.94)/(3.33) = 23.41 \text{ rad/s}^2$$

Example 18.10 A homogeneous steel bar AB of length l is suspended horizontally as shown in Fig. 18.12. If the string at B is snapped, determine the reaction at the hinge point A at that instant. In addition, determine the velocity of centre of mass of the bar when it swings to the vertical position.

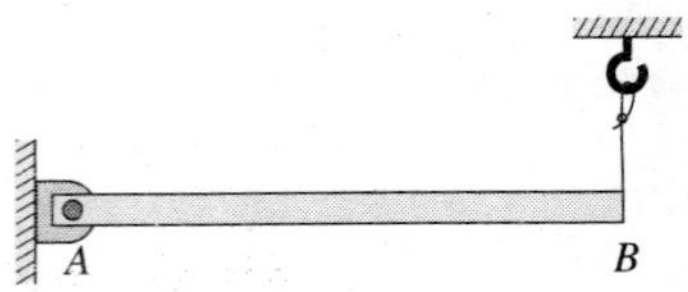

Fig. 18.12

Solution The free-body diagram of the bar is shown in Fig. 18.12(a). It should be noted that as the string is cut, the tension at B would cease to exist. Hence, the forces acting on it are its weight mg acting at its midpoint and reactions A_x and A_y at the hinge point A. The net external moment acting on it causing it to rotate about the hinge is equal to the moment of the forces acting on the bar about point A, i.e., $M_A = mg(l/2)$. Hence, writing the kinetic equation for angular motion,

$$M_A = I_A\alpha$$

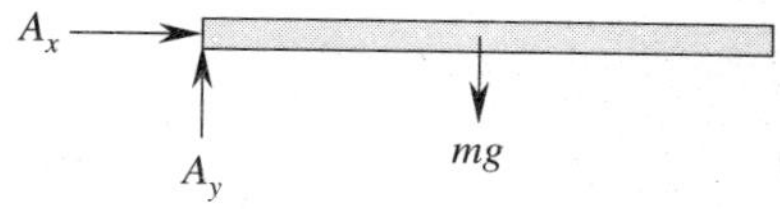

Fig. 18.12(a)

Since the mass moment of inertia of the bar about an axis passing through the hinge is $I_A = ml^2/3$,

$$mg\frac{l}{2} = \frac{ml^2}{3}\alpha$$

$$= \frac{2ml}{3}\left[\frac{l}{2}\alpha\right] = \frac{2ml}{3}a$$

where a is the linear acceleration of the midpoint and it is given as

$$a = \frac{3}{4}g$$

When the bar swings to the vertical position, the centre of mass would have traversed a vertical distance of $l/2$. Therefore, applying the kinematic equation of motion of the centre of mass, we have

$$v^2 = v_o^2 + 2as$$

$$= 0 + 2\left[\frac{3}{4}g\right]\left[\frac{l}{2}\right]$$

Therefore, $$v_{\text{cm}} = \sqrt{\frac{3gl}{4}}$$

At the instant that the string is cut, the linear acceleration of the midpoint is a in the vertically downward direction. Note that the normal component of acceleration is zero as the bar starts from rest. Hence, applying the kinetic equation for tangential motion,

$$\sum F_y = ma_y \Rightarrow$$

$$mg - A_y = ma = m\frac{3}{4}g$$

$$\Rightarrow \qquad A_y = \frac{mg}{4}$$

Example 18.11 In the above problem, suppose the bar is hinged at C such that $AC = l/4$. Determine the accelerations of ends A and B, and the reaction at the hinge point C.

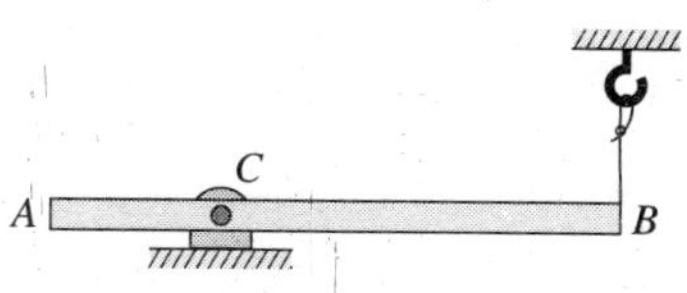

Fig. 18.13

Solution The free-body diagram of the bar is shown in Fig. 18.13(a). The net external moment acting on it causing it to rotate about the hinge is equal to the moment of the forces acting on the bar about point C, i.e., $M_C = mg(l/4)$. Hence, writing the kinetic equation for angular motion,

$$M_C = I_C \alpha$$

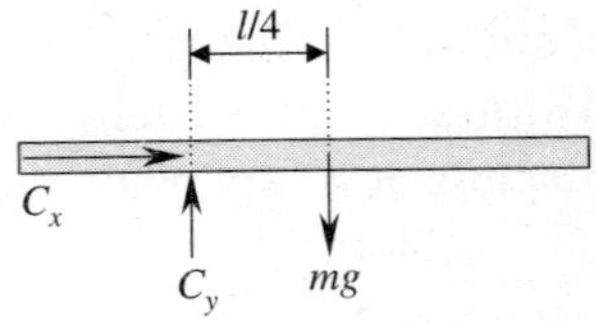

Fig. 18.13(a)

Since the mass moment of inertia of the bar about an axis passing through the hinge is $I_C = ml^2/12 + m(l/4)^2 = 7ml^2/48$,

$$\frac{mgl}{4} = \frac{7ml^2}{48}\alpha$$

$$\Rightarrow \quad \alpha = \frac{12g}{7l}$$

Therefore, the tangential accelerations of ends A and B are respectively given as

$$a_A = r_A\alpha = \left[\frac{l}{4}\right]\left[\frac{12g}{7l}\right] = \frac{3g}{7} \text{ vertically upwards}$$

and

$$a_B = r_B\alpha = \left[\frac{3l}{4}\right]\left[\frac{12g}{7l}\right] = \frac{9g}{7} \text{ vertically downwards}$$

At the instant that the string is cut, the linear acceleration of the centre of mass is in the vertically downward direction. Note that the normal component of acceleration, i.e., a_x is zero as the bar starts from rest. Hence, applying the kinetic equation for tangential motion,

$$\sum F_y = ma_y \Rightarrow$$

$$mg - C_y = ma_{\text{cm}} = m\left[\frac{l}{4}\right]\left[\frac{12g}{7l}\right] = \frac{3mg}{7}$$

$$\Rightarrow \quad C_y = \frac{4mg}{7}$$

Example 18.12 A wooden metre stick AB of 300g mass hangs vertically as shown in Fig. 18.14. If a horizontal force of 2 N is applied at a point that is 20 cm from the bottom end B, determine (i) the angular acceleration of the stick, (ii) the components of reaction at the hinge at A. In addition, determine the point of application of the horizontal force at which the horizontal component of the reaction at A is zero.

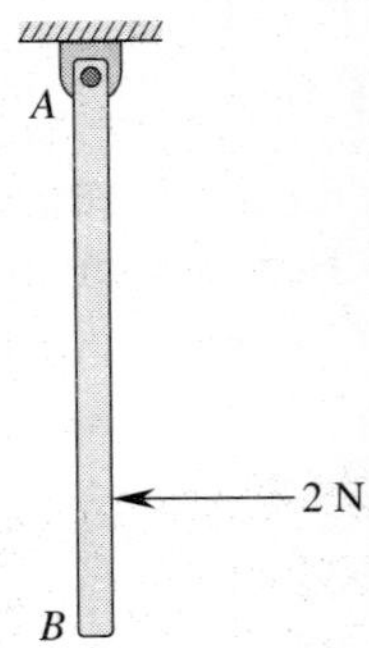

Fig. 18.14

Solution The free-body diagram of the stick is shown in Fig. 18.14(a). Since the externally applied force does not pass through the point of suspension A, it alone contributes to the externally applied moment tending to rotate the stick about A.

(i) *The angular acceleration of the stick*

Applying the kinetic equation for rotational motion, we have

$$M_A = I_A\alpha$$

We know that the moment of the applied force about A is $M_A = (2)(0.8) = 1.6$ N.m and the mass moment of inertia of the stick about A is $I_A = ml^2/3$. Therefore, we can write

$$1.6 = \frac{(0.3)(1)^2}{3}\alpha$$

$$\Rightarrow \quad \alpha = 16 \text{ rad/s}^2$$

Therefore, the tangential acceleration of the centre of mass is given as

$$(a_{cm})_x = r_{cm}\,\alpha = (0.5)(16) = 8 \text{ m/s}^2$$

Note that as the stick is initially at rest, its normal component of acceleration, i.e. $(a_{cm})_y$ is zero.

(ii) *The components of reaction at the hinge at A*

Applying the kinetic equations for translational motion, assuming all the forces are acting at the centre of mass, we have

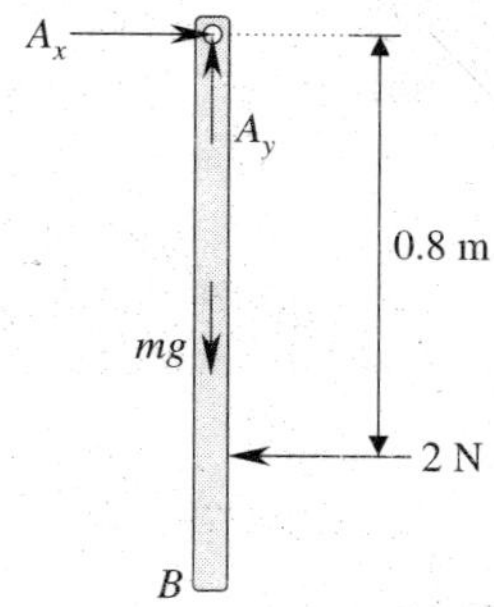

Fig. 18.14(a)

$$\sum F_x = m(a_{cm})_x \Rightarrow$$

$$2 - A_x = m(a_{cm})_x$$

$$2 - A_x = (0.3)(8)$$

$$\Rightarrow \qquad A_x = -0.4 \text{ N}$$

$$\sum F_y = m(a_{cm})_y \Rightarrow$$

$$A_y - mg = 0$$

$$\Rightarrow \qquad A_y = (0.3)(9.81) = 2.943 \text{ N}$$

(iii) The point of application of the horizontal force at which the horizontal component of the reaction at A is zero is determined by equating A_x to zero. Then the horizontal component of acceleration of centre of mass is given as

$$2 - 0 = (0.3)(a_{cm})_x$$

$$\Rightarrow \qquad (a_{cm})_x = 6.67 \text{ m/s}^2$$

Hence, the angular acceleration of the stick is given as

$$\alpha = (a_{cm})_x / r_{cm} = 13.34 \text{ rad/s}^2$$

Therefore, applying the kinetic equation of rotation, we have

$$M_A = I_A \alpha$$

$$(2)(h) = \frac{(0.3)(1)^2}{3}(13.34)$$

$$\Rightarrow \qquad h = 0.667 \text{ m}$$

18.5.4 General Plane Motion of a Rigid Body

The general plane motion of a rigid body can be analyzed by determining the resultant of the forces acting on the body and applying the two kinetic equations of translational motion. We can determine the components of acceleration of centre of mass using

$$\sum F_x = m(a_{cm})_x \qquad (18.21')$$

and

$$\sum F_y = m(a_{cm})_y \qquad (18.22')$$

Similarly, determining the moment resultant of the forces about the centre of mass G and applying the kinetic equation of motion for fixed-axis rotation, we can determine the angular acceleration of the body.

$$\sum M_G = \bar{I}\alpha \tag{18.33'}$$

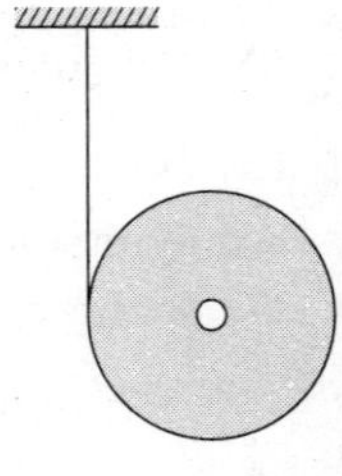

Fig. 18.15

Example 18.13 A string is wound several times around a solid cylinder of 2 kg mass. The free end of the string is fixed to the ceiling and the cylinder is released from rest. Determine its velocity after it has fallen through a height of 2 m. In addition, determine the tension in the string.

Solution The free-body diagram of the cylinder is shown in Fig. 18.15(a). The forces acting on it are its own weight acting at its centre of mass and tension T in the string. As these forces are non-concurrent, they not only cause translational motion but also cause it to rotate. We can replace the system of forces by an equivalent force at the centre of mass and resultant moment about the centre of mass and apply the kinetic equations of motion for general plane motion. Applying the kinetic equation of motion along vertical direction,

$$\sum F_y = ma_y \Rightarrow$$

$$mg - T = ma \tag{a}$$

and kinetic equation for rotational motion,

$$\sum M_z = \bar{I}\alpha \Rightarrow$$

$$Tr = \frac{mr^2}{2}\alpha = \frac{mr}{2}(r\alpha)$$

$$\Rightarrow \qquad T = \frac{m}{2}a \tag{b}$$

Note that there is no force acting along the X-direction. By adding equations (a) and (b), we can eliminate the tension T and hence,

$$mg = ma + \frac{m}{2}a = \frac{3}{2}ma$$

$$\Rightarrow \qquad a = \frac{2}{3}g$$

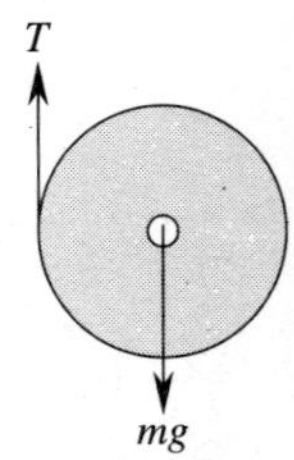

Fig. 18.15(a)

Using the kinematic equation of motion,

$$v^2 = v_o^2 + 2as$$

$$= 0 + 2\left[\frac{2}{3}g\right](2)$$

$$\therefore \qquad v = 5.11 \text{ m/s}$$

From the equation (b), we get the tension in the string as

$$T = \frac{m}{2}\frac{2}{3}g = \frac{mg}{3} = \frac{2\times 9.81}{3} = 6.54 \text{ N}$$

Example 18.14 A solid cylinder rolls down an incline without slipping. Determine the acceleration of the mass centre. In addition, determine its velocity after rolling down through a vertical height h.

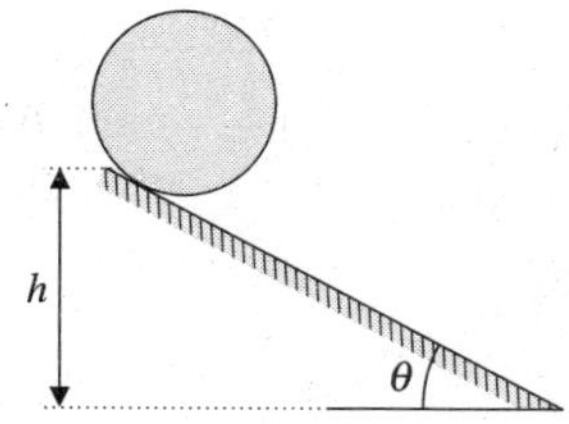

Fig. 18.16

Solution The free-body diagram of the cylinder is shown in Fig. 18.16(a). The forces acting on it are the components of its weight $mg \sin\theta$ and $mg \cos\theta$, normal reaction N and force of friction f acting upward. Applying the kinetic equations for translational motion along the inclined plane and normal to the plane,

$$\sum F_y = ma_y \Rightarrow$$

$$N - mg\cos\theta = 0 \quad \text{(a)}$$

$$\therefore \quad N = mg\cos\theta$$

$$\sum F_x = ma_x \Rightarrow$$

$$mg\sin\theta - f = ma \quad \text{(b)}$$

Similarly, applying the kinetic equation for rotational motion,

$$\sum M_z = \bar{I}\alpha \Rightarrow$$

[The lines of action of all the other forces except the force of friction pass through the centre of mass and hence they do not contribute to the moment.]

$$fr = \frac{mr^2}{2}\alpha = \frac{mr}{2}(r\alpha)$$

$$\Rightarrow \quad f = \frac{m}{2}a \quad \text{(c)}$$

Substituting the value of f in the equation (b), we get

$$mg\sin\theta - \frac{m}{2}a = ma$$

$$mg\sin\theta = \frac{3}{2}ma$$

or

$$a = \frac{2}{3}g\sin\theta$$

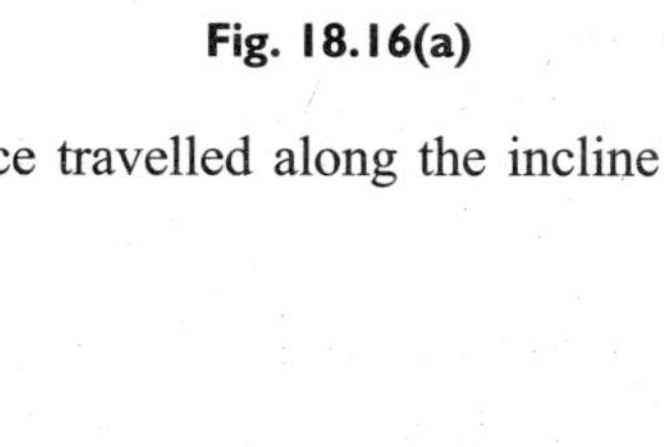

Fig. 18.16(a)

Using the kinematic equation of motion and noting that the distance travelled along the incline is $h/\sin\theta$,

$$v^2 = v_o^2 + 2as$$

$$= 0 + 2\left[\frac{2g\sin\theta}{3}\right]\frac{h}{\sin\theta}$$

$$= \frac{4}{3}gh$$

$$\therefore \quad v = \sqrt{\frac{4}{3}gh}$$

We see that the speed of the centre of mass of the rolling cylinder is less than the speed had it slid down a *frictionless* incline, where $v = \sqrt{2gh}$.

Substituting the value of acceleration in the equation (c), we get

$$f = \frac{m}{2} a = \frac{m}{2}\frac{2}{3} g \sin\theta = \frac{mg \sin\theta}{3}$$

This is the minimum force of static friction required for rolling.

Example 18.15 Solve the above problem, if (i) a sphere is released from rest, and (ii) an annular ring is released. Assume that they roll without slipping. If a cylinder, sphere and an annular ring are released from rest at the same instant from the top of an incline, state which of the three would reach the bottom of the incline at the earliest.

Solution

Sphere

Since the mass moment of inertia of a sphere of mass m and radius r is $(2/5)mr^2$, the kinetic equation for rotational motion can be written as

$$fr = \frac{2}{5} mr^2 \alpha = \frac{2}{5} mr\,a$$

$$\Rightarrow \quad f = \frac{2}{5} ma$$

The other equations (a) and (b) remaining the same, substituting the value of f in the equation (b), we have

$$mg \sin\theta - \frac{2}{5} ma = ma$$

$$\Rightarrow \quad mg \sin\theta = \frac{7}{5} ma$$

$$\text{or} \quad a = \frac{5}{7} g \sin\theta$$

Annular ring

Since the mass moment of inertia of an annular ring of mass m and radius r is mr^2, the kinetic equation for rotational motion can be written as

$$fr = mr^2 \alpha = mr\,a$$

$$\Rightarrow \quad f = ma$$

Substituting the value of f in the equation (b),

$$mg \sin\theta - ma = ma$$

$$\Rightarrow \quad mg \sin\theta = 2ma$$

$$\text{or} \quad a = \frac{1}{2} g \sin\theta$$

The accelerations of the three bodies are listed below:

$$a_{\text{cylinder}} = \frac{2}{3}g\sin\theta = 0.67g\sin\theta$$

$$a_{\text{sphere}} = \frac{5}{7}g\sin\theta = 0.71g\sin\theta$$

$$a_{\text{ring}} = \frac{1}{2}g\sin\theta = 0.5g\sin\theta$$

Comparing the three values of accelerations respectively for cylinder, sphere and ring, we see that the sphere has greater acceleration and hence it reaches the bottom of the incline at the earliest.

Example 18.16 A sphere and a cylinder are released from the top of an incline of 10 m at the same instant. Determine which of the two reaches the bottom of the incline first and what is the position of the other body at that instant.

Solution From the conclusion in the previous example, we can note that the sphere reaches the bottom of the incline first as it has greater acceleration than that of the cylinder. Writing the kinematic equations of motion respectively for the sphere and the cylinder,

$$s_1 = (v_{1o})t_1 + (1/2)a_1t_1^2 \quad \text{(a)}$$

$$s_2 = (v_{2o})t_2 + (1/2)a_2t_2^2 \quad \text{(b)}$$

Since both the bodies start from rest, their initial velocities are zero. In addition, as we are interested in the position of the bodies at the same instant, their times of motion are same equal to t. Hence, we can write

$$s_1 = (1/2)a_1t^2 \quad \text{(c)}$$

$$s_2 = (1/2)a_2t^2 \quad \text{(d)}$$

Dividing the equation (c) by the equation (d), we can eliminate t and hence we have

$$\frac{s_1}{s_2} = \frac{a_1}{a_2} = \frac{(5/7)g\sin\theta}{(2/3)g\sin\theta} = \frac{15}{14}$$

When the sphere has travelled a distance of 10 m down the incline, the distance travelled by the cylinder is then given as

$$s_2 = \frac{14}{15}s_1 = \frac{14}{15}(10) = 9.33\text{ m}$$

Therefore, when the sphere reaches the bottom of the incline, the cylinder is at a distance of 0.67 m behind the sphere.

Example 18.17 A solid cylinder and an annular ring of equal masses m and radii r are connected at their centres by a bar as shown in Fig. 18.17. If released from rest, determine the acceleration of the system and the force in the connecting bar. Assume the bodies roll without slipping and neglect the weight of the connecting bar.

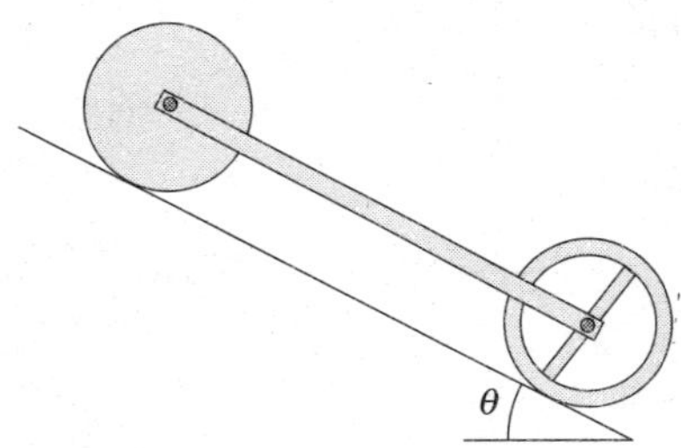

Fig. 18.17

Solution The free-body diagrams of the two bodies are shown in Fig. 18.17(a). It should be noted that as the cylinder and the hoop are connected at their centres, the acceleration of their centres of mass are same.

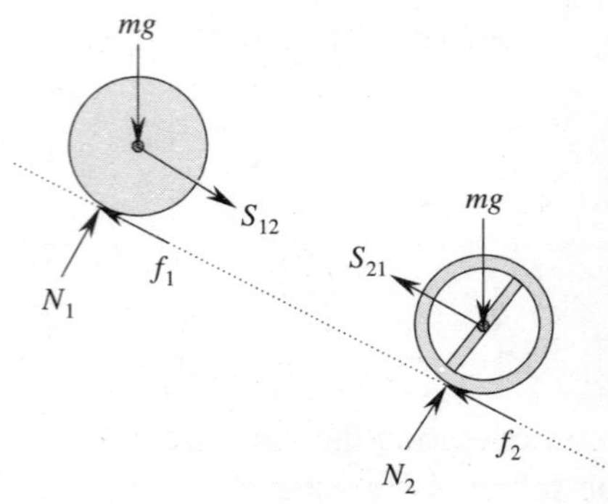

Fig. 18.17(a)

Cylinder

Applying the kinetic equations for translational motion along the inclined plane and normal to the plane,

$\sum F_y = ma_y \Rightarrow$

$$N_1 - mg\cos\theta = 0$$

$$\therefore \quad N_1 = mg\cos\theta$$

$\sum F_x = ma_x \Rightarrow$

$$S_{12} + mg\sin\theta - f_1 = ma \tag{a}$$

Similarly, applying the kinetic equation for rotational motion,

$\sum M_z = \bar{I}\alpha \Rightarrow$

$$f_1 r = \frac{mr^2}{2}\alpha$$

$$\therefore \quad f_1 = \frac{m}{2}a \tag{b}$$

Substituting the value of f_1 in the equation (a), we get

$$S_{12} + mg\sin\theta = \frac{3}{2}ma \tag{c}$$

Annular ring

Similarly, writing the equations of motion for annular ring,

$\sum F_y = ma_y \Rightarrow$

$$N_2 - mg\cos\theta = 0$$

$$\therefore \quad N_2 = mg\cos\theta$$

$\sum F_x = ma_x \Rightarrow$

$$mg\sin\theta - f_2 - S_{21} = ma \tag{d}$$

and $\sum M_z = \bar{I}\alpha \Rightarrow$

$$f_2 r = mr^2\alpha$$

$$\Rightarrow \quad f_2 = ma \tag{e}$$

Substituting the value of f_2 in the equation (d), we get

$$mg\sin\theta - S_{21} = 2ma \tag{f}$$

Adding equations (c) and (f), noting that $S_{12} = S_{21}$,

$$a = \frac{4}{7} g \sin\theta$$

Substituting the value of a in either of the equations (c) or (f), we get

$$S_{12} = S_{21} = \frac{-1}{7} mg \sin\theta$$

The negative sign indicates that the bar is under compression.

Example 18.18 A horizontal force P is applied on a cylinder as shown in Fig. 18.18. Determine the acceleration of centre of mass of the cylinder assuming there is no slipping.

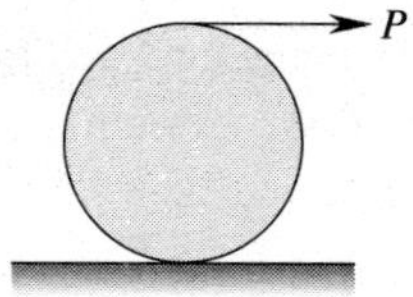

Fig. 18.18

Solution Taking moment of the externally applied force about the contact point, we have

$$M_A = P(2r) = 2Pr$$

Therefore, applying the kinetic equation of rotation,

$$M_A = I_A \alpha \quad \text{(a)}$$

The moment of inertia of the cylinder about a perpendicular axis passing through the contact point is

$$I_A = \bar{I} + mr^2$$

$$= \frac{mr^2}{2} + mr^2 = \frac{3}{2} mr^2$$

Substituting the values in the equation (a), we have

$$2Pr = \frac{3}{2} mr^2 \alpha$$

$$\Rightarrow \quad \alpha = \frac{4P}{3mr}$$

Therefore, the acceleration of centre of mass is given as

$$a = r\alpha = \frac{4P}{3m}$$

18.6 WORK–ENERGY METHOD

Just as we solved the kinetics of particles by an alternative method, namely, work–energy method, we will solve kinetics of rigid bodies using work–energy method. We already know the advantages of this method over the kinetic equations of motion in that the quantities involved, namely, work and energy are *scalars*, and that *displacement* and *velocity* could be determined without knowing the acceleration.

18.6.1 Work Done by Forces Acting on a Rigid Body

Consider a force $\vec{F}$ acting on a particle in the rigid body. If the body is in pure rotational motion, then the particle we have considered is representative of all the particles in the body as the angular displacement, velocity and acceleration are *same* for every particle in the body.

In a small interval of time dt, the particle would have undergone an infinitesimal angular displacement $d\theta$ and hence we can write

$$ds = rd\theta \tag{18.35}$$

Therefore, work done by the force is given as scalar product of force and displacement vectors, i.e.,

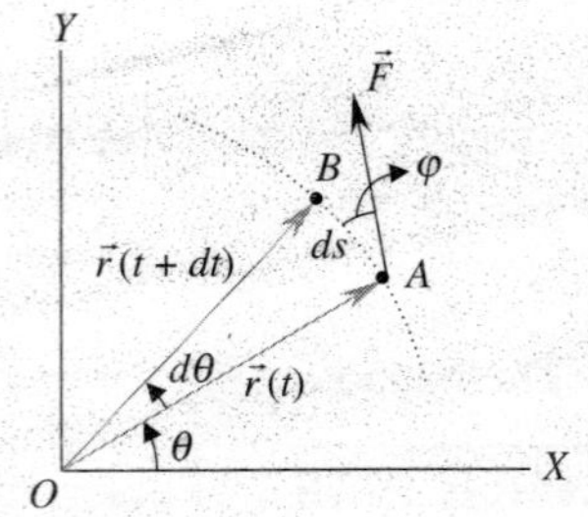

Fig. 18.19 Particle undergoing angular displacement

$$\begin{aligned} dW &= \vec{F}.d\vec{s} \\ &= F(ds)\cos\varphi \\ &= F(rd\theta)\cos\varphi \\ &= (Fr\cos\varphi)d\theta \end{aligned} \tag{18.36}$$

Similarly, the rotational effect of $\vec{F}$, i.e., moment about the origin causes angular displacement and we can determine the work done in causing this angular displacement as follows:

We know that the magnitude of moment of the applied force about the origin O is

$$M = Fr\sin(90° - \varphi) = Fr\cos\varphi \tag{18.37}$$

Therefore, work done can be written as

$$dW = Md\theta \tag{18.38}$$

Hence, the total work done in displacing from the point 1 to the point 2 is obtained by integrating between limits:

$$W_{1\text{-}2} = \int_{\theta_1}^{\theta_2} Md\theta$$

If the applied moment is constant then

$$W_{1\text{-}2} = M(\theta_2 - \theta_1) \tag{18.39}$$

Instantaneous power is defined as the rate of change of work done. Mathematically,

$$P = \frac{dW}{dt} = M\frac{d\theta}{dt} = M\omega \tag{18.40}$$

18.6.2 Kinetic Energy of a Rigid Body

As we derived the kinetic equations of motion first for a system of particles and then applied them to kinetics of rigid bodies, we will proceed in a similar manner to derive the mathematical expression for kinetic energy of a rigid body also. Consider a system of n particles. A particle in motion is said to possess energy called kinetic energy. If m_i is the mass of the ith particle and $\vec{v}_i$ is its velocity, its kinetic energy is expressed as

$$(\text{K.E})_i = \frac{1}{2} m_i v_i^2 \tag{18.41}$$

Since energy is a scalar, the total kinetic energy of the system of particles is given by the algebraic sum of kinetic energies of individual particles. Hence,

$$\text{K.E} = \frac{1}{2} \sum_{i=1}^{n} m_i v_i^2 \tag{18.42}$$

Since v_i^2 can be expressed as a scalar product of $\vec{v}_i . \vec{v}_i$, we can also write

$$\text{K.E} = \frac{1}{2} \sum_{i=1}^{n} (m_i \vec{v}_i . \vec{v}_i) \tag{18.43}$$

Since the absolute velocity of the particle $\vec{v}_i$ can be expressed as

$$\vec{v}_i = \vec{v}_{\text{cm}} + \vec{v}_i' \tag{18.44}$$

we have,

$$\text{K.E} = \frac{1}{2} \sum_{i=1}^{n} m_i (\vec{v}_{\text{cm}} + \vec{v}_i').(\vec{v}_{\text{cm}} + \vec{v}_i')$$

$$= \frac{1}{2}\left(\sum_{i=1}^{n} m_i\right) v_{\text{cm}}^2 + \vec{v}_{\text{cm}}.\left(\sum_{i=1}^{n} m_i \vec{v}_i'\right) + \frac{1}{2}\left(\sum_{i=1}^{n} m_i v_i'^2\right)$$

We know that $\sum_{i=1}^{n} m_i \vec{v}_i'$ is equal to $m\vec{v}'_{\text{cm}}$, which is zero for a centroidal frame of reference. Hence, we can write

$$\text{K.E} = \frac{1}{2} m v_{\text{cm}}^2 + \frac{1}{2} \sum_{i=1}^{n} m_i v_i'^2 \tag{18.45}$$

Suppose the body is rigid and it is under fixed-axis rotation. Then every particle in the body has the same angular velocity ω about the centroidal frame of reference. Hence, the above equation can be written as

$$\text{K.E} = \frac{1}{2} m v_{\text{cm}}^2 + \frac{1}{2} \sum_{i=1}^{n} \Delta m_i r_i'^2 \omega^2$$

$$= \frac{1}{2} m v_{\text{cm}}^2 + \omega^2 \frac{1}{2} \sum_{i=1}^{n} \Delta m_i r_i'^2$$

$$\therefore \quad \text{K.E} = \frac{1}{2} m v_{\text{cm}}^2 + \frac{1}{2} \bar{I} \omega^2 \tag{18.46}$$

Thus, the total kinetic energy of a rigid body can be expressed as a sum of translational energy of its centre of mass and kinetic energy of rotation about its centre of mass.

18.6.3 Work–Energy Principle for a Rigid Body

The work–energy principle for a rigid body states that the change in kinetic energy of the rigid body during any displacement is equal to the work done by the net force acting on it.

$$W = (\text{K.E})_{\text{final}} - (\text{K.E})_{\text{initial}}$$
$$= \text{change in K.E} \tag{18.47}$$

18.6.4 Conservation of Mechanical Energy

When a rigid body moves under the action of a conservative force such as gravitational force, the total mechanical energy is conserved for any position in the force field, i.e.,

$$(\text{P.E}) + (\text{K.E}) = \text{constant} \tag{18.48}$$

or

$$(\text{P.E})_1 + (\text{K.E})_1 = (\text{P.E})_2 + (\text{K.E})_2 \tag{18.49}$$

Example 18.19 The speed of a flywheel rotating at 200 rpm is uniformly increased to 300 rpm in 5 seconds. Determine the work done by the driving torque and the increase in kinetic energy during this time. What do you infer from the result? Take mass of the flywheel as 25 kg and its radius of gyration as 20 cm.

Solution Initial speed of the flywheel,

$$\omega_o = 200 \text{ rpm} = 6.67\pi \text{ rad/s}$$

and its final speed,

$$\omega = 300 \text{ rpm} = 10\pi \text{ rad/s}$$

Mass moment of inertia of the flywheel about its centroidal axis is

$$\bar{I} = mk^2 = (25)(0.2)^2 = 1 \text{ kg.m}^2$$

Since the angular acceleration is uniform, we can use the kinematic equation

$$\omega = \omega_o + \alpha t$$

$$\Rightarrow \quad \alpha = \frac{\omega - \omega_o}{t}$$

$$= \frac{10\pi - 6.67\pi}{5} = 2.09 \text{ rad/s}^2$$

In addition, the angular displacement during this time is obtained as

$$\omega^2 = \omega_o^2 + 2\alpha\theta$$

$$\Rightarrow \quad \theta = \frac{\omega^2 - \omega_o^2}{2\alpha}$$

$$= \frac{(10\pi)^2 - (6.67\pi)^2}{2(2.09)} = 131.07 \text{ rad}$$

Since the angular acceleration is constant, the driving torque is constant and hence applying the kinetic equation of motion about fixed axis,

$$M = \bar{I}\alpha$$
$$= (1)(2.09) = 2.09 \text{ N.m}$$

Therefore, work done by the driving torque is given as

$$W = M(\theta_2 - \theta_1)$$
$$= (2.09)(131.07) = 273.94 \text{ J}$$

The increase in kinetic energy is given as

$$\Delta(\text{K.E}) = (\text{K.E})_f - (\text{K.E})_i$$
$$= \frac{1}{2}\bar{I}\omega^2 - \frac{1}{2}\bar{I}\omega_o^2$$
$$= \frac{1}{2}\bar{I}(\omega^2 - \omega_o^2)$$
$$= \frac{1}{2}(1)[(10\pi)^2 - (6.67\pi)^2] = 273.94 \text{ J}$$

Thus, we see that the work–energy principle holds.

Example 18.20 A flywheel of 15 kg mass and 20 cm radius of gyration is directly coupled to an electric motor, which can develop 10 kW power when rotating at a speed of 1200 rpm. Determine the driving torque to maintain this speed. If power is switched off and the flywheel comes to rest in 20 seconds, determine the uniform retarding torque on the flywheel.

Solution The initial speed of the flywheel is ω_o = 1200 rpm = 40π rad/s

The instantaneous power delivered to the flywheel is given as

$$P = M\omega$$

Therefore, the driving torque acting on the flywheel to maintain this speed is given as

$$M = \frac{P}{\omega} = \frac{10 \times 10^3}{40\pi} = 79.58 \text{ N.m}$$

When power is switched off, the final speed is zero. Hence, using the kinematic equation,

$$\omega = \omega_o + \alpha t$$
$$0 = 40\pi + \alpha(20)$$
$$\Rightarrow \qquad \alpha = -6.28 \text{ rad/s}^2$$

Therefore, the retarding torque acting on the flywheel is given as

$$M = \bar{I}\alpha = (mk^2)(\alpha)$$
$$= 15 \times (0.2)^2 (-6.28) = -3.77 \text{ N.m}$$

Example 18.21 In Example 18.3, determine the velocity of the block when it has fallen through a height of 2 m, using the work–energy method.

Solution As the block falls down by a height of 2 m, there is a loss in its potential energy. By the principle of conservation of energy, we know that this loss in potential energy is equal to the increase in kinetic energy of the system consisting of the block and the pulley.

Loss in potential energy = $m_1 gh$

Increase in kinetic energy of the system = $\frac{1}{2}m_1 v^2 + \frac{1}{2}\bar{I}\omega^2$

Equating the two, we have

$$m_1 gh = \frac{1}{2}m_1 v^2 + \frac{1}{2}\bar{I}\omega^2$$

$$= \frac{1}{2}m_1 v^2 + \frac{1}{2}\frac{mr^2}{2}\omega^2$$

$$= \frac{1}{2}m_1 v^2 + \frac{1}{4}mv^2 \quad \text{[since } v = r\omega\text{]}$$

$$= \frac{1}{4}(2m_1 + m)v^2$$

$$\Rightarrow \quad v^2 = \frac{4m_1 gh}{2m_1 + m}$$

Substituting the values, we have

$$v^2 = \frac{4(30)(2)}{2(30/9.81) + 20} = 9.2$$

$$\therefore \quad v = 3.03 \text{ m/s}$$

Example 18.22 In the Example 18.10, determine the velocity of the bar when it reaches the vertical position using the work–energy method. In addition, determine the vertical reaction at A at the instant the bar reaches the vertical position.

Solution In this example, we observe that the bar undergoes fixed axis rotation about hinge point A. Therefore, applying the principle of conservation of energy,

loss in P.E = gain in kinetic energy

$$mgh = \frac{1}{2}I_A\omega^2$$

When the bar swings to the vertical position, we see that the height through which the centre of mass (at which the entire weight, mg assumed to be concentrated) displaces is $l/2$. Hence,

$$mg\frac{l}{2} = \frac{1}{2}\frac{ml^2}{3}\omega^2$$

$$\Rightarrow \quad \omega = \sqrt{\frac{3g}{l}}$$

Therefore, the velocity of the centre of mass of the bar is given as

$$v_{cm} = r_{cm}\,\omega$$

$$= \frac{l}{2}\sqrt{\frac{3g}{l}} = \sqrt{\frac{3gl}{4}}$$

To determine the vertical reaction at A at the instant the bar reaches the vertical position, we must apply the kinetic equation of motion in the vertical direction,

$$\sum F_y = ma_y \Rightarrow$$

$$mg - A_y = ma_r = -mr\omega^2$$

$$\Rightarrow \qquad mg - A_y = -m\frac{l}{2}\frac{3g}{l}$$

Therefore, $\qquad A_y = 2.5mg$

Example 18.23 In Example 18.11, determine the angular velocity of the bar when it swings to the vertical position.

Solution Applying the principle of conservation of energy,

loss in P.E = gain in kinetic energy

$$mgh = \frac{1}{2}I_C\omega^2$$

When the bar swings to the vertical position, we see that the height through which the centre of mass (at which the entire weight, *mg* assumed to be concentrated) displaces is $l/4$. Hence,

$$mg\frac{l}{4} = \frac{1}{2}\left[\frac{ml^2}{12} + m(l/4)^2\right]\omega^2$$

$$\Rightarrow \qquad \omega = \sqrt{\frac{24g}{7l}}$$

Example 18.24 In Example 18.11, determine the location of the hinge point C from the centre of mass such that the angular velocity of the bar when it swings to the vertical position is maximum.

Solution Let a be the distance of the hinge point C from the centre of mass then applying the principle of conservation of energy as before, we have

$$mga = \frac{1}{2}\left[\frac{ml^2}{12} + ma^2\right]\omega^2$$

$$\Rightarrow \qquad \omega^2 = \frac{24ga}{l^2 + 12a^2}$$

We know from calculus that the angular velocity ω is maximum when $d\omega/da = 0$. Hence, differentiating the above equation with respect to the variable a, we have

$$2\omega\frac{d\omega}{da}=\frac{(l^2+12a^2)24g-24ga(24a)}{(l^2+12a^2)^2}=\frac{24g(l^2-12a^2)}{(l^2+12a^2)^2}$$

Taking $d\omega/da = 0$, we get

$$a=\frac{l}{\sqrt{12}}$$

Example 18.25 A homogeneous steel bar AB of length l hinged at end A and end B attached to a string is held inclined at an angle of 30° to the horizontal as shown in Fig. 18.20. If the string at B is snapped, determine the reaction at the hinge point A at the instant the bar passes through the horizontal position.

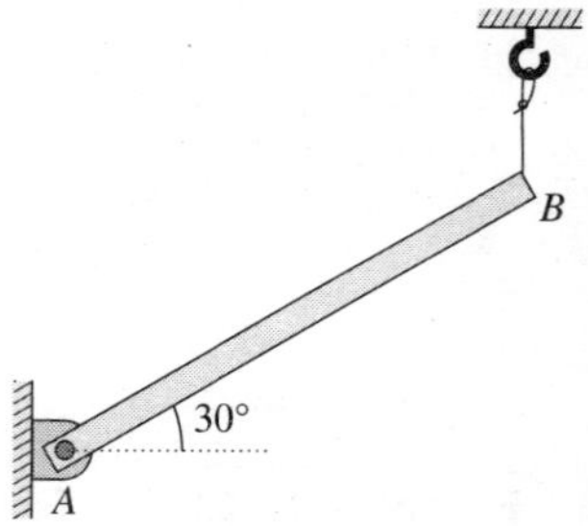

Fig. 18.20

Solution When the bar swings to the horizontal position, we see that the height through which the centre of mass displaces is $\frac{l}{2}\sin 30°$.

Applying the principle of conservation of energy,

loss in P.E = gain in kinetic energy

$$mg\frac{l}{2}\sin 30° = \frac{1}{2}\frac{ml^2}{3}\omega^2$$

$$\Rightarrow \qquad \omega=\sqrt{\frac{3g}{2l}}$$

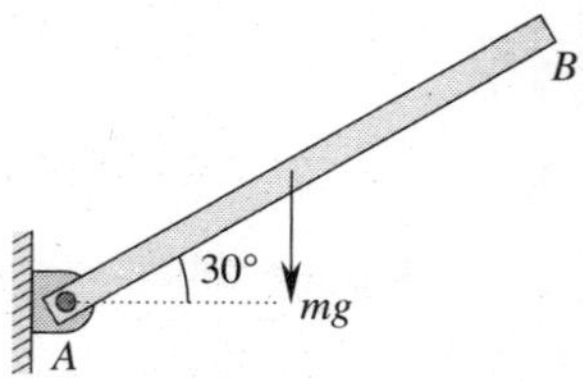

Fig. 18.20(a)

Therefore, the radial acceleration of the centre of mass of the bar at the horizontal position is given as

$$a_x = r_{cm}\,\omega^2=\frac{l}{2}\frac{3g}{2l}=\frac{3g}{4}$$

To determine the instantaneous tangential acceleration of the centre of mass in the horizontal position, we apply the kinetic equation of motion for rotation,

$\sum M_z = \bar{I}\alpha \Rightarrow$

$$mg\frac{l}{2}=\frac{ml^2}{3}\alpha$$

$$\Rightarrow \qquad \alpha=\frac{3g}{2l}$$

Therefore, the tangential acceleration of the centre of mass is given as

$$a_y = r_{cm}\,\alpha=\frac{l}{2}\frac{3g}{2l}=\frac{3g}{4}$$

To determine the reaction at A, we apply the kinetic equations of translational motion,

$\sum F_y = ma_y \Rightarrow$

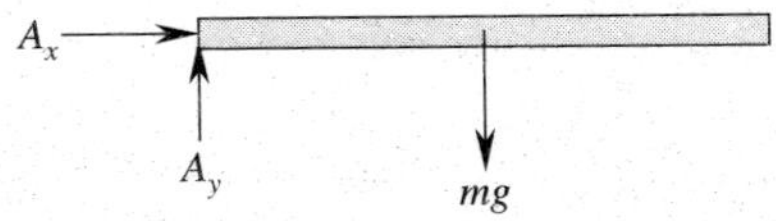

Fig. 18.20(b)

$$mg - A_y = ma_t = \frac{3mg}{4}$$

Therefore, $$A_y = \frac{mg}{4}$$

$\sum F_x = ma_x \Rightarrow$

$$A_x = mr_{cm}\omega^2$$

$$= m\frac{l}{2}\frac{3g}{2l}$$

$\therefore$ $$A_x = \frac{3mg}{4}$$

Example 18.26 In Example 18.14, determine the velocity of the cylinder using work–energy method.

Solution Here, as the cylinder undergoes general plane motion, we see that the kinetic energy of the body is the sum of translational and rotational kinetic energies. Applying the work–energy principle,

Work done = change in kinetic energy

$$mgh = \frac{1}{2}mv_{cm}^2 + \frac{1}{2}\bar{I}\omega^2$$

$$= \frac{1}{2}mv_{cm}^2 + \frac{1}{2}\frac{mr^2}{2}\omega^2$$

$$= \frac{1}{2}mv_{cm}^2 + \frac{1}{4}mv_{cm}^2 \quad [\text{since } v_{cm} = r\omega]$$

$$= \frac{3}{4}mv_{cm}^2$$

$\Rightarrow$ $$v_{cm} = \sqrt{\frac{4}{3}gh}$$

Example 18.27 In the Example 18.13, determine the velocity of the cylinder using work–energy method after it has fallen through a height of 2 m.

Solution Applying the work–energy principle,

Work done = change in kinetic energy

$$mgh = \frac{1}{2}mv_{cm}^2 + \frac{1}{2}\bar{I}\omega^2$$

$$= \frac{1}{2}mv_{cm}^2 + \frac{1}{2}\frac{mr^2}{2}\omega^2$$

$$= \frac{1}{2}mv_{cm}^2 + \frac{1}{4}mv_{cm}^2 \quad [\text{since } v_{cm} = r\omega]$$

$$= \frac{3}{4}mv_{cm}^2$$

$$\Rightarrow \qquad v_{cm} = \sqrt{\frac{4}{3}gh} = \sqrt{\frac{4}{3}\times 9.81\times 2} = 5.11 \text{ m/s}$$

Example 18.28 A cylinder of 10 kg mass and 10 cm radius rolls without slipping on a horizontal surface. If it moves with a velocity of 3 m/s, determine (i) its kinetic energy of translation, (ii) kinetic energy of centroidal rotation.

Solution

(i) The translational velocity of the centre of mass is given as 3 m/s. Hence, its kinetic energy of translation is given as

$$\text{K.E} = \frac{1}{2}mv_{cm}^2$$

$$= \frac{1}{2}(10)(3)^2 = 45 \text{ J}$$

(ii) Since the cylinder is rolling without slipping, we know that its angular velocity is given as

$$\omega = v/r = 3/0.1 = 30 \text{ rad/s}$$

Therefore, its kinetic energy due to centroidal rotation is given as

$$\text{K.E} = \frac{1}{2}\bar{I}\omega^2$$

$$= \frac{1}{2}\frac{mr^2}{2}\omega^2$$

$$= \frac{1}{2}\frac{(10)(0.1)^2}{2}(30)^2 = 22.5 \text{ J}$$

Example 18.29 A constant force of 100 N is applied as shown tangentially on a cylinder at rest, whose mass is 50 kg and radius is 10 cm, for a distance of 5 m. Determine the angular velocity of the cylinder and the velocity of its centre of mass. Assume that there is no slip.

Solution Since the applied force is horizontal and the displacement is in the direction of the force, the work done by the force in causing a displacement s is given as

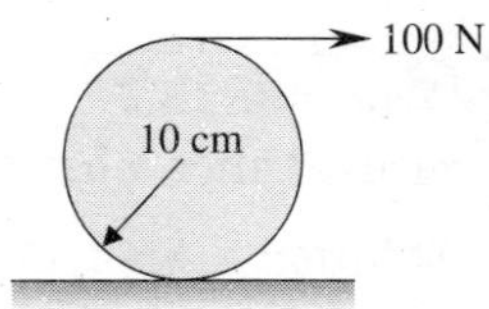

Fig. 18.21

$$W = Fs$$

Applying the work–energy principle,

Work done = change in kinetic energy

$$Fs = \frac{1}{2}mv^2 + \frac{1}{2}\bar{I}\omega^2$$

$$= \frac{1}{2} mr^2 \omega^2 + \frac{1}{2} \frac{mr^2}{2} \omega^2$$

$$= \frac{3}{4} mr^2 \omega^2$$

$$\Rightarrow \qquad \omega^2 = \frac{4Fs}{3mr^2} = \frac{4(100)(5)}{3(50)(0.1)^2} = 1333.33$$

Therefore, $\omega = 36.51$ rad/s

Therefore, velocity of the centre of mass is given as

$$v_{cm} = r\omega$$

$$= (0.1)(36.51) = 3.651 \text{ m/s}$$

SUMMARY

The *kinematics* of rigid bodies deals with the geometry of motion of rigid bodies, while the *kinetics* of rigid bodies relates the cause of motion with the resulting motion. For analyzing the motion of a rigid body, it is assumed to be made up of a large number of particles, which are closely spaced and fixed relative to one another.

Translational Motion of a System of Particles

For a system of particles, we can write Newton's equation of motion as

$$\sum \vec{F}_{ext} = \sum m_i \vec{a}_i$$

where m_i is the mass of the ith particle and $\vec{a}_i$ is its acceleration, $\vec{F}_i$ is the external force acting on it. This equation can also be expressed in terms of the motion of centre of mass of the system as

$$\sum \vec{F}_{ext} = m\vec{a}_{cm}$$

The centre of mass of a system of particles moves in such a way as if the total mass of the system were concentrated at that point and all the external forces were acting at that point.

Rotational Motion of a System of Particles

The rotational motion of a system of particles can be expressed as

$$\sum \vec{M}_O = \sum (\vec{r}_i \times \vec{F}_i) = \sum (\vec{r}_i \times m_i \vec{a}_i)$$

where $\sum \vec{M}_O$ is the resultant moment of the system of forces about the origin O of the fixed reference frame.

If the rotational motion with respect to centre of mass is considered, we have

$$\sum \vec{M}_G = \sum (\vec{r}_i' \times m_i \vec{a}_i')$$

General Plane Motion of a Rigid Body

When a system of non-concurrent forces acts on a rigid body, we can replace them by a resultant force $\sum \vec{F}_i$ acting at the centre of mass together with the resultant moment $\sum \vec{M}_G$. The resultant force causes the translational motion of the rigid body and hence we can write

$$\sum \vec{F}_i = \sum \vec{F}_{\text{ext}} = m\vec{a}_{\text{cm}}$$

or in scalar forms

$$\sum F_x = m(a_{\text{cm}})_x \quad \text{and} \quad \sum F_y = m(a_{\text{cm}})_y$$

Similarly, the resultant moment causes the rotational motion of the rigid body and hence we can write

$$\sum \vec{M}_G = \sum (\vec{r}_i' \times \vec{F}_i) = \sum (\vec{r}_i' \times \Delta m_i \vec{a}_i')$$

which can also be expressed in compact form as

$$\sum \vec{M}_G = \frac{d\vec{H}_G}{dt}$$

where $\vec{H}_G$ is the *angular momentum* of the rigid body. Thus, moment resultant of the forces is equal to the rate of change of angular momentum. The magnitude of angular momentum can be expressed as

$$H_G = \bar{I}\omega$$

Hence, $$\sum \vec{M}_G = \bar{I}\alpha$$

Centroidal Rotation of a Rigid Body

For pure rotation of a rigid body about a perpendicular axis passing through the centre of mass, the kinetic equation of motion is

$$\sum M_G = \bar{I}\alpha$$

Non-centroidal Rotation of a Rigid Body

When a rigid body rotates about a perpendicular axis passing through a point O other than the centre of mass, the kinetic equations of motion are

$$\sum M_O = I_O\alpha = (\bar{I} + md^2)\alpha$$

$$\sum F_x = m(a_{\text{cm}})_x$$

and $$\sum F_y = m(a_{\text{cm}})_y$$

Work–Energy Method

Just as we solved the kinetics of particles by an alternative method, namely, work–energy method, we can also solve kinetics of rigid bodies using work–energy method. The advantages of this method over the kinetic equations of motion are that the quantities involved, namely, work and energy are scalars, and that displacement and velocity could be determined without knowing the acceleration.

Work Done by Forces Acting on a Rigid Body

The work done by a force acting on a particle in a rigid body in causing a displacement from the point 1 to the point 2 is given as

$$W_{1\text{-}2} = \int_{\theta_1}^{\theta_2} M d\theta$$

$$W_{1\text{-}2} = M(\theta_2 - \theta_1), \text{ for a constant applied moment } M$$

Instantaneous power is defined as the rate of change of work done. Mathematically,

$$P = \frac{dW}{dt} = M\frac{d\theta}{dt} = M\omega$$

Kinetic Energy of a Rigid Body

The total kinetic energy of a rigid body can be expressed as a sum of translational energy of its centre of mass and kinetic energy of rotation about its centre of mass. Mathematically,

$$\text{K.E} = \frac{1}{2}mv_{\text{cm}}^2 + \frac{1}{2}\bar{I}\omega^2$$

Work–Energy Principle for a Rigid Body

The work–energy principle for a rigid body states that the change in kinetic energy of the rigid body during any displacement is equal to the work done by the net force acting on it.

$$W = (\text{K.E})_{\text{final}} - (\text{K.E})_{\text{initial}}$$

$$= \text{change in K.E}$$

Conservation of Mechanical Energy

When a rigid body moves under the action of a conservative force such as gravitational force, the total mechanical energy is conserved for any position in the force field, i.e.,

$$(\text{P.E}) + (\text{K.E}) = \text{constant}$$

or

$$(\text{P.E})_1 + (\text{K.E})_1 = (\text{P.E})_2 + (\text{K.E})_2$$

EXERCISES

Objective-type Questions

1. A rigid body is made up of
   (a) a large number of particles
   (b) particles closely spaced
   (c) particles that remain at fixed distances relative to one another
   (d) all of these
2. The motion of the centre of mass represents the ________ motion of a rigid body.
   (a) translational (b) rotational
   (c) fixed axis rotational (d) general plane

3. The angular momentum of a rigid body is expressed mathematically as
   (a) $\bar{I}\alpha$ (b) $\bar{I}\omega$ (c) $\bar{I}\omega^2$ (d) mv
4. The rotational analog of Newton's equation of motion is
   (a) $\bar{I}\alpha$ (b) $\bar{I}\omega$ (c) $\bar{I}\omega^2$ (d) ma
5. The kinetic equation for general plane motion of a rigid body is
   (a) $F_x = m(a_{cm})_x, F_y = m(a_{cm})_y$ (b) $\sum M_G = \bar{I}\alpha$
   (c) $\sum M_O = I_O\alpha$ (d) $F_x = m(a_{cm})_x, F_y = m(a_{cm})_y, \sum M_G = \bar{I}\alpha$
6. Instantaneous power in fixed axis rotation is expressed mathematically as
   (a) $\bar{I}\alpha$ (b) $\bar{I}\omega$ (c) $M\alpha$ (d) $M\omega$

Answers

1. (d) **2.** (a) **3.** (b) **4.** (a) **5.** (d) **6.** (d)

Short-answer Questions

1. Distinguish between kinematics and kinetics of rigid body motion.
2. Explain how the sum of internal forces in a system of particles reduces to zero.
3. Describe the motion of the centre of mass in a system of particles.
4. Explain how the sum of moment of internal forces in a system of particles reduces to zero.
5. State the kinetic equations of motion in a (i) centroidal rotation, (ii) non-centroidal rotation, and (iii) general plane motion.
6. Define linear momentum and angular momentum. Express them in mathematical forms.
7. Derive the rotational analog of Newton's second law of motion.
8. Define work done by a force acting on a rigid body.
9. Define instantaneous power.
10. Express the kinetic energy of a rigid body as a sum of translational motion of its centre of mass and kinetic energy of rotation about its centre of mass.
11. State the work–energy principle and conservation of mechanical energy for a rigid-body motion.

Numerical Problems

18.1 A flywheel of 15 kg mass and 25 cm radius of gyration rotates at a constant angular speed of 1200 rpm. When the power supply is switched off, if it coasts to rest in 15 seconds, determine retarding torque due to friction in the bearings assuming it to be uniform.

Ans. 7.86 N.m

18.2 The water jet is shut off when a hydraulic turbine is rotating at a speed of 3000 rpm. If the friction in the bearings is uniform exerting a retarding torque of 20 N.m, determine the number of revolutions made by the rotor before coming to rest and the time taken to come to rest. The mass of the rotor is 20 kg and the radius of gyration is 15 cm.

Ans. 176.73 rev, 7.07 s

18.3 A wheel rotates freely about a horizontal axis. If a vertical force P of 50 N magnitude is applied to one of the spokes as shown in Fig. E.18.3, determine the resulting angular acceleration of the wheel. Take mass of the wheel as 3 kg and radius of gyration as 35 cm.

Ans. 28.86 rad/s^2

18.4 In the previous problem, if the hub of 8 cm diameter about which the wheel rotates offers a resistance of 20 N, determine the resulting angular acceleration of the wheel.

Ans. 26.68 rad/s^2

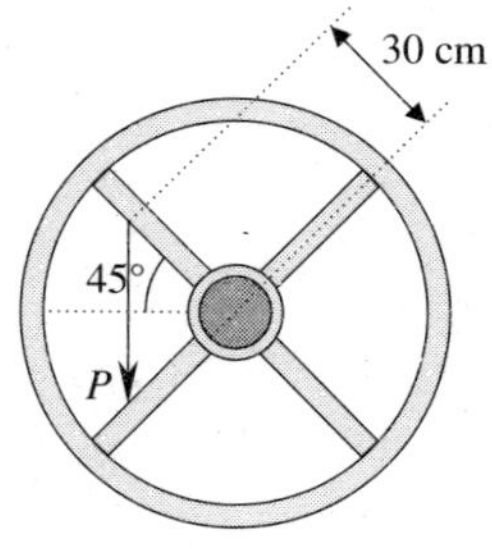

Fig. E.18.3, E.18.4

18.5 Water is drawn from a well using a bucket-and-pulley arrangement. If a constant force P is applied at the free end of the rope, determine the acceleration of the system. The pulley can be approximated to a cylinder of mass M and radius R, and the mass of the bucket together with water is m.

Ans. $\dfrac{2(P - mg)}{2m + M}$

18.6 An equilateral triangular plate revolves round about one of its sides, which is horizontal and fixed, under the action of gravity. If the plate is initially horizontal, compare the pressure on the axis with the weight of the plate.

Ans. 1:3

18.7 Find the ratio of tension in the string AB before and after the string BC is cut. Refer Fig. E.18.7.

Ans. 4/3

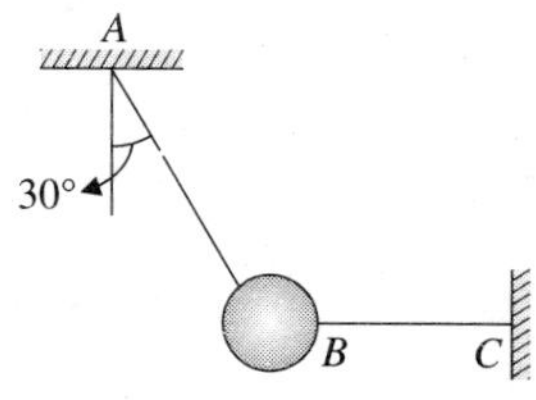

Fig. E.18.7

18.8 A block of mass m_1 is lifted up by a string passing over two pulleys of same mass m_2 and radius of gyration r by applying a constant pull P at the free end of the string. Determine the acceleration of the block. Refer Fig. E.18.8.

Ans. $\dfrac{P - m_1 g}{m_1 + m_2}$

18.9 The stepped pulley system shown in Fig. E.18.9, when released from rest, determine the acceleration of the blocks, angular acceleration of the pulley and tension in the strings connecting the blocks. The mass of the pulley is 10 kg and its radius of gyration is 22 cm. Take $m_1 = 40$ kg, $m_2 = 25$ kg, $r_1 = 20$ cm and $r_2 = 25$ cm.

Ans. $a_1 = 0.94$ m/s^2, $a_2 = 1.18$ m/s^2, $\alpha = 4.71$ rad/s^2, $T_1 = 354.7$ N and $T_2 = 274.7$ N

18.10 A block of 10 kg mass slides down a smooth inclined plane as shown in Fig. E.18.10. The other end of the string connecting the block is wound several times around a pulley of 10 kg mass and of 10 cm radius free to rotate about the horizontal axis. Determine the acceleration of the block and the tension in the string. Assume the pulley to be a perfect cylinder.

Ans. 3.27 m/s^2, 16.35 N

18.11 In the previous problem, suppose the other end of the string passes over the pulley supporting a block of 4 kg mass, determine the acceleration of the blocks and the tensions in the different portions of the string. Refer Fig. E.18.11.

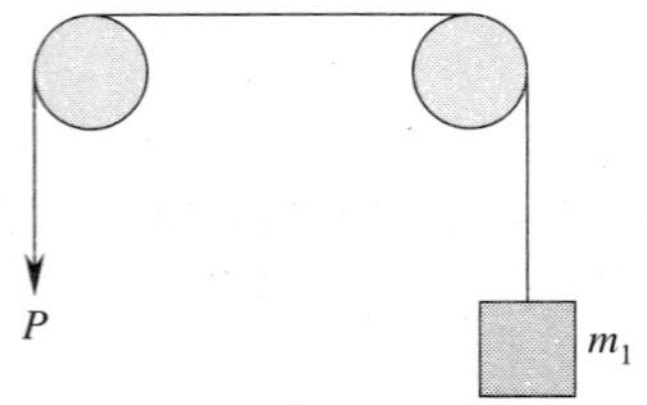

Fig. E.18.8

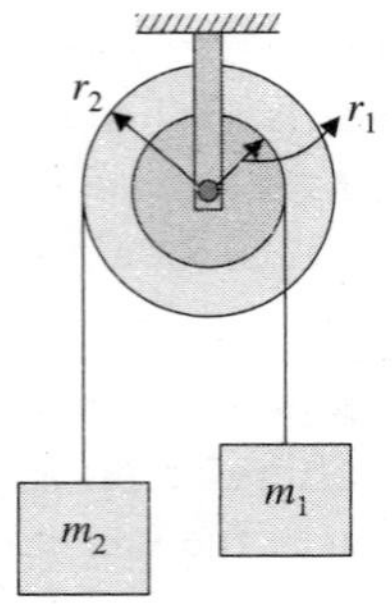

Fig. E.18.9

Ans. 0.516 m/s^2 down the incline, 43.9 N in the string supporting the 10 kg block, 41.3 N in the string supporting the 4 kg block.

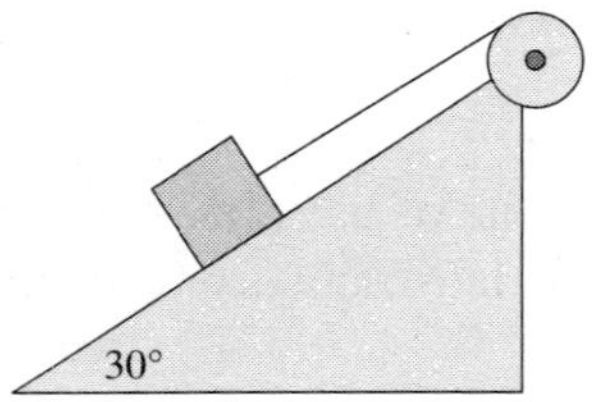

Fig. E.18.10

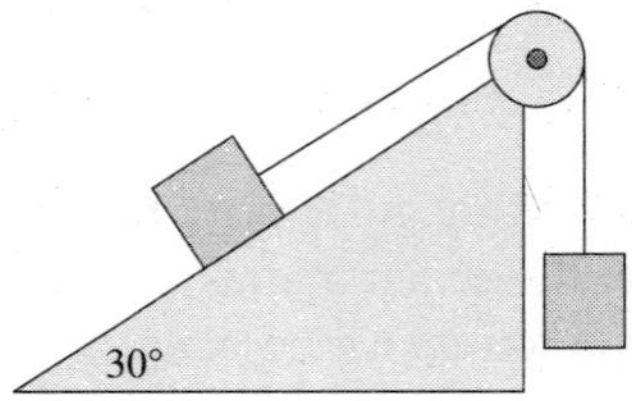

Fig. E.18.11

18.12 A slender rod *AB* of 20 kg mass is pivoted at its centre as shown in Fig. E.18.12. If a force of 100 N is gently applied at end *B* in the position shown, determine the resulting acceleration. In addition, determine its rotational kinetic energy when it reaches the horizontal position.

Ans. 8.66 rad/s^2, 68 J

18.13 A jet of water issued from a nozzle strikes normally a smooth flat plate of 2 kg mass and 50 cm height hinged as shown in Fig. E.18.13. The water after striking the plate at its centre leaves parallel to the plate. If the jet diameter is 6 mm and water moves at 15 m/s, determine the acceleration of the centre of mass and the reaction at the hinge point. In addition, determine the point at which the jet should strike such that the horizontal component of reaction at the hinge is zero.

Ans. $a_{cm} = 2.4$ m/s^2 $A_x = 1.6$ N $\leftarrow$, $A_y = 19.6$ N $\uparrow$, 0.33 m from the hinge

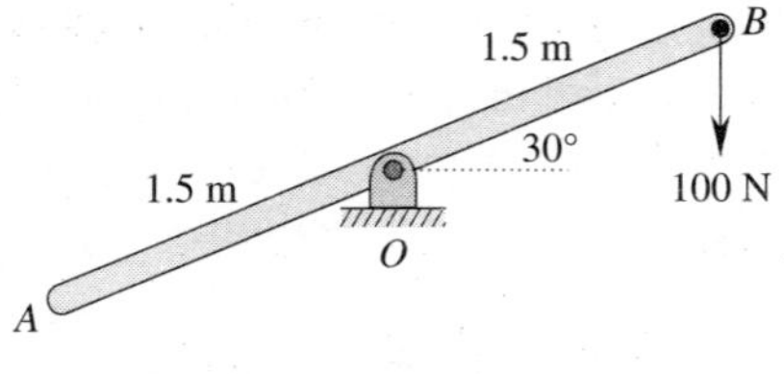

Fig. E.18.12

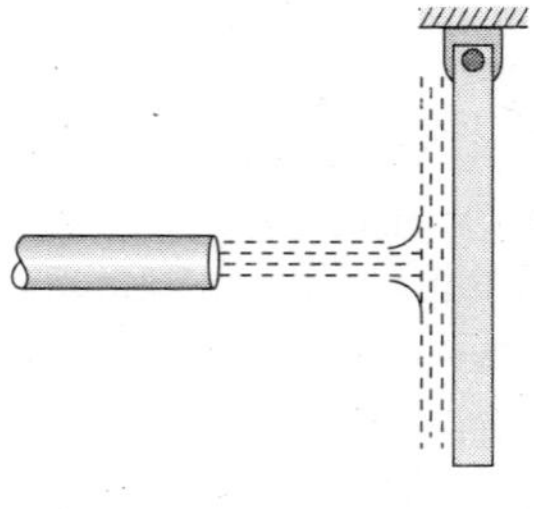

Fig. E.18.13

18.14 A compound pendulum consists of a circular metallic plate suspended by a thin metallic strip as shown in Fig. E.18.14. If their respective masses are 0.4 kg and 0.2 kg, determine the angular acceleration of the pendulum at the instant it is released from the position shown.

Ans. 4.53 rad/s

18.15 A block of mass $m_1 = 20$ kg resting on a frictionless horizontal plane is attached to a string, which passes over a pulley (solid cylinder) of 10 kg mass and 10 cm radius. The other end of the string supports a block of mass $m_2 = 25$ kg. If released from rest, determine the acceleration of the system and the tension in each part of the string. Refer Fig. E.18.15.

Ans. $a = g/2$ m/s^2; 98.1 N in the portion of the string attached to m_1; 122.6 N in the portion of the string attached to m_2

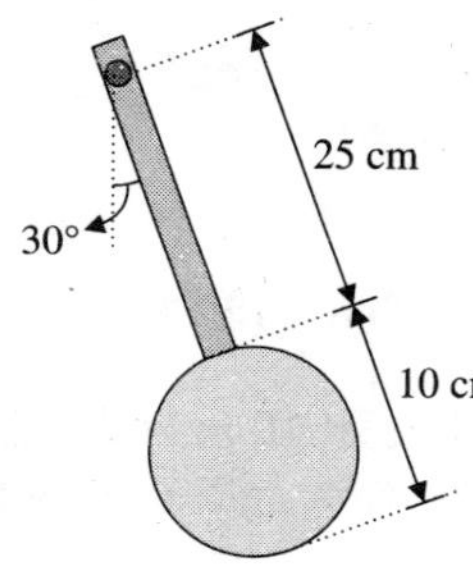

Fig. E.18.14

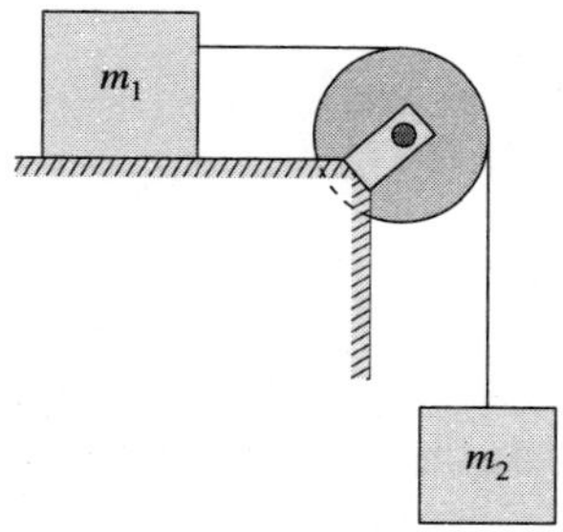

Fig. E.18.15

18.16 A thin annular ring of mass m and radius r is hinged at a point as shown in Fig. E.18.16. Determine the angular acceleration and the reaction at the hinge if released from a position such that the diameter passing through the hinge is horizontal.

Ans. g/r clockwise; $A_x = A_y = 0$

18.17 A thin homogeneous semicircular plate *ABC* of 500*g* mass and 10 cm radius is held such that the diameter *AB* is horizontal and released. Refer Fig. E.18.17. Determine the angular acceleration of the plate and the reactions at A at the instant it is released.

Ans. 65.4 rad/s^2 clockwise; $A_x = 1.39$ N $\rightarrow$, $A_y = 1.62$ N $\uparrow$

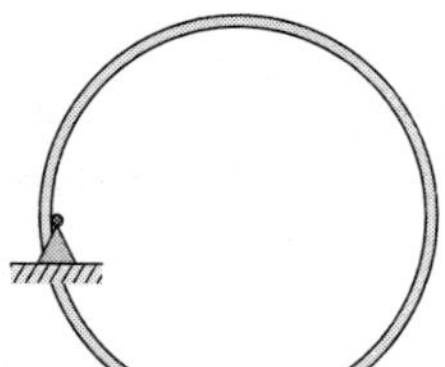

Fig. E.18.16

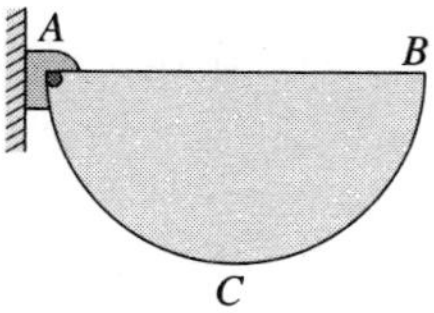

Fig. E.18.17

18.18 A homogeneous bar *AB* of mass m and length l hinged at A is released from rest from the position shown in Fig. E.18.18. Determine the acceleration of centre of mass of the bar and the reaction at A.

Ans. $0.375g$, $0.625mg$

18.19 A sphere and a cylinder of equal masses m and radii r roll down an inclined plane in contact with each other as shown in Fig. E.18.19. Determine the acceleration of the two bodies assuming that they roll without slipping and the contact force between them.

Ans. $a = \frac{20}{29} g \sin\theta, R = \frac{1}{29} mg \sin\theta$

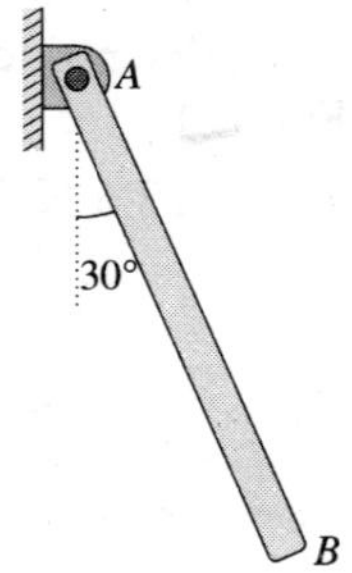

Fig. E.18.18

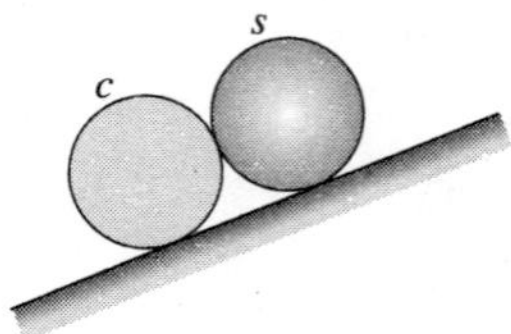

Fig. E.18.19

18.20 A sphere and a hoop are released from the top of an incline of 10 m at the same instant. Determine which of the two reaches the bottom of the incline first and what is the position of the other body at that instant.

Ans. The sphere reaches the bottom of the incline first. The hoop is at a distance of 3 m from the bottom of the incline.

18.21 A cylinder and a hoop are released from the top of an incline of 10 m at the same instant. Determine which of the two reaches the bottom of the incline first and what is the position of the other body at that instant.

Ans. The cylinder reaches the bottom of the incline first. The hoop is at a distance of 2.5 m from the bottom of the incline.

18.22 A sphere, a cylinder and a hoop are released from rest along an inclined plane in parallel paths. If all three reach the bottom of the incline the same time, determine the initial distance between them.

Ans. sphere: cylinder: hoop = 1: 0.067: 0.3

18.23 In Example 18.17, if the cylinder and the hoop are connected in the reverse direction, that is the hoop is trailing behind, determine the acceleration of the system and the force in the connecting bar.

Ans. $\frac{4}{7} g \sin\theta, \frac{1}{7} mg \sin\theta$ (tension)

18.24 A car when moving at a speed of 60 kmph, one of its rear wheels gets detached due to loose bolting. Determine the work required to be done on the wheel to stop it. Its mass is 4 kg and diameter is 40 cm. Its radius of gyration is 15 cm.

Ans. 868.4 J

18.25 The speed of a flywheel rotating at 200 rpm is increased to 300 rpm in 6 seconds. Determine the work done by the driving torque and the increase in its kinetic energy. The mass of the flywheel is 25 kg and radius of gyration is 20 cm.

Ans. 274.2 J, 274.2 J

18.26 Determine the angular velocity of the earth assuming it to be a perfect sphere revolving about the north and south poles. If the radius of the earth is 6370 km and its mass is 6×10^{24} kg, determine its angular momentum and rotational kinetic energy.

Ans. 7.1×10^{33} kg.m^2/s; 2.6×10^{29} J

18.27. A block of 2 kg mass is suspended from a stepped cylinder as shown in Fig. E.18.27. If the system is released from rest, determine the velocity of the block after it has fallen through a distance of 3 m. The mass of the cylinder is 3 kg and its radius of gyration is 10 cm.

Ans. 3.38 m/s

18.28. Two circular disks mounted on frictionless bearings are in contact with each other as shown in Fig. E.18.28. The mass and radius of the larger disk are respectively 5 kg and 10 cm and that of the smaller disk are respectively 2 kg and 8 cm. If a clockwise moment of 2 N.m is applied to the larger disk, determine the angular velocity of each if the larger disk makes one complete revolution.

Ans. 8.7 rad/s, 10.9 rad/s

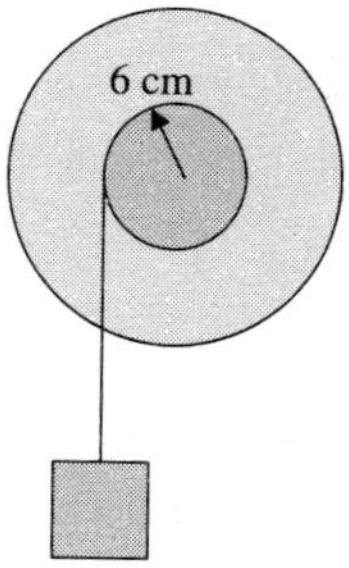

Fig. E.18.27

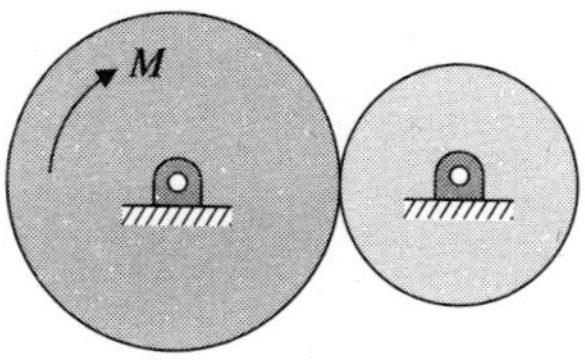

Fig. E.18.28

18.29. A horizontal force of 100 N magnitude is applied on a cylinder of 10 kg mass and of 20 cm diameter about its centre of mass. Refer Fig. E.18.29. Determine its velocity after the cylinder makes 5 complete revolutions without slipping.

Ans. 6.5 m/s

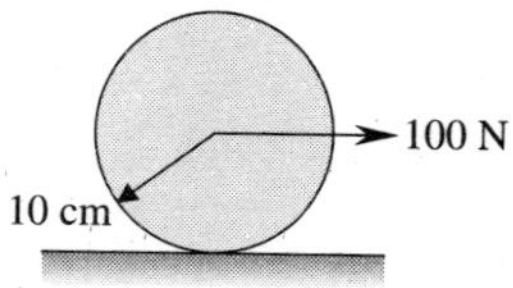

Fig. 18.29

19 Vibrations

19.1 INTRODUCTION

So far, in dynamics, we have analyzed *linear* and *angular* motions of *particles* and *rigid bodies*. In this chapter, we will discuss a *special* case of motion, which occurs commonly in real structures such as bridges, towers, transmission cables and in mechanical systems when they are acted on by external forces such as wind, water waves, earthquake ground motions, human induced and others.

Water waves, because of their unconstrained nature, move from one place to another under the action of wind. The motion of waves as we observe in oceans occurs *repeatedly* after *regular* intervals of time. Such types of motions, which occur after *equal* intervals of time, are termed **periodic** motions.

On the other hand, real structures such as transmission cables, diving boards in swimming pools, bridges, tall buildings, etc., and even mechanical systems such as simple pendulums, string of a musical instrument, etc., due to their *attachments* with the surroundings do not displace from one place to another upon the action of external forces. Instead, they move back and forth over the *same path* periodically. Such types of periodic motions, which trace the same path in a *cyclic* manner, are termed **vibratory** or **oscillatory** motions.

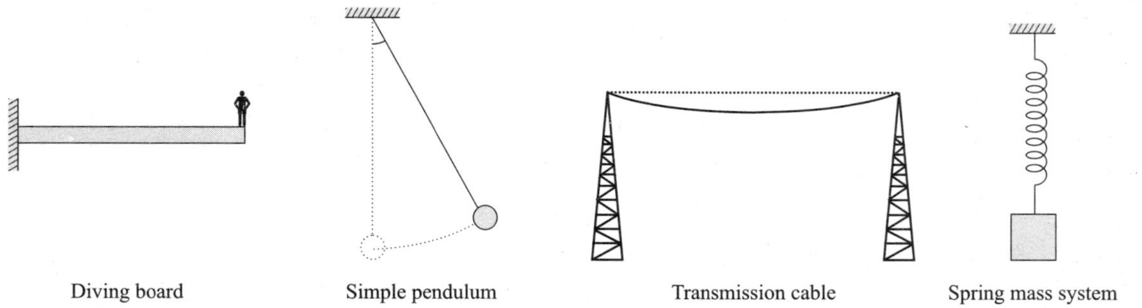

Fig. 19.1 Structures and mechanical systems executing vibratory motions

Throughout our study, we have assumed the bodies to be *rigid*, that is, they *do not* deform under the action of external forces. This however is an ideal condition, as in reality, bodies do deform to some extent under the action of external forces. Hence, when an external force acts on a mechanical system or a structure, which is in **equilibrium**, it causes **elastic deformation** of the system or structure, resulting in *displacement* of the members from the equilibrium position.

If this force is removed, due to elastic property of the material and in some cases due to force of gravity, the system or structure tries to return to its *equilibrium* position. The restoring force acting on the system or structure is not a constant one but of varying nature and it depends upon the displacement from the equilibrium position. Hence, the acceleration (which is directly proportional to the force) of the system or structure also varies with the displacement. As a result, the system accelerates to the equilibrium position reaching a *maximum velocity* at the equilibrium position. The restoring force vanishes at the equilibrium position; but due to the increased velocity, which the system attains at the equilibrium position, it overshoots and begins to move to the other side of the equilibrium position. Again, the restoring force comes into play tending to bring it to the equilibrium position and this way the system or structure *oscillates* back and forth about the equilibrium position.

Generally, we observe these vibrations to *die* out after some time. This happens mainly due to resistance offered by the surroundings on the structure. Hence, such vibrations are termed **damped free vibrations**. However, for elementary analytical purposes, we assume all resistance to vibration to be eliminated and such vibrations are termed **undamped free vibrations**. Further, in some cases, the external force may *continue* to act on the structure periodically. Such vibrations are termed **forced vibrations**.

Like friction, vibration in structures is an unavoidable evil. It causes discomfort to the passengers when a vehicle moves over a bump on a road or panic to the occupants in a building, which is subjected to earthquake ground motion, or it may even cause collapse of structures. There has been much work done to reduce such vibrations. Though vibration cannot be eliminated completely, it can be suppressed to a greater extent using what are called **dampers**.

On the other hand, vibration is used in a positive sense in some of the instruments such as vibrating machines used in compacting concrete, drilling machines used in making boreholes, mine explosions, electric shaver, and so on. Hence, vibrational analysis is essential to understand such behaviour of structures and mechanical systems. There are books devoted exclusively to this topic and getting into all details is beyond the scope of this book. However, we will discuss the basics of vibrational analysis, which can pave the way for further study along this line.

In the next two sections, we will introduce the student to **simple harmonic motion**. This is a special type of *rectilinear* motion and most of the structures and mechanical systems execute such motion under *small* displacements from the equilibrium position. In Sections 19.4–19.6, we will discuss simple mechanical systems, which execute simple harmonic motion; in Sections 19.7–19.10, we will discuss the motion of various pendulums and in the last few sections, analytical modeling of structures and their responses to external disturbances.

19.2 SIMPLE HARMONIC MOTION (SHM)

Simple harmonic motion, abbreviated to SHM, is a special case of **rectilinear** motion with **variable** acceleration, in which the acceleration of the particle is proportional to the displacement from the origin and is always directed towards the origin.

Consider a particle executing SHM along a rectilinear path, say X-axis. Let it cross the origin O with a velocity v directed towards the positive X-axis [refer Fig. 19.2(i)]. The acceleration of the particle at the origin is zero, since by definition the acceleration is proportional to the displacement from the origin. However, the acceleration increases with increase in displacement and as this acceleration is directed opposite to the displacement, the velocity of the particle goes on decreasing till the particle reaches a momentary rest at the position *A* [refer Fig. 19.2(iii)]. At this position, the acceleration of the particle reaches a maximum value due to maximum displacement from the origin.

Because of the maximum acceleration at this extreme position *A* which is directed towards the origin, the particle begins to move towards the negative direction with increasing velocity, reaching a maximum value at the origin *O*. At this position, its acceleration becomes zero as the displacement from the mean position is zero [refer Fig. 19.2(v)]. With this velocity, it continues to move to the left of the origin. Again, the acceleration being proportional to the displacement and directed towards the origin retards the speed of the particle. Hence, the particle comes to a momentary rest at position *B* [refer Fig. 19.2(vii)]. As the acceleration is maximum at this position and directed towards the origin, the particle now begins to move from *B* to *O* with increasing velocity reaching a maximum velocity at *O*. This cycle of to and fro motion is repeated. As a result, the particle oscillates between the extreme positions *A* and *B* with the origin *O* as the mean position.

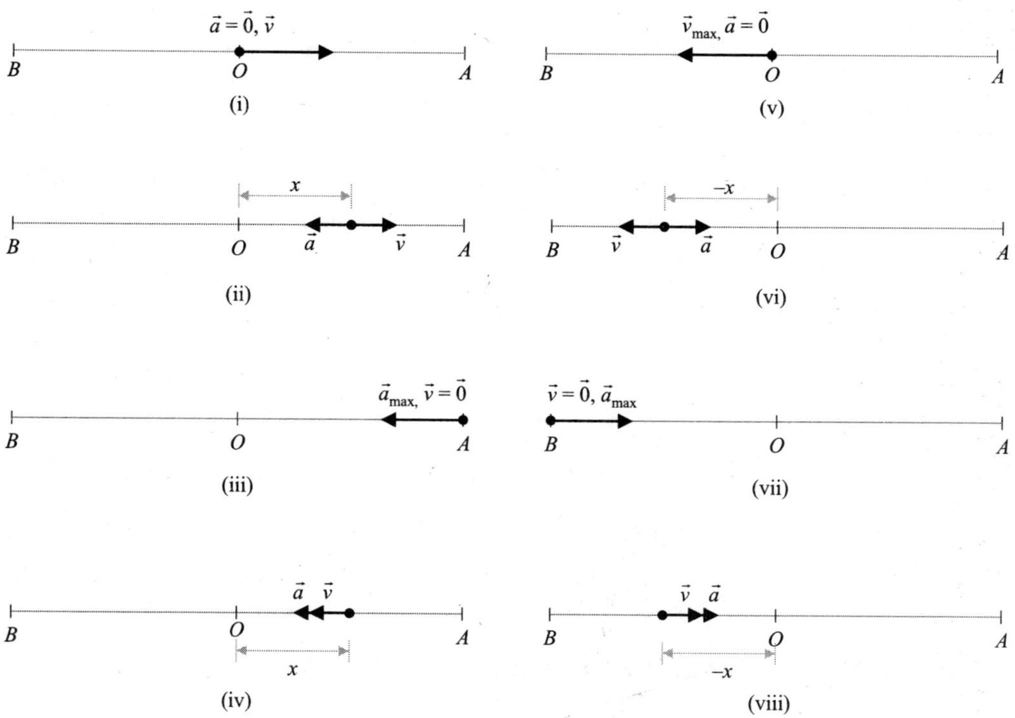

Fig. 19.2 Acceleration and velocity vectors of a particle executing SHM

In Chapter 12, we saw that to describe the motion of a particle completely, we must specify its displacement, velocity and acceleration at any instant of time. Let *P* be the position of the particle at a given instant of time, where its displacement from the origin is x and its velocity at that instant be v directed along the positive x-direction.

As the particle is executing SHM, by definition we know that the acceleration of the particle is proportional to the displacement and directed opposite to it as shown in Fig. 19.3. Mathematically, this can be written as

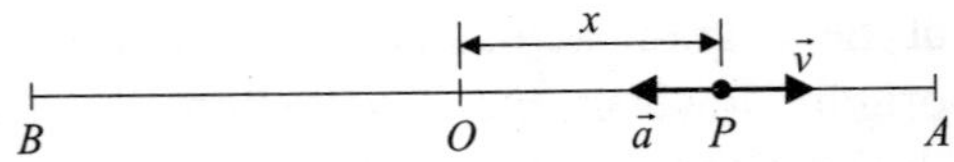

Fig. 19.3 A particle executing SHM

$$a \propto -x \tag{19.1}$$

Note that as the motion is rectilinear, we have avoided the vector notation taking into account that the vectors pointing along the $+X$ axis are taken as *positive* and pointing along $-X$ axis are taken as *negative*. Introducing a constant of proportionality ω^2 (whose meaning we will explain later), we have

$$a = -\omega^2 x \tag{19.2}$$

By chain rule of differentiation, we can write acceleration as

$$a = \frac{dv}{dt} = \frac{dv}{dx}\frac{dx}{dt} = v\frac{dv}{dx}$$

Hence,

$$v\frac{dv}{dx} = -\omega^2 x \tag{19.3}$$

Upon integration, we have

$$\frac{v^2}{2} = -\omega^2 \frac{x^2}{2} + C_1 \tag{19.4}$$

where C_1 is the constant of integration, whose value can be determined from the conditions at the extreme positions. If $\overline{OA} = \overline{OB} = A$ in Fig. 19.3, then we know that

$$v = 0 \text{ at } x = \pm A \tag{19.5}$$

Substituting this condition in the Eq. (19.4), we get

$$C_1 = \frac{\omega^2}{2}A^2 \tag{19.6}$$

Therefore,

$$\frac{v^2}{2} = -\frac{\omega^2 x^2}{2} + \frac{\omega^2 A^2}{2}$$

$$\Rightarrow \quad v = \pm\,\omega\sqrt{A^2 - x^2} \tag{19.7}$$

The positive value of v indicates the velocity of the particle when the motion is directed towards $+X$ axis and the negative value indicates the velocity when the motion is directed towards $-X$ axis.

19.2.1 Motion of the Particle as Functions of Time

Considering the positive value of velocity

$$v = \frac{dx}{dt} = \omega\sqrt{A^2 - x^2}$$

Upon rearranging

$$\frac{dx}{\sqrt{A^2 - x^2}} = \omega\, dt$$

and integrating, we have

$$\sin^{-1}\left[\frac{x}{A}\right] = \omega t + \varphi \tag{19.8}$$

where φ is a constant of integration. Therefore,

$$x = A \sin(\omega t + \varphi) \tag{19.9}$$

The values of φ depend upon the initial conditions. If the motion starts from O at $t = 0$, i.e., $x = 0$, then

$$0 = A \sin[\omega(0) + \varphi]$$

$$\Rightarrow \quad \varphi = 0 \tag{19.10}$$

Therefore, $$x = A \sin \omega t \tag{19.11}$$

In some cases, we may come across instances where the motion starts from the extreme position. In such cases $x = A$ at $t = 0$, then

$$A = A \sin[\omega(0) + \varphi]$$

$$\Rightarrow \quad \varphi = \frac{\pi}{2} \tag{19.12}$$

Therefore, $$x = A \sin\left[\omega t + \frac{\pi}{2}\right] = A \cos \omega t \tag{19.13}$$

Similarly, we can also consider the negative value of velocity and get the expression for displacement.

Upon differentiating the expressions for displacement (19.11 and 19.13) with respect to time successively, we can get the velocity and acceleration of the particle as functions of time. Thus, we have seen that the motion can be expressed as functions of *displacement* [Eqs 19.2 and 19.7] as well as *time* and these are summarized below.

Table 19.1 SHM as functions of displacement and time

	As function of displacement	*As function of time*	
		t = 0, at mean position	*t = 0, at extreme position*
Displacement	x	$x = A \sin \omega t$	$x = A \cos \omega t$
Velocity	$v = \pm\omega\sqrt{A^2 - x^2}$	$v = A\omega \cos \omega t$	$v = -A\omega \sin \omega t$
Acceleration	$a = -\omega^2 x$	$a = -A\omega^2 \sin \omega t$	$a = -A\omega^2 \cos \omega t$

We know that sine and cosine functions are also termed harmonic functions. Hence, any periodic motion is termed **harmonic** motion. In addition, if the acceleration of the particle is also proportional to the displacement and directed opposite to it then it is termed **simple harmonic** motion.

The maximum displacement A from the mean position is termed **amplitude** of the motion. The motion of the particle in Fig. 19.2 from O to A and A to B and from B to O is termed as a complete **cycle** or **oscillation**. The constant of proportionality ω in Eq. (19.2) is termed **natural circular frequency**. Its SI unit is radian per second. As the simple harmonic motion is periodic, it repeats itself after equal intervals of time. The time taken for one complete cycle of motion is termed the **time period**. From the properties of harmonic functions (sine and cosine), we see that the displacement can be written as

$$x = A \sin \omega t = A \sin [\omega t + 2\pi] = A \sin \omega \left[t + \frac{2\pi}{\omega} \right] \tag{19.14}$$

Similarly, velocity can also be written as

$$v = A\omega \cos \omega t = A\omega \cos [\omega t + 2\pi] = A\omega \cos \omega \left[t + \frac{2\pi}{\omega} \right] \tag{19.15}$$

We see that the displacement and velocity remain the same after a time of $\frac{2\pi}{\omega}$ s. Thus, we can generalize and say that the particle has the same position, velocity and direction after an interval of

$$\frac{2\pi}{\omega}, \frac{4\pi}{\omega}, \ldots$$

Hence, we can say that the period of motion is

$$T = \frac{2\pi}{\omega} \tag{19.16}$$

Its unit is **second** and it should be noted that the period is independent of the amplitude of motion. The number of cycles the particle completes in unit time is defined as **frequency** and its SI unit is Hertz (Hz). Mathematically, it can also be expressed as the inverse of time period. Hence,

$$f = \frac{1}{T} = \frac{\omega}{2\pi} \tag{19.17}$$

19.2.2 General Solution of SHM

Equations 19.11 and 19.13 can be combined together and a *general* solution can be written as

$$x = a_1 \cos \omega t + a_2 \sin \omega t \tag{19.18}$$

where a_1 and a_2 are arbitrary constants whose values can be determined from the initial conditions. Upon differentiation, we get the expression for velocity as

$$v = -a_1 \omega \sin \omega t + a_2 \omega \cos \omega t \tag{19.19}$$

For instance, if at $t = 0$, $x = A$ and $v = 0$ then substituting these values in the above equations, we get

$$x/_{t=0} = A \Rightarrow a_1 = A$$

and

$$v/_{t=0} = 0 \Rightarrow a_2 = 0$$

Therefore,

$$x = A \cos \omega t \tag{19.20}$$

Similarly, if at $t = 0$, $x = 0$ and $v = v_o$, then

$$x/_{t=0} = 0 \Rightarrow a_1 = 0$$

and

$$v/_{t=0} = v_o \Rightarrow a_2 = v_o/\omega$$

Therefore,

$$x = [v_o/\omega] \sin \omega t = A \sin \omega t \tag{19.21}$$

Thus, we see that we get the same expressions as before. Hence, the general solution of simple harmonic motion can be written as

$$x = a_1 \cos \omega t + a_2 \sin \omega t \qquad (19.18')$$

19.3 GRAPHICAL REPRESENTATION OF SHM

Consider a particle moving in a *circular* path of radius r with *constant* angular speed, ω in the clockwise direction. At time $t = 0$, let the particle be at position A and after an interval of time t, let it be at P. In this time t, the radius of the particle would have swept through an angle $\theta = \omega t$. The velocity and acceleration vectors are constant in their magnitudes equal to $r\omega$ and $r\omega^2$ respectively but their directions vary with time. The instantaneous directions at point P are as shown in Fig. 19.4(b), where velocity is along the tangential direction and acceleration is radially inward.

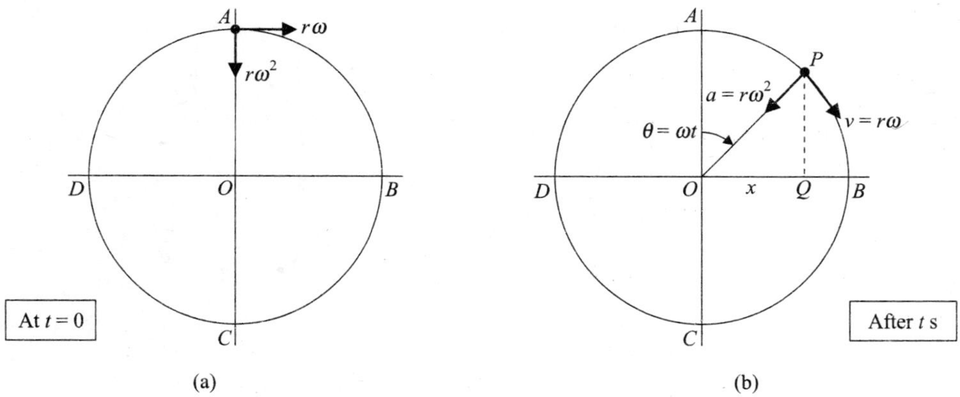

Fig. 19.4

Let us project this particle on the X-axis and let Q be that position. Then the position of Q with respect to origin is

$$\overline{OQ} = x = r \sin \theta = r \sin \omega t \qquad (19.22)$$

The velocity and acceleration of Q are given respectively by the x-components of velocity and acceleration of the particle. Therefore,

$$v_Q = (v_P)_x = r\omega \cos \theta = r\omega \cos \omega t \qquad (19.23)$$

and

$$a_Q = (a_P)_x = -r\omega^2 \sin \theta = -r\omega^2 \sin \omega t \qquad (19.24)$$

The negative sign indicates that the acceleration of Q is pointed along the negative X direction. We readily see that these expressions are same as that derived for SHM. Thus, we can conclude that when a particle moves in a circular path with constant angular velocity, its projection on any diameter will execute simple harmonic motion.

Particle Starting its Motion from Origin When a particle starts its motion from origin, i.e., at $t = 0$, $x = 0$, we know that the equations of motion are

$$x = A \sin \omega t \qquad (19.25)$$

$$v = A\omega \cos \omega t \qquad (19.26)$$

$$a = -A\omega^2 \sin \omega t \tag{19.27}$$

Since these are sine and cosine functions, their graphical representations are shown below for up to three cycles.

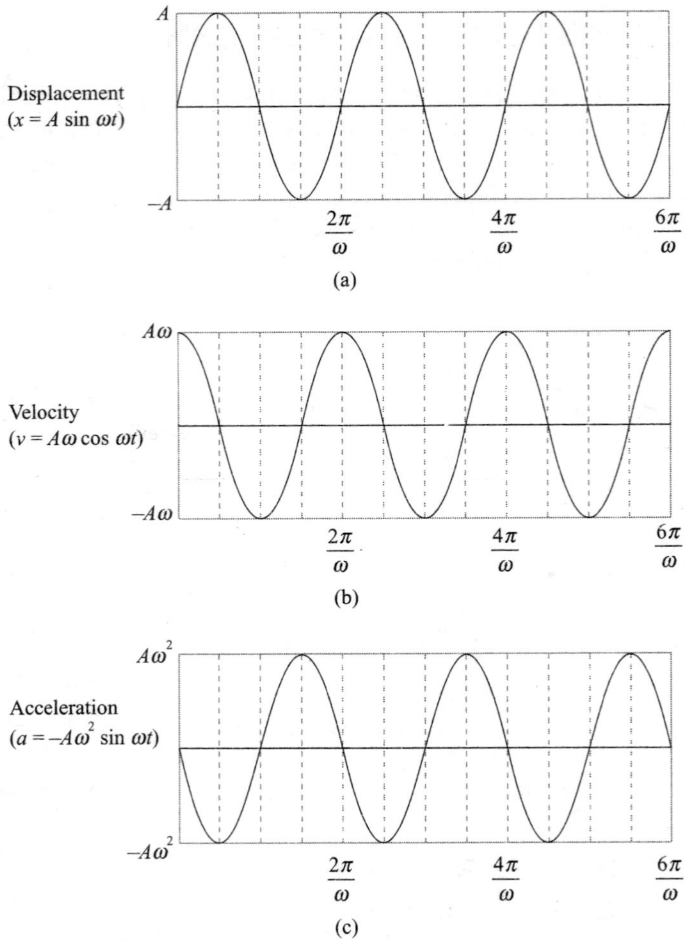

Fig. 19.5

Particle Starting its Motion from Extreme Position Similarly, it can be shown that if the particle starts its motion from the extreme position, i.e., at time $t = 0$ then the displacement, velocity and acceleration are given as expected:

$$x = A \cos \omega t \tag{19.28}$$

$$v = -A\omega \sin \omega t \tag{19.29}$$

$$a = -A\omega^2 \cos \omega t \tag{19.30}$$

The corresponding graphical representations are as shown below for up to three cycles:

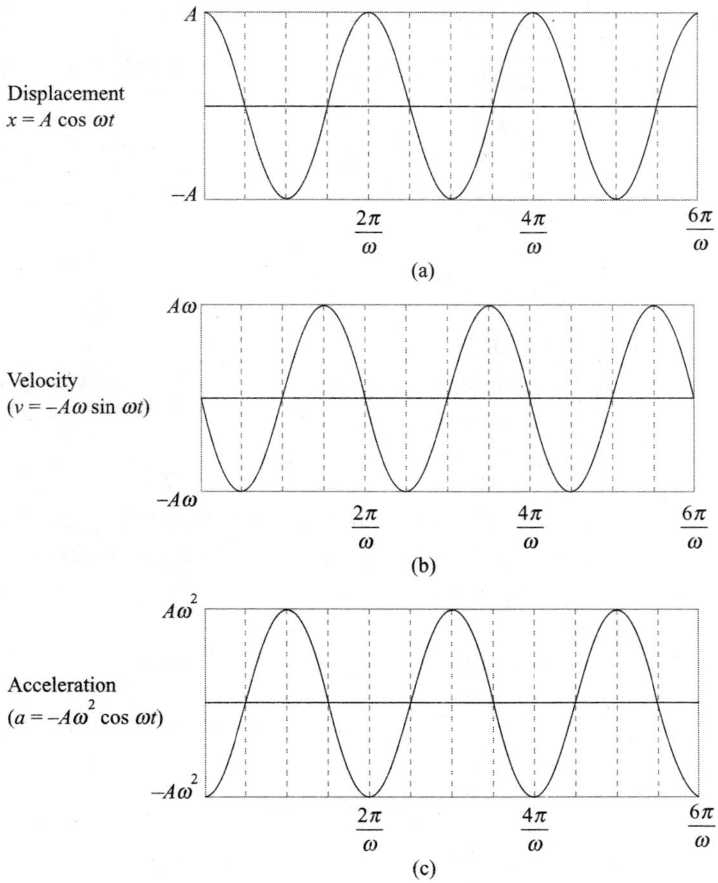

Fig. 19.6

Example 19.1 A particle executing SHM has a time period of 4 seconds. If the velocity of the particle at a distance of 10 cm from the extreme right position is 20 cm/s, determine the amplitude, maximum velocity and maximum acceleration.

Solution We know that time period of motion is given as

$$T = \frac{2\pi}{\omega}$$

$$\Rightarrow \qquad \omega = \frac{2\pi}{T} = \frac{2\pi}{4} = \frac{\pi}{2} \text{ rad/s}$$

Let P be the instantaneous position of the particle. Since the position of the particle from the extreme right position is given as 10 cm, its position from the mean position is

$$x = (A - 10) \text{ cm}$$

where A is the amplitude of motion. The velocity of the particle as a function of displacement is given as

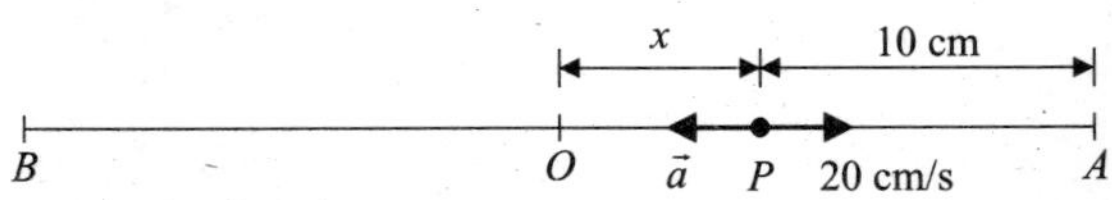

Fig. 19.7

$$v = \omega\sqrt{A^2 - x^2}$$
$$= \omega\sqrt{A^2 - (A-10)^2}$$
$$= \omega\sqrt{20A - 100}$$
$$20 = \frac{\pi}{2}\sqrt{20A - 100}$$
$$\Rightarrow \quad \left(\frac{40}{\pi}\right)^2 = 20A - 100$$
$$\therefore \quad A = 13.11 \text{ cm}$$

The maximum velocity and acceleration are given as

$$v_{max} = A\omega = 20.59 \text{ cm/s}$$
$$a_{max} = A\omega^2 = 32.35 \text{ cm/s}^2$$

Example 19.2 A particle executes SHM with a time period of 1s. It has half the maximum velocity when it is at a distance of $1/\sqrt{3}$ m from the mean position. Find the acceleration at that instant and the time taken to reach that point from the mean position. In addition, determine the amplitude, maximum velocity and maximum acceleration.

Solution Given that, time period of motion is 1 second. We know that

$$T = \frac{2\pi}{\omega}$$
$$\Rightarrow \quad \omega = \frac{2\pi}{T} = \frac{2\pi}{1} = 2\pi \text{ rad/s}$$

Velocity of the particle in terms of displacement is given as

$$v = \omega\sqrt{A^2 - x^2}$$

Given at $\quad x = \frac{1}{\sqrt{3}}$ m, $v = \frac{1}{2}v_{max} = \frac{1}{2}A\omega$

$$\therefore \quad \frac{1}{2}A\omega = \omega\sqrt{A^2 - \left(\frac{1}{\sqrt{3}}\right)^2}$$
$$\frac{1}{2}A = \sqrt{A^2 - \frac{1}{3}}$$
$$\Rightarrow \quad A = \frac{2}{3} \text{ m}$$

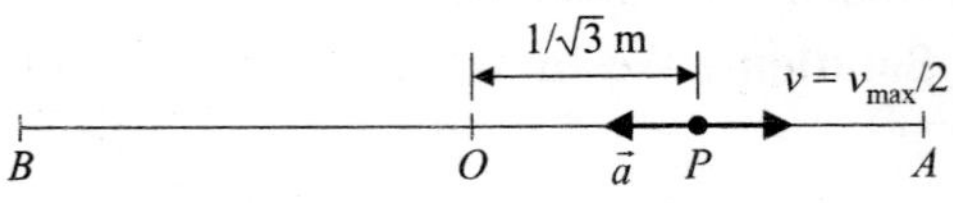

Fig. 19.8

Therefore, maximum velocity and acceleration of the particle are given as

$$v_{max} = A\omega = \frac{2}{3}2\pi = \frac{4\pi}{3} \text{ m/s}$$
$$a_{max} = A\omega^2 = \frac{2}{3}(2\pi)^2 = \frac{8\pi^2}{3} \text{ m/s}^2$$

Acceleration of the particle at that point is given as

$$a = -\omega^2 x$$

$$= -(2\pi)^2 \frac{1}{\sqrt{3}} = -22.79 \text{ m/s}^2$$

The displacement of the particle at any instant of time from mean position is given as

$$x = A \sin \omega t$$

$$\frac{1}{\sqrt{3}} = \frac{2}{3} \sin (2\pi t)$$

$$\Rightarrow \quad 2\pi t = \sin^{-1}\left[\frac{\sqrt{3}}{2}\right] = 60° = \frac{\pi}{3} \text{ rad}$$

$$\therefore \quad t = \frac{1}{6} \text{ s}$$

Example 19.3 A particle executing SHM passes through two points P and Q that are 20 cm apart with the same velocity in a time interval of 4 seconds. If the particle returns to Q after another 2 seconds, find the period and amplitude of motion.

Solution Since the particle passes through points P and Q with the same velocity, the two points should be on either side of equilibrium position O at equal distances. Therefore, $OP = OQ = 10$ cm. Hence, the time taken to travel from mean position to P or Q is 1/2(4) = 2 s. Since the time taken to come back to Q is 2 seconds, the time taken to travel from Q to the extreme position B is 1/2(2) = 1 s. Hence, the total time to travel quarter of a cycle, i.e., from the origin O to the extreme position is 2 + 1 = 3 s. Therefore, time period for the complete oscillation is

$$T = 4 \times 3 = 12 \text{ s}$$

However, we know,

$$T = \frac{2\pi}{\omega}$$

$$\therefore \quad \omega = \frac{\pi}{6} \text{ rad/s}$$

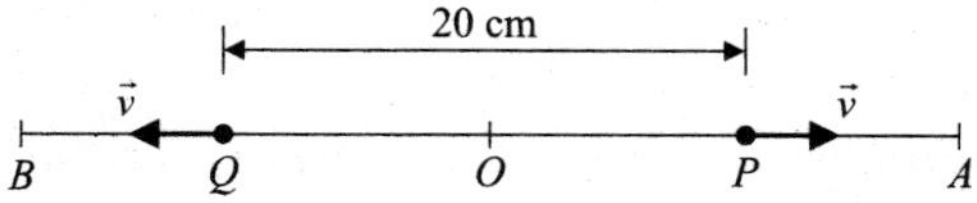

Fig. 19.9

The displacement of the particle as a function of time is

$$x = A \sin \omega t$$

Since the particle takes 2 seconds to travel from O to P,

$$10 = A \sin\left(\frac{\pi}{6} 2\right)$$

$$\therefore \quad A = 11.55 \text{ cm}$$

Example 19.4 A particle executing SHM has a maximum velocity of 10 m/s and a maximum acceleration of 20 m/s^2. Determine (i) period and amplitude of motion, (ii) displacement, velocity and acceleration of the particle at $t = 2$ s from the mean position, (iii) displacement, velocity and acceleration of the particle at $t = 0.5$ s from the extreme right position, (iv) velocity and acceleration at a point 3 m to the right of mean position and the time taken to reach it.

Solution

Given $v_{max} = A\omega = 10$ m/s and (a)

$a_{max} = A\omega^2 = 20$ m/s^2 (b)

(i) *Period and amplitude of motion*

Solving for ω and A from the above two equations (a) and (b), we get

$$\omega = \frac{20}{10} = 2 \text{ rad/s}$$

and $A = 5$ m

Therefore, time period is given as

$$T = \frac{2\pi}{\omega} = \frac{2\pi}{2} = \pi \text{ s}$$

(ii) *Displacement, velocity and acceleration of the particle at t = 2 s from the mean position*

Since time is measured from the *mean position*, we know,

$$x = A \sin \omega t$$
$$= 5 \sin 2t$$

Upon differentiation, we get

$$v = \frac{dx}{dt} = 10 \cos 2t$$

and

$$a = \frac{d^2x}{dt^2} = -20 \sin 2t$$

Therefore, at $t = 2$ s, we have

$$x = 5 \sin (2 \times 2) = -3.78 \text{ m}$$
$$v = 10 \cos (2 \times 2) = -6.54 \text{ m/s}$$
$$a = -20 \sin (2 \times 2) = 15.14 \text{ m/s}^2$$

(iii) *Displacement, velocity and acceleration of the particle at t = 0.5 s from the extreme right position*

We know that the displacement from the extreme right position is given as

$$x = A \cos \omega t$$
$$= 5 \cos 2t$$

Upon differentiation, we get

$$v = \frac{dx}{dt} = -10 \sin 2t$$

and

$$a = \frac{d^2x}{dt^2} = -20 \cos 2t$$

Therefore, at $t = 0.5$ s, we have

$$x = 5 \cos (2 \times 0.5) = 2.7 \text{ m}$$
$$v = -10 \sin (2 \times 0.5) = -8.41 \text{ m/s}$$
$$a = -20 \cos (2 \times 0.5) = -10.81 \text{ m/s}^2$$

(iv) Velocity and acceleration at a point 3 m to the right of mean position and the time taken to reach it

The velocity of the particle as a function of displacement is given as

$$v = \omega\sqrt{A^2 - x^2}$$
$$= 2\sqrt{5^2 - 3^2} = 8 \text{ m/s}$$

The acceleration of the particle as a function of displacement is given as

$$a = -\omega^2 x$$
$$= -(2)^2 (3) = -12 \text{ m/s}^2$$

The displacement of the particle as a function of time is given as

$$x = A \sin \omega t$$

Therefore, $3 = 5 \sin (2t)$

$\Rightarrow$ $$2t = \sin^{-1}\left(\frac{3}{5}\right)$$

$\therefore$ $$t = 0.322 \text{ s}$$

Example 19.5 A horizontal platform executes simple harmonic oscillations horizontally; it oscillates over a distance of 2 m and makes 15 complete oscillations per minute. Determine the least value of coefficient of friction between the platform and a heavy block placed on it preventing the block from slipping.

Solution Since the platform moves through a distance of 2 m, the amplitude of oscillation is

$$A = \frac{1}{2}(2) = 1 \text{ m}$$

Frequency of oscillation is,

$$f = \frac{15}{60} = \frac{1}{4} \text{ rev/s}$$

We know that $$f = \frac{\omega}{2\pi}$$

$\therefore$ $$\omega = \frac{\pi}{2} \text{ rad/s}$$

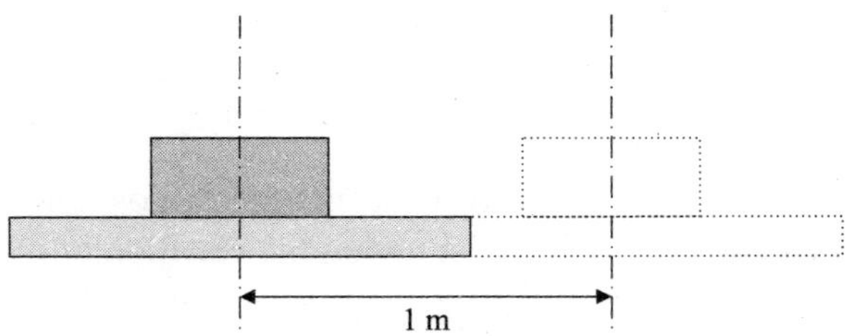

Fig. 19.10

Maximum acceleration of the platform is given as

$$a_{\max} = A\omega^2$$
$$= 1\left(\frac{\pi}{2}\right)^2 = \frac{\pi^2}{4} \text{ m/s}^2$$

We know that this maximum acceleration occurs at the extreme position of the platform. The frictional resistance between the platform and the block should be such as to overcome this maximum acceleration to prevent the block from slipping. Hence,

$$ma_{\max} = \mu mg$$

$\Rightarrow$ $$\mu = \frac{a_{\max}}{g} = \frac{\pi^2}{4g} = \frac{\pi^2}{4 \times 9.81} = 0.252$$

Example 19.6 A horizontal platform executes simple harmonic oscillations vertically. The period of oscillation is 2 seconds. Determine the greatest amplitude it can have so that a block of 1 kg mass resting on it may always be in contact with it. In addition, determine the greatest and least pressures exerted by the block on the platform.

Solution Given that, the period of oscillation T is 2 s. We know that

$$T = \frac{2\pi}{\omega}$$

$$\Rightarrow \qquad \omega = \frac{2\pi}{T} = \pi \text{ rad/s}$$

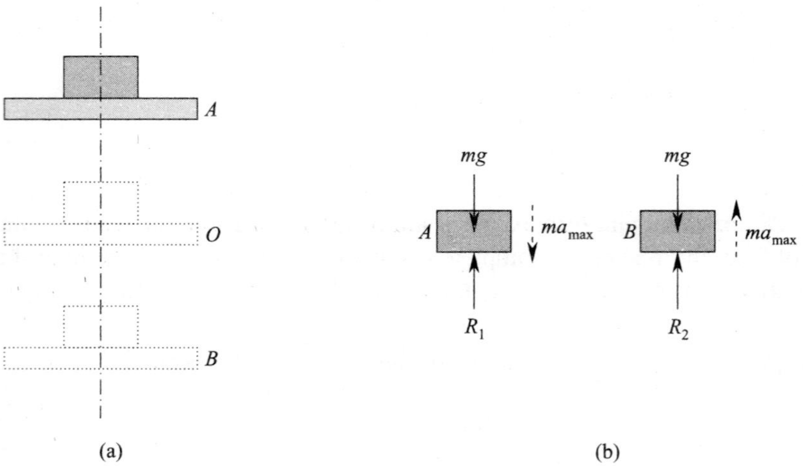

Fig. 19.11

When the platform is at its highest position A, its acceleration is acting downwards towards the mean position. Hence, applying the equation of motion,

$$R_1 - mg = -ma_{max}$$

$$\Rightarrow \qquad R_1 = m(g - a_{max}) \qquad \text{(a)}$$

When the platform is at its lowest position B, its acceleration is acting upwards towards the mean position. Hence, applying the equation of motion,

$$R_2 - mg = ma_{max}$$

$$\Rightarrow \qquad R_2 = m(g + a_{max}) \qquad \text{(b)}$$

From equations (a) and (b), we see that R_1 is less than R_2. Hence, the block will lose contact at the highest position A. Thus, making $R_1 = 0$, we get

$$0 = m(g - a_{max})$$

$$\Rightarrow \qquad a_{max} = g$$

But we know that maximum acceleration is $A\omega^2$. Therefore,

$$A\omega^2 = g$$

$$\Rightarrow \qquad A = \frac{g}{\pi^2} = \frac{9.81}{\pi^2} \approx 1 \text{ m}$$

By Newton's third law of motion, we know that the normal reaction exerted by the block on the platform will be equal in magnitude and opposite in direction to the reaction exerted by the platform on the block. Hence, from equation (a),

least pressure, $R_1 = 0$

and from equation (*b*),

$$\text{greatest pressure, } R_2 = m[g + a_{\max}]$$
$$= m[g + g] = 19.62 \text{ N}$$

19.4 MECHANICAL SYSTEMS EXECUTING SHM

So far, we have discussed the *kinematics* of simple harmonic motion, that is, motion without considering the forces causing it. From now on, we will analyze the *kinetics* of simple harmonic motion. Here we will consider simple mechanical systems, which we normally encounter. When they are displaced from their equilibrium positions, they execute simple harmonic motion for smaller displacements. We also derive expressions for the periods of motions in each case. In Section 19.5, we will discuss motion of a body attached to a spring; in Section 19.6 motion of a body attached to an elastic string and in Sections 19.7–19.9 three different types of pendulums whose motions are simple harmonic when their displacements from the equilibrium positions are smaller.

19.5 MOTION OF A BODY ATTACHED TO A SPRING

Consider a body of mass m suspended by a spring of spring constant k. The mass will extend the spring by an amount e from the unstretched position in the vertical direction. Within the elastic limit, the extension e in the equilibrium position is obtained from Hooke's law as

$$mg = ke \tag{19.31}$$

$$\Rightarrow \qquad e = \frac{mg}{k} \tag{19.32}$$

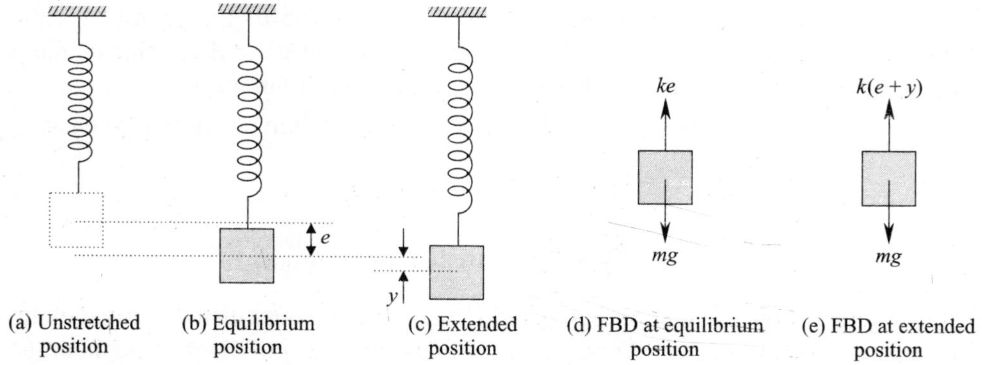

Fig. 19.12

Suppose the mass is pulled down by a vertical distance y from the equilibrium position and released, then the net force acting on the mass tending to restore it to the equilibrium position is [refer Fig. 19.12(e)]:

$$k(e+y)-mg \tag{19.33}$$

From Eq. (19.31), we know $mg = ke$. Hence, the net restoring force becomes,

$$ky \tag{19.34}$$

Writing the equation of motion of the block considering the downward displacement of the block from the equilibrium position as *positive* and that the net force acts upwards in the $-Y$ direction,

$$m\frac{d^2y}{dt^2} = -ky \tag{19.35}$$

$$\therefore \quad \frac{d^2y}{dt^2} = -\left[\frac{k}{m}\right]y \tag{19.36}$$

From the above equation, we see that the acceleration is *directly proportional* to the displacement and *oppositely* directed, which is a characteristic of simple harmonic motion. Hence, the motion of the block is *simple harmonic* with natural frequency

$$\omega = \sqrt{\frac{k}{m}}$$

Therefore, its time period of motion is given as

$$T = \frac{2\pi}{\omega} = 2\pi\sqrt{\frac{m}{k}} \tag{19.37}$$

From the Eq. 19.31, we know $m/k = e/g$. Hence, time period can also be written as

$$T = 2\pi\sqrt{\frac{e}{g}} \tag{19.38}$$

It can be seen from the above two equations that the time period of the motion is dependent upon the mass of the block and the stiffness of the spring or the extension of the spring in the equilibrium position.

Example 19.7 A block of 3 kg mass is suspended from a spring, whose spring constant is 300 N/m. If the block is displaced by 5 cm from its equilibrium position and released, determine (i) the period of vibration, (ii) amplitude of vibration, and (iii) velocity at the equilibrium position.

Solution The extension of the spring, when the mass is in equilibrium position is given as

$$e = \frac{mg}{k}$$

$$= \frac{3\times 9.81}{300} = 9.81\times 10^{-2}\text{ m (or) } 9.81\text{ cm}$$

(i) As the displacement from the equilibrium position (i.e., 5 cm) is less than the initial elongation of 9.81 cm of the spring at the equilibrium position, the body will execute SHM throughout its motion. Therefore, the period of vibration is given as

$$T = 2\pi\sqrt{\frac{e}{g}}$$

$$= 2\pi\sqrt{\frac{9.81 \times 10^{-2}}{9.81}} = 0.63 \text{ s}$$

(ii) Since the mass is pulled down by 5 cm from the equilibrium position and released, the amplitude of vibration is 5 cm.

(iii) We know that the angular frequency is given as

$$\omega = \sqrt{\frac{g}{e}}$$

$$= \sqrt{\frac{9.81}{9.81 \times 10^{-2}}} = 10 \text{ rad/s}$$

Since velocity is maximum at the equilibrium position, velocity at the equilibrium position is given as

$$v = A\omega$$

$$= 5 \times 10 = 50 \text{ cm/s.}$$

Example 19.8 A mass of 5 kg extends a spring vertically by 15 cm. The same spring supports a mass of 2 kg attached at its lower end. Determine the frequency of oscillation when displaced vertically from its equilibrium position.

Solution We know by Hooke's law,

$$mg = ke$$

Given that when $m = 5$ kg elongation $e = 15$ cm. Therefore, when $m = 2$ kg, the elongation is given as $\frac{2}{5} \times 15 = 6$ cm. Therefore, frequency of oscillation is given as

$$f = \frac{\omega}{2\pi}$$

$$= \frac{1}{2\pi}\sqrt{\frac{g}{e}} = \frac{1}{2\pi}\sqrt{\frac{9.81}{0.06}} = 2.04 \text{ Hz}$$

Example 19.9 A body of mass m when hung from a spring vibrates with a period T. What mass should be hung so that the period of vibration is reduced by half?

Solution We know that the time period of motion is given as

$$T = 2\pi\sqrt{\frac{e}{g}} = 2\pi\sqrt{\frac{m}{k}} \qquad \text{(a)}$$

Let T' be the time period for the second case when suspended mass is m'. Therefore,

$$T' = 2\pi\sqrt{\frac{m'}{k}} \qquad \text{(b)}$$

Dividing equation (a) by equation (b), we get

$$\frac{T}{T'} = \sqrt{\frac{m}{m'}}$$

Note that as the same spring is used, the spring constant is same for both the cases.

$$\therefore \quad \frac{1}{1/2} = \sqrt{\frac{m}{m'}}$$

$$\Rightarrow \quad m' = \frac{m}{4}$$

Hence, *one-fourth* of the original mass would reduce the time period by *half*.

Example 19.10 The maximum velocity of a spring mass system with 2 kg mass oscillating vertically is 1 m/s and the oscillation is over a space of 0.5 m. Determine (i) amplitude (ii) frequency (iii) elongation of the spring in the equilibrium position, and (iv) spring constant.

Solution

(i) Amplitude

Since the oscillation is over a space of 0.5 m, the amplitude of oscillation is equal to half the value, i.e.,

$$A = \frac{1}{2}(0.5) = 0.25 \text{ m}$$

(ii) Frequency

$$v_{max} = 1 \text{ m/s}$$

$$\omega = \frac{v_{max}}{A} = \frac{1}{0.25} = 4 \text{ rad/s}$$

Therefore, frequency of oscillation is given as

$$f = \frac{\omega}{2\pi} = \frac{4}{2\pi} = \frac{2}{\pi} \text{ Hz}$$

(iii) Elongation of the spring in the equilibrium position

We know $\omega = \sqrt{\frac{g}{e}}$

$$\Rightarrow \quad e = \frac{g}{\omega^2} = \frac{9.81}{4^2} = 0.613 \text{ m}$$

(iv) Spring constant

$$mg = ke$$

$$\Rightarrow \quad k = \frac{mg}{e} = \frac{2 \times 9.81}{0.613} = 32 \text{ N/m}$$

19.5.1 Springs in Series and in Parallel

Springs in parallel Consider a body of mass m suspended by two springs of equal lengths and spring constants k_1 and k_2 arranged in parallel as shown in Fig. 19.13. Since the lengths of the springs are equal,

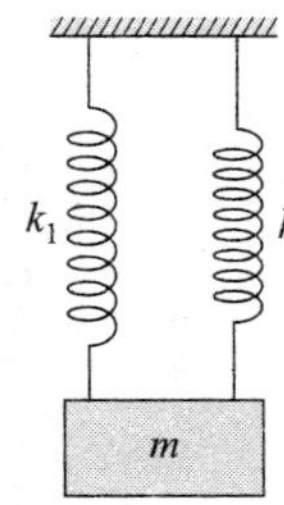

Fig. 19.13 Springs in parallel

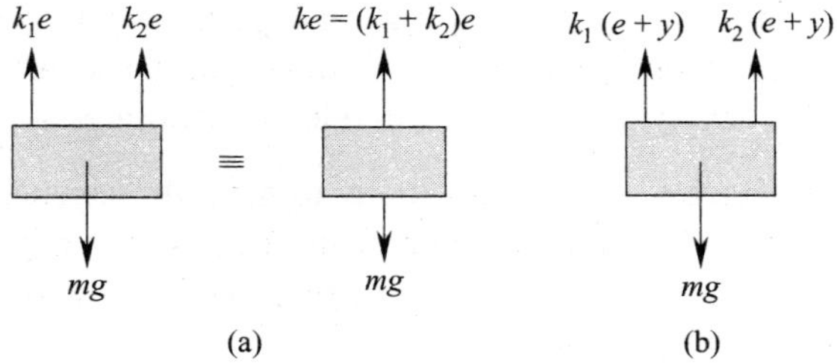

Fig. 19.14

it can be seen that the extension of the spring will be same in both the springs. Let e be the extension of both the springs.

The free-body diagram of the mass in the equilibrium position is shown in Fig. 19.14(a). Then

$$mg = k_1e + k_2e$$
$$= (k_1 + k_2)e \tag{19.39}$$

Suppose we replace the springs by an equivalent spring of stiffness k such that the elongation under the same load is e then we can see that the equivalent spring constant is given as

$$k = k_1 + k_2$$

If the body is given a small downward displacement y vertically downwards then the forces acting on the body tending to bring it to the equilibrium position are as shown in Fig. 19.14(b). Hence, the net restoring force acting on it tending to bring it to the equilibrium position is given as

$$k_1(e + y) + k_2(e + y) - mg$$
$$= (k_1 + k_2)e - mg + (k_1 + k_2)y$$

Substituting the Eq. (19.39) in the above expression, it can be further simplified as

$$(k_1 + k_2)y \tag{19.40}$$

Since this force acts in the direction opposite to the direction of displacement of the body, the equation of motion can be written as

$$m\frac{d^2y}{dt^2} = -(k_1 + k_2)y \tag{19.41}$$

$$\Rightarrow \quad \frac{d^2y}{dt^2} = -\left[\frac{(k_1 + k_2)}{m}\right]y \tag{19.42}$$

Since the acceleration is *proportional* to the displacement and *oppositely* directed, we conclude that the motion is *simple harmonic* and hence the period of vibration is given as

$$T = \frac{2\pi}{\omega} = 2\pi\sqrt{\frac{m}{(k_1 + k_2)}} \tag{19.43}$$

Though this expression was derived for *two* springs arranged in *parallel*, it is true for any number of springs in parallel. If n numbers of springs are arranged in parallel then its equivalent spring constant k is $k = k_1 + k_2 + \cdots + k_n$. Therefore, the time period is given as

$$T = 2\pi\sqrt{\frac{m}{k}} = 2\pi\sqrt{\frac{m}{(k_1 + k_2 + \ldots + k_n)}} \tag{19.44}$$

The arrangement shown in Fig. 19.15 is also equivalent to two springs arranged in parallel. The only difference is that the sense of reaction is opposite in the two springs, tension in one, and compression in the other.

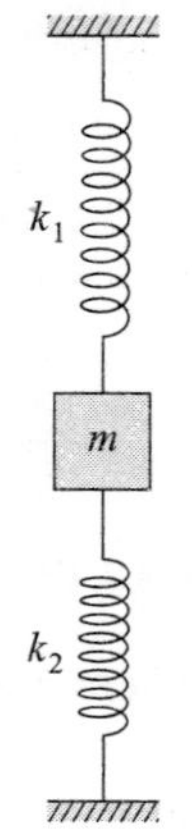

Fig. 19.15 An alternative representation of springs in parallel

Springs in Series Consider a body of mass m suspended by two springs of spring constants k_1 and k_2 arranged in series as shown in Fig. 19.16. If the springs are assumed to be *massless*, then the force acting on each spring tending to extend it is mg.

Hence, the extension produced in each spring is given as

$$mg = k_1 e_1 \qquad\qquad mg = k_2 e_2$$

$$\Rightarrow \quad e_1 = \frac{mg}{k_1} \qquad\qquad \Rightarrow \quad e_2 = \frac{mg}{k_2} \tag{19.45}$$

Hence, the net downward displacement of mass m is

$$e = e_1 + e_2$$

$$= \frac{mg}{k_1} + \frac{mg}{k_2}$$

$$= mg\left[\frac{k_1 + k_2}{k_1 k_2}\right]$$

Suppose we replace the springs by an equivalent spring of stiffness k such that the elongation under the same load is e. Then we can see that the equivalent spring constant is given as

$$k = \left(\frac{k_1 k_2}{k_1 + k_2}\right) \tag{19.46}$$

If the mass is given a small vertical displacement then period of vibration is given as

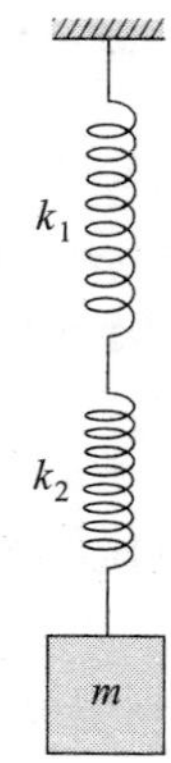

Fig. 19.16 Springs arranged in series

$$T = 2\pi\sqrt{\frac{e}{g}} = 2\pi\sqrt{\frac{mg}{kg}}$$

$$= 2\pi\sqrt{m\left[\frac{k_1 + k_2}{k_1 k_2}\right]} \tag{19.47}$$

Though this was derived for two springs in series, it can be extended to any number of springs arranged in series. If n numbers of springs are arranged in series then the equivalent spring constant is $\frac{1}{k} = \frac{1}{k_1} + \frac{1}{k_2} + \dots + \frac{1}{k_n}$ and hence the time period is given as

$$T = 2\pi\sqrt{\frac{m}{k}} \tag{19.48}$$

Example 19.11 A block of 12 kg mass is suspended by two springs with spring constants 2 kN/m and 3 kN/m arranged in series. The block is pulled down by 8 cm and released. Determine the period of oscillation, maximum velocity and maximum acceleration of the block.

Solution Since the springs are arranged in series, they can be replaced by an equivalent spring, whose spring constant k can be determined as follows

$$\frac{1}{k} = \frac{1}{2} + \frac{1}{3}$$

$$\Rightarrow \quad k = 1.2 \text{ kN/m}$$

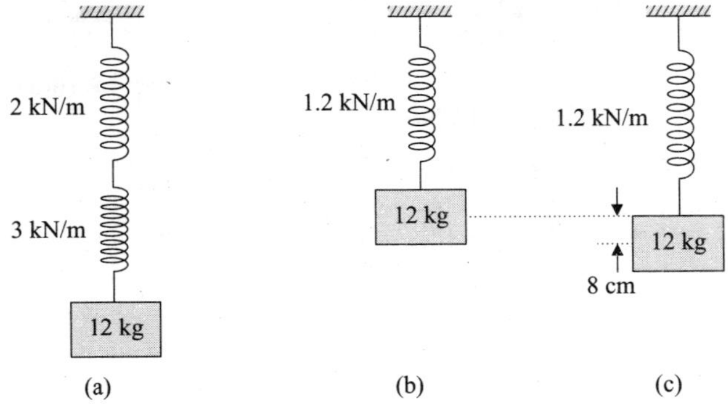

Fig. 19.17

Therefore, elongation of the spring in the equilibrium position is obtained as

$$e = \frac{mg}{k} = \frac{12 \times 9.81}{1.2 \times 10^3} = 0.098 \text{ m (or) } 9.8 \text{ cm}$$

When the block is pulled down by 8 cm (which is less than the elongation of the spring in the equilibrium position) and released, the block will execute simple harmonic motion with amplitude of 8 cm. Therefore, the natural frequency is given as

$$\omega = \sqrt{\frac{k}{m}} = \sqrt{\frac{1.2 \times 10^3}{12}} = 10 \text{ rad/s}$$

Hence, the period of oscillation is given as

$$T = \frac{2\pi}{\omega} = \frac{2\pi}{10} = 0.628 \text{ s}$$

Therefore, the maximum velocity and maximum acceleration of the block are given as

$$v_{max} = A\omega = 0.08 \times 10 = 0.8 \text{ m/s}$$

$$a_{max} = A\omega^2 = 0.08 \times 10^2 = 8 \text{ m/s}^2$$

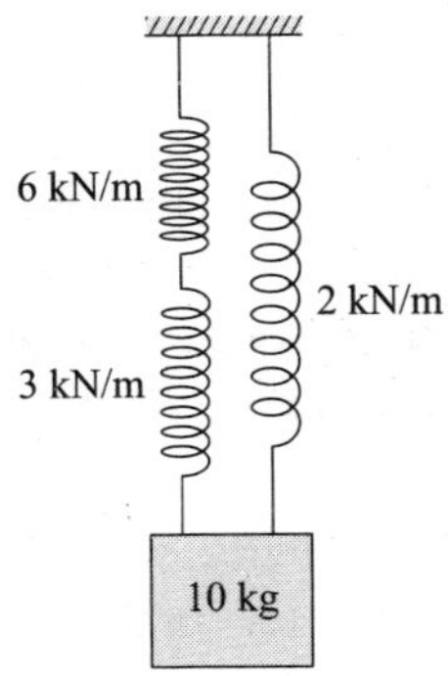

Fig. 19.18

Example 19.12 A 10 kg block is suspended by a system of three springs as shown. The respective spring constants are also shown in Fig. 19.18. Determine the period of oscillation for small displacement from the equilibrium position.

Solution The equivalent spring constant k' of the two springs in series is given as

$$\frac{1}{k'} = \frac{1}{6} + \frac{1}{3}$$

$\Rightarrow$ $$k' = 2 \text{ kN/m}$$

Hence, the spring system reduces to that as shown in Fig. 19.18(a).

These two springs in parallel can further be replaced by an equivalent spring as shown in Fig. 19.18(b) with spring constant k given as

$$k = 2 + 2 = 4 \text{ kN/m}$$

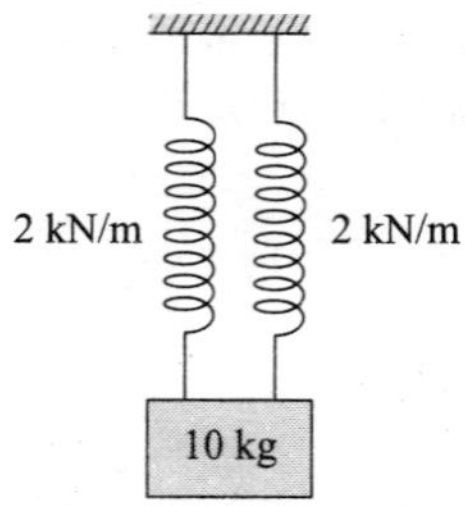

Fig. 19.18(a)

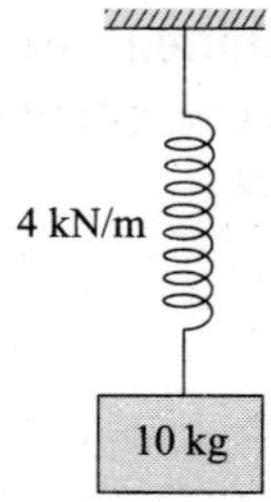

Fig. 19.18(b)

Hence, the period of vibration is given as

$$T = \frac{2\pi}{\omega} = 2\pi\sqrt{\frac{m}{k}}$$

$$= 2\pi\sqrt{\frac{10}{4 \times 10^3}} = 0.314 \text{ s}$$

19.5.2 Horizontal Motion of a Block Attached to a Spring

Consider now a block of mass m attached to a spring of spring constant k and free to move *horizontally* on a frictionless surface as shown.

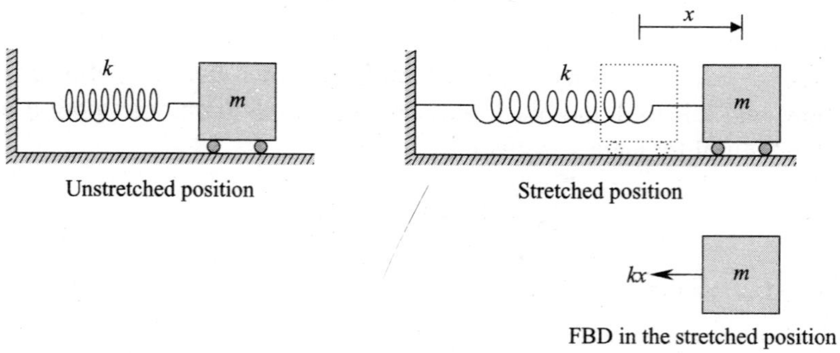

Fig. 19.19

If the block is pulled from its unstretched position and released, it will oscillate back and forth about the mean or unstretched position. Suppose x be the position of the block at an instant of time from the unstretched position, then the restoring force acting on the block tending to bring it to the unstretched or mean position is

$$kx \tag{19.49}$$

Hence, we can write the horizontal motion of the block taking into consideration that the restoring force acts in the direction opposite to that of the displacement as

$$m\frac{d^2x}{dt^2} = -kx \tag{19.50}$$

or $$\frac{d^2x}{dt^2} = -\left[\frac{k}{m}\right]x \tag{19.51}$$

We can see that the acceleration is proportional to the displacement and directed opposite to it. Thus, we can conclude that the motion is simple harmonic with natural frequency

$$\omega = \sqrt{\frac{k}{m}} \tag{19.52}$$

Hence, the period of motion is

$$T = \frac{2\pi}{\omega} = 2\pi\sqrt{\frac{m}{k}} \tag{19.53}$$

The expressions derived above are same as that discussed in Section 19.5 for a spring mass system with the mass oscillating *vertically*. Thus, we see that the motion of the block is similar whether it oscillates vertically or horizontally.

19.6 MOTION OF A BODY ATTACHED TO AN ELASTIC STRING

Consider a body of mass m suspended by an elastic string of natural or unstretched length l and cross-sectional area A as shown in Fig. 19.20. Due to elasticity of the string, the mass will extend the string and let the extension at the equilibrium position be Δ.

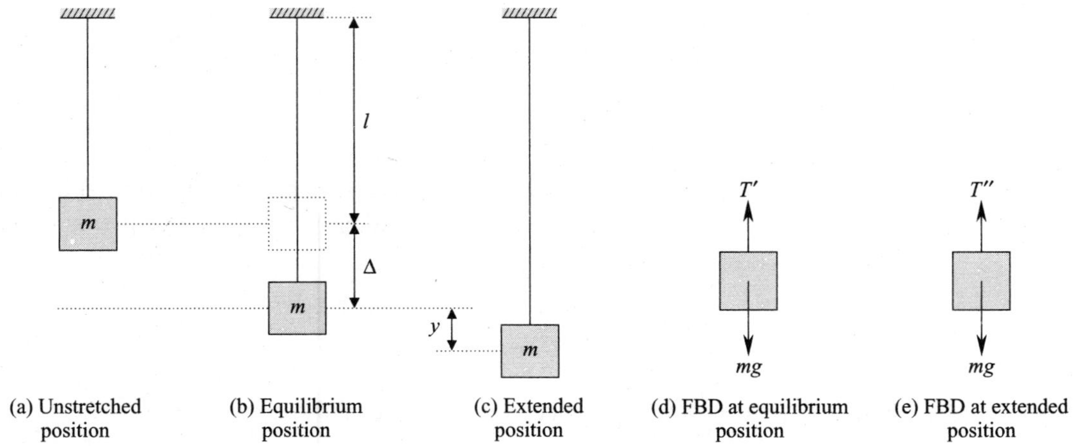

Fig. 19.20

Hooke's law states that within the elastic limit, stress is proportional to strain, i.e.,

$$\text{stress } [\sigma] \propto \text{strain } [\varepsilon] \tag{19.54}$$

Introducing a constant of proportionality E, we can write

$$\sigma = E\varepsilon \tag{19.55}$$

where E is called the **elastic** or **Young's modulus** of material of the string. The force acting on the string at the equilibrium position is mg; hence stress, that is force per unit area is given as mg/A. Its strain is given as change in length per unit original length, i.e., Δ/l. Therefore, we can write the Eq. 19.55 as

$$\frac{mg}{A} = E\frac{\Delta}{l} \tag{19.56}$$

The tension in the string at the equilibrium position [refer Figs 19.20 (b) and (d)] is

$$T' = mg = \left[\frac{AE}{l}\right]\Delta \tag{19.57}$$

Since the terms within the bracket on the right-hand side are constants for a particular string, we see that the tension is proportional to the extension. Suppose the body is pulled down by a vertical distance y from the equilibrium position [refer Fig. 19.20 (c) and (e)], then the tension in the string is

$$T'' = \frac{AE}{l}(\Delta + y) \tag{19.58}$$

Therefore, restoring force acting on the mass after it is pulled down by y and released is

$$T'' - mg = \frac{AE}{l}(\Delta + y) - mg$$

Substituting the Eq. (19.57) in the above equation, we get the net restoring force as

$$\frac{AE}{l}y \tag{19.59}$$

Writing the equation of motion of the block taking into consideration that the downward displacement of the block from the equilibrium position as positive

$$m\frac{d^2y}{dt^2} = -\frac{AE}{l}y \tag{19.60}$$

or

$$\frac{d^2y}{dt^2} = -\left[\frac{AE}{ml}\right]y \tag{19.61}$$

From the above equation, we see that the acceleration is directly proportional to the displacement and oppositely directed. Hence, the motion of the block is simple harmonic and its time period is given as

$$T = \frac{2\pi}{\omega} = 2\pi\sqrt{\frac{ml}{AE}} \tag{19.62}$$

From the Eq. 19.56, we know $\frac{ml}{AE} = \frac{\Delta}{g}$ Hence, time period can also be written as

$$T = 2\pi\sqrt{\frac{\Delta}{g}} \tag{19.63}$$

It can be seen from the above equation that like in a spring mass system, for small displacements from the equilibrium position, the period of motion is dependent on the extension of the string in the equilibrium position.

Example 19.13 A block of 3 kg mass is suspended by an elastic string of length 1 m and 4 mm in diameter. The extension of the string is observed to be 25 cm. Determine the elastic modulus of the string. Describe the motion of the mass and determine period of motion if the (i) mass is released from the unstretched position, (ii) mass is pulled down by 20 cm from equilibrium position and released, (iii) mass is pulled down by 30 cm from equilibrium position and released.

Solution *Elastic modulus of the string*

We know that the elastic modulus of material of the string is given as

$$E = \frac{mgl}{A\Delta}$$

$$= \frac{3 \times 9.81 \times 1}{\pi(0.002)^2 (0.25)} = 93.68 \times 10^5 \text{ N/m}^2$$

(i) Mass is released from the unstretched position

As the mass is *released* from the unstretched position P, its velocity at that point is zero. It begins to increase its velocity and reaches its maximum value at the equilibrium or mean position Q. With this velocity, it begins to move further down and this motion is restrained by the tension in the string. As a result, the mass begins to decelerate and reaches a momentary rest at position R. Due to the restoring force acting on the mass, it begins to move upwards with increasing velocity and thus it oscillates between P and R.

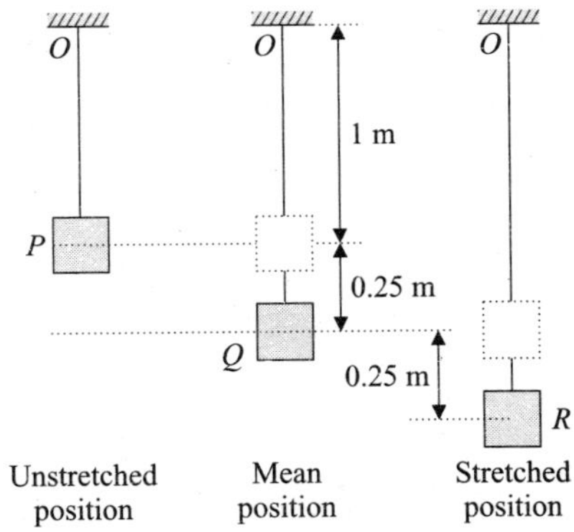

Fig. 19.21

From the Eq. 19.60, we can write the equation of motion of the mass for a displacement of y from mean position as

$$m\frac{d^2y}{dt^2} = -\frac{AE}{l}y$$

$$\Rightarrow \quad \frac{d^2y}{dt^2} = -\frac{AE}{ml}y$$

Hence, the motion is simple harmonic. Its amplitude is 0.25 m as the extension at the mean position is 0.25 m and natural frequency ω is given as

$$\omega = \sqrt{\frac{AE}{ml}} = \sqrt{\frac{g}{\Delta}} = 6.26 \text{ rad/s}$$

Hence, time period is given as

$$T = \frac{2\pi}{\omega} = \frac{2\pi}{6.26} = 1 \text{ s}$$

The maximum velocity at the mean position Q is given as

$$v_{\max} = A\omega = 0.25 \times 6.26 = 1.57 \text{ m/s}$$

(ii) Mass is pulled down by 20 cm and released

As this distance is less than the extension of the string in the equilibrium position, the block will execute simple harmonic motion with 20 cm as amplitude. The time period will be same as in the previous case as it is independent of amplitude; whereas the velocity is dependent upon amplitude. Hence,

$$v_{\max} = A\omega = 0.2 \times 6.26 = 1.25 \text{ m/s}$$

(iii) Mass is pulled down by 30 cm and released

When the mass is pulled down by 30 cm and released (position R), the velocity of the mass at that position is zero. As the pulled distance is greater than the extension (25 cm) of the string in the equilibrium position, the block will execute simple harmonic motion with 30 cm as amplitude up to the unstretched position. The time period will be same as the previous two cases as it is independent of amplitude; whereas the velocity is dependent upon amplitude.

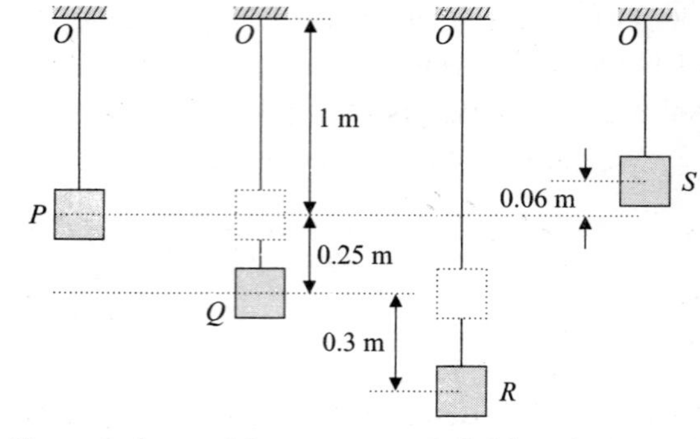

Fig. 19.22

The position of the block with respect to the mean position, till it reaches the unstretched position P is given as

$$x = A \cos \omega t \qquad \text{[note that the motion starts from extreme position]}$$

$$-0.25 = 0.3 \cos [6.26t] \qquad \text{[note that } x \text{ is taken as positive downwards]}$$

$$\therefore \quad t = 0.41 \text{ s} \qquad \text{(a)}$$

The velocity of the block when it reaches the unstretched position is

$$|v| = \omega\sqrt{A^2 - x^2}$$

$$= 6.26\sqrt{0.3^2 - (-0.25)^2} = 1.04 \text{ m/s}$$

With this velocity it moves upwards purely under gravity up to a distance given as

$$\frac{v^2}{2g} = \frac{(1.04)^2}{2 \times 9.81} = 0.055 \text{ m}$$

The time taken to reach this height is given as

$$t = \frac{v}{g} = \frac{1.04}{9.81} = 0.11 \text{ s} \qquad \text{(b)}$$

Hence, the time taken to move from position R to S (half the oscillation) is

$$0.41 + 0.11 = 0.52 \text{ s}$$

Hence, the time period of motion for complete cycle is

$$T = 2 \times 0.52 = 1.04 \text{ s}$$

Example 19.14 A string of length l is stretched to twice its length retaining elasticity and both its ends are fixed. A particle of mass m is attached to it at its centre. Determine the period of oscillation of the mass for small displacements of the mass (i) along the string, and (ii) perpendicular to the string.

Solution

(i) When the particle is displaced along the string

Since the string is stretched to twice its initial length, we see that at the equilibrium position

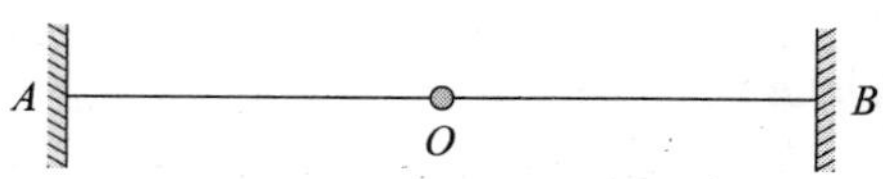

Fig. 19.23

$$\overline{OA} = \overline{OB} = l$$

Let the particle be displaced through a small distance x from its equilibrium position. The tensions in the string on both sides of the particle are shown in the free-body diagram [Fig. 19.23(b)].

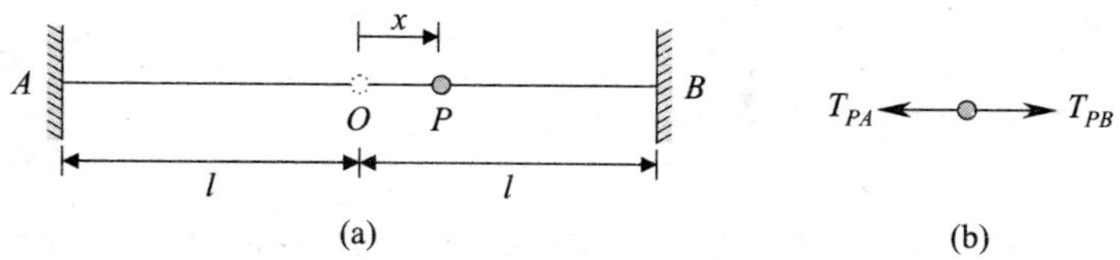

Fig. 19.23

From the Fig. 19.23(a) we see that

$$\overline{PA} = l + x \quad \text{and} \quad \overline{PB} = l - x$$

The elongation of portion of the string on the left-hand side of the particle is

Final length – initial length $= (l + x) - l/2 = l/2 + x$

Note that the original unstretched length of string on the left-hand side as well as that on the right-hand side is $l/2$.

Similarly, elongation of the string on the right-hand side of the particle is

$$l - x - l/2 = l/2 - x$$

From the Eq. 19.57, we know that the tension in the string is proportional to the extension. Hence, the tensions on the left and right-hand sides of the particle are respectively

$$T_{PA} = \frac{AE}{l/2}[l/2 + x]$$

and

$$T_{PB} = \frac{AE}{l/2}[l/2 - x]$$

Hence, the net tension acting on the particle tending to restore it to the equilibrium position is

$$T_{PA} - T_{PB} = \frac{AE}{l/2}[l/2 + x - (l/2 - x)] = \frac{AE}{l/2}2x$$

Therefore, writing the equation of motion of the particle taking into consideration that the restoring force acts in the direction opposite to that of the displacement,

$$m\frac{d^2x}{dt^2} = -\frac{AE}{l/2}2x$$

or

$$\frac{d^2x}{dt^2} = -\left[\frac{4AE}{ml}\right]x$$

We see that the acceleration is proportional to the displacement x and directed opposite to it. Hence, the motion is simple harmonic with natural frequency, $\omega = \sqrt{\dfrac{4AE}{ml}}$. Therefore, the period of oscillation is

$$T = \frac{2\pi}{\omega} = 2\pi\sqrt{\frac{ml}{4AE}} = \pi\sqrt{\frac{ml}{AE}}$$

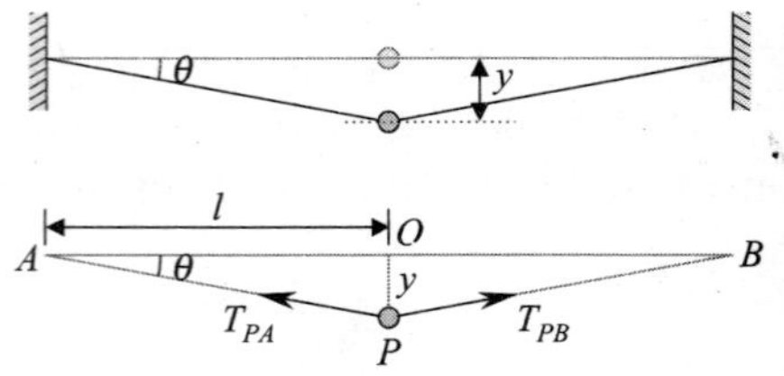

Fig. 19.24

(ii) When the particle is displaced perpendicular to the string

Let the particle be displaced through a small distance y from the equilibrium position. Due to symmetry, the tensions in the string on both sides of the particle will be equal, i.e.,

$$T_{PA} = T_{PB} = T$$

Hence, the net restoring force acting on the particle tending to bring it to the equilibrium position is

$$2T\sin\theta$$

From the figure, we see that

$$\overline{PA} = \overline{PB} = \sqrt{l^2 + y^2} = l\sqrt{1 + \frac{y^2}{l^2}} \approx l$$

by neglecting higher order terms. Hence, extension of PA or PB is

$$l - l/2 = l/2$$

We know that tension depends upon the extension of the string. Therefore,

$$T_{PA} = T_{PB} = T = \frac{AE}{l/2}[l/2] = AE$$

Also, $$\sin\theta = \frac{OP}{PA} = \frac{y}{l}$$

Hence, the net restoring force acting on the particle is

$$\frac{2(AE)y}{l}$$

Therefore, writing the equation of motion of the particle taking into consideration that the restoring force acts in the direction opposite to that of the displacement,

$$m\frac{d^2y}{dt^2} = -\frac{2AE}{l}y$$

or $$\frac{d^2y}{dt^2} = -\left[\frac{2AE}{ml}\right]y$$

We see that the acceleration is proportional to the displacement y and directed opposite to it. Hence, the motion is simple harmonic with natural frequency, $\omega = \sqrt{\frac{2AE}{ml}}$. Therefore, the period of oscillation is

$$T = \frac{2\pi}{\omega} = 2\pi\sqrt{\frac{ml}{2AE}} = \pi\sqrt{\frac{2ml}{AE}}$$

19.7 SIMPLE PENDULUM

A *simple pendulum* is defined as a small but heavy mass of bob suspended from a fixed support by a weightless, inextensible string. The bob is allowed to oscillate in an arc in a vertical plane under the influence of gravity.

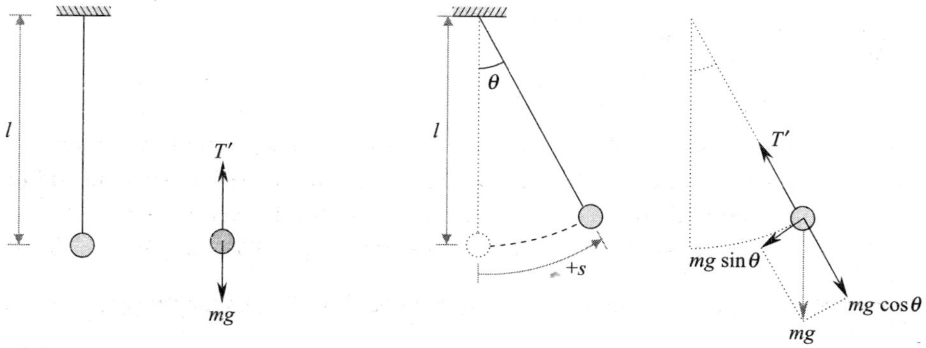

(a) Equilibrium position of simple pendulum (b) Simple pendulum

Fig. 19.25

Consider a simple pendulum with mass of bob *m* and length of string '*l*.' In the equilibrium position, the weight of the bob *mg* is balanced by the tension *T′* in the string [refer Fig. 19.25(a)]. If the bob is given a small displacement from its equilibrium position, it begins to *oscillate* about the equilibrium position in a vertical plane under the influence of gravity. To determine the period of such oscillatory motion, let us consider the free-body diagram of the bob shown in Fig. 19.25(b), where the weight *mg* acting vertically downwards and tension *T′* acting along the string are shown. It will be convenient to resolve the weight into components $mg \sin\theta$ and $mg \cos\theta$ along tangential to the arc and normal to the arc respectively. The normal component $mg \cos\theta$ is balanced by the tension in the string as there is no motion in the radial direction. The unbalanced component $mg \sin\theta$ is the restoring force acting on the bob tending to restore it to the equilibrium position. Hence, writing the equation of motion along the tangential direction, taking the displacement *s* as positive in the direction shown, we have

$$ma_t = -mg \sin\theta$$

$$\Rightarrow \quad a_t = -g \sin\theta \tag{19.64}$$

If we consider small displacements, then $\sin\theta$ can be approximated to θ. Hence, we can write

$$a_t = -g\theta \tag{19.65}$$

From the figure, we see that

$$\theta = \frac{\text{arc length}}{\text{radius}} = \frac{s}{l}$$

Therefore,

$$a_t = -g\theta = -g\,\frac{s}{l} = -\left[\frac{g}{l}\right]s \tag{19.66}$$

From which we see that the acceleration is proportional to the displacement s and directed towards the mean position. Thus we can conclude that the motion is simple harmonic in the angular direction, where

$$\omega^2 = \frac{g}{l} \tag{19.67}$$

Hence, the period of motion is given as

$$T = \frac{2\pi}{\omega} = 2\pi\sqrt{\frac{l}{g}} \tag{19.68}$$

From the above expression, we see that the time period is independent of the mass of the bob, but dependent on the length of the string and acceleration due to gravity at that place. In addition, the period is independent of the amplitude of motion. It is for this reason it is used in clocks to maintain the time period. It is also useful in determining the acceleration due to gravity at a particular location.

Example 19.15 Determine the period of a simple pendulum whose length is 0.5 m at a place where $g = 9.81$ m/s^2.

Solution The time period of a simple pendulum is given as

$$T = 2\pi\sqrt{\frac{l}{g}}$$

$$= 2\pi\sqrt{\frac{0.5}{9.81}} = 1.42 \text{ s}$$

Example 19.16 The period of a simple pendulum whose length is 0.3 m is found to be 1.1 s. Determine the acceleration due to gravity at that place.

Solution We know that the period of simple pendulum is given as

$$T = 2\pi\sqrt{\frac{l}{g}}$$

$$\Rightarrow \quad g = \left(\frac{2\pi}{T}\right)^2 l$$

$$= \left(\frac{2\pi}{1.1}\right)^2 \times 0.3 = 9.79 \text{ m/s}^2$$

Example 19.17 If the time period of a simple pendulum has to be 1 s, determine the length of the pendulum required at a place where $g = 9.81$ m/s^2.

Solution The time period of a simple pendulum is given as

$$T = 2\pi\sqrt{\frac{l}{g}}$$

$$\therefore \qquad 1 = 2\pi\sqrt{\frac{l}{9.81}}$$

$$\Rightarrow \qquad l = \frac{9.81}{4\pi^2} = 0.248 \text{ m}$$

Example 19.18 A simple pendulum of length l is suspended in an elevator to oscillate freely in a vertical plane. Determine the time period of oscillation when the elevator is (a) accelerating upwards with an acceleration of a m/s^2, (*b*) accelerating downwards at a m/s^2 and (c) is falling freely.

Solution The free-body diagram of the pendulum is shown in Fig. 19.26. The forces acting on the bob are its weight and tension in the string.

Fig. 19.26

(i) When the elevator is accelerating upwards with an acceleration of a m/s^2

Writing the equation of motion in the vertical direction,

$$T' - mg = ma$$

$$\Rightarrow \qquad T' = m(g + a)$$

Thus, we can see that the effective acceleration of the bob when the elevator is accelerating upwards is $(g + a)$ m/s^2. Therefore, the period of motion of the pendulum is given as

$$T = 2\pi\sqrt{\frac{l}{g + a}}$$

(ii) When the elevator is accelerating downwards with an acceleration of a m/s^2

Writing the equation of motion in the vertical direction,

$$mg - T' = ma$$

$$\Rightarrow \qquad T' = m(g - a)$$

Thus, we can see that the effective acceleration of the bob when the elevator is accelerating downwards is $(g - a)$ m/s^2. Therefore, the period of motion of the pendulum is given as

$$T = 2\pi\sqrt{\frac{l}{g - a}}$$

(c) When the elevator is falling freely

Writing the equation of motion in the vertical direction,

$$mg - T' = mg$$

$$\Rightarrow \qquad T' = 0$$

Hence, the net acceleration of the bob is zero. Therefore, the period of motion of the pendulum is *infinity*, that is, it does not oscillate.

19.7.1 Seconds Pendulum

A seconds pendulum is one which takes *two seconds* to execute *one* complete oscillation, or *a second* to execute *half* an oscillation. The time taken to complete half an oscillation is known as a **beat**. Hence, a seconds pendulum is said to beat a second.

Example 19.19 A pendulum clock beats seconds at a place where $g = 9.81$ m/s^2. If it is brought to a place where $g = 9.8$ m/s^2, how much does it gain or lose per day?

Solution As the clock beats seconds at the former place, its time period T_1 for *half the oscillation* is 1 s.

$$T_1 = 1 \text{ s} \quad \text{at} \quad g_1 = 9.81 \text{ m/s}^2$$
$$T_2 = ? \quad \text{at} \quad g_2 = 9.8 \text{ m/s}^2$$

We know that the time period for half an oscillation is

$$T = \pi\sqrt{\frac{l}{g}}$$

Since the length of the pendulum remains constant at both the places, we have

$$\frac{T_1}{T_2} = \sqrt{\frac{g_2}{g_1}}$$

$$\therefore \quad T_2 = T_1\sqrt{\frac{g_1}{g_2}} = \sqrt{\frac{9.81}{9.8}}$$

In the latter place in T_2 s, the clock records 1 s.

Therefore, in 1 s, the clock records $\frac{1}{T_2}$ s $= \sqrt{\frac{9.8}{9.81}}$ s

Since the difference between the numerator and the denominator is very small, the above expression can be approximated as

$$= \left[\frac{9.8}{9.81}\right]^{1/2} = \left[1 - \frac{0.01}{9.81}\right]^{1/2} \approx 1 - \frac{1}{2}\left[\frac{0.01}{9.81}\right]$$

$$\therefore \quad \text{Loss of time in 1 s} = \frac{1}{2}\left[\frac{0.01}{9.81}\right] = 5.09684 \times 10^{-4} \text{ s}$$

$$\therefore \quad \text{Loss of time in a day} = (5.09684 \times 10^{-4})(86400) = 44.04 \text{ s}$$

[Note that 1 day = 24 hr = 86400 s]

Example 19.20 A clock, which keeps correct time when its pendulum beats seconds, was found to be losing 5 minutes per day. By what amount should the length be altered to correct the clock if the length of the seconds pendulum is 0.5 m?

Solution *Time period of pendulum*

Given that, the clock loses 5 minutes per day. Hence, in 1 second it loses

$$\frac{5 \times 60}{86400} \text{ s}$$

Therefore, the time recorded by the clock in 1 s is

$$\left[1 - \frac{300}{86400}\right] \text{s} = \frac{86100}{86400} = \frac{861}{864} \text{ s}$$

Therefore, the time period of the pendulum when it records 1 s is $\frac{864}{861}$ s. As the time period of the pendulum is greater than 1, the actual length should be greater than 0.5 m, by an amount say x metres.

$$T_1 = \frac{864}{861}\ \text{s} \qquad \text{for} \qquad l = (0.5 + x)\ \text{m}$$

$$T_2 = 1\ \text{s} \qquad \text{for} \qquad l = (0.5)\ \text{m}$$

We know that the time period is directly proportional to the length for a given value of g. Hence,

$$\frac{864}{861} = \sqrt{\frac{0.5 + x}{0.5}}$$

$$1 + \frac{3}{861} = 1 + \frac{1}{2}\frac{x}{0.5}$$

$$\Rightarrow \qquad x = \frac{3}{861} = 3.48\ \text{mm}$$

Hence, the length of the pendulum must be diminished by 3.48 mm.

19.8 COMPOUND PENDULUM

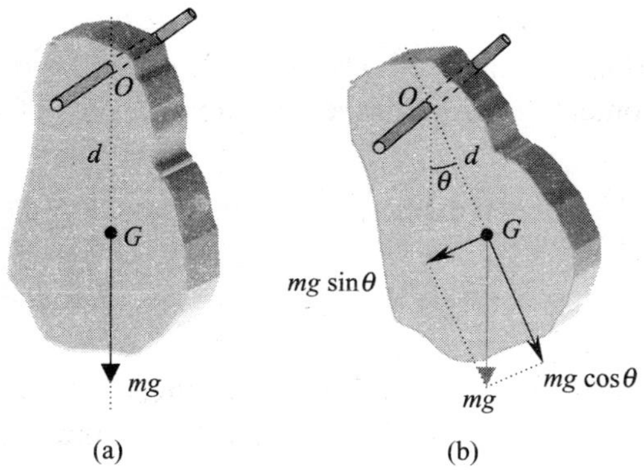

Fig. 19.27

A compound pendulum is a rigid body mounted in such a way that it can oscillate freely in a vertical plane about a horizontal axis passing through the body.

Figure 19.27(a) shows a rigid body of general shape mounted to oscillate freely about a horizontal axis through O. Point O is called **point of support** or **suspension**. When the body is in equilibrium, the centre of gravity G of the body lies vertically below the point of support O. Let d be the distance of centre of gravity from the axis through O. If the body is displaced from its equilibrium position, it tends to return to the equilibrium position. However, when it reaches its equilibrium position, it possesses some velocity and hence it overshoots and displaces angularly on the other side. As a result, the body oscillates about the equilibrium position in a vertical plane. From the Fig. 19.27(b), we see that the restoring moment is provided by the tangential component of weight, i.e., $mg \sin \theta$.

Restoring moment acting on the body tending to bring it to the equilibrium position is

$$M = mg \sin \theta d \tag{19.69}$$

Note that the normal component $mg \cos \theta$ does not contribute to the moment as its line of action passes through the point of support O.

Writing the equation of angular motion considering angular displacement θ as positive when in anticlockwise direction and moment as negative when in clockwise direction,

$$I\alpha = I\frac{d^2\theta}{dt^2} = -mg \sin \theta d \tag{19.70}$$

For small displacements, $\sin\theta \approx \theta$. Therefore,

$$I\frac{d^2\theta}{dt^2} = -mgd\theta$$

$$\therefore \quad \frac{d^2\theta}{dt^2} = -\left(\frac{mgd}{I}\right)\theta \qquad (19.71)$$

From which we see that angular acceleration is proportional to angular displacement and oppositely directed. Thus, the above equation represents angular simple harmonic motion, where natural frequency is given as

$$\omega = \sqrt{\frac{mgd}{I}} \qquad (19.72)$$

Therefore, time period of oscillation is given as

$$T = \frac{2\pi}{\omega} = 2\pi\sqrt{\frac{I}{mgd}} \qquad (19.73)$$

19.8.1 Length of Equivalent Simple Pendulum

As a special case, consider the mass to be concentrated as a point mass m suspended by a weightless string of length 'l'. The moment of inertia of the point mass about its own axis is zero. Hence, the moment of inertia of the mass about the point of suspension 'O' is given as

$$I = md^2 = ml^2 \qquad (19.74)$$

Thus, the expression for time period in the Eq. (19.73) reduces to

$$T = 2\pi\sqrt{\frac{ml^2}{mgl}} = 2\pi\sqrt{\frac{l}{g}} \qquad (19.75)$$

which is same as the expression already derived for the time period of a simple pendulum. Thus, from the Eqs 19.73 and 19.75, we see that the time period of a compound pendulum is same as that of a simple pendulum of length, $L_e = l = \dfrac{I}{md}$. This length is called **length of equivalent** simple pendulum.

Corollary Further, we know that the moment of inertia of the body about the axis of rotation is

$$I = \bar{I} + md^2 \qquad (19.76)$$

$$\therefore \quad L_e = \frac{I}{md} = \frac{\bar{I} + md^2}{md} = \frac{\bar{I}}{md} + d \qquad (19.77)$$

From the above expression, we see that the equivalent length is greater than d. Hence, it lies below the centre of gravity as shown in Fig. 19.28(a). This point C is known as **point of oscillation**. Therefore,

$$\overline{OG} = d,\ \overline{OC} = L_e \text{ and } \overline{GC} = L_e - d = \frac{\bar{I}}{md} \qquad (19.78)$$

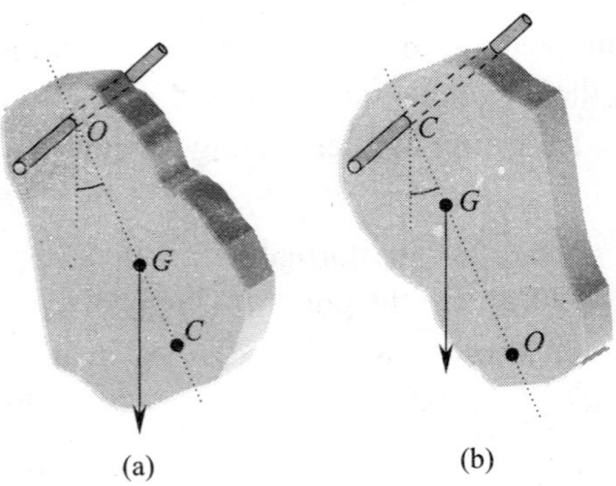

Fig. 19.28

Substituting the value of I from the Eq. (19.76) in the Eq. (19.73), we get

$$T = 2\pi\sqrt{\frac{I}{mgd}} = 2\pi\sqrt{\frac{\frac{\bar{I}}{md} + d}{g}} \tag{19.79}$$

We see from the above expression that the time period depends only upon value of d, i.e., distance of the axis of suspension from the centre of gravity of the body. Note that all other terms are constant for a given body. Suppose the body is allowed to oscillate about a horizontal axis passing through the point of oscillation C as shown in Fig. 19.28(b), then the distance of the axis of support from the centre of gravity is $\overline{CG} = L_e - d = \frac{\bar{I}}{md}$. Hence,

$$T = 2\pi\sqrt{\frac{\frac{\bar{I}}{m(L_e - d)} + (L_e - d)}{g}}$$

$$= 2\pi\sqrt{\frac{\frac{\bar{I}}{md} + d}{g}} \tag{19.80}$$

which is same as the Eq. 19.79. Hence, we see that the time period of motion remains the same when the point of suspension and point of oscillation are interchanged. This principle is used in the design of certain equipments like cricket bats. When a batsman bats a ball, we can assume that the bat is swinging about an axis (his hands), and there is a point at which if he strikes the ball, he requires less effort to do it and this point happens to be the point of *oscillation*, also called the point of *percussion*.

Example 19.21 A slender rod of length L is pivoted at a distance d from the centre of mass, so as to rotate freely about a horizontal axis. For small angular displacements, determine the period of oscillation. Also, determine for what value d will the period be minimum.

Solution *Period of oscillation for small angular displacements*

We know that the moment of inertia of a slender rod about centriodal perpendicular axis Z–Z is

$$\bar{I}_{zz} = \frac{1}{12}ml^2$$

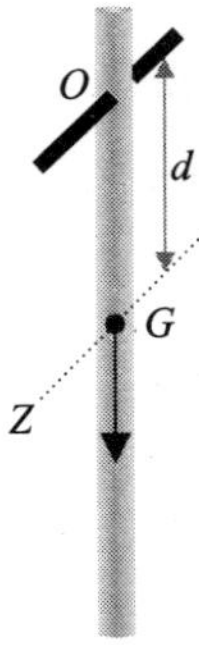

Fig. 19.29

Therefore, moment of inertia of the rod about the perpendicular axis through the pivot is

$$I = \bar{I}_{zz} + md^2$$

$$= \frac{1}{12}ml^2 + md^2$$

Therefore, time period of oscillation for small angular displacement is

$$T = 2\pi\sqrt{\frac{I}{mgd}}$$

$$= 2\pi\sqrt{\frac{\frac{1}{12}ml^2 + md^2}{mgd}}$$

$$= 2\pi\sqrt{\frac{l^2}{12gd} + \frac{d}{g}} \tag{a}$$

Condition for minimum period

Squaring both sides and rearranging,

$$\frac{T^2}{4\pi^2} = \frac{l^2}{12gd} + \frac{d}{g}$$

Taking l and g as constants and differentiating time period T with respect to d

$$\frac{2T}{4\pi^2}\frac{dT}{dd} = -\frac{l^2}{12gd^2} + \frac{1}{g}$$

From calculus, we know that time period can be minimum when $\frac{dT}{dd} = 0$. Therefore,

$$-\frac{l^2}{12gd^2} + \frac{1}{g} = 0$$

$$\Rightarrow \qquad d = \frac{l}{\sqrt{12}} \tag{b}$$

Substituting this value in the equation (a), we get the minimum time period.

$$T = 2\pi\sqrt{\frac{l^2}{12gd} + \frac{d}{g}} = 2\pi\sqrt{\frac{\sqrt{12}l^2}{12gl} + \frac{l}{\sqrt{12}g}} = 2\pi\sqrt{\frac{4}{\sqrt{12}}\left(\frac{l}{g}\right)}$$

$$= 2\pi\sqrt{0.577\frac{l}{g}} = 4.77\sqrt{\frac{l}{g}}$$

Example 19.22 A thin annular ring of 50 cm diameter and 800 g mass is mounted on a knife-edge as shown. Determine the period of oscillation for small displacements. Also, determine the length of equivalent simple pendulum.

Solution Mass moment of inertia of the annular ring about the perpendicular axis passing through the centroid is

$$\bar{I}_{zz} = mr^2$$

Therefore, by transfer theorem, moment of inertia about the perpendicular axis through knife-edge is given as

$$I_{zz} = \bar{I}_{zz} + mr^2$$
$$= 2mr^2$$
$$= 2 \times 0.8 \times (0.25)^2 = 0.1 \text{ kg.m}^2$$

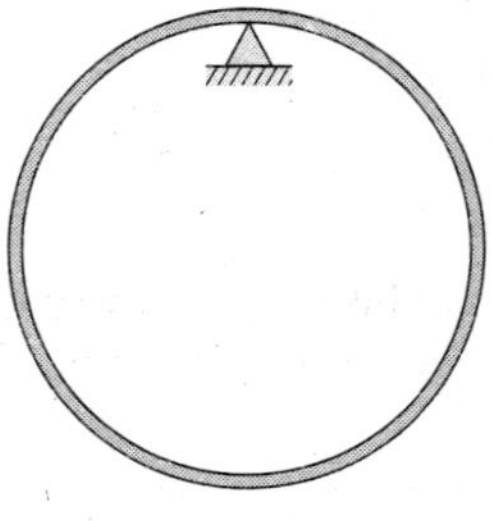

Fig. 19.30

Time period of oscillation for small displacement is given as

$$T = 2\pi\sqrt{\frac{I_{zz}}{mgd}} = 2\pi\sqrt{\frac{0.1}{0.8 \times 9.81 \times 0.25}} = 1.42 \text{ s}$$

Length of equivalent simple pendulum is given as

$$L_e = \frac{I_{zz}}{md} = \frac{0.1}{0.8 \times 0.25} = 0.5 \text{ m}$$

Example 19.23 A compound pendulum consists of a circular metallic plate suspended by a thin metallic strip as shown. If their respective masses are 0.4 kg and 0.2 kg, determine the frequency of oscillation for small displacements.

Solution *Given data*

Mass of thin strip, $m_1 = 0.2$ kg

Length of strip, $l = 0.25$ m

Mass of circular plate, $m_2 = 0.4$ kg

Radius of plate, $r = 0.05$ m

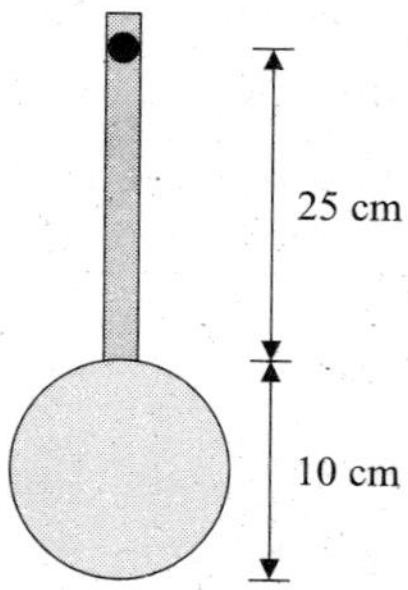

Fig. 19.31(a)

Location of centre of mass from the pivot is given as

$$d = \frac{m_1 d_1 + m_2 d_2}{m_1 + m_2}$$

$$= \frac{0.2 \times (0.25/2) + 0.4 \times (0.25 + 0.05)}{0.2 + 0.4} = 0.242 \text{ m}$$

Mass moment of inertia of the composite body

Thin strip: $(I_{zz})_1 = (\bar{I}_{zz})_1 + m_1 d_1^2$

$$= \frac{m_1 l^2}{12} + m_1 (l/2)^2$$

$$= \frac{m_1 l^2}{3}$$

$$= \frac{0.2 \times (0.25)^2}{3}$$

$$= 4.167 \times 10^{-3} \text{ kg.m}^2$$

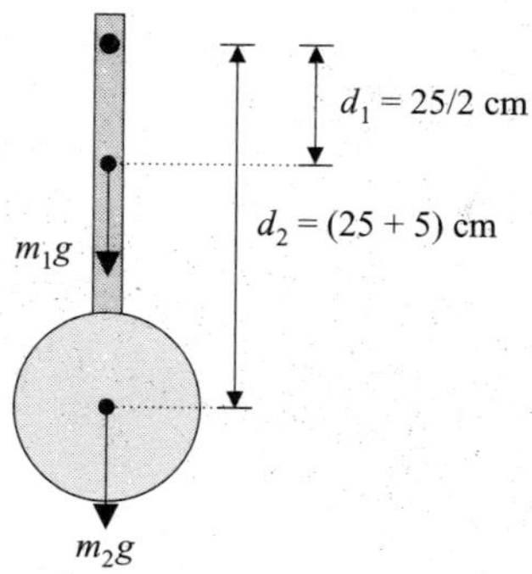

Fig. 19.31(b)

Circular plate: $(I_{zz})_2 = (\bar{I}_{zz})_2 + m_2 d_2^2$

$$= \frac{m_2 r^2}{2} + m_2 d_2^2$$

$$= \frac{0.4 \times (0.05)^2}{2} + [0.4 \times (0.05 + 0.25)^2] = 36.5 \times 10^{-3} \text{ kg.m}^2$$

Therefore, moment of inertia of the composite section is given as

$$(I_{zz}) = (I_{zz})_1 + (I_{zz})_2$$
$$= 40.667 \times 10^{-3}\ \text{kg.m}^2$$

Therefore, frequency of oscillation is given as

$$f = \frac{\omega}{2\pi} = \frac{1}{2\pi}\sqrt{\frac{mgd}{I}}$$
$$= \frac{1}{2\pi}\sqrt{\frac{(0.2+0.4)(9.81)(0.242)}{40.667\times 10^{-3}}} = 0.942\ \text{Hz}$$

Example 19.24 A metallic disk of 1 m diameter is mounted at its outer edge as shown. Determine the time period for small oscillations. Also, determine the length of equivalent simple pendulum.

Solution Let the mass of the disc be m and its radius be r. Then the moment of inertia of the disc about perpendicular axis through the centroid is

$$\bar{I}_{zz} = \frac{mr^2}{2}$$

Therefore, by transfer formula, moment of inertia about perpendicular axis through pivot is given as

$$I_{zz} = \bar{I}_{zz} + mr^2$$
$$= \frac{3}{2}mr^2$$

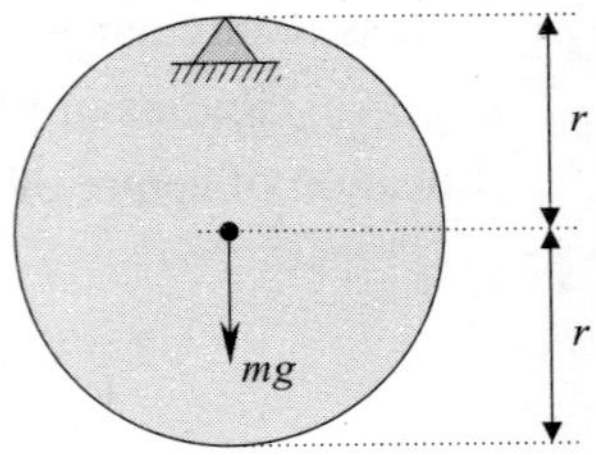

Fig. 19.32

Distance of the centre of mass from the pivot is $d = r$. Therefore, time period for small oscillations is given as

$$T = 2\pi\sqrt{\frac{I}{mgd}}$$
$$= 2\pi\sqrt{\frac{(3/2)mr^2}{mgr}}$$
$$= 2\pi\sqrt{\frac{1.5r}{g}} = 2\pi\sqrt{\frac{1.5\times 0.5}{9.81}} = 1.74\ \text{s}$$

From the expression for T, we see that length of equivalent simple pendulum is

$$L_e = 1.5r = 1.5\times 0.5 = 0.75\ \text{m}$$

19.9 TORSIONAL PENDULUM

A torsional pendulum consists of a solid body suspended by a wire and free to rotate in a horizontal plane about an axis coinciding with the wire. The wire is rigidly attached to the body in such a way that its axis passes through centre of gravity of the body to avoid eccentricity during oscillation.

Consider a disk suspended by a wire as shown in Fig. 19.33. If the disk is displaced from the equilibrium position by giving an angular displacement θ about an axis coinciding with the wire then the wire gets twisted through the same angle θ. Due to resistive property of material of the wire, it opposes this twisting and tends to regain its original position. However, when it reaches its original position, it possesses some velocity and hence it overshoots and displaces angularly on the other side. As a result, the disk oscillates about the equilibrium position in a horizontal plane about the axis of the wire.

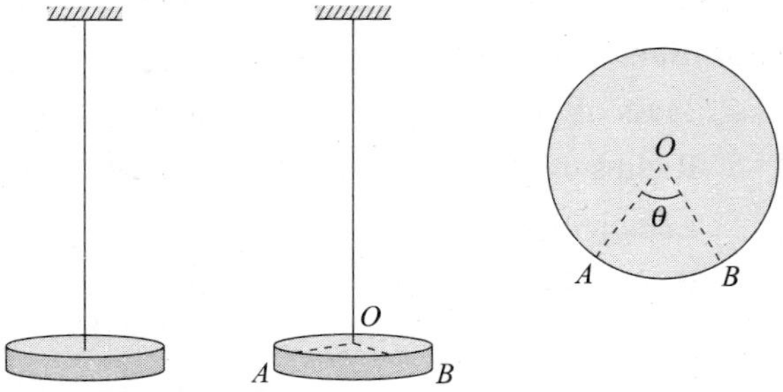

Fig. 19.33 Torsional pendulum

From the theory of torsion based on principles of solid mechanics, it can be shown that the angle of twist of a material with uniform circular cross section subjected to an external torque is

$$\theta = \frac{ML}{I_p G} \tag{19.81}$$

where M is torque applied, L is length of the wire, I_p is polar moment of inertia of the cross section and G is shear modulus of material of the wire. This twisting torque of the wire is the external force acting on the disk tending to restore it to the equilibrium position. When displaced through an angle θ, the torque is given as

$$M = \frac{I_p G}{L}\theta \tag{19.82}$$

Thus, writing the equation of angular rotation of disk, taking into account that the restoring torque acts in the direction opposite to that of the displacement,

$$I\frac{d^2\theta}{dt^2} = -M = -\frac{I_p G}{L}\theta$$

or

$$\frac{d^2\theta}{dt^2} = -\left(\frac{I_p G}{IL}\right)\theta \tag{19.83}$$

This shows that the angular acceleration is proportional to angular displacement and oppositely directed. Hence, we can say that the motion is simple harmonic with natural angular frequency

$$\omega = \sqrt{\frac{I_p G}{IL}} \tag{19.84}$$

Therefore, time period for the oscillation is given as

$$T = \frac{2\pi}{\omega} = 2\pi\sqrt{\frac{IL}{I_p G}} \tag{19.85}$$

Torsional pendulum is particularly useful in determining mass moment of inertia of any body about the axis of rotation by suspending it about its centre of gravity and measuring the period of oscillation.

Example 19.25 A circular metallic disk of 15 cm diameter and 1.5 kg mass is suspended such as to oscillate in a horizontal plane by a metallic wire passing through its centre and whose other end is clamped. The wire is 1 m long and 2 mm in diameter. If the shear modulus of the wire is 0.75×10^5 N/mm², determine the period of oscillation of the disk.

Solution *Given data*

Mass of disk, $m = 1.5$ kg

Radius of disk, $r = 7.5 \text{ cm} = 7.5 \times 10^{-2}$ m

Length of wire, $L = 1$ m

Radius of wire, $r = 1 \text{ mm} = 1 \times 10^{-3}$ m

Shear modulus of wire, $G = 0.75 \times 10^5 \text{ N/mm}^2 = 0.75 \times 10^{11} \text{ N/m}^2$

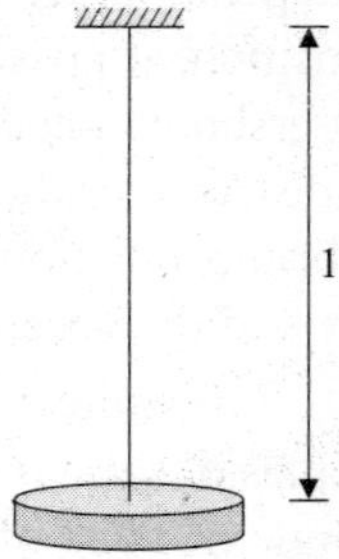

Fig. 19.34

Mass moment of inertia of the disk about an axis coinciding with the wire is

$$I = \frac{mr^2}{2} = \frac{1.5 \times (0.075)^2}{2} = 4.219 \times 10^{-3} \text{ kg.m}^2$$

Polar moment of inertia of the wire,

$$I_P = \frac{\pi r^4}{2} = \frac{\pi \times (1 \times 10^{-3})^4}{2} = 1.571 \times 10^{-12} \text{ m}^4$$

Therefore, time period for the oscillation is obtained as

$$T = 2\pi \sqrt{\frac{IL}{I_p G}} = 2\pi \sqrt{\frac{4.219 \times 10^{-3} \times 1}{1.571 \times 10^{-12} \times 0.75 \times 10^{11}}} = 1.19 \text{ s}$$

Example 19.26 A thin rod of 200 g mass and 0.2 m length is suspended such as to oscillate in a horizontal plane by a metallic wire passing through its centre of gravity. The wire is 1 m long and 1 mm in diameter. If its shear modulus is 0.9×10^5 N/mm², determine the period of oscillation of the rod.

Solution *Given data*

Mass of rod, $m = 200 \text{ gm} = 0.2$ kg

Length of rod, $l = 0.2$ m

Length of wire, $L = 1$ m

Radius of wire, $r = 0.5 \text{ mm} = 0.5 \times 10^{-3}$ m

Shear modulus of wire, $G = 0.9 \times 10^5 \text{ N/mm}^2 = 0.9 \times 10^{11} \text{ N/m}^2$

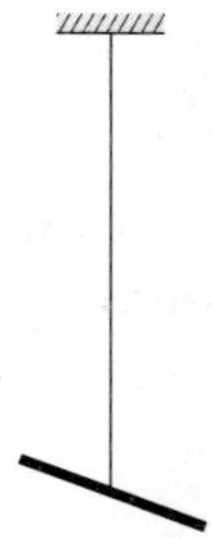
Fig. 19.35

Mass moment of inertia of rod about an axis coinciding with wire is given as

$$I = \frac{ml^2}{12} = \frac{0.2 \times (0.2)^2}{12} = 6.667 \times 10^{-4} \text{ kg.m}^2$$

Polar moment of inertia of wire is given as

$$I_p = \frac{\pi r^4}{2} = \frac{\pi \times (0.5 \times 10^{-3})^4}{2} = 9.817 \times 10^{-14} \text{ m}^4$$

Therefore, time period for the oscillation is given as

$$T = 2\pi\sqrt{\frac{IL}{I_pG}} = 2\pi\sqrt{\frac{6.667\times10^{-4}\times1}{9.817\times10^{-14}\times0.9\times10^{11}}} = 1.73 \text{ s}$$

Example 19.27 A thin rod formed into the shape of an I is suspended horizontally by a 1 m long and 2 mm diameter metallic wire passing through its centre of gravity. The shear modulus of the wire is 0.75×10^5 N/mm^2. If the period of oscillation is 1.5 s, determine the mass moment of inertia of the rod about the axis coinciding with the wire.

Solution *Given data*

Length of wire, $L = 1$ m

Radius of wire, $r = 1 \text{ mm} = 1 \times 10^{-3}$ m

Shear modulus of wire, $G = 0.75 \times 10^5 \text{ N/mm}^2 = 0.75 \times 10^{11} \text{ N/m}^2$

Time period of oscillation, $T = 1.5$ s

Polar moment of inertia of wire is,

$$I_p = \frac{\pi r^4}{2} = \frac{\pi\times(1\times10^{-3})^4}{2} = 1.571 \times 10^{-12} \text{ m}^4$$

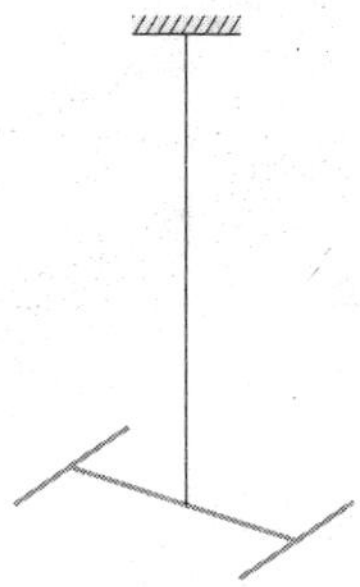

Fig. 19.36

Time period for the oscillation is given as

$$T = 2\pi\sqrt{\frac{IL}{I_pG}}$$

$$\therefore \quad 1.5 = 2\pi\sqrt{\frac{(I)(1)}{(1.571\times10^{-12})(0.75\times10^{11})}}$$

$$\Rightarrow \quad I = 6.715 \times 10^{-3} \text{ kg.m}^2$$

19.10 MISCELLANEOUS TYPES OF OSCILLATIONS

In this section, we will consider miscellaneous types of oscillations not covered in the previous sections. Consider a U-tube differential manometer used for measuring difference in pressures as shown in Fig. 19.37. Initially, the level of liquid in the two limbs will remain same. When there is a difference in pressure between two points to which the manometer is connected then the level of liquid in the two limbs will be different. It will rise in one limb and fall in the other. If this pressure is removed then the liquid will oscillate about the mean position. The time period of such an oscillation can be determined as follows.

Let a be the cross-sectional area of the manometer limb and L be the total length of the column of manometric liquid of density ρ. Then the total mass of manometric liquid undergoing motion is

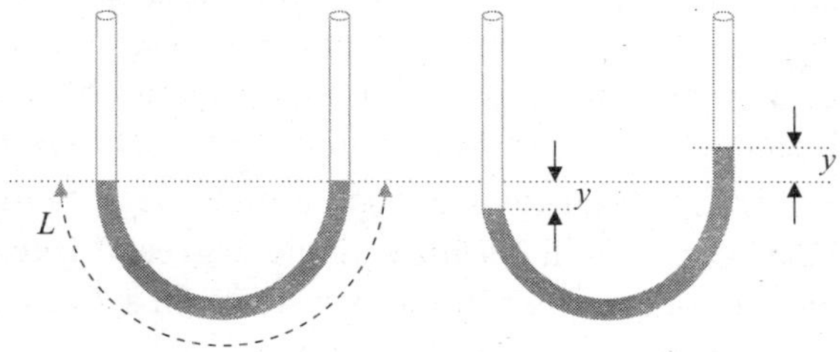

Fig. 19.37 U-tube differential manometer

$$m = \text{density} \times \text{volume}$$
$$= \rho(aL) \tag{19.86}$$

If the column of liquid is displaced vertically by a distance of y, then the restoring force acting on the liquid mass tending to bring it to the equilibrium position is the weight of the excess liquid above the level of liquid on the other limb, i.e., weight of liquid of height $2y$. Therefore,

$$\text{restoring force} = \text{mass} \times \text{acceleration due to gravity}$$
$$= \rho(a \times 2y)g \tag{19.87}$$

Taking into account that the restoring force acts in the direction opposite to the direction of positive displacement, the equation of motion can be written as

$$m\frac{d^2y}{dt^2} = -\rho(a)(2y)g$$

$$\Rightarrow \quad \frac{d^2y}{dt^2} = \frac{\rho(a)(2y)g}{m}$$

$$= -\frac{\rho(a)(2y)g}{\rho aL} = -\left[\frac{2g}{L}\right]y \tag{19.88}$$

We can readily see from the above expression that the acceleration is proportional to the displacement and directed opposite to it. Hence, the motion is simple harmonic, where $\omega^2 = \dfrac{2g}{L}$. Therefore, the period of motion is given as

$$T = \frac{2\pi}{\omega} = 2\pi\sqrt{\frac{L}{2g}} \tag{19.89}$$

19.11 VIBRATIONS OF STRUCTURES

So far, we have analyzed the vibration of simple mechanical systems. In this section, we will discuss the vibration of real structures. Unlike rigid bodies, structures because of their elastic nature, can take different shapes (called vibration modes) during their vibrational motion. To describe such motion of the entire structure, we need to specify the motion of all the particles in the body.

We know that a body is made up of *infinite* number of particles. In addition, each particle has *three* degrees of freedom, as their displacement can be expressed by the components along the X, Y and Z directions. Hence, while solving problems of this nature, we must analyze structures with *infinite* degrees of freedom and hence *infinite* number of unknowns to be solved. These types of problems are difficult to solve even with modern computers. Hence, we make certain idealizations, in which we assume the body to be made up of a large but *finite* number of particles and hence *finite* degrees of freedom. Such types of structures are termed **multi-degrees of freedom** (MDOF) systems.

Even the analysis of such a multi-degrees of freedom system is beyond the scope of our study. However, we will discuss a **single degree of freedom** (SDOF) system, which will give an insight into the multi-degrees of freedom systems. In a single degree of freedom system, we assume the motion of the structure to be in *one* direction only and hence, one time-dependent coordinate describes the vibration of the whole structure. In this, the entire mass of the structure is assumed to be a concentrated mass which is connected to a spring of spring constant equal to the stiffness of the structure. The mass is allowed to move in only *one* direction. Such an idealization is called **spring mass system**, whose motion we have already discussed in Section 19.5. There we discussed that in a spring mass system, the mass can have motion vertically or horizontally. For convenience, we choose only the horizontal motion.

When a structure is given an initial displacement from the equilibrium position and the external force is removed, the resulting vibration is termed **free vibration**. The motion of a diving board after the diver has dived is an example of free vibration. Further, we observe these vibrations to die out after some time. This is mainly due to the resistance offered by the surroundings on the structure. Hence, such vibrations are termed **damped free vibrations**. However, for analytical purposes, we assume all resistance to vibration to be eliminated and such vibrations are termed **undamped free vibrations**. Further, in some cases, the external force may continue to act on the structure periodically. Such vibrations are termed **forced vibrations**. The vibration of a structure caused by a rotating machine is an example of a forced vibration. We will discuss each of these in the following sections and describe the motion of the structure.

19.11.1 Undamped Free Vibration

As said in the previous section, we will assume the entire mass of the structure to be concentrated and it is connected to a spring of spring constant equal to the stiffness of the structure. The mass is also allowed to move freely in the horizontal direction without friction.

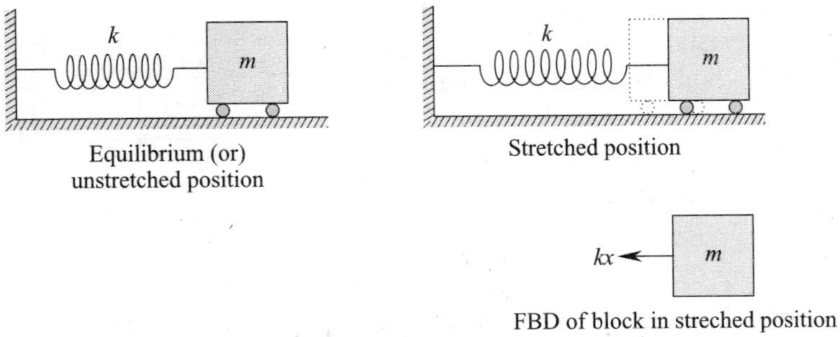

Fig. 19.38

The motion of such a block was already discussed in Section 19.5.2 and they are repeated here for continuation sake. The horizontal motion of the block is given as

$$m\frac{d^2x}{dt^2} = -kx$$

From calculus, we can write the above expression in a simplified form as

$$m\ddot{x} = -kx$$

where $\ddot{x}$ represents the second derivative of x with respect to time. Hence,

$$\ddot{x} = -\left[\frac{k}{m}\right]x$$

or $$\ddot{x} = -\omega^2 x \tag{19.51' \& 52'}$$

Since the acceleration is proportional to the displacement and directed opposite to it, the motion is simple harmonic, whose period is given as

$$T = \frac{2\pi}{\omega} = 2\pi\sqrt{\frac{m}{k}} \tag{19.53}$$

Equation 19.52' is a differential equation and its general solution was derived in Section 19.2.2 as

$$x = a_1 \cos \omega t + a_2 \sin \omega t \tag{19.18'}$$

where a_1 and a_2 are arbitrary constants, whose values can be determined from the initial conditions.

19.11.2 Damped Free Vibration

The motion of the structure, if resistance to motion is eliminated, will go on infinitely according to the Eq. (19.18') discussed in the previous section. However, we observe in nature that the vibrations of bodies, after an initial disturbance, to die out after some time. For instance, a violin string or a tuning fork stops vibrating after a lapse of time. This is due to damping effect and such vibrations are termed damped free vibrations.

Damping is generally a good thing for the structures. The higher the damping, the faster is the decay of free vibration. Damping is mainly caused by friction in the attachments or resistive nature of the material or air resistance. In the analytical model that we consider, we must include a damping mechanism to account for this damping effect. Normally, a dashpot mechanism, like oil damper shown in Fig. 19.39(a) or a symbolical one shown in Fig. 19.39(b) is used.

It is a piston mechanism moving in a viscous fluid. As the frictional resistance offered by any fluid medium to a moving body is directly proportional to the velocity of the body and oppositely directed, in damping mechanism also we can write the frictional resistance mathematically as

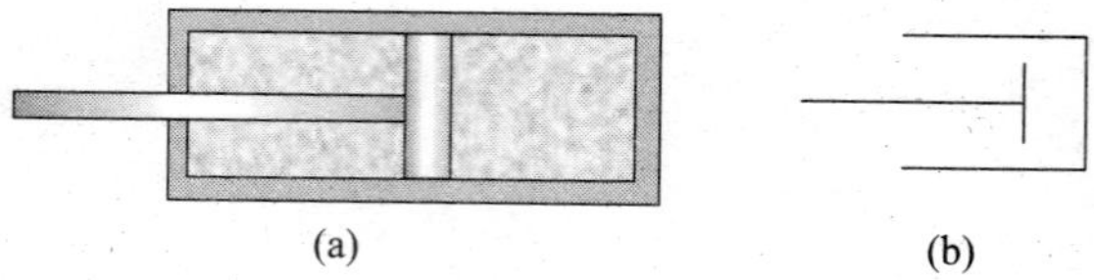

Fig. 19.39 Dashpot mechanism

$$F \alpha - \dot{x}$$

or $$F = -c\dot{x} \tag{19.90}$$

where c is a constant of proportionality, called **coefficient of friction** or **dashpot constant**. Its SI unit is given as N.s/m.

Consider a spring mass system together with damping mechanism as shown in Fig. 19.40. Suppose the body is displaced from its equilibrium position, the forces acting on it tending to restore it to the equilibrium position are kx and $c\dot{x}$ as shown in the free-body diagram.

Hence, writing the equation of motion of the block:

$$m\ddot{x} = -c\dot{x} - kx \tag{19.91}$$

where $\dot{x}$ and $\ddot{x}$ represent respectively the first and second derivatives of x with respect to time. On rearranging,

$$m\ddot{x} + c\dot{x} + kx = 0 \tag{19.92}$$

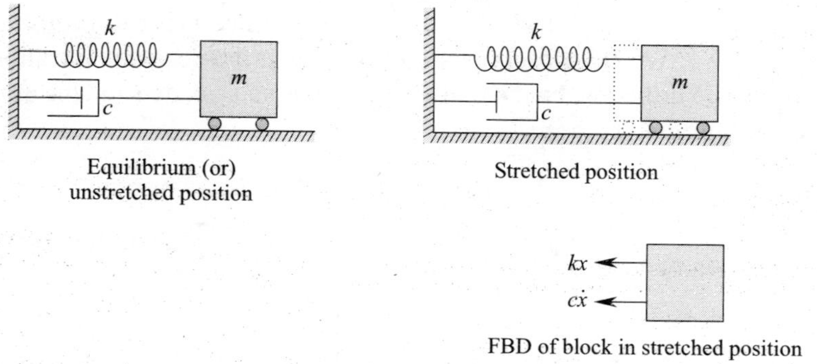

Fig. 19.40

Dividing throughout by m, we have

$$\ddot{x} + \frac{c}{m}\dot{x} + \frac{k}{m}x = 0 \tag{19.93}$$

As before, if we take $\frac{k}{m} = \omega_0^2$ and introduce a new dimensionless parameter, $\xi = \frac{c}{2\sqrt{km}}$, the above equation can be written as

$$\ddot{x} + 2\xi\omega_0\dot{x} + \omega_0^2 x = 0 \tag{19.94}$$

This is a linear second-order differential equation. Since its right-hand term is zero, it is called **homogeneous differential equation** and the corresponding solution is called **homogeneous solution.** The general solution of the above equation depends on the value of ξ, the damping ratio, which indicates the level of damping.

Case I When $\xi < 1$

From calculus, we can determine the general solution as

$$x(t) = x_h(t) = e^{-\xi\omega_0 t}[a_1 \cos \omega_d t + a_2 \sin \omega_d t] \tag{19.95}$$

where x_h indicates homogeneous solution. The term ω_d in the above expression is called **damped natural frequency** and it is defined as $\omega_d = \omega_0\sqrt{1-\xi^2}$.

Case II When $\xi = 1$

The general solution is given as

$$x(t) = x_h(t) = e^{-\xi\omega_0 t}[a_1 + a_2 t] \tag{19.96}$$

Case III When $\xi > 1$

The general solution is given as

$$x(t) = x_h(t) = e^{-\xi\omega_0 t}[a_1 \cosh \omega'_d t + a_2 \sinh \omega'_d t] \tag{19.97}$$

where $\omega'_d = \omega_0\sqrt{\xi^2 - 1}$

In the above equations, a_1 and a_2 are arbitrary constants, whose values can be determined from the initial conditions. The term $e^{-\xi\omega_0 t}$ indicates that the solutions have an exponential decay. Responses to initial velocity $[\dot{x}(0) \neq 0, x(0) = 0]$ for different values of ξ are shown below in Fig. 19.41. The parameter ξ indicates the level of damping. The condition $\xi = 1$ is called **critical damping** because no real vibration occurs when $\xi \geq 1$. We also observe that for $0 < \xi < 1$, the decay of vibration increases as the damping ratio ξ increases. Further, when $\xi < 0$, the vibration is self-excited and the amplitude of motion keeps on increasing with time.

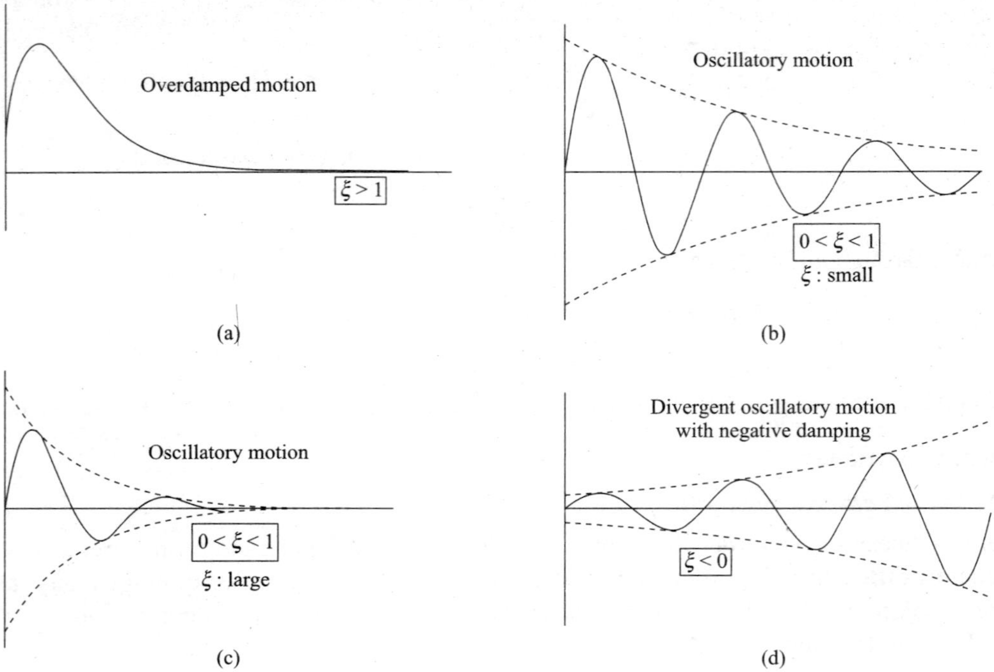

Fig. 19.41

19.11.3 Damped Forced Vibration

When the external force continues to act on the structure periodically, the resulting motion is termed **forced vibration**. The external forces could be human induced, machine induced, vehicle induced or caused by wind or water waves and earthquake or underground explosion.

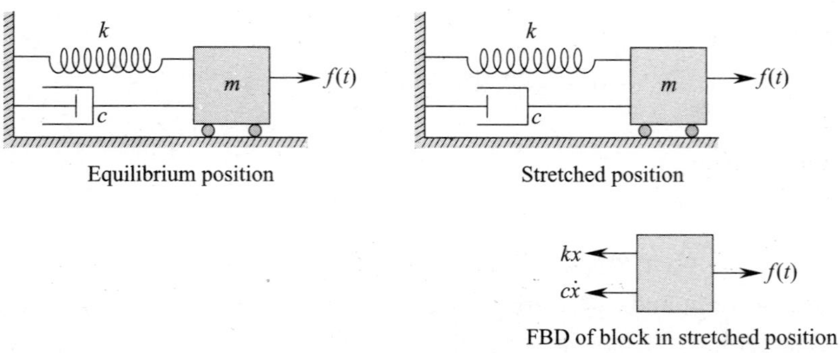

Fig. 19.42

When the block is displaced from the equilibrium position, the forces acting on it are shown in the free-body diagram. Writing the equation of motion of the block,

$$f(t) - kx - c\dot{x} = m\ddot{x} \tag{19.98}$$

On rearranging,

$$m\ddot{x} + c\dot{x} + kx = f(t) \tag{19.99}$$

The solution of the above differential equation can be obtained in two steps: first, we consider a homogeneous differential equation, which is obtained by making the right-hand term in the above equation as zero.

$$m\ddot{x} + c\dot{x} + kx = 0 \tag{19.92'}$$

The solution of the above homogeneous equation is known as **homogeneous solution**, $x_h(t)$, which we already discussed in the previous section. Next, we consider the non-homogeneous equation,

$$m\ddot{x} + c\dot{x} + kx = f(t)$$

whose solution if we take as $x_p(t)$, called the **particular solution** then the solution of forced vibration is given by superposition principle as

$$x(t) = x_h(t) + x_p(t) \tag{19.100}$$

Computation of Particular Solution There are different types of time-dependent external forces that can act on a structure causing vibration. They are harmonic force caused by rotating machines, periodic force caused by water waves, non-periodic force caused by turbulent wind and transient force caused by ground motion, explosion, etc. There are different approaches to solve different types of external forces, but we will concentrate mainly on harmonic force.

Dividing the Eq. 19.99 throughout by m and considering the same parameters ξ and ω_0 as before:

$$\ddot{x} + 2\xi\omega_0\dot{x} + \omega_0^2 x = \frac{f(t)}{m} \tag{19.101}$$

The damped natural frequency is given as $\omega_d = \omega_0\sqrt{1-\xi^2}$. Here we consider only Case (I), where $\xi < 1$, as most structures have a damping ratio less than one.

Consider a harmonic force with unit amplitude. We know that both sine and cosine functions are harmonic in nature. Their responses we will analyze below:

Let $$f(t) = 1.\sin\omega t$$

$$\therefore \quad \ddot{x} + 2\xi\omega_0\dot{x} + \omega_0^2 x = \frac{\sin\omega t}{m} \tag{19.102}$$

From background of calculus, we can assume a particular solution in the form

$$x_p(t) = a\sin\omega t \tag{19.103}$$

Substituting this value in the Eq. 19.102

$$-a\omega^2\sin\omega t + 2\xi\omega_0 a\omega\cos\omega t + \omega_0^2 a\sin\omega t = \frac{\sin\omega t}{m}$$

or

$$a\sin\omega t[\omega_0^2 - \omega^2] + 2a\xi\omega_0\omega\cos\omega t = \frac{\sin\omega t}{m} \tag{19.104}$$

Similarly, let $f(t) = 1.\cos \omega t$

and its particular solution will be in the form

$$x_p(t) = a \cos \omega t \tag{19.105}$$

Substituting this in the original equation,

$$-a\omega^2 \cos \omega t - 2\xi\omega_0 a\omega \sin \omega t + \omega_0^2 a \cos \omega t = \frac{\cos \omega t}{m}$$

$$a \cos \omega t [\omega_0^2 - \omega^2] - 2a\xi\omega_0 \omega \sin \omega t = \frac{\cos \omega t}{m} \tag{19.106}$$

Squaring Eqs (19.104 and 106) and adding

$$a^2 [\omega_0^2 - \omega^2]^2 + 4a^2(\xi\omega_0 \omega)^2 = \frac{1}{m^2}$$

$$\therefore \quad a = \frac{1}{m} \frac{1}{\sqrt{(\omega_0^2 - \omega^2)^2 + (2\xi\omega_0\omega)^2}} \tag{19.107}$$

Introducing parameters, $\beta = \dfrac{\omega}{\omega_0}$ and $\omega_0 = \sqrt{\dfrac{k}{m}}$, we can write

$$a = \frac{1}{k} \frac{1}{\sqrt{(1-\beta^2)^2 + (2\xi\beta)^2}} \tag{19.108}$$

When $\omega = 0$ then $\beta = 0$ and $a = \dfrac{1}{k}$ which is termed as **static response** (displacement). The remaining part of the expression is termed as **dynamic response** with respect to static response.

We consider

$$M_0 = \frac{1}{\sqrt{(1-\beta^2)^2 + (2\xi\beta)^2}} \tag{19.109}$$

which is termed **magnification factor** and from its expression, it can be seen that it is a *dimensionless* quantity. The variation of M_0 with respect to β is shown in Fig. 19.43. When $\beta = 1$, i.e., when the natural frequency of vibration $[\omega_0]$ of the structure is equal to the frequency $[\omega]$ of the external harmonic force, we see that the magnification factor $[M_0]$ takes a peak value. This condition is called **resonance**. Further, we see that as the damping ratio $\xi \to 0$, the magnification factor tends to infinity.

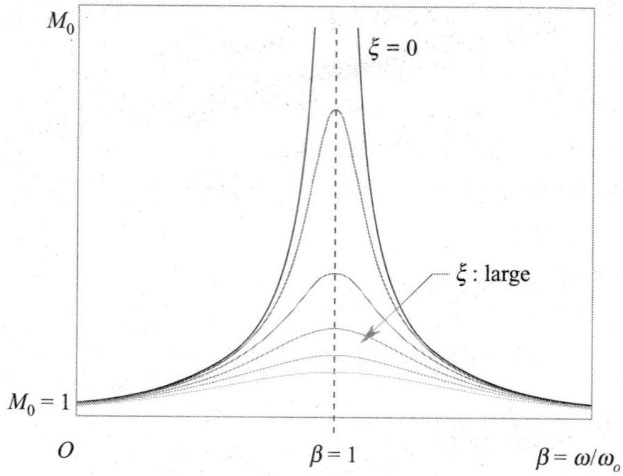

Fig. 19.43

$\beta = 1 \quad M_0 = \dfrac{1}{2\xi}$ (0.02 ≈ 0.05) (response)

$\beta << 1, \quad M_0 \approx 1$ (static)

$\beta >> 1, \quad M_0 \approx 0$

The bigger the damping, the faster the decay of free vibration, and the smaller the steady-state amplitude of resonant forced vibration.

SUMMARY

Any motion that repeats itself after equal intervals of time is termed *periodic* motion. The motion of water waves in seas under the action of wind can be cited as an example. As water waves are free to move, they displace from one place to the other. On the other hand, real civil structures such as transmission cable, diving board in swimming pool, bridge, tall towers, etc., and even mechanical systems such as simple pendulum, string of a musical instrument due to their attachment with the surroundings do not displace from one place to the other upon the action of external forces. Instead, they move back and forth over the *same path*. Such types of periodic motions, which trace the same path in a cyclic manner, are termed *vibratory* or *oscillatory* motions.

For all analytical purposes, we assume all resistance to vibration to be eliminated and such vibrations are termed *undamped free vibrations*. However, this is an ideal condition as we normally observe these vibrations to die out after some time and such vibrations are termed *damped free vibrations*. If the external force continues to act on the structure periodically then the resulting vibrations are termed *forced vibrations*.

Simple Harmonic Motion

Simple harmonic motion is a special case of *rectilinear* motion with *variable* acceleration, in which the acceleration of the particle is proportional to the displacement from the origin and is always directed towards the origin. Most of the structures and mechanical systems are observed to execute this motion under small displacements from the equilibrium position.

The equations of motion as functions of time and displacement for ready reference are shown below:

	As function of displacement	*As function of time*	
		t = 0, at mean position	*t = 0, at extreme position*
Displacement	x	$x = A \sin \omega t$	$x = A \cos \omega t$
Velocity	$v = \pm\omega\sqrt{A^2 - x^2}$	$v = A\omega \cos \omega t$	$v = -A\omega \sin \omega t$
Acceleration	$a = -\omega^2 x$	$a = -A\omega^2 \sin \omega t$	$a = -A\omega^2 \cos \omega t$

By combining the motions starting from mean position and from extreme position, the general solution of simple harmonic motion can be written as

$$x = a_1 \cos \omega t + a_2 \sin \omega t$$

Amplitude of motion is defined as the maximum displacement from the mean position. The *period* of motion, that is the time taken for one complete cycle of motion is expressed mathematically as

$$T = \frac{2\pi}{\omega}$$

and its *frequency*, that is, the number of cycles completed in unit time is expressed as

$$f = \frac{1}{T} = \frac{\omega}{2\pi}$$

Motion of a Body Attached to a Spring

When a body of mass m is attached to a spring of spring constant k and allowed to oscillate vertically or horizontally, its period of motion is given as

$$T = 2\pi\sqrt{\frac{m}{k}} = 2\pi\sqrt{\frac{e}{g}}$$

where e is extension of the spring.

When more than one spring is connected in *parallel*, its equivalent spring constant is

$$k = k_1 + k_2 + \dots + k_n$$

and if they are connected in *series*,

$$\frac{1}{k} = \frac{1}{k_1} + \frac{1}{k_2} + \dots + \frac{1}{k_n}$$

Motion of a Body Attached to an Elastic String

When a body of mass m attached to an elastic string of length l and cross-sectional area A is allowed to oscillate vertically, its period of motion for small displacements from the equilibrium position is given as

$$T = 2\pi\sqrt{\frac{ml}{AE}} = 2\pi\sqrt{\frac{\Delta}{g}}$$

Simple Pendulum

A simple pendulum is defined as a small but heavy mass of bob suspended by a weightless, inextensible string. For small displacements from the equilibrium position, the motion of the mass is observed to be angular simple harmonic. Its time period is given as

$$T = \frac{2\pi}{\omega} = 2\pi\sqrt{\frac{l}{g}}$$

A seconds pendulum is one which takes *two seconds* to execute *one* complete oscillation or *a second* to execute *half* an oscillation. The time taken to complete half an oscillation is known as a *beat*. Hence, a seconds pendulum is said to beat a second.

Compound Pendulum

A compound pendulum is any rigid body mounted in such a way that it can oscillate freely in a vertical plane about a horizontal axis passing through the body. Its period of motion for small displacement from the equilibrium position is given as

$$T = \frac{2\pi}{\omega} = 2\pi\sqrt{\frac{I}{mgd}}$$

The length of an equivalent simple pendulum is the length of the simple pendulum, which would have the same time period as the compound pendulum. It is given as

$$l = L_e = \frac{I}{md}$$

Torsional Pendulum

A torsional pendulum consists of a solid body suspended by a wire and free to rotate in a horizontal plane about an axis coinciding with the wire. For small displacements from the equilibrium position, its time period is given as

$$T = \frac{2\pi}{\omega} = 2\pi\sqrt{\frac{IL}{I_p G}}$$

Vibrations of Structures

The vibration of any structure can be idealized as a spring mass system, which is allowed to move freely when we consider undamped vibrations; a damping mechanism is added when we consider damped vibrations.

The motion of undamped structure is simple harmonic and its displacement is

$$x = a_1 \cos \omega t + a_2 \sin \omega t$$

while the motion of damped structure is harmonic subjected to exponential decay. The displacement is dependent on the value of ξ, damping ratio.

$$x(t) = x_h(t) = e^{-\xi\omega_0 t}[a_1 \cos \omega_d t + a_2 \sin \omega_d t] \qquad \text{when } \xi < 1$$

$$x(t) = x_h(t) = e^{-\xi\omega_0 t}[a_1 + a_2 t] \qquad \text{when } \xi = 1$$

$$x(t) = x_h(t) = e^{-\xi\omega_0 t}[a_1 \cos h\omega_d' t + a_2 \sin h\omega_d' t] \qquad \text{when } \xi > 1$$

Damped Forced Vibration

When the external force continues to act on the structure periodically, the resulting motion is termed **forced vibration**. The external forces could be human induced, machine induced, vehicle induced or caused by wind or water waves and earthquake or underground explosion. The solution of forced vibration is given by superposition principle as the sum of homogeneous solution and particular solution:

$$x(t) = x_h(t) + x_p(t)$$

EXERCISES

Objective-type Questions

1. A body is said to undergo free vibration when

(a) it vibrates in free space

(b) it vibrates freely with no force acting on it

(c) the force causing the initial displacement is removed

(d) it vibrates freely with no resistive force acting on it

2. Which of the following is an ideal motion?
   (a) Periodic motion
   (b) Damped free vibration
   (c) Undamped free vibration
   (d) Forced vibration
3. SHM is a
   (a) rectilinear motion with uniform acceleration
   (b) rectilinear motion with variable acceleration
   (c) curvilinear motion with variable acceleration
   (d) circular motion with variable angular speed
4. Which of the following is NOT a simple harmonic motion?
   (a) A particle moving in a circular path with constant angular speed
   (b) A simple pendulum undergoing very small displacements
   (c) A compound pendulum undergoing very small displacements
   (d) A car moving on a straight road
5. The SI unit of frequency of vibration is
   (a) s (b) rad/s (c) rad/s^2 (d) Hz
6. When a particle is executing simple harmonic motion, the velocity is maximum at
   (a) the mean position
   (b) extreme position
   (c) midway between extreme and mean positions
   (d) one-third distance from mean position
7. When a particle is executing simple harmonic motion, the acceleration is maximum at
   (a) the mean position
   (b) extreme position
   (c) midway between extreme and mean positions
   (d) one-third distance from mean position
8. When two springs of spring constants k_1 and k_2 are arranged in series then their equivalent spring constant is
   (a) $k_1 + k_2$ (b) $k_1 - k_2$ (c) $\dfrac{k_1 k_2}{k_1 + k_2}$ (d) $\dfrac{k_1 + k_2}{k_1 k_2}$
9. When two springs of spring constants k_1 and k_2 are arranged in parallel then their equivalent spring constant is
   (a) $k_1 + k_2$ (b) $k_1 - k_2$ (c) $\dfrac{k_1 k_2}{k_1 + k_2}$ (d) $\dfrac{k_1 + k_2}{k_1 k_2}$

Answers

1. (c) **2.** (c) **3.** (b) **4.** (d) **5.** (d) **6.** (a) **7.** (b) **8.** (c)
9. (a)

Short-answer Questions

1. Define periodic motion by giving examples.
2. Define oscillatory or vibratory motion by giving examples.
3. Distinguish between free vibration and forced vibration. Give suitable examples for each.
4. Distinguish between undamped and damped vibration.
5. Vibration is an unavoidable evil. Discuss.
6. Is it possible to eliminate vibration completely? What are used to suppress vibrations?
7. Distinguish between harmonic and simple harmonic motions. Can all periodic motions be considered simple harmonic?
8. State the mathematical expression of simple harmonic motion.
9. How does the motion of a particle executing simple harmonic motion differ, when it starts from the mean position and when it starts from the extreme position?
10. Define one cycle, amplitude, natural frequency, time period and frequency in vibratory motion.
11. Prove that the motion of a mass suspended from a spring is simple harmonic for small displacements from the equilibrium position and derive the expression for the period of motion.
12. Derive the equivalent spring constant for springs arranged in (i) parallel, and (ii) in series.
13. Prove that the motion of a mass suspended from an elastic string is simple harmonic for small displacements from the equilibrium position and derive the expression for the period of motion.
14. Define a simple pendulum and derive an expression to determine its period of motion for small displacements from equilibrium position.
15. Define a seconds pendulum.
16. Define a compound pendulum and derive an expression to determine its period of motion for small displacements from equilibrium position.
17. Distinguish between centre of oscillation and centre of suspension.
18. Define a torsional pendulum and derive an expression to determine its period of motion for small displacements from equilibrium position.
19. Define multi-degrees of freedom system.
20. Describe the dashpot mechanism used in analytical modeling.
21. Explain the determination of homogeneous and particular solutions for the vibration of a structure subjected to harmonic force.
22. When does resonance of the structure occur?

Numerical Problems

19.1 A particle moving with SHM has a time period of 4 seconds and an amplitude of 1 m. Determine the maximum velocity and acceleration of the particle.

Ans. $\frac{\pi}{2}$ m/s, $\frac{\pi^2}{4}$ m/s^2

19.2 A particle executing SHM makes 50 complete revolutions per minute. If its acceleration at the extreme position is 20 m/s^2, determine (i) the time period and the amplitude of oscillation, (ii) the velocity of the particle at the extreme position, mean position and halfway between mean and extreme positions, (iii) the position of the particle from the mean position at which the velocity is half the maximum velocity.

Ans. (i) 1.2 s, 0.73 m, (ii) 0, 3.83 m/s, 3.31 m/s, (iii) 0.63 m

19.3 A particle executing SHM has a time period of 2 seconds. If it starts from rest at a distance of 25 cm from the mean position, determine the maximum velocity and the velocity when it has traveled 10 cm.

Ans. 25π cm/s, 20π cm/s

19.4 If the displacement of a particle at any instant of time is given as $x = a \cos \omega t + b \sin \omega t$, show that the particle executes simple harmonic motion.

19.5 The velocities of a particle moving with SHM are 16 m/s and 12 m/s respectively when at distances of 6 m and 8 m from the mean position. Determine the period of motion, amplitude, maximum velocity and acceleration.

Ans. π s, 10 m, 20 m/s, 40 m/s^2

19.6 A particle executing SHM has a maximum velocity of 6 m/s and maximum acceleration of 12 m/s^2. Determine (i) the period and the amplitude of motion, (ii) displacement, velocity and acceleration of the particle at $t = 2$ s starting from the mean position, (iii) velocity and acceleration of the particle at a point, 1.5 m to the left of the mean position and the time taken to reach it from the mean position.

Ans. (i) π s, 3 m, (ii) –2.27 m, –3.92 m/s, 9.08 m/s^2, (iii) –5.2 m/s, 6 m/s^2, 0.262 s

19.7 A particle executing SHM has an acceleration of 30 m/s^2 at a point 1.2 m from the origin. Determine the time period for the oscillation.

Ans. 1.26 s

19.8 The equation of motion of a particle executing SHM is $x = 5 \sin\left(2\pi t + \dfrac{\pi}{4}\right)$, where x and t are in metres and seconds respectively. Determine the (i) displacement, (ii) velocity, (iii) acceleration of the particle at $t = 2$ s. Also, determine the natural frequency $[\omega]$ and the period of vibration $[T]$.

Ans. 3.54 m, –22.21 m/s, 139.6 m/s^2, 2π rad/s, 1 s

19.9 The equation of motion of a particle executing SHM is $x = 4 \cos\left(\pi t + \dfrac{\pi}{3}\right)$. Determine the (i) displacement, (ii) velocity, (iii) acceleration of the particle at $t = 1$ s. Also, determine the natural frequency $[\omega]$ and the period of vibration $[T]$.

Ans. –2 m, 10.9 m/s, 19.7 m/s^2, π rad/s, 2 s

19.10 A horizontal platform executes simple harmonic oscillations vertically; it moves through a distance of 1.5 m and makes 30 complete oscillations per minute. Determine whether a block of 1 kg mass placed on it will stay in contact with the platform throughout the motion or not. Also, determine the greatest and least pressures exerted by the block on the platform.

Ans. Will stay in contact, 17.2 N, 2.4 N

19.11 A horizontal platform executes simple harmonic oscillations vertically; it moves through a distance of 1 m. If a block is resting on it, what should be the minimum number of oscillations per minute so that the block is thrown off?

Ans. 42.3 rev/min

19.12 A horizontal platform executes simple harmonic oscillations horizontally; it makes 15 oscillations per minute. What maximum amplitude can it reach if a book placed on the platform would not slip? Take coefficient of static friction between contact surfaces as 0.2.

Ans. 0.8 m

19.13 A horizontal platform executes simple harmonic oscillations vertically; its velocity at the mean position is 4 m/s. If a block is resting on it, what should be the minimum number of oscillations per minute so that the block is thrown off?

Ans. 23.4 osc./min

19.14 A particle executes SHM on a line *OAB*, between the two points *A* and *B*; it has velocity v at the mean position between *A* and *B*. If the distances of *A* and *B* from the origin *O* are a and b respectively then show that the time period for oscillation is $\dfrac{\pi(b-a)}{v}$.

19.15 The period of oscillation of a spring mass system in the vertical direction is 0.5 s. Determine (i) its frequency of vibration, (ii) elongation of the spring in the equilibrium position, (iii) its maximum velocity if the system oscillates over a space of 1 m.

Ans. (i) 2 Hz, (ii) 6.2 cm, (iii) 2π m/s

19.16 The periods of vibration of two spring mass systems with same mass m are T_1 and T_2. Prove that the difference between their spring constants is $4\pi^2 m\left[\dfrac{T_2^2 - T_1^2}{T_1^2 T_2^2}\right]$.

19.17 A body of 4 kg mass when suspended from a spring extends it by 10 cm. If a body of mass 1.5 kg is suspended from the same spring, determine the elongation of the spring. If it is pulled by 1 cm from its equilibrium position and released, determine the period of vibration, amplitude and maximum velocity.

Ans. 3.75 cm, 0.39 s, 1 cm and 16.17 cm/s

19.18 A spring mass system is observed to make 50 complete oscillations in 1 minute. Determine the period of vibration and the extension of the spring in equilibrium position.

Ans. 1.2 s, 0.36 m

19.19 A block of 15 kg mass is suspended by two springs arranged in series as shown in Fig. E.19.19. The block is pulled down by 5 cm and released. Determine the period of oscillation, maximum velocity and maximum acceleration of the block.

Ans. 0.52 s, 0.6 m/s, 7.2 m/s^2

19.20 A block of 20 kg mass is suspended by two springs of equal lengths arranged in parallel as shown in Fig. E.19.20. The respective spring constants are also shown in the figure. The block is pulled down by 2 cm and released. Determine the period of oscillation, maximum velocity and maximum acceleration of the block.

Ans. 0.1π s, 0.4 m/s, 8 m/s^2

19.21 A 15 kg block is suspended by a system of three springs as shown in Fig. E.19.21. The respective spring constants are also shown in the figure. Determine the period of oscillation for small displacement from the equilibrium position.

Ans. 0.444 s

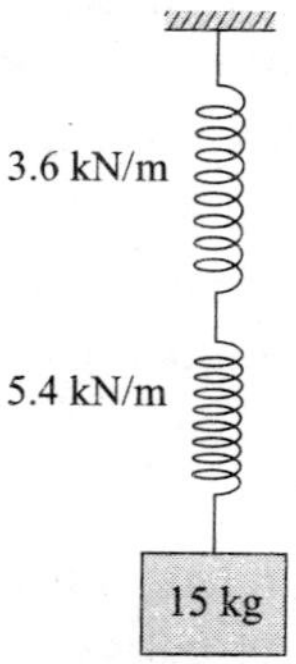

Fig. E.19.19

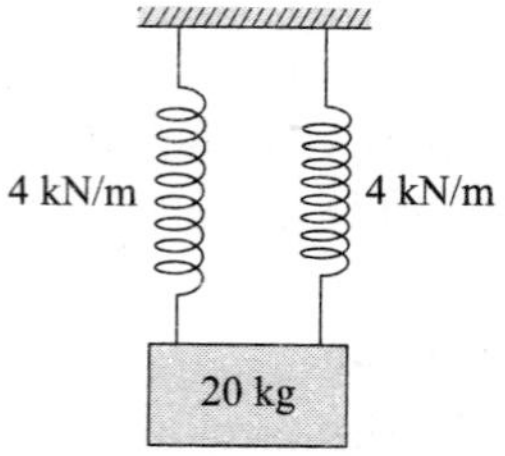

Fig. E.19.20

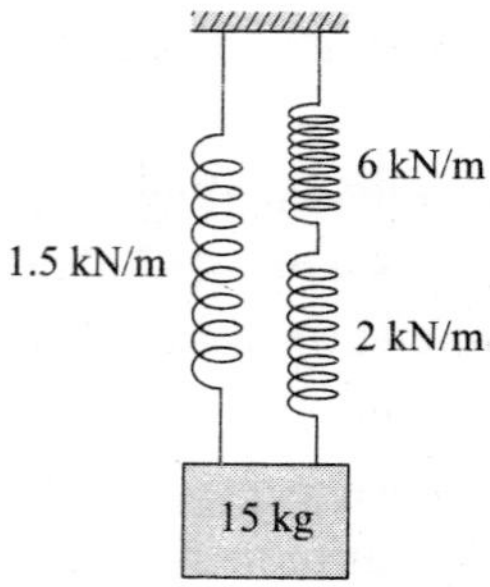

Fig. E.19.21

19.22 If the time period of oscillation is the same for both the spring mass systems shown in Fig. E.19.22, determine the unknown spring constant k.

Ans. 12 kN/m

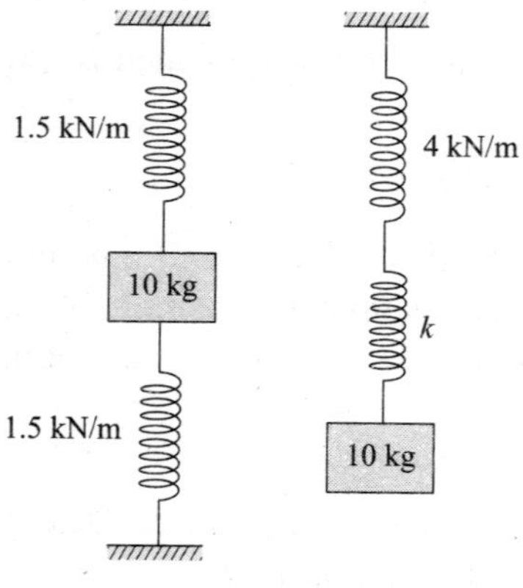

Fig. E.19.22

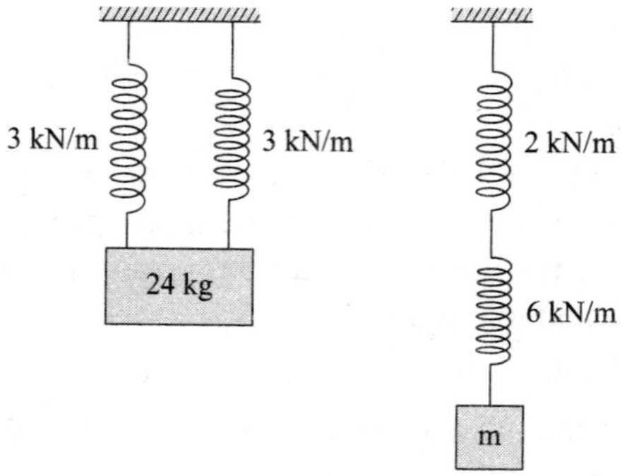

Fig. E.19.23

19.23 Determine the mass of the block in the second system shown in Fig. E.19.23, such that the time period is the same for both the systems.

Ans. 6 kg

19.24 A body of 800 g mass oscillates horizontally as shown in Figs E.19.24. Determine the period of motion in each case. The spring constants are respectively 6 kN/m and 4 kN/m.

Ans. 1.78 s, 3.63 s, 1.78 s

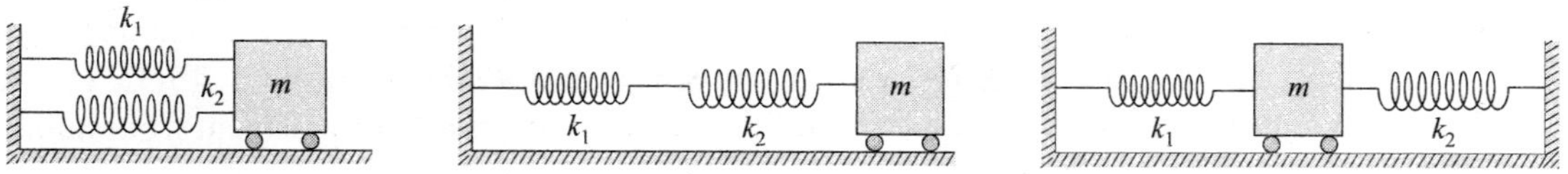

Fig. E.19.24

19.25 A mass of 4 kg is suspended by an elastic string of 1 m length and 3 mm diameter. The extension of the string is observed to be 15 cm. Determine the elastic modulus of the string. Describe the motion of the mass and determine period of motion if the (i) mass is released from the unstretched position, (ii) mass is pulled down by 10 cm from equilibrium position and released, (iii) mass is pulled down by 20 cm from the equilibrium position and released.

Ans. 370.09×10^5 N/m^2, (i) $T = 0.78$ s, $v_{max} = 1.21$ m/s (ii) $T = 0.78$ s, $v_{max} = 0.81$ m/s, (iii) $T = 0.82$ s

19.26 A simple pendulum with a string length of 36 cm is observed to make 50 complete oscillations in 1 minute. Determine the acceleration due to gravity at that place.

Ans. 9.87 m/s^2

19.27 If the period of a simple pendulum is 1.0 s, determine the length of the pendulum at a place where $g = 9.81$ m/s^2.

Ans. 0.248 m

19.28 A pendulum clock beats seconds at a place where $g = 9.81$ m/s^2. If it is brought to a place where $g = 9.82$ m/s^2, how much does it gain or lose per day?

Ans. Gains 44.04 s in a day

19.29 A pendulum clock beats seconds at a place where $g = 9.81$ m/s^2. If it is brought to a place where $g = 9.8$ m/s^2, by how much should the length be altered so that it again beats seconds?

Ans. Diminished by 0.76 mm

19.30 A slender massless rod AB of length l carries two equal point masses m at its extreme end B and at its midpoint. If the rod is supported at A by a horizontal axis, determine the period of oscillation for small displacements from equilibrium position.

Ans. $2\pi\sqrt{\dfrac{5l}{6g}}$

19.31 A thin square plate of side a and mass m is suspended at one of its corner by a horizontal axis. Determine the length of equivalent simple pendulum.

Ans. $\dfrac{\sqrt{8}}{3}a$

19.32 A thin rectangular plate of breadth a and height $2a$ is suspended about its shorter edge by a horizontal axis. Determine the period of oscillation for small displacements from equilibrium position. Also, determine the length of equivalent simple pendulum.

Ans. $2\pi\sqrt{\dfrac{4a}{3g}}, \dfrac{4}{3}a$

19.33 A cylinder of base radius R, height H and mass M is suspended by a horizontal axis (i) about a diameter, and (ii) about an axis parallel to its central axis as shown in Fig. E.19.33. Determine the ratio of the periods of oscillation for the two cases for small displacement from quilibrium position.

Ans. $\left[\dfrac{3R^2 + 4H^2}{9RH}\right]$

19.34 A slender rod formed into an isosceles triangle is mounted so as to oscillate about a horizontal axis as shown in Fig. E.19.34. If its period of oscillation is 0.77 s then determine its mass moment of inertia about the axis of oscillation. Mass per unit length is 1 kg/m.

Ans. 4.8×10^{-3} kg.m^2

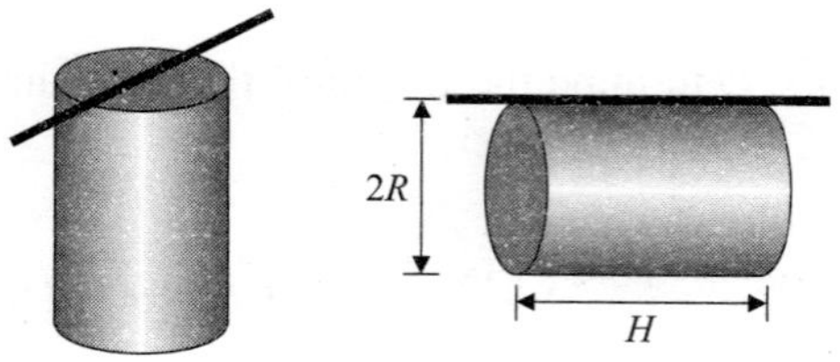

Fig. E.19.33

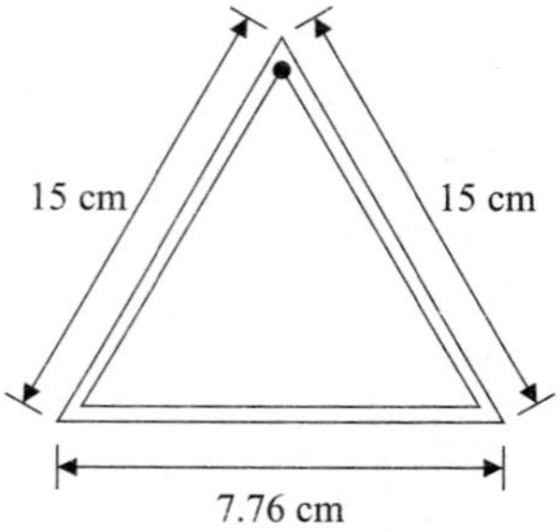

Fig. E.19.34

19.35 An equilateral triangular plate of 0.25 m side and 0.5 kg mass is mounted at one of its corners so as to oscillate about a horizontal axis. If its period of oscillation is 0.85 s then determine its mass moment of inertia about the axis of oscillation.

Ans. 6.5×10^{-3} kg.m^2

19.36 A short steel cylinder of 5 cm diameter and 10 cm height is suspended by a metallic wire coinciding with the axis of the cylinder. Refer Fig. E.19.36. The wire is 1 m long and 2 mm in diameter and its shear modulus is 0.8×10^5 N/mm^2. Determine the period of oscillation of the cylinder. Mass density of steel is 7850 kg/m^3.

Ans. 0.39 s

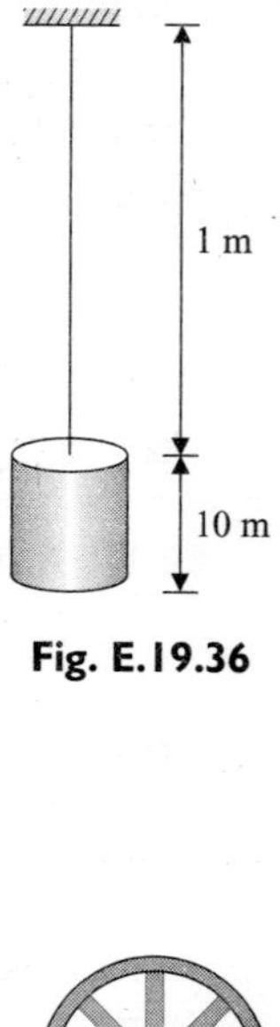

Fig. E.19.36

19.37 A circular wheel is suspended as shown in Fig. E.19.37 such as to oscillate in a horizontal plane by a metallic wire passing through its centre of gravity. The wire is 1 m long and 1 mm in diameter and its shear modulus is 0.9×10^5 N/mm^2. If the period of oscillation of the wheel is 1.2 s, determine the mass moment of inertia of the wheel.

Ans. 3.22×10^{-4} kg.m^2

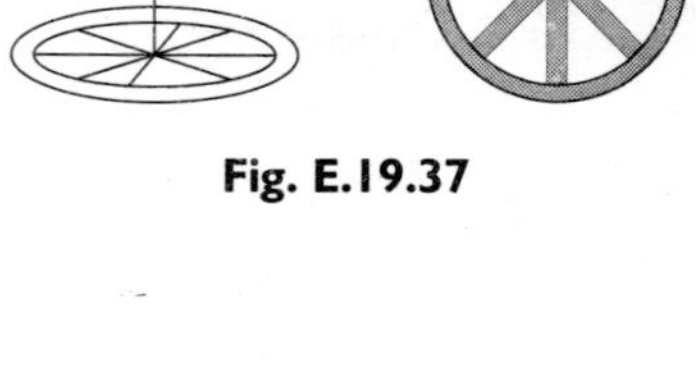

Fig. E.19.37

Index

D

E

T